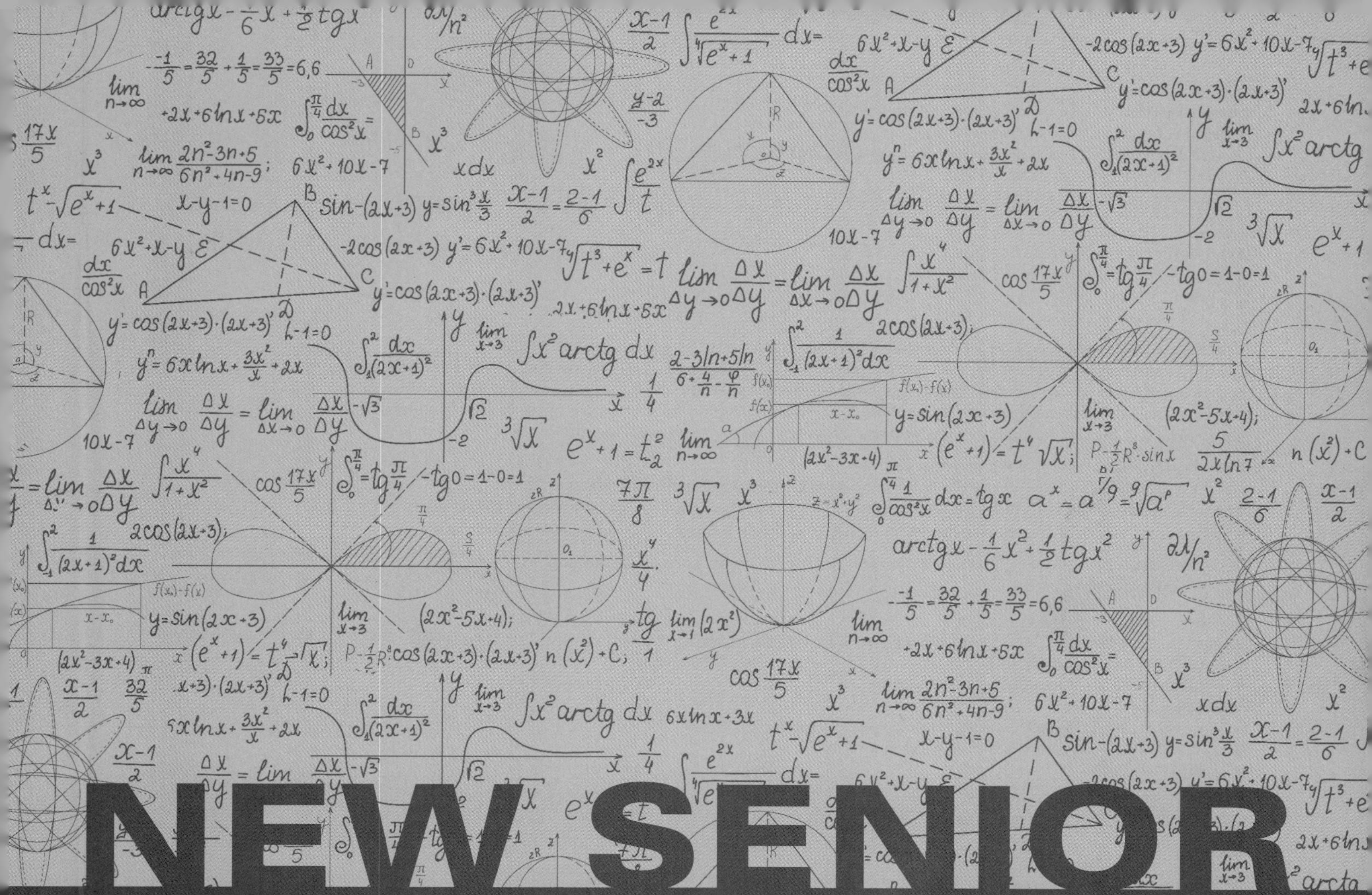

NEW SENIOR MATHEMATICS

ADVANCED

FOR YEARS 11 & 12

THIRD EDITION

J.B. FITZPATRICK
BOB AUS

Pearson Australia
(a division of Pearson Australia Group Pty Ltd)
459–471 Church Street, Level 1, Building B, Richmond, Victoria 3121
PO Box 23360, Melbourne, Victoria 8012
www.pearson.com.au

First published 1983 by Pearson Australia
2027 2026 2025 2024
1 2 3 4 5

Publishers: Julian Lumb, Tim Carruthers
Project Managers: Michelle Thomas, Shelly Wang, Beth Zeme, Anubhuti Harsh
Production Editor: Casey McGrath
Editors: Maryanne Park, Marcia Bascombe, Sam Trafford
Proofreader: Laura Rentsch
Designers: Jay Urbanic, Anne Donald, iEnergizer Aptara Limited
Rights and Permissions Editors: Sian Human, Amirah Fatin
Illustrator: Diacritech
Printed in China, GCC/01 (2024)

National Library of Australia Cataloguing-in-Publication entry:

A catalogue record for this book is available from the National Library of Australia

ISBN 978 1 4886 1829 1
Pearson Australia Group Pty Ltd ABN 40 004 245 943

Acknowledgements
We would like to thank the following teachers for their reviews:
Jennie Bucco, David Coffey, John Hamilton, Gareth Thompson, Roger Walter.

Attributions
COVER: Marina Sun/Shutterstock
Australian Bureau of Statistics (ABS): Licensed under CC BY 2.5 AU, p. 534.
Fairfax Photo Sales: Rodolfo Pazos, Conrad Walters, The Age, December 29, 2012, p. 288–9.
Nine.com.au: http://livescores.ninemsn.com.au/football/epl/ladder.html, p. 557b.
National Basketball League: www.nbl.com.au, p. 557t.
NSW Education Standards Authority (NESA): Mathematics Advanced Stage 6 Syllabus 2017 © NSW Education Standards Authority for and on behalf of the Crown in right of the State of New South Wales, 2017, pp.609–10.
Shutterstock: Laborant, p. 225; Valeriy Lebedev, p. 234; vilax, p. 279.

INTRODUCTION AND DEDICATION

J.B. Fitzpatrick

It is interesting to wonder whether J. B. Fitzpatrick ('Bernie') realised in 1983 just how popular his book *New Senior Mathematics* would be. That first edition of *New Senior Mathematics* was to remain in print for almost 30 years. It has stood the test of time thanks to the quality, rigour and variety of its questions, its accuracy and its high mathematical standards.

As Fitzpatrick wrote in 1983: 'Mathematics, like many other things, is best learnt by doing. A student begins to appreciate the power of mathematics when he or she has achieved a mastery of basic techniques, not after reading lengthy explanations … The emphasis throughout the book is on the understanding of mathematical concepts'. (Introduction, *New Senior Mathematics* 1984.)

J. B. Fitzpatrick passed away in 2008. Fitzpatrick was a respected author, teacher and figurehead of mathematics education.

Bob Aus

Bob Aus taught in New South Wales high schools for 40 years, retiring in 2007. During that time Bob taught all courses from years 7 to 12 up to Level 1 / 4-unit / Extension 2. He has marked HSC examination papers and has been involved in the Standards Setting Process as Judge and Chief Judge for the three calculus-based courses over four years. He has also completed review work for the NSW Board of Studies and represented NSW at a week-long review and standards setting of the upper level course from each state prior to the development of the Australian National Curriculum for senior students.

Bob spent time as Regional Vocational Education Consultant in the North Coast Region and Mathematics Consultant in the Hunter Region. When he retired he was Head Teacher Mathematics at Merewether High School and enjoyed teaching an Extension 2 class with 24 students.

Bob's first publication was in 1983 and he has been involved with writing a range of textbooks and study guides since then, including revising and updating the *New Senior Mathematics* series 2nd Edition in 2013.

Bob has presented talks on the three calculus-based courses throughout the state. He has co-written the Years 6–9 Mathematics Syllabus for the Abu Dhabi Education Authority, as well as managing the writing project for support material for this course. He also wrote the years 10–12 syllabus for their calculus-based course.

This third edition of *New Senior Mathematics* updates the series for the new Stage 6 courses in New South Wales to be implemented in Year 11, 2019.

NEW SENIOR MATHEMATICS

THIRD EDITION

New Senior Mathematics Advanced for Years 11 & 12 is part of a new edition of the well-known mathematics series for New South Wales. The series has been updated to address all requirements of the new Stage 6 syllabus. We have maintained our focus on mathematical rigour and challenging student questions, while providing new opportunities for students to consolidate their understanding of concepts and ideas with the aid of digital resources and activities.

Student Book

The first three chapters of the student book contain revision material that provides the necessary foundation for the development of senior mathematics concepts. In the new edition you'll also find:

- Content built on a rigorous, academic approach that promotes excellence and prepares students for higher education
- A simple convenient approach with Year 11 and 12 content in one book for Advanced and Extension
- Digital technology activities that promote deeper understanding allowing students to make connections, and visualise and manipulate data in real time.

Student Worked Solutions

The *New Senior Mathematics for Years 11 & 12 Student Worked Solutions* contain the fully worked solutions for every second question in *New Senior Mathematics for Years 11 & 12*.

Reader+

Reader+, our next generation eBook, features content and digital activities, with technology such as graphing software and spreadsheets, to help students engage on their devices.

Teacher Support includes materials such as practice exams, question banks, investigation assignments, and fully worked solutions to cover all internal and external assessment items to save you time.

Pearson Places

FEATURES OF THE THIRD EDITION STUDENT BOOK/READER⁺

YEAR LEVELS

Year levels are indicated on each page for easy identification of Year 11 & 12 content.

MAKING CONNECTIONS

This eBook feature provides teachers and students with a visual interactive of specific mathematics concepts or ideas to aid students in their conceptual understanding.

MAKING CONNECTIONS

EXPLORE FURTHER

This eBook feature provides an opportunity for students to consolidate their understanding of concepts and ideas with the aid of technology, and answer a small number of questions to deepen their understanding and broaden their skill base. These activities should take approximately 5–15 minutes to complete.

EXPLORE FURTHER

CHAPTER REVIEW

Each chapter contains a comprehensive review of chapter content.

CHAPTER REVIEW

SUMMARY PAGES

A comprehensive course summary is provided at the end of the book.

SUMMARY

CONTENTS

Sections 20.4 and 20.5 are available as online resources only.

CHAPTER 1
Algebraic techniques

1.1 SIMPLIFYING ALGEBRAIC EXPRESSIONS

When adding and subtracting algebraic expressions, you can only combine like terms (that is, algebraic parts that have the same pronumerals). Be careful that when subtracting terms with brackets, the subtraction must be applied as a negative to each term inside the brackets (as in part **(c)** of the example below).

Example 1

Simplify each expression by collecting the like terms.

(a) $3x + 2y + 5x - 6y$ **(b)** $x^2 + 2x - x + 3x^2$ **(c)** $2(3a - 4b) - 3(a - 5b)$

Solution

(a) $3x + 2y + 5x - 6y$
$= 3x + 5x + 2y - 6y$
$= 8x - 4y$

(b) $x^2 + 2x - x + 3x^2$
$= x^2 + 3x^2 + 2x - x$
$= 4x^2 + x$

(c) $2(3a - 4b) - 3(a - 5b)$
$= 6a - 8b - 3a + 15b$
$= 6a - 3a - 8b + 15b$
$= 3a + 7b$

EXERCISE 1.1 SIMPLIFYING ALGEBRAIC EXPRESSIONS

Simplify each expression by collecting the like terms.

1 $3x + 5 + 7x + 10$ **2** $7x - 3 + 3x - 2$ **3** $4a + b - a - 4b$

4 $6ab + 3ab + 5a + 4a$ **5** $3xy + 2xy - yx$ **6** $3a^2b - 3ab^2 + 2a^2b$

7 $2x^2y + 3x^2y^2 - x^2y + 3x^2y^2$ **8** $3abc + 5bca - 2cba$ **9** $12mn + 3m - 6mn - m$

10 $x^2 - 3x + 2x + 4x^2$ **11** $2x^2 + 5y^2 - 4x^2$

Simplify each expression by expanding the brackets and then collecting the like terms.

12 $5a - 3(a + b)$ **13** $4(2x - y) - 6x$ **14** $8m - 5(2m - 3n)$

15 $3(2x + 5y) + 4(x - y)$ **16** $5(2x + 3) - 5(x + 7)$ **17** $6(2a + 3b) + 3(a - b)$

18 $5a(a + 2) - 3a(a + 1)$ **19** $5x(x - 2y) + 3x(2x - y)$ **20** $2a + 3b - (a - b)$

21 $x + 5y - (3x + 2y)$ **22** $5x(2x + 1) - (x^2 + x)$ **23** $15(x - 2) + 4(3x - 3)$

24 $3x(x - 2) - 4(x - 1)$ **25** $3(x^2 + 5x - 1) - (2x^2 + x - 2)$ **26** $5x + 2y - 3 - (x - 7y + 9)$

27 The expression $a(a + 1) - 3(2a + 1)$ simplifies to:

A $a^2 - 7a - 3$ **B** $a^2 - 5a + 3$ **C** $a^2 - 5a - 3$ **D** $a^2 - 7a + 3$

28 The expression $3(m^2 - m) - 2(m^2 + 2m + 5)$ simplifies to:

A $5m^2 - 7m - 10$ **B** $m^2 - 7m - 10$ **C** $m^2 + m - 10$ **D** $m^2 - 7m + 10$

1.2 SUBSTITUTION IN FORMULAE

Example 2

If $V = \pi r^2 h$, find:

(a) V when $r = 3.5$, $h = 5$

(b) r when $V = 275$, $h = 14$

Solution

(a) $V = \pi \times 3.5^2 \times 5$

$= 61.25\pi$ (exact value)

$= 192.4$ correct to one decimal place

(b) $r^2 = \dfrac{V}{\pi h}$

$r^2 = \dfrac{275}{14 \times \pi}$

$r = \sqrt{\dfrac{275}{14\pi}}$ (exact value)

$r = 2.5$ correct to one decimal place

EXERCISE 1.2 SUBSTITUTION IN FORMULAE

Use the value of π on your calculator. Give your answer correct to one decimal place when needed.

1. If $P = 2(l + b)$, find the value of P when $l = 20$, $b = 12$.
2. If $E = IR$, find E when $I = 2.4$, $R = 40$.
3. If $F = ma$, find F when $m = 50$, $a = 0.2$.
4. If $F = \dfrac{9C}{5} + 32$, find: (a) F when $C = 60$ (b) C when $F = 41$.
5. If $A = \pi r^2$, find A when $r = 3.5$.
6. If $V = \pi r^2 h$, find V when $r = 4.2$, $h = 10$.
7. If $E = mc^2$, find: (a) E when $m = 10$, $c = 1.6$ (b) c when $E = 13.5$, $m = 1.5$.
8. If $v = u + at$, find v when $u = 20$, $a = 1.8$, $t = 10$.
9. If $s = ut + \dfrac{1}{2}at^2$, find: (a) s when $u = 5$, $a = 6$, $t = 2.4$ (b) a when $s = 50$, $t = 2.5$, $u = 10$.
10. If $v^2 = u^2 + 2as$, find v when $u = 12$, $a = 2$, $s = 20.25$.
11. If $s = \dfrac{1}{2}(u + v)t$, find s when $u = 2.6$, $v = 3.2$, $t = 2.5$.
12. If $S = 2\pi rh$, find S when $r = 2.5$, $h = 3.5$.
13. If $r = \sqrt{\dfrac{A}{\pi}}$, find: (a) r when $A = 154$ (b) A when $r = 1.75$.
14. If $E = \dfrac{m}{2}\left(v^2 - u^2\right)$, find E when $m = 4$, $v = 4$, $u = 2$.
15. If $t = a + (n - 1)d$, find: (a) t when $a = 3.8$, $n = 20$, $d = 0.2$ (b) n when $a = 5.6$, $d = 5$, $t = 25.6$.
16. If $F = \dfrac{m(v - u)}{t}$, find F when $m = 20$, $v = 4$, $u = 2$, $t = 6$.
17. If $t = ar^5$, find t when $a = 64$, $r = 0.5$.
18. If $S = \dfrac{a\left(r^3 - 1\right)}{r - 1}$, find S when $a = 5$, $r = 3$.

19 If $A = \pi(R^2 - r^2)$, find A when $R = 5.6$, $r = 1.4$.

20 If $V = \pi(R^2 - r^2)h$, find V when $R = 0.9$, $r = 0.2$, $h = 1.5$.

21 If $V = \frac{1}{3}\pi r^2 h$, find V when $r = 3$, $h = 3.5$.

22 If $P = \sqrt{\frac{2R - V}{5}}$, find: **(a)** P when $R = 50$, $V = 20$ **(b)** V when $P = 0.2$, $R = 20$.

23 If $W = \frac{1}{3}d\pi r^2 h$, find W when $d = 3$, $r = \frac{7}{11}$, $h = \frac{11}{14}$.

24 If $f = \frac{vu}{v+u}$, find: **(a)** f when $v = 20$, $u = 25$ **(b)** v when $f = 20$, $u = 25$.

25 If $A = P\left(1 + \frac{r}{100}\right)^n$, find A when $P = 1000$, $r = 10$, $n = 2$.

1.3 BASIC POLYNOMIALS

Common terms

A **monomial** is an expression that contains only **one** term, e.g. $5x$, x^2, $2ab$, $5a^2b^3$.

A **binomial** is an expression that contains **two** terms added or subtracted, e.g. $x + y$, $3a - 2b$, $x^2 + 1$, $3y - 4$.

A **trinomial** is an expression that contains **three** terms added or subtracted, e.g. $x^2 - 5x + 6$, $x + y - 4$, $4x^2 - 2xy + y^2$, $m + n - p$.

A **quadratic trinomial** is a trinomial of the form $ax^2 + bx + c$ (where $a \neq 0$, $b \neq 0$, $c \neq 0$); a is the coefficient of x^2, b is the coefficient of x, and c is the constant term.

Standard results

$(x + m)(x + n) = x^2 + (m + n)x + mn$

$(a + b)^2 = a^2 + 2ab + b^2$

$(a - b)^2 = a^2 - 2ab + b^2$

$(a - b)(a + b) = a^2 - b^2$

In each of these results, the expression on the left-hand side has been **expanded** to obtain the expression on the right.

If we start with the expression on the right-hand side, then we can **factorise** it to obtain the (usually) shorter form on the left.

Example 3

Expand and simplify each expression.

(a) $(x + 2)(x + 3)$ **(b)** $(3x - 2)(2x + 3)$ **(c)** $(2y + 5)^2$

(d) $(3x - 4)(3x + 4)$ **(e)** $(x + 2)(x^2 - 5x + 6)$ **(f)** $(x - 1)(x + 2)(x + 3)$

Solution

(a) $(x + 2)(x + 3)$
$= x(x + 3) + 2(x + 3)$
$= x^2 + 3x + 2x + 6$
$= x^2 + 5x + 6$

(b) $(3x - 2)(2x + 3)$
$= 6x^2 + 9x - 4x - 6$
$= 6x^2 + 5x - 6$

(c) $(2y + 5)^2$
$= 4y^2 + 20y + 25$

(d) $(3x - 4)(3x + 4)$
$= 9x^2 - 16$

(e) $(x + 2)(x^2 - 5x + 6)$
$= x(x^2 - 5x + 6) + 2(x^2 - 5x + 6)$
$= x^3 - 5x^2 + 6x + 2x^2 - 10x + 12$
$= x^3 - 3x^2 - 4x + 12$

(f) $(x - 1)(x + 2)(x + 3)$
$= (x - 1)(x^2 + 5x + 6)$
$= x^3 + 5x^2 + 6x - x^2 - 5x - 6$
$= x^3 + 4x^2 + x - 6$

EXERCISE 1.3 BASIC POLYNOMIALS

Write the expansion of the following.

1 $(x+5)(x+1)$
2 $(x-2)(x-3)$
3 $(a-3)(a+4)$
4 $(x-2)^2$
5 $(y+7)^2$
6 $(2x+3)(x+5)$
7 $(3x-4)(x-2)$
8 $(3m+7)(2m-1)$
9 $(3x+2)(3x+2)$
10 $(2p-9)(2p+9)$
11 $(3x+2)(2x+3)$
12 $(4p-5)^2$
13 $(3x+4)^2$
14 $(x-3)(2x^2+3x+1)$
15 $(3x^2-5x+2)(2x-4)$
16 $x(x-2)(x+2)$
17 $(x-1)(x-1)(x-2)$
18 $2(x-1)(x-2)(x-3)$
19 $(x^2+5)(x^2-2x-3)$
20 $(x-2)(x+2)(x+2)$
21 $(x^2-y^2)^3$

22 The correct expansion of $\left(\sqrt{x}+\sqrt{y}\right)^2$ is:
A $x^2+2xy+y^2$ B $x+y+2xy$ C $x+y+2\sqrt{xy}$ D $x^2+y^2+2\sqrt{xy}$

23 Indicate whether each answer is a correct or incorrect factorisation of $x^2-4xy+4y^2$.
(a) $(x+2y)^2$ (b) $(2y-x)^2$ (c) $(2x-y)^2$ (d) $(x-2y)^2$

1.4 FACTORISING BY GROUPING IN PAIRS

This method is used when there are four terms in the expression.

Example 4

Factorise:

(a) $bx+by+cx+cy$ (b) $m^2-mn-2m+2n$

Solution

	(a)	(b)
Group in pairs:	$\underbrace{bx+by}+\underbrace{cx+cy}$	$\underbrace{m^2-mn}-\underbrace{2m+2n}$
Take out common factor:	$=b(x+y)+c(x+y)$	$=m(m-n)-2(m-n)$
Take out common factor:	$=(x+y)(b+c)$	$=(m-2)(m-n)$

EXERCISE 1.4 FACTORISING BY GROUPING IN PAIRS

Factorise:

1 $a(x+2)+b(x+2)$
2 $3a(2b-3c)-m(2b-3c)$
3 $p(a+b)+q(a+b)-r(a+b)$
4 $x^2(2x-1)+4(2x-1)$
5 $ax+4a+bx+4b$
6 $x^2-xy+xz-yz$
7 $2xy+2xz+y+z$
8 $a^2-ab-ac+bc$
9 $10y-25y^2+4x-10xy$
10 $a^3+3a^2b+ab^2+3b^3$
11 $ac-2bc-2ad+4bd$
12 $3xy-6y+7x-14$
13 $x^2-2xy-xz+2yz$
14 $a^3-a^2b-ab+b^2$
15 $2mn+2mp+pn^2+p^2n$
16 $x^3+3x^2+4x+12$
17 $p^2q-pq^2+5p-5q$
18 m^2p+m^2+np+n

19 $x^2y + x^2 + y + 1$ **20** $ab - 3a - 4b + 12$ **21** $2x - 6y - xy + 3y^2$

22 When $3m^2 - 3mn - m + n$ is factorised, the answer is:

A $(3m-1)(m-n)$ **B** $(3m-n)(m-1)$ **C** $(3m-1)(m+n)$ **D** $(3m+1)(m-n)$

23 Indicate whether each answer is a correct or incorrect factorisation of $2x^3 - 2x^2 - 2x + 2$.

(a) $2(x+1)(x+1)(x-1)$ **(b)** $2(x+1)(x-1)^2$ **(c)** $2(x+1)(x-1)(x-1)$ **(d)** $2(x-1)(x+1)^2$

1.5 STANDARD FACTORISATIONS

Factorising using the difference of two squares

Remember the difference of two squares: $a^2 - b^2 = (a-b)(a+b)$

Example 5

Factorise:

(a) $a^2 - 25$ **(b)** $9x^2 - 49$ **(c)** $(x+1)^2 - (y-1)^2$ **(d)** $a^3 - a^2b - ab^2 + b^3$

Solution

(a) $a^2 - 25 = a^2 - 5^2$
$= (a-5)(a+5)$

(b) $9x^2 - 49 = (3x)^2 - 7^2$
$= (3x-7)(3x+7)$

(c) $(x+1)^2 - (y-1)^2$
$= [(x+1) - (y-1)][(x+1) + (y-1)]$
$= (x - y + 2)(x + y)$

(d) $a^3 - a^2b - ab^2 + b^3$
$= a^2(a-b) - b^2(a-b)$
$= (a^2 - b^2)(a-b)$
$= (a-b)(a+b)(a-b)$
$= (a-b)^2(a+b)$

Sum and difference of two cubes

Two important identities are: $a^3 + b^3 = (a+b)(a^2 - ab + b^2)$
$a^3 - b^3 = (a-b)(a^2 + ab + b^2)$

The identities can be verified by expanding the right-hand side.

Example 6

Factorise:

(a) $x^3 - 8$ **(b)** $27y^3 + 64x^3$ **(c)** $(x+2)^3 + y^3$ **(d)** $x^2y^3 - z^2y^3 - x^2w^3 + z^2w^3$

Solution

(a) $x^3 - 8 = x^3 - 2^3$
$= (x-2)(x^2 + 2x + 4)$

(b) $27y^3 + 64x^3 = (3y)^3 + (4x)^3$
$= (3y + 4x)(9y^2 - 12xy + 16x^2)$

(c) $(x+2)^3 + y^3$
$= (x + 2 + y)[(x+2)^2 - (x+2)y + y^2]$
$= (x + 2 + y)(x^2 + 4x + 4 - xy - 2y + y^2)$

(d) $x^2y^3 - z^2y^3 - x^2w^3 + z^2w^3$
$= y^3(x^2 - z^2) - w^3(x^2 - z^2)$
$= (x^2 - z^2)(y^3 - w^3)$
$= (x-z)(x+z)(y-w)(y^2 + yw + w^2)$

EXERCISE 1.5 STANDARD FACTORISATIONS

Factorise:

1 m^2-1 2 x^2-16 3 $64-m^2$ 4 $9a^2-25$

5 $x^2-0.36$ 6 $a^2b^2-c^2$ 7 $9x^2-4y^2$ 8 $(x+1)^2-9$

9 $x^2-y^2z^2$ 10 $\frac{a^2}{25}-1$ 11 $p^2-\frac{1}{4}$ 12 $\frac{x^2}{4}-\frac{1}{9}$

13 $(a+2)^2-4$ 14 $x^2-(y+z)^2$ 15 99^2-1 16 523^2-477^2

17 a^3b-ab^3 18 $12a^3-3ab^2$ 19 $3x^2y-27y$ 20 $(x+y)^2-4$

21 $a^2-(a-b)^2$ 22 $x^3-x^2y-9x+9y$ 23 $x^3+3x^2-4x-12$ 24 $p^2q-p^2-16q+16$

25 a^2x-x 26 $48a^2-75b^2$ 27 $(1+h)^2-1$ 28 $\frac{x^2}{25}-y^2$

29 When $(p+2)^2-(p-2)^2$ is factorised, the answer is:

A $2p^2+8$ B $-8p$ C $2p^2-8$ D $8p$

30 Indicate whether each answer is a correct or incorrect factorisation of $\frac{a^2}{b^2}-\frac{b^2}{a^2}$.

(a) $\left(\frac{a}{b}-\frac{b}{a}\right)\left(\frac{a}{b}+\frac{b}{a}\right)$ (b) $\left(\frac{a}{b}-\frac{b}{a}\right)\left(\frac{b}{a}+\frac{a}{b}\right)$ (c) $\left(\frac{a}{b}-\frac{b}{a}\right)\left(\frac{a}{b}-\frac{b}{a}\right)$ (d) $\frac{(a-b)(a+b)(a^2+b^2)}{a^2b^2}$

31 y^3-125 32 z^3+1 33 $8p^3+27$ 34 $216-a^3$

35 $(x+5)^3+(x-2)^3$ 36 $(2x+3)^3-(x-4)^3$ 37 b^6-a^6 38 $64a^3+8b^3$

39 $\frac{4}{3}\pi R^3-\frac{4}{3}\pi r^3$ 40 $p^7x^4-p^4x^7$ 41 x^6+y^6 42 $\frac{8}{a^3}-\frac{27}{b^3}$

43 $a^3m^3+a^3n^3-b^3n^3-b^3m^3$ 44 $4x^5-9x^3-4x^2+9$ 45 $(x+h)^3-x^3$

46 $a^3+(a-b)^3$ 47 $(a+b)^3-(a-b)^3$ 48 $(2x+1)^3-(2x-1)^3$

49 $8-(2-x)^3$ 50 $a^5b^4-a^2b$ 51 $2(x-y)^3+54$

52 When $1000p^3-q^6$ is factorised, the answer is:

A $(10p-q)(100p^2+10pq+q^2)$ B $(10p-q^2)(100p^2+10pq^2+q^4)$

C $(10p+q)(100p^2-10pq+q^2)$ D $(10p+q^2)(100p^2-10pq^2+q^4)$

53 When $(2x+1)^3+(2x-1)^3$ is factorised, the answer is:

A $2(12x^2+1)$ B $4x(12x^2+1)$ C $2(4x^2+3)$ D $4x(4x^2+3)$

1.6 FACTORISING QUADRATIC TRINOMIALS

To factorise quadratic trinomials, you must remember how to expand binomial products and then work backwards. We know the following:

- $(x+m)(x+n)=x^2+(m+n)x+mn=x^2+$ (sum of m and n)$x+$ (the product of m and n)
- $(x-m)(x-n)=x^2-(m+n)x+mn=x^2+$ (sum of $-m$ and $-n$)$x+$ (the product of $-m$ and $-n$)
- $(x+m)(x-n)=x^2+(m-n)x-mn=x^2+$ (sum of m and $-n$)$x+$ (the product of m and $-n$)

To factorise x^2+5x+6 you must write it in the form $(x+m)(x+n)$, where $m+n=5$ and $mn=6$. This means you must find two numbers whose sum is 5 and whose product is 6.

Example 7

Factorise:

(a) x^2+5x+6 (b) $x^2-7x+10$ (c) x^2+x-12 (d) x^2-6x+9 (e) $x^2-5x-24$

Solution

(a) x^2+5x+6

Write: $x^2+5x+6=(x+m)(x+n)$

Look for numbers m and n whose sum is 5 and whose product is 6.

List possible factors of 6 and check the sum: $6\times 1=6$ $\quad 6+1=7$

$3\times 2=6$ $\quad 3+2=5$

Hence: $x^2+5x+6=(x+3)(x+2)$

The information could also be set out using the cross method:

$$\begin{matrix} x & \diagdown\!\!\!\diagup & \not{6} & 3 \\ x & & \not{1} & 2 \end{matrix}$$

The correct pair will give $5x$ when multiplied across.

(b) $x^2-7x+10$

Write: $x^2-7x+10=(x+m)(x+n)$

Look for numbers m and n whose sum is -7 and whose product is 10.

Since the sum is negative and the product is positive, both the numbers are negative.

List possible factors of 10 and check the sum: $-10\times(-1)=10$ $\quad -10+(-1)=-11$

$-5\times(-2)=10$ $\quad -5+(-2)=-7$

Hence: $x^2-7x+10=(x-5)(x-2)$

(c) x^2+x-12

Write: $x^2+x-12=(x+m)(x+n)$

Look for numbers m and n whose sum is 1 and whose product is -12.

Since the sum is positive and the product is negative, the numbers have different signs and the larger number is positive.

List possible factors of -12 and check the sum: $12\times(-1)=-12$ $\quad 12+(-1)=11$

$6\times(-2)=-12$ $\quad 6+(-2)=4$

$4\times(-3)=-12$ $\quad 4+(-3)=1$

Hence: $x^2+x-12=(x+4)(x-3)$

Using the cross method:

$$\begin{matrix} x & \diagdown\!\!\!\diagup & \not{12} & \not{6} & 4 \\ x & & \not{-1} & \not{-2} & -3 \end{matrix}$$

The correct pair will give x when multiplied across.

(d) x^2-6x+9

Write: $x^2-6x+9=(x+m)(x+n)$

Look for numbers m and n whose sum is -6 and whose product is 9.

Since the sum is negative and the product is positive, both the numbers are negative.

List possible factors of 9 and check the sum: $-9\times(-1)=9$ $\quad -9+(-1)=-10$

$-3\times(-3)=9$ $\quad -3+(-3)=-6$

Hence: $x^2-6x+9=(x-3)(x-3)=(x-3)^2$

(e) $x^2 - 5x - 24$

Write: $x^2 - 5x - 24 = (x + m)(x + n)$

Look for numbers m and n whose sum is -5 and whose product is -24.

Since the sum is negative and the product is negative, the numbers have different signs and the smaller number is positive.

List possible factors of -24 and check the sum:

$-24 \times 1 = -24$	$-24 + 1 = -23$
$-12 \times 2 = -24$	$-12 + 2 = -10$
$-6 \times 4 = -24$	$-6 + 4 = -2$
$-8 \times 3 = -24$	$-8 + 3 = -5$

Hence: $x^2 - 5x - 24 = (x - 8)(x + 3)$

x	╳	~~−24~~	~~−12~~	~~−6~~	-8
x		~~1~~	~~2~~	~~4~~	3

The correct pair will give $-5x$ when multiplied across.

With practice, you will be able to write the factors simply by looking at the sum and product.

MAKING CONNECTIONS

Factorising quadratic trinomials

Use technology to check the factorisation of quadratic trinomials.

EXERCISE 1.6 FACTORISING QUADRATIC TRINOMIALS

Factorise the following quadratic trinomials:

1 $x^2 + 4x + 3$ **2** $x^2 + 10x + 21$ **3** $x^2 + 11x + 24$ **4** $a^2 + 12a + 32$

5 $m^2 + 9m + 20$ **6** $x^2 + 13x + 12$ **7** $x^2 + 8x + 12$ **8** $x^2 - 7x + 12$

9 $x^2 - 13x + 12$ **10** $x^2 - 8x + 12$ **11** $p^2 + 2p - 15$ **12** $p^2 + 14p - 15$

13 $p^2 - 2p - 15$ **14** $p^2 - 14p - 15$ **15** $x^2 - 2x - 35$ **16** $x^2 - 3x - 10$

17 $x^2 + 17x + 72$ **18** $a^2 - 4a - 12$ **19** $x^2 - 7x + 6$ **20** $x^2 - x - 72$

21 $x^2 + 6x - 72$ **22** $x^2 - 21x - 72$ **23** $a^2 + 13a + 30$ **24** $x^2 - x - 42$

25 $x^2 - 19x - 42$ **26** $x^2 + 19x - 42$

27 The factors of $x^2 - 11x - 42$ are:

A $(x + 14)(x - 3)$ **B** $(x - 7)(x + 6)$ **C** $(x - 6)(x + 7)$ **D** $(x - 14)(x + 3)$

28 Indicate whether each answer is a correct or incorrect factorisation of $x^2 - 8x + 7$.

(a) $(x + 1)(x - 7)$ **(b)** $(1 - x)(7 - x)$ **(c)** $(x - 1)(x + 7)$ **(d)** $(x - 1)(x - 7)$

1.7 FACTORISING NON-MONIC TRINOMIALS

When the x^2 term in the quadratic has a coefficient other than 1, finding the factors becomes more difficult because there are more possibilities. You must find factors of the coefficient of x^2 as well as the factors of the constant term and get the correct pairs together. You could use trial and error, but the cross method is easier because it keeps the information more organised.

Example 8

Factorise $6x^2 + 19x + 10$.

Solution

Write: $6x^2 + 19x + 10 = (ax + m)(bx + n) = abx^2 + (an + bm)x + mn$

This gives: $ab = 6$, $an + bm = 19$, $mn = 10$

List the factors of 6: 6, 1 or 3, 2

List the factors of 10: 10, 1 or 5, 2

List possible binomial factors:

$(6x + 10)(x + 1)$ $(6x + 1)(x + 10)$ $(6x + 5)(x + 2)$ $(6x + 2)(x + 5)$

$(3x + 10)(2x + 1)$ $(3x + 1)(2x + 10)$ $(3x + 5)(2x + 2)$ $(3x + 2)(2x + 5)$

You can expand the binomial factors to see which answer gives the original quadratic trinomial.

Before doing this, you can eliminate any possibility that has a common factor, because there is no common factor in the original quadratic trinomial. This means you can eliminate the answers containing $(6x + 10)$, $(6x + 2)$, $(2x + 10)$ and $(2x + 2)$, because they all have a common factor of 2 (and the original quadratic trinomial does not).

Expand the others:

$(6x + 1)(x + 10) = 6x^2 + 61x + 10$ Not correct

$(6x + 5)(x + 2) = 6x^2 + 17x + 10$ Not correct

$(3x + 10)(2x + 1) = 6x^2 + 23x + 10$ Not correct

$(3x + 2)(2x + 5) = 6x^2 + 19x + 10$ Correct

Hence: $6x^2 + 19x + 10 = (3x + 2)(2x + 5)$

Alternatively, using the cross method:

$6x$	X	10	1	5	2	or	$3x$	X	10	1	5	2
x		1	10	2	5		$2x$		1	10	2	5

The correct pair will give $19x$ when multiplied across.

Hence: $6x^2 + 19x + 10 = (3x + 2)(2x + 5)$

Example 9

Factorise $4x^2 - 4x - 15$.

Solution

The factors of $4x^2$ are either $4x$ and x, or $2x$ and $2x$.

The factors of -15 are -15 and 1, 15 and -1, -5 and 3, or 5 and -3.

Set up the cross method:

$4x$	X	−15	15	−5	5	or	$4x$	X	1	−1	3	−3
x		1	−1	3	−3		x		−15	15	−5	5

or

$2x$	X	−15	15	−5	5	or	$2x$	X	1	−1	3	−3
$2x$		1	−1	3	−3		$2x$		−15	15	−5	5

The only combination that gives $-4x$ is:

$2x$	X	−5
$2x$		3

Hence: $4x^2 - 4x - 15 = (2x - 5)(2x + 3)$

Example 10

Factorise $3x^2 + 8x - 16$.

Solution

The factors of $3x^2$ are $3x$ and x.

The factors of -16 are -16 and 1; 16 and -1; -8 and 2; 8 and -2; -4 and 4; or 4 and -4.

Set up the cross method:

$3x$	X	-16	1	16	-1	-8	2	-2	8	-4	4
x		1	-16	-1	16	2	-8	8	-2	4	-4

The only combination that gives $8x$ is:

$3x$	X	-4
x		4

Hence: $3x^2 + 8x - 16 = (3x - 4)(x + 4)$

EXERCISE 1.7 FACTORISING NON-MONIC TRINOMIALS

Factorise:

1 $2x^2 + 3x + 1$ **2** $3x^2 + 11x - 4$ **3** $2x^2 + 7x + 6$ **4** $4a^2 + 13a + 3$

5 $3a^2 - 5a + 2$ **6** $8x^2 - 14x + 3$ **7** $13c^2 - 7c - 6$ **8** $8x^2 + 14x + 5$

9 $3x^2 - 17x + 10$ **10** $6a^2 - 13a - 63$ **11** $3x^2 - 11x - 4$ **12** $10x^2 - 11x - 8$

13 $2x^2 + 3x - 2$ **14** $4x^2 - 12x + 9$ **15** $9x^2 - 12x + 4$ **16** $2x^2 - 9x + 10$

17 $6x^2 - 85x + 14$ **18** $y^2 - 2y - 3$ **19** $12y^2 + 14y - 6$ **20** $6x^2 - 25x + 14$

21 $6x^2 - 29x + 28$ **22** $6x^2 - 19x + 14$ **23** $6x^2 - 20x + 14$ **24** $8x^2 + 2x - 3$

25 $6p^2 + 25p + 21$ **26** $10a^2 - 11a - 6$ **27** $12y^2 + 28y - 5$ **28** $24x^2 - 59x + 36$

29 $15x^2 - 19x + 6$ **30** $3x^2 - 2x - 1$ **31** $9x^2 + 9x - 10$ **32** $2x^2 - 9x + 4$

33 The factors of $8x^2 - 6x - 9$ are:

A $(4x - 3)(2x + 3)$ **B** $(8x - 9)(x + 1)$ **C** $(4x + 3)(2x - 3)$ **D** $(4x - 9)(2x + 1)$

34 Indicate whether each answer is a correct or incorrect factorisation of $4x^2 + 12x + 9$.

(a) $(2x + 3)(2x + 3)$ **(b)** $(3 + 2x)(3 + 2x)$ **(c)** $(2x + 3)^2$ **(d)** $(2x - 3)^2$

1.8 MIXED FACTORISATIONS

To factorise the expressions in the next exercises, you will need to use **one or more** of the techniques learned so far:

- remove a common factor
- group in pairs and remove common factors
- difference of two squares
- quadratic trinomials.

EXERCISE 1.8 MIXED FACTORISATIONS

Factorise completely:

1 $x^2 - 3x$ **2** $2a^3 - 8a$ **3** $x^2 - 9$ **4** $x^2 - 8x - 9$

5 $3x^2y - 12y^3$ **6** $5x^3y - 20xy^3$ **7** $1 - (b + c)^2$ **8** $10x^2 + 9x - 1$

9 $(a+b)^2-b^2$ **10** $6x^2-24$ **11** a^2-a-42 **12** $a(m+n)-b(m+n)$

13 $2x^3+14x^2-16x$ **14** $3a^3+24a^2+21a$ **15** $(x+2y)^2-4$ **16** $ab^2+abc+abd$

17 x^2-36y^2 **18** $x(y-z)+y(y-z)$ **19** $4x^2-28x-480$ **20** $bx^2-14bxy+49by^2$

21 $6y^3+3y^2-3y$ **22** $6y^3+26y^2+8y$ **23** $15a^2-60$ **24** $9mn-25m^3n^3$

25 $5a^2x-125x$ **26** $(x+y)^2-(x-y)^2$ **27** $5t^3+5t^2-360t$ **28** $m^2-mn+6m-6n$

29 $x^2(x+3)-4(x+3)$ **30** $mx^2-xy+ly-mlx$

31 The factors of $4-(x+1)^2$ are:

A $(3-x)(5+x)$ **B** $(2-x)(2+x)$ **C** $(1+x)(3-x)$ **D** $(1-x)(3+x)$

For questions **32** to **39**, write an algebraic expression in factorised form for the shaded area in each of the figures.

32

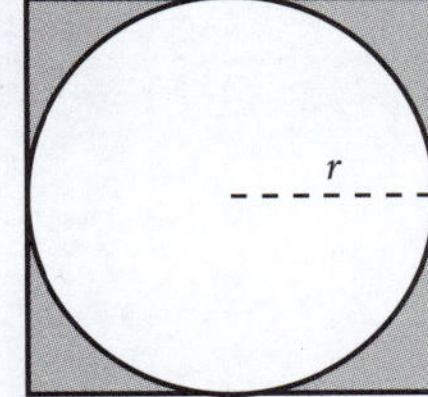

33

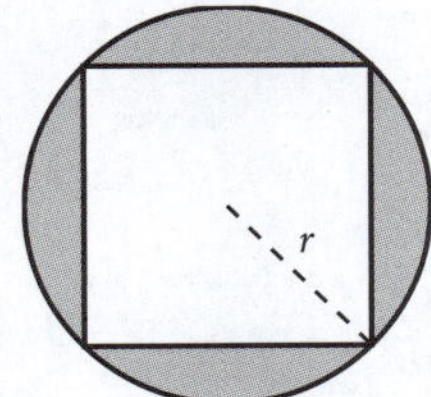

34

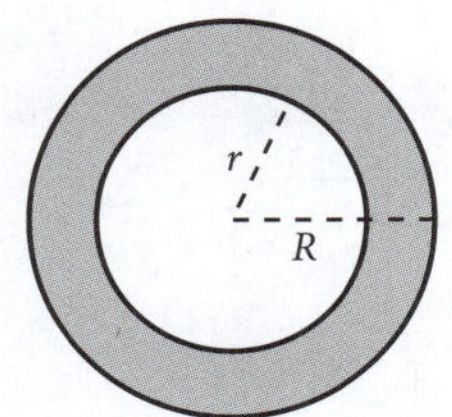

35

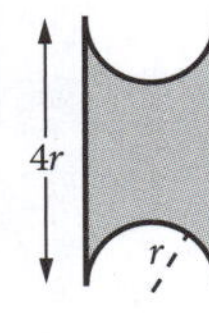

36

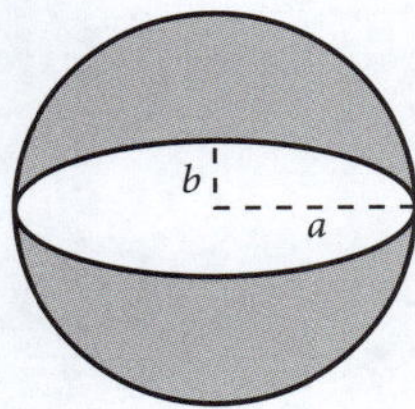

37

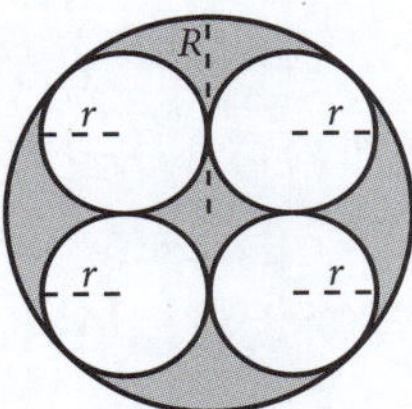

38

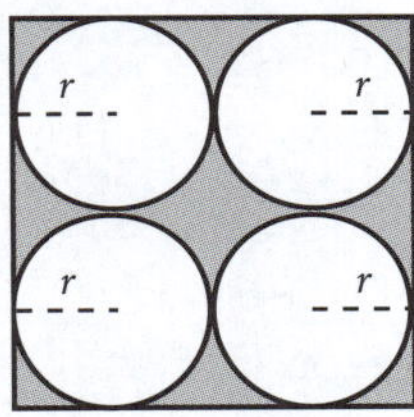

39

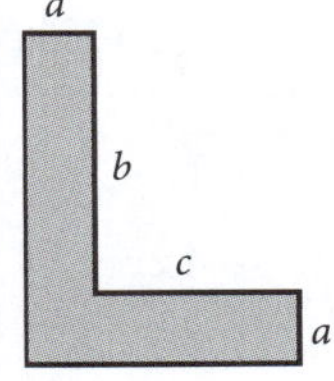

40 Indicate whether each answer is a correct or incorrect expression for the shaded area in question **39**.

(a) $(a+b)(a+c)-bc$ **(b)** $a^2+ab+ac$ **(c)** $(a+b)(a+c)$ **(d)** $a(a+b+c)$

1.9 ALGEBRAIC FRACTIONS

To **simplify** algebraic fractions:

1 Factorise the numerator and denominator.

2 Cancel any common factors.

Example 11

Simplify:

(a) $\dfrac{9x+6}{3x+2}$ **(b)** $\dfrac{15a^2-5ab}{10ab}$ **(c)** $\dfrac{9x^2-y^2}{6xy-2y^2}$

Solution

(a) $\dfrac{9x+6}{3x+2}=\dfrac{3(\cancel{3x+2})}{\cancel{3x+2}}$

$=3$

(b) $\dfrac{15a^2-5ab}{10ab}=\dfrac{5a(3a-b)}{10ab}$

$=\dfrac{3a-b}{2b}$

(c) $\dfrac{9x^2-y^2}{6xy-2y^2}=\dfrac{(\cancel{3x-y})(3x+y)}{2y(\cancel{3x-y})}$

$=\dfrac{3x+y}{2y}$

Example 12

Simplify: $\dfrac{x^2-5x+6}{x^2-9}\times\dfrac{x^2+3x}{x^2-x-2}$

Solution

$$\frac{x^2-5x+6}{x^2-9}\times\frac{x^2+3x}{x^2-x-2}=\frac{\cancel{(x-2)}\,\cancel{(x-3)}}{\cancel{(x-3)}\,\cancel{(x+3)}}\times\frac{x\,\cancel{(x+3)}}{(x+1)\cancel{(x-2)}}$$

$$=\frac{x}{x+1}$$

EXERCISE 1.9 ALGEBRAIC FRACTIONS

Simplify:

1 $\dfrac{8a-4b}{4}$ **2** $\dfrac{15x+10y}{15}$ **3** $\dfrac{14x-7y}{2x-y}$ **4** $\dfrac{8x^2-4xy}{8xy}$

5 $\dfrac{12ab-6b^2}{9ab}$ **6** $\dfrac{8x+2}{4x+1}$ **7** $\dfrac{3a-5b}{3a^2-5ab}$ **8** $\dfrac{m+m^2}{m}$

9 $\dfrac{mn-n^2}{n}$ **10** $\dfrac{p^2q-pq^2}{pq}$ **11** $\dfrac{x^2+xy}{2x}$ **12** $\dfrac{2rs-12r}{r^2+rs}$

13 $\dfrac{x^2-y^2}{(x+y)^2}$ **14** $\dfrac{k^2+k}{k+1}$ **15** $\dfrac{x^2-9}{x^2+3x}$ **16** $\dfrac{15a^2-5ab}{3ab-b^2}$

17 $\dfrac{4x^2-4xy}{x^2-y^2}$ **18** $\dfrac{x^2-7x+6}{x^2-36}$ **19** $\dfrac{a^2+ab}{ab+b^2}$ **20** $\dfrac{a^2-b^2}{a^2+ab}$

21 $\dfrac{x^2-1}{x^2-5x+4}$ **22** $\dfrac{x^2-6x+8}{x^2-x-2}$ **23** $\dfrac{x^2+3x+2}{x^2-4}$ **24** $\dfrac{x^2-5x+6}{x^2+x-12}$

25 $\dfrac{x^2+4x+4}{x^2-3x-10}$ **26** $\dfrac{4x^3y-16xy}{x^2+2x-8}$ **27** $\dfrac{12a+9}{15}\times\dfrac{5}{4a+3}$ **28** $\dfrac{3x^2-xy}{xy}\times\dfrac{x^2y}{3xy-y^2}$

29 $\dfrac{m^2+m-2}{m^2-m}$ simplifies to:

A $\dfrac{(m-2)(m+1)}{m(m-1)}$ **B** 2 **C** $\dfrac{m+2}{m}$ **D** $\dfrac{m-2}{m}$

30 $\dfrac{2a^2-3ab}{ab-b^2}\times\dfrac{2a^2-2ab}{4a-6b}$ **31** $\dfrac{15x^2-5xy}{10xy}\div\dfrac{3x-y}{2y}$ **32** $\dfrac{12x^2-4x}{3x^2-x}\div\dfrac{10x^2y}{5x^2y^2}$

33 $\dfrac{x^2-2x-3}{x^2-4x-5}\times\dfrac{x^2-25}{(x-3)(x+5)}$ **34** $\dfrac{(a+2b)(a-b)}{a^2-4b^2}\times\dfrac{a^2-3ab+2b^2}{ab-b^2}$

35 Simplify $\dfrac{m^2-9}{m^2-m-12}\div\dfrac{m^2-3m}{m^2-9m+20}$. Indicate whether each answer is correct or incorrect.

(a) $\dfrac{m(m-3)^2}{(m-4)(m-5)}$ **(b)** $\dfrac{3}{4}$ **(c)** $\dfrac{m+5}{m}$ **(d)** $\dfrac{m-5}{m}$

Simplify:

36 $\dfrac{x^3+y^3}{x^2-y^2}$ **37** $\dfrac{8x^2+4x+2}{8x^3-1}$ **38** $\dfrac{2x+2y}{x^3-y^3}\times\dfrac{x^2-2xy+y^2}{x^2-y^2}$

39 $\dfrac{(x+h)^3-x^3}{h}$ **40** $\dfrac{x^3-(x-y)^3}{x^2-(x-y)^2}$

1.10 ADDING AND SUBTRACTING ALGEBRAIC FRACTIONS

When adding and subtracting fractions, first rewrite each fraction with the same (common) denominator, then add or subtract the numerators.

Example 13

Express as a single fraction:

(a) $\frac{3}{5}-\frac{2}{7}$ (b) $\frac{3x}{5}-\frac{2x+1}{7}$ (c) $\frac{3}{5x}+\frac{2}{7x}$ (d) $\frac{2x-y}{3}-\frac{x-y}{6}$

Solution

(a) $\frac{3}{5}-\frac{2}{7}$

$=\frac{21}{35}-\frac{10}{35}$

$=\frac{11}{35}$

(b) $\frac{3x}{5}-\frac{2x+1}{7}$

$=\frac{21x}{35}-\frac{5(2x+1)}{35}$

$=\frac{21x-10x-5}{35}$

$=\frac{11x-5}{35}$

(c) $\frac{3}{5x}+\frac{2}{7x}$

$=\frac{21}{35x}+\frac{10}{35x}$

$=\frac{31}{35x}$

(d) $\frac{2x-y}{3}-\frac{x-y}{6}$

$=\frac{2(2x-y)}{6}-\frac{(x-y)}{6}$

$=\frac{4x-2y-x+y}{6}$

$=\frac{3x-y}{6}$

Harder algebraic fractions

More-complex algebraic fractions require you to factorise the denominator before you find the common denominator. Write each fraction with the common denominator before you add or subtract the numerators.

Example 14

Express as a single fraction:

(a) $\frac{3}{x^2-4}+\frac{1}{x-2}$ (b) $\frac{1}{x-y}-\frac{1}{x+y}$ (c) $\frac{3}{x^2+2x}-\frac{2}{x^2-4}$ (d) $\frac{1}{x^2-5x+6}-\frac{1}{x^2+2x-8}$

Solution

(a) $\frac{3}{x^2-4}+\frac{1}{x-2}$

$=\frac{3}{(x-2)(x+2)}+\frac{1}{x-2}$

$=\frac{3}{(x-2)(x+2)}+\frac{x+2}{(x-2)(x+2)}$

$=\frac{x+5}{(x-2)(x+2)}$

(b) $\frac{1}{x-y}-\frac{1}{x+y}$

$=\frac{x+y}{(x-y)(x+y)}-\frac{x-y}{(x-y)(x+y)}$

$=\frac{x+y-(x-y)}{(x-y)(x+y)}$

$=\frac{2y}{(x-y)(x+y)}$

(c) $\frac{3}{x^2+2x}-\frac{2}{x^2-4}$

$=\frac{3}{x(x+2)}-\frac{2}{(x-2)(x+2)}$

$=\frac{3(x-2)}{x(x+2)(x-2)}-\frac{2x}{x(x-2)(x+2)}$

$=\frac{3x-6-2x}{x(x+2)(x-2)}$

$=\frac{x-6}{x(x+2)(x-2)}$

(d) $\frac{1}{x^2-5x+6}-\frac{1}{x^2+2x-8}$

$=\frac{1}{(x-2)(x-3)}-\frac{1}{(x-2)(x+4)}$

$=\frac{x+4}{(x-2)(x-3)(x+4)}-\frac{x-3}{(x-2)(x-3)(x+4)}$

$=\frac{x+4-(x-3)}{(x-2)(x-3)(x+4)}$

$=\frac{x+4-x+3}{(x-2)(x-3)(x+4)}$

$=\frac{7}{(x-2)(x-3)(x+4)}$

EXERCISE 1.10 ADDING AND SUBTRACTING ALGEBRAIC FRACTIONS

Simplify:

1 $\frac{x}{5}-\frac{x}{6}$ **2** $\frac{3x}{8}+\frac{x}{2}$ **3** $\frac{a}{3}+\frac{4a}{5}-\frac{a}{6}$ **4** $\frac{y}{2}+\frac{2y}{3}-\frac{y}{4}$

5 $\frac{a+2}{5}-\frac{a-1}{3}$ **6** $\frac{2x-y}{3}-\frac{x-3y}{6}$ **7** $\frac{3x+2}{6}-\frac{x+1}{4}$ **8** $\frac{3m-2n}{5}+\frac{m+n}{10}$

9 $\frac{x}{2}+\frac{y}{4}-\frac{x+y}{3}$ **10** $\frac{a-2b}{6}-\frac{2a+b}{9}$ **11** $\frac{3(a+b)}{4}-\frac{a-b}{6}$ **12** $\frac{1}{x}-\frac{2}{3x}$

13 $\frac{3}{a}+\frac{1}{a^2}$ **14** $\frac{1}{ab}-\frac{2}{b}$ **15** $\frac{m}{n}-\frac{n}{m}$ **16** $\frac{4}{xy}+\frac{3}{yz}$

17 $\frac{5}{a^2b}-\frac{2}{ab^2}$ **18** $\frac{a+1}{6a}+\frac{a-4}{2a}$ **19** $\frac{1}{x+1}+\frac{2}{3}$ **20** $\frac{1}{x}+\frac{2}{x}-\frac{1}{x^2}$

21 Simplify $\frac{1}{ab}+\frac{a}{bc}$. Which of the following is correct?

A $\frac{1+a}{abc}$ B $\frac{c+a^2}{abc}$ C $\frac{1+a}{ab^2c}$ D $\frac{c+a^2}{ab^2c}$

For questions **22** to **29**, write the lowest common multiple (LCM).

22 $(x-3)$ and $(x+3)$ **23** x and $(x-2)$ **24** $(2x-4)$ and $(3x-6)$

25 (x^2-4x) and $(x-4)$ **26** (x^2-4x) and (x^2-16) **27** $(x+2)$ and (x^2+4x+4)

28 $(x-y)$, $(x+y)$, and (x^2-y^2) **29** (x^2-y^2), (x^2+xy), $(xy-y^2)$

30 The lowest common multiple of x, $x+3$ and x^2-9 is:

A $x(x-3)(x+3)$ B $x(x-3)(x^2-9)$ C $x(x+3)(x^2-9)$ D $x(x-3)^2$

Express each of the following as a single fraction.

31 $\frac{1}{a-b}+\frac{1}{a+b}$ **32** $\frac{3}{x-y}-\frac{2}{x+y}$ **33** $\frac{x}{x-y}+\frac{y}{x-y}$

34 $\frac{x}{x-y}+\frac{y}{x+y}$ **35** $\frac{3a-b}{a^2-b^2}+\frac{1}{a-b}$ **36** $\frac{1}{x^2-4}-\frac{1}{x+2}$

37 $\frac{x}{x^2-y^2}-\frac{y}{x^2-y^2}$ **38** $\frac{3}{(x-2)^2}+\frac{2}{x-2}$ **39** $\frac{1}{x^2-4x+3}-\frac{1}{x^2-1}$

40 $\frac{3}{x-2}+\frac{1}{x+3}$ **41** $\frac{1}{x+y}-\frac{1}{x-y}$ **42** $\frac{5}{2a+6}+\frac{a}{a^2-9}$

43 $\frac{1}{x-5}-\frac{1}{x+5}+\frac{x+10}{x^2-25}$ **44** $\frac{6}{3x-2}-\frac{8}{4x+1}$ **45** $\frac{7a}{3a-4}-\frac{5a}{2a-3}$

46 $\frac{y}{x^2-xy}+\frac{1}{x}$ **47** $\frac{1}{x+4}+\frac{x}{x^2-16}$ **48** $\frac{1}{x+2}+\frac{1}{x-2}+\frac{4}{x^2-4}$

49 $\frac{1}{a+3}+\frac{a+4}{a^2+5a+6}$ **50** $\frac{3}{x^2-4}-\frac{2}{x^2-3x+2}$ **51** $\frac{x+1}{x-1}-\frac{x-1}{x+1}$

52 $\frac{5}{x-2}+\frac{3}{x^2-4}$ **53** $\frac{3x}{x^2-16}-\frac{2}{x+4}$ **54** $\frac{5}{3x}-\frac{2}{x^2-5x}$

55 Simplify $\frac{5}{x+1}-\frac{3}{x-2}$. Indicate whether each answer is correct or incorrect.

(a) $\frac{2x-7}{(x+1)(x-2)}$ (b) $\frac{2x-13}{(x+1)(x-2)}$ (c) $\frac{2}{(x+1)(x-2)}$ (d) $\frac{2x-1}{(x+1)(x-2)}$

1.11 REAL NUMBERS AND SURDS

Rational numbers may be expressed in the form $\frac{a}{b}$ where a and b are integers and $b \neq 0$. Rational numbers include fractions, terminating decimals and recurring decimals. Integers are rational numbers with a denominator of one, for example $5 = \frac{5}{1}$.

Irrational numbers are numbers that cannot be represented as a fraction, such as $\sqrt{2}$, $\sqrt[3]{15}$, $2\sqrt{7}$ or π. They are not rational. Irrational numbers that are roots, such as $\sqrt{2}$, $\sqrt[3]{15}$ and $2\sqrt{7}$, are called surds. Some other irrational numbers, such as π, are called transcendental numbers.

The following diagram shows how to construct exact lengths for $\sqrt{2}$, $\sqrt{3}$, $\sqrt{4}$, $\sqrt{5}$ and $\sqrt{6}$, starting with a right-angled triangle with side lengths of one unit. Each triangle's hypotenuse is an exact value, written as a surd.

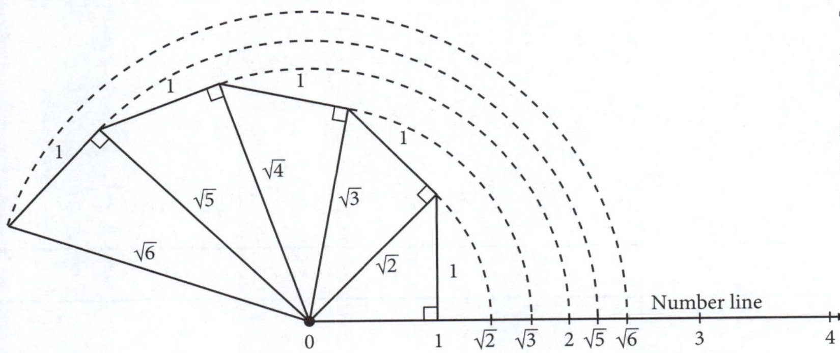

The diagram shows that $\sqrt{4} = 2$, which is a rational number. The square root of a perfect square is always a rational number. Similarly, $\sqrt[3]{8} = 2$ is rational. Surds that simplify to a rational number do not represent irrational numbers.

The sets of all rational and irrational numbers together form a larger set called the set of **real numbers**.

The real numbers can be represented by all the points on the real number line.

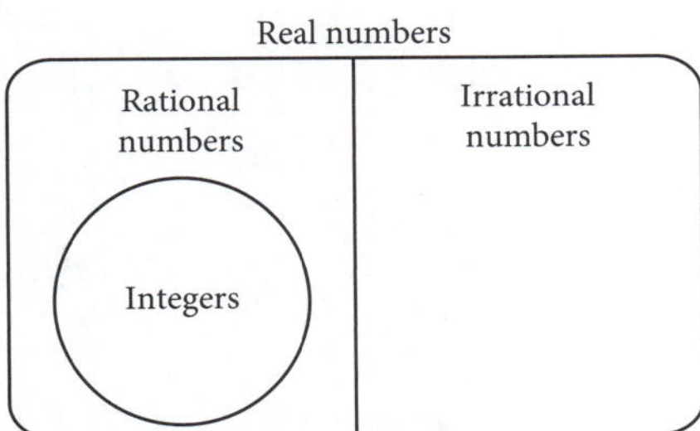

Operations with surds—basic rules

Surds have their own set of simplifying rules for the four operations of arithmetic. These rules are used to simplify expressions involving surds.

If a and b are positive numbers, then:

(a) $\sqrt{ab} = \sqrt{a} \times \sqrt{b}$ **(b)** $\sqrt{\frac{a}{b}} = \frac{\sqrt{a}}{\sqrt{b}}$ **(c)** $\sqrt{a^2} = a$

(d) $\sqrt{a} \times \sqrt{b} = \sqrt{ab}$ **(e)** $\frac{\sqrt{a}}{\sqrt{b}} = \sqrt{\frac{a}{b}}$ **(f)** $\sqrt{a} \times \sqrt{a} = a$

(g) $\sqrt{a} \div \sqrt{b} = \frac{\sqrt{a}}{\sqrt{b}} = \frac{\sqrt{a} \times \sqrt{b}}{\sqrt{b} \times \sqrt{b}} = \frac{\sqrt{ab}}{b}$ to write the expression with a rational denominator.

Surds can also be simplified sometimes by identifying a factor in the surd that is a perfect square and then taking that factor's square root. (For example, see part **(a)** in Example **15**.) You might also multiply two surds and find that the answer has a perfect square factor.

You should know the perfect squares so that you can look for them as factors in surds. When a surd contains a perfect square factor, you can use rules **(a)**, **(c)** and **(g)** above to simplify the expression.

Example 15

Simplify each expression.

(a) $\sqrt{12}$ (b) $\sqrt{80}$ (c) $3\sqrt{28}$ (d) $\dfrac{\sqrt{175}}{5}$

(e) $\sqrt{6}\times\sqrt{10}$ (f) $\dfrac{\sqrt{80}}{\sqrt{5}}$ (g) $\sqrt{3}\times\sqrt{12}$ (h) $4\sqrt{2}\times\sqrt{24}$

Solution

(a) Look for a factor of 12 that is a perfect square.
$12 = 4 \times 3$
$\therefore \sqrt{12}=\sqrt{4\times 3}=\sqrt{4}\times\sqrt{3}=2\sqrt{3}$

(b) Look for a factor of 80 that is a perfect square.
$80 = 16 \times 5$
$\therefore \sqrt{80}=\sqrt{16\times 5}=4\sqrt{5}$

(c) $3\sqrt{28}=3\sqrt{4}\times\sqrt{7}=6\sqrt{7}$

(d) $\dfrac{\sqrt{175}}{5}=\dfrac{\sqrt{25\times 7}}{5}=\dfrac{5\sqrt{7}}{5}=\sqrt{7}$

(e) $\sqrt{6}\times\sqrt{10}=\sqrt{60}=\sqrt{4\times 15}=2\sqrt{15}$

(f) $\dfrac{\sqrt{80}}{\sqrt{5}}=\sqrt{\dfrac{80}{5}}=\sqrt{16}=4$

(g) $\sqrt{3}\times\sqrt{12}=\sqrt{3}\times 2\sqrt{3}=2\times 3=6$

(h) $4\sqrt{2}\times\sqrt{24}=4\sqrt{2}\times 2\sqrt{6}=8\sqrt{12}=16\sqrt{3}$

Example 16

Simplify each expression, writing your answer with a rational denominator (or no denominator).

(a) $\dfrac{6}{\sqrt{3}}$ (b) $\dfrac{\sqrt{60}}{3\sqrt{6}}$ (c) $\dfrac{3\sqrt{5}}{\sqrt{15}}$

Solution

(a) $\dfrac{6}{\sqrt{3}}=\dfrac{6}{\sqrt{3}}\times\dfrac{\sqrt{3}}{\sqrt{3}}$
$=\dfrac{6\sqrt{3}}{3}$
$=2\sqrt{3}$

(b) $\dfrac{\sqrt{60}}{3\sqrt{6}}=\dfrac{\sqrt{6}\times\sqrt{10}}{3\sqrt{6}}$
$=\dfrac{\sqrt{10}}{3}$

(c) $\dfrac{3\sqrt{5}}{\sqrt{15}}=\dfrac{3\sqrt{5}}{\sqrt{3}\times\sqrt{5}}$
$=\dfrac{3}{\sqrt{3}}$
$=\dfrac{\sqrt{3}\times\sqrt{3}}{\sqrt{3}}$
$=\sqrt{3}$

EXERCISE 1.11 REAL NUMBERS AND SURDS

Simplify all the expressions in this exercise, writing each answer with a rational denominator (or no denominator).

1 $\sqrt{8}$ **2** $\sqrt{20}$ **3** $\sqrt{27}$ **4** $\sqrt{32}$

5 When simplified, $\sqrt{40}$ equals:

A $4\sqrt{5}$ **B** $\sqrt{10}$ **C** $2\sqrt{10}$ **D** 20

6 $\sqrt{45}$ **7** $\sqrt{72}$ **8** $\sqrt{84}$ **9** $\sqrt{98}$ **10** $\sqrt{108}$ **11** $\sqrt{125}$

12 $\sqrt{162}$ **13** $\sqrt{200}$ **14** $5\sqrt{128}$ **15** $4\sqrt{800}$ **16** $2\sqrt{150}$ **17** $3\sqrt{52}$

18 When simplified, $7\sqrt{245}$ equals:

A $7\sqrt{5}$ **B** $35\sqrt{7}$ **C** $35\sqrt{5}$ **D** $49\sqrt{5}$

19 $\dfrac{\sqrt{320}}{2}$ **20** $\dfrac{\sqrt{175}}{5}$ **21** $\dfrac{4\sqrt{243}}{3}$ **22** $\dfrac{\sqrt{90}}{9}$

23 When simplified, $\frac{\sqrt{50}}{\sqrt{5}}$ equals:

A $\sqrt{2}$ **B** $\sqrt{10}$ **C** $5\sqrt{2}$ **D** 10

24 Simplify $\frac{2\sqrt{7}}{\sqrt{35}}$. Indicate whether each answer below is correct or incorrect.

(a) $\frac{2}{\sqrt{5}}$ **(b)** $\frac{14}{\sqrt{5}}$ **(c)** $\frac{2\sqrt{5}}{5}$ **(d)** 10

25 $\sqrt{3}\times\sqrt{5}$ **26** $\sqrt{8}\times\sqrt{2}$ **27** $\sqrt{6}\times\sqrt{2}$ **28** $\sqrt{6}\times\sqrt{10}$ **29** $2\sqrt{3}\times4\sqrt{2}$

30 $\sqrt{8}\times2\sqrt{2}$ **31** $2\sqrt{5}\times5\sqrt{2}$ **32** $4\sqrt{6}\times2\sqrt{3}$ **33** $2\sqrt{5}\times3\sqrt{7}$ **34** $4\sqrt{3}\times\sqrt{18}$

35 $2\sqrt{8}\times\sqrt{12}$ **36** $4\sqrt{5}\times\sqrt{20}$ **37** $\frac{\sqrt{3}}{\sqrt{2}}$ **38** $\frac{2}{\sqrt{3}}$ **39** $\frac{\sqrt{28}}{\sqrt{7}}$

40 $\frac{7}{\sqrt{7}}$ **41** $\frac{\sqrt{80}}{\sqrt{5}}$ **42** $\frac{3\sqrt{2}}{\sqrt{6}}$ **43** $\frac{7\sqrt{2}}{\sqrt{98}}$ **44** $\frac{2\sqrt{18}}{\sqrt{8}}$

1.12 ADDING AND SUBTRACTING SURDS

Surds of the same kind can be added and subtracted just like pronumerals, using the distributive law to collect like terms: $ab + ac = a(b + c)$

You can only add and subtract like algebraic terms, such as $2a + 3a - a = 4a$. Similarly, you can only add and subtract like surds, such as $2\sqrt{2}+3\sqrt{2}=5\sqrt{2}$. It may be necessary to first simplify the surd terms by removing perfect square factors.

Example 17

Simplify each expression by collecting like terms.

(a) $2\sqrt{3}+5\sqrt{3}$ **(b)** $4\sqrt{10}-\sqrt{10}$ **(c)** $3\sqrt{6}+4\sqrt{6}-\sqrt{5}$ **(d)** $3\sqrt{5}+4\sqrt{7}+2\sqrt{5}-6\sqrt{7}$

Solution

(a) $2\sqrt{3}+5\sqrt{3}=(2+5)\sqrt{3}=7\sqrt{3}$ **(b)** $4\sqrt{10}-\sqrt{10}=(4-1)\sqrt{10}=3\sqrt{10}$

(c) $3\sqrt{6}+4\sqrt{6}-\sqrt{5}=7\sqrt{6}-\sqrt{5}$ **(d)** $3\sqrt{5}+4\sqrt{7}+2\sqrt{5}-6\sqrt{7}=5\sqrt{5}-2\sqrt{7}$

In parts **(c)** and **(d)** the different surds cannot be combined into a single term.

Example 18

Simplify:

(a) $\sqrt{8}-\sqrt{18}+\sqrt{50}$ **(b)** $5\sqrt{3}+\sqrt{20}-2\sqrt{12}+\sqrt{45}$

Solution

Where possible, simplify each term before attempting to add or subtract the surds.

(a) $\sqrt{8}-\sqrt{18}+\sqrt{50}$
$=\sqrt{4\times2}-\sqrt{9\times2}+\sqrt{25\times2}$
$=2\sqrt{2}-3\sqrt{2}+5\sqrt{2}$
$=4\sqrt{2}$

(b) $5\sqrt{3}+\sqrt{20}-2\sqrt{12}+\sqrt{45}$
$=5\sqrt{3}+2\sqrt{5}-4\sqrt{3}+3\sqrt{5}$
$=\sqrt{3}+5\sqrt{5}$

EXERCISE 1.12 ADDING AND SUBTRACTING SURDS

Simplify the expressions in this exercise.

1 $\sqrt{3}+2\sqrt{3}+4\sqrt{3}$ **2** $5\sqrt{7}-2\sqrt{7}+4\sqrt{7}$ **3** $3\sqrt{5}+5\sqrt{5}-2\sqrt{5}$

4 $4\sqrt{2}-\sqrt{3}+4\sqrt{3}-\sqrt{2}$ **5** $\sqrt{5}+\sqrt{2}+3\sqrt{5}-6\sqrt{2}$ **6** $\sqrt{8}-\sqrt{2}$

7 When $\sqrt{27}-\sqrt{18}+\sqrt{3}$ is simplified, the answer is:

A $4\sqrt{3}-3\sqrt{2}$ **B** $\sqrt{3}$ **C** $3+\sqrt{3}$ **D** $2\sqrt{3}$

8 $\sqrt{20}+\sqrt{5}$ **9** $\sqrt{18}+\sqrt{32}-\sqrt{2}$ **10** $\sqrt{27}+2\sqrt{48}-4\sqrt{3}$

11 $\sqrt{12}+\sqrt{3}+\sqrt{48}$ **12** $2\sqrt{50}-3\sqrt{18}$ **13** $\sqrt{7}+\sqrt{28}-\sqrt{63}$

14 Simplify $\sqrt{5}+\sqrt{2}-\sqrt{45}+\sqrt{8}$. Some steps in this simplification are given below. Indicate whether each statement is a correct or incorrect step.

(a) $\sqrt{10}-\sqrt{45}$ **(b)** $-\sqrt{35}$ **(c)** $\sqrt{5}+\sqrt{2}-3\sqrt{5}+2\sqrt{2}$ **(d)** $3\sqrt{2}-2\sqrt{5}$

15 $6\sqrt{5}+4\sqrt{7}-2\sqrt{5}$ **16** $5\sqrt{3}+\sqrt{27}-\sqrt{45}$ **17** $\sqrt{20}+\sqrt{5}+\sqrt{18}$

18 $3\sqrt{15}+\sqrt{60}-\sqrt{40}$ **19** $4\sqrt{7}-\sqrt{28}+\sqrt{63}$ **20** $2\sqrt{50}-3\sqrt{18}+\sqrt{3}$

21 $\sqrt{3}+3\sqrt{75}-\sqrt{48}$ **22** $\sqrt{150}-\sqrt{200}$ **23** $\sqrt{6}+\sqrt{24}+\sqrt{54}$

24 $5\sqrt{7}+3\sqrt{5}-2\sqrt{28}$ **25** $5\sqrt{45}-2\sqrt{32}$ **26** $\sqrt{98}-2\sqrt{20}-\sqrt{12}$

27 $\sqrt{125}-5\sqrt{2}+\sqrt{50}$ **28** $7\sqrt{3}-2\sqrt{2}+\sqrt{12}+\sqrt{8}$ **29** $\sqrt{150}-\sqrt{96}-\sqrt{24}$

30 $6\sqrt{2}+\sqrt{12}-2\sqrt{3}-3\sqrt{8}$

1.13 THE DISTRIBUTIVE LAW

The distributive law can be used with surds to expand expressions with a binomial factor: $a(b + c) = ab + ac$

Example 19

Expand and simplify:

(a) $\sqrt{6}\left(\sqrt{2}+2\sqrt{3}\right)$ **(b)** $\left(\sqrt{3}+\sqrt{2}\right)\left(\sqrt{5}+\sqrt{6}\right)$ **(c)** $\left(\sqrt{5}-\sqrt{3}\right)\left(\sqrt{5}+\sqrt{3}\right)$ **(d)** $\left(\sqrt{5}+\sqrt{3}\right)\left(\sqrt{5}+\sqrt{3}\right)$

Solution

(a) $\sqrt{6}\left(\sqrt{2}+2\sqrt{3}\right)$

$=\sqrt{6}\times\sqrt{2}+\sqrt{6}\times2\sqrt{3}$

$=\sqrt{12}+2\sqrt{18}$

$=2\sqrt{3}+6\sqrt{2}$

(b) $\left(\sqrt{3}+\sqrt{2}\right)\left(\sqrt{5}+\sqrt{6}\right)$

$=\sqrt{3}\left(\sqrt{5}+\sqrt{6}\right)+\sqrt{2}\left(\sqrt{5}+\sqrt{6}\right)$

$=\sqrt{15}+3\sqrt{2}+\sqrt{10}+2\sqrt{3}$

(c) $\left(\sqrt{5}-\sqrt{3}\right)\left(\sqrt{5}+\sqrt{3}\right)$

$=\left(\sqrt{5}\right)^2-\left(\sqrt{3}\right)^2$

$=5-3$

$=2$

This is similar to $(a - b)(a + b) = a^2 - b^2$.

(d) $\left(\sqrt{5}+\sqrt{3}\right)\left(\sqrt{5}+\sqrt{3}\right)$

$=\sqrt{5}\left(\sqrt{5}+\sqrt{3}\right)+\sqrt{3}\left(\sqrt{5}+\sqrt{3}\right)$

$=5+2\sqrt{15}+3$

$=8+2\sqrt{15}$

This is similar to $(a + b)^2 = a^2 + 2ab + b^2$.

EXERCISE 1.13 THE DISTRIBUTIVE LAW

Expand and simplify the expressions in this exercise.

1 $\sqrt{5}\left(\sqrt{2}+\sqrt{3}\right)$ **2** $\sqrt{5}\left(\sqrt{5}+\sqrt{2}\right)$ **3** $\sqrt{2}\left(\sqrt{2}+\sqrt{8}\right)$

4 $\sqrt{3}\left(\sqrt{2}-\sqrt{6}\right)$ **5** $\sqrt{6}\left(\sqrt{3}-2\right)$ **6** $7\left(2\sqrt{5}-1\right)$

7 When simplified, $\sqrt{2}\left(\sqrt{32}-\sqrt{8}\right)$ equals:

A $8\sqrt{3}$ **B** 4 **C** $2\sqrt{2}$ **D** $8-2\sqrt{2}$

8 $3\sqrt{2}\left(2\sqrt{6}-\sqrt{5}\right)$ **9** $\sqrt{a}\left(\sqrt{a}+\sqrt{b}\right)$ **10** $\sqrt{x}\left(\sqrt{x}-\sqrt{y}\right)$

11 $\left(\sqrt{5}+\sqrt{3}\right)\left(\sqrt{7}-\sqrt{2}\right)$ **12** $\left(\sqrt{2}+\sqrt{7}\right)\left(\sqrt{3}+2\sqrt{2}\right)$ **13** $\left(\sqrt{3}-1\right)\left(\sqrt{2}+3\right)$

14 $\left(\sqrt{5}+2\right)\left(2\sqrt{5}+3\right)$ **15** $\left(2\sqrt{3}-5\right)\left(2\sqrt{3}+3\right)$ **16** $\left(\sqrt{3}-\sqrt{2}\left(2\sqrt{3}-\sqrt{2}\right)\right)$

17 $\left(2\sqrt{5}-\sqrt{2}\right)\left(2\sqrt{5}+3\right)$ **18** $\left(2\sqrt{2}-\sqrt{6}\right)\left(2\sqrt{3}-1\right)$ **19** $\left(\sqrt{3}+1\right)^2$

20 $\left(\sqrt{5}-\sqrt{2}\right)^2$ **21** $\left(2\sqrt{6}+\sqrt{3}\right)^2$ **22** $\left(2\sqrt{2}-1\right)\left(2\sqrt{2}+1\right)$

23 $\left(2\sqrt{6}-\sqrt{3}\right)\left(2\sqrt{6}+\sqrt{3}\right)$ **24** $\left(\sqrt{11}-\sqrt{7}\right)\left(\sqrt{11}+\sqrt{7}\right)$ **25** $\left(\sqrt{7}-2\right)\left(\sqrt{7}+2\right)$

26 When simplified, $\left(3\sqrt{7}-2\right)^2$ equals:

A 59 **B** $67+12\sqrt{7}$ **C** $23-12\sqrt{7}$ **D** $67-12\sqrt{7}$

27 $\left(\sqrt{5}-\sqrt{3}\right)^2$ **28** $\left(\sqrt{11}+\sqrt{7}\right)^2$ **29** $\left(2\sqrt{3}-1\right)\left(\sqrt{3}+2\right)$

30 $\left(\sqrt{11}-\sqrt{10}\right)\left(\sqrt{11}+\sqrt{10}\right)$ **31** $\left(\sqrt{6}-\sqrt{5}\right)\left(\sqrt{6}+\sqrt{5}\right)$ **32** $\left(2\sqrt{2}+\sqrt{3}\right)^2$

33 $\left(3\sqrt{5}-2\sqrt{2}\right)\left(3\sqrt{5}+2\sqrt{2}\right)$ **34** $\left(\sqrt{5}+2\sqrt{2}\right)\left(\sqrt{6}-1\right)$ **35** $\left(2\sqrt{6}-\sqrt{3}\right)\left(\sqrt{6}+3\sqrt{3}\right)$

36 Expand and simplify $\left(4\sqrt{3}+1\right)\left(2\sqrt{3}-3\right)$. Some steps in this simplification are given below. Indicate whether each statement is a correct or incorrect step.

(a) $72-12\sqrt{3}+2\sqrt{3}-3$ **(b)** $24-12\sqrt{3}+2\sqrt{3}-3$ **(c)** $21-10\sqrt{3}$ **(d)** $27-10\sqrt{3}$

37 $\left(5\sqrt{2}-4\right)\left(5\sqrt{2}+4\right)$ **38** $\left(2\sqrt{7}+3\sqrt{6}\right)^2$ **39** $\left(2\sqrt{15}+\sqrt{5}\right)\left(\sqrt{15}-3\sqrt{5}\right)$

40 $\left(2\sqrt{2}+3\sqrt{3}\right)^2$ **41** $\left(2\sqrt{3}-1\right)\left(2\sqrt{3}+1\right)$ **42** $\left(2\sqrt{3}-1\right)^2$

1.14 RATIONALISING DENOMINATORS

The expressions $\sqrt{a}\times\sqrt{a}$ and $\left(\sqrt{a}-\sqrt{b}\right)\left(\sqrt{a}+\sqrt{b}\right)$ both have rational answers (that is, answers that do not involve surds): $\sqrt{a}\times\sqrt{a}=a$ and $\left(\sqrt{a}-\sqrt{b}\right)\left(\sqrt{a}+\sqrt{b}\right)=\left(\sqrt{a}\right)^2-\left(\sqrt{b}\right)^2=a-b$. For example:

- $\sqrt{3}\times\sqrt{3}=3$
- $\left(\sqrt{7}-\sqrt{2}\right)\left(\sqrt{7}+\sqrt{2}\right)=7-2=5$

The expansion $\left(\sqrt{a}-\sqrt{b}\right)\left(\sqrt{a}+\sqrt{b}\right)=a-b$ is known as the 'difference of two squares'. You have seen this previously as $(x-y)(x+y)=x^2-y^2$. By letting $x=\sqrt{a}$ and $y=\sqrt{b}$, we have obtained a process to convert any binomial surd into a rational number.

When you have a surd expression in the denominator of a fraction, it is normal to make the denominator into a rational number. This process is called **rationalising the denominator**.

Remember: to change a fraction without changing its value, multiply the numerator and the denominator by the same amount.

The expressions $\sqrt{a}-\sqrt{b}$ and $\sqrt{a}+\sqrt{b}$ are called **conjugate** surds, with each expression being the conjugate of the other. To convert any surd into a rational number, multiply the surd by its conjugate.

Example 20

Express with a rational denominator:

(a) $\frac{1}{\sqrt{2}}$ (b) $\frac{\sqrt{7}}{\sqrt{3}}$ (c) $\frac{1}{\sqrt{3}+\sqrt{2}}$ (d) $\frac{\sqrt{3}}{2\sqrt{7}-\sqrt{3}}$

Solution

(a) $\frac{1}{\sqrt{2}}$: multiply by $\frac{\sqrt{2}}{\sqrt{2}}$

$$=\frac{1}{\sqrt{2}}\times\frac{\sqrt{2}}{\sqrt{2}}$$

$$=\frac{\sqrt{2}}{2}$$

(b) $\frac{\sqrt{7}}{\sqrt{3}}$: multiply by $\frac{\sqrt{3}}{\sqrt{3}}$

$$=\frac{\sqrt{7}}{\sqrt{3}}\times\frac{\sqrt{3}}{\sqrt{3}}$$

$$=\frac{\sqrt{21}}{3}$$

In parts **(c)** and **(d)** the denominator is a binomial surd, so you have to multiply both numerator and denominator by the conjugate of the denominator.

(c) $$\frac{1}{\sqrt{3}+\sqrt{2}}=\frac{1}{\sqrt{3}+\sqrt{2}}\times\frac{\sqrt{3}-\sqrt{2}}{\sqrt{3}-\sqrt{2}}$$

$$=\frac{\sqrt{3}-\sqrt{2}}{3-2}$$

$$=\sqrt{3}-\sqrt{2}$$

(d) $$\frac{\sqrt{3}}{2\sqrt{7}-\sqrt{3}}=\frac{\sqrt{3}}{2\sqrt{7}-\sqrt{3}}\times\frac{2\sqrt{7}+\sqrt{3}}{2\sqrt{7}+\sqrt{3}}$$

$$=\frac{2\sqrt{21}+3}{4\times7-3}$$

$$=\frac{2\sqrt{21}+3}{25}$$

Example 21

Express $\frac{\sqrt{2}}{2\sqrt{2}+1}+\frac{2}{\sqrt{3}+1}$ as a single fraction with a rational denominator.

Solution

You can add the two fractions by finding a common denominator and then rationalising the result, or by first rationalising each denominator and then adding the resulting fractions. The latter is usually the easier method, unless the denominators happen to be conjugates.

$$\frac{\sqrt{2}}{2\sqrt{2}+1}+\frac{2}{\sqrt{3}+1}=\frac{\sqrt{2}}{2\sqrt{2}+1}\times\frac{2\sqrt{2}-1}{2\sqrt{2}-1}+\frac{2}{\sqrt{3}+1}\times\frac{\sqrt{3}-1}{\sqrt{3}-1}$$
$$=\frac{4-\sqrt{2}}{4\times2-1}+\frac{2\left(\sqrt{3}-1\right)}{3-1}$$
$$=\frac{4-\sqrt{2}}{7}+\frac{2\left(\sqrt{3}-1\right)}{2}$$
$$=\frac{4-\sqrt{2}}{7}+\frac{\sqrt{3}-1}{1}$$
$$=\frac{4-\sqrt{2}+7\left(\sqrt{3}-1\right)}{7}$$
$$=\frac{7\sqrt{3}-\sqrt{2}-3}{7}$$

EXERCISE 1.14 RATIONALISING DENOMINATORS

For questions **1** to **27**, express each fraction with a rational denominator.

1 $\frac{2}{\sqrt{3}}$ **2** $\frac{\sqrt{5}}{\sqrt{3}}$ **3** $\frac{3\sqrt{5}}{\sqrt{15}}$ **4** $\frac{1}{\sqrt{3}-\sqrt{2}}$

5 $\frac{1}{2\sqrt{7}+\sqrt{6}}$ **6** $\frac{1}{\sqrt{5}+2}$ **7** $\frac{1}{2\sqrt{5}-3\sqrt{2}}$ **8** $\frac{3\sqrt{2}}{\sqrt{5}-\sqrt{3}}$

9 $\frac{\sqrt{6}}{2\sqrt{5}-3\sqrt{2}}$ **10** $\frac{2\sqrt{5}}{3\sqrt{11}-2\sqrt{8}}$ **11** $\frac{\sqrt{3}+2}{\sqrt{3}-2}$ **12** $\frac{4\sqrt{2}+3\sqrt{5}}{2\sqrt{5}-\sqrt{2}}$

13 $\frac{\sqrt{7}-2\sqrt{5}}{3\sqrt{5}-2\sqrt{2}}$ **14** $\frac{3\sqrt{2}+2\sqrt{3}}{3\sqrt{2}-2\sqrt{3}}$ **15** $\frac{5\sqrt{3}+3\sqrt{5}}{5\sqrt{5}-3\sqrt{3}}$ **16** $\frac{2\sqrt{3}}{3\sqrt{3}-2}$

17 $\frac{3\sqrt{6}}{2\sqrt{2}+\sqrt{3}}$ **18** $\frac{5\sqrt{11}+3}{3\sqrt{11}-2}$ **19** $\frac{\sqrt{2}}{2\sqrt{2}-2\sqrt{3}}$ **20** $\frac{3\sqrt{3}}{2\sqrt{3}+\sqrt{2}}$

21 $\frac{\sqrt{3}}{\sqrt{24}-\sqrt{48}}$ **22** $\frac{\sqrt{2}-1}{\sqrt{2}+1}$ **23** $\frac{2\sqrt{5}-\sqrt{2}}{2\sqrt{5}+\sqrt{2}}$ **24** $\frac{\sqrt{6}+2\sqrt{3}}{2\sqrt{6}-\sqrt{3}}$

25 $\frac{2\sqrt{7}-3\sqrt{2}}{2\sqrt{7}-\sqrt{2}}$ **26** $\frac{\sqrt{5}+\sqrt{3}}{2\sqrt{10}-\sqrt{6}}$ **27** $\frac{2\sqrt{2}-5\sqrt{3}}{3\sqrt{6}-\sqrt{15}}$

28 If $x=\frac{3-2\sqrt{2}}{3+2\sqrt{2}}$ then the value of $x+\frac{1}{x}$ is:

A 34 **B** $24\sqrt{2}$ **C** $34-24\sqrt{2}$ **D** 2

29 If $x=\sqrt{3}+1$, find the value of $x^2-\frac{1}{x^2}$.

30 If $x=3\sqrt{2}+1$, find the value of $\frac{x^2-2x}{x-1}$.

31 If $x=\sqrt{5}-2$, find the value of $\frac{x^2+2x}{x+3}$.

32 Show that $x=2\sqrt{2}-3$ is one solution of the equation $x^2+6x+1=0$.

33 Which values of x satisfy $x^2+4x+2=0$? Indicate whether each answer is correct or incorrect.

(a) $x=2-\sqrt{2}$ **(b)** $x=\sqrt{2}-2$ **(c)** $x=\sqrt{2}+2$ **(d)** $x=-2-\sqrt{2}$

34 Show that $x = \sqrt{5} - 1$ is one solution of the equation $x^3 + 3x^2 - 2x - 4 = 0$.

35 Show that $x = \dfrac{1}{\sqrt{3} - 1}$ is one solution of the equation $2x^2 - 2x - 1 = 0$.

For questions **36** to **44**, express as a single fraction with a rational denominator:

36 $\dfrac{1}{2\sqrt{3} - 1} + \dfrac{3}{\sqrt{3} + 1}$

37 $\dfrac{\sqrt{5} + \sqrt{2}}{\sqrt{5} - \sqrt{2}} - \dfrac{\sqrt{5} - \sqrt{2}}{\sqrt{5} + \sqrt{2}}$

38 $\dfrac{\sqrt{3} - \sqrt{2}}{\sqrt{3} + \sqrt{2}} \times \dfrac{2\sqrt{2} - \sqrt{3}}{2\sqrt{2} + \sqrt{3}}$

39 $\dfrac{\sqrt{5} + \sqrt{2}}{\sqrt{5} - \sqrt{2}} \times \dfrac{2\sqrt{5} - 3\sqrt{2}}{\sqrt{5} + 2\sqrt{2}}$

40 $\dfrac{2\sqrt{5} + 1}{2\sqrt{5} - 1} - \dfrac{\sqrt{5} - 1}{2\sqrt{5} - 3}$

41 $\dfrac{2\sqrt{3}}{\sqrt{6} - \sqrt{3}} - \dfrac{\sqrt{3}}{2\sqrt{6} + 3\sqrt{3}}$

42 $\dfrac{\sqrt{3} - 1}{\sqrt{3} + 2} - \dfrac{\sqrt{5} - \sqrt{3}}{2\sqrt{5} + \sqrt{3}}$

43 $\dfrac{2\sqrt{5}}{\sqrt{10} - \sqrt{15}} - \dfrac{3\sqrt{7}}{\sqrt{35} - \sqrt{14}}$

44 $\dfrac{1}{x - 1} + \dfrac{1}{x + 1} - \dfrac{2}{x^2 - 1}$, where $x = 2\sqrt{3} + 1$.

CHAPTER REVIEW 1

1 Expand and simplify:

(a) $(2x - 5)^2$ (b) $(x + 3)(x - 7)$ (c) $(2y + 1)(3y + 4)$

(d) $(5x - 4)(5x + 4)$ (e) $(2x - y)(x^2 - xy + y^2)$

2 Factorise:

(a) $81 - 4a^2$ (b) $10xy - 5y$ (c) $5x^2y - 10xy^2 - 5xy$ (d) $a^2 - 18a + 56$

(e) $16x^2 - 1$ (f) $3x^2 + 4x - 7$ (g) $a^2 - b^2 + 2a - 2b$ (h) $3x + 7x^2 - 6x^3$

(i) $8a^3 - 27$ (j) $8 - (x + h)^3$ (k) $8y^2 - 6y - 9$ (l) $x^4 - 8x$

3 If $a = -3$ and $b = 4$, evaluate $\sqrt{a^2 + b^2}$.

4 Simplify:

(a) $\dfrac{3x^3}{4a^2} \times \dfrac{ay - a}{xy^2} \div \dfrac{3y - 3}{4ay^2}$ (b) $\dfrac{5x}{3} - \dfrac{2x + 3}{4} + \dfrac{x}{6}$ (c) $\dfrac{12m^2 - 4n}{3m^2 - n} \div \dfrac{10m^2n}{5m^2n^2}$

(d) $\dfrac{3x + 4}{x^2} - \dfrac{5}{x}$ (e) $\dfrac{3}{x + 1} + \dfrac{1}{x^2 + 2x + 1}$ (f) $\dfrac{2x^2 - 3xy}{xy - y^2} \div \dfrac{4x - 6y}{2x^2 - 2xy}$

(g) $\dfrac{a^3 + b^3}{a^2 - b^2}$ (h) $\dfrac{2}{m^2 - 4} - \dfrac{1}{m^2 - 3m + 2}$ (i) $\dfrac{1}{x^2 - 9x + 20} + \dfrac{1}{x^2 - 11x + 30}$

5 Expand and simplify:

(a) $\left(a^{\frac{1}{3}} - b^{\frac{1}{3}}\right)\left(a^{\frac{2}{3}} + a^{\frac{1}{3}}b^{\frac{1}{3}} + b^{\frac{2}{3}}\right)$ (b) $\left(a^{\frac{1}{2}} - b^{-\frac{1}{2}}\right)^2$

6 Find the exact value of $\dfrac{a^4b}{c^4}$ where $a = \left(\dfrac{2}{3}\right)^2$, $b = \left(\dfrac{8}{3}\right)^7$, $c = \left(\dfrac{4}{3}\right)^4$.

7 If $E = \dfrac{m}{2}\left(v^2 - u^2\right)$, find:

(a) E when $m = 13$, $v = 17$, $u = 15$ (b) v when $E = 98$, $m = 4$, $u = 24$, if $v > 0$

8 Simplify:

(a) $\sqrt{12} + 4\sqrt{3} + \sqrt{27} - 3\sqrt{54}$ (b) $\dfrac{2}{\sqrt{5}} + \sqrt{20} + \dfrac{8}{\sqrt{80}}$

9 (a) $\dfrac{\sqrt{5} - \sqrt{2}}{\sqrt{5} + \sqrt{2}}$ (b) $\dfrac{3^{\frac{5}{4}} \times 15^{\frac{3}{4}}}{5^{-\frac{3}{4}} \times 45^{\frac{1}{2}}}$ (A challenge!)

10 Expand and simplify $\left(3\sqrt{5}-2\sqrt{3}\right)^2$.

11 The correct expansion of $\left(\sqrt{5}+2\sqrt{2}\right)\left(\sqrt{6}-\sqrt{5}\right)$ is:

A $\sqrt{30}+2\sqrt{2}-5$ **B** $\sqrt{6}+2\sqrt{12}-2\sqrt{10}$ **C** $4\sqrt{3}-\sqrt{10}-5$ **D** $\sqrt{30}+4\sqrt{3}-2\sqrt{10}-5$

12 Express $\dfrac{\sqrt{2}-1}{2\sqrt{2}-1}$ with a rational denominator.

13 Find the exact value of x^4-2x^2+1 when $x=3\sqrt{2}$.

14 Show by substitution that $x=\dfrac{\sqrt{5}-1}{2}$ is a root of the equation $x^3+3x^2+x-2=0$.

15 Simplify:

(a) $1-\dfrac{1}{1-\left(-\frac{1}{4}\right)}$ (b) $\dfrac{8^2+6^2-4^2}{2\times 8\times 6}$ (c) $H\pi\left(\dfrac{2R}{3}\right)^2-\dfrac{H}{R}\pi\left(\dfrac{2R}{3}\right)^3$ (d) $\dfrac{2x(x-1)-\left(x^2+3\right)\times 1}{(x-1)^2}$

(e) $\dfrac{1}{36}\times\dfrac{\left(\frac{35}{36}\right)^2}{1-\left(\frac{35}{36}\right)^2}$ (f) $2+\dfrac{t^2}{t+1}-\left(1+\dfrac{1}{t+1}\right)$ (g) $2\left(\dfrac{8}{3}+2-\left(\dfrac{1}{3}+1\right)\right)$

(h) Simplify, factorising your answer as much as possible: $\left(\sqrt{(x+1)^2+(y-0)^2}\right)^2+\left(\sqrt{(x-3)^2+(y-0)^2}\right)^2=40$

16 Factorise fully:

(a) $2x^2-8$ (b) $6x^2-216$

17 Evaluate:

(a) $100000\times 1.006^{120}-780\left(\dfrac{1.006^{120}-1}{0.006}\right)$

(b) correct to 2 decimal places, $M=\dfrac{500000\times 1.005^{360}\times 0.005}{1\left(1.005^{360}-1\right)}$

(c) correct to 2 decimal places, $A_{20}=500000\times 1.005^{240}-2998\times\dfrac{1.005^{240}-1}{0.005}$

(d) $\$500\times\dfrac{1.003^{60}\left(\left(\frac{1.01}{1.003}\right)^{60}-1\right)}{\frac{1.01}{1.003}-1}$

18 Make y the subject of the formula in $\dfrac{H}{H-y}=\dfrac{R}{x}$.

19 If $\dfrac{1}{\sqrt{n}+\sqrt{n+1}}=\sqrt{n+1}-\sqrt{n}$ then find the value of $\dfrac{1}{\sqrt{1}+\sqrt{2}}+\dfrac{1}{\sqrt{2}+\sqrt{3}}+\dfrac{1}{\sqrt{3}+\sqrt{4}}+\ldots+\dfrac{1}{\sqrt{99}+\sqrt{100}}$.

CHAPTER 2
Trigonometry

2.1 REVIEW OF RIGHT-ANGLED TRIANGLES

Trigonometric ratio definitions

From your study of the trigonometry of a right-angled triangle, you should already know the definitions of the trigonometric ratios for sine, cosine and tangent.

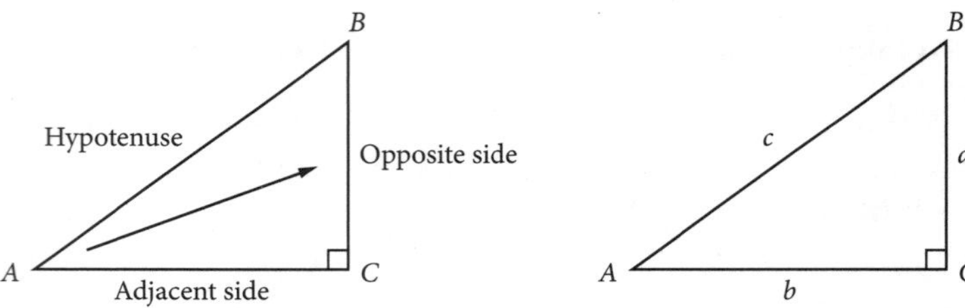

$$\sin A = \frac{\text{Opposite side}}{\text{Hypotenuse}} = \frac{a}{c} \qquad \cos A = \frac{\text{Adjacent side}}{\text{Hypotenuse}} = \frac{b}{c} \qquad \tan A = \frac{\text{Opposite side}}{\text{Adjacent side}} = \frac{a}{b}$$

To remember these ratios you can use the mnemonic **SOHCAHTOA**, which stands for **S**in is **O**pposite over **H**ypotenuse, **C**os is **A**djacent over **H**ypotenuse, **T**an is **O**pposite over **A**djacent.

These trigonometric ratios describe the relationship between the angles and sides of a right-angled triangle, while Pythagoras' theorem $c^2 = a^2 + b^2$ describes the relationship between the side lengths without reference to the angles.

Example 1

Use the trigonometric ratios to find the value of each variable given on the diagram, giving the lengths correct to 2 decimal places.

Hence, calculate the area of the quadrilateral.

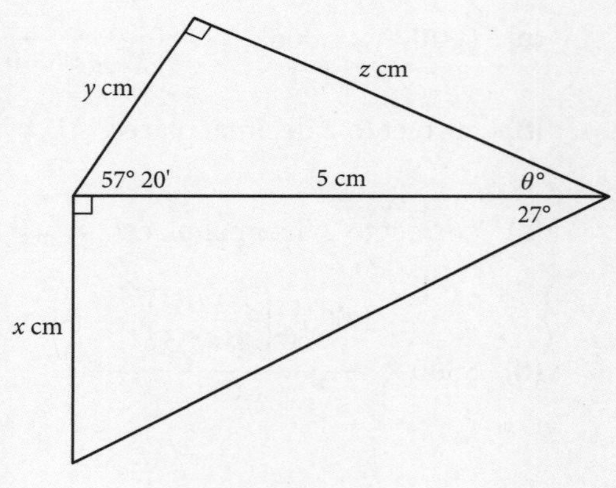

Solution

$\frac{x}{5} = \tan 27°$

$x = 5 \tan 27°$

$x = 2.55$ cm

$\frac{y}{5} = \cos 57°20'$

$y = 5 \cos 57°20'$

$y = 2.699 \approx 2.70$ cm

$\frac{z}{5} = \sin 57°20'$

$z = 5 \sin 57°20'$

$z = 4.21$ cm

$\theta = 180 - (90 + 57°20')$

$\theta = 32°40'$

$$\begin{aligned}\text{Area of quadrilateral} &= \frac{1}{2}(5x + yz) \\ &= \frac{1}{2}(5 \times 2.55 + 2.70 \times 4.21) \\ &= 12.06 \text{ cm}^2\end{aligned}$$

EXERCISE 2.1 REVIEW OF RIGHT-ANGLED TRIANGLES

1 Which is the correct expression for the value of x in the following diagram?

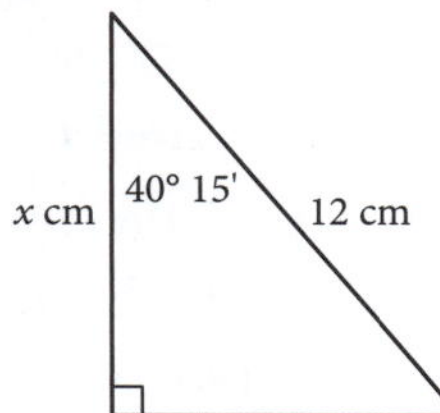

A $x = 12 \sin 40° 15'$
B $x = 12 \tan 40° 15'$
C $x = 12 \cos 40° 15'$
D $x = \dfrac{12}{\sin 49°15'}$

2 Find the value of x in each diagram. Where necessary give your answer correct to 2 decimal places.

(a) 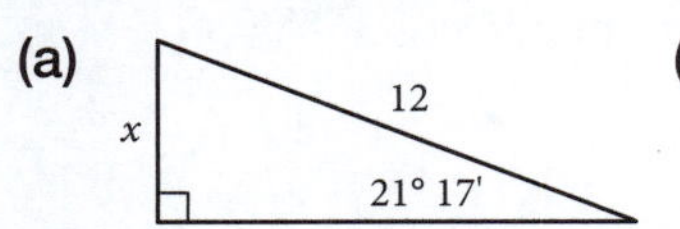

(b) 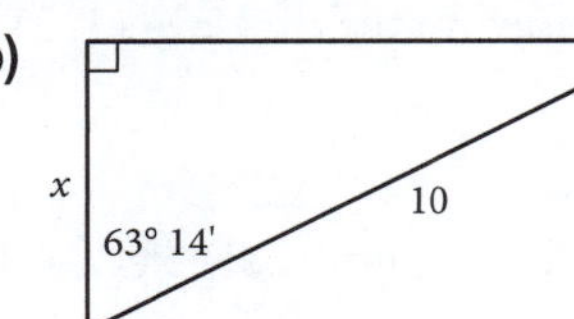

(c) 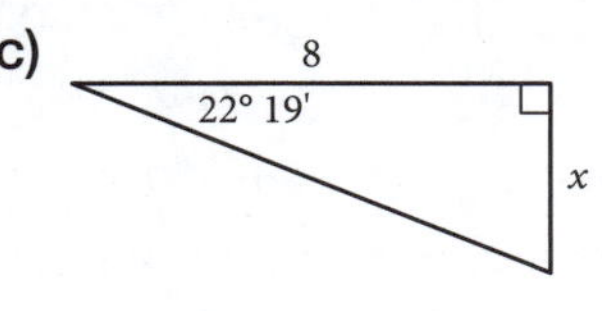

(d) 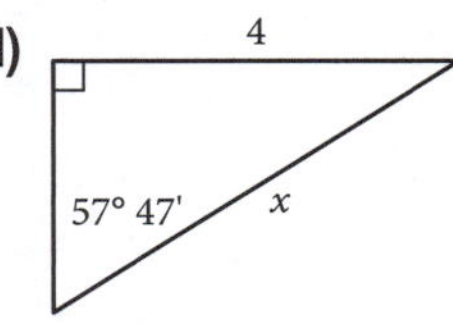

3 Find the value of θ in each diagram. Where necessary give your answer in degrees and minutes.

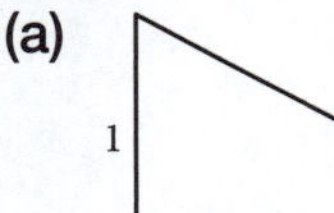

(a) 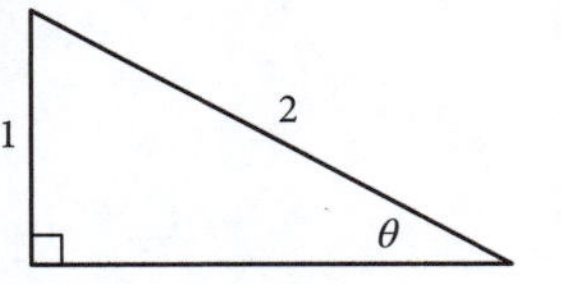

(b) 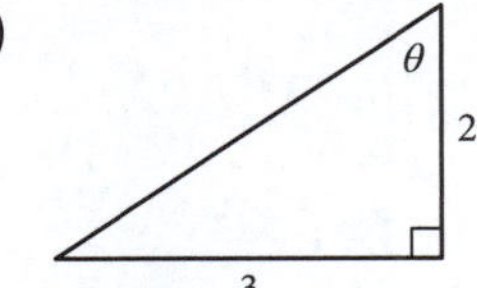

(c) 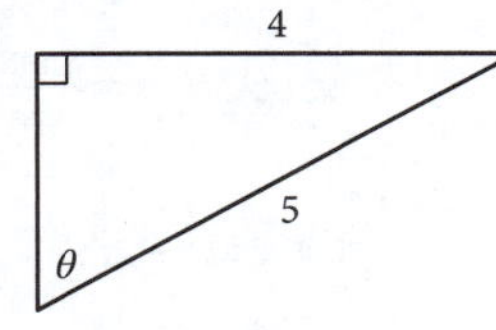

(d) 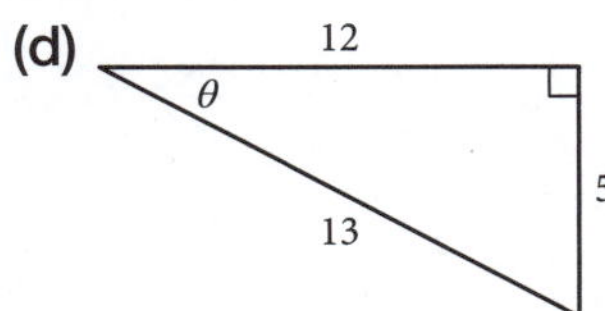

4 Use trigonometry to find the values of x and y, giving your answer correct to 1 decimal place.

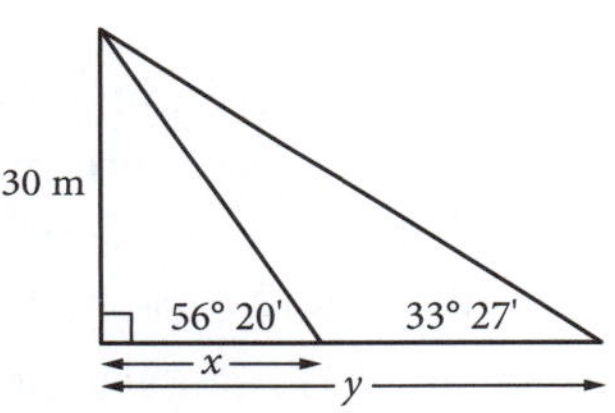

2.2 ANGLES OF ANY MAGNITUDE

The unit circle

Consider a circle of unit radius whose centre is at the origin. The equation of the circle is $x^2 + y^2 = r^2$ and $r = 1$.

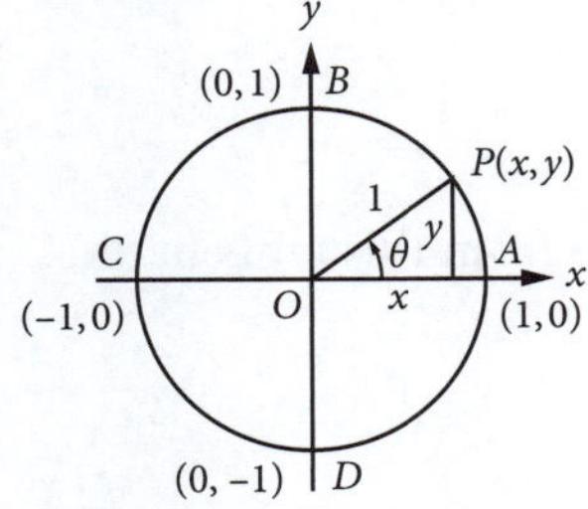

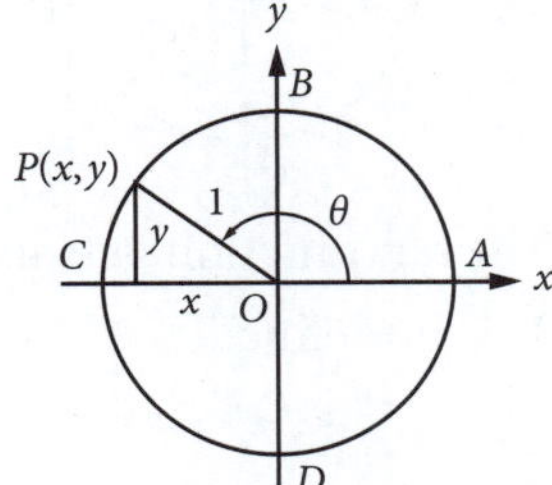

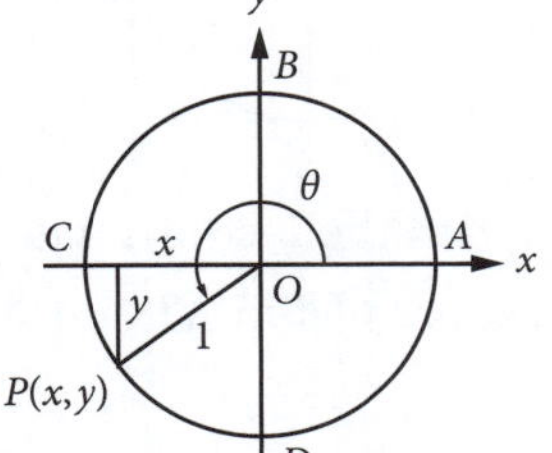

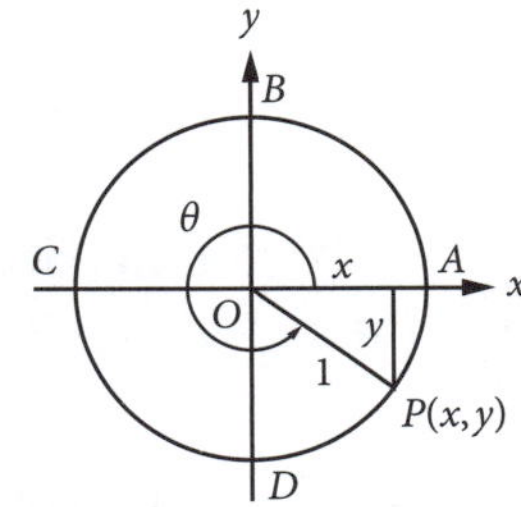

Take any point P on the circumference of this unit circle whose coordinates are (x, y). Consider the point P as starting at A and rotating in an anticlockwise direction, taking various positions around the circumference as shown in the diagrams above. In each position $\angle AOP = \theta$.

You can define **cosine** (cos) and **sine** (sin) as:

$$\cos\theta = \frac{\text{Adjacent}}{\text{Hypotenuse}} = \frac{x}{1} = x = x\text{-coordinate of } P \quad \text{(the abscissa)}$$

$$\sin\theta = \frac{\text{Opposite}}{\text{Hypotenuse}} = \frac{y}{1} = y = y\text{-coordinate of } P \quad \text{(the ordinate)}$$

The definitions of sine and cosine apply to angle θ of any magnitude. Also note:

- Because $-1 \leq x \leq 1$, $\cos\theta$ is from -1 to 1.
- Because $-1 \leq y \leq 1$, $\sin\theta$ is from -1 to 1.

Referring to the unit circle diagram, if the point P is:

(a) at A, $\theta = 0°$, the coordinates of A are $(1, 0)$ and hence $\cos 0° = 1$, $\sin 0° = 0$

(b) at B, $\theta = 90°$, the coordinates of B are $(0, 1)$ and hence $\cos 90° = 0$, $\sin 90° = 1$

(c) at C, $\theta = 180°$, the coordinates of C are $(-1, 0)$ and hence $\cos 180° = -1$, $\sin 180° = 0$

(d) at D, $\theta = 270°$, the coordinates of D are $(0, -1)$ and hence $\cos 270° = 0$, $\sin 270° = -1$

(e) all the way around to A again, $\theta = 360°$, the coordinates of A are $(1, 0)$ and hence $\cos 360° = 1$, $\sin 360° = 0$.

Note that $\cos 360° = \cos 0° = 1$, and $\sin 360° = \sin 0° = 0$.

There are four other trigonometric ratios: tangent (tan), cotangent (cot), secant (sec) and cosecant (cosec, or sometimes csc). These can all be defined in terms of cos and sin. The functions $\cot\theta$, $\sec\theta$ and $\operatorname{cosec}\theta$ are the reciprocals of $\tan\theta$, $\cos\theta$ and $\sin\theta$ respectively.

$$\tan\theta = \frac{y}{x} = \frac{\sin\theta}{\cos\theta},\ \cos\theta \neq 0 \qquad \cot\theta = \frac{x}{y} = \frac{\cos\theta}{\sin\theta} = \frac{1}{\tan\theta},\ \sin\theta \neq 0$$

$$\sec\theta = \frac{1}{x} = \frac{1}{\cos\theta},\ \cos\theta \neq 0 \qquad \operatorname{cosec}\theta = \frac{1}{y} = \frac{1}{\sin\theta},\ \sin\theta \neq 0$$

Because sin and cos appear as denominators, there are restrictions on the values of θ in these functions. The functions tan and sec are undefined when $\cos\theta = 0$, which is when $\theta = 90°, 270°, 450°, \ldots$ (i.e. $90° + n180°$, where n is any integer); cot and cosec are undefined when $\sin\theta = 0$, which is when $\theta = 0°, 180°, 360°, \ldots$ (i.e. $n180°$, where n is any integer).

The tangent ratio can also be considered as a ratio without reference to sin or cos.

At the point $A(1, 0)$ where the unit circle cuts the x-axis, draw a tangent line AT.

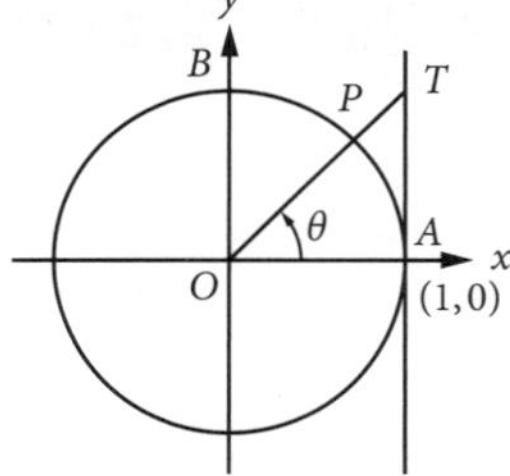

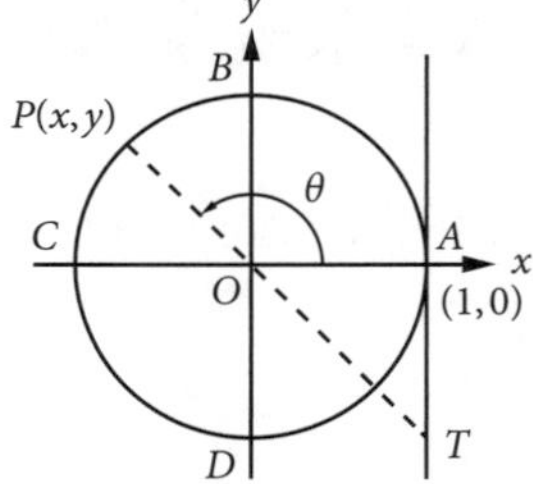

If $\angle AOP = \theta$, you can define $\tan\theta$ as the y-coordinate of T. (The tan function gets its name from this tangent line.) If P is at B, $\theta = 90°$, OP (or OB) is parallel to AT and so $\tan 90°$ is undefined.

Symmetry properties of trigonometric ratios

The coordinate axes divide the circle into four quarters, called **quadrants**.

First quadrant $0° < \theta < 90°$

Consider the point $P(a, b)$, where P lies between A and B.

Both the x- and y-coordinates of P are positive numbers, so all the ratios are positive.

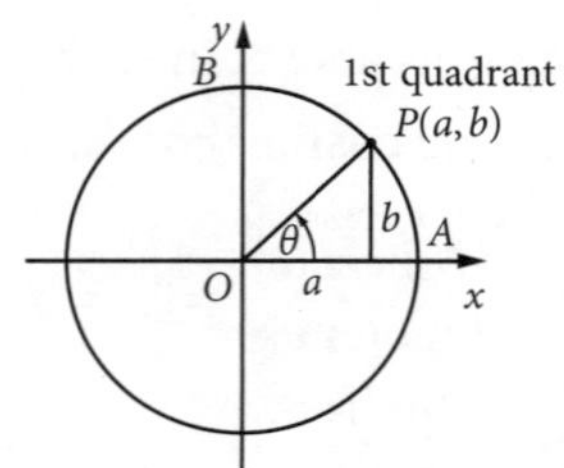

Second quadrant $90^\circ < \theta < 180^\circ$

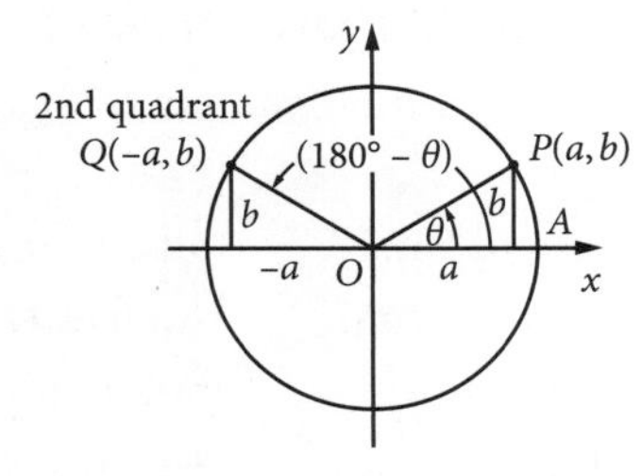

Consider the point $P(a, b)$ in the first quadrant such that $\angle AOP = \theta$, and a point Q in the second quadrant such that $\angle AOQ = 180^\circ - \theta$.

By symmetry, the coordinates of Q are $(-a, b)$. Hence:

$$\cos(180^\circ - \theta) = -a = -\cos\theta$$

$$\sin(180^\circ - \theta) = b = \sin\theta$$

$$\tan(180^\circ - \theta) = \frac{b}{-a} = -\tan\theta$$

Because the triangles are congruent, we can see that for every angle in the second quadrant there is a corresponding angle in the first quadrant whose sine, cosine and tangent ratios are numerically the same. You can find this angle by subtracting the second quadrant angle from 180°. (For example, $180^\circ - 40^\circ = 140^\circ$, so the angle in the first quadrant corresponding to 140° is 40°.) Because the x-coordinate is negative in the second quadrant, the values of cos and tan are now negative; sin remains positive. For example:

$$\cos 140^\circ = \cos(180^\circ - 40^\circ) = -\cos 40^\circ \approx -0.7660$$

$$\sin 140^\circ = \sin(180^\circ - 40^\circ) = \sin 40^\circ \approx 0.6428$$

$$\tan 140^\circ = \tan(180^\circ - 40^\circ) = -\tan 40^\circ \approx -0.8391$$

Third quadrant $180^\circ < \theta < 270^\circ$

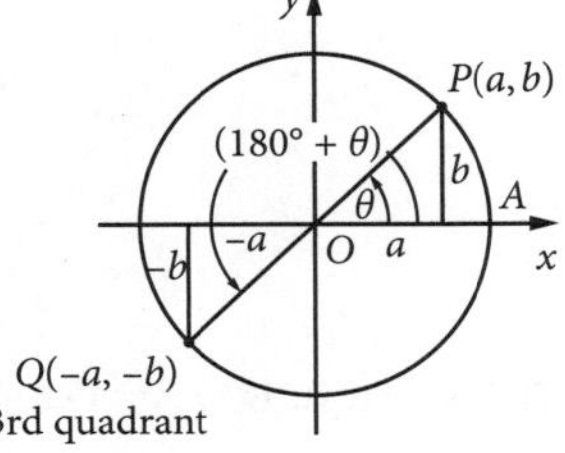

Consider the point $P(a, b)$ in the first quadrant such that $\angle AOP = \theta$, and a point Q in the third quadrant such that $\angle AOQ = 180^\circ + \theta$.

By symmetry, the coordinates of Q are $(-a, -b)$. Hence:

$$\cos(180^\circ + \theta) = -a = -\cos\theta$$

$$\sin(180^\circ + \theta) = -b = -\sin\theta$$

$$\tan(180^\circ + \theta) = \frac{-b}{-a} = \tan\theta$$

For every angle in the third quadrant there is a corresponding angle in the first quadrant whose sine, cosine and tangent ratios are numerically the same. You can find this angle by subtracting 180° from the third quadrant angle. (For example, $220^\circ - 180^\circ = 40^\circ$ and $220^\circ = 180^\circ + 40^\circ$, so the angle in the first quadrant corresponding to 220° is 40°.) Because in the third quadrant $x < 0$ and $y < 0$, only tan is positive, while sin and cos are both negative. For example:

$$\cos 220^\circ = \cos(180^\circ + 40^\circ) = -\cos 40^\circ \approx -0.7660$$

$$\sin 220^\circ = \sin(180^\circ + 40^\circ) = -\sin 40^\circ \approx -0.6428$$

$$\tan 220^\circ = \tan(180^\circ + 40^\circ) = \tan 40^\circ \approx 0.8391$$

Fourth quadrant $270^\circ < \theta < 360^\circ$

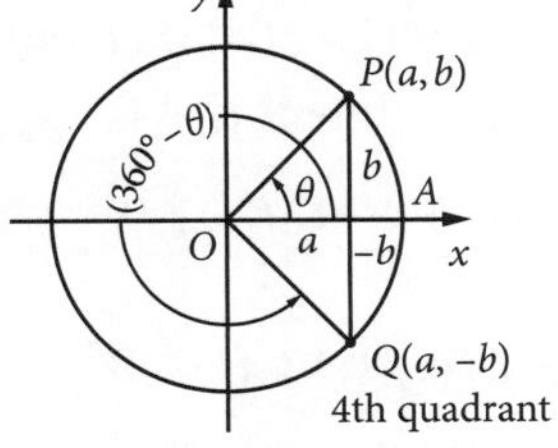

Consider the point $P(a, b)$ in the first quadrant such that $\angle AOP = \theta$, and a point Q in the fourth quadrant such that $\angle AOQ = 360^\circ - \theta$.

By symmetry, the coordinates of Q are $(a, -b)$. Hence:

$$\cos(360^\circ - \theta) = a = \cos\theta$$

$$\sin(360^\circ - \theta) = -b = -\sin\theta$$

$$\tan(360^\circ - \theta) = \frac{-b}{a} = -\tan\theta$$

For every angle in the fourth quadrant there is a corresponding angle in the first quadrant whose sine, cosine and tangent ratios are numerically the same. You can find this angle by subtracting the fourth quadrant angle from 360°. (For example, $320^\circ = 360^\circ - 40^\circ$, so the angle in the first quadrant corresponding to 320° is 40°.) In the fourth quadrant only cos is positive. For example:

$$\cos 320^\circ = \cos(360^\circ - 40^\circ) = \cos 40^\circ \approx 0.7660$$

$$\sin 320^\circ = \sin(360^\circ - 40^\circ) = -\sin 40^\circ \approx -0.6428$$

$$\tan 320^\circ = \tan(360^\circ - 40^\circ) = -\tan 40^\circ \approx -0.8391$$

Sign of the trigonometric ratios

The sign of $\cos\theta$, $\sin\theta$ and $\tan\theta$ for the first four quadrants can be summarised as follows:

- First quadrant: all are positive (A)
- Second quadrant: $\sin\theta$ only is positive (S)
- Third quadrant: $\tan\theta$ only is positive (T)
- Fourth quadrant: $\cos\theta$ only is positive (C)

S A
T C

The trigonometric ratio positive signs in the different quadrants can be remembered with the mnemonic **ASTC**: **A**ll **S**tations **T**o **C**entral.

Complementary angles: θ and $(90° - \theta)$

Consider the point $P(a, b)$ on the unit circle such that $\angle AOP = \theta$, and a point Q such that $\angle AOQ = (90° - \theta)$.

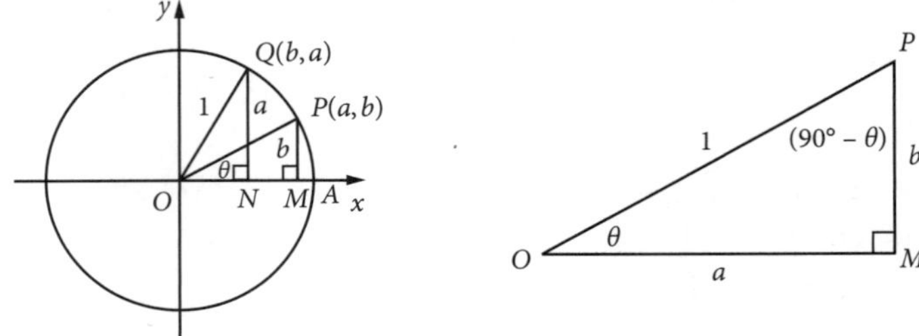

From congruent triangles, the coordinates of Q are (b, a) because $ON = PM = b$ and $QN = OM = a$. Hence:

$$\sin(90° - \theta) = a = \cos\theta \qquad \cos(90° - \theta) = b = \sin\theta$$

$$\tan(90° - \theta) = \frac{a}{b} = \cot\theta \qquad \cot(90° - \theta) = \frac{b}{a} = \tan\theta$$

$$\sec(90° - \theta) = \frac{1}{b} = \operatorname{cosec}\theta \qquad \operatorname{cosec}(90° - \theta) = \frac{1}{a} = \sec\theta$$

These relationships are said to be **complementary**. This is why the prefix 'co-' is in the words cosine, cosecant and cotangent.

The ratios sine and cosine, tangent and cotangent, secant and cosecant are complementary pairs. For example:

$\sin 50° = \cos(90° - 50°) = \cos 40°$ $\qquad$ $\tan 75° = \cot(90° - 75°) = \cot 15°$

$\sec 80° = \operatorname{cosec}(90° - 10°) = \operatorname{cosec} 10°$ $\qquad$ $\cos 60° = \sin(90° - 60°) = \sin 30°$

Use your calculator to verify these results.

Negative angles: $\theta < 0°$

So far we have only considered $\theta > 0°$. If we start from the point A and rotate anticlockwise to P, then $\theta > 0°$. However, if we rotate clockwise to Q so that $\angle AOQ = \angle AOP$, then $\theta < 0°$.

Hence, by symmetry:

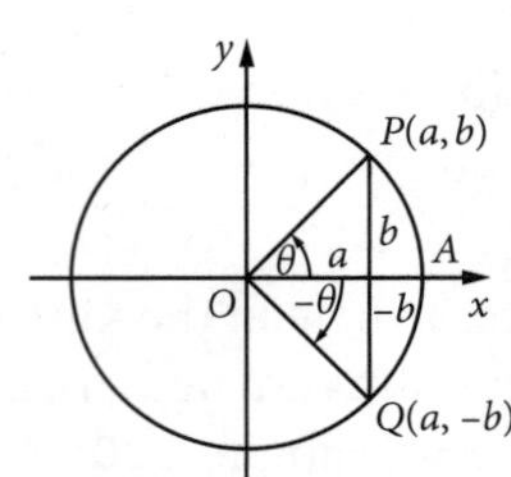

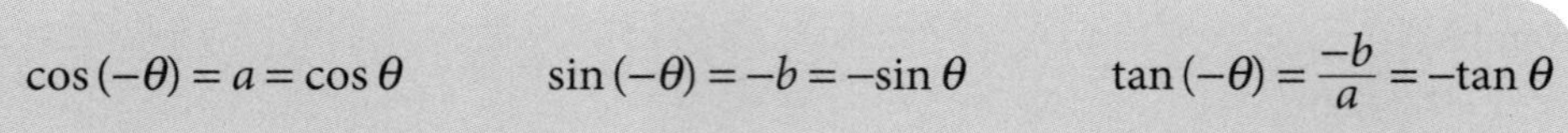

$$\cos(-\theta) = a = \cos\theta \qquad \sin(-\theta) = -b = -\sin\theta \qquad \tan(-\theta) = \frac{-b}{a} = -\tan\theta$$

For example:

$\cos(-40°) = \cos 40° \approx 0.7660$

$\tan(-25°) = -\tan 25° \approx -0.4663$

$\sin(-70°) = -\sin 70° \approx -0.9397$

$\cos(-160°) = \cos 160° = -\cos 20° \approx -0.9397$

$\tan(-245°) = -\tan 245° = -\tan 65° \approx -2.1445$

$\sin(-210°) = -\sin 210° = -(-\sin 30°) = \sin 30° = 0.5$

$\cot(-135°) = -\cot 135° = \cot 45° = \tan 45° = \frac{1}{\sqrt{2}} \approx 0.7071$

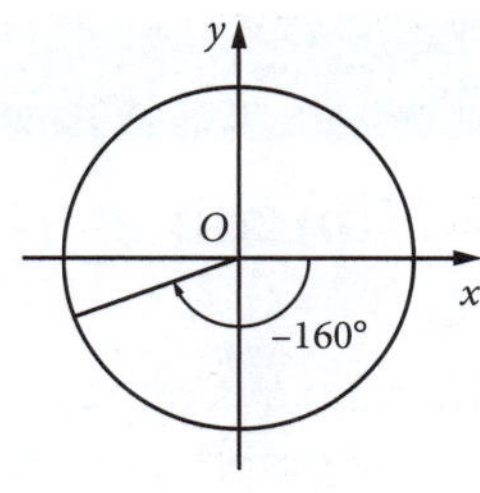

EXERCISE 2.2 ANGLES OF ANY MAGNITUDE

1 In which of the four quadrants is the following true?

(a) $\sin\theta > 0$ **(b)** $\tan\theta < 0$ **(c)** $\cos\theta < 0$ **(d)** $\sin\theta < 0$ and $\tan\theta < 0$
(e) $\sin\theta > 0$ and $\cos\theta < 0$ **(f)** $\cos\theta < 0$ and $\tan\theta > 0$ **(g)** $\cos\theta > 0$ and $\tan\theta > 0$

2 State the quadrant of each angle.

(a) 72° **(b)** 114° **(c)** 95° **(d)** 200° **(e)** 321°
(f) 183° **(g)** 83° **(h)** 216° **(i)** 300° **(j)** 155°

3 Express each of the following as a trigonometric ratio of angle A.

(a) $\sin(180° - A)$ **(b)** $\cos(90° - A)$ **(c)** $\tan(360° - A)$
(d) $\cos(180° + A)$ **(e)** $\sin(360° - A)$ **(f)** $\cot(90° - A)$

4 Use a calculator to evaluate $\sin\theta$, $\cos\theta$, $\tan\theta$, $\operatorname{cosec}\theta$, $\sec\theta$, and $\cot\theta$ for each of the following values of θ, writing each answer correct to 4 decimal places:

(a) (i) 125° **(ii)** 152° **(iii)** 117° **(b) (i)** 205° **(ii)** 217° **(iii)** 251°
(c) (i) 282° **(ii)** 301° **(iii)** 342° **(d) (i)** −25° **(ii)** −122° **(iii)** −215°

5 If $\sin\alpha = 0.2$, write the value of:

(a) $\sin(180° - \alpha)$ **(b)** $\sin(360° - \alpha)$ **(c)** $\sin(-\alpha)$
(d) $\cos(90° - \alpha)$ **(e)** $\sin(180° + \alpha)$ **(f)** $\operatorname{cosec}\alpha$

6 If $\tan\theta = t$, express in terms of t:

(a) $\cot\theta$ **(b)** $\cot(90° - \theta)$ **(c)** $\tan(180° - \theta)$
(d) $\tan(360° - \theta)$ **(e)** $\cot(180° - \theta)$ **(f)** $\tan(180° + \theta)$

7 If $\cos A = c$, express in terms of c:

(a) $\sec A$ **(b)** $\cos(-A)$ **(c)** $\cos(180° - A)$
(d) $\cos(360° - A)$ **(e)** $\sec(-A)$ **(f)** $\cos(180° + A)$

8 Use a calculator to evaluate, correct to 4 decimal places:

(a) $\tan 305°$ **(b)** $\sin 212°$ **(c)** $\cos(-140°)$
(d) $\sin(-160°)$ **(e)** $\cot 42°$ **(f)** $\cos 260°$

9 If θ is an angle in the second quadrant, state whether the following are positive or negative:

(a) $\cos(180° - \theta)$ **(b)** $\tan(180° - \theta)$ **(c)** $\sin(90° - \theta)$
(d) $\sin(360° - \theta)$ **(e)** $\cos(180° + \theta)$ **(f)** $\tan(90° - \theta)$

10 If $90° < \theta < 180°$, use a unit circle diagram to show that:

(a) $\cos(180° + \theta) = -\cos\theta$ **(b)** $\sin(360° - \theta) = -\sin\theta$

11 Which expression is equal to $\sin(360° + \theta)$?

A $\cos\theta$ **B** $-\sin\theta$ **C** $-\cos\theta$ **D** $\sin\theta$

12 Write 'correct' or 'incorrect' for each answer: $\cos\theta = \ldots$

(a) $\cos(360° - \theta)$ **(b)** $\sin(180° - \theta)$ **(c)** $-\cos(180° + \theta)$ **(d)** $\sin(90° - \theta)$

2.3 TRIGONOMETRIC GRAPHS

Graphs of trigonometric functions

Sine (sin) and cosine (cos)

In the figure below, the graph of $y = \sin\theta$ is represented by the continuous curve, and the graph of $y = \cos\theta$ is represented by the broken curve. The graphs are drawn in the domain $-360° \leq \theta \leq 360°$. For a domain of real numbers, the graphs continue similarly in both directions: for example, the graph for the domain $360° \leq \theta \leq 720°$ is the same as the graph for the domain $0° \leq \theta \leq 360°$, as $720° = 360° + 360°$.

As θ increases, the values of $\sin\theta$ and $\cos\theta$ repeat after each interval or **period** of 360°. The functions sine and cosine are therefore called **periodic functions**, with the period being 360°. We saw earlier that the point $P(x, y)$ corresponds to angles θ, $360° + \theta$, $720° + \theta$ and so on, and hence $\sin(360° + \theta) = \sin\theta$ and $\cos(360° + \theta) = \cos\theta$.

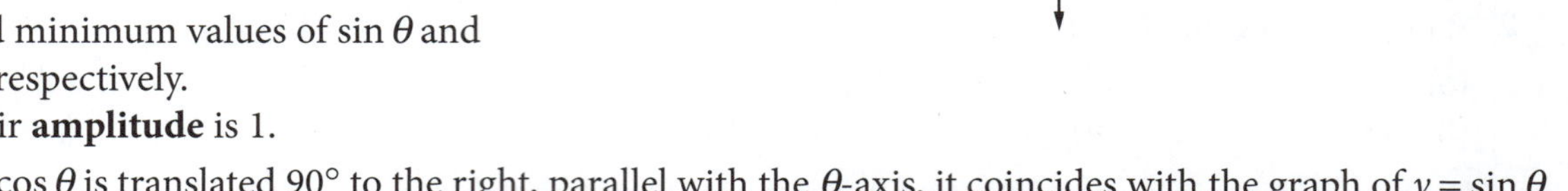

The maximum and minimum values of $\sin\theta$ and $\cos\theta$ are 1 and –1 respectively.
In other words their **amplitude** is 1.

If the graph of $y = \cos\theta$ is translated 90° to the right, parallel with the θ-axis, it coincides with the graph of $y = \sin\theta$. This can be seen to follow from the fact that $\cos\theta = \sin(90° + \theta)$.

MAKING CONNECTIONS

Transforming a sine graph into a cosine graph

Move the slider to see how the translated sine graph matches the cosine graph.

Tangent (tan)

The graph of $y = \tan\theta$ can be drawn by plotting points from a table of values or by using graphing software. But you can determine features of the graph simply from the definition of tan:

- Because $\tan\theta = \dfrac{\sin\theta}{\cos\theta}$, the graph of $y = \tan\theta$ will not exist for values of θ that make $\cos\theta = 0$. In the domain $0° \leq \theta \leq 360°$, $\tan\theta$ will be undefined when $\theta = 90°, 270°$.
- If $\sin\theta = 0$ then $\tan\theta = 0$, so in the domain $0° \leq \theta \leq 360°$ you know that $\tan\theta = 0$ when $\theta = 0°, 180°, 360°$.
- If $\sin\theta = \cos\theta$ then $\tan\theta = 1$. In the domain $0° \leq \theta \leq 360°$, $\tan\theta = 1$ when $\theta = 45°, 225°$; $\tan\theta = -1$ when $\theta = 135°, 315°$.

For $0° \leq \theta \leq 360°$, the graph of $y = \tan\theta$ is:

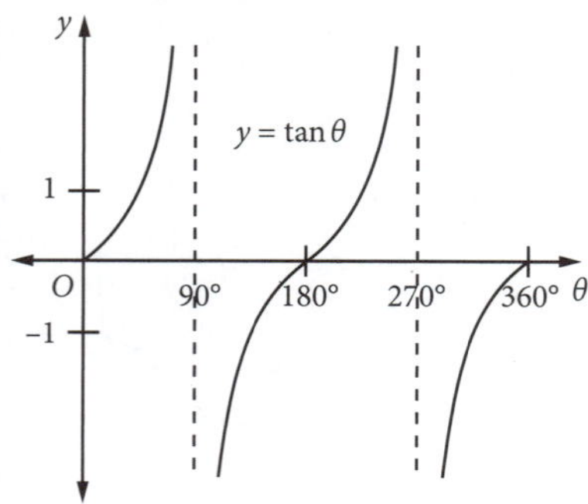

As θ increases, the values of $\tan\theta$ repeat after each interval or period of 180°. The curve has no greatest or least value, so it does not have an amplitude: $y = \tan\theta$ is **discontinuous** at $\theta = 90°, 270°$. The dotted vertical lines drawn at these values are called **asymptotes**.

Cotangent (cot)

The graph of $y = \cot\theta$ can be drawn by plotting points from a table of values or by using graphing software, but you can determine some features from its definition:

- Because $\cot\theta = \dfrac{1}{\tan\theta} = \dfrac{\cos\theta}{\sin\theta}$, the graph of $y = \cot\theta$ will not exist for values of θ that make $\tan\theta = 0$ or $\sin\theta = 0$. In the domain $0° \le \theta \le 360°$, $\cot\theta$ will be undefined when $\theta = 0°, 180°, 360°$.
- If $\cos\theta = 0$ then $\cot\theta = 0$. In the domain $0° \le \theta \le 360°$, $\cot\theta = 0$ when $\theta = 90°, 270°$.
- If $\sin\theta = \cos\theta$ then $\cot\theta = 1$. In the domain $0° \le \theta \le 360°$, $\cot\theta = 1$ when $\theta = 45°, 225°$; $\cot\theta = -1$ when $\theta = 135°, 315°$.

For $0° \le \theta \le 360°$, the graph of $y = \cot\theta$ is:

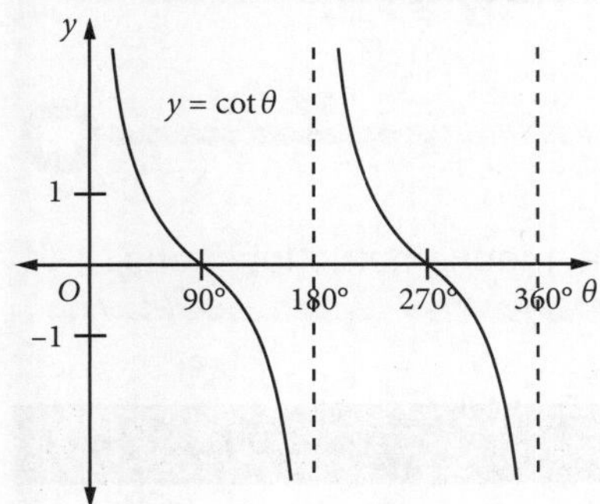

As θ increases, the values of $\cot\theta$ repeat after an interval or period of 180°. The curve has no greatest or least value, so it does not have an amplitude: $y = \cot\theta$ is discontinuous at $\theta = 0°, 180°, 360°$. The dotted vertical lines drawn at these values are asymptotes.

Because $\cot\theta = \tan(90° - \theta)$ and $\cot\theta = \dfrac{1}{\tan\theta}$, these two ratios are both complementary ratios and reciprocal ratios.

Cosecant (cosec)

The graph of $y = \operatorname{cosec}\theta$ can be drawn by plotting points from a table of values or by using graphing software, but you can determine some features from its definition:

- Because $\operatorname{cosec}\theta = \dfrac{1}{\sin\theta}$, the graph of $y = \operatorname{cosec}\theta$ will not exist for values of θ that make $\sin\theta = 0$. In the domain $0° \le \theta \le 360°$, $\operatorname{cosec}\theta$ will be undefined when $\theta = 0°, 180°, 360°$.
- If $\sin\theta = 1$ then $\operatorname{cosec}\theta = 1$. In the domain $0° \le \theta \le 360°$, $\operatorname{cosec}\theta = 1$ when $\theta = 90°$; $\operatorname{cosec}\theta = -1$ when $\theta = 270°$.

For $0° \le \theta \le 360°$, the graph of $y = \operatorname{cosec}\theta$ is:

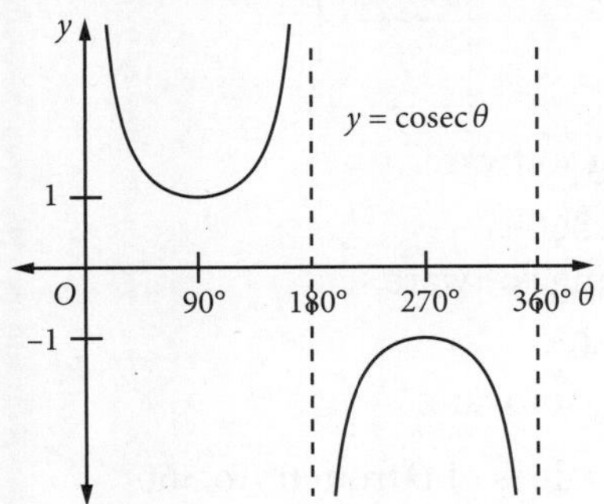

As θ increases, the values of $\operatorname{cosec}\theta$ repeat after an interval or period of 360°. The curve has no greatest or least value, so it does not have an amplitude, but the range is $\operatorname{cosec}\theta \le -1$ or $\operatorname{cosec}\theta \ge 1$. Also, $y = \operatorname{cosec}\theta$ is discontinuous at $\theta = 0°, 180°, 360°$. The dotted vertical lines drawn at these values are asymptotes.

Secant (sec):

The graph of $y = \sec\theta$ can be drawn by plotting points from a table of values or by using graphing software, but you can determine some features from its definition:

- Because $\sec\theta = \dfrac{1}{\cos\theta}$, the graph of $y = \sec\theta$ will not exist for values of θ that make $\cos\theta = 0$. In the domain $0° \le \theta \le 360°$, $\sec\theta$ will be undefined when $\theta = 90°, 270°$.
- If $\cos\theta = 1$ then $\sec\theta = 1$. In the domain $0° \le \theta \le 360°$, $\sec\theta = 1$ when $\theta = 0°, 360°$; $\sec\theta = -1$ when $\theta = 180°$.

For $0° \le \theta \le 360°$, the graph of $y = \sec\theta$ is:

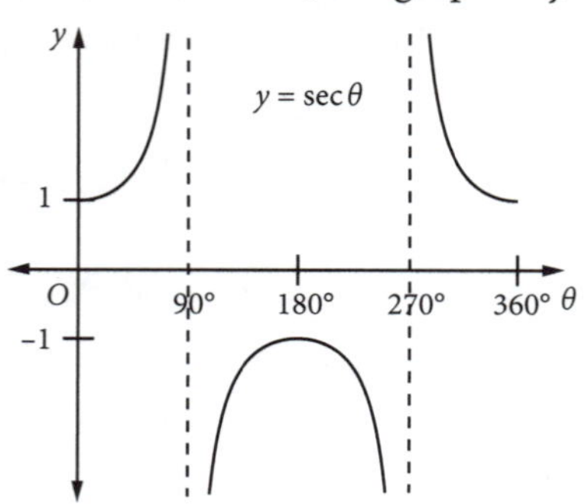

As θ increases, the values of $\sec\theta$ repeat after an interval or period of 360°. The curve has no greatest or least value, so it does not have an amplitude, but the range is $\sec\theta \le -1$ or $\sec\theta \ge 1$. Also, $y = \sec\theta$ is discontinuous at $\theta = 90°, 270°$. The dotted vertical lines drawn at these values are asymptotes.

Graphing software is an ideal way to investigate these trigonometric curves.

EXPLORE FURTHER

Reciprocal trigonometric functions

Use technology to explore the relationship between trigonometric and reciprocal trigonometric functions.

EXERCISE 2.3 TRIGONOMETRIC GRAPHS

Draw these graphs using a table of values where given and appropriate graphing software.

1 **(a)** Complete the following table and use it to draw the graph of $y = \sin\theta$ for values of θ from 0° to 360°.

θ	0°	30°	90°	150°	180°	210°	270°	330°	360°
$\sin\theta$									

(b) Draw the graph of $y = \sin\theta$ for values of θ from 0° to 360° using graphing software.

2 **(a)** Complete the following table and use it to draw the graph of $y = \cos\theta$ for values of θ from 0° to 360°.

θ	0°	60°	90°	120°	180°	240°	270°	300°	360°
$\cos\theta$									

(b) Draw the graph of $y = \cos\theta$ for values of θ from 0° to 360° using graphing software.

3 **(a)** Use your graph in question **1** to draw the graph of $y = \operatorname{cosec}\theta$ for $0° \le \theta \le 360°$.
(b) Draw the graph of $y = \operatorname{cosec}\theta$ for values of θ from 0° to 360° using graphing software.

4 **(a)** Use your graph in question **2** to draw the graph of $y = \sec\theta$ for $0° \le \theta \le 360°$.
(b) Draw the graph of $y = \sec\theta$ for values of θ from 0° to 360° using graphing software.

5 **(a)** Complete the following table and use it to draw the graph of $y = \tan\theta$ for values of θ from 0° to 360°.

θ	0°	30°	90°	150°	180°	210°	270°	330°	360°
$\tan\theta$									

(b) Draw the graph of $y = \tan\theta$ for values of θ from 0° to 360° using graphing software.

6 **(a)** Use your graph in question **5** to draw the graph of $y = \cot\theta$ for $0° \le \theta \le 360°$.
(b) Draw the graph of $y = \cot\theta$ for values of θ from 0° to 360° using graphing software.

7 Repeat questions **1** to **6** using a domain of $-180° \le \theta \le 180°$.

2.4 EXACT VALUES OF TRIGONOMETRIC RATIOS

The exact values of sin, cos and tan for 30°, 60°, and 45° can be found from the following diagrams.

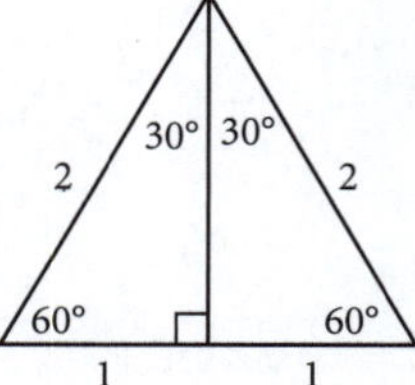

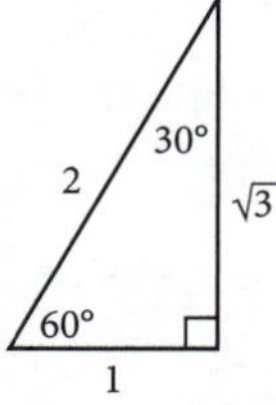

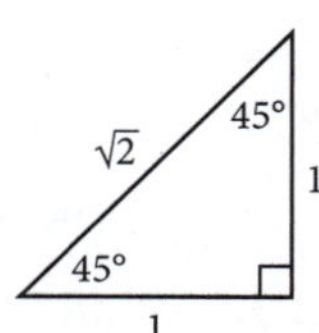

You can use Pythagoras' theorem to calculate the side lengths of the right-angled triangles and relate the lengths to the trigonometric values, as summarised in the following table.

θ	30°	45°	60°
$\sin\theta$	$\frac{1}{2}$	$\frac{1}{\sqrt{2}}$	$\frac{\sqrt{3}}{2}$
$\cos\theta$	$\frac{\sqrt{3}}{2}$	$\frac{1}{\sqrt{2}}$	$\frac{1}{2}$
$\tan\theta$	$\frac{1}{\sqrt{3}}$	1	$\sqrt{3}$

You can either remember how to calculate these values from the triangles or remember the table of values. It is also easy to obtain the exact values for the reciprocal functions cosec, sec and cot by taking the reciprocal of each value above. These are given in the following table.

θ	30°	45°	60°
$\operatorname{cosec}\theta$	2	$\sqrt{2}$	$\frac{2}{\sqrt{3}}$
$\sec\theta$	$\frac{2}{\sqrt{3}}$	$\sqrt{2}$	2
$\cot\theta$	$\sqrt{3}$	1	$\frac{1}{\sqrt{3}}$

Example 2

Calculate the exact length of the sides in each triangle.

(a)

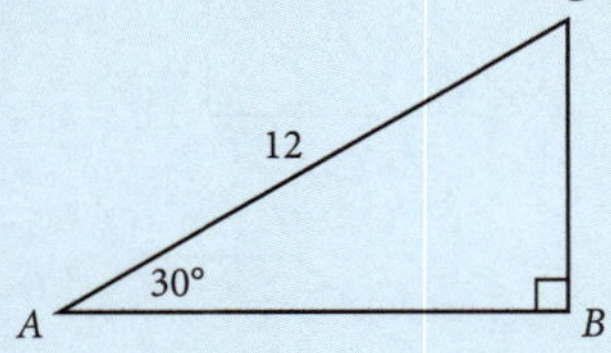

(b)

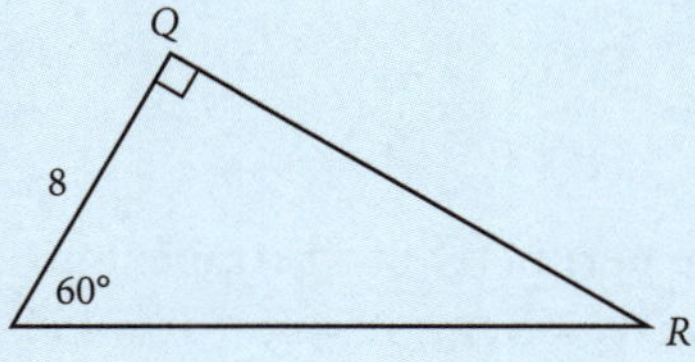

Solution

(a) $\sin 30° = \frac{BC}{12}$

$BC = 12\sin 30°$

$= 12 \times \frac{1}{2}$

$= 6$

$\cos 30° = \frac{AB}{12}$

$AB = 12\cos 30°$

$= 12 \times \frac{\sqrt{3}}{2}$

$= 6\sqrt{3}$

(b) $\tan 60° = \frac{QR}{8}$

$QR = 8\tan 60°$

$= 8\sqrt{3}$

$PR^2 = QR^2 + PQ^2$

$= (8\sqrt{3})^2 + 8^2$

$= 192 + 64 = 256$

$PR = \sqrt{256} = 16$

EXERCISE 2.4 EXACT VALUES OF TRIGONOMETRIC RATIOS

Give the exact answer to each of the following, expressing lengths in simplest surd form where necessary. (Do not use a calculator.)

1 In ΔABC, $B = 90°$, $A = 30°$, $AC = 20$ cm. Calculate the lengths of:
(a) BC (b) AB

2 In ΔABC, $C = 90°$, $A = 45°$, $BC = 10$ cm. The lengths of AC and AB are:

A $AC = 10\sqrt{2}$ cm, $AB = 10$ cm **B** $AC = 10$ cm, $AB = 10\sqrt{2}$ cm
C $AC = 10$ cm, $AB = 5\sqrt{2}$ cm **D** $AC = 10$ cm, $AB = 10$ cm

3 A vertical pole of height 15 m stands on level ground and a straight wire 30 m long joins the top of the pole to a point on the ground. Find:
(a) the distance of this point on the ground from the foot of the pole
(b) the angle the wire makes with the ground.

4 A ladder 10 m long, standing on level ground, leans against a vertical wall and makes an angle of 60° with the ground. Calculate:
(a) how high up the wall the ladder reaches
(b) the distance of the foot of the ladder from the wall.

5 In ΔABC, $AB = 12$ cm, $AC = 8$ cm, $A = 60°$. CF is drawn perpendicular to AB to meet AB at F. Calculate the length of:
(a) AF
(b) FC
(c) BC

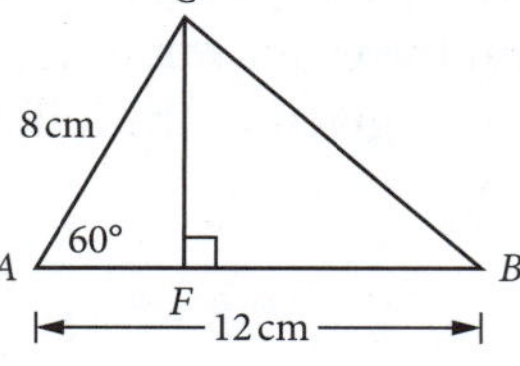

6 For the given diagram, calculate the length of:
(a) AC
(b) DC
(c) AB
(d) BC

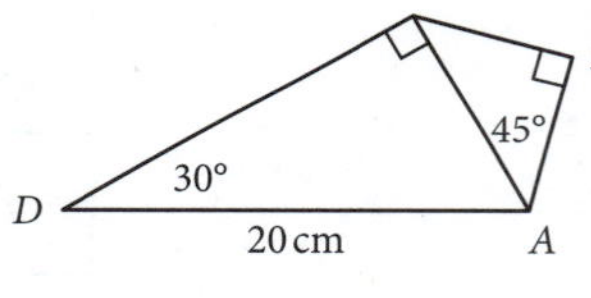

7 In the diagram, $AC = 12$ cm. Calculate the length of:
(a) AB
(b) BC
(c) DC
(d) AD

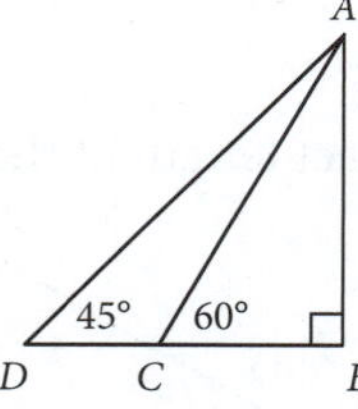

8 Calculate the perimeter of the trapezium $ABCD$ given that $AD = 16$ cm, $DC = 8$ cm, $A = 30°$, $B = 45°$.

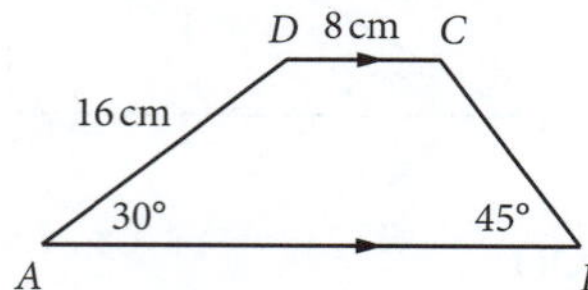

9 A stepladder stands on horizontal ground with its feet 2 m apart. If the angle formed by the legs is 60°, how high above the ground is the top of the ladder?

10 The magnitude of the angle formed by the diagonal of a rectangle and one of its longer sides is 30°. Find the dimensions of the rectangle if the length of the diagonal is 60 cm.

11 In the diagram, $AC = 16\,\text{cm}$, $\angle ABD = 90°$, $\angle ACB = 60°$ and $\angle ADB = 30°$. Indicate whether each answer is correct or incorrect.

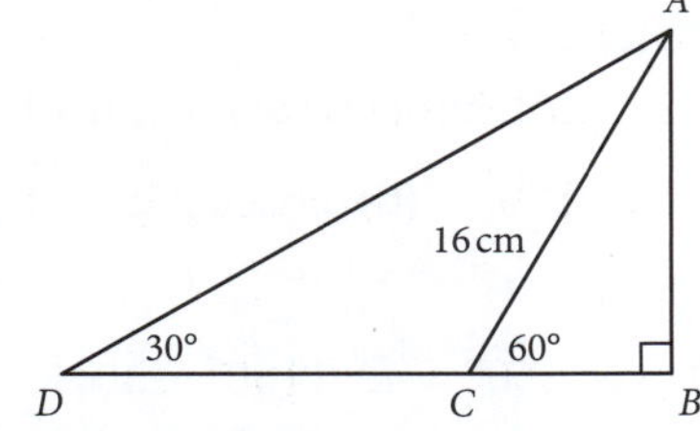

(a) $\angle DAC = 30°$ (b) $DC = 16\,\text{cm}$

(c) $AB = 8\,\text{cm}$ (d) $CB = 8\,\text{cm}$

2.5 MORE TRIGONOMETRIC EXACT VALUES

You have already found exact values of trigonometric ratios for the first quadrant angles of 30°, 45° and 60°. You can also find exact values for 0° and 90°.

Using the unit circle:

- $\theta = 0°$ when P is at A
- $\theta = 90°$ when P is at B

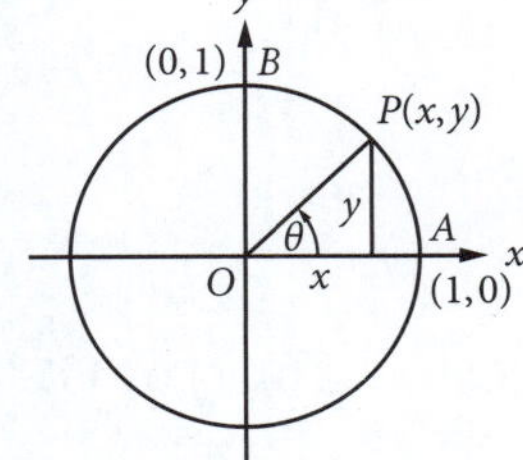

The table of exact values now becomes:

θ	0°	30°	45°	60°	90°
$\sin\theta$	0	$\frac{1}{2}$	$\frac{1}{\sqrt{2}}$	$\frac{\sqrt{3}}{2}$	1
$\cos\theta$	1	$\frac{\sqrt{3}}{2}$	$\frac{1}{\sqrt{2}}$	$\frac{1}{2}$	0
$\tan\theta$	0	$\frac{1}{\sqrt{3}}$	1	$\sqrt{3}$	undefined

You should learn these results for sin, cos and tan. Once learnt, they can easily be used to find exact values for the reciprocal functions cosec, sec and cot.

Example 3

Find the exact value of each expression.

(a) $\sin 150°$ (b) $\cos 225°$ (c) $\tan 240°$ (d) $\sin 270°$ (e) $\cos(-300°)$ (f) $\cos 405°$

Solution

(a) $\sin 150° = \sin(180° - 30°) = \sin 30° = \frac{1}{2}$

(b) $\cos 225° = \cos(180° + 45°) = -\cos 45° = -\frac{1}{\sqrt{2}}$

(c) $\tan 240° = \tan(180° + 60°) = \tan 60° = \sqrt{3}$

(d) $\sin 270° = \sin(180° + 90°) = -\sin 90° = -1$

(e) $\cos(-300°) = \cos 300° = \cos(360° - 60°)$
$= \cos 60° = \frac{1}{2}$

(f) $\cos 405° = \cos(360° + 45°) = \cos 45° = \frac{1}{\sqrt{2}}$

Example 4

Find the exact value of each expression.

(a) sec 150° **(b)** cosec 225° **(c)** cot 240° **(d)** cosec 270° **(e)** sec (−300°) **(f)** cot 405°

Solution

(a) $\sec 150° = \sec(180° - 30°)$

$= -\sec 30°$ (1st quadrant)

$= -\dfrac{2}{\sqrt{3}}$

(b) $\operatorname{cosec} 225° = \operatorname{cosec}(180° + 45°)$

$= -\operatorname{cosec} 45°$ (3rd quadrant)

$= -\sqrt{2}$

(c) $\cot 240° = \cot(180° + 60°)$

$= \cot 60°$ (3rd quadrant)

$= \dfrac{1}{\sqrt{3}}$

(d) $\operatorname{cosec} 270° = \dfrac{1}{y}$ (unit circle definition)

$= \dfrac{1}{-1}$ (In the unit circle an angle of 270° corresponds to a position of (0, −1).)

$= -1$

(e) $\sec(-300°) = \sec(-300° + 360°)$

$= \sec 60°$

$= 2$

(f) $\cot 405° = \cot(405° - 360°)$

$= \cot 45°$

$= 1$

Example 5

Find all values of θ, for $0° \le \theta \le 360°$, for which:

(a) $\cos\theta = \dfrac{1}{2}$ **(b)** $\sin\theta = -\dfrac{1}{\sqrt{2}}$ **(c)** $\tan\theta = 1$ **(d)** $\cos 2\theta = -\dfrac{\sqrt{3}}{2}$ **(e)** $\sin\theta = -1$

Solution

(a) $\cos\theta > 0$, so θ is in the 1st and 4th quadrants.

$\cos\theta = \dfrac{1}{2}$: $\theta = 60°, 360° - 60°$

$\theta = 60°, 300°$

(b) $\sin\theta < 0$, so θ is in the 3rd and 4th quadrants.

$\sin\theta = -\dfrac{1}{\sqrt{2}}$: $\theta = 180° + 45°, 360° - 45°$

$\theta = 225°, 315°$

(c) $\tan\theta > 0$, so θ is in the 1st and 3rd quadrants.

$\tan\theta = 1$: $\theta = 45°, 180° + 45°$

$\theta = 45°, 225°$

(d) $0° \le \theta \le 360°$, so $0° \le 2\theta \le 720°$. Because $\cos 2\theta < 0$, 2θ is in the 2nd and 3rd quadrants.

$\cos 2\theta = -\dfrac{\sqrt{3}}{2}$: $2\theta = 180° - 30°, 180° + 30°, 540° - 30°, 540° + 30°$

$2\theta = 150°, 210°, 510°, 570°$

$\theta = 75°, 105°, 255°, 285°$

(e) $\sin\theta < 0$, so θ is in the 3rd and 4th quadrants.

$\sin\theta = -1$: $\theta = 180° + 90°, 360° - 90°$

$\theta = 270°$

Example 6

Find all values of θ, for $0° \le \theta \le 360°$, for which:

(a) $\sec\theta = 2$ (b) $\operatorname{cosec}\theta = -\sqrt{2}$ (c) $\cot\theta = 1$ (d) $\sec 2\theta = -\dfrac{2}{\sqrt{3}}$ (e) $\operatorname{cosec}\theta = -1$

Solution

(a) $\sec\theta > 0$, so θ is in the 1st and 4th quadrants.

$\sec 60° = 2$: $\theta = 60°, 360° - 60°$

$\theta = 60°, 300°$

(b) $\operatorname{cosec}\theta < 0$, so θ is in the 3rd and 4th quadrants.

$\operatorname{cosec} 45° = \sqrt{2}$: $\theta = 180° + 45°, 360° - 45°$

$\theta = 225°, 315°$

(c) $\cot\theta > 0$, so θ is in the 1st and 3rd quadrants.

$\cot 45° = 1$: $\theta = 45°, 180° + 45°$

$\theta = 45°, 225°$

(d) $0° \le \theta \le 360°$ so $0° \le 2\theta \le 720°$. Since $\sec 2\theta < 0$, then 2θ is in the 2nd and 3rd quadrants.

$\sec 30° = \dfrac{2}{\sqrt{3}}$: $2\theta = 180° - 30°, 180° + 30°, 540° - 30°, 540° + 30°$

$2\theta = 150°, 210°, 510°, 570°$

$\theta = 75°, 105°, 255°, 285°$

(e) $\operatorname{cosec}\theta = -1$ so $\dfrac{1}{y} = -1$ (unit circle definition)

$y = -1$

The position is $(0, -1)$

$\theta = 270°$

EXERCISE 2.5 MORE TRIGONOMETRIC EXACT VALUES

For questions **1** to **4**, write the exact value without using a calculator.

1 (a) $\sin 90°$ (b) $\cos 120°$ (c) $\tan 150°$ (d) $\cos 180°$ (e) $\sin 120°$

2 (a) $\sin 180°$ (b) $\cos 210°$ (c) $\tan 225°$ (d) $\cos 240°$ (e) $\tan 180°$

3 (a) $\sin 270°$ (b) $\tan 300°$ (c) $\tan 315°$ (d) $\cos 330°$ (e) $\sin 300°$

4 (a) $\sin 360°$ (b) $\cos 390°$ (c) $\tan 405°$ (d) $\cos 450°$ (e) $\sin 420°$

5 The exact value of $\sin 210°$ is:

A $-\dfrac{1}{2}$ B $\dfrac{\sqrt{3}}{2}$ C $\dfrac{2}{\sqrt{3}}$ D 0

6 Write 'correct' or 'incorrect' for each answer: $-1 = \ldots$

(a) $\cos 180°$ (b) $\sin 45°$ (c) $\sin 270°$ (d) $\tan 495°$

For questions **7** to **18**, find all the correct values of θ for $0° < \theta < 360°$.

7 $\sin\theta = -\dfrac{\sqrt{3}}{2}$ **8** $\tan\theta = -1$ **9** $\cos\theta = -1$ **10** $\sin\theta = \cos\theta$

11 $\sin\theta = 0$ **12** $2\cos\theta + 1 = 0$ **13** $2\sin\theta = \sqrt{3}$ **14** $\sin\theta + \sqrt{3}\cos\theta = 0$

15 $\sin 2\theta = 0.5$ **16** $\cos\dfrac{\theta}{2} = \dfrac{1}{2}$ **17** $\tan 2\theta = 1$ **18** $\sin\dfrac{\theta}{2} = -\dfrac{1}{\sqrt{2}}$

19 Use a calculator to evaluate cosec θ, sec θ, and cot θ for each of the following values of θ, writing each answer correct to 4 decimal places:

(a) (i) $125°$ (ii) $152°$ (iii) $117°$ (b) (i) $205°$ (ii) $217°$ (iii) $251°$
(c) (i) $282°$ (ii) $301°$ (iii) $342°$ (d) (i) $-25°$ (ii) $-122°$ (iii) $-215°$

20 If $\cos A = c$, express in terms of c:

(a) $\sec A$ (b) $\sec(180° - A)$ (c) $\sec(-A)$ (d) $\sec(180° + A)$

21 Use a calculator to evaluate, correct to 4 decimal places:

(a) cosec $305°$ (b) cot $212°$ (c) sec $(-140°)$ (d) cot $42°$

22 Which expression is equal to cosec $(360° + \theta)$?

A $\cos\theta$ B $\sin\theta$ C $\sec\theta$ D cosec θ

23 Write 'correct' or 'incorrect' for each answer: $\sec\theta = \ldots$

(a) $\sec(360° - \theta)$ (b) cosec $(180° - \theta)$ (c) $\sec(180° + \theta)$ (d) cosec $(90° - \theta)$

24 Write the exact value without using a calculator.

(a) cosec $90°$ (b) sec $150°$ (c) cot $240°$ (d) cosec $180°$ (e) sec $300°$
(f) cot $315°$ (g) sec $(-30°)$ (h) cosec $510°$ (i) cot $(-150°)$ (j) sec $120°$ + cosec $150°$

25 The exact value of cosec $300°$ is:

A $-\frac{\sqrt{3}}{2}$ B $\frac{\sqrt{3}}{2}$ C $-\frac{2}{\sqrt{3}}$ D $\frac{2}{\sqrt{3}}$

26 Solve for $0° \le \theta \le 360°$.

(a) cosec $\theta = -\frac{2}{\sqrt{3}}$ (b) $\cot\theta = \sqrt{3}$ (c) $\sec\theta = -1$ (d) cosec $\theta = -\sec\theta$

(e) cosec $\theta = 0$ (f) $\sec\theta = 2$ (g) $\cot 2\theta = \frac{1}{\sqrt{3}}$ (h) $\sec\frac{\theta}{2} = \frac{2}{\sqrt{3}}$

2.6 DIRECTION AND BEARING

Navigators and surveyors measure direction by reference to the points of the compass: north, south, east and west. Directions are indicated in terms of the number of degrees east or west of north or south, known as compass bearings, or are measured **clockwise from north** and written in standard three-figure notation, known as true bearings:

N 30° E or 030° means the direction is 30° east of north.

A **bearing** is a direction angle that indicates the direction of one point relative to another point. In this diagram, the bearing of B from A is N 70° E or 070°. (In bearings, the word 'from' indicates the starting point.) The bearing of A from B is S 70° W or 250°.

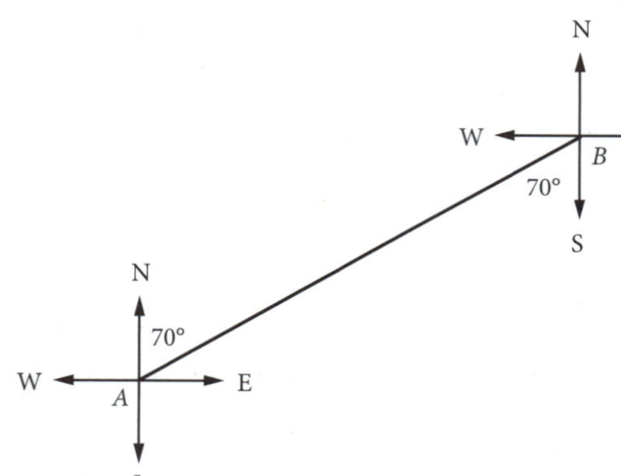

All bearings are in a horizontal plane.

MAKING CONNECTIONS

Bearings

Move the compass point to view the relationship between compass bearings and true bearings.

Example 7

Two yachts sail in a straight line away from a buoy B. One sails 12 km in the direction 038° and the other sails 16 km in the direction 128°.

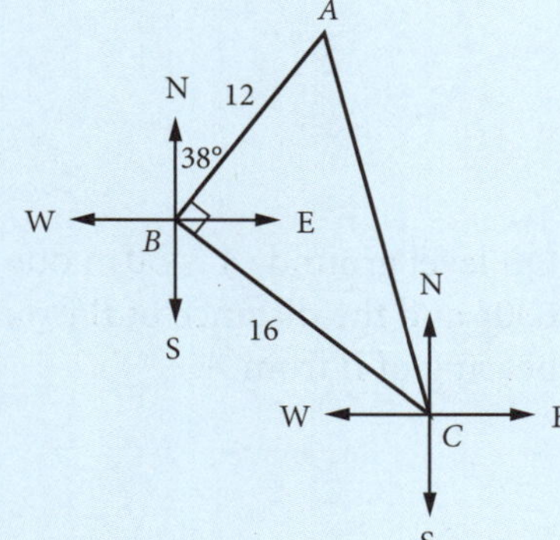

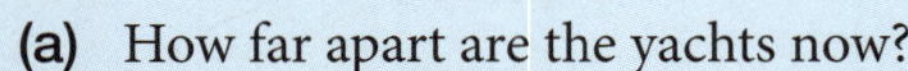

(a) How far apart are the yachts now?

(b) What is the bearing of the first yacht, as seen from the second yacht?

Solution

(a) $\angle ABC = 128° - 38° = 90°$

Use Pythagoras' theorem for ΔABC: $AC^2 = 12^2 + 16^2 = 400$

$AC = \sqrt{400} = 20$

The two yachts are 20 km apart.

(b) To find the bearing of A from C you first must calculate the size of $\angle ACN$.

In ΔABC: $\tan(\angle ACB) = \frac{12}{16} = 0.75$

$\angle ACB = 36° 52'$

From angle sum at B: $\angle EBC = 38°$

$BE \parallel CE$, alternate angles give: $\angle BCW = 38°$

$\angle ACN = 90° - (\angle ACB + \angle BCW)$

$= 90° - (36° 52' + 38°) = 15° 8'$

The bearing of A from C is N 15° 8′ W or 344° 52′.

The bearing of C from A would be S 15° 8′ E or 164° 52′.

When working with bearings, draw a diagram to show all the given information. Use the diagram to calculate missing angles.

EXERCISE 2.6 DIRECTION AND BEARING

1 Two towns A and B are 15 km apart, with B due west of A. (For compass directions, 'due' means 'exactly'.) Town C is due south of B and 12 km away. Which diagram represents this information correctly?

A B 15 km A, 12 km, C

B A 15 km B, 12 km, C

C B 15 km A, 12 km, C

D C, 12 km, B 15 km A

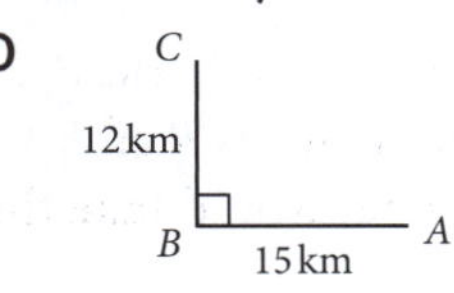

2 The bearing of B from A is 120°, the bearing of C from A is NE and the bearing of C from B is N. Find the bearing of:

(a) A from C

(b) A from B

(c) B from C

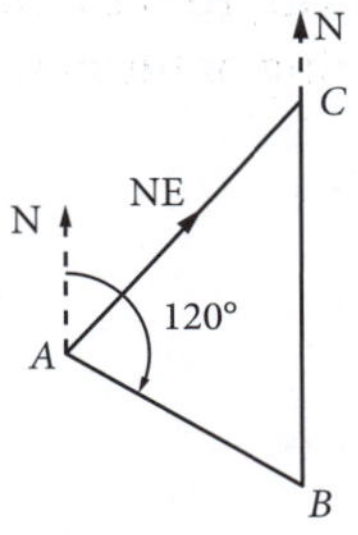

3 Two towns A and B are 15 km apart, with B due west of A. Town C is due south of B and 12 km away. Calculate the distance and bearing of A from C.

4 Ahmed cycles 15 km due north, then 12 km due east and finally 20 km due south. What are his distance and bearing from his original position?

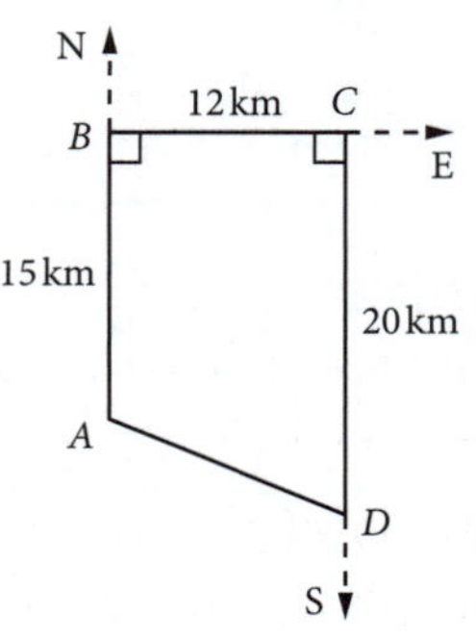

5 On level ground, A is 50 m due east of O. The bearing of B from O is 030° and the distance of B from O is also 50 m. Find the distance and bearing of B from A.

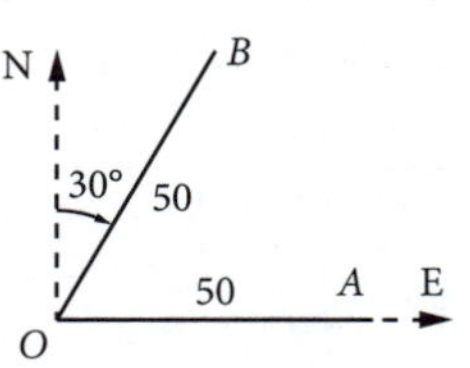

6 A is 5 km from a lighthouse, L, on a bearing N 37° W. B is 12 km from the same lighthouse on a bearing of S 53° W. Find the distance and bearing of B from A.

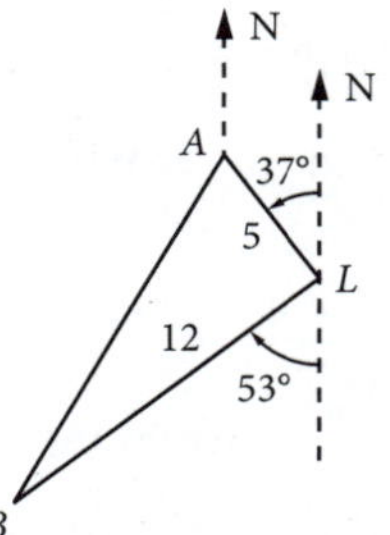

7 Karen and David set out from home at the same time. Karen cycles due north at $15\,\text{km}\,\text{h}^{-1}$ and David cycles due east at $20\,\text{km}\,\text{h}^{-1}$. Find:

(a) how far apart they are after 1 hour
(b) after how many minutes they are 10 km apart
(c) the bearing of Karen from David at any time.

8 A and B are two lighthouses, A being 20 km due north of B. The bearing of a ship is 145° from A and 055° from B. Calculate the distance of each lighthouse from the ship.

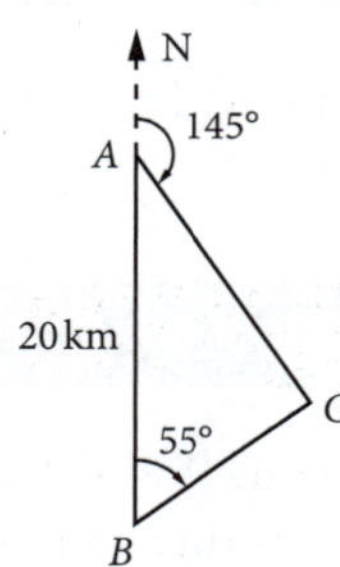

9 Two ports A and B are such that B is due west of A. A is due north of a ship C. The ship is on a course N 32° W and reaches B after travelling for 3 hours at $25\,\text{km}\,\text{h}^{-1}$. Calculate the distance between the two ports and the time it would take the ship to reach A from C.

10 A hiker walks 15 km from camp in the direction S 36° 52′ W and then walks 7 km due west. What are the distance and bearing of his new position from the camp?

11 A ship sails for 20 km on a course S 20° W and then 25 km on a course S 25° W. Calculate:

(a) how far south the ship now is from its original position
(b) how far west the ship now is from its original position
(c) the bearing of the ship now from its original position.

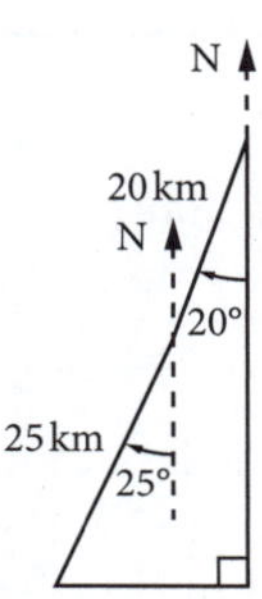

2.7 ANGLES OF ELEVATION AND DEPRESSION

(a)

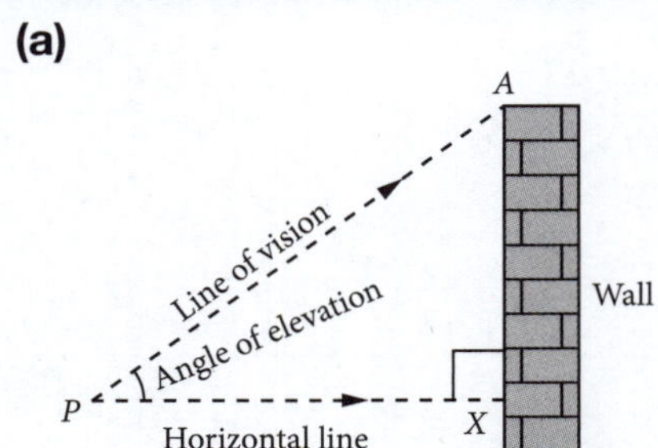

(b)

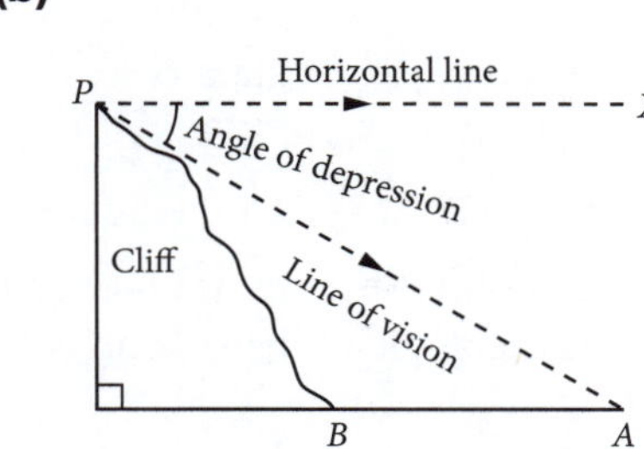

In **(a)** above, if you look up at A from the point P then the **angle of elevation** of A from P is the angle between the horizontal line PX and the line of sight PA. $\angle APX$ is an angle of elevation.

The point P is the eye of the observer. A could be, for example, a point on top of a wall.

In **(b)** above, if you look down at A from the point P then the **angle of depression** of A from P is the angle between the horizontal line PX and the line of sight PA. $\angle APX$ is an angle of depression.

The point P is the eye of the observer. The observer could be, for example, at the top of a cliff looking down on a boat at A in the water below.

If we look up from A to P, then $\angle PAB$ is an angle of elevation. $\angle APX = \angle BAP$ because alternate angles between parallel lines are equal, so the angle of depression from P is the same as the angle of elevation from A.

Angles of elevation and depression are in the same vertical plane.

Example 8

A point A is level with the foot of a vertical pole and 25 m away from it. The angle of elevation from point A to the top of the pole P is 40°. Calculate:

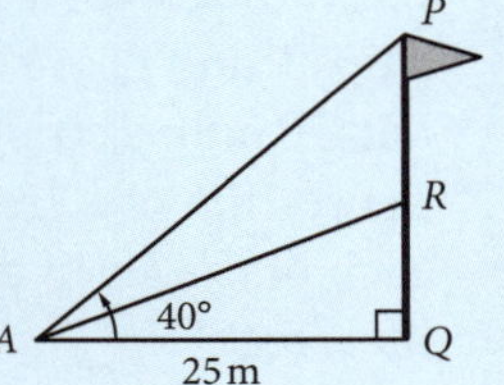

(a) the height of the pole, in metres to the nearest centimetre

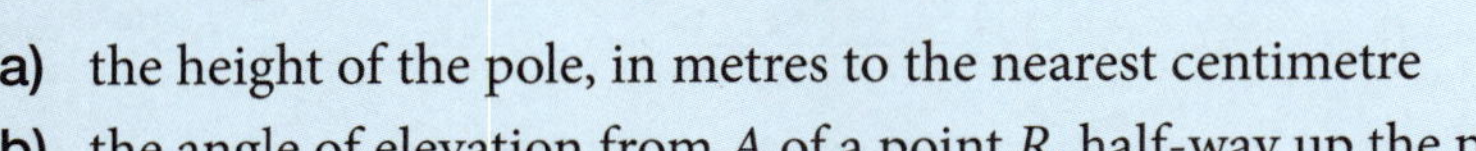

(b) the angle of elevation from A of a point R, half-way up the pole, to the nearest minute.

Solution

In the diagram, PQ is the vertical pole and $\angle PAQ$ is the angle of elevation of P from A.

(a) In ΔPAQ: $\tan 40° = \dfrac{PQ}{25}$

$PQ = 25 \times \tan 40° = 20.98$

The height of the pole is 20.98 m.

(b) $PQ = 2RQ$: $RQ = 10.49$

In ΔRAQ: $\tan(\angle RAQ) = \dfrac{10.49}{25} = 0.4196$

$\angle RAQ = 22° 46'$

The angle of elevation of R from A is $22° 46'$.

Example 9

An observer in a lighthouse, 100 m above sea level, is watching a ship sailing towards the lighthouse. The angle of depression of the ship from the observer is 15°.

(a) How far is the ship from the lighthouse?

(b) Sometime later, the angle of depression is measured to be 25°. How far has the ship travelled in this time?

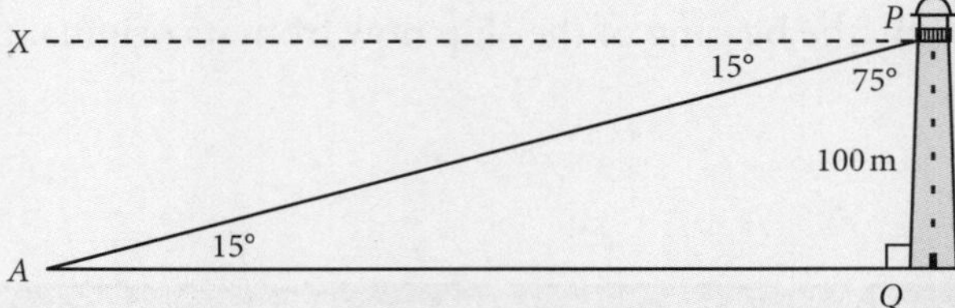

Solution

(a) In ΔPAQ:

$$\tan 15° = \frac{100}{AQ} \qquad \textbf{OR} \qquad \tan 75° = \frac{AQ}{100}$$

$$AQ = \frac{100}{\tan 15°} \qquad\qquad AQ = 100 \times \tan 75°$$

$$AQ = 373.2 \qquad\qquad AQ = 373.2$$

The ship is 273.2 m from the lighthouse.

(b) In the diagram, the ship has moved from A to B. $\angle BPX$ is the angle of depression when the ship is at B.

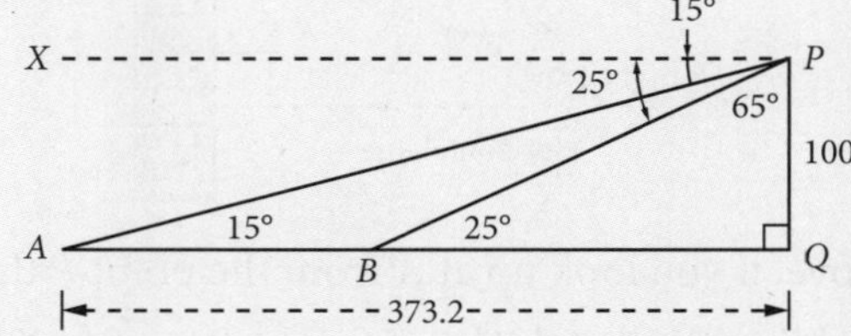

In ΔBPQ:

$$\tan 25° = \frac{100}{BQ} \qquad \textbf{OR} \qquad \tan 65° = \frac{BQ}{100}$$

$$BQ = \frac{100}{\tan 25°} \qquad\qquad BQ = 100 \times \tan 65°$$

$$BQ = 214.5 \qquad\qquad BQ = 214.5$$

$$\begin{aligned} AB &= AQ - BQ \\ &= 373.2 - 214.5 \\ &= 158.7 \end{aligned}$$

The ship has travelled 158.7 m.

EXERCISE 2.7 ANGLES OF ELEVATION AND DEPRESSION

1 From the top of a cliff, T, a walker sees two ships P and Q at horizontal distances of 50 m and 70 m in a straight line. The angle of depression of P from T is 25°. Indicate whether each of the given diagrams is correct or incorrect.

(a)

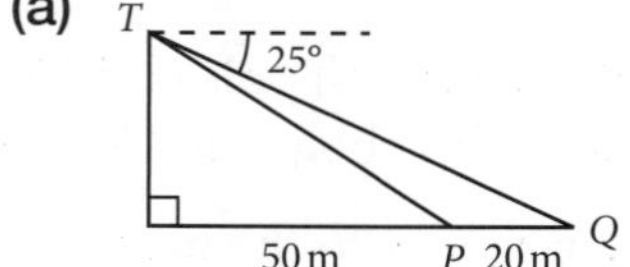

(b)

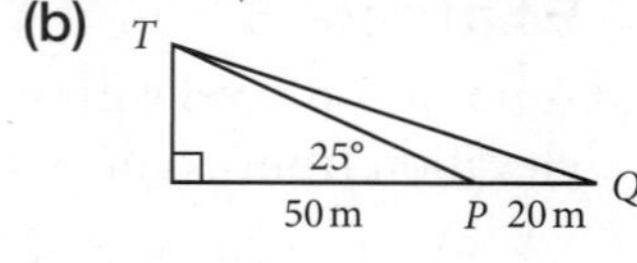

(c)

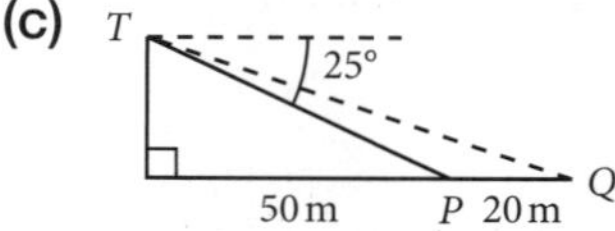

(d)

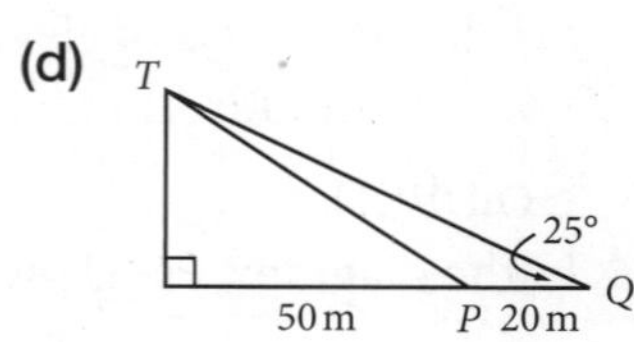

2 A person 2 m tall is standing on the ground and looking up at the top of a building. If the person is 18 m from the building and the angle of elevation of the top of the building is 30°, calculate the height of the building.

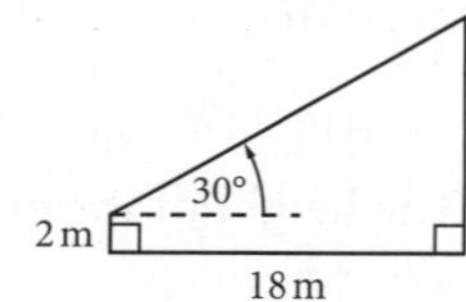

3 An aircraft flying in a horizontal straight line at an altitude of 2000 m passes directly over an observer on the ground. One minute later, the observer finds that the angle of elevation of the plane is 13° 24′. Calculate:

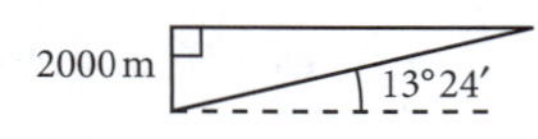

(a) the distance flown by the aircraft in that time

(b) the speed of the aircraft in km h^{-1}.

4 The diagram represents a vertical flagpole AB on top of a building. From a point P on the ground, the angle of elevation of A is 36° 52′. $PD = 55$ m, $CB = 5$ m, $AB = 12$ m. Calculate:

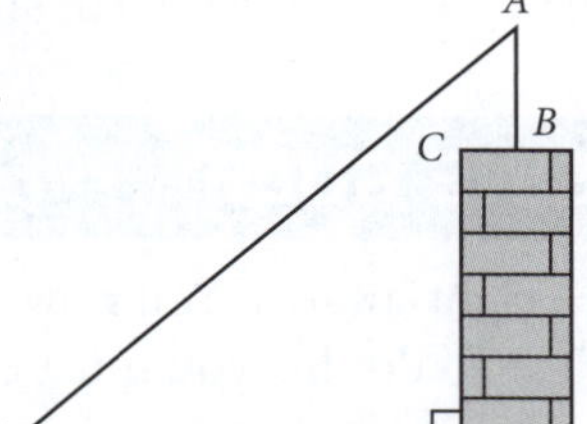

(a) the height of A above the ground

(b) the distance from A to P

(c) the angle of elevation of A from C.

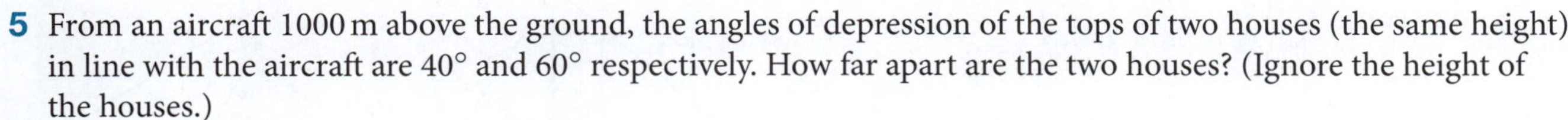

5 From an aircraft 1000 m above the ground, the angles of depression of the tops of two houses (the same height) in line with the aircraft are 40° and 60° respectively. How far apart are the two houses? (Ignore the height of the houses.)

6 AB and CD are two vertical buildings with their bases A and C on level ground. The height of AB is 50 m. The angle of elevation of B as seen from C is 20° and that of D as seen from A is 35°. Calculate:

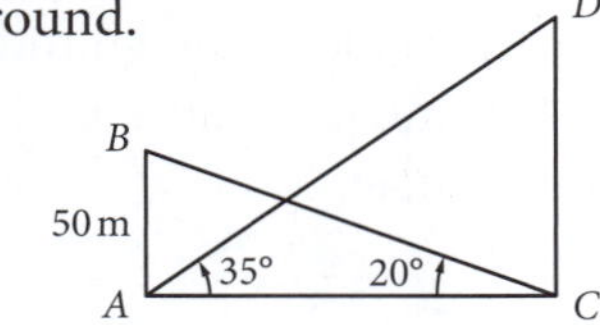

(a) the horizontal distance between the buildings

(b) the height of CD

(c) the angle of elevation of D as seen from B.

7 From the top of a cliff, T, an observer sees two ships P and Q in line with the observer and at horizontal distances of 50 m and 70 m. The angle of depression of P from T is 25°. Calculate:

(a) the vertical height of the cliff

(b) the angle of elevation of T from Q.

8 From a point 5 m above the ground, the angle of elevation of the top of a wall is 32° and the angle of depression of the bottom of the wall is 21°. Find:

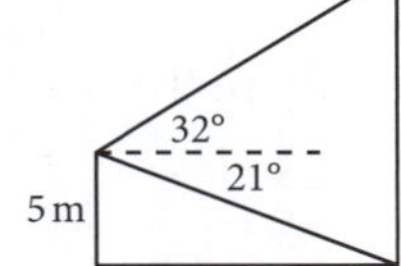

(a) the horizontal distance from the point of observation to the wall

(b) the height of the wall, correct to the nearest metre.

9 From a point A on the ground, the angle of elevation of the top of a tower is 38° and the angle of elevation of the top of a vertical flagpole on top of the tower is 41°. A is 80 m from the foot of the tower. The ground between A and the tower is horizontal. Calculate the length of the flagpole.

10 A city building is 45 m high. From the top of this building, the angle of depression of an object O on the wall of a building opposite is 50°. The width of the street is 20 m. Find:

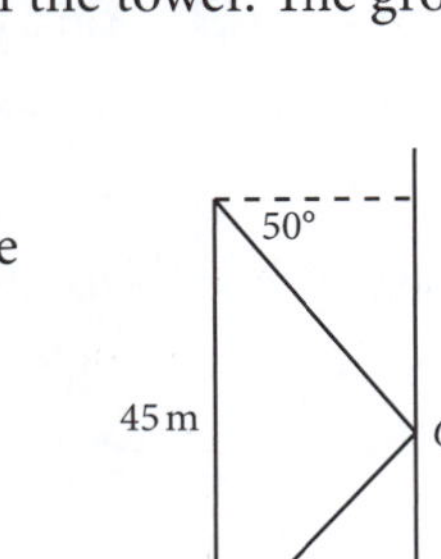

(a) the height of O above street level

(b) the angle of elevation of O from the foot of the first building.

11 Two buildings of unequal height stand at a distance apart on horizontal ground. The taller building is 60 m high and from its top an observer finds that lines of sight to the bottom and the top of the shorter building are at angles of depression of 25° and 10° respectively. Calculate:

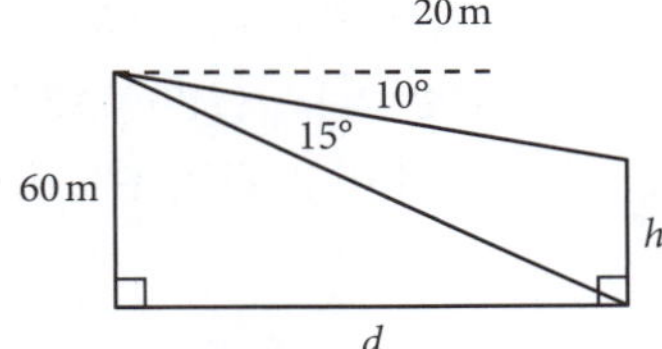

(a) the horizontal distance apart of the buildings **(b)** the height of the shorter building.

12 From the top of a lighthouse 75 m above sea level, the angles of depression of two buoys due north of the lighthouse are 60° and 30° respectively. Find, in simplest surd form:

(a) the distance of each buoy from the lighthouse (b) the distance between the two buoys.

13 From a point P on horizontal ground, the angle of elevation of the top of a building 40 m high is 30°. From a point Q on the same horizontal level as P and in line with the foot of the building, the angle of elevation is 60°. Calculate the distance PQ in simplest surd form.

2.8 THE SINE RULE

Not all triangles are right-angled. You now consider triangles that have either three acute angles or one obtuse angle and two acute angles. To do this you must establish two rules: the **sine rule** and the **cosine rule**.

The sine rule: $\dfrac{a}{\sin A} = \dfrac{b}{\sin B} = \dfrac{c}{\sin C}$

In triangle ABC, a, b and c are the lengths of the sides opposite the angles of magnitudes A, B and C respectively. A, B and C are also used to label the vertices of the triangle.

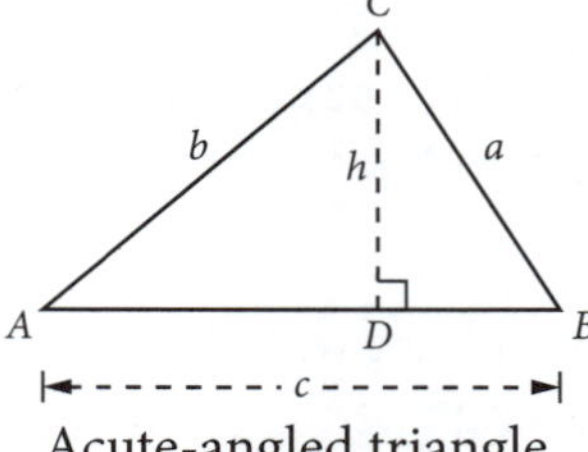

Acute-angled triangle

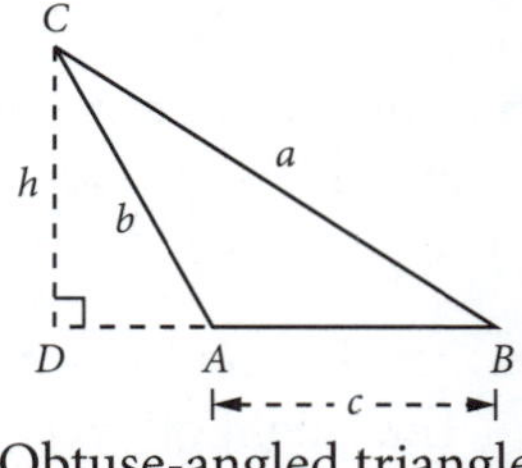

Obtuse-angled triangle

Proof

Let h be the length of the perpendicular from C to AB. The foot of this perpendicular lies on AB if ΔABC is acute-angled (diagram above left), or on BA produced if ΔABC is obtuse-angled (diagram above right).

In the right-angled triangles ΔACD and ΔBCD:

Acute-angled triangle:

$$h = b\sin A = a\sin B$$

So: $b\sin A = a\sin B$

Hence: $\dfrac{a}{\sin A} = \dfrac{b}{\sin B}$.

Obtuse-angled triangle:

$$h = b\sin(180° - A) = a\sin B$$

But: $\sin(180° - A) = \sin A$

So: $b\sin A = a\sin B$

Similarly, by drawing a perpendicular from A to BC it can be seen that $\dfrac{b}{\sin B} = \dfrac{c}{\sin C}$.

Hence $\dfrac{a}{\sin A} = \dfrac{b}{\sin B} = \dfrac{c}{\sin C}$.

The sine rule may also be written as $\dfrac{\sin A}{\sin B} = \dfrac{a}{b}$: the ratio of the sines = the ratio of the corresponding sides.

The sine rule can be used with a triangle when you are given:

(a) the size of two angles and the length of one side, **OR**

(b) the lengths of two sides and the size of an angle opposite one of these sides.

Important result

$$\sin(180° - \theta) = \sin\theta$$

Hence: $\sin 150° = \sin(180° - 30°) = \sin 30° = 0.5$

Solving an equation like $\sin\theta = 0.5$ gives two possible answers: $\theta = 30°$ or $(180 - 30)° = 150°$

Example 10

In ΔABC, given $A = 45°$, $B = 30°$ and $BC = 5$ cm, calculate the size of C, b and c.

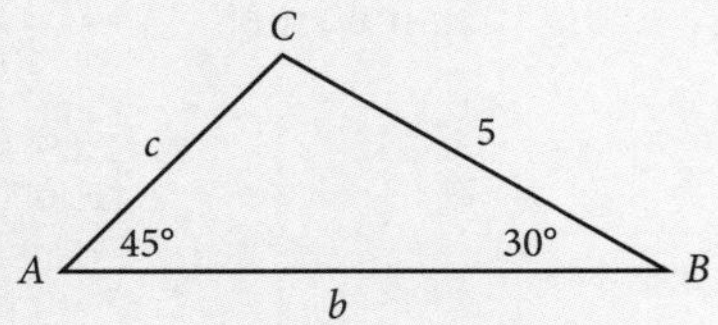

Solution

Because $A + B + C = 180°$, it follows that $C = 180° - (45° + 30°) = 105°$.

Using the sine rule:

$$\frac{a}{\sin A} = \frac{b}{\sin B} \qquad \frac{c}{\sin C} = \frac{a}{\sin A}$$

$$\frac{5}{\sin 45°} = \frac{b}{\sin 30°} \qquad \frac{c}{\sin 105°} = \frac{5}{\sin 45°}$$

$$b = \frac{5 \sin 30°}{\sin 45°} \qquad c = \frac{5 \sin 105°}{\sin 45°}$$

$$b = 5 \times \frac{1}{2} \times \frac{\sqrt{2}}{1} \qquad c = 6.8\text{ cm}$$

The exact value of b is $\frac{5\sqrt{2}}{2}$ cm, which is 3.5 cm correct to one decimal place.

The ambiguous case

It is sometimes possible to find two different solutions that use the same values for the two sides of a triangle and an angle opposite one of these sides. This situation is investigated in the examples below.

Example 11

Solve the triangle ABC given $A = 52°$, $BC = 12$ cm and $AC = 8$ cm.

(In this context, to 'solve the triangle' is to find the sizes of all unknown angles and sides.)

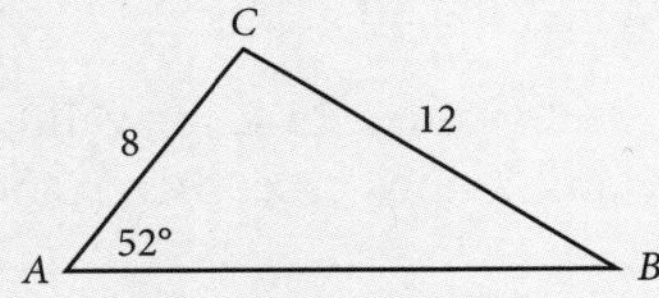

Solution

Find B by writing the sine rule in the form with $\sin\theta$ in the numerator.

Sine rule:

$$\frac{\sin A}{a} = \frac{\sin B}{b}$$

$$\frac{\sin 52°}{12} = \frac{\sin B}{8}$$

$$\sin B = \frac{8 \sin 52°}{12}$$

If needed, write as a decimal: $\sin B = 0.5253$

Because $\sin(180° - \theta) = \sin\theta$ you must consider both the acute angle and its supplement as possible solutions.

$B = 31°41'$ or $180° - 31°41'$

$B = 31°41'$ or $148°19'$

Check whether $B = 148°19'$ is a possible answer. The angle sum of a triangle is 180°:

$A + B = 52° + 148°19' = 200°19' > 180°$

Thus the only value of B is $31°41'$.

Find C using A and B:

$C = 180° - (52° + 31°41')$

$C = 96°19'$

Find c using the sine rule: $\frac{c}{\sin C} = \frac{a}{\sin A}$

$$\frac{c}{\sin 96°19'} = \frac{12}{\sin 52°}$$

$$c = \frac{12 \sin 96°19'}{\sin 52°}$$

$$c = 15.14 \quad \text{(correct to 2 decimal places)}$$

$\therefore B = 31°41', C = 96°19', c = 15.14\,\text{cm}$

Example 12

In ΔABC, given $A = 40°$, $BC = 20$ cm and $AC = 30$ cm, find the sizes of the other angles.

Solution

Draw a diagram:

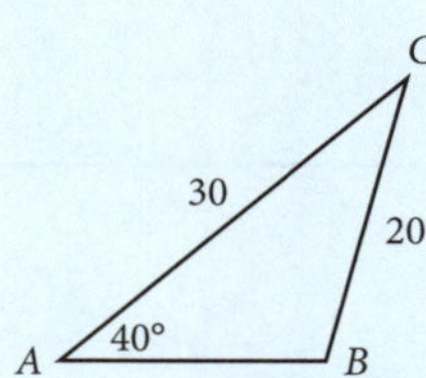

By the sine rule:

$$\frac{\sin A}{a} = \frac{\sin B}{b}$$

$$\frac{\sin 40°}{20} = \frac{\sin B}{30}$$

$$\sin B = \frac{30 \sin 40°}{20}$$

$$B = 74°37' \quad \text{or} \quad 105°23'$$

Test to see whether these values for B give a valid value for C:

When $B = 74°37'$: $C = 180° - (40° + 74°37') = 65°23'$

When $B = 105°23'$: $C = 180° - (40° + 105°23') = 34°37'$

Both these values for C are valid, so there are two possible triangles for the given information. This is an example of the ambiguous case: there is not enough information for you to find a unique solution.

In general, for a triangle ABC if you are given two sides and a non-included angle (that is, values of a, b and A), then one of four situations can occur.

Case 1: If $a < b \sin A$ then no triangle can be constructed.

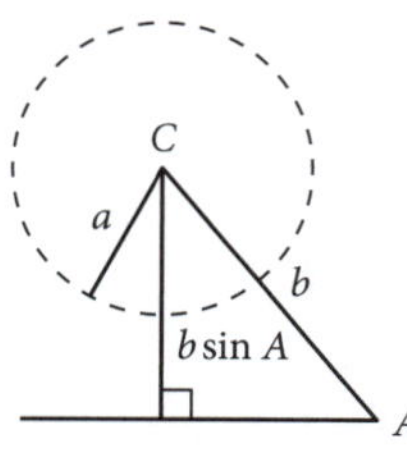

Case 2: If $a = b \sin A$ then one triangle exists and it is right-angled at B.

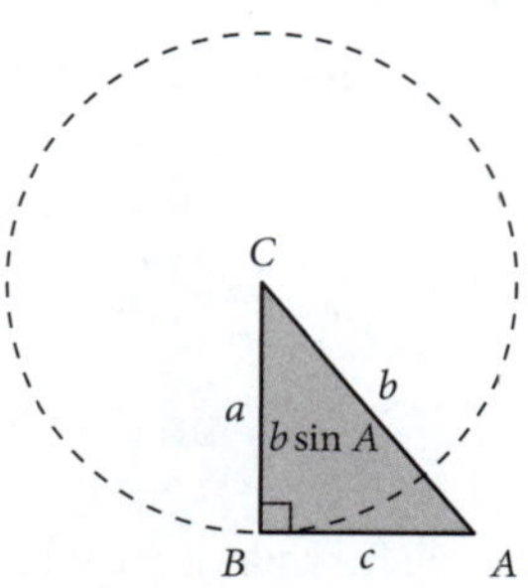

Case 3: If $a > b$, there is a unique solution as B must be an acute angle.

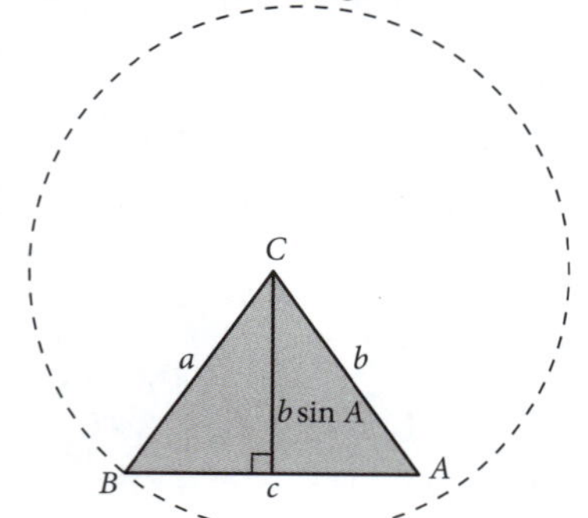

Case 4: If $b > a > b \sin A$, there are two distinct triangles. This is the ambiguous case.

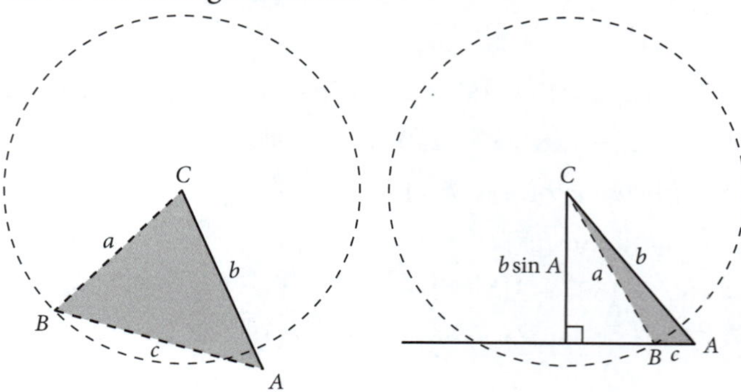

The ambiguous case arises only when the given angle is opposite the shorter of the two given sides. Angle B may be either acute or obtuse. The two possible sizes for angle B are supplementary to each other.

MAKING CONNECTIONS

The ambiguous case of the sine rule

Use technology to explore the ambiguous case of the sine rule.

EXERCISE 2.8 THE SINE RULE

1 In a triangle ABC, $a = 8$, $A = 30°$ and $B = 75°$. The correct diagram is:

A
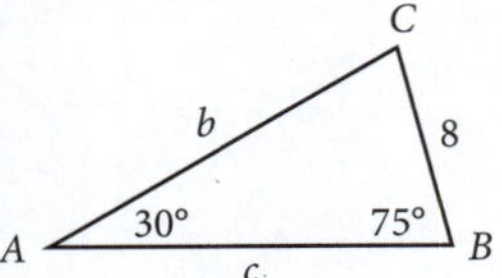

B
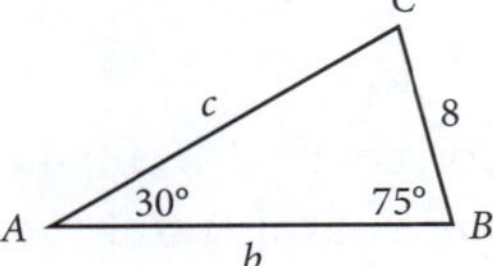

C
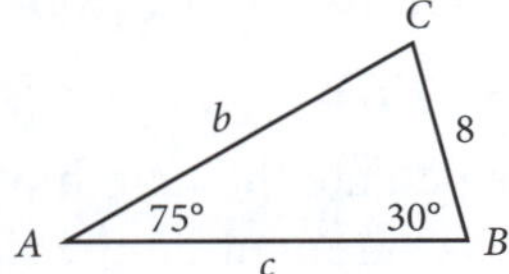

D
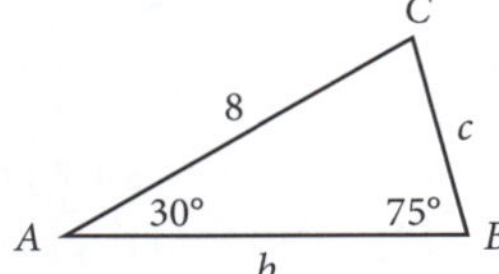

2 In ΔABC, $b = 10$, $B = 45°$ and $C = 120°$. Calculate the size of A, a and c.

3 In ΔPQR, $P = 65°$, $Q = 70°$ and $PR = 25$ cm. Calculate the length of PQ.

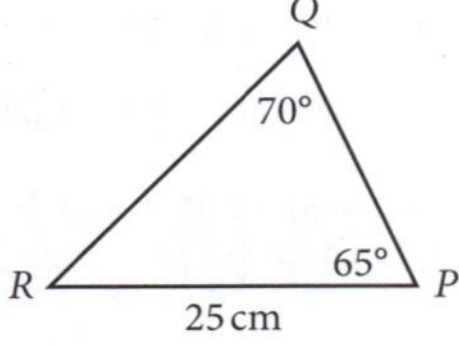

4 In ΔABC, $a = 5$, $A = 60°$ and $B = 45°$. Find the exact value of b.

5 In ΔABC, $\sin B = \frac{4}{5}$, $a = 6$ and $b = 9$. Find $\sin A$.

6 In ΔABC, $B = 2A$, $b = 5$ and $a = 3$. Show that $5 \sin A = 3 \sin 2A$.

7 Find the length of the longest side of a triangle ABC in which $A = 36° 52'$, $B = 30°$ and $AC = 8$ cm.

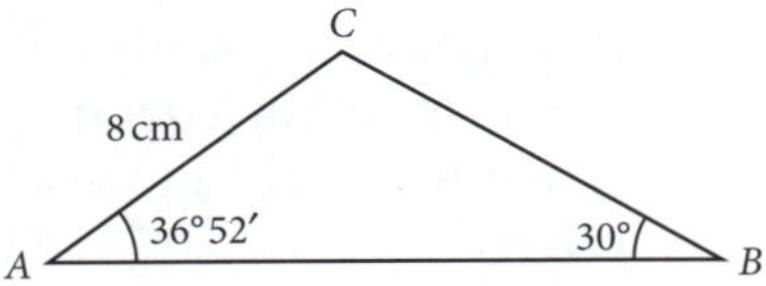

8 In a triangle ABC, $A = 42°$, $B = 28°$ and $BC = 6$ cm. Indicate whether each statement is correct or incorrect.

(a) $\frac{AB}{\sin 110°} = \frac{6}{\sin 28°}$ (b) $\frac{AB}{\sin 110°} = \frac{6}{\sin 42°}$ (c) $\frac{AB}{\sin 100°} = \frac{6}{\sin 42°}$ (d) $\frac{AB}{\sin 70°} = \frac{6}{\sin 42°}$

9 In a triangle ABC, $a = 16$, $b = 12$ and $\sin A = 0.4$. Calculate:

(a) $\sin B$ (b) the length of the perpendicular from C to AB (the altitude).

10 A wooden stake S is 13 m from a point A on a straight fence. SA makes an angle of 20° with the fence. If a goat is tethered to S by a 10 m rope, where on the fence is the closest point to A that the goat can reach?

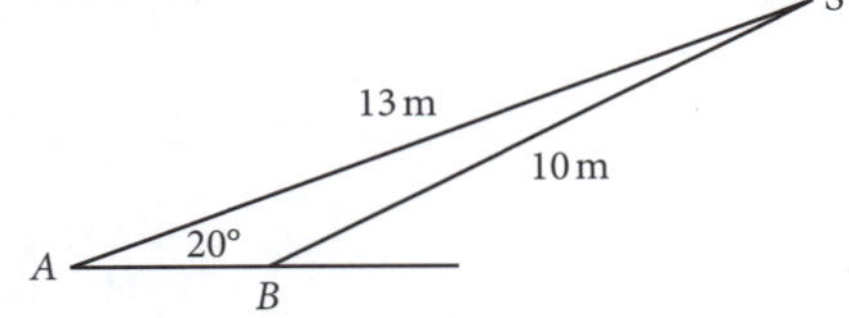

11 In a triangle ABC, $AC = 30$ cm, $AB = 44$ cm and $B = 37°$. Find two possible values of angles A and C.

12 In a triangle PQR, $PQ = 20$ cm, $QR = 22$ cm and $R = 15°$. Find two possible values for $\angle RPQ$ and PR.

13 In a triangle ABC, $A = 36° 52'$, $B = 30°$ and the perpendicular distance from C to AB is 3 cm. Calculate the perimeter of ΔABC.

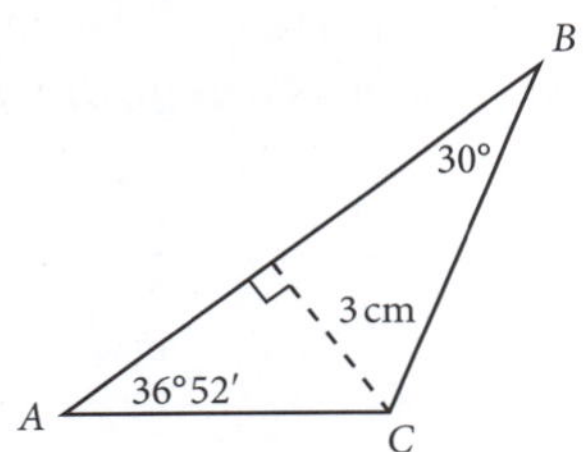

14 In ΔABC, the lengths of BC and AC are in the ratio $2:1$ and $\sin B = \frac{1}{4}$. Calculate:

(a) $\sin B$ **(b)** two possible sizes for A.

15 In ΔABC, $B = 25°$, $C = 55°$, $BC = 5$ m. Calculate:

(a) the length of AC
(b) the length of AX, where X is the foot of the perpendicular line from A to BC.

16 In ΔPQR, $PR = 3$ cm, $P = 40°$, $Q = 60°$. Calculate:

(a) the perimeter of ΔPQR
(b) the length of the perpendicular line from R to PQ.

17 In $\angle KLM$, $L = 55°$, $L = 35°$, $KL = 10$ cm.

(a) Use the sine rule to calculate the length of KM correct to 1 decimal place.
(b) What is the size of K? Comment on the significance of this result.

18 In ΔXYZ, $XY = 5$ m, $X = 68°$, $Y = 82°$. Calculate the length of XZ.

19 An aircraft flies 400 km from point A to point B on a course 040°. It then flies on a course 160° from B to C, 500 km from A. Calculate:

(a) the distance BC **(b)** the bearing of C from A.

20 The bearing of a ship from a lighthouse A is N 75° E. The ship's bearing from a second lighthouse B, 44 km south of A, is N 40° E. Find the distance of the ship from B.

21 Three points A, B and C lie on a horizontal plane. B is 2000 m due south of A. C is on a bearing 145° from A and 052° from B. Calculate the distance of C from both A and B.

22 Two points A and B on the same bank of a river are 50 m apart. C is a point on the other bank. $A = 80°$, $B = 70°$. Calculate the width of the river, assuming the river is straight and its width is constant.

23 O, A and B are three points in a straight line (in that order). The bearings of A and B from O are 020°. From a point P, 4 km from O in a direction NW, the bearings of A and B are 112° and 064° respectively. Calculate the distance from A to B.

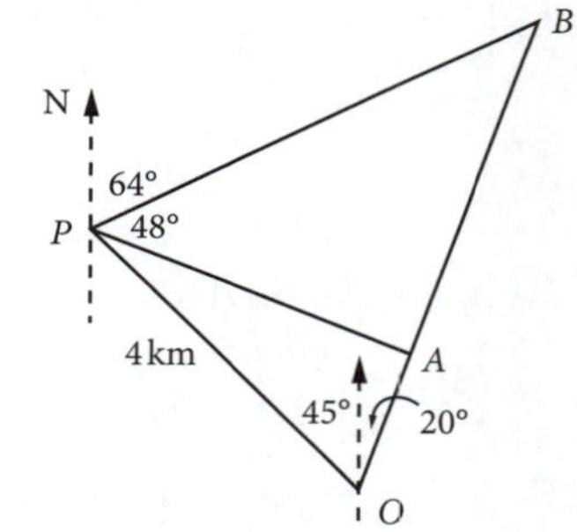

24 In ΔABC, AD bisects angle A. Use the sine rule to prove that $\frac{AB}{AC} = \frac{BD}{DC}$.

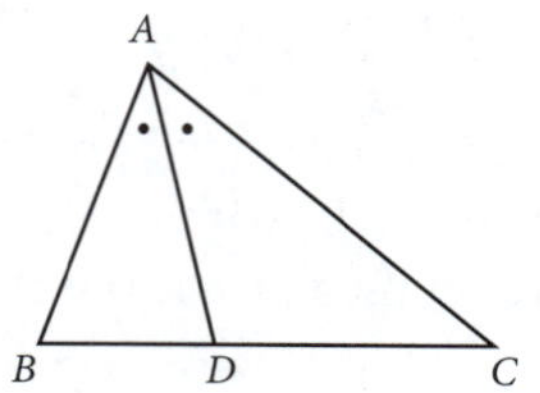

25 In ΔABC, $a = 11$ cm and $B = 53°$.

What can you say about the value of b if:

(a) two distinct triangles can be drawn to fit the information
(b) only one triangle can be drawn to fit the information
(c) no triangles can be drawn to fit the information.

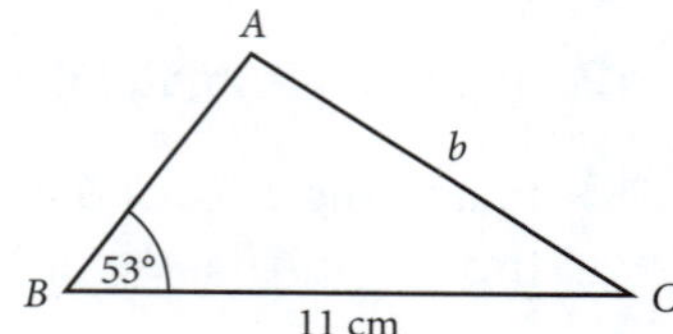

26 In ΔABC, $a = 6$ cm, $b = 7$ cm, and either A or B is given. In which case is there certain to be a unique triangle?

27 In ΔABC, $C = 110°$, $b = 9$ and the value of c is given. If it is possible to find a unique value for a, what can you say about the value of c?

28 In ΔABC, $a = 6.5$ cm, $b = 8.5$ cm, $A = 40°$. Find the size of B.

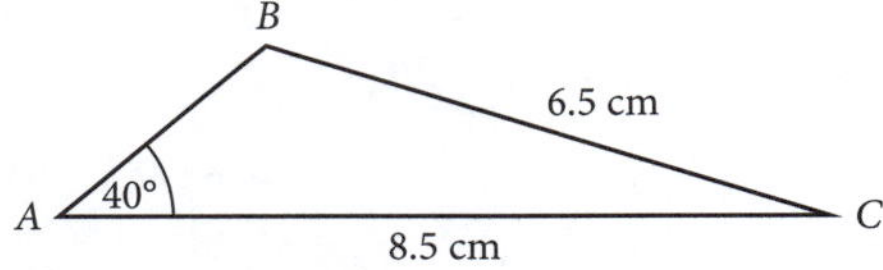

29 In ΔABC, $b = 12$ cm, $c = 17$ cm, $B = 40°$. Find the size of C.

30 In ΔABC, $a = 5.4$ cm, $b = 6.7$ cm and $A = 53°40'$. Find the size of B.

31 In ΔABC, $a = 7.8$ cm, $c = 8.3$ cm and $C = 62°30'$. Calculate the size of the other sides and angles in this triangle.

2.9 THE COSINE RULE

For ΔABC, the cosine rule states $a^2 = b^2 + c^2 - 2bc\cos A$. In words:

- The square of one side of a triangle is equal to the sum of the squares of the other two sides minus twice the product of those two sides and the cosine of the included angle.

To find a side, use:

$$a^2 = b^2 + c^2 - 2bc\cos A$$

$$b^2 = a^2 + c^2 - 2ac\cos B$$

$$c^2 = a^2 + b^2 - 2ab\cos C$$

To find an angle, use:

$$\cos A = \frac{b^2 + c^2 - a^2}{2bc}$$

$$\cos B = \frac{a^2 + c^2 - b^2}{2ac}$$

$$\cos C = \frac{a^2 + b^2 - c^2}{2ab}$$

Proof

From C, draw the perpendicular to meet AB at D (for acute-angled triangles) or to meet BA produced at D (for obtuse-angled triangles).

Let $CD = h$ and $AD = x$-units.

Apply Pythagoras' theorem to ΔBCD:

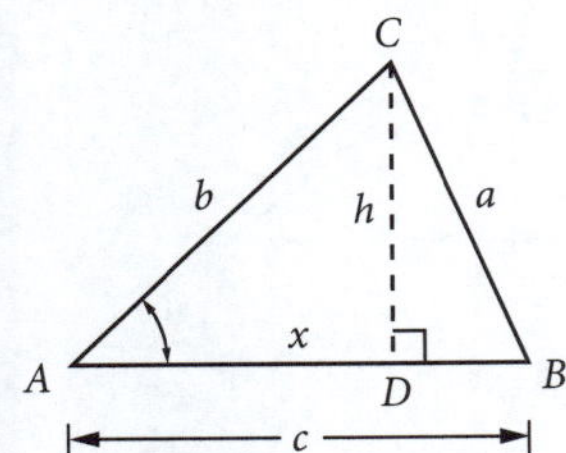

Acute-angled triangle

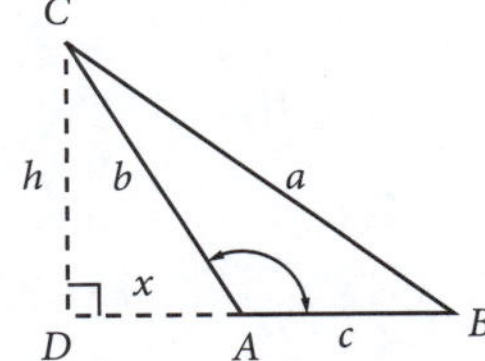

Obtuse-angled triangle

Acute-angled triangle:

$$a^2 = h^2 + (c - x)^2$$

$$a^2 = h^2 + c^2 - 2cx + x^2$$

But: $h^2 + x^2 = b^2$

And: $x = b\cos A$

$$\therefore a^2 = b^2 + c^2 - 2bc\cos A$$

Obtuse-angled triangle:

$$a^2 = h^2 + (c + x)^2$$

$$a^2 = h^2 + c^2 + 2cx + x^2$$

But: $h^2 + x^2 = b^2$

And: $x = b\cos(180° - A)$

$x = -b\cos A$

$$\therefore a^2 = b^2 + c^2 - 2bc\cos A$$

The cosine rule can be used with a triangle when you are given:

(a) the lengths of three sides, **OR**

(b) the lengths of two sides and the size of the included angle.

If $A = 90°$ then $\cos A = \cos 90° = 0$. The size of A determines the proportions of the sides:

- If $A < 90°$ then $a^2 < b^2 + c^2$
- If $A = 90°$ then $a^2 = b^2 + c^2$ (Pythagoras' theorem)
- If $A > 90°$ then $a^2 > b^2 + c^2$

Important result

$$\cos(180° - \theta) = -\cos\theta$$

Hence: $\cos 120° = \cos(180° - 60°) = -\cos 60° = -0.5$

Solving an equation like $\cos\theta = -0.5$ gives $\theta = (180 - 60)° = 120°$.

Example 13

In triangle XYZ, $XY = 7$ cm, $YZ = 6$ cm and $Y = 60°$. Calculate the exact length of XZ.

Solution

Show the information on a diagram.

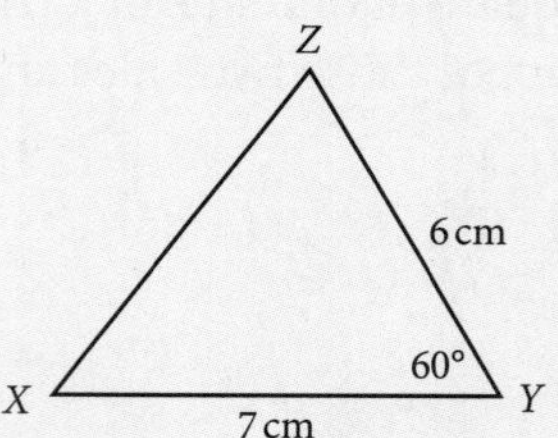

Two sides and the included angle are given.

Use the cosine rule: $y^2 = x^2 + z^2 - 2xz\cos Y$

$$y^2 = 6^2 + 7^2 - 2 \times 6 \times 7 \times \cos 60°$$

$$y^2 = 36 + 49 - 84 \times 0.5$$

$$y^2 = 43$$

$$y = \sqrt{43}$$

$\therefore XZ = \sqrt{43}$ cm

Example 14

Find the size of the largest angle of a triangle with side lengths 3 cm, 5 cm and 7 cm.

Solution

Show the information on a diagram.

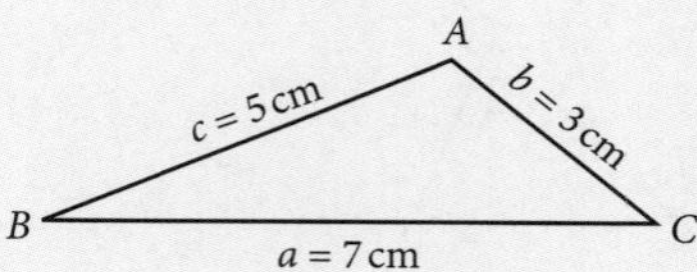

The largest angle is opposite the longest side. Find the size of A.

Use the cosine rule: $\cos A = \dfrac{b^2 + c^2 - a^2}{2bc}$

$$\cos A = \frac{3^2 + 5^2 - 7^2}{2 \times 3 \times 5}$$

$$= \frac{9 + 25 - 49}{30}$$

$$= -\frac{15}{30} = -\frac{1}{2}$$

Because $\cos A < 0$, A is obtuse: $A = 180° - 60° = 120°$

The largest angle in the triangle is 120°.

Example 15

In ΔABC, $a = 10$, $c = 5$ and $B = 36° 52'$. Calculate:

(a) The length b **(b)** the size of angle C

Solution

Show the information on a diagram.

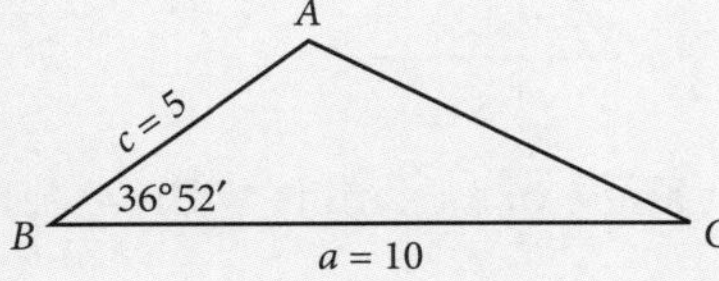

(a) Two sides and the included angle are known.

Use the cosine rule: $b^2 = a^2 + c^2 - 2ac\cos B$

$$b^2 = 10^2 + 5^2 - 2 \times 10 \times 5 \times \cos 36° 52'$$

$$b^2 = 125 - 100\cos 36° 52'$$

$$b^2 = 44.9966$$

$$b = 6.71$$

(b) Because you now know the length of the three sides and the size of one angle, you can use either the sine rule or the cosine rule to find the required angle. The sine rule is usually quicker and easier to use.

Use the sine rule: $\dfrac{\sin C}{c} = \dfrac{\sin B}{b}$

$$\frac{\sin C}{5} = \frac{\sin 36° 52'}{6.71}$$

$$\sin C = \frac{5\sin 36° 52'}{6.71} = 0.4471$$

$$C = 26° 33'$$

Example 16

Two people set out from point P at the same time. One travels at $20\,\text{km h}^{-1}$ along a straight road in the direction 032°. The other travels at $25\,\text{km h}^{-1}$ along another straight road in the direction 132°. Find their distance apart after 3 hours.

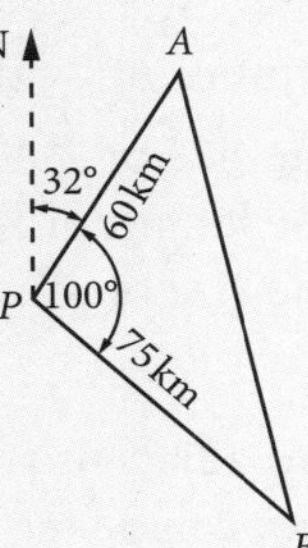

Solution

After 3 hours, one person is at A, 60 km from P, and the other is at B, 75 km from P, as shown above.

$$\angle APB = 132° - 32° = 100°$$

Two sides and the included angle are known. Use the cosine rule:

$$AB^2 = 60^2 + 75^2 - 2 \times 60 \times 75 \times \cos 100°$$

$$AB^2 = 10\,787.8$$

$$AB = 103.9$$

The two people are 103.9 km apart.

EXERCISE 2.9 THE COSINE RULE

1 In ΔABC, $b = 4$ cm, $c = 5$ cm, $A = 53°8'$. The correct diagram for this information is:

A

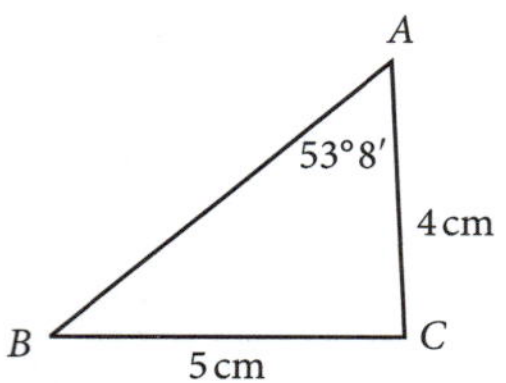

B

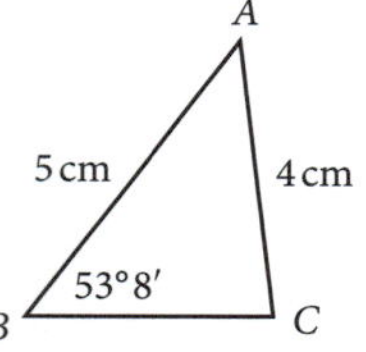

C

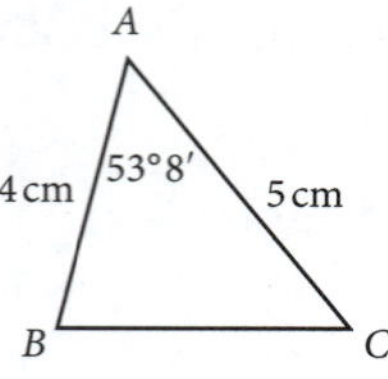

D

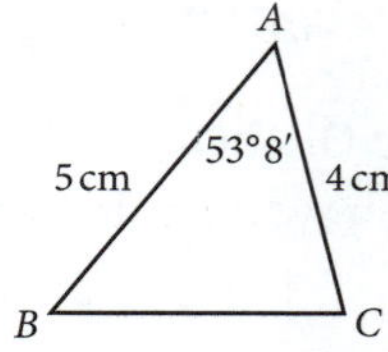

2 Calculate the cosine of the smallest angle of a triangle with side lengths 9 cm, 11 cm and 13 cm.

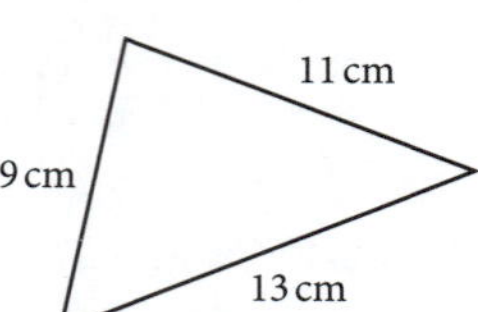

3 In ΔABC, $A = 36°52'$, $b = 7$ and $c = 10$. Calculate: **(a)** a **(b)** B.

4 Two adjacent sides of a parallelogram have lengths of 12 cm and 15 cm and the included angle is 50°. Calculate the lengths of the diagonals.

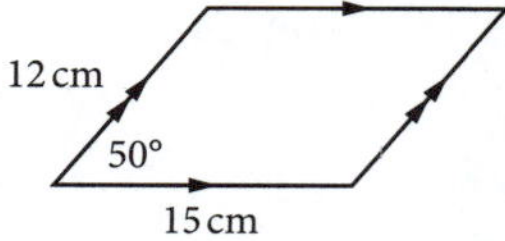

5 In ΔABC, $b = 4$ cm, $c = 5$ cm, $A = 53°8'$. Calculate the perimeter of the triangle, correct to the nearest centimetre.

6 Two sides of a triangle have lengths of 3.2 cm and 4.8 cm and the included angle is 65°. Calculate the length of the third side, correct to one decimal place. Indicate whether each statement is a correct or incorrect step in the working.

(a) $x = \dfrac{3.2\sin 65°}{4.8}$ **(b)** $x = \dfrac{4.8\sin 65°}{3.2}$ **(c)** $x^2 = 3.2^2 + 4.8^2 - 30.72\cos 65°$ **(d)** $x = 4.5$ cm

7 In ΔABC, $B = 126°52'$, $a = 12$ and $c = 15$. Find: **(a)** b **(b)** A.

8 The lengths of the sides of a triangle are in the ratio 5 : 6 : 9. Find the size of the largest angle.

9 In ΔABC, $BC = 8$ cm, $AC = 9$ cm and $AB = 7$ cm. If D is the midpoint of BC, calculate:

(a) the size of $\angle ABC$
(b) the length of AD
(c) the size of $\angle DAC$.

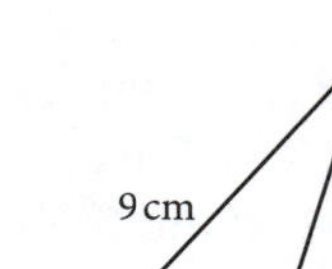

10 The two adjacent sides of a parallelogram have lengths of 8 cm and 10 cm. If the length of the longer diagonal is 14 cm, calculate:

(a) the size of the angles of the parallelogram **(b)** the length of the other diagonal.

11 In ΔPQR, $q = 12$, $r = 5$ and $\angle QPR = 108°$. Calculate: **(a)** p **(b)** $\angle PQR$.

12 In ΔABC, $BC = 11$ cm, $AC = 5$ cm and $AB = 8$ cm. P is a point on BC such that $BP = 4$ cm. Indicate whether each statement is correct or incorrect.

(a) $AP = 5$ cm **(b)** $\cos(\angle ABC) = \dfrac{10}{11}$ **(c)** $AP^2 = \dfrac{240}{11}$ **(d)** $\cos(\angle ACB) = \dfrac{41}{55}$

13 Two cars A and B depart from the same position. A travels along a straight road due east at 30 km h^{-1}. B departs 15 min after A and travels along another straight road in a north-east direction at 40 km h^{-1}. How far apart are they 15 min after B departs?

14 A, B and C are three towns such that B is 20 km from A in a direction 330° and C is 30 km from A in a direction 203° 8′. Find the distance from B to C.

15 P and Q are two towns 50 km apart with Q due east of P. A third town R, to the north of the line joining P and Q, is 70 km from P and 30 km from Q. Find the bearing of R from: **(a)** Q **(b)** P.

16 A lighthouse is 10 km north-west of a ship travelling due west at 16 km h^{-1}.

(a) How far is the ship from the lighthouse, 45 minutes later?
(b) What is the bearing of the lighthouse from the ship then?

17 P, A, B and C are four points (in that order) on a straight road that runs up a hill and makes a constant angle of 10° with the horizontal. A flagpole of height h m stands at P. From A and B, the top of the flagpole has elevations of 30° and 5° respectively above the horizontal.

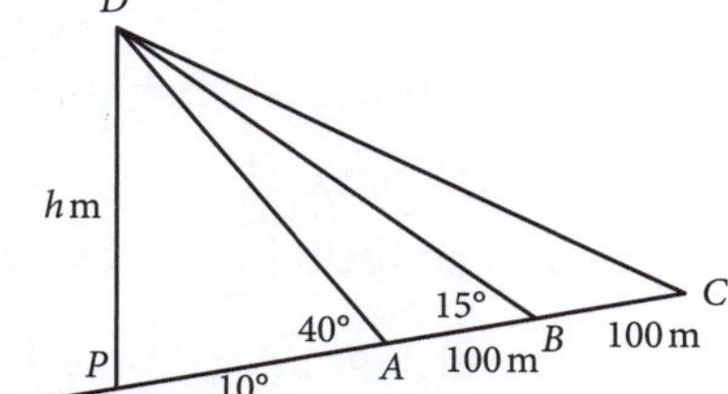

(a) If AB is 100 m, what is the height of the flagpole?
(b) If BC is also 100 m long, what is the elevation of the top of the flagpole from C?

18 From a point P, a person observes that the angle of elevation of the top of a cliff A is 40°. After walking 100 m towards A along a straight road inclined upwards at an angle of 15° to the horizontal, the angle of elevation of A is observed to be 50°. Find the vertical height of A above P.

19 The captain of a ship at D sailing on a bearing of 065° observes two lighthouses, A and B, in a line due north. After travelling 4 km to C, she notes that one lighthouse is on a bearing of 285° and the other 315°. Calculate the distance between the lighthouses.

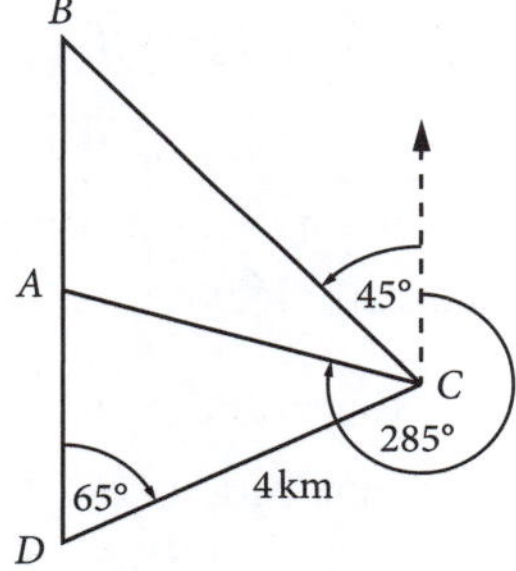

20 Two cars leave a point A at the same time. One car averages 80 km h^{-1} along a straight road in the direction 025°. The other car averages 90 km h^{-1} along a straight road in the direction 135°. How far apart are they after 3 hours?

2.10 AREA OF A TRIANGLE

The formula for the area of a triangle is the same for both acute-angled and obtuse-angled triangles.

$$\text{Area of } \Delta ABC = \frac{\text{base} \times \text{altitude}}{2} = \frac{1}{2} \times \text{base} \times \text{perpendicular height}$$

In each triangle the altitude is of length h.

$$\therefore \text{ Area of } \Delta ABC = \frac{ch}{2}$$

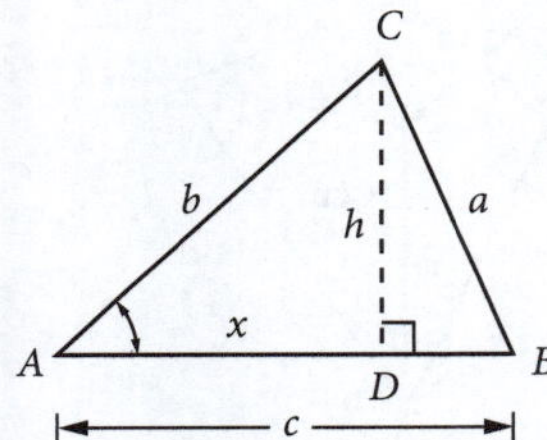

Acute-angled triangle

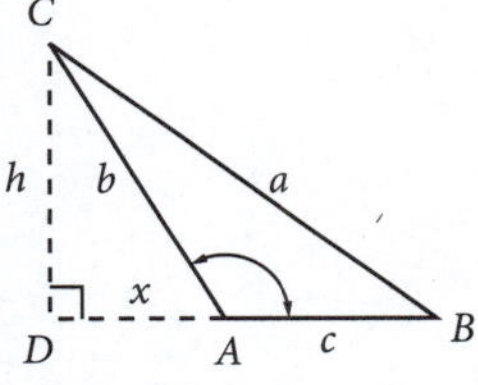

Obtuse-angled triangle

Acute-angled triangle:

$h = b \sin A$

Obtuse-angled triangle:

$h = b \sin (180° - A) = b \sin A$

Hence: Area of $\Delta ABC =$ Area of $\Delta ABC = \frac{1}{2}bc\sin A = \frac{1}{2}ab\sin C = \frac{1}{2}ac\sin B$

= half the product of two sides with the sine of the angle between them.

Example 17

The adjacent sides of a parallelogram are 12 cm and 15 cm and the angle between them is 50°. Calculate the area of the parallelogram correct to the nearest square centimetre.

Solution

Show the information on a diagram.

Draw the shorter diagonal.

The area of the parallelogram is twice the area of the triangle.

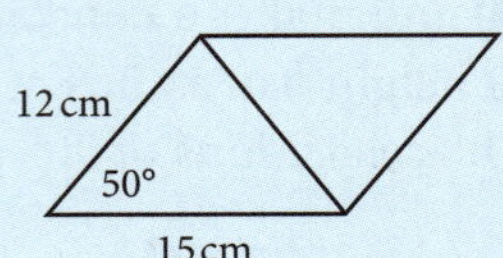

$$\text{Area of parallelogram} = 2 \times \frac{12 \times 15 \sin 50^\circ}{2}$$
$$= 180 \sin 50^\circ = 138\text{ cm}^2$$

EXERCISE 2.10 AREA OF A TRIANGLE

1 In ΔABC, $A = 36^\circ 52'$, $b = 7$ cm and $c = 10$ cm. Calculate:
   (a) a (b) B (c) the area of ΔABC.

2 In ΔABC, $b = 4$ cm, $c = 5$ cm, $A = 53^\circ 8'$. Calculate the area of the triangle, correct to the nearest square centimetre.

3 In ΔABC, $BC = 11$ cm, $AC = 5$ cm and $AB = 8$ cm. Calculate:
   (a) the magnitude of A (b) the length of the perpendicular from A to BC
   (c) the area of ΔABC.

4 The area of ΔABC if $a = 6$ cm, $b = 7$ cm and $c = 11$ cm is nearest to:
   A 6 cm^2 **B** 19 cm^2 **C** 20 cm^2 **D** 22 cm^2

5 ABC is a triangle in which $AC = 7$ cm. A circle, centre B and radius BC, cuts AB internally at D. $AD = 5$ cm, $DC = 4$ cm. Calculate:
   (a) the length of BC
   (b) the area of ΔABC.

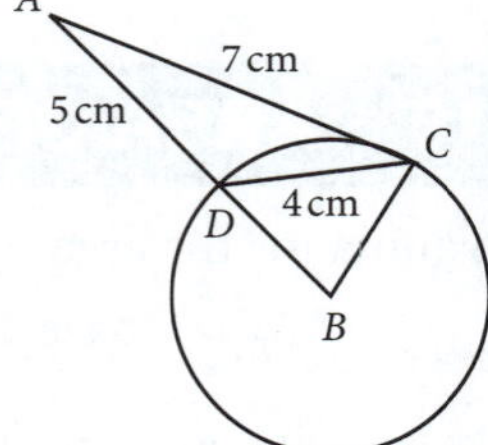

6 P, A, B and C are four points in a plane such that angles BPA and CPA are obtuse and on opposite sides of PA. $PA = 8$ cm, $BP = 10$ cm, $PC = 12$ cm, $AB = 14$ cm and $AC = 18$ cm. Calculate:
   (a) the length of BC
   (b) the area of ΔABC.

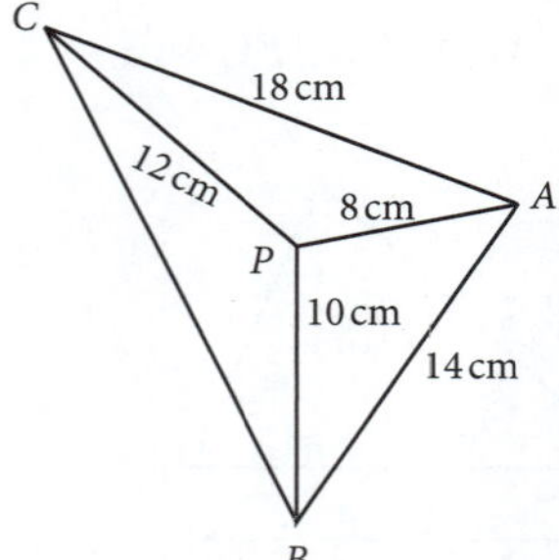

7 The equal sides AB and AC of an isosceles triangle ABC are each 5 cm and $BC = 4$ cm. D is a point on AC such that $DC = 1$ cm. Calculate:

(a) the size of A **(b)** the length of BD **(c)** the area of ΔABC.

8 $ABCD$ is a parallelogram in which $AB = 8$ cm, $BC = 5$ cm and $\angle DAB = 60°$. Calculate:

(a) the length of the diagonals AC and BD **(b)** the area of $ABCD$.

9 Two sides of a triangular field are 60 m and 50 m and the included angle is 140°. Indicate whether each statement is correct or incorrect.

(a) Third side = 103.4 m **(b)** Area = 964.2 m^2 **(c)** Third side = 38.8 m **(d)** Area = 1149 m^2

10 The sides of a triangular field have lengths 80 m, 90 m and 100 m. Calculate the area of the field.

2.11 APPLIED TRIGONOMETRY

This section uses the skills from earlier in this chapter to solve harder two-dimensional problems and problems in three dimensions. In three-dimensional problems it is important that you break the 3D diagram into its 2D parts.

Example 18

In this diagram, $AC = 15$ cm, $AB = 7$ cm, $DB = 10$ cm and $\angle DBA = \alpha$. Find the perpendicular distance CE of C from DA in terms of α and evaluate it for $\alpha = 25°$.

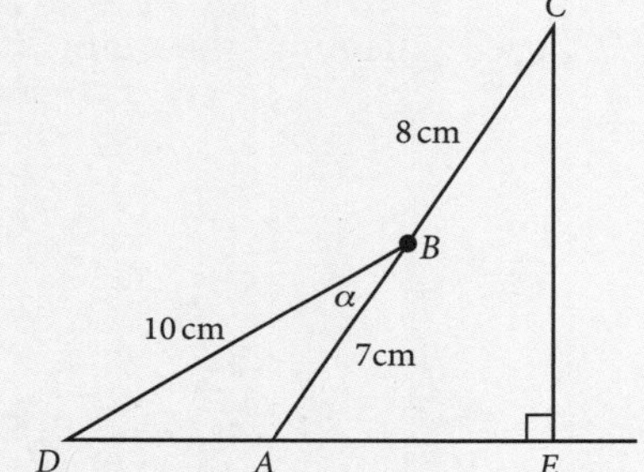

Solution

Use the cosine rule in ΔDBA: $DA^2 = 10^2 + 7^2 - 2 \times 10 \times 7\cos\alpha$

$DA^2 = 149 - 140\cos\alpha$

$\therefore \quad DA = \sqrt{149 - 140\cos\alpha}$ [1]

Use the sine rule in ΔDBA: $\dfrac{10}{\sin(\angle DAB)} = \dfrac{DA}{\sin\alpha}$

$\therefore \quad \sin(\angle DAB) = \dfrac{10\sin\alpha}{DA}$ [2]

In ΔCAE: $CE = 15\sin(\angle CAE)$ [3]

Now: $\angle DAB = 180° - \angle CAE$

Hence: $\sin(\angle DAB) = \sin(\angle CAE)$

Using [1], [2] and [3]: $CE = \dfrac{150\sin\alpha}{\sqrt{149 - 140\cos\alpha}}$

For $\alpha = 25°$: $CE = \dfrac{150\sin 25°}{\sqrt{149 - 140\cos 25°}}$

$= 13.48$

Perpendicular distance = 13.48 cm

Example 19

From a point P due south of a vertical tower, the angle of elevation of the top of the tower is 20°; from a point Q due east of the tower, the angle of elevation is 35°. The distance from P to Q is 40 metres. Find the height of the tower.

Solution

(Trigonometric values are rounded to 3 decimal places.)

Draw a 3D diagram to represent this information. Let the height of the tower be x m.

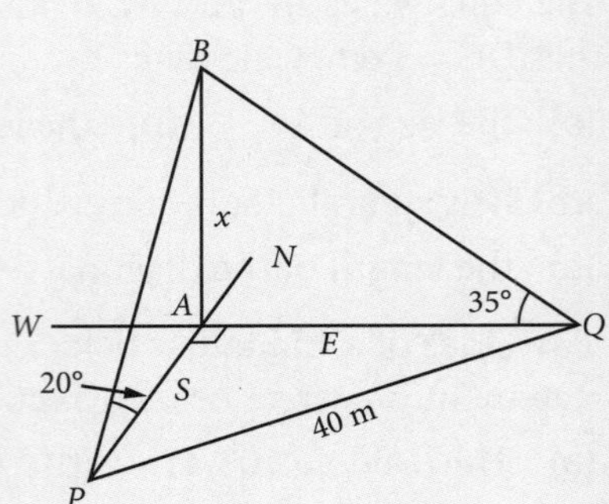

There are three right-angled triangles: ΔPAB and ΔQAB in the vertical plane, and ΔPAQ in the horizontal plane. Draw each right-angled triangle separately as you use it.

In ΔPAB: $\tan 20° = \frac{x}{PA}$ **OR** $\tan 70° = \frac{PA}{x}$

$\therefore \quad PA = \frac{x}{\tan 20°} \qquad PA = x\tan 70°$

$\therefore \quad PA = 2.747x \qquad PA = 2.747x$

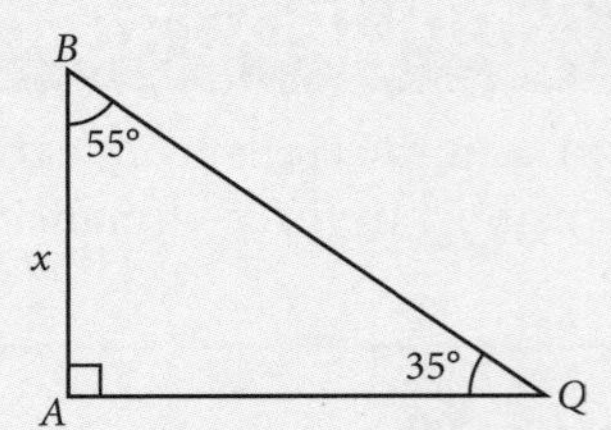

In ΔQAB: $\tan 35° = \frac{x}{QA}$ **OR** $\tan 55° = \frac{QA}{x}$

$\therefore \quad QA = \frac{x}{\tan 35°} \qquad QA = x\tan 55°$

$\therefore \quad QA = 1.428x \qquad QA = 1.428x$

In ΔPAQ, use Pythagoras' theorem:

$$(2.747x)^2 + (1.428x)^2 = 40^2$$
$$7.546x^2 + 2.039x^2 = 1600$$
$$9.585x^2 = 1600$$
$$x = \sqrt{\frac{1600}{9.585}}$$
$$x = 12.920$$

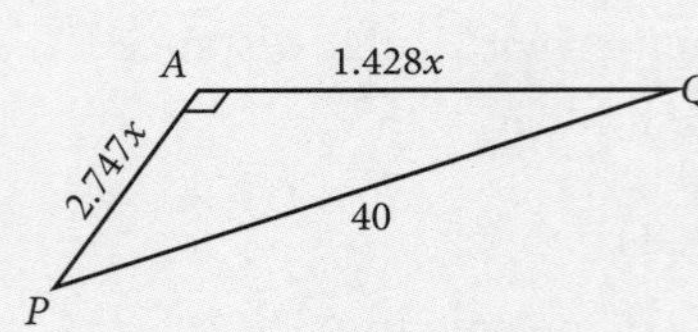

The height of the tower is 12.92 m.

To avoid errors due to rounding, it is better if you can avoid using your calculator until the last step. Without rounding, the calcuation is:

$$(x\tan 70°)^2 + (x\tan 55°)^2 = 1600$$
$$x^2(\tan^2 70° + \tan^2 55°) = 1600$$
$$x^2 = \frac{1600}{\tan^2 70° + \tan^2 55°}$$
$$x = \frac{40}{\sqrt{\tan^2 70° + \tan^2 55°}} \qquad x = 12.918 \approx 12.92\text{ m}$$

EXERCISE 2.11 APPLIED TRIGONOMETRY

1 P, A, B, C are four points in a plane such that the angles BPA and CPA are obtuse and on opposite sides of PA. $PA = 8$ cm, $BP = 10$ cm, $PC = 12$ cm, $AB = 14$ cm and $AC = 18$ cm. Calculate the length of BC and the area of the triangle ABC.

2 P, A, B, C are four points (in order) on a straight road that runs up a hill at a constant angle of 10° to the horizontal. A flagpole with a height of h m stands at P. From A and B, the top of the flagpole has elevations of 30° and 5° respectively above the horizontal. If AB is 100 m long, what is the height of the flagpole? If BC is also 100 m long, what is the angle of elevation of the top of the flagpole from C?

3 From a point P, a person observes that the angle of elevation of the top of a cliff A is 40°. After walking 100 m towards A along a straight road that inclines upwards at an angle of 15° to the horizontal, the angle of elevation of A is observed to be 50°. Find the vertical height of A above P.

4 A ship sailing in a direction 065° observes two lighthouses in a line due north. After the ship has travelled 4 km, one of the lighthouses is on a bearing of 285° and the other 315°. The distance between the lighthouses is given by which two of the following expressions?

A $4\left(\frac{\sin 75°}{\sin 45°} - \frac{\sin 40°}{\sin 70°}\right)$ **B** $4\left(\frac{\sin 70°}{\sin 45°} - \frac{\sin 40°}{\sin 75°}\right)$ **C** $\frac{2\sin 65°}{\sin 75° \sin 45°}$ **D** $\frac{2\sin 75°}{\sin 65° \sin 45°}$

5 A and B are the bases of two vertical towers each 30 m high. B is due north of A. From a point P that is due east of A and in the same horizontal plane, the angles of elevation of the tops of the towers A and B are 45° and 36° 52′ respectively. Calculate the distance:

(a) from P to the bases of the two towers **(b)** between the two towers.

6 From a point P, an observer finds that the angle of elevation of the top of a vertical tower is $\alpha°$. After walking x metres horizontally towards the base of the tower, the observer finds the angle of elevation of the top of the tower is now $\beta°$. If the height of the tower is h metres, show that $x = h(\cot \alpha° - \cot \beta°)$.

7 A flagpole 5 m high stands on top of a vertical tower. From a point on the ground, the angles of elevation of the top and bottom of the flagpole are 68° and 62° respectively. Show that the height of the tower (h metres) is given by $h = \frac{5\tan 62°}{\tan 68° - \tan 62°}$.

8 An open rectangular tank a units deep and b units wide holds water and is tilted so that the base BC makes an angle θ with the horizontal. When BC is returned to the horizontal, show that the depth of the water is $\frac{a^2 \cot\theta}{2b}$ units.

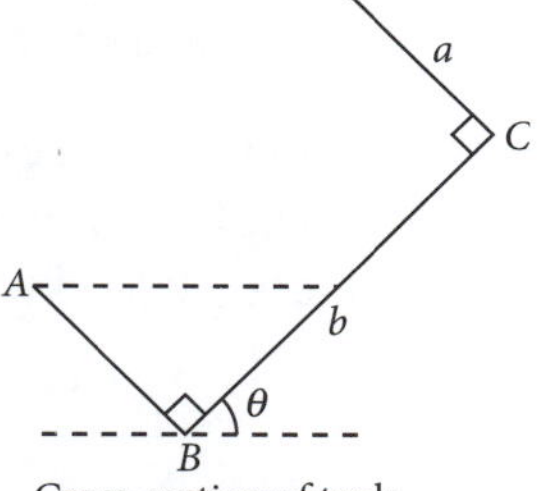

Cross-section of tank

9 A conical vessel with vertex at O and semi-vertical angle α is suspended from a point A on the rim of the base. G is a point on the axis OC of the cone such that $OG = \frac{2}{3} OC$. If the vessel rests with G vertically below A, show that the acute angle β that AO makes with the vertical is given by $\tan\beta = \frac{2\tan\alpha}{1 + 3\tan^2\alpha}$.

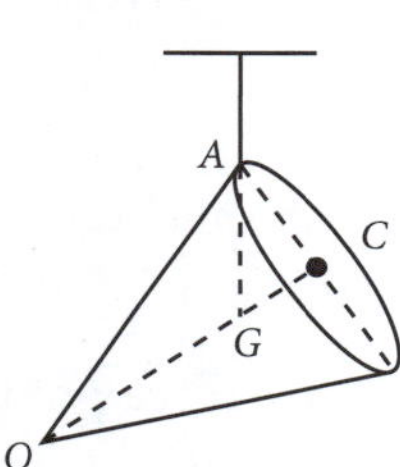

10 In ΔABC, $AC = 7$ cm. A circle with centre B and radius BC cuts AB internally at D. $AD = 5$ cm, $DC = 4$ cm. The length of BC is:

A 10 cm **B** 5 cm **C** 7 cm **D** impossible to find

11 A and B are two towers, with B 4 km due east of A. A flagpole C has true bearings from A and B that are $\alpha°$ east of north and $\alpha°$ west of north respectively. A second flagpole D has true bearings from A and B that are $(\alpha + \beta)°$ east of north and $(\alpha - \beta)°$ west of north respectively. Draw a sketch to indicate the positions of A, B, C and D. Assuming that A, B, C and D are on level ground and that $\alpha = 25$, $\beta = 10$, find the distance between C and D.

12 A, B, C, D are four points equally spaced on the circumference of a horizontal circle of radius 3 cm. P is a point above the circle such that $PA = PD = 4$ cm and $PB = PC = 6$ cm. Calculate the size of $\angle APC$.

13 The angle of elevation of the top of a building from a point P due east of it is 40°, and from a point Q due south of P is 30°. If the distance from P to Q is 20 metres, find the height of the building.

14 A railway line runs south-west between two railway stations A and B. Two spires 3 km apart are both directly north-west of A. From B, the bearings of the spires are N 7.5° E and N 37.5° E. Show that $AB = 6\sin 52.5° \cos 7.5°$. If a train takes 4 minutes to travel from A to B, find its average speed.

15 Two buildings of equal height are 40 m apart. At a point on the horizontal line joining the buildings' bases, the angles of elevation of the tops of the buildings are 47° and 28°. Show that the height h of the buildings is given by $h = \frac{40\tan 47° \tan 28°}{\tan 47° + \tan 28°}$.

16 An observer's eye is 2 m above the ground. A vertical pole fixed in the ground subtends an angle of 45° at the observer's eye. The angle of depression of the base of the flagpole from the eye is 15°. Show that the height of the flagpole is $2(1 + \tan 30° \tan 75°)$ metres.

17 From a point A, two points B and C are in line in a direction 049°. From a point D that is 100 m from A in a direction 139°, B is in a direction 352° and C is in a direction 022°. Calculate the distance between B and C.

18 From a point A, the angle of elevation of the top of a tower due north of it is 20°. From point B, due east of the tower, the angle of elevation is 18°. A and B are 100 m apart. Show that the height h of the tower is given by $h = \dfrac{100}{\sqrt{\tan^2 72° + \tan^2 70°}}$.

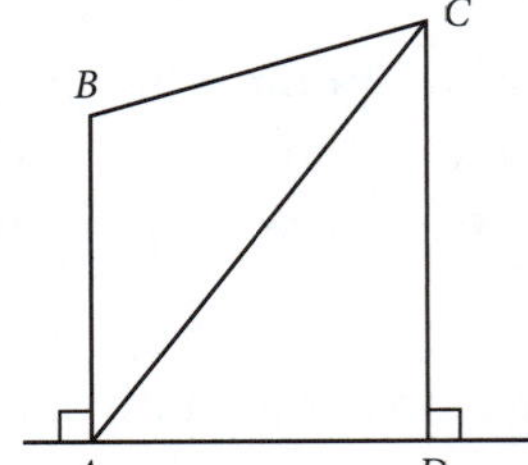

19 In the diagram shown, $AB = 25$ cm, $BC = 27$ cm, $AC = 40$ cm. Calculate the length of AD.

20 $ABCD$ is a cyclic quadrilateral in which $AB = 5$ cm, $BC = 6$ cm, $CD = 7$ cm, $AD = 8$ cm. Show that $\cos(\angle ADC) = \frac{13}{43}$.

21 A and B represent successive positions of an aircraft flying horizontally at an altitude of 2000 metres. From a point P the angles of elevation of A and B are 15° and 25° respectively. If $\angle DPC = 60°$, calculate the distance AB.

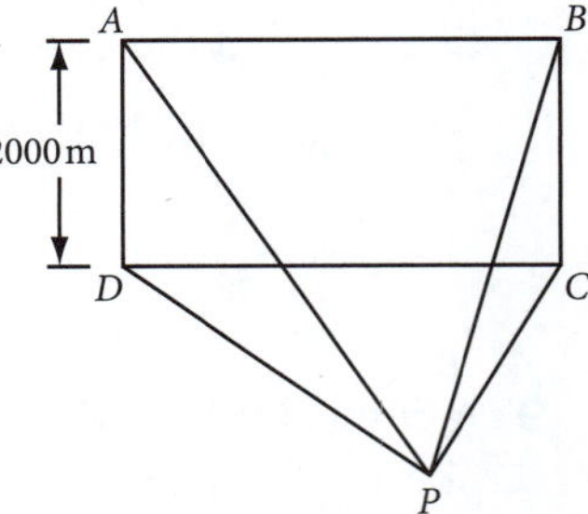

22 The elevation of a hill at a point P due east of it is 42° and at a point Q due south of it is 22°. If the distance from P to Q is 400 metres, find the height of the hill.

23 O, A, B are three points in order in a straight line. The bearings of A and B from O are both 020° T. From a point P that is 4 km from O in a NW direction, the bearings of A and B are 112° T and 064° T. Calculate the distance from A to B.

24 From a point P an observer finds that the angle of elevation of the top of a vertical tower is $\alpha°$. After walking x m horizontally towards the foot of the tower, the observer finds that the angle of elevation is $\beta°$. If the height of the tower is h m, prove that $h = \dfrac{x \sin\alpha° \sin\beta°}{\sin(\beta - \alpha)°}$.

25 Using the sine rule, show that $\dfrac{x}{y} = \dfrac{\sqrt{3}}{2}$ and also that $\dfrac{\text{Area } \Delta ABC}{\text{Area } \Delta ADC} = \dfrac{\sqrt{3}}{2}$.

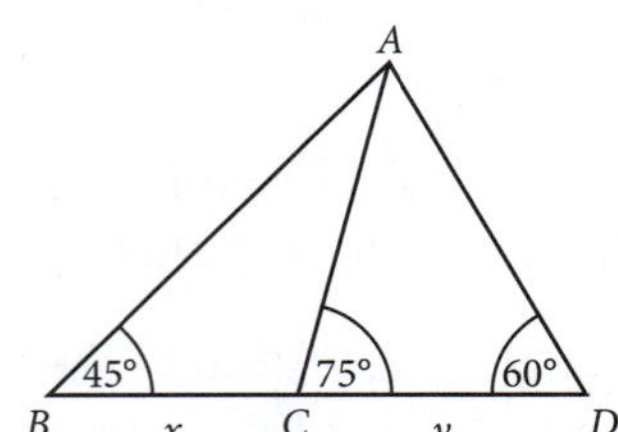

26 A side of a hill can be regarded as a plane making an angle of 20° with the horizontal.

(a) From a point A at the base of the hill, I walk 100 m up the hill along a straight line of steepest slope to the point B. Find the vertical height of B above A.

(b) A tower 40 m tall is constructed at B. Find the angle subtended by the tower at my eyes, which are 1.5 m above the ground, when I stand at A.

(c) A straight road is constructed on the hillside with a slope of 1 in 10, i.e. with a 1 m vertical rise for every 10 m horizontally forward. Find the angle this road must make with the line of steepest slope on the hillside.

27 A tower PQ of height h metres has an angle of elevation of 13° at a point A due east of it. From another point B, the bearing of the tower is 050° T and the angle of elevation is 10°. The points A and B are 1000 metres apart on the same level as the tower's base Q.

(a) Show that $\angle AQB = 140°$.
(b) Consider the triangle APQ and show that $AQ = h \tan 77°$.
(c) Find a similar expression for BQ.
(d) Use the cosine rule in the triangle AQB to calculate h to the nearest metre.

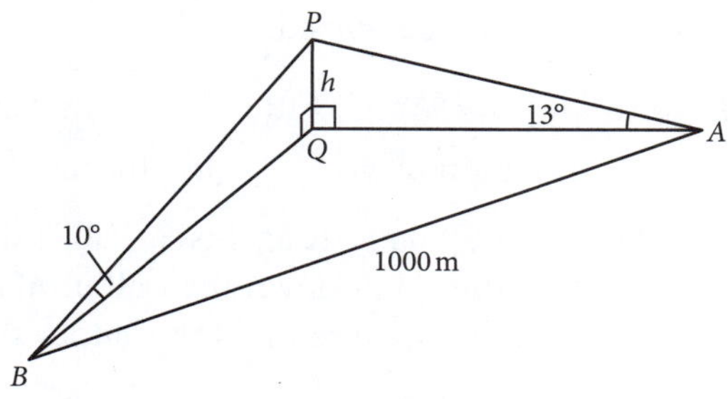

28 Two observers P and Q, who are 2000 metres apart, take the bearing and elevation of a balloon B at the same instant. The observer P finds the bearing to be 018° T and the angle of elevation to be 22°, while the observer Q finds the bearing to be 345° T and the angle of elevation to be 19°. Find the height BC of the balloon to the nearest metre.

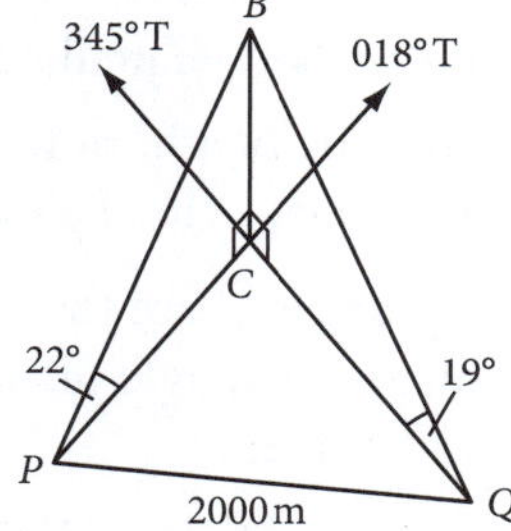

29 AB is a tower. CD is a road. A person walking along the road observes the tower AB from C at a bearing of 038° T and with an angle of elevation of 11°. After travelling 3000 m along the road to D, the person observes the tower from D at a bearing of 320° T and an angle of elevation of 8°. By letting $CB = x$, $BD = y$ and $AB = h$:

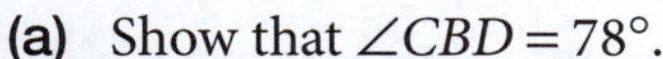

(a) Show that $\angle CBD = 78°$.
(b) Express x in terms of h.
(c) Express y in terms of h.
(d) Using the cosine rule with ΔCBD, find the value of h correct to one decimal place.

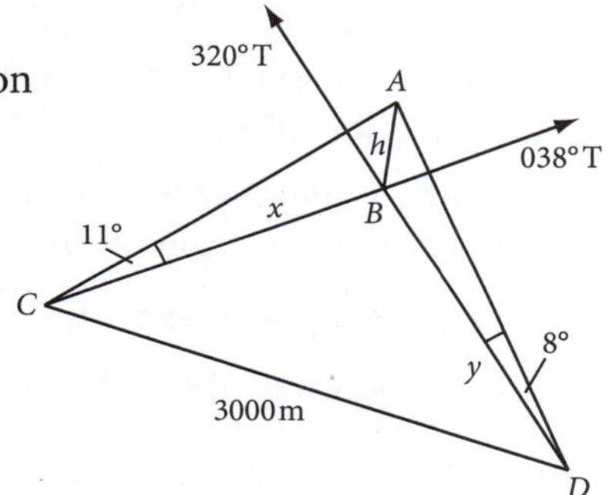

CHAPTER REVIEW 2

1 Simplify:

(a) $\cos(90° - \theta)$	**(b)** $\sin(270° - \theta)$	**(c)** $\cos(90° + \theta)$
(d) $\tan(\theta - 180°)$	**(e)** $\tan(180° - \theta)$	**(f)** $\sin(\theta + 180°)$

2 Write the exact value.

(a) $\tan 315°$	**(b)** $\sin 225°$	**(c)** $\cos 180°$
(d) $\tan 360°$	**(e)** $\sin 60°$	**(f)** $\cos 210°$

3 Simplify:

(a) $\dfrac{\cos\theta}{\sin(90° - \theta)}$ **(b)** $\cos(90° + \theta) + \sin\theta$

4 If $\tan\theta = \dfrac{3}{5}$ and $180° < \theta < 270°$, write the exact value of: **(a)** $\sin\theta$ **(b)** $\cos\theta$

5 If $\sin\alpha = 0.6$ and $0° < \alpha < 90°$, write the exact value of:

(a) $\sin(180° - \alpha)$	**(b)** $\cos(90° - \alpha)$	**(c)** $\cos(180° + \alpha)$
(d) $\tan\alpha$	**(e)** $\tan(180° - \alpha)$	**(f)** $\sin(360° - \alpha)$

6 If $\tan\theta = t$, express in terms of t:

(a) $\tan(90° - \theta)$	**(b)** $\tan(180° + \theta)$	**(c)** $\cot(180° - \theta)$
(d) $\tan(360° - \theta)$	**(e)** $\tan(-\theta)$	**(f)** $\tan(90° + \theta)$

7 Calculate the cosine of the smallest angle of the triangle with side lengths 5 cm, 6 cm and 7 cm.

8 Find the size of the largest angle of the triangle with side lengths 5 cm, 6 cm and 8 cm. Hence, show that the triangle is obtuse-angled.

9 In ΔABC, $B = 53°$, $C = 48°$, $AC = 8$ cm. Calculate:

(a) the length of BC (b) the area of ΔABC.

10 A ladder 8 m long rests against a wall and its foot makes an angle of 60° with the horizontal ground. The top of the ladder then slips down the wall until its foot makes an angle of 45° with the ground. Find, in simplest surd form, how far down the wall the ladder slips.

11 From a point A, level with the foot of a vertical pole and 30 m from it, the angle of elevation of the top of the pole is 40°. Calculate:

(a) the height of the pole (b) the direct distance from A to the top of the pole
(c) the angle of elevation from A of a point half-way up the pole.

12 AB and CD are two vertical buildings with their bases at A and at C on horizontal ground. The height of AB is 30 m. The angle of elevation of B as seen from C is 25° and the angle of elevation of D as seen from A is 40°. Calculate:

(a) the horizontal distance between the buildings (b) the height of CD
(c) the angle of depression of B as seen from D.

13 Two yachts sail in a straight line from a buoy B. One sails 10 km in the direction 040° and the other sails 20 km in the direction 160°.

(a) How far apart are the yachts?
(b) What is the bearing of the first yacht as seen from the second yacht?

14 (a) Find a simplified expression for r given that $r^2 = (100 - 50t)^2 + (80t)^2 - 4(100 - 50t) \times 80t \times \cos 60°$.
(b) Find the value of r to the nearest whole number when $t = \dfrac{30}{43}$.

15 The elevation of a hill at a place K due east of the hill is 38°; at a place L, due south of K, the elevation of the hill is 26°. If the distance from K to L is 500 metres, calculate the height of the hill to the nearest metre.

CHAPTER 3
Further algebraic techniques

3.1 LINEAR EQUATIONS IN ONE VARIABLE

To **solve** a linear equation in one variable is to find the value of the variable that will make the equation true.

Example 1

Solve:

(a) $5x - 3 = 2x + 9$

(b) $5(x - 3) - 3(x - 2) = 19$

Solution

(a)
$$\begin{aligned} 5x - 3 &= 2x + 9 \\ 5x - 3 - 2x &= 9 \\ 3x - 3 &= 9 \\ 3x &= 12 \\ x &= 4 \end{aligned}$$

(b)
$$\begin{aligned} 5(x - 3) - 3(x - 2) &= 19 \\ 5x - 15 - 3x + 6 &= 19 \\ 2x - 9 &= 19 \\ 2x &= 28 \\ x &= 14 \end{aligned}$$

EXERCISE 3.1 LINEAR EQUATIONS IN ONE VARIABLE

Solve:

1 $5x = 3x + 9$

2 $4(2x - 7) = 3x - 5$

3 $3(x + 3) = 3(9 - 3x)$

4 $x - 7 = 3 - x$

5 $5(2a + 1) = 6(a - 5)$

6 $3(5x - 1) - 2(4 - x) = 6$

7 $4 - 3x = x + 9$

8 $3x - 2 = 5x$

9 $5(2x - 1) = 2(x - 5) - 3$

10 $4(3a + 2) - 6(3 - a) = 8$

11 $5 - 9y = 10 - 11y$

12 $4(3m - 1) = 11 + 2m$

13 $2c - (4 - c) = 5 - c$

14 $m + 8 = 5(m - 1) - 2m$

15 $4(x + 5) - (x - 1) = 9$

16 $18 - 3(a - 2) = 2(a + 2)$

17 $3(2x + 5) = 4 - (3x - 2)$

18 $5(c - 7) = 3(3c + 8)$

19 $6(2x - 3) = 8 - 2(3x + 1)$

20 $a + 6 = 3 - a$

3.2 LINEAR EQUATIONS INVOLVING FRACTIONS

For equations involving fractions, write the terms with a common denominator and then solve the equation given by the numerators.

If the equation has only one term on each side of the equals sign, you can multiply each numerator by the other term's denominator to achieve the same result (see Example **2(d)**).

Example 2

Solve:

(a) $\frac{x}{5}+\frac{4x}{3}=23$ (b) $\frac{2x+1}{4}-\frac{2x-3}{6}=\frac{7}{3}$ (c) $\frac{3}{x-4}=\frac{2}{x-2}$ (d) $\frac{3a-2}{2a-1}=\frac{3a+1}{2a+3}$

Solution

(a)
$$\frac{x}{5}+\frac{4x}{3}=23$$
$$\frac{3x}{15}+\frac{5\times 4x}{15}=23$$
$$\frac{23x}{15}=23$$
$$23x=15\times 23$$
$$x=15$$

(b)
$$\frac{2x+1}{4}-\frac{2x-3}{6}=\frac{7}{3}$$
$$\frac{3(2x+1)}{12}-\frac{2(2x-3)}{12}=\frac{28}{12}$$
$$6x+3-4x+6=28$$
$$2x+9=28$$
$$2x=19$$
$$x=9.5$$

(c)
$$\frac{3}{x-4}=\frac{2}{x-2},\ x\neq 4,\ x\neq 2$$
$$\frac{3(x-2)}{(x-4)(x-2)}=\frac{2(x-4)}{(x-2)(x-4)}$$
$$3(x-2)=2(x-4)$$
$$3x-6=2x-8$$
$$3x-2x=6-8$$
$$x=-2$$

(d)
$$\frac{3a-2}{2a-1}=\frac{3a+1}{2a+3},\ a\neq 0.5,\ a\neq -1.5$$
$$\frac{3a-2}{2a-1}=\frac{3a+1}{2a+3}$$
$$(3a-2)(2a+3)=(3a+1)(2a-1)$$
$$6a^2+5a-6=6a^2-a-1$$
$$6a=5$$
$$a=\frac{5}{6}$$

Solutions may have restrictions, which can be easy to miss. In **(c)**, $x\neq 4$ and $x\neq 2$, because these values would make the denominators zero. In **(d)**, $a\neq 0.5$ and $a\neq -1.5$.

EXERCISE 3.2 LINEAR EQUATIONS INVOLVING FRACTIONS

Solve:

1 $\frac{2x}{3}+\frac{3x}{4}=34$

2 $\frac{3x}{4}-\frac{2x}{5}=14$

3 $\frac{3x}{5}=\frac{4x}{3}-22$

4 $\frac{2x+5}{3}=\frac{x+2}{7}+4$

5 $\frac{5x-3}{2}=x+2$

6 $\frac{x+3}{3}-\frac{2x-1}{6}=x+1$

7 $\frac{2x-1}{8}=\frac{3x+1}{4}$

8 $\frac{2x-1}{3}-5=\frac{x}{6}$

9 $\frac{2(2a+1)}{3}=\frac{5(a-2)}{2}$

10 $\frac{5y+1}{4}=6-\frac{2y}{3}$

11 $\frac{x}{3}+\frac{x-7}{4}=\frac{x+1}{6}$

12 $\frac{x+4}{2}-\frac{3-4x}{4}=\frac{5-x}{8}$

13 $\frac{3(x-2)}{5}=\frac{2(x-1)}{3}-\frac{2}{5}$

14 $\frac{x-4}{x+2}=5$

15 $\frac{5}{x}+\frac{3}{2x}=2$

16 $\frac{7}{x}+2=\frac{3}{x}$

17 $\frac{3}{p+1}=\frac{2}{2p+1}$

18 $\frac{3}{x+2}=\frac{5}{2x-1}$

19 $\frac{5}{a+3}=2$

20 $\frac{y+3}{y+2}=\frac{y+1}{y+4}$

21 $\frac{x-2}{x+3}=\frac{x+4}{x-5}$

22 Solve $\frac{2x+3}{3x-1}=\frac{2x-5}{3x+4}$. Which of the following is correct?

A $\frac{1}{3}$ **B** $-\frac{4}{3}$ **C** $-\frac{7}{34}$ **D** $-\frac{1}{2}$

Solve:

23 $\frac{2m-5}{m} = \frac{2m}{m+3}$

24 $\frac{2}{x+1} + \frac{3}{x+9} = 0$

25 $\frac{5}{y} = \frac{2}{y-7}$

26 $\frac{3}{x-2} - \frac{2}{x+2} = \frac{1}{x^2-4}$

27 $\frac{4}{a} - \frac{3}{a+2} = \frac{1}{a(a+2)}$

28 $\frac{1}{x-1} + \frac{1}{2x-1} = 0$

29 $\frac{1}{x+2} + \frac{1}{x-3} = \frac{1}{(x+2)(x-3)}$

30 $\frac{1}{x+1} + \frac{1}{x+2} = \frac{1}{x^2+3x+2}$

31 $\frac{1}{x-3} + \frac{1}{x+3} = \frac{1}{x^2-9}$

32 Solve $\frac{1}{1-t} + \frac{1}{1+t} = \frac{2t}{1-t^2}$. Indicate whether each answer is correct or incorrect.

(a) -1 (b) 0 (c) 1 (d) 2

3.3 SIMPLE LINEAR INEQUALITIES

The solution to an inequality is usually a range of numbers described by another inequality, related to the inequality in the question. The solution is also usually a range of real numbers, unless another set of numbers is specified in the question (e.g. integers).

Rules for inequalities

If both sides of an inequality are multiplied or divided by a negative number, then the direction of the inequality is reversed.

If $a > b$, then: $a + c > b + c$
$a - c > b - c$
$ac > bc$ if $c > 0$
$ac < bc$ if $c < 0$
$\frac{a}{c} > \frac{b}{c}$ if $c > 0$
$\frac{a}{c} < \frac{b}{c}$ if $c < 0$

If $a < b$, then: $a + c < b + c$
$a - c < b - c$
$ac < bc$ if $c > 0$
$ac > bc$ if $c < 0$
$\frac{a}{c} < \frac{b}{c}$ if $c > 0$
$\frac{a}{c} > \frac{b}{c}$ if $c < 0$

On number lines:

- $a > b$ means that a is to the right of b.
- $a < b$ means that a is to the left of b.
- $x \geq 2$ is shown by a solid circle over 2 and an arrow to the right.
- $x > 2$ is shown by an empty circle over 2 and an arrow to the right.

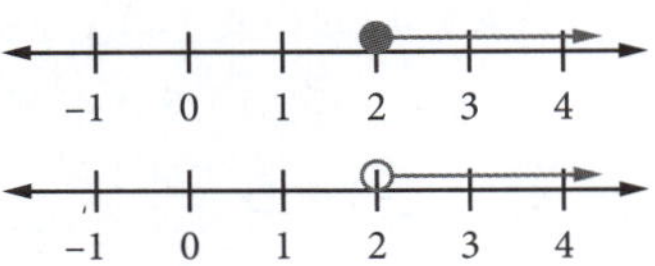

This is demonstrated in the following examples.

Example 3

Solve $2x + 3 \geq 9$ and show the solution on a number line.

Solution

$2x + 3 \geq 9$
$2x \geq 6$
$x \geq 3$

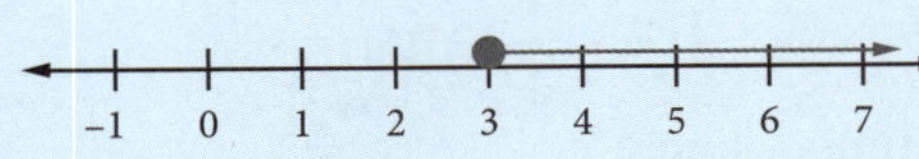

Example 4

Solve $35 - 3x > 19 - x$ and show the solution on a number line, for the conditions:

(a) x is a real number **(b)** x is an integer **(c)** x is not negative

Solution

$35 - 3x > 19 - x$

$35 > 19 + 2x$	Add $3x$ to both sides
$16 > 2x$	Subtract 19 from both sides
$8 > x$	Divide both sides by 2
$x < 8$	Rewrite starting with x

(a) For real numbers:
Solution is $x < 8$

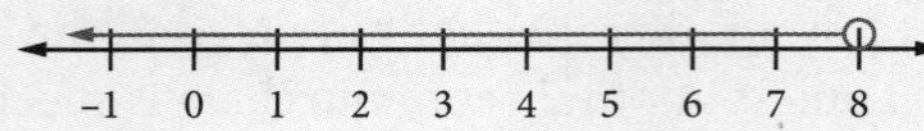

(b) For integers:
Solution is $x = 7, 6, 5, \ldots$
(all integers to the left on the number line)

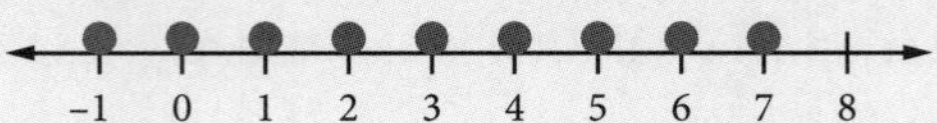

(c) For non-negative numbers, $x \geq 0$:
Solution is $0 \leq x < 8$

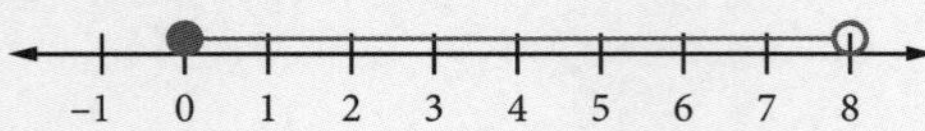

Example 5

Solve $\frac{x-1}{5} \leq \frac{x+1}{3}$ and show the solution on a number line.

Solution

$\frac{x-1}{5} \leq \frac{x+1}{3}$	
$15 \times \frac{(x-1)}{5} \leq 15 \times \frac{(x+1)}{3}$	Multiply both sides by 15
$3(x-1) \leq 5(x+1)$	
$3x - 3 \leq 5x + 5$	Expand both sides
$-3 \leq 2x + 5$	Subtract $3x$ from both sides
$-8 \leq 2x$	Subtract 5 from both sides
$-4 \leq x$	Divide both sides by 2
$x \geq -4$	Rewrite starting with x

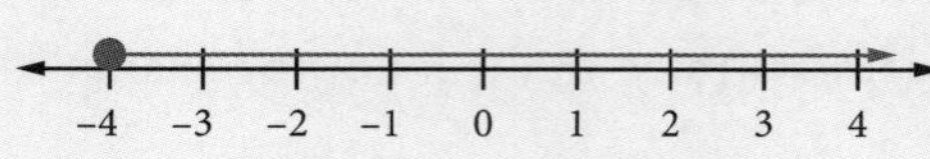

Example 6

Solve $-5 < 2x - 3 \leq 7$ and show the solution on a number line.

Solution

This is really two separate inequalities, $-5 < 2x - 3$ and $2x - 3 \leq 7$, solved simultaneously.

Solve them together by applying the same steps to each part.

$-5 < 2x - 3 \le 7$		or	$2x - 3 > -5$	$2x - 3 \le 7$
$-5 + 3 < 2x \le 7 + 3$	Add 3 to each part		$2x > -5 + 3$	$2x \le 7 + 3$
$-2 < 2x \le 10$	Simplify		$2x > -2$	$2x \le 10$
$-1 < x \le 5$	Divide each part by 2		$x > -1$	$x \le 5$

$-1 < x \le 5$

The solution shows that x is greater than -1 but less than or equal to 5.

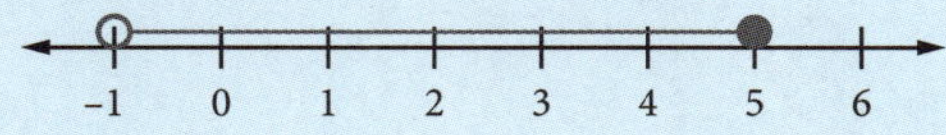

EXERCISE 3.3 SIMPLE LINEAR INEQUALITIES

In questions **1** to **15**, solve each inequality and show the solution on a number line.

1 $5x - 2 \ge 13$ **2** $2x - 2 < 0$ **3** $2x + 5 < -5$ **4** $6x + 2 \ge -10$

5 $3x > 2x + 12$ **6** $3(x + 1) \ge 9$ **7** $7x < 3(2x + 1)$ **8** $2(x - 6) \ge 8$

9 $\frac{x}{3} - \frac{x}{4} > 1$ **10** $\frac{3x}{2} - \frac{x}{3} < \frac{3}{2}$ **11** $3x - 1 \ge x + 2$ **12** $2x + 7 < 3x + 10$

13 $\frac{x-1}{3} > -1$ **14** $\frac{x+3}{5} < 7$ **15** $x - 1 > 5$, x is an integer

16 Solve $4x < x + 15$ for x as a positive integer. The solution on a number line is:

A 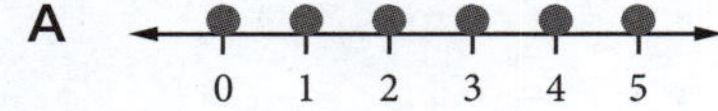

B 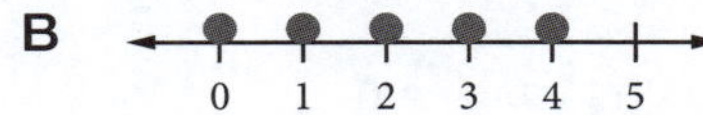

C 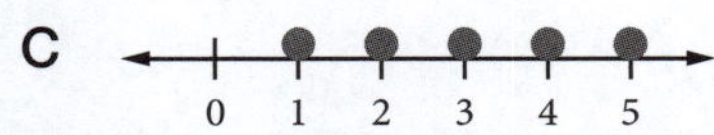

D 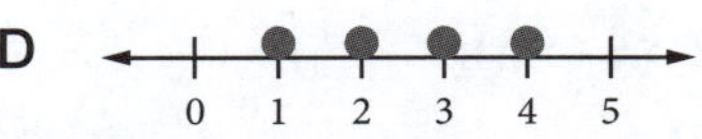

In questions **17** to **23**, solve each inequality and show the solution on a number line.

17 $\frac{3x}{5} - \frac{2x}{3} > -2$ **18** $\frac{7x}{3} < 3 + \frac{4x}{3}$ **19** $\frac{x-5}{2} > \frac{5x-3}{6}$ **20** $\frac{5x-3}{2} < x + 2$

21 $-5 < x + 4 < 1$ **22** $22 \le 5x - 3 \le 32$ **23** $-3 \le \frac{2x-1}{3} < 3$

24 Solve simultaneously $x - 2 > -2$ and $x - 3 \le 0$. Indicate whether each answer is correct or incorrect.

(a) $0 \le x \le 3$ **(b)** $0 < x \le 3$ **(c)** $0 \le x \le 3$ **(d)** $x > 0$ or $x \le 3$

25 If 5 is subtracted from a certain positive integer, the result is greater than 5 but less than 12. What values can this integer take? (Let the positive integer be x, so that $5 < x - 5 < 12$.)

26 If a certain number is divided by 2, the result is greater than 4 but less than 8. What values can this number take?

27 The sum of two consecutive positive integers is no more than 35. What are the possible values of these integers?

28 A committee consists of 3 more women than men. The total number of committee members is at least 7 but not more than 15. How many women could be on the committee?

29 When a certain number is added to 5 and the sum is divided by 5, the result is not more than if this same number is added to 13 and the sum is divided by 9. What is the largest value this number can take?

30 The base length of an isosceles triangle is an integer (in cm) and is 4 cm less than the sum of the two equal sides. The perimeter is an integer (in cm) less than 80 cm. What are the possible base lengths?

31 The sum of three consecutive integers is greater than 7 and less than 25. Find all possible values of the smallest of these integers.

32 The side lengths of a triangle are 8 cm, 10 cm and x cm. What are the possible values of x?

3.4 QUADRATIC EQUATIONS

An equation of the form $ax^2 + bx + c = 0$, $a \neq 0$ is called a quadratic equation in x. The values of x that make this equation true are called the solutions or the roots of the equation.

A quadratic equation can be solved by factorising the quadratic expression and making the factored expression equal to zero:

If $AB = 0$, then $A = 0$ or $B = 0$ or $A = B = 0$. (This is the **null factor law**.)

Example 7

Solve:

(a) $x(x-2)=0$ (b) $(x+1)(x-4)=0$

Solution

(a) $x(x-2)=0$

$x=0$ or $x-2=0$

$x=0$ or $x=2$

(b) $(x+1)(x-4)=0$

$x+1=0$ or $x-4=0$

$x=-1$ or $x=4$

EXERCISE 3.4 QUADRATIC EQUATIONS

Solve:

1 $x(x-5)=0$

2 $(x-2)(x-3)=0$

3 $x(2x+1)=0$

4 $(x-7)(2x+5)=0$

5 $3x(2x-9)=0$

6 $-5x(x+1)=0$

7 $(x-a)(x-b)=0$

8 $(x-3a)(x+2b)=0$

9 $(x-2)(x+2)=0$

10 $(2x-11)(2x+11)=0$

11 $(x-1)^2=0$

12 $(2x+3)^2=0$

3.5 QUADRATIC EQUATIONS WITHOUT A LINEAR TERM

To solve $ax^2 + c = 0$, $c < 0$, remember the difference of two squares: $a^2 - b^2 = (a-b)(a+b)$.

Example 8

Solve:

(a) $x^2-4=0$ (b) $2x^2-18=0$ (c) $9x^2=25$ (d) $(x-2)^2=9$

Solution

(a) $x^2-4=0$

$(x-2)(x+2)=0$

$x-2=0$ or $x+2=0$

$x=2$ or $x=-2$

(b) $2x^2-18=0$

$2(x^2-9)=0$

$2(x-3)(x+3)=0$

$x=3$ or $x=-3$

(c) $9x^2=25$

$9x^2-25=0$

$(3x-5)(3x+5)=0$

$3x-5=0$ or $3x+5=0$

$x=\frac{5}{3}$ or $x=-\frac{5}{3}$

(d) $(x-2)^2=9$

$(x-2)^2-9=0$

$(x-2-3)(x-2+3)=0$

$(x-5)(x+1)=0$

$x=5$ or $x=-1$

Example 9

Solve:

(a) $x^2 - 6 = 0$ (b) $3x^2 = 15$

Solution

(a) $x^2 - 6 = 0$

Use $6 = \left(\sqrt{6}\right)^2$

$x^2 - \left(\sqrt{6}\right)^2 = 0$

$\left(x - \sqrt{6}\right)\left(x + \sqrt{6}\right) = 0$

$x = \sqrt{6}$ or $x = -\sqrt{6}$

or $x^2 = 6$

$x = \pm\sqrt{6}$

(b) $3x^2 = 15$

$3(x^2 - 5) = 0$

$3\left(x - \sqrt{5}\right)\left(x + \sqrt{5}\right) = 0$

$x = \sqrt{5}$ or $x = -\sqrt{5}$

or $x^2 = 5$

$x = \pm\sqrt{5}$

MAKING CONNECTIONS

Quadratic equations without a linear term

Use technology to investigate solutions to $ax^2 - c = 0$

EXERCISE 3.5 QUADRATIC EQUATIONS WITHOUT A LINEAR TERM

Solve:

1 $x^2 - 1 = 0$ **2** $x^2 - 25 = 0$ **3** $x^2 - 49 = 0$ **4** $x^2 = 16$

5 The roots of the equation $x^2 - 0.25 = 0$ are:

A $x = 0.5$ B $x = -0.5$ C $x = \pm 0.5$ D $x = \pm\frac{1}{16}$

Solve:

6 $4x^2 = 9$ **7** $x^2 - 6\frac{1}{4} = 0$ **8** $4x^2 - 25 = 0$ **9** $x^2 - 2\frac{1}{4} = 0$

10 $5x^2 - 5 = 0$ **11** $16x^2 - 1 = 0$ **12** $16 - x^2 = 0$ **13** $25x^2 = 49$

14 For the equation $9(x-1)^2 - 36 = 0$, indicate whether each statement is correct or incorrect.

(a) $(x-1)^2 = 4$ (b) $x = 1 \pm \sqrt{4}$ (c) $x = -3, 1$ (d) $x = -1, 3$

Solve:

15 $(x-2)^2 = 16$ **16** $5x^2 - 245 = 0$ **17** $7x^2 = 63$ **18** $(5x-1)^2 = 16$

19 $(x+3)^2 - 4 = 0$ **20** $x^2 - 5 = 0$ **21** $x^2 = 2$ **22** $(x+1)^2 = 8$

3.6 QUADRATIC EQUATIONS WITHOUT A CONSTANT TERM

To solve $ax^2 + bx = 0$: the equation has no constant term, so the common factor is x or a multiple of x.

Example 10

Solve:

(a) $x^2 - 3x = 0$ (b) $4x^2 = 8x$

Solution

(a)
$$x^2 - 3x = 0$$
$$x(x-3) = 0$$
$$x = 0 \quad \text{or} \quad x = 3$$

(b)
$$4x^2 = 8x$$
$$4x^2 - 8x = 0$$
$$4x(x-2) = 0 \qquad \text{Divide by 4, do \textbf{not} divide by } x$$
$$x = 0 \quad \text{or} \quad x = 2$$

EXERCISE 3.6 QUADRATIC EQUATIONS WITHOUT A CONSTANT TERM

Solve:

1 $x^2 - 6x = 0$ **2** $x^2 - 5x = 0$ **3** $x^2 + 5x = 0$ **4** $x^2 + 10x = 0$

5 The solution to $x^2 = 4x$ is:

A $x = 0$ **B** $x = 4$ **C** $x = 0, 2$ **D** $x = 0, 4$

Solve:

6 $2x^2 - 5x = 0$ **7** $x^2 = 7x$ **8** $3x^2 - 21x = 0$ **9** $2x^2 + 20x = 0$

10 $6x^2 = 24x$ **11** $2x^2 - x = 0$ **12** $5x^2 + x = 0$ **13** $3x^2 = -9x$

14 $12x^2 - 5x = 0$ **15** $15x - x^2 = 0$

3.7 GENERAL QUADRATIC EQUATIONS

To solve $ax^2 + bx + c = 0$, factorise the trinomial if possible.

Example 11

Solve:

(a) $x^2 - 5x + 6 = 0$ (b) $2x^2 = x + 3$ (c) $x(x-2) = 3$ (d) $(3x+4)(x-3) = 16$

Solution

(a)
$$x^2 - 5x + 6 = 0$$
$$(x-2)(x-3) = 0$$
$$x = 2 \quad \text{or} \quad x = 3$$

(b)
$$2x^2 = x + 3$$
$$2x^2 - x - 3 = 0$$
$$(2x-3)(x+1) = 0$$
$$x = 1.5 \quad \text{or} \quad x = -1$$

(c)
$$x(x-2) = 3$$
$$x^2 - 2x - 3 = 0$$
$$(x-3)(x+1) = 0$$
$$x = 3 \quad \text{or} \quad x = -1$$

(d)
$$(3x+4)(x-3) = 16$$
$$3x^2 - 5x - 12 = 16$$
$$3x^2 - 5x - 28 = 0$$
$$(3x+7)(x-4) = 0$$
$$x = -2\frac{1}{3} \quad \text{or} \quad x = 4$$

EXERCISE 3.7 GENERAL QUADRATIC EQUATIONS

Solve:

1 $x^2 - 3x + 2 = 0$ **2** $x^2 - 6x + 5 = 0$ **3** $x^2 - 2x - 8 = 0$ **4** $x^2 - 4x + 3 = 0$

5 $x^2 - 6x + 9 = 0$ **6** $x^2 - 5x + 4 = 0$ **7** $x^2 + 9x + 8 = 0$ **8** $9x^2 + 4x - 5 = 0$

9 The roots of the equation $5x^2 + 7x - 12 = 0$ are:

A $x = -1, 2.4$ **B** $x = -2.4, 1$ **C** $x = -1.2, 1$ **D** $x = -0.8, 3$

Solve:

10 $x^2 + 4x - 12 = 0$ **11** $5x^2 - 11x + 2 = 0$ **12** $4x^2 - 12x - 7 = 0$ **13** $2x^2 - x - 10 = 0$

14 $x^2 + 10x + 25 = 0$ **15** $x^2 + 5x + 4 = 0$ **16** $4x^2 - 8x - 21 = 0$ **17** $3x^2 - 28x + 25 = 0$

18 $x^2 - 8x + 16 = 0$ **19** $5x^2 + 26x + 24 = 0$ **20** $3x^2 - 41x + 60 = 0$ **21** $30 - 7x - x^2 = 0$

22 $5x^2 = 8x - 3$ **23** $x(2x - 11) = 6$ **24** $x(x + 5) = 6$ **25** $x(3x + 19) = 72$

26 $x^2 + 15 = 8x$ **27** $(x - 2)(2x + 5) = 2x + 5$ **28** $12 - 4x - x^2 = 0$ **29** $(2x + 1)^2 = 4$

30 $(x + 1)^2 = 4x$ **31** $(x + 6)^2 = x + 6$ **32** $6x^2 = 10 - 11x$ **33** $7x^2 = 2(17x - 12)$

3.8 COMPLETING THE SQUARE

In the expansion $(x + 6)^2 = x^2 + 12x + 36$, the constant term 36 is half the coefficient of x squared: $\left(\frac{12}{2}\right)^2 = 36$.

In the expansion $(x - 5)^2 = x^2 - 10x + 25$, the constant term 25 is half the coefficient of x squared: $\left(\frac{-10}{2}\right)^2 = 25$.

This can be shown for the general expansion $(x - a)^2 = x^2 - 2ax + a^2$ as $\left(\frac{-2a}{2}\right)^2 = a^2$.

Thus, an expression like $x^2 + 6x$ can be made into a perfect square by adding $\left(\frac{6}{2}\right)^2 = 9$ to obtain $x^2 + 6x + 9 = (x + 3)^2$.

Example 12

What must be added to each expression to complete the square?

(a) $x^2 + 8x$ **(b)** $x^2 - 3x$

Solution

(a) $x^2 + 8x$
- Half of 8 is 4 (This is the value that will go in the brackets.)
- The square of 4 is 16
- Hence 16 must be added

Check: $x^2 + 8x + 16 = (x + 4)^2$

(b) $x^2 - 3x$
- Half of -3 is $-\frac{3}{2}$
- The square of $-\frac{3}{2}$ is $\frac{9}{4}$
- Hence $\frac{9}{4}$ must be added

Check: $x^2 - 3x + \frac{9}{4} = \left(x - \frac{3}{2}\right)^2$

EXERCISE 3.8 COMPLETING THE SQUARE

In questions **1–4** and **6–13**, write the number to be added to complete the square.

1 $x^2 + 4x$ **2** $x^2 - 6x$ **3** $x^2 + 14x$ **4** $x^2 + 2x$

5 The number added to $x^2 - 12x$ to complete the square is:

A 36 **B** 144 **C** -36 **D** -144

6 $x^2 - x$ **7** $x^2 + 5x$ **8** $x^2 + 3x$ **9** $x^2 - 7x$

10 $x^2 + x$ **11** $x^2 + 2ax$ **12** $x^2 - 2bx$ **13** $x^2 + cx$

14 The square is completed for the expression $x^2 - ax$. Indicate whether the following statements would be correct or incorrect.

(a) $x^2 - ax + \frac{a^2}{2}$ (b) $x^2 - ax + \frac{a^2}{4}$ (c) $\left(x - \frac{a}{2}\right)^2$ (d) $x^2 - ax + a^2$

3.9 SOLVING QUADRATIC EQUATIONS BY COMPLETING THE SQUARE

Example 13

Complete the square to solve:

(a) $x^2 + 4x - 5 = 0$ (b) $x^2 - 5x + 6 = 0$ (c) $x^2 = 8x$

Solution

(a) $x^2 + 4x - 5 = 0$

$x^2 + 4x = 5$ Add 5 to both sides

$x^2 + 4x + 4 = 5 + 4$ Add 4 to both sides to complete the square

$(x + 2)^2 = 9$ Factorise

$x + 2 = \pm 3$ Take the square roots of each side

$x + 2 = 3$ or $x + 2 = -3$

$x = 1$ or $x = -5$

(b) $x^2 - 5x + 6 = 0$

$x^2 - 5x = -6$ Subtract 6 from both sides

$x^2 - 5x + \frac{25}{4} = -6 + \frac{25}{4}$ Add $\frac{25}{4}$ to both sides to complete the square

$\left(x - \frac{5}{2}\right)^2 = \frac{1}{4}$ Factorise

$x - \frac{5}{2} = \pm\frac{1}{2}$ Take the square roots of each side

$x - \frac{5}{2} = \frac{1}{2}$ or $x - \frac{5}{2} = -\frac{1}{2}$

$x = 3$ or $x = 2$

OR, instead of taking the square roots of each side, rewrite the expression as:

$\left(x - \frac{5}{2}\right)^2 - \frac{1}{4} = 0$

Then, factorise using the difference of two squares:

$\left(x - \frac{5}{2} - \frac{1}{2}\right)\left(x - \frac{5}{2} + \frac{1}{2}\right) = 0$

$(x - 3)(x - 2) = 0$

$x = 3$ or $x = 2$

(c) $x^2 = 8x$

$x^2 - 8x = 0$ Subtract $8x$ from both sides

$x^2 - 8x + 16 = 16$ Add 16 to both sides to complete the square

$(x - 4)^2 = 16$ Factorise

$x - 4 = \pm 4$ Take the square roots of each side

$x - 4 = 4$ or $x - 4 = -4$

$x = 8$ or $x = 0$

EXERCISE 3.9 SOLVING QUADRATIC EQUATIONS BY COMPLETING THE SQUARE

Complete the square to solve the quadratic equations in this exercise.

1 $x^2 - 6x + 5 = 0$ **2** $x^2 - 2x - 8 = 0$ **3** $x^2 + 4x - 5 = 0$ **4** $x^2 + 4x = 12$

5 When solving $x^2 - 10x + 24 = 0$ by completing the square, the line of working after completing the square could be:

A $x^2 - 10x + 16 = -8$ B $x^2 - 10x + 25 = 0$ C $x^2 - 10x + 100 = 76$ D $x^2 - 10x + 25 = 1$

6 $x^2 - 4x = 21$ **7** $x^2 - 26x + 25 = 0$ **8** $x^2 - 3x + 2 = 0$ **9** $x^2 + x - 12 = 0$

10 $x^2 - 5x + 4 = 0$ **11** $x^2 + 7x = 30$ **12** $x^2 - 11x = 12$ **13** $x^2 - 3x - 10 = 0$

14 $x^2 = 7x - 10$ **15** $x^2 + x = 72$ **16** $x^2 - 10x - 11 = 0$ **17** $x^2 - 10x = 0$

18 $x^2 = 3x$ is solved by completing the square. Indicate whether the following steps in the working are correct or incorrect.

(a) $x^2 - 3x + \frac{9}{4} = \frac{9}{4}$ (b) $\left(x - \frac{3}{2}\right)^2 = \frac{3}{2}$ (c) $x - \frac{3}{2} = \pm\frac{3}{2}$ (d) $x = 0, 3$

3.10 QUADRATIC EQUATIONS WITH NON-RATIONAL SOLUTIONS

All the questions in Exercise 3.9 could have been solved by factorising. Practice in completing the square will also enable you to solve equations that do not have rational factors.

Example 14

Complete the square to solve:

(a) $x^2 + 2x - 5 = 0$ (b) $x^2 = 4x + 8$ (c) $x^2 - 5x + 2 = 0$

Solution

(a) $x^2 + 2x - 5 = 0$

$x^2 + 2x = 5$	Move constant to RHS
$x^2 + 2x + 1 = 5 + 1$	Add 1^2 to complete the square
$(x + 1)^2 = 6$	Factorise
$x + 1 = \pm\sqrt{6}$	Take square roots of both sides
$x = -1 + \sqrt{6}$ or $x = -1 - \sqrt{6}$	Exact answers
$x = 1.45$ or $x = -3.45$	Answers correct to 2 decimal places

(b) $x^2 = 4x + 8$

$x^2 - 4x = 8$	Rewrite with only the constant on RHS
$x^2 - 4x + 4 = 8 + 4$	Add 2^2 to complete the square
$(x - 2)^2 = 12$	Factorise
$x - 2 = \pm\sqrt{12}$	Take square roots of both sides
$x = 2 + 2\sqrt{3}$ or $x = 2 - 2\sqrt{3}$	Exact answers
$x = 5.46$ or $x = -1.46$	Answers correct to 2 decimal places

(c) $x^2 - 5x + 2 = 0$

$x^2 - 5x = -2$	Move constant to RHS
$x^2 - 5x + \frac{25}{4} = -2 + \frac{25}{4}$	Add $\left(\frac{5}{2}\right)^2$ to complete the square
$\left(x - \frac{5}{2}\right)^2 = \frac{17}{4}$	Factorise
$x - \frac{5}{2} = \pm\frac{\sqrt{17}}{2}$	Take square roots of both sides
$x = \frac{5}{2} + \frac{\sqrt{17}}{2}$ or $x = \frac{5}{2} - \frac{\sqrt{17}}{2}$	Exact answers
$x = 4.56$ or $x = 0.44$	Answers correct to 2 decimal places

EXERCISE 3.10 QUADRATIC EQUATIONS WITH NON-RATIONAL SOLUTIONS

Complete the square to solve the following quadratic equations, giving answers to even-numbered questions in surd form and answers to odd-numbered questions correct to two decimal places.

1 $x^2 - 2x - 4 = 0$ **2** $x^2 + 4x - 4 = 0$ **3** $x^2 - x - 5 = 0$ **4** $x^2 - 6x + 2 = 0$

5 $x^2 - 5x + 1 = 0$ **6** $x^2 + 2x - 2 = 0$ **7** $x^2 = 6x - 4$ **8** $x^2 + x - 1 = 0$

9 $x^2 - 6x - 5 = 0$ **10** $x^2 + 4x = 1$ **11** $x^2 = 2x + 5$ **12** $x^2 + 3x - 6 = 0$

3.11 COMPLETING THE SQUARE FOR NON-MONIC EQUATIONS

Completing the square is more difficult when the coefficient of x^2 is not 1. You can overcome this problem by first dividing every term by the coefficient of x^2.

Example 15

Complete the square to solve $2x^2 - 3x - 3 = 0$.

Solution

$2x^2 - 3x - 3 = 0$

$2x^2 - 3x = 3$	Move constant to RHS
$x^2 - \frac{3x}{2} = \frac{3}{2}$	Divide by coefficient of x^2
$x^2 - \frac{3x}{2} + \frac{9}{16} = \frac{3}{2} + \frac{9}{16}$	Add $\left(\frac{3}{4}\right)^2$ to complete the square
$\left(x - \frac{3}{4}\right)^2 = \frac{33}{16}$	Factorise
$x - \frac{3}{4} = \pm\frac{\sqrt{33}}{4}$	Take square roots of both sides
$x = \frac{3}{4} + \frac{\sqrt{33}}{4} = \frac{3+\sqrt{33}}{4}$ or $x = \frac{3}{4} - \frac{\sqrt{33}}{4} = \frac{3-\sqrt{33}}{4}$	Exact answers
$x = 2.19$ or $x = -0.69$	Answers correct to two decimal places

EXERCISE 3.11 COMPLETING THE SQUARE FOR NON-MONIC EQUATIONS

Complete the square to solve the following quadratic equations, giving answers to even-numbered questions in surd form and answers to odd-numbered questions correct to two decimal places.

1 $2x^2 - x - 5 = 0$ **2** $2x^2 + 6x - 5 = 0$ **3** $2x^2 + x - 2 = 0$ **4** $2x^2 + 3x - 1 = 0$

5 $3x^2 - 5x - 1 = 0$ **6** $3x^2 + 4x = 5$ **7** $3x^2 - 2x = 4$ **8** $2x^2 - 6x + 1 = 0$

9 $3x^2 = 7x + 3$ **10** $4x^2 + 4x - 5 = 0$ **11** $2x^2 - 5x = 9$ **12** $3x^2 - 2x - 2 = 0$

13 $2x^2 + x = 4$ **14** $3x^2 - 8x + 3 = 0$ **15** $6x^2 - 10x + 3 = 0$

3.12 THE QUADRATIC FORMULA

The equation $ax^2 + bx + c = 0$, $a \neq 0$ is the general quadratic equation.

If you solve this equation by completing the square, you obtain the **quadratic formula**, which is a solution that must be true for all quadratic equations. This formula enables us to solve quadratic equations even when the factors are not obvious.

$$ax^2 + bx + c = 0$$

$$ax^2 + bx = -c \quad \text{Move constant to RHS}$$

$$x^2 + \frac{b}{a}x = -\frac{c}{a} \quad \text{Divide by coefficient of } x^2$$

$$x^2 + \frac{b}{a}x + \left(\frac{b}{2a}\right)^2 = \frac{b^2}{4a^2} - \frac{c}{a} \quad \text{Add } \left(\frac{b}{2a}\right)^2 \text{ to complete the square}$$

$$\left(x + \frac{b}{2a}\right)^2 = \frac{b^2 - 4ac}{4a^2} \quad \text{Factorise}$$

$$x + \frac{b}{2a} = \pm\frac{\sqrt{b^2 - 4ac}}{2a} \quad \text{Take square roots of both sides}$$

$$x = -\frac{b}{2a} \pm \frac{\sqrt{b^2 - 4ac}}{2a}$$

$$x = \frac{-b \pm \sqrt{b^2 - 4ac}}{2a} \quad \text{The quadratic formula}$$

Example 16

Use the quadratic formula to solve the following quadratic equations.

(a) $x^2 + 8x + 12 = 0$ (b) $x^2 - 3x - 2 = 0$ (c) $2x^2 - 4x + 1 = 0$ (d) $4x^2 + 5x - 2 = 0$

Solution

(a) $x^2 + 8x + 12 = 0$

$$x = \frac{-b \pm \sqrt{b^2 - 4ac}}{2a}, \; a = 1, b = 8, c = 12$$

$$x = \frac{-8 \pm \sqrt{8^2 - 4 \times 1 \times 12}}{2}$$

$$= \frac{-8 \pm \sqrt{64 - 48}}{2}$$

$$= \frac{-8 \pm \sqrt{16}}{2}$$

$$= \frac{-8 + 4}{2} \quad \text{or} \quad \frac{-8 - 4}{2}$$

$$= -2 \quad \text{or} \quad -6$$

It would have been faster to use factors to solve **(a)**: $x^2 + 8x + 12 = (x + 2)(x + 6)$

(b) $x^2 - 3x - 2 = 0$

$$x = \frac{-b \pm \sqrt{b^2 - 4ac}}{2a}, \; a = 1, b = -3, c = -2$$

$$x = \frac{-(-3) \pm \sqrt{(-3)^2 - 4 \times 1 \times (-2)}}{2}$$

$$= \frac{3 \pm \sqrt{9 + 8}}{2}$$

$$= \frac{3 \pm \sqrt{17}}{2}$$

$$= \frac{3 + \sqrt{17}}{2} \quad \text{or} \quad \frac{3 - \sqrt{17}}{2} \quad \text{(Exact solution)}$$

$$= 3.56 \quad \text{or} \quad -0.56 \quad \text{(Correct to 2 d.p.)}$$

(c) $2x^2 - 4x + 1 = 0$

$$x = \frac{-b \pm \sqrt{b^2 - 4ac}}{2a}, \; a = 2, b = -4, c = 1$$

$$x = \frac{-(-4) \pm \sqrt{(-4)^2 - 4 \times 2 \times 1}}{4}$$

$$= \frac{4 \pm \sqrt{16 - 8}}{4}$$

$$= \frac{4 \pm \sqrt{8}}{4}$$

$$= \frac{4 \pm 2\sqrt{2}}{4}$$

$$= \frac{2 \pm \sqrt{2}}{2} \quad \text{(Exact solution)}$$

$$= 1.71 \quad \text{or} \quad 0.29 \quad \text{(Correct to 2 d.p.)}$$

(d) $4x^2 + 5x - 2 = 0$

$$x = \frac{-b \pm \sqrt{b^2 - 4ac}}{2a}, \; a = 4, b = 5, c = -2$$

$$x = \frac{-5 \pm \sqrt{5^2 - 4 \times 4 \times (-2)}}{8}$$

$$= \frac{-5 \pm \sqrt{25 + 32}}{8}$$

$$= \frac{-5 \pm \sqrt{57}}{8}$$

$$= \frac{-5 + \sqrt{57}}{8} \quad \text{or} \quad \frac{-5 - \sqrt{57}}{8} \quad \text{(Exact solution)}$$

$$= 0.32 \quad \text{or} \quad -1.57 \quad \text{(Correct to 2 d.p.)}$$

EXERCISE 3.12 THE QUADRATIC FORMULA

Solve the following quadratic equations, giving the answers to the even-numbered questions in surd form, and to the odd-numbered questions correct to two decimal places (if necessary).

1 $x^2 + 6x + 5 = 0$ **2** $x^2 + 2x - 8 = 0$ **3** $x^2 - 6x - 7 = 0$ **4** $x^2 - 7x + 10 = 0$

5 $x^2 + 2x - 1 = 0$ **6** $x^2 - 6x + 4 = 0$ **7** $x^2 - 2x - 5 = 0$ **8** $x^2 + 5x - 1 = 0$

9 $x^2 - 2x - 9 = 0$ **10** $x^2 + 4x + 2 = 0$ **11** $x^2 - 15x + 56 = 0$ **12** $x^2 + 2x - 15 = 0$

13 $2x^2 + 5x + 1 = 0$ **14** $2x^2 - 8x + 3 = 0$ **15** $2x^2 + 3x + 1 = 0$ **16** $2x^2 - 3x = 0$

17 $3x^2 + 2x - 2 = 0$ **18** $2x^2 + 3x - 5 = 0$ **19** $x^2 + 6x + 1 = 0$ **20** $x^2 - 8x + 16 = 0$

21 $2x^2 - x - 3 = 0$ **22** $7x^2 - 7x - 2 = 0$ **23** $4x^2 - 9x + 4 = 0$ **24** $3x^2 - 11x - 4 = 0$

25 $2x^2 + x = 3$ **26** $x(x + 3) = 2$ **27** $2x^2 + 6x + 1 = 0$ **28** $2x^2 - 6x = 3$

29 $x^2 = 2x + 2$ **30** $x^2 = 6x + 2$ **31** $2x^2 = 3x + 4$ **32** $2x^2 + 10x + 5 = 0$

33 $x^2 + 17x = 60$ **34** $3x^2 + 9x + 5 = 0$ **35** $3x^2 - 15 = 0$ **36** $x(x + 1) = 1$

3.13 PROBLEMS INVOLVING QUADRATIC EQUATIONS

Example 17

One side of a rectangle is 2 cm longer than the other side. The area of the rectangle is 120 cm^2. What are the dimensions of the rectangle?

Solution

Let one side length be x cm.

The other side length is $(x + 2)$ cm.

Draw a diagram to show this information.

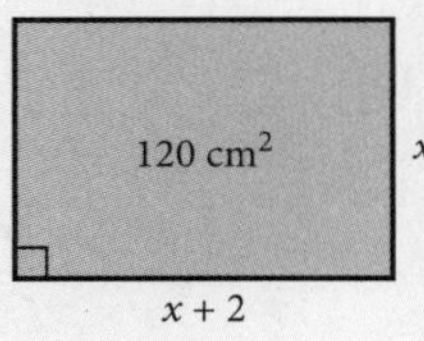

The area of the rectangle is 120 cm^2, so:

$$x(x + 2) = 120$$
$$x^2 + 2x = 120$$
$$x^2 + 2x - 120 = 0$$
$$(x + 12)(x - 10) = 0$$
$$x + 12 = 0 \quad \text{or} \quad x - 10 = 0$$
$$x = -12 \quad \text{or} \quad x = 10$$

Because x represents the side length of a rectangle, $x > 0$. This means the only possible solution is $x = 10$.

Therefore the dimensions of the rectangle are 10 cm by 12 cm.

We should have written the original equation as $x(x + 2) = 120$, $x > 0$, to remember that x must represent a positive length.

Also, don't forget that when solving $AB = 0$, either $A = 0$ or $B = 0$ or $A = B = 0$, but zero is not always a valid solution to the problem. This is one reason why we don't always use all solutions to the quadratic equation in practical problems.

Example 18

The height h metres of a stone, t seconds after being thrown straight up, is given by the equation $h = 30t - 5t^2$. When is the stone at a height of 40 metres?

Solution

$$h = 30t - 5t^2$$

For $h = 40$: $\quad 40 = 30t - 5t^2$

$$5t^2 - 30t + 40 = 0$$
$$5(t^2 - 6t + 8) = 0$$
$$5(t - 2)(t - 4) = 0$$
$$t = 2 \quad \text{or} \quad t = 4$$

On the way up, the stone reaches a height of 40 m after 2 seconds; on the way down, it comes back to a height of 40 m at 4 seconds. In this problem both answers make sense.

EXERCISE 3.13 PROBLEMS INVOLVING QUADRATIC EQUATIONS

1 In each diagram, all measurements are in centimetres and the area of the shaded region is given. Find the value of x in each case.

(a) 40 cm^2 $(x - 1)$ $(x + 2)$

(b)

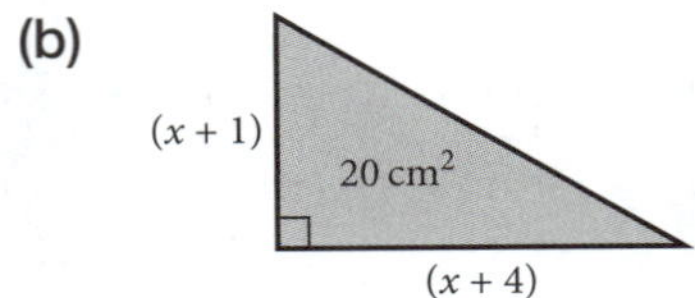

(c)

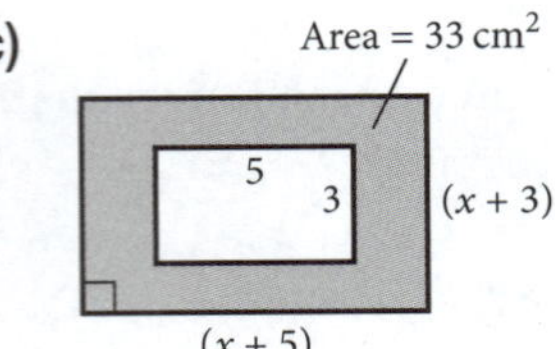

(d) Area = 44 cm^2 x $(x + 3)$

(e)

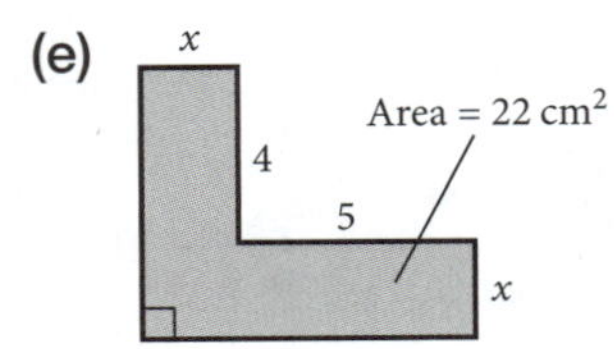

2 For the information given on the triangle, which statement is correct?

A $x(x - 1) = 2.5$ **B** $2x - 1 = 5$
C $x^2 = 12$ **D** $x^2 - x - 12 = 0$

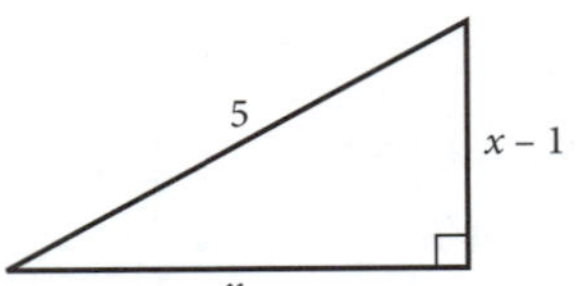

3 Use Pythagoras' theorem to find the value of x, given that all measurements are in centimetres.

(a) 13 x $x + 7$

(b)

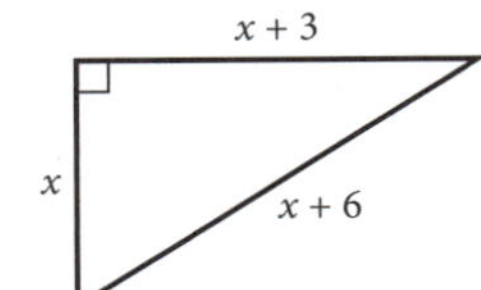

(c)

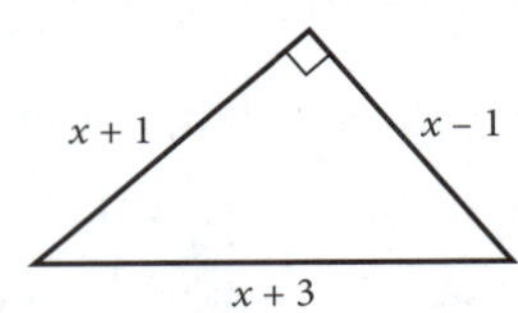

4 The sum of a certain positive number and the square of that number is 12. What is the number?

5 The product of two numbers is 88. If one of the numbers is 3 more than the other, what are the numbers?

6 The product of two consecutive numbers is 72. What are the numbers?

7 The height h metres of a stone, t seconds after being thrown straight up, is given by $h = 40t - 5t^2$. At what times is the stone at a height of: **(a)** 60 m **(b)** 80 m?

8 The sum of the square of a positive number and four times the number is 60. What is the number?

9 A rectangular swimming pool, 12 m by 8 m, is surrounded by a concrete path of uniform width. If the area of the path alone is 224 m^2, find its width.

10 A carpet is placed in a room measuring 6 m by 4 m, leaving an uncarpeted border of uniform width around it. If the area of the carpet is 8 m^2, find the width of the border.

11 A picture on a wall measures 24 cm by 20 cm. It is surrounded by a frame of uniform width whose area, not including the picture, is 416 cm^2. What is the width of the frame?

12 A rectangle is 8 cm longer than its breadth. If the area of the rectangle is 48 cm^2, what are its dimensions?

13 In a right-angled triangle, one of the sides adjacent to the right angle is 4 cm longer than the other side. If the area of the triangle is 96 cm^2, find the length of each of the three sides.

14 The perimeter of a rectangle is 40 cm and its area is 84 cm^2.

(a) If the breadth of the rectangle is x cm, express the length in terms of x.
(b) Write the area of the rectangle in terms of x.
(c) Form a quadratic equation in x and solve it to find the length and breadth.

CHAPTER REVIEW 3

1 Solve:

(a) $5a - 6 = 4(2a + 3)$ (b) $3(8a - 2) - 3(2a + 4) = 0$ (c) $8(x + 2) - 3(x + 5) = 2(x - 2)$

2 Solve:

(a) $\frac{x}{5} = \frac{3}{20}$ (b) $\frac{3x - 1}{5} = \frac{x}{20}$ (c) $\frac{x - 2}{x + 3} = \frac{3}{5}$

3 Solve, showing your solution on a number line:

(a) $\frac{3x - 2}{5} > 2$ (b) $-8 \le 3x - 2 < 16$ (c) $|x - 1| > 1$

4 Solve:

(a) $x^2 = 4$ (b) $x^2 = 4x$ (c) $x^2 = 4x - 4$
(d) $(x^2 - 3x)^2 = 16$ (e) $(x^2 - 3x - 10)(x^2 - 3x - 4) = 0$ (f) $6x^2 + 7x - 3 = 0$

5 The hypotenuse of a right-angled triangle is $(x + 1)$ cm in length and the other two sides are x cm and $(x - 7)$ cm. Form an equation and solve it to find the length of each side.

6 Solve the quadratic equation $2x^2 - x - 5 = 0$, giving your solutions:

(a) in simplest surd form (b) correct to 2 decimal places.

7 Find the solutions of the equation $\frac{x}{x+1} - \frac{1}{x+2} = 3$ as surds.

8 Expand and simplify $(2x - y)(x^2 - xy + y^2)$.

9 Simplify: $\frac{2}{m^2 - 4} - \frac{1}{m^2 - 3m + 2}$

10 If $n \ge 0$, solve $262\,500 = \frac{n}{2}\left(66\,000 + 1500(n - 1)\right)$.

11 The perimeter of a rectangle is 18 cm and its area is 20 cm^2.

(a) If the length is x cm, express the breadth in terms of x.
(b) Write the area in terms of x.
(c) Form a quadratic equation in x and solve it to find the length and breadth.

12 If $n \ge 0$, solve $200 = \frac{n}{2}(6 + 2(n - 1))$, rounding your answer to the nearest integer.

13 Solve $12x^3 + 12x^2 - 24x = 0$.

14 Solve $\frac{a - x}{x} = \frac{x}{b - x}$ for x.

15 If $x > 0$, solve $22^2 = x^2 + 20^2 - 40x\cos 60°$, giving your answer to the nearest integer.

16 Solve $x(x^2 + 5) = 6x^2$.

CHAPTER 4
Functions

4.1 FUNCTIONS AND RELATIONS

A function is a type of mathematical object that precisely describes a relationship between variables.

Definitions

- A real function f of a real variable x assigns to each element x of a given set of real numbers **exactly one** real number y, which is called the value of the function f at x. The dependence of y on f and x is made explicit by the notation $f(x)$, which means the value of f at x. This can be written as $y = f(x)$.
- The set of real numbers x on which f is defined is called the **domain** of f, while the set of values $f(x)$ obtained as x varies over the domain of f is called the **range** or **image** of f.
  (In other words, the domain may be thought of as the function's potential 'input', while the range is the 'output'.)
- The variable x is called the **independent variable**, as it may be chosen freely from the domain of f, while y is called the **dependent variable**, as its value depends on the particular value chosen for x.
- A **function** may also be defined as a set of ordered pairs with the special property that no two pairs have the same first element (x value).

Finding the value of a function

Consider a function defined by the rule $f(x) = 2x - 7$. What is the value of the function when $x = 4$, $x = 0$ and $x = -1$? In the past you would have written the two equations $x = 4$, $y = 8 - 7 = 1$. With function notation, you can simply write $f(4) = 8 - 7 = 1$. Thus:

- $f(4) = 1$
- $f(0) = 0 - 7 = -7$
- $f(-1) = -2 - 7 = -9$

The function notation $f(x)$ allows you to write, in a single statement, the value of the independent variable as well as the corresponding dependent variable.

Example 1

Plot the following points on a number plane: $(-1, 2)$, $(0, 1)$, $(1, 0)$, $(2, 3)$, $(3, -2)$, $(4, 0)$.

Does this set of points represent a function? Write its domain and range.

Solution

The graph shows that for each first value, x, there is only one second value, y.

This means the set of points is a function:

Domain = $\{-1, 0, 1, 2, 3, 4\}$

Range = $\{-2, 0, 1, 2, 3\}$

Note that 0 is the y value for two points, but it only needs to be listed once in the range.

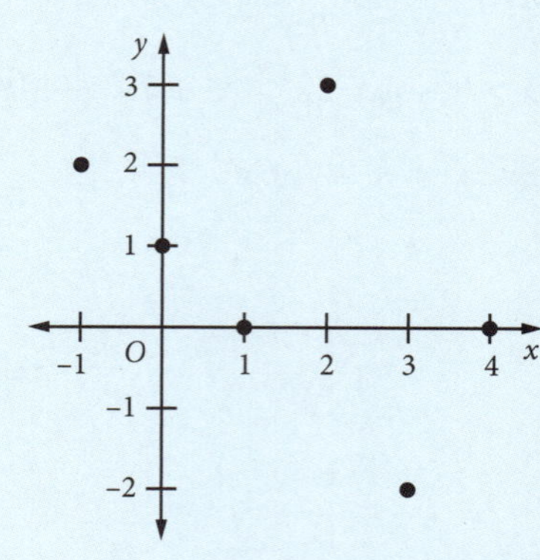

Example 2

Determine whether each graph represents a function. Write the domain and range for each.

(a)

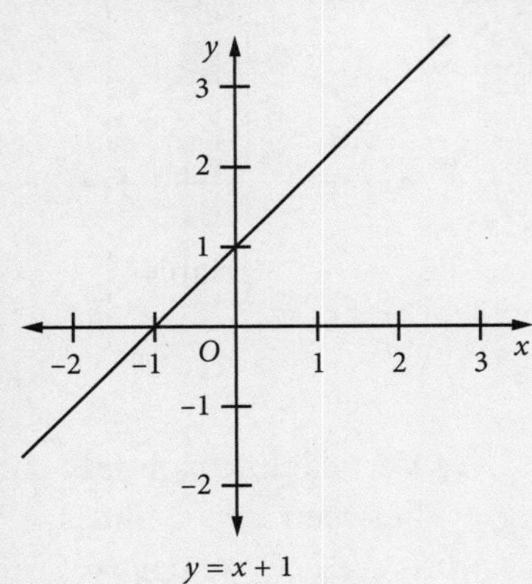

(b)

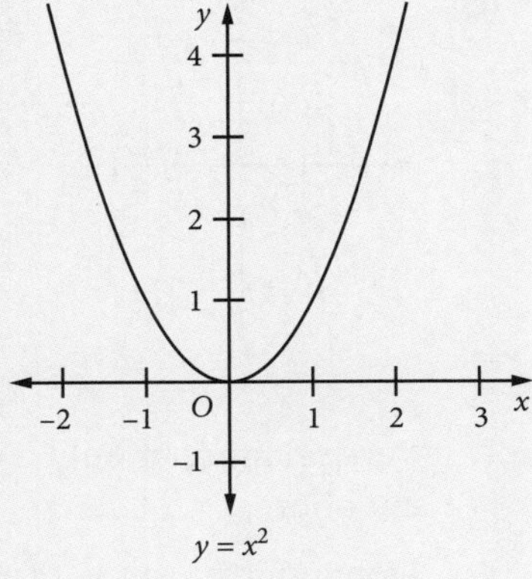

(c)

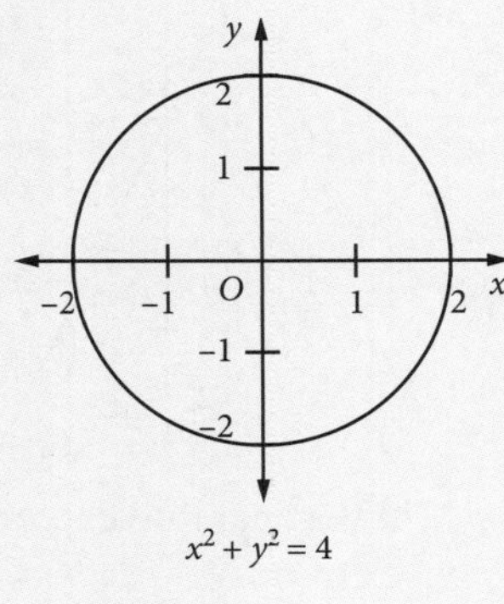

Solution

(a) Function

Domain: real numbers

Range: real numbers

Hence we can write $f(x) = x + 1$

(b) Function

Domain: real numbers

Range: real numbers, $y \geq 0$

Hence we can write $g(x) = x^2$

(c) Not a function

Domain: real numbers, $-2 \leq x \leq 2$

Range: real numbers, $-2 \leq x \leq 2$

In parts **(a)** and **(b)** the functions are different, so they have been given different labels, *f* and *g*.

Vertical line test

A simple way to determine whether a graph represents a function is to draw vertical lines. If each vertical line cuts the graph only once, then the graph represents a function. If any of the vertical lines cut the graph more than once, then the graph does not represent a function. The *x* values for which vertical lines do not cut the graph are not in the domain.

Example 3

Use the vertical line test to show that each graph represents a function. Write the domain and range for each.

(a) $f(x) = 2 - 2x$

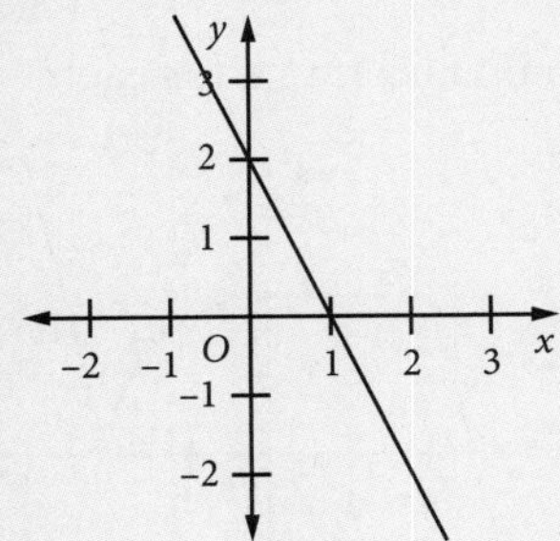

(b) $g(x) = x^3$

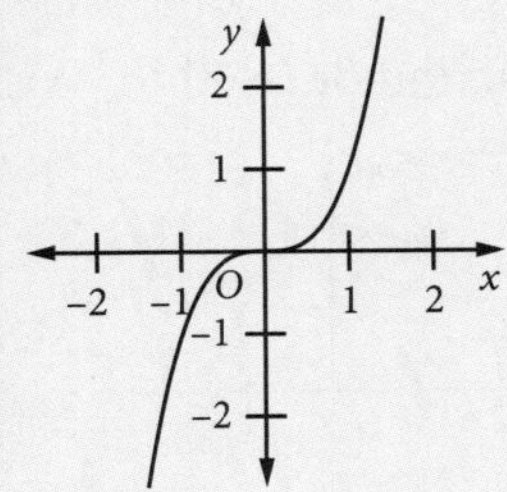

(c) $h(x) = \sqrt{4 - x^2}$

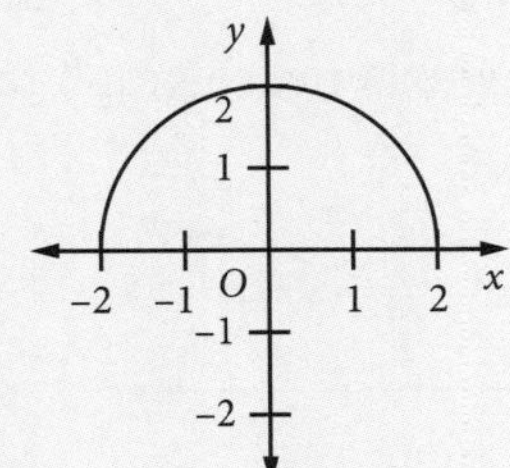

Solution

Draw several vertical lines on each graph.

(a)

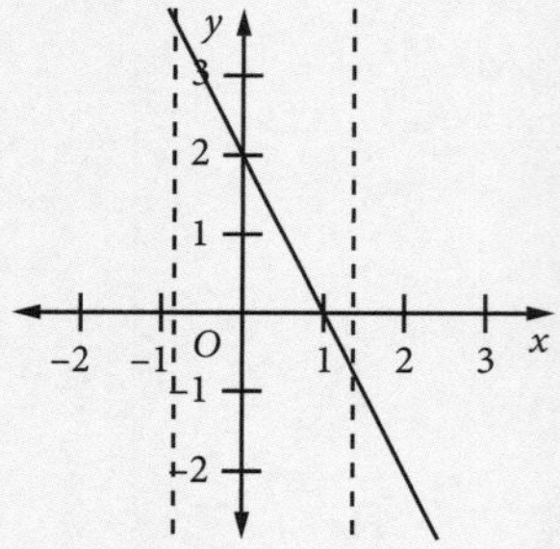

Vertical line can only cut once.

Domain: real values of x

Range: real values of y

(b)

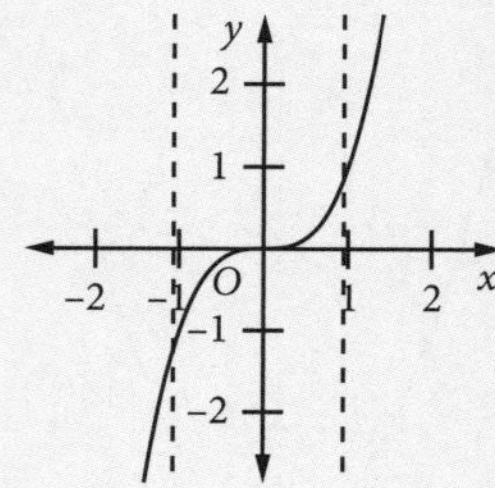

Vertical line can only cut once.

Domain: real values of x

Range: real values of y

(c)

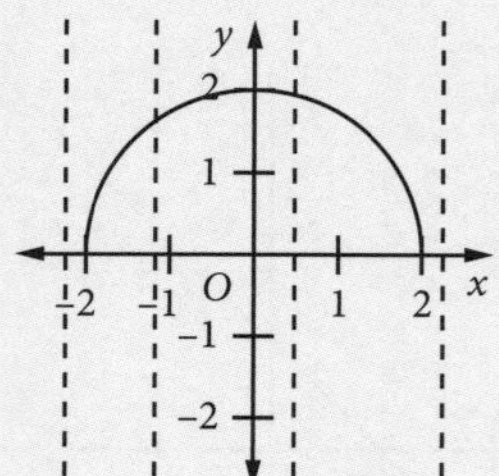

Vertical line can only cut once, between $x = -2$ and $x = 2$.

Domain: $-2 \le x \le 2$

Range: $0 \le y \le 2$

One-to-one functions

In a one-to-one function, every element in the range of the function corresponds to exactly one element in the domain.

Horizontal line test

If a horizontal line intersects the graph of a function more than once then the function is not a one-to-one function. The graph of the parabola $y = x^2$ is an example of a function that can be intersected by a horizontal line.

Relations

The simplest definition of a relation is that it is a multi-valued function: it is a set of ordered pairs, usually defined by some rule, but each first member of the ordered pairs can have more than one second member. If the semicircle in Example **2(c)** were a complete circle instead, then it would not be a function, but it would be a relation.

The language of functions also applies to relations, so the terms independent variable, dependent variable, domain and range have the same meaning.

Example 4

Determine whether each graph represents a function or relation. Write the domain and range for each.

(a)

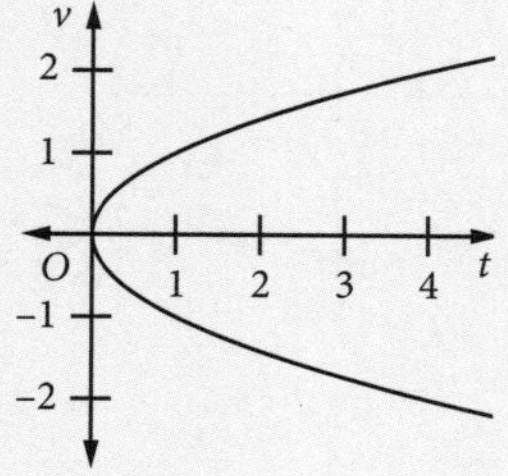

(b)

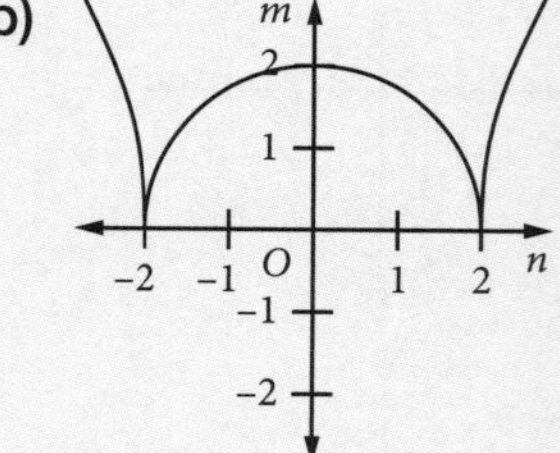

(c)

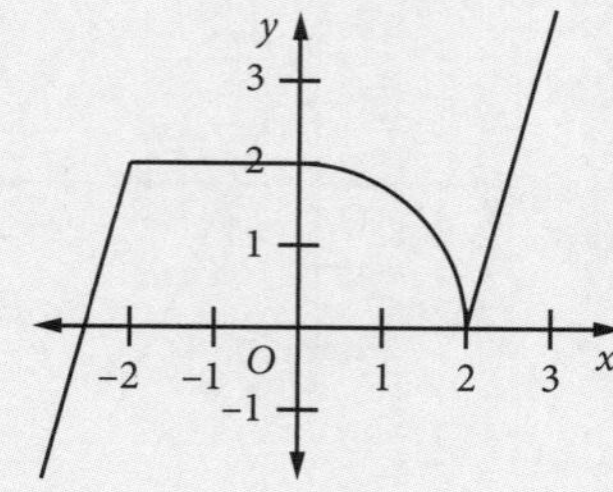

Solution

(a) Relation

Independent variable is t

Domain: real t, $t \ge 0$

Range: real numbers

(b) Function

Independent variable is n

Domain: real numbers

Range: real m, $m \ge 0$

(c) Function

Independent variable is x

Domain: real numbers

Range: real numbers

Types of functions and relations

Sometimes other terms are used to define a function or relation in terms of a 'mapping' between two sets of numbers.

Consider the sets $X = \{1, 2, 3, 4, 5, 6\}$ and $Y = \{5, 11, 14, 17, 23, 27\}$.

One-to-one functions

Make up the set of ordered pairs $\{(1, 5), (2, 11), (3, 14), (4, 17), (5, 23), (6, 27)\}$. This is a one-to-one mapping of set X onto set Y as each element of set X is paired with a different element of set Y. The resulting set of ordered pairs form a function as shown in the diagram.

This function would pass the vertical line test.

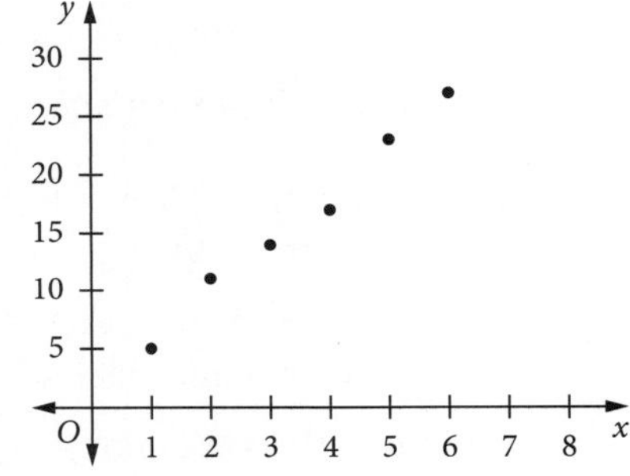

Many-to-one functions

Make up the set of ordered pairs $\{(1, 11), (2, 5), (3, 14), (4, 11), (5, 5), (6, 17)\}$. This is a many-to-one mapping of set X onto set Y as more than one element of set X is paired with an element of set Y. The resulting set of ordered pairs form a function as shown in the diagram.

This function would pass the vertical line test, but it would not pass the horizontal line test as some y values are paired with more than one x value. This is why it is called 'many (x values) to one (y value)'.

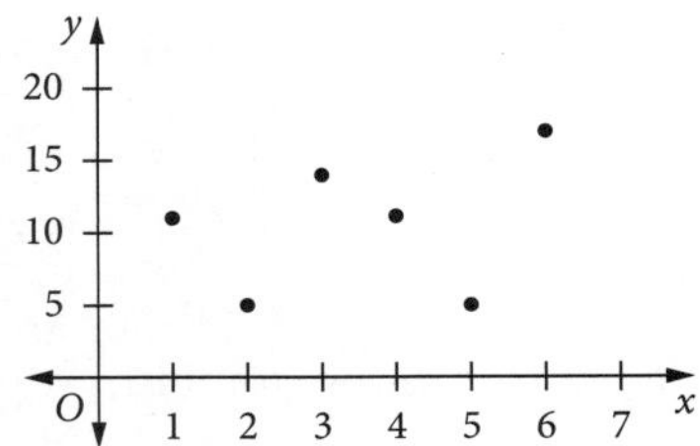

One-to-many relations

Make up the set of ordered pairs $\{(1, 5), (2, 11), (3, 14), (4, 17), (2, 23), (4, 27)\}$. This is a one-to-many mapping of set X onto set Y as some elements of set X are paired with more than one element of set Y. The resulting set of ordered pairs is not a function, but a relation as shown in the diagram.

This set would not pass the vertical line test, but it would pass the horizontal line test as no y values occur more than once. This is why it is called 'one (y value) to many (x values)'.

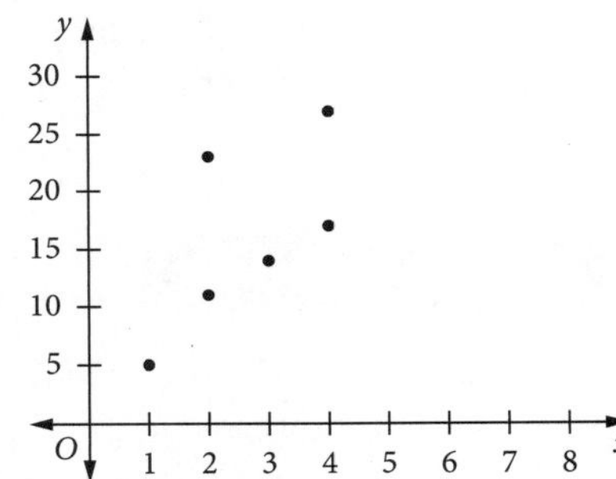

Many-to-many relations

Make up the set of ordered pairs $\{(1, 5), (2, 5), (2, 14), (4, 14), (5, 23), (6, 17)\}$. This is a many-to-many mapping of set X onto set Y as some elements of set X are paired with more than one element of set Y and some elements of set Y are paired with more than one element of set X. The resulting set of ordered pairs form a relation as shown in the diagram.

This set would not pass the vertical line test nor the horizontal line test. Some y values are paired with more than one x value and some x values are paired with more than one y value. This is why it is called 'many (y values) to many (x values)'.

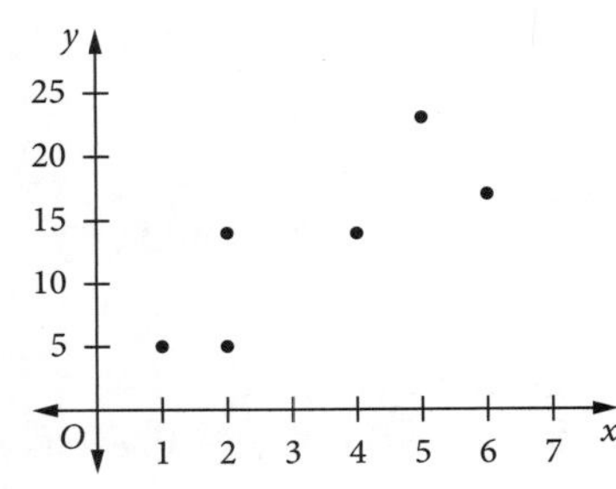

Function rules

When a function rule f is given and a domain is not specified, it is assumed that the domain of the function is the set of real numbers for which $f(x)$ defines a real number range. To find the domain, the solution of an inequality may be needed.

When the domain of f is all values of x over an interval, the graph of $y = f(x)$ is called the **curve** $y = f(x)$ and a part of the curve between two points is called an **arc**.

Example 5

State the largest possible domain for the function defined by the given rule. What is the range of each function?

(a) $f(x) = x^2$ **(b)** $f(x) = \frac{1}{x}$ **(c)** $f(x) = \sqrt{x}$ **(d)** $f(x) = \sqrt{4 - x^2}$ **(e)** $f(x) = \frac{x}{x^2 - 1}$

Solution

(a) Any real number squared is also a real number, so the domain of $f(x) = x^2$ is all real numbers.
Any real number squared is never negative, so the range of the function is all positive real numbers and zero.

(b) Fractions are not defined for a denominator of zero, so $\frac{1}{x}$ is defined for all values of x except $x = 0$. Thus the domain of $f(x) = \frac{1}{x}$ is all real numbers except $x = 0$. You can write $f(x) = \frac{1}{x}, x \neq 0$.
Because the numerator of $f(x)$ is never zero, we have $f(x) \neq 0$. The reciprocal of every non-zero real number is another non-zero real number, so the range of the function is all real numbers except zero.

(c) Only the square roots of non-negative numbers are real, so the domain of $f(x) = \sqrt{x}$ is real x, $x \geq 0$.
The square root of zero is zero and the square root of a positive real number is another positive real number, so the range of the function is all positive real numbers and zero: $f(x) \geq 0$.

(d) For the value of $f(x) = \sqrt{4 - x^2}$ to be real, $4 - x^2 \geq 0$, so $-2 \leq x \leq 2$. Therefore the domain of $f(x) = \sqrt{4 - x^2}$ is real x, $-2 \leq x \leq 2$ (or $|x| \leq 2$).
When $x = 0$, the value of the function is $f(0) = 2$; also, $f(2) = 0$ and $f(-2) = 0$. For all other values of x in the domain, $0 < f(x) < 2$, so the range of the function is the real numbers $0 \leq f(x) \leq 2$.

(e) The function is not defined when the denominator is zero, i.e. when $x^2 - 1 = 0$. This is true for $x = \pm 1$, so the domain of $f(x) = \frac{x}{x^2 - 1}$ is real x, $x \neq \pm 1$.
$f(0) = 0$, and for all values of x in the domain the function exists. The range of the function is the set of real numbers.

EXPLORE FURTHER

Investigating domain and range

Use technology to investigate the domain and range of a set of functions.

A function may be defined over its domain by several different rules.

Example 6

A function is defined as $f(x) = \begin{cases} x^2, & x \leq 1 \\ 2 - x, & x > 1 \end{cases}$. Find the domain and range of this function.

Solution

When $x \leq 1$, $f(x) = x^2$ exists for all real values of x.

When $x > 1$, $f(x) = 2 - x$ exists for all real values of x.

Thus the domain of the function is all real x.

When $x \leq 1$, $f(1) = 1$, $f(0) = 0$, $f(-1) = 1$ and $f(x) > 0$ when $x \neq 0$.

When $x > 1$, $f(1.01) = 0.99$. $f(2) = 0$, $f(3) = -1$ and for $x > 2$, $f(x) < 0$.

Thus the range of the function is all real numbers.

A sketch of the function shows this information more clearly.

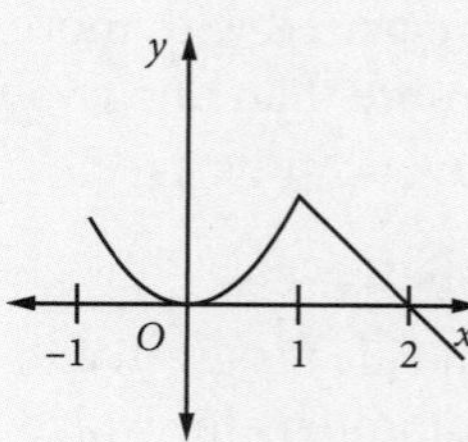

Interval notation

Parentheses ( ) are used to indicate open intervals while square brackets [ ] are used to indicate closed intervals.

A combination of these two symbols, ( ] or [ ), is used to indicate intervals that are open at one end and closed at the other end.

For example:

- The domain $1 < x < 3$ is an open interval and may be written $(1, 2)$ or $x \in (1, 2)$.
- The range $-3 \leq y \leq 5$ is a closed interval and may be written $[-3, 5]$ or $y \in [-3, 5]$.
- $2 \leq x < 7$ may be written as $[2, 7)$ or $x \in [2, 7)$.
- $0 < y \leq 17$ may be written as $(0, 17]$ or $y \in (0, 17]$.
- The statement $x \in (-\infty, 0)$ represents the interval $-\infty < x < 0$ or just $x < 0$.

When graphing an interval on a number line, the interval is shown as a line with a circle at each end: a solid circle for a closed interval, or a hollow circle for an open interval.

Open interval

$(1, 3)$ represents the interval $1 < x < 3$ and when graphed on the number line is shown as:

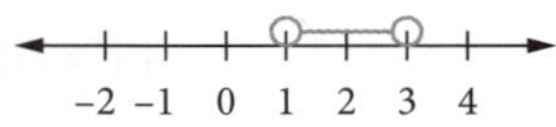

Closed interval

$[-2, 3]$ represents the interval $-2 \leq x \leq 3$ and when graphed on the number line is shown as:

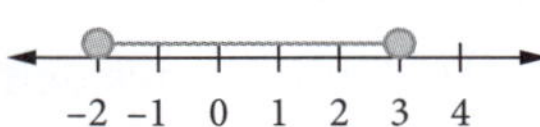

Half-open intervals

$(0, 4]$ represents the interval $0 < x \leq 4$ and when graphed on the number line is shown as:

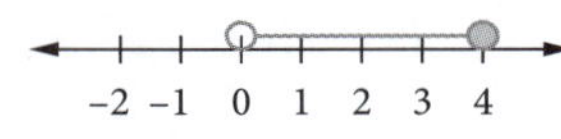

$[-1, 3)$ represents the interval $-1 \leq x < 3$ and when graphed on the number line is shown as:

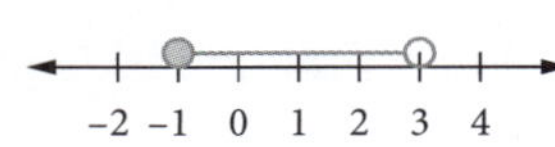

EXERCISE 4.1 FUNCTIONS AND RELATIONS

1 Indicate whether each set of points is a function or relation. In each case write the largest possible domain and range.

(a) $\{(1,1), (2,2), (3,5), (3,7)\}$ **(b)** $\{(-3,1), (3,1), (8,7), (9,-2), (11,6)\}$
(c) $(x, 5)$ for all real x **(d)** $(2, y)$ for all real y **(e)** $x^2 + y^2 = 9$
(f) $y = \sqrt{16 - x^2}$ **(g)** $y = x + 2$ **(h)** $y = 2 - x^2$

2 Which statement is correct about the function $f(t) = t^2 - 9$?

A Domain is $t \geq 9$ **B** Range is $f(t) \leq -9$ **C** Range is $f(t) \geq -9$ **D** Domain is $t \geq 0$

3 For each of the following functions, sketch the graph and state the largest possible domain and range.

(a) $f(x) = 9 - x$ **(b)** $g(x) = 9 - x^2$ **(c)** $h(x) = -\sqrt{4 - x^2}$ **(d)** $f(t) = \sqrt{t}$

4 Consider the function defined by $f(x) = 3x - 6$ for all x.

(a) Find the value of $f(1), f(-2), f(a)$. **(b)** For what values of a is $f(a) = a$?
(c) For what values of x is $f(x) > x$? **(d)** Sketch the graph of f.

5 Consider the function $f(x) = x^2 - 1$, where x is any real number.

(a) Find $f(3), f(-3)$. **(b)** Find $f(a), f(b), f(a + b)$.
(c) Is $f(a) + f(b) = f(a + b)$ true for all a and b? **(d)** State the range of f.

6 For the function $g(x) = \sqrt{x}, x \geq 0$, state whether each statement is correct or incorrect.

(a) $g(1) = 1$ **(b)** $g(9) = -3$ **(c)** $g(x^2) = x$ **(d)** $g(x + 2) = \sqrt{x + 2}$

7 Sketch each function and write the domain and range.

(a) $f(x)=\begin{cases} x+1 & \text{for } x\geq 1 \\ 2 & \text{for } x<1 \end{cases}$ (b) $f(x)=\begin{cases} \dfrac{1}{x+2} & \text{for } x\neq -2 \\ 4 & \text{for } x=-2 \end{cases}$

(c) $f(x)=\begin{cases} 2x & \text{for } x\geq 0 \\ -2x & \text{for } x<0 \end{cases}$ (d) $f(x)=\begin{cases} x & x\in(1,\infty) \\ 1 & x\in[-1,1] \\ -x & x\in(-\infty,-1) \end{cases}$

8 State the largest possible domain and range for:

(a) $f(x)=\sqrt{x-2}$ (b) $f(x)=\sqrt{3-x}$ (c) $f(x)=\sqrt{x^2-9}$

(d) $g(x)=\frac{1}{x}$ (e) $h(t)=t^3$ (f) $g(k)=5-k^2$

9 A function is defined by the rule $f(x)=\begin{cases} 1-x, & x\in(-\infty,-2] \\ -1, & x\in(-1,1) \\ x+1, & x\in[1,\infty) \end{cases}$

Find the values (if they exist) for:

(a) $f(1)$ (b) $f(-1)$ (c) $f(0)$ (d) $f(2)+f(-2)$

10 A function is defined by the rule $f(x)=\begin{cases} \frac{1}{x}, & x<0 \\ x, & x\geq 0 \end{cases}$

Find:

(a) $f(0)$ (b) $f(2)$ (c) $f(-2)$ (d) $f(a^2)$

11 A function is defined by the rule $f(x)=\begin{cases} x-1, & x<1 \\ \sqrt{x-1}, & x\geq 1 \end{cases}$

Find:

(a) $f(1)$ (b) $f(-1)$ (c) $f(10)+f(-10)$ (d) $f(5)$

4.2 SKETCHING BASIC FUNCTIONS

You should be able to sketch simple linear functions from either the gradient-intercept form or the general form of the equation. You also need to be able to quickly and neatly sketch the power functions, such as $f(x)=x^2, f(x)=-x^2, f(x)=x^3, f(x)=-x^3, f(x)=x^4, f(x)=-x^4, f(x)=\frac{1}{x}$ and $f(x)=\frac{1}{x^2}$. You should already be familiar with many of these.

The following examples should also be graphed using graphing software. You may need to rewrite each equation in the form $y=f(x)$.

Example 7

Sketch each straight line. State the gradient and both axis intercepts of each.

(a) $y=2x+1$ (b) $2x+3y-6=0$ (c) $y=4-x$

Solution

(a) $y = 2x + 1$

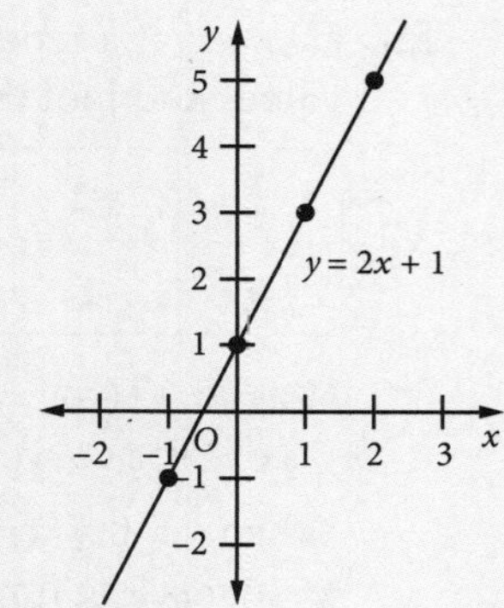

Find the value of y for three different values of x: $(0, 1)$, $(2, 5)$, $(-1, -1)$

Plot these points on the number plane. Join them to obtain the line.

OR

From the form of the equation, recognise that the y-intercept is 1 and the gradient is 2.

Because the line passes through $(0, 1)$ it also passes through $(1, 1 + 2 = 3)$ and $(2, 3 + 2 = 5)$. This is because the gradient is 2, which means that as x increases by 1, y increases by 2. Plot and join the points.

Gradient = 2, x-intercept = –0.5, y-intercept = 1.

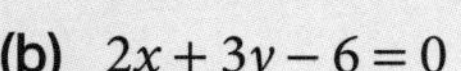

(b) $2x + 3y - 6 = 0$

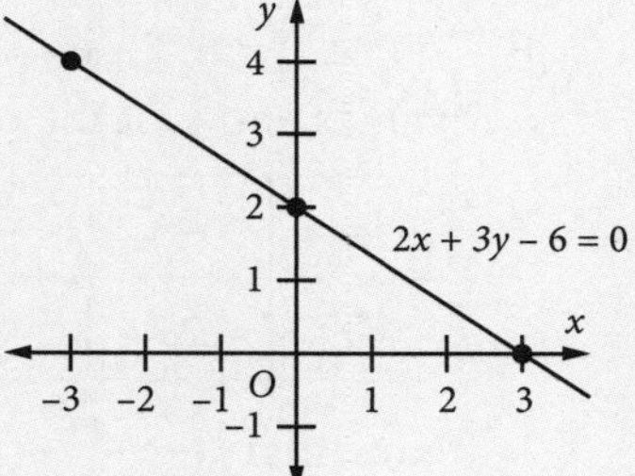

Find the value of y for three different values of x: $(0, 2)$, $(3, 0)$, $(-3, 4)$

Plot these points on the number plane. Join them to obtain the line.

OR

Rewrite the equation in the gradient-intercept form: $y = \frac{-2x}{3} + 2$

The gradient is a fraction, so this is not so convenient.

OR

Rewrite the equation by putting the constant term on the RHS of the equation and dividing by 6: the equation becomes $\frac{x}{3} + \frac{y}{2} = 1$.

This shows that the x-intercept is 3 and the y-intercept is 2. Draw a line through these intercept points to obtain the graph.

Because the line falls as x increases, the gradient is negative.

$$\text{Gradient} = -\frac{y\text{-intercept}}{x\text{-intercept}} = -\frac{2}{3}, \; x\text{-intercept} = 2, \; y\text{-intercept} = 3.$$

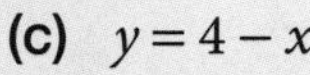

(c) $y = 4 - x$

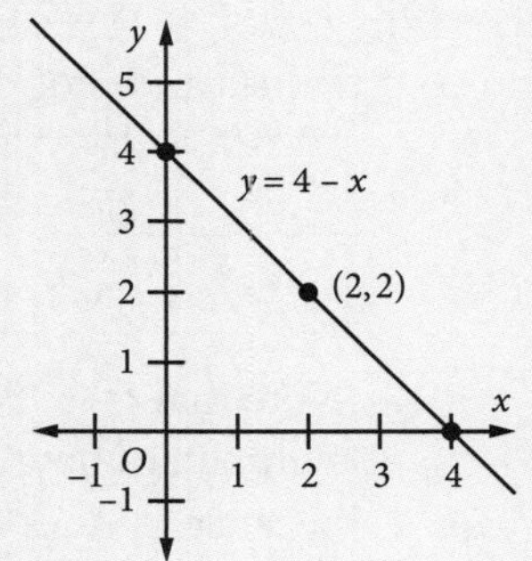

From the equation:

gradient = −1, x-intercept = 4, y-intercept = 4.

Use this information to sketch the graph.

If you use this method, you should also find the coordinates of a third point to check that you haven't made a mistake, e.g. $(2, 2)$.

Example 8

Sketch each function, showing any intercepts on the coordinate axes.

(a) $f(x) = x^2$ **(b)** $f(x) = -x^3$ **(c)** $f(x) = \frac{1}{x}$

Solution

(a) $f(x)=x^2$ is a type of curve called a parabola. Create a table of values and plot the points.

x	−2	−1	0	1	2
$f(x)$	4	1	0	1	4

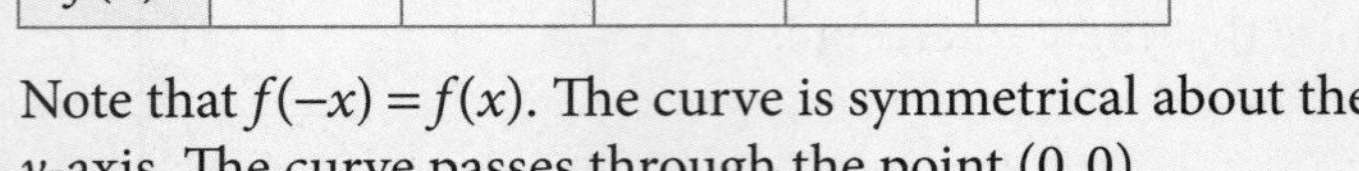

Note that $f(-x)=f(x)$. The curve is symmetrical about the y-axis. The curve passes through the point $(0,0)$.

When $x>0$, $f(x)$ increases as x increases, so we say that $f(x)$ is an increasing function for $x>0$.

When $x<0$, $f(x)$ decreases as x increases, so we say that $f(x)$ is a decreasing function for $x<0$.

(b) $f(x)=-x^3$ is a type of curve called a cubic. Create a table of values and plot the points.

x	−3	−2	−1	0	1	2	3
$f(x)$	27	8	1	0	−1	−8	−27

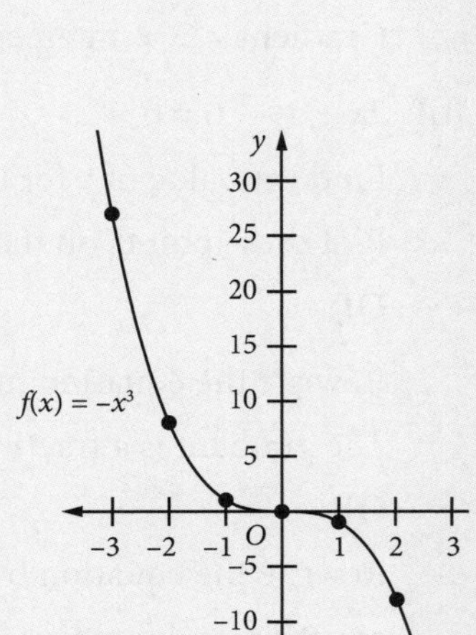

Note that $f(-x)=-f(x)$. The curve has rotational or point symmetry about the origin.

The curve passes through the point $(0,0)$.

As x increases over the domain, the value of $f(x)$ decreases, so $f(x)$ is a decreasing function over its domain.

(c) $f(x)=\frac{1}{x}$ is a type of curve called a hyperbola. Create a table of values and plot the points.

x	−2	−1	−0.5	0	0.5	1	2
$f(x)$	−0.5	−1	−2	undefined	2	1	0.5

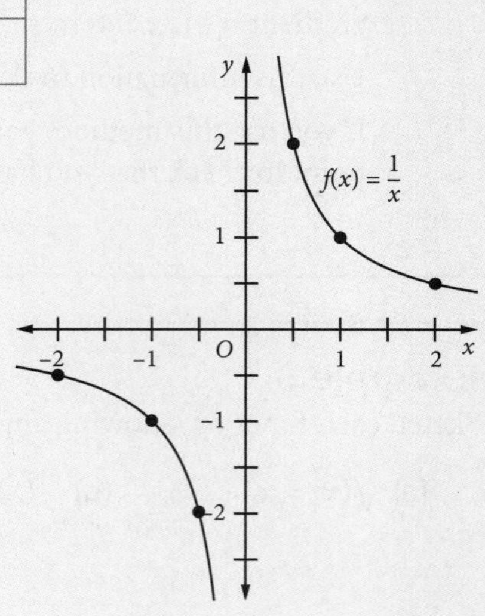

Note that $f(-x)=-f(x)$. The curve has rotational or point symmetry about the origin. Also note that $f(0)$ is undefined because $\frac{1}{0}$ does not exist.

The curve does not cut either axis.

As $x\to\pm\infty$, $f(x)\to 0$ and as $f(x)\to\pm\infty$, $x\to 0$.

The line $f(x)=0$ is called a horizontal asymptote.

The line $x=0$ is called a vertical asymptote.

When $x<0$, $f(x)$ decreases as x increases, so we say that $f(x)$ is a decreasing function for $x<0$.

When $x>0$, $f(x)$ decreases as x increases, so we say that $f(x)$ is a decreasing function for $x>0$.

Thus $f(x)$ is a decreasing function over each part of its domain.

Odd and even functions

An **odd function** has the property that $f(-x) = -f(x)$. For example:

$$\begin{aligned} \text{If} \quad & f(x) = x^3 \\ \text{then} \quad & f(-x) = (-x)^3 \\ & \quad = -x^3 \\ & \quad = -f(x) \end{aligned}$$

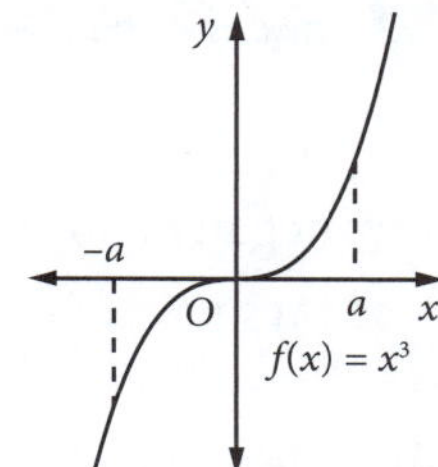

Hence $f(x) = x^3$ is an odd function.

Because $f(x)$ and $f(-x)$ are opposite in sign, the graph of f for $x \leq 0$ can be obtained by rotating the graph for $x \geq 0$ through an angle of 180° about the origin.

An **even function** has the property that $f(-x) = f(x)$. For example:

$$\begin{aligned} \text{If} \quad & f(x) = x^2 \\ \text{then} \quad & f(-x) = (-x)^2 \\ & \quad = x^2 \\ & \quad = f(x) \end{aligned}$$

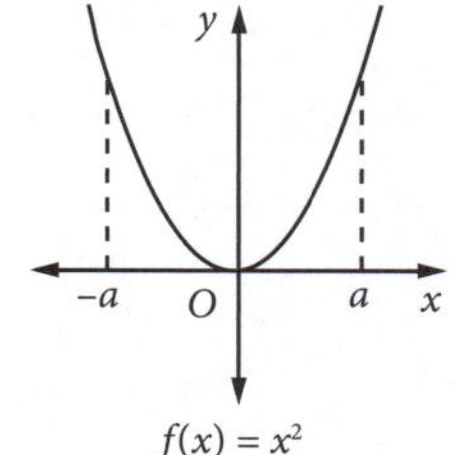

Hence $f(x) = x^2$ is an even function.

The graph of an even function is symmetrical about the y-axis. The graph for $x \leq 0$ can be obtained by reflecting the graph for $x \geq 0$ in the y-axis.

Note that the statement $f(-a) = f(a)$ implies that the function is defined at both $x = a$ and $x = -a$. The function $f(x) = x^2$, $x > 0$ is **not** an even function, because $f(-a)$ is not defined.

The properties of odd and even functions are useful when sketching the curves for these functions. After drawing a curve for $x \geq 0$, the other half of the curve can be drawn immediately from the odd or even symmetrical properties. Disappointingly, however, most functions are neither even nor odd.

EXPLORE FURTHER

Odd and even functions

Use graphing software to determine if a function is odd or even.

EXERCISE 4.2 SKETCHING BASIC FUNCTIONS

1 Sketch each function. State the gradient and the x- and y-intercepts for each.

(a) $y = 3x + 1$ (b) $3x + 2y - 6 = 0$ (c) $y = 4 - 2x$ (d) $y = x - 1$ (e) $4x - y - 8 = 0$
(f) $y = -x$ (g) $y = 3$ (h) $x = 4$ (i) $x + 2y + 5 = 0$

2 For each part of question **1**, determine whether the function is increasing, decreasing or neither. What do you notice about the gradient in each case?

3 Sketch each function, showing any intercepts on the coordinate axes. State the domain for which each function is increasing.

(a) $f(x) = x$ (b) $f(x) = -x^2$ (c) $f(x) = x^3$ (d) $f(x) = -\frac{1}{x}$ (e) $f(x) = x^4$ (f) $f(x) = \frac{1}{x^2}$

4 For each part of question **3**, determine whether the function is odd, even or neither.

5 If $f(x) = 2x^2$ and $g(x) = \frac{3}{x}$, which of the following statements is correct?

A f and g are both odd functions
B f is an even function and g is an odd function
C f and g are both even functions
D f and g are neither even nor odd functions

6 For $y = 2x - 3$, indicate whether each statement is correct or incorrect.

(a) gradient = 2 (b) x-intercept = 3 (c) y is an increasing function (d) y-intercept = 3

4.3 SQUARE ROOTS AND ABSOLUTE VALUE

Square roots

If $a \geq 0$, then $\sqrt{a}$ is a non-negative number such that $\left(\sqrt{a}\right)^2 = a$. $\sqrt{a}$ is called the positive square root of a.

An equation like $x^2 = 9$ always has two solutions: the positive and negative square roots of 9. In this case, they are $x = \sqrt{9} = 3$ and $x = -\sqrt{9} = -3$.

The important fact is that the value of $\sqrt{a}$ will always be either positive or zero (non-negative) if $a \geq 0$.

What meaning can be given to $\sqrt{x^2}$?

If $x = 2$, then $\sqrt{x^2} = \sqrt{(2)^2} = \sqrt{4} = 2$.

If $x = -2$, then $\sqrt{x^2} = \sqrt{(-2)^2} = \sqrt{4} = 2$.

If $x = 0$, then $\sqrt{x^2} = \sqrt{0^2} = \sqrt{0} = 0$.

This means:

$$\begin{aligned} \sqrt{x^2} &= x && \text{if } x > 0 \\ &= -x && \text{if } x < 0 \\ &= 0 && \text{if } x = 0 \end{aligned}$$

Here $-x$ means the opposite sign of x, hence $-x > 0$.

Absolute value

The absolute value (also called 'modulus') of a real number x is written $|x|$. It is the non-negative number that defines the magnitude of the given number.

Thus $|3| = 3$, $|-3| = 3$ and $|0| = 0$.

This means:

$$\begin{aligned} |x| &= x && \text{if } x > 0 \\ &= -x && \text{if } x < 0 \\ &= 0 && \text{if } x = 0 \end{aligned}$$

This is identical to $\sqrt{x^2}$, so it leads to another definition of absolute value: $|x| = \sqrt{x^2}$.

Because x is a real number, it can be represented by a point on the number line, and $|x|$ is the distance of the point x from the origin. Distance is always positive, so $|x| > 0$ for all $x \neq 0$.

The expression $|x| = 2$ thus describes a distance of 2 from the origin. The two points that are 2 units distant from 0 are $x = 0 + 2 = 2$ and $x = 0 - 2 = -2$.

Similarly, the expression $|x - a| = 4$ is saying that the distance between the points x and a is 4 units. For example, the values of x for which $|x - 2| = 4$ are $x = 2 + 4 = 6$ and $x = 2 - 4 = -2$. This can be shown on a number line:

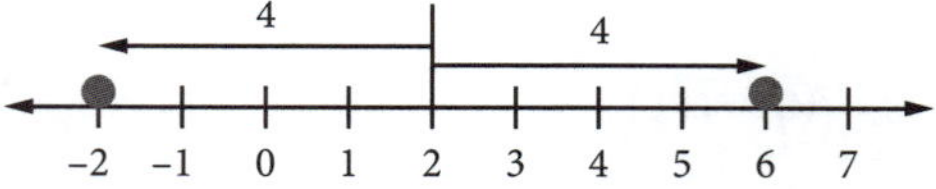

In general:

$$\begin{aligned} |x - y| &= x - y && \text{if } x > y \\ &= y - x && \text{if } x < y \\ &= 0 && \text{if } x = y \end{aligned}$$

Thus $|5 - 3| = |3 - 5| = 2$.

Important results

1 $|xy| = |x| \times |y|$

2 $|x + y| \le |x| + |y|$ (the 'triangle inequality')
and $|x + y| = |x| + |y|$ if and only if x and y are either zero or have the same sign.

These results can be checked with specific values of x and y. Test them for $x = -7$ and $y = 3$.

EXPLORE FURTHER

Absolute value and addition

Use a spreadsheet to explore the result $|a + b| \le |a| + |b|$.

Example 9

On a number line, show the values of x for which:

(a) $|x| > 1$ (b) $|x| \le 2$

Solution

(a) $|x| > 1$

$x > 1$ or $-x > 1$

$x > 1$ or $x < -1$

$x < -1$ or $x > 1$

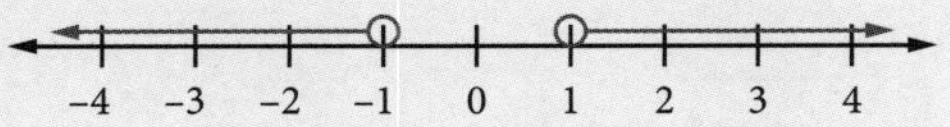

(b) $|x| \le 2$

$x \le 2$ or $-x \le 2$

$x \le 2$ or $x \ge -2$

$-2 \le x \le 2$

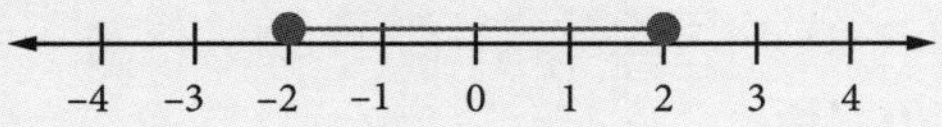

When the circle is filled in, the point is included, as it is in part **(b)**.

Example 10

Solve for x:

(a) $|2x - 1| = 3$ (b) $|3x + 2| = 1$ (c) $|2x - 1| \ge 3$ (d) $|3x + 2| < 1$

Solution

(a) $|2x - 1| = 3$

$2x - 1 = 3$ or $-(2x - 1) = 3$

$2x - 1 = 3$ or $2x - 1 = -3$

$2x = 4$ or $2x = -2$

$x = 2$ or $x = -1$

(b) $|3x + 2| = 1$

$3x + 2 = 1$ or $3x + 2 = -1$

$3x = -1$ or $3x = -3$

$x = -\frac{1}{3}$ or $x = -1$

(c) $|2x - 1| \ge 3$

$2x - 1 \ge 3$ or $-(2x - 1) \ge 3$

(d) $|3x + 2| < 1$

$3x + 2 < 1$ or $-(3x + 2) < 1$

Multiplying both sides of an inequality by −1 reverses the direction of the inequality, so:

$$2x-1\geq 3 \quad \text{or} \quad 2x-1\leq -3$$
$$2x\geq 4 \quad \text{or} \quad 2x\leq -2$$
$$x\geq 2 \quad \text{or} \quad x\leq -1$$

$$3x+2<1 \quad \text{or} \quad 3x+2>-1$$
$$3x<-1 \quad \text{or} \quad 3x>-3$$
$$x<-\frac{1}{3} \quad \text{or} \quad x>-1$$
$$-1<x<-\frac{1}{3}$$

In **(b)** above, the first line of working has been left out. When you are confident solving absolute value equations, you can do this too. However, beware that skipping the first line of working in problems like parts **(c)** and **(d)** could easily lead to wrong inequality signs.

Example 11

Write expressions to give meaning to the following:

(a) $|2x-3|$ **(b)** $\sqrt{(y-2)^2}$ **(c)** $\sqrt{(2y-x)^2}$

Solution

(a)
$$|2x-3| = 2x-3 \quad \text{if } 2x-3>0\text{, i.e. if } x>\frac{3}{2}$$
$$= 3-2x \quad \text{if } 2x-3<0\text{, i.e. if } x>\frac{3}{2}$$
$$= 0 \quad \text{if } 2x-3=0\text{, i.e. if } x=\frac{3}{2}$$

or

$$|2x-3| = 2x-3 \quad \text{if } x\geq\frac{3}{2}$$
$$= 3-2x \quad \text{if } x<\frac{3}{2}$$

(b)
$$\sqrt{(y-2)^2} = |y-2|$$
$$= y-2 \quad \text{if } y>2$$
$$= 2-y \quad \text{if } y<2$$
$$= 0 \quad \text{if } y=2$$

(c)
$$\sqrt{(2y-x)^2} = |2y-x|$$
$$= 2y-x \quad \text{if } y>\frac{x}{2}$$
$$= x-2y \quad \text{if } y<\frac{x}{2}$$
$$= 0 \quad \text{if } y=\frac{x}{2}$$

Algebraic denominators

A fraction cannot have zero as a denominator. If we write a fraction with an algebraic expression as the denominator, then that fraction is undefined when that algebraic expression is zero.

For example, $\frac{1}{x}$ is undefined when $x=0$, so we say that the expression is defined for all real x, $x\neq 0$.

The expression $\frac{x-1}{x(x-1)}$ is undefined for $x=0$ or 1. The expression is defined for all real x, $x\neq 0, 1$.

When an expression is undefined for some values, you must write the restrictions in the answer.

Example 12

Simplify these expressions:

(a) $\frac{|x|}{x}$ for $x\neq 0$ **(b)** $\frac{|x-2|}{x^2-4}$ for $x\neq -2, 2$ **(c)** $\frac{\sqrt{(x-1)^2}}{x-1}$ for x between 0 and 1

Solution

(a) $\frac{|x|}{x}$ for $x \neq 0$

$\frac{|x|}{x} = \frac{x}{x} = 1$ if $x > 0$

$\frac{|x|}{x} = \frac{-x}{x} = -1$ if $x < 0$

(b) $\frac{|x-2|}{x^2-4}$ for $x \neq -2, 2$

$\frac{|x-2|}{x^2-4} = \frac{x-2}{(x-2)(x+2)} = \frac{1}{x+2}$ if $x > 2$

$\frac{|x-2|}{x^2-4} = \frac{-(x-2)}{(x-2)(x+2)} = \frac{-1}{x+2}$ if $x < 2, x \neq -2$

(c) $\frac{\sqrt{(x-1)^2}}{x-1}$ for x between 0 and 1

$\frac{\sqrt{(x-1)^2}}{x-1} = \frac{-(x-1)}{x-1} = -1$ because $0 < x < 1$ means that $(x - 1) < 0$, but the square root of its square must be positive.

EXERCISE 4.3 SQUARE ROOTS AND ABSOLUTE VALUE

Write expressions for the following:

1 $\sqrt{81}$ **2** $\sqrt{(-2.5)^2}$ **3** $|6x-4|$ **4** $|\sqrt{3}-2|$

5 $|x+y|$ **6** $|x| + |y|$ **7** $\sqrt{x^2} + x$ **8** $|x-5| + |x+5|$

9 $\sqrt{16} + \sqrt{(-4)^2}$ **10** $\sqrt{(2x+3)^2}$ **11** $\sqrt{9-6x+x^2}$ when $x > 3$

12 Solve for x:

(a) $|x-2| = 3$ (b) $|x+3| = 7$ (c) $|4-x| = 5$ (d) $|x+7| = 2$

(e) $|x-6| = 0$ (f) $|x-5| = 1$ (g) $|x+1| = 0$ (h) $|10+x| = 3$

(i) $|2x+1| = 2$ (j) $|2x-5| = 3$ (k) $|5x+1| = 4$ (l) $|3x-4| = 5$

(m) $|3x+1| = 0$ (n) $|6x+1| = 7$ (o) $|4x-1| = 0$ (p) $|2x-9| = 13$

13 The solution to $|2+x| = 5$ is:

A $x = -3, 7$ B $x = -7, -3$ C $x = -7, 3$ D $x = 3, 7$

14 Solve:

(a) $|x-1| < 3$ (b) $|y+2| > 4$ (c) $|t-6| \leq 2$ (d) $|x+4| \geq 2$

(e) $|m-5| \geq 0$ (f) $|3-x| \leq 5$ (g) $|y+1| \leq 0$ (h) $|7+x| < 3$

(i) $|2x+1| > 3$ (j) $|3z-5| < 1$ (k) $|4x+3| \geq 5$ (l) $|3t-2| < 5$

(m) $|3y+1| < 0$ (n) $|5x+4| > 9$ (o) $|1-2x| > 0$ (p) $|2x-7| \geq 11$

Solve questions **15** to **32** and show your solution on a number line.

15 $|x+1| > 1$ **16** $|y-4| < 3$ **17** $|x+2| < 4$ **18** $|y-2| > 3$

19 $|t-3| \leq 2$ **20** $|x+2| \geq 1$ **21** $|3-m| \leq 2$ **22** $|3+x| \geq 3$

23 $|3t-1| < 4$ **24** $|2x+5| < 3$ **25** $|1-2x| > 3$ **26** $|2+4x| \geq 6$

27 $|x-1| < -2$ **28** $|2x-3| \leq 5$ **29** $|3x+2| < 2$ **30** $|x^2-1| \leq 4$

31 $|x+3|>1$ and $|2x+5|<3$

32 $|2x+5|<3$ or $|2+4x|\geq 6$

33 $|x^2-1|\leq 3$ and $|x^2+1|\geq 0$ are solved and shown on a number line. Indicate whether each of the following is a correct or incorrect part of the solution.

(a) $-3\leq x^2-1\leq 3 \quad$ and $\quad x^2+1\geq 0$

(b) $0\leq x\leq 2$

(c) $-2\leq x\leq 2$

(d) 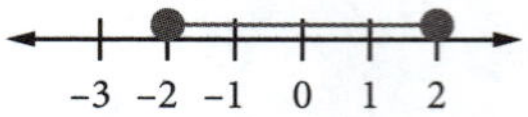 [number line: −3 −2 −1 0 1 2, closed dots at −2 and 2]

Simplify the following expressions, stating the values of x for which your answers apply.

34 $\dfrac{\sqrt{x^2}}{|x|}$

35 $\dfrac{\sqrt{x^2}}{x}$

36 $\sqrt{\dfrac{(x-4)^2}{x^2}}$

37 $\dfrac{1-x}{|1-x|}$

38 $\sqrt{x^2-10x+25}$

39 $\dfrac{|x^2-1|}{x-1}$

40 For the following values of x and y, verify that **(i)** $|xy|=|x|\times|y|$ and **(ii)** $|x+y|\leq|x|+|y|$.

(a) $x=5, y=2$

(b) $x=3, y=-2$

(c) $x=-6, y=8$

(d) $x=-4, y=-3$

4.4 ABSOLUTE VALUE FUNCTIONS

You can now consider $f(x)=|x|$, which is called the absolute value function (or the numerical value function).

From our earlier definition of absolute value, you have: $f(x)=|x|=\begin{cases} x & x\in(0,\infty) \\ -x & x\in(-\infty,0)\end{cases}$

Or, the alternative definition: $\quad f(x)=|x|=\sqrt{x^2}$ for all real x

The domain of $f(x)=|x|$ is all real x. The range of $f(x)=|x|$ is non-negative real numbers (i.e. $f(x)$ is zero or a positive real number).

Example 13

Find the domain and range for each function. Sketch each function.

(a) $f(x)=\sqrt{x^2}$

(b) $f(x)=|2x-1|$

(c) $f(x)=x+|x|$

(d) $f(x)=|x^2-4|$

Solution

(a) $f(x)=\sqrt{x^2}=|x|=\begin{cases} x & \text{for } x\geq 0 \\ -x & \text{for } x<0\end{cases}$

Domain: real x.

Range: $f(x)$ is zero or a positive real number.

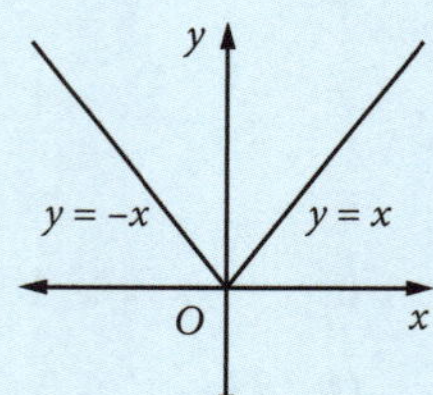

(b) $f(x)=|2x-1|$

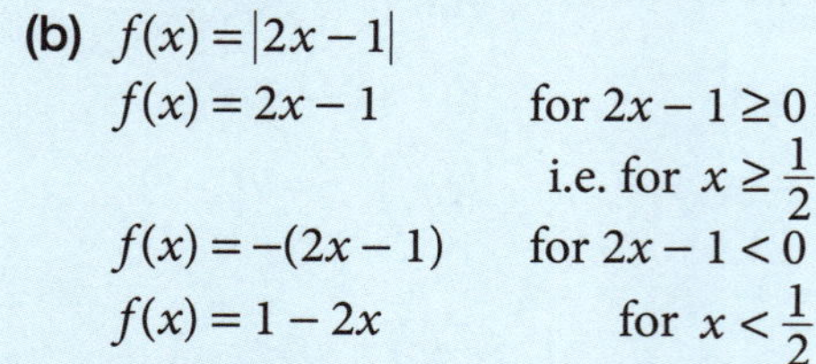

$f(x)=2x-1 \quad$ for $2x-1\geq 0$

i.e. for $x\geq\frac{1}{2}$

$f(x)=-(2x-1) \quad$ for $2x-1<0$

$f(x)=1-2x \quad$ for $x<\frac{1}{2}$

Domain: real x.

Range: $f(x)$ is zero or a positive real number.

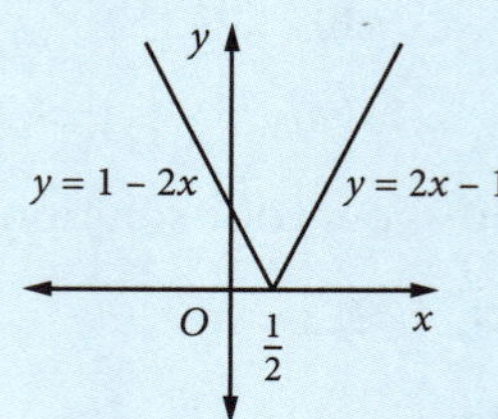

(c) $f(x) = x + |x|$

$f(x) = x + x \qquad \text{for } x \geq 0$

$f(x) = 2x \qquad \text{for } x \geq 0$

$f(x) = x - x \qquad \text{for } x < 0$

$f(x) = 0 \qquad \text{for } x < 0$

Domain: real x.

Range: $f(x)$ is zero or a positive real number.

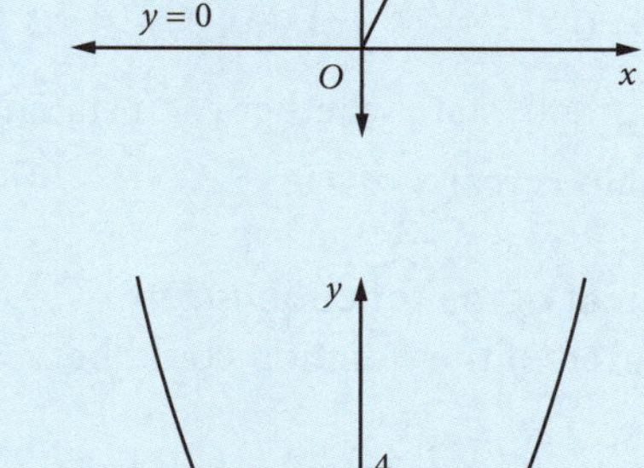

(d) $f(x) = |x^2 - 4|$

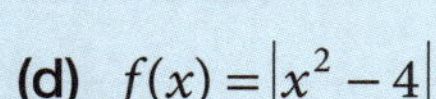

$f(x) = x^2 - 4 \qquad \text{for } x^2 - 4 \geq 0$

$\qquad \text{i.e. for } |x| \geq 2$

$f(x) = -(x^2 - 4) \qquad \text{for } x^2 - 4 < 0$

$f(x) = 4 - x^2 \qquad \text{for } |x| < 2$

Domain: real x.

Range: $f(x)$ is zero or a positive real number.

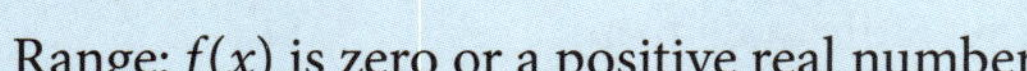

This sketch can be obtained by first sketching the graph of $f(x) = x^2 - 4$. Take the part of that curve that is below the x-axis and reflect it above the x-axis.

EXERCISE 4.4 ABSOLUTE VALUE FUNCTIONS

1 Sketch the graphs of the following absolute value functions defined for all x and state the range in each case.

(a) $f(x) = |x - 4|$ **(b)** $g(x) = |x| - 2$ **(c)** $h(x) = |x + 1|$ **(d)** $f(x) = \sqrt{(x+2)^2}$

(e) $h(x) = |3x - 6|$ **(f)** $f(x) = |4 - 2x|$ **(g)** $g(x) = 4 - |2x|$ **(h)** $f(x) = 2x + |x|$

2 Which diagram is the correct sketch of $y = |3x - 2|$?

A

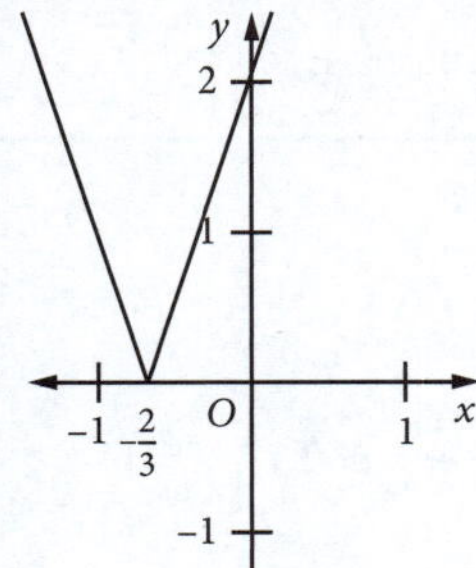

B

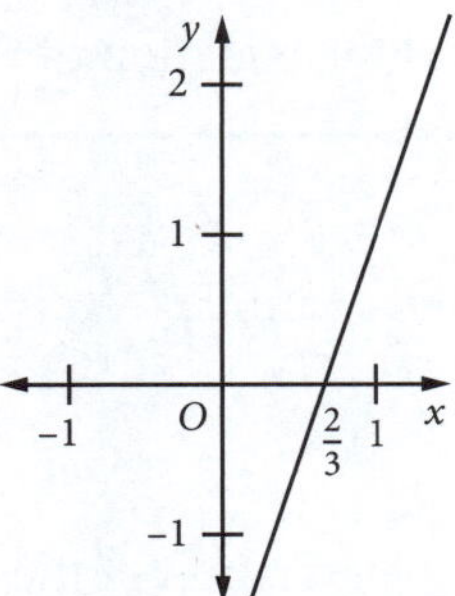

C

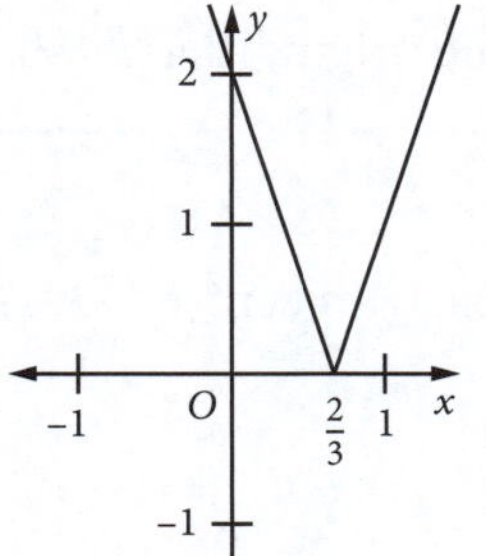

D

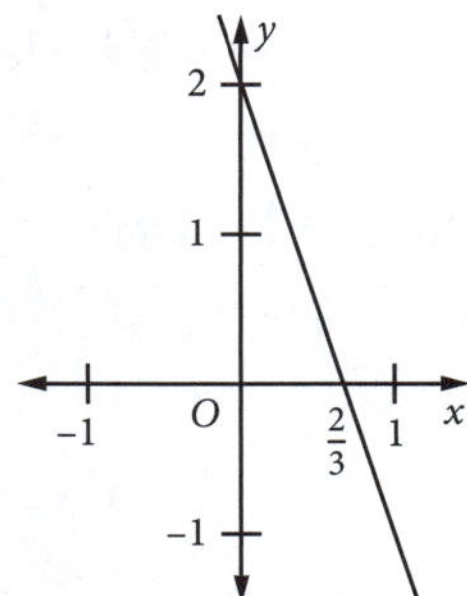

3 State the largest possible domain for:

(a) $f(x) = \sqrt{x-2} + \sqrt{3-x}$ **(b)** $f(x) = \dfrac{x}{|x|}$

4 State whether the following functions are odd, even or neither, defined on their largest possible domain.

(a) $f(x) = x$ **(b)** $f(x) = x + 1$ **(c)** $f(x) = |x|$

(d) $f(x) = x^3 + x$ **(e)** $f(x) = 4 - x^2$ **(f)** $f(x) = (x-2)^2$

(g) $f(x) = \sqrt{4 - x^2}$ **(h)** $f(x) = \dfrac{x}{x^2 - 1}$ **(i)** $f(x) = x^2 + x$

5 Find the largest possible range for the following functions:

(a) $f(x) = (x-3)^2$ **(b)** $f(x) = x + |x|$ **(c)** $f(x) = \sqrt{16 - x^2}$ **(d)** $f(x) = 16 - x^2$

6 Sketch the graph of $f(x) = |2x - 5|$. On the graph, indicate the values of x for which $f(x) = 3$.

7 Sketch the relation $|x| + |y| = 1$. Is this relation a function? State the greatest possible domain and the range.

8 The relation $|x| + |y| = 2$ is sketched and the equations of the four boundary lines are written. State whether each of the following statements is a correct or incorrect equation of a boundary.

(a) $x + y - 2 = 0$ **(b)** $x + y + 2 = 0$ **(c)** $x - y - 2 = 0$ **(d)** $x - y + 2 = 0$

9 For the given graph, state whether each statement is correct or incorrect.

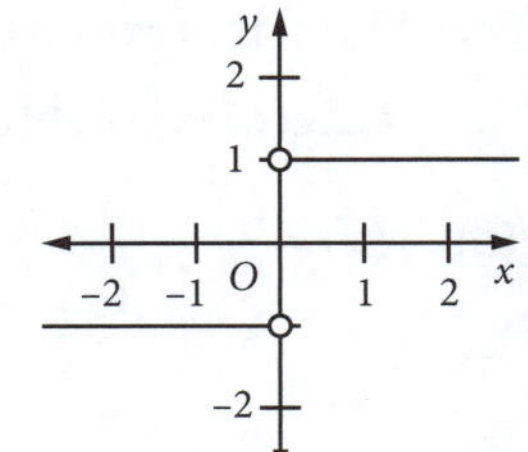

(a) The domain is real x, $x \neq 0$.
(b) The range is real y, $-1 < y < 1$.
(c) The gradient of the function is zero.
(d) The equation of the function could be $y = \frac{|x|}{x}$.

4.5 CIRCLES

A **circle** can be defined as the set of all points P in a plane at a given distance from a fixed point in the plane. The fixed point is the centre of the circle and the given distance is the **radius**.

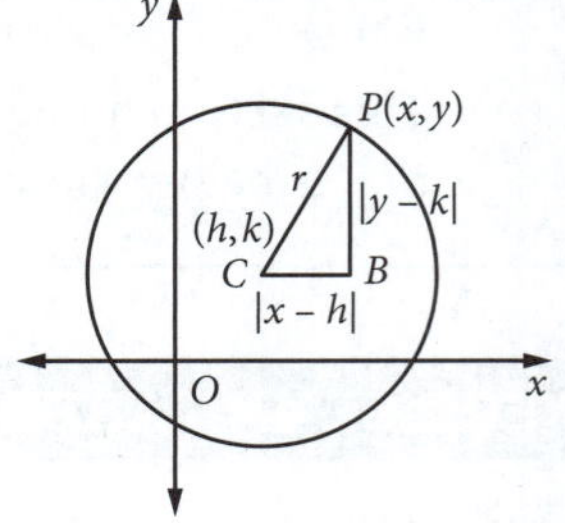

Consider the circle of radius r units with its centre at $C(h, k)$. If P is a point with coordinates (x, y) on the circumference of this circle, then the distance of P from C is r units.

Applying Pythagoras' theorem to triangle CBP in the diagram gives:

$$BC^2 + BP^2 = CP^2$$

$$(x - h)^2 + (y - k)^2 = r^2$$

Thus the equation of the circle is given by $(x - h)^2 + (y - k)^2 = r^2$, with the values for x and y restricted:

- The set of values for x is given by $h - r \leq x \leq h + r$.
- The set of values for y is given by $k - r \leq y \leq k + r$.

If the centre of the circle is at the origin, then $h = 0$, $k = 0$ and the equation of the circle is $x^2 + y^2 = r^2$.

Example 14

Find the equation of the circle with centre $(-3, 4)$ and radius 6 units.

Solution

Use the result: $(x - h)^2 + (y - k)^2 = r^2$

Substitute $(-3, 4)$, $r = 6$: $(x + 3)^2 + (y - 4)^2 = 36$ is the equation of the circle.

Example 15

Find the coordinates of the centre and the length of the radius for the circle whose equation is $x^2 + y^2 - 4x + 10y + 14 = 0$.

Solution

Rewrite equation: $x^2 - 4x + y^2 + 10y = -14$

Complete the square for x and y: $x^2 - 4x + 4 + y^2 + 10y + 25 = -14 + 4 + 25$

Factorise: $(x - 2)^2 + (y + 5)^2 = 15$

The circle has its centre at $(2, -5)$ and has a radius of $\sqrt{15}$ units.

Example 16

Find the equation of the circle with centre (3, 4) that passes through the point (−1, 1).

Solution

Use the result: $(x-h)^2+(y-k)^2=r^2$

Centre is (3, 4): $(x-3)^2+(y-4)^2=r^2$

(−1, 1) satisfies equation: $(-4)^2+(-3)^2=r^2$

$$r^2=25$$

Equation of circle is $(x-3)^2+(y-4)^2=25$.

Example 17

The diagram shows the graph of a circle with centre (1, 2) that passes through the point (4, 6). Find the equation of the circle:

(a) in the form $(x-h)^2+(y-k)^2=r^2$

(b) in general form.

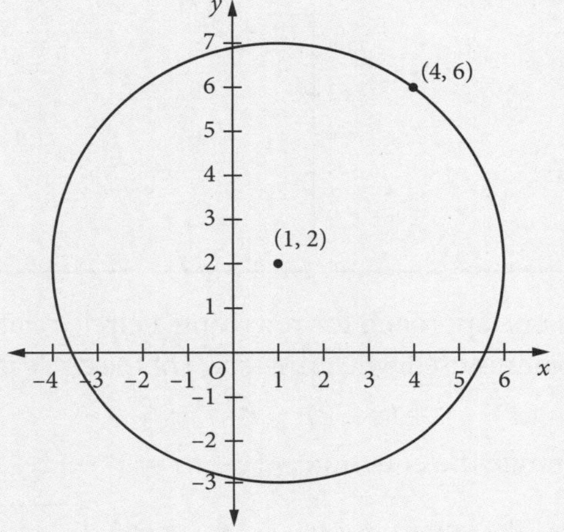

Solution

(a) $C(1,2)$, $P(4,6)$

Radius of circle, $r=CP=\sqrt{(4-1)^2+(6-2)^2}$

$=\sqrt{3^2+4^2}$

$=5$

Equation of the circle is: $(x-1)^2+(y-2)^2=25$

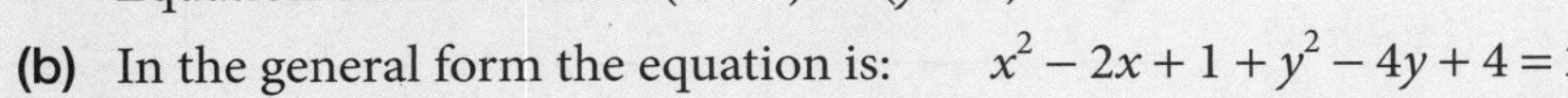

(b) In the general form the equation is: $x^2-2x+1+y^2-4y+4=25$

$$x^2+y^2-2x-4y-20=0$$

Semicircles

The equation $x^2+y^2=r^2$ can be written so that instead of being a relation, it becomes two functions that each represent a semicircle.

Rearrange the equation: $y^2=r^2-x^2$

Take square roots: $y=\pm\sqrt{r^2-x^2}$

Thus the circle can be represented by two functions, $y=\sqrt{r^2-x^2}$ and $y=-\sqrt{r^2-x^2}$. Both these functions have the same domain, $x\in[-r,r]$.

- $y=\sqrt{r^2-x^2}$ represents a semicircle in the upper half plane. The range is $y\in[0,r]$.
- $y=-\sqrt{r^2-x^2}$ represents a semicircle in the lower half plane. The range is $y\in[-r,0]$.

Example 18

Given the equation $x^2+y^2=25$, find the equations of the two functions that represent the semicircles that make up this circle. Sketch their graphs on separate diagrams.

Solution

$x^2 + y^2 = 25$: $\quad y^2 = 25 - x^2$

$$y = \pm\sqrt{25 - x^2}$$

The equations of the semicircles are $y = \sqrt{25 - x^2}$ and $y = -\sqrt{25 - x^2}$.

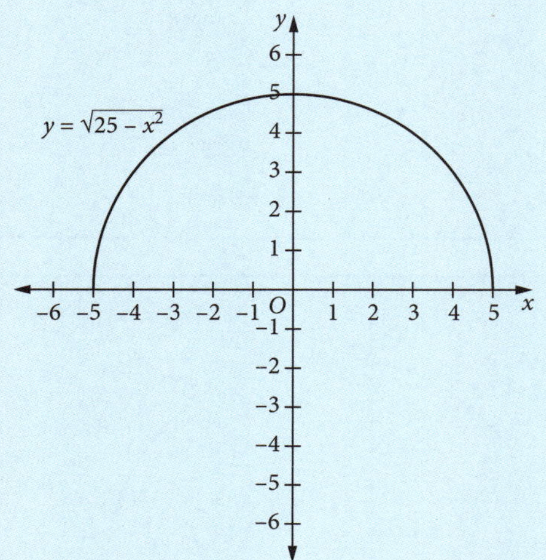

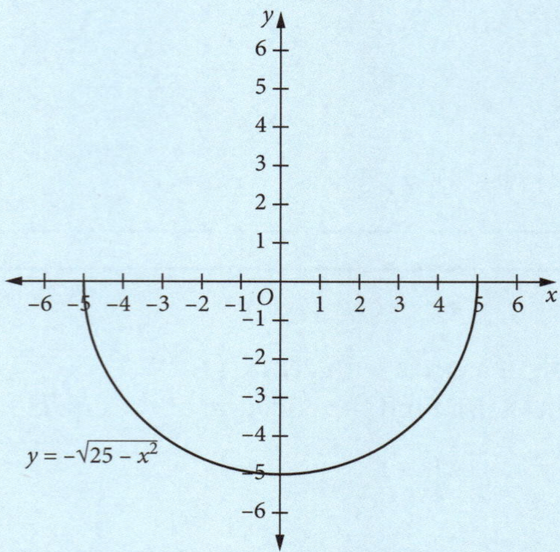

The same approach for rewriting a circle relation as two functions is used when the centre of the circle is not the origin.

Given $(x - h)^2 + (y - k)^2 = r^2$:

Rearrange the equation: $\quad (y - k)^2 = r^2 - (x - h)^2$

Take square roots: $\quad y - k = \pm\sqrt{r^2 - (x - h)^2}$

$$y = k \pm \sqrt{r^2 - (x - h)^2}$$

The equations of the semicircles are $y = k + \sqrt{r^2 - (x - h)^2}$ and $y = k - \sqrt{r^2 - (x - h)^2}$.

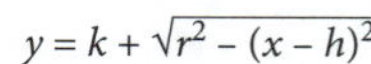

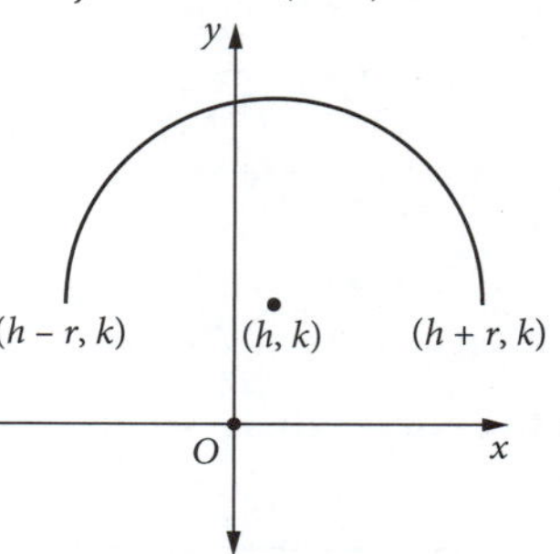

Upper semicircle

$y = k - \sqrt{r^2 - (x - h)^2}$

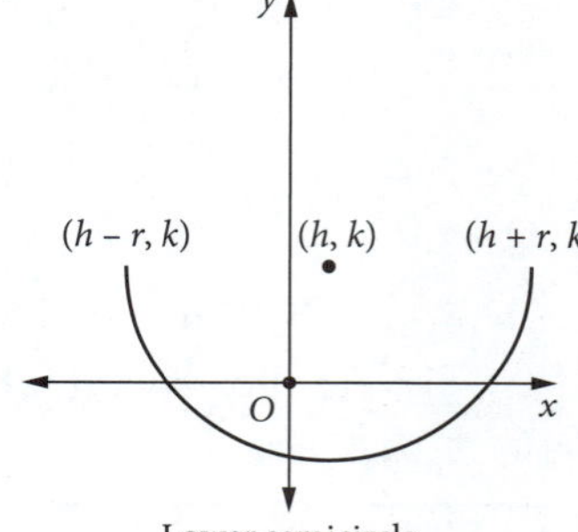

Lower semicircle

MAKING CONNECTIONS

Graphing circles on the Cartesian plane

Move the sliders to explore transformations of the graph of $(x - h)^2 + (y - k)^2 = r^2$.

EXERCISE 4.5 CIRCLES

1 Find the equation of each of the following circles.

(a) centre $(3, 2)$, radius 4 units **(b)** centre $(-1, -4)$, radius 3 units **(c)** centre $(3, -3)$, radius $\sqrt{5}$ units

(d) centre $\left(-2, \frac{5}{2}\right)$, radius $\frac{7}{2}$ units **(e)** centre $\left(0, -\frac{3}{2}\right)$, radius 4 units **(f)** centre $(4, 0)$, radius 3 units

2 The equation of the circle with centre $(-4, 4)$ and radius 6 units is:

A $(x - 4)^2 + (y - 4)^2 = 36$ **B** $(x + 4)^2 + (y - 4)^2 = 36$ **C** $(x - 4)^2 + (y + 4)^2 = 36$ **D** $x^2 + y^2 = 36$

3 Find the equation for each of the following circles.

(a) centre $(3, 2)$ and passing through the point $(5, -5)$
(b) centre $(-1, 4)$ and passing through the origin
(c) centre $(0, 0)$ and passing through the point $(-3, 4)$

4 Find the coordinates of the centre and the length of the radius for the following circles.

(a) $x^2+y^2-6x+4y-3=0$
(b) $x^2+y^2+4x+2y-4=0$
(c) $(x-3)^2+y^2=3$
(d) $(x+a)^2+(y-b)^2=8$
(e) $x^2+y^2-5x+3y-1=0$
(f) $x^2+y^2+4x+2y-5=0$
(g) $2x^2+2y^2-8x+5y+3=0$
(h) $3x^2+3y^2+9x-4y-24=0$

5 Using the fact that the centre of a circle is the midpoint of a diameter, find the equation of the circle with the diameter endpoints given.

(a) $(3,4)$ and $(9,-6)$
(b) $(0,0)$ and $(5,-3)$
(c) $(5,8)$ and $(-2,3)$

6 For the equation $x^2+y^2-6x+2y+10=0$, indicate whether each statement is correct or incorrect.

(a) centre $(3,-1)$, radius $=1$
(b) centre $(-3,1)$, radius $=0$
(c) centre $(3,-1)$, radius $=2\sqrt{5}$
(d) centre $(3,-1)$, radius $=0$

7 For the circle with equation $x^2+y^2+6x-8y=0$:

(a) find the coordinates of the centre and the length of the radius
(b) show that the origin is on the circle
(c) find the equation of the diameter that passes through the origin.

8 For each circle relation, find and sketch the equation of the semicircle in the upper half plane. State the range of each function.

(a) $x^2+y^2=4$
(b) $x^2+y^2=36$
(c) $x^2+y^2=5$
(d) $x^2+y^2=\frac{25}{9}$

9 For each circle relation, find and sketch the equation of the semicircle in the lower half plane. State the range of each function.

(a) $x^2+y^2=4$
(b) $x^2+y^2=36$
(c) $x^2+y^2=5$
(d) $x^2+y^2=\frac{25}{9}$

10 The equation of a circle is given by $x^2+y^2+2x-6y+1=0$.

(a) What are the coordinates of the centre and the radius of this circle?
(b) What is the equation of the upper semicircle of this circle?
(c) Sketch the graph of your equation in **(b)**.
(d) State the domain and range of your function in **(b)**.

11 The equation of a circle is given by $x^2+y^2-8y=0$.

(a) What are the coordinates of the centre and the radius of this circle?
(b) What is the equation of the lower semicircle of this circle?
(c) Sketch the graph of your equation in **(b)**.
(d) State the domain and range of your function in **(b)**.

12 Find the equation of the circle that touches the x-axis at $(4,0)$ and the y-axis at $(0,4)$.

13 Show that the point $(4,-3)$ is not on the circle $x^2+y^2-5x+3y+2=0$. Determine whether the point is inside or outside the circle.

14 Determine whether the origin is inside or outside the circle $x^2+y^2-4x-y+1=0$.

15 (a) Find the equation of a circle with a radius of 5 units and its centre at the point $(-1,2)$.
(b) What is the length of the intercept cut off by this circle on the x-axis?
(c) Find the length of the tangent to this circle from the point $(4,6)$.

(**Note**: A tangent line to a circle is perpendicular to a radius that is drawn to the point of contact.)

16 The equation of a circle is $x^2+y^2+4x-2y-20=0$. Find:

(a) the length of the tangent to this circle from the point $(5,2)$
(b) the length of the intercept on the y-axis.

17 A diameter intersects a circle at the points $(6,-4)$ and $(-2,6)$.

(a) Find the centre and radius of the circle.
(b) What is the length of the circle's tangent from the point $(-5,5)$?

18 The coordinates of two points A and B are $(-1, 3)$ and $(5, 7)$. Find:

(a) the coordinates of the midpoint of AB

(b) the equation of the circle of which AB is a diameter

(c) the coordinates of the intersection points of the circle with the y-axis.

19 **(a)** Find the coordinates of the centre and the length of the radius for the circle $x^2 + y^2 - 4x - 8y - 5 = 0$.

(b) The point $(3, 2)$ is the midpoint of a chord of this circle. Find the distance of the chord from the centre and the length of the chord.

20 Given the coordinates of the centre and a point on each circle, find the equation of each circle:

(i) in the form $(x-h)^2 + (y-k)^2 = r^2$

(ii) in general form.

(a)

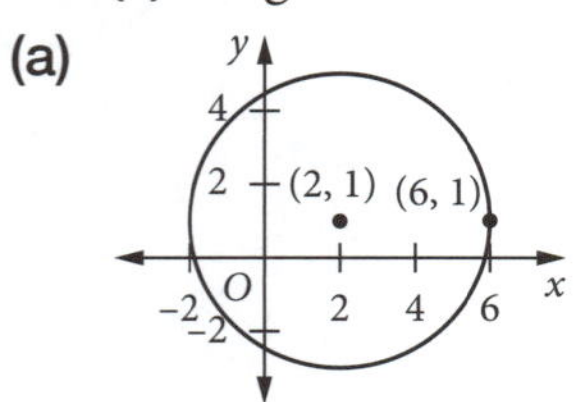

(b)

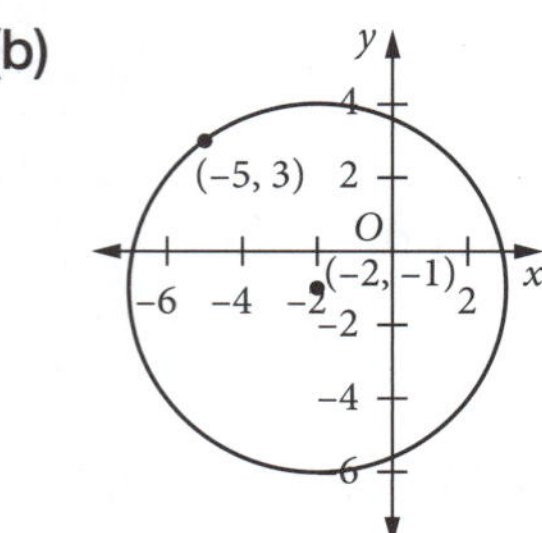

(c)

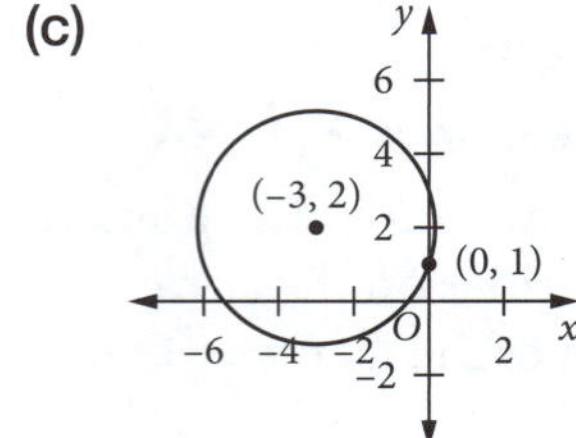

(d)

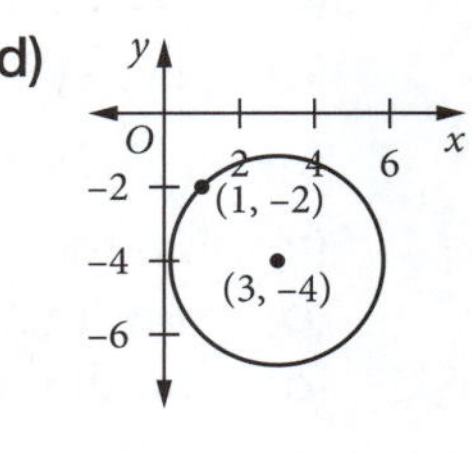

4.6 CUBIC POLYNOMIALS

So far you have considered the linear polynomial, $ax + b$ and the quadratic polynomial $ax^2 + bx + c$. The graph of $y = ax + b$ is a straight line and the graph of $y = ax^2 + bx + c$ is a parabola. The equation $ax + b = 0$ has one root whilst the equation $ax^2 + bx + c = 0$ has at most two real roots, $a \neq 0$ in each case.

These functions are continuous: they have no gaps in them over their domain. Simply put, this means that they could be drawn without taking your pen off the paper. In particular, all polynomial functions are continuous.

The general polynomial of the form $a_n x^n + a_{n-1}x^{n-1} + a_{n-2}x^{n-2} + \ldots + a_0$ is a polynomial of degree n, n a positive integer, $a_n \neq 0$, with coefficients $a_n, a_{n-1}, a_{n-2}, \ldots, a_0$. a_0 is called the constant term.

The cubic polynomial is of the form $ax^3 + bx^2 + cx + d$. The cubic function is easy to sketch without technology when it can be written in any of the forms $y = kx^3$, $y = k(x-b)^3 + c$ or $y = k(x-a)(x-b)(x-c)$ where a, b, c and k are constants, $k \neq 0$.

A cubic polynomial is of degree 3, that is the largest power of x is x^3. In $ax^3 + bx^2 + cx + d$, ax^3 is called the leading term and a is called the coefficient of the leading term. b and c are coefficients and d is called the constant term.

Example 19

Draw the graph of $y = (x+2)^3$. On your graph, draw the lines $y = 1$ and $y = -8$. Use the graph to solve the equations:

(a) $(x+2)^3 = 0$ **(b)** $(x+2)^3 = -8$ **(c)** $(x+2)^3 = 1$

Solution

(a) $(x+2)^3 = 0$ where the graph cuts the x-axis, so $x = -2$ is the solution.

(b) $(x+2)^3 = -8$ where the graph cuts the line $y = -8$, so $x = -4$ is the solution.

(c) $(x+2)^3 = 1$ where the graph cuts the line $y = 1$, so $x = -1$ is the solution.

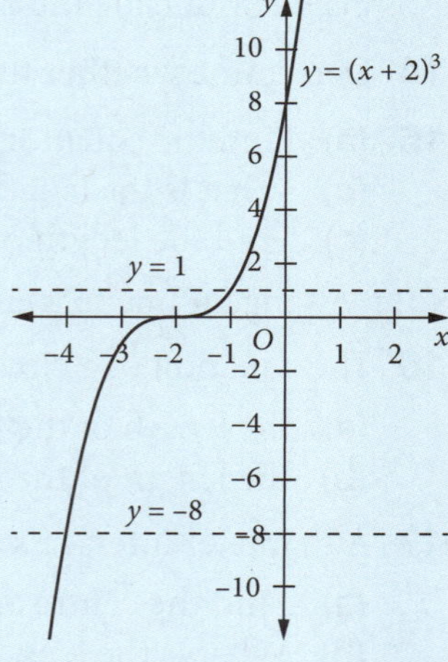

These equations could also have been solved algebraically. In each case, take cube root of both sides:

(a) $(x+2)^3 = 0$
$x + 2 = 0$
$x = -2$

(b) $(x+2)^3 = -8$
$x + 2 = -2$
$x = -4$

(c) $(x+2)^3 = 1$
$x + 2 = 1$
$x = -1$

You may know the terms 'turning point', 'local maximum' and 'local minimum'.

The term **turning point** describes itself. It is the point where the curve turns around, i.e. stops rising and starts to fall or stops falling and starts to rise. The curve in Example 20 has a turning point between $x = -1$ and $x = 0$ and another between $x = 1$ and $x = 2$. The first is a maximum turning point (or **local maximum**); the second is a minimum turning point (or **local minimum**).

At a maximum turning point a curve is concave down; at a minimum turning point a curve is concave up. Between two points like this, there must be a point where the concavity changes. This is called a **point of inflection** and will be discussed more fully later in the course.

In Example **19**, the graph of $y = (x + 2)^3$ does not have any turning points as it is always sloping up to the right. For this curve, $x = -2$ gives the point of inflection as it is clear that the concavity changes from concave down to concave up at the point $(-2, 0)$.

Example 20

Draw the graph of $y = (x + 1)(x - 1)(x - 2)$, using technology if necessary. On your graph, draw the lines $y = 2$ and $y = -4$. Use the graph to solve the following equations.

(a) $(x + 1)(x - 1)(x - 2) = 0$ **(b)** $(x + 1)(x - 1)(x - 2) = 2$ **(c)** $(x + 1)(x - 1)(x - 2) = -4$

Solution

(a) $(x + 1)(x - 1)(x - 2) = 0$ where the graph cuts the x-axis, so $x = -1, 1, 2$

(b) $(x + 1)(x - 1)(x - 2) = 2$ where the graph cuts the line $y = 2$, so $x = -0.414, 0, 2.414$ (3 d.p.)

(c) $(x + 1)(x - 1)(x - 2) = -4$ where the graph cuts the line $y = -4$, so $x = -1.468$ (3 d.p.)

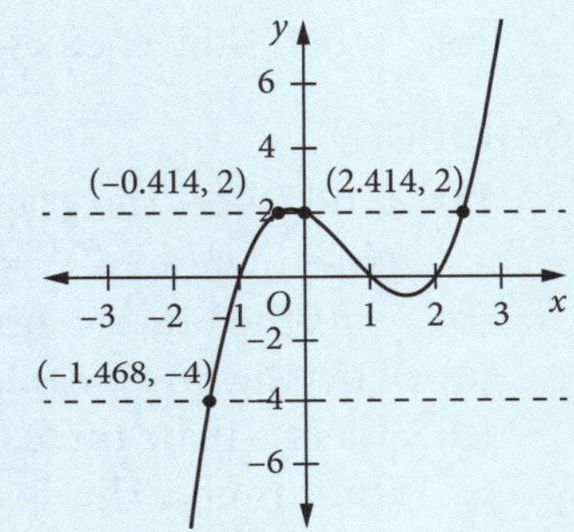

These equations could also have been solved algebraically. It is easy to solve **(a)** $(x + 1)(x - 1)(x - 2) = 0$ algebraically because it is in factored form.

To solve the other two equations algebraically, you need to expand the left-hand side and collect like terms. Part **(b)** is then solvable as the constant term disappears, but in part **(c)** the result is still a cubic and not factorised. Using graphing software is often the only easy way to solve a cubic equation like this.

MAKING CONNECTIONS

Solving cubic polynomial equations graphically

Use technology to solve cubic polynomial equations graphically.

Example **20** shows how horizontal lines $y = c$ drawn on a cubic graph can be used to see how many solutions there will be. If the line $y = c$ is drawn on the graph in Example **20**, then the graph shows that the equation $(x + 1)(x - 1)(x - 2) = c$ would always have at least one solution. If $y = c$ touches the curve at the local maximum or minimum value, it will cut the curve again so that the equation $(x + 1)(x - 1)(x - 2) = c$ will have two distinct solutions.
If $y = c$ cuts the curve between the local maximum or minimum values, the equation $(x + 1)(x - 1)(x - 2) = c$ will have three distinct solutions. If $y = c$ cuts the curve above the local maximum or below the local minimum value, the equation $(x + 1)(x - 1)(x - 2) = c$ will only have one solution.

Example 21

Draw the graph of $y = x^3 - 2x^2 - 3x + 4$. By drawing appropriate lines on your graph, use this to solve:

(a) $x^3 - 2x^2 - 3x + 4 = 0$ **(b)** $x^3 - 2x^2 - 3x + 4 = 4$ **(c)** $x^3 - 2x^2 - 3x + 4 = -3$

Solution

Use graphing software to obtain this graph. Software will also give the points of intersection.

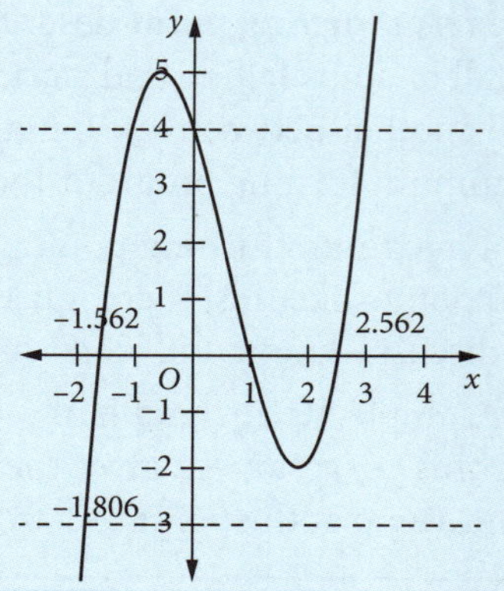

(a) Use the x-axis (for the line $y = 0$) and find the intersections: $x = -1.562, 1, 2.562$. This means the equation has a factor of $(x - 1)$.

(b) Draw $y = 4$ and find the intersections: $x = -1, 0, 3$. This means that the equation can be factorised as $x(x + 1)(x - 3) = 0$.

(c) Draw $y = -3$ and find the intersections: $x = -1.806$

Example 22

Draw the graph of $y = (x + 2)^2(1 - x)$. By drawing appropriate lines on your graph, use this to solve the following:

(a) $(x + 2)^2(1 - x) = 0$ **(b)** $(x + 2)^2(1 - x) = 4$ **(c)** $(x + 2)^2(1 - x) = -2$

(d) For what values of c will the equation $(x + 2)^2(1 - x) = c$ have three distinct roots?

(e) What is the coefficient of x^3 when $(x + 2)^2(1 - x)$ is expanded?

Solution

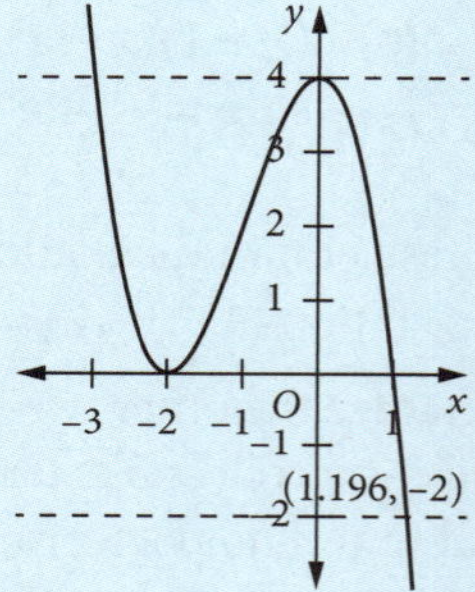

(a) Find where the graph cuts the x-axis: $x = -2, 2$.
Note that $(x + 2)^2 = 0$ gives a single value of x.

(b) Cuts the line $y = 4$: $x = -3, 0$.

(c) Cuts the line $y = -2$: $x = 1.196 \approx 1.2$

(d) There will be three distinct roots where a horizontal line cuts the graph three times. This happens only between the local minimum and local maximum, so: $0 < c < 4$

(e) $(x + 2)^2(1 - x) = -x^3 - 3x^2 + 4$. The coefficient of x^3 is -1.

Note that as $x \to \infty$, $y \to -\infty$ and that the coefficient of x^3 is -1.

Properties of the cubic polynomial

For the polynomial $y = ax^3 + bx^2 + cx + d$, $a \neq 0$:

- For $a > 0$, if $x \to \infty$ then $y \to ax^3$ and $y \to \infty$
  if $x \to -\infty$ then $y \to -ax^3$ and $y \to -\infty$
- For $a < 0$, if $x \to \infty$ then $y \to ax^3$ and $y \to -\infty$
  if $x \to -\infty$ then $y \to -ax^3$ and $y \to \infty$
- The graph will cut the x-axis at least once;
  $ax^3 + bx^2 + cx + d = 0$ for at least one value of x.

Cubic equations

The general cubic function is written $y = ax^3 + bx^2 + cx + d$, $a \neq 0$. In this course you will look at the simpler versions of the cubic function given by $y = k(x - b)^3 + c$, $k \neq 0$ and $y = k(x - a)(x - b)(x - c)$, $k \neq 0$.

You can show by expanding and collecting like terms that:

$$(a + b)^3 = a^3 + 3a^2b + 3ab^2 + b^3$$

$$(a - b)^3 = a^3 - 3a^2b + 3ab^2 - b^3$$

Note that these results are quite different from the factorisation of $a^3 - b^3$ and $a^3 + b^3$ studied in Chapter 1.

Using the expansion of $(x - b)^3$ you can show that $a(x - b)^3 + c$ becomes an expression of the form $ax^3 + bx^2 + cx + d$, $a \neq 0$.

Solving equations of the type $k(x - b)^3 + c = 0$

Example 23

Solve the equation $2(x - 1)^3 + 5 = 0$.

Solution

Rearrange the equation:

$$2(x - 1)^3 + 5 = 0$$
$$2(x - 1)^3 = -5$$
$$(x - 1)^3 = -\frac{5}{2}$$

Take the cube root of both sides:
(You can always find the odd root of a negative number.)

$$x - 1 = -\sqrt[3]{\frac{5}{2}}$$
$$x = 1 - \sqrt[3]{\frac{5}{2}}$$

To put this in standard form required for surds (that is, to rationalise the denominator), multiply the numerator and denominator of the fraction by $\sqrt[3]{4}$:

$$x = 1 - \frac{\sqrt[3]{5}}{\sqrt[3]{2}} \times \frac{\sqrt[3]{4}}{\sqrt[3]{4}}$$
$$x = 1 - \frac{\sqrt[3]{20}}{\sqrt[3]{8}}$$
$$x = 1 - \frac{\sqrt[3]{20}}{2}$$

Even though the equation in Example 23 is cubic, it only has one root. This becomes obvious when you consider the following graphs.

$y = x^3$

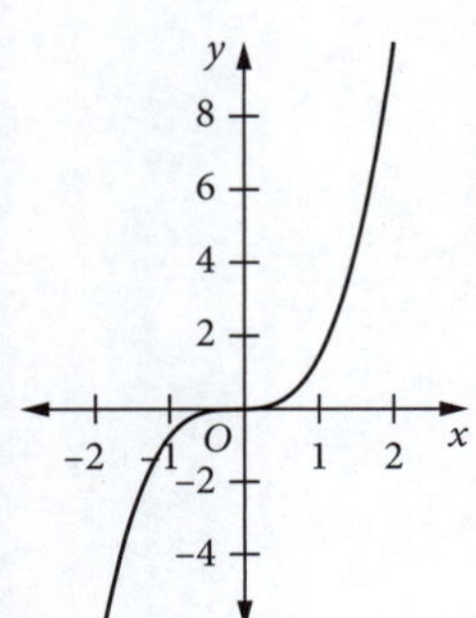

$y = (x - 1)^3$

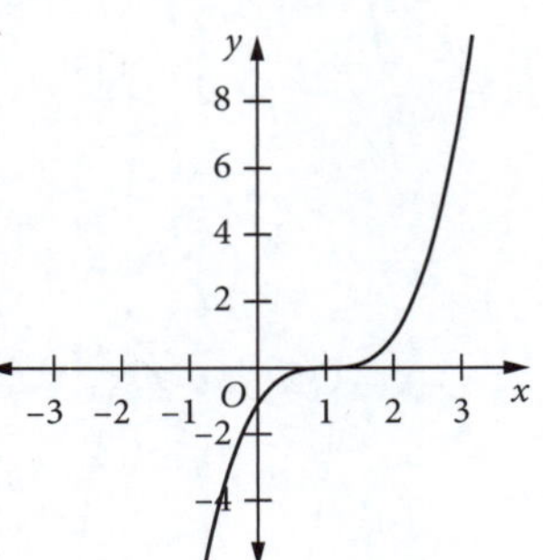

This is the graph of $y = x^3$ moved 1 unit to the right.

$y = 2(x - 1)^3$

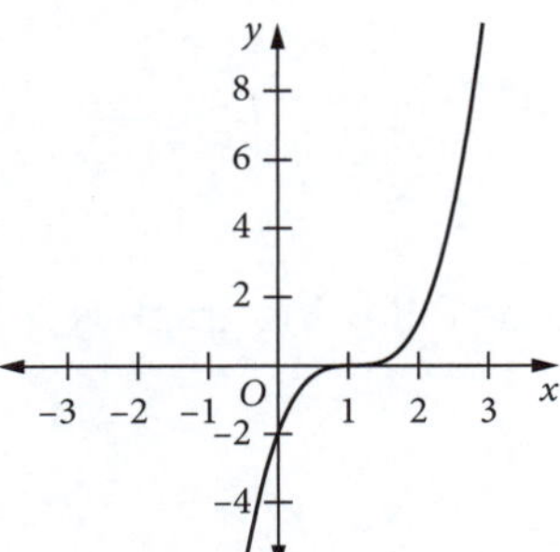

This is the graph of $y = (x - 1)^3$ stretched vertically by a factor of 2.

$y = 2(x - 1)^3 + 5$

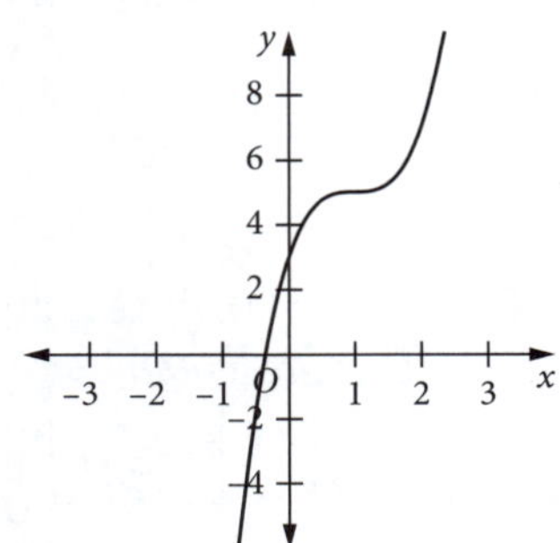

This is the graph of $y = (x - 1)^3$ stretched vertically by a factor of 2 and moved upwards 5 units.

Graphically, it is easy to see the solution to $x^3 = 0$, $(x - 1)^3 = 0$ and $2(x - 1)^3 = 0$. It is not so easy to see the solution to $2(x - 1)^3 + 5 = 0$. An approximate solution may be found using graphing software and clicking on the point of intersection of the graph and the x-axis, and reading off the x-value at this point.

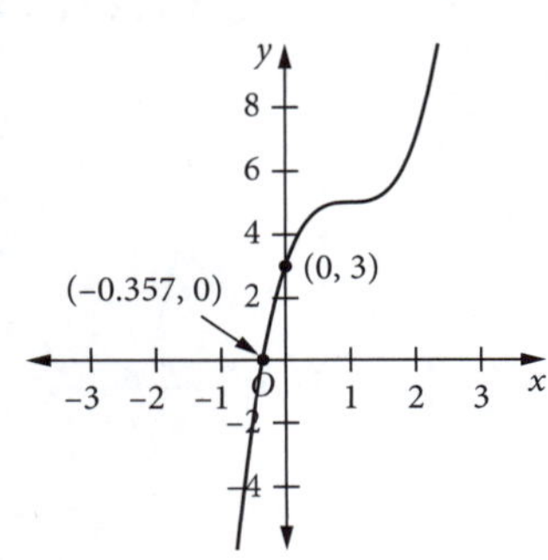

The diagram shows that the graph cuts the x-axis at $(-0.357, 0)$ so the solution to $2(x - 1)^3 + 5 = 0$ is $x = -0.357$. This is a good approximation to the exact value of the root found in Example 23.

In graphing each of these functions notice how the curve flattens out at $x = 0$ for $y = x^3$ and at $x = 1$ for the other graphs. It would appear that the slope of the curve is zero at this point. It is also worth noting that at all other points on the curve the slope is positive. The significance of these observations will become clear later, after you have learnt how to find the derivative of a function.

EXERCISE 4.6 CUBIC POLYNOMIALS

1 Draw the graph of $y = (x + 2)^3$. On your graph, draw the lines $y = 1$, $y = 8$ and $y = 2$. Use this to solve the equations:

(a) $(x + 2)^3 = 0$ (b) $(x + 2)^3 = 1$ (c) $(x + 2)^3 = 8$ (d) $(x + 2)^3 = 2$

2 Solve the following equations algebraically:

(a) $(x + 2)^3 = 0$ (b) $(x + 2)^3 = 1$ (c) $(x + 2)^3 = 8$
(d) $(x + 2)^3 = 2$ (e) $(x - 1)^3 = 4$ (f) $3(x - 4)^3 = 5$

3 Draw the graph of $y = (x - 1)(x + 1)(x + 2)$. On your graph, draw the lines $y = -2$ and $y = 4$. Use this to solve the equations:

(a) $(x - 1)(x + 1)(x + 2) = 0$ (b) $(x - 1)(x + 1)(x + 2) = -2$ (c) $(x - 1)(x + 1)(x + 2) = 4$

4 Solve the following equations algebraically: (a) $(x - 1)(x + 1)(x + 2) = 0$ (b) $(x - 1)(x + 1)(x + 2) = -2$

5 The line $y = c$ is drawn on the graph in question **3**. For what values of c will the equation $(x - 1)(x + 1)(x + 2) = c$ have three distinct roots? Give your answer to one decimal place.

6 Draw the graph of $y = (x - 2)^2(1 - x)$. By drawing appropriate lines on your graph, use this to solve the following:

(a) $(x - 2)^2(1 - x) = 0$ (b) $(x - 2)^2(1 - x) = 4$ (c) $(x - 2)^2(1 - x) = -2$
(d) For what values of c will the equation $(x - 2)^2(1 - x) = c$ have three distinct roots?
(e) What is the coefficient of x^3 when $(x - 2)^2(1 - x)$ is expanded?

7 For $a, b, c > 0$, the equation of the following graph is best represented by:

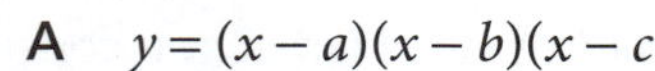

A $y = (x - a)(x - b)(x - c)$
B $y = (x - a)(x + b)(x + c)$
C $y = (x - a)(x - b)(x + c)$
D $y = (x + a)(x - b)(x - c)$

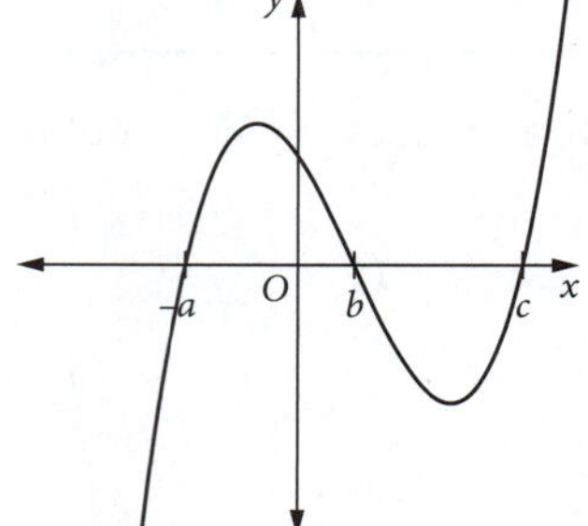

8 For $a, b > 0$, the equation of the following graph is best represented by:

A $y = (x + a)(x - b)^2$
B $y = (x + a)^2(b - x)$
C $y = (a - x)(x + b)^2$
D $y = (x - a)^2(x + b)$

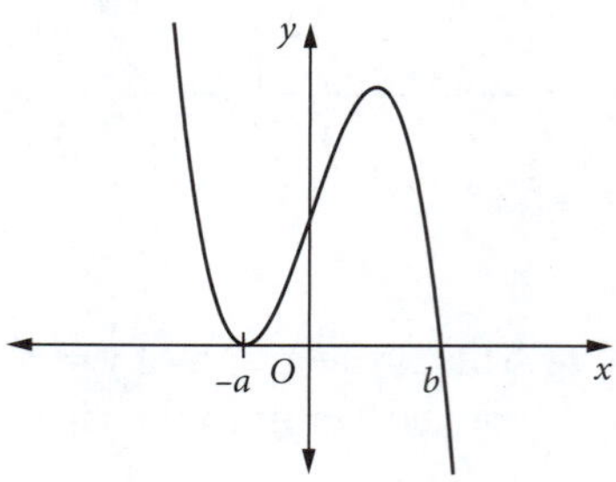

9 If $y = (x + 2)(x + 1)(2 - x)$, which of the following is the graph of this function?

A

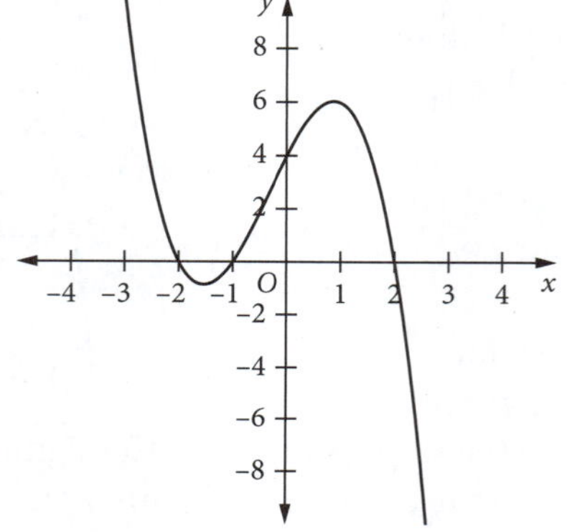

B

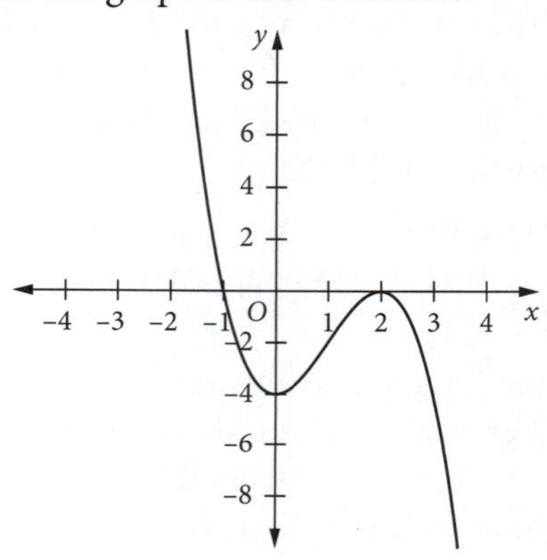

C

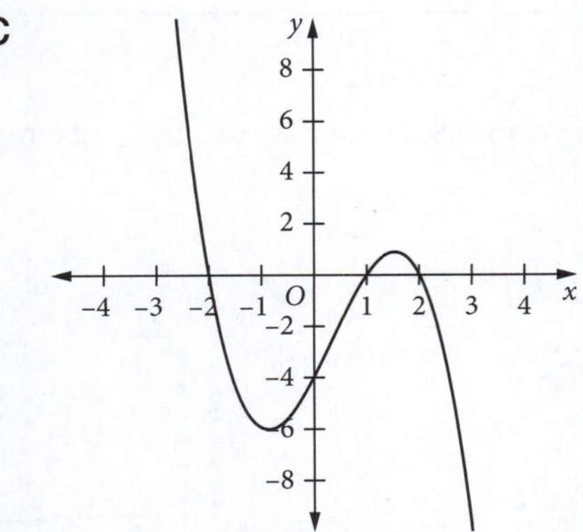

D

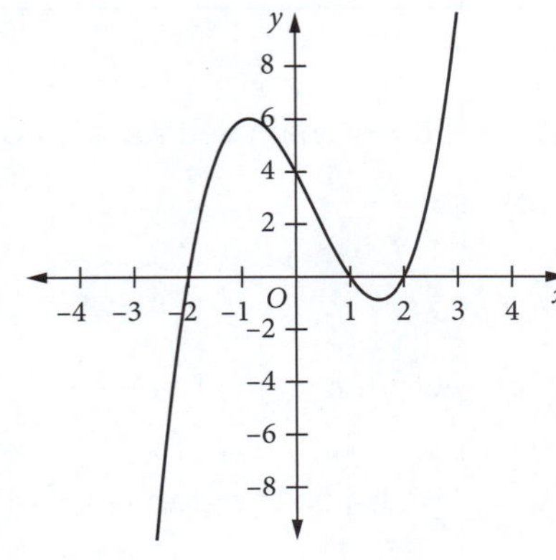

10 Find the equation of the cubic function that cuts the x-axis at $x = -1, 2, 3$ and has a y-intercept of 6.

11 Find the values of a and b if the curve $y = x(x - a)(x + b) + 4$ passes through the points $(1, 0)$ and $(-2, 12)$.

4.7 THE EQUATION $y = \frac{k}{x}$ AND INVERSE VARIATION

Earlier you have seen the link between direct variation and the equation of the straight line $y = kx$, which passes through the origin. By definition, two variables are in direct variation if one is a constant multiple of the other. This means if one variable increases then the other variable also increases at the same rate. This can also be written as $\frac{y}{x} = k$: the ratio between the two variables is a constant.

Similarly, the equation $xy = k$ can be written as $y = \frac{k}{x}$. In this situation, as x increases then y decreases; or, as x decreases then y increases. A change in one variable produces the opposite change in the other variable. This is called **inverse variation** or inverse proportion. This is expressed by saying that 'y is inversely proportional to x, where k is the constant of proportion (or variation)'.

Inverse variation has many applications in science. For example, in physics, Boyle's law states that 'at constant temperature a fixed mass of gas occupies a volume inversely proportional to the pressure exerted on it'. This is written as the formula $V = \frac{k}{P}$ or $PV = k$, where k is a constant.

The equation $y = \frac{k}{x}$

The equation $y = \frac{k}{x}$ is the same as $xy = k$, the rectangular hyperbola.

Consider the graph of $y = \frac{4}{x}$:

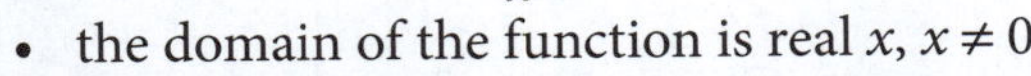

- the domain of the function is real x, $x \neq 0$
- the range is real y, $y \neq 0$
- the line $x = 0$ is a vertical asymptote
- the line $y = 0$ is a horizontal asymptote.

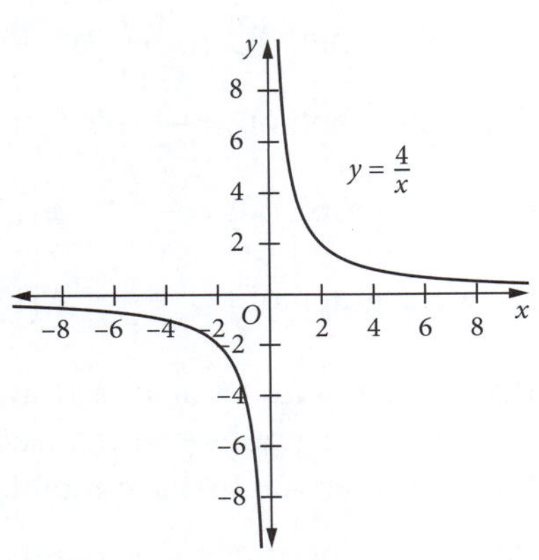

An **asymptote** is a line that the curve approaches but never meets.

In both branches of the curve, as x increases then y decreases.

Notice that this graph passes through the points $(2, 2)$ and $(-2, -2)$.

In general, the graph of $y = \frac{k}{x}$ will pass through the points $\left(\sqrt{k}, \sqrt{k}\right)$ and $\left(-\sqrt{k}, -\sqrt{k}\right)$.

Example 24

Sketch the following graphs. Give the domain and range of each function and state the equations of any asymptotes.

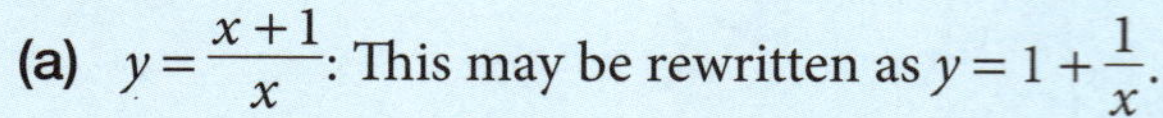

(a) $y = \dfrac{x+1}{x}$ **(b)** $y = \dfrac{x}{x+1}$

Solution

(a) $y = \dfrac{x+1}{x}$: This may be rewritten as $y = 1 + \dfrac{1}{x}$.

Thus the graph of $y = \dfrac{x+1}{x}$ is just the graph of $y = \dfrac{1}{x}$ moved up 1 unit.

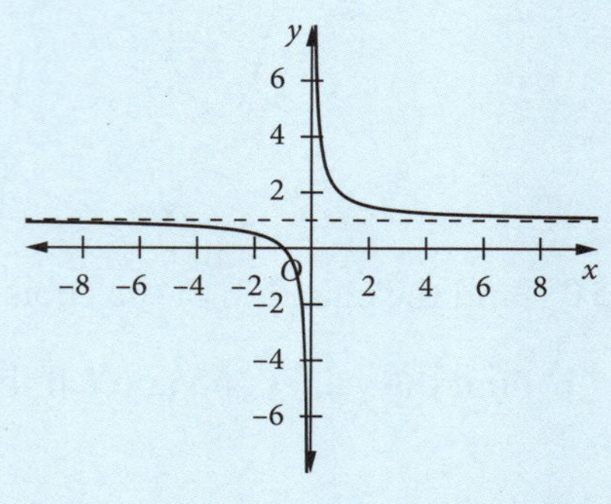

Vertical asymptote is $x = 0$.
Horizontal asymptote is $y = 1$.
Domain is real x, $x \neq 0$.
Range is real y, $y \neq 1$.

(b) $y = \dfrac{x}{x+1}$: This may be rewritten as $y = \dfrac{x+1-1}{x+1} = 1 - \dfrac{1}{x+1}$.
Vertical asymptote is $x = -1$.
Horizontal asymptote is $y = 1$.
Domain is real x, $x \neq -1$.
Range is real y, $y \neq 1$.

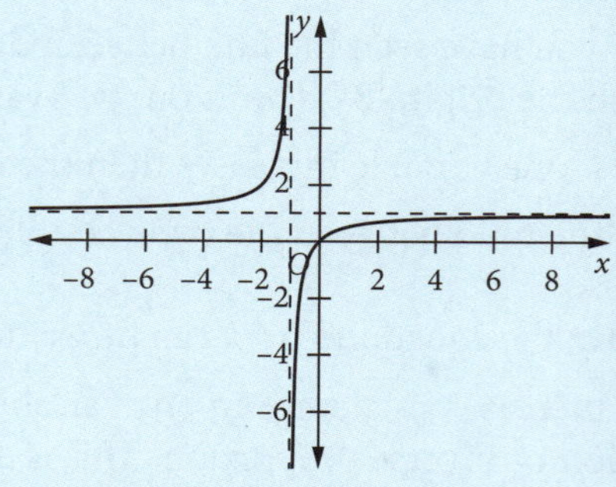

Note: The equations in this example do not represent inverse variation as their graphs do not have the coordinate axes as their asymptotes.

EXERCISE 4.7 THE EQUATION $y = \frac{k}{x}$ AND INVERSE VARIATION

1 Draw the graph of $y = \dfrac{9}{x}$. Write the equations of its asymptotes.

2 Draw the graph of $y = \dfrac{2}{x}$. Write the equations of its asymptotes.

3 Draw the graph of $y = \dfrac{x+2}{x}$. Write the equations of its asymptotes, the domain and the range.

4 Draw the graph of $y = \dfrac{x}{x-2}$. Write the equations of its asymptotes, the domain and the range.

5 **(a)** On the same set of axes, draw the graphs of $y = 2 + \dfrac{1}{x}$ and $y = 2 - \dfrac{1}{x}$.
(b) Do these graphs ever intersect?
(c) Comment on their asymptotes.

6 **(a)** In an experiment it is found that at a temperature of 100°C, 2 litres of argon gas is at a pressure of 15.28 atmospheres. If this gas obeys Boyle's law, $PV = k$, where V is in litres and P is in atmospheres, then find the value of k.
(b) If the volume was expanded to 4 litres with the temperature held at 100°C, then what would be the expected pressure?
(c) If the pressure was increased to 90 atmospheres with the temperature held at 100°C, then what would be the expected volume?

4.8 WORKING WITH FUNCTIONS

Sum and difference of functions

The sum or difference of two functions is found by adding or subtracting any corresponding terms of the two functions, where this is possible. Where one of the functions has a restrictive domain, this domain must apply for the sum or difference function.

For a sum $h(x) = f(x) + g(x)$ or for a difference $k(x) = f(x) - g(x)$, any simplification of the resulting functions depends on the original functions.

Example 25

If $f(x) = x - 5$, $g(x) = x^2 + 3$, $h(x) = \sqrt{x+4}$ and $k(x) = x^3 - 2x^2 + 6$, find expressions for each of the following functions, stating the domain and range in each case.

(a) $f(x) + g(x)$ **(b)** $f(x) - g(x)$ **(c)** $f(x) + h(x)$ **(d)** $k(x) - g(x)$ **(e)** $g(x) - h(x)$

Solution

(a) $f(x) + g(x) = x - 5 + x^2 + 3$
$= x^2 + x - 2$

Domain is the set of real numbers.

For the range you need to find the least value of the quadratic expression.

Solve: $x^2 + x - 2 = 0$

$(x - 1)(x + 2) = 0$

so $x = -2$ or 1

Least value occurs when: $x = \frac{-2+1}{2} = -\frac{1}{2}$

$x = -\frac{1}{2}$, $f(x) + g(x) = \frac{1}{4} - \frac{1}{2} - 2$

$= -2\frac{3}{4}$

The range is $y \geq -2\frac{3}{4}$.

(b) $f(x) - g(x) = x - 5 - (x^2 + 3)$
$= -x^2 + x - 8$

Domain is the set of real numbers.

For the range you need to find the greatest value of the quadratic expression. Since there are no obvious factors of the quadratic expression, use the axis of symmetry of the parabola, $x = -\frac{b}{2a}$ to find where the greatest value occurs.

$x = -\frac{1}{(-2)} = \frac{1}{2}$

$x = \frac{1}{2}$, $f(x) - g(x) = -\frac{1}{4} + \frac{1}{2} - 8$

$= -7\frac{3}{4}$

The range is $y \leq -7\frac{3}{4}$.

(c) $f(x) + h(x) = x - 5 + \sqrt{x+4}$

No further simplification of this expression is possible. The domain of the new function will be the same as the domain of $h(x)$ as this is the more restrictive. Domain is $x \geq -4$.

The least value of $h(x)$ occurs when: $x = -4$

so the least value of $f(x) + h(x) = -4 - 5 + 0$
$= -9$

The range is $y \geq -9$.

(d) $k(x) - g(x) = x^3 - 2x^2 + 6 - (x^2 + 3)$
$= x^3 - 3x^2 + 3$

Domain is the set of real numbers.
Since $k(x) - g(x)$ is a polynomial of degree 3, the range is the set of real numbers.

(e) $g(x) - h(x) = x^2 + 3 - \sqrt{x+4}$

The domain of the new function will be the same as the domain of $h(x)$ as this is the more restrictive. Domain is $x \geq -4$.
The least value of $x^2 + 3$ occurs at $x = 0$ and is 3.

The least value of $\sqrt{x+4}$ occurs when $x = -4$ and is 0.

When $x = 0$, $g(x) - h(x) = 0 + 3 - 2 = 1$ and this appears to be the least value of $g(x) - h(x)$.
The only way to check this assertion at this stage is to draw a graph of the function.

The graph shows that the least value of $x^2 + 3 - \sqrt{x+4}$ occurs when $x = 0.123$.

The range of $x^2 + 3 - \sqrt{x+4}$ is then $y \geq 0.985$.
The initial answer was a good estimate.

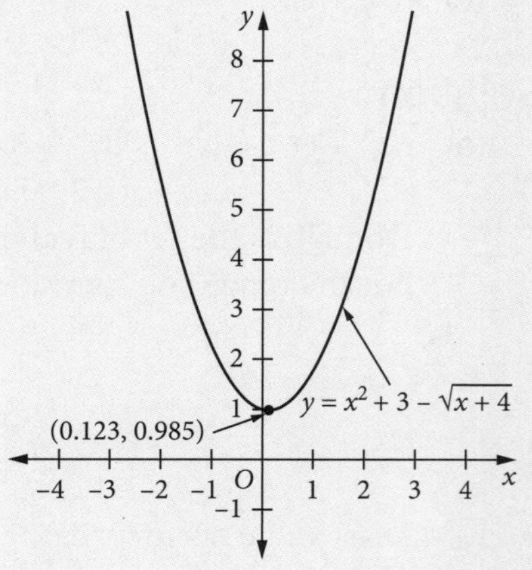

Product and quotient of functions

The product of two functions, which may be written as $f(x) \times g(x)$ or $f(x) \cdot g(x)$, will have its domain determined by the more restrictive of the two original domains.

The quotient of two functions, $\frac{f(x)}{g(x)}$, will have its domain further restricted by the values of x for which $g(x) = 0$, because at those values $\frac{f(x)}{g(x)}$ is undefined.

Example 26

Given $f(x) = x - 3$ and $g(x) = x$, find expressions for each of the following functions, stating the domain and range in each case.

(a) $f(x) \cdot g(x)$ (b) $\frac{f(x)}{g(x)}$ (c) $\frac{g(x)}{f(x)}$

Solution

(a) $f(x) \cdot g(x) = (x-3) \times x$
$= x^2 - 3x$

Domain: Real x as the quadratic expression exists for all values of x.
Since $f(x) \cdot g(x)$ is a quadratic polynomial, find the axis of symmetry:

$$x = -\frac{-3}{2} = \frac{3}{2}$$

$$x = \frac{3}{2}: f(x) \cdot g(x) = \frac{9}{4} - \frac{9}{2}$$
$$= -2\frac{1}{4}$$

This is the least value of the function as it is a concave up parabola.

Range is the set of real numbers $\geq -2\frac{1}{4}$.

(b) $\dfrac{f(x)}{g(x)} = \dfrac{x-3}{x} = \dfrac{x}{x} - \dfrac{3}{x} = 1 - \dfrac{3}{x}$

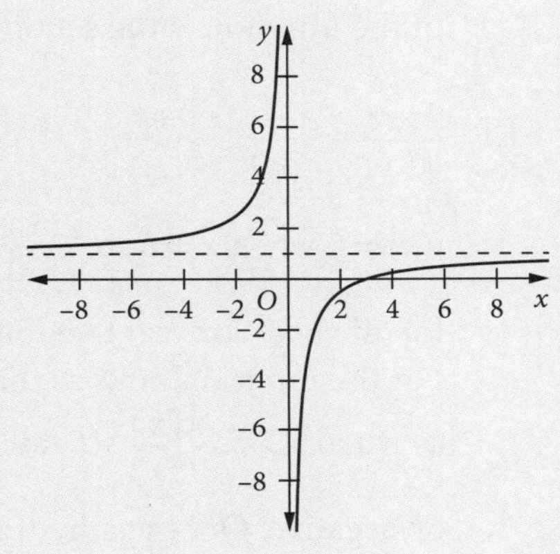

Domain: Real x, $x \neq 0$.

As x gets larger, $x \to \infty$, $1 - \dfrac{3}{x}$ gets smaller and approaches 1 from below.

As x gets smaller, $x \to -\infty$, $1 - \dfrac{3}{x}$ gets smaller and approaches 1 from above.

Range is the set of real numbers except 1.

(c)

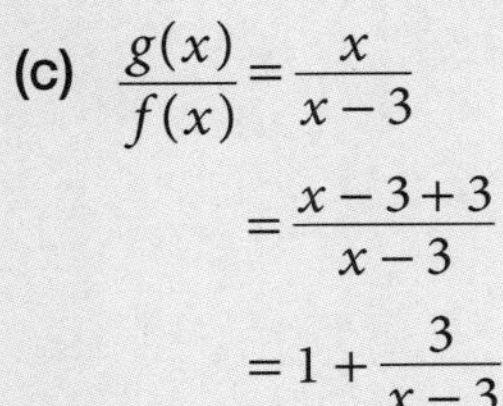

$$\frac{g(x)}{f(x)} = \frac{x}{x-3}$$
$$= \frac{x-3+3}{x-3}$$
$$= 1 + \frac{3}{x-3}$$

The calculation here shows a way to simplify algebraic fractions when the power of the numerator is the same or greater than the denominator: create a factor of the denominator in the numerator, such as $(x-3)$ here, by adding and subtracting the same number (here $-3+3$). Because $-3+3$ cancels to zero, the expression has not really changed, but the term $(x-3)$ that appears can be divided to simplify.

Domain: Real x, $x \neq 3$.

For $x > 3$, as x gets larger, $x \to \infty$, $1 + \dfrac{3}{x-3}$ gets smaller and approaches 1 from above.

For $x < 3$, as x gets smaller, $x \to -\infty$, $1 + \dfrac{3}{x-3}$ gets larger and approaches 1 from below.

Range is the set of real numbers except 1.

Example 27

If $f(x) = x - 5$, $g(x) = x^2 + 3$, $h(x) = \sqrt{x+4}$ and $k(x) = x^3 - 2x^2 + 6$, find expressions for each of the following functions, stating the domain and range in each case.

(a) $f(x) \cdot g(x)$
(b) $g(x) \cdot h(x)$
(c) $\dfrac{f(x)}{g(x)}$
(d) $\dfrac{k(x)}{f(x)}$
(e) $\dfrac{h(x)}{f(x)}$

Solution

(a) $f(x) \cdot g(x) = (x-5)(x^2+3)$
$= x^3 - 5x^2 + 3x - 15$

Both $f(x)$ and $g(x)$ are defined for all real x.

The domain of $f(x) \cdot g(x)$ is all real x.

$f(x) \cdot g(x) = x^3 - 5x^2 + 3x - 15$ is a cubic polynomial.

The range is all real y.

(b) $g(x) \cdot h(x) = (x^2+3)\sqrt{x+4}$

$\sqrt{x+4}$ is only defined for $x \geq -4$.

The domain of $g(x) \cdot h(x)$ is real $x \geq -4$.

Since $x^2 + 3 \geq 3$ and $\sqrt{x+4} \geq 0$, the product $(x^2+3)\sqrt{x+4} \geq 0$.

The range is the set of non-negative real numbers, or $g(x) \cdot h(x) \geq 0$.

(c) $\dfrac{f(x)}{g(x)} = \dfrac{x-5}{x^2+3}$

$x^2 + 3 \geq 3$ so the denominator is never zero. The domain of $\dfrac{f(x)}{g(x)}$ is real x.

The sign of $\dfrac{f(x)}{g(x)}$ is the same as the sign of $f(x)$ and since the denominator is never zero the range of the function is the set of real numbers.

(d) $\dfrac{k(x)}{f(x)} = \dfrac{x^3 - 2x^2 + 6}{x-5}$

$f(x) = 0$ when $x = 5$ so $\dfrac{k(x)}{f(x)}$ is undefined at $x = 5$. The numerator exists for all values of x so the domain is real x, $x \neq 5$.

For the range, it looks as though the answer is the set of real numbers as when $x > 5$, $\dfrac{k(x)}{f(x)} > 0$, and when $x < 5$ then $\dfrac{k(x)}{f(x)}$ can be positive or negative. Once again, drawing a graph of the function can confirm these answers.

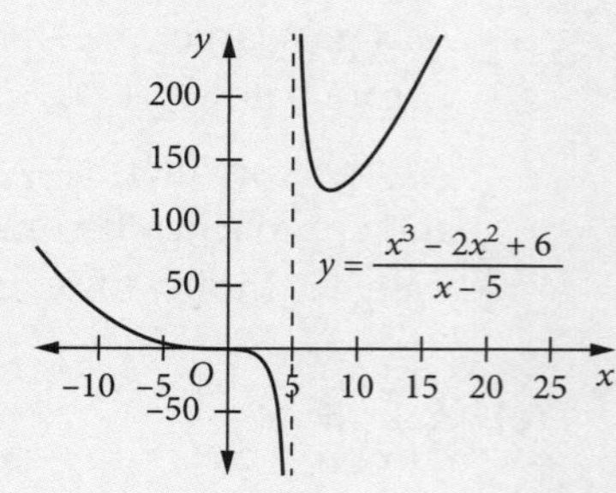

(e) $\dfrac{h(x)}{f(x)} = \dfrac{\sqrt{x+4}}{x-5}$

$f(x) = 0$ when $x = 5$ so $\dfrac{h(x)}{f(x)}$ is undefined at $x = 5$.

The numerator is defined for $x \geq -4$.

The domain of $\dfrac{h(x)}{f(x)}$ is $-4 \leq x < 5$ and $x > 5$.

The numerator is always positive or zero, the denominator can be positive or negative so the range of the function is the set of real numbers since $\dfrac{h(x)}{f(x)} = 0$ when $x = -4$.

The graph confirms this answer.

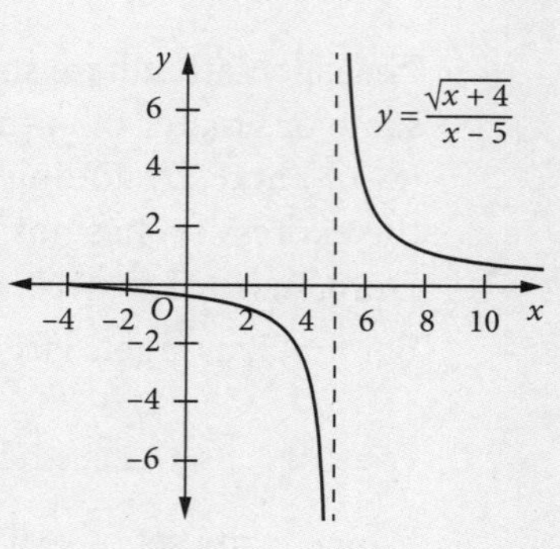

Using graphing software is a very effective way of finding domain and range when the functions are not straight polynomial functions.

Composite functions

$f(x) = (3x+1)^2$ is a function in terms of x.

If $g(x) = 3x + 1$ then it is possible to write $f(x) = (g(x))^2$. A 'function of a function' like this is called a **composite function.**

It is sometimes written $f \circ g$ so if $f(x) = (g(x))^2$ and $g(x) = 3x + 1$ then $(f \circ g)(x) = (3x+1)^2$.

The order of the functions is important, because $(g \circ f)(x) = 3(3x+1)^2 + 1$ which is a very different function to $(f \circ g)(x)$.

In a composite function, the output of one function $g(x)$ has become the input of the other function: $f(g(x))$ or $(f \circ g)(x)$.

Example 28

If $f(x) = \sqrt{x+3}$, $g(x) = x^3 + 2$ and $h(x) = \dfrac{1}{x^2 - 1}$, find the expressions for:

(a) $f(g(x))$ (b) $f(h(x))$ (c) $g(h(x))$ (d) $g(f(x))$ (e) $h(f(x))$ (f) $h(g(x))$

Solution

(a) $f(g(x)) = \sqrt{(x^3+2)+3} = \sqrt{x^3+5}$

(b) $f(h(x)) = \sqrt{\dfrac{1}{x^2-1} + 3} = \sqrt{\dfrac{1+3x^2-3}{x^2-1}} = \sqrt{\dfrac{3x^2-2}{x^2-1}}$

(c) $g(h(x)) = \left(\dfrac{1}{x^2-1}\right)^3 + 2 = \dfrac{1}{(x^2-1)^3} + 2$

(d) $g(f(x)) = \left(\sqrt{x+3}\right)^3 + 2 = (x+3)^{\frac{3}{2}} + 2$

(e) $h(f(x)) = \dfrac{1}{\left(\sqrt{x+3}\right)^2 - 1} = \dfrac{1}{x+3-1} = \dfrac{1}{x+2}$

(f) $h(g(x)) = \dfrac{1}{(x^3+2)^2 - 1} = \dfrac{1}{x^6 + 4x^3 + 3}$

With composite functions, you must be careful when finding the domain and range.

To find the domain of $f(x) = (3x+1)^2$ you must first find the domain of $g(x) = 3x + 1$.

As $g(x)$ is defined for all x, then $f(x)$ is also defined for all x, because you can square any number. The range of $f(x)$ is $f(x) \geq 0$.

Example 29

Find the domain and range of $f(x)$ given the following.

(a) $f(x) = \sqrt{x-4}$ (b) $f(x) = \sqrt{x^3+8}$ (c) $f(x) = \dfrac{1}{x^2-1}$

Solution

(a) $f(x) = \sqrt{x-4}$: You can only take the square root of a non-negative number,
$x - 4 \geq 0$ gives $x \geq 4$.
Domain is $x \geq 4$
Range is $f(x) \geq 0$

(b) $f(x) = \sqrt{x^3+8}$: You can only take the square root of a non-negative number,
$x^3 + 8 \geq 0$ gives $x^3 \geq -8$ so $x \geq -2$ is the domain of $f(x)$.
Range is $f(x) \geq 0$

(c) $f(x) = \dfrac{1}{x-1}$: The denominator $x - 1 = 0$ at $x = 1$ so $f(x)$ is undefined for $x = 1$.
The domain of $f(x)$ is real x, $x \neq 1$.
Where $x > 1$ then $f(x) > 0$
Where $x < 1$ then $f(x) < 0$
$f(x)$ is never zero so the range of $f(x)$ is all real numbers, $f(x) \neq 0$.

EXERCISE 4.8 WORKING WITH FUNCTIONS

1 If $f(x) = x^2 + 7$ and $g(x) = 5 - 2x$, then the correct expression for $f(x) + g(x)$ is:

A $x^2 + 2x + 12$ B $x^2 - 2x + 12$ C $x^2 - 2x + 2$ D $x^2 + 2x + 2$

2 If $f(x) = x^2 + 7$ and $g(x) = 5 - 2x$, then the correct expression for $f(x) \cdot g(x)$ is:

A $2x^3 - 5x^2 + 14x - 35$ B $x^2 - 2x + 12$ C $-2x^3 - 5x^2 + 14x + 35$ D $-2x^3 + 5x^2 - 14x + 35$

3 If $f(x) = x^2 + 7$ and $g(x) = 5 - 2x$, then the correct expression for $f(g(x))$ is:

A $4x^2 - 20x + 32$ B $-2x^2 - 9$ C $4x^2 + 32$ D $x^2 + 2x + 2$

4 If $f(x) = x + 4$, $g(x) = x^2 - 6$, find expressions for each of the following functions, stating the domain and range in each case.

(a) $f(x) + g(x)$ (b) $f(x) - g(x)$

5 If $f(x) = x^2 + 4$, $g(x) = x^3 - 3x^2 + 2x + 6$, find expressions for each of the following functions, stating the domain and range in each case.

(a) $f(x) + g(x)$ (b) $f(x) - g(x)$

6 If $f(x) = x$, $g(x) = x + 4$, find expressions for each of the following functions, stating the domain and range in each case. Use technology to sketch the new function.

(a) $f(x) \cdot g(x)$ (b) $\dfrac{g(x)}{f(x)}$ (c) $\dfrac{f(x)}{g(x)}$

7 If $f(x) = x - 1$, $g(x) = x + 1$, find expressions for each of the following functions, stating the domain and range in each case. Use technology to sketch the new function.

(a) $f(x) \cdot g(x)$ (b) $\dfrac{f(x)}{g(x)}$

8 If $f(x) = x$, $g(x) = x^2 + 4$, find expressions for each of the following functions, stating the domain and range in each case. Use technology to sketch the new function.

(a) $f(x) \cdot g(x)$ (b) $\dfrac{f(x)}{g(x)}$ (c) $\dfrac{g(x)}{f(x)}$

9 If $f(x) = x + 6$, $g(x) = x^2 - 9$, find expressions for each of the following functions, stating the domain and range in each case. Use technology to sketch the new function.

(a) $f(x) \cdot g(x)$ (b) $\dfrac{g(x)}{f(x)}$

10 If $f(x) = x + 4$, $g(x) = x^2 - 16$, find expressions for each of the following functions, stating the domain and range in each case. Use technology to sketch the new function.

(a) $f(x) \cdot g(x)$ (b) $\dfrac{g(x)}{f(x)}$ (c) $\dfrac{f(x)}{g(x)}$

11 If $f(x) = x + 3$, $g(x) = x^2 - 5$, $h(x) = \sqrt{x-4}$ and $k(x) = x^3 + 2x^2 - 6$, find expressions for each of the following functions, stating the domain and range in each case. Use technology to sketch the new function.

(a) $f(x) + g(x)$ (b) $f(x) - g(x)$ (c) $f(x) + h(x)$ (d) $k(x) - g(x)$ (e) $g(x) - h(x)$

12 If $f(x) = x + 3$, $g(x) = x^2 - 5$, $h(x) = \sqrt{x-4}$ and $k(x) = x^3 + 2x^2 - 6$, find expressions for each of the following functions, stating the domain and range in each case. Use technology to sketch the new function.

(a) $f(x) \cdot g(x)$ (b) $g(x) \cdot h(x)$ (c) $\dfrac{f(x)}{g(x)}$ (d) $\dfrac{k(x)}{f(x)}$ (e) $\dfrac{h(x)}{f(x)}$

CHAPTER REVIEW 4

1 State the largest possible domain for the following functions:

(a) $f(x)=\sqrt{x-1}$ (b) $f(x)=\dfrac{1}{x^2-4}$ (c) $f(x)=\sqrt{25-x^2}$ (d) $f(x)=|x|$

2 Sketch the graph of each function given in question **1**.

3 If $g(x)=x^4-x^2+1$, show that $g(x)$ is an even function.

4 Draw the graph of $y=1-|x|$.

5 Draw the graph of $y=x-|x|$ and state the domain of the function.

6 State the largest possible domain for each function:

(a) $f(x)=\sqrt{x+3}+\sqrt{2-x}$ (b) $g(x)=\dfrac{x}{|x-1|}$

7 Is the function $y=x^3-1$ even, odd or neither?

8 Draw the graph of $y=|x|+1$.

9 The equation of a circle is $x^2+y^2-2x-2y-23=0$.

(a) Find the circle's centre and radius.
(b) Calculate the distance from the point $(7,-2)$ to the centre of the circle.
(c) Explain why the point $(7,-2)$ is outside the circle.
(d) Use Pythagoras' theorem to find the length of the tangent to the circle from the point $(7,-2)$. (Note that tangent $\perp$ radius drawn to point of contact.)

10 Show algebraically that the line $y=x-4$ is a tangent to the circle $x^2+y^2=8$ and find the coordinates of the point of contact.

11 Solve: (a) $|x+7|=11$ (b) $|3x-4|\geq 5$

12 On the graph of $y=(x-2)(x-1)(x+1)$, which of the following lines would you need to draw on this graph in order to solve $(x-2)(x-1)(x+1)+3=0$?

A $y=-1$ B $y=-3$ C $y=1$ D $y=3$

13 Solve algebraically: (a) $(x-3)^3=-8$ (b) $(x+5)^3=4$ (c) $(x-2)^3=81$

14 Find the equation of the cubic function that cuts the x-axis at $x=-3, -2, 1$ and has a y-intercept of -12.

15 What are the equations of the asymptotes of the graph of $y=\dfrac{x}{x+3}$?

A $x=-3, y=-1$ B $x=3, y=-1$ C $x=-3, y=1$ D $x=3, y=1$

16 (a) On the same diagram, draw the graphs of $y=\dfrac{x+1}{x}$ and $y=\dfrac{x-2}{x}$.

(b) Do these graphs have the same asymptotes?
(c) Will these graphs ever intersect? Why?

CHAPTER 5
Equations and functions

In the seventeenth century, mathematics leapt forward with two great advances: analytical geometry and calculus. René Descartes (1596–1650) was one of the mathematicians most responsible for the creation of analytical geometry. He did this by setting up a coordinate system and applying algebra to geometry. The name of Descartes is still used today as an adjective in terms such as Cartesian plane, Cartesian coordinates and Cartesian axes. The Cartesian plane is also referred to as the number plane and the x-y plane.

5.1 GRADIENT OF A STRAIGHT LINE

$A(x_1, y_1)$ and $B(x_2, y_2)$ are two points on the number plane, with B to the right of A.

The **gradient** (or slope) m of AB is defined as:

$$m = \frac{y_2 - y_1}{x_2 - x_1} \qquad \text{if } x_1 \neq x_2$$

If $x_1 = x_2$, the gradient is undefined and the line is vertical (that is, parallel to the y-axis).

If $y_1 = y_2$, the gradient is zero and the line is horizontal (that is, parallel to the x-axis).

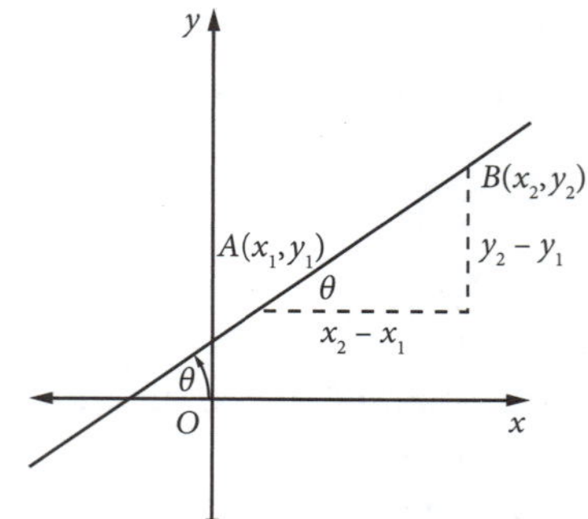

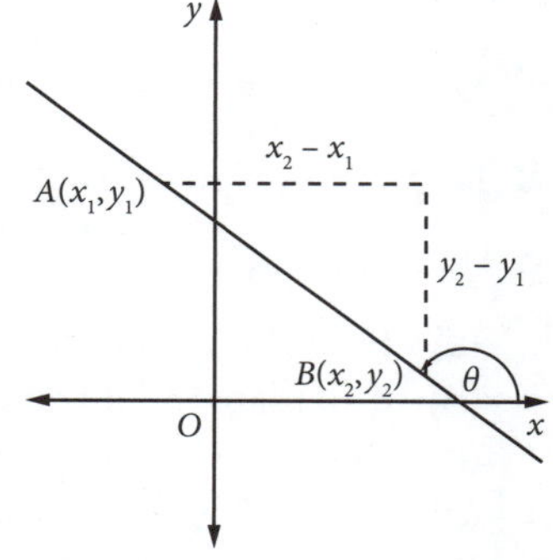

Angle of inclination, θ

The angle that a line makes with the positive direction of the x-axis is called the **angle of inclination**. It is usually denoted by the symbol θ where $0° \leq \theta < 180°$. The gradient m is also equal to $\tan\theta$, hence:

$$\tan\theta = \frac{y_2 - y_1}{x_2 - x_1} \qquad \text{if } x_1 \neq x_2$$

In the figure on the left above, the line rises from left to right and makes an acute angle θ with the x-axis. The values $(x_2 - x_1)$ and $(y_2 - y_1)$ are both positive numbers, so the gradient is positive.

In the figure on the right above, the line falls from left to right and makes an obtuse angle θ with the x-axis. The value $(x_2 - x_1)$ is a positive number, but $(y_2 - y_1)$ is negative, so the gradient is negative.

MAKING CONNECTIONS

Gradient

Drag the points on the Cartesian plane to explore the calculation of the gradient and angle of inclination.

Example 1

Find the gradient of the line joining the given points and calculate the size of the angle of inclination.

(a) $(2, 3)$ and $(4, 7)$ **(b)** $(-3, 5)$ and $(2, -1)$

Solution

(a) $(2, 3)$ and $(4, 7)$

$$m = \frac{7-3}{4-2} = 2$$

$\tan\theta = 2$

$\theta = 63°\,26'$

(b) $(-3, 5)$ and $(2, -1)$

$$m = \frac{-1-5}{2+3} = -1.2$$

$\tan\theta = -1.2$

$\theta = 180° - 50°\,12' = 129°\,48'$

Parallel lines

Parallel lines make angles of equal magnitude with the positive direction of the x-axis, i.e. their angle of inclination is the same.

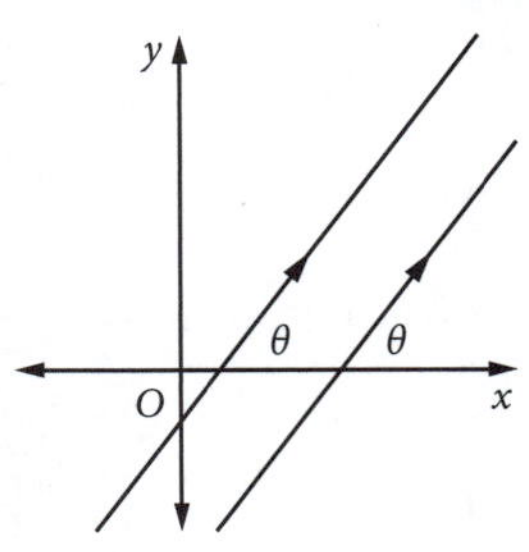

> Two lines are parallel if their gradients are equal, i.e. $m_1 = m_2$.

Perpendicular lines

In the diagram, OP and OQ are perpendicular to each other, with gradients m_1 and m_2 respectively. $\Delta ORP \equiv \Delta QSO$ (SSS).

By definition: $\tan(\angle POR) = \dfrac{b}{a} = m_1$

$$\tan(\angle SOQ) = \frac{a}{-b} = m_2$$

$$m_1 m_2 = \frac{b}{a} \times \frac{a}{-b} = -1$$

i.e. $m_2 = \dfrac{-1}{m_1}$, provided that $m_1, m_2 \neq 0$.

> Two lines are perpendicular if the product of their gradients is -1, i.e. $m_1 m_2 = -1$.
>
> This can also be written as $m_2 = \dfrac{-1}{m_1}$. Each gradient is the negative reciprocal of the other.

The converse of this statement says:

> If $m_1 \times m_2 = -1$, then the lines with gradient m_1 and m_2 are perpendicular.

These two statements may be written as a single statement:

> Two lines with gradient m_1 and m_2 are perpendicular if and only if $m_1 m_2 = -1$.

Example 2

The coordinates of the vertices of a triangle ABC are $(-2, 1)$, $(3, 2)$ and $(4, -3)$ respectively. Show that AB is perpendicular to BC.

Solution

Gradient of $AB = \dfrac{2-1}{3+2} = \dfrac{1}{5}$ Gradient of $BC = \dfrac{-3-2}{4-3} = -5$

Because $\dfrac{1}{5} \times -5 = -1$, AB is perpendicular to BC. ($AB \perp BC$).

Example 3

Given the points $P(1, 3)$, $Q(5, 2)$, $R(6, 6)$, $S(2, 7)$, show that $PQRS$ is a rectangle.

Solution

Gradient of $PQ = \frac{3-2}{1-5} = \frac{-1}{4}$ Gradient of $QR = \frac{2-6}{5-6} = 4$

Gradient of $RS = \frac{6-7}{6-2} = \frac{-1}{4}$ Gradient of $SP = \frac{7-3}{2-1} = 4$

Gradient of PQ = Gradient of RS, so $PQ \parallel RS$

Gradient of QR = Gradient of SP, so $QR \parallel SP$

$\therefore$ $PQRS$ is a parallelogram.

But: Gradient of $PQ \times$ Gradient of $QR = \frac{-1}{4} \times 4 = -1$

$\therefore$ $PQ \perp QR$ so $PQRS$ is a parallelogram with at least one angle of 90° i.e. a rectangle.

MAKING CONNECTIONS

Parallel and perpendicular lines

Move the sliders to explore the relationship between the gradients of two lines.

EXERCISE 5.1 GRADIENT OF A STRAIGHT LINE

1 Find the gradient of the line containing the given points.

(a) $(2, 4), (0, 6)$ **(b)** $(-3, -1), (-5, 6)$ **(c)** $(-2, 2), (-6, 2)$ **(d)** $(2, -3), (-3, 2)$
(e) $(a, b), (b, a)$ **(f)** $(3b, 2c), (2b, -3c)$ **(g)** $(-1, -5), (-4, -3)$ **(h)** $(0, 5), (-8, -7)$

2 Calculate the angle of inclination of the line joining the given points.

(a) $(-4, -2), (4, 6)$ **(b)** $(0, 5), (-2, 4)$ **(c)** $(-5, 6), (3, 3)$
(d) $(4, 5), (-2, -4)$ **(e)** $(2a, b), (2b, a)$ **(f)** $(b, c), (c, b)$

3 Calculate the gradients of the lines joining the points $(-2, 3)$ and $(4, -2)$ and the points $(-1, 7)$ and $(-7, 12)$. These two lines are:

A parallel **B** perpendicular **C** intersecting **D** coincident

4 For each of the following, show that $ABCD$ is a parallelogram.

(a) $A(0, 0), B(3, 0), C(5, 5), D(2, 5)$ **(b)** $A(-3, -1), B(4, 1), C(8, 5), D(1, 3)$
(c) $A(-1, 4), B(4, 6), C(2, 7), D(-3, 5)$ **(d)** $A(-2, -3), B(6, 2), C(8, 7), D(0, 2)$

5 For each of the following, show that $PQRS$ is a trapezium.

(a) $P(-1, 2), Q(3, 4), R(8, -1), S(-4, -7)$ **(b)** $P(-2, 3), Q(3, 7), R(9, -5), S(2, -5)$

6 Show that the points $(-2, 0)$, $(2, 12)$ and $(-5, -9)$ are collinear.

7 For the points $A(2a, b)$, $B(a, 2b)$ and $C(-a, 4b)$, indicate whether each statement is correct or incorrect.

(a) $AB = BC$ **(b)** $AB + BC = AC$ **(c)** $AB \perp BC$ **(d)** A, B and C are collinear

8 For each of the following, show that ABC is a right-angled triangle.

(a) $A(2, -3), B(5, 2), C(-3, 0)$ **(b)** $A(-1, 2), B(3, 4), C(7, -4)$

9 For each of the following, show that $PQRS$ is a rectangle.

(a) $P(1, 4), Q(5, 2), R(2, -4), S(-2, -2)$ **(b)** $P(4, -7), Q(8, -4), R(2, 4), S(-2, 1)$

10 $A(-8, 6)$, $B(2, 4)$, $C(5, -7)$ and $D(-5, -3)$ are the vertices of a quadrilateral. Prove that the diagonals of the quadrilateral are perpendicular.

5.2 EQUATION OF A STRAIGHT LINE

Point-gradient form

Given the gradient of a straight line and a point on the line, you can find the equation of the line.

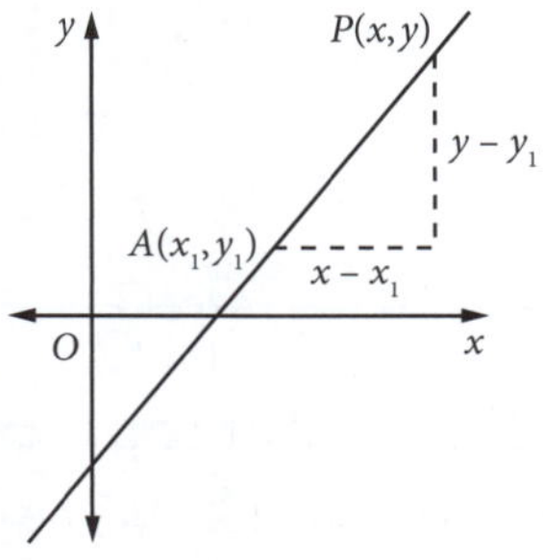

Let $A(x_1, y_1)$ be a given point and $P(x, y)$ be any point on a line with gradient m.

Thus: $\dfrac{y - y_1}{x - x_1} = m$

$\therefore y - y_1 = m(x - x_1)$

The equation of the line AP must be: $y - y_1 = m(x - x_1)$

This is known as the **point-gradient form**, because it is obtained using a point on the line and its gradient.

Example 4

Find the equation of the straight line through the point $(2, 3)$ with a gradient of $\frac{3}{4}$.

Solution

Use the point-gradient form: $y - y_1 = m(x - x_1)$

Substitute information: $y - 3 = \frac{3}{4}(x - 2)$

Multiply through by 4: $4(y - 3) = 3(x - 2)$

$4y - 12 = 3x - 6$

$3x - 4y + 6 = 0$ is the equation of the line.

Two-point form

Given the coordinates of two points, you can find the equation of the line that passes through those points.

Let $A(x_1, y_1)$ and $B(x_2, y_2)$ be the given points and $P(x, y)$ be any point on the line.

Gradient of $AB = \dfrac{y_2 - y_1}{x_2 - x_1}$ Gradient of $AP = \dfrac{y - y_1}{x - x_1}$

Because these gradients are the same, you have:

$$\frac{y - y_1}{x - x_1} = \frac{y_2 - y_1}{x_2 - x_1}$$

This must be the equation of the line AB. This equation can also be written as:

$$\frac{y - y_1}{y_2 - y_1} = \frac{x - x_1}{x_2 - x_1} \quad \text{or} \quad y - y_1 = \frac{y_2 - y_1}{x_2 - x_1}(x - x_1)$$

You can use the form that you find the easiest to remember. This is known as the **two-point form**, because it is obtained using two points on the line.

Example 5

Find the equation of the straight line through the points $(-3, 4)$ and $(2, -6)$.

Solution

Use the two-point form: $\dfrac{y - y_1}{y_2 - y_1} = \dfrac{x - x_1}{x_2 - x_1}$

Points are $(-3, 4), (2, -6)$:

$$\frac{y-4}{-6-4} = \frac{x-(-3)}{2-(-3)}$$

$$\frac{y-4}{-10} = \frac{x+3}{5}$$

$$y-4 = -2(x+3)$$

$$y-4 = -2x-6$$

$2x + y + 2 = 0$ is the equation of the line.

Gradient-intercept form of the straight line

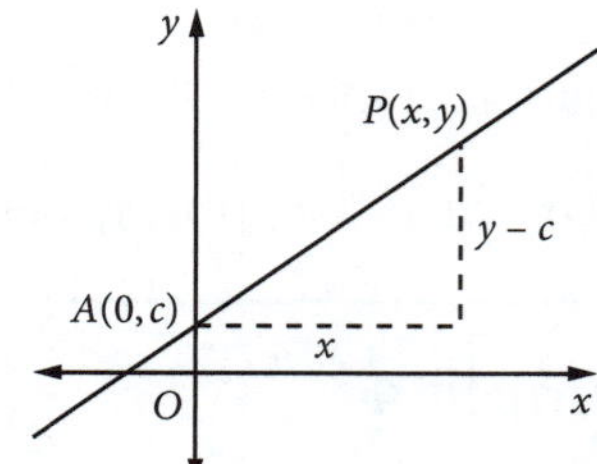

Let the fixed point $A(0, c)$ be the point where the line AP cuts the y-axis.
Let $P(x, y)$ be any other point on the line.

Thus: $m = \dfrac{y-c}{x-0}$

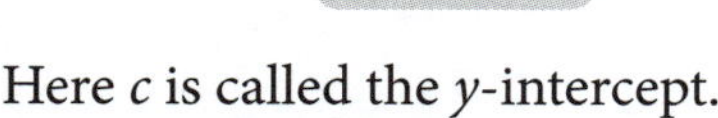

$\therefore$ $y = mx + c$

Here c is called the y-intercept.

Example 6

For the line of the equation $3x + 2y - 5 = 0$, find:

(a) the gradient **(b)** the angle at which the line crosses the x-axis **(c)** the y-intercept.

Solution

(a) $3x + 2y - 5 = 0$

$$2y = -3x + 5$$

$$y = -\frac{3}{2}x + \frac{5}{2}$$

$$m = -\frac{3}{2} = -1.5$$

(b) $\tan\theta = -1.5$

$$\theta = 180° - 56°\,19'$$

$$\theta = 123°\,41'$$

(c) Using the form $y = mx + c$ gives the y-intercept as:

$$c = \frac{5}{2} = 2.5$$

When $c = 0$, the line $y = mx + c$ becomes $y = mx$ and passes through the origin.

If $x > 0$ and $y > 0$, $y = mx$ can be written as $\dfrac{y}{x} = m$. This is an example of direct variation as m is a constant.

As x increases then y increases at the same rate.

As y increases then x increases at the same rate.

This rate of increase is m. It is interesting to note that this rate is also the gradient of the straight line.

MAKING CONNECTIONS

Equation of a straight line

Move the sliders to view the effect of changing m and c on the graph $y = mx + c$.

Line parallel to the x-axis

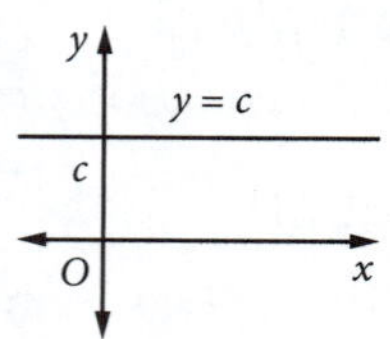

Any line parallel to the x-axis has zero gradient, or $m = 0$.

The gradient-intercept form of the straight line gives:

$y = 0x + c$ or $y = c$

Thus $y = c$ is the equation of the straight line parallel to the x-axis passing through the point $(0, c)$.

Line parallel to the *y*-axis

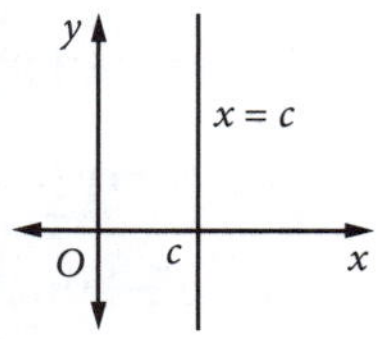

Any line parallel to the y-axis has a gradient that is undefined.

Any line parallel to the y-axis is a vertical line, so its equation must be of the form $x = c$.

Another way of looking at this situation is to recall that any fraction with a zero denominator is undefined. If the denominator in $m = \frac{y - y_1}{x - x_1}$ is zero, so that $x - x_1 = 0$, then the gradient is undefined.

This gives $x = x_1$ as the equation of the line through the point (x_1, y_1), which means that x is a constant. This can be written as $x = c$.

General form of the straight line equation

The equation of a straight line is a 'first degree' or linear equation. It can be written in many forms, but it is called the **general form** when written as $ax + by + c = 0$, where a, b and c are integers:

> $ax + by + c = 0$
>
> a is the coefficient of x, b is the coefficient of y and c is the constant term.

Rewriting the general form in the gradient-intercept form:

$$y = \frac{-a}{b}x - \frac{c}{b}, \qquad b \neq 0$$

Hence the gradient $= -\frac{a}{b}$ and the y-intercept $= -\frac{c}{b}$.

Useful results:

1. If $a = 0$, the line is parallel to the x-axis.
2. If $b = 0$, the line is parallel to the y-axis.
3. If $c = 0$, the line passes through the origin.

Parallel and perpendicular lines

Consider the lines with equations:

$$ax + by + c = 0 \qquad [1]$$

$$a_1x + b_1y + c_1 = 0 \qquad [2]$$

Writing these in gradient-intercept form:

$$y = \frac{-a}{b}x - \frac{c}{b} \qquad [1]$$

$$y = \frac{-a_1}{b_1}x - \frac{c_1}{b_1} \qquad [2]$$

These lines have the same gradient when $\frac{a}{b} = \frac{a_1}{b_1}$, so the lines are parallel when $\frac{a}{b} = \frac{a_1}{b_1}$. The product of the gradients is -1 when $\frac{a}{b} \times \frac{a_1}{b_1} = -1$ or $aa_1 = -bb_1$, so the lines are perpendicular when $aa_1 = -bb_1$.

Useful results:

1. The equation of a line parallel to $ax + by + c = 0$ is $ax + by + d = 0$, where $c \neq d$.
2. The equation of a line perpendicular to $ax + by + c = 0$ is $bx - ay + d = 0$.

Example 7

Find the equation of the line through the point $(2, -3)$ that is:

(a) parallel to the line with equation $3x + 4y - 5 = 0$

(b) perpendicular to the line with equation $3x + 4y - 5 = 0$.

Solution

For the line equation $3x + 4y - 5 = 0$, $a = 3$ and $b = 4$.

(a) **Method 1:**

Parallel lines, so $m = -\frac{3}{4}$; point is $(2, -3)$.

$\therefore$ equation is: $y + 3 = -\frac{3}{4}(x - 2)$

$$4y + 12 = -3x + 6$$

$$3x + 4y + 6 = 0$$

Method 2:

The equation of the line parallel to $3x + 4y - 5 = 0$ is of the form $3x + 4y + c = 0$.

Substitute the values of the point $(2, -3)$ to find c:

$$6 - 12 + c = 0$$

$$c = 6$$

$\therefore$ equation is $3x + 4y + 6 = 0$

(b) **Method 1:**

Perpendicular lines, so $m = -1 \div -\frac{3}{4} = \frac{4}{3}$; point is $(2, -3)$.

$\therefore$ equation is: $y + 3 = \frac{4}{3}(x - 2)$

$$3y + 9 = 4x - 8$$

$$4x - 3y - 17 = 0$$

Method 2:

The equation of the line perpendicular to $3x + 4y - 5 = 0$ is of the form $4x + 3y + c = 0$.

Substitute the values of the point $(2, -3)$ to find c:

$$8 + 9 + c = 0$$

$$c = -17$$

$\therefore$ equation is $4x + 3y - 17 = 0$

EXERCISE 5.2 EQUATION OF A STRAIGHT LINE

1 Find the equation of the straight line with:

(a) gradient $\frac{3}{4}$, passing through $(-6, 5)$ (b) gradient $-\frac{1}{2}$, passing through $(4, -3)$

2 Find the equation of the straight line passing through:

(a) $(3, 3)$ and $(-4, -5)$ (b) $(2, -8)$ and $(7, 2)$

3 Find the equation of the straight line passing through:

(a) $(6, 6)$ with an angle of inclination of $45°$

(b) $(-2, 3)$ with an angle of inclination of $53°8'$ $\left(\tan 53°8' \approx \frac{4}{3}\right)$

(c) $(-5, -2)$ with an angle of inclination of $135°$

(d) $(-7, 4)$ with an angle of inclination of $143°8'$ $\left(\tan 143°8' \approx -\frac{3}{4}\right)$

4 Find the equation of the straight line parallel to the x-axis and passing through the point $(5, 2)$.

5 Find the equation of the straight line parallel to the y-axis and passing through the point $(-2, -4)$.

6 The equation of the straight line with x-intercept 2 and y-intercept -5 is:

A $2x - 5y - 10 = 0$ B $5x - 2y - 10 = 0$ C $2x - 5y + 10 = 0$ D $5x - 2y + 10 = 0$

7 Find the equation of the straight line with x-intercept -3 and y-intercept -2.

8 Write each equation in the form $y = mx + c$ and find the gradient of each line.

(a) $2x + 3y = 4$ (b) $3x - 2y = 7$ (c) $2y = 6 - 3x$ (d) $5y - 2x = 8$

9 Indicate whether each statement is correct or incorrect for the line $2x + 3y - 12 = 0$.

(a) $m = -\frac{2}{3}$ (b) x-intercept $= 6$ (c) y-intercept $= -4$ (d) passes through $(3, 2)$

10 (a) Show that $(2,3)$ is on the line $2x+3y-13=0$.
(b) If $(-1,2)$ is on the line $ax-4y+11=0$, find a.

11 Find the equation of the line containing the point $(2,-3)$ that is:
(a) parallel to the line $3x+2y-6=0$ (b) perpendicular to the line $3x+2y-6=0$

12 Find the equation of the line passing through the origin that is:
(a) parallel to the line $4x-5y+3=0$ (b) perpendicular to the line $4x-5y+3=0$

13 The coordinates of two points A and B are $(0,-2)$ and $(3,0)$ respectively. The x-coordinate of a point C on the line AB is 6. Find:
(a) the equation of AB (b) the angle of inclination of AB
(c) the y-coordinate of C (d) the equation of the line through C that is perpendicular to AB.

14 Show that the line with equation $2x-y=5$ is parallel to the line joining the points $(-1,5)$ and $(1,9)$.

15 $ABCD$ is a parallelogram. The coordinates of A, B and C are $(-1,4)$, $(4,6)$ and $(2,7)$ respectively. Find:
(a) the equation of CD (b) the equation of AD (c) the coordinates of D.

16 $ABCD$ is a rectangle. The coordinates of A and B are $(1,4)$ and $(5,2)$ respectively. The x-coordinate of D is -2. Find:
(a) the equation of AB (b) the equation of AD (c) the y-coordinate of D
(d) the equation of BC (e) the equation of DC (f) the coordinates of C.

17 $A(0,0)$, $B(2,1)$ and $C(1,5)$ are the vertices of a triangle ABC. The triangle is rotated anticlockwise about the point A, through a right angle in the x–y plane. Find the equation of the image lines of:
(a) the line AC (b) the line AB (c) the line BC.

5.3 INTERSECTION OF TWO LINES

The point of intersection of two lines is found by solving the equations of the lines simultaneously.

You will look at finding the equation of a line that passes through the intersection point of two other lines and also satisfies one other condition. You will learn two different methods of doing this.

Example 8

Find the equation of the line through the point $A(4,-2)$ that also passes through the intersection point B of the lines with equations $4x+2y+2=0$ and $3x+5y-9=0$.

Solution

Method 1:

To find B, solve simultaneously:

$$4x+2y+2=0 \quad [1]$$
$$3x+5y-9=0 \quad [2]$$

$3\times[1]$: $\quad 12x+6y+6=0 \quad [3]$

$4\times[2]$: $\quad 12x+20y-36=0 \quad [4]$

$[3]-[4]$: $\quad -14y+42=0$

$$y=3$$

Substitute into [1]: $\quad 4x+6+2=0$

$$x=-2$$

∴ Intersection point solution: $B(-2,3)$

Gradient of AB: $m = \frac{3+2}{-2-4} = -\frac{5}{6}$

Equation of AB: $y-(-2) = -\frac{5}{6}(x-4)$

$$6(y+2) = -5(x-4)$$

$$6y+12 = -5x+20$$

The equation of AB is $5x+6y-8=0$.

Method 2:

Any line passing through the intersection point of $4x+2y+2=0$ and $3x+5y-9=0$ has the equation:

$$4x+2y+2+k(3x+5y-9)=0$$

(You can check this statement by substituting some values for k and then seeing if the point B, found above, satisfies the new equation.)

The line $4x+2y+2+k(3x+5y-9)=0$ is to pass through the point $A(4,-2)$, so substitute the coordinate values of A to find k: $16-4+2+k(12-10-9)=0$

$$14-7k=0$$

$$k=2$$

Substitute $k=2$ into the line equation: $4x+2y+2+2(3x+5y-9)=0$

$$10x+12y-16=0$$

The equation of AB is $5x+6y-8=0$.

Two straight lines—only three possibilities

Two straight lines on the number plane must either:

(i) intersect at a point or **(ii)** be parallel or **(iii)** coincide.

This means that when you solve a pair of simultaneous linear equations, you will always get either:

(i) a unique solution (as the lines intersect)

or **(ii)** no solution (as the lines are parallel)

or **(iii)** an infinite number of solutions (as the lines coincide).

You can determine which of these three situations is the case without actually solving the equations.

For the pair of equations $a_1x+b_1y+c_1=0$ and $a_2x+b_2y+c_2=0$, you have:

(i) intersecting lines if $\frac{a_1}{a_2} \neq \frac{b_1}{b_2}$ (the gradients are not equal)

(ii) parallel lines if $\frac{a_1}{a_2} = \frac{b_1}{b_2} \neq \frac{c_1}{c_2}$ (the gradients are equal, but the lines are different)

(iii) coincident lines if $\frac{a_1}{a_2} = \frac{b_1}{b_2} = \frac{c_1}{c_2}$ (the equations are equivalent).

Example 9

Consider the simultaneous equations. Discuss their solution.

(a) $3x+2y-5=0$
$6x+4y-16=0$

(b) $3x+2y-5=0$
$6x+4y-10=0$

Solution

(a) The two lines have the same gradient, $-\frac{6}{4} = -\frac{3}{2}$, but $\frac{3}{6} = \frac{2}{4} \neq \frac{5}{16}$, so the lines are parallel (not coincident) and do not intersect. Therefore there is no solution to the pair of equations.

Algebraically:	$3x + 2y - 5 = 0$	[1]
	$6x + 4y - 16 = 0$	[2]
$2 \times [1]$:	$6x + 4y - 10 = 0$	[3]
$[2] - [3]$:	$-6 = 0$	

This is obviously false, so there is no solution to the pair of simultaneous equations.

(b) The two lines have the same gradient, $-\frac{6}{4} = -\frac{3}{2}$, and $\frac{3}{6} = \frac{2}{4} = \frac{5}{10}$, so the lines are coincident. Therefore the lines intersect at all points on the line $3x + 2y - 5 = 0$. Some of the points of intersection include $(1,1)$, $(0,2.5)$, $(-1,4)$.

EXPLORE FURTHER

Two straight lines—only three possibilities

Use graphing software to explore the possibilities for two straight lines on the number plane.

EXERCISE 5.3 INTERSECTION OF TWO LINES

1 The coordinates of the intersection point of the lines $x + 2y - 3 = 0$ and $2x - 2y - 6 = 0$ are:

A $(0,3)$ **B** $(0,-3)$ **C** $(3,0)$ **D** $(-3,0)$

2 Find the equation of the line that contains the intersection point of the lines $2x + 5y - 19 = 0$ and $3x - 4y + 6 = 0$ and is parallel to the line with equation $4x - y - 8 = 0$.

3 The coordinates of the vertices A, B and C of a triangle are $(-1,3)$, $(2,5)$ and $(1,-1)$ respectively. Find:

(a) the equation of the perpendicular line from A to BC
(b) the equation of the line through B, parallel to this perpendicular line.

4 Consider the equations $x + y + 1 = 0$, $y = 2 - x$, $3y = 2x + 1$ and $2x - 3y + 6 = 0$.

(a) Show that these equations represent the sides of a parallelogram.
(b) Find the coordinates of the vertices of the parallelogram.
(c) Find the equations of the diagonals of the parallelogram.

5 Find the equation of the straight line that contains the intersection point of the lines $3x + 2y - 12 = 0$ and $5x - y - 7 = 0$ and that:

(a) passes through the point $(-4,-5)$ **(b)** is parallel to the line $2x - y + 4 = 0$
(c) is perpendicular to the line $y = 5$.

6 $ABCD$ is a parallelogram. The coordinates of A, B and C are $(-1,4)$, $(4,6)$ and $(2,7)$ respectively. Indicate whether each statement is correct or incorrect.

(a) AD is $x + 2y - 7 = 0$ **(b)** CD is $2x - 5y + 31 = 0$
(c) gradient of $AB = 0.5$ **(d)** D is $(-3,5)$

7 $ABCD$ is a quadrilateral. The coordinates of A, B and C are $(-8,6)$, $(2,4)$ and $(5,-7)$ respectively. If the diagonals are perpendicular and DC is parallel to the x-axis, find:

(a) the coordinates of D **(b)** the coordinates of the intersection point of the diagonals.

8 The equations of the sides of a triangle ABC are as follows:

AB: $5x + y - 10 = 0$; BC: $3x - 2y - 6 = 0$; CA: $x - 5y + 24 = 0$

(a) Show that angle BAC is a right angle.

(b) Find the coordinates of the foot of the perpendicular from A to BC.

9 Without actually solving the simultaneous equations, state whether the following pairs of lines intersect, are parallel or coincide.

(a) $2x - 3y - 8 = 0$
$4x - 6y - 16 = 0$

(b) $x + 3y + 7 = 0$
$2x + 7y + 16 = 0$

(c) $6x - 5y - 24 = 0$
$9x - 4y - 22 = 0$

(d) $x + y - 7 = 0$
$x + y - 8 = 0$

5.4 SIMULTANEOUS EQUATIONS

A **linear equation** in two variables x and y is an equation in which both the pronumerals are of the **first degree**, e.g. $x + y = 6$. There are an infinite number of values for x and y that satisfy an equation like this, e.g. $(0, 6)$, $(1, 5)$, $(2, 4)$, $(4, 2)$, $(-8, 14)$. Graphically, the points that satisfy a linear equation all lie on the same straight line.

Consider another equation $x - y = 2$, for which the graph is also a straight line containing an infinite number of points, e.g. $(0, -2)$, $(1, -1)$, $(2, 0)$, $(4, 2)$, $(-10, -12)$.

From the two lists of points, you can see that the ordered pair $(4, 2)$ satisfies both equations. Hence $x = 4$, $y = 2$ is said to be the solution of the simultaneous linear equations $x + y = 6$ and $x - y = 2$. Graphically, $x = 4$, $y = 2$ are the coordinates of the point of intersection of the two lines.

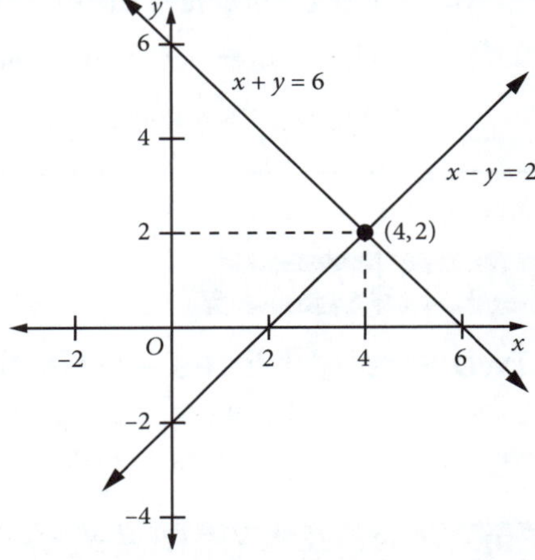

Algebraic solution of simultaneous equations

The most common algebraic methods of solving a simultaneous pair of linear equations in two variables are **elimination** and **substitution**.

In the elimination method, you can eliminate one of the variables x or y by adding or subtracting the equations, as long as the coefficients of one of the variables are equal in both equations.

> Solving simultaneous equations by elimination:
> - If the coefficients have the same size but opposite sign, eliminate by addition.
> - If the coefficients are equal, eliminate by subtraction.

In the substitution method, rewrite one of the equations to make it equal to one of the variables, then substitute this into the other equation.

Example 10

Solve the simultaneous equations $x + y = 6$ and $x - y = 2$.

Solution

$x + y = 6$ [1]

$x - y = 2$ [2]

Elimination method

The coefficients of y have the same size but opposite sign, so eliminate y by adding the equations [1] and [2].

[1] + [2]: $2x = 8$

$x = 4$

Substitute $x = 4$ into [1]: $4 + y = 6$

$y = 2$

Check this solution by substituting it into the other equation, [2].

LHS = 4 − 2 = 2 = RHS

∴ Solution is $x = 4, y = 2$.

Substitution method

Rewrite [1] in the form $y = \ldots$ to make a new equation, then substitute into [2].

Rewrite [1]: $y = 6 - x$ [3]

Substitute [3] into [2]: $x - (6 - x) = 2$

$x - 6 + x = 2$

$2x = 8$

$x = 4$

Substitute $x = 4$ into [3]: $y = 6 - 4 = 2$

∴ Solution is $x = 4, y = 2$.

Example 11

Solve the simultaneous equations $3x + 2y = 10$ and $4x + 3y = 13$.

Solution

$3x + 2y = 10$ [1]

$4x + 3y = 13$ [2]

Elimination method

To use the elimination method, make the coefficients of y both the same by multiplying [1] by three and multiplying [2] by two.

3 × [1]: $9x + 6y = 30$ [3]

2 × [2]: $8x + 6y = 26$ [4]

[3] − [4]: $x = 4$

Substitute $x = 4$ into [1]: $12 + 2y = 10$

$2y = -2$

$y = -1$

∴ Solution is $x = 4, y = -1$.

Substitution method

Rewrite [1]: $y = \dfrac{10 - 3x}{2}$ [3]

Substitute [3] into [2]: $4x + \dfrac{3(10 - 3x)}{2} = 13$

$8x + 30 - 9x = 26$

$-x = -4$

$x = 4$

Substitute $x = 4$ into [3]: $y = \dfrac{10 - 12}{2} = \dfrac{-2}{2} = -1$

∴ Solution is $x = 4, y = -1$.

Example 12

Solve the simultaneous equations $\dfrac{x}{6} + \dfrac{y}{4} = 1$ and $\dfrac{3x}{4} - \dfrac{x - y}{2} = \dfrac{7}{4}$.

Solution

First, it is best to get rid of the fractions in each equation by multiplying each term by the LCM (lowest common multiple) of the denominators.

$\dfrac{x}{6} + \dfrac{y}{4} = 1$, LCM of 6 and 4 is 12: $2x + 3y = 12$ [1]

$\dfrac{3x}{4} - \dfrac{x - y}{2} = \dfrac{7}{4}$, LCM of 4 and 2 is 4: $3x - 2(x - y) = 7$

$x + 2y = 7$ [2]

$2\times[2]$:	$2x+4y=14$	[3]
$[1]-[3]$:	$-y=-2$	
	$y=2$	
Substitute into [2]:	$x+4=7$	
	$x=3$	

$\therefore$ Solution is $x=3$, $y=2$.

Check your solutions by substituting them back into the original equations.

EXPLORE FURTHER

Solving simultaneous equations

Use technology to solve simultaneous equations graphically.

EXERCISE 5.4 SIMULTANEOUS EQUATIONS

Solve the simultaneous equations in this exercise. (Try to solve some using graphing software.)

1 $x+7y=5$
$x-7y=-9$

2 $x+5y=34$
$x-5y=-6$

3 $4x-5y=30$
$4x-2y=24$

4 $3x-y=5$
$5x+3y=-8$

5 $2m+3n=-4$
$3m+2n=-6$

6 $-2x+7y=4$
$-3x+5y=-5$

7 $x+5y=-13$
$2x-y=7$

8 $5x+2y=9$
$9x-7y=-5$

9 The solution to the simultaneous equations $2a-3b=5$ and $2a-5b=-1$ is:

A $a=-7, b=3$ **B** $a=7, b=-3$ **C** $a=7, b=3$ **D** $a=3, b=7$

10 $2x+5y=16$
$10x-3y=-4$

11 $2x+3y=14$
$4x-5y=17$

12 $5m-6n=12$
$2m+9n=20$

13 $y=4x-2$
$y=-3x+5$

14 $x-4=4(y+2)$
$3(x-2)=2y+20$

15 $3(x-y)+8y=42$
$9x-(x-2y)=78$

16 $2x-\dfrac{y}{4}=5$
$x+\dfrac{3y}{4}=-1$

17 $\dfrac{3a-2b}{2}=9$
$\dfrac{6a-b}{5}=9$

18 $\dfrac{x-3}{2}=\dfrac{2y+1}{3}$
$\dfrac{3x-1}{5}-\dfrac{2y+1}{2}=1$

19 $3(x-y)-8(x+y)=7$
$2(x+y)+5(x-y)=-65$

20 $2(3a-b)=3(a+b)$
$3(a-4b)+46=5a$

21 $5(2x-y)=7x+1$
$3(3x+y)=5(x-y+12)$

5.5 PROBLEM SOLVING WITH SIMULTANEOUS EQUATIONS

Example 13

Three books and five pens cost \$5.55. A book costs 10 cents more than ten pens. Find the cost of a book and a pen.

Solution

Let a book cost x cents and a pen cost y cents.

3 books and 5 pens cost 555 cents:	$3x + 5y = 555$	[1]
1 book costs 10 cents more than 10 pens:	$x = 10y + 10$	[2]
Substitute [2] into [1]:	$3(10y + 10) + 5y = 555$	
	$30y + 30 + 5y = 555$	
	$35y = 525$	
	$y = 15$	
Substitute back into [2]:	$x = 150 + 10$	
	$x = 160$	

Hence, a book costs \$1.60 and a pen costs 15 cents, so a book and a pen together cost \$1.75.

Example 14

$y = mx + c$ is the equation of a straight line that passes through the points $(1, 8)$ and $(-2, -1)$. Form a pair of simultaneous equations in m and c and solve them to find m and b. Find the equation of the line.

Solution

	$y = mx + c$	
At $(1, 8)$:	$8 = m + c$	[1]
At $(-2, -1)$:	$-1 = -2m + c$	[2]
[1] – [2]:	$9 = 3m$	
	$m = 3$	
Substitute into [1]:	$8 = 3 + c$	
	$c = 5$	

The equation of the line is $y = 3x + 5$.

Break-even analysis

One application of simultaneous equations is to find the break-even point in manufacturing and sales.

A business can determine a formula for the cost to produce a given number of items – the cost function. This consists of the fixed costs plus a cost per item produced. They can also determine a revenue function, which is the sale price multiplied by the number of units sold.

Where these two functions intersect is the break-even point. For sales above the break-even point the revenue function is above the cost function so a profit is made. For sales below the break-even point the revenue function is below the cost function and a loss is made.

Example 15

Georgia decides to turn her jewellery-making hobby into a small business. She spends \$120 on equipment and estimates that it costs her \$8 to manufacture each item. She plans to sell the items for \$20 each.

(a) If C is her total cost and R her total revenue, in dollars, set up the cost and revenue functions for the sale of x items.

(b) Determine her break-even point.

(c) Determine her profit if she sells: **(i)** 6 items **(ii)** 50 items.

Solution

(a) Georgia has a fixed cost of \$120 and a variable cost of \$8 per item. If she produces x items, then $C = 8x + 120$.

Her selling price is \$20 per item. The revenue function is $R = 20x$.

(b) At the break-even point, $R = C$.

$20x = 8x + 120$

$12x = 120$

$x = \frac{120}{12}$

$x = 10$

Her break-even point is 10 items.

(c) Profit, P, is given by: $P = R - C$

$$P = 20x - (8x + 120)$$

$$P = 12x - 120$$

(i) $x = 6$: $P = 72 - 120 = -48$
She makes a loss of \$48.

(ii) $x = 50$: $P = 600 - 120 = 480$
She makes a profit of \$480.

The information in this example may be represented graphically, with C and R on the vertical axis and x on the horizontal axis. The point of intersection of the two graphs is the break-even point, which you can see is at $x = 10$.

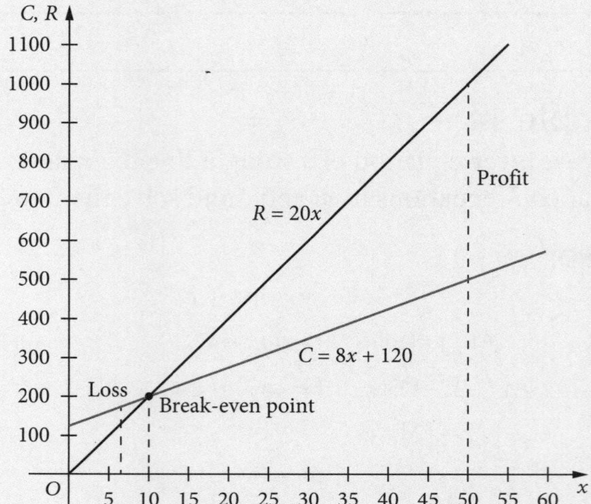

The revenue line is below the cost line when $x = 6$ so a loss occurs at that point. The revenue line is above the cost line when $x = 50$ so a profit occurs at that point.

Georgia incurs a loss when $0 \le x < 10$, breaks even when $x = 10$ and makes a profit when $x > 10$.

The break-even point will change if any of the fixed values in the equation are changed: for example, the fixed cost of production, the cost of producing each unit or the selling price. The effect of these changes can be seen quickly using graphing software.

In this example, if the cost to produce each unit rose to \$9, then the cost function would become $C = 9x + 120$. The following graph shows the change in the break-even point.

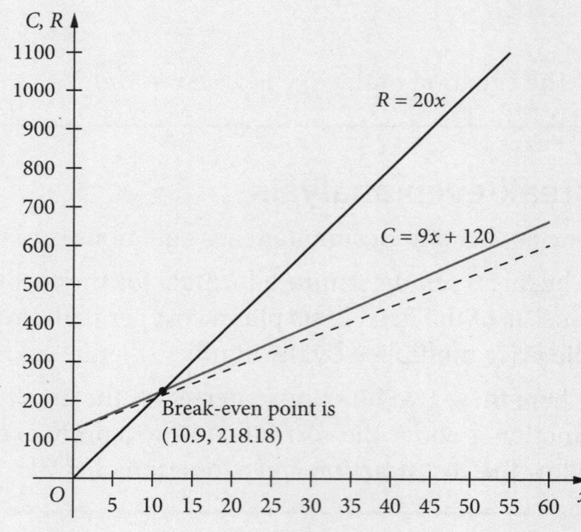

EXERCISE 5.5 PROBLEM SOLVING WITH SIMULTANEOUS EQUATIONS

1 There are 450 students at Newton High School. If there are 50 more boys than girls, how many boys and girls are there at the school?

2 A contractor has 8 trucks. Some trucks carry a load of 10 tonnes and the other trucks carry a load of 5 tonnes. When all 8 trucks are filled, they contain a total load of 70 tonnes. How many of each size of truck does the contractor own?

3 A father is 7 times as old as his daughter. In 5 years' time he will be 4 times as old as his daughter. If he is now x years old and his daughter is y years old, which set of equations correctly describes their ages?

A $x - 7y = 0,\ x - 4y + 5 = 0$ B $x + 7y = 0,\ x - 4y - 15 = 0$
C $x - 7y = 0,\ x - 4y - 15 = 0$ D $x + 7y = 0,\ x - 4y + 5 = 0$

4 John's mother is now 5 times as old as John. Three years ago, she was 9 times his age. What are their ages now?

5 Tickets to a movie cost \$15 for adults and \$12 for children. If 1000 people paid to see a movie and the total money paid was \$13 800, how many adults and how many children were there?

6 The straight line $ax + by = 12$ passes through the points $(2, 2)$ and $(-4, 5)$. Form two equations in a and b, then solve them to find the equation of the line.

7 Let x be the numerator and y be the denominator of a fraction. The denominator is 5 more than the numerator. If 2 is subtracted from both the numerator and the denominator, the denominator is then twice the numerator. What is the fraction?

8 Find two numbers such that if 18 is added to the first number it becomes twice the second number, and if 6 is added to the second number it becomes three times the first number.

9 The weekly wages of 5 carpenters and 3 apprentices total \$5480 while the wages of 3 carpenters and 5 apprentices total \$4440. Find the weekly wages of a carpenter and of an apprentice.

10 The equation of a straight line is given as $y = mx + c$. By forming a pair of simultaneous equations, find the equation of the line if it passes through the points given.

(a) $(4, 1)$ and $(-1, -9)$ **(b)** $(0, 4)$ and $(1, 0)$ **(c)** $(2, -1.5)$ and $(-4, -6)$ **(d)** $(2, 4)$ and $(-6, 8)$

11 For each set of cost and revenue functions **(a)**–**(d)**, find:

(i) the break-even point
(ii) the revenue at the break-even point
(iii) the profit function, P.

(a) $C = 5x + 200,\ R = 15x$ **(b)** $C = 0.5x + 100,\ R = 1.5x$
(c) $C = 15x + 3000,\ R = 45x$ **(d)** $C = 0.3x + 5000,\ R = 1.1x$

12 Julian buys a coffee cart for \$15 000. His repayments work out to be \$80 per day for the first year. He calculates that it will cost him \$2 per cup of coffee for the ingredients. He sells coffee at \$4 per cup and he sells x cups of coffee per day.

(a) Write the cost function C and revenue function R.
(b) What is his break-even point?
(c) Write the profit function P.
(d) What is his profit if he sells 100 cups per day?

13 Maya runs a market stall at the weekends, selling paintings. It costs her \$90 per day for the site. It costs her on average \$4 per painting that she sells and she sells them for an average price of \$10. If she sells x articles each day, find:

(a) the cost function C and revenue function R
(b) her break-even point
(c) the profit she will make if she sells 40 paintings in a day.
(d) One weekend, the weather is fine on the Saturday and rainy on the Sunday. On Saturday Maya sells 30 paintings, but on Sunday she only sells 10 paintings. What profit (or loss) does she make for the weekend?

5.6 SOLVING SIMULTANEOUS EQUATIONS—LINEAR AND SECOND DEGREE

Equations like $y = x^2 - 5x + 6$ (a parabola when graphed), $x^2 + y^2 = 4$ (a circle when graphed) and $xy = 6$ (a rectangular hyperbola when graphed) may be intersected by a straight line. To find the intersection points, you can solve the pair of simultaneous equations.

For non-linear simultaneous equations like these, the substitution method is usually the best. You might also use graphing software to find the solutions graphically.

Example 16

Solve the simultaneous equations $y = x^2$ and $y = x + 2$.

Solution

Graphically:

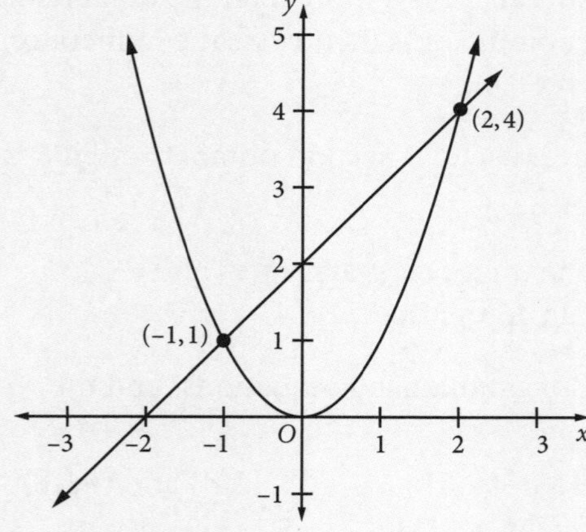

Algebraically:

$y = x^2$ [1]

$y = x + 2$ [2]

Substitute [2] into [1]: $x + 2 = x^2$

$x^2 - x - 2 = 0$

$(x - 2)(x + 1) = 0$

$x = 2 \quad \text{or} \quad -1$

Substitute into [2]: $y = 4 \quad \text{or} \quad 1$

$\therefore$ Solutions are $x = 2, y = 4$ or $x = -1, y = 1$.

Example 17

Solve the simultaneous equations $x^2 + y^2 = 25$ and $x + y = 5$.

Solution

Graphically:

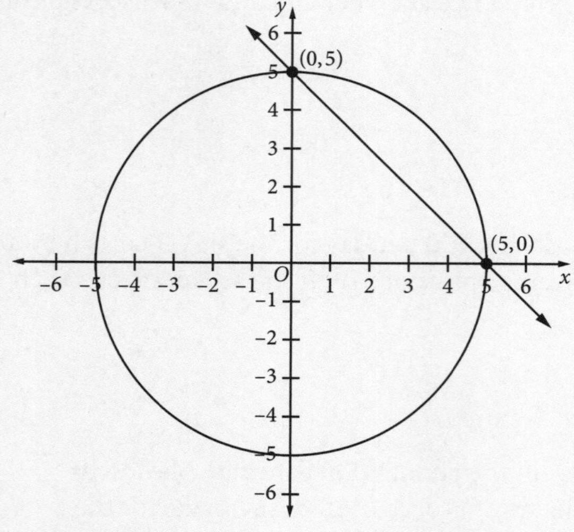

Algebraically:

$x^2 + y^2 = 25$ [1]

$x + y = 5$ [2]

Rewrite [2]: $y = 5 - x$ [3]

Substitute [3] into [1]: $x^2 + (5 - x)^2 = 25$

$x^2 + 25 - 10x + x^2 = 25$

$2x^2 - 10x = 0$

$2x(x - 5) = 0$

$x = 0, x = 5$

Substitute into [3]: $y = 5, y = 0$

$\therefore$ Solutions are $x = 0, y = 5$ or $x = 5, y = 0$.

EXERCISE 5.6 SOLVING SIMULTANEOUS EQUATIONS—LINEAR AND SECOND DEGREE

Solve the following simultaneous equations. (Try to solve some using graphing software.)

1 $y = 5x + 6$
$y = x^2$

2 $y = 3x - 2$
$y = x^2$

3 $y = x + 5$
$y = x^2 - 3x$

4 $x + y = 15$
$y = x^2 - 6x + 1$

5 $y = x - 3$
$xy = 10$

6 $y - 2x = 1$
$x^2 + y^2 = 10$

7 $x + y = 5$
$x^2 + y^2 = 13$

8 $2x - y = 2$
$y = x^2 - x - 2$

9 $x - y = 1$
$xy = 2$

10 $y = 2x - 5$
$y = x^2 - 4x + 4$

11 $y = 2x - 6$
$x^2 - xy + 2y^2 = 16$

12 $x + y = 5$
$3x^2 + xy - y^2 = 29$

13 $y - 2x + 1 = 0$
$3y^2 - y - 2x^2 = 0$

14 $y - 4x - 8 = 0$
$y = 4 - x^2$

15 $x - y + 3 = 0$
$xy = 10$

16 $3x + y = 11$
$2x^2 - xy - y = 10$

17 $x + y = 2$
$x^2 + y^2 = 2$

18 $x = 2y - 1$
$3x^2 = x + 2y^2$

19 $x + 2y = -8$
$xy = 8$

20 $y = x + 9$
$y = x^2 - x - 6$

5.7 SOLVING SIMULTANEOUS EQUATIONS—LINEAR AND SECOND DEGREE IN THE GENERAL FORM

Example 18

Solve the simultaneous equations $3x + 2y = 4$ and $x^2 + xy - y^2 = 1$.

Solution

$$3x + 2y = 4 \qquad [1]$$

$$x^2 + xy - y^2 = 1 \qquad [2]$$

Rewrite [1]: $2y = 4 - 3x$

$$y = \frac{4 - 3x}{2} \qquad [3]$$

Substitute [3] into [2]: $x^2 + \frac{x(4 - 3x)}{2} - \frac{(4 - 3x)^2}{4} = 1$

Multiply by 4: $4x^2 + 8x - 6x^2 - (16 - 24x + 9x^2) = 4$

$$8x - 2x^2 - 16 + 24x - 9x^2 = 4$$

$$11x^2 - 32x + 20 = 0$$

$$(11x - 10)(x - 2) = 0$$

$$x = \frac{10}{11} \text{ or } 2$$

Substitute into [3]: $y = \frac{4 - \frac{30}{11}}{2} = \frac{44 - 30}{22} = \frac{7}{11}$ or $y = \frac{4 - 6}{2} = -1$

When rewriting equation [1], you could have instead written $x = \frac{4 - 2y}{3}$ and substituted for x in equation [2].

EXERCISE 5.7 SOLVING SIMULTANEOUS EQUATIONS— LINEAR AND SECOND DEGREE IN THE GENERAL FORM

Solve the following simultaneous equations. (Try to solve some using graphing software.)

1 $2x - 5y = 3$
$x^2 - 2y^2 - 3x = 2$

2 $3y - 4x = 0$
$x^2 + y^2 = 25$

3 $x + 2y = 3$
$xy + 2x + y = 4$

4 $3x - 2y = 2$
$x^2 - xy + y^2 = 21$

5 $3x + 2y = 1$
$xy + y^2 - y = 8$

6 $2x = 3y + 1$
$xy + x + y = 23$

5.8 QUADRATIC FUNCTIONS

Definitions

$ax^2 + bx + c$, a polynomial of the second degree, is called a quadratic polynomial or a quadratic expression.

A function defined by the rule $y = ax^2 + bx + c$, where a, b and c are constants, $a \neq 0$, is called a **quadratic function**. The domain of this function is the set of real numbers (unless otherwise stated or implied).

When $y = 0$ the quadratic function becomes the quadratic equation $ax^2 + bx + c = 0$. The values of x that satisfy this equation are called the roots of the equation. They are also called the zeros of the quadratic polynomial, as they are the values that make the polynomial zero.

Example 19

Give the domain of each function.

(a) $y = 2x^2 - 3x - 5$ (b) $y = t^2 + 2t$ (c) $y = 4p^2 - 8$

(d) $A = s^2$, where s is the side length of a square

Solution

(a) Real numbers for x (b) Real numbers for t (c) Real numbers for p

(d) Positive real numbers for s (because side length cannot be negative or zero)

Graph of a quadratic function

The graph of a quadratic function has the characteristic shape of a **parabola**. The parabola has a turning point at its vertex, where the function has a minimum value if $a > 0$ or a maximum value if $a < 0$.

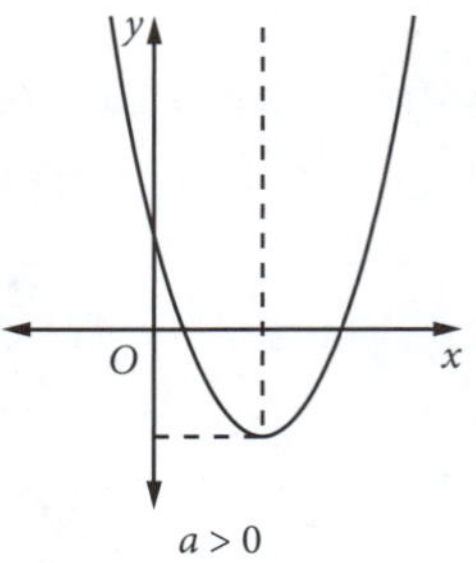

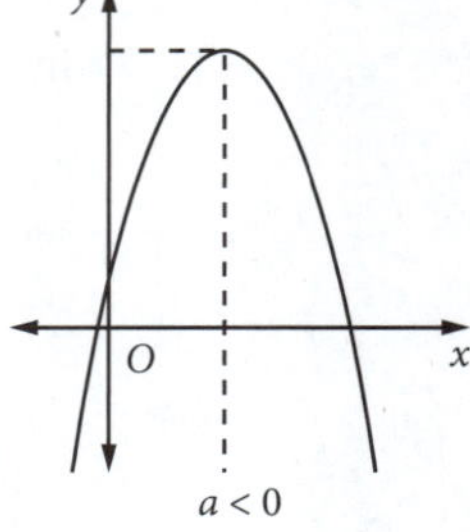

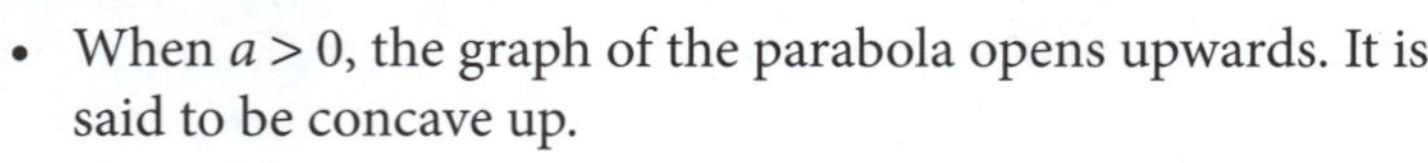

- When $a > 0$, the graph of the parabola opens upwards. It is said to be concave up.
- When $a < 0$, the graph of the parabola opens downwards. It is said to be concave down.

Thus a straight line has no concavity.

- On either side of the minimum turning point of a parabola the curve is concave up.
- On either side of the maximum turning point of a parabola the curve is concave down.

The graph of the quadratic function $y = ax^2 + bx + c$ will cut the y-axis when $x = 0$. Thus the y-intercept is c.

The graph of the quadratic function $y = ax^2 + bx + c$ will cut the x-axis when $y = 0$. Thus the x-intercepts are the roots of the quadratic equation $ax^2 + bx + c = 0$, for which there may be none, one or two roots. The number of roots can be predicted using the **discriminant**, which is explained in the next section.

Maximum or minimum value of a quadratic function

All quadratic polynomials can be expressed in the form $a(x+B)^2+C$, where a, B and C are constants, $a \neq 0$.

In this form, you are effectively completing the square on x. Because $(x+B)^2$ is a perfect square, it is non-negative for all real values of x, i.e. $(x+B)^2 \geq 0$. Thus the smallest possible value of $(x+B)^2$ is zero, when $x=-B$.

(i) If $a>0$, then the **minimum** value of $a(x+B)^2+C$ is C.

(ii) If $a<0$, then the **maximum** value of $a(x+B)^2+C$ is C.

Example 20

Express $2x^2-4x-5$ in the form $a(x+B)^2+C$ and hence state its minimum value.

Solution

Expression $= 2x^2-4x-5$

Take out the factor of 2 from terms in x: $= 2(x^2-2x)-5$

Complete the square on x: $= 2(x^2-2x+1)-5-2\times 1$

$= 2(x-1)^2-7$

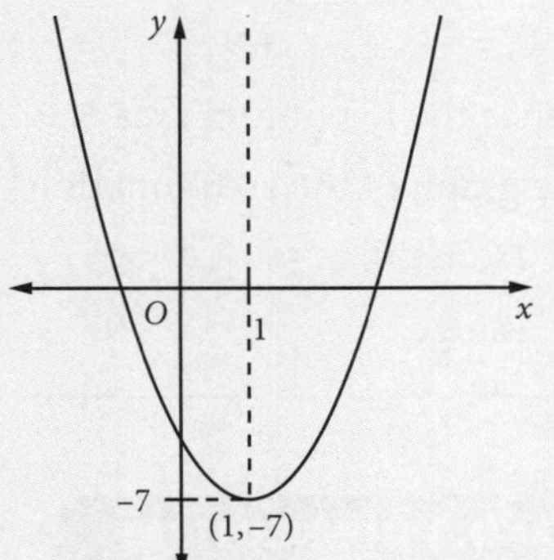

Hence the minimum value of $2x^2-4x-5$ is -7, when $x=1$.

The graph of the function $y=2x^2-4x-5$ is shown in the diagram.

The range of the function is $y \geq -7$.

The graph is symmetrical about the line $x=1$, which is the axis of symmetry of a parabola that has its vertex at $(1,-7)$.

Example 21

Find the maximum value of $-3x^2-12x-7$ by the method of completing the square.

Solution

Expression $= -3x^2-12x-7$

Take out the factor of 3 from terms in x: $= -3(x^2+4x)-7$

Complete the square on x: $= -3(x^2+4x+4)-7+3\times 4$

$= -3(x+2)^2+5$

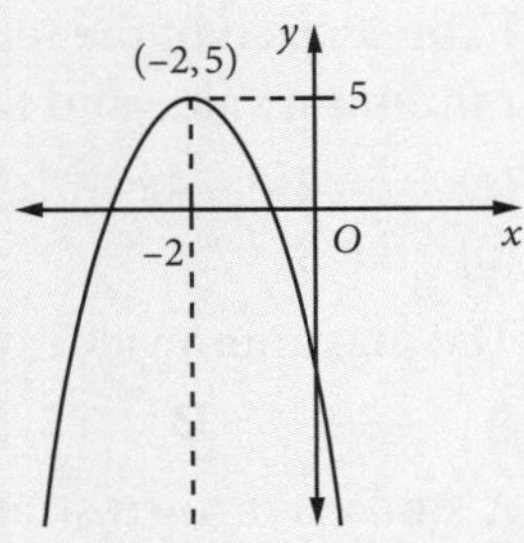

Hence the maximum value of $-3x^2-12x-7$ is 5 when $x=-2$.

The graph of the function $y=-3x^2-12x-7$ is shown in the diagram.

The range of the function is $y \leq -5$.

The graph is symmetrical about the line $x=-2$, which is the axis of symmetry of a parabola that has its vertex at $(-2,5)$.

Turning point of a parabola

From the examples above, it can be seen that the turning point (minimum or maximum value) of the quadratic function $y=ax^2+bx+c$ will occur at $x=-\dfrac{b}{2a}$.

Thus the turning point of $f(x)=ax^2+bx+c$ has the coordinates $\left(-\dfrac{b}{2a}, f\left(-\dfrac{b}{2a}\right)\right)$.

- If $a<0$ this will be a **maximum** turning point, so the **maximum value** of the function is $f\left(-\dfrac{b}{2a}\right)$.
- If $a>0$ this will be a **minimum** turning point, so the **minimum value** of the function is $f\left(-\dfrac{b}{2a}\right)$.

Later in the course, you will be able to find the turning point using calculus and the fact that $\dfrac{dy}{dx}=0$.

Example 22

A piece of wire of length 12 cm is bent in the shape of a rectangle. Find the maximum area of the rectangle.

Solution

Let one dimension of the rectangle be x cm.

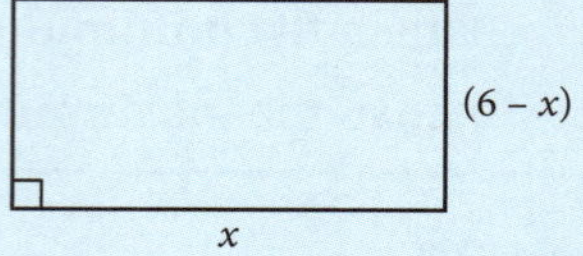

Hence the other dimension will be $(12 \div 2 - x) = (6 - x)$ cm.

For any given x in the domain $0 < x < 6$, the area function is given by:

$$\begin{aligned} y &= x(6-x) \qquad \text{for } 0 < x < 6 \\ &= 6x - x^2 \\ &= -(x^2 - 6x + 9) + 9 \\ &= -(x-3)^2 + 9 \end{aligned}$$

Hence the maximum area is 9 cm^2, when $x = 3$ and the rectangle is a square.

The graph of the area function is shown in the diagram.

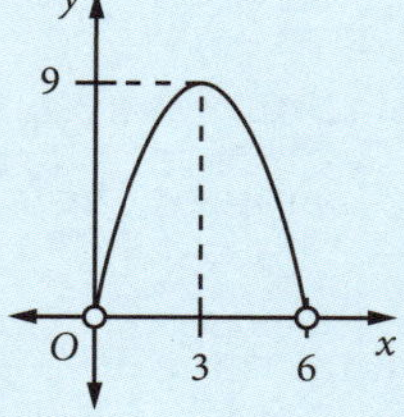

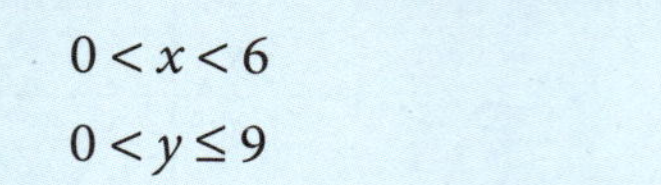

Domain: $0 < x < 6$

Range: $0 < y \leq 9$

MAKING CONNECTIONS

Graphing quadratic functions

Move the sliders to view the effect of changing a, B and C on the graph of $y = a(x + B)^2 + C$.

EXERCISE 5.8 QUADRATIC FUNCTIONS

1 The minimum value of $x^2 - 2x + 6$ is:

A -1 **B** 1 **C** 5 **D** 6

2 Express each of the following functions in the form $y = a(x + B)^2 + C$ and hence find the maximum or minimum value and the range.

(a) $y = 2x^2 - 4x$ **(b)** $y = -2x^2 + 8x - 3$ **(c)** $y = 7 + 16x - 4x^2$ **(d)** $y = 4x^2 + 8x - 7$

(e) $y = 8 - 2x^2$ **(f)** $y = 7 - 2x - x^2$ **(g)** $y = 2x^2 - 6x$ **(h)** $y = 6 - 10x - 5x^2$

3 The maximum value of $-x^2 + x + 11$ is:

A -0.5 **B** 0.5 **C** 10.75 **D** 11.25

4 A stone is thrown straight up from the ground. The height above the ground, $h(t)$ metres, is a function of time, t seconds $(t \geq 0)$, according to the rule $h(t) = 20t - 5t^2$. Find:

(a) the domain of h **(b)** the greatest height reached.

5 The equation of the path of a cricket ball thrown at an angle of 45° to the horizontal is $y = x - \frac{x^2}{50}$, where x m and y m are the horizontal distance travelled and the vertical height reached, respectively. Calculate the greatest vertical height reached and the horizontal distance travelled.

6 The sum of two numbers is 20. Find the numbers and their product if their product is a maximum.

7 A piece of wire 60 cm long is bent in the shape of a rectangle. Find the maximum area of the rectangle.

8 A farmer wants to make a rectangular enclosure using her existing fence as one side. If she has 20 metres of fencing material available to make the other three sides, find the area and the dimensions of the largest enclosure that can be formed.

9 A piece of wire 6 metres long is cut into two parts. One part is used to form a square and the other part is used to form a rectangle whose length is three times its breadth. Find the lengths of the two parts if the sum of the two areas is a minimum.

10 A large open area is to have a section surrounded by a rectangular fence. This rectangle is then divided into six smaller rectangles, using one dividing fence parallel to its length and two fences parallel to its width. If the total length of fencing available is 1200 m, find the maximum possible area.

11 A machine comes in two parts, which weigh x kg and b kg respectively. The cost c of the machine (in dollars) is given by $c = 2x + b$. The earning capacity y of the machine is given by $y = x(x + b)$. If c has the fixed value 10, express y as a function of x and hence find the value of x for which y is a maximum. Find the maximum value of y.

12 $ABCD$ is a square of unit length. Points E and F are on the sides AB and AD respectively such that $AE = AF = x$. Show that the area y of the quadrilateral $CDFE$ is given by $y = \frac{1}{2}(1 + x - x^2)$. What is this quadrilateral's greatest possible area?

5.9 PARABOLAS AND DISCRIMINANTS

Discriminants

In Chapter 3 you solved the general quadratic equation $ax^2 + bx + c = 0$, $a \neq 0$, by 'completing the square' to obtain the quadratic formula:

$$ax^2 + bx + c = 0$$

Move constant to RHS: $$ax^2 + bx = -c$$

Divide by coefficient of x^2: $$x^2 + \frac{b}{a}x = -\frac{c}{a}$$

Add $\left(\frac{b}{2a}\right)^2$ to complete the square: $$x^2 + \frac{b}{a}x + \left(\frac{b}{2a}\right)^2 = \frac{b^2}{4a^2} - \frac{c}{a}$$

Factorise: $$\left(x + \frac{b}{2a}\right)^2 = \frac{b^2 - 4ac}{4a^2}$$

Take square roots of both sides: $$x + \frac{b}{2a} = \frac{\pm\sqrt{b^2 - 4ac}}{2a}$$

$$x = \frac{-b \pm \sqrt{b^2 - 4ac}}{2a}$$

The symbol Δ (delta) is sometimes used for the expression $b^2 - 4ac$.
This is called the **discriminant**: $\Delta = b^2 - 4ac$.

The roots of the quadratic equation can be written as $x = \frac{-b + \sqrt{\Delta}}{2a}$ or $x = \frac{-b - \sqrt{\Delta}}{2a}$.

Case 1 If $\Delta > 0$, $\sqrt{\Delta}$ is a real number and so the roots are two real numbers.
The equation has two unequal roots or two different roots.

Case 2 If $\Delta = 0$, $\sqrt{\Delta} = 0$ and so the roots are only one real number: $-\frac{b}{2a}$.
The equation has only one root or two equal roots.

Case 3 If Δ is a perfect square, $\sqrt{\Delta}$ is a rational number and so the roots are two rational numbers.
The equation has two unequal, rational roots.

Case 4 If $\Delta < 0$, $\sqrt{\Delta}$ does not exist as a real number and so there are no real roots.
The equation has no real roots.

Intersection of the parabola with the x-axis

The roots of the equation $ax^2 + bx + c = 0$ are the x values of the points of intersection of the graph of the parabola $y = ax^2 + bx + c = 0$ with the x-axis. In other words, at the points where the parabola cuts the x-axis, $y = 0$.

Thus, the parabola $y = ax^2 + bx + c = 0$:

- cuts the x-axis at two distinct points if $\Delta > 0$
- touches the x-axis at one point only (or two coincident points) if $\Delta = 0$
- does not touch the x-axis if $\Delta < 0$.

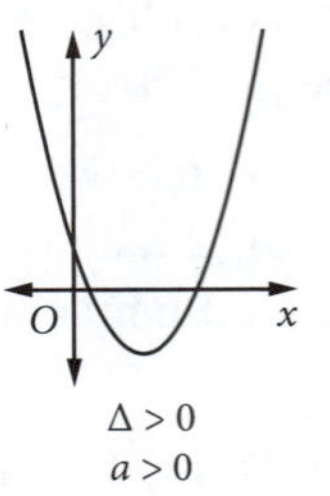

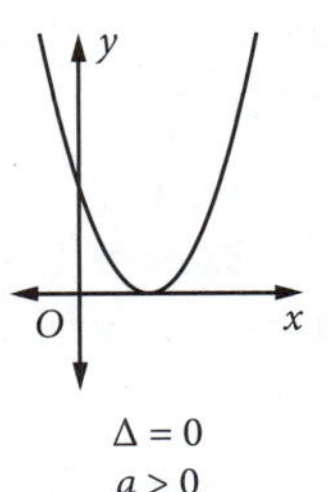

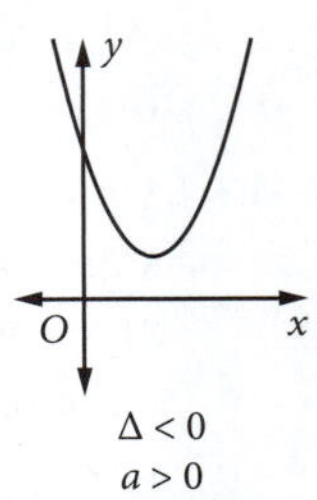

For the parabolas shown, the coefficient of x^2 is positive, so the vertex is at the bottom and the curve opens upwards.

When $a > 0$ and $\Delta < 0$, as in the third diagram shown, the graph of $f(x) = ax^2 + bx + c = 0$ is always above the x-axis, i.e. $f(x) > 0$ for all values of x. This f is called a **positive definite** function because it is always positive.

For these parabolas the coefficient of x^2 is negative, so the vertex is at the top and the curve opens downwards.

When $a < 0$ and $\Delta < 0$, as in the third diagram at right, the graph of $f(x) = ax^2 + bx + c = 0$ is always below the x-axis, i.e. $f(x) < 0$ for all values of x. This f is called a **negative definite** function because it is always negative.

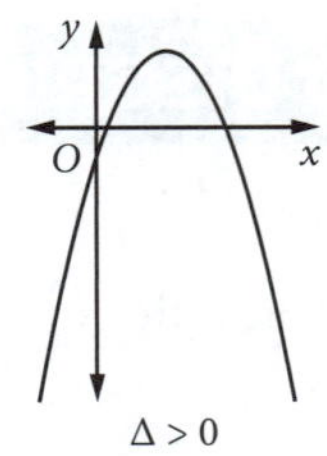

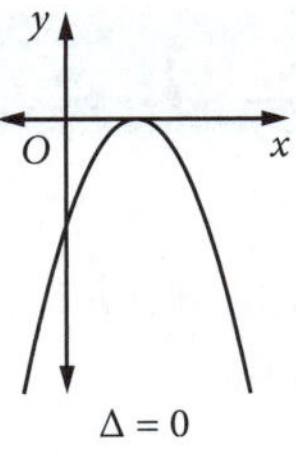

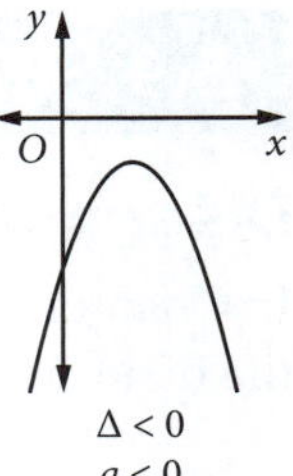

If a quadratic function is neither positive definite nor negative definite, it is said to be **indefinite**.

Example 23

Write the discriminant of each of the following quadratic equations and hence state whether the corresponding parabola cuts the x-axis at two, one or no points.

(a) $x^2 - 5x + 6 = 0$ (b) $x^2 - 4x + 4 = 0$ (c) $2x^2 - 3x + 3 = 0$

Solution

(a) $x^2 - 5x + 6 = 0$

$\Delta = b^2 - 4ac$

$\Delta = 25 - 24 = 1$

$\Delta > 0$

Two real roots, cuts x-axis at two distinct points

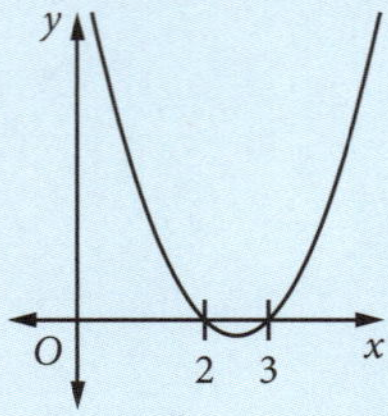

$y = x^2 - 5x + 6$

$y = (x - 2)(x - 3)$

Cuts at $(2, 0)$, $(3, 0)$

(b) $x^2 - 4x + 4 = 0$

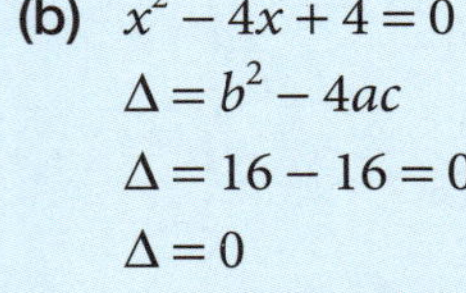

$\Delta = b^2 - 4ac$

$\Delta = 16 - 16 = 0$

$\Delta = 0$

One real root, touches x-axis at one point

$y = x^2 - 4x + 4$

$y = (x - 2)^2$

Touches at $(2, 0)$

(c) $2x^2 - 3x + 3 = 0$

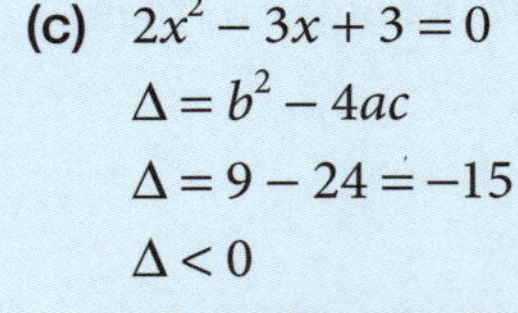

$\Delta = b^2 - 4ac$

$\Delta = 9 - 24 = -15$

$\Delta < 0$

No real roots, does not cut x-axis

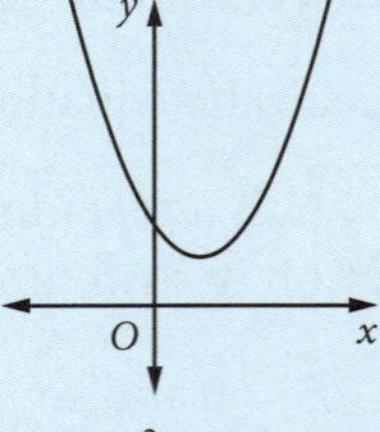

$y = 2x^2 - 3x + 3$

Cannot factorise with real numbers, does not cut or touch x-axis.

Positive definite function, $a = 2$

The discriminant tells you whether the parabola cuts, touches, or does not cut the x-axis. To find the point at which this occurs, you need to solve the quadratic equation. Thus the discriminant tells you whether you can factorise the quadratic expression.

Example 24

Solve the following equations after first finding the discriminant.

(a) $3x^2 - 11x - 4 = 0$ (b) $4x^2 - 20x + 25 = 0$ (c) $x^2 = 2x + 5$ (d) $2x^2 + x + 3 = 0$

Solution

(a) $3x^2 - 11x - 4 = 0$: $\Delta = 121 + 48 = 169$

Because the discriminant is a perfect square, $3x^2 - 11x - 4$ has rational factors.

$$3x^2 - 11x - 4 = 0$$
$$(3x + 1)(x - 4) = 0$$
$$x = -\frac{1}{3} \quad \text{or} \quad 4$$

(b) $4x^2 - 20x + 25 = 0$: $\Delta = 400 - 400 = 0$

Because the discriminant is zero, $4x^2 - 20x + 25$ has equal rational factors.

$$4x^2 - 20x + 25 = 0$$
$$(2x - 5)^2 = 0$$
$$x = 2.5$$

(c) $x^2 - 2x - 5 = 0$: $\Delta = 4 + 20 = 24$

Because $\Delta > 0$ but it is not a perfect square, the equation has two unequal, irrational roots, so it is best solved using the quadratic formula.

$$x^2 - 2x - 5 = 0$$
$$x = \frac{2 \pm \sqrt{24}}{2} = \frac{2 \pm 2\sqrt{6}}{2} = 1 \pm \sqrt{6}$$

(d) $2x^2 + x + 3 = 0$: $\Delta = 1 - 24 = -23$

Because $\Delta < 0$, there are no real roots.

EXPLORE FURTHER

Investigating the discriminant

Use technology to investigate the relationship between the discriminant and the parabola.

EXERCISE 5.9 PARABOLAS AND DISCRIMINANTS

1 Select the equation for which $\Delta > 0$.

A $x^2 + 2x + 3 = 0$ B $x^2 + 6x + 9 = 0$ C $2x^2 - 4x + 5 = 0$ D $x^2 - 4x - 5 = 0$

2 Calculate the discriminant for each of the following equations and hence state whether the equations have two, one or no real roots.

(a) $x^2 + 6x + 2 = 0$ (b) $2x^2 + 3x + 4 = 0$ (c) $4x^2 - 12x + 9 = 0$
(d) $-3x^2 + 2x - 1 = 0$ (e) $2x^2 = 3x + 7$

3 For which curve can you say that $a > 0$ and $\Delta < 0$?

A
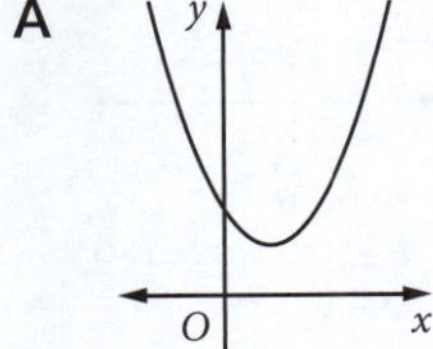

B
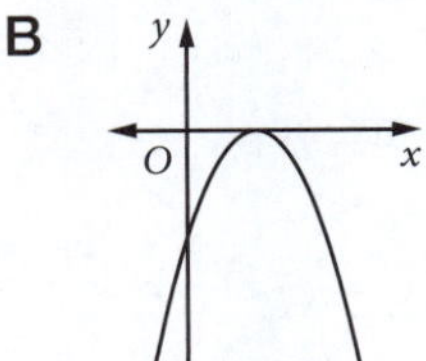

C
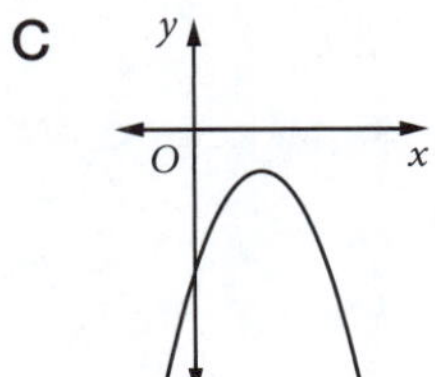

D
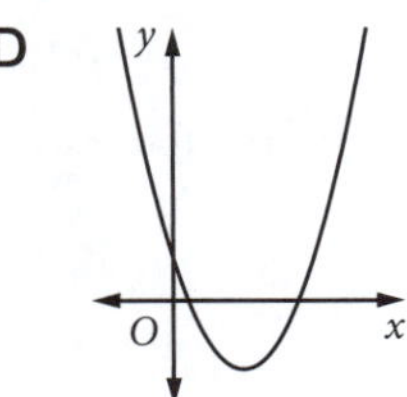

4 Without sketching the graphs of each function, determine whether or not they cross the x-axis.

(a) $y = x^2 - 5x + 2$ (b) $y = -4x^2 + 2x - 1$ (c) $y = x^2 - 6x + 9$
(d) $y = 8 - 3x - 2x^2$ (e) $y = 3x^2 + 2x + 5$ (f) $y = -x^2 - x - 1$

5 Calculate the discriminant for each of the following equations. Use this information to decide on the best solution method, then solve each equation.

(a) $x^2 + 2x - 15 = 0$ (b) $x^2 - 9x - 5 = 0$ (c) $12x^2 = 25x - 12$
(d) $4x^2 - 12x + 9 = 0$ (e) $7x^2 = 63$ (f) $x^2 - 6x = 0$
(g) $(x + 1)^2 = 4x$ (h) $x^2 = 4(x - 24)$ (i) $2x^2 - x = 25$
(j) $3x^2 = 2x + 2$ (k) $3x^2 - 7x - 3 = 0$ (l) $(x + 6)^2 = x + 6$
(m) $4x^2 = 9x - 4$ (n) $9x^2 + 24x + 16 = 0$ (o) $3x^2 + 4x = 5$
(p) $2x^2 + x - 4 = 0$ (q) $x^2 = 15x - 56$ (r) $x(2x - 3) = 0$
(s) $2x^2 + 5x + 1 = 0$ (t) $5x^2 - 7x + 2 = 0$ (u) $(x - 1)(x - 2) + (x + 1)(x + 3) = 0$

5.10 FURTHER EXAMPLES INVOLVING DISCRIMINANTS

Example 25

Find the values of m for which the equation $x^2 + (m - 2)x + 4 = 0$ has:

(a) one root (b) two roots (c) no roots.

Solution

Consider Δ, the discriminant.

(a) For one root only, $\Delta = 0$:

$$\begin{aligned}\Delta &= (m-2)^2 - 16\\ &= m^2 - 4m - 12\\ &= (m-6)(m+2)\end{aligned}$$

$\Delta = 0$ when $m = 6$ or -2

(b) For two roots, $\Delta > 0$:

i.e. $(m - 6)(m + 2) > 0$

The graph of $\Delta = m^2 - 4m - 12$ cuts the m-axis at $m = 6$ and $m = -2$.

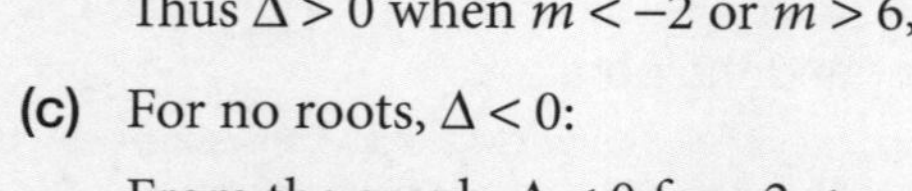

Thus $\Delta > 0$ when $m < -2$ or $m > 6$, as shown by the graph.

(c) For no roots, $\Delta < 0$:

From the graph, $\Delta < 0$ for $-2 < m < 6$.

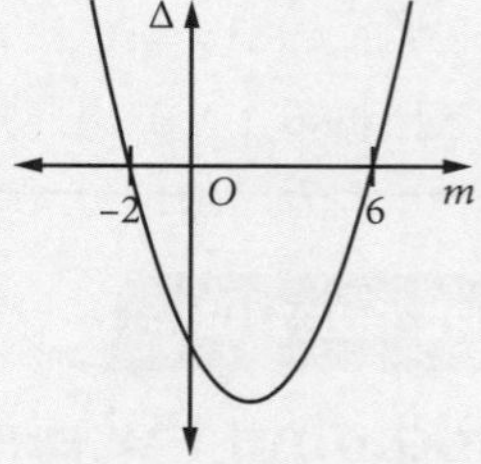

$\Delta = m^2 - 4m - 12$

Example 26

Prove that the equation $x^2 + (k - 3)x - k = 0$ has real roots for all values of k.

Solution

For real roots, $\Delta \geq 0$:

$$\begin{aligned}\Delta &= (k-3)^2 + 4k\\ &= k^2 - 2k + 9\end{aligned}$$

The graph of $\Delta = k^2 - 2k + 9$ does not cross the k-axis and the coefficient of k^2 is positive, so $\Delta > 0$ for all values of k.

Hence the equation $x^2 + (k - 3)x - k = 0$ has real roots for all values of k.

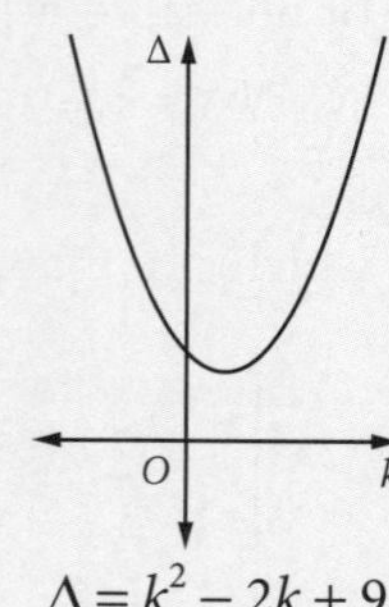

$\Delta = k^2 - 2k + 9$

Alternatively:

$$\begin{aligned}\Delta &= k^2 - 2k + 9 \\ &= k^2 - 2k + 1 + 8 \\ &= (k-1)^2 + 8 \\ &= \text{a perfect square} + \text{a positive number}\end{aligned}$$

Hence $\Delta \geq 8$ for all k, so the given equation has real roots for all k.

Example 27

Show that the roots of the equation $x^2 + 2x - (m^2 - 1) = 0$ are rational for all rational values of m.

Solution

For rational roots, Δ is a perfect square: $\Delta = 4 + 4(m^2 - 1)$
$= 4m^2$

Hence Δ is a perfect square if m is a rational number.

$\therefore$ the equation $x^2 + 2x - (m^2 - 1) = 0$ has rational roots for all rational values of m.

EXERCISE 5.10 FURTHER EXAMPLES INVOLVING DISCRIMINANTS

1 Find the values of k for which the following quadratic equations have: **(i)** one root **(ii)** two roots.
(a) $x^2 - 3x + k = 0$ **(b)** $x^2 + kx + 3 = 0$
(c) $x^2 + (k-1)x - (2k+1) = 0$ **(d)** $(k-1)x^2 + (k+1)x = 1 - k$

2 The quadratic equation $(2k-3)x^2 + (k+1)x - 1 = 0$ has two roots when:
A $k > 1$ or $k < -11$ **B** $-11 < k < 1$ **C** $-11 \leq k \leq 1$ **D** $k \geq 1$ or $k \leq -11$

3 For what values of m does the quadratic equation $(5m-3)x^2 - 4mx + m + 1 = 0$ have only one root?

4 For what values of m are the roots of the following equations real?
(a) $x^2 + 2x + m^2 - 1 = 0$ **(b)** $(m-1)x^2 + (m+1)x + m - 1 = 0$ **(c)** $x^2 + 2mx + 2(m+12) = 0$

5 If the real roots of the equation $x^2 - 2mx + 3 = 0$ are a and b, find the values of m for which:
(a) $a = b$ **(b)** $a \neq b$

6 Show that the roots of the equation $4(m+1)x^2 - 4(m-1)x - 3 = 0$, for $m \neq -1$, are real for all real m.

7 Show that the roots of the equation $(3m-5)x^2 - 3m^2x + 5m^2 = 0$, for $m \neq \frac{5}{3}$, are rational if m is rational.

8 Show that the equation $x^2 - (2a+b)x + ab = 0$ has real roots for all values of a and b.

9 Find the values of p for which the equation $2x^2 - 4x + p = 0$ has:
(a) one root (i.e. two equal roots) **(b)** two distinct roots.

10 For what values of m does the equation $x^2 - 2mx + 8m - 15 = 0$ have: **(a)** one root **(b)** two roots?

11 Show that the roots of the equation $ax^2 - (a+b)x + b = 0$, for $a \neq 0$, are rational for all rational values of a and b.

12 For what values of m does the quadratic equation $x^2 + mx + (m+1)^2 = 0$ have two roots?

5.11 SOLUTION SET OF SIMULTANEOUS EQUATIONS

To find the points of intersection of a straight line and a parabola, you can algebraically calculate the solution set of a system consisting of the equation of the line and the equation of the parabola.

A straight line may cut a parabola at two distinct points, touch it at one point or not intersect it at all. If the line is a tangent to the parabola then the solution set has only one ordered pair (a 'double root').

Example 28

Find the points of intersection of the line $y = x + 6$ and the parabola $y = x^2 - x - 2$.

Solution

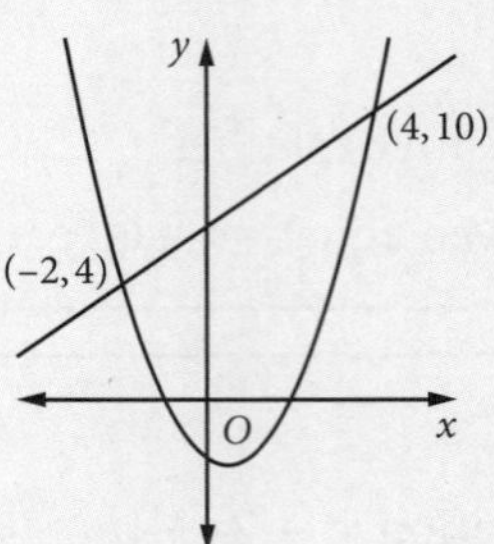

For the points of intersection:

$$\begin{aligned} x^2 - x - 2 &= x + 6 \\ x^2 - 2x - 8 &= 0 \\ (x-4)(x+2) &= 0 \\ x &= 4 \quad \text{or} \quad -2 \\ y &= 10 \quad \text{or} \quad 4 \end{aligned}$$

They intersect at the points $(4, 10)$ and $(-2, 4)$.

Example 29

Prove that the line $x - y + 1 = 0$ is a tangent to the parabola $y = x^2 - 3x + 5$.

Solution

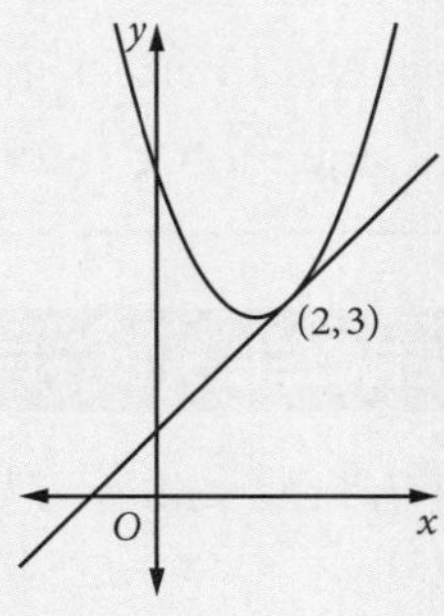

For the points of intersection:

$$\begin{aligned} x^2 - 3x + 5 &= x + 1 \\ x^2 - 4x + 4 &= 0 \\ (x-2)^2 &= 0 \\ x &= 2,\ y = 3 \end{aligned}$$

This is a double root, so the straight line is a tangent to the parabola at the point $(2, 3)$

Example 30

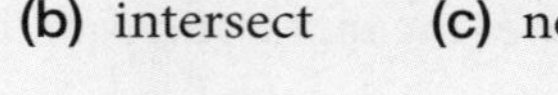

For what values of m does the line $y = mx - 6$ **(a)** touch **(b)** intersect **(c)** not intersect the parabola $y = x^2 - 2x + 3$?

Solution

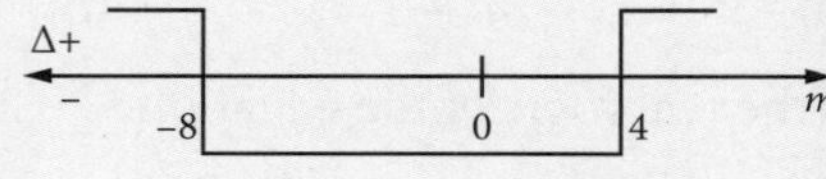

For the points of intersection:

$$\begin{aligned} x^2 - 2x + 3 &= mx - 6 \\ x^2 - (m+2)x + 9 &= 0 \end{aligned}$$

(a) Touch when $\Delta = 0$:

$$\begin{aligned} \Delta &= b^2 - 4ac \\ &= (m+2)^2 - 36 \\ &= m^2 + 4m - 32 \\ &= (m+8)(m-4) \end{aligned}$$

Will touch when $m = -8$ or 4

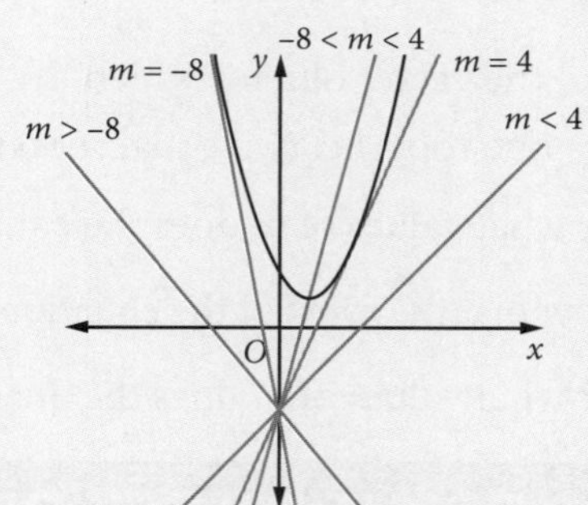

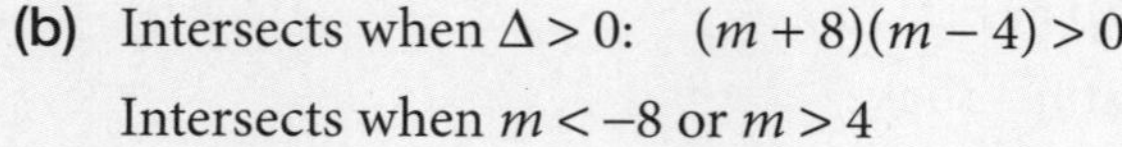

(b) Intersects when $\Delta > 0$: $(m+8)(m-4) > 0$

Intersects when $m < -8$ or $m > 4$

(c) Does not intersect when $\Delta < 0$: $(m+8)(m-4) < 0$

Does not intersect for $-8 < m < 4$.

Note: The principles involved in the three examples above apply to the intersection of a straight line and any curve whose defining equation generates an equation of the second degree when solved simultaneously with a linear equation (e.g. a quadratic, a circle, or a rational function $y=\frac{a}{x+b}$).

EXPLORE FURTHER

The intersection of a parabola and a straight line

Use technology to explore the points of intersection of parabolas with straight lines.

EXERCISE 5.11 SOLUTION SET OF SIMULTANEOUS EQUATIONS

1 Find algebraically the solution set of each system of equations.

(a) $y=2x^2-3x+4$ and $y=12-3x$
(b) $y=2-x-3x^2$ and $y=2-7x$
(c) $y=x^2+3x-2$ and $x+y=3$
(d) $x-y=1$ and $x^2+xy=6$

2 Find algebraically the coordinates of the intersection points of:

(a) the straight line $y=x-3$ and the circle $x^2+y^2=9$
(b) the straight line $y=2x-1$ and the parabola $y=x^2-3x+5$
(c) the straight line $y=3-2x$ and the parabola $y=(x-2)^2$

3 For what value of c is the line $y=2x+c$ a tangent to the parabola $y=x^2-x-2$?

A $c=-4.25$ **B** $c=0.25$ **C** $c=-2.25$ **D** $c=4.25$

4 For what value of c is the line $y=x+c$ a tangent to the circle $x^2+y^2=4$?

5 For what value of a is the line $y=ax$ a tangent to the circle $x^2+y^2+20x-10y+100=0$?

6 For what value of m is the line $y=mx$ a tangent to the parabola $y=x^2-8x+25$?

7 For what value of m does the line $y=mx-6$ (a) touch (b) intersect (c) not intersect the parabola $y=x^2-2x+3$?

8 For what value of m does the line $y=mx+5$ (a) touch (b) intersect (c) not intersect the parabola $y=3+5x-2x^2$?

9 For what value of m does the line $y=mx-12$ (a) touch (b) intersect (c) not intersect the parabola $y=2x^2-x-10$?

10 For what value of a does the line $y=ax$ intersect the curve with equation $y=\frac{2}{x-3}$?

11 For what value of a does the line $y=ax$ not meet the rectangular hyperbola $y=\frac{3}{x-2}$?

12 Find the equations of the two lines that contain the point $(1,3)$ and are tangent to the parabola $y=x^2-2x+5$.

13 Prove that the parabolas $y=2x^2-6x+7$ and $y=x^2-2x+3$ touch each other and find the coordinates of the point of contact.

CHAPTER REVIEW 5

1 Given $A(1,5)$, $B(2,4)$, $C(-2,3)$ and $D(3,-2)$, find:

(a) the gradients of AB, AC, AD, BC, BD, CD
(b) which lines are parallel
(c) the lengths of AB, CD, AD
(d) the midpoints of AD, BC, BD.

2 $P(2,3)$, $Q(6,-1)$, $R(-4,-5)$ are the vertices of ΔPQR. M is the midpoint of PQ and N is the midpoint of PR.

(a) Find the coordinates of M and N.
(b) Show that $MN \parallel QR$.
(c) Calculate the length of QR.
(d) Show that the length of MN is half the length of QR.

3 Find the perpendicular distance from the point $(2,3)$ to the line $6x+8y-5=0$.

4 Find the equation of the line that is perpendicular to the line $3x+4y=5$ and also passes through the midpoint of the line segment joining the points $(3,-2)$ and $(5,8)$.

5 Find the coordinates of the intersection points of the lines $5x-3y-19=0$, $3x+5y+9=0$ and $x-4y+3=0$. Show that these intersection points are the vertices of a right-angled triangle.

6 Solve these simultaneous equations: $3(x-3)=2(2y+1)$
$2(3x-1)=5(2y+1)+10$

7 By the method of completing the square, find:

(a) the maximum value of $5+4x-x^2$ (b) the minimum value of $2x^2+6x-9$.

8 Solve these simultaneous equations: $x+y-9=0$
$y=x^2+4x+3$

9 Solve the following equations.

(a) $(x^2-x)^2-5(x^2-x)+6=0$ (b) $x^4-4x^2-45=0$ (c) $3^{2x}-12(3^x)+27=0$

10 Without actually solving the following quadratic equations, determine whether each equation has two, one or no roots.

(a) $x^2-6x+5=0$ (b) $2x^2-3x-7=0$ (c) $x^2-20x+100=0$ (d) $3x^2+4x-1=0$

11 Show that the roots of the equation $mx^2-(m+n)x+n=0$ are rational for all rational values of m and n.

12 For what values of k does the quadratic equation $x^2-5x+(k-1)=0$ have:

(a) two roots (b) one root (c) no roots?

13 If $2x^2-9x+9\equiv(ax-b)(x-b)$ for all values of x, find the values of a and b.

14 Find the coordinates of the points of intersection of the line $y=2x-3$ and the parabola $y=x^2-4x+5$.

15 For what values of m does the line $y=mx-5$ (a) touch (b) intersect (c) not intersect the parabola $y=x^2-5x+4$?

16 A firm sells its product for \$20 per unit.

(a) If x units are produced, write down the revenue function.
(b) In manufacturing this product, the fixed costs are \$10 000 and the variable costs are \$10 per unit produced. Hence write down the cost function.
(c) What is the break-even point for this firm?
(d) How much profit is made when 3000 units are sold?

17 The sum of two numbers is 24. Find the numbers and their product if the product is a maximum.

18 A farmer has 24 metres of fencing material and wants to make three enclosures of equal area using the side of the barn as one side of the enclosures as shown in the diagram.

Side of barn

If the total area of the three enclosures is to be as large as possible, find the dimensions of each enclosure and the total area enclosed.

CHAPTER 6
Further trigonometry

6.1 RADIAN MEASURE OF AN ANGLE

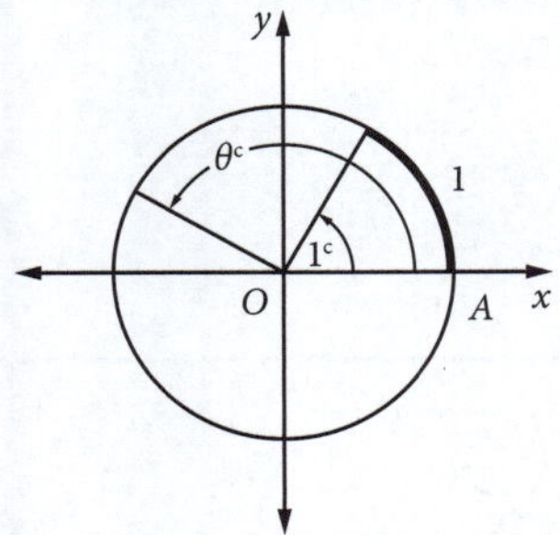

Later in this course, when you look at the calculus of trigonometric functions, you will see that in many situations degrees are not a useful measure of angle size. Instead of using degrees, it is often better to use **radians**.

Consider a unit circle. Starting at point A and moving around the circle in an anticlockwise direction, mark an arc of length 1 unit.

The angle that this arc subtends at the circle's centre is defined to have a magnitude of 1 radian. If you mark any arc of length θ, then the angle that this arc subtends at the circle's centre has a magnitude of θ radians. In this way, the length of the arc is the same number as the size of the angle in radians.

In other words: 1 radian is the angle subtended by an arc of 1 unit in a circle of radius 1 unit.

You can write an angle of θ radians as θ^c, but the 'c' is usually left off so that the angle is written simply as θ. If an angle has no degree symbol °, you should assume that it is measured in radians.

The circumference of a circle of radius r units is given by $C = 2\pi r$. Thus the circumference of the unit circle is 2π and half the circumference is π. An arc of length π therefore subtends a straight angle of size π^c at the centre of the circle. You already know that the size of a straight angle is 180°, so $\pi^c = 180°$:

$$1^c = \left(\frac{180}{\pi}\right)^\circ \approx 57°\,18' \qquad \text{and} \qquad 1° = \left(\frac{\pi}{180}\right)^c \approx 0.017\,45^c$$

So: $$\theta^c = \left(\frac{180\theta}{\pi}\right)^\circ \qquad \text{and} \qquad \theta° = \left(\frac{\pi\theta}{180}\right)^c$$

The relations $\pi^c = 180°$ and $1° = \left(\frac{\pi}{180}\right)^c$ enable you to convert radians to degrees and degrees to radians.

Thus, for example: $\frac{\pi}{2}^c = 90°$, $\frac{\pi}{3}^c = 60°$, $\frac{4\pi}{5}^c = \frac{4}{5}\times 180° = 144°$, $\frac{7\pi}{6}^c = \frac{7}{6}\times 180° = 210°$.

Note that angles in radians are not always given in terms of π. It is usually necessary to use a calculator and multiply by $\frac{180}{\pi}$ to convert radians to degrees.

MAKING CONNECTIONS

Radians and degrees

Move the point around the unit circle to see the degree and radian measures of the angle.

Example 1

Convert the following to degrees: **(a)** 2.4^c **(b)** 4.6^c

Solution

(a) $$2.4^c = 2.4 \times \frac{180°}{\pi} = 137.5099\ldots° \approx 137°\,31'$$

(b) $$4.6^c = 4.6 \times \frac{180°}{\pi} = 263.561\ldots° \approx 263°\,34'$$

You can use a calculator to find $\sin 2.4^c$, $\cos 4.6^c$ or any other calculation using radians without converting to degrees. Most calculators will be set to calculate trigonometric functions using degrees, but they will also have a key labelled as **MODE**, **DRG** or something similar that can be used to switch between the use of degrees (often displayed as 'DEG'), radians ('RAD') and sometimes also gradians ('GRAD', which is an obscure unit equal to $\frac{9}{10}$ of a degree). If you are using calculator software on a computer, you may need to change the trigonometry calculation mode through menu settings.

Example 2

Convert to radians: **(a)** 127° **(b)** 214° 35′

Solution

(a) $127° = 127 \times \frac{\pi^c}{180}$
≈ 2.217 radians

(b) $214° 35' = 214.58 \times \frac{\pi^c}{180}$
≈ 3.745 radians

Common conversions

It is important to be able to easily convert between angles in degrees and angles in radians for the common angles that give exact values in trigonometric functions.

Degrees	0°	30°	45°	60°	90°	120°	135°	150°	180°	210°	225°	240°	270°	300°	315°	330°	360°
Radians	0	$\frac{\pi}{6}$	$\frac{\pi}{4}$	$\frac{\pi}{3}$	$\frac{\pi}{2}$	$\frac{2\pi}{3}$	$\frac{3\pi}{4}$	$\frac{5\pi}{6}$	π	$\frac{7\pi}{6}$	$\frac{5\pi}{4}$	$\frac{4\pi}{3}$	$\frac{3\pi}{2}$	$\frac{5\pi}{3}$	$\frac{7\pi}{4}$	$\frac{11\pi}{6}$	2π

Adding 360° to an angle in degrees is the same as adding 2π to an angle in radians. Both 360° and 2π each represent a full turn.

EXERCISE 6.1 RADIAN MEASURE OF AN ANGLE

1 Express the following angles in radians, in terms of π.
(a) 30° **(b)** 225° **(c)** 72° **(d)** 210° **(e)** 315° **(f)** 112° 30′ **(g)** 330° **(h)** 144°

2 Express, in degrees, the angles whose radian measures are:
(a) $\frac{3\pi}{4}$ **(b)** $\frac{7\pi}{8}$ **(c)** $\frac{6\pi}{5}$ **(d)** $\frac{3\pi}{2}$ **(e)** $\frac{5\pi}{9}$ **(f)** $\frac{11\pi}{12}$ **(g)** 1.8π **(h)** $\frac{11\pi}{8}$

3 Use a calculator to express, in degrees and minutes, the angles whose radian measures are:
(a) 0.5 **(b)** 1.82 **(c)** 3 **(d)** 4.26 **(e)** 2.72 **(f)** 3.426 **(g)** 5.24 **(h)** 4.782

4 Use a calculator to find the radian equivalent of each angle, giving your answer correct to 4 decimal places.
(a) 42° **(b)** 74° **(c)** 105° **(d)** 164° **(e)** 48° 9′ **(f)** 220° **(g)** 138° 12′ **(h)** 72° 8′

5 Indicate whether each statement is correct or is incorrect. 315° is:
(a) $\frac{5\pi}{4}$ **(b)** $\frac{7\pi}{4}$ **(c)** 5.4978 **(d)** 3.9270

6 Convert each angle into radians, in terms of π.
(a) 120° **(b)** 315° **(c)** 210° **(d)** 135° **(e)** 390° **(f)** 405° **(g)** 480° **(h)** 720°
(i) −30° **(j)** −135° **(k)** 450° **(l)** −180° **(m)** 15° **(n)** 22.5° **(o)** 345° **(p)** −67.5°

7 Convert each angle to degrees.
(a) $\frac{5\pi}{6}$ **(b)** $\frac{5\pi}{3}$ **(c)** $\frac{5\pi}{4}$ **(d)** 3π **(e)** $\frac{9\pi}{4}$ **(f)** $\frac{7\pi}{2}$ **(g)** $\frac{10\pi}{3}$ **(h)** $-\pi$
(i) $-\frac{3\pi}{2}$ **(j)** $-\frac{7\pi}{6}$ **(k)** $-\frac{3\pi}{4}$ **(l)** -2π **(m)** $\frac{\pi}{8}$ **(n)** $\frac{5\pi}{12}$ **(o)** $\frac{9\pi}{8}$ **(p)** $-\frac{7\pi}{12}$

6.2 ARC LENGTH AND SECTOR AREA OF A CIRCLE

Arc length

Let the arc AB subtend an angle of θ^c at the centre of a circle of radius r units.
Let the length of the arc $AB = \ell$.
Let $\angle POQ = 1^c$, hence the length of arc $PQ = r$.

Compare arcs: $\dfrac{\text{arc } AB}{\text{arc } PQ} = \dfrac{\angle AOB}{\angle POQ}$

Hence $\dfrac{\ell}{r} = \dfrac{\theta}{1}$, so:

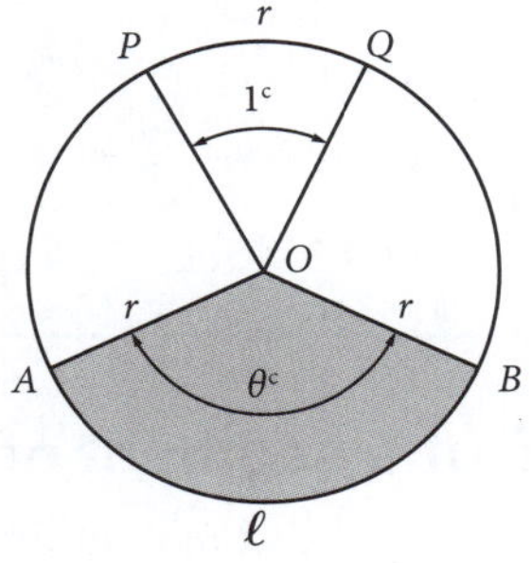

Arc length: $\ell = r\theta$

Example 3

Find the length of an arc of a circle of radius 12 cm, if the arc subtends an angle at the centre of:

(a) 1.5 radians **(b)** 40°

Solution

(a) $r = 12,\ \theta = 1.5$

$\ell = r\theta$
$= 12 \times 1.5$
$= 18$

Arc length is 18 cm.

(b) $r = 12,\ \theta = 40 \times \dfrac{\pi}{180} = \dfrac{2\pi}{9}$

$\ell = r\theta$

$\ell = 12 \times \dfrac{2\pi}{9} = \dfrac{8\pi}{3}$

$= 8.378$

Arc length is 8.4 cm (correct to 1 decimal place).

Area of a sector

Consider a unit circle with a sector AOB.

Comparing the shaded area AOB to the area of the whole circle gives:

$$\frac{\text{Area of sector } AOB}{\text{Area of circle}} = \frac{\angle AOB}{2\pi}$$

$$\frac{\text{Area of sector } AOB}{\pi r^2} = \frac{\theta}{2\pi}$$

Area of sector $AOB = \frac{1}{2}r^2\theta$ $\qquad A = \frac{1}{2}r^2\theta$

A useful variation of this formula is $A = \frac{1}{2}r^2\theta = \frac{1}{2} \times r\theta \times r = \frac{1}{2}\ell r$.
Thus if you know the arc length and the radius, you can find the area of the sector without knowing the angle.

Example 4

The length of an arc AB of a circle with centre O and radius 10 cm is 8 cm. Calculate:

(a) $\angle AOB$ in degrees and minutes **(b)** the area of the sector AOB.

Solution

(a) $r = 10, \ell = 8$

$\ell = r\theta$

$8 = 10\theta$

$\theta = 0.8^c$

$\theta = 0.8 \times \dfrac{180°}{\pi}$

$\theta = 45° 50'$

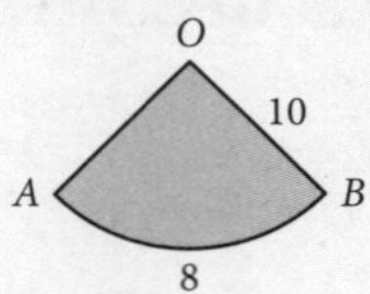

(b) $r = 10, \ell = 8, \theta = 0.8$

$A = \dfrac{1}{2}r^2\theta$

$= \dfrac{1}{2} \times 100 \times 0.8$

$= 40$

OR

$A = \dfrac{1}{2}\ell r$

$= \dfrac{1}{2} \times 8 \times 10$

$= 40$

Area of sector is $40\,\text{cm}^2$.

Area of a segment of a circle

This is an application combining the formula for the area of a sector with the formula for the area of a triangle (see Section 2.10 Area of a triangle).

The chord AB cuts the circle into two segments: minor segment ACB (the shaded portion at right) and major segment ADB.

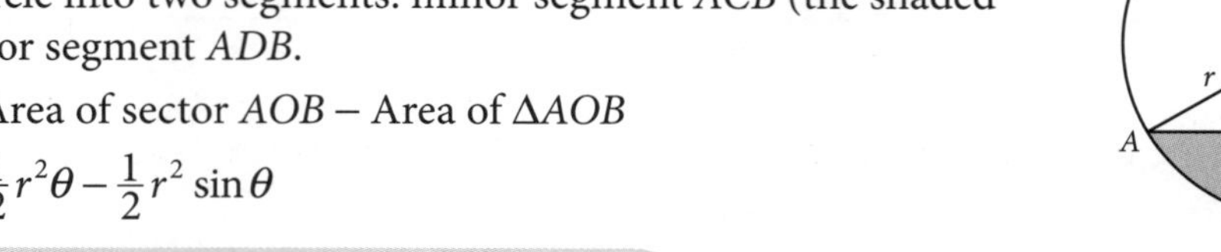

Area of segment ACB = Area of sector AOB – Area of ΔAOB

$$= \frac{1}{2}r^2\theta - \frac{1}{2}r^2 \sin\theta$$

Area of segment $= \dfrac{1}{2}r^2(\theta - \sin\theta)$, where θ is in radians.

Example 5

An arc AB subtends an angle of $126° 52'$ at the centre O of a circle with radius 6 cm. Calculate:

(a) the length of the arc AB, correct to 1 decimal place

(b) the length of the chord AB, correct to 1 decimal place

(c) the area of the sector AOB, to the nearest cm^2

(d) the area of the minor segment cut by the chord AB, to the nearest cm^2.

Solution

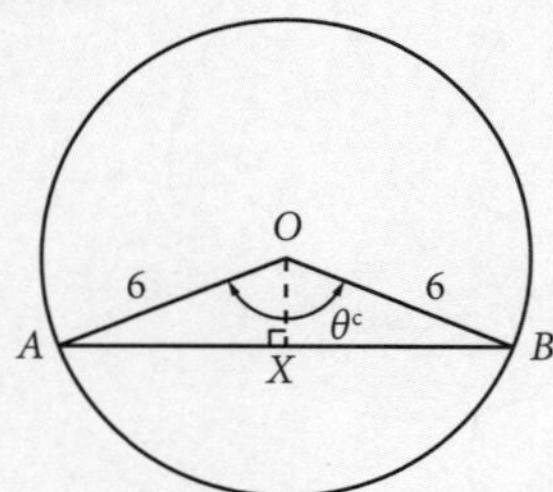

(a) $\theta = 126° 52', r = 6$

Use your calculator to convert $126° 52'$ to $126.866...°$, then:

$\theta = 126.867 \times \dfrac{\pi}{180} = 2.214^c$

$\ell = r\theta$

$\ell = 6 \times 2.214$

$= 13.284$

Length of arc $AB = 13.3\,\text{cm}$

(b) $AB = 2AX = 2 \times 6 \sin 63° 26'$

$= 10.73$

Length of chord $AB = 10.7\,\text{cm}$

(c) Area of sector $AOB = \dfrac{1}{2}r^2\theta$

$= \dfrac{1}{2} \times 36 \times 2.214$

$= 39.85$

Area of sector $AOB = 40\,\text{cm}^2$

(d) Area of minor segment $= \dfrac{1}{2}r^2(\theta - \sin\theta)$

$= \dfrac{1}{2} \times 36(2.214 - \sin 2.214)$ [Remember to make sure that your calculator is set to **radians**.]

$= 25.4$

Area of minor segment $= 25\,\text{cm}^2$

EXERCISE 6.2 ARC LENGTH AND SECTOR AREA OF A CIRCLE

1 Calculate, to the nearest minute of a degree, the size of the angle subtended at the centre of a circle of radius 8 cm by an arc whose length is 15 cm.

2 Find the length of an arc of a circle of radius 15 cm if the arc subtends an angle of 70° at the centre.

3 An arc of a circle subtends an angle of 100° at the centre. If the radius of the circle is 12 cm, then the length of the arc in terms of π is:

A 40π cm   **B** $\frac{108}{5\pi}$ cm   **C** $\frac{20\pi}{3}$ cm   **D** $\frac{108\pi}{5}$ cm

4 The area of a sector AOB of a circle with centre O, radius 20 cm is 240 cm^2. Calculate:

(a) the size of $\angle AOB$, in radians   **(b)** the length of the arc AB **(c)** the length of the chord AB.

5 A chord PQ, 24 cm long, is 5 cm from the centre of a circle. Calculate the length of the arc PQ correct to 1 decimal place.

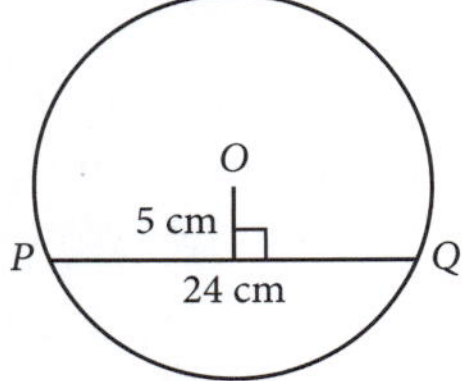

6 A point P is 8 cm from the centre of a circle of radius 5 cm. Find the length of the major arc between the points where the circle touches the tangents that are drawn from P to the circle.

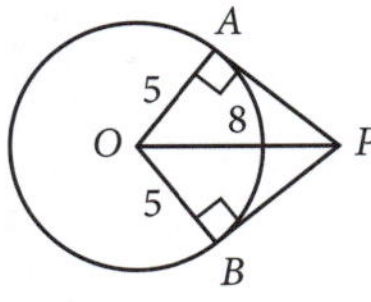

7 In a circle with centre O, the length of the chord AB is 16 cm. If the radius of the circle is 10 cm, calculate:

(a) the size of $\angle AOB$ **(b)** the length of the minor arc AB

(c) the area of the minor segment formed by the chord AB.

8 The length of the minute hand of a clock is 20 cm. Calculate:

(a) the length of the arc that the tip of the hand travels along in 16 minutes

(b) the shortest distance between the initial and final positions of the tip of the hand.

9 An arc AB subtends an angle of 60° at the centre O of a circle of radius 15 cm. Indicate whether each statement below is correct or incorrect.

(a) Arc $AB = 5\pi$ cm   **(b)** Arc $AB = 15$ cm

(c) Chord $AB = 15$ cm   **(d)** Area of sector $AOB = \frac{75\pi}{2}$ cm^2

10 From a circular piece of metal 6 cm in diameter, a sector of angle 30° is removed. Find the area that remains, expressing your answer in terms of π.

11 A chord subtends an angle of 140° at the centre of a circle of radius 16 cm. Find the difference in length between the chord and the arc.

12 For the diagram, calculate (in terms of π):

(a) the area of the shaded region between the two circular arcs

(b) the perimeter of the shaded region.

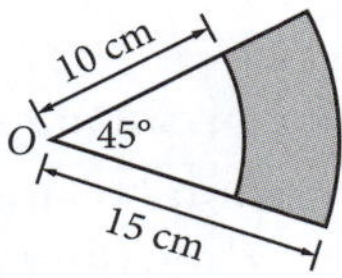

13 The minute hand of a clock is 3 cm long. What area is swept out by the hand in a time interval of 40 minutes? Express your answer in terms of π.

14 A circular metal plate is cut into two segments along a chord equal in length to the radius. What is the ratio of the areas of the two segments?

15 A sheep, grazing in a paddock, is tethered to a stake by a rope 20 m long. If the stake is 10 m from a fence, find the area (to the nearest m^2) over which the sheep can graze.

16 A sector of angle 160° is cut out of a circular piece of thin cardboard of radius 10 cm. The cut edges of this sector are brought together to shape the cardboard into a cone. Find the circumference of the circular base of the cone.

17 A piece of land is in the shape of a circular sector of radius 25 metres and angle 30°. Find the piece of land's: **(a)** perimeter **(b)** area.

18 A pendulum 40 cm long swings through an angle of 25°. Find:

(a) the length of the arc

(b) the shortest distance between the extreme positions of the bob (bottom of the pendulum).

19 Two circles, each of radius 10 cm, have their centres 16 cm apart. Calculate the area common to the two circles.

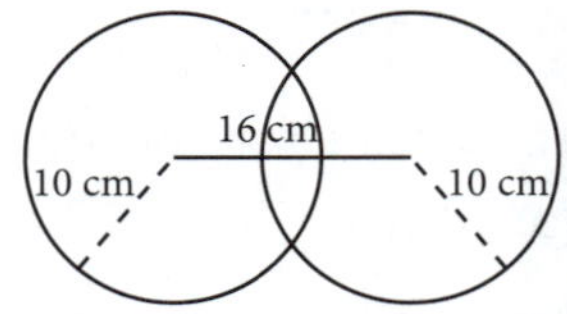

20 Three circles are drawn, each with a radius of 5 cm. Their centres are at the vertices of an equilateral triangle with sides of length 10 cm. Calculate the area that is between the circles and inside the triangle.

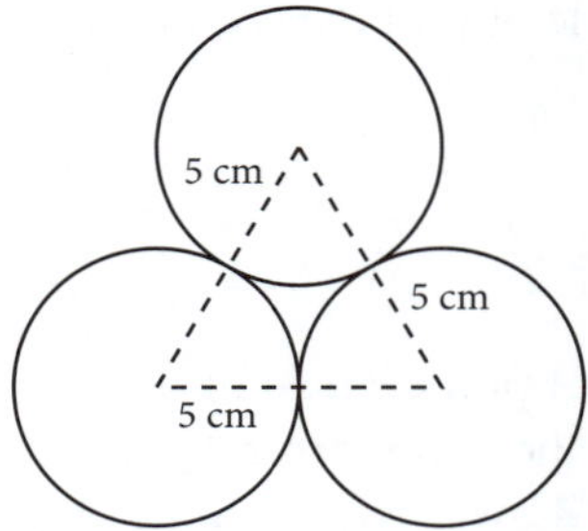

21 The points K, L, N and M are on the circumference of a circle with centre O and radius r. $KL = KM = r\sqrt{3}$.

(a) Use the cosine rule, or otherwise, to find $\angle LKO$.

(b) Calculate the area of ΔKLM.

(c) Write the length of LM.

(d) By finding the area of the circle, or otherwise, calculate the area of the segment LMN.

(e) A circle with centre K is drawn through the points L and M. Calculate the area between the arc LNM and this new arc joining L and M.

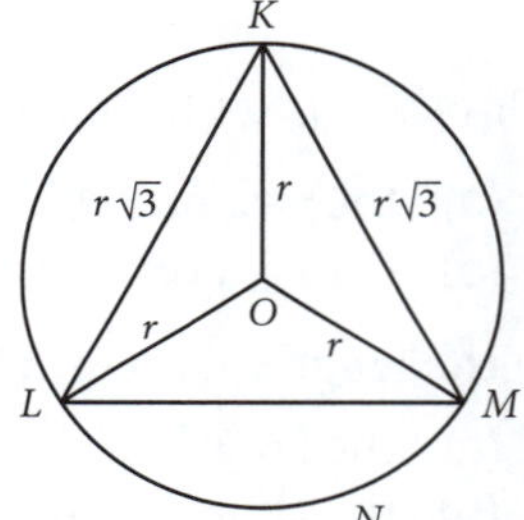

6.3 ANGLES OF ANY MAGNITUDE—RADIANS

This topic was introduced in Chapter 2 (Section 2.2 Angles of any magnitude) using degrees. You can now investigate the same topic using radians.

Definitions

You can define circular or trigonometric functions as functions of real numbers. For this purpose, consider a circle of unit radius defined by the equation $x^2 + y^2 = 1$.

Starting from the point A, you can mark an arc of length θ for any real number θ. The arc AP subtends an angle of θ^c at the centre of the circle. In this way, the same number θ measures the length of the arc as well as the number of radians in $\angle AOP$. For $\theta > 0$, you mark the arc in an anticlockwise direction; for $\theta < 0$ you mark the arc in a clockwise direction. If $P(\theta)$ is the endpoint of this arc, with coordinates (x, y), then you can define the functions cosine and sine as:

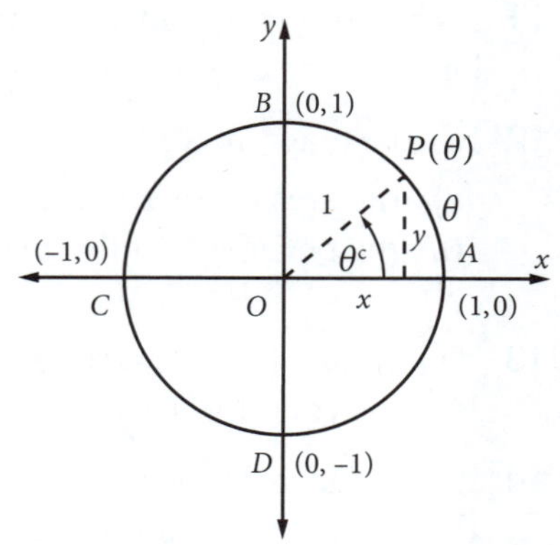

$\cos\theta = x$ $\sin\theta = y$

Both these functions have the set of real numbers as their domain. Because $-1 \leq x \leq 1$ and $-1 \leq y \leq 1$, the range of each function is from -1 to 1 inclusive. Considering the unit circle diagram above and our definition of x and y:

- At A, $\theta = 0$; the coordinates of A are $(1, 0)$ and so $\cos 0 = 1$ and $\sin 0 = 0$.
- At B, $\theta = \frac{\pi}{2}$; the coordinates of B are $(0, 1)$ and so $\cos \frac{\pi}{2} = 0$ and $\sin \frac{\pi}{2} = 1$.
- At C, $\theta = \pi$; the coordinates of C are $(-1, 0)$ and so $\cos \pi = -1$ and $\sin \pi = 0$.
- At D, $\theta = \frac{3\pi}{2}$; the coordinates of D are $(0, -1)$ and so $\cos \frac{3\pi}{2} = 0$ and $\sin \frac{3\pi}{2} = -1$.
- At A again, $\theta = 2\pi$; the coordinates of A are $(1, 0)$ and so $\cos 2\pi = 1$ and $\sin 2\pi = 0$.
- The circumference of the unit circle is 2π.

Notice that $\cos 2\pi = \cos 0 = 1$ and $\sin 2\pi = \sin 0 = 0$. In fact, the points $P(\theta)$, $P(2\pi + \theta)$, $P(4\pi + \theta)$ etc. all coincide. In general, $P(\theta)$ and $P(2k\pi + \theta)$ coincide, where k is any integer. This allows us to write $\cos(2\pi + \theta) = \cos\theta$ and $\sin(2\pi + \theta) = \sin\theta$. Functions whose values recur at regular intervals like this (in this case, with a period of 2π) are called **periodic functions**.

Aside from sine and cosine, there are four other trigonometric functions: tangent, cotangent, secant and cosecant (abbreviated as tan, cot, sec and cosec, respectively) that can all be defined in terms of cos and sin:

$$\tan\theta = \frac{y}{x} = \frac{\sin\theta}{\cos\theta}, \quad \cos\theta \neq 0 \qquad \sec\theta = \frac{1}{x} = \frac{1}{\cos\theta}, \quad \cos\theta \neq 0$$

$$\cot\theta = \frac{x}{y} = \frac{\cos\theta}{\sin\theta}, \quad \sin\theta \neq 0 \qquad \operatorname{cosec}\theta = \frac{1}{y} = \frac{1}{\sin\theta}, \quad \sin\theta \neq 0$$

Thus there are restrictions on the domains of these functions. The functions tan and sec are not defined for $\cos\theta = 0$, i.e. for $\theta = 0, \frac{\pi}{2}, \frac{3\pi}{3}, \frac{5\pi}{2}, \ldots$; cot and cosec are not defined for $\sin\theta = 0$, i.e. for $\theta = 0, \pi, 2\pi, \ldots$

Also note that cot, sec and cosec are the reciprocal functions of tan, cos and sin respectively.

Symmetry properties of trigonometric functions

The results established in Chapter 2 Trigonometry for $90° - \theta$, $180° \pm \theta$, $360° \pm \theta$ and $-\theta$ are still true when you change from degrees to radians measure.

Degrees	Radians
$\sin(90° - \theta) = \cos\theta$	$\sin\left(\frac{\pi}{2} - \theta\right) = \cos\theta$
$\cos(90° - \theta) = \sin\theta$	$\cos\left(\frac{\pi}{2} - \theta\right) = \sin\theta$
$\tan(90° - \theta) = \cot\theta$	$\tan\left(\frac{\pi}{2} - \theta\right) = \cot\theta$
$\sin(180° - \theta) = \sin\theta$	$\sin(\pi - \theta) = \sin\theta$
$\cos(180° - \theta) = -\cos\theta$	$\cos(\pi - \theta) = -\cos\theta$
$\tan(180° - \theta) = -\tan\theta$	$\tan(\pi - \theta) = -\tan\theta$
$\sin(180° + \theta) = -\sin\theta$	$\sin(\pi + \theta) = -\sin\theta$
$\cos(180° + \theta) = -\cos\theta$	$\cos(\pi + \theta) = -\cos\theta$
$\tan(180° + \theta) = \tan\theta$	$\tan(\pi + \theta) = \tan\theta$
$\sin(360° - \theta) = -\sin\theta$	$\sin(2\pi - \theta) = -\sin\theta$
$\cos(360° - \theta) = \cos\theta$	$\cos(2\pi - \theta) = \cos\theta$
$\tan(360° - \theta) = -\tan\theta$	$\tan(2\pi - \theta) = -\tan\theta$
$\sin(360° + \theta) = \sin\theta$	$\sin(2\pi + \theta) = \sin\theta$
$\cos(360° + \theta) = \cos\theta$	$\cos(2\pi + \theta) = \cos\theta$
$\tan(360° + \theta) = \tan\theta$	$\tan(2\pi + \theta) = \tan\theta$

Sign of the trigonometric ratios

The sign of sin, cos and tan for the first four quadrants can be summarised as follows:

- First quadrant: All are positive (A)
- Second quadrant: sin only is positive (S)
- Third quadrant: tan only is positive (T)
- Fourth quadrant: cos only is positive (C)

You may find this easier to remember using the mnemonic **ASTC**: '**A**ll **S**tations **T**o **C**entral'.

Exact values in radians

θ	0	$\frac{\pi}{6}$	$\frac{\pi}{4}$	$\frac{\pi}{3}$	$\frac{\pi}{2}$
$\sin\theta$	0	$\frac{1}{2}$	$\frac{1}{\sqrt{2}}$	$\frac{\sqrt{3}}{2}$	1
$\cos\theta$	1	$\frac{\sqrt{3}}{2}$	$\frac{1}{\sqrt{2}}$	$\frac{1}{2}$	0
$\tan\theta$	0	$\frac{1}{\sqrt{3}}$	1	$\sqrt{3}$	undefined
$\operatorname{cosec}\theta$	undefined	2	$\sqrt{2}$	$\frac{2}{\sqrt{3}}$	1
$\sec\theta$	1	$\frac{2}{\sqrt{3}}$	$\sqrt{2}$	2	undefined
$\cot\theta$	undefined	$\sqrt{3}$	1	$\frac{1}{\sqrt{3}}$	0

Example 6

Find the exact values of: **(a)** $\sin\frac{5\pi}{6}$ **(b)** $\cos\frac{5\pi}{4}$ **(c)** $\tan\frac{4\pi}{3}$ **(d)** $\sin\frac{3\pi}{2}$ **(e)** $\cos\left(-\frac{5\pi}{3}\right)$ **(f)** $\sec\frac{2\pi}{3}$ **(g)** $\operatorname{cosec}\frac{11\pi}{6}$

Solution

(a) $\sin\frac{5\pi}{6} = \sin\left(\pi - \frac{\pi}{6}\right) = \sin\frac{\pi}{6} = \frac{1}{2}$

(b) $\cos\frac{5\pi}{4} = \cos\left(\pi + \frac{\pi}{4}\right) = -\cos\frac{\pi}{4} = -\frac{1}{\sqrt{2}}$

(c) $\tan\frac{4\pi}{3} = \tan\left(\pi + \frac{\pi}{3}\right) = \tan\frac{\pi}{3} = \sqrt{3}$

(d) $\sin\frac{3\pi}{2} = \sin\left(\pi + \frac{\pi}{2}\right) = -\sin\frac{\pi}{2} = -1$

(e) $\cos\left(-\frac{5\pi}{3}\right) = \cos\frac{5\pi}{3} = \cos\left(2\pi - \frac{\pi}{3}\right) = \cos\frac{\pi}{3} = \frac{1}{2}$

(f) $\sec\frac{2\pi}{3} = \sec\left(\pi - \frac{\pi}{3}\right) = -\sec\frac{\pi}{3} = -\frac{1}{\cos\frac{\pi}{3}} = -2$

(g) $\operatorname{cosec}\frac{11\pi}{6} = \operatorname{cosec}\left(2\pi - \frac{\pi}{6}\right) = -\operatorname{cosec}\frac{\pi}{6} = -\frac{1}{\sin\frac{\pi}{6}} = -2$

Example 7

Find all values of θ, $0 < \theta < 2\pi$, for which:

(a) $\cos\theta = \frac{1}{2}$ (b) $\sin\theta = -\frac{1}{\sqrt{2}}$ (c) $\tan\theta = 1$ (d) $\sec\theta = -\frac{2}{\sqrt{3}}$ (e) $\sin\theta = -1$

Solution

(a) $\cos\theta > 0$, so θ is in the 1st and 4th quadrants: $\cos\theta = \frac{1}{2}$: $\theta = \frac{\pi}{3}, 2\pi - \frac{\pi}{3}$

$\theta = \frac{\pi}{3}, \frac{5\pi}{3}$

(b) $\sin\theta < 0$, so θ is in the 3rd and 4th quadrants: $\sin\theta = -\frac{1}{\sqrt{2}}$: $\theta = \pi + \frac{\pi}{4}, 2\pi - \frac{\pi}{4}$

$\theta = \frac{5\pi}{4}, \frac{7\pi}{4}$

(c) $\tan\theta > 0$, so θ is in the 1st and 3rd quadrants: $\tan\theta = 1$: $\theta = \frac{\pi}{4}, \pi + \frac{\pi}{4}$

$\theta = \frac{\pi}{4}, \frac{5\pi}{4}$

(d) $\sec\theta < 0$, so θ is in the 2nd and 3rd quadrants: $\sec\theta = -\frac{2}{\sqrt{3}}$, $\cos\theta = -\frac{\sqrt{3}}{2}$: $\theta = \pi - \frac{\pi}{6}, \pi + \frac{\pi}{6}$

$\theta = \frac{5\pi}{6}, \frac{7\pi}{6}$

(e) $\sin\theta < 0$, so θ is in the 3rd and 4th quadrants: $\sin\theta = -1$: $\theta = \pi + \frac{\pi}{2}, 2\pi - \frac{\pi}{2}$

$\theta = \frac{3\pi}{2}$

EXERCISE 6.3 ANGLES OF ANY MAGNITUDE—RADIANS

1 Express each of the following as a simpler trigonometric function.

(a) $\sin(\pi - x)$ (b) $\cos\left(\frac{\pi}{2} - x\right)$ (c) $\tan(2\pi - x)$

(d) $\cos(\pi + x)$ (e) $\sin(2\pi - x)$ (f) $\cot\left(\frac{\pi}{2} - x\right)$

2 For any angle θ, $\cos(\pi - \theta)$ is equal to:

A $-\cos\theta$ B $\cos\theta$ C $\sin\theta$ D $-\sin\theta$

3 Indicate whether each statement is correct or incorrect.

(a) $\cos\left(\frac{\pi}{2} - \theta\right) = \cos\theta$ (b) $\cos(2\pi - \theta) = \cos\theta$ (c) $\sin(\pi + \theta) = \sin\theta$ (d) $\sin(2\pi - \theta) = -\sin\theta$

4 If $\sin x = 0.2$, write the value of:

(a) $\sin(\pi - x)$ (b) $\sin(2\pi - x)$ (c) $\sin(-x)$ (d) $\cos\left(\frac{\pi}{2} - x\right)$ (e) $\sin(\pi + x)$ (f) $\operatorname{cosec} x$

5 If $\tan\theta = t$, express the following in terms of t:

(a) $\cot\theta$ (b) $\cot\left(\frac{\pi}{2} - \theta\right)$ (c) $\tan(\pi - \theta)$ (d) $\tan(2\pi - \theta)$ (e) $\cot(\pi - \theta)$ (f) $\tan(\pi + \theta)$

6 If $\cos x = c$, express the following in terms of c:

(a) $\sec x$ (b) $\cos(-x)$ (c) $\cos(\pi - x)$ (d) $\cos(2\pi - x)$ (e) $\sec(-x)$ (f) $\cos(\pi + x)$

7 Write the exact value of:

(a) $\sin\frac{\pi}{2}$ (b) $\cos\frac{2\pi}{3}$ (c) $\tan\frac{5\pi}{6}$ (d) $\cos\pi$

(e) $\sec\frac{3\pi}{4}$ (f) $\cot\frac{5\pi}{6}$ (g) $\operatorname{cosec}\frac{\pi}{2}$ (h) $\sin\frac{2\pi}{3}$

8 Write the exact value of:

(a) $\sin\pi$ (b) $\cos\frac{7\pi}{6}$ (c) $\tan\frac{5\pi}{4}$ (d) $\cot\frac{4\pi}{3}$

(e) $\cos\frac{4\pi}{3}$ (f) $\sec\frac{5\pi}{4}$ (g) $\tan\pi$ (h) $\sin\frac{7\pi}{6}$

9 Write the exact value of:

(a) $\sin\frac{3\pi}{2}$ (b) $\tan\frac{5\pi}{3}$ (c) $\operatorname{cosec}\frac{11\pi}{6}$ (d) $\tan\frac{7\pi}{4}$

(e) $\cot\frac{7\pi}{4}$ (f) $\cos\frac{11\pi}{6}$ (g) $\sin\frac{5\pi}{3}$ (h) $\operatorname{cosec}\frac{5\pi}{3}$

10 Write the exact value of:

(a) $\sin 2\pi$ (b) $\sin\frac{13\pi}{6}$ (c) $\tan\frac{9\pi}{4}$ (d) $\cot\frac{7\pi}{3}$

(e) $\sec\frac{13\pi}{6}$ (f) $\cos\frac{5\pi}{2}$ (g) $\sin\frac{7\pi}{3}$ (h) $\tan\frac{11\pi}{4}$

11 For $\frac{\pi}{2} < x < \pi$, use a diagram of a unit circle to show that:

(a) $\cos(\pi + x) = -\cos x$ (b) $\sin(2\pi - x) = -\sin x$

12 If θ is an angle in the 2nd quadrant, state whether the following are positive or negative:

(a) $\cos(\pi - \theta)$ (b) $\tan(\pi + \theta)$ (c) $\sin\left(\frac{\pi}{2} - \theta\right)$

(d) $\sin(2\pi - \theta)$ (e) $\cos(\pi + \theta)$ (f) $\tan\left(\frac{\pi}{2} - \theta\right)$

13 Solve, for $0 < x < 2\pi$:

(a) $\sin x = -\frac{\sqrt{3}}{2}$ (b) $\tan x = -1$ (c) $\cos x = -1$ (d) $\cot x = \sqrt{3}$

(e) $\sec x = -\sqrt{2}$ (f) $\sin x = \cos x$ (g) $\sin x = 0$ (h) $2\cos x + 1 = 0$

(i) $2\sin x = \sqrt{3}$ (j) $\sin x + \sqrt{3}\cos x = 0$ (k) $\operatorname{cosec} x = \sec x$

14 Using a diagram, find equivalent expressions for:

(a) $\cos\left(\frac{3\pi}{2} + x\right)$ (b) $\tan\left(\frac{3\pi}{2} - x\right)$ (c) $\sin\left(\frac{3\pi}{2} - x\right)$

6.4 GRAPHS OF TRIGONOMETRIC FUNCTIONS USING RADIANS

In the diagram below, the graph of $y = \sin x$ is the unbroken curve and the graph of $y = \cos x$ is the dashed curve. The graphs are drawn in the domain $-2\pi \le x \le 2\pi$. For a domain of real numbers, the graphs continue to repeat in both directions, so the graphs for the domains $-2\pi \le x \le 0$, $0 \le x \le 2\pi$ and $2\pi \le \theta \le 4\pi$ are the same.

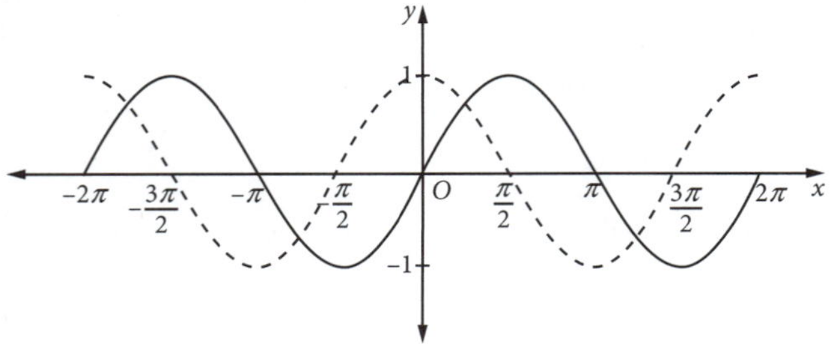

As x increases, the values of $\sin x$ and $\cos x$ repeat themselves after an interval or **period** of 2π. Sine and cosine are therefore called **periodic functions**. This means that the points $P(x)$, $P(2\pi + x)$, $P(4\pi + x)$ on the unit circle all coincide and hence $\sin(2\pi + x) = \sin x$, $\cos(2\pi + x) = \cos x$ and so on.

The maximum and minimum values of $\sin x$ and $\cos x$ are 1 and -1 respectively, so we say that their **amplitude** is 1.

MAKING CONNECTIONS

Trigonometric functions and the unit circle with radians

Move the point around the unit circle to explore the construction of the sine and cosine graphs.

If the graph of $y = \cos x$ is translated $\frac{\pi}{2}$ units to the right, parallel with the x-axis, it coincides with the graph of $y = \sin x$. You can check that this follows from $\cos x = \sin\left(\frac{\pi}{2} + x\right)$. You can also check by sketching graphs by hand or by using graphing software.

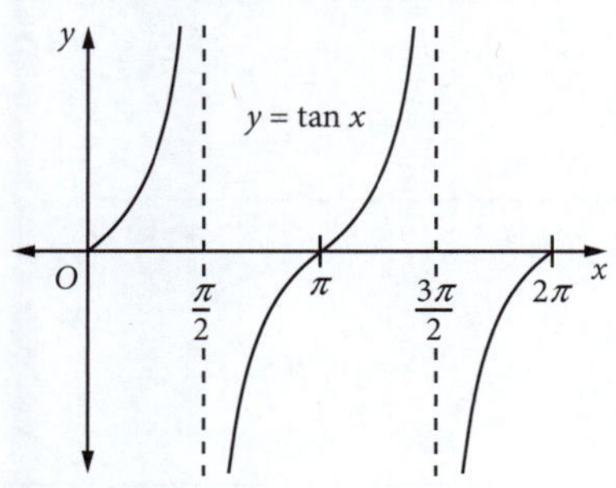

The diagram shows the graph of $y = \tan x$ for $0 \le x \le 2\pi$. As x increases, the values of $\tan x$ repeat after an interval or **period** of π. Because $\tan x = \frac{\sin x}{\cos x}$, the graph of $y = \tan x$ does not exist when $\cos x = 0$, so the tan function is undefined at $x = \frac{\pi}{2}, \frac{3\pi}{2}, \ldots$

The domain of $\tan x$ is all real x except for $x = \frac{(2n-1)\pi}{2}$, where n is an integer (i.e. odd multiples of $\frac{\pi}{2}$). The range of $\tan x$ is all real y.

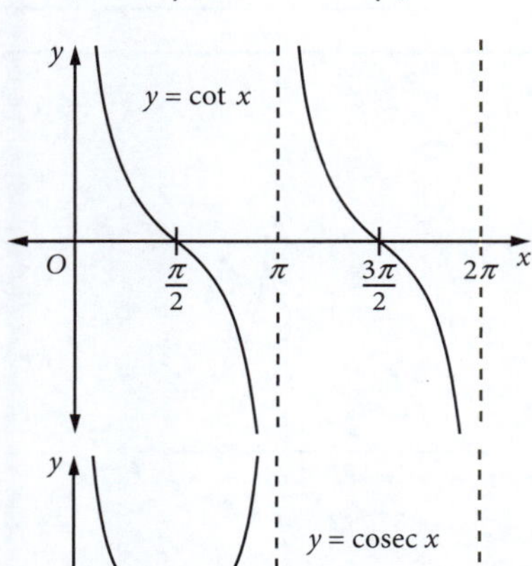

The diagram shows the graph of $y = \cot x$ for $0 \le x \le 2\pi$. As x increases, the values of $\cot x$ repeat after an interval or **period** of π.

The cot function is undefined at $x = 0, \pi, 2\pi, \ldots$

The domain of $\cot x$ is all real x except for $x = n\pi$, where n is an integer (i.e. whole multiples of π). The range of $\cot x$ is all real y.

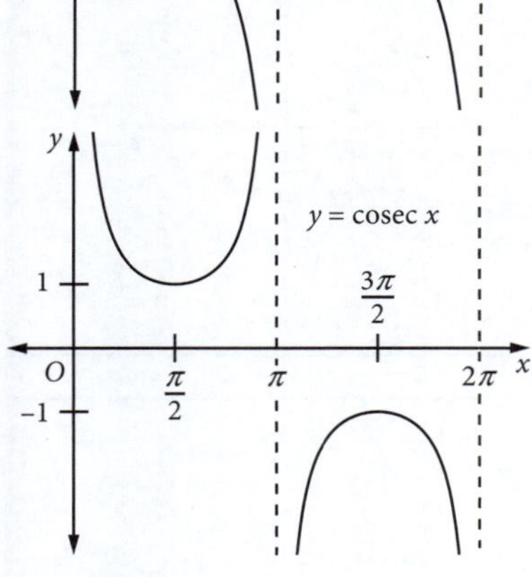

The diagram shows the graph of $y = \operatorname{cosec} x$ for $0 \le x \le 2\pi$. As x increases, the values of $\operatorname{cosec} x$ repeat after an interval or **period** of 2π.

The cosec function is undefined at $x = 0, \pi, 2\pi, \ldots$

The domain of $\operatorname{cosec} x$ is all real x except for $x = n\pi$, where n is an integer (i.e. whole multiples of π). The range of $\operatorname{cosec} x$ is $|y| \ge 1$.

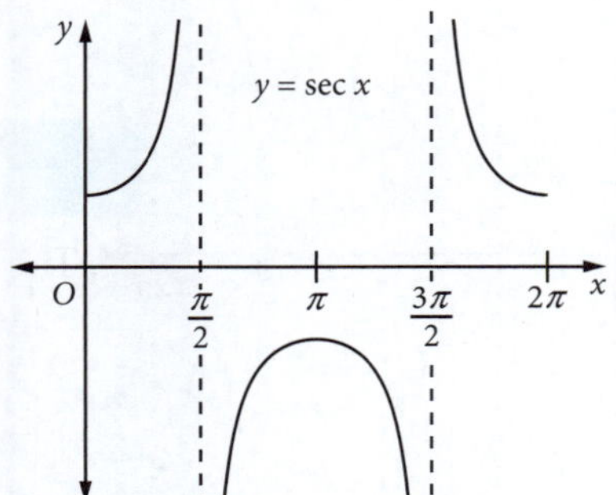

The diagram shows the graph of $y = \sec x$ for $0 \le x \le 2\pi$. As x increases, the values of $\sec x$ repeat after an interval or **period** of 2π.

The sec function is undefined at $x = \frac{\pi}{2}, \frac{3\pi}{2}, \ldots$

The domain of $\sec x$ is all real x except for $x = \frac{(2n-1)\pi}{2}$, where n is an integer (i.e. odd multiples of $\frac{\pi}{2}$).

The range of $\sec x$ is $|y| \ge 1$.

The graphs of the six circular functions can be drawn as pairs of complementary functions:

- sine and cosine—because $\cos x = \sin\left(\frac{\pi}{2} - x\right)$ and $\sin x = \cos\left(\frac{\pi}{2} - x\right)$
- tangent and cotangent—because $\cot x = \tan\left(\frac{\pi}{2} - x\right)$ and $\tan x = \cot\left(\frac{\pi}{2} - x\right)$
- cosecant and secant—because $\operatorname{cosec} x = \sec\left(\frac{\pi}{2} - x\right)$ and $\sec x = \operatorname{cosec}\left(\frac{\pi}{2} - x\right)$.

Reciprocal functions

The trigonometric functions can also be considered as pairs of reciprocal functions: sine and cosecant, cosine and secant, and tangent and cotangent. Because these can be defined according to $\operatorname{cosec} x = \frac{1}{\sin x}$, $\sec x = \frac{1}{\cos x}$ and $\cot x = \frac{1}{\tan x}$, you now have another way of sketching these trigonometric functions. For example, to draw $y = \operatorname{cosec} x$, first draw $y = \sin x$ and then take the reciprocal of each y value to obtain the graph of $y = \operatorname{cosec} x$.

Example 8

Use the graph of $y = \sin x$ to sketch the graph of $y = \operatorname{cosec} x$ for $0 \leq x \leq 2\pi$.

Solution

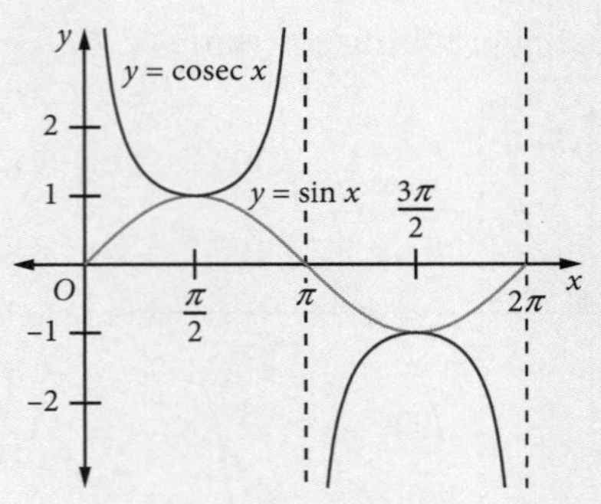

For $y = 0$, $\frac{1}{y}$ is undefined. For $y = 1$, $\frac{1}{y} = 1$, so the curves intersect at this point. For $y = -1$, $\frac{1}{y} = -1$, so the curves also intersect at this point.

For $0 < y < 1$, $\frac{1}{y} > 1$. Because $\sin\frac{\pi}{6} = \frac{1}{2}$, $\operatorname{cosec}\frac{\pi}{6} = 2$.

For $-1 < y < 0$, $\frac{1}{y} < -1$. Because $\sin\frac{7\pi}{6} = -\frac{1}{2}$, $\operatorname{cosec}\frac{7\pi}{6} = -2$.

Example 9

Use the graph of $y = \cos x$ to sketch the graph of $y = \sec x$ for $0 \leq x \leq 2\pi$.

Solution

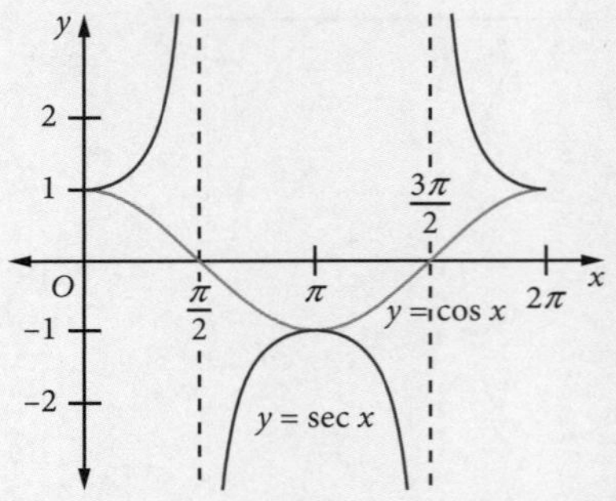

For $y = 0$, $\frac{1}{y}$ is undefined. For $y = 1$, $\frac{1}{y} = 1$, so the curves intersect at this point. For $y = -1$, $\frac{1}{y} = -1$, so the curves also intersect at this point.

For $0 < y < 1$, $\frac{1}{y} > 1$. Because $\cos\frac{\pi}{3} = \frac{1}{2}$, $\sec\frac{\pi}{3} = 2$.

For $-1 < y < 0$, $\frac{1}{y} < -1$. Because $\cos\frac{2\pi}{3} = -\frac{1}{2}$, $\sec\frac{2\pi}{3} = -2$.

Example 10

Use the graph of $y = \tan x$ to sketch the graph of $y = \cot x$ for $0 \leq x \leq 2\pi$.

Solution

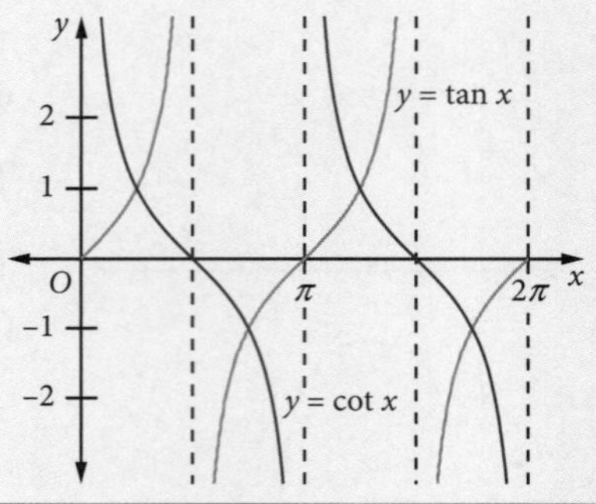

For $y = 0$, $\frac{1}{y}$ is undefined. For $y = 1$, $\frac{1}{y} = 1$, so the curves intersect at this point. For $y = -1$, $\frac{1}{y} = -1$, so the curves also intersect at this point.

For $0 < y < 1$, $\frac{1}{y} > 1$. Because $\tan\frac{\pi}{4} = \frac{1}{\sqrt{2}}$, $\cot\frac{\pi}{4} = \sqrt{2}$.

For $-1 < y < 0$, $\frac{1}{y} < -1$. Because $\tan\frac{3\pi}{4} = -\frac{1}{\sqrt{2}}$, $\cot\frac{3\pi}{4} = -\sqrt{2}$.

To investigate these graphs further, use graphing software to draw the following pairs of graphs on the same axes:

- $y = \sin x$ and $y = \operatorname{cosec} x$

Select the scale for each axis carefully. Start with $0 \leq x \leq 7$ and $-5 \leq y \leq 5$ and investigate what happens as you change the dimensions of the graph window.

Remember that as $y \to 0$, $\frac{1}{y} \to \infty$ and so the function will be undefined. The only functions that have an amplitude are $\sin x$ and $\cos x$. Also note that the period of a reciprocal function is the same as the original function.

EXERCISE 6.4 GRAPHS OF TRIGONOMETRIC FUNCTIONS USING RADIANS

1 Sketch the graph of $y = \sin x$, $-\pi \le x \le \pi$. On the same axes, sketch the graph of:

(a) $y = -\sin x$ (b) $y = 1 - \sin x$ (c) $y = -\cos x$

2 Sketch the graph of each of the following, stating the period and amplitude of the function:

(a) $y = \sin \pi x$, $-1 \le x \le 1$ (b) $y = \cos 2\pi x$, $0 \le x \le 2$ (c) $y = \tan \frac{\pi x}{2}$, $0 \le x \le 2$

3 Sketch the graph of $y = \tan x$, $-\pi \le x \le \pi$. On the same axes, sketch the graph of:

(a) $y = 2\tan x$ (b) $y = -2\tan x$ (c) $y = 2 - 2\tan x$

6.5 TRIGONOMETRIC IDENTITIES AND PROOFS

Because the equation of the unit circle is: $x^2 + y^2 = 1$

and, by definition: $x = \cos\theta$, $y = \sin\theta$

it follows that:

$$\cos^2\theta + \sin^2\theta = 1 \qquad [1]$$
$$\cos^2\theta = 1 - \sin^2\theta$$
$$\sin^2\theta = 1 - \cos^2\theta$$

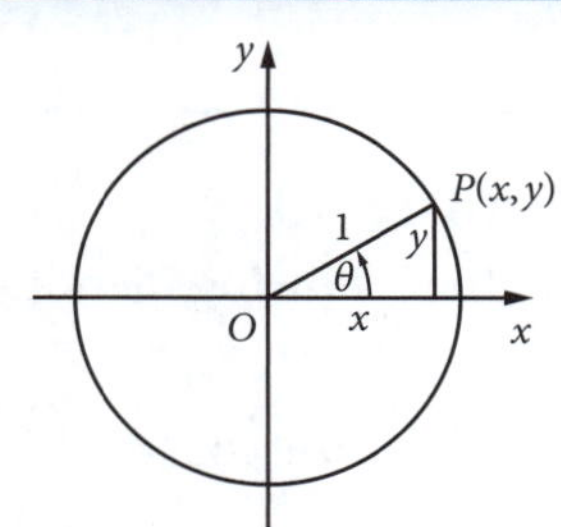

Dividing [1] by $\cos^2\theta$: $$1 + \tan^2\theta = \sec^2\theta \qquad [2]$$

Dividing [1] by $\sin^2\theta$: $$\cot^2\theta + 1 = \operatorname{cosec}^2\theta \qquad [3]$$

Statements [1], [2] and [3] are called **identities** as they are true for all values of θ.

Example 11

If $\sin\theta = \frac{3}{5}$ and $\frac{\pi}{2} < \theta < \pi$, find the exact values of $\cos\theta$ and $\tan\theta$.

Solution

Using: $\cos^2\theta + \sin^2\theta = 1$

$\sin\theta = \frac{3}{5}$: $\cos^2\theta + \frac{9}{25} = 1$

$$\cos^2\theta = \frac{16}{25}$$

So: $\cos\theta = \pm\frac{4}{5}$; because it is given that $\frac{\pi}{2} < \theta < \pi$, we have $\cos\theta = -\frac{4}{5}$

$$\tan\theta = \frac{\sin\theta}{\cos\theta} = \frac{\frac{3}{5}}{-\frac{4}{5}}$$

$$\tan\theta = -\frac{3}{4}$$

Alternatively: in a right-angled triangle with a side length of 3 and a hypotenuse length of 5, you can calculate that the third side length is 4. You also know from $\frac{\pi}{2} < \theta < \pi$ that the angle is in the second quadrant, so the x value will be negative. Hence from $\sin\theta = \frac{3}{5}$ you can obtain $\cos\theta = -\frac{4}{5}$ and $\tan\theta = -\frac{3}{4}$.

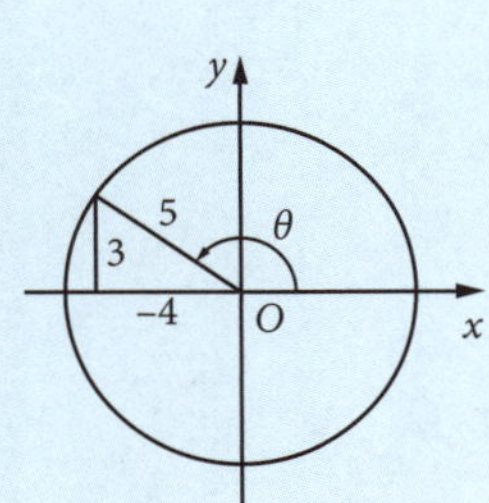

Example 12

Simplify:

(a) $\dfrac{1-\cos^2\theta}{1-\sin^2\theta}$ (b) $1+\tan^2(\frac{\pi}{2}-\theta)$ (c) $\sqrt{a^2+x^2}$, for $x = a\tan\theta$ and $0<\theta<\frac{\pi}{2}$

Solution

(a) $\dfrac{1-\cos^2\theta}{1-\sin^2\theta}$

$=\dfrac{\sin^2\theta}{\cos^2\theta}$

$=\tan^2\theta$

(b) $1+\tan^2(\frac{\pi}{2}-\theta)$

$=1+\cot^2\theta$

$=\operatorname{cosec}^2\theta$

(c) $\sqrt{a^2+x^2}$

$=\sqrt{a^2+a^2\tan^2\theta}$

$=\sqrt{a^2\left(1+\tan^2\theta\right)}$

$=\sqrt{a^2\sec^2\theta}$

$=a\sec\theta$

Example 13

Prove the following identities.

(a) $\dfrac{1-\sin^2\theta}{\sin^2\theta+\cos^2\theta}=\cos^2\theta$ (b) $\dfrac{\cos\theta}{1-\sin\theta}-\tan\theta=\sec\theta$ (c) $\tan\theta\sin\theta+\cos\theta=\sec\theta$

Solution

(a) LHS $=\dfrac{1-\sin^2\theta}{\sin^2\theta+\cos^2\theta}$ Use $\cos^2\theta=1-\sin^2\theta$ and $\sin^2\theta+\cos^2\theta=1$

$=\dfrac{\cos^2\theta}{1}$

$=\cos^2\theta$

= RHS

(b) LHS $=\dfrac{\cos\theta}{1-\sin\theta}-\tan\theta$ Replace $\tan\theta$ with $\dfrac{\sin\theta}{\cos\theta}$

$=\dfrac{\cos\theta}{1-\sin\theta}-\dfrac{\sin\theta}{\cos\theta}$ Find common denominator

$=\dfrac{\cos^2\theta-\sin\theta(1-\sin\theta)}{(1-\sin\theta)\cos\theta}$ Expand

$=\dfrac{\cos^2\theta-\sin\theta+\sin^2\theta}{(1-\sin\theta)\cos\theta}$ Use $\sin^2\theta+\cos^2\theta=1$

$=\dfrac{1-\sin\theta}{(1-\sin\theta)\cos\theta}$ Cancel common factor

$=\dfrac{1}{\cos\theta}$ for $\sin\theta\neq 1$

$=\sec\theta$

= RHS

(c) LHS $= \tan\theta\sin\theta + \cos\theta$

$= \dfrac{\sin\theta}{\cos\theta} \times \sin\theta + \cos\theta$ Replace $\tan\theta$ with $\dfrac{\sin\theta}{\cos\theta}$

$= \dfrac{\sin^2\theta}{\cos\theta} + \dfrac{\cos\theta}{1}$ Find common denominator

$= \dfrac{\sin^2\theta + \cos^2\theta}{\cos\theta}$

$= \dfrac{1}{\cos\theta}$ Use $\sin^2\theta + \cos^2\theta = 1$

$= \sec\theta$

$=$ RHS

EXERCISE 6.5 TRIGONOMETRIC IDENTITIES AND PROOFS

1 If $\tan\theta = \frac{7}{24}$ and $\pi < \theta < \frac{3\pi}{2}$, find the exact value of:
(a) $\sin\theta$ (b) $\cos\theta$

2 If $\sin\theta = \frac{5}{13}$ and θ is acute, indicate whether each statement is correct or incorrect.
(a) $\cos\theta = \dfrac{12}{13}$ (b) $\sec\theta = \dfrac{13}{5}$ (c) $\tan\theta = \dfrac{5}{12}$ (d) $\cot\theta = \dfrac{13}{12}$

3 If $\tan\theta = -\frac{4}{3}$ and $\frac{\pi}{2} < \theta < \pi$, find the exact value of:
(a) $\cot\theta$ (b) $\sin\theta$ (c) $\cos\theta$

4 If $\cos u = \frac{2}{3}$ and u is not in the first quadrant, then $\dfrac{\cos u - 2\cot u}{\tan u - 3\sin u}$ simplifies to:
A $\dfrac{4(\sqrt{5}+6)}{15}$ B $\dfrac{5-2\sqrt{5}}{9}$ C $\dfrac{2(5-\sqrt{5})}{21}$ D $\dfrac{14}{3(4\sqrt{5}-5)}$

5 Simplify:
(a) $\dfrac{\sin^2\theta + \cos^2\theta}{\tan^2\theta}$ (b) $\dfrac{\sin^2\theta}{1-\sin^2\theta}$ (c) $\dfrac{2\cot\alpha}{1+\cot^2\alpha}$ (d) $(\sec^2\theta - 1)\tan(\frac{\pi}{2} - \theta)$
(e) $\dfrac{\sin A}{\cos A} + \dfrac{\cos A}{\sin A}$ (f) $\sin^3\theta + \sin\theta\cos^2\theta$ (g) $\operatorname{cosec}^2\theta\sin^2\theta$ (h) $1 - \sin^2(\pi + \theta)$

6 Simplify:
(a) $\dfrac{x^2}{\sqrt{x^2 - a^2}}$ for $x = a\sec\theta$ (b) $\sqrt{a^2 - x^2}$ for $x = a\cos\theta$

7 If $\sec\alpha = -\frac{17}{8}$ and $\frac{\pi}{2} < \alpha < \pi$, find the exact value of:
(a) $\cos\alpha$ (b) $\sin\alpha$ (c) $\tan\alpha$

8 If $\pi < \theta < \frac{3\pi}{2}$ and $\sin\theta = p$, express $\cos\theta$ in terms of p.

9 Find the exact value of $\sec\theta$ if $\tan\theta = 0.6$ and θ is not in the first quadrant.

10 If $\sin\theta = x$, express $\dfrac{1-\cos^2\theta}{\sec^2\theta}$ in terms of x.

11 If $a\sin^2\theta + b\cos^2\theta = c$, express $\sin\theta$ and $\cos\theta$ in terms of a, b and c.

12 If $x = a\sin\theta$ and $y = b\cos\theta$, find an expression from the two equations that does not involve θ.

13 If $\tan^2\theta + 2\sec^2\theta = 5$, find the value of $\sin^2\theta$.

14 Simplify:

(a) $(1+\tan^2 u)(1-\sin^2 u)$ (b) $\dfrac{1}{1-\sin V}+\dfrac{1}{1+\sin V}$ (c) $\dfrac{\sin\theta}{1+\cos\theta}+\dfrac{1+\cos\theta}{\sin\theta}$

(d) $\dfrac{1-\sin\theta}{1+\cos\theta}\times\dfrac{1+\sin\theta}{1-\cos\theta}$ (e) $\sin\theta\cos(\frac{\pi}{2}-\theta)+\cos\theta\sin(\frac{\pi}{2}-\theta)$ (f) $1-\dfrac{\sin A\cos A}{\tan A}$

(g) $2\cos^2\frac{\pi}{6}-1$ (h) $1-\sin\theta\cos(\frac{\pi}{2}-\theta)$

Prove the following identities.

15 $(1-\tan x)^2+(1+\tan x)^2=2\sec^2 x$

16 $(\cot t+\operatorname{cosec} t)^2=\dfrac{1+\cos t}{1-\cos t}$

17 $\sin^2\alpha\cos^2\beta-\cos^2\alpha\sin^2\beta=\sin^2\alpha-\sin^2\beta$

18 $\sec\theta+\tan\theta=\dfrac{1+\sin\theta}{\cos\theta}$

19 $\sin^2\theta\tan\theta+\cos^2\theta\cot\theta+2\sin\theta\cos\theta=\tan\theta+\cot\theta$

20 $\sin\theta(1+\tan\theta)+\cos\theta(1+\cot\theta)=\dfrac{\sin\theta+\cos\theta}{\sin\theta\cos\theta}$

21 $\tan\theta(1-\cot^2\theta)+\cot\theta(1-\tan^2\theta)=0$

22 $\dfrac{\cot\theta\cos\theta}{\cot\theta+\cos\theta}=\dfrac{\cos\theta}{1+\sin\theta}$

23 $\dfrac{1+\cot\theta}{\operatorname{cosec}\theta}-\dfrac{\sec\theta}{\tan\theta+\cot\theta}=\cos\theta$

24 $\dfrac{(\cos t\cot t-\sin t\tan t)\sin t\cos t}{\cos t-\sin t}=1+\sin t\cos t$

6.6 SOLVING TRIGONOMETRIC EQUATIONS

You will solve some simple trigonometric equations where the values are given in degrees. You will then solve equations where the answers are given in radians and learn how the question tells you the form of the answer.

To solve trigonometric equations you can use the same techniques that you have used for algebraic equations. You will also use trigonometric identities to help solve trigonometric equations.

Example 14

Solve each equation for $0°\le x\le 180°$.

(a) $3+2\sin x=5\sin x$ (b) $\sin x=2\cos x$ (c) $6\cos 2x=3\sin 30°$

Solution

(a)
$$3+2\sin x=5\sin x$$
$$3=3\sin x$$
$$\sin x=1$$
$$x=90°$$

(b)
$$\sin x=2\cos x$$
$$\frac{\sin x}{\cos x}=2$$
$$\tan x=2$$
$$x=63.43°$$
$$x=63°\,26'$$

(c)
$$6\cos 2x=3\sin 30°$$
$$\cos 2x=\frac{3\sin 30°}{6}$$
$$\cos 2x=0.25$$
$$2x=75.52°,\ 360°-75.52°$$
$$2x=75.52°,\ 284.48°$$
$$x=37.76°,\ 142.24°$$
$$x=37°\,46',\ 142°\,14'$$

Looking at the solution to **(c)** above, you might ask, 'Why two answers?' Consider that the value of x is given as between 0° and 180°, so the value of $2x$ is between 0° and $2\times 180°=360°$. There are thus four quadrants in which the cosine function might have solutions, but you also know that the solutions must be positive. This means that the answers can be in the first and the fourth quadrants.

Example 15

Solve each equation for $0° \le x \le 360°$.

(a) $\sec^2 x = 2$ (b) $2\sin x - \tan x = 0$ (c) $\sin x - \cos x = 0$

Solution

(a) $\sec^2 x = 2$

$\sec x = \pm\sqrt{2}$

This gives values of x in all 4 quadrants.

In the first quadrant, $\sec x = \sqrt{2}$ gives $x = 45°$.

$\therefore\ x = 45°, 180° - 45°, 180° + 45°, 360° - 45°$

$x = 45°, 135°, 225°, 315°$

(b) $2\sin x - \tan x = 0$

$2\sin x - \dfrac{\sin x}{\cos x} = 0$

$\sin x\left(\dfrac{2\cos x - 1}{\cos x}\right) = 0$

$\sin x(2\cos x - 1) = 0, \cos x \ne 0$

$\sin x = 0, \cos x = 0.5$

$x = 0°, 180°, 360°$ or $x = 60°, 300°$

(c) $\sin x - \cos x = 0$

$\sin x = \cos x$

$\tan x = 1$

$x = 45°, 180° + 45°$

$x = 45°, 225°$

The following equations give their domain in terms of π, so the answers are to be given in radians.

You will need to remember the exact values for the common trigonometric ratios, or how to calculate them. You also need to know the signs of the values in the four quadrants (remember 'ASTC').

Example 16

Solve each equation for $0 \le x \le 2\pi$.

(a) $\cos x = \dfrac{\sqrt{3}}{2}$ (b) $\sin^2 x + 2\sin x + \cos^2 x = 0$ (c) $\sec x = -2$

Solution

(a)

Solve: $\cos x = \dfrac{\sqrt{3}}{2}$

From the sign, angles are in 1st and 4th quadrants: $x = \dfrac{\pi}{6}, 2\pi - \dfrac{\pi}{6}$

$x = \dfrac{\pi}{6}, \dfrac{11\pi}{6}$

(b)

Solve: $\sin^2 x + 2\sin x + \cos^2 x = 0$

Rearrange equation: $\sin^2 x + \cos^2 x + 2\sin x = 0$

Use the identity $\sin^2 x + \cos^2 x = 1$: $1 + 2\sin x = 0$

$2\sin x = -1$

$\sin x = -\dfrac{1}{2}$

From the sign, angles are in 3rd and 4th quadrants: $x = \pi + \dfrac{\pi}{6}, 2\pi - \dfrac{\pi}{6}$

$x = \dfrac{7\pi}{6}, \dfrac{11\pi}{6}$

(c)

Solve: $\sec x = -2$

Change to $\cos x$: $\dfrac{1}{\cos x} = -2$

$\cos x = -\dfrac{1}{2}$

From the sign, angles are in 2nd and 3rd quadrants: $x = \pi - \dfrac{\pi}{3}, \pi + \dfrac{\pi}{3}$

$x = \dfrac{2\pi}{3}, \dfrac{4\pi}{3}$

Example 17

Solve, for $-\pi \le \theta \le \pi$: **(a)** $2\sin^2\theta = 1$ **(b)** $3\sin\theta = 2\cos\theta$ **(c)** $\sin^2\theta + 2\sin\theta + 1 = 0$

Solution

(a)

Solve: $2\sin^2\theta = 1$

Rearrange: $\sin^2\theta = \dfrac{1}{2}$

Square root of each side: $\sin\theta = \pm\dfrac{1}{\sqrt{2}}$

From the sign, angles are in all quadrants: $\theta = -\dfrac{3\pi}{4}, -\dfrac{\pi}{4}, \dfrac{\pi}{4}, \dfrac{3\pi}{4}$ as $-\pi \le \theta \le \pi$

(b)

Solve: $3\sin\theta = 2\cos\theta$

Divide both sides by $3\cos\theta$: $\dfrac{\sin\theta}{\cos\theta} = \dfrac{2}{3}$

But $\tan\theta = \dfrac{\sin\theta}{\cos\theta}$: $\tan\theta = \dfrac{2}{3}$

From the sign, angles are in 1st and 3rd quadrants: $\theta = 0.5880, -\pi + 0.5880$ as $-\pi \le \theta \le \pi$

$\theta = -2.5536, 0.5880$ (to 4 d.p.)

(c)

Solve: $\sin^2\theta + 2\sin\theta + 1 = 0$

Factorise: $(\sin\theta + 1)^2 = 0$

Square root of each side: $\sin\theta + 1 = 0$

Rearrange: $\sin\theta = -1$

From the sign, angles are in 3rd and 4th quadrants: $\theta = -\dfrac{\pi}{2}$ as $-\pi \le \theta \le \pi$

Example 18

Solve the equation $\sin\theta = \dfrac{4\sin\frac{\pi}{9}}{5}$ if θ is an angle in a triangle. Give your answer correct to 2 decimal places.

Solution

Evaluate the expression to find θ: $\theta = 0.277$

Because θ is an angle in a triangle, consider also: $\theta = \pi - 0.277 = 2.868$

The original equation can be rearranged into $\dfrac{4}{\sin\theta} = \dfrac{5}{\sin\frac{\pi}{9}}$, which is a statement of the sine rule (see Section 2.8 The sine rule, page 47). This implies that $\frac{\pi}{9}$ is also an angle of the triangle so the second value is too large (because $2.868 + \frac{\pi}{9} > \pi$ will not fit), hence: $\theta = 0.28$ (to 2 d.p.)

Example 19

Solve, for $0 \le x \le \pi$: **(a)** $\cos\frac{x}{2} = \frac{1}{\sqrt{2}}$ **(b)** $\sin 2x = -\frac{1}{2}$ **(c)** $\sec^2 2x - 2\tan 2x = 0$

Solution

(a)

Solve: $\cos\frac{x}{2} = \frac{1}{\sqrt{2}}$

$0 \le x \le \pi$ so $0 \le \frac{x}{2} \le \frac{\pi}{2}$: $\frac{x}{2} = \frac{\pi}{4}$

$x = \frac{\pi}{2}$

(b)

Solve: $\sin 2x = -\frac{1}{2}$

$0 \le x \le \pi$ so $0 \le 2x \le 2\pi$: $2x = \pi + \frac{\pi}{6}, 2\pi - \frac{\pi}{6}$

$2x = \frac{7\pi}{6}, \frac{11\pi}{6}$

$x = \frac{7\pi}{12}, \frac{11\pi}{12}$

(c)

Solve: $\sec^2 2x - 2\tan 2x = 0$

Use identity $\sec^2 2x = 1 + \tan^2 2x$: $1 + \tan^2 2x - 2\tan 2x = 0$

Rearrange: $\tan^2 2x - 2\tan 2x + 1 = 0$

Factorise: $(\tan 2x - 1)^2 = 0$

$\tan 2x = 1$

$0 \le x \le \pi$ so $0 \le 2x \le 2\pi$: $2x = \frac{\pi}{4}, \pi + \frac{\pi}{4}$

$x = \frac{\pi}{8}, \frac{5\pi}{8}$

Example 20

Solve for $-\pi \le \theta \le \pi$: **(a)** $\sec^2 2\theta - 2\tan 2\theta = 0$ **(b)** $\sin^2\theta + 2\sin\theta + \cos^2\theta = 0$

Solution

(a)

Solve: $\sec^2 2\theta - 2\tan 2\theta = 0$

Use $\sec^2 x = 1 + \tan^2 x$: $1 + \tan^2 2\theta - 2\tan 2\theta = 0$

Rearrange: $\tan^2 2\theta - 2\tan 2\theta + 1 = 0$

Factorise: $(\tan 2\theta - 1)^2 = 0$

$\tan 2\theta - 1 = 0$

$\tan 2\theta = 1$

$2\theta = -2\pi + \frac{\pi}{4}, -\pi + \frac{\pi}{4}, \frac{\pi}{4}, \pi + \frac{\pi}{4}$, as $-2\pi \le 2\theta \le 2\pi$.

$2\theta = -\frac{7\pi}{4}, -\frac{3\pi}{4}, \frac{\pi}{4}, \frac{5\pi}{4}$

$\theta = -\frac{7\pi}{8}, -\frac{3\pi}{8}, \frac{\pi}{8}, \frac{5\pi}{8}$

(b) $\sin^2\theta + 2\sin\theta + \cos^2\theta = 0$

Use $\sin^2\theta + \cos^2\theta = 1$: $1 + 2\sin\theta = 0$

$$2\sin\theta = -1$$

$$\sin\theta = -\frac{1}{2}$$

$$\theta = -\pi + \frac{\pi}{6}, -\frac{\pi}{6}$$

$$\theta = -\frac{5\pi}{6}, -\frac{\pi}{6}$$

Example 21

Solve the equation $\operatorname{cosec}^2 x - 2\cot x = 4$, for $0 \le x \le 2\pi$.

Solution

Use $\operatorname{cosec}^2 x = 1 + \cot^2 x$: $\quad 1 + \cot^2 x - 2\cot x = 4$

Rearrange: $\quad \cot^2 x - 2\cot x - 3 = 0$

Factorise: $\quad (\cot x - 3)(\cot x + 1) = 0$

$\cot x = 3$ is the same as $\tan x = \frac{1}{3}$: $\quad x = 0.32, \pi + 0.32$

$x = 0.32, 3.46$

$\cot x = -1$ is the same as $\tan x = -1$: $\quad x = \pi - \frac{\pi}{4}, 2\pi - \frac{\pi}{4}$

$x = \frac{3\pi}{4}, \frac{7\pi}{4}$

The solutions in order are: $\quad x = 0.32, \frac{3\pi}{4}, 3.46, \frac{7\pi}{4}$

EXERCISE 6.6 SOLVING TRIGONOMETRIC EQUATIONS

Give answers correct to 3 decimal places where necessary.

1 Solve each equation for $0° \le x \le 180°$.

(a) $3 + 2\cos x = 5\cos x$ (b) $\sin x = 3\cos x$ (c) $6\sin 2x = 3\cos 30°$
(d) $4 - 3\tan x = \tan x$ (e) $3\sin x = \cos x$ (f) $\sin 2x = \sin 30°$

2 The solution to $\sin x + \cos x = 0$ for $0° \le x \le 180°$ is:

A $x = -45°$ **B** $x = 45°$ **C** $x = 90°$ **D** $x = 135°$

3 Solve each equation for $0° \le x \le 360°$.

(a) $\operatorname{cosec}^2 x = 2$ (b) $\sin^2 x = 1$ (c) $\tan^2 x = 3$
(d) $3 + 4\sin x = 9\sin x$ (e) $\tan^2 x - \tan x = 0$ (f) $\sin 2x = \cos 2x$

4 (a) On the same diagram draw $y = \sin x$ and $y = \frac{1}{2}$ for $0° \le x \le 360°$. Use your diagram to solve the equation $\sin x = \frac{1}{2}$ for $0° \le x \le 360°$.

(b) What line would you need to draw to solve the equation $\sin x = -\frac{1}{2}$? What are the solutions to this equation for $0° \le x \le 360°$?

5 Solve, for $0 \le \theta \le 2\pi$:

(a) $\sin\theta = \frac{-1}{\sqrt{2}}$ (b) $\sec\theta = \frac{2}{\sqrt{3}}$ (c) $\cot\theta = 1$
(d) $\sin^2\theta - 2\cos\theta + \cos^2\theta = 0$ (e) $\sin^2\theta + \cos\theta - 1 = 0$ (f) $\sec^2\theta - 2\tan\theta = 0$
(g) $4\sin^2\theta = 3$ (h) $\sin\theta = \cos\theta$ (i) $\cos^2\theta - 2\cos\theta + 1 = 0$
(j) $\tan^2\theta = 3$ (k) $\sin\theta + \cos\theta = 0$ (l) $2\cos^2\theta - \cos\theta = 0$

6 What is the solution to $\operatorname{cosec} x = 3$, for $0 < x < 2\pi$?

A $x = 0.340, 2.802$ **B** $x = 19°28', 160°32'$ **C** $x = 0.340$ **D** $x = 19°28'$

7 Solve, for $-\pi \le \theta \le \pi$:

(a) $1 + \sin\theta = \cos^2\theta$ (b) $2\cos^2\theta + \sin\theta = 1$ (c) $4\cos\theta = \sec\theta$
(d) $2\sin^2\theta - 2 = 3\cos\theta$ (e) $4\sin\theta = \operatorname{cosec}\theta$ (f) $2\sec^2\theta = 5\tan\theta$

8 Solve, for $0 \le x \le 2\pi$:

(a) $\tan x = 2\sin x$ (b) $5\cos^2 x + 2\sin x - 2 = 0$ (c) $3\sec^2 x - 5\tan x = 5$
(d) $6\tan x = 5\operatorname{cosec} x$ (e) $7\sin x\cos x + \cos^2 x = 1$ (f) $\sin^2 x + 2\sin x\cos x = 3\cos^2 x$

9 Solve the equation $\operatorname{cosec}^2 x + \cot^2 x = 3$, for $0 \le x \le \pi$. Indicate whether each statement below is a correct or incorrect step of the solution.

(a) $2\cot^2 x = 2$ (b) $2\cot^2 x = 4$ (c) $\cot x = \pm 1$ (d) $x = \pm\frac{\pi}{4}$

10 Solve, for $0 \le x \le 2\pi$: (a) $\sin\frac{x}{2} = \frac{\sqrt{3}}{2}$ (b) $\tan\frac{x}{2} = -\frac{1}{\sqrt{3}}$ (c) $\cos\frac{x}{3} = 1$

11 Solve, for $0 \le x \le \pi$: (a) $\cos 2x = 0$ (b) $\tan 3x = \sqrt{3}$ (c) $\sin 4x = \frac{1}{\sqrt{2}}$

12 Solve for $0 \le x \le \pi$: (a) $\sin\frac{x}{3} = 0.4$ (b) $\cos 3x = -0.7$ (c) $\tan\frac{x}{4} = 1.5$

13 What is the solution to $\operatorname{cosec} x = 3$, for $0 < x < 2\pi$?

A $x = 0.340, 2.802$ **B** $x = 19°28', 160°32'$ **C** $x = 0.340$ **D** $x = 19°28'$

14 Solve for $0 < x < 2\pi$: (a) $5\cos^2 x + 8\sin x - 8 = 0$ (b) $6\tan x = 5\cot x$

15 Simplify:

(a) $1 + \tan^2(\frac{\pi}{2} - \theta)$ (b) $1 - \cos^2(\pi + \theta)$ (c) $\sin\theta\cos(\frac{\pi}{2} - \theta) + \cos\theta\sin(\frac{\pi}{2} - \theta)$

(d) $\cos^2\frac{\pi}{6} - 1$ (e) $1 - \sin\theta\cos(\frac{\pi}{2} - \theta)$

16 Solve for values between 0 and 2π inclusive:

(a) $\tan^2 x + \tan x = 0$ (b) $\sin^2 x - \sin x\cos x = 0$ (c) $2\sin^2\theta + \cos\theta = 1$
(d) $3\cos^2\theta - 2\sin\theta = 2$ (e) $(2\cos x + 1)(\sin x - 1) = 0$ (f) $5\sec^2 x + 2\tan x = 8$

17 If $0 \le \theta \le 2\pi$, then the solution to $3\sin^2\theta - 4\cos\theta + 1 = 0$ is (approximately or exactly):

A $\theta = 0.841, 5.442$ **B** $\theta = 1.969, 4.315$ **C** $\theta = 2.301, 3.983$ **D** $\theta = \frac{2\pi}{3}, \frac{4\pi}{3}$

18 Solve for $0 \le \theta \le 2\pi$:

(a) $3\tan^3\theta - 3\tan^2\theta - \tan\theta + 1 = 0$ (b) $\cos^3\theta - 2\cos^2\theta + \cos\theta = 0$

CHAPTER REVIEW 6

1 The minute hand of a clock is 1.2 metres long. How far does the tip move in 40 minutes?

2 An arc AB of length 6 cm subtends an angle of 56° at the centre of a circle. Calculate:

(a) the length of the radius (b) the length of the chord AB.

3 Find the exact value of $\sec^2\frac{\pi}{4} + \operatorname{cosec}^2\frac{\pi}{4}$.

4 Solve, for $-\pi \le x \le \pi$, $6\cos^2 x - 5\cos x + 1 = 0$.

5 An arc AB of a sector of a circle is $\frac{\pi}{4}$ metres long and subtends an angle of 60° at the centre, O, of the circle.
Calculate: **(a)** the length of the radius
(b) the area of the sector AOB (correct to 1 decimal place)
(c) the length of the chord AB (correct to 1 decimal place).

6 Find all values of θ between 0 and 2π for which:
(a) $\sin\theta = -0.5$ **(b)** $\cos\theta = 0$ **(c)** $\tan\theta = -1$
(d) $\sec\theta = \frac{2}{\sqrt{3}}$ **(e)** $\cot\theta = \sqrt{3}$ **(f)** $\operatorname{cosec}\theta = \sqrt{2}$

7 Draw the graph of $y = \cos 2x$ for $-\pi \le x \le \pi$. On the same set of axes, draw the graph of $y = -\frac{x}{2}$. Use your graphs to solve the equation $\cos 2x = -\frac{x}{2}$.

8 An arc AB of a circle, centre C, is 20 cm long.
(a) If the chord AB is 15 cm long and subtends an angle of x radians at C, show that $8\sin\frac{x}{2} = 3x$.
(b) Solve the equation in part **(a)** graphically, giving your answer correct to 1 decimal place.

9 Express in radians, in terms of π: **(a)** 45° **(b)** 240° **(c)** 160° **(d)** −210°

10 Express in radians, correct to 4 decimal places: **(a)** 65° **(b)** 281° **(c)** −100° **(d)** −326°

11 Express in degrees the angles whose radian measures are: **(a)** $\frac{4\pi}{5}$ **(b)** $\frac{7\pi}{6}$ **(c)** $\frac{23\pi}{12}$ **(d)** $-\frac{3\pi}{2}$

12 Express in degrees and minutes, to the nearest minute, the angles whose radian measures are:
(a) 2.6 **(b)** −1.4 **(c)** 0.341 **(d)** −3

13 Simplify: **(a)** $\sin(\pi + x)$ **(b)** $\cos(2\pi - x)$ **(c)** $\tan(\pi - x)$

14 Write the exact value of: **(a)** $\cos\pi$ **(b)** $\tan\frac{7\pi}{6}$ **(c)** $\sin\frac{3\pi}{4}$ **(d)** $\cos\frac{5\pi}{3}$

15 Solve for $0 < x < 2\pi$: **(a)** $\sin x = -\frac{1}{2}$ **(b)** $\sin x = \sqrt{3}\cos x$ **(c)** $\sqrt{2}\cos x + 1 = 0$

16 Simplify: **(a)** $\sin\left(\frac{\pi}{2} - x\right)$ **(b)** $\cos\left(\frac{3\pi}{2} + x\right)$ **(c)** $\tan\left(\frac{\pi}{2} + x\right)$

CHAPTER 7
Introduction to differentiation

The study of calculus is rich in history and controversy. Isaac Newton (1642–1727) and Gottfried Wilhelm Leibniz (1646–1716) are both credited with the original development of calculus, but there was much argument at the time over who was first. Their approaches were quite different, although ultimately equivalent, and today the notation that we use comes from both sources. Important contributions to calculus were also made by mathematicians such as John Wallis, Isaac Barrow, Guillaume de l'Hôpital and later Augustin-Louis Cauchy. The work of mathematicians on differential calculus and integral calculus is the basis of much mathematics still used in a huge number of applications today.

In this chapter, you will learn about differentiation—what it is, and how it can be used to find rates of change in a variety of situations.

7.1 CONTINUITY AND GRADIENTS OF TANGENTS

Converting between fractions, decimals and percentages without using a calculator is an important skill. A calculator cannot always simplify algebraic fractions, so you are expected to be proficient at this skill.

Continuity

You have drawn many different graphs in this course so far, but all of them can be divided into two special types:

- Graphs such as $f(x) = x, f(x) = x^3, f(x) = \sin x$ and $f(x) = |x|$ are **continuous** curves: they do not have any gaps or jumps in them.
- Graphs such as $f(x) = \frac{1}{x}$, $f(x) = \tan x$, $f(x) = \frac{|x|}{x}$ and $f(x) = \frac{1}{x^2}$ are not continuous ('**discontinuous**') functions, because they all have gaps or jumps.

A simple way to describe a continuous function is that it can be drawn without lifting your pencil from the paper. Continuous functions like $f(x) = x, f(x) = x^3, f(x) = \sin x$ are smooth graphs, but $f(x) = |x|$ is not a smooth graph because the slope of its curve changes suddenly at $x = 0$ to create a sharp corner.

You should use appropriate software to graph all of these functions so that you can remember what they look like.

Example 1

For each graph, decide whether it is **(i)** continuous or discontinuous, and, if continuous, whether it is **(ii)** smooth or not smooth.

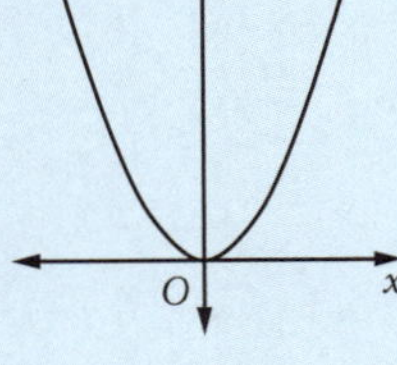

(c)

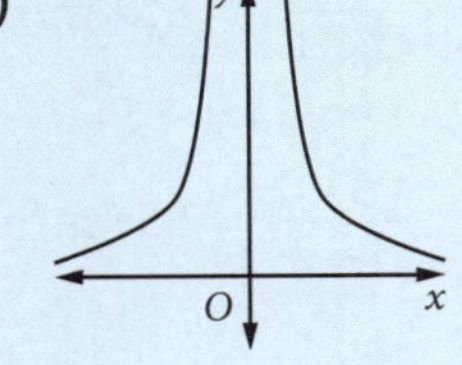

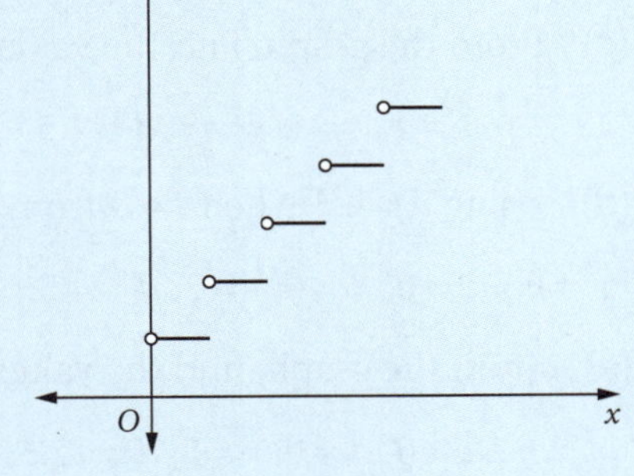

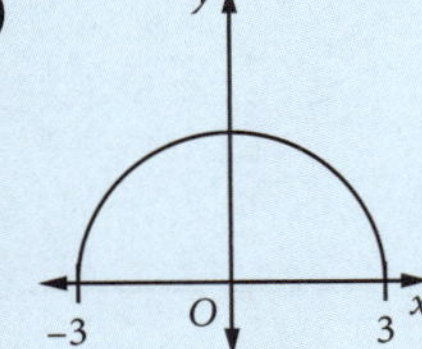

(d)

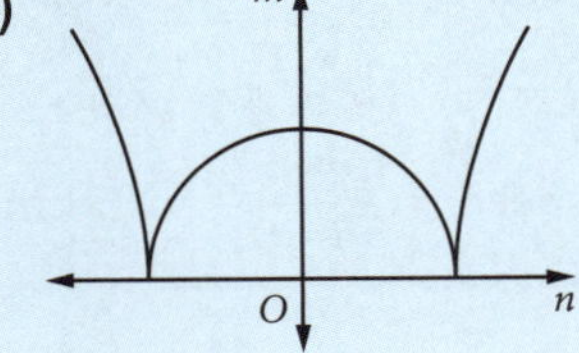

Solution

(a) (i) Continuous
(ii) Smooth

(b) (i) Continuous for $|x| \le 3$
(ii) Smooth

(c) (i) Discontinuous

(d) (i) Continuous
(ii) Not smooth

(e) (i) Discontinuous

Each branch of **(c)**, if considered on its own, is a smooth continuous curve. In **(e)**, if the open circles had not been at the start of each horizontal line then the graph would have represented a relation, not a function.

Gradient and rates of change

Smooth curves change gradually as x (the independent variable) increases. The rate of this change between two points on the curve can be calculated as the **average rate of change** over that interval. If the function increases as x increases, then the rate of change is positive. If the function decreases as x increases, then the rate of change is negative.

Now consider the gradient of a line. The gradient is positive if y increases as x increases (i.e. the line moves from left to right). The gradient is negative if y decreases as x increases. Thus you can see that in general the gradient is the rate of change.

Example 2

The graph shows the distance d in kilometres that a car travels in t minutes. The function is given by the equation $d = \frac{t^2}{2}$.

The average speed of the car is the rate of change of the distance travelled over a time interval. Find the average speed over the:

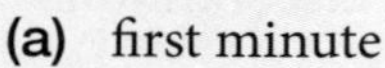

(a) first minute
(b) second minute
(c) third minute
(d) first three minutes
(e) first five minutes.

Solution

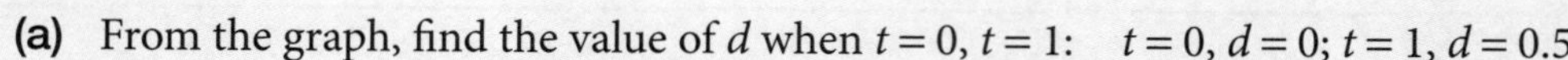

(a) From the graph, find the value of d when $t = 0$, $t = 1$: $t = 0, d = 0; t = 1, d = 0.5$

$$\text{Average speed} = \frac{\text{distance covered}}{\text{time taken}} = \frac{0.5 - 0}{1 - 0} = 0.5\,\text{km/min}$$

(b) From the graph, find the value of d when $t = 2$: $t = 2, d = 2$

$t = 1, d = 0.5; t = 2, d = 2$: Average speed $= \frac{2 - 0.5}{2 - 1} = 1.5\,\text{km/min}$

(c) From the graph, find the value of d when $t = 3$: $t = 3, d = 4.5$

$t = 2, d = 2; t = 3, d = 4.5$: Average speed $= \frac{4.5 - 2}{3 - 2} = 2.5\,\text{km/min}$

(d) Using $d = 4.5$ when $t = 3$ from part **(c)**

For $t = 0, d = 0; t = 3, d = 4.5$: Average speed $= \frac{4.5 - 0}{3 - 0} = 1.5\,\text{km/min}$

(e) From the graph, find the value of d when $t = 5$: $t = 5, d = 12.5$

$t = 0, d = 0; t = 5, d = 12.5$: Average speed $= \frac{12.5 - 0}{5 - 0} = 2.5\,\text{km/min}$

In Example **2** you were actually finding the gradient of each secant. A **secant** is a line joining two points on a curve. This is shown on the following graph.

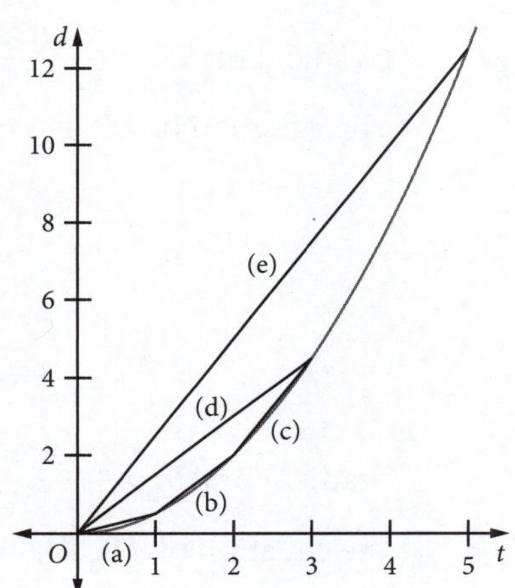

Average speed over the:

(a) first minute
(b) second minute
(c) third minute
(d) first three minutes
(e) first five minutes.

The slope of a secant gives the average rate of change of the function between the two values of its domain.

In the diagram, the slope of the secant AB is:

$$\frac{f(x+h)-f(x)}{(x+h)-x} = \frac{f(x+h)-f(x)}{h}$$

This is the average rate of change of $f(x)$ between x and $(x+h)$.

The position of B varies according to the value of h.

Let $A(x_1, y_1)$ be a fixed point on a curve. Take any other point $B(x, y)$ on the curve and join the points to obtain the secant AB. Let the point B get closer to the point A. This means that h is getting smaller as the two points get closer together. As the secant gets smaller, it becomes more like the tangent to the curve at the fixed point. The tangent can be considered to be the limiting position of the secant.

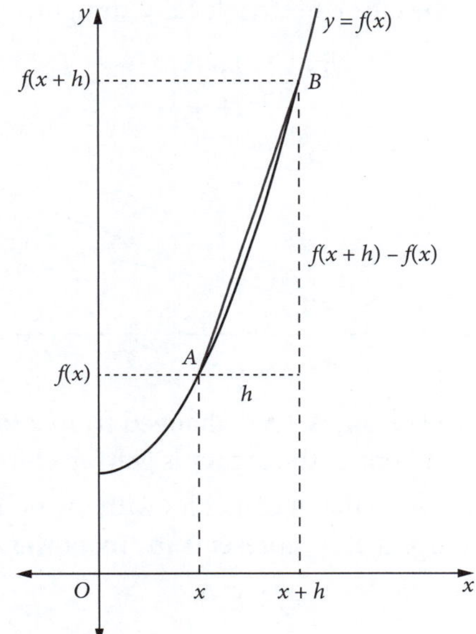

Consider the following graphs in which $A(1, 0.5)$ is a fixed point and B moves from the point $B(3, 4.5)$ to $B_1(2, 2)$ and then to $B_2(1.5, 1.125)$. By the third graph, it is hard to distinguish the secant AB from the curve.

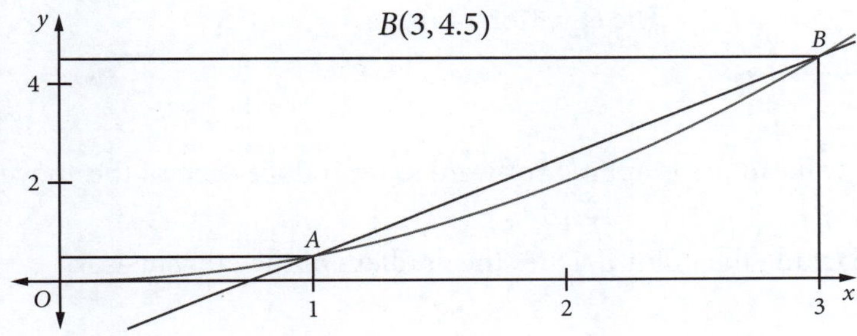

Gradient of $AB = \frac{4.5-0.5}{3-1} = 2$

The equation of AB is $y = 2x - 1.5$

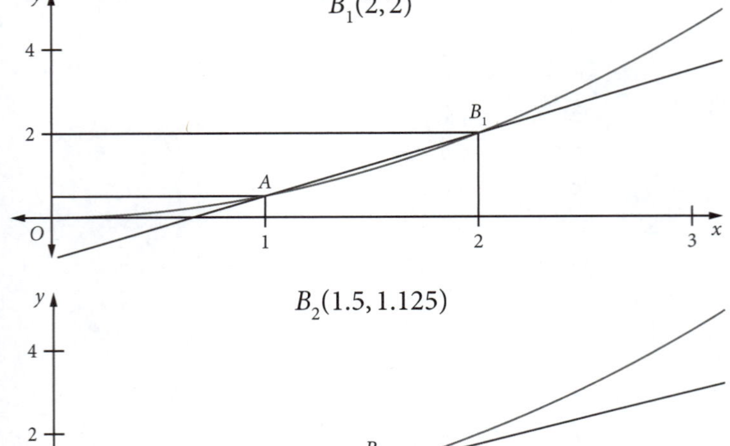

Gradient of $AB_1 = \dfrac{2 - 0.5}{2 - 1} = 1.5$

The equation of AB_1 is $y = 1.5x - 1$

Gradient of $AB_2 = \dfrac{1.125 - 0.5}{1.5 - 1} = 1.25$

The equation of AB_2 is $y = 1.25x - 0.75$

Let B move to $B_3(1.1, 0.605)$, as shown in the next graph. Here the secant and the curve appear to be the same curve. You can say that the curve here is 'locally straight'.

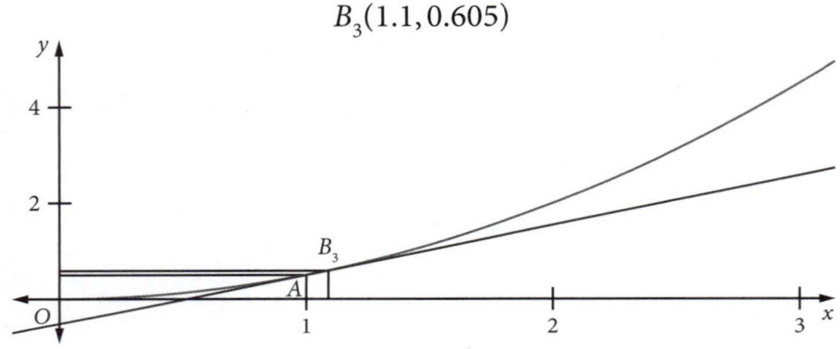

Gradient of $AB_3 = \dfrac{0.605 - 0.5}{1.1 - 1} = 1.05$

The equation of AB_3 is $y = 1.05x - 0.55$

The gradient of the secant AB has changed from 2 to 1.05 as B has moved closer to A. This suggests that as B gets closer to A, the gradient of the secant is getting closer to 1.

Let the point B move so that it coincides with the point A: the secant has now become the tangent at A. Importantly, the gradient of this tangent at A represents the **instantaneous rate of change** of the curve at the point A.

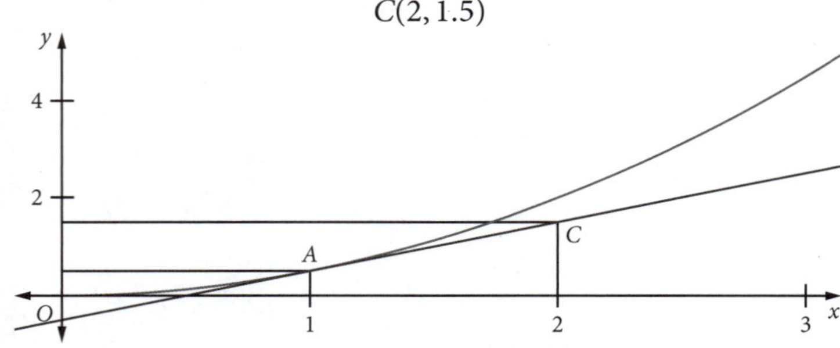

The tangent at A appears to pass through the point $C(2, 1.5)$.

Gradient of $AC = 1$

The equation of AC is $y = x - 0.5$

Given that the equation of the above curve is $y = \frac{1}{2}x^2$, you can use graphing software to reproduce each of the above diagrams. Investigate the point $B_4(1.05, 0.55125)$.

Repeat this investigation process for the point A approaching the point B. Does the gradient of the tangent at B appear to be 3?

MAKING CONNECTIONS

The secant and the rate of change

Move the slider to see how the secant approximates the gradient of the tangent to a curve.

Example 3

The graph of $y = \frac{1}{2}x^2$ is drawn for $-3 \le x \le 3$. Tangent lines are drawn at the points where $x = -2, -1, 0, 1, 2, 3$, as shown on the graph.

From the graph, find the slope of each of these tangents and use it to complete the following table:

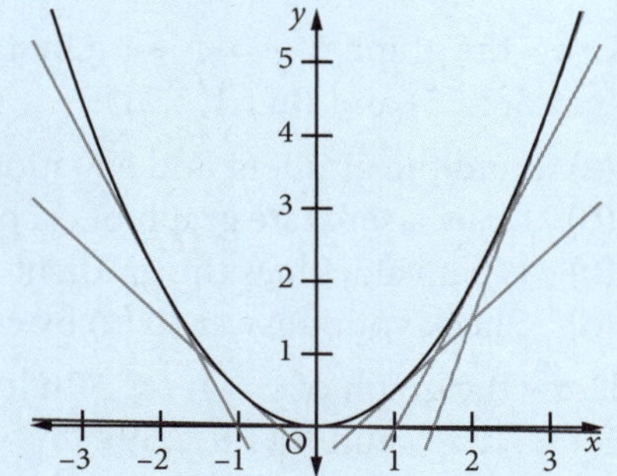

x	−2	−1	0	1	2	3
m						

Plot these points on a separate number plane and write the equation of the curve that passes through them.

Solution

x	−2	−1	0	1	2	3
m	−2	−1	0	1	2	3

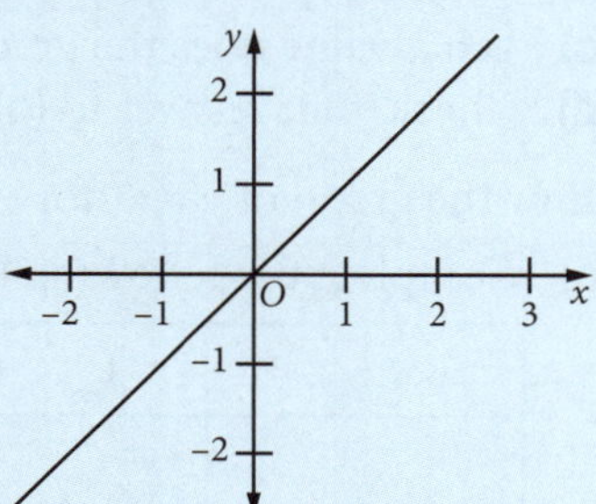

The equation of the line through these points is $y = x$.

This equation represents the gradient function for the curve, because it gives the gradient of the tangent to the curve at each point on the curve.

EXERCISE 7.1 CONTINUITY AND GRADIENTS OF TANGENTS

1 Indicate whether each graph is continuous or discontinuous.

(a)
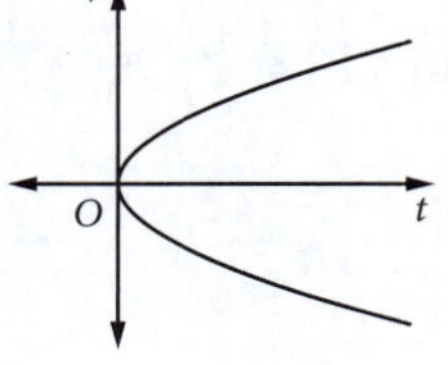

(b)
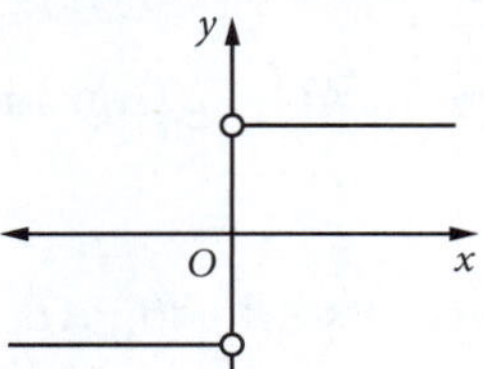

(c)
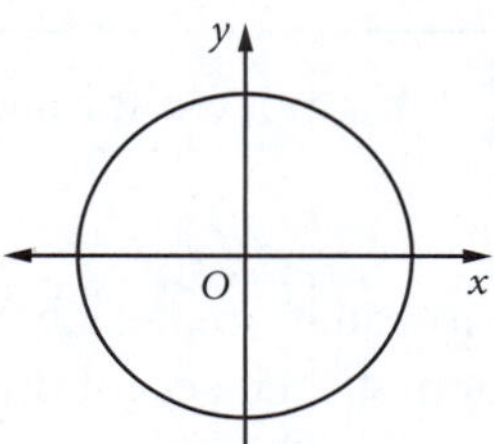

(d)
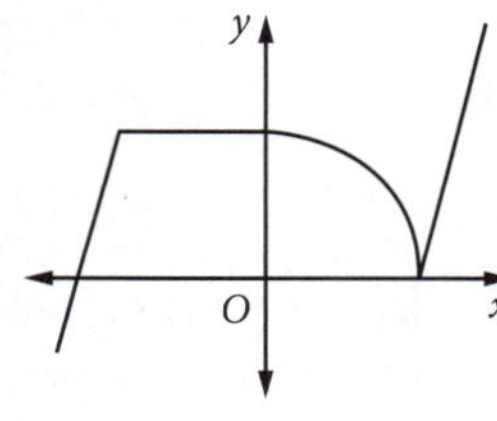

(e)
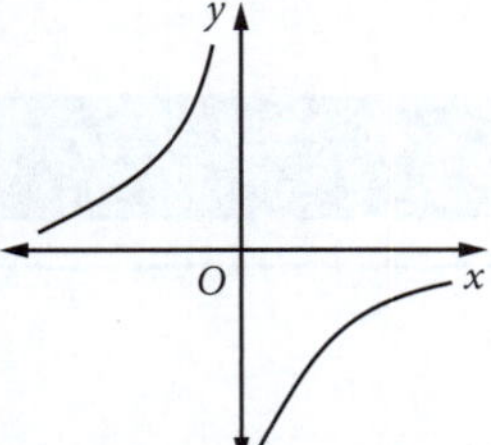

(f)
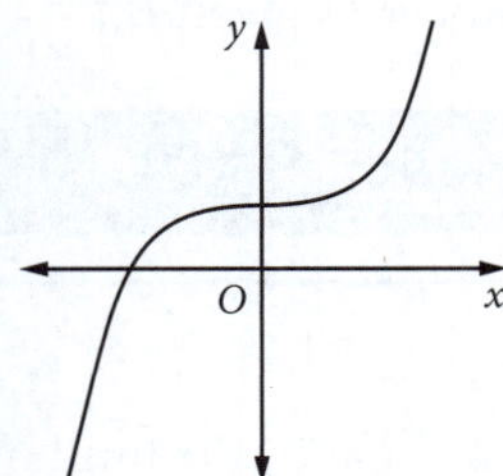

(g)
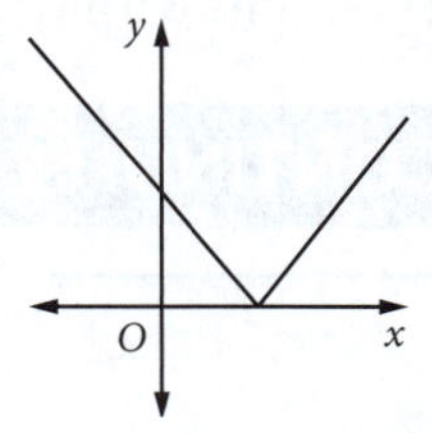

(h)
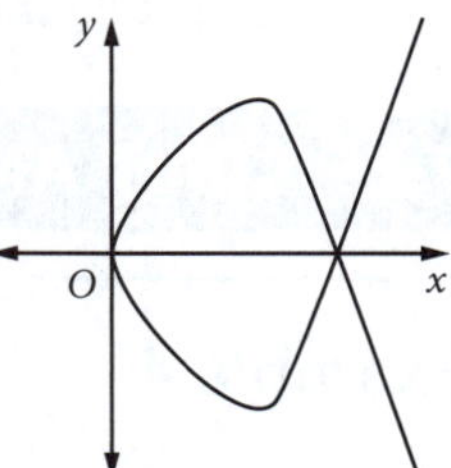

2 In question **1**, which curves could be called smooth?

3 Which graph represents a smooth continuous curve?

A
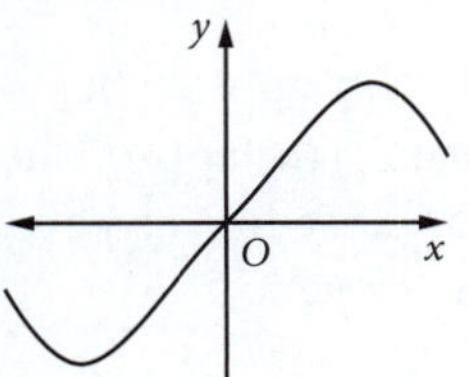

B
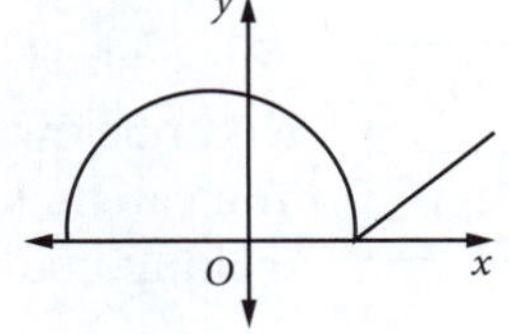

C
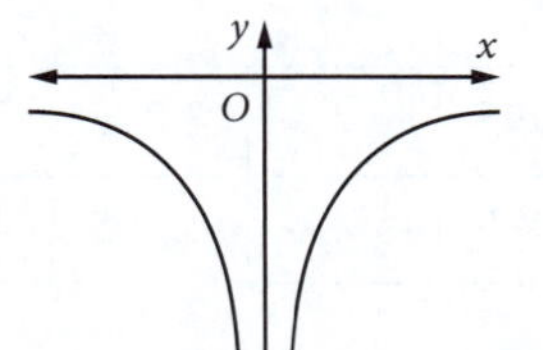

D
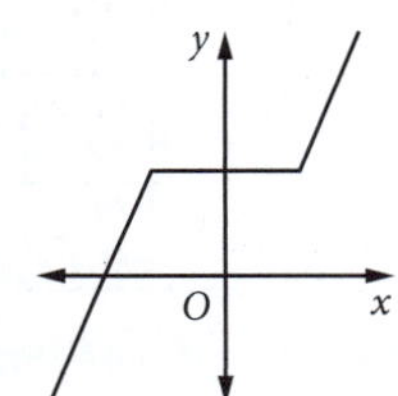

The following questions must be completed using graphing software.

4 Draw the graph of $y = x^2 + 1$. On it mark the point $A(1, 2)$. Consider the points $B(2.5, 7.25)$, $B_1(2, 5)$, $B_2(1.5, 3.25)$ and $B_3(1.1, 2.21)$.
   (a) Find the gradient and equation of each of the secants AB, AB_1, AB_2 and AB_3.
   (b) Draw a separate graph of each secant on the curve.
   (c) What value does the gradient of the secant appear to approach as B gets closer to A?
   (d) Check your answer to **(c)** by considering the secant AB_4, given $B_4(0.9, 1.81)$.

5 Draw the graph of $y = 4 - x^2$. On it mark the point $A(2, 0)$. Consider the points $B(0.5, 3.75)$, $B_1(1, 3)$, $B_2(1.5, 1.75)$ and $B_3(1.9, 0.39)$.
   (a) Find the gradient and equation of each of the secants AB, AB_1, AB_2 and AB_3.
   (b) Draw a separate graph of each secant on the curve.
   (c) What value does the gradient of the secant appear to approach as B gets closer to A?
   (d) Check your answer to **(c)** by considering the secant AB_4, given $B_4(2.1, -0.41)$.

6 Draw the graph of $y = x^2$ for $-3 \le x \le 3$. Draw tangent lines at the points where $x = -2, -1, 0, 1, 2, 3$.
   (a) Complete the following table, where m is the gradient of the tangent at each point x.

x	-2	-1	0	1	2	3
m						

   (b) Plot these points on a number plane.
   (c) Write the equation of the curve in **(b)**.

7 Draw the graph of $y = x^3$ for $-2.5 \le x \le 2.5$. Draw tangent lines at the points where $x = -2, -1, 0, 1, 2$.
   (a) Complete the following table, where m is the gradient of the tangent at each point x.

x	-2	-1	0	1	2
m					

   (b) Plot these points on a number plane.
   (c) Write the equation of the curve in **(b)**.

8 Draw the graph of $y = \frac{1}{x}$ for $x > 0$. On it mark the point $A(1, 1)$. Consider the points $B\left(3, \frac{1}{3}\right)$, $B_1\left(2, \frac{1}{2}\right)$, $B_2\left(\frac{3}{2}, \frac{2}{3}\right)$ and $B_3\left(\frac{11}{10}, \frac{10}{11}\right)$.
   (a) Find the gradient and equation of each of the secants AB, AB_1, AB_2 and AB_3.
   (b) Draw a separate graph of each secant on the curve.
   (c) What value does the gradient of the secant appear to approach as B gets closer to A?
   (d) Check your answer by considering the secant AB_4, given $B_4\left(\frac{9}{10}, \frac{10}{9}\right)$.

7.2 LIMIT AND CONTINUITY

Example 4

For the function f, where $f(x) = x + 2$, find $\lim_{x \to 2}(x + 2)$ (that is, the limit of the function as x approaches 2).

Solution

The domain of this function is the set of real numbers. The table shows $f(x)$ for values of x near (or 'in the neighbourhood of') 2.

x	1.95	1.99	1.995	$\to 2 \leftarrow$	2.005	2.01	2.05
$f(x)$	3.95	3.99	3.995	$\to 4 \leftarrow$	4.005	4.01	4.05

This table shows that as x approaches 2 from either below or above 2, $f(x)$ approaches 4. You can make $f(x)$ as close to 4 as you like by making x sufficiently close to 2.

Hence: $\lim_{x \to 2}(x + 2) = 4$

Notice that in this example, $f(2) = 4$. Thus $\lim_{x \to 2}(x + 2) = f(2)$.

In this case you say that the function is **continuous** at $x = 2$.

Example 5

For the function f where $f(x)=\dfrac{x^2-4}{x-2}$, find $\lim\limits_{x\to 2}\dfrac{x^2-4}{x-2}$.

Solution

If $x = 2$, this function is not defined (as the denominator is zero).

If $x \neq 2$: $\quad \dfrac{x^2-4}{x-2}=\dfrac{(x-2)(x+2)}{x-2}=x+2$

$$\therefore \lim_{x\to 2}\frac{x^2-4}{x-2}=\lim_{x\to 2}(x+2)=4$$

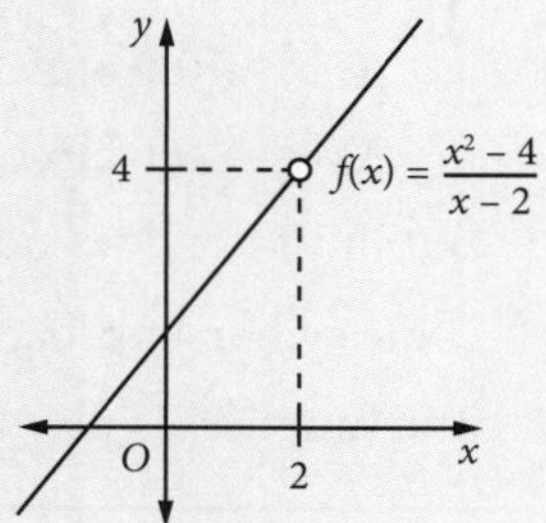

If $x \neq 2$ then $f(x)=\dfrac{x^2-4}{x-2}$ is equivalent to $f(x) = x + 2$, so its graph is a straight line with a missing point at $x = 2$.

In this case you say that the function is **discontinuous** at $x = 2$:

$$\lim_{x\to 2}\frac{x^2-4}{x-2}\neq f(2)$$

Example 6

Investigate the following functions:

(a) $f(x)=\begin{cases} x+1 & \text{when } x\geq 1 \\ 3 & \text{when } x<1 \end{cases}$ (b) $f(x)=\dfrac{|x|}{x},\ x\neq 0$

Solution

(a) Draw the graph of the function.

$x \geq 1$: $\quad f(x) = x + 1$

$\lim\limits_{x\to 1}(x+1)=2$

Thus as $x \to 1$ from above, $f(x) \to 2$ from above.

You can write this as $\lim\limits_{x\to 1^+} f(x)=2$

This can be confirmed by evaluating $f(1.01), f(1.001)$ etc.

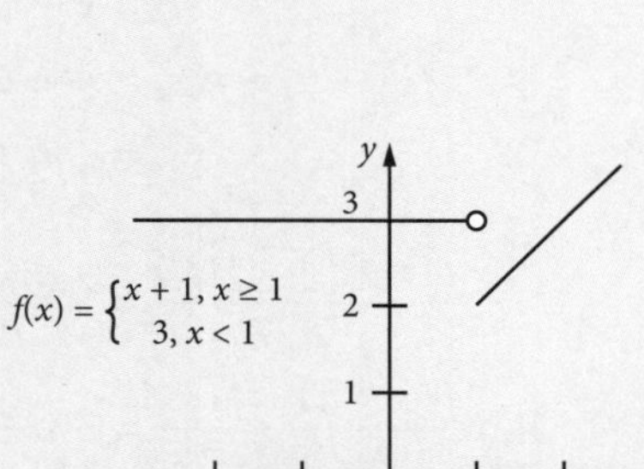

$x < 1$: $\quad f(x) = 3$

This means that the value of $f(x)$ is always 3, for all $x < 1$.

Hence $\lim\limits_{x\to 1^-} f(x)=3$

Because $\lim\limits_{x\to 1^+} f(x)\neq \lim\limits_{x\to 1^-} f(x)$ the function is discontinuous at $x = 1$.

(b) $f(x)=\begin{cases}\dfrac{x}{x}=1 & \text{when } x>0\\ \dfrac{0}{0} & \text{(which is undefined) if } x=0\\ -\dfrac{x}{x}=-1 & \text{when } x<0\end{cases}$

Draw the graph of the function.

The function is defined for all values of x except $x = 0$.

As $x \to 0$ from above, $f(x) \to 1$

As $x \to 0$ from below, $f(x) \to -1$

But $\lim\limits_{x\to 0} f(x)$ does not exist, because it approaches different values when approached from above or from below.

The function is discontinuous at $x = 0$.

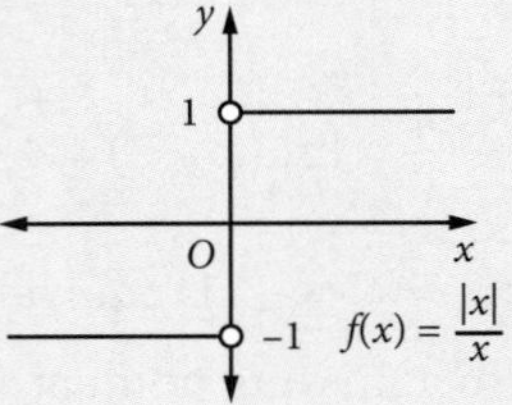

Example 7

Find $\lim\limits_{h\to 0}\dfrac{f(x+h)-f(x)}{h}$ where: **(a)** $f(x)=x^2$ **(b)** $f(x)=6+5x-2x^2$

Solution

(a)

$$\begin{aligned}f(x)&=x^2\\ f(x+h)&=(x+h)^2\\ &=x^2+2xh+h^2\\ f(x+h)-f(x)&=x^2+2xh+h^2-x^2\\ &=2xh+h^2\end{aligned}$$

$$\begin{aligned}\frac{f(x+h)-f(x)}{h}&=\frac{2xh+h^2}{h}\\ \lim_{h\to 0}\frac{f(x+h)-f(x)}{h}&=\lim_{h\to 0}\frac{2xh+h^2}{h}\\ &=\lim_{h\to 0}\frac{h(2x+h)}{h}\\ &=\lim_{h\to 0}(2x+h) \qquad \text{for } h\neq 0\\ &=2x\end{aligned}$$

(b)

$$\begin{aligned}f(x)&=6+5x-2x^2\\ f(x+h)&=6+5(x+h)-2(x+h)^2\\ &=6+5x+5h-2x^2-4xh-2h^2\\ f(x+h)-f(x)&=6+5x+5h-2x^2-4xh-2h^2-(6+5x-2x^2)\\ &=5h-4xh-2h^2\\ \frac{f(x+h)-f(x)}{h}&=\frac{5h-4xh-2h^2}{h}\\ \lim_{h\to 0}\frac{f(x+h)-f(x)}{h}&=\lim_{h\to 0}\frac{5h-4xh-2h^2}{h}\\ &=\lim_{h\to 0}(5-4x-2h) \qquad \text{for } h\neq 0\\ &=5-4x\end{aligned}$$

Limit theorems

These theorems on limits of functions will be stated without their proofs.

Theorem 1

For the constant function f, where $f(x) = c$: $\lim\limits_{x \to a} f(x) = c$

Theorem 2

If $\lim\limits_{x \to a} f(x) = L$ and $\lim\limits_{x \to a} g(x) = M$, then: $\lim\limits_{x \to a}\left(f(x) \pm g(x)\right) = \lim\limits_{x \to a} f(x) \pm \lim\limits_{x \to a} g(x)$

$$= L \pm M$$

The limit of a sum = the sum of the limits.

The limit of a difference = the difference of the limits.

e.g. $\lim\limits_{x \to 2}\left(x^2 - 3x + 5\right) = \lim\limits_{x \to 2}\left(x^2\right) - \lim\limits_{x \to 2}(3x) + \lim\limits_{x \to 2} 5$

$$= 4 - 6 + 5 = 3$$

Theorem 3

If $\lim\limits_{x \to a} f(x) = L$ and $\lim\limits_{x \to a} g(x) = M$, then: $\lim\limits_{x \to a}\left(f(x) \times g(x)\right) = \lim\limits_{x \to a} f(x) \times \lim\limits_{x \to a} g(x)$

$$= L \times M$$

The limit of a product = the product of the limits.

e.g. $\lim\limits_{x \to -1}\left\{2x\left(x^2 - 4\right)\right\} = \lim\limits_{x \to -1}(2x) \times \lim\limits_{x \to -1}\left(x^2 - 4\right)$

$$= (-2) \times (-3) = 6$$

Theorem 4

If $\lim\limits_{x \to a} f(x) = L$ and $\lim\limits_{x \to a} g(x) = M$, then: $\lim\limits_{x \to a}\left(\dfrac{f(x)}{g(x)}\right) = \dfrac{\lim\limits_{x \to a} f(x)}{\lim\limits_{x \to a} g(x)}$

$$= \frac{L}{M} \quad \text{for } M \neq 0$$

The limit of a quotient = the quotient of the limits.

e.g. $\lim\limits_{x \to 3} \dfrac{x^2 + 2}{x + 1} = \dfrac{\lim\limits_{x \to 3}\left(x^2 + 2\right)}{\lim\limits_{x \to 3}(x + 1)}$

$$= \frac{11}{4}$$

Continuity

If $f(x)$ and $g(x)$ are both continuous at $x = a$, then the new functions $f(x) \pm g(x)$, $f(x) \times g(x)$ and $\dfrac{f(x)}{g(x)}$ are also continuous at $x = a$. This is a direct consequence of the limit theorems.

Continuity at a point—formal definition

A function f that is defined in some region that includes $x = c$ is said to be continuous at c if:

(a) the function has a definite value $f(c)$ at c, **AND**

(b) as $x \to c$, $f(x) \to f(c)$ as a limit, i.e. $\lim\limits_{x \to c} f(x) = f(c)$.

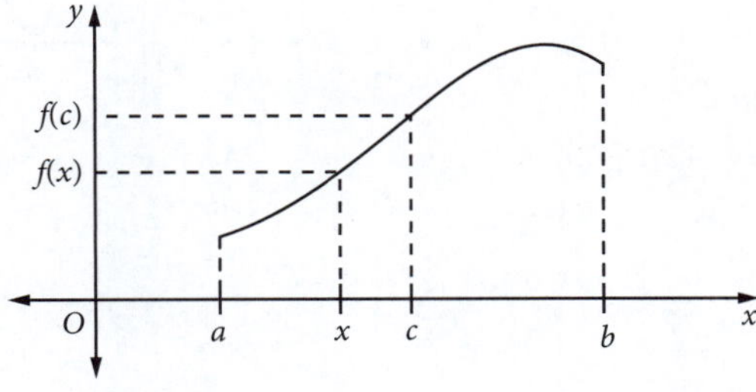

A function f is said to be continuous in an (open or closed) interval if it is continuous at all points of that interval. For a closed interval $[a, b]$, continuity at a and b implies that $\lim_{x\to a^+} f(x) = f(a)$ and $\lim_{x\to b^-} f(x) = f(b)$.

In simple language, a function is continuous in the interval $[a, b]$ if its graph can be drawn from $x = a$ to $x = b$ without lifting your pencil off the paper. (See the diagram above.) Thus all polynomial functions are continuous.

EXERCISE 7.2 LIMIT AND CONTINUITY

1 Evaluate each limit.

(a) $\lim_{x\to 3}(3x)$ **(b)** $\lim_{x\to -1}\left(x^2+4x\right)$ **(c)** $\lim_{x\to 3}\left(9-x^2\right)$ **(d)** $\lim_{x\to -2}\left(x^2-2x+1\right)$

(e) $\lim_{x\to -4} x^2(x+2)$ **(f)** $\lim_{h\to 2}\left(h^2-4h+4\right)$ **(g)** $\lim_{a\to -1}(a+3)(a-4)$ **(h)** $\lim_{x\to 3}\left(\frac{x^2-5}{x+2}\right)$

2 The value of $\lim_{x\to -3}\frac{(x+5)(x+3)}{x+3} = \ldots$

A -2 **B** 0 **C** 2 **D** indeterminate

3 Evaluate the following limits.

(a) $\lim_{x\to 3} 1$ **(b)** $\lim_{x\to 0}\left(\frac{x^2+5x}{x}\right)$ **(c)** $\lim_{x\to -2}\left(\frac{x^3+8}{x+2}\right)$

(d) $\lim_{x\to 3}\left(\frac{x^2-5x+6}{x-3}\right)$ **(e)** $\lim_{x\to 3}\left(\frac{3x}{x+3}\right)$ **(f)** $\lim_{x\to 5}\left(\frac{x-5}{2x^2-9x-5}\right)$

(g) $\lim_{x\to 1}\left(\frac{x-1}{x^2+x-2}\right)$ **(h)** $\lim_{x\to 4}\left(\frac{x-1}{x^2+x-2}\right)$ **(i)** $\lim_{x\to 1}\left(\frac{x^3-1}{x-1}\right)$

4 Evaluate each limit.

(a) $\lim_{x\to 0} f(x)$ where $f(x) = \begin{cases} x^2+1 & \text{for } x \ge 0 \\ 1 & \text{for } x < 0 \end{cases}$ **(b)** $\lim_{x\to 1} f(x)$ where $f(x) = \begin{cases} 2x & \text{for } x \ge 1 \\ -2x+4 & \text{for } x < 1 \end{cases}$

5 Evaluate $\lim_{h\to 0}\frac{f(x+h)-f(x)}{h}$ where:

(a) $f(x) = x^2 - 1$ **(b)** $f(x) = 2x^2 - 3x + 2$ **(c)** $f(x) = x^3$ **(d)** $f(x) = x(6-x)$

6 Show that the following limits do not exist:

(a) $\lim_{x\to 0}\frac{1}{x}$ **(b)** $\lim_{x\to 0} f(x)$ where $f(x) = \begin{cases} 0 & \text{for } x < 0 \\ 1 & \text{for } x > 0 \end{cases}$

(c) $\lim_{x\to 0} f(x)$ where $f(x) = \begin{cases} x & \text{for } x > 0 \\ x+1 & \text{for } x < 0 \end{cases}$ **(d)** $\lim_{x\to 0} f(x)$ where $f(x) = \begin{cases} x^2+1 & \text{for } x > 0 \\ 2 & \text{for } x < 0 \end{cases}$

7 The function whose graph is shown is:

A discontinuous at $x = 0$
B continuous for all x
C discontinuous at $x = 2$
D continuous for all $x > 0$ but discontinuous at $x = 1$

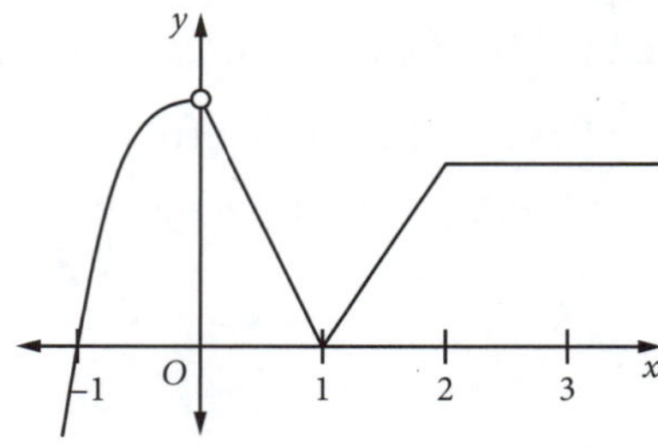

7.3 GRADIENT OF A CURVE

Two expressions have been used to find the gradient of a secant, or the average gradient of the interval joining two points on a curve. This can also be called the average rate of change of the function represented by the curve.

The average gradient of the secant joining the points $(c, f(c))$ and $(x, f(x))$ on the curve $y = f(x)$ is given by $\frac{f(x)-f(c)}{x-c}$. (To visualise this, you may want to refer to the graph of the secant in Section 7.1 on page 165.)

The average gradient of the secant joining the points $(a, f(a))$ and $(a + h, f(a + h))$ on the curve $y = f(x)$ is given by $\frac{f(a+h)-f(a)}{h}$. This could also be called the average rate of change of the function $f(x)$ on the interval $[a, a + h]$.

The earlier calculations have shown that as h gets smaller, the gradient of the secant becomes closer to the gradient of the tangent at the point $(a, f(a))$.

Consider the point $P(2,4)$ on the curve $f(x) = x^2$. Let Q be a point near P on the same curve. Consider the gradient of the secant PQ for different positions of Q near P, as shown in the following tables:

x-coordinate of Q	Gradient of secant PQ
2.1	$\frac{2.1^2-4}{2.1-2}=4.1$
2.01	$\frac{2.01^2-4}{2.01-2}=4.01$
2.001	$\frac{2.001^2-4}{2.001-2}=4.001$
2.0001	$\frac{2.0001^2-4}{2.0001-2}=4.0001$

x-coordinate of Q	Gradient of secant PQ
1.9	$\frac{1.9^2-4}{1.9-2}=3.9$
1.99	$\frac{1.99^2-4}{1.99-2}=3.99$
1.999	$\frac{1.999^2-4}{1.999-2}=3.999$
1.9999	$\frac{1.9999^2-4}{1.9999-2}=3.9999$

It appears from the tables that as Q gets closer to P (that is, as x gets closer to 2), the gradient of the secant gets closer to 4. In general:

Let Q be the point $(x, f(x))$ and P be the fixed point $(2,4)$ on $f(x) = x^2$. The gradient of $PQ = \frac{f(x)-4}{x-2}$.

The information from our table for the function $f(x) = x^2$ indicates that as x approaches 2, the value of the gradient $\frac{f(x)-4}{x-2}$ approaches 4.

This sentence can be written in a new notation as $\lim_{x\to 2}\frac{f(x)-4}{x-2}=4$. This is read as 'the limit as x approaches 2 of function x minus 4 divided by x minus 2 is equal to 4'.

If you let Q be the point $(2 + h, f(2 + h))$ and P be the fixed point $(2, 4)$,

then the gradient of $PQ = \frac{f(2+h)-f(2)}{2+h-2} = \frac{f(2+h)-f(2)}{h}$.

Now as h approaches zero, the value of the gradient $\frac{f(2+h)-f(2)}{h}$ approaches 4.

This sentence can be written as $\lim_{h\to 0}\frac{f(2+h)-f(2)}{h}=4$. This is read as 'the limit as h approaches zero of function $(2 + h)$ minus function 2 divided by h is 4'.

The gradient of a curve

Following the example above: if you let your points be $P(c, f(c))$ and $Q(x, f(x))$, then the gradient of the secant PQ is given by the general expression $\frac{f(x)-f(c)}{x-c}$. This expression does not exist when $x = c$ because the denominator becomes zero.

From our calculations above you have seen that the limiting value of the gradient of the secant is defined to be the gradient of the tangent at P, and is simply called the gradient of the curve at P.

The gradient of the curve $y = f(x)$ at the point $P(c, f(c))$ is defined as the limiting value $\lim_{x \to c} \frac{f(x)-f(c)}{x-c} = f'(c)$, provided that this limit exists.

Thus $f'(c)$ is the slope of the tangent to the curve $y = f(x)$ at the point $x = c$.
This is also called the **derivative** of $f(x)$ at $x = c$.

Gradient and derivative notation

For the function $f(x)$, the gradient function or the derived function is given by the notation $f'(x)$. However, there are also other common notations used for this derivative function.

Let $x = c + h$, where h may be positive or negative. Thus $x - c = h$, so as $x \to c$ then $h \to 0$.

This gives $f'(c) = \lim_{h \to 0} \frac{f(c+h)-f(c)}{h}$.

Also, if $\delta x = h$ and $\delta y = f(c+h) - f(c)$ when $y = f(x)$, then you can write $f'(c) = \lim_{\Delta x \to 0} \frac{\delta y}{\delta x} = \frac{dy}{dx}$.

If in $f'(c) = \lim_{h \to 0} \frac{f(c+h)-f(c)}{h}$ we replace c by x, then you obtain $f'(x) = \lim_{h \to 0} \frac{f(x+h)-f(x)}{h}$.

This now gives us a definition of the derivative at any point (x, y) on the curve $y = f(x)$.

The gradient at any point (x, y) on the curve $y = f(x)$ can be represented by any of the following notations: $f'(x)$, $\frac{dy}{dx}$, $\frac{d}{dx}(f(x))$, y'.

EXERCISE 7.3 GRADIENT OF A CURVE

1 $P(1, 1)$ is a point on the curve $f(x) = x^2$. Complete the following table of values.

x-coordinate of Q	0.9	0.99	0.999	1.001	1.01	1.1
Gradient of secant PQ						

Use the table to find $\lim_{x \to 1} \frac{f(x)-f(1)}{x-1}$.

2 $P(1, 5)$ is a point on the curve $f(x) = 2x + 3$. Complete the following table of values.

x-coordinate of Q	0.9	0.99	0.999	1.001	1.01	1.1
Gradient of secant PQ						

Use the table to find $\lim_{x \to 1} \frac{f(x)-f(1)}{x-1}$.

Could you have predicted this answer from the function? Why?

3 $P(3,9)$ is a point on the curve $f(x)=x^2$. Complete the following table of values.

x-coordinate of Q	2.9	2.99	2.999	3.001	3.01	3.1
Gradient of secant PQ						

Use the table to find $\lim\limits_{x\to 3}\dfrac{f(x)-f(3)}{x-3}$.

4 If $g(x)=x^2$, then $\dfrac{g(3.999)-g(4)}{3.999-4}$ is closest to:

A -8 **B** 0.008 **C** 0.001 **D** 8

5 $P(1,1)$ is a point on the curve $f(x)=x^3$. Complete the following table of values.

x-coordinate of Q	0.9	0.99	0.999	1.001	1.01	1.1
Gradient of secant PQ						

Use the table to find $\lim\limits_{x\to 1}\dfrac{f(x)-f(1)}{x-1}$.

6 $P\left(2,\frac{1}{2}\right)$ is a point on the curve $f(x)=\frac{1}{x}$. Complete the following table of values.

x-coordinate of Q	1.9	1.99	1.999	2.001	2.01	2.1
Gradient of secant PQ						

Use the table to find $\lim\limits_{x\to 2}\dfrac{f(x)-f(2)}{x-2}$.

7 $P(-2,4)$ is a point on the curve $f(x)=x^2$. Complete the following table of values.

x-coordinate of Q	−2.1	−2.01	−2.001	−1.999	−1.99	−1.9
Gradient of secant PQ						

Use the table to find $\lim\limits_{x\to -2}\dfrac{f(x)-f(-2)}{x+2}$.

8 For each of the functions given below, find $\dfrac{f(x+h)-f(x)}{h}$, $h\neq 0$:

(a) $f(x)=2x^2$ **(b)** $f(x)=2x^2+x$ **(c)** $f(x)=4x-x^2$ **(d)** $f(x)=x^3$

9 For each of the functions given below, find $\lim\limits_{h\to 0}\dfrac{f(x+h)-f(x)}{h}$, $h\neq 0$:

(a) $f(x)=5x+1$ **(b)** $f(x)=x^2$ **(c)** $f(x)=3x^2+7x$ **(d)** $f(x)=x^3+2x$

7.4 FINDING THE DERIVATIVE FROM FIRST PRINCIPLES

This process is also called 'differentiation from first principles'. You will use this to show how the general result for the derivative of x^n (powers of x) is obtained.

Example 8

Use the result $f'(x)=\lim\limits_{h\to 0}\dfrac{f(x+h)-f(x)}{h}$ to find the expression for $f'(x)$, if:

(a) $f(x)=x$ **(b)** $f(x)=x^2$ **(c)** $f(x)=x^3$

Solution

(a) $f(x) = x$:

$$f'(x) = \lim_{h\to 0}\frac{f(x+h)-f(x)}{h}$$
$$f'(x) = \lim_{h\to 0}\frac{x+h-x}{h}$$
$$= \lim_{h\to 0} 1$$
$$= 1$$

(b) $f(x) = x^2$:

$$f'(x) = \lim_{h\to 0}\frac{f(x+h)-f(x)}{h}$$
$$= \lim_{h\to 0}\frac{(x+h)^2-x^2}{h}$$
$$= \lim_{h\to 0}\frac{((x+h)-x)((x+h)+x)}{h}$$
$$= \lim_{h\to 0}\frac{h(2x+h)}{h}$$
$$= \lim_{h\to 0}(2x+h)$$

As $h \to 0$, $2x + h \to 2x$, so: $f'(x) = 2x$

(c) $f(x) = x^3$:

$$f'(x) = \lim_{h\to 0}\frac{f(x+h)-f(x)}{h}$$
$$f'(x) = \lim_{h\to 0}\frac{(x+h)^3-x^3}{h}$$

Now $(x+h)^3 = (x+h)(x+h)^2 = x^3 + 3x^2h + 3xh^2 + h^3$

$$f'(x) = \lim_{h\to 0}\frac{x^3+3x^2h+3xh^2+h^3-x^3}{h}$$
$$= \lim_{h\to 0}\frac{h\left(3x^2+3xh+h^2\right)}{h}$$
$$= \lim_{h\to 0}\left(3x^2+3xh+h^2\right)$$
$$= 3x^2$$

OR

Use $a^3 - b^3 = (a-b)(a^2+ab+b^2)$: $f'(x) = \lim_{h\to 0}\frac{((x+h)-x)\left((x+h)^2+x(x+h)+x^2\right)}{h}$

$$= \lim_{h\to 0}\frac{h\left((x+h)^2+x(x+h)+x^2\right)}{h}$$
$$= \lim_{h\to 0}\left((x+h)^2+x(x+h)+x^2\right)$$

As $h \to 0$, $x + h \to x$, so: $f'(x) = x^2 + x^2 + x^2 = 3x^2$

These examples seem to suggest that there might be a rule for finding derivatives without resorting to first principles every time.

Important result

$$a^n - b^n = (a-b)\left(a^{n-1} + a^{n-2}b + a^{n-3}b^2 + \ldots + ab^{n-2} + b^{n-1}\right)$$

This factorisation can be shown to be true by expanding the right-hand side and collecting like terms.

If $f(x) = x^n$, then the above factorisation can be used to find the derivative as follows:

$$f(x) = x^n: \qquad f'(x) = \lim_{h\to 0}\frac{f(x+h)-f(x)}{h}$$

$$= \lim_{h\to 0}\frac{(x+h)^n - x^n}{h}$$

$$= \lim_{h\to 0}\frac{(x+h-x)\left((x+h)^{n-1} + (x+h)^{n-2}x + (x+h)^{n-3}x^2 + \ldots + (x+h)x^{n-2} + x^{n-1}\right)}{h}$$

$$= \lim_{h\to 0}\frac{h\left((x+h)^{n-1} + (x+h)^{n-2}x + (x+h)^{n-3}x^2 + \ldots + (x+h)x^{n-2} + x^{n-1}\right)}{h}$$

$$= \lim_{h\to 0}\left((x+h)^{n-1} + (x+h)^{n-2}x + (x+h)^{n-3}x^2 + \ldots + (x+h)x^{n-2} + x^{n-1}\right)$$

$$= \underbrace{x^{n-1} + x^{n-1} + x^{n-1} + \ldots + x^{n-1}}_{n \text{ factors}}$$

$$= nx^{n-1}$$

This process can be described in words: 'To differentiate a power of x, multiply the power by x to the power reduced by one.'

EXERCISE 7.4 FINDING THE DERIVATIVE FROM FIRST PRINCIPLES

1 For the graph of $f(x) = 6x - 2x^2$:

(a) find the gradient of the secant joining the points whose x-coordinates are 1 and $1 + h$ respectively

(b) deduce the gradient of the curve at $x = 1$.

2 Use the result $f'(x) = \lim_{h\to 0}\frac{f(x+h)-f(x)}{h}$ to find:

(a) $f'(-2)$ when $f(x) = x^2$ (b) $f'(-1)$ when $f(x) = x^3$

3 $P(1,1)$ and $Q(2,8)$ are points on the curve $f(x) = x^3$. Indicate whether each statement is correct or incorrect.

(a) Gradient of $PQ = 7$ (b) $f'(2) = \lim_{x\to 1}\frac{x^3-8}{x-2}$ (c) $f'(1) = \lim_{x\to 1}\frac{x^3-1}{x-1}$ (d) $f'(x) = 3x^2$

4 If $f(x) = 3 - 2x + 4x^2$, then $\lim_{h\to 0}\frac{f(1+h)-f(1)}{h} = \ldots$

A -10 **B** 0 **C** 5 **D** 6

5 For the function $f(x) = 2x^2 - 4x$, find the following:

(a) $\lim_{h\to 0}\frac{f(3+h)-f(3)}{h}$ (b) $\lim_{h\to 0}\frac{f(x+h)-f(x)}{h}$

Interpret your results geometrically.

6 Find $\lim_{h\to 0}\frac{f(x+h)-f(x)}{h}$ for the following:

(a) $f(x) = 4x^2 - 1$ (b) $f(x) = \frac{x^2}{2} - 2x - 3$ (c) $f(x) = x^3 - 2x^2$

7.5 CONDITIONS FOR DIFFERENTIABILITY

If $f(x)$ possesses a derivative $f'(x)$ for each x belonging to the domain of f, then $f(x)$ is called a **differentiable function**.

The statement '$f(x)$ is differentiable' means '$f(x)$ has a derivative at each point of its domain'.

Example 9

Investigate the continuity and differentiability of the given graphs.

(a)

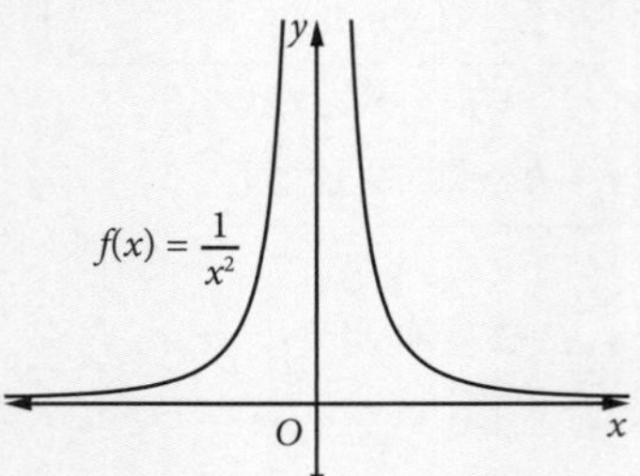

(b)

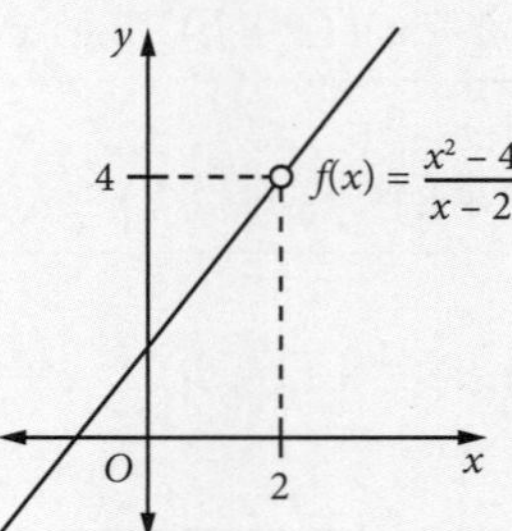

(c)

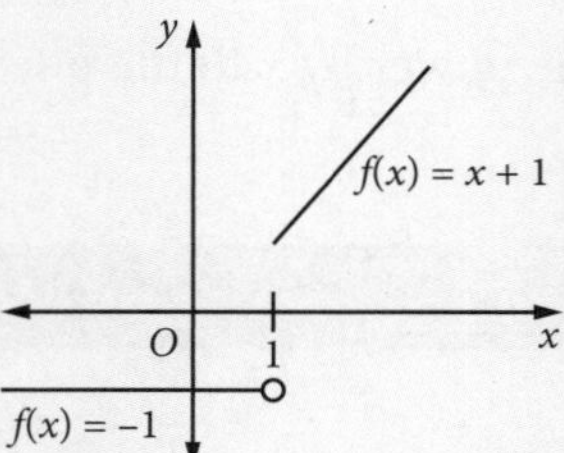

(d)

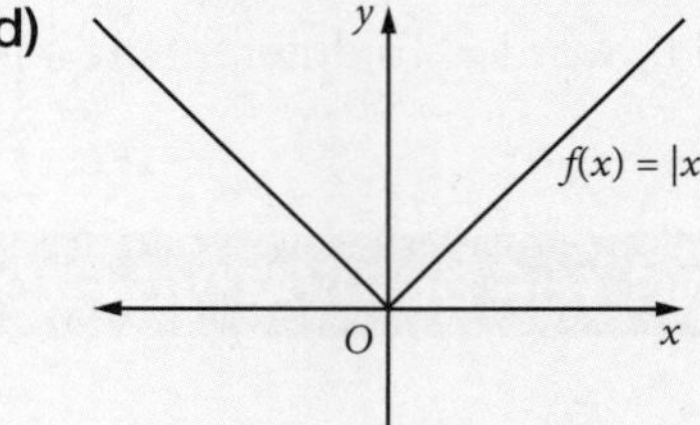

Solution

(a) $f(x)=\dfrac{1}{x^2}, x\neq 0$

$f'(0)$ cannot be found because $f(0)$ is undefined.
Not differentiable at $x=0$. The function is discontinuous at $x=0$.

(b) $f(x)=\dfrac{x^2-4}{x-2}, x\neq 2$

$f'(2)$ cannot be found because $f(2)$ is undefined.
Not differentiable at $x=2$. The function is discontinuous at $x=2$.

(c) $f(x)=\begin{cases} x+1 & \text{for } x\geq 1 \\ -1 & \text{for } x<1 \end{cases}$

$f'(1)$ cannot be found because the left-hand derivative does not exist at $x=1$, even though $f(1)$ is defined.
Not differentiable at $x=1$. The function is discontinuous at $x=1$.

(d) $f(x)=|x|$

$f'(0)$ cannot be found because the left-hand derivative is -1 and the right-hand derivative is 1.
Not differentiable at $x=0$. The function is continuous at $x=0$.

EXERCISE 7.5 CONDITIONS FOR DIFFERENTIABILITY

In each case write the value of $f'(a)$, if it exists, and sketch the graph of f.

1 (a) $f(x)=x-2, a=2$ (b) $f(x)=x^2-4, a=2$ (c) $f(x)=\dfrac{x^2-4}{x+2}, a=2$ (d) $f(x)=\dfrac{x^2-2x}{x-2}, a=2$

2 $f(x)=\begin{cases} x, & x\leq 1 \\ (x-2)^2, & x>1 \end{cases}$ for $a=1, a=3$

3 $f(x)=|x-2|$ for $a=2, a=4$

4 $f(x)=\begin{cases} x^2, & x \le 0 \\ x+1, & x > 0 \end{cases}$ for $a=0, a=-1$

5 $f(x)=\begin{cases} x^2-2x, & x \le 2 \\ x-2, & x > 2 \end{cases}$ for $a=2, a=3$

6 $f(x)=\begin{cases} x, & x > 3 \\ 3, & -3 \le x \le 3 \\ x+1, & x < -3 \end{cases}$ for $a=3, a=-3$

7 For the points at $x=a$, b and c, comment on the continuity and differentiability of the function given in the diagram:

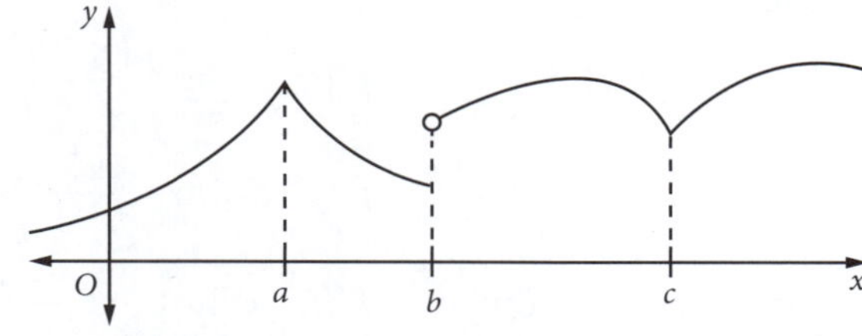

7.6 STANDARD DERIVATIVES

There are several different notations for differentiation, the processes used for the derivative of a function, or finding the derived function.

If $y=f(x)$ then the derivative may written as $\frac{dy}{dx}$, $f'(x)$, y' or $\frac{d}{dx}[f(x)]$ where the definition of each of these expressions is given by $\frac{dy}{dx}=f'(x)=y'=\frac{d}{dx}[f(x)]=\lim\limits_{h\to 0}\frac{f(x+h)+f(x)}{h}$.

$\frac{dy}{dx}$ is read as dee y dee x. $f'(x)$ is read as f dash x. $\frac{d}{dx}[f(x)]$ is read as dee dee x of f of x. y' is read as y dash or y prime.
In each case you are differentiating the given function with respect to x.

You have shown that if $y=x^n$, where n is a positive integer, then $\frac{dy}{dx}=nx^{n-1}$. In fact this result is true for all values of n.

Derivative of $\sqrt{x}$

If $f(x)=\sqrt{x}$, then $f'(x)=\frac{1}{2\sqrt{x}}$.

Proof

$$f'(x)=\lim_{h\to 0}\frac{f(x+h)-f(x)}{h}$$

$f(x)=\sqrt{x}$:
$$f'(x)=\lim_{h\to 0}\frac{\sqrt{x+h}-\sqrt{x}}{h}$$

Rationalise the numerator:
$$f'(x)=\lim_{h\to 0}\left(\frac{\sqrt{x+h}-\sqrt{x}}{h}\times\frac{\sqrt{x+h}+\sqrt{x}}{\sqrt{x+h}+\sqrt{x}}\right)$$
$$=\lim_{h\to 0}\frac{(x+h)-x}{h\left(\sqrt{x+h}+\sqrt{x}\right)}$$
$$=\lim_{h\to 0}\frac{h}{h\left(\sqrt{x+h}+\sqrt{x}\right)}$$
$$=\lim_{h\to 0}\frac{1}{\sqrt{x+h}+\sqrt{x}}$$
$$=\frac{1}{\sqrt{x}+\sqrt{x}}=\frac{1}{2\sqrt{x}}$$

Now $\sqrt{x}=x^{\frac{1}{2}}$, so if $y=x^{\frac{1}{2}}$ then you have shown that $\frac{dy}{dx}=\frac{1}{2\sqrt{x}}=\frac{1}{2}x^{-\frac{1}{2}}$.

Hence the rule for the derivative of x^n works for $n=\frac{1}{2}$.

Derivative of $\frac{1}{x}$

If $f(x) = \frac{1}{x}$, then $f'(x) = -\frac{1}{x^2}$.

Proof

$f(x) = \frac{1}{x}$: $\quad f'(x) = \lim_{h \to 0} \frac{f(x+h) - f(x)}{h}$

$$f'(x) = \lim_{h \to 0} \frac{\frac{1}{x+h} - \frac{1}{x}}{h}$$

Simplify the fraction: $\quad f'(x) = \lim_{h \to 0} \left(\frac{1}{h} \times \frac{x - (x+h)}{(x+h)x} \right)$

$$= \lim_{h \to 0} \left(\frac{1}{h} \times \frac{-h}{(x+h)x} \right)$$

$$= \lim_{h \to 0} \left(\frac{-1}{(x+h)x} \right)$$

$$= \frac{-1}{x \times x} = -\frac{1}{x^2}$$

Now $\frac{1}{x} = x^{-1}$, so if $y = x^{-1}$ then you have shown that $\frac{dy}{dx} = -\frac{1}{x^2} = -1x^{-2}$.

Hence the rule for the derivative of x^n works for $n = -1$.

Derivative of c

If $f(x) = c$, a constant, then $f'(x) = 0$.

Proof

$$f'(x) = \lim_{h \to 0} \frac{f(x+h) - f(x)}{h}$$

$f(x) = c$: $\quad f'(x) = \lim_{h \to 0} \frac{c - c}{h} = 0$

Now $c = cx^0$, so if $y = cx^0$ then you have shown that $\frac{dy}{dx} = 0x^{-1} = 0$.

Hence the rule for the derivative of x^n works for $n = 0$.

Derivative of x^n

If $f(x) = x^n$, then $f'(x) = nx^{n-1}$, where n is any real number.

This can also be written as $\frac{d}{dx}\left(x^n\right) = nx^{n-1}$ or $\frac{dy}{dx} = nx^{n-1}$.

Example 10

Consider $f(x) = cx^n$, where c is a constant. Find $f'(x)$.

Solution

$$f'(x) = \lim_{h \to 0} \frac{f(x+h) - f(x)}{h}$$

$f(x) = cx^n$: $\quad f'(x) = \lim_{h \to 0} \frac{c(x+h)^n - cx^n}{h}$

$$= c \lim_{h \to 0} \frac{(x+h)^n - x^n}{h}$$

$$= c \times nx^{n-1}$$

Derivative of $cg(x)$

If $y = cu$, where c is a constant and $u = g(x)$, then $\frac{dy}{dx} = c\frac{du}{dx}$.

Proof

If $u = g(x)$, then: $\frac{du}{dx} = \lim_{h \to 0} \frac{g(x+h) - g(x)}{h}$

Then:
$$\begin{aligned}\frac{dy}{dx} &= \lim_{h \to 0} \frac{cg(x+h) - cg(x)}{h} \\ &= \lim_{h \to 0} c\left(\frac{g(x+h) - g(x)}{h}\right) \\ &= c\lim_{h \to 0}\left(\frac{g(x+h) - g(x)}{h}\right) \\ &= cg'(x) = c\frac{du}{dx}\end{aligned}$$

Example 11

If $f(x) = x^2 + x$, find $f'(x)$.

Solution

$$f'(x) = \lim_{h \to 0} \frac{f(x+h) - f(x)}{h}$$

$f(x) = x^2 + x$:
$$\begin{aligned}f'(x) &= \lim_{h \to 0} \frac{(x+h)^2 + (x+h) - (x^2 + x)}{h} \\ &= \lim_{h \to 0} \frac{x^2 + 2hx + h^2 + x + h - x^2 - x}{h} \\ &= \lim_{h \to 0} \frac{2hx + h^2 + h}{h} \\ &= \lim_{h \to 0} \frac{h(2x + 1 + h)}{h} \\ &= \lim_{h \to 0}(2x + 1 + h) \\ &= 2x + 1\end{aligned}$$

Derivative of $f(x) \pm g(x)$

If $y = u \pm v$ where $u = f(x)$ and $v = g(x)$, then $\frac{dy}{dx} = \frac{du}{dx} \pm \frac{dv}{dx}$.

Proof

If $u = f(x)$, then: $\frac{du}{dx} = \lim_{h \to 0} \frac{f(x+h) - f(x)}{h}$ and $\frac{dv}{dx} = \lim_{h \to 0} \frac{g(x+h) - g(x)}{h}$

Then for $y = u + v$:
$$\begin{aligned}\frac{dy}{dx} &= \lim_{h \to 0} \frac{\big(f(x+h) + g(x+h)\big) - \big(f(x) + g(x)\big)}{h} \\ &= \lim_{h \to 0} \frac{\big(f(x+h) - f(x)\big) + \big(g(x+h) - g(x)\big)}{h} \\ &= \lim_{h \to 0}\left(\frac{\big(f(x+h) - f(x)\big)}{h} + \frac{\big(g(x+h) - g(x)\big)}{h}\right) \\ &= \lim_{h \to 0}\left(\frac{\big(f(x+h) - f(x)\big)}{h}\right) + \lim_{h \to 0}\left(\frac{\big(g(x+h) - g(x)\big)}{h}\right) \\ &= \frac{du}{dx} + \frac{dv}{dx}\end{aligned}$$

The result follows similarly for $y = u - v$.

Summary of important results so far

1 If $y = x^n$, then $\frac{dy}{dx} = nx^{n-1}$, where n is any real number.

2 If $y = cu$, where c is a constant and $u = g(x)$, then $\frac{dy}{dx} = c\frac{du}{dx}$.

3 If $y = u \pm v$ where $u = f(x)$ and $v = g(x)$, then $\frac{dy}{dx} = \frac{du}{dx} \pm \frac{dv}{dx}$.

4 If $y = \sqrt{x}$ then $\frac{dy}{dx} = \frac{1}{2\sqrt{x}}$.

5 If $y = \frac{1}{x}$ then $\frac{dy}{dx} = -\frac{1}{x^2}$.

6 From results 1 and 3, it follows that if $y = x^n + x^{n-1} + x^{n-2} + \ldots + x$, then $\frac{dy}{dx} = nx^{n-1} + (n-1)x^{n-2} + (n-2)x^{n-3} + \ldots + 1$.

Example 12

Find $\frac{dy}{dx}$ for each function.

(a) $y = x^4 + x^2 + x$ (b) $y = 5x^3 - 4x + 3$ (c) $y = x^5 + 6x^4 - 7x$

Solution

(a) $y = x^4 + x^2 + x$

$\frac{dy}{dx} = 4x^3 + 2x + 1$

(b) $y = 5x^3 - 4x + 3$

$\frac{dy}{dx} = 5 \times 3x^2 - 4$

$= 15x^2 - 4$

(c) $y = x^5 + 6x^4 - 7x$

$\frac{dy}{dx} = 5x^4 + 6 \times 4x^3 - 7$

$= 5x^4 + 24x^3 - 7$

Example 13

Find $\frac{dy}{dx}$ for each function.

(a) $y = \frac{1}{x^2}$ (b) $y = 3x^2 + \sqrt{x}$ (c) $y = 4 - \frac{1}{x} + \sqrt[3]{x}$

Solution

(a) $y = x^{-2}$

$\frac{dy}{dx} = -2x^{-3}$

$= -\frac{2}{x^3}$

(b) $y = 3x^2 + x^{\frac{1}{2}}$

$\frac{dy}{dx} = 3 \times 2x + \frac{1}{2}x^{-\frac{1}{2}}$

$= 6x + \frac{1}{2\sqrt{x}}$

(c) $y = 4 - x^{-1} + x^{\frac{1}{3}}$

$\frac{dy}{dx} = 0 - \left(-1x^{-2}\right) + \frac{1}{3}x^{-\frac{2}{3}}$

$= \frac{1}{x^2} + \frac{1}{3\sqrt[3]{x^2}}$

Example 14

If $y = x^3 - 6x$, find: (a) $\frac{dy}{dx}$ (b) the gradient when $x = -1$

(c) the coordinates of the points at which the gradient is 6.

Solution

(a) $y = x^3 - 6x$: $\frac{dy}{dx} = 3x^2 - 6$ (b) $x = -1$: $\frac{dy}{dx} = 3 - 6 = -3$

(c) $\frac{dy}{dx} = 6$:

$$\begin{aligned} 3x^2 - 6 &= 6 \\ 3x^2 &= 12 \\ x^2 &= 4 \\ x &= \pm 2 \end{aligned}$$

$x = 2, y = -4; x = -2, y = 4$. Points are $(2, -4)$ and $(-2, 4)$.

Example 15

If $f(x) = x^4 - 3x^3 + 2x$, find: (a) $f'(x)$ (b) $f'(a)$ (c) $f'(2)$ (d) $f'(-2)$

Solution

(a) $f(x) = x^4 - 3x^3 + 2x$: $f'(x) = 4x^3 - 9x^2 + 2$ (b) $f'(a) = 4a^3 - 9a^2 + 2$

(c) $f'(2) = 4 \times 8 - 9 \times 4 + 2 = 32 - 36 + 2 = -2$ (d) $f'(-2) = 4 \times (-8) - 9 \times 4 + 2 = -32 - 36 + 2 = -66$

Practical application of differentiation

In business, manufacturers create a function called the cost function based on the expenses involved in producing their product. This function may be written $c = f(x)$ where x is the number of units produced.

The marginal cost is given by the derivative, $\frac{dc}{dx}$ and represents the change in the total cost of producing one extra unit.

Example 16

The cost function, in dollars, for a manufacturer is given by $c = 0.4x^2 + 3x + 750$, where x is the number of items produced in a week. What is the marginal cost when $x = 50$?

Solution

$$c = 0.4x^2 + 3x + 750$$

$$\frac{dc}{dx} = 0.8x + 3$$

When $x = 50$: $\frac{dc}{dx} = 0.8 \times 50 + 3 = 43$

It costs $43 for an extra item to be produced when 50 items are produced in a week.

Families of curves

Example 17

(a) For each of the functions find their derivative: **(i)** $f(x) = x^2$ **(ii)** $g(x) = x^2 + 2$ **(iii)** $h(x) = x^2 - 3$. Comment on your answers.

(b) On the same diagram, graph each of the functions in part **(a)** and well as the derivative function obtained in each part.

Solution

(a) **(i)** $f(x) = x^2$: $f'(x) = 2x$ **(ii)** $g(x) = x^2 + 2$: $g'(x) = 2x$ **(iii)** $h(x) - x^2 - 3$: $h'(x) = 2x$

The derivative of each function is the same, $2x$.

(b)

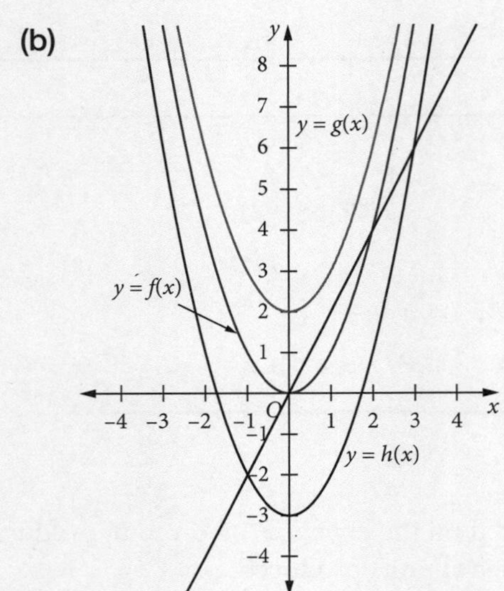

This example shows that functions that only differ by a constant will have the same derivative.

Thus when $x = 0$ on each function, the gradient of the tangent at this point on each function is 0. This is called a horizontal tangent and the point at which it occurs is called stationary point.

When $x > 0$ the gradient of each function is positive; for example, when $x = 1$ the gradient of the tangent for each function is 2. For $x > 0$, as x increases the value of the gradient remains positive so each function is called an increasing function.

When $x < 0$ the gradient of each function is negative; for example, when $x = -1$ the gradient of the tangent for each function is -2. For $x < 0$, as x increases the value of the gradient remains negative so each function is called a decreasing function.

This shows that for a given derivative, $f'(x)$, there are many functions that will give this derivative, each one differing by a constant. For example, $f(x)$, $f(x) + 1$, $f(x) - 2$, $f(x) + c$, where c is a constant, will all give $f'(x)$ as their derivative.

Properties of the derivative

1. If $f(x)$ is a polynomial then the degree of $f'(x)$ will be one less than the degree of $f(x)$
2. $f'(x)$ is also a function.
3. Functions that only differ by a constant will have the same derivative.
4. If the derivative is a constant then the original function increases or decreases at a constant rate.
5. If the derivative is itself a function of the variable, then the original function will increase and/or decrease at a variable rate.

Example 18

On the same diagram, draw the graph of each function and its derivative.

(a) $f(x) = 2x$

(b) $g(x) = x^3$

(c) $h(x) = x^3 - 3x$

(d) Describe the rate of change of each function.

Solution

The derived function is shown as a dashed curve.

(a) $f(x) = 2x$: $f'(x) = 2$

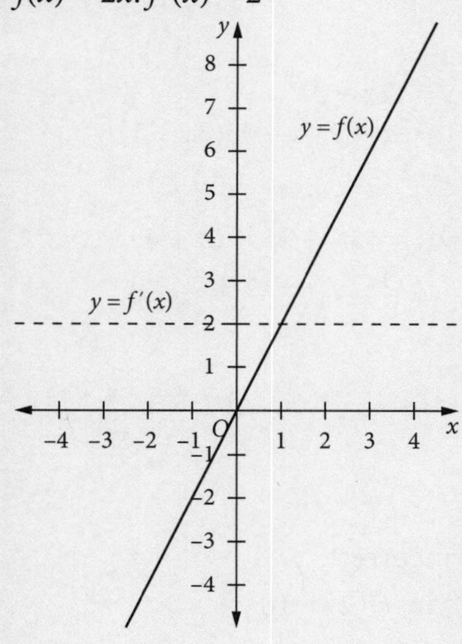

(b) $g(x) = x^3$: $g'(x) = 3x^2$

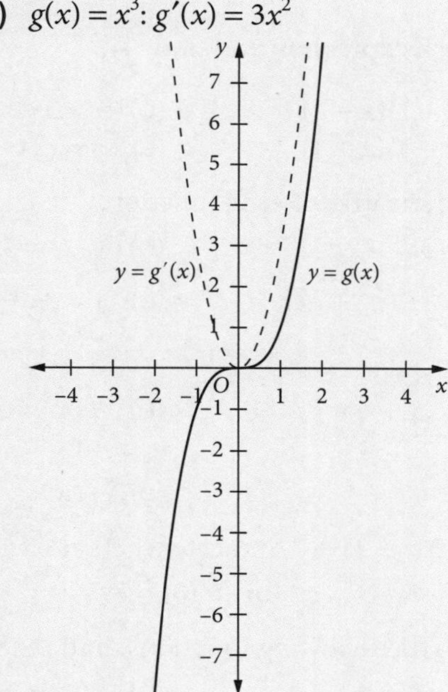

(c) $h(x) = x^3 - 3x$: $h'(x) = 3x^2 - 3$

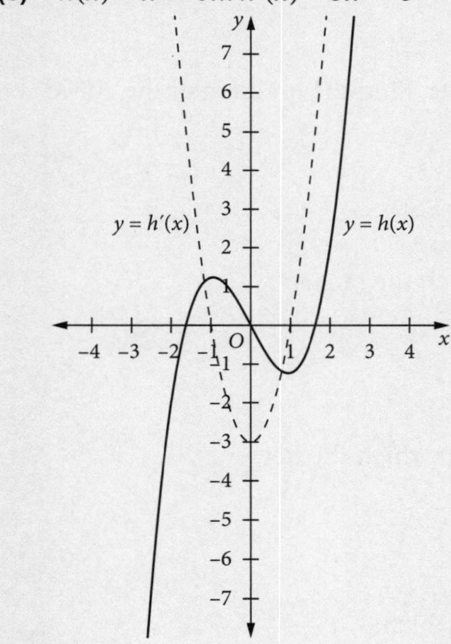

(d) In **(a)**, $f(x) = 2x$ is a straight line which has a constant gradient. This is verified by the gradient function $f'(x) = 2$.

In **(b)** $g(x) = x^3$ is a cubic function whose gradient function, $g'(x) = 3x^2$, is zero when $x = 0$, but positive elsewhere. This means that $g(x)$ is an increasing function. When $x > 0$, then as x increases $g(x)$ also increases at an increasing rate. $x = 0$ is a stationary point on $g(x)$.

In **(c)**, $h(x) = x^3 - 3x$ is a cubic function whose gradient function, $h'(x) = 3x^2 - 3$, is zero when $x = \pm 1$. Stationary points occur when $x = \pm 1$. For $|x| > 1$ then $h'(x) > 0$ and $h(x)$ increases as x increases over those domains. For $-1 < x < 1$, $h'(x) < 0$ and $h(x)$ decreases as x increases over that domain.

EXERCISE 7.6 STANDARD DERIVATIVES

1 Find the derivative of:

(a) $y = 3x^2 + 2x - 1$ **(b)** $y = 4x - 3x^2$ **(c)** $y = 7x - 4x^2$

(d) $y = x^4 + x^2 + 1$ **(e)** $y = x^5 - x^3 + x$ **(f)** $v = t^3 + 4t^2 - 2t + 5$

2 Find the derivative of:

(a) $y = x^{\frac{3}{2}}$ **(b)** $y = \dfrac{2}{x}$ **(c)** $y = 2\sqrt{x}$ **(d)** $v = \sqrt[3]{t^2}$ **(e)** $h(m) = \dfrac{1}{m^3}$ **(f)** $f(x) = \dfrac{1}{\sqrt{x}}$

3 If $g(x) = x^{\frac{4}{3}}$, then $g'(x) = \ldots$

A $g'(x) = 4x^{\frac{1}{3}}$ **B** $g'(x) = \dfrac{4}{3x^{\frac{1}{3}}}$ **C** $g'(x) = x^{\frac{1}{3}}$ **D** $g'(x) = \frac{4}{3}x^{\frac{1}{3}}$

4 Expand each expression and find $\dfrac{dy}{dx}$.

(a) $y = (x-1)(x+2)$ **(b)** $y = 3x(x^2-2)$ **(c)** $y = (2x-3)^2$
(d) $y = (x-4)(x+4)$ **(e)** $y = (2x-3)^3$ **(f)** $y = (x-2)(x+1)(3x+1)$

5 Write the derivative of each function.

(a) $f(x) = 3x^2 - 2x^3$ **(b)** $g(x) = 4x(1-x^3)$ **(c)** $v(t) = 3t^3 - 2t^2 + 5t + 4$
(d) $y = ax^3 + bx^2 + cx + d$ **(e)** $y = 3x + 4$ **(f)** $y = \dfrac{3x^3 - 2x^2 + x}{x}, x \neq 0$

6 Find $f'(x)$.

(a) $f(x) = x + \sqrt{x}$ **(b)** $f(x) = x^2 + \dfrac{1}{x}$ **(c)** $f(x) = x^2 + x + 1 + \dfrac{1}{x} + \dfrac{1}{x^2}$
(d) $f(x) = x^{\frac{2}{3}} + x^{\frac{1}{3}}$ **(e)** $f(x) = \left(x - \dfrac{1}{x}\right)^2$ **(f)** $f(x) = x\sqrt{x}$

7 For $f(x) = 3x^2 - 2x + 7$, indicate whether each statement is correct or incorrect.

(a) $f'(x) = 6x - 2$ **(b)** $f'(0) = 7$ **(c)** $f(1) = 8$ **(d)** $f'(2) = 10$

8 For each of the following functions, find the value of x for which $f'(x) = 0$.

(a) $f(x) = x^2 - 4$ **(b)** $f(x) = 2x^3 - 6x$ **(c)** $f(x) = x^3 - 4x^2$

9 Find the gradient of the curve $y = x^2 - x - 6$ at the points where $y = 0$.

10 If $f(x) = x^3 - x^2 - 6x + 1$, find the values of x for which $f'(x) = -5$.

11 Show that the graph of $y = x^2 + 4x - 12$ crosses the x-axis at two points. Find the gradient of the curve at these points.

12 For the graph of $f(x) = (x-1)^2$, find the values of x for which:

(a) $f(x) = 0$ **(b)** $f'(x) = 0$ **(c)** $f'(x) = -1$

13 Find the coordinates of the points on the curve $y = x^2 - 5x + 6$ at which the tangent:

(a) makes an angle of 45° with the x-axis
(b) is parallel to the line with equation $3x + y - 4 = 0$
(c) is perpendicular to the line with equation $2y - x + 3 = 0$.

14 Find the x values of the points on the curve $y = \frac{1}{3}x^3 - \frac{3}{2}x^2 + 2x + 1$ at which the tangent:

(a) is parallel to the x-axis
(b) makes an angle θ with the x-axis such that $\tan\theta = 2$
(c) is parallel to the line $y - 6x - 1 = 0$.

15 Find the coordinates of the points on the parabola $y = x^2 - 2x - 8$ at which:

(a) the gradient is zero **(b)** the tangent is parallel to the line $2x + y = 7$.

16 The cost function, in dollars, for a manufacturer is given by $c = 0.6x^2 + 4x + 650$, where x is the number of items produced in a week. What is the marginal cost when $x = 40$?

17 The profit function, in dollars, for a manufacturer is given by the function $P = 6x - \dfrac{x^2}{2} - 10$, where x is the number of items produced in a day up to a maximum of 6 items.

(a) If the break-even point is when the profit is zero, what is the break-even point for this manufacturer?
(b) Find $\dfrac{dP}{dx}$.
(c) For what values of x is $\dfrac{dP}{dx} > 0$?

18 The value, in dollars, of a machine is given by $v = 40\,000 - 4000t$, where t is in years and $0 \le t \le 10$. How fast is the value falling with respect to time when $t = 2$, $t = 5$ or at any time, t, in its domain?

19 **(a)** Find $\frac{dy}{dx}$ for each of the following functions:
(i) $y = x^2 + 3x$ **(ii)** $y = x^2 + 3x + 5$ **(iii)** $y = x^2 + 3x - 7$
(b) What do you notice about each of the answers in **(a)**?

20 **(a)** Given $f(x) = x^2 + 3$, find $f'(x)$.
(b) On the same diagram sketch the graph of $y = f(x)$ and $y = f'(x)$.

21 **(a)** Given $f(x) = x^3 - 1$, find $f'(x)$.
(b) On the same diagram sketch the graph of $y = f(x)$ and $y = f'(x)$.

22 **(a)** Given $f(x) = x^3 + x^2 + x$, find $f'(x)$.
(b) On the same diagram sketch the graph of $y = f(x)$ and $y = f'(x)$.

23 **(a)** If $g'(x) = 3x^2$, give the equations for 3 possible functions for $g(x)$.
(b) On the same diagram, graph $g'(x)$ and one of your functions for $g(x)$.

7.7 THE PRODUCT RULE

In previous exercises you have differentiated the product of two functions by first expanding the product, then differentiating term by term. Luckily there is a rule, known as the **product rule**, that allows you to differentiate products without expanding first.

The product rule:

- If u and v are functions of x, and $y = uv$, then $\frac{d}{dx}(uv) = v\frac{du}{dx} + u\frac{dv}{dx}$.

The product rule can also be written as $\frac{d}{dx}(uv) = u\frac{dv}{dx} + v\frac{du}{dx}$. You may find it easier to remember this in the form $\frac{d}{dx}(uv) = v\frac{du}{dx} + u\frac{dv}{dx}$, because this is consistent with the form of the quotient rule to come later.

Example 19

If $y = x^2(3x + 4)$, find $\frac{dy}{dx}$:

(a) by expanding the expression and differentiating term by term

(b) by using the product rule.

Solution

(a) $y = 3x^3 + 4x^2$: $\frac{dy}{dx} = 9x^2 + 8x$

(b) $y = x^2(3x + 4)$: $u = x^2$, $v = 3x + 4$

$\frac{du}{dx} = 2x$, $\frac{dv}{dx} = 3$

Product rule: $\frac{d}{dx}(uv) = v\frac{du}{dx} + u\frac{dv}{dx}$

$$\frac{d}{dx}(uv) = (3x + 4) \times 2x + x^2 \times 3$$
$$= 6x^2 + 8x + 3x^2$$
$$= 9x^2 + 8x$$

The answers are the same. This example verifies the product rule.

Proof of the product rule

If $f(x)=uv$, $u=g(x)$ and $v=k(x)$ are functions of x, then:

$$f'(x)=\lim_{h\to 0}\frac{f(x+h)-f(x)}{h} \qquad g'(x)=\lim_{h\to 0}\frac{g(x+h)-g(x)}{h} \qquad k'(x)=\lim_{h\to 0}\frac{k(x+h)-k(x)}{h}$$

Now $f(x)=g(x)k(x)$ and $f(x+h)=g(x+h)k(x+h)$, so:

$$\begin{aligned}
f'(x)&=\lim_{h\to 0}\frac{g(x+h)k(x+h)-g(x)k(x)}{h}\\
&=\lim_{h\to 0}\left\{\frac{g(x+h)k(x+h)-g(x)k(x+h)+g(x)k(x+h)-g(x)k(x)}{h}\right\}\\
&=\lim_{h\to 0}\left\{\frac{g(x+h)k(x+h)-g(x)k(x+h)}{h}+\frac{g(x)k(x+h)-g(x)k(x)}{h}\right\}\\
&=\lim_{h\to 0}\left\{\frac{k(x+h)\big(g(x+h)-g(x)\big)}{h}\right\}+\lim_{h\to 0}\left\{\frac{g(x)\big(k(x+h)-g(x)\big)}{h}\right\}\\
&=k(x)\lim_{h\to 0}\left\{\frac{g(x+h)-g(x)}{h}\right\}+g(x)\lim_{h\to 0}\left\{\frac{k(x+h)-g(x)}{h}\right\}\\
&=k(x)g'(x)+g(x)k'(x)
\end{aligned}$$

Hence $\dfrac{d}{dx}(uv)=v\dfrac{du}{dx}+u\dfrac{dv}{dx}$.

Example 20

Use the product rule to differentiate $y=(3x+2)(2x^2-3x+4)$.

Solution

$y=(3x+2)(2x^2-3x+4)$: $\qquad u=3x+2 \qquad v=2x^2-3x+4$

$\qquad \dfrac{du}{dx}=3 \qquad \dfrac{dv}{dx}=4x-3$

Product rule: $\dfrac{d}{dx}(uv)=v\dfrac{du}{dx}+u\dfrac{dv}{dx}$

$$\begin{aligned}
\frac{dy}{dx}&=(2x^2-3x+4)\times 3+(3x+2)\times(4x-3)\\
&=6x^2-9x+12+12x^2-9x+8x-6\\
&=18x^2-10x+6
\end{aligned}$$

You can check this answer by expanding the function in the question and then differentiating term by term.

EXERCISE 7.7 THE PRODUCT RULE

1 Use the product rule to find the derivative of each function.

(a) $y=(x-2)(6x+7)$ **(b)** $f(x)=(2x+1)(x+3)$ **(c)** $y=(3x+4)(x^2-2x)$

(d) $g(x)=(x-1)(x^2-3x)$ **(e)** $y=(2x^2-5x)(x-2)$ **(f)** $f(x)=(x^2-4x)(x^2+3)$

(g) $y=(x-1)(3x+5)$ **(h)** $f(x)=(x^2-5x)(2x+3)$ **(i)** $g(x)=(4x-1)(5x^2-7)$

2 If $g(x)=(3x-1)(3x^2+1)$ then $g'(x)=\ldots$

A $g'(x)=27x^2-6x+3$ **B** $g'(x)=18x$

C $g'(x)=3+6x-9x^2$ **D** $g'(x)=9x^3-3x^2+3x-1$

3 For $f(x) = (x^3 - 1)(2x + 3)$, indicate whether each statement is correct or incorrect.

(a) $f'(x) = 6x^2$ (b) $f'(x) = 8x^3 + 9x^2 - 2$ (c) $f'(0) = -2$ (d) $f(0) = -3$

4 For $g(x) = (x^2 + 5x)(x^3 + x^2 + 1)$, find: (a) $g'(x)$ (b) $g'(1)$ (c) $g'(-2)$

5 Find $\frac{dy}{dx}$ for: (a) $y = \sqrt{x}(x-1)$ (b) $y = x\left(\sqrt{x} - 1\right)$ (c) $y = x\left(\frac{1}{x} + 1\right)$

(d) $y = \left(3\sqrt{x} + 1\right)(x^2 + 4)$ (e) $y = \left(x + \frac{1}{x}\right)\left(x - \frac{1}{x}\right)$ (f) $y = \left(x^2 + \frac{2}{x}\right)\left(1 + \sqrt{x}\right)$

7.8 THE CHAIN RULE

The chain rule is also known as the 'composite function rule' or the 'function-of-a-function rule'. This rule allows you to differentiate functions such as $f(x) = (x^2 - 5)^7$ without having to expand.

Example 21

Differentiate each function after expanding the parentheses. Factorise your answer.

(a) $f(x) = (x^2 - 5)^2$ (b) $g(x) = (x^2 - 5)^3$

Solution

(a)
$$\begin{aligned} f(x) &= (x^2 - 5)^2 = x^4 - 10x^2 + 25 \\ f'(x) &= 4x^3 - 20x \\ &= 4x(x^2 - 5) \\ &= 2x \times 2(x^2 - 5) \end{aligned}$$

(b)
$$\begin{aligned} g(x) &= (x^4 - 10x^2 + 25)(x^2 - 5) \\ &= x^6 - 15x^4 + 75x^2 - 125 \\ g'(x) &= 6x^5 - 60x^3 + 150x \\ &= 6x(x^4 - 10x^2 + 25) \\ &= 2x \times 3(x^2 - 5)^2 \end{aligned}$$

There is a pattern in the answers to Example **21**. In each case, $2x$ is the derivative of x^2 and the second part looks a bit like $\frac{d}{dx}\left(u^n\right) = nu^{n-1}$ where $u = h(x)$.

Let $u = (x^2 - 5)$ so that $\frac{du}{dx} = 2x$. This allows you to write:

$$\begin{aligned} f(x) &= u^2 & g(x) &= u^3 \\ f'(x) &= 2x \times 2u & g'(x) &= 2x \times 3u^2 \\ f'(x) &= 2u \times \frac{du}{dx} & g'(x) &= 3u^2 \times \frac{du}{dx} \end{aligned}$$

If $y = f(x) = [h(x)]^2$, where $u = h(x)$, then you have $y = u^2$.

Hence, the chain rule: If $f(x) = u(h(x))$ then $f'(x) = u'(h(x))\, h'(x)$

OR $f'(x) = \frac{d}{du}(f(x)) \times \frac{du}{dx}$ for any differentiable function f.

$$\frac{dy}{dx} = \frac{dy}{du} \times \frac{du}{dx}$$

You can check this rule using Example **21(b)**:

$y = (x^2 - 5)^3$: $y = u^3$ where $u = x^2 - 5$

$$\frac{dy}{du} = 3u^2 \text{ and } \frac{du}{dx} = 2x$$

Chain rule:
$$\begin{aligned} \frac{dy}{dx} &= \frac{dy}{du} \times \frac{du}{dx} \\ \frac{dy}{dx} &= 3u^2 \times 2x \\ &= 3\left(x^2 - 5\right)^2 \times 2x = 6x\left(x^2 - 5\right)^2 \end{aligned}$$

This is the same as the derivative obtained by expanding the function and differentiating term by term.

Proof of the chain rule

The chain rule: if $y = f(u)$ and $u = g(x)$ then $\frac{dy}{dx} = \frac{dy}{du} \times \frac{du}{dx}$.

Using first principles, you can write: $$\frac{du}{dx} = \lim_{h\to0}\left(\frac{g(x+h)-g(x)}{h}\right)$$

and $$\frac{dy}{du} = \lim_{k\to0}\left(\frac{f(u+k)-f(u)}{k}\right) = \lim_{k\to0}\left(\frac{f(g(x)+k)-f(g(x))}{k}\right).$$

If $k = g(x+h) - g(x)$ then $g(x+h) = g(x) + k$.

If $h \to 0$ then $g(x+h) \to g(x)$ and thus $k \to 0$.

Using first principles, you can write:
$$\begin{aligned}\frac{dy}{dx} &= \lim_{h\to0}\left(\frac{f(g(x+h))-f(g(x))}{h}\right)\\ &= \lim_{h\to0}\left(\frac{f(g(x+h))-f(g(x))}{g(x+h)-g(x)} \times \frac{g(x+h)-g(x)}{h}\right)\\ &= \lim_{h\to0}\left(\frac{f(g(x+h))-f(g(x))}{g(x+h)-g(x)}\right) \times \lim_{h\to0}\left(\frac{g(x+h)-g(x)}{h}\right)\\ &= \lim_{k\to0}\left(\frac{f(g(x)+k)-f(g(x))}{k}\right) \times \lim_{h\to0}\left(\frac{g(x+h)-g(x)}{h}\right)\\ &= \frac{dy}{du} \times \frac{du}{dx}\end{aligned}$$

Example 22

Use the chain rule to find the derivative of each function.

(a) $f(x) = (2x+1)^2$ **(b)** $g(x) = (2x^2 - 3x + 1)^4$ **(c)** $y = \sqrt{4 - x^2}$

Solution

(a) Let $u = 2x + 1$ so $\frac{du}{dx} = 2$

$f(x) = u^2$ so $\frac{d}{du}(f(x)) = 2u$

Hence: $$\begin{aligned}f'(x) &= \frac{d}{du}(f(x)) \times \frac{du}{dx}\\ &= 2u \times 2\\ &= 4(2x+1)\end{aligned}$$

(b) Let $u = 2x^2 - 3x + 1$ so $\frac{du}{dx} = 4x - 3$

$g(x) = u^4$ so $\frac{d}{du}(g(x)) = 4u^3$

Hence: $$\begin{aligned}g'(x) &= \frac{d}{du}(g(x)) \times \frac{du}{dx}\\ &= 4u^3 \times (4x-3)\\ &= 4(4x-3)\left(2x^2-3x+1\right)^3\end{aligned}$$

(c) Let $u = 4 - x^2$ so $\frac{du}{dx} = -2x$

$y = \sqrt{u}$ so $\frac{dy}{du} = \frac{1}{2\sqrt{u}}$

Hence: $$\begin{aligned}\frac{dy}{dx} &= \frac{dy}{du} \times \frac{du}{dx}\\ &= \frac{1}{2\sqrt{u}} \times (-2x)\\ &= \frac{-x}{\sqrt{4-x^2}}\end{aligned}$$

Example 23

Find $f'(t)$ if $f(t)=t\sqrt{t^2-9}$.

Solution

This example requires both the product rule and the chain rule.

$$u=t \qquad \frac{du}{dt}=1$$

Chain rule: $v=\sqrt{t^2-9}$

$$\frac{dv}{dt}=\frac{1}{2}\left(t^2-9\right)^{-\frac{1}{2}}\times 2t$$

$$=\frac{t}{\sqrt{t^2-9}}$$

Product rule: $\frac{d}{dt}(uv)=v\frac{du}{dt}+u\frac{dv}{dt}$:

$$f'(t)=\sqrt{t^2-9}\times 1+t\times\frac{t}{\sqrt{t^2-9}}$$

$$=\sqrt{t^2-9}+\frac{t^2}{\sqrt{t^2-9}}$$

Write with common denominator:

$$=\frac{t^2-9+t^2}{\sqrt{t^2-9}}$$

$$=\frac{2t^2-9}{\sqrt{t^2-9}}$$

EXERCISE 7.8 THE CHAIN RULE

1 Use the chain rule to differentiate:

(a) $y=(x^2-4)^5$ (b) $f(x)=\sqrt{5x-1}$ (c) $y=(x^3-3x)^4$

(d) $y=(2x+5)^{-1}$ (e) $f(t)=\sqrt[3]{t^2+3t}$ (f) $g(x)=(2x^2+5x-4)^4$

(g) $f(x)=\sqrt{x^2-2x}$ (h) $f(t)=(t^2+4)^{-2}$ (i) $y=\sqrt{25-x^2}$

2 If $f(x)=(x^3-1)^5$ then $f'(x)=\ldots$

A $3x^2(x^3-1)^4$ B $15x^2(x^3-1)^4$ C $5x^2(x^3-1)^4$ D $15x^2$

3 Find the derivative of each function.

(a) $y=\sqrt{x^2-4}$ (b) $f(x)=\left(x^2+1\right)^{\frac{1}{2}}$ (c) $y=(1+2x)^{-1}$

(d) $v=(m^2+25)^2$ (e) $y=(2x-1)^5$ (f) $g(t)=\sqrt[3]{t+1}$

(g) $f(x)=(3x^2-2x-1)^4$ (h) $f(t)=(t^2+4)^{-2}$ (i) $y=\left(x-\frac{1}{x}\right)^4$

4 For $g(x)=x^2+5x+\sqrt[3]{x^2-4}$, indicate whether each statement is correct or incorrect.

(a) $g'(x)=\frac{2x}{3}\left(x^2-4\right)^{-\frac{2}{3}}$ (b) $g'(x)=4x$

(c) $g'(x)=2x+5+\frac{2x}{3}\left(x^2-4\right)^{-\frac{2}{3}}$ (d) $g'(x)=2x+5+\frac{2x}{3\left(x^2-4\right)^{\frac{2}{3}}}$

5 Find the derivative of each function.

(a) $y=(x-3)(3x+4)^6$ (b) $f(x)=x^2\sqrt{1-x^2}$ (c) $h(t)=t^3+(4-t)^4$

(d) $y=\sqrt{1-x}+\sqrt{1+x}$ (e) $y=t^2-1+(1+t)^{-1}$ (f) $g(x)=x\left(1+\frac{1}{x}\right)^2$

7.9 THE QUOTIENT RULE

This rule is a special application of the product rule.

Consider: If u and v are functions of x, and $y = u \times \frac{1}{v}$, $v \neq 0$, then by the product rule

$$\frac{d}{dx}\left(u \times \frac{1}{v}\right) = \frac{1}{v}\frac{du}{dx} + u\frac{d}{dx}\left(\frac{1}{v}\right)$$

Hence:

$$\frac{d}{dx}\left(\frac{u}{v}\right) = \frac{1}{v}\frac{du}{dx} + u\frac{d}{dx}\left(v^{-1}\right)$$

$$= \frac{1}{v}\frac{du}{dx} + u \times \frac{-1}{v^2} \times \frac{dv}{dx} \quad \text{using the chain rule}$$

$$= \frac{1}{v}\frac{du}{dx} - \frac{u}{v^2}\frac{dv}{dx}$$

$$= \frac{v\frac{du}{dx} - u\frac{dv}{dx}}{v^2}$$

Hence, the quotient rule:

- If u and v are functions of x, and $y = \frac{u}{v}$, $v \neq 0$, then $\frac{dy}{dx} = \frac{v\frac{du}{dx} - u\frac{dv}{dx}}{v^2}$.

Example 24

Use the quotient rule to differentiate each function: **(a)** $y = \frac{2x+1}{4x-3}$ **(b)** $f(t) = \frac{t}{1+t^2}$

Solution

(a) $y = \frac{2x+1}{4x-3}$: $\quad u = 2x+1 \quad v = 4x-3$

$\frac{du}{dx} = 2 \quad \frac{dv}{dx} = 4$

Quotient rule: $$\frac{dy}{dx} = \frac{v\frac{du}{dx} - u\frac{dv}{dx}}{v^2}$$

$$\frac{dy}{dx} = \frac{(4x-3)\times 2 - (2x+1)\times 4}{(4x-3)^2}$$

$$= \frac{8x-6-8x-4}{(4x-3)^2}$$

$$= \frac{-10}{(4x-3)^2}$$

Because the function is only defined for $x \neq \frac{3}{4}$, the same restriction applies for the derivative.

(b) $f(t) = \frac{t}{1+t^2}$: $\quad u = t \quad v = 1+t^2$

$\frac{du}{dt} = 1 \quad \frac{dv}{dt} = 2t$

Quotient rule: $$f'(t) = \frac{v\frac{du}{dt} - u\frac{dv}{dt}}{v^2}$$

$$f'(t) = \frac{\left(1+t^2\right)\times 1 - t \times 2t}{\left(1+t^2\right)^2}$$

$$= \frac{1+t^2-2t^2}{\left(1+t^2\right)^2}$$

$$= \frac{1-t^2}{\left(1+t^2\right)^2}$$

Example 25

Find the gradient of the tangent to the curve $y=\dfrac{\sqrt{4-x^2}}{x+1}$ at the point where $x=0$.

Solution

To find $\dfrac{dy}{dx}$ you have to use both the chain rule and the quotient rule.

$y=\dfrac{\sqrt{4-x^2}}{x+1}$: $\quad u=\sqrt{4-x^2} \quad\quad v=x+1$

$$\frac{du}{dx}=\frac{1}{2}\left(4-x^2\right)^{-\frac{1}{2}}\times(-2x) \quad\quad \frac{dv}{dx}=1$$

$$=\frac{-x}{\sqrt{4-x^2}}$$

Quotient rule: $\dfrac{dy}{dx}=\dfrac{v\dfrac{du}{dx}-u\dfrac{dv}{dx}}{v^2}$

$$\frac{dy}{dx}=\frac{(x+1)\times\dfrac{-x}{\sqrt{4-x^2}}-\sqrt{4-x^2}\times 1}{(x+1)^2}$$

$$=\frac{-1}{(x+1)^2}\left(\frac{x(x+1)}{\sqrt{4-x^2}}+\sqrt{4-x^2}\right)$$

$$=\frac{-1}{(x+1)^2}\left(\frac{x^2+x+4-x^2}{\sqrt{4-x^2}}\right)$$

$$=\frac{-(x+4)}{(x+1)^2\sqrt{4-x^2}}$$

Where $x=0$: $\dfrac{dy}{dx}=\dfrac{-4}{1\times 2}=-2$

EXERCISE 7.9 THE QUOTIENT RULE

1 Use the quotient rule to differentiate each function.

(a) $y=\dfrac{x-1}{x+1}$ (b) $f(x)=\dfrac{3x-7}{4x+5}$ (c) $g(t)=\dfrac{2t+5}{t+2}$

(d) $f(x)=\dfrac{1}{x^2-5x+6}$ (e) $h(n)=\dfrac{n^2+3n+4}{2n-1}$ (f) $y=\dfrac{x}{x-3}$

(g) $y=\dfrac{4x^2}{2x+5}$ (h) $y=\dfrac{4x^2-2}{x^2+5}$ (i) $v(x)=\dfrac{x+1}{x^3-1}$

2 If $y=\dfrac{x+1}{x^2+1}$ then $\dfrac{dy}{dx}=\ldots$

A $\dfrac{1-2x-x^2}{\left(x^2+1\right)^2}$ B $\dfrac{3x^2+2x+1}{x^4+1}$ C $\dfrac{1-2x-x^2}{x^4+1}$ D $\dfrac{3x^2+2x+1}{\left(x^2+1\right)^2}$

3 Differentiate each function with respect to x.

(a) $y=\dfrac{\sqrt{x+1}}{x}$ (b) $f(x)=\dfrac{(x+1)^2}{x}$ (c) $y=\dfrac{x}{(x+1)^2}$

(d) $y=\dfrac{(2x+1)^3}{\left(3-x^2\right)^2}$ (e) $f(x)=\dfrac{x}{\sqrt{x+1}}$ (f) $y=\left(\dfrac{x+1}{x}\right)^2$

4 $f(x)=\dfrac{\sqrt{x}}{x^2+1}$. Four steps in finding the simplest form of $f'(x)$ are given. Indicate whether each step is correct or incorrect.

A $f'(x)=\dfrac{\left(x^2+1\right)\times\dfrac{1}{2\sqrt{x}}-\sqrt{x}\times 2x}{\left(x^2+1\right)^2}$

B $f'(x)=\dfrac{\dfrac{x^2+1}{2\sqrt{x}}-2x\sqrt{x}}{\left(x^2+1\right)^2}$

C $f'(x)=\dfrac{x^2+1-4x^2}{2\sqrt{x}\left(x^2+1\right)^2}$

D $f'(x)=\dfrac{5x^2+1}{2\sqrt{x}\left(x^2+1\right)^2}$

5 Find the derivative of each function.

(a) $y=(x^2-4)(x+2)$

(b) $f(x)=4x^{\frac{5}{2}}-2x^{\frac{3}{2}}+6x^{\frac{1}{2}}$

(c) $y=\sqrt{5x-1}$

(d) $g(t)=t^{\frac{2}{3}}+(2t-1)^3$

(e) $y=\dfrac{x+2}{x+5}$

(f) $f(x)=\sqrt{x^2+2}$

(g) $y=x^2-5x+\sqrt{x-2}$

(h) $f(z)=\dfrac{z+4}{2z-1}$

(i) $y=\dfrac{1}{x^2-2x+2}$

(j) $f(x)=(x^2-7x)(x+1)$

(k) $y=(x^2+1)\sqrt{x}$

(l) $g(x)=\dfrac{\sqrt{x+1}}{\sqrt{x^2+1}}$

6 Show that the gradient of the tangent to the curve $y=\dfrac{x}{x^2+1}$ is zero twice, at $x=-1$ and $x=1$.

7.10 TANGENTS AND NORMALS TO A CURVE

Differentiation gives the gradient function of a curve. This function can be evaluated at a point on the curve to obtain the gradient of the curve's **tangent** at that point. The point-gradient form of the straight line can then be used to find the equation of that tangent.

A **normal** at a point on a curve is the line that is perpendicular to the tangent at that point. Because the lines are perpendicular, the gradient of the normal can be obtained from the gradient of the tangent at that point.

MAKING CONNECTIONS

Tangent and normal to a curve

Move the sliders to explore the relationship between the tangent and normal to a curve.

Example 26

Find the equation of the tangent and the normal to the curve $y=x^2-4x+4$ at the point on the curve where $x=3$.

Solution

$x=3,\ y=9-12+4=1$ so the point is $(3, 1)$

$y=x^2-4x+4$: $\quad \dfrac{dy}{dx}=2x-4$

$x=3$: $\quad \dfrac{dy}{dx}=6-4=2$

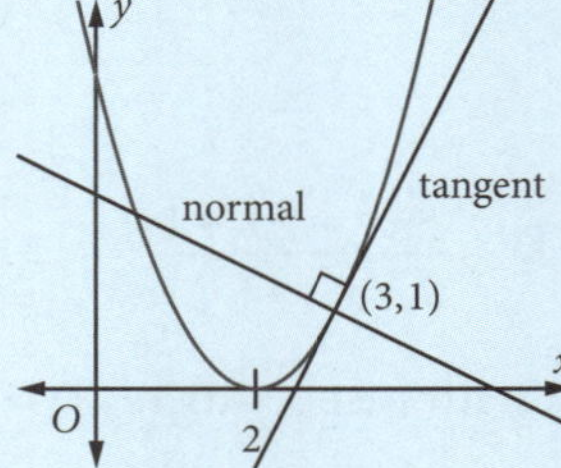

Equation of the tangent (from point-gradient form): $y-1=2(x-3)$

$$y=2x-5$$

Gradient of normal $=\dfrac{-1}{\text{gradient of tangent}}=-\dfrac{1}{2}$ (Remember $m_1\times m_2=-1$ for perpendicular lines)

Equation of the normal (from point-gradient form): $y-1=-\frac{1}{2}(x-3$

$$2y-2=-x+3$$

$$x+2y-5=0$$

Example 27

Find the equation of the tangent and normal to the curve $y=\sqrt{x}$ at the point on the curve where $x=4$.

Solution

Draw a sketch.

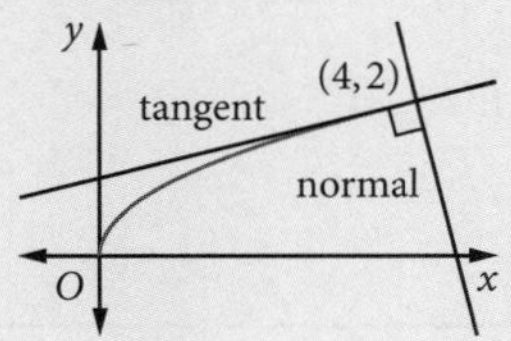

Point is at $x=4$, $y=2$: $y=x^{\frac{1}{2}}$: $\dfrac{dy}{dx}=\dfrac{1}{2}x^{-\frac{1}{2}}=\dfrac{1}{2\sqrt{x}}$

$x=4$: $\dfrac{dy}{dx}=\dfrac{1}{2\sqrt{4}}=\dfrac{1}{4}$

For $(4, 2)$, $m=\frac{1}{4}$, tangent is: $y-2=\frac{1}{4}(x-4)$

$$4y-8=x-4$$

The equation of the tangent is $x-4y+4=0$.

For $(4, 2)$, $m=-4$, normal is: $y-2=-4(x-4)$

The equation of the normal is $4x+y-18=0$.

Example 28

Find the equation of the tangent to the curve $y=\dfrac{x^3}{3}-x^2-x+1$ at the points on the curve where the tangent is parallel to the line $7x-y+5=0$.

Solution

$y=\dfrac{x^3}{3}-x^2-x+1$: $\dfrac{dy}{dx}=x^2-2x-1$

Gradient of $7x-y+5=0$: $m=7$

Solve: $x^2-2x-1=7$

$$x^2-2x-8=0$$

$$(x-4)(x+2)=0$$

$$x=-2, 4$$

At $x=2$, $y=-\frac{11}{3}$; at $x=4$, $y=\frac{7}{3}$.

At $\left(-2,-\frac{11}{3}\right)$, tangent is: $y+\frac{11}{3}=7(x+2)$

$$3y+11=21x+42$$

$$21x-3y+31=0$$

At $\left(4,\frac{7}{3}\right)$, tangent is: $y-\frac{7}{3}=7(x-4)$

$$3y-7=21x-84$$

$$21x-3y-77=0$$

Example 29

Find the points of intersection of the parabolas $y=x^2-2x$ and $y=4x-x^2$. Sketch the parabolas and find the angle between the parabolas at their point of intersection in the first quadrant.

Solution

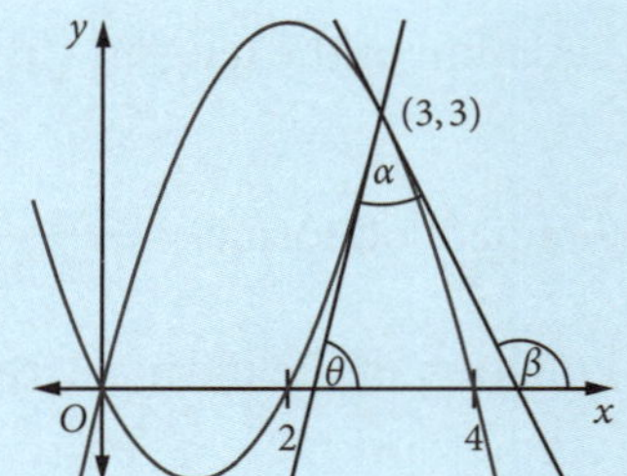

$y = x^2 - 2x$, $y = 4x - x^2$, intersection at:

$$x^2 - 2x = 4x - x^2$$
$$2x^2 - 6x = 0$$
$$2x(x-3) = 0$$
$$x = 0, 3 \quad y = 0, 3$$

Curves intersect at $(0, 0)$ and $(3, 3)$. Intersection point in the first quadrant is $(3, 3)$. The angle between the curves at this point is the angle between their tangents at this point.

For $y = x^2 - 2x$: $\quad \dfrac{dy}{dx} = 2x - 2$

At $(3, 3)$: $\quad \dfrac{dy}{dx} = 6 - 2 = 4 \qquad$ Hence $\tan\theta = 4$, so $\theta = 75°58'$

For $y = 4x - x^2$: $\quad \dfrac{dy}{dx} = 4 - 2x$

At $(3, 3)$: $\quad \dfrac{dy}{dx} = 4 - 6 = -2 \qquad$ Hence $\tan\beta = -2$, so $\beta = 180° - 63°26' = 116°34'$

Now $\beta = \alpha + \theta$ (exterior angle of a triangle), so $\alpha = 116°34' - 75°58' = 40°36'$.

EXERCISE 7.10 TANGENTS AND NORMALS TO A CURVE

1 Find the equations of the tangent and normal to the curve $y = x^2$ at $(2, 4)$.

2 Find the equations of the tangent and normal to the curve $y = 3x - x^2$ at $(0, 0)$.

3 The equation of the tangent to the curve $y = x^2 - 5x + 6$ at $(3, 0)$ is:

A $y = x - 3$ **B** $y = x + 3$ **C** $y = -x - 3$ **D** $y = -x + 3$

4 Find the equations of the tangent and normal to the curve $y = 2x^2 + 3x - 4$ where $x = 0$.

5 Find the equations of the tangent and normal to the curve $y = 2x^2 - 4x + 1$ where the gradient is 4.

6 Find the equations of the tangent and normal to the curve $y = \frac{1}{x}$ at the point where $x = -2$. Indicate whether each statement below is a correct or incorrect step in answering this question.

(a) $\dfrac{dy}{dx} = \dfrac{1}{x^2}$

(b) At $\left(-2, -\frac{1}{2}\right)$, $\dfrac{dy}{dx} = -\dfrac{1}{4}$

(c) Equation of tangent is $x + 4y + 4 = 0$

(d) Equation of normal is $8x - 2y + 15 = 0$

7 Find the equations of the tangent and normal to the curve $y = 3 - x - x^2$ at the point where the curve crosses the y-axis.

8 Find the equations of the tangent and normal to the curve $y = 3x^3 - 7x^2 + 2x$ at the point where $x = 2$.

9 The straight line $y = x + 2$ cuts the parabola $y = \dfrac{x^2}{2} - 2$ at two points P and Q. Find the coordinates of P and Q. Also find the equations of the tangents to the parabola at P and Q and the coordinates of the point of intersection of these tangents.

10 Find the equations of the tangents to the parabola $y = 4x - 3x^2$ at the points where the parabola cuts the x-axis.

11 Find the equations of the tangent and normal to the parabola $y = 2x^2 - 4x + 1$ at the point of zero gradient.

12 Prove that the parabolas $y = 2x^2 - 6x + 5$ and $y = x^2 - 2x + 1$ touch at a point and find the equation of their common tangent.

13 The line $y = x + 4$ cuts the parabola $y = x^2 - 2x$ at two points A and B. Find the size of the angles that the tangents to the curve at A and B make with the x-axis.

14 The line $y = x + 1$ cuts the parabola $y = x^2 - x - 2$ at two points P and Q. Find:

(a) the coordinates of P and Q

(b) the equations of the tangents at P and Q

(c) the coordinates of the intersection point of these two tangents.

15 Find the equations of the tangents to the parabola $y = x^2 - 2x + 4$ at the points where:

(a) the gradient is zero

(b) the tangent is parallel to the line $y = 2x + 1$.

16 Find the equations of the tangent and normal to the parabola $y = 2x^2 - 5x + 1$ at the point on the parabola where the gradient is 3.

17 The normal to the curve $y = (x + 2)^2$ at the point $A(-3, 1)$ meets the curve again at B. Find

(a) the equation of the normal

(b) the coordinates of B

(c) the angle at B between the curve and the chord AB.

18 The line $y = x - 2$ cuts the curve $y = x^3(x - 2)$ at two points A and B. Calculate the angles that the tangents to the curve at A and B make with the x-axis and hence find the angle between the tangents.

19 Find the equations of the tangents to the curve $y = (x^2 - 1)(x - 2)$ at the points where the curve crosses the x-axis.

20 Find the coordinates of the points on the curve $y = x^2(2x - 3)$ at which the tangent is parallel to:

(a) the line $y = 12x - 1$

(b) the x-axis.

7.11 THE GRADIENT AS A RATE OF CHANGE

Consider the linear function $y = f(x) = 3x + 2$. An increase of h in the value of x leads to an increase of $3h$ in the value of y. For any value of h, $h \neq 0$, y increases by an amount equal to three times the increase in x.

You can say that 'the rate of change of y with respect to x is 3'. The 3 is the gradient of the line. Because the gradient in this case is constant, y increases at a constant rate.

(This should remind you of differentiation from first principles, where the gradient of the function $= \dfrac{f(x+h) - f(x)}{h}$.)

Consider the function $y = f(x) = 2x^2$ at the point $(1, 2)$:

$$f(x) = 2x^2 \qquad f(1) = 2 \qquad f(1 + h) = 2(1 + h)^2 = 2 + 4h + 2h^2$$

$$k = f(1 + h) - f(1) = 2(1 + h)^2 - 2 = 4h + 2h^2$$

$$\frac{k}{h} = 4 + 2h,\ h \neq 0$$

Thus when x changes by an amount h, from 1 to $1 + h$, $f(x)$ changes by an amount $k = 4h + 2h^2$; that is, the value of the function changes by an amount that is $(4 + 2h)$ times the amount of the change in x.

The 'average rate of change of y with respect to x' $= 4 + 2h$. This rate varies with h.

As $h \to 0$, $4 + 2h \to 4$, i.e. $f'(1) = 4$. The number 4 in this case gives you:

(a) the gradient of the tangent at $x = 1$

(b) the rate of change of y with respect to x at $x = 1$.

This concept of the derivative as a 'rate of change' is very important in differential calculus. For example, there are many practical situations in which the change in a physical quantity depends on time:

- If a vessel is being filled with water, the volume V of the water in the vessel is a function of time. $\dfrac{dV}{dt}$ is the rate at which the volume changes, which may or may not be a constant rate.
- The population P of a town may increase or decrease with time. $\dfrac{dP}{dt}$ is the rate of change of population with respect to time.

- As a spherical balloon is being inflated, $\frac{dV}{dt}$ is the instantaneous rate of change of its volume with respect to time. Its spherical radius is also increasing, so $\frac{dr}{dt}$ is the rate of change of the radius with respect to time. Because a sphere's volume is given by $V = \frac{4}{3}\pi r^3$, we have $\frac{dV}{dr} = 4\pi r^2$, where $\frac{dV}{dr}$ measures the rate of change of volume with respect to the radius.

Note: When dealing with time t, you will only consider $t \geq 0$, so you will use the term **initially** to mean 'at time $t = 0$'.

Rates of change have many applications in physics and chemistry. Boyle's law for gases states that 'the pressure P of a given mass of gas varies inversely with the volume V', so that P is a function of V where:

$$P = \frac{k}{V} = kV^{-1}$$

$$\frac{dP}{dV} = -\frac{k}{V^2} = \text{rate of change of pressure with respect to volume.}$$

The negative sign indicates that an increase in V leads to a decrease in P.

Proportion

If two variables are in **direct proportion**, this means that their absolute values vary in the same direction: as the magnitude of the independent variable increases, the magnitude of the dependent variable will increase proportionally (and vice versa). Mathematically, the two variables are linear functions of each other (e.g. $y = kx$).

Inverse proportion means that the absolute values of the variables vary in opposite directions: as the magnitude of the independent variable increases, the magnitude of the dependent variable will decrease proportionally (and vice versa). Mathematically, the two variables are the inverse (reciprocal) of each other $\left(\text{e.g. } y = \frac{k}{x}\right)$.

Example 30

The volume V litres of water in a tank is given by $V = 4t + 30$, where t is in seconds.

(a) How much water is in the tank initially? **(b)** At what rate is water flowing into the tank?

Solution

(a) 'Initially' means $t = 0$, as we are dealing with time $t \geq 0$.

When $t = 0$: $V = 30$

Initially there is 30 L of water in the tank.

(b) If $V = 4t + 30$: $\frac{dV}{dt} = 4$

Water is flowing into the tank at a constant rate of 4 litres per second (L s^{-1}).

EXERCISE 7.11 THE GRADIENT AS A RATE OF CHANGE

1 A trench is being dug by a team of labourers who remove V cubic metres of soil in t minutes, where $V = 10t - \frac{t^2}{20}$.

(a) State the domain of the function, i.e. the values of t during which soil is being removed.
(b) At what rate is the soil being removed at the end of 40 minutes?
(c) Are the labourers working at a constant rate?
(d) What is their initial rate of work, i.e. when $t = 0$?
(e) At what time are they removing soil at the rate of 5 m^3 per minute?

2 When some concentrated chemical solutions are allowed to evaporate slowly, crystals are formed. The surface area of a particular crystal is given by $A = 0.6t^2$, where A is in mm^2 and t is in days. The rate at which the surface is increasing after 4 days is:

A $9.6\text{ mm}^2/\text{day}$ **B** $12.8\text{ mm}^2/\text{day}$ **C** $0.6\text{ mm}^2/\text{day}$ **D** $4.8\text{ mm}^2/\text{day}$

3 A cube of ice has an edge length of 10 cm. It melts so that its volume decreases at a constant rate and the block remains a cube. If the edge length measures 5 cm after 70 minutes, find:

(a) the rate at which the volume decreases **(b)** the volume at any time t.

4 A water tank is being emptied. The quantity Q litres of water remaining in the tank at any time t minutes after it starts to empty is given by $Q(t) = 1000(20 - t)^2$, $t \geq 0$.

(a) At what rate is the tank being emptied at any time t?
(b) How much time does it take to empty the tank? (When is $V = 0$?)
(c) At what time is the water flowing out at a rate of 20 000 litres per minute?
(d) What is the average rate at which the water flows out in the first 5 minutes?

5 The following table shows the temperature T degrees Celsius (°C) of water in a vessel, initially at 100°C, after t minutes.

t (min)	0	5	10	15	20	25	30
T (°C)	100	85	74	64	56	48	44

Plot the graph of this data to show the relation between temperature and time. From the graph, estimate the rate at which the temperature is falling:

(a) after 15 minutes **(b)** when the temperature is 80°C.

6 A machine manufactures items at a variable rate given by $\frac{dQ}{dt} = 2t + 1$, $t \geq 0$, where Q is the number of items manufactured in a time t minutes.

At what rate is the machine working: **(a)** initially **(b)** after 10 minutes?

7 If the area of a circle is given by $A = \pi r^2$, show that the rate of change of the area with respect to the radius, $\frac{dA}{dr}$, is proportional to the radius. Find this rate when the radius is 2 cm.

8 A right circular cylinder of volume V has height h and radius of its base r. Find:

(a) the rate of change of volume with respect to height, if the radius of the base is constant
(b) the rate of change of volume with respect to the radius of the base, if the height is constant.

9 The pressure P and volume V of a given mass of gas at constant temperature are connected by the formula $PV = 500$. Find $\frac{dP}{dV}$ when $V = 10$.

10 Using the straight-line method of depreciation, it is found that after t years have elapsed the value V of a certain machine is given by $V = 40\,000 - 5000t$, where $0 \leq t \leq 8$.

(a) Find $\frac{dV}{dt}$ and interpret your result. **(b)** What is the value of the machine after 3 years?

11 The revenue function for a particular manufacturer is $R = x\left(15 - \frac{x}{30}\right)$, where x is the number of units of the product sold. If the marginal revenue is given by $\frac{dR}{dx}$, find the marginal revenue when:

(a) $x = 6$ **(b)** $x = 15$ **(c)** $x = 225$

7.12 VELOCITY AND ACCELERATION AS A RATE OF CHANGE

Particle is the term used for a body that behaves such that all forces acting on the body can be regarded as acting through a single point. This means that you can represent the body as a single point, regardless of its actual size and shape. This definition of a particle means that quite large bodies, e.g. trains, can still be classified as 'particles' provided this condition applies.

Displacement

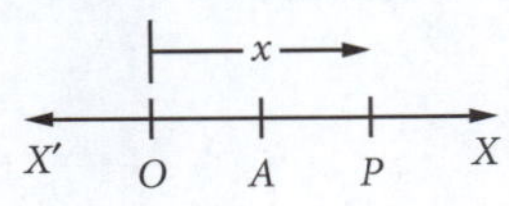

Consider a particle, which can be represented by a point P, moving in a straight line $X'OX$. The **displacement** x is the particle's position relative to the fixed point O. It may be a positive or negative number, according to whether P is to the right or left of O. The origin of the motion is not necessarily at O, so when $t = 0$, P may be (for example) at the point A.

Displacement is defined as the position relative to a starting point. It can be positive or negative.
Displacement does not necessarily represent the total distance travelled.
Unlike displacement, distance is always a positive quantity.

Velocity

Consider the equation $x = f(t)$ that gives the position coordinate x of a particle moving in a straight line at time t.

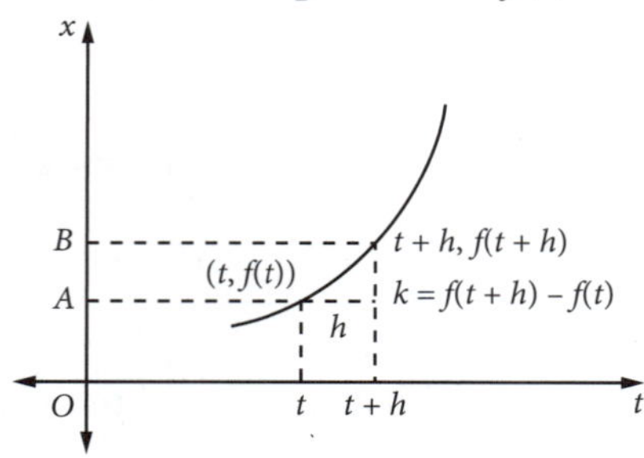

At time t, the particle is at A, and at time $(t + h)$ the particle is at B, as shown in the diagram. Thus in the small time interval h the particle has changed its position by an amount $k = f(t + h) - f(t)$.

The average **velocity** in this time interval $= \dfrac{k}{h} = \dfrac{f(t+h) - f(t)}{h}$, $h \neq 0$.

The instantaneous velocity of the particle at time t is defined by $\lim\limits_{h \to 0} \dfrac{f(t+h) - f(t)}{h}$. It may be denoted by $v(t)$, $f'(t)$, $\dfrac{dx}{dt}$ or $\dot{x}$.

Velocity is defined as the rate of change of position (i.e. of displacement) with respect to time, or as the time rate of change of position in a given direction.

$$v(t) = f'(t) = \frac{dx}{dt} = \dot{x} = \lim_{h \to 0} \frac{f(t+h) - f(t)}{h}$$

Velocity can be positive or negative, depending on the direction of travel.
Speed is the magnitude of the velocity and is always positive.

Example 31

Consider the equation $x = 5 + 4t - t^2$, which defines the displacement x metres from O at time t seconds (for $t \geq 0$) of a particle moving in a straight line.

(a) Find the velocity function. **(b)** Discuss the sign of the velocity over $0 \leq t \leq 6$.

Solution

(a) For $x = 5 + 4t - t^2$: $v = \dfrac{dx}{dt} = 4 - 2t$

(b)

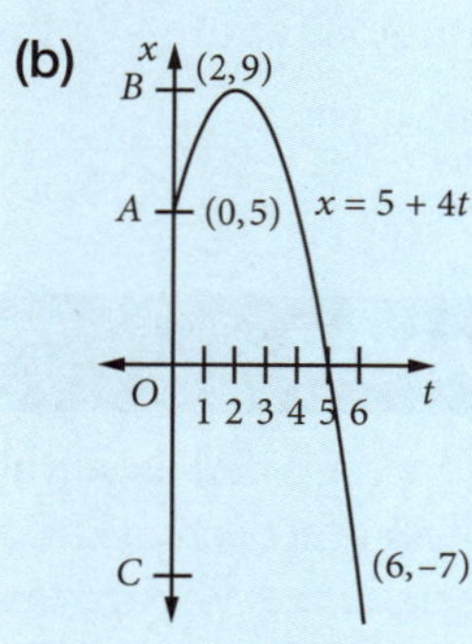

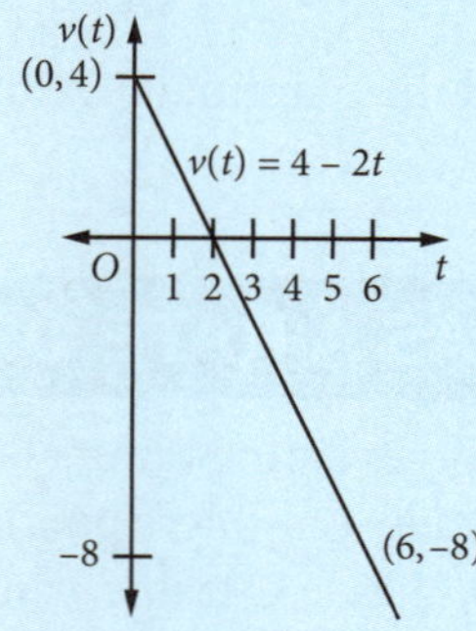

The diagram at left is the graph of the displacement function. The diagram at right is the graph of the velocity function.

- When $t = 0$, $v = 4$ and the particle is at point A.
- When $t = 2$, $v = 0$ and the particle is at rest at B. For the first 2 seconds of motion the particle is moving in a positive direction with a positive velocity.
- When $t = 3$, $v = -2$. This negative velocity means that the particle is moving in a negative direction (i.e. 'backwards') with a speed of $2\,\text{m}\,\text{s}^{-1}$.
- When $t = 6$, $v = -8$. The particle is moving with a speed of $8\,\text{m}\,\text{s}^{-1}$ in the negative direction and is at the point C.

Over $0 \le t \le 6$, the particle moves from A to B to C. Its final displacement from O is -7 m. The total distance travelled is 20 m, which includes the distance from A to B (4 m) plus the distance from B to C (16 m).

After the particle was at rest (at $t = 2$ seconds), it reversed its direction and was then moving in the opposite direction.

Also, when $t = 3$, $v = -2$ and $x = 8$. The particle has moved from a displacement of 9 m to a displacement of 8 m. The displacement has decreased, so its rate of change (i.e. velocity) is negative.

Acceleration

Acceleration is defined as the rate of change of velocity with respect to time. Acceleration, like velocity, can be positive or negative. Positive acceleration indicates that the velocity is increasing, while negative acceleration indicates that the velocity is decreasing, which is often called deceleration or retardation.

(Note that 'increasing velocity' is not necessarily 'faster speed', it only means acceleration in the direction of positive displacement.)

If you denote the velocity by $v(t)$, then the average acceleration over the time interval from t to $(t+h)$ is $\dfrac{v(t+h)-v(t)}{h}$.

The instantaneous acceleration at time t is defined by $\lim\limits_{h\to 0}\dfrac{v(t+h)-v(t)}{h}$. It may be denoted by $v'(t)$, $a(t)$, $f''(t)$, $\dfrac{dv}{dt}$, $\dfrac{d^2x}{dt^2}$ or $\ddot{x}$:

$$a(t) = v'(t) = \frac{d^2x}{dt^2} = \ddot{x} = \lim_{h\to 0}\frac{v(t+h)-v(t)}{h}$$

The notations for the acceleration involving double differentiation are introduced here for completeness. They will be formally introduced in Chapter 14.

Example 32

A particle is moving in a straight line such that its displacement x m from a fixed point O on the line at time t seconds (for $t \ge 0$) is given by $x = t^3 - 12t + 16$. Find:

(a) the particle's initial displacement, velocity and acceleration

(b) the time when its velocity is zero, and its displacement and acceleration at this time.

Solution

(a) $x = t^3 - 12t + 16$

$v = \dfrac{dx}{dt} = 3t^2 - 12$

$a = \dfrac{dv}{dt} = 6t$

When $t = 0$: $x = 16$, $v = -12$, $a = 0$

Initially the particle is 16 m from O, moving with a velocity of -12 m s^{-1} and with zero acceleration.

(b) $v = 0$: $3t^2 - 12 = 0$

Factorise: $3(t-2)(t+2) = 0$

$\therefore t = 2, -2$

But $t \ge 0$, so the particle is at rest at 2 seconds.

When $t = 2$: $x = 8 - 24 + 16 = 0$, so the particle is at O.

$a = 12$, so acceleration is 12 m s^{-2}.

It is worth looking at the graphs of the functions for x, v and a:

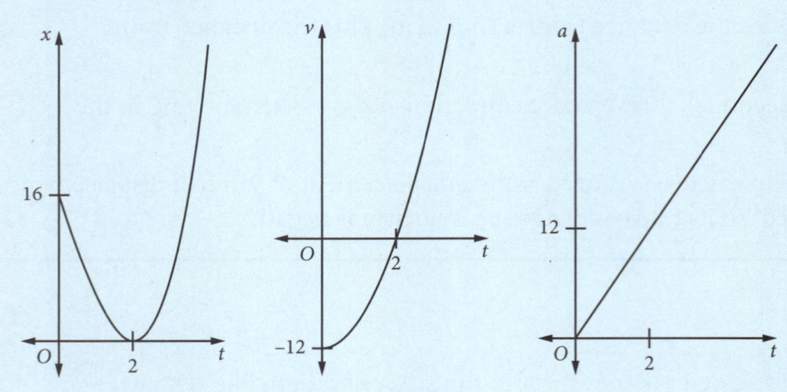

From the graphs we can see that the acceleration, after initially being zero, is positive and increasing at a constant rate. As $a > 0$, the velocity is increasing at all times. During the first two seconds the particle moves from 16 metres on the positive side of O back to O, with an increasing negative velocity. After 2 seconds the particle is at rest at O ($v = 0$). It then moves back in a positive direction with an increasing velocity.

Summary of important motion terms

'initially':	$t = 0$	'at the origin':	$x = 0$
'at rest':	$v = 0$	'velocity is constant':	$a = 0$

Units and symbols

Physical quantity	Unit	Symbol
Time	s	t
Displacement	cm, m	x (or s in Physics)
Velocity	cm s^{-1}, m s^{-1}	$v, \frac{dx}{dt}, \dot{x}$
Acceleration	cm s^{-2}, m s^{-2}	$a, \frac{dv}{dt}, \frac{d^2x}{dt^2}, \ddot{x}$

Note that 's' is the abbreviation for second, 'cm' for centimetre and 'm' for metre.
Constant acceleration due to gravity $= 9.8\text{ m s}^{-2}$ ($\approx 10\text{ m s}^{-2}$)

EXERCISE 7.12 VELOCITY AND ACCELERATION AS A RATE OF CHANGE

1 A particle moves in a straight line so that its displacement x m from a fixed point O on the line at any time t seconds ($t \geq 0$) is given by $x = t^2 - 5t + 6$. Find:

(a) its initial displacement **(b)** its initial velocity
(c) when it first passes through O and with what velocity
(d) when it passes through O the second time and with what velocity
(e) when and where its velocity is zero.

2 The displacement x m at time t seconds ($t \geq 0$) of a particle moving in a straight line is given by $x = 2t^3 - t^2 + 4t + 1$. Its acceleration is given by:

A $a = 2t^3 - t^2 + 4t + 1$ **B** $a = 6t^2 - 2t + 4$ **C** $a = 12t - 2$ **D** $a = 12$

3 The displacement x m at time t seconds $(t \geq 0)$ of a particle moving in a straight line is given by $x = t^2 - 5t + 4$.

(a) At what time is its velocity zero? **(b)** What is the acceleration at this time?
(c) What is the distance travelled in the first 4 seconds? **(d)** At what time is the velocity 8 m s^{-1}?

4 A point moving in a straight line is distant x m from the origin O at time t, where $x = 2t^3 - 15t^2 + 36t$.

(a) Find the velocity and acceleration at any time t.
(b) Find the initial velocity and acceleration. **(c)** At what times is the velocity zero?
(d) At what time is the acceleration zero? Find the velocity and position at this time.
(e) During what interval of time is the velocity negative?

5 The displacement x m at time t seconds of a particle moving in a straight line is given by $x = 2t^3 - 9t^2 + 12t + 6$. Find:

(a) when its acceleration is zero, and its velocity at this time
(b) when its velocity is zero, and its acceleration at this time.

6 Two bodies move along a straight path, starting at the same time, so that their displacement x m from a fixed point O at any time t is given by $x = t + 6$ and $x = t^2 + 4$ respectively. At what times are they:

(a) together **(b)** travelling with the same velocity?

7 The velocity $v\text{ m s}^{-1}$ at time t seconds $(t \geq 0)$ of a body moving in a straight line is given by $v = 6t^2 + 6t - 12$. Its initial displacement is 7 m from O. Find:

(a) the acceleration when the velocity is zero **(b)** the initial velocity and acceleration.

8 Two cars A and B travel along a straight road in the same direction. Their respective distances x km from a fixed point O at any time t hours are given by the following rules:

A: $x = 50t - 20t^2$ B: $x = 80t^2 + 20t$

(a) Calculate each car's speed at the point O.
(b) At what time are the cars travelling at the same speed?
(c) Both cars reach a point Q at the same time. Calculate the distance from O to Q.
(d) A third car, travelling at uniform speed, is 2 km ahead of A and B when they pass the point O. If this car arrives at Q at the same time as A and B, find a rule connecting x and t for it.

9 A particle moves in a straight line so that its displacement $x(t)$ from a fixed point in the line at time $t \geq 0$ is given by $x(t) = 3 + 4t - 5\sqrt{t^2 + 4}$. Find the particle's displacement when it comes to rest.

10 A particle is moving so that, for $0 < t < 1$, its velocity is positive and its acceleration is negative. Which graph could represent the displacement function of this particle?

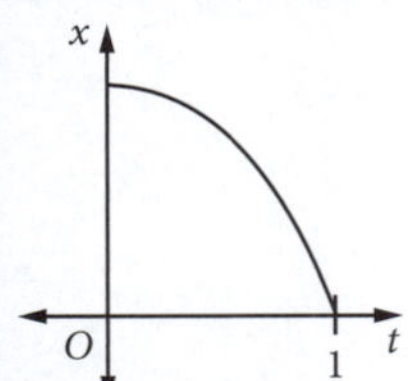

B
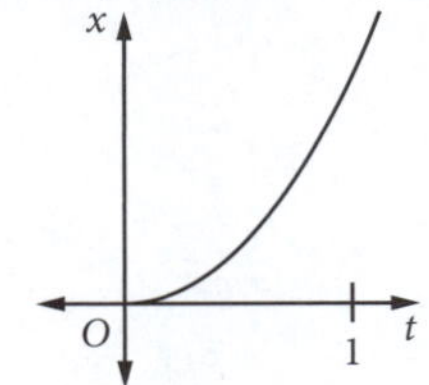

C
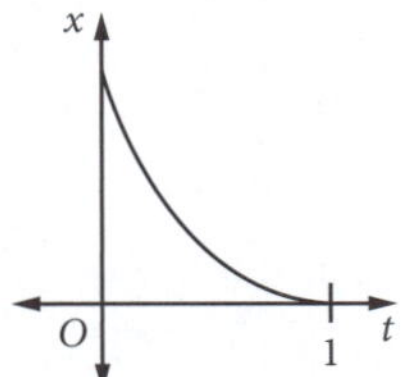

D
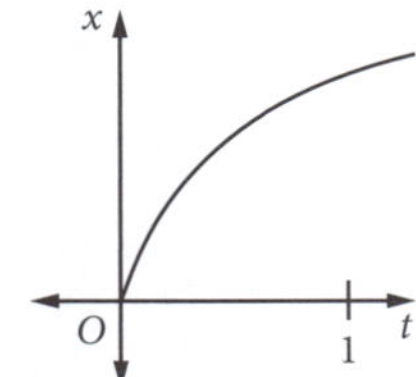

CHAPTER REVIEW 7

1 Find the following limits. **(a)** $\lim\limits_{x \to \frac{1}{2}} \dfrac{1 - 4x^2}{1 - 2x}$ **(b)** $\lim\limits_{x \to 3} \dfrac{x^3 - 27}{x - 3}$

2 Evaluate: **(a)** $\lim\limits_{h \to 0} \dfrac{2x^2h + 3h}{h}$ **(b)** $\lim\limits_{h \to 0} \dfrac{(2+h)^2 - 4}{h}$ **(c)** $\lim\limits_{h \to 0} \dfrac{(1+h)^3 - 1}{h}$

3 Find $\dfrac{f(x+h) - f(x)}{h}$, $h \neq 0$, for $f(x) = 2x^2 - 3x$.

4 For $f(x) = x^2 + 6x + 8$, find:

(a) $f(2)$ **(b)** $f'(2)$ **(c)** $f'(c)$ **(d)** the value of c for which $f'(c) = -2$

5 Find $f'(x)$ for $f(x)=\sqrt{2x-1}$.

6 Given $y=(x^2-4)(3x^2-2x+1)^5$, find $\frac{dy}{dx}$.

7 Find the derivative of each function.

(a) $y=(x-2)^3$ (b) $f(x)=(x^2+x^3)^5$ (c) $v=\sqrt{25-t^2}$

(d) $y=(x+x^{-1})^2$ (e) $g(x)=\dfrac{1}{(x+4)^2}$ (f) $y=\dfrac{x-2}{x}$

(g) $f(x)=(x-1)^6(x+2)$ (h) $y=x^2+3x+\sqrt{x-2}$ (i) $f(x)=\dfrac{1}{x^2-2}$

(j) $u=\dfrac{2m-7}{2m+3}$ (k) $y=\dfrac{1+x^3}{x^2}$ (l) $h(t)=(t-3)\sqrt{t-3}$

8 Differentiate with respect to x: (a) x^4 (b) $x(x-1)$ (c) x^3-2x

9 Find the equation of the tangent to the parabola $y=4x-x^2$ at the point where the gradient is -2.

10 Find the equation of the tangents to the curve $y=2x^2(4-x)$ at the point where the curve intersects the x-axis.

11 A body starts from rest and moves in a straight line so that its velocity v m s^{-1} after t seconds is given by $v=2t+6t^2$. Calculate:

(a) its acceleration at the end of the first second

(b) its displacement after 5 seconds, given that the body is initially at zero displacement.

12 A particle is moving along the x-axis and is initially at the origin. Its velocity v metres per second at time t seconds is given by $v=\dfrac{2t}{9+t^2}$.

(a) What is the initial velocity of the particle?

(b) Find an expression for the acceleration of the particle.

(c) When is the acceleration zero?

(d) What is the maximum velocity attained by the particle and when does it occur?

CHAPTER 8
Exponential and logarithmic functions

8.1 INDEX LAWS WITH INTEGERS AS INDICES

You have frequently used the following index laws, where a and b are real numbers and m and n are positive integers. These rules can now be extended to include m and n as any real numbers.

Index laws

$a^m \times a^n = a^{m+n}$	$(ab)^n = a^n \times b^n$	$a^{-n} = \frac{1}{a^n}$ and hence $a^{-1} = \frac{1}{a}$
$\frac{a^m}{a^n} = a^{m-n}$	$a^1 = a$	$a^{\frac{1}{n}} = \sqrt[n]{a}$
$\left(a^m\right)^n = a^{mn}$	$a^0 = 1$	$a^{\frac{m}{n}} = \left(\sqrt[n]{a}\right)^m = \sqrt[n]{a^m}$

Example 1

Simplify, writing the answers with positive indices:

(a) 4^{-2} (b) $\left(\frac{1}{2}\right)^{-3}$ (c) $\frac{3^{-2} \times 6^3 \times 12^{-2}}{9^{-3} \times 2^{-1}}$

Solution

(a) 4^{-2}

$= \frac{1}{4^2}$

$= \frac{1}{\left(2^2\right)^2}$

$= \frac{1}{2^4}$

(b) $\left(\frac{1}{2}\right)^{-3}$

$= \frac{1^{-3}}{2^{-3}}$

$= 2^3$

(c) $\frac{3^{-2} \times 6^3 \times 12^{-2}}{9^{-3} \times 2^{-1}}$

$= \frac{3^{-2} \times 2^3 \times 3^3 \times 3^{-2} \times 4^{-2}}{3^{-6} \times 2^{-1}}$

$= \frac{3^{-1} \times 2^3 \times 2^{-4}}{3^{-6} \times 2^{-1}}$

$= 3^5$

Example 2

Simplify $\frac{x^{-1}+1}{x^{-1}-x}$, writing the answer with positive indices.

Solution

First simplify the numerator and denominator:

$$x^{-1}+1 = \frac{1}{x}+1 = \frac{1+x}{x}$$

$$x^{-1}-x = \frac{1}{x}-x = \frac{1-x^2}{x}$$

$$\therefore \frac{x^{-1}+1}{x^{-1}-x} = \frac{\frac{1+x}{x}}{\frac{1-x^2}{x}} = \frac{1+x}{x} \times \frac{x}{1-x^2}$$

$$= \frac{1+x}{(1+x)(1-x)}, \quad x \neq 0$$

$$= \frac{1}{1-x}, \quad x \neq -1, 0$$

EXERCISE 8.1 INDEX LAWS WITH INTEGERS AS INDICES

1 Simplify, writing your answer with positive indices.

(a) $2^3 \times 4^2 \times 8^2$ (b) $a^3b^{-2} \times (a^2b^2)^4$ (c) $m^2p^3 \times (m^3n^2)^3 \times (p^{-1})^2$ (d) $2^n \times 2^{2n} \times 2^{3n}$

(e) $\dfrac{(x^2y^3)^4 \times (xy)^{-2}}{xy}$ (f) $\dfrac{(2m^2n)^3}{(mn^3)^2 \times (4m^2)^2}$ (g) $\dfrac{5x^5y^2 \times 3(xy^3)^2}{15x^{-2}y}$ (h) $\dfrac{(a^2b)^2 \times (ab)^4}{(a^2b)^3}$

2 $\dfrac{(-2xy)^2 \times 2(x^2y^{-1})^3}{8(xy)^{-3}}$ simplifies to: **A** $\dfrac{x^{11}y^2}{2}$ **B** x^8y^2 **C** $-x^{11}y^2$ **D** $x^{11}y^2$

3 Simplify:

(a) $m^2n^2p^{-2} \times (mnp^2)^{-3}$ (b) $\dfrac{p^2q^3r^{-3}}{p^3q^{-1}r}$ (c) $\dfrac{x^2yz^{-2} \times 2(x^2y^{-1}z)^2}{xyz}$ (d) $2^3 \times (2^n)^2 \times 2^{-n}$

4 Write the following as negative powers of 2.

(a) $\frac{1}{4}$ (b) $\frac{1}{16}$ (c) $\frac{1}{32}$ (d) 0.125 (e) $\frac{1}{64}$ (f) $\frac{1}{128}$ (g) 0.25 (h) 8^{-3}

5 Write the following as powers of 10.

(a) 100 (b) 10 (c) 1 (d) 0.1 (e) 0.01 (f) 0.001 (g) $\frac{1}{1000}$ (h) $\frac{1}{100000}$

6 Simplify, writing your answer with positive indices.

(a) $\dfrac{3^{2n} \times 25^{2n-1}}{15^{n-1}}$ (b) $(x^{-1}+y^{-1})(x^{-1}-y^{-1})$ (c) $\dfrac{2^n \times 4^{n+1}}{8^{n-2}}$ (d) $(x^{-2}+x^{-1})^2$

(e) $\dfrac{x-5+6x^{-1}}{1-2x^{-1}}$ (f) $\dfrac{x-4x^{-1}}{1+2x^{-1}}$ (g) $4^{-2} \times 6^3 \times 8^4 \times 12^{-2}$ (h) $\dfrac{15^{n+1} \times 25 \times 5^{3n-4}}{9^{n-1} \times 25^{n-2}}$

7 For n as a positive integer, decide whether each statement is correct or incorrect.

(a) $(-1)^n = 1$ when n is even (b) $(-1)^n = -1$ when n is odd
(c) $(-1)^n = -1$ when n is even (d) $(-1)^n = 1$ when n is odd

8 Expand and simplify the following, expressing the results with positive indices.

(a) $(a^{-1}+b)(a^{-1}-b)$ (b) $(x^{-1}+y)(x+y^{-1})$ (c) $(x^{-2}+y^{-2})(x^{-2}-y^{-2})$

(d) $(a^2-2b^{-1})(a^{-2}-b)$ (e) $\dfrac{a^{-1}+b^{-1}}{a+b}$ (f) $\dfrac{y^{-1}+y}{1+y^2}$

8.2 INDEX LAWS WITH FRACTIONAL INDICES

Example 3

Simplify:

(a) $32^{\frac{2}{5}}$ (b) $125^{-\frac{2}{3}}$ (c) $x^{\frac{5}{2}} \times x^{-\frac{3}{4}}$

Solution

(a) $32^{\frac{2}{5}} = (2^5)^{\frac{2}{5}} = 2^2 = 4$ or $32^{\frac{2}{5}} = (\sqrt[5]{32})^2 = 2^2 = 4$

(b) $125^{-\frac{2}{3}} = (5^3)^{-\frac{2}{3}} = 5^{-2} = \frac{1}{5^2} = \frac{1}{25}$ or $125^{-\frac{2}{3}} = (\sqrt[3]{125})^{-2} = 5^{-2} = \frac{1}{25}$

(c) $x^{\frac{5}{2}} \times x^{-\frac{3}{4}} = x^{\frac{5}{2}-\frac{3}{4}} = x^{\frac{7}{4}}$

Example 4

Simplify:

(a) $\dfrac{5^{\frac{1}{4}} \times \sqrt{10} \times \sqrt[4]{2}}{20^{\frac{3}{4}}}$ (b) $\dfrac{3^{n-2} \times 9^{n+1}}{81^{n-1}}$ (c) $\left(x^{\frac{1}{2}} - x^{-\frac{1}{2}}\right)^2$

Solution

(a) $\dfrac{5^{\frac{1}{4}} \times \sqrt{10} \times \sqrt[4]{2}}{20^{\frac{3}{4}}} = \dfrac{5^{\frac{1}{4}} \times (2 \times 5)^{\frac{1}{2}} \times 2^{\frac{1}{4}}}{\left(2^2 \times 5\right)^{\frac{3}{4}}}$

$= \dfrac{5^{\frac{1}{4}} \times 2^{\frac{1}{2}} \times 5^{\frac{1}{2}} \times 2^{\frac{1}{4}}}{2^{\frac{3}{2}} \times 5^{\frac{3}{4}}}$

$= \dfrac{1}{2^{\frac{3}{4}}}$

(b) $\dfrac{3^{n-2} \times 9^{n+1}}{81^{n-1}} = \dfrac{3^{n-2} \times \left(3^2\right)^{n+1}}{\left(3^4\right)^{n-1}}$

$= \dfrac{3^{n-2} \times 3^{2n+3}}{3^{4n-4}}$

$= \dfrac{3^{3n+1}}{3^{4n-4}}$

$= 3^{4-n}$ or $\dfrac{1}{3^{n-4}}$

(c) $\left(x^{\frac{1}{2}} - x^{-\frac{1}{2}}\right)^2 = \left(x^{\frac{1}{2}}\right)^2 - 2x^{\frac{1}{2}}x^{-\frac{1}{2}} + \left(x^{-\frac{1}{2}}\right)^2$

$= x - 2 + x^{-1}$

$= x - 2 + \dfrac{1}{x}$

or $\left(x^{\frac{1}{2}} - x^{-\frac{1}{2}}\right)^2 = \left(\sqrt{x} - \dfrac{1}{\sqrt{x}}\right)^2$

$= \left(\dfrac{x-1}{\sqrt{x}}\right)^2$

$= \dfrac{(x-1)^2}{x} = x - 2 + \dfrac{1}{x}$

EXERCISE 8.2 INDEX LAWS WITH FRACTIONAL INDICES

1 Evaluate the following:

(a) $64^{\frac{2}{3}}$ (b) $49^{-\frac{1}{2}}$ (c) $\left(9^3\right)^{\frac{1}{2}}$ (d) $\left(\dfrac{1}{3}\right)^{-2}$

(e) $2^{\frac{2}{3}} \times 4^{\frac{1}{6}}$ (f) $\left(\dfrac{1}{16}\right)^{-\frac{3}{2}}$ (g) $\sqrt[3]{27} \times \sqrt[5]{32}$ (h) $\sqrt{6\frac{1}{4}} \times \sqrt[3]{8}$

2 The simplest correct expression for $\sqrt[5]{8} \times \sqrt[5]{4}$ is:

A $2^{\frac{3}{5}} \times 2^{\frac{2}{5}}$ B 32 C 2 D $\sqrt[5]{32}$

3 Express each of the following as simply as possible using index notation.

(a) $\sqrt[4]{36}$ (b) $\sqrt[8]{32}$ (c) $\sqrt[3]{4} \times \sqrt[6]{16}$ (d) $\sqrt{3} \times \sqrt[3]{81}$

4 Simplify the following, writing your answer with positive indices.

(a) $x^{\frac{2}{3}} \times x^{\frac{3}{2}}$ (b) $\left(a^{-1}b\right)^2 \times \left(\dfrac{1}{b^{-2}}\right)^{\frac{1}{2}}$ (c) $\left(x^{\frac{1}{2}}\right)^2 - \left(x^{-2}\right)^{\frac{1}{2}}$ (d) $\left(x^{\frac{1}{3}}\right)^2 \times \left(x^{-1}y^3\right)^{-1} \times x^{-\frac{5}{3}}y^2$

(e) $\left(x^{\frac{1}{2}} + y^{\frac{1}{2}}\right)\left(x^{\frac{1}{2}} - y^{\frac{1}{2}}\right)$ (f) $\sqrt[6]{x^2y^3} \times \dfrac{x^{\frac{1}{3}}}{y^{-\frac{1}{2}}}$ (g) $\dfrac{54^{\frac{1}{4}}}{6^{\frac{3}{4}} \times 12^{-\frac{1}{2}}}$ (h) $\dfrac{\left(x^{m+1}\right)^n \times x^{m+n}}{\left(x^m\right)^{n+1} \times x^{2n}}$

5 Simplify $\left(a^2b^{-1}\right)^{-2} \div \left(a^{-1}b^2\right)$. Indicate whether each expression below is a correct or incorrect step in the simplification.

(a) $a^4b^2 \times a^2b^4$ (b) $a^{-4}b^2 \times \dfrac{1}{a^{-2}b^4}$ (c) $a^{-2}b^{-2}$ (d) $\dfrac{1}{a^2b^2}$

8.3 SOLVING EQUATIONS WITH EXPONENTS

Example 5

Solve the equations:

(a) $3^x = 27$ (b) $5^{2x} = 125^{\frac{1}{2}}$ (c) $\left(3^x - 1\right)\left(2^{2x} - \frac{1}{16}\right) = 0$

Solution

(a) $3^x = 27$

$3^x = 3^3$

$x = 3$

(b) $5^{2x} = 125^{\frac{1}{2}}$

$5^{2x} = \left(5^3\right)^{\frac{1}{2}}$

$5^{2x} = 5^{\frac{3}{2}}$

$2x = \frac{3}{2}$

$x = \frac{3}{4}$

(c) $\left(3^x - 1\right)\left(2^{2x} - \frac{1}{16}\right) = 0$

$\left(3^x - 1\right)\left(2^{2x} - 2^{-4}\right) = 0$

$3^x - 1 = 0$ or $2^{2x} - 2^{-4} = 0$

$3^x = 1 = 3^0$ $\quad$ $2^{2x} = 2^{-4}$

$x = 0$ $\quad$ $2x = -4$

$x = 0$ $\quad$ $x = -2$

EXERCISE 8.3 SOLVING EQUATIONS WITH EXPONENTS

1 Solve:

(a) $2^x = 8$ (b) $3^{x-1} = 27$ (c) $x^3 = -125$ (d) $x^{-2} = 81$

(e) $\frac{2^{x-3}}{4^{1-x}} = 1$ (f) $4^x = 32$ (g) $9^x = 27$ (h) $3^x + 5 \times 3^x = 54$

(i) $\frac{3 \times 5^x - 1}{5^x + 2} = 2$ (j) $3^x = \frac{1}{9}$ (k) $2^{-x} = \frac{1}{64}$ (l) $5^x = \frac{1}{125}$

2 The solution to the equation $9^x = \frac{1}{3}$ is:

A $x = \frac{1}{2}$ B $x = 2$ C $x = -\frac{1}{2}$ D $x = -2$

3 Solve:

(a) $2^x = \frac{1}{8}$ (b) $a^{x-3} = 1$ (c) $4^x = \frac{1}{2}$ (d) $2^x \times 4^x \times 8^x = 2^{-3}$

(e) $3^x \times 2^x = 1$ (f) $5^x = \frac{1}{125}$ (g) $16^x = 128$ (h) $5^x = 125$

(i) $8^{-x} = \frac{1}{32}$ (j) $\left(2^x - 1\right)\left(3^x - \frac{1}{9}\right) = 0$ (k) $(3^x - 9)(5^x - 1) = 0$ (l) $3^{2x+1} = \frac{1}{27}$

8.4 LOGARITHMS

Logarithms (often called 'logs') were once used mainly as a way to simplify difficult calculations, but electronic calculators and computers have made this technique obsolete. However, logarithms are still useful! You will study logarithmic functions in detail later, but at this stage you need to learn and understand the laws of logarithms. These laws are still needed to solve some equations where the variable is in the exponent (index).

Consider the statement $2^3 = 8$. This is equivalent to the statement $\log_2 8 = 3$. These two statements are different ways to write the same result. Given one statement, you can always write the other. The statement $\log_2 8 = 3$ is read as 'the logarithm of 8 to the base 2 is equal to 3'.

The following pairs of statements are all equivalent:

$3^2 = 9 \Leftrightarrow \log_3 9 = 2$ $\qquad$ $a^0 = 1 \Leftrightarrow \log_a 1 = 0$

$10^4 = 10\,000 \Leftrightarrow \log_{10} 10\,000 = 4$ $\qquad$ $a^x = y \Leftrightarrow \log_a y = x$

In general, if $a > 0$, $a \neq 1$, then the statements $a^x = y$ and $\log_a y = x$ are equivalent. Thus you can see that a logarithm is equivalent to an **index** or **exponent**.

Example 6

Without using a calculator, find the value of:

(a) $\log_2 16$ **(b)** $\log_5 125$

Solution

(a) Let
$$\begin{aligned} \log_2 16 &= x \\ 16 &= 2^x \\ 2^4 &= 2^x \\ x &= 4 \\ \log_2 16 &= 4 \end{aligned}$$

(b) Let
$$\begin{aligned} \log_5 125 &= x \\ 125 &= 5^x \\ 5^3 &= 5^x \\ x &= 3 \\ \log_5 125 &= 3 \end{aligned}$$

Each logarithm has been evaluated by converting it to the equivalent index (exponential) function.

You can obtain the laws of logarithms from the equivalent index laws.

Index laws Let $a^x = m$ and $a^y = n$	**Logarithm laws** Let $\log_a m = x$ and $\log_a n = y$	Logarithm law examples:
$a^x \times a^y = a^{x+y} \rightarrow mn = a^{x+y}$	$\log_a (mn) = x + y = \log_a m + \log_a n$	$\log_a 15 = \log_a 3 + \log_a 5$
$\frac{a^x}{a^y} = a^{x-y} \rightarrow \frac{m}{n} = a^{x-y}$	$\log_a \left(\frac{m}{n}\right) = x - y = \log_a m - \log_a n$	$\log_a \left(\frac{17}{5}\right) = \log_a 17 - \log_a 5$
$\frac{1}{a^x} = a^{-x} \rightarrow \frac{1}{m} = a^{-x}$	$\log_a \left(\frac{1}{m}\right) = -x = -\log_a m$	$\log_a \left(\frac{1}{2}\right) = -\log_a 2$
$\left(a^x\right)^p = a^{xp} \rightarrow m^p = a^{xp}$	$\log_a m^p = px = p\log_a m$	$\log_a 81 = \log_a (3^4) = 4\log_a 3$
$a^1 = a$	$\log_a a = 1$	$\log_{10} 10 = 1$
$a^0 = 1$	$\log_a 1 = 0$	

Note that these laws are all for $a > 0$, and so you also have $m > 0$ and $n > 0$.

For $m > 0$ you can also see that $\log_a m > 0$; for $m = 1$, $\log_a 1 = 0$; and for $0 < m < 1$, $\log_a m < 0$. This means that m is never negative (because $a^x \geq 0$). Therefore you cannot find the logarithm of a negative number.

On most calculators you will notice two keys, **log** and **ln**. These are both logarithm keys. The **log** is the logarithm to base 10, also called the 'common logarithm'; **ln** is the logarithm to base e and is called the 'natural logarithm' (or occasionally the 'Naperian' or 'Napierian' logarithm).

Logarithms to different bases are useful in different situations, but base 10 and base e are often the most useful because of the properties of the numbers 10 and e, as you will see. It is useful to be able to convert all logarithms to a standard base. Luckily, this is possible by using the 'change of base' rule.

Change of base rule

Let $\log_a n = y$, so $n = a^y$.

Take logarithms to base b of both sides: $\log_b n = \log_b a^y$

Using the logarithm law for exponents: $\log_b n = y\log_b a$

So: $y = \frac{\log_b n}{\log_b a}$

Hence: $\log_a n = \frac{\log_b n}{\log_b a}$

This rule shows that the logarithm of a number to a given base is equal to the logarithm of the number to a new base divided by the logarithm of the old base to the new base.

Logarithm laws

$\log_a(mn) = \log_a m + \log_a n$ $\qquad \log_a a = 1$ $\qquad \log_a n = \dfrac{\log_b n}{\log_b a}$

$\log_a\left(\dfrac{m}{n}\right) = \log_a m - \log_a n$ $\qquad \log_a 1 = 0$

$\log_a m^p = p\log_a m$ $\qquad \log_a\left(\dfrac{1}{m}\right) = -\log_a m$

Example 7

Simplify:

(a) $\log_{10} 20 + \log_{10} 5$ **(b)** $\log_a 4 + \log_a 3 + \log_a 2$ **(c)** $\log_2 20 - \log_2 5$

Solution

(a) $\log_{10} 20 + \log_{10} 5$
$= \log_{10} 100$
$= \log_{10} 10^2$
$= 2\log_{10} 10$
$= 2$

(b) $\log_a 4 + \log_a 3 + \log_a 2$
$= \log_a (4 \times 3 \times 2)$
$= \log_a 24$

(c) $\log_2 20 - \log_2 5$
$= \log_2\left(\dfrac{20}{5}\right)$
$= \log_2 4$
$= \log_2 2^2$
$= 2$

Example 8

Simplify:

(a) $\log_{10} 5 + \log_{10} 4 - \log_{10} 2$ **(b)** $3\log_{10} 2 + \log_{10} 18 - 2\log_{10}\left(\dfrac{6}{5}\right)$

Solution

(a) $\log_{10} 5 + \log_{10} 4 - \log_{10} 2$
$= \log_{10}(5 \times 4) - \log_{10} 2$
$= \log_{10}\left(\dfrac{20}{2}\right)$
$= \log_{10} 10 = 1$

(b) $3\log_{10} 2 + \log_{10} 18 - 2\log_{10}\left(\dfrac{6}{5}\right)$
$= \log_{10} 8 + \log_{10} 18 - \log_{10}\left(\dfrac{36}{25}\right)$
$= \log_{10}\left(8 \times 18 \times \dfrac{25}{36}\right)$
$= \log_{10} 100 = 2$

Example 9

Evaluate $\log_2 9$ using base 10 logarithms.

Solution

Use the change of base rule: $\log_2 9 = \dfrac{\log_{10} 9}{\log_{10} 2}$

Use the **log** key on your calculator: $\log_2 9 = 3.1699$ (correct to 4 decimal places)

EXERCISE 8.4 LOGARITHMS

1 Simplify:

(a) $\log_3 9$ (b) $\log_9 3$ (c) $\log_2 128$ (d) $\log_a a$

(e) $\log_3 \frac{1}{27}$ (f) $\log_4 0.25$ (g) $\log_{\sqrt{3}} 243$ (h) $\log_3 \sqrt{3}$

2 $\log_8 512$ is equal to: **A** 3 **B** 5 **C** 7 **D** 9

3 Simplify: (a) $\log_5 625$ (b) $\log_9 243$ (c) $\log_a a^3$ (d) $\log_{10} 0.0001$

4 Simplify the following:

(a) $\log_2 16 + \log_2 8$ (b) $\log_{10} 2 + \log_{10} 5$ (c) $(\log_2 16)(\log_2 4)$ (d) $\log_3 54 - \log_3 18$

(e) $\dfrac{\log_a 8}{\log_a 2}$ (f) $\log_a 5 + \log_a \frac{1}{5}$ (g) $\log_2 18 - 2\log_2 3$ (h) $\log_3 81 \times \log_5 125$

5 Simplify $\log_{10} 125 + \log_{10} 32 - \log_{10} 4$ and state whether each of the following expressions is a correct or incorrect step in the simplification.

(a) $\log_{10} 1000$ (b) $\log_{10} 16\,000$ (c) $3\log_{10} 5 + 3\log_{10} 2$ (d) 3

6 Simplify the following:

(a) $\frac{1}{2}\log_{10} 16 + 2\log_{10} 5$ (b) $\log_2 (2^x)$ (c) $10^{\log_{10} 3}$ (d) $\dfrac{\log_{10} 25}{\log_{10} 5}$

(e) $\log_{10} 125 + \log_{10} 25 + \log_{10} 5$ (f) $\dfrac{\log\left(x^3\right)}{\log x}$ (g) $\log_{10} \dfrac{1+\sqrt{5}}{2} + \log_{10} \dfrac{3+\sqrt{5}}{2}$ (h) $\dfrac{\log x}{\log \sqrt{x}}$

7 If $x = \log_{10} 2$ and $y = \log_{10} 3$, express the following in terms of x and y:

(a) $\log_{10} \frac{2}{3}$ (b) $\log_{10} \frac{4}{9}$ (c) $\log_{10} 15$ (d) $\log_{10} 54$ (e) $\log_{10} 5.4$ (f) $\log_{10} 75$

(g) $\log_{10} 150$ (h) $\log_{10} 0.27$ (i) $\log_{10} 4.5$ (j) $\log_{10} 0.6$ (k) $\log_{10} 81$ (l) $\log_{10} 1.8$

8 Use the change of base rule to evaluate each expression, giving your answer correct to 3 decimal places:

(a) $\log_2 5$ (b) $\log_3 12$ (c) $\log_5 20$ (d) $\log_4 3$ (e) $\log_3 16$ (f) $\log_6 4$ (g) $\log_5 3$ (h) $\log_2 10$

8.5 SOLVING EQUATIONS WITH LOGARITHMS

In the previous section you investigated the logarithm laws and the change of base rule. You will now see how to use these techniques in an algebraic setting to solve more difficult equations.

Note that when the notation 'log' is written without a base, then by convention you should assume it represents $\log_{10}$, the common logarithm.

Example 10

Solve the equations.

(a) $\log_{10} x = \log_{10} 9 + \log_{10} 3$ (b) $3\log_{10} x + 4 = 7\log_{10} x$

Solution

(a)

$\log_{10} x = \log_{10} 9 + \log_{10} 3$

Use $\log_a m + \log_a n = \log_a (mn)$: $\log_{10} x = \log_{10} 27$

If $\log a = \log b$ then $a = b$: $x = 27$

(b)

$3\log_{10} x + 4 = 7\log_{10} x$

Collect like terms: $4 = 4\log_{10} x$

Simplify: $\log_{10} x = 1$

If $\log_a n = y$ then $n = a^y$: $x = 10^1 = 10$

Example 11

For what value of x is $\log_2(x+1) - \log_2(x-1) = 3$ true?

Solution

For $\log x$ to exist, $x > 0$. Hence the equation requires $x + 1 > 0$ and so $x > -1$.

The equation also requires $x - 1 > 0$, and so $x > 1$.

Therefore, for a solution to exist for this equation it requires $x > 1$.

Use $\log_a m - \log_a n = \log_a\left(\frac{m}{n}\right)$: $\quad \log_2\left(\frac{x+1}{x-1}\right) = 3$

If $\log_a n = y$ then $n = a^y$: $\quad \frac{x+1}{x-1} = 2^3$

$$x + 1 = 8x - 8$$

Solve equation: $\quad x = 1\frac{2}{7}$

As $1\frac{2}{7} > 1$, this is a valid solution to the equation.

Remember that 'log' written without a base should be assumed to mean $\log_{10}$.

Example 12

Solve, giving answers correct to 3 decimal places: **(a)** $2^x = 7$ **(b)** $3^{x+1} = 12$

Solution

(a)

$$2^x = 7$$

Take logs to base 10: $\quad \log_{10} 2^x = \log_{10} 7$

$$x \log_{10} 2 = \log_{10} 7$$

$$x = \frac{\log_{10} 7}{\log_{10} 2}$$

$$x = 2.807 \quad \text{(3 d.p.)}$$

(b)

$$3^{x+1} = 12$$

Take logs to base 10: $\quad \log_{10} 3^{(x+1)} = \log_{10} 12$

$$(x+1)\log_{10} 3 = \log_{10} 12$$

$$x + 1 = \frac{\log_{10} 12}{\log_{10} 3}$$

$$x = \frac{\log_{10} 12}{\log_{10} 3} - 1$$

$$x = 1.262 \quad \text{(3 d.p.)}$$

Example 13

Solve the inequalities: **(a)** $2^x > 9$ **(b)** $0.4^x < 0.3$

Solution

(a)

$$2^x > 9$$
$$x > \log_2 9$$
Change of base rule: $x > \dfrac{\log_{10} 9}{\log_{10} 2}$
$$x > 3.17 \quad \text{(2 d.p.)}$$

(b)

$$0.4^x < 0.3$$
$$x \log 0.4 < \log 0.3$$
$$x < \frac{\log_{10} 0.3}{\log_{10} 0.4}$$

Although this step looks correct, it contains an error. This is because $\log_{10} 0.4 < 0$, so you have divided by a negative quantity without reversing the direction of the inequality sign.

Correct result: $x > \dfrac{\log_{10} 0.3}{\log_{10} 0.4}$
$$x > 1.31 \quad \text{(2 d.p.)}$$

Be very careful when using logarithms with inequalities. Remember that if $0 < m < 1$ and $a > 1$, then $\log_a m < 0$, so when you divide by that logarithm you must reverse the direction of the inequality.

Example 14

How many years does it take for \$2000 to grow to \$3000, at 7% p.a. compound interest?

Note that 'p.a.' = 'per annum' = per year. The compound interest formula is $A = P\left(1 + \frac{r}{100}\right)^n$, where P is the initial money invested and A is the amount that P grows to after n periods of time (years) with interest applied at r% per period. Compound interest is investigated further in Chapter 18.

Solution

$A = 2000, r = 7, P = 3000.$ Find n: $2000 \times 1.07^n = 3000$
$$1.07^n = \frac{3}{2}$$
Take logs: $n = \log_{1.07} 1.5$ **OR** $n \log_{10} 1.07 = \log_{10} 1.5$

Use change of base rule **OR** rearrange: $n = \dfrac{\log_{10} 1.5}{\log_{10} 1.07}$
$$n \approx 5.993 \text{ years}$$

The value of n is just under 6 years. Assuming that interest is added at the end of each year, it will take 6 years.

EXERCISE 8.5 SOLVING EQUATIONS WITH LOGARITHMS

1 Solve for x:

(a) $\log_3 x = 3$ (b) $\log_x 81 = 2$ (c) $\log_6 x = 3$ (d) $\log_x 343 = 3$

(e) $\log_5 x = -3$ (f) $\log_3 81 = x$ (g) $\log_x \frac{1}{64} = -3$ (h) $\log_9 x = 0.25$

(i) $\log_3 27\sqrt{3} = x$ (j) $\log_7 x = 2.5$ (k) $\log_2 (\log_2 x) = 2$ (l) $\log_2 x = \log_2 8 + \log_4 8$

2 Without using a calculator, solve each equation:

(a) $\log_{10} x = \log_{10} 4 + \log_{10} 2$ (b) $\log_{10} x = \log_{10} 4 - \log_{10} 2$ (c) $\log_{10} x = \dfrac{\log_{10} 4}{\log_{10} 2}$

(d) $\log_{10} x = \frac{1}{2}\log_{10}\left(\frac{1}{4}\right)$ (e) $2\log_{10} x + 3 = \log_{10}(x^5)$ (f) $\log_{10} x^2 = 2$

3 Find a relation between x and y that does not involve logarithms.

(a) $\log x + \log y = \log(x+y)$
(b) $2\log_{10} y - 3\log_{10} x = 2$
(c) $2\log_3 y - 3\log_3 x = 2$
(d) $2\log_{10} y + 3\log_{10} x = \log_{10}(2x)$
(e) $\log_5 y = 2 + \log_5 x$
(f) $\log y = \log 5 + 3\log x$
(g) $2\log x + 3\log y = 0$
(h) $\log_{10}(1+y) - \log_{10}(1-y) = x$

4 Solve:

(a) $\log_{10} 2 + \log_{10} 5 + \log_{10} x - \log_{10} 3 = 2$
(b) $2\log_{10} x + 3 = 5\log_{10} x$
(c) $\log_{10} 2 + 5\log_{10} x - \log_{10} 5 - \log_{10}(x^3) = \log_{10} 40$
(d) $\log_{10} x = 4\log_{10} 2 - 2\log_{10} x$
(e) $\log_{10} x - \log_{10}(x-1) = 1$
(f) $\log_{10} x = 2\log_{10} 3 + \log_{10} 5 - \log_{10} 2 - 1$

5 Solve $2^{-x} = 5$. Indicate whether each statement below is a correct or incorrect step in the solution.

(a) $x = \dfrac{\log 5}{\log 2}$
(b) $x = \log_2\left(\dfrac{1}{5}\right)$
(c) $x = \dfrac{-\log 5}{\log 2}$
(d) $x = -2.32$ (2 d.p.)

6 Solve, correct to 3 decimal places:

(a) $2^x = 7$
(b) $3^x = 18$
(c) $5^x = 2$
(d) $0.4^x = 2$
(e) $6^x = 21$
(f) $3^{-x} = 0.1$
(g) $5^x = 16$
(h) $4^x = 5$

7 Find the values of x (to 2 decimal places) for which:

(a) $5^x > 2$
(b) $1.6^x \geq 0.5$
(c) $3^x < 0.2$
(d) $3^{-x} > 27$
(e) $2^x \geq 5$
(f) $0.25^{-x} < 1.5$
(g) $0.8^x < 3$
(h) $0.7^x \geq 0.3$

8 If $y = a10^{bx}$, then:

A $x = \log_{10}\dfrac{y}{ab}$
B $x = \dfrac{1}{b}\log_{10}\dfrac{y}{a}$
C $y = \dfrac{1}{b}\log_{10}\dfrac{x}{a}$
D $x = \dfrac{1}{a}\log_{10}\dfrac{y}{b}$

9 If $\log_{10} A = bt + \log_{10} P$, express A in terms of the other symbols.

10 If $\log y = \log a + n\log x$, find an expression for y.

11 If $y = \dfrac{\log x}{\log 2}$, express x in terms of y.

12 If $x = a^2\sqrt{b^3 c}$, express $\log x$ in terms of $\log a$, $\log b$ and $\log c$.

13 If $\log x = 0.6$ and $\log y = 0.2$, evaluate $\log\left(\dfrac{x^2}{\sqrt{y}}\right)$.

14 If $y = ae^{4t}$, express t in terms of a and y.

15 If $\log_b a = p$ and $c = a^2$, find the following in terms of p: (a) $\log_b c$ (b) $\log_c b$

16 If $\log_a 2 = \log_b 16$, show that $b = a^4$.

17 \$5000 is invested at 7% p.a. compound interest. How long does it take for this money to:

(a) double in value
(b) grow to \$20 000
(c) grow to \$30 000?

18 \$5000 is invested at 6% p.a. compound interest. If the interest is calculated monthly, how long does it take for this money to:

(a) double in value
(b) grow to \$20 000
(c) grow to \$30 000?

19 Marika and Joe deposit \$4000 in an account that pays 9% p.a. compound interest, to be withdrawn when it has grown to \$20 000. If the interest is calculated monthly, for how many whole months must they leave the money in the account?

20 A company is considering a merger. Currently the company earnings are \$5 per share and these earnings are growing by 5% p.a. It is predicted that after the merger the earnings will drop to \$4 per share, but will then grow by 7% p.a.

(a) Show that if no merger occurs, the earnings per share after n years are given by $y_n = 5 \times 1.05^n$.
(b) Show that if the merger occurs, the earnings per share after n years are given by $z_n = 4 \times 1.07^n$.
(c) If the merger goes ahead, how many years will it take for a shareholder to be better off?

8.6 EXPONENTIAL FUNCTIONS

An exponential function is a function $f(x) = a^x$, where the base a is a positive real number other than 1. Its domain is the set of real numbers and its range is the set of positive real numbers.

The exponential function can be used to model many real-life situations. The compound interest formula is an example of this. You can also use exponential functions to model population growth and radioactive decay.

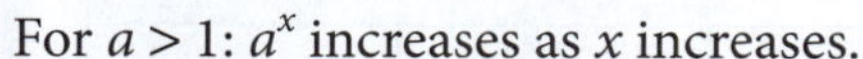

For $a > 1$: a^x increases as x increases.

As $x \to -\infty$, $a^x \to 0$ from above.

For all values of x, $a^x > 0$.

For all values of $x < 0$, $0 < a^x < 1$.

At $x = 0$, $a^0 = 1$.

$y = a^x$ cuts the y-axis at $(0, 1)$ for all values of a.

For all values of $x > 0$, $a^x > 1$.

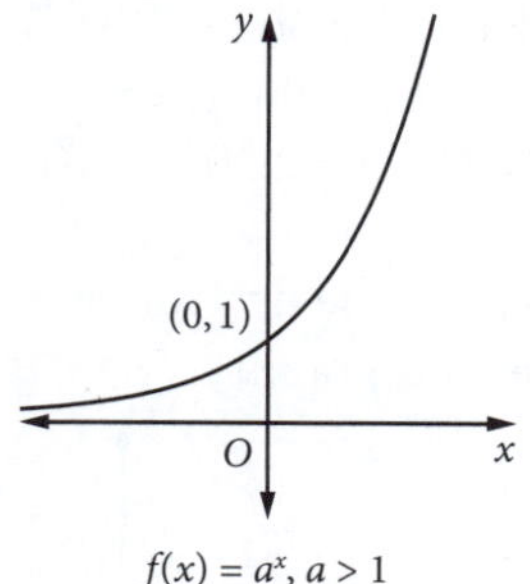

$f(x) = a^x$, $a > 1$

For $0 < a < 1$: a^x decreases as x increases.

As $x \to \infty$, $a^x \to 0$ from above.

For all values of x, $a^x > 0$.

For all values of $x < 0$, $a^x > 1$.

When $x = 0$, $a^0 = 1$.

$y = a^x$ cuts the y-axis at $(0, 1)$ for all values of a.

For all values of $x > 0$, $0 < a^x < 1$.

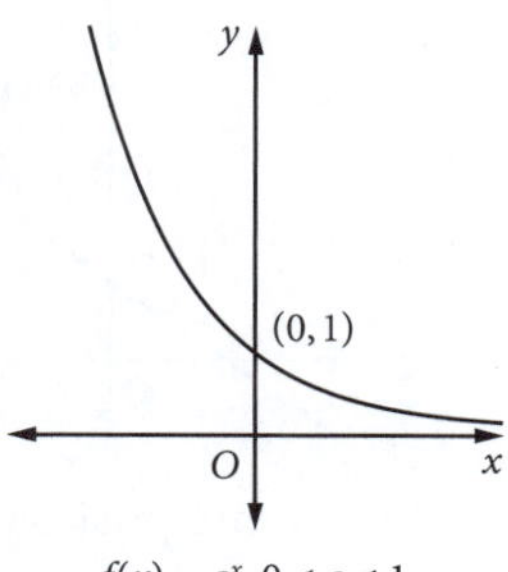

$f(x) = a^x$, $0 < a < 1$

For $a = \frac{1}{2}$, $f(x) = \left(\frac{1}{2}\right)^x = \left(2^{-1}\right)^x = 2^{-x}$, so $f(x) = a^x$ for $0 < a < 1$ is the same as $f(x) = a^{-x}$ for $a > 1$. Hence the graphs now represent $f(x) = a^x$ and $f(x) = a^{-x}$ for $a > 1$:

From these graphs you can see that the gradient of $f(x) = a^x$ is always positive (i.e. $f'(x) > 0$) and the gradient of $f(x) = a^{-x}$ is always negative (i.e. $f'(x) < 0$). You cannot find the gradient functions yet, but the graphs tell us their signs. Also note that both graphs are concave upwards.

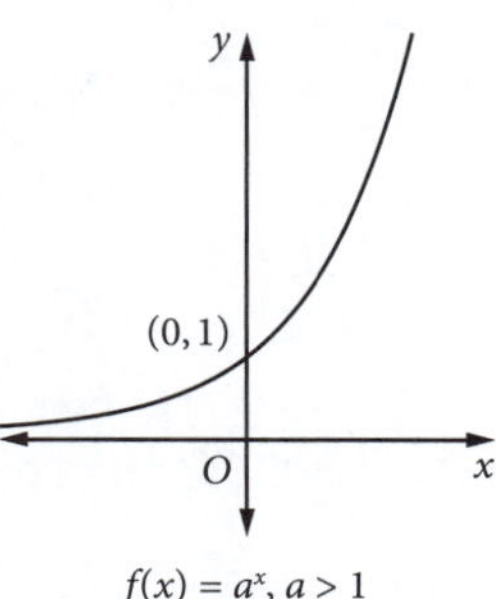

$f(x) = a^x$, $a > 1$

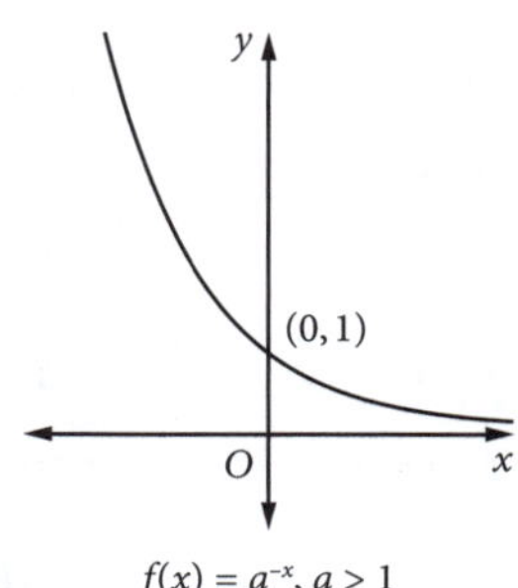

$f(x) = a^{-x}$, $a > 1$

Gradient of exponential functions

The diagram shows the graph of $y = 10^x$.

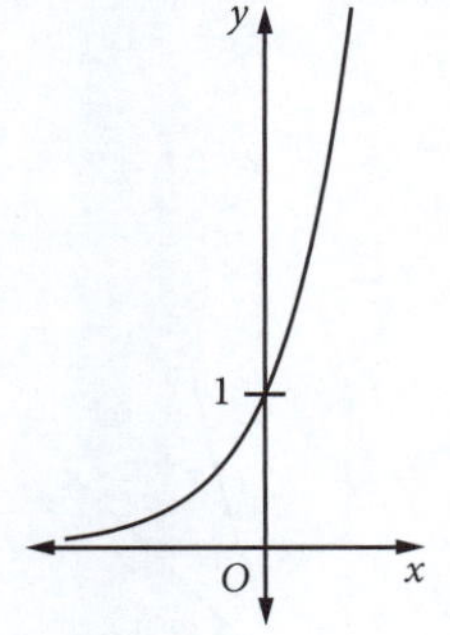

To find the gradient at any point P on the curve, you need to go back to our first-principles definition of the gradient function (see Chapter 7).

Let $f(x) = 10^x$ so $f(x + h) = 10^{x+h}$. From the definition of the gradient function:

$$
\begin{aligned}
f'(x) &= \lim_{h \to 0} \frac{f(x+h) - f(x)}{h} \\
&= \lim_{h \to 0} \frac{10^{x+h} - 10^x}{h} \\
&= \lim_{h \to 0} \frac{10^x\left(10^h - 1\right)}{h} \\
&= 10^x \lim_{h \to 0} \frac{10^h - 1}{h} \qquad \text{because } 10^x \text{ is independent of } h.
\end{aligned}
$$

You can use a spreadsheet or a calculator to investigate the values of $\frac{10^h - 1}{h}$ as $h \to 0$:

h	$\frac{10^h - 1}{h}$	h	$\frac{10^h - 1}{h}$
0.1	2.5893	−0.1	2.0567
0.05	2.4404	−0.05	2.1750
0.01	2.3293	−0.01	2.2763
0.001	2.3052	−0.001	2.2999
0.0001	2.3029	−0.0001	2.3023

As $h \to 0$ from either positive or negative values, the value of $\frac{10^h - 1}{h}$ approaches a limit. The value of this limit is between 2.3029 and 2.3023, so a good approximation to 4 decimal places is 2.3026. This limit is the gradient at the point where the graph of $y = 10^x$ crosses the y-axis. Thus $f'(x) = 2.3026 \times 10^x$.

You can repeat this investigation of the derivative for $f(x) = 2^x$ and $f(x) = 3^x$, evaluating $\frac{2^h - 1}{h}$ and $\frac{3^h - 1}{h}$ as $h \to 0$, as shown in the following table and graphs.

$f(x)$	$f'(x)$
2^x	0.6931×2^x
3^x	1.0986×3^x
10^x	2.3026×10^x

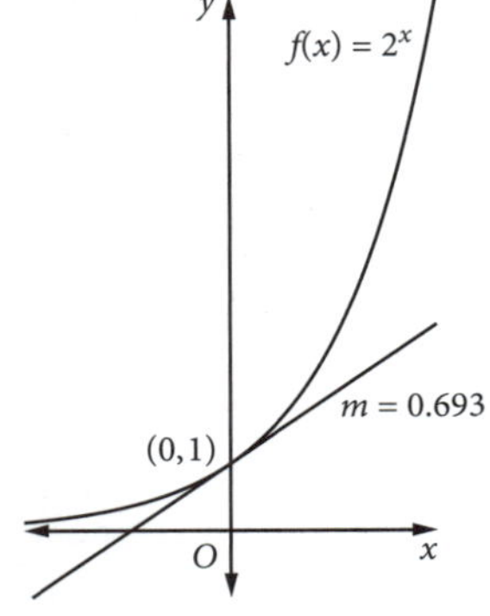

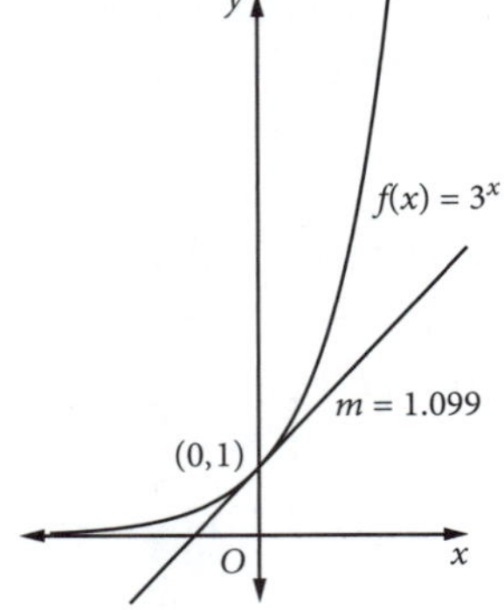

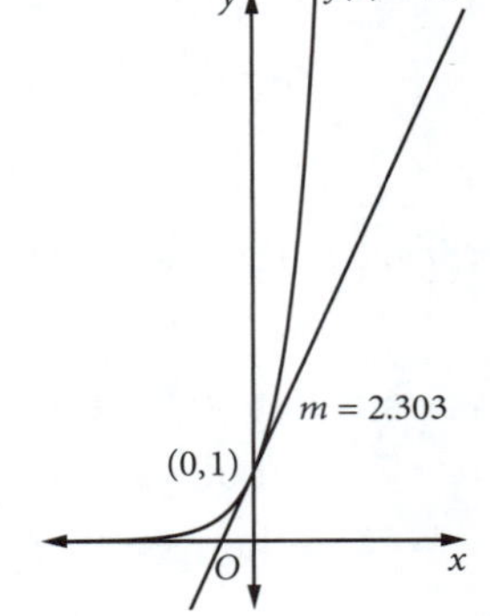

In general, you have $\frac{d}{dx}\left(a^x\right) = ka^x$ for some positive integer k, which is the gradient of the function a^x at $(0, 1)$. If you could find a value for a that gave $k = 1$, then that function would be its own derivative. When $a = 2$, $k = 0.693$ and when $a = 3$, $k = 1.099$, so you are looking for a value of a between 2 and 3. Denote this special base by e, where $2 < e < 3$ such that you can write $\frac{d}{dx}\left(e^x\right) = e^x$.

Evaluation of e

Let $e = 10^b$, so that $e^x = 10^{bx}$. So: $\frac{d}{dx}\left(e^x\right) = \frac{d}{dx}\left(10^{bx}\right)$

Now let $y = 10^{bx} = 10^u$, where $u = bx$.

You showed earlier that $\frac{d}{dx}\left(10^x\right) = 2.3026 \times 10^x$, so $\frac{d}{du}\left(10^u\right) = 2.3026 \times 10^u$.

Using the chain rule $\frac{dy}{dx} = \frac{dy}{du} \times \frac{du}{dx}$, so $\frac{dy}{dx} = 2.3026 \times 10^u \times b$ because $\frac{du}{dx} = b$.

Hence: $\frac{dy}{dx} = 2.3026b \times 10^{bx}$ replacing u, as $u = bx$

Thus: $\frac{d}{dx}\left(10^{bx}\right) = 2.3026be^x$ replacing y, as $y = 10^{bx} = e^x$

So: $\frac{d}{dx}\left(e^x\right) = 2.3026be^x$

But the original definition of e requires that $\frac{d}{dx}\left(e^x\right) = e^x$, so you can now write that $e^x = 2.3026be^x$.

Solve this equation for b: $b = \frac{1}{2.3026} = 0.4343$ (4 decimal places)

Hence: $e = 10^{0.4343} = 2.7183$ (4 decimal places)

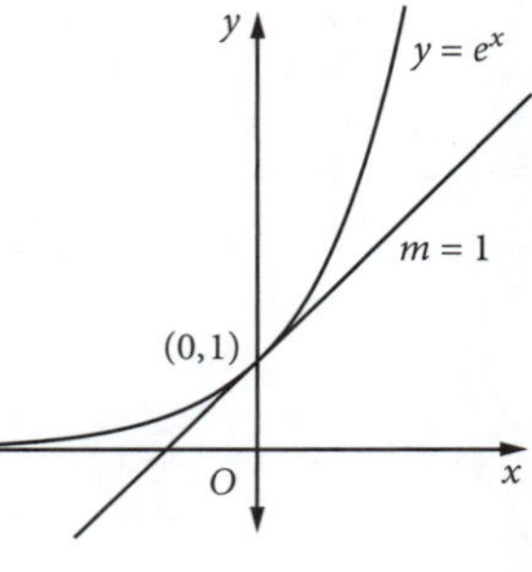

This number e is known as **Euler's number** after mathematician Leonhard Euler, who invented its notation. After π, e is one of the most famous transcendental numbers and scientific constants. (For more on transcendental numbers, see Section 12.3.)

Here the graph of $y = e^x$ is drawn, as well as the tangent to the curve at $(0, 1)$. The gradient of this tangent is 1, because at $(0, 1)$ you have $\frac{dy}{dx} = e^0 = 1$.

Formal derivation of e

Consider the function $f(x) = e^x$, where e is some positive number. Differentiating from first principles, you have:

$$f'(x) = \lim_{h \to 0} \frac{f(x+h) - f(x)}{h}$$

$$= \lim_{h \to 0} \frac{e^{x+h} - e^x}{h}$$

$$= \lim_{h \to 0} \frac{e^x\left(e^h - 1\right)}{h}$$

$$= e^x \lim_{h \to 0} \frac{e^h - 1}{h} \quad \text{because } e^x \text{ is independent of } h.$$

Hence: $f'(x) = f(x) \times \lim_{h \to 0} \frac{e^h - 1}{h}$

Define the value of e such that $\frac{d}{dx}\left(e^x\right) = e^x$, or $f'(x) = f(x)$.

Thus you require $\lim_{h \to 0}\left(\frac{e^h - 1}{h}\right) = 1$, which requires that $\frac{e^h - 1}{h} = 1$ for a very small value of h.

Hence: $e^h - 1 = h$ for very small h.

Recall that if n is very large ($n \to \infty$) then the value of $\frac{1}{n}$ is very small $\left(\frac{1}{n} \to 0\right)$.

Thus for very small h, $e^h - 1 = h$ can be written as $e^{\frac{1}{n}} - 1 = \frac{1}{n}$, where n is very large.

Hence: $e^{\frac{1}{n}} = \frac{1}{n} + 1$

Raise both sides to the power n: $\left(e^{\frac{1}{n}}\right)^n = \left(\frac{1}{n} + 1\right)^n$

So: $e = \left(1 + \frac{1}{n}\right)^n$ for $n \to \infty$.

Thus you have found the value of e for which the function $f(x) = e^x$ has the property that $\frac{d}{dx}\left(e^x\right) = e^x$.

More formally, you could write: $e = \lim_{n \to \infty}\left(1 + \frac{1}{n}\right)^n$

You can use this equation to calculate values of e for increasing values of n using a calculator or a spreadsheet. Some values are given in the following table.

n	100	1000	10^4	10^5	10^6	10^9
e	2.704 814	2.716 923	2.718 146	2.718 268	2.718 280	2.718 282

Thus $e = 2.718\,28$ is an approximation for e correct to 5 decimal places. Most calculators will give you an approximation for e correct to 9 decimal places.

Derivative of e^{kx}, k a constant

To find the derivative of $y = e^{kx}$, you first write $y = e^{kx} = e^u$, where $u = kx$.

Use the chain rule: $\frac{dy}{dx} = \frac{dy}{du} \times \frac{du}{dx}$

$$\frac{dy}{dx} = e^u \times k = k e^x$$

$$\therefore \frac{d}{dx}\left(e^{kx}\right) = ke^{kx}$$

Thus the function $f(x) = e^{kx}$ is a function whose derivative is proportional to the value of the function itself.

That is: $f'(x) = kf(x)$, or $\frac{dy}{dx} = ky$ where $y = e^{kx}$.

Consider some values for k, for example:

$k=2$: $\frac{d}{dx}\left(e^{2x}\right)=2e^{2x}$

$k=-3$: $\frac{d}{dx}\left(e^{-3x}\right)=-3e^{-3x}$

$k=\frac{1}{2}$: $\frac{d}{dx}\left(e^{\frac{x}{2}}\right)=\frac{1}{2}e^{\frac{x}{2}}$

Derivative of e^{ax+b}, a and b constants

Again, you first write $y=e^{ax+b}=e^u$, where $u=ax+b$.

Chain rule: $\frac{dy}{dx}=\frac{dy}{du}\times\frac{du}{dx}$

$\frac{dy}{dx}=e^u\times a=ae^{ax+b}$

So: $\frac{dy}{dx}=ay$, where $y=e^{ax+b}$

$\therefore \frac{d}{dx}\left(e^{ax+b}\right)=ae^{ax+b}$

Derivative of $e^{f(x)}$

$y=e^{f(x)}=e^u$, where $u=f(x)$

Chain rule: $\frac{dy}{dx}=\frac{dy}{du}\times\frac{du}{dx}$

$\frac{dy}{dx}=e^u\times f'(x)=f'(x)e^{f(x)}$

So: $\frac{dy}{dx}=f'(x)e^{f(x)}$

Summary of derivatives involving e^x

$\frac{d}{dx}\left(e^x\right)=e^x$ $\quad\frac{d}{dx}\left(e^{kx}\right)=ke^{x}$ $\quad\frac{d}{dx}\left(e^{ax+b}\right)=ae^{ax+b}$ $\quad\frac{d}{dx}\left(e^{f(x)}\right)=f'(x)e^{f(x)}$

Example 15

Differentiate: **(a)** $y=e^{3x-4}$ **(b)** e^{x^2} **(c)** $e^{x+\sqrt{x}}$

Solution

(a) $\frac{dy}{dx}=3e^{3x-4}$

(b) Let $y=e^{x^2}=e^u$ where $u=x^2$.

Chain rule: $\frac{dy}{dx}=\frac{dy}{du}\times\frac{du}{dx}$

$\frac{dy}{dx}=e^u\times 2x$

$=2xe^{x^2}$

(c) Let $y=e^{x+\sqrt{x}}=e^u$ where $u=x+\sqrt{x}$.

$\frac{dy}{dx}=\frac{dy}{du}\times\frac{du}{dx}$

$\frac{dy}{dx}=e^u\times\left(1+\frac{1}{2\sqrt{x}}\right)$

$=\left(1+\frac{1}{2\sqrt{x}}\right)e^{x+\sqrt{x}}$

Example 16

Differentiate: **(a)** $(2x^2+1)e^{3x}$ **(b)** $\frac{e^x}{x}$

Solution

(a) Let $y = (2x^2 + 1)e^{3x} = uv$ where $u = 2x^2 + 1$ and $v = e^{3x}$.

Product rule: $\dfrac{dy}{dx} = v\dfrac{du}{dx} + u\dfrac{dv}{dx}$

$$\frac{dy}{dx} = e^{3x} \times 4x + (2x^2 + 1) \times 3e^{3x}$$
$$= (6x^2 + 4x + 3)e^{3x}$$

(b) Let $y = \dfrac{e^x}{x} = \dfrac{u}{v}$ where $u = e^x$ and $v = x$.

Quotient rule: $\dfrac{dy}{dx} = \dfrac{v\dfrac{du}{dx} - u\dfrac{dv}{dx}}{v^2}$

$$\frac{dy}{dx} = \frac{x \times e^x - e^x \times 1}{x^2}$$
$$= \frac{e^x(x-1)}{x^2}$$

EXERCISE 8.6 EXPONENTIAL FUNCTIONS

1 Write the derivative of:

(a) e^{4x} (b) $2e^{\frac{x}{2}}$ (c) $e^{4x} - e^{3x}$ (d) $2e^{3x} + e^{-x}$ (e) $4e^{3x} - e^{-2x}$

(f) $e^{3.2x} - e^{1.6x}$ (g) $3e^x - 2e^{-x}$ (h) $4e^{2x} + \frac{1}{2}e^{-2x}$ (i) $e^{2x}(e^x - e^{-x})$

2 If $y = e^{x^2}$ then $\dfrac{dy}{dx}$ is: **A** $x^2e^{x^2}$ **B** $2xe^{x^2}$ **C** x^2e^{2x} **D** $2xe^{2x}$

3 Differentiate:

(a) x^2e^{3x} (b) $(2x+1)e^{-x}$ (c) $(x^2 + x + 1)e^{2x}$ (d) xe^{-2x}

(e) $e^{-x}x^3$ (f) $x^3 - xe^{2x}$ (g) $x^2 - x^3e^{2x}$ (h) $\dfrac{e^x}{x^2}$

(i) $\dfrac{e^{3x}}{x}$ (j) $\dfrac{x^3}{e^x}$ (k) $\dfrac{e^{4x}}{x-1}$ (l) $\dfrac{e^x}{\sqrt{x}}$

4 Differentiate:

(a) e^{2x+3} (b) e^{x^2-2x} (c) $3e^{-x^2}$ (d) $2e^{3x-1}$ (e) $e^{3x-1} + e^{4x+2}$ (f) $\sqrt{x}e^{-x}$ (g) $3e^{2x^2}$ (h) $3e^{2x-1}$ (i) xe^{x^2}

5 If $y = e^{2x} + e^{8x}$, indicate whether each statement below is correct or incorrect.

(a) $\dfrac{dy}{dx} = 2e^{2x} + 8e^{8x}$ (b) $\dfrac{dy}{dx} = 10e^x$ (c) $\dfrac{d^2y}{dx^2} = 4e^{2x} + 64e^{8x}$ (d) $\dfrac{d^2y}{dx^2} - 10\dfrac{dy}{dx} + 16y = 0$

6 If $x = (1+t)e^{5t}$, prove that $\dfrac{d^2x}{dt^2} - 10\dfrac{dx}{dt} + 25x = 0$.

7 Find the equation of the tangent to the curve $y = e^x$ at the point where it crosses the y-axis.

8 Find the equation of the tangent to the curve $y = e^{-x}$ at the point where it crosses the y-axis.

9 Find the equation of the tangent to the curve $y = e^{2x}$ at the point where $x = 1$. Find also the coordinates of the points where the tangent intersects: (a) the x-axis (b) the y-axis.

10 Write the equation of the tangent and the normal to the curve $y = 2 + e^{-x}$ at the point where $x = 0$.

11 After n years, the value V of a principal of P dollars that is invested at a rate of r% per year (with r expressed as a decimal) and compounded continuously is given by $V = Pe^{rn}$. Show that $\dfrac{dV}{dn} = Vr$.

12 The expression $y = 500(1 - e^{-0.2t})$ represents the daily output of y units on day t of a production run. Find the instantaneous rate of change of the output y with respect to t.

8.7 SOME APPLICATIONS OF EXPONENTIAL FUNCTIONS

Note: This section is enhancement material and may be skipped as these concepts are developed in Chapter 14

Example 17

For the function $f(t) = 2te^{-0.5t}$, find the value of t for which $f(t)$ has a maximum and hence calculate the maximum value. Sketch the graph of $f(t)$.

Solution

$f(t) = 2te^{-0.5t}$

$f'(t) = 2\left(e^{-0.5t} + t \times \left(-\frac{1}{2}\right)e^{-0.5t}\right)$ using the product rule

$f'(t) = e^{-0.5t}(2 - t)$

For stationary points, $f'(t) = 0$: $e^{-0.5t}(2 - t) = 0$

But $e^{-0.5t} > 0$ for all t, so $t = 2$ is the only solution and $f(2) = \frac{4}{e}$

For $t < 2$: $f'(t) > 0$

For $t > 2$: $f'(t) < 0$

Gradient changes from positive to negative as x increases, so $\left(2, \frac{4}{e}\right)$ is a maximum turning point.

The maximum value of the function is $\frac{4}{e} \approx 1.472$

$f(t) = 0$ at $t = 0$ as $e^{-0.5t} > 0$ for all t.

$t < 0, f(t) < 0$ $\quad$ $t < 2, f'(t) > 0$

$t > 0, f(t) > 0$ $\quad$ $t > 2, f'(t) < 0$

$t \to \infty, f(t) \to 0$ from above

$t = 0$ is horizontal asymptote

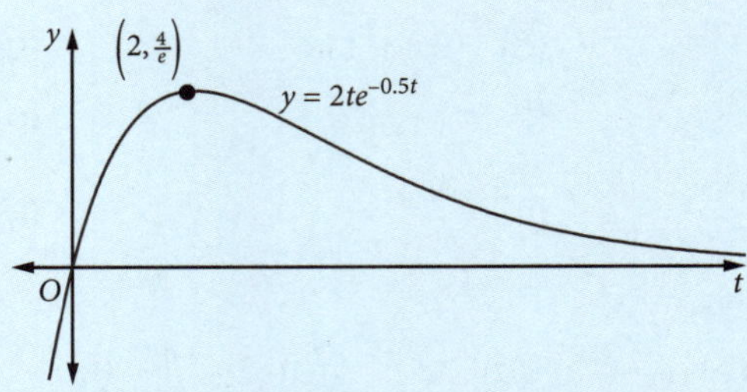

EXERCISE 8.7 SOME APPLICATIONS OF EXPONENTIAL FUNCTIONS

1 Find the minimum value of $(x - 2)e^x$.

2 Find the coordinates of the turning point of the curve $y = xe^{-0.5x}$ and state whether it is a maximum or minimum. Find also the values of x for which:

(a) $y > 0$ **(b)** $\frac{dy}{dx} > 0$

3 Consider the function defined by the rule $f(x) = 3 - e^{-x}, x \geq 0$.

(a) Find the value of $f(0)$ and $f'(0)$.

(b) Show that $f'(x) > 0$ for all values of x in the domain.

(c) What is the value of $\lim_{x\to\infty} f(x)$?

(d) Sketch the graph of $f(x)$.

4 Consider the function defined by $f(x) = e^{-x^2}$ for all values of x.

(a) Find $f'(x)$.

(b) Find the values of x for which: **(i)** $f'(x) = 0$ **(ii)** $f'(x) > 0$ **(iii)** $f'(x) < 0$

(c) Sketch the graph of the function.

5 The concentration of a certain drug in the blood at a time t hours after taking the dose is x units, where $x = 0.3te^{-1.1t}$.

(a) Determine the maximum concentration and the time at which this is reached.

(b) Plot the graph of $x = 0.3te^{-1.1t}$ for $t = 0, 0.1, 0.5, 1, 2, 3$ using graph paper or graphing software.

(c) This drug kills germs only while its concentration is at least 0.06 units. From the graph, find the length of time during which the drug will kill germs.

6 For $y = e^t + 4e^{-t}$, find the minimum value of y. Indicate whether each of the statements below is a correct or incorrect step in solving this problem.

(a) $y' = e^t - 4e^{-t}$ **(b)** Stationary point when $e^t = \pm 2$

(c) $y'' = e^t + 4e^{-t}$ **(d)** Minimum value is 4

7 Sketch the graph of $f(t) = \dfrac{5}{2+3e^{-t}}$, $t \geq 0$.

(a) Show that $f'(t) > 0$ for all values of t in the domain. **(b)** Find $\lim\limits_{t\to\infty} f(t)$.

(c) State the range of the function.

8 The rectangle $PQRS$ has two vertices on the x-axis and two on the curve $y = e^{-x^2}$. Find:

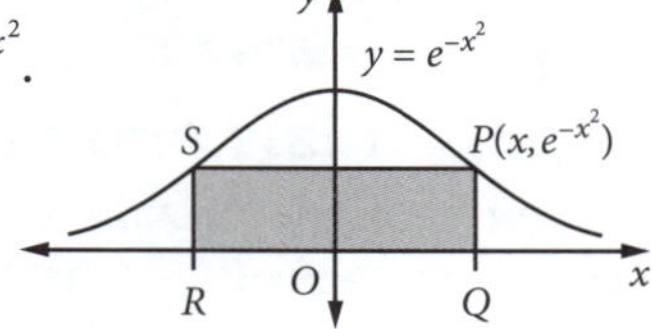

(a) the value of x for which the rectangle has a maximum area

(b) the maximum area of the rectangle.

8.8 NATURAL LOGARITHMS

You have previously looked at the relationship between the expressions $y = a^x$ and $x = \log_a y$. You have also defined the exponential function $y = e^x$ so that $x = \log_e y$. This logarithm to the base e is called the **natural logarithm**, or sometimes the 'Naperian' or 'Napierian' logarithm after John Napier, the Scottish mathematician who introduced logarithms in the 1600s. The function $\log_e x$ is sometimes written as $\ln x$, and calculators usually have a natural logarithm key labelled **ln**.

The diagram shows the graphs of $g(x) = e^x$ and $f(x) = \log_e x$.

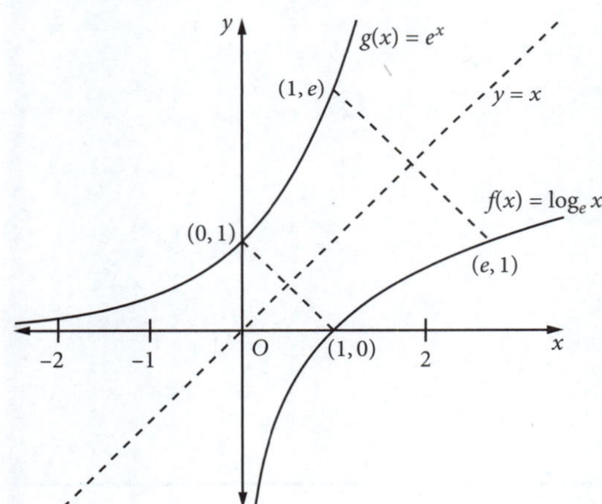

These two graphs are symmetrical about the line $y = x$. You can say that $f(x)$ is the **reflection** of $g(x)$ in the line $y = x$. To 'reflect' a curve of $y = e^x$ in the line $y = x$ means that you interchange (swap) the x and y values, so the equation of the reflection becomes $x = e^y$. Because you know how the exponential and logarithmic functions are linked, you can rewrite $x = e^y$ as $y = \log_e x$.

Because of this property, $y = e^x$ and $y = \log_e x$ are known as **inverse functions**.

On the graph of $g(x) = e^x$ the point $(1, e)$ shows that $e^1 = e$. On the graph of $f(x) = \log_e x$ the point $(e, 1)$ shows that $\log_e e = 1$.

Hence you can write that $e^{\log_e e} = e$ and in general that $e^{\log_e x} = x$.

You can verify this result by taking the point $(2, \log_e 2)$ on $y = f(x)$ and checking whether the point $(\log_e 2, 2)$ is on $y = g(x)$ using graphing software or your calculator.

Important properties of e^x and $\log_e x$

If $y = \log_e x$ then you can write:	$e^y = e^{\log_e x}$	or	$e^y = e^{\ln x}$
But $e^y = x$ so:	$x = e^{\log_e x}$	or	$x = e^{\ln x}$
Hence:	$e^{\log_e x} = x$	or	$e^{\ln x} = x$
Similarly, if $y = e^x$ then:	$\log_e y = \log_e e^x$	or	$\ln y = \ln e^x$
But $\log_e y = x$, so:	$x = \log_e e^x$	or	$x = \ln e^x$
Hence:	$\log_e e^x = x$	or	$\ln e^x = x$

In general, you can write $a^{\log_a x} = x$ for all real x, and $\log_a a^x = x$ for all $x > 0$.

Note that the operations 'square' and 'square root' are similarly inverse operations (for positive values), because when repeated after each other they return to the starting value: $\sqrt{x^2} = x$ and $\left(\sqrt{x}\right)^2 = x$. This is the same as for the exponential function and the natural logarithm function: $e^{\log_e x} = x$ and $\log_e e^x = x$.

Summary of important results—exponentials and logarithms

- If $y = e^x$ then $\log_e y = x$ or $\ln y = x$
- If $y = \log_e x$ or $y = \ln x$ then $e^y = x$
- $e^{\log_e x} = x$ or $e^{\ln x} = x$
- $\log_e e^x = x$ or $\ln e^x = x$
- $e^0 = 1$ and $\log_e 1 = 0$
- $e^x > 0$ for all x
- Domain of $y = \log_e x$ is $x > 0$

EXPLORE FURTHER

Exponentials and natural logarithms

Use technology to explore how exponentials and natural logarithms are inverse functions of each other.

Example 18

Solve $\log_e(x+3) + \log_e(x-4) = 1$.

Solution

State the domain for possible solutions: $(x+3) > 0$ and $(x-4) > 0$, so $x > 4$

Write the LHS as a single log expression: $\log_e(x+3)(x-4) = 1$

Hence: $x^2 - x - 12 = e$

Use the quadratic formula: $x^2 - x - 12 - e = 0$

$$x = \frac{1 \pm \sqrt{1 + 48 + 4e}}{2}$$

$$= \frac{1 \pm \sqrt{4e + 49}}{2}$$

$$\approx 4.37, -3.37$$

Check domain for valid solutions: $x > 4$ for the first solution only.

The only solution is: $x = \dfrac{1 + \sqrt{4e + 49}}{2}$

EXERCISE 8.8 NATURAL LOGARITHMS

1 Solve for x: **(a)** $e^{2x+3} = e^5$ **(b)** $e^{4x+7} = e^{5x-3}$ **(c)** $e^x = 5$

2 Solve for x: **(a)** $\log_e e^{3x+5} = 2$ **(b)** $e^{\frac{x}{4}} = 3$ **(c)** $5e^{4x} = 8$

3 Solve for x:

(a) $\log_e(x+5) = \log_e 3$ **(b)** $\log_e(x-1) + \log_e(x+1) = 3\log_e 2$ **(c)** $\log_e(x-3) - \log_e(x+1) = 2\log_e 3$

4 Solve for x:

(a) $\log_e(x+2) = 3$ **(b)** $\log_e(2x-2) = 4$ **(c)** $\log_e(x+2) - \log_e(x-2) = 1$

5 Solve for x: **(a)** $\log_e(x+2e) + \log_e(x-2e) = 2$ **(b)** $\log_e x + \log_e(x-2e^2) = 2$

6 Solve for x:

(a) $\log_e x + \log_e(x+5) = \log_e(x+2) + \log_e 6$
(b) $\log_e x - \log_e(x+5) = \log_e(x-4) - \log_e(x+2)$

7 Which expression is equivalent to $3 + \log_e x$?

A $\log_e 3x$ **B** $\log_e(e^3 + x)$ **C** $3\log_e(ex)$ **D** $\log_e(e^3 x)$

8.9 GRAPHS OF EXPONENTIAL AND LOGARITHMIC FUNCTIONS

There are many transformations of exponential and logarithmic functions of the types $f(x) = ke^{ax+b} + c$ and $g(x) = k\log_e(ax+b)+c$.

In each case, adding c provides a vertical translation. Multiplying by k extends the range of the function (a dilation) by a factor of $|k|$. Reflection in the x-axis occurs if k is negative.

Example 19

Draw the graphs of $f(x) = e^{2x}$, $g(x) = e^{2x+1}$ and $h(x) = e^{2x} + 1$. Describe the properties of these graphs. Find the coordinates of any points of intersection by solving the appropriate equations.

Solution

$f(x) = e^{2x}$ $\quad g(x) = e^{2x+1}$ $\quad h(x) = e^{2x} + 1$

$f'(x) = 2e^{2x}$ $\quad g'(x) = 2e^{2x+1}$ $\quad h'(x) = 2e^{2x}$ $\quad f'(x) = h'(x)$

All three functions increase as x increases.
All three functions have a positive slope which gets steeper as x increases.
$g(x)$ is $f(x)$ translated 0.5 units to the left. $h(x)$ is $f(x)$ translated 1 unit up.
The three curves have the same shape; only their position is different.
As $x \to -\infty$, $f(x) = e^{2x} \to 0$ from above; $g(x) = e^{2x+1} \to 0$ from above; and $h(x) = e^{2x+1} \to 1$ from above.
$f(0) = 1$ so $f(x)$ cuts the y-axis at (0, 1).
$g(0) = e$ so $g(x)$ cuts the y-axis at (0, e).
$h(0) = 2$ so $h(x)$ cuts the y-axis at (0, 2).

Intersection of $g(x)$ and $h(x)$:

$$e^{2x+1} = e^{2x} + 1$$
$$e \times e^{2x} - e^{2x} = 1$$
$$e^{2x}(e-1) = 1$$
$$e^{2x} = \frac{1}{e-1}$$
$$x = \frac{1}{2}\log_e\left(\frac{1}{e-1}\right)$$

Hence: $$h(x) = e^{2\left(\frac{1}{2}\log_e\left(\frac{1}{e-1}\right)\right)} + 1$$
$$= \frac{1}{e-1} + 1 = \frac{e}{e-1}$$

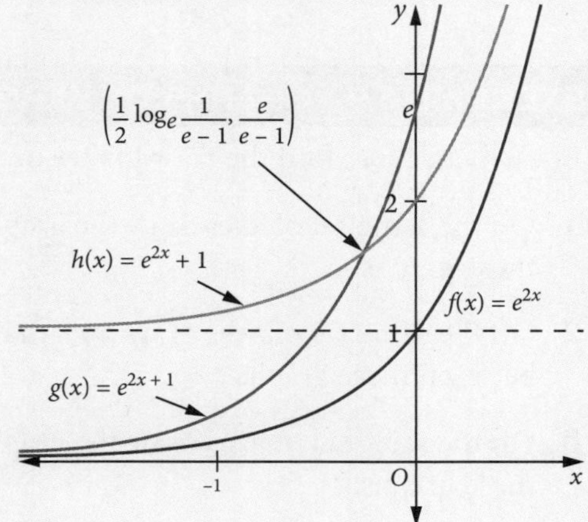

Point of intersection is $\left(\frac{1}{2}\log_e\left(\frac{1}{e-1}\right), \frac{e}{e-1}\right)$ or about (−0.27, 1.58).

In the example above if the functions were $j(x) = e^{-2x}$, $k(x) = e^{-2x+1}$ and $m(x) = e^{-2x} + 1$, then the curves would be the reflection of $f(x)$, $g(x)$ and $h(x)$ in the y-axis. The graphs of $f(x)$ and $g(x)$ would approach zero as x increases and $h(x)$ would approach 1 as x increases. The curves would still cut the y-axis at the same points.

Also note: $j'(x) = -2e^{-2x}$, $k'(x) = -2e^{-2x+1}$, $m'(x) = -2e^{-2x}$

Hence: $j'(x) = m'(x)$

MAKING CONNECTIONS

Graph of the exponential function

Move the sliders to explore the effects of k, a, b, c, on the graph of $y = ke^{ax+b} + c$.

Example 20

Draw the graphs of $f(x) = \log_e x$, $g(x) = \log_e (2x)$ and $h(x) = \log_e (2x + 1)$. Describe the properties of these graphs. Find the coordinates of any points of intersection by solving the appropriate equations.

Solution

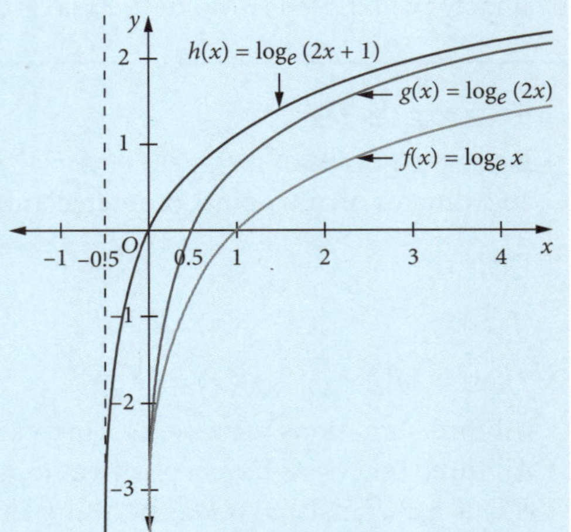

$f(x) = \log_e x$, so $x > 0$.
$f(1) = 0$ so $f(x)$ cuts the x-axis at $(1, 0)$.
As $x \to 0$ from the right $f(x) \to -\infty$ so $x = 0$ is an asymptote.

$g(x) = \log_e (2x) = \log_e 2 + \log_e x = \log_e 2 + f(x)$
$g(x)$ is just $f(x)$ translated up by $\log_e 2$.
$g(0.5) = 0$ so $g(x)$ cuts the x-axis at $(0.5, 0)$.

$h(x) = \log_e (2x + 1)$ so $x > -0.5$.
$h(0) = 0$ so $h(x)$ cuts the axes at $(0, 0)$.
As $x \to 0$ from the right $h(x) \to -\infty$ so $x = -0.5$ is an asymptote.

All three functions increase as x increases.
All have a positive slope which flattens out as x increases.
The graphs do not intersect.

MAKING CONNECTIONS

Graph of the logarithmic function

Move the sliders to explore the effects of k, a, b, c, on the graph of $y = k\log_e (ax + b) + c$.

EXERCISE 8.9 GRAPHS OF EXPONENTIAL AND LOGARITHMIC FUNCTIONS

In each question, find the coordinates of any points of intersection by solving the appropriate equations.

1 On the same set of axes draw the graphs of $f(x) = e^x$, $g(x) = e^{-x}$ and $h(x) = -e^x$. Describe the properties of these graphs.

2 On the same set of axes draw the graphs of $f(x) = e^x$, $g(x) = 3e^x$ and $h(x) = e^{3x}$. Comment on the differences between these graphs.

3 On the same set of axes draw the graphs of $f(x) = e^x$, $g(x) = e^x + 1$ and $h(x) = e^x - 1$. Describe the properties of these graphs.

4 On the same set of axes draw the graphs of $f(x) = \log_e x$, $g(x) = \log_e (3x)$ and $h(x) = \log_e (x + 3)$. Describe the properties of these graphs.

5 On the same set of axes draw the graphs of $f(x) = \log_e (x - 2)$, $g(x) = 2\log_e x$ and $h(x) = \log_e (x + 2)$. Comment on the differences between these graphs.

6 On the same set of axes draw the graphs of $f(x) = \log_e (2x)$, $g(x) = \log_e x^2$ and $h(x) = \log_e \left(\frac{x}{2}\right)$. Comment on the differences between these graphs.

7 On the same set of axes draw the graphs of $f(x) = \log_e (-x)$, $g(x) = e^{-x}$ and $h(x) = -x$. Describe the properties of these graphs.

8 On the same set of axes draw the graphs of $f(x) = e^x$, $g(x) = e^{-x}$ and $h(x) = e^x + e^{-x}$. Describe the properties of these graphs.

8.10 LOGARITHMS IN THE REAL WORLD

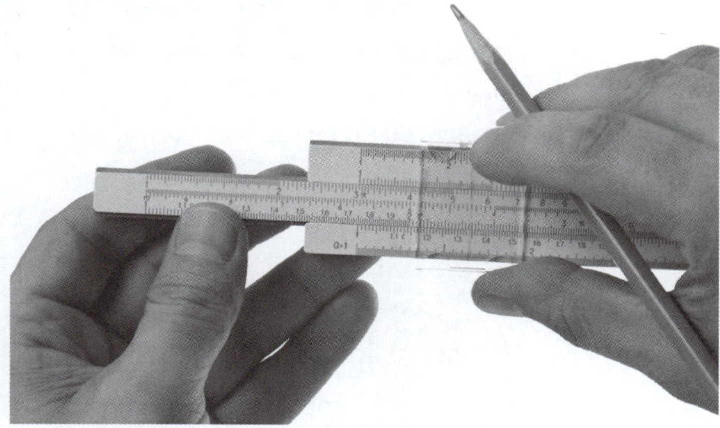

A range of logarithmic scales are used in science and engineering to measure natural phenomena. When phenomena involve large differences in values, logarithms of values can be used to reduce the numbers to a size that is meaningful to non-scientific people so that useful comparisons can be made. Sometimes the common logarithm base 10 is used; at other times base 2 is more useful.

Logarithmic scales, rather than linear scales, are also used to graph experimental results. If this produces a straight line graph, then this indicates a logarithmic mathematical relationship.

Before calculators were invented, tables of logarithms and devices called 'slide rules' (see the picture at right) were used to perform calculations involving multiplication, division and corresponding operations, e.g. raising to a power or finding the square root. The slide rule had a logarithmic scale and used the fact that you can find the product of two numbers by adding their indices or logarithms.

Decibels (dB)

The decibel was originally used to measure sound levels, but now is also widely used as a measurement unit in electronics, signals and communications. It is based on a logarithmic scale. The difference in intensity or 'loudness' between two sounds, or between two sources of power P_1 and P_2, is defined to be $10\log_{10}\left(\frac{P_2}{P_1}\right)$ dB. The absolute measurement of the intensity or loudness of a sound is given by $L = 10\log_{10}\left(\frac{P}{P_0}\right)$, where P_0 is the reference value and has an intensity of 0 dB.

This is the quietest noise that can be (just) heard by the typical human ear. It has an intensity of 10^{-12} W/m^2 (watts per square metre). The intensity of any other sound measured in these units is

$$\begin{aligned}10\log_{10}\left(\frac{P}{P_0}\right) &= 10\log_{10}P - 10\log_{10}P_0\\ &= 10\log_{10}P - 0\\ &= 10\log_{10}P\end{aligned}$$

Sounds quieter than P_0 have a negative number of decibels.

Because it is a logarithmic scale, every increase of 10 dB is 10 times more powerful. So if 0 dB the smallest audible sound, then 10 dB is a sound 10 times more powerful than that, 20 dB is a sound 100 times more powerful, and 30 dB is 1000 times more powerful.

Sound intensities for some common events, in dB:

- near total silence: 0 dB
- a normal conversation: 60 dB
- a lawnmower: 90 dB
- a rock concert or a jet engine: 120 dB
- a gunshot or a firecracker: 140 dB and above.

You should not be exposed to a continuous noise level of 85 dB for more than 8 hours, or else you risk hearing damage. For each 3 dB increase, the length of time that you can be exposed is halved. This means that if the noise level in a nightclub is 100 dB, then you should only stay 15 minutes. Of course, the further away from the noise source you are, the lower the sound intensity.

The decibel scale reduces very large numbers to numbers that people can relate to.

Example 21

(a) **(i)** How many times more intense is a sound P_2 which is 30 dB louder than another sound P_1?

(ii) How many times more intense is a sound P_2 which is 14 dB louder than another sound P_3?

(b) How much louder is:

(i) a sound of 50 dB than the smallest audible sound

(ii) a sound of 47 dB than a sound of 27 dB

(iii) a sound of 102 dB than a sound of 64 dB?

Solution

(a) **(i)**

$$10 \log_{10}\left(\frac{P_2}{P_1}\right) = 30$$

$$\log_{10}\left(\frac{P_2}{P_1}\right) = 3$$

$$\frac{P_2}{P_1} = 10^3 = 1000$$

P_2 is 1000 times as loud as P_1.

(ii)

$$10 \log_{10}\left(\frac{P_2}{P_3}\right) = 14$$

$$\log_{10}\left(\frac{P_2}{P_3}\right) = 1.4$$

$$\frac{P_2}{P_3} = 10^{1.4} \approx 25$$

P_2 is about 25 times as loud as P_3.

(b) **(i)** The smallest audible sound is 0 dB

50 dB − 0 dB = 50 dB

$$10 \log_{10}\left(\frac{P}{P_0}\right) = 50$$

$$\log_{10}\left(\frac{P}{P_0}\right) = 5$$

$$\frac{P}{P_0} = 10^5$$

$$P = 10^5\, P_0$$

The sound is 100 000 times as loud.

(ii) 47 dB − 27 dB = 20 dB

$$10 \log_{10}\left(\frac{P}{P_0}\right) = 20$$

$$\log_{10}\left(\frac{P}{P_0}\right) = 2t$$

$$\frac{P}{P_0} = 10^2$$

$$P = 100\, P_0$$

The sound is 100 times as loud.

(iii) 102 dB − 64 dB = 38 dB

$$10 \log_{10}\left(\frac{P}{P_0}\right) = 38$$

$$\log_{10}\left(\frac{P}{P_0}\right) = 3.8$$

$$\frac{P}{P_0} = 10^{3.8}$$

$$P \approx 6300\, P_0$$

The sound is 6300 times as loud.

Richter scale

The Richter scale is a logarithmic scale that is used to compare the magnitude of earthquakes. It was developed in the 1930s in California by Charles Richter. Using the Richter scale, the magnitude of an earthquake is expressed as a number (without units) determined by the logarithm of the amplitude of waves recorded by seismographs, which are devices that directly measure the seismic vibrations of the earth. Adjustments are made for the variations in distance between the various seismographs and the epicentre of the earthquake.

The Richter scale formula is $M_L = \log_{10}\left(\frac{A}{A_0}\right)$, where M_L is the measurement value of the Richter scale, A is the amplitude of the wave recorded by the seismograph and A_0 is the reference value that corresponds to a zero-level, earthquake, i.e. no earthquake. Because the scale is logarithmic, every increase of 1 on the scale corresponds to an earthquake that is 10 times 'stronger'. This means that an earthquake that measures 6.3 on the Richter scale is 10 times as strong as an earthquake measuring 5.3. This corresponds to a tenfold increase in the wave amplitude, but this actually means it releases approximately 32 times more energy.

The difference on the Richter scale between the intensity (amplitude) of two earthquakes is $\log_{10}\left(\frac{A_2}{A_1}\right)$.

pH

The pH level of a solution (where the word 'solution' here means water that has something dissolved in it) is a measure of how acidic or basic the solution is. The more acidic a solution, the more positive hydrogen ions are produced for chemical reactions; the more basic a solution, the more negative hydroxide ions are produced for chemical reactions. Acids and bases are important because they are involved in many common chemical reactions that involve water and other useful compounds. pH levels vary from 0 to 14, with 7 being neutral (neither acidic nor basic); a pH of less than 7 indicates acidity, whereas a pH greater than 7 indicates basicity.

The definition of pH is given by $\text{pH} = -\log_{10}[\text{H}^+]$, where $[\text{H}^+]$ is the concentration of the H^+ (hydrogen) ions measured in units of mol/L.

In pure water, $[\text{H}^+] = 1.0 \times 10^{-7}$ mol/L, so the pH is $-\log_{10}[\text{H}^+] = -\log_{10}(1.0 \times 10^{-7}) = 7$.

Octaves

In music, there are 8 notes in each octave, named using the letters from A to G. An octave is any sequence of 8 notes starting and ending at the same note, e.g. C D E F G A B C.

Every musical note is a sound wave that has a particular vibration frequency. 'Higher'-sounding notes have higher frequency values, while 'lower' (or 'deeper') notes have lower frequency values.

The note at the high end of any octave has double the frequency of the note at the lower end. For every octave higher, the frequency doubles, so for two octaves higher the frequency would quadruple; three octaves higher, the frequency would increase by a factor of 8; and so on. This suggests using logarithms to the base 2 as a measure of how much higher one note is than another.

So, $\log_2\left(\frac{f_2}{f_1}\right)$ is a measure of how many octaves higher a note of frequency f_2 is than a note of frequency f_1.

For example, if two notes have frequencies of 120 Hz and 1920 Hz, then the difference can be measured as $\log_2\left(\frac{1920}{120}\right) = \log_2 16 = 4$, meaning that one note is 4 octaves higher than the other.

EXERCISE 8.10 LOGARITHMS IN THE REAL WORLD

1 What is the value of $10\log_{10}\left(\frac{P_2}{P_1}\right)$ when: **(a)** $P_2 = P_1$ **(b)** $P_2 = 100\,000P_1$

2 How many times louder is:

(a) a sound which is 20 dB louder than another sound
(b) a 75 dB sound than a 35 dB sound
(c) a 79 dB sound than a 72 dB sound?

3 If one sound is twice as loud as another, how many more decibels is its intensity?

4 An earthquake measuring 8.7 on the Richter scale is followed by one that measures 6.5 on the Richter scale. How many times stronger is the first earthquake than the second?

5 The energy released by an earthquake, E, can be given by $\log_{10} E = 11.8 + 1.5\,M_L$, where M_L is the measurement of its magnitude on the Richter scale. Calculate the energy released by both of the earthquakes in Question **4** and state how many times more energy is released by the first earthquake than by the second.

6 Calculate, correct to one decimal place, the pH level of a solution where the concentration of H^+ (hydrogen) ions is 2.3×10^{-5} mol/L. Is this an acidic or a basic solution?

7 The frequency of the note A3, the A below middle C, is 220 Hz. Another note has a frequency of 1760 Hz.

(a) How many octaves higher than A3 is this note? **(b)** What is this note?

8 Why can you not use the decibel scale to measure a sound of zero intensity, i.e. no sound at all?

CHAPTER REVIEW 8

1 Find the values of x for which the following are true.

(a) $7^{x+2} = 343$ **(b)** $4^{x-2} < 128$ **(c)** $3^x \geq 12$

2 Simplify:

(a) $\log_3 18 + 2\log_3 9 - \log_3 54$ **(b)** $\log_a(xy^2) + \log_a(yz^2) - \log_a(xz^2)$

(c) $\log_{10}\frac{6+4\sqrt{6}}{5} + \log_{10}\frac{2\sqrt{6}-3}{2}$

(d) $2\log(x+1) - \log(x-1) - 2\log(y+1) + \log(y-1)$, given $x = 5$ and $y = 2$

3 Express y in terms of x: **(a)**$\log_a y = x$ **(b)** $\log 10\,y = 2 + \log 10\,x - \log 10\,(x^2)$

4 Differentiate with respect to x:

(a) $(x^2 + 2x)e^x$ **(b)** $(x^2 + 3x)e^{-3x}$ **(c)** $e^{\sqrt{x}}$

5 Simplify the following:

(a) $\frac{10^3 \times 5^{-2} \times 18^{-1}}{8^{-4} \times 5^{-2}}$ **(b)** $\frac{3^n \times 9^{n+1}}{27^n}$ **(c)** $\frac{a^{n-1} \times b^3}{a^{n-3} \times b^2}$ **(d)** $x^{-\frac{3}{2}} \times x^{\frac{1}{4}}$ **(e)** $\frac{\sqrt{6} \times \sqrt[3]{3} \times 2^{\frac{1}{2}}}{12^{\frac{5}{4}}}$

6 Solve for x:

(a) $(3x-1)\left(2x - \frac{1}{8}\right)$ **(b)** $23x + 1 = \frac{1}{32}$ **(c)** $e^{4x+1} = \frac{1}{e^3}$

7 Use the change of base rule to give each expression as a single term involving natural logarithms.

(a) $\log_{10} 5$ **(b)** $3 + \log_3 6$

8 Solve for x:

(a) $\log_e(x+2) = \log_e(2x)$ **(b)** $\log_e(2x+3) = \log_e x^2$ **(c)** $\log_e x^2 = \log_e\left(\frac{x}{3}\right)$

9 Solve for x: **(a)** $2e^x = e^{3x}$ **(b)** $2e^{x+1} = e^{2x}$ **(c)** $2e^{-x} = e^{3x-4}$

10 \$4000 is invested at 3% p.a. compound interest. How long does it take for this money to:

(a) double in value **(b)** grow to \$10 000 **(c)** grow to \$80 000?

11 On the same set of axes sketch the functions $f(x) = \log_e(3x)$, $g(x) = \log_e(3x) + 1$ and $h(x) = \log_e(3x+1)$. Describe the properties of these graphs, including asymptotes, intercepts on axes and points of intersection.

12 If $\theta = \theta_0 e^{-kt}$, show that $\frac{d\theta}{dt} = -k\theta$.

13 A car is worth \$10 000 when new. After t years, the value of the car (in dollars) is given by the formula $V = Ae^{-0.2t}$.

(a) Find the value of A. **(b)** Find the value of the car after 6 years.

(c) Find the rate in dollars per year at which the car's value is depreciating, when:

(i) $t = 6$ **(ii)** $V = 5000$

(d) How much time will it take until the car is worth only \$1000?

14 Solve:

(a) $370\,000 \times 1.005^n = 2998 \times \left(\frac{1.005^n - 1}{0.005}\right)$, giving your answer correct to 2 decimal places.

(b) $48\,500 \times 1.006^n - 780\left(\frac{1.006^n - 1}{0.006}\right) = 0$, giving your answer to the next integer.

CHAPTER 9
Probability

9.1 INTRODUCTION TO PROBABILITY

People often make statements like the following:

- It will **probably** rain today.
- I have a **good chance** of passing my exams.
- There is a **50-50 chance** that heads will come up when I toss a coin.
- I have a **slight chance** of being dux of the class.

In each case there is some doubt about the outcome, but the degree of doubt is different. **Probability** is about this doubt: it is the study of events that may or may not happen, rather than of events that will happen or that have already happened.

If you toss a coin, the result must be heads (H) or tails (T). There are only two possible outcomes, H or T, and each outcome is **equally likely**. You say that the probability of heads is $\frac{1}{2}$, and you can write this as:

$$P(\text{H}) = \frac{1}{2}$$

This probability be expressed in words in a variety of ways:

- There is a 50-50 chance that heads will turn up.
- There is a 1-in-2 chance that heads will turn up.
- The odds that heads will turn up are 1 to 1.
- It is an even-money bet that heads will turn up.

Before beginning a game of football, cricket, basketball etc., the team captains usually toss a coin. Why?

Equally likely events are events where each outcome has the same chance of occurring.

Coin experiment

Toss a coin 10 times, 50 times, 100 times, 200 times and 500 times, counting the number of heads tossed each time. (This can be simulated using a calculator or spreadsheet.) Complete the following table:

Number of coin tosses	10	50	100	200	500
Number of heads					
Relative frequency (number of heads / number of throws)					

If you toss the coin 10 times, how many heads would you expect? If you toss the coin 50 times, how many heads would you expect? Would you be surprised if you didn't get exactly the number that you expected?

The relative frequency of each experiment gives you the actual probability of that outcome occurring. Is it as close to 0.5 as you expected? Did the relative frequency get closer to 0.5 as you increased the number of throws?

MAKING CONNECTIONS

Coin experiment

Use a spreadsheet to simulate the coin experiment. Select the number of coin tosses and investigate the results.

An American soldier, while a prisoner of war in World War II, passed the time by tossing a coin 1000 times. He performed this experiment ten times and obtained the following numbers of heads: 502, 511, 497, 529, 504, 476, 507, 528, 504, 529. Are these numbers surprising?

Die

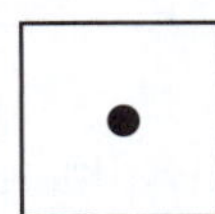 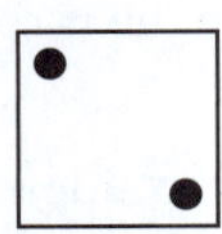 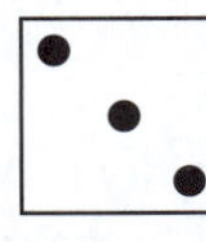 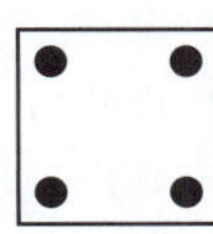 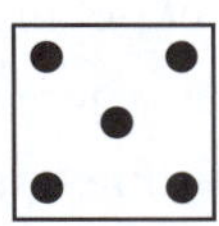 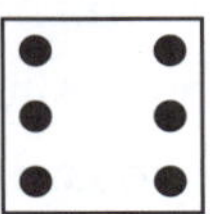

Faces

A die (plural: dice) is a cube with six faces numbered 1 to 6, as shown above. If a die is rolled on a flat surface, one of the six numbers must appear on the top. (For example, the number 5 appears on top of the die in the picture above.)

- Each number from 1 to 6 is **equally likely** to turn up. Why?
- What is the probability that a 5 will appear on top?

There are six possible outcomes. Of these six outcomes, the outcome that you are interested in is a 5. This is called the **favourable** outcome. The probability of a 5 appearing is $\frac{1}{6}$, so the probability of a favourable outcome is written as is $P(5) = \frac{1}{6}$.

This can be expressed in a variety of ways:

- There is a 1-in-6 chance that a 5 will turn up.
- The odds are 5 to 1 against a 5 turning up. This means that for every 1 favourable outcome there are 5 unfavourable outcomes.

The probability of an outcome is defined as the ratio of the number of favourable outcomes to the number of possible outcomes, assuming that the outcomes are equally likely. Thus, the probability $P(A)$ of a particular result A is:

$$P(A) = \frac{\text{number of favourable outcomes}}{\text{number of possible outcomes}} \quad \textbf{OR} \quad P(\text{Event}) = \frac{\text{number of ways Event can occur}}{\text{number of possible outcomes}}$$

Dice experiment

Roll a die 6 times, 60 times, 120 times, 300 times and 600 times, counting the number of times a 5 (or another particular number) is rolled. (This can be simulated using a calculator or spreadsheet.) Complete the following table:

Number of rolls of the die	6	60	120	240	300	600
Number of times 5 appears						
Relative frequency (number of 5s / number of rolls)						

If you roll a die 6 times, how many 5s would you expect? If you roll a die 60 times, how many 5s would you expect? Would you be surprised if you didn't get exactly the number you expected?

The relative frequency of each experiment gives you the actual probability of that outcome occurring. Is it as close to $\frac{1}{6}$ as you expected? Did the relative frequency get closer to $\frac{1}{6}$ (≈ 0.1667) as you increased the number of throws?

MAKING CONNECTIONS

Dice experiment

Use a spreadsheet to simulate the dice-roll experiment. Select the number of rolls and investigate the results.

From the definition of probability as the number of favourable outcomes divided by the number of possible outcomes, you can see that if the number of favourable outcomes is equal to the number of possible outcomes then the probability is 1. If the probability is 1, then the event is **certain** to happen. For example:

- It is certain that the Sun will rise in the east tomorrow.
- The probability of obtaining a number less than 7 when a die is rolled $= \frac{6}{6} = 1$ (a certainty).

List some other situations where the probability is 1.

If there are no favourable outcomes, then the probability is 0. For example, many athletes can run very fast, but no athlete can run at the speed of light. The probability that an athlete will run faster than light is 0. Similarly, the probability that a 7 will appear when a normal die is rolled $= \frac{0}{6} = 0$.

Probabilities must always be from 0 to 1 inclusive:

- $0 \le P(\text{Event}) \le 1$
- If $P(\text{Event}) = 0$, the event is impossible.
- If $P(\text{Event}) = 1$, the event is certain.

To win Lotto®, you must pick six winning numbers out of the numbers 1 to 45. The number of different ways that you can choose 6 numbers from 45 numbers is 8 145 060, but only one of these ways is favourable (i.e. will win):

$P(\text{6 winning numbers}) = \frac{1}{8145060}, \approx 0.000\,000\,1$, which is rather small.

As another example, the following table shows the number of live births and the number of male and female babies born in Australia over the three-year period 2000 to 2002.

Year	Live births	Boys	Girls	Proportion of boys
2000	249 636	128 190	121 446	0.514
2001	246 394	126 298	120 096	0.513
2002	250 988	128 623	122 365	0.512

With rare exceptions, statistics over the years indicate a slight excess of male over female births.

$P(\text{newborn is male}) = \frac{\text{number of boys born}}{\text{total number of births}} \approx 0.51$

Insurance companies prepare 'life tables', which show the number of Australian people per 100 000 who are statistically expected to be alive at any given age. For example, a table like this could show that 92 515 people out of every 100 000 people born are expected to be still alive after 60 years:

$P(\text{an Australian will live until age 60}) = \frac{92515}{100000} = 0.93$

The following scale shows the probabilities of the events discussed.

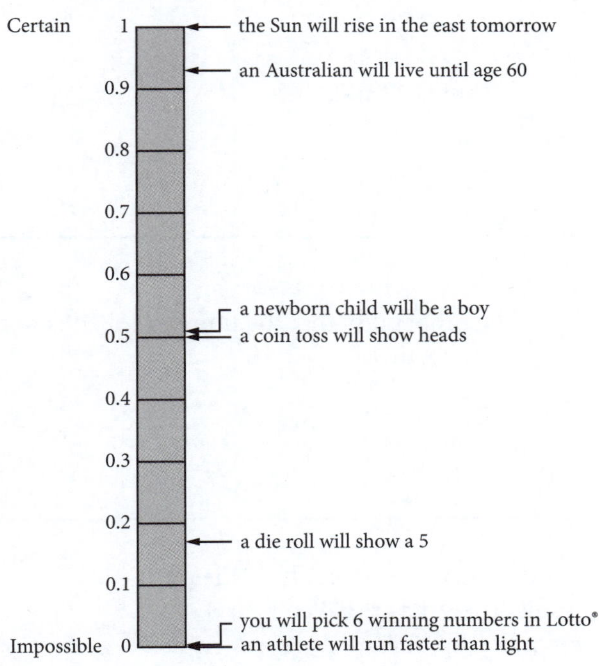

Example 1

A die is rolled on a flat surface. What is the probability that the number appearing is:

(a) odd

(b) divisible by 3

(c) odd **or** divisible by 3

(d) odd **and** divisible by 3?

Solution

The six possible outcomes from rolling a die are equally likely.

(a) There are three odd numbers (1, 3, 5), so there are 3 favourable outcomes.

$$P(\text{odd}) = \frac{3}{6} = \frac{1}{2}$$

(b) There are two numbers divisible by 3 (3, 6), so there are 2 favourable outcomes.

$$P(\text{divisible by 3}) = \frac{2}{6} = \frac{1}{3}$$

(c) 1, 3, 5, 6 are favourable, so there are 4 favourable outcomes. Note carefully the word 'or' in the question.

$$P(\text{odd or divisible by 3}) = \frac{4}{6} = \frac{2}{3}$$

(d) 3 is the only number that is odd and also divisible by 3, so there is only 1 favourable outcome.

$$P(\text{odd and divisible by 3}) = \frac{1}{6}$$

Complementary events

The **complement of event** A is the event 'A **does not occur**', and can be denoted by $\overline{A}$.

If an experiment has n possible outcomes, m of which are associated with event A, then $(n - m)$ outcomes are associated with event $\overline{A}$, and so:

$$P(A) = \frac{m}{n} \qquad P(\overline{A}) = \frac{n-m}{n} = \frac{n}{n} - \frac{m}{n} = 1 - \frac{m}{n} = 1 - P(A)$$

Thus:

$$P(A)+P(\bar{A})=1$$

In general, P(Event does not occur) = 1 − P(Event occurs), or:

$$P(\bar{A})=1-P(A)$$

Example 2

A die is rolled on a flat surface. What is the probability that the number rolled is **not** 4?

Solution

$P(4) = \frac{1}{6}$ $\qquad$ $P(A) = \frac{1}{6}$

$\therefore P(\text{not } 4) = 1 - \frac{1}{6} = \frac{5}{6}$ $\qquad$ $P(\bar{A}) = 1 - \frac{1}{6} = \frac{5}{6}$

A standard deck (or pack) of playing cards consists of 52 cards, made up of 4 sets of 13 cards in groups called a suit. The suits are called hearts, diamonds, spades and clubs and have symbols on them that represent the name of the suit. The hearts have red hearts on them, the diamonds have red diamonds on them, the spades have a black shape called a spade on them and the clubs have a black shape like a clover leaf on them.

This means that two of the suits have red coloured symbols on them and the other two of the suits have black coloured symbols on them. Hence half the deck is red cards and half the deck is black cards.

In each suit the cards are numbered 2, 3, 4, 5, 6, 7, 8, 9 ,10, J, Q, K, A, where J stands for Jack, Q stands for Queen, K stands for King and A stands for Ace.

The jacks, queens and kings are called court cards or picture cards.

Example 3

From a standard deck of 52 playing cards, one card is drawn at random. What is the probability that it is:

(a) a diamond $\qquad$ **(b)** not a spade $\qquad$ **(c)** an ace

(d) a diamond **or** an ace $\qquad$ **(e)** a diamond **and** an ace $\qquad$ **(f)** a heart **and** the ace of clubs?

Solution

There are 52 possible outcomes.

(a) There are 13 diamonds: $\quad P(\text{diamond}) = \frac{13}{52} = \frac{1}{4}$

(b) There are 13 spades, so there are 39 cards that are not spades: $\quad P(\text{not spade}) = \frac{39}{52} = \frac{3}{4}$

Or, using complementary events: $\quad P(\text{not spade}) = 1 - P(\text{spade}) = 1 - \frac{1}{4} = \frac{3}{4}$

(c) There are 4 aces: $\quad P(\text{ace}) = \frac{4}{52} = \frac{1}{13}$

(d) There are 13 diamonds and 4 aces, but one ace is also a diamond, so there are 13 + 3 = 16 favourable outcomes: $\quad P(\text{diamond } \textbf{or} \text{ ace}) = \frac{16}{52} = \frac{4}{13}$

(e) The ace of diamonds is the only card that is a diamond and also an ace: $\quad P(\text{diamond } \textbf{and} \text{ ace}) = \frac{1}{52}$

(f) There is no card that is a heart and also the ace of clubs: $\quad P(\text{heart } \textbf{and} \text{ ace of clubs}) = 0$

EXERCISE 9.1 INTRODUCTION TO PROBABILITY

1 Select 20 lines of text from a page in a book. Count the number of times each letter of the alphabet appears. Would you expect each of the 26 letters of the alphabet to appear approximately the same number of times? If a letter in your 20 lines is selected at random, would you say that the probability that the letter is 'e' is the same as the probability that it is 'z'?

2 Select a page in a telephone directory (or search at random on a telephone directory website) and count the number of times that each digit 0, 1, 2, 3, … 9 appears as the last digit of a phone number. Prepare a table showing your results.
   (a) Did you expect results like these?
   (b) Would similar results have occurred if the first digit of each number was recorded? Why?
   (c) What would have happened if the first letter of each person's name was recorded?

3 What is wrong with the following arguments?
   (a) The probability that a child born in Australia is born in New South Wales is $\frac{1}{7}$.
   (b) Because there are fewer road accidents in Hobart than in Sydney, it is safer to drive in Hobart.

4 Which statement has a probability closest to 1?
   A The cost of buying a house will increase this year.
   B You will win the lottery.
   C A person will be born in Sydney in the next hour.
   D It will rain in Sydney tomorrow.

5 Perform the following experiment:
From a bag containing 7 black marbles and 3 white marbles, withdraw a marble, note its colour and then replace it in the bag. Shake the marbles and then withdraw a marble again. Repeat this 10, 50, 100, 150 and 200 times to complete the following table.

Number of withdrawals	10	50	100	150	200
Number of black marbles withdrawn					
Relative frequency (number of black / number of withdrawals)					

How many black marbles do you expect to get in each case?
Did the relative frequency get closer to 0.7 as you increased the number of withdrawals?

6 Perform the following experiment:
From a standard deck of 52 playing cards, withdraw a card and note whether it is a diamond, heart, spade or club. Replace the card in the deck, shuffle the cards and then withdraw another card. Repeat this 20, 40, 60, 80 and 100 times to complete the following table.

Number of withdrawals	20	40	60	80	100
Number of hearts withdrawn					
Relative frequency (number of hearts / number of withdrawals)					

How many hearts do you expect to get in each case?
Did the relative frequency get closer to 0.25 as you increased the number of withdrawals?

7 A box contains 50 matches, of which 40 are 'live' (ready to be struck) and the remainder 'dead' (used). A match is selected at random. Indicate whether each statement below is correct or incorrect.
   (a) $P(\text{live}) = \frac{4}{5}$
   (b) $P(\text{dead}) = \frac{4}{5}$
   (c) $P(\text{live}) = \frac{1}{5}$
   (d) $P(\text{dead}) = \frac{1}{5}$

8 A carton contains a dozen eggs, of which three are brown and the rest are white. An egg is chosen at random from the carton. What is the probability that it is:

(a) brown (b) white?

9 A bag contains 2 white marbles, 2 black marbles and 1 red marble. A marble is selected from the bag. What is the probability that it is:

(a) red (b) black (c) not white
(d) red or black (e) not black, nor white (f) red or white?

10 A set of 20 cards is numbered 1, 2, 3, ... 20. A card is drawn at random from the set. What is the probability that the number on the card is divisible by:

(a) 3 (b) 5 (c) 3 or 5 or both (d) 3 and 5?

11 A standard six-sided die is rolled on a flat surface. What is the probability that the number appearing is:

(a) greater than 2 (b) greater than 3 but less than 5
(c) odd (d) odd and divisible by 3?

12 A set of 10 cards is numbered 3, 4, 5, ... 12. A card is drawn at random. What is the probability that the number on the card is:

(a) even (b) odd (c) odd or even (d) greater than 7 (e) divisible by 3
(f) even and divisible by 3 (g) divisible by 5 (h) odd or divisible by 5, but not both?

13 A letter is chosen at random from the letters of the word SUNDAY. What is the probability that it is:

(a) a vowel (b) a consonant (c) D or A?

14 A card is drawn at random from a standard deck of 52 playing cards. What is the probability that it is:

(a) a heart (b) a king (c) a heart or a king
(d) a heart **and** a king (e) the queen of diamonds?

15 A number is chosen at random from the numbers 7, 8, 9, ... 15. What is the probability that the number is:

(a) odd (b) odd and divisible by 5 (c) odd but not divisible by 5
(d) a multiple of 2 (e) a multiple of 3 (f) a multiple of both 2 and 3
(g) a multiple of 2 or 3 (h) a multiple of 2 or 3, but not both?

16 A bag contains 5 blue marbles, 3 red marbles and 2 green marbles. One marble is drawn at random from the bag. What is the probability that it is:

(a) blue (b) red (c) not green (d) blue or green (e) not red, nor blue?

17 A box contains marbles of the same size but different colours: red, white and blue. If a marble is drawn at random, the probability that it is red is the same as the probability that it is white and twice the probability that it is blue.

(a) What is the smallest number of marbles that the box could have?
(b) If a marble is chosen at random, what is the probability that it is:
(i) red (ii) white (iii) red or white (iv) not blue?

18 A bag contains 10 balls, of which 6 are red, 3 are yellow and 1 is white. A ball is drawn at random. What is the probability that it is:

(a) red (b) not yellow (c) white or yellow (d) neither red nor white (e) not red?

19 A bag contains 12 discs, of which 7 are black, 3 are white and 2 are red. A black disc is withdrawn. What is the probability that the next disc withdrawn is:

(a) black (b) not white (c) neither black nor white?

20 A survey of a certain district showed that 6% of families there have 1 child, 38% have 2 children, 42% have 3 children, and 10% have more than 3 children. A family from the district is selected at random. What is the probability that the family will have:

(a) some children (b) no children
(c) at least 2 children (d) not more than 2 children?

21 The diagram shows a circular disc divided into four sections A, B, C and D. A pointer is free to spin, pivoted at the centre O. If the pointer is spun, find the probability that the end of the pointer stops inside:

(a) A **(b)** B **(c)** A or B

(d) not C **(e)** B, C or D

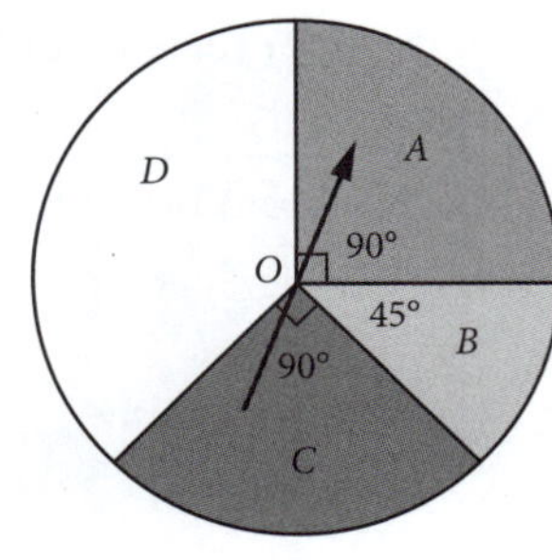

22 A horse is grazing inside an enclosed rectangular paddock 50 m by 20 m and is free to move anywhere inside the paddock. Assuming that the horse's position is random, what is the probability that at any given time the horse is:

(a) more than 5 m from the fence

(b) less than 5 m from the fence

(c) not more than 5 m from a corner?

D 50 m C
20 m
A B

23 A French roulette wheel has 37 compartments around its rim. One of these is numbered 0 and is coloured green (shown as grey at right). The others are numbered 1 to 36, with half of these 36 compartments coloured red (shown as white at right) and the other half coloured black. The wheel is spun in one direction and a small ball is rolled in the opposite direction. The chances of the ball falling into any of the 37 compartments are equally likely. Find the probability that the ball will land on:

(a) black **(b)** an odd number

(c) a multiple of 3 **(d)** any number from 1 to 12

(e) 15

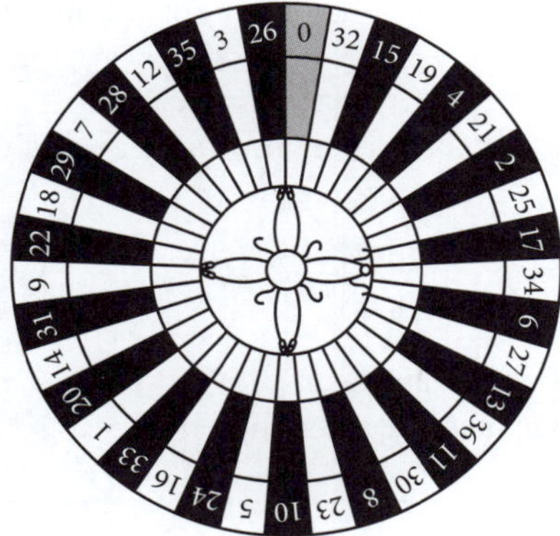

9.2 VENN DIAGRAMS

Venn diagrams are a visual way to represent sets of data. You can name the sets with letters and talk about various properties of the sets. Venn diagrams are an efficient tool to be used when analysing data to find the probability of events.

If $S = \{1, 2, 3, 4, \ldots 25\}$ (i.e. whole numbers from 1 to 25), $D = \{1, 3, 5, \ldots 25\}$ (i.e. odd numbers from 1 to 25), $F = \{3, 6, 9, \ldots 24\}$ (i.e. multiples of 3 between 1 and 25), then $n(S) = 25$, $n(D) = 13$ and $n(F) = 8$, where the notation $n(S)$ means the number of members of set S. This information can be shown on a Venn diagram as follows:

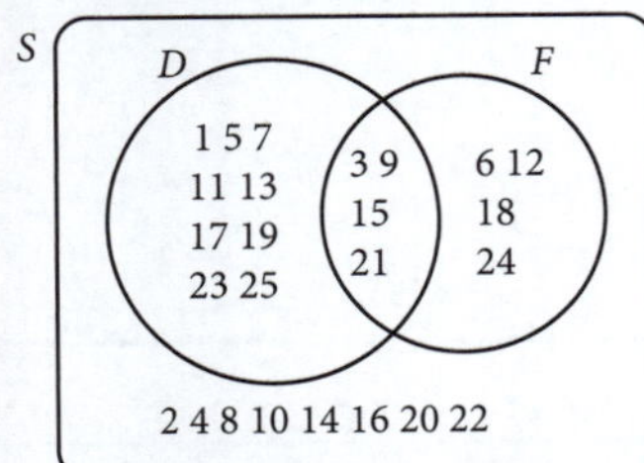

The outer group contains all the members of set S; the inner group on the left contains all the members of set D; the inner group on the right contains all the members of set F. The area where the inner groups overlap contains the members that are in both sets D and F. The region outside the groups of D and F contain all the members of set S that are in neither set D nor set F.

To enter information onto a Venn diagram like this it is best to first find the members common to both sets D and F, then put the rest of the data into D and F. The remaining members of S that were in neither D nor F can then be put into the region outside D and F.

The set made by combining sets D and F is called the **union** of the two sets, written $D \cup F$. It consists of all the elements that are in D or in F or in both D and F:

$D \cup F = \{1, 3, 5, 6, 7, 9, 11, 12, 13, 15, 17, 18, 19, 21, 23, 24, 25\}$

The set made up of the elements common to sets D and F is called the **intersection** of the two sets, written $D \cap F$:

$D \cap F = \{3, 9, 15, 21\}$

The intersection contains numbers that belong to both sets, i.e. the odd multiples of 3 between 1 and 25.

Now $n(D) = 13$, $n(F) = 8$, $n(D \cup F) = 17$, $n(D \cap F) = 4$. Hence:

$$n(D \cup F) = n(D) + n(F) - n(D \cap F)$$

Because sets D and F contain some common elements, the two sets are not mutually exclusive. Also, sets D and F are each **subsets** of set S, because all their elements are in set S.

Note that sets do not always overlap. If $A = \{\text{letters of the alphabet}\}$, $V = \{\text{vowels}\}$ and $C = \{\text{consonants}\}$, this information can be shown on a Venn diagram:

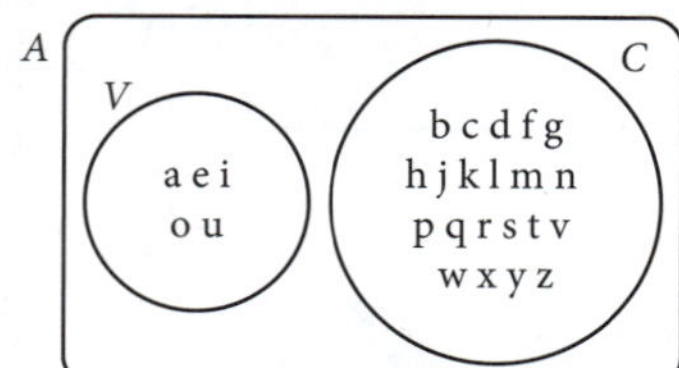

Sets V and C have no members in common, so they do not overlap. Hence $V \cap C = \{\}$, where $\{\}$ is called the empty set.

All members of set A are in either set V or set C. Because of this, sets V and C are called **mutually exclusive** or **disjoint**.

Now $n(V) = 5$, $n(C) = 21$, $n(V \cup C) = 26$, $n(V \cap C) = 0$.

Hence, for disjoint or mutually exclusive sets:

$$n(V \cup C) = n(V) + n(C)$$

Sets V and C are each subsets of set A, because all their elements are in set A. They are also called **complementary** sets, because together they make up set A with no overlapping.

Example 4

A is the set of possible sums from the numbers showing when two dice are rolled. B is the set of odd sums and C is the set of sums that are more than 7. Show this information on a Venn diagram and find:

(a) $B \cup C$ (b) $B \cap C$

Solution

List the elements in each set:

$A = \{2, 3, 4, 5, 6, 7, 8, 9, 10, 11, 12\}$,

$B = \{3, 5, 7, 9, 11\}$, $C = \{8, 9, 10, 11, 12\}$

Elements common to B and C: $\{9, 11\}$

(a) $B \cup C = \{3, 5, 7, 8, 9, 10, 11, 12\}$

(b) $B \cap C = \{9, 11\}$

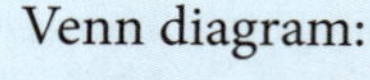

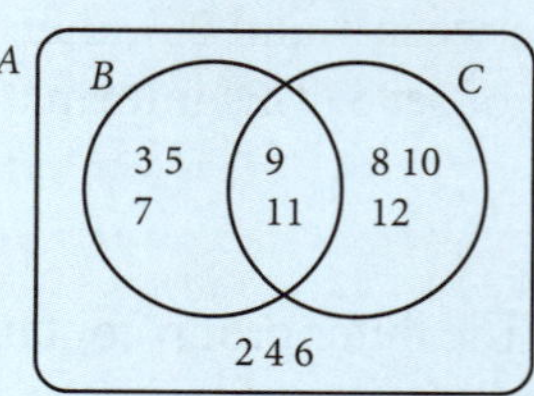

Example 5

A die is rolled on a flat surface. A is the set of even results, B is the set of odd results and C is the set of results less than 4. Show this information on a Venn diagram and find:

(a) $A \cup C$ (b) $C \cup B$ (c) $A \cup B \cup C$ (d) $A \cap B$ (e) Which two sets are mutually exclusive?

Solution

List the sets: $A = \{2, 4, 6\}$, $B = \{1, 3, 5\}$, $C = \{1, 2, 3\}$

There are 3 intersecting sets, so 3 groups needed.

Common elements between sets: 2 is common to A and C

1, 3 are common to B and C

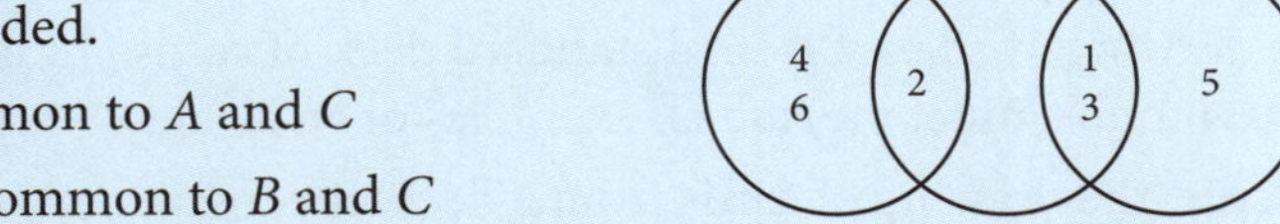

(a) $A \cup C = \{1, 2, 3, 4, 6\}$ **(b)** $C \cup B = \{1, 2, 3, 5\}$ **(c)** $A \cup B \cup C = \{1, 2, 3, 4, 5, 6\}$

(d) $A \cap B = \{\}$ This set is called the empty set as it does not contain any elements.

(e) Since the intersection of sets A and B is the empty set then these two sets are mutually exclusive.

EXERCISE 9.2 VENN DIAGRAMS

1 For the Venn diagram shown, the set S is given by:

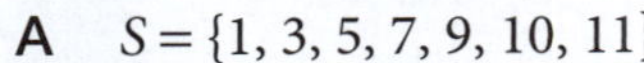

A $S = \{1, 3, 5, 7, 9, 10, 11\}$ **B** $S = \{1, 2, 3, 4, 5, 6, 7, 8, 9, 10, 11\}$
C $S = \{2, 4, 6, 8, 9, 10, 11\}$ **D** $S = \{9, 10, 11\}$

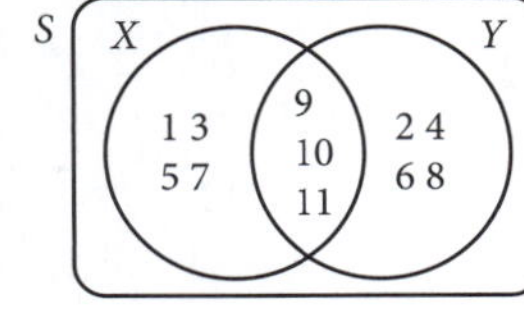

To show the information given in questions **2** to **5**, draw a Venn diagram for each.

2 S is the set of possible sums from the numbers showing when two dice are rolled, D is the set of even sums, F is the set of sums less than 6. Find: **(a)** $D \cup F$ **(b)** $D \cap F$

3 A coin is tossed and a die is rolled. S is the set of all possible results obtained, H is the set of all results containing heads, G is the set of all results containing an odd number. Find:
(a) $H \cup G$ **(b)** $n(H \cap G)$

4 S is the set of all possible outcomes on a French roulette wheel, B is the set of black numbers, R is the set of red numbers. (For a description of a French roulette wheel, see Exercise 9.1 question **23**.) Find:
(a) $B \cup R$ **(b)** $B \cap R$ **(c)** $n(B \cup R)$ **(d)** $n(B \cap R)$ **(e)** $n(S)$

5 $S = \{1, 2, 3, \ldots 25\}$, $E = \{\text{even numbers in set } S\}$, $P = \{\text{perfect squares in set } S\}$. Find:
(a) $n(S)$ **(b)** $n(E)$ **(c)** $n(P)$ **(d)** $n(E \cap P)$ **(e)** $n(E \cup P)$

6 In a group of 25 students, 18 study French, 12 study German and 5 study neither French nor German. Find how many students:
(a) study only French **(b)** study only German **(c)** study only one language.

7 $S = \{1, 2, 3, \ldots 20\}$, $K = \{\text{multiples of 3 in set } S\}$, $L = \{\text{multiples of 6 in set } S\}$. Find:
(a) $K \cup L$ **(b)** $K \cap L$ **(c)** $n(K \cap L)$

8 A die is rolled on a flat surface. A is the set of even results, B is the set of odd results, C is the set of results more than 4. Find: **(a)** $A \cap B$ **(b)** $A \cup B$ **(c)** $B \cap C$ **(d)** $n(A \cap C)$

9 Two dice are rolled. Of the results, $E = \{\text{even sums}\}$, $F = \{\text{odd sums}\}$, $G = \{\text{sums where one die shows a 6}\}$. Find: **(a)** $n(E \cap F)$ **(b)** $n(F \cap G)$ **(c)** $n(E \cup F)$ **(d)** $n(E \cup G)$

10 For the Venn diagram shown, indicate whether each statement is correct or incorrect.
(a) $S = \{1, 2, 3, \ldots 12\}$ **(b)** $X \cup Y = \{1, 2, 3, 4, 5, 6, 7, 8, 9, 11, 12\}$
(c) $X \cap Y = \{1, 11, 12\}$ **(d)** $S = \{1, 2, 3, \ldots 14\}$

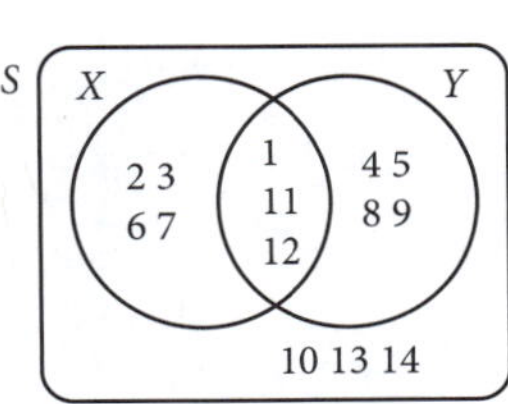

9.3 FINITE SAMPLE SPACES

A set of favourable outcomes can be considered an **event**. The rolling of even numbers with a die (3 favourable outcomes), the drawing of a heart from a standard deck of cards (13 favourable outcomes) or the tossing of heads with a coin (1 favourable outcome) are all examples of an event.

The rolling of a die, the drawing of cards from a deck, the tossing of a coin—that is, activities that will produce various events—are called **experiments** or **trials**.

In modern probability theory, all possible outcomes of an experiment are considered as points in a space, called a **sample space** S. If S contains a finite number of points n and if the outcomes of an experiment are equally likely, then you can assign each point (called a **sample point**) a probability of $\frac{1}{n}$. The sum of the probabilities of all the sample points is therefore 1.

If S is the sample space and A is a set of points drawn from the sample space (i.e. A is an event), then:

$$P(A) = \frac{\text{number of sample points in } A}{\text{number of sample points in } S} = \frac{n(A)}{n(S)}$$
$$= \frac{\text{number of outcomes corresponding to } A}{\text{total number of possible outcomes}}$$

If $S = \{a, b, c, \ldots x, y, z\}$ then $n(S) = 26$.

Let $A = \{a, e, i, o, u\}$ so $n(A) = 5$: $P(A) = \frac{n(A)}{n(S)} = \frac{5}{26}$

Thus the probability of selecting a vowel from S is $\frac{5}{26}$.

Mutually exclusive events

If two or more events cannot occur simultaneously then the events are said to be **mutually exclusive** or **disjoint**. In the language of the sample space, the events have no points in common.

If a coin is tossed, either heads or tails may turn up, but both events cannot occur at the same time. If a card is drawn at random from a deck, it may be a heart or the ace of spades or some other card, but if the card is a heart then this excludes the possibility of it being the ace of spades. The two events 'heart' and 'ace of spades' are mutually exclusive. However, the events 'heart' and 'ace' are not mutually exclusive. Why?

Example 6

If one card is drawn at random from a standard deck of 52 playing cards, what is the probability that it is a heart or the ace of spades?

Solution

These two events are mutually exclusive, because they can't happen together.

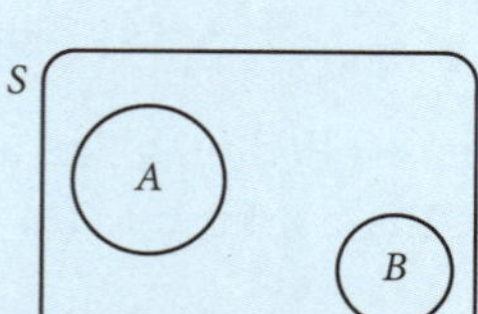

$S = \{\text{playing cards}\}$ $\quad n(S) = 52$

$A = \{\text{heart}\}$ $\quad n(A) = 13$

$B = \{\text{ace of spades}\}$ $\quad n(B) = 1$

$A \cup B = A \text{ or } B = \{\text{heart or ace of spades}\}$

$n(A \text{ or } B) = 14$

$P(A \text{ or } B) = \frac{n(A \text{ or } B)}{n(S)} = \frac{14}{52} = \frac{7}{26}$

A and B are disjoint sets, so they don't intersect.

Now $P(A) = \frac{13}{52}$, $P(B) = \frac{1}{52}$ and so $P(A) + P(B) = \frac{13}{52} + \frac{1}{52} = \frac{14}{52} = \frac{7}{26}$

Because the sets A and B have no points in common and thus are mutually exclusive, then:

$P(A \text{ or } B) = P(A) + P(B)$

Example 7

From a set of 15 cards numbered 1 to 15, one card is drawn at random. What is the probability that it is a multiple of 3 or 5 or both?

Solution

$S = \{1, 2, 3, \ldots 15\}$ $\quad n(S) = 15$

$A = \{3, 6, 9, 12, 15\}$ $\quad n(A) = 5$

$B = \{5, 10, 15\}$ $\quad n(B) = 3$

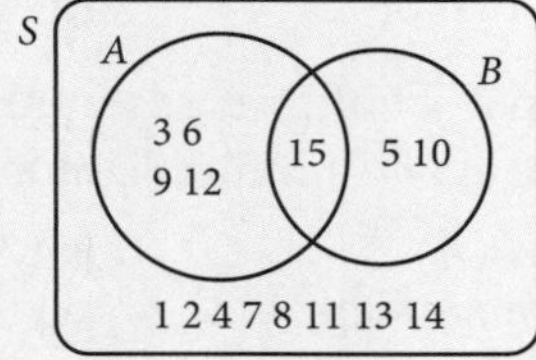

One card, 15, is common to both sets A and B:

$A \cap B = A \text{ and } B = \{15\}$ $\quad n(A \text{ and } B) = 1$

$A \cup B = A \text{ or } B = \{3, 5, 6, 9, 10, 12, 15\}$ $\quad n(A \text{ or } B) = 7$

$\therefore P(A \text{ or } B) = \dfrac{n(A \text{ or } B)}{n(S)} = \dfrac{7}{15}$

The notation 'A and B' means the elements common to both sets (in Example **7** above, the multiples of 3 and 5). This is also known as the intersection of the two sets.

The notation 'A or B' means the elements in either set (in the example above, the multiples of 3 or 5 or both). This is also known as the union of the two sets.

You have $n(A \text{ or } B) = n(A) + n(B) - n(A \text{ and } B)$, because the sets are not mutually exclusive. Thus:

$P(A \text{ or } B) = P(A) + P(B) - P(A \text{ and } B)$

This formula also works for mutually exclusive sets, where $P(A \text{ and } B) = 0$ and so the result becomes:
$P(A \text{ or } B) = P(A) + P(B)$

Example 8

A card is drawn at random from a standard deck of 52 playing cards. R is the event 'drawing a red card,' H is the event 'drawing the ace of hearts,' A is the event 'drawing an ace.' Find:

(a) $\overline{R}$ **(b)** $\overline{A}$ **(c)** $P(R)$ **(d)** $P(\overline{R})$
(e) $P(R \text{ or } A)$ **(f)** $P(R \text{ or } H)$ **(g)** $P(R \text{ and } A)$ **(h)** $P(\overline{R} \text{ or } H)$

Solution

(a) $\overline{R} = \{\text{not red}\} = \{\text{black}\}$ **(b)** $\overline{A} = \{\text{not an ace}\}$ **(c)** $n(R) = 26, \quad P(R) = \frac{26}{52} = \frac{1}{2}$

(d) $n(\overline{R}) = 26, \quad P(\overline{R}) = \frac{26}{52} = \frac{1}{2}$ **OR** $P(\overline{R}) = 1 - P(R) = 1 - \frac{1}{2} = \frac{1}{2}$

(e) $n(R \cup A) = 26 + 4 - 2 = 28, \quad P(R \text{ or } A) = \dfrac{n(R \cup A)}{52} = \dfrac{28}{52} = \dfrac{7}{13}$

(f) $n(R \cup H) = n(R) = 26$ because the ace of hearts is a red card.

$P(R \text{ or } H) = \dfrac{n(R \cup H)}{52} = \dfrac{26}{52} = \dfrac{1}{2}$

(g) $n(R \cap A) = 2$ because there are 2 red aces.

$P(R \text{ and } A) = \dfrac{n(R \cap A)}{52} = \dfrac{2}{52} = \dfrac{1}{26}$

(h) $\overline{R} \cup H = \{\text{black or ace of hearts}\}, \ P(\overline{R} \text{ or } H) = \dfrac{n(\overline{R} \cup H)}{52} = \dfrac{27}{52}$

EXERCISE 9.3 FINITE SAMPLE SPACES

1 A card is drawn at random from a standard deck of 52 playing cards. A is the event 'drawing a heart', B the event 'drawing the ace of clubs', C is the event 'drawing an ace'. Find the following:

(a) $P(A)$ (b) $P(B)$ (c) $P(C)$ (d) $P(A \text{ or } B)$ (e) $P(A \text{ or } C)$
(f) $P(B \text{ or } C)$ (g) Which pairs of events A and B, A and C, or B and C are mutually exclusive?

2 From a set of 17 cards numbered 1, 2, 3, … 17, one card is selected at random. A is the event 'a multiple of 3', B is the event 'a multiple of 8', C is the event 'a multiple of 5'. Find the following:

(a) $P(A)$ (b) $P(B)$ (c) $P(C)$ (d) $P(\bar{A})$ (e) $P(\bar{B})$
(f) $P(A \text{ or } B)$ (g) $P(A \text{ or } C)$ (h) $P(\bar{A} \text{ or } B)$ (i) $P(A \text{ or } \bar{C})$ (j) $P(\bar{B} \text{ or } \bar{C})$
(k) Which pairs of the events A, B and C are mutually exclusive?

3 A standard die is rolled on a flat surface. A is the event 'an even number', B is the event 'an odd number', C is the event 'a multiple of 5'. Which of the following statements is true?

A $P(A \text{ or } B) = 0.5$ **B** $P(\bar{A} \text{ or } B) = 1$ **C** $P(A \text{ or } C) = 0.5$ **D** $P(\bar{B} \text{ or } \bar{C}) = \frac{5}{6}$

4 A container holds 8 red marbles, 7 white marbles and 5 black marbles. One marble is drawn at random from the container. Indicate whether each statement below is correct or incorrect.

(a) $P(\text{red or black}) = \frac{3}{4}$ (b) $P(\text{not white}) = \frac{1}{4}$
(c) $P(\text{neither black nor white}) = \frac{2}{5}$ (d) $P(\text{not red}) = \frac{3}{5}$

5 A coin is tossed three times. A is the event 'at least two heads', B is the event 'head, tail, head', C is the event 'not more than one head'. Find:

(a) $P(A)$ (b) $P(B)$ (c) $P(C)$ (d) $P(A \text{ or } B)$ (e) $P(A \text{ or } C)$ (f) $P(B \text{ or } C)$
(g) $P(A \text{ or } B \text{ or } C)$ (h) Which pairs of the events A, B and C are mutually exclusive?

6 An integer is chosen at random from the first 50 positive integers. A is the event 'divisible by 2', B is the event 'divisible by 3', C is the event 'divisible by 5'. Find:

(a) $P(A \text{ and } B)$ (b) $P(A \text{ and } C)$ (c) $P(B \text{ and } C)$ (d) $P(A \text{ and } B \text{ and } C)$
(e) $P(A \text{ or } B)$ (f) $P(A \text{ or } C)$ (g) $P(B \text{ or } C)$ (h) $P(A \text{ or } B \text{ or } C)$

7 A die is rolled twice and the numbers rolled are added together. A is the event 'the sum is 4 or more', B is the event 'the sum is less than 6', C is the event 'the two numbers rolled are the same'. Find:

(a) $P(A \text{ and } B)$ (b) $P(A \text{ and } C)$ (c) $P(B \text{ and } C)$ (d) $P(A \text{ and } B \text{ and } C)$
(e) $P(A \text{ or } B)$ (f) $P(A \text{ or } C)$ (g) $P(B \text{ or } C)$ (h) $P(A \text{ or } B \text{ or } C)$

8 A die is rolled twice and the numbers rolled are added together. A is the event 'the sum is greater than 5', B is the event 'the sum is less than 8'. Find:

(a) $P(A)$ (b) $P(B)$ (c) $P(A \text{ and } B)$ (d) $P(A \text{ or } B)$.
(e) Hence show that $P(A \text{ or } B) = P(A) + P(B) - P(A \text{ and } B)$. Are A and B mutually exclusive?

9 If $P(A) + P(B) = 1$, does it follow that one of these events must occur?

10 $P(A \text{ or } B) = \frac{15}{16}$, $P(A \text{ and } B) = \frac{1}{8}$, $P(A) = \frac{3}{8}$. Find $P(B)$.

11 In a group of 50 students, 30 study Mathematics, 25 study Physics and 20 study both Mathematics and Physics. One student is selected at random from the group. What is the probability that this student studies:

(a) Mathematics but not Physics (b) Physics but not Mathematics
(c) neither Physics nor Mathematics?

12 From a group of 100 students, 50 study History, 30 study English and 20 study both. If a student is selected at random from the group, what is the probability that the student studies:

(a) at least one of these subjects (b) History but not English
(c) History, given that the student also studies English?

9.4 SUCCESSIVE OUTCOMES

It is important to be able to organise information so that probabilities can be calculated easily, especially when there are multiple events occurring successively. A **tree diagram** is one of the most common ways to do this.

Example 9

A coin is tossed three times. Use a tree diagram to show the possible outcomes.

Solution

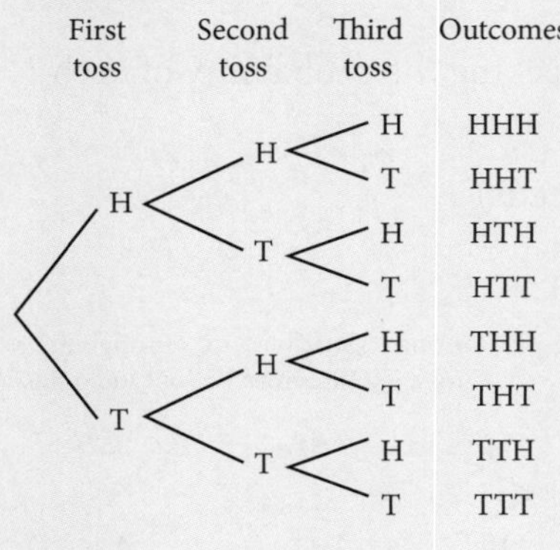

Each time a coin is tossed, it can land as either a head (H) or a tail (T).

There are 8 possible outcomes ($2 \times 2 \times 2 = 2^3$) for this experiment.

How you describe the outcomes depends on whether or not the order of the events is important.

If the order is important, then the outcomes HHT, HTH and THH are three different outcomes. If order is not important, then these three results represent the same single outcome: 2 heads and 1 tail.

When order is not important, there are only 4 possible outcomes for this experiment: 3 heads, 2 heads and 1 tail, 1 head and 2 tails, or 3 tails.

As well as listing outcomes, a tree diagram can be used to show the probability of each outcome in a repeated experiment. The fundamental counting principle (or 'multiplication principle'—see below) means that you can multiply the successive probabilities along the branches to obtain the probability of a final outcome.

Example 10

At a school sports day Joely runs in the 100 m, 200 m and 400 m races. In terms of win (W) and loss (L), there are two possible outcomes for each race.

(a) How many possibilities are there for the three races?

(b) What are the probabilities of Joely winning 3, 2, 1 or no races?

Solution

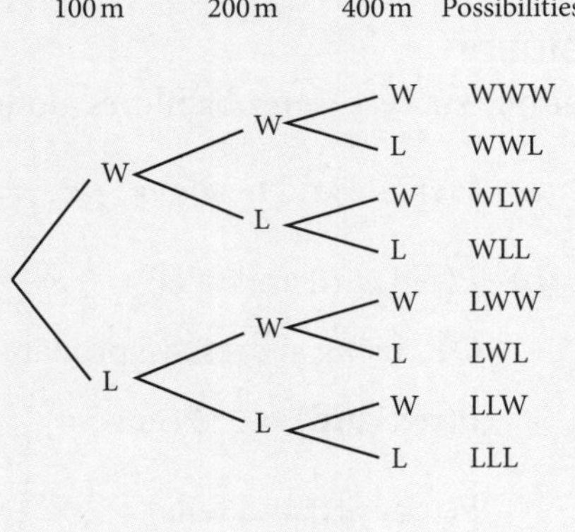

(a) There are 8 possible outcomes ($2 \times 2 \times 2$). WWW represents a win in each of the three races; WWL, for example, represents a win in 100 m and 200 m and a loss in the 400 m.

(b) The question does not say that Joely has an equally likely chance of winning each race, so the question cannot be answered due to lack of information.

However, you can find the probabilities if you assume that she has a 50-50 chance of winning each race, in which case:

$P(\text{wins 3 races}) = \frac{1}{8}$ $\quad P(\text{wins 2 races}) = \frac{3}{8}$

$P(\text{wins 1 race}) = \frac{3}{8}$ $\quad P(\text{wins 0 races}) = \frac{1}{8}$

Multiplication principle

Tree diagrams can be used to calculate probabilities. To obtain the probability of a final outcome, you can multiply the probabilities of the successive intermediate outcomes along the branches. For example:

> Number of ways a two-part outcome can occur = $m \times n$
> (where m = number of ways the first part can occur,
> n = number of ways the second part can occur)

Example 11

A coin is tossed twice. Draw a tree diagram to show the possible outcomes, marking the probability of each outcome on each branch. Use your tree diagram to find:

(a) $P(2 \text{ heads})$ **(b)** $P(1 \text{ head})$ **(c)** $P(0 \text{ heads})$

Solution

$P(\text{H}) = P(\text{T}) = 0.5$

First toss	Second toss	Possible Outcomes	Probability of Outcomes
0.5 H	0.5 H	HH	0.25
	0.5 T	HT	0.25
0.5 T	0.5 H	TH	0.25
	0.5 T	TT	0.25

Find each final probability by multiplying the successive probabilities along each branch of the tree diagram.

(a) $P(2 \text{ heads}) = P(\text{H}) \times P(\text{H}) = 0.5 \times 0.5 = 0.25$

(b) There are 2 branches that have the outcome of 1 head, so their probabilities must be added together.

$$P(1 \text{ head}) = P(\text{H}) \times P(\text{T}) + P(\text{T}) \times P(\text{H})$$
$$= 0.5 \times 0.5 + 0.5 \times 0.5$$
$$= 0.5$$

(c) $P(0 \text{ heads}) = P(\text{T}) \times P(\text{T}) = 0.5 \times 0.5 = 0.25$

Example 12

A card is drawn from a standard deck of 52 playing cards and a coin is tossed. In this example, you are only concerned with the suit of the card (i.e. whether hearts, diamonds, clubs or spades). Draw a tree diagram to show the possible outcomes, marking the probability of each outcome on each branch. Use your tree diagram to find:

(a) $P(\text{spade and heads})$ **(b)** $P(\text{red card and tails})$ **(c)** $P(\text{not a club and heads})$

Solution

Multiply successive probabilities along the branches.

(a) $P(\text{spade and heads}) = \frac{1}{4} \times \frac{1}{2} = \frac{1}{8}$

(b) $P(\text{red card and tails}) = \frac{1}{4} \times \frac{1}{2} + \frac{1}{4} \times \frac{1}{2} = \frac{1}{4}$

OR, using successive probabilities:

$P(\text{red card}) = \frac{1}{2}$, $P(\text{tails}) = \frac{1}{2}$

$P(\text{red card and tails}) = \frac{1}{2} \times \frac{1}{2} = \frac{1}{4}$

(c) $P(\text{not a club and heads}) = 3 \times \frac{1}{4} \times \frac{1}{2} = \frac{3}{8}$

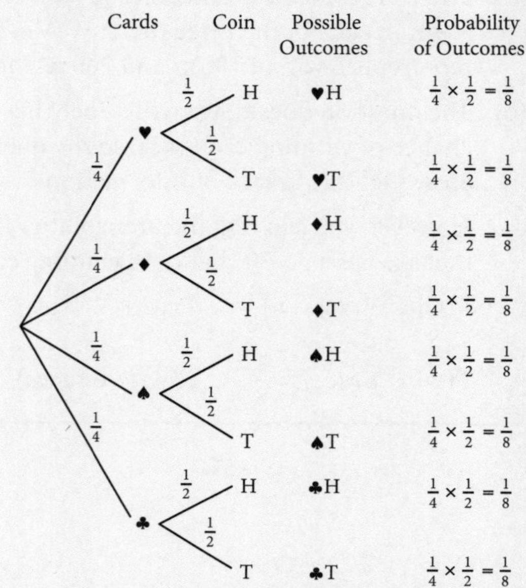

EXERCISE 9.4 SUCCESSIVE OUTCOMES

1 A coin is tossed and a die is rolled. Draw a tree diagram to show the outcomes, marking the probability of the outcome on each branch. If each outcome is equally likely, find the probability of:

(a) a head and an even number **(b)** a tail and a number greater than 4.

2 Two cubes each have two faces painted red, another two faces painted white and the remaining two faces painted blue. Both cubes are rolled on a table like dice. Draw a tree diagram to show the possible outcomes of the colours that are rolled. If each outcome is equally likely, what is the probability that:

(a) both colours rolled are the same **(b)** one colour rolled is red and the other is white?

3 A committee consisting of one boy and one girl is to be selected from three girls (Anna, Kate and Sue-Lin) and four boys (Brent, Jimmy, Mustafa and Rajiv). What is the probability that the committee consists of Kate and Rajiv?

4 A restaurant has a set menu that offers a choice of three different entrées, three different main courses and three different desserts. The entrées are vegetable soup, spring rolls and calamari; the main courses are roast beef, roast eggplant and spaghetti bolognese; the desserts are cheesecake, ice-cream and fruit salad. Draw a tree diagram to show all the menu choices that include one item from each course.

(a) If you must choose one item from each course, how many different meal choices are there?
If each choice is equally likely, find the probability of selecting:
(b) vegetable soup, roast beef, and either cheesecake or ice-cream
(c) spring rolls, not roast beef, and fruit salad.

5 A die is rolled twice and the numbers rolled are added together. Which of the following events has a probability of $\frac{1}{18}$?

A both numbers rolled are the same **B** the sum is 5 **C** the sum is 3 **D** the sum is 12

6 A coin is tossed three times. Illustrate on a tree diagram the eight possible outcomes. Use the tree diagram to find the probability of tossing:

(a) 3 tails **(b)** 1 head **(c)** 2 heads **(d)** 3 heads

7 Three students, Kristian, Soung-ho and Amélie, work independently on a mathematics problem. Each student has a 50-50 chance of success. Draw a tree diagram and use it to indicate whether each of the following statements is correct or incorrect.

(a) $P(\text{all successful}) = \frac{1}{2}$ **(b)** $P(\text{only Kristian and Amélie successful}) = \frac{1}{8}$

(c) $P(\text{none successful}) = \frac{3}{8}$ **(d)** $P(\text{at least one is successful}) = \frac{7}{8}$

8 One cube has 4 red faces and 2 white faces, another has 3 red and 3 white faces, and a third cube has 2 red and 4 white faces. The three cubes are rolled like dice.

(a) Draw a tree diagram to show all possible outcomes and the associated probabilities.
What is the probability that:
(b) 3 red faces are rolled **(c)** only one cube rolls a white face
(d) more red faces than white faces are rolled?

9 A 50-cent coin, a 20-cent coin and a 10-cent coin are tossed. What is the probability of:

(a) 3 heads **(b)** a head only with the 50-cent coin **(c)** at least 2 tails?

10 A card is drawn at random from a standard deck of 52 playing cards. It is replaced, the deck is shuffled and a card is drawn again. What is the probability that:

(a) both cards are spades **(b)** neither card is a heart **(c)** both cards are the same suit?

9.5 INDEPENDENT EVENTS

Two events are **independent** if the outcome of each event is not affected by the outcome of the other event. This means that the probability of both events together is the same as the probability of both events separately, and so their probabilities can be multiplied together as if they are separate successive outcomes (as shown in the previous section).

In other words:

If events A and B are independent, then:	$P(A \text{ and } B) = P(A) \times P(B)$
Or, writing $P(A \text{ and } B)$ as the 'product' of A and B:	$P(AB) = P(A) \times P(B)$

Many of the questions in Exercise 9.4 involved independent events.

Example 13

If two dice are rolled, what is the probability of rolling an even number with the first die and a 3 or a 5 with the second die?

Solution

The sample space consists of $6 \times 6 = 36$ points, as listed below. Each sample point is represented for convenience below as (**a**, b), where **a** is the number on the first die and b is the number on the second die:

(1, 1)	(2, 1)	(3, 1)	(4, 1)	(5, 1)	(6, 1)
(1, 2)	(2, 2)	(3, 2)	(4, 2)	(5, 2)	(6, 2)
(1, 3)	(2, 3)	(3, 3)	(4, 3)	(5, 3)	(6, 3)
(1, 4)	(2, 4)	(3, 4)	(4, 4)	(5, 4)	(6, 4)
(1, 5)	(2, 5)	(3, 5)	(4, 5)	(5, 5)	(6, 5)
(1, 6)	(2, 6)	(3, 6)	(4, 6)	(5, 6)	(6, 6)

There are 18 points that correspond to event A, an even number with the first die and any number with the second die. There are 12 points that correspond to event B, any number with the first die and a 3 or a 5 with the second die.

Thus: $P(A) = \frac{18}{36} = \frac{1}{2}$ and $P(B) = \frac{12}{36} = \frac{1}{3}$.

The intersection of A and B is the set of 6 points above that are shaded.

$A \cap B = \{(2, 3), (2, 5), (4, 3), (4, 5), (6, 3), (6, 5)\}$

Hence: $P(A \text{ and } B) = \frac{6}{36} = \frac{1}{6}$

And: $P(A) \times P(B) = \frac{1}{2} \times \frac{1}{3} = \frac{1}{6} = P(A \text{ and } B)$

This result conforms to our definition of independent events, as well as to the everyday meaning of the word 'independent'. Clearly, whichever number the first die happens to roll will have absolutely no influence on the second die's roll.

Alternatively:

This example could also be solved by a careful observation of the 36 sample points. Observe that $\frac{1}{2}$ of the outcomes contain an even number with the first die; then note that $\frac{1}{3}$ of this $\frac{1}{2}$ (that is, $\frac{1}{6}$) contain an even number with the first die **and** a 3 or a 5 with the second die.

We can think of rolling two dice as a two-stage process, which can be represented by a tree diagram with two sets of branches. For convenience, we label an even number on the first die E, an odd number on the first die O, a 3 or a 5 with the second die a success S, and 'not a 3 or a 5' a failure F.

1st die	2nd die	Possible Outcomes	Probability
E ($\frac{1}{2}$)	S ($\frac{1}{3}$)	ES	$\frac{1}{2} \times \frac{1}{3} = \frac{1}{6}$
	F ($\frac{2}{3}$)	EF	$\frac{1}{2} \times \frac{2}{3} = \frac{2}{6}$
O ($\frac{1}{2}$)	S ($\frac{1}{3}$)	OS	$\frac{1}{2} \times \frac{1}{3} = \frac{1}{6}$
	F ($\frac{2}{3}$)	OF	$\frac{1}{2} \times \frac{2}{3} = \frac{2}{6}$

The tree diagram shows the four possible outcomes, but they are not all equally likely. An even number E with the first die and a 3 or a 5 with the second die is given by ES with probability $\frac{1}{2} \times \frac{1}{3} = \frac{1}{6}$.

Example 14

A fair coin is tossed 3 times. What is the probability of:

(a) 2 heads and 1 tail, in any order **(b)** at least 1 head?

Solution

The experiment of tossing a coin 3 times can be represented by a tree diagram with 3 sets of branches. As usual, H = head and T = tail:

First toss	Second toss	Third toss	Possible Outcomes	Probability
H ($\frac{1}{2}$)	H ($\frac{1}{2}$)	H ($\frac{1}{2}$)	HHH	$\frac{1}{2} \times \frac{1}{2} \times \frac{1}{2} = \frac{1}{8}$
		T ($\frac{1}{2}$)	HHT	$\frac{1}{2} \times \frac{1}{2} \times \frac{1}{2} = \frac{1}{8}$
	T ($\frac{1}{2}$)	H ($\frac{1}{2}$)	HTH	$\frac{1}{2} \times \frac{1}{2} \times \frac{1}{2} = \frac{1}{8}$
		T ($\frac{1}{2}$)	HTT	$\frac{1}{2} \times \frac{1}{2} \times \frac{1}{2} = \frac{1}{8}$
T ($\frac{1}{2}$)	H ($\frac{1}{2}$)	H ($\frac{1}{2}$)	THH	$\frac{1}{2} \times \frac{1}{2} \times \frac{1}{2} = \frac{1}{8}$
		T ($\frac{1}{2}$)	THT	$\frac{1}{2} \times \frac{1}{2} \times \frac{1}{2} = \frac{1}{8}$
	T ($\frac{1}{2}$)	H ($\frac{1}{2}$)	TTH	$\frac{1}{2} \times \frac{1}{2} \times \frac{1}{2} = \frac{1}{8}$
		T ($\frac{1}{2}$)	TTT	$\frac{1}{2} \times \frac{1}{2} \times \frac{1}{2} = \frac{1}{8}$

Hence $S = \{(HHH), (HHT), (HTH), (HTT), (THH), (THT), (TTH), (TTT)\}$

$\therefore n(S) = 8$ and each of the 8 outcomes is equally likely.

(a) From observation of the 8 sample points, we can pick out the points corresponding to the event A, '2 heads and 1 tail': $A = \{(HHT), (HTH), (THH)\}$

$$\therefore P(\text{2 heads and 1 tail}) = \frac{1}{8} + \frac{1}{8} + \frac{1}{8} = \frac{3}{8}$$

(b) Each of the 8 outcomes except (TTT) has at least 1 head: $\therefore P(\text{at least 1 head}) = 1 - P(\text{0 heads})$

$$= 1 - \frac{1}{8} = \frac{7}{8}$$

Example 15

Two friends, Suzy and Dimitri, often play golf and tennis with each other. Overall, Suzy tends to win 3 rounds of golf out of every 5 and 1 game of tennis out of every 4. If Suzy and Dimitri play one round of golf and one game of tennis, find the probability that Suzy:

(a) wins both
(b) loses both
(c) wins only the round of golf
(d) wins either the golf or the tennis but not both.

Solution

Assume that no game ends in a draw.

Let A be the event 'Suzy wins golf', so $\bar{A}$ is the event 'Suzy loses golf'. From the overall probabilities:

$$P(A)=\frac{3}{5} \text{ and } P(\bar{A})=\frac{2}{5}$$

Let B be the event 'Suzy wins tennis', so $\bar{B}$ is the event 'Suzy loses tennis'.

$$P(B)=\frac{1}{4} \text{ and } P(\bar{B})=\frac{3}{4}$$

Assuming that A and B are independent events (which seems reasonable), then:

(a) $P(A \text{ and } B) = P(A) \times P(B)$

$$= \frac{3}{5} \times \frac{1}{4}$$
$$= \frac{3}{20}$$

(b) $P(\bar{A} \text{ and } \bar{B}) = P(\bar{A}) \times P(\bar{B})$

$$= \frac{2}{5} \times \frac{3}{4}$$
$$= \frac{3}{10}$$

(c) $P(A \text{ and } \bar{B}) = P(A) \times P(\bar{B})$

$$= \frac{3}{5} \times \frac{3}{4}$$
$$= \frac{9}{20}$$

(d) For this outcome, Suzy either **(i)** wins golf and loses tennis, or **(ii)** loses golf and wins tennis.

(i) $P(A \text{ and } \bar{B}) = P(A) \times P(\bar{B})$

$$= \frac{3}{5} \times \frac{3}{4}$$
$$= \frac{9}{20}$$

(ii) $P(\bar{A} \text{ and } B) = P(\bar{A}) \times P(B)$

$$= \frac{2}{5} \times \frac{1}{4}$$
$$= \frac{2}{20}$$

(i) and **(ii)** are mutually exclusive, so $P\left(\left(A \text{ and } \bar{B}\right) \text{ or } \left(\bar{A} \text{ and } B\right)\right) = \frac{9}{20} + \frac{2}{20} = \frac{11}{20}$.

This problem can also be solved using the tree diagram method. There are four possible outcomes (not all equally likely); if we denote a win by Suzy as W and a loss as L, then the possible outcomes are WW, WL, LW and LL. Their respective probabilities can be illustrated by the following tree diagram.

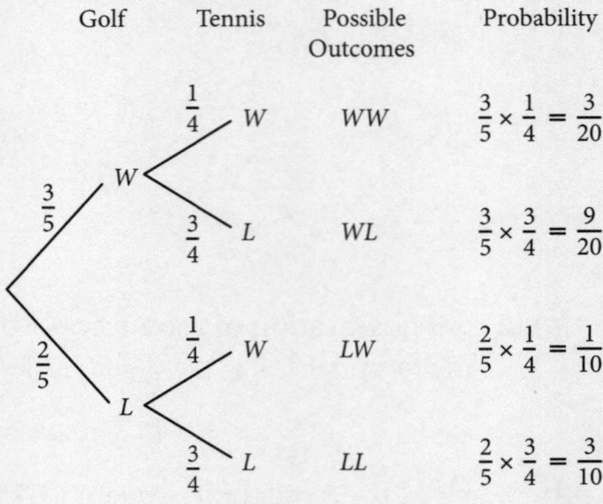

Example 16

A die is rolled, a coin is tossed and a letter is taken from the set {a, b, c}. Find the probability of each event described below.

(a) Event A: roll 3 with the die
(b) Event B: a head with the coin
(c) Event C: 'b' from the set {a, b, c}
(d) A and B and C.

Solution

Events A, B and C are independent.

(a) $P(A) = \frac{1}{6}$
(b) $P(B) = \frac{1}{2}$
(c) $P(C) = \frac{1}{3}$
(d)

$$\begin{aligned} P(A \text{ and } B \text{ and } C) &= P(A) \times P(B) \times P(C) \\ &= \frac{1}{6} \times \frac{1}{2} \times \frac{1}{3} \\ &= \frac{1}{36} \end{aligned}$$

When finding the probability of an event, given that you know another event has occurred, the notation | is used to mean 'given'. Hence P(A|B) means 'the probability of event A, given the situation that event B has occurred'.

It is important to realise that:

> If events A and B are independent, then $P(A \mid B) = P(A)$ and $P(B \mid A) = P(B)$.

EXERCISE 9.5 INDEPENDENT EVENTS

1 You toss a coin and roll a die. What is the probability of:

(a) a head with the coin
(b) a number greater than 4 with the die
(c) a head and a number greater than 4?

2 In a class of 25 boys and 15 girls, 8 boys and 7 girls wear glasses. One student from the class is selected. Indicate whether each statement below is correct or incorrect.

(a) $P(\text{boy}) = \frac{5}{8}$
(b) $P(\text{wears glasses}) = \frac{3}{8}$
(c) $P(\text{boy and wears glasses}) = \frac{1}{5}$
(d) $P(\text{girl and does not wear glasses}) = \frac{7}{40}$

3 A coin is tossed and a die is rolled. The probability of 'a head and a number greater than 4' or 'a tail and a number not exceeding 3' is:

A $\frac{1}{24}$
B $\frac{1}{6}$
C $\frac{1}{4}$
D $\frac{5}{12}$

4 A red die and a blue die are rolled on a table. Find the probability of each event:

(a) the same number with each die
(b) the sum of the numbers exceeds 9
(c) a 3 with the red die and a 4 with the blue die
(d) an odd number with the red die and an even number with the blue die
(e) the sum of the numbers is less than 2.

5 A coin is tossed three times. A is the event 'at least two tails'; B is the event 'three heads or three tails'; C is the event 'at least one tail'. Which of the following are independent?

(a) A and B
(b) A and C
(c) B and C

6 A die is rolled and a number is selected at random from the set {1, 2, 3, 4, 5}. The die number and the selected number are added together to find the score. Find the probability of each event:

(a) A: an even number is selected from the set
(b) B: the die rolls an odd number
(c) C: the score exceeds 9
(d) D: the score is 10

7 A die is loaded so that the probability of rolling a 6 is $\frac{3}{10}$, the probability of a 5 is $\frac{3}{10}$, and each of the other numbers is equally likely. If the die is rolled twice, find the probability of:

(a) two 6s **(b)** no 6s **(c)** at least one 6 **(d)** the sum of the two numbers being six.

8 In a large school, 25% of the students ride bicycles to school and 40% of the students have fair hair. One student is selected at random. What is the probability that the student:

(a) has fair hair and rides a bicycle to school
(b) does not have fair hair and does not ride a bicycle to school
(c) has fair hair but does not ride a bicycle to school
(d) rides a bicycle to school but does not have fair hair?

9 For a certain species of bird, there is a chance of 4 in 5 that a fledgling will survive the first month after birth. From a brood of 3 fledglings, what is the probability that:

(a) all will survive **(b)** none will survive **(c)** at least one will survive?

10 One card is drawn at random from a standard deck of 52 playing cards. The card is replaced in the deck and the deck is shuffled. A second card is then drawn. What is the probability that:

(a) both cards are diamonds **(b)** neither card is a diamond
(c) only one of the cards is a diamond **(d)** only the first card is a diamond
(e) only the second card is a diamond **(f)** at least one of the cards is a diamond?

11 Cube A has 4 red faces and 2 white faces; cube B has 3 red faces and 3 white faces; cube C has 2 red faces and 4 white faces. The three cubes are rolled like dice. What is the probability of rolling:

(a) 3 red faces **(b)** 3 white faces **(c)** red with A and B, white with C
(d) red with A, white with B and C **(e)** at least one red face?

12 An athlete competes in races over 100 m, 200 m and 400 m, and she estimates her chances of winning as $\frac{1}{2}$, $\frac{1}{3}$ and $\frac{1}{4}$ respectively. Using these probabilities, calculate the probability that:

(a) she wins all three races **(b)** she loses all three races
(c) she wins the 100 m race and loses the others **(d)** she wins the 400 m race and loses the others
(e) she wins the 100 m race and the 200 m race but loses the 400 m race.

13 A container holds 2 blue, 3 black and 5 white balls. A ball is withdrawn and then replaced in the container. This is repeated three times. Find the probability of each event:

(a) 3 white balls **(b)** a black ball in the first two drawings, but not in the third
(c) white, black, blue (in that order) **(d)** white, black, white (in that order)
(e) not more than two white balls **(f)** a white or a black ball each time.

14 A coin is tossed four times. What is the probability of:

(a) 4 heads **(b)** 4 tails **(c)** head, tail, head, tail (in that order)
(d) heads in the first 3 tosses but not in the fourth **(e)** heads in any one of the 4 tosses?

15 A student estimates that the chances of passing English, Mathematics and Physics are respectively $\frac{9}{10}$, $\frac{4}{5}$ and $\frac{3}{4}$. Estimate the probability of the events:

(a) passes English only **(b)** passes English and Mathematics only
(c) passes all three subjects **(d)** fails every subject.

16 A survey finds that, on average, 2 out of every 3 people interviewed are in favour of a certain proposal. If a random group of 3 people are interviewed, what is the probability that:

(a) all will be in favour of the proposal **(b)** none will be in favour
(c) the first and third people interviewed will be in favour, but not the second?

17 One container holds 2 red cubes and 4 blue cubes and a second container holds 4 red cubes and 3 blue cubes. One cube is selected at random from each of the two containers. What is the probability that one of the cubes is red?

18 The probability that a certain woman will be alive in 20 years is $\frac{2}{3}$ and the probability that a certain man will be alive is $\frac{3}{5}$. What is the probability that in 20 years' time:

(a) both will be alive **(b)** only one will be alive **(c)** at least one will be alive?

19 There are three containers A, B and C. A contains 3 black and 2 white cubes; B contains 3 black and 1 white cube; C contains 3 black and 3 white cubes.

(a) A container is chosen at random and from it a cube is chosen at random. Illustrate this two-stage process with a probability tree diagram.

(b) What is the probability that the cube is black?

20 On average, a student misses her bus to school once every eight weeks (where there are five days per school week). Find the probability that she:

(a) does not miss the bus on any one morning **(b)** catches the bus on two successive mornings

(c) catches the bus each morning for a week (5 days) **(d)** misses the bus on at least one morning in a week.

21 To open a locked safe requires a correct 3-digit combination. If a combination is chosen at random, calculate the probability of:

(a) succeeding at the first attempt **(b)** failing at the first attempt.

22 The first race at Randwick has 13 horses running and the second race has 16 horses. Assuming that all horses have an equal chance of winning, calculate the probability of predicting:

(a) a double (i.e. a winner in each race)

(b) a quinella (i.e. the two horses that come first and second, in either order) in the first race

(c) a quinella in the second race **(d)** a quinella in both races.

23 To gain a driver licence in NSW, you must pass both a touch-screen computer-based test and a practical driving test. Statistics show that 70% of learners pass the computer-based test on the first attempt; of those who fail, 90% pass on the second attempt. Also, 60% of learners pass their first practical test and 80% pass their second practical test. The computer-based and practical tests are independent. Calculate the probability of:

(a) passing the computer-based test on the second attempt

(b) passing the computer-based test after no more than two attempts

(c) needing to take a third computer-based test

(d) passing the practical test on the second attempt

(e) receiving a licence after taking one practical test and two computer-based tests.

9.6 DEPENDENT EVENTS

Two events are **dependent** if the outcome of one event is affected by the outcome of the other event. This means that the probability of both events together is *not* the same as the probability of both events separately. Their probabilities cannot be simply multiplied together as if they are separate successive outcomes (as for independent events).

In other words:

If events A and B are dependent, then: $P(A \text{ and } B) \neq P(A) \times P(B)$

Example 17

Two dice are rolled. A is the event '5 with the first die' and B is the event 'sum of the numbers on the two dice exceeds 10'.

(a) Find $P(A \text{ and } B)$ (also written $P(A \cap B)$).

(b) Are A and B independent?

(c) Find the probability that the sum of the numbers exceeds 10, given that the first die rolls a 5.

Solution

(a) Rolling two dice has 36 possible outcomes.

$A = \{(5, 1), (5, 2), (5, 3), (5, 4), (5, 5), (5, 6)\}$, so $P(A) = \frac{6}{36} = \frac{1}{6}$

$B = \{(5, 6), (6, 5), (6, 6)\}$, so $P(B) = \frac{3}{36} = \frac{1}{12}$

A and $B = A \cap B = \{5, 6)\}$, so $P(A \text{ and } B) = \frac{1}{36}$

Note that in this case $P(A \text{ and } B) \neq P(A) \times P(B)$.

(b) Because $P(A \text{ and } B) \neq P(A) \times P(B)$, events A and B are not independent. You say they are **dependent**.

(c) You are asked to find the probability of event B given that event A has happened. This can be written as $P(B \mid A)$.

Because A has occurred, the only possible events in this case are (5, 1), (5, 2), (5, 3), (5, 4), (5, 5), (5, 6). Of these 6 outcomes, only one is favourable: (5, 6).

Hence $P(B \mid A) = \frac{1}{6}$.

For dependent events:

$$P(A \cap B) = P(A) \times P(B \mid A)$$

where $P(B \mid A)$ is the probability of B occurring given that A has already occurred.

This statement can also be written as: $P(B \mid A) = \frac{P(A \cap B)}{P(A)}$, $P(A) \neq 0$

Sampling without replacement from a small population

Example 18

From a set of 5 cards numbered 1, 2, 3, 4, 5, two cards are selected at random **without replacement** (i.e. without putting the first card back before taking the second). What is the probability that both cards are odd-numbered cards?

Solution

Let A be the event 'odd number first card' and B the event 'odd number second card'. You must find $P(A \text{ and } B)$. First: $P(A) = \frac{3}{5}$

If the first card is odd, then this leaves 2 odd cards in the remaining 4 cards, hence: $P(B \mid A) = \frac{2}{4}$

Thus: $P(A \text{ and } B) = P(A) \times P(B \mid A) = \frac{3}{5} \times \frac{2}{4} = \frac{3}{10}$

It may have been easier to say $P(\text{odd, odd}) = \frac{3}{5} \times \frac{2}{4} = \frac{3}{10}$.

Example 19

A carton contains 10 batteries, 2 of which are defective. A sample of 3 batteries is drawn at random from the carton. Find the probability that not more than one battery is defective, if the sampling is done:

(a) without replacement **(b)** with replacement.

Solution

(a) Let D denote 'defective' and N 'non-defective'. You can represent the outcomes using a tree diagram:

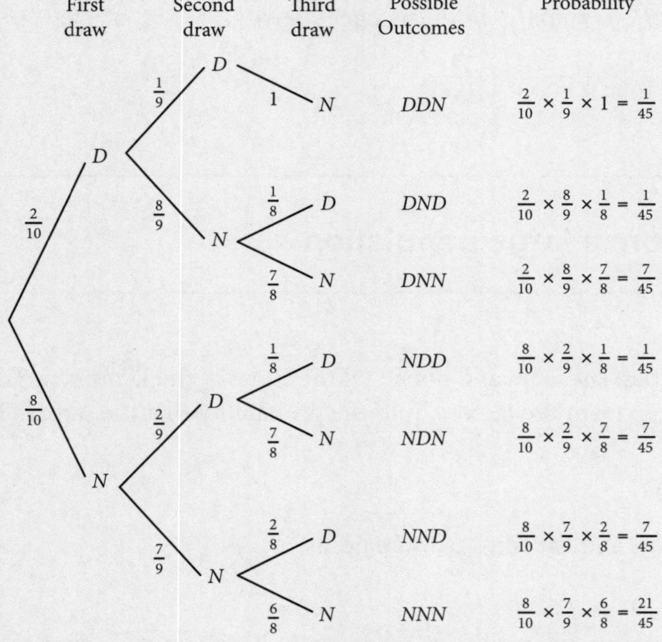

There is no branch for DDD because there are only 2 defective batteries in the carton.

The statement 'not more than 1 defective' is the same as '0 defective or 1 defective'.

$$\therefore \text{ Required probability} = P(DNN) + P(NDN) + P(NND) + P(NNN)$$
$$= \frac{7}{45} + \frac{7}{45} + \frac{7}{45} + \frac{21}{45} = \frac{42}{45} = \frac{14}{15}$$

(b) Because each battery is replaced in the carton before the next withdrawal is made, the probability of a defective battery being withdrawn remains the same each time: $\frac{2}{10}$ for each withdrawal. You can again represent the outcomes using a tree diagram (note the difference compared to part **(a)**):

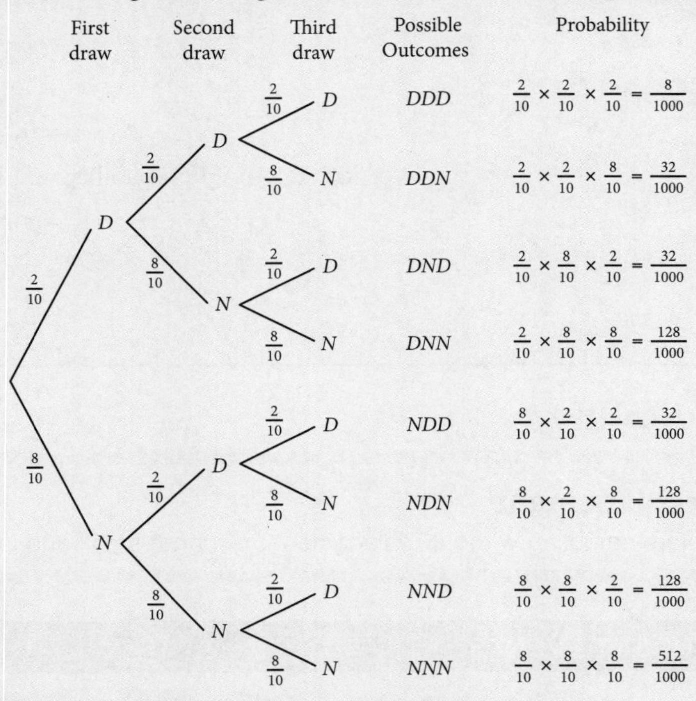

$$\therefore \text{ Required probability} = P(DNN) + P(NDN) + P(NND) + P(NNN)$$
$$= \frac{128}{1000} + \frac{128}{1000} + \frac{128}{1000} + \frac{512}{1000} = 0.896$$

Alternatively:

$$\begin{aligned}\text{Required probability} &= 1 - P(2 \text{ defectives} + 3 \text{ defectives})\\ &= 1 - [P(DDN) + P(DND) + P(NDD) + P(DDD)]\\ &= 1 - \left[\frac{32}{1000} + \frac{32}{1000} + \frac{32}{1000} + \frac{8}{1000}\right]\\ &= 0.896\end{aligned}$$

Sampling without replacement from a large population

Example 20

A shoe manufacturer makes two types of shoes: sneakers and boots. Of the shoes at the factory, 60% are sneakers. If a random sample of 3 pairs of shoes is taken from the factory, find the probability that the sample is such that:

(a) there are exactly 2 pairs of sneakers

(b) there is at least 1 pair of boots

(c) the first two pairs chosen are sneakers and the third pair are boots.

Solution

Let $p = 0.6$ and $q = 0.4$ be the respective probabilities of choosing a pair of sneakers and a pair of boots.

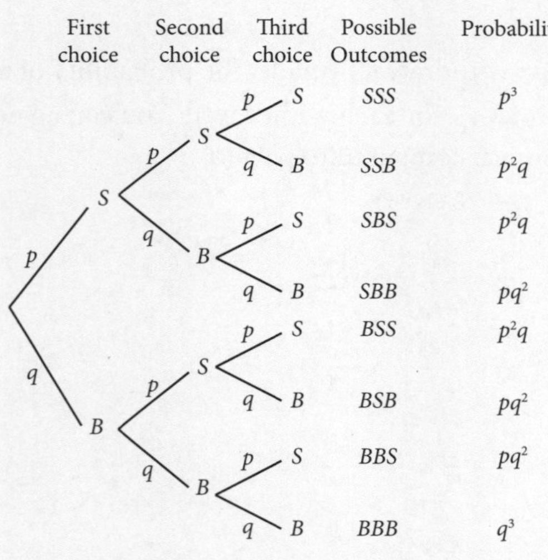

(a) $$\begin{aligned}\text{Required probability} &= P(SSB) + P(SBS) + P(BSS)\\ &= 3p^2q\\ &= 3 \times 0.6^2 \times 0.4\\ &= 0.432\end{aligned}$$

(b) $$\begin{aligned}\text{Required probability} &= 1 - P(0 \text{ boots})\\ &= 1 - P(SSS)\\ &= 1 - p^3\\ &= 1 - 0.6^3\\ &= 0.784\end{aligned}$$

(c) $$\begin{aligned}\text{Required probability} &= P(SSB)\\ &= p^2q\\ &= 0.6^2 \times 0.4\\ &= 0.144\end{aligned}$$

EXPLORE FURTHER

Sampling without replacement

Use a spreadsheet to explore sampling without replacement from both small and large populations.

EXERCISE 9.6 DEPENDENT EVENTS

1 A box contains 5 black cubes and 3 red cubes. Two cubes are drawn at random from the box. Find the probability that:

(a) both cubes are black **(b)** both cubes are the same colour

(c) both cubes are different colours.

2 A box contains 6 green balls and 4 white balls. A batch of two balls is drawn at random from the box. The probability that the two balls are the same colour is:

A $\frac{1}{3}$ **B** $\frac{7}{15}$ **C** $\frac{2}{15}$ **D** $\frac{2}{45}$

3 An angler has caught 15 fish, of which 3 are undersized. A random sample of 3 fish is taken without replacement by an inspector. The angler is fined if one or more of the fish in the sample is undersized. What is the probability that the angler is fined?

4 A carton contains a dozen eggs, 3 of which have a double yolk. If 3 eggs are taken to make a cake, find the probability that all 3 eggs will have double yolks.

5 In a group of 9 people, 3 have brown hair and 6 have black hair. A random sample of 2 people is selected from the group. What is the probability that:

(a) the first person selected has black hair **(b)** both people selected have black hair
(c) one person has brown hair and the other has black hair?

6 From a standard deck of 52 playing cards, two cards are selected at random without replacement. Indicate whether each statement below is correct or incorrect.

(a) $P(\text{both cards are diamonds}) = \frac{1}{17}$ **(b)** $P(\text{both cards are the same suit}) = \frac{1}{17}$

(c) $P(\text{one card is a spade and the other is a club}) = \frac{13}{102}$ **(d)** $P(\text{both cards are different suits}) = \frac{13}{17}$

7 A container holds 3 white balls, 4 red balls and 5 black balls. Two balls are drawn at random without replacement. What is the probability that the balls are:

(a) both white **(b)** both red **(c)** both black?

8 From 7 teachers and 5 pupils, a random selection of 2 people is made. What is the probability that:

(a) they are both teachers **(b)** they are both pupils **(c)** one is a teacher and the other is a pupil?
(d) How can the answer to **(c)** be deduced from the answers to **(a)** and **(b)**?

9 From the letters of the word 'PROMISE', three letters are chosen at random, one at a time. What is the probability that the three letters are:

(a) vowels **(b)** consonants **(c)** vowel, consonant, vowel (in that order)?

10 A punter correctly picked first and second in a race of 10 horses. What is the probability of this, if all the horses were equally likely to win?

11 In a raffle, 20 tickets are sold and there are 2 prizes. If you buy 5 tickets, what is the probability that you win at least one of the prizes?

12 Group A contains 10 gorillas and 5 chimpanzees. Group B contains 4 gorillas and 6 chimpanzees. Two apes are selected at random from the groups. What is the probability that:

(a) they are both gorillas, if they are selected from Group A
(b) they are both chimpanzees, if they are selected from Group B
(c) one is a gorilla, the other is a chimpanzee, if one is selected from each group?

13 A box of 10 chocolates contains 4 hard-centred and 6 soft-centred chocolates. If two chocolates are selected at random, what is the probability that:

(a) they both have hard centres **(b)** they both have soft centres
(c) one has a soft centre and the other has a hard centre?

14 In a lottery game, three numbers are selected at random from 1, 2, 3, 4, … 40. Find the probability that the three selected numbers are all even.

15 A sample of three items is selected at random without replacement from a batch of ten items, four of which are defective. Find the probability that there is at most one defective item in the sample.

16 In a race with nine runners, if every possible order of finishing is equally likely, find the probability of picking the runners who come first, second and third in the correct order.

17 Three cards are drawn at random from a standard deck of 52 playing cards. What is the probability that 3 aces are drawn, if the drawing is done:

(a) without replacement **(b)** with replacement?

18 A bag contains a large number of five-cent coins and ten-cent coins in a ratio of 2 to 3. If three coins are randomly selected from the bag, find the probability that:

(a) two of them are five-cent coins **(b)** at least two of them are five-cent coins
(c) not more than two of them are five-cent coins.

19 A manufacturer finds that 10% of the products made in a factory are defective. If three products are taken at random, what is the probability that:

(a) all are defective **(b)** none are defective **(c)** more are defective than are non-defective?

20 A hand of three cards dealt from a standard deck of 52 playing cards contains the ace of clubs. What is the probability of this happening?

21 In a raffle, 30 tickets are sold and there are two prizes. What is the probability that a person who buys five tickets wins:

(a) neither prize **(b)** both prizes **(c)** at least one prize?

22 From a set of 10 cards numbered 1 to 10, two cards are drawn at random without replacement. What is the probability that:

(a) both numbers are even **(b)** one is even and the other is odd
(c) the sum of the two numbers is 12?

23 It is known that 7 out of 10 students from a certain school will go on to university. If a group of 3 students is chosen at random from this school, find the probability that:

(a) all will go on to university **(b)** some will go on to university.

24 An archer finds that her ratio of success to failure in hitting a bullseye is 9 to 1. If 3 arrows are shot, what is the probability of:

(a) 3 successes **(b)** at least 2 successes **(c)** not more than 1 success?

25 On average, at a particular beach it rains on 2 days out of every 7. Find the probability that on a given weekend it will rain:

(a) on both days **(b)** on at least one day.

26 A container holds a number of cubes: 60% are white and the remainder are black. Two cubes are randomly selected without replacement. What is the probability that:

(a) they are the same colour **(b)** they are different colours?

27 In a large flock of birds, 50% are red and 50% are green. If two birds are selected at random, what is the probability that:

(a) they are both red **(b)** at least one is green?

28 Container X holds 1 white cube and 2 black cubes. Container Y holds 2 white cubes and 1 black cube. A container is selected at random and from it two cubes are selected without replacement. Draw a tree diagram to represent this three-stage process and find the probability that both cubes drawn are:

(a) the same colour **(b)** different colours.

29 A and B play a 'set' of tennis: when a player wins two games, the set is won. If A has probability 0.6 of winning any one game, what is the probability that A wins the set? Construct a tree diagram.

30 Of two coins A and B, A is a fair coin while B is loaded with a probability of 0.6 for heads. A coin is chosen at random and tossed twice. What is the probability of tossing two heads?

CHAPTER REVIEW 9

1 The combination lock on a safe has three concentric circular discs, each showing the digits 0 to 9. Only one combination of digits will open the safe. What is the probability of opening the safe at your first attempt if you do not know the combination?

2 One student has a pencil whose cross-section is square and whose faces are coloured black, white, green and red. Another student has a pencil whose cross-section is hexagonal (six-sided) and whose faces are coloured black, white, green, red, yellow and orange. Both pencils are rolled on a flat surface and the colours appearing uppermost are noted. Find the probability of each event:

(a) both colours uppermost are the same
(b) both colours uppermost are different
(c) black is uppermost on at least one of the pencils
(d) neither black nor white appears uppermost.

3 Four names beginning with B and five names beginning with G are in a hat. If two names are randomly drawn without replacement, find the probability that:

(a) two G names are drawn out **(b)** at least one B name is drawn out.

4 A jar contains red buttons and white buttons in a ratio of 3 : 2. If three buttons are chosen at random from the jar, find the probability that:

(a) exactly two are red **(b)** not more than one is white.

5 A box contains four balls marked 1, 2, 3, 4. A spinner has an equal chance of spinning R, G or W. A ball is selected at random from the box and the spinner is spun. Illustrate on a tree diagram the 12 possible outcomes.
Find the probability of each outcome:

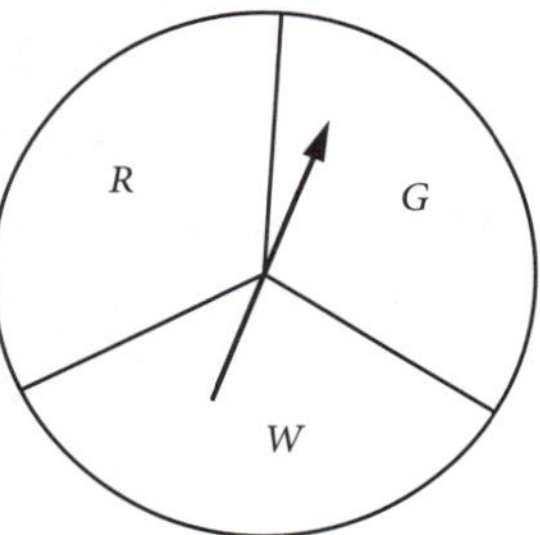

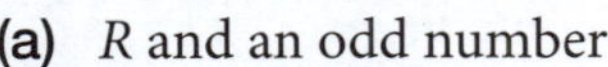

(a) R and an odd number
(b) W and a number greater than 2, or R and an even number.

6 In a group of 25 students, 18 study Biology, 12 study Chemistry and 5 study neither Biology nor Chemistry. If a student is chosen at random, what is the probability that the student studies:

(a) Biology only **(b)** Chemistry only
(c) Biology or Chemistry or both **(d)** both Biology and Chemistry?

7 A coin is tossed and a die is rolled. What is the probability of 'a head and a number greater than 3' or 'a tail and a number not exceeding 4'?

8 Consider two spinners. One spinner shows the letters A, B, C, D and E, and the other spinner shows the digits 1, 2, 3, 4 and 5. When the spinners are spun, it is equally likely that they will stop on any letter or number. Find the probability of each event:

(a) B and an even number **(b)** C or D, and an odd number
(c) E and an even number, or C and a number greater than 3
(d) a consonant and an odd number, or a consonant and a number greater than 2.

9 A die is rolled three times. What is the probability of:

(a) 3 sixes **(b)** 0 sixes **(c)** 3 odd numbers **(d)** 3 even numbers
(e) a six in the first two tosses only **(f)** a six, not a six, a six (in that order)?

10 Trinh and Oscar play three tennis matches. Trinh's chance of winning any one match is $\frac{2}{3}$. What is the probability that Trinh:

(a) wins all three matches **(b)** loses all three matches
(c) wins the first and third but loses the second **(d)** loses the first and wins the other two?

11 A man finds that he is late for work on 10% of occasions if he was on time the previous day, but late on 20% of occasions if he was late the previous day. Given that he was on time on Monday and worked on Tuesday, what is the probability that he is on time on Wednesday? Illustrate using a tree diagram.

12 A certain factory has three machines A, B and C, which manufacture 25%, 35% and 40% respectively of the factory's products. Of the machines' products, 5%, 4% and 2% respectively are defective. A product from the machines is selected at random. What is the probability that:

(a) it was manufactured by A and is defective **(b)** it was manufactured by B and is not defective?

13 From a group of 5 boys and 6 girls, two are selected at random for a class committee. What is the probability that a boy and a girl are selected?

14 A carton contains 10 electric lights, 3 of which are defective. Two are drawn at random. What is the probability that:

(a) the first drawn is defective **(b)** both are defective
(c) neither is defective **(d)** exactly one is defective?

15 Three cards are drawn at random from a standard deck of 52 playing cards. What is the probability that they are all from the same suit?

16 A bag contains 6 red balls and 4 white balls. A random sample of 3 is withdrawn. What is the probability that the balls are the same colour, if the sampling is done:

(a) without replacement **(b)** with replacement?

17 Three cards are taken without replacement from a standard deck of 52 playing cards. What is the probability of taking:

(a) exactly 3 hearts **(b)** exactly 3 aces **(c)** at least 1 heart?

18 A manufacturer of metal pistons finds that on average 20% of pistons are rejected because they are either oversize or undersize. What is the probability that a batch of three pistons will contain:

(a) no more than two rejects **(b)** at least two rejects?

19 In an assortment of bananas, the ratio of ripe bananas to unripe bananas is 3 to 5. If 3 bananas are chosen at random, find the probability that:

(a) exactly 2 will be ripe **(b)** at least 1 will be unripe.

20 In an opinion poll, the ratio of those in favour to those against a particular proposal was 7 to 3. If three randomly chosen people are interviewed about the proposal, what is the probability that:

(a) all will be in favour **(b)** the majority will be in favour
(c) not more than two will be against the proposal?

21 A container holds 10 cubes: 6 are white and the remainder are black. Two cubes are randomly selected without replacement. What is the probability that:

(a) they are the same colour **(b)** they are different colours?

22 In a group of 10 birds, 5 are red and 5 are green. If two birds are selected at random, what is the probability that:

(a) they are both red **(b)** at least one is green?

CHAPTER 10
Discrete probability distributions

10.1 DISCRETE RANDOM VARIABLES

When performing a sampling procedure, a number of different outcomes are expected. For example, when rolling a normal die a large number of times, you can expect to observe some of each of the values from {1, 2, 3, 4, 5, 6}. The outcome can vary between rolls. X, the observed outcome, is a **random variable**. In particular, X is a **discrete random variable** because the list of possible outcomes is countable. Discrete random variables are often associated with number or size.

On the other hand, if the list of possible outcomes is not countable, the variable is a **continuous random variable**. For example, when measuring the heights of a sample of people, although you might expect the measurements to fall within a range (say, 140 cm to 190 cm), each individual value is dependent only on the degree of accuracy of the measuring instrument.

Continuous random variables are often associated with height, mass and time. Further work will be done with continuous random variables in a later chapter.

Discrete variables and whole numbers

A discrete random variable is not restricted to taking on whole number values; the important criterion is that the number of outcomes must be countable. For example, shoe sizes using the British measuring system increase in half sizes such as 7, $7\frac{1}{2}$, 8, . . ., but since the total number of different sizes can be counted, the variable is discrete.

Statistical models

Statistical modelling is a process used to predict real-world events. A **statistical model** consists of equations that are based on assumptions and simplifications that produce a mathematical result. The predictions of the model can then be tested against some real-world data. Since it is unlikely that a model will be completely accurate, it is often modified to make it better fit the data. As an example, one might suggest that adults have the same arm span as their height. However, after collecting a large quantity of data, arm span might be found to be closer to 97% of a person's height.

Notation

Capital letters (e.g. X and Y) are used for random variables, and their corresponding lower-case letters (e.g. x and y) are used for the values that the random variable takes. Subscripts distinguish the various possible values of X. For example, we denote $P(X = x_2)$ as the probability that the variable X takes the value x_2.

By varying the number of times a coin is tossed in the activity below, you will see how the probability of obtaining a fixed number of heads changes.

MAKING CONNECTIONS

Tossing three coins

Use a spreadsheet to simulate the results of tossing three coins n times.

In some cases, the variable is not the actual outcome but rather a value assigned to an outcome. Consider an experiment where three coins are tossed. Let X stand for the number of heads obtained. There are 8 possible outcomes.

Outcome observed	HHH	HHT	HTH	THH	HTT	THT	TTH	TTT
Value of X	3	2	2	2	1	1	1	0

So, X can be any of the values from {0, 1, 2, 3}. Assuming that each of the 8 observed outcomes are equally likely, which is a reasonable assumption using fair coins, a probability table can be created for the variable X.

x	0	1	2	3
$P(X = x)$	$\frac{1}{8}$	$\frac{3}{8}$	$\frac{3}{8}$	$\frac{1}{8}$

Example 1

Four coins are tossed. What is the probability of obtaining exactly two heads?

Solution

Write out the sample space for the experiment.

(There are two options for each coin, so the sample space will consist of $2 \times 2 \times 2 \times 2 = 16$ options.)

$$\left\{\begin{matrix} \text{HHHH}, & \text{HHHT}, & \text{HHTH}, & \text{HHTT} \\ \text{HTHH}, & \text{HTHT}, & \text{HTTH}, & \text{HTTT} \\ \text{THHH}, & \text{THHT}, & \text{THTH}, & \text{THTT} \\ \text{TTHH}, & \text{TTHT}, & \text{TTTH}, & \text{TTTT} \end{matrix}\right\}$$

Define the variable to be used: Let X = the number of heads obtained, so $X = 2$.

From the sample space write out the events which correspond to the situation.

(Here there are six events.)

$$\left\{\begin{matrix} \text{HHTT}, & \text{HTHT}, & \text{HTTH} \\ \text{THHT}, & \text{THTH}, & \text{TTHH} \end{matrix}\right\}$$

State the probability for each successful event:

$$P(\text{HHTT}) = \frac{1}{2} \times \frac{1}{2} \times \frac{1}{2} \times \frac{1}{2} = \frac{1}{16}$$

and:

$$P(\text{HHTT}) = P(\text{HTHT}) = P(\text{HTTH}) = P(\text{THHT}) = P(\text{THTH}) = P(\text{TTHH}) = \frac{1}{16}$$

Calculate the required probability by adding together the individual probabilities for each outcome.

$$P(X = 2) = \frac{6}{16} = \frac{3}{8}$$

When the outcomes are equally likely you can use P(Event):

$$P(\text{Event}) = \frac{\text{number of successful outcomes}}{\text{total number of outcomes}}$$

Finding sample size

A more interesting question would be to find the sample size necessary to meet a particular condition. In many of these questions it is useful to use the complementary event. Recall that $P(\bar{A}) = 1 - P(A)$.

Example 2

A netball player scores a goal from 75% of her shots. How many shots at goal would she need to have at least a 95% chance of scoring at least one goal?

Solution

Use complementary events: P(scoring at least one goal) = 1 – P(scoring no goals)
You require: P(scoring no goals) < 0.05
Systematically vary the number of misses to find the required value: let M be the event that she misses the goal.

$$\begin{aligned} P(M) &= 0.25 \\ P(MM) &= 0.25 \times 0.25 \\ &= 0.0625 \\ P(MMM) &= 0.25 \times 0.25 \times 0.25 \\ &= 0.015\,625 \\ &< 0.05 \end{aligned}$$

The netball player would need to take at least three shots at goal to be at least 95% certain of scoring at least one goal.

Discrete probability distributions

Consider an experiment where three coins are tossed. The table of probabilities for each outcome forms what is known as a discrete probability distribution.

x	0	1	2	3
$P(X = x)$	$\frac{1}{8}$	$\frac{3}{8}$	$\frac{3}{8}$	$\frac{1}{8}$

It lists each of the x values possible and the associated probabilities. Strictly speaking you should indicate that the probability for any other value of x is zero, but in cases such as this it is obvious so you do not need to state it explicitly.

Some conditions need to be met before you can say you are dealing with a discrete probability distribution.

For a discrete probability distribution, then: $0 \leq P(X) \leq 1$ for all values of x,

and $\sum P(X = x) = 1$ (where $\sum$ stands for 'the sum of').

If either of these conditions is not met, then you do not have a discrete probability distribution.

Example 3

Does the following table represent a discrete probability distribution?

x	1	3	5	7
$P(X = x)$	$\frac{1}{5}$	$\frac{1}{10}$	$\frac{3}{10}$	$\frac{2}{5}$

Solution

Are all the probabilities between 0 and 1, inclusive; i.e. is $0 \leq P(X) \leq 1$? Yes.
Do the probabilities add up to 1?

$$\begin{aligned} \sum P(X = x) &= \frac{1}{5} + \frac{1}{10} + \frac{3}{10} + \frac{2}{5} = \frac{2}{10} + \frac{1}{10} + \frac{3}{10} + \frac{4}{10} \\ &= \frac{10}{10} \\ &= 1 \end{aligned}$$

Both conditions have been met, so the table of data represents a discrete probability distribution.

The probabilities can be represented as fractions, decimals, or percentages.

Example 4

The table of data below represents a discrete probability distribution.

x	3	4	5	6	7
$P(X=x)$	0.14	k	0.36	0.21	0.13

Find the value of k.

Solution

Add up the given probabilities: $0.14 + k + 0.36 + 0.21 + 0.13 = 0.84 + k$
As the sum of probabilities = 1, then: $0.84 + k = 1$
Solve for k: $k = 1 - 0.84$
$k = 0.16$

Example 5

The following discrete probability distribution represents a five-sector spinner where the areas of the sectors are not equal.

x	1	2	3	4	5
$P(X=x)$	k	$\frac{k}{2}$	$\frac{k}{3}$	$\frac{k}{4}$	0.25

Which of the values for k makes this a discrete probability distribution?

A 0.48 **B** 0.1875 **C** 0.25 **D** 0.36

Solution

Add up the terms involving the unknown: $k + \frac{k}{2} + \frac{k}{3} + \frac{k}{4} = \frac{25k}{12}$

As the sum of probabilities = 1, then: $\frac{25k}{12} + 0.25 = 1$

$$\frac{25k}{12} = \frac{3}{4}$$

$$k = \frac{9}{25} = 0.36$$

The answer is D.
Note that the other options came from common mistakes that students make.
A ignores the probability of 0.25 and just solves $\frac{25k}{12} = 1$.
B assumes that there were no fractions and the equation to be solved is $4k = 0.75$
C assumes that there were no fractions and the equation to be solved is $4k = 1$

As with many areas of mathematics, you may find it useful to draw a graph as a pictorial representation of the distribution.

Consider the following table of values:

x	0	1	2	3	4
$P(X=x)$	0.25	0.1	0.3	0.15	0.2

The graph of the distribution is shown in two different ways: a bar graph on the left and a dot graph on the right.

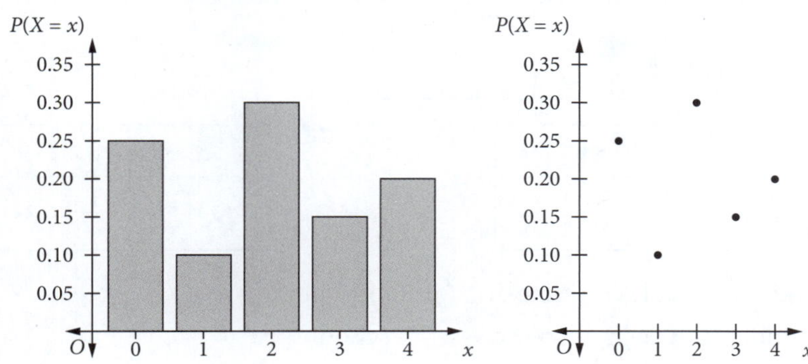

There is another notation that can be used for describing a discrete probability distribution that is particularly useful if the possible values all have the same probability. To describe the distribution for rolling a normal six-sided die,

you can write: $P(X=x)=\begin{cases}\frac{1}{6} & \text{if } x\in\{1,2,3,4,5,6\}\\ 0 & \text{for all other values of } x\end{cases}$

Here the probability attached to every value is explicitly stated, not just values with non-zero probability.

EXERCISE 10.1 DISCRETE RANDOM VARIABLES

1 Find the probability of each of the following events. Express your answers as fractions in simplest form.

- **(a)** The probability of obtaining exactly one head if three coins are tossed.
- **(b)** The probability of obtaining at least one head if three coins are tossed.
- **(c)** The probability of obtaining exactly two even numbers if two dice are rolled.
- **(d)** The probability of obtaining exactly one odd number if two dice are rolled.
- **(e)** The probability of obtaining a pair of numbers that are the same if two dice are rolled.

2 Wayne is a soccer player who takes the penalty kicks awarded to his team. He has a 60% chance of scoring from the penalty spot. For Wayne to be 95% certain of scoring at least one goal, how many penalty kicks would he need to take?

3 Do the following tables represent discrete probability distributions? Give a reason for your answer.

(a)

x	0	1	2	3	4
$P(X=x)$	$\frac{1}{3}$	$\frac{1}{3}$	$\frac{1}{9}$	$\frac{1}{9}$	$\frac{1}{9}$

(b)

x	2	3	4	5	6
$P(X=x)$	$\frac{1}{5}$	$\frac{1}{4}$	$\frac{1}{2}$	$\frac{1}{6}$	$\frac{1}{3}$

(c)

x	−1	0	1
$P(X=x)$	$\frac{2}{3}$	$\frac{1}{6}$	$\frac{1}{6}$

(d)

x	1	2	3	4	5
$P(X=x)$	$\frac{2}{5}$	$\frac{1}{5}$	$\frac{2}{5}$	$\frac{1}{5}$	$-\frac{1}{5}$

(e)

x	1	1.5	2	2.5	3
$P(X=x)$	20%	15%	30%	18%	17%

(f)

x	$1\frac{1}{2}$	$1\frac{3}{4}$	2	$2\frac{1}{4}$	$2\frac{3}{4}$
$P(X=x)$	0.24	0.16	0	0.38	0.22

4 For each of the following examples find the value of k that makes each table represent a discrete probability distribution.

(a)

x	0	1	2	3	4	5
$P(X=x)$	$\frac{1}{18}$	$\frac{1}{6}$	$\frac{1}{9}$	$\frac{5}{18}$	k	$\frac{3}{18}$

(b)

x	0	1	2	3	4	5
$P(X=x)$	$\frac{1}{8}$	$\frac{5}{24}$	k	$\frac{7}{24}$	$\frac{1}{24}$	$\frac{1}{6}$

(c)

x	5	6	7	8	9
$P(X=x)$	$\frac{1}{9}$	k	$\frac{k}{2}$	$\frac{4}{9}$	$\frac{1}{9}$

(d)

x	8	9	10	11	12
$P(X=x)$	$\frac{1}{6}$	$3k$	$\frac{1}{3}$	k	$\frac{1}{6}$

5 Are the following random variables discrete or continuous data?

(a) The number of students in each of your classes at school.
(b) The height of teachers at your school.
(c) The sizes of the shirts worn by each of the students in your mathematics class.
(d) The neck circumference of each of the members of the Australian netball team.
(e) The number of tosses of a coin before a head is observed.
(f) The time it takes for a process worker to complete 50 items.
(g) The distance jumped in the long jump by the competitors at the sports carnival.
(h) The number of red lights you stop at, per day, on the way to school over the course of a month.
(i) The number of whole lessons missed, per student, this year by members of your English class.
(j) The retail price in cents per litre charged for petrol over the course of a year.
(k) The number of people in the queue when you enter the bank each Friday for a year.
(l) The number of passionfruit collected from each vine in your orchard this season.

6 You have 10 cards. Five of the cards are hearts, three are diamonds and two are spades. You draw two cards, with replacement, from the pack. Let X be the number of diamonds drawn.

(a) What are the only values that X can take?
A 0 only **B** 0 and 1 only **C** 0, 1, and 2 only **D** 1 and 2 only

(b) Complete the table to show the probability distribution of X. Express the probabilities in decimal form.

x	0	1	2
$P(X=x)$			

(c) What is the probability of drawing exactly one red card from the pack?
(d) What is the probability of drawing at least one red card from the pack?

7 The following table represents a discrete probability distribution.
The value of k is:

x	3	5	7	9
$P(X=x)$	$\frac{1}{5}$	$\frac{2}{15}$	$\frac{1}{3}$	k

A $\frac{2}{3}$ **B** 0 **C** 1 **D** $\frac{1}{3}$

8 Which value of k makes the following table a discrete probability distribution?

x	1	2	3	4
$P(X=x)$	$k+0.1$	$k-0.1$	$k+0.55$	$k-0.15$

A 0.6 **B** 0.15 **C** 0.9 **D** 0.1

9 A board game uses a spinner divided equally into eight sections, each of which has a different number from 1 to 8 written on it. Before a player can put a piece on the board they must spin a 6, 7 or 8.

(a) What is the probability of spinning a 6, 7 or 8?
(b) What is the probability of not rolling a 6, 7 or 8 on the first 3 attempts? Give your answer in decimal form, correct to 2 decimal places.
(c) What is the probability that a player has not been able to start after the sixth attempt? Give your answer in decimal form, correct to 2 decimal places.
(d) Part of the game design process is to ensure it is not too difficult for a player to start. How many attempts does a player need to be at least 97% certain that they can start to play the game?

10 An apartment complex is built with the units having two, three or four bedrooms in the ratio 3 : 6 : 1. One of the units is chosen at random.

(a) Complete the table to show the probability distribution of the number of bedrooms the unit contains. Write the probabilities in decimal form.

x	2	3	4
$P(X=x)$			

(b) What type of graph would best represent the data?

11 Two six-sided dice are rolled. Let X be the total of the two dice.

(a) What are the lowest and highest totals possible?

(b) Complete the table to show the probability distribution of X. Express the probabilities as fractions in simplest form.

x	2	3	4	5	6	7	8	9	10	11	12
$P(X=x)$											

(c) Find $P(X \geq 6)$. **(d)** Find $P(X < 10)$. **(e)** Find $P(4 \leq X \leq 10)$.

12 Two coins are tossed. T stands for the number of tails obtained.

(a) Complete the table to show the probability distribution of T. Express the probabilities in decimal form.

t	0	1	2
$P(T=t)$			

(b) What type of graph would best represent this distribution? Draw the graph.

13 A six-sided die is rolled, and X is the cube of the number showing.

(a) Complete the table to show the probability distribution of X. Express the probabilities as fractions in simplest form.

x	1	8	27	64	125	216
$P(X=x)$						

(b) Find $P(X < 100)$.

14 In a game a coin is tossed and a six-sided die is rolled. If the coin shows tails then X is the score on the die. If the coin shows heads then X is the square of the score on the die.

(a) Draw a table to show the probability distribution of X. Express the probabilities as fractions in simplest form.

(b) Find $P(X \leq 12)$. **(c)** Find $P(X \geq 16)$. **(d)** Find $P(X \leq 9 \mid \text{coin showed heads})$.

15 On the way to work Enzo must pass through three sets of traffic lights. The probability that he will stop at any particular set of lights is $\frac{3}{5}$.

(a) Assuming the traffic lights are independent of each other, construct a table to show the probability distribution of X, the number of traffic lights at which Enzo stops.

(b) What is the probability that, on a particular day, Enzo stops at:

(i) exactly two sets of traffic lights

(ii) no more than two sets of traffic lights

(iii) at least two sets of traffic lights

(iv) fewer than two sets of traffic lights?

(c) What do you notice about the answers to **(b)(iii)** and **(iv)**? Why does this occur?

16 A die is loaded so that the probability of obtaining each number is as shown in the table.

x	1	2	3	4	5	6
$P(X=x)$	$\frac{1}{9}$	$\frac{2}{9}$	$\frac{1}{9}$	$\frac{2}{9}$	$\frac{1}{9}$	$\frac{2}{9}$

The die is rolled twice and Y is the sum of the two results.

(a) Draw a table to show the probability distribution of Y. Express the probabilities as fractions in simplest form.

(b) Find $P(Y \geq 4)$. **(c)** Find $P(Y < 7)$. **(d)** Find $P(3 \leq Y \leq 9)$. **(e)** Find $P(Y \geq 6 \mid Y \leq 10)$.

17 Sharmela is an interior designer who uses up to six colours in her designs. She sometimes chooses colours randomly for a unique effect. The number of colours Sharmela uses in her designs is a discrete random variable X with a probability distribution formula $P(X=x) = \frac{k}{x}$.

(a) Calculate the value of k.

(b) Draw a table to show the probability distribution of X. Express the probabilities as fractions in simplest form.

(c) Add a third column to the table to display the probabilities in decimal form, correct to 2 decimal places, and then draw a dot plot of the distribution.

18 Analyse each of the following goal-scoring situations.

(a) Lou has a 71% chance of scoring a goal when she kicks a penalty kick for her football team. How many penalty kicks would she need to take to have a 95% chance of scoring at least one goal?

(b) Buddy plays full-forward in an AFL team. He has a 68% chance of scoring a goal when taking a kick from 45 m out from goal. How many shots at goal would he need to have so that he has a 95% chance of scoring at least one goal?

(c) What is the largest whole number percentage that your chance of scoring needs to be so that you would need more than three shots to have a 95% chance of scoring at least once?

19 Alexander is the manager of a shoe shop. The shop sells shoes only in the following sizes: 7, 8, 9, 10, 11. To stock the shop according to market demand, Alexander records the sizes of shoes he sells in a month. The table shows his results.

Shoe size	7	8	9	10	11
Number of pairs	24	a	72	a	8

(a) If Alexander sold a total of 200 pairs of shoes and the random variable X represents the shoe size sold, create a discrete probability distribution table for the sale of shoes given their size, by first calculating the value of a.

(b) What is the probability that the next customer who buys a pair of shoes will buy a pair of shoes of size less than 10 but greater than 7?

(c) Calculate the probability that the next two pairs of shoes that Alexander sells are a size 10 followed by a size 11. Assume the events are independent.

(d) Alexander has noticed that he made a mistake when he recorded the sales over the given month. In one sale he sold two pairs of shoes whose sizes added up to 18. Mistakenly he recorded two pairs of shoes of size 9, but when he checked the receipt again he realised that the two pairs were of sizes 7 and 11.

(i) Explain how this mistake changes the probabilities in the probability distribution table. Write down the revised values, correct to 3 decimal places.

(ii) Calculate the probability that the next two pairs of shoes that Alexander sells are a size 10 followed by a size 11 using the revised values. Assume the events are independent.

10.2 EXPECTED VALUE, VARIANCE AND STANDARD DEVIATION OF DISCRETE PROBABILITY DISTRIBUTIONS

In the past, you have calculated the mean of a set of values as a measure of central tendency for that set. Another way of describing the mean is to call it the **expected value**. The study of expected value was first related to gambling, where being able to decide if a game is fair is certainly an advantage. The following example shows how this can be done.

Consider a game in which you roll a die and receive as your prize the number of dollars showing on the die. The following table shows the possible outcomes.

Number showing	1	2	3	4	5	6
Probability	$\frac{1}{6}$	$\frac{1}{6}$	$\frac{1}{6}$	$\frac{1}{6}$	$\frac{1}{6}$	$\frac{1}{6}$
Gain ($)	1	2	3	4	5	6

If you were running the game you would need to decide how much to charge players for a turn. So, you need to know the expected gain for a player. Since a player can expect to win \$1 on $\frac{1}{6}$ of the rolls, \$2 on $\frac{1}{6}$ of the rolls, etc., for any particular roll of the die the expected gain would be:

$$\$1 \times \frac{1}{6} + \$2 \times \frac{1}{6} + \$3 \times \frac{1}{6} + \$4 \times \frac{1}{6} + \$5 \times \frac{1}{6} + \$6 \times \frac{1}{6}$$

$$= \$\left(\frac{1}{6} + \frac{2}{6} + \frac{3}{6} + \frac{4}{6} + \frac{5}{6} + \frac{6}{6}\right)$$

$$= \$\frac{21}{6}$$

$$= \$3.50$$

This means you would expect the player to gain, on average, \$3.50 on each roll of the die. You should note that this average outcome could not happen itself, since there is no result which will return \$3.50 to the player. As the game operator, you could charge \$3.50 to play, in which case the game would be considered fair, or you could charge more than \$3.50 if you wanted to make a profit. It is important to note that gambling games in businesses like casinos are designed so that the casino always has a slight advantage to ensure they make a profit.

This example leads to the definition for the expected value of a discrete random variable.

Expected value

The expected value or mean, $E(X)$ or μ (mu), of a discrete random variable, X, is the sum of each possible value multiplied by its probability.

In symbolic form:

Let X take the values $x_1, x_2, x_3, \ldots, x_i, \ldots x_n$ with associated probabilities $p_1, p_2, p_3, \ldots, p_i, \ldots p_n$.

Then:

$$E(X) = x_1p_1 + x_2p_2 + x_3p_3 + \ldots + x_ip_i + x_np_n = \sum_{i=1}^{n} x_i p_i$$

This sum can also be expressed as $\sum_{x=1}^{n} x\, P(X = x)$.

Remember that Σ is the Greek letter sigma and stands for 'the sum of'.

This is similar to the formula used to find the mean of a data set that has been expressed in a frequency table.

$$\text{Therefore: mean} = \frac{\Sigma xf}{\Sigma f} = \frac{105}{30} = 3.5$$

x	f	xf
1	4	4
2	7	14
3	3	9
4	6	24
5	6	30
6	4	24
	$\Sigma f = 30$	$\Sigma xf = 105$

Summation notation

There are several different notations used for summations of this type.

You can leave out the index from the summation sign when it is a sum over all possible values.

Another notation is $\sum_x$ which means to use all possible values of x. This form is especially useful when the values of x are not consecutive.

Example 6

Consider the following probability distribution table.
Find the expected value of the probability distribution.

x	4	8	12	16
$P(X = x)$	$\frac{3}{8}$	$\frac{1}{4}$	$\frac{1}{4}$	$\frac{1}{8}$

Solution

Use the rule: $E(X) = \sum x_i p_i$

$$= \left(4\times\frac{3}{8}\right)+\left(8\times\frac{1}{4}\right)+\left(12\times\frac{1}{4}\right)+\left(16\times\frac{1}{8}\right)$$

$$= \frac{3}{2}+2+3+2$$

$$= 8\frac{1}{2}$$

Hence $\sum x_i p_i = E(X) = 8\frac{1}{2}$

Example 7

The following table represents a probability distribution.
The expected value $E(X) = 3.4$.

x	1	2	3	4	5	6
$P(X = x)$	0.1	a	0.3	0.2	0.2	b

Find the values of a and b.

Solution

Write an equation using the fact that the probabilities must add to 1: $0.1 + a + 0.3 + 0.2 + 0.2 + b = 1$

$$a + b = 0.2$$

Write an equation using $E(X)$: $E(X) = 3.4$

$$0.1 + 2a + 0.9 + 0.8 + 1 + 6b = 3.4$$

$$2a + 6b = 0.6$$

$$a + 3b = 0.3$$

Solve the equations simultaneously: $a + b = 0.2$ [1]

$$a + 3b = 0.3 \quad [2]$$

[2] − [1]: $2b = 0.1$

$$b = 0.05$$

Substitute into [1]: $a + 0.05 = 0.2$

$$a = 0.15$$

Hence $a = 0.15$, $b = 0.05$

Sometimes you want to know information about the random variable for which you have the probability distribution. In these cases, apply the function rule to the values the variable can take.

Example 8

The following table represents a probability distribution.
Find the expected value of $3X - 2$.

x	2	3	4	5
$P(X = x)$	0.4	0.15	0.25	0.2

Solution

Apply the function to each value the variable can take. (This has no effect on the probability of each outcome): $f(x) = 3x - 2$

$$f(2) = 3(2) - 2$$
$$= 4$$
$$f(3) = 3(3) - 2$$
$$= 7$$
$$f(4) = 3(4) - 2$$
$$= 10$$
$$f(5) = 3(5) - 2$$
$$= 13$$

Rewrite the probability distribution table replacing the x values with the newly calculated values for $f(x)$:
Multiply $f(x_i)$ by p_i and add the products:
(This is just a modified form of $E(X) = \sum x_i p_i$ where $f(x_i)$ replaces x_i.)

x	4	7	10	13
$P[f(X) = f(x)]$	0.4	0.15	0.25	0.2

$$E(3X - 2) = 4 \times 0.4 + 7 \times 0.15 + 10 \times 0.25 + 13 \times 0.2$$
$$= 1.6 + 1.05 + 2.5 + 2.6$$
$$= 7.75$$

Hence $E(3X - 2) = 7.75$

From the previous example you can conclude $E(aX + b) = aE(X) + b$.

The following activity provides a verification of this result in a non-probability context. However, the ideas are the same for the probability application.

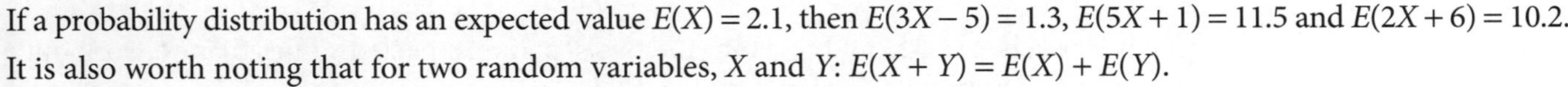

MAKING CONNECTIONS

Expected value $E(X)$

Use a spreadsheet to see how $E(aX + b) = aE(X) + b$.

If a probability distribution has an expected value $E(X) = 2.1$, then $E(3X - 5) = 1.3$, $E(5X + 1) = 11.5$ and $E(2X + 6) = 10.2$.

It is also worth noting that for two random variables, X and Y: $E(X + Y) = E(X) + E(Y)$.

Example 9

Consider the situation where two spinners are spun.
One spinner, X, has the following probability distribution:

x	2	3	4	5
$P(X = x)$	0.3	0.25	0.15	0.3

The other, Y, has the following probability distribution:
Show that $E(X + Y) = E(X) + E(Y)$.

y	1	2	3	4
$P(Y = y)$	0.2	0.3	0.4	0.1

Solution

The first step is to find the value of $E(X)$:

$$E(X) = 2 \times 0.3 + 3 \times 0.25 + 4 \times 0.15 + 5 \times 0.3$$
$$= 0.6 + 0.75 + 0.6 + 1.5$$
$$= 3.45$$

Now find the value of $E(Y)$:

$$E(Y) = 1 \times 0.2 + 2 \times 0.3 + 3 \times 0.4 + 4 \times 0.1$$
$$= 0.2 + 0.6 + 1.2 + 0.4$$
$$= 2.4$$

Then add these together to find the value of $E(X) + E(Y)$:

$$E(X) + E(Y) = 3.45 + 2.4$$
$$= 5.85$$

Draw a table that shows the sample space, and associated probabilities, for $X + Y$:

$Y\downarrow$	$X\rightarrow$	2	3	4	5
	1	3 : 0.06	4 : 0.05	5 : 0.03	6 : 0.06
	2	4 : 0.09	5 : 0.075	6 : 0.045	7 : 0.09
	3	5 : 0.12	6 : 0.1	7 : 0.06	8 : 0.12
	4	6 : 0.03	7 : 0.025	8 : 0.015	9 : 0.03

This table can then be used to draw up the probability distribution of $X + Y$:

$x+y$	3	4	5	6	7	8	9
$P(X+Y)$	0.06	0.14	0.225	0.235	0.175	0.135	0.03

Use the values in this probability distribution table to find $E(X + Y)$:

$$\begin{aligned} E(X+Y) &= 3\times 0.06 + 4\times 0.14 + 5\times 0.225 + 6\times 0.235 + 7\times 0.175 + 8\times 0.135 + 9\times 0.03 \\ &= 0.18 + 0.56 + 1.125 + 1.41 + 1.225 + 1.08 + 0.27 \\ &= 5.85 \end{aligned}$$

The values of $E(X + Y) = E(X) + E(Y)$ are equal, therefore: $E(X + Y) = E(X) + E(Y)$.

You should be able to see that it is much easier to add together the two separate expected values than to work out the distribution for the addition and then find its expected value.

Variance

The study of statistics includes **measures of central tendency** (mean, median, mode) and **measures of spread** (range, interquartile range, variance, standard deviation). You have already seen a measure of central tendency, the expected value or mean, $E(X)$ or μ. Consider now a measure of spread, the **variance**.

The variance is usually denoted by σ^2 ('sigma squared') or by Var(X). In simple terms, the variance describes how far the values of a data set are spread out. The larger the variance, the more spread out the data.

The variance is defined as the expected value of the square of the difference from the mean $\text{Var}(x) = E(x-\mu)^2$

For a finite probability distribution, to find the variance first subtract the mean μ from each value of x_i and square the result. Multiply that value by the corresponding probability p_i. The variance is the sum of these products. The further away from the mean the observed values, the greater the value of the variance.

$$\begin{aligned} \text{Var}(X) &= \sigma^2 \\ &= \Sigma(x_i-\mu)^2\, p_i \text{ where } \mu = E(X) \end{aligned}$$

As $E(X) = \Sigma x_i p_i$, and recognising $(x_i-\mu)^2$ as a function of X: $\text{Var}(X) = E(X-\mu)^2$

This version of the variance is not always convenient, especially if $E(X)$ is already known.

Expand the perfect square:	$E(X-\mu)^2 = E(X^2 - 2X\mu + \mu^2)$
Use the property that $E(X+Y) = E(X) + E(Y)$:	$= E(X^2) - E(2X\mu) + E(\mu^2)$
As μ is a constant:	$= E(X^2) - 2\mu E(X) + E(\mu^2)$
And as $E(X) = \mu$:	$= E(X^2) - 2\mu^2 + \mu^2$
	$= E(X^2) - \mu^2$
And again, as $\mu = E(X)$:	$= E(X^2) - [E(X)]^2$

$$\begin{aligned}\text{Var}(X) &= \sigma^2 \\ &= E(X-\mu)^2 \\ &= \sum(x_i-\mu)^2 p_i \\ &= E(X^2) - [E(X)]^2\end{aligned}$$

Use whichever format is most convenient for the situation.

Example 10

For the following probability distribution:

(a) find the expected value

(b) find the variance.

x	1	2	3	4	5	6
$P(X=x)$	0.2	0.15	0.35	0.05	0.1	0.15

Solution

(a) Use $E(X) = \sum x_i p_i$:

$$\begin{aligned}E(X) &= 1\times 0.2 + 2\times 0.15 + 3\times 0.35 + 4\times 0.05 + 5\times 0.1 + 6\times 0.15 \\ &= 0.2 + 0.3 + 1.05 + 0.2 + 0.5 + 0.9 \\ &= 3.15\end{aligned}$$

(b) Rewrite the probability distribution table, replacing the x values with the $f(x) = x^2$ values.

x^2	1	4	9	16	25	36
$P(X=x)$	0.2	0.15	0.35	0.05	0.1	0.15

Find $E(X^2)$ by applying the rule $E[f(X)] = \sum f(x_i)p_i$:

$$\begin{aligned}E(X^2) &= 1\times 0.2 + 4\times 0.15 + 9\times 0.35 + 16\times 0.05 + 25\times 0.1 + 36\times 0.15 \\ &= 0.2 + 0.6 + 3.15 + 0.8 + 2.5 + 5.4 \\ &= 12.65\end{aligned}$$

Calculate $[E(X)]^2$:

$$\begin{aligned}E(X) &= 3.15 \\ [E(X)]^2 &= (3.15)^2 \\ &= 9.9225\end{aligned}$$

Calculate $\text{Var}(X) = E(X^2) - [E(X)]^2$:

$$\begin{aligned}\text{Var}(X) &= 12.65 - 9.9225 \\ &= 2.7275\end{aligned}$$

You could also use a vertical table structure to find the value of $E(X^2)$. For the probability distribution used in this example you would have:

x	$P(X=x)$	x^2	$x^2P(X=x)$
1	0.2	1	0.2
2	0.15	4	0.6
3	0.35	9	3.15
4	0.05	16	0.8
5	0.1	25	2.5
6	0.15	36	5.4

$\sum x^2 P(X=x) = 12.65$

Var(X) must be positive. If you obtain a negative value, then there is an error.

Standard deviation

The measure of spread used most frequently is the standard deviation, measured in the same units as the variable itself. Fortunately, this value is simple to calculate if you know the variance.

The standard deviation of X is written as σ.

$$\text{Recall that } \text{Var}(X) = \sigma^2$$

$$\text{Then the standard deviation of } X,\ \sigma = \sqrt{\sigma^2} = \sqrt{\text{Var}(X)}$$

Only the positive square root of the variance is useful, because σ is a measure of spread, which is not dependent on direction.

Example 11

The variable X has the following probability distribution:

x	−1	0	1	2	3
$P(X = x)$	0.2	0.15	0.25	0.3	0.1

Find the standard deviation correct to 2 decimal places.

Solution

Rewrite the table and add columns for $xP(X = x)$, x^2, $x^2P(X = x)$:

x	$P(X = x)$	$x\,P(X = x)$	x^2	$x^2P(X = x)$
−1	0.2	−0.2	1	0.2
0	0.15	0	0	0
1	0.25	0.25	1	0.25
2	0.3	0.6	4	1.2
3	0.1	0.3	9	0.9
Σ	1	0.95		2.55

From the table: $E(X) = 0.95$ $\quad E(X^2) = 2.55$

Hence $[E(X)]^2 = 0.9025$

Calculate Var(X).

$$\begin{aligned}\text{Var}(X) &= E(X^2) - [E(X)]^2\\ &= 2.55 - 0.9025\\ &= 1.6475\end{aligned}$$

Calculate σ using $\sigma = \sqrt{\text{Var}(X)}$

$$\begin{aligned}\sigma &= \sqrt{1.6475}\\ &= 1.28 \text{ (to 2 decimal places)}\end{aligned}$$

Just as there is a relationship between a function and the expected value based on the probability distribution, there is a relationship between a function and the standard deviation.

The connection that exists between the standard deviation of a discrete probability distribution and the standard deviation of a function based on that discrete probability distribution is summarised as:

$\sigma(aX+b)=|a|\sigma(X)$

MAKING CONNECTIONS

Standard deviation $\sigma(X)$

Use a spreadsheet to see how $\sigma(aX + b) = |a|\sigma(X)$.

One useful characteristic of the standard deviation is that, for many variables, about 95% of their values will lie within two standard deviations of the mean. The following example illustrates this.

Example 12

The following table represents the probability distribution of Y, the number of consultations per hour by a dentist.

y	0	1	2	3	4	5	6	7
$P(Y=y)$	0.015	0.05	0.255	0.321	0.179	0.155	0.016	0.009

For this distribution it has been calculated that $E(Y)=3.173$ and $\sigma=1.283$.
Find the probability of the values being within two standard deviations of the expected value, i.e. $P(\mu-2\sigma\leq Y\leq\mu+2\sigma)$, and comment on the result.

Solution

Calculate $\mu-2\sigma$: $\mu-2\sigma=3.173-2\times1.283$
$=0.607$

Calculate $\mu+2\sigma$: $\mu+2\sigma=3.173+2\times1.283$
$=5.739$

Rewrite the probability interval using the values just calculated: $P(0.607\leq Y\leq5.739)$

Calculate the probability from the table. (Remember, these are discrete values):

$$\begin{aligned}&P(0.607\leq Y\leq5.739)\\&=P(Y=1)+P(Y=2)+P(Y=3)+P(Y=4)+P(Y=5)\\&=0.05+0.255+0.321+0.179+0.155\\&=0.96\end{aligned}$$

About 95% (96%) of the values lie within two standard deviations of the expected value.

EXERCISE 10.2 EXPECTED VALUE, VARIANCE AND STANDARD DEVIATION OF DISCRETE PROBABILITY DISTRIBUTIONS

1 Find $E(X)$ for each of the following probability distributions.

(a)

x	1	3	5	7	9
$P(X=x)$	0.2	0.3	0.25	0.15	0.1

(b)

x	−1	0	1	2	3
$P(X=x)$	0.4	0.15	0.2	0.05	0.2

(c)

x	2	3	4	5	8
$P(X=x)$	$\frac{1}{9}$	$\frac{1}{9}$	$\frac{1}{3}$	$\frac{1}{6}$	$\frac{5}{18}$

(d)

x	−2	−1	1	3	5
$P(X=x)$	$\frac{1}{4}$	$\frac{1}{6}$	$\frac{1}{12}$	$\frac{1}{3}$	$\frac{1}{6}$

2 Use the given value of $E(X)$ to solve for the unknowns.

(a) Find the values of a and b in the following probability distribution, given that $E(X) = 3.8$.

x	2	3	4	5	6
$P(X=x)$	0.4	0.1	a	0.1	b

(b) Find the values of i and j in the following probability distribution, given that $E(X) = 0.15$.

x	−3	−2	−1	0	1
$P(X=x)$	0.05	i	0.05	j	0.55

(c) Find the values of i and j in the following probability distribution, given that $E(X) = 6\frac{2}{5}$.

x	2	4	6	8	10
$P(X=x)$	$\frac{3}{20}$	$\frac{5}{20}$	i	$\frac{7}{20}$	j

(d) Find the values of a and b in the following probability distribution, given that $E(X) = 1\frac{4}{15}$

x	−3	−1	1	3	5
$P(X=x)$	a	$\frac{1}{5}$	$\frac{1}{15}$	b	$\frac{1}{5}$

3 For the following probability distribution, find:

x	0	1	2	3	4
$P(X=x)$	0.25	0.3	0.2	0.15	0.1

(a) $E(X)$ (b) $E(2X-3)$ (c) $E(X^2-5)$ (d) $E(X^2+X-4)$

4 For the following probability distributions find the expected value and variance.

(a)

x	1	2	3	4	5	6
$P(X=x)$	0.1	0.3	0.25	0.05	0.15	0.15

(b)

x	5	6	7	8	9
$P(X=x)$	0.15	0.35	0.1	0.25	0.15

5 Find the standard deviation of the variable X which has the following probability distribution. Give your answer correct to 2 decimal places.

x	−2	−1	0	1	2	3
$P(X=x)$	0.1	0.15	0.3	0.15	0.2	0.1

6 For the following probability distribution, find:

y	−3	−2	−1	0	1	2
$P(Y=y)$	0.02	0.03	0.25	0.35	0.3	0.05

(a) the standard deviation of Y (b) $P(\mu-2\sigma \le Y \le \mu+2\sigma)$.

7 In each of the following probability distributions, find the values of k and $E(X)$.

(a)

x	1	2	3	4	5
$P(X=x)$	k	$2k$	$3k$	$12k$	$6k$

(b)

x	−3	−1	1	3	5
$P(X=x)$	$12k$	$2k$	$3k$	k	$2k$

8 For the following probability distribution, find:

(a) $E(W)$ (b) $E(3W-4)$

(c) $E(2W+5)$ (d) $E(W^2-7)$

w	2	4	6	8
$P(W=w)$	$\frac{5}{16}$	$\frac{1}{8}$	$\frac{3}{8}$	$\frac{3}{16}$

9 Variable X has the following probability distribution:

x	−2	−1	0	1
$P(X=x)$	0.3	0.4	0.1	0.2

What is the value of $E(X^2-4)$?

A 0.2 **B** −5.4 **C** 2.2 **D** −2.2

10 Variable Y has the following probability distribution:

y	0	1	2	3
$P(Y=y)$	$\frac{1}{9}$	$\frac{7}{18}$	$\frac{1}{3}$	$\frac{1}{6}$

What is the value of $\text{Var}(Y)$?

A $1\frac{5}{9}$ **B** $1\frac{153}{162}$ **C** $\frac{65}{81}$ **D** $2\frac{5}{38}$

11 Eric just got a new job selling cars. He is offered a choice of two salary packages. In the first package he receives a weekly retainer of \$200 and an additional \$650 for every car sold. In the second package his retainer would be \$400, but he would only receive \$400 for every car sold. Past sales patterns indicate that the probability distribution for the number of cars sold per week is as follows:

Number of vehicles	0	1	2	3	4	5
Probability	0.45	0.35	0.1	0.05	0.04	0.01

Which salary package would Eric be better off taking?

12 A biased die has the following probability distribution:

d	1	2	3	4	5	6
$P(D=d)$	$\frac{1}{12}$	$\frac{1}{6}$	$\frac{1}{4}$	$\frac{1}{4}$	$\frac{1}{6}$	$\frac{1}{12}$

(a) If you rolled this die twenty times, what would be the expected (mean) value of the data, stated in mixed number form?

(b) State the range of $\mu \pm 2\sigma$ correct to 2 decimal places.

(c) Does the range of $\mu \pm 2\sigma$ cover all possible results when rolling the die? Should any of the values be considered unusual?

13 A random variable, T, has the following probability distribution:

t	$w-3$	$w-2$	$w-1$	w	$w+1$
$P(T=t)$	0.2	0.5	0.1	0.05	0.15

(a) Given that $E(T) = 8.45$, find the value of w.

(b) Find $\text{Var}(T)$. (c) Find the standard deviation of T correct to 2 decimal places.

(d) Find $\text{Var}(2T-6)$ correct to 2 decimal places.

(e) Find $\text{Var}(5-3T)$ correct to 2 decimal places.

14 The probability distribution of G is given by:

$$P(G=g) = \begin{cases} k(6-g) & \text{if } g \in \{0,1,2,3,4\} \\ 0 & \text{for all other values of } g \end{cases}$$

Find the following values: (a) k (b) $E(G)$ (c) $\text{Var}(G)$ (d) σ

15 Two spinners are spun. One spinner, X, has the following probability distribution:

x	1	2	3	4
$P(X=x)$	0.2	0.2	0.1	0.5

The other spinner, Y, has the following probability distribution:

y	2	3	4	5
$P(Y = y)$	0.2	0.3	0.4	0.1

Show that $E(X + Y) = E(X) + E(Y)$.

16 Enrico runs a game of chance at the Sydney Show. The player pays a fee to play, and draws a card at random from a normal pack of 52 playing cards. If the card is black, the player gets \$1 and their fee is refunded. If the card is a diamond, the player gets \$5 and their fee is refunded.

(a) Find Enrico's loss for the following conditions:
   (i) the card the player chooses is black
   (ii) the card the player chooses is a diamond.
(b) Describe the event for which Enrico keeps the game fee.
(c) Let Enrico's profit for the event described in part **(b)** be \$$p$, and draw up a probability distribution table that shows the three possible outcomes from Enrico's point of view.
(d) Find the value of p so that the expected value for the distribution is 0.
(e) If Enrico wants to make a profit, what is the minimum whole dollar amount he should charge to play the game?

17 Dubravko and Erina often play a best-of-three-sets match of tennis. From past experience they know that the probability that Dubravko will win a set is $\frac{3}{5}$.

(a) Draw a tree diagram to show the possible outcomes for their three-set match.
(b) Use the results from your tree diagram to draw a probability distribution table for their matches. Let X stand for the number of sets played.
(c) Find the expected number of sets they play in a match.
(d) Comment on your answer to part **(c)** in real-life terms.

18 A spinner has nine equal sections, of which five are yellow, three are blue and one is red. If the spinner lands on yellow, you receive \$1. If it lands on blue you receive \$3 and if it lands on red you receive \$5. Let X stand for the amount of money you receive.

(a) Draw up a probability distribution table for this game.
(b) What is the expected value of X?
(c) If the game is to be fair, how much should you pay to play?
(d) Comment on your answer to part **(c)** in real-life terms.

19 The game of 'Take a Chance' requires the player to roll three dice, of which one is blue, one is red and one is white. If a 1 shows on the blue die the player receives \$1, if a 1 shows on the red die the player receives \$2 and if a 1 shows on the white die the player receives \$5. In all other circumstances the player receives nothing. A player can receive more than one prize.

(a) Find the probability that the player receives the following amounts:
   (i) \$1 **(ii)** \$2 **(iii)** \$3 **(iv)** \$5 **(v)** \$6 **(vi)** \$7 **(vii)** \$8 **(viii)** \$0
(b) What is the expected return on this game?
(c) How much should the operator charge to play if the game is fair?

20 The number, Y, appearing on a spinner showing the numbers 1, 2, 3, 4, has a probability distribution as follows:

y	1	2	3	4
$P(Y = y)$	0.1	0.2	0.3	0.4

For another spinner, the number appearing, X, has a probability distribution as follows:

x	0	1	2	3	4
$P(X = x)$	0.6	0.1	0.1	0.1	0.1

(a) Find $E(Y)$. **(b)** Find $E(X)$.
(c) The numbers appearing on each of the two spinners are added together to give the variable M.
   (i) Draw a table to show the probability distribution of M.
   (ii) Find $E(M)$.
(d) What connection is there between $E(X)$, $E(Y)$ and $E(M)$?

10.3 THE UNIFORM DISTRIBUTION

There are some discrete distributions that have special properties worthy of separate investigation—**uniform distribution** and binomial distribution. In this section, you will look at uniform distribution.

A discrete probability distribution is said to be uniform if all values of the random variable are equally likely.

A common example of a uniform distribution is the random variable, X, that is the value of the face showing when a normal, six-sided die is rolled.

$$P(X=x)=\begin{cases}\dfrac{1}{6} & \text{if } x\in\{1,2,3,4,5,6\}\\ 0 & \text{for all other values of } x\end{cases}$$

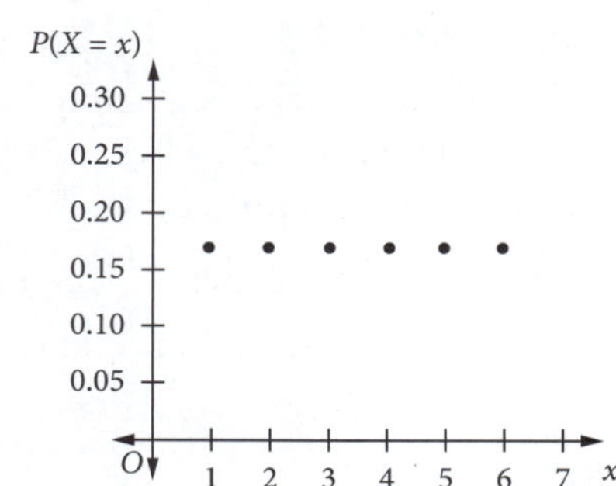

The graph of this distribution is shown on the right:

The expected value is: $E(X)=1\times\frac{1}{6}+2\times\frac{1}{6}+3\times\frac{1}{6}+4\times\frac{1}{6}+5\times\frac{1}{6}+6\times\frac{1}{6}$

$$=\frac{21}{6}$$

$$=3\frac{1}{2}$$

This can also be determined from the symmetry of the graph of the distribution.

The variance of the distribution is given by $\text{Var}(X)=E(X^2)-[E(X)]^2$:

$$\text{Var}(X)=\frac{1}{6}(1+4+9+16+25+36)-\left(\frac{7}{2}\right)^2$$

$$=\frac{35}{12}$$

In this text a uniform distribution will be assumed to be based on the first n positive integers. This is so some general expressions can be established. The general expression for a discrete uniform distribution with n values is:

$$P(X=x)=\begin{cases}\dfrac{1}{n} & \text{if } x\in\{1,2,\ldots n\}\\ 0 & \text{for all other values of } x\end{cases}$$

If you use the fact that the sum of the first n positive integers $1+2+\ldots+n=\frac{n}{2}(n+1)$, then the expected value is given by:

$$E(X)=1\times\frac{1}{n}+2\times\frac{1}{n}+3\times\frac{1}{n}+\ldots+n\times\frac{1}{n}$$

$$=\frac{1}{n}(1+2+3+\ldots+n)$$

$$=\frac{1}{n}\times\frac{n}{2}(n+1)$$

$$=\frac{n+1}{2}$$

The variance of X can be obtained by evaluating $\text{Var}(X)=E(X^2)-[E(X)]^2$ to get: $\text{Var}(X)=\dfrac{n^2-1}{12}$

Technology can be used to establish the expected value and variance of a discrete uniform distribution with n values. The probability function for this distribution is $P(X=x)=\frac{1}{n}$ for $x\in\{1, 2, 3, \ldots, n\}$ and zero otherwise.

The general form for the mean of a finite discrete distribution is: $E(X)=x_1p_1+x_2p_2+\ldots+x_np_n=\sum_{i=1}^{n}x_ip_i$

For the uniform distribution with n values this becomes: $\sum_{i=1}^{n}x_ip_i=\sum_{i=1}^{n}i\frac{1}{n}=\sum_{i=1}^{n}\frac{i}{n}$

EXPLORE FURTHER

Expected value and variance of a uniform distribution

Use a spreadsheet to find the expected value and variance of the first n natural numbers.

For a discrete uniform probability distribution with n values, 1 to n:

$$E(X) = \frac{n+1}{2} \qquad \text{Var}(X) = \frac{(n+1)(n-1)}{12} = \frac{n^2-1}{12}$$

Example 13

A roulette wheel in the United States usually has 38 equal-sized spaces showing the numbers 1 to 36 as well as 0 and 00. When the wheel is spun, a ball will land in one of the 38 spaces at random. For this question assume that 0 and 00 represent the 37th and 38th possible outcomes.

(a) Find the mean of the number of the space the ball lands in.

(b) Find the variance of the number of the space the ball lands in.

Solution

(a) Use the rule for $E(X)$:

$$E(X) = \frac{n+1}{2} = \frac{38+1}{2} = \frac{39}{2} = 19\frac{1}{2}$$

(b) Use the rule for Var (X):

$$\text{Var}(X) = \frac{n^2-1}{12} = \frac{38^2-1}{12} = \frac{1443}{12} = 120\frac{1}{4}$$

This section has focused on uniform distributions where the values of x are 1 to n. However, it is still quite easy to find the expected value and variance for other uniform distributions where the values the distribution takes are consecutive numbers. This is because such a distribution is a lateral shift of the uniform distribution where x takes the values 1 to n, so the rules $E(X + b) = E(X) + b$ and $\text{Var}(X + b) = \text{Var}(X)$ can be applied.

EXERCISE 10.3 THE UNIFORM DISTRIBUTION

1 A number from 1 to 16 is chosen at random. The random variable, R, represents the value chosen. Find the following values:

(a) $E(R)$ **(b)** $\text{Var}(R)$

2 A cleaner has nine similar-looking keys on a key chain. He tries them in turn, until he finds the one that opens the lock.

(a) What is the expected number of attempts for the cleaner to open the lock?

(b) What is the variance of the number of attempts?

3 A die in the shape of a tetrahedron (a solid with four triangular faces) is rolled. The four faces are numbered 1 to 4. Let F be a random variable that represents the value that is face down on the table.

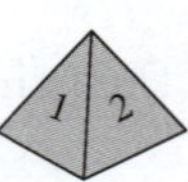

(a) What is the expected value, $E(F)$? **(b)** What is the variance, $\text{Var}(F)$?

4 A spinner is equally divided into n segments and each segment contains a value from 1 to n.

(a) If the expected value is 11.5, then n is equal to which of the following values?

A 20 **B** 21 **C** 22 **D** 23

(b) If the variance is 14 then n is equal to which of the following values?

A 9 **B** 10 **C** 12 **D** 13

5 Consider the random variable X that has the following probability distribution:

x	1	2	3	4	5
$P(X=x)$	0.2	0.2	0.2	0.2	0.2

(a) Find the following values:

(i) $E(X)$ **(ii)** $\text{Var}(X)$

(b) Now consider the random variable Y that has the following probability distribution:

y	2	3	4	5	6
$P(Y=y)$	0.2	0.2	0.2	0.2	0.2

Find, from first principles, the following values:

(i) $E(Y)$ **(ii)** $\text{Var}(Y)$

(c) What can you say about the values of $E(X)$ and $E(Y)$?

(d) What can you say about the values of $\text{Var}(X)$ and $\text{Var}(Y)$?

(e) Now consider the random variable Z that has the following probability distribution:

z	6	7	8	9	10
$P(Z=Z)$	0.2	0.2	0.2	0.2	0.2

Without doing any additional calculations, determine the following values:

(i) $E(Z)$ **(ii)** $\text{Var}(Z)$

6 You enter a room that contains a digital clock, such as the one shown. Let T represent the minutes shown on the clock.

(a) Show, using symmetry, that $E(T) = 29.5$.

(b) Now calculate $\text{Var}(T)$ and $\sigma(T)$, correct to 3 decimal places.

(c) If the numbers shown were 1 to 60, instead of 0 to 59, how would this effect $\text{Var}(T)$ and $\sigma(T)$?

7 A six-sided die numbered 1 to 6 is rolled twice and the values obtained are added together.

(a) Construct a probability distribution table for this event, using S to represent the variable.

(b) Draw a bar chart to illustrate the distribution.

(c) How would you describe this distribution?

(d) Calculate the following values:

(i) $E(S)$ **(ii)** $\text{Var}(S)$

(e) Recall that for a single roll of such a die: $E(X) = 3\frac{1}{2}$ and $\text{Var}(X) = 2\frac{11}{12}$. Comment on the relationship between the values for $E(X)$, $\text{Var}(X)$, $E(S)$ and $\text{Var}(S)$.

(f) Consider a spinner that has four equally-sized sections labelled 1 to 4. Let V be the value the spinner stops on. Calculate the following values:

(i) $E(V)$ **(ii)** $\text{Var}(V)$

(g) The spinner is now spun twice. Let T be the sum of the two values obtained. Calculate the following values:

(i) $E(T)$ **(ii)** $\text{Var}(T)$

(h) Comment on the relationship between the values for $E(V)$, $\text{Var}(V)$, $E(T)$ and $\text{Var}(T)$.

(i) Can you now make a general comment on the values obtained for the mean (expected value) and variance when two identical uniform variables, with values between 1 and n, are added?

10.4 DISCRETE DISTRIBUTIONS IN PRACTICAL SITUATIONS

Example 14

Two standard dice are rolled and the variable X represents the number of sixes obtained.

Find the expected number of sixes obtained.

Solution

Write the sample space:

$$\left\{\begin{matrix}(1,1),(1,2),(1,3),(1,4),(1,5),(1,6)\\(2,1),(2,2),(2,3),(2,4),(2,5),(2,6)\\(3,1),(3,2),(3,3),(3,4),(3,5),(3,6)\\(4,1),(4,2),(4,3),(4,4),(4,5),(4,6)\\(5,1),(5,2),(5,3),(5,4),(5,5),(5,6)\\(6,1),(6,2),(6,3),(6,4),(6,5),(6,6)\end{matrix}\right\}$$

There are 36 outcomes in the sample space.
Draw a table showing the probability distribution of the random variable.
Leave the probabilities in fraction form with the same denominator to make calculations easier:

x	0	1	2
$P(X=x)$	$\frac{25}{36}$	$\frac{10}{36}$	$\frac{1}{36}$

Calculate $E(X)$ from first principles. Express the final answer as a fraction in simplest form:

$$\begin{aligned}E(X) &= 0\times\frac{25}{36}+1\times\frac{10}{36}+2\times\frac{1}{36}\\&=\frac{12}{36}\\&=\frac{1}{3}\end{aligned}$$

The expected number (or average number) of sixes in two rolls of a standard die is $\frac{1}{3}$.

Example 15

One of the games at the local sporting club's 'Vegas Night' involves rolling a standard six-sided die. If an even number is shown, there is no game charge and the player wins the number of dollars shown on the face of the die. If an odd number is shown, the cost of playing is the number of dollars shown on the face of the die. Let Z stand for the number of dollars received by the player.

(a) Draw up a probability distribution table.
(b) What type of distribution is this?
(c) Find $E(Z)$.
(d) Is the game fair?

Solution

(a) A win could be represented by a positive value and a loss by a negative value.
Draw up the probability distribution table:

z	−1	2	−3	4	−5	6
$P(Z=z)$	$\frac{1}{6}$	$\frac{1}{6}$	$\frac{1}{6}$	$\frac{1}{6}$	$\frac{1}{6}$	$\frac{1}{6}$

(b) The probabilities are all equal, so this is a uniform distribution.

(c) The rule for $E(Z)$ cannot be used since the values are not (1, 2, 3, …, n). Express the final answer as a fraction in simplest form:

$$\begin{aligned}E(Z) &= -1\times\frac{1}{6}+2\times\frac{1}{6}-3\times\frac{1}{6}+4\times\frac{1}{6}-5\times\frac{1}{6}+6\times\frac{1}{6}\\&=\frac{3}{6}\\&=\frac{1}{2}\end{aligned}$$

(d) The game is not fair. It is actually in favour of the player since the expected return is \$0.50.

EXPLORE FURTHER

Mean, variance and standard deviation of a discrete distribution

Use a spreadsheet to find the mean, variance and standard deviation for a discrete probability distribution.

EXERCISE 10.4 DISCRETE DISTRIBUTIONS IN PRACTICAL SITUATIONS

1 A die, labelled 1 to 6, is rolled until the total of the scores is 4 or greater. Answer each of the following, giving all answers as exact values, in simplest fraction form.

(a) Find the probability distribution of the number of rolls X required to achieve this total.
(b) Find the expected number of rolls required.
(c) Find the variance for the number of rolls required.

2 One of the games at the local sporting club's 'Vegas Night' involves rolling a standard six-sided die. If a non-prime number is shown, there is no game charge and the player wins the number of dollars shown on the face of the die. If a prime number is shown, the cost of playing is the number of dollars shown on the face of the die. Let Z stand for the number of dollars received by the player.

(a) Draw a table to show the probability distribution of the variable.
(b) Is this best described as a uniform or a non-uniform distribution?
(c) Find the value of $E(Z)$.
(d) Is this game fair? If not, is it biased in favour of the operator or the player?

3 Four cards are labelled from 1 to 4. Two cards are dealt at random, without replacement. Let X represent the larger of the two numbers shown on the cards.

(a) How many values can X take?
(b) Find $P(X = 2)$.

4 For the discrete random variable X, the probability distribution is given by:

$$P(X = x) = \begin{cases} kx, & x = 1, 2, 3, 4 \\ k(9 - x), & x = 5, 6, 7, 8 \end{cases}$$

(a) Find the value of k.
(b) Complete the following table to show the probability distribution of X.

x	1	2	3	4	5	6	7	8
$P(X = x)$								

(c) Find the value of $E(X)$.
(d) Find the value of $\text{Var}(X)$.

5 The ratio boys : girls in a particular town was found to be 11:10, where the gender of one child in the family is independent of the gender of any other child in the family, and all the children are either boys or girls.

(a) State all possible combinations of boys and/or girls for a family with three children.
(b) What is the probability that a family with three children will have at least one boy?
(c) What proportion of families with exactly 4 children will have at least 3 girls?
(d) What proportion of families with exactly 4 children will have 2 girls and 2 boys?

6 Tomino has written the following as his answer for the probability distribution of the random variable X:

$$P(X = x) = \begin{cases} \dfrac{4 - x}{5} & \text{for } x = 1, 2, 3, 4, 5 \\ 0 & \text{otherwise} \end{cases}$$

Explain to Tomino why this cannot be correct.

7 The discrete random variable X has the probability distribution shown in the following table.

x	1	2	3
$P(X=x)$	$\frac{1}{4}$	$\frac{1}{2}$	$\frac{1}{4}$

(a) Find $E(X)$.
(b) Find $\text{Var}(X)$.
(c) A second random variable Y has the same distribution as X, and the two variables are independent. Draw a table to show the probability distribution of $X + Y$.
(d) Find $E(X + Y)$.
(e) Find $\text{Var}(X + Y)$.
(f) How is the value of $E(X + Y)$ related to the values of $E(X)$ and $E(Y)$?
(g) How is the value of $\text{Var}(X + Y)$ related to the values of $\text{Var}(X)$ and $\text{Var}(Y)$?

8 A standard six-sided die is rolled twice.

(a) How many ordered pairs make up the sample space?
(b) Assign Z to be the maximum number in each ordered pair. Draw a table to show the probability distribution of Z. Write your answers in fraction form using a denominator of 36.
(c) Now assign Y to be the minimum number in each ordered pair. Draw a table to show the probability distribution of Y. Write your answers in fraction form using a denominator of 36.
(d) How is the distribution of Z related to the distribution of Y?
(e) Find the expected value for each of the following:
(i) Z **(ii)** Y
(f) How far from the greatest value of Z is $E(Z)$?
(g) How far from the least value of Y is $E(Y)$?
(h) Comment on your answers to part **(f)** and part **(g)**.
(i) Given $\text{Var}(Z) = 1\frac{926}{1296}$, what can you say about $\text{Var}(Y)$?

9 Stephan has made a game in which the probability of randomly picking a number from 0 to 5 is given by the probability distribution shown in the following table. Answer each of the following, giving all answers correct to 3 decimal places.

X	0	1	2	3	4	5
$P(X=x)$	0.002	0.076	0.293	0.268	a	0.098

(a) Calculate the expected value for this random variable.
(b) Leanne made a game similar to Stephan's, but the probability of randomly picking a number from 0 to 5 is given by the following probability distribution.

Y	0	1	2	3	4	5
$P(Y=y)$	0.005	0.029	0.047	0.219	0.386	0.314

Calculate the expected value for this random variable.
(c) If one value in Stephan's game and one value in Leanne's game are chosen at random, calculate the probability that:
(i) they are the same value **(ii)** they are different **(iii)** their sum is greater than 8.

10 The discrete random variable X has the following probability distribution.

x	1	2	3	4
$P(X=x)$	0.25	0.1	0.45	0.2

(a) Find $E(X)$.
(b) Verify that $E(2X) = 2E(X)$.
(c) Find $\text{Var}(X)$.
(d) Find the following values:
(i) $\text{Var}(2X)$ **(ii)** $\text{Var}(3X)$
(e) State the relationship between $\text{Var}(X)$ and $\text{Var}(kX)$.

11 The probability distribution table of a random variable X is shown. Answer each of the following, giving all answers correct to 3 decimal places where necessary.

X	n	$n+1$	$n+2$	$n+3$
$P(X=x)$	0.80	0.12	0.05	0.03

(a) Show that $E(X) = n + 0.31$.
(b) If two independent values of X are chosen at random, calculate the probability of choosing two consecutive values.
(c) If two independent values of X are chosen at random, calculate the probability that the sum of the two values is even.
(d) If four independent values of X are chosen at random, calculate the probability that they are one of each type.

CHAPTER REVIEW 10

1 In each of the following find the value of k that makes the table a discrete probability distribution. In each case express the value in simplest fraction form.

(a)

x	1	2	3	4	5	6
$P(X=x)$	$\frac{1}{8}$	$\frac{1}{16}$	k	$\frac{1}{4}$	$\frac{1}{16}$	$\frac{3}{16}$

(b)

x	5	6	7	8	9	10
$P(X=x)$	$\frac{1}{12}$	k	$\frac{1}{6}$	$\frac{1}{3}$	$2k$	$\frac{1}{4}$

(c)

x	−1	0	1	2	3
$P(X=x)$	k	$2k$	$3k$	$4k$	$5k$

2 Two six-sided dice, numbered 3 to 8, are rolled. Let X be the total of the two dice.

(a) Complete the following table that shows the probability distribution of X. Express the probabilities as fractions in simplest form.

x	6	7	8	9	10	11	12	13	14	15	16
$P(X=x)$		$\frac{1}{18}$	$\frac{1}{12}$		$\frac{5}{36}$			$\frac{1}{9}$			$\frac{1}{36}$

(b) Find $P(X \geq 11)$. **(c)** Find $P(X < 15)$. **(d)** Find $P(7 \leq X \leq 11)$.

3 A spinner showing the numbers 1, 2, 3 and 4 is spun twice. Let X be the total of the two numbers obtained.

(a) Complete the table using fractions in simplest form.

x	2	3	4	5	6	7	8
$P(X=x)$							

(b) Use your table to find each of the following values:
(i) $E(X)$ **(ii)** $\mathrm{Var}(X)$

4 A random variable, X, has the following probability distribution.

x	−2	−1	0	1	2
$P(X=x)$	0.15	0.2	0.1	0.35	0.2

Find the following, rounding your answers to 4 decimal places where necessary.

(a) $E(X)$ **(b)** $\mathrm{Var}(X)$ **(c)** $E(3X-2)$ **(d)** $E(X^2-2X)$ **(e)** $\mathrm{Var}(3X-2)$

5 A game at the local festival requires a player to toss two coins. If two heads are obtained, the player receives $2. Otherwise, the player loses the cost of the game. For the game to be fair, what should it cost in dollars?

6 Technology is used to select a random number from 1 to 20 inclusive. The variable, R, represents the value chosen. Find the following values:

(a) $E(R)$ (b) $\text{Var}(R)$

7 Each of the numbers from 1 to n appearing on an n-sided spinner has an equal chance of appearing. If the variable S describes this situation, and $\text{Var}(S) = 18\frac{2}{3}$, what is the value of n?

8 A coin is biased in such a way that $P(\text{H}) = 3 \times P(\text{T})$. The coin is tossed 100 times. Let X stand for the number of tails obtained. Find the value of $E(X)$.

9 When Yehudi and Carlos play racquetball, the probability that Yehudi wins a point is 0.35.

(a) How many points would you expect Yehudi to win from the first 15 points? Give your answer to the nearest whole number of points.

(b) Choose the correct terms in the following statement.
If Yehudi won 10 out of the first 15 points I would [not be/be slightly/be very] surprised as the number is [about the same as/just above/well above] the expected number.

10 A box contains 10 items, of which three are defective. A sample of three items is selected at random from the box. Let Y represent the number of defective items selected.

(a) Complete the table to show the probability distribution of the variable. State the probabilities in simplest fraction form.

(b) Find the expected number of defective items, $E(Y)$.

y	0	1	2	3
$P(Y=y)$				

11 Which one of the following random variables is not associated with a discrete scale?

A The number of dim sims sold at the local fish and chip shop per day
B Your height, to the nearest centimetre, measured monthly for two years
C The time taken to fill your bath each night for one month
D The number of goals scored in each netball match by a team throughout the season

12 A jar contains seven white marbles, three green marbles and two blue marbles. Two marbles are drawn, with replacement, from the jar. What is the probability of drawing exactly one white marble?

A $\frac{7}{12} \times \frac{7}{12}$ B $\left(\frac{7}{12} \times \frac{5}{11}\right) + \left(\frac{5}{12} \times \frac{7}{11}\right)$ C $2 \times \frac{7}{12} \times \frac{5}{12}$ D $\frac{7}{12} \times \frac{5}{12}$

13 What is the value of t in the following probability distribution table?

A 10 B 0.1 C 6 D 0.6

x	0	1	2	3
$P(X=x)$	t	$2t$	$3t$	$4t$

14 Two six-sided dice are rolled. Let X be the total shown on the two dice.

(a) What is $P(X \geq 10)$?

A $\frac{11}{12}$ B $\frac{1}{6}$ C $\frac{1}{36}$ D $\frac{1}{12}$

(b) What is $P(5 < X \leq 9)$?

A $\frac{4}{9}$ B $\frac{1}{3}$ C $\frac{2}{3}$ D $\frac{5}{9}$

(c) What is the expected value, $E(X)$?

A 7 B 3 C 6.5 D 7.5

15 A probability distribution table is shown.

x	1	2	3	4	5
$P(X = x)$	0.2	0.1	0.35	0.05	0.3

(a) What is the expected value, $E(X)$?
A 0.2 **B** 1.34 **C** 1.8 **D** 3.15

(b) What is the expected value, $E(2X - 3)$?
A 3 **B** 3.3 **C** 3.5 **D** 3.7

(c) What is the variance, Var(X)?
A 2.13 **B** 1.46 **C** 8.81 **D** 2.97

16 For the random variable X it is known that $E(X) = 2.3$. If the expected value $E(X^2) = 6.2$, what is the standard deviation σ?

A 0.8281 **B** 0.91 **C** 0.95 **D** 2.3

17 A variable Y has the probability distribution shown in the following table.

y	−2	−1	0	1
$P(Y = y)$	0.4	0.3	0.2	0.1

What is the variance, Var(Y)?

A 1 **B** −1 **C** 2.2 **D** 1.1

18 A spinner is equally divided into n segments. Each segment contains a different value from 1 to n

(a) If the expected value is 26.5, what is the value of n?
A 26 **B** 51 **C** 52 **D** 53

(b) If the expected value is 18, what is the value of n?
A 9 **B** 10 **C** 34 **D** 35

(c) If the expected value is 101, what is the value of n?
A 201 **B** 200 **C** 52 **D** 51

19 Audrey is playing a popular quest-style game on her games console. It randomly generates whole numbers in the range 3 to 8 each time she hits the *Play* button. She cannot start her quest until the total of the scores is 7 or greater. Give exact answers, in simplest fraction form, for this question.

(a) Find the probability distribution of the number of times Audrey hits *Play*, given the variable X, required to achieve this total.

(b) Find the expected number of hits of *Play* required.

(c) Find the variance for the number of hits of *Play* required. (If you cannot get the exact answer for this part write the answer as a decimal correct to 3 decimal places.)

20 The discrete random variable X can take only the values 1, 2, 3, 4, 5 and 6. The probability distribution of X is described by the following statements:

$P(X = 1) = P(X = 3) = P(X = 5) = a$
$P(X = 2) = P(X = 4) = P(X = 6) = b$
$a = 3b$

(a) Find the values of a and b.

(b) Draw up a probability distribution table for X.

(c) Show that $E(X) = 3\frac{1}{4}$.

(d) Show that $\text{Var}(X) = 2\frac{41}{48}$.

(e) Find the probability that the sum of two independent observations from this distribution is greater than 9.

21 Freya rolls a normal six-sided die marked 1 to 6. If she obtains a 4 she rolls the die a second time, and in this case her score is the sum of 4 and the second number obtained. If she does not obtain a 4, her score is the number rolled. Freya has at most two rolls of the die. Let Z be the random variable representing Freya's score.

(a) Draw a table showing the probability distribution of Z.

(b) What is the expected value, $E(Z)$?

(c) What is $P(Z > 6)$?

(d) What is $P(Z < 7)$?

(e) What is $P(Z > 4 | Z \leq 8)$?

22 Use mathematical reasoning to explain why the function given by $P(X = x) = \dfrac{2x - 3}{15}$ is a probability distribution for $x \in \{3, 4, 5\}$ but it is not a probability distribution for $x \in \{1, 2, 3, 4, 5\}$.

23 Many calculators can produce random numbers as whole-number values within a specified range.

(a) If the technology produced truly random numbers, what type of distribution would you expect the results to follow?

(b) Listed below are 96 random numbers (as displayed by a calculator) in the range 1 to 6, inclusive.

$$\left\{\begin{array}{cccccccccccc} 3 & 6 & 2 & 5 & 6 & 2 & 3 & 1 & 6 & 1 & 1 & 4 \\ 6 & 6 & 2 & 2 & 1 & 1 & 5 & 1 & 3 & 2 & 6 & 1 \\ 6 & 4 & 2 & 6 & 5 & 2 & 2 & 6 & 6 & 1 & 1 & 5 \\ 6 & 4 & 2 & 1 & 6 & 2 & 3 & 1 & 2 & 6 & 3 & 4 \\ 4 & 1 & 1 & 1 & 5 & 5 & 6 & 3 & 1 & 3 & 6 & 5 \\ 4 & 4 & 6 & 5 & 5 & 4 & 3 & 2 & 4 & 6 & 1 & 3 \\ 3 & 2 & 5 & 6 & 3 & 3 & 2 & 4 & 4 & 5 & 5 & 3 \\ 6 & 4 & 3 & 3 & 5 & 6 & 2 & 5 & 3 & 3 & 2 & 1 \end{array}\right\}$$

Create a probability distribution table based on these results. To make analysis easier, write the probabilities with a denominator of 96 and use X as the variable.

(c) What do these results suggest about whether or not the numbers are truly random? Refer to the size of the sample space in your answer.

(d) Calculate $E(X)$ and $\text{Var}(X)$ for the sample of 96 random values. Compare them to the theoretical values for the underlying distribution.

(e) Now use technology to generate a similar set of 96 numbers in the range 1 to 6. Calculate $E(X)$ and $\text{Var}(X)$ for the sample of 96 random values that you generated.

(f) Compare the three sets of statistics that you have now calculated.

(g) What do you think would happen if you generated 5000 random numbers in the range 1 to 6?

24 A standard six-sided die is rolled until an even number shows or five odd numbers in a row have shown.

(a) Draw up a probability table showing the number of rolls and the associated probabilities.

(b) Find the expected number of rolls. Give your answer correct to 1 decimal place.

(c) Consider tossing a coin where you will stop as soon as the coin lands on tails or if five heads in a row appear. Explain how you can easily state the expected number of tosses of the coin.

25 Raduhas developed a website where subscribers can pick up to six motivational videos per day to watch. The number of each video chosen is a discrete random variable X with a probability distribution formula $P(X = x) = k(30 - (x - 1)^2)$, where $x \in \{1, 2, 3, 4, 5\}$.

(a) Calculate the value of k.

(b) Display all probabilities as fractions in simplest form in a probability distribution table.

(c) Draw a scatterplot of the distribution.

26 Karina and Achim print and sell business cards in packs of 500. They sell a maximum of 4 packs per hour and incur an average of \$16 running costs per hour. The average price for a pack of 500 business cards is \$20. The probability distribution table shown represents the random variable X, the number of packs they sell in any given hour.

X	0	1	2	3	4
$P(X = x)$	0.012	0.097	0.138	0.325	0.428

(a) Calculate the number of packs of business cards they can expect to print and sell in any given hour.

(b) Determine whether Karina and Achim incur a loss or make a profit, and the amount on average, in any given hour.

27 Four scatterplots are shown.

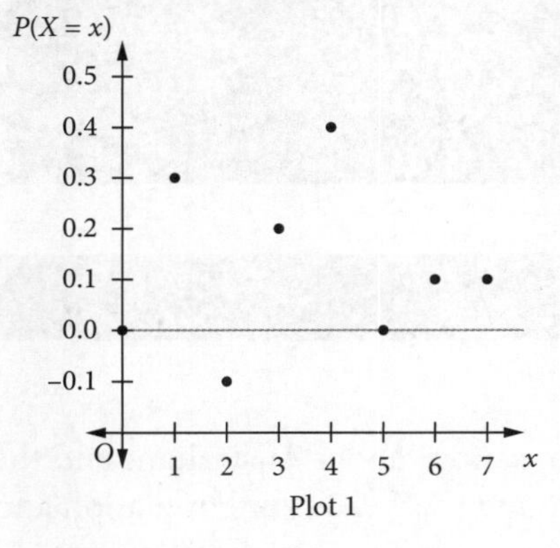

Plot 1

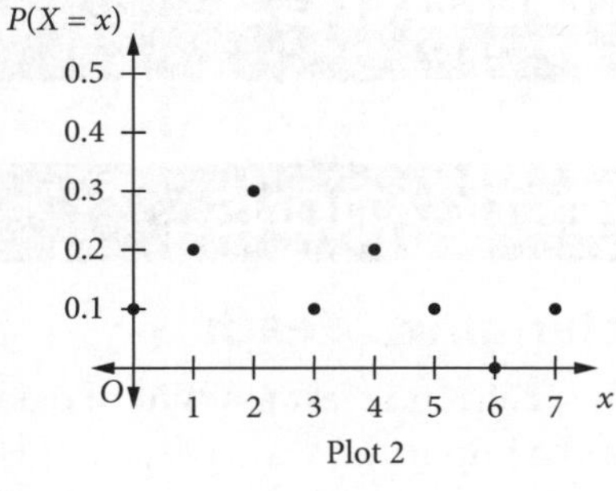

Plot 2

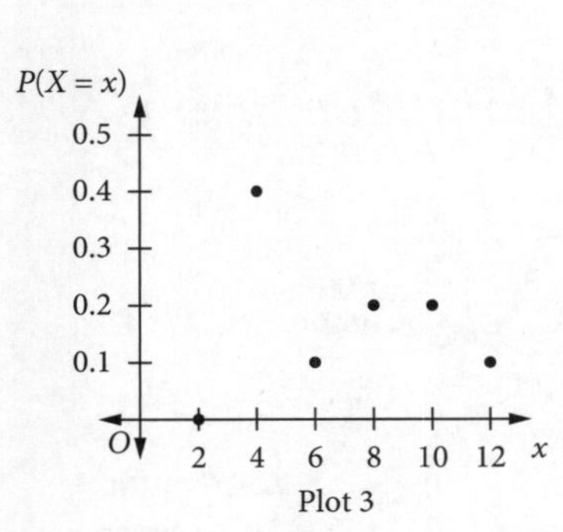

Plot 3

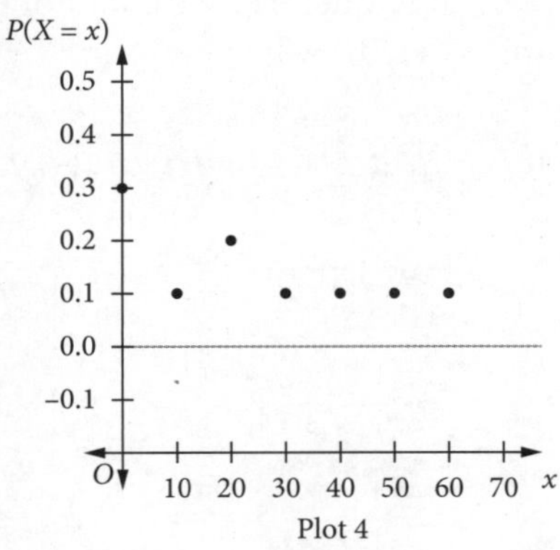

Plot 4

(a) Establish which of the four scatterplots represents a probability distribution function. Explain your answer using mathematical reasoning.

(b) Calculate the mean and variance of the plots that represent probability distribution functions.

CHAPTER 11
Descriptive statistics

11.1 STATISTICAL INVESTIGATIONS

Statistics reveal how life determines death

The Australian Bureau of Statistics (ABS) collects a vast amount of data (information) about Australians and their daily lives. The largest and one of the most important ways they do this is through the Census of Population and Housing, which is conducted every five years. It is a descriptive count of everyone who is in Australia on one particular night, what they do, and how and where they live.

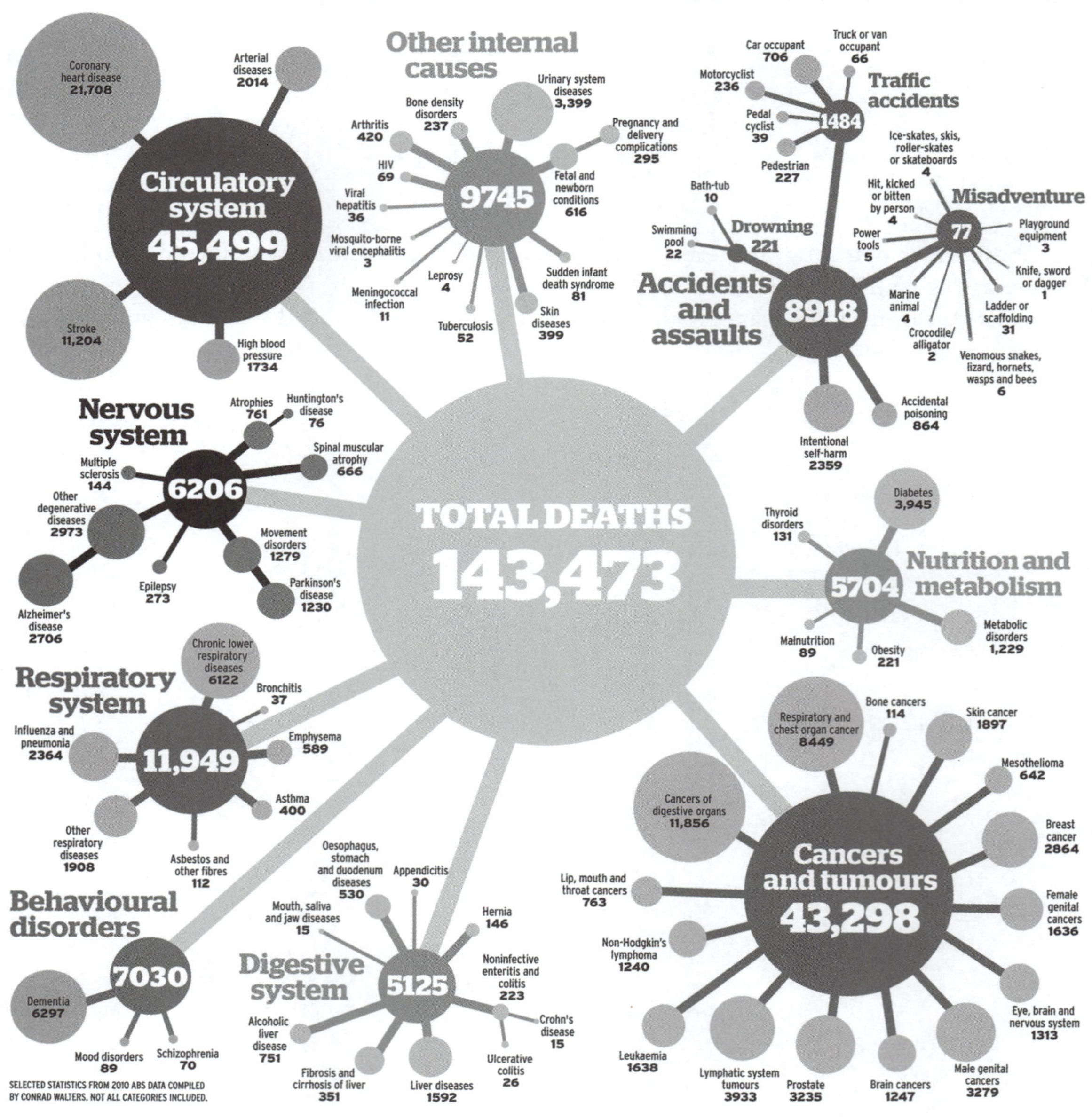

Because a census includes every member of a target population, it is a huge and costly undertaking, which is why the ABS Census is only conducted every 5 years. Surveys that involve samples of the whole population are used more frequently to predict trends for the whole population.

Information collected by the ABS can be used to help us understand about aspects of the lives of Australians—and their deaths. For example, the graphic on the previous page presents information about deaths in Australia in the year 2010 and is based on ABS statistics.

In a good statistical diagram, the data should be clearly displayed. For example, if you take a closer look at the information shown for the nervous system, you can see 761 deaths are attributed to atrophies, 76 of which are the specific atrophy called Huntington's disease.

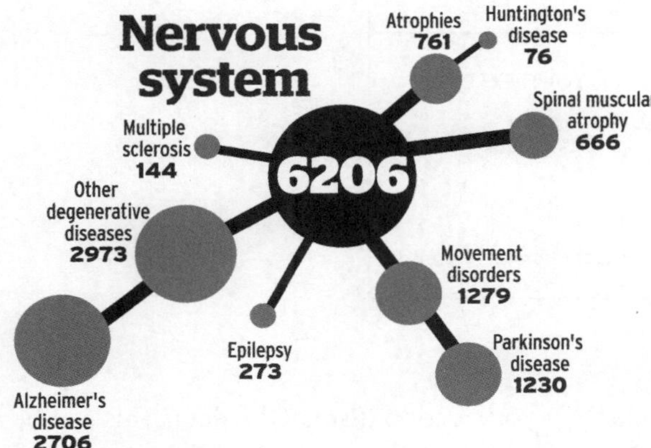

As well as just presenting data, newspaper (and other) articles and reports could include interpretation of the data or opinions that are based on the data, such as this quote from Mike Daube, the director of Curtin University's Public Health Advocacy Institute:

> 'Because of their later uptake of smoking, we are only now starting to see the declines in female mortality. But in 20 years we will be wondering why we did so little on obesity. Diabetes and all these problems caused by obesity, I think they are going to be looking catastrophic. The crisis then is not going to be in emergency departments but in chronic conditions.'

'Statistics reveal how life determines death', by Amy Corderoy, *The Age*, 29 December 2014.

His opinion may be valid due to his expertise, but it is not statistical fact based on the data.

Conducting a statistical investigation

A statistical investigation should typically include four phases:

Phase 1: Identify the problem and set a statistical question

Phase 2: Collect statistical data

Phase 3: Analyse the data

Phase 4: Interpret and communicate the results.

To help understand these phases, consider how the phases apply in a simple example. Imagine that you are conducting a small investigation into some of your classmates' characteristics. You will be collecting information on the following:

- gender
- left-handed or right-handed
- arm span (to nearest centimetre)
- position in family (eldest, middle or youngest child).

Phase 1: Identify the problem and set a statistical question.

The problem is to discover whether connections exist between pairs of these characteristics for the students in your class. For example, you could compare students based on whether they are right-handed or left-handed and their position in the family.

Phase 2: Collect the statistical data.

As a data collection device you could use a simple card as shown.

GENDER: FEMALE	ARM SPAN: 153 cm Stretch out your arms and measure from finger tip to finger tip
(LEFT-HANDED) RIGHT-HANDED Circle the correct response	YOUNGEST (MIDDLE) ELDEST Circle the correct response

Phase 3: Analyse the data.

After you have collected data, to analyse it you need to display the data. This can be done in a table, a bar graph, a dot plot, a stem-and-leaf plot or a histogram. (These display tools will be discussed in more detail later.) Two-way tables could be used to pair up the variables as shown below. Note that e.g. 150–<160 means values including 150 but less than 160.

Arm span (cm)	Male	Female
150–<160	2	4
160–<170	3	5
170–<180	3	2
180–<190	2	1
190–<200	2	0

Place in family	Left-handed	Right-handed
Youngest	3	8
Middle	2	4
Eldest	1	6

Phase 4: Interpret and communicate the results.

For the final part of the process you will need to write some conclusions about the data. Some examples based on the sample data recorded above could be:

- The median arm span for males was greater than the median arm span for females.
- A greater percentage of right-handed students were the eldest in the family compared with left-handed students.

You could also draw some appropriate statistical graphs to help explain the connections you have found.

EXERCISE 11.1 STATISTICAL INVESTIGATIONS

1 Answer the following questions. Use the graphic and the information from the section 'Statistics reveal how life determines death' on page 288.

(a) For which year are these statistics compiled?

(b) What percentage of deaths, to 1 decimal place, is attributed to each of the following causes?

(i) accidents and assaults **(ii)** cancers and tumours **(iii)** respiratory system

2 Look at the following graph that relates to Tennant Creek in the Northern Territory.

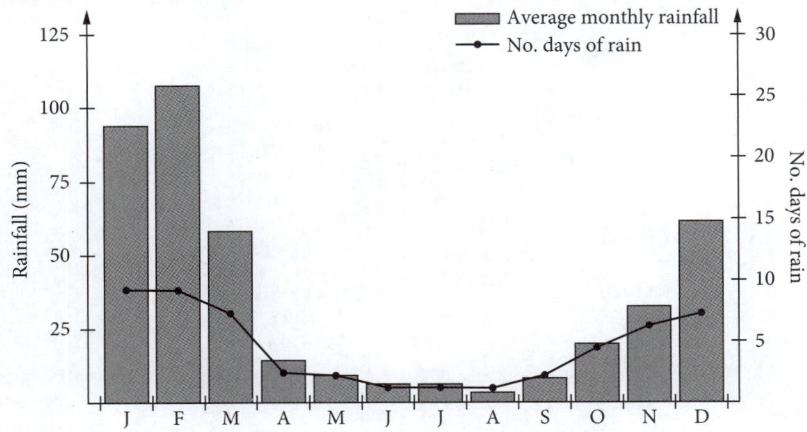

(a) Why does this graph have a vertical axis on both sides with a different scale on each?
(b) Which month has the highest average monthly rainfall?
(c) Which month has the most days of rain?

Answer the following as true or false. If false, rewrite the statement to make it true.

(d) The average daily rainfall in January must be more than in February because the same number of rainy days produces more rainfall in January than in February.
(e) The days of rain in August are rare and when it rains the rainfall is virtually non-existent.
(f) The production of this graph most likely falls into the identification phase of the statistical investigation process.

3 Your class is conducting a statistical investigation. You are in the group that is interviewing students about their study plans when they finish school. In which part of the investigation process are you involved?

A identification of a problem and the setting of a statistical question
B collection of the statistical data
C analysis of the data
D interpretation and communication of the results

4 Your class is conducting a statistical investigation. You are in the group that is presenting the findings to the school council. In which part of the investigation process are you involved?

A identification of a problem and setting of a statistical question
B collection of the statistical data
C analysis of the data
D interpretation and communication of the results

5 Look at the following graphic showing information about child adoption into the USA from other countries.

(a) If asked to interpret the numbers in the graph you would need to read the article that was published with the graph. True or false? If false, rewrite the statement to make it true.
(b) What percentage of the adoptions, correct to 1 decimal place, is from Russia?
(c) Looking at the reported distribution by gender, how many boys and girls were adopted?
(d) Why is the total number of boys and girls different from the number in the centre of the main graphic?
(e) What percentage, correct to 1 decimal place, of the distribution by gender (reported) adoptions were female?

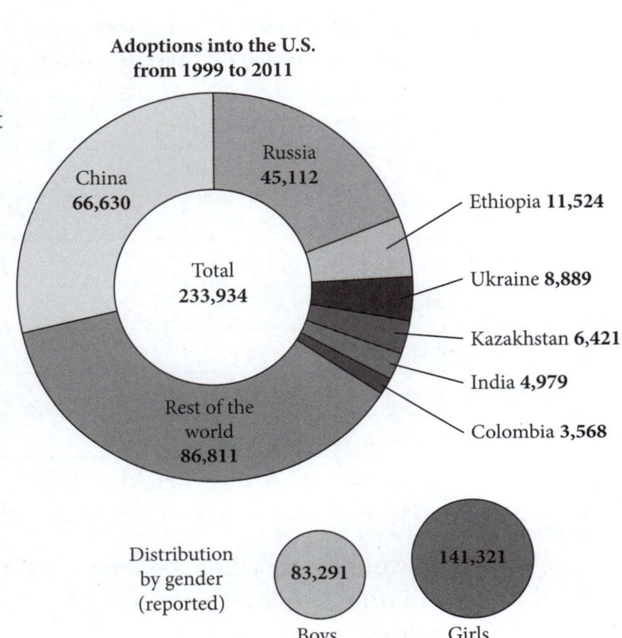

6 Look at this extract from the cause of death graphic.

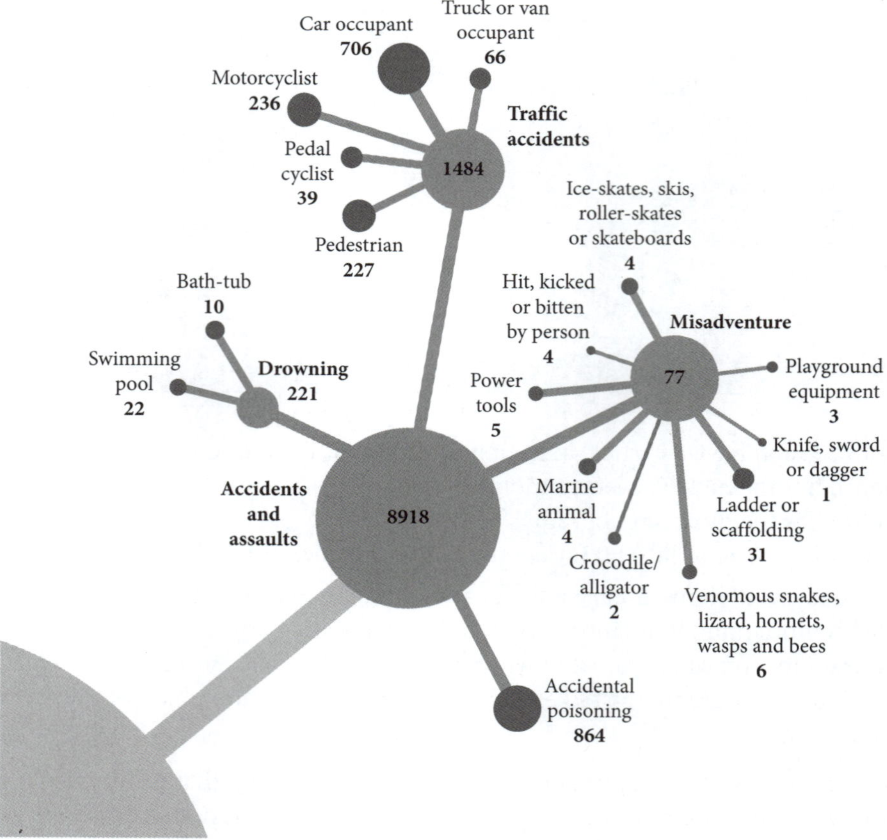

(a) Of the deaths in the 'Accidents and assaults' category, what percentage, correct to 1 decimal place, is linked to drowning?

(b) There are two sub-branches from the drowning branch. What do they add to?

(c) Based on the data in this graphic, where is it reasonable to say that most cases of drowning occur?

A In swimming pools **B** In oceans, rivers, lakes ponds and other water sources

C Among young people **D** In bath-tubs **E** At home

(d) How many deaths related to traffic accidents are not detailed in one of the sub-categories?

7 The following table shows the numbers of people presenting to hospital emergency departments with injuries caused by exercise equipment, and the exercise equipment that caused their injuries.

Total injuries	795
Treadmill	296
Weights	166
Punching bag	119
Exercise bike	116
Swiss/gym ball	48
Other exercise equipment	27
Elliptical/cross-trainer	9
Medicine/fitness ball; rowing machine; balance/wobble board; step machine/Stairmaster; yoga/exercise mat; resistance board	Less than 5 each

(a) What is your first reaction to seeing these figures?

(b) What information is missing from the table that would enable you to make a judgement on the severity of the injuries?

(c) The article accompanying the table indicates that these figures are for Victoria over the period 2008–2012, and that in 2011–2012 alone the tally was 178. Does this additional information change the reaction you had in parts (a) and (b)?

(d) The article indicates that these figures include 227 children under the age of four. Write a paragraph that could have been included in this article that highlights this particular figure related to children.

8 Conduct the investigation outlined from page 289 above regarding four characteristics of students. You will need at least 50 respondents to make this process worthwhile. You may need a class discussion on how to choose this sample. Collate the results and use them to answer questions such as those listed below. (This may be best done as a whole class task.)

(a) What percentage of respondents is male?

(b) What percentage of respondents are left-handed males?

(c) What percentage of respondents are left-handed males who are the youngest in their family?

(d) What percentage of respondents are left-handed males who are the youngest in their family and have an arm span less than 155 cm?

(e) What collation procedures did you find useful in answering these questions?

(f) Would the percentages you found in parts (a)–(d) be the same for every sample of students who respond to the survey questions?

11.2 TYPES OF DATA

Categorical and numerical data

Data is information that will most often be raw facts that can be collected and/or measured. There are two basic types of data:

- **Categorical data** can be grouped into categories. Examples include hair colour and position in the family.
- **Numerical data** has values that can be counted or measured. Examples include height and number of brothers and sisters.

There are two types of categorical data:

- **Nominal data** has no special order associated with the possible responses. Examples include hair colour.
- **Ordinal data** does have some sort of order associated with it. Examples include position in family.

There are two types of numerical data:

- **Discrete data** has a countable number of numerical values, but they don't have to be whole numbers. Examples include the number of brothers and sisters, or the age of your teacher.
- **Continuous data** has an infinite number of possibilities within a particular range. This type of data is measured using some sort of instrument. The accuracy with which we record the data may vary but does not alter the fact that there are an infinite number of possible values. Examples include the heights of a group of people.

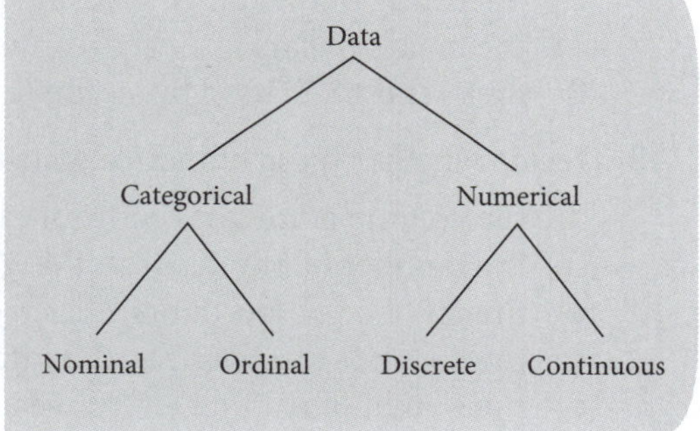

Example 1

Decide whether these data samples are ordinal or nominal (categorical), or discrete or continuous (numerical).

(a) The number of students present in your mathematics class for each lesson over a 3-week period.

(b) The favourite music style for each student in your mathematics class.

(c) The eye colour of each member of your class.

Solution

Identify whether the data is categorical or numerical. Does the data collected involve numbers? If yes, the data is numerical, otherwise it is categorical.

(a) The data is clearly numerical.
Will the data be collected by counting (discrete) or measuring (continuous)?
The data can only be whole numbers. The data is discrete.

(b) As numbers are not involved, the data is categorical. Answers might include categories such as: classical, rock, rap, jazz.
Is the order of the data important? If yes, the data is ordinal, otherwise it is nominal.
The data is nominal.

(c) As numbers are not involved, the data is categorical.
Is the order of the data important?
No, so the data is nominal.

EXERCISE 11.2 TYPES OF DATA

1 Decide whether these data examples are categorical or numerical.

(a) the amount of money carried by each student in your mathematics class on a particular day
(b) favourite chocolate truffle flavours, selected from a sample provided, for 100 people chosen at random in a shopping centre
(c) the hair colour of the teachers at your school
(d) the ages, in years, of the parents of the students in your class
(e) the individual diameter of each orange in a fruit bowl
(f) the size of running shoes worn by each of the competitors in the 100 m final at the last Olympic Games

2 Decide whether these categorical data examples are nominal or ordinal.

(a) the favourite radio station of each member of your extended family
(b) the degree of support for the new jumper design of your local sporting team, as recorded by the members of the team
(c) the starting letter or digit of the numberplate for each vehicle in the staff car park at your school
(d) the listing of the top 10 grossing movies for the week
(e) the makes of cars in a used-car lot
(f) the brand of oil used by a sample of masseurs

3 Decide whether these numerical data examples are continuous or discrete.

(a) the volume of milk taken from each cow in a herd of dairy cattle
(b) the number of brands of clothing available in a shopping centre
(c) the number of spectators at each of the Sydney Swifts home games
(d) the distance thrown by each of the competitors in the Olympic shot put for women
(e) the weight of fruit taken from each individual tree in an orchard
(f) the distance between consecutive stations on the exercise circuit in the local park

4 Decide whether these examples are nominal, ordinal, discrete or continuous.

(a) the strength of agreement with a number of statements, recorded using a scale of 1 to 5, where 1 represents strongly disagree and 5 represents strongly agree.
(b) the height of the high tide at Sydney Heads for 30 consecutive days.
(c) the number of brothers and sisters of each member of the teaching staff at your school.

5 Imagine you were setting up a series of surveys to collect various data sets. For each scenario below:
- (i) Write a survey question to ask.
- (ii) What type of data you will collect and why?
  - (a) You are trying to find out the most popular brand of soft drink.
  - (b) You are investigating the height of basketball players in a local league.
  - (c) You are investigating the degree of popularity of singers from a list of ten artists you have named.
  - (d) You are trying to determine the level of support for the question: *Should state governments be scrapped?*
  - (e) You are investigating which brand of dog food a person is most likely to buy.
  - (f) You are trying to find the ages, in years, of the players at the local netball club on their last birthday.

6 The measurement of time is supposed to result in continuous data. However, all Olympic records that involve time appear to produce discrete data. How can this be?

7 Surveys often use the numbers 1 to 5 to represent various levels of support for statements that are made. Often 1 stands for strongly disagree and 5 stands for strongly agree. The responses would appear to be numerical, but are they really? Discuss.

11.3 DISPLAYING DATA

In this section, you will look at the following data displays to represent sets of data:

- frequency table: for categorical or numerical data
- bar graph: for categorical data
- Pareto chart
- two-way table
- divided bar graph
- dot plot: for discrete numerical or categorical data
- histogram: for continuous numerical data
- stem-and-leaf plot: for discrete numerical data.

Frequency tables

A **frequency table** is used to summarise data. It can show frequency values for individual data values or grouped data. When grouped, the data may represent discrete data or continuous data.

Individual data

Number of TVs	Frequency
0	1
1	4
2	9
3	7
4	3

Grouped discrete data

Number of CDs	Frequency
0–9	3
10–19	5
20–29	6
30–39	2
40–49	1

Grouped continuous data

Distance travelled (km)	Frequency
0–<1	3
1–<2	5
2–<3	6
3–<4	2
4–<5	1

For the continuous data example, 1–<2 means responses from 1 km to less than 2 km. So, this would include distances such as 1.2 km, 1.23 km, 1.235 km and so on to 1.999… km. A distance of 2.0 km is recorded in the next group.

EXPLORE FURTHER

Frequency tables

Explore how to create frequency tables using a spreadsheet to display frequency, relative frequency and percentage frequency.

Bar graphs

A **bar graph** is used to display categorical (nominal) information. Values are simply read from the graph, using the frequency scale. Bar graphs can have vertical or horizontal bars, which are spaced out evenly and a gap is left between each of the categories. When vertical bars are used, the graph is sometimes called a column graph.

Example 2

Draw a bar graph to represent the following data collected as a result of a survey asking for the respondents' favourite code of football.

AFL	AFL	NRL	AFL	AFL	Soccer	AFL	NRL	NRL	AFL	Soccer
Soccer	NRL	Union	AFL	Union	NRL	Union	AFL	NRL	NRL	Union
Union	AFL	AFL	NRL	NRL	Union	AFL	Soccer	Union	NRL	NRL
AFL	Soccer	Soccer	AFL	Union	NRL	AFL	Union	NRL	AFL	Soccer

Solution

Construct a frequency table for the following data.

Football code	Frequency
AFL	15
NRL	13
Union	9
Soccer	7

Draw and label the bar graph, remembering to leave a gap between each of the categories.

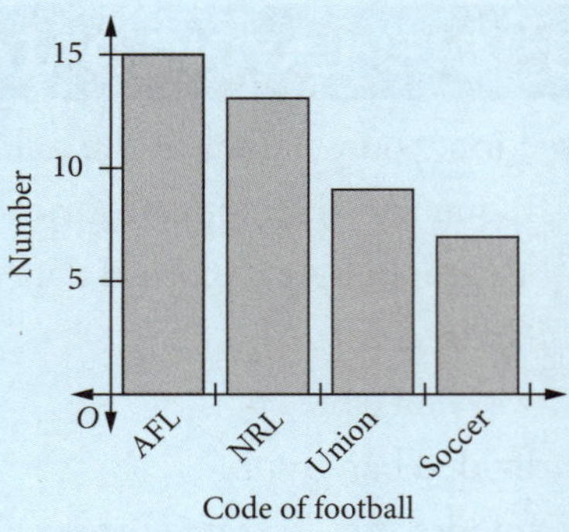

EXPLORE FURTHER

Bar graphs

Explore how to create, read and compare bar graphs using a spreadsheet.

Pareto charts

A Pareto chart is a type of chart that contains both bars and a line graph, where individual values are represented in descending order by the bars and the cumulative total is represented by the line graph. It is named after Vilfredo Pareto, an influential Italian sociologist and economist (although he originally trained as an engineer), who is also known for his work in politics and philosophy.

The left axis of the Pareto chart gives the frequency, cost or another unit of measure for the information shown in the bar graph, whilst the right axis shows the cumulative percentage total. By having the bars in descending order, the most frequent items are shown first.

The purpose of a Pareto chart is to distinguish the 'vital few' from the 'trivial many'. For example, if you are trying to analyse various different problems to decide which problems are the most important, then a Pareto chart can helpfully show which problems are having the greatest impact. The cumulative line will show how much of the total will be solved by fixing the few most frequent problems shown in the bar graph.

Example 3

An online seller of clothing summarised the complaints received in a month in the following table.

Type of complaint	Number of complaints
Problems completing the order online	50
Difficulty accessing website	30
Cancelled order	9
Wrong article sent	5
Overcharging for delivery	4
Late delivery	2
Total	**100**

(a) Copy this table and add percentage and cumulative percentage columns.

(b) Construct a Pareto chart for this information.

(c) Which problems account for 80% of the complaints?

Solution

(a)

Complaint	Number of complaints	Percentage	Cumulative percentage
Problems completing the order online	50	50	50
Difficulty accessing website	30	30	80
Cancelled order	9	9	89
Wrong article sent	5	5	94
Overcharging for delivery	4	4	98
Late delivery	2	2	100
	100	**100**	

(b)

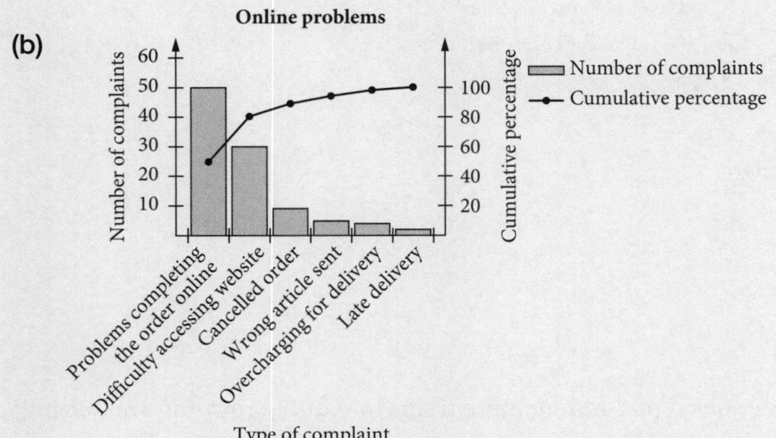

(c) 80% of the complaints occur in the categories 'Problems completing the order online' and 'Difficulty accessing the website'. These are the two areas that you would concentrate on fixing. If drawing a Pareto chart using spreadsheet software, it will help set the gridlines from the 'Cumulative percentage' axis as this will then clearly show the 80% line.

Example 4

A suggestion box is left at a train station for 5 days. The complaints received are summarised in the following table.

Type of complaint	Number of complaints
Crowded platform	1100
Lateness	1655
Crowded train	3290
Lack of toilets on platform	55
Dirty carriages	100
Total	**6200**

(a) Arrange the number of complaints in descending order and add percentage and cumulative percentage columns.

(b) Construct a Pareto chart for this information.

(c) What should the train company do to make the customers happier?

Solution

(a)

Type of complaint	Frequency	Percentage	Cumulative percentage
Crowded train	3290	53.1	53.1
Lateness	1655	26.7	79.8
Crowded platform	1100	17.7	97.5
Dirty carriages	100	1.6	99.1
Lack of toilets on platform	55	0.9	100
	6200	**100**	

(b)

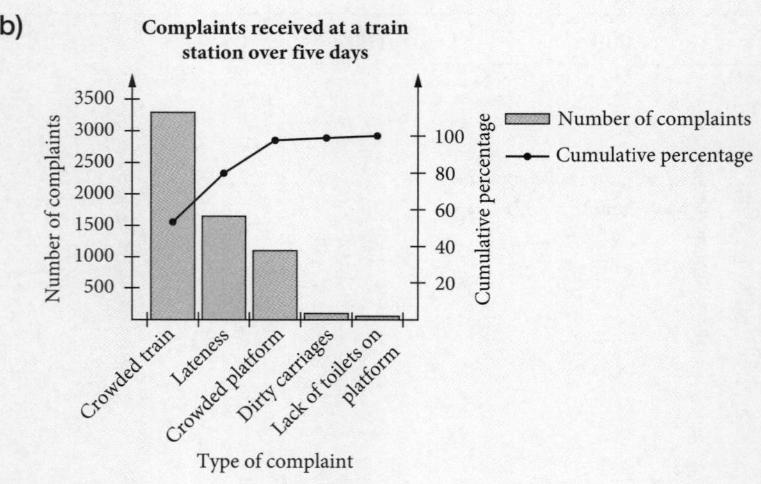

(c) Most customers' problems would be solved by running more trains to reduce crowding and having them run on time.

Two-way tables

Two-way tables allow more information to be obtained from the table. As well as the information in each cell, you can also read information from each row or column.

Example 5

The foreign language studied by the students in a class are listed in the following table.

	Japanese	French	German	Total
Girls	6	5	4	15
Boys	2	7	6	15
Total	8	12	10	30

(a) How many students are in the class?

(b) How many girls are studying Japanese or German?

(c) How many students are studying French?

(d) What proportion of the students are not studying German?

Solution

(a) 30 students (the total number of girls and boys)

(b) Studying Japanese or German = $6 + 4 = 10$

(c) Studying French = $5 + 7 = 12$

(d) Studying German = 10

Not studying German = $30 - 10 = 20$

Proportion not studying German = $\frac{20}{30} = \frac{2}{3}$

Divided bar graphs

A **divided bar graph** or **segmented bar graph** is a rectangle divided into lengths according to the proportion of each group or category. Sometimes you can convert the values to percentages, rounded appropriately, to enable a simpler scale to be used.

Example 6

Draw a divided bar graph to represent the number of grams of various components in a particular brand of muesli bars.

Total per serving (g)	Protein (g)	Fat (g)	Carbohydrate (g)	Dietary fibre (g)	Other (g)
35	2.3	4.9	20.9	2.6	4.3

Solution

State the total grams per serving: There are 35 g in total.

Find the factors of the total (this will give convenient lengths for the rectangular whole):

$35 = 1 \times 35, 5 \times 7$

Select an appropriate length (approximately 10–20 cm where possible). Determine the length of each 'piece' by dividing the amount present by the number of grams per centimetre. Where necessary, state your answers to 1 decimal place:

Length: 7 cm

$$\frac{35}{7} = 5\,\text{g/cm (each centimetre represents 5 g)}$$

Protein: $\frac{2.3}{5} = 0.46\,\text{cm}$ Dietary fibre: $\frac{2.6}{5} = 0.52\,\text{cm}$

Fat: $\frac{4.9}{5} = 0.98\,\text{cm}$ Other: $\frac{4.3}{5} = 0.86\,\text{cm}$

Carbohydrate: $\frac{20.9}{5} = 4.18\,\text{cm}$

Draw a rectangle of a suitable length and divide it into pieces according to the category values. Include a key.

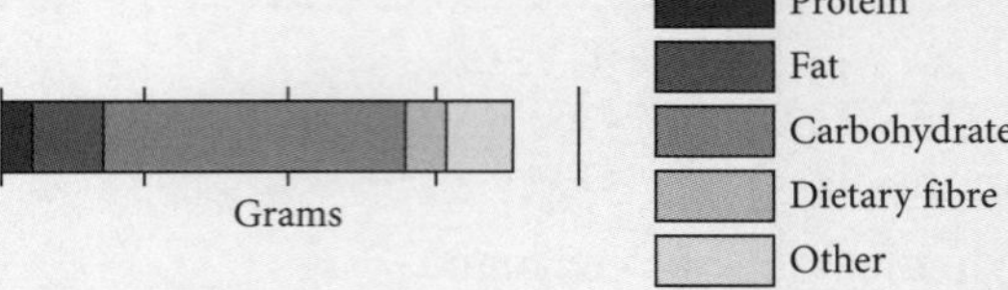

Dot plots

A **dot plot** can be used with either discrete numerical or categorical data. The categories or values are written along the horizontal axis, and then dots are arranged vertically to represent the number of each category. In this chapter you will only use dot plots for numerical data because categorical data can be better represented using a bar graph.

Example 7

Draw a dot plot to represent the following discrete data set.

Score	Frequency
7	4
8	8
9	6
10	2
11	1
Total	**21**

Solution

Draw a horizontal line marked with the values from the 'Score' column:

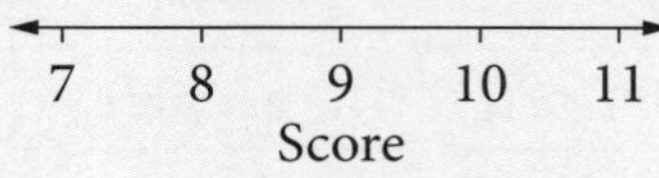

For each score add the number of dots represented by the 'Frequency' value.

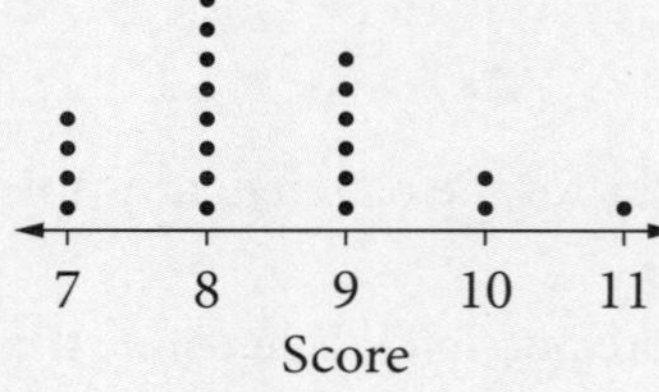

A dot plot is useful for identifying symmetry in a data distribution and for identifying extreme values that don't seem to fit with the other values. These values are called **outliers**.

A dot plot should not be used if there are too many values. It would be best to limit this type of display to no more than 10 columns. A dot plot can be used when the data is in its raw state. This means the data has not already been collated into a frequency table. In cases like this you should go through the data set once, marking dots in the appropriate places as you go.

A dot plot is also not very useful if there are high frequencies involved. As each value needs to be represented by its own dot you don't want to be drawing a dot plot where the individual frequencies are greater than 10.

Histograms

A **histogram** is used to display continuous numerical data. A histogram looks similar to a bar graph, but there are no gaps between the bars. Also, the first column is usually placed one half-column width from the vertical axis.

Example 8

Draw a histogram of the following data that represents the haemoglobin level for a sample of students.

Haemoglobin level	Frequency
9.0–<10	1
10.0–<11	2
11.0–<12	5
12.0–<13	2
13.0–<14	7
14.0–<15	5
15.0–<16	2
16.0–<17	1
Total	**25**

Solution

Find the minimum and maximum values to determine the range to be marked on the axes:

The minimum is greater than or equal to 9.

The maximum is less than 17.

The frequency is from 1 to 7.

Draw a set of axes marking 'Haemoglobin level' on the horizontal axis and 'Frequency' on the vertical axis. Draw a break at the start of the axis to indicate that part of the horizontal axis has been left out and complete the diagram.

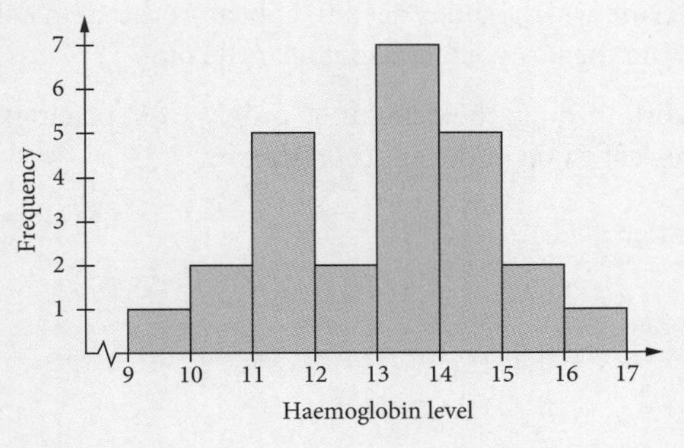

Stem-and-leaf plots

A **stem-and-leaf plot** (or **stem plot**) is a useful way to display grouped discrete data. It does not lose any of the information, unlike for example a histogram (which does not show individual values).

A stem-and-leaf plot consists of two parts: a stem and a leaf. Typically, the leaf contains the last digit of the number and the stem contains all of the other digits. For example, the leaf may represent the 'ones'—single-digit numbers 1 to 9—and the stem will represent the rest of the number (tens, hundreds, thousands etc.). Use spaces, not commas, to separate each entry.

The values in the leaf section should be ordered. There should also be a key that explains the sort of values contained in the plot. The key will use a vertical line to show how the numbers are divided into stem and leaf.

The stem is usually based on a class interval of 10 but is also often based on 5. If based on 5, you can use subscript letters L (meaning *low*) and H (meaning *high*) to indicate where values belong. So, for a stem of 6, leaf values of 0, 1, 2, 3 and 4 would be placed alongside 6_L and leaf values of 5, 6, 7, 8 and 9 would be placed alongside 6_H, as in the following example that shows the data set 61, 61, 62, 64, 66, 68, 69, 69:

Stem	Leaf
6_L	1 1 2 4
6_H	6 8 9 9

Key: 6|4 = 64

A stem-and-leaf plot can also be used to record decimal values. For example, the following shows the times recorded for competitors in the 100 m run at an athletics carnival.

Stem	Leaf
10	9 9
11	0 1 4 5 6 8 9
12	0 1 1 1 2 4 5 6 9
13	0 0 1 1 2 2 2 2 3 4 6 7

Key: 10|9 = 10.9

Example 9

A class obtained the following marks, out of 100, on a mathematics test.

42, 55, 46, 78, 83, 85, 99, 78, 83, 66, 50, 49, 72, 84, 92, 81, 60, 56, 57, 82, 78, 57, 93, 58

Show the data in a stem-and-leaf plot with a class interval of 10.

Solution

Decide which values need to appear in the stem column. List them down the page in ascending order. Write the key you are using near the plot.

Work through the data in the order given, placing the leaf value in the appropriate row.

Stem	Leaf
4	2 6 9
5	5 0 6 7 7 8
6	6 0
7	8 8 2 8
8	3 5 3 4 1 2
9	9 2 3

Key: 5|0 = 50

Redraw the plot with the leaf values in order.

Stem	Leaf
4	2 6 9
5	0 5 6 7 7 8
6	0 6
7	2 8 8 8
8	1 2 3 3 4 5
9	2 3 9

Key: 5|0 = 50

EXERCISE 11.3 DISPLAYING DATA

1 Use this data set for the following questions.

red, blue, green, blue, red, yellow, blue, green, red, green, yellow, blue, red, green, green, yellow, blue, red, green blue, yellow, red, green, yellow, blue, red, red, blue, yellow, yellow.

(a) Complete a frequency table and draw a bar graph to represent the set of categorical (nominal) data.

Colour	Frequency
Red	
Blue	
Green	
Yellow	

(b) Which two colours each have a frequency of 7?

2 Use this data set for the following questions.

Holden, BMW, Mazda, BMW, Holden, Ford, Ford, Ford, Holden, BMW, Mazda, Mazda, Holden, Ford, Holden, Ford, Ford, Ford, Mazda, BMW, Holden, Ford, Ford, Ford.

(a) Draw a bar graph to represent the set of categorical (nominal) data.
(b) Which car maker has a frequency of 6, and which has a frequency of 10?

3 Use this data set for the following questions.

blonde, brown, blonde, blonde, brown, red, red, blonde, brown, brown, brown, red, red, brown, blonde, red, blonde, blonde, brown, brown, red, blonde, brown, blonde, brown.

(a) Complete a frequency table and draw a bar graph to represent the set of categorical (nominal) data.

Colour	Frequency
Blonde	
Brown	
Red	

(b) What is the least frequently observed hair colour?

4 A survey conducted by a bus company received the results as summarised in the following table.

Type of complaint	Number of complaints
Punctuality	120
Cleanliness	30
Slow trip	20
Lack of seats	80
Cost of trip	15
Attitude of the driver	5
Total	**270**

(a) Find the percentage for each complaint and reorder the table staring with the largest number of complaints. Then complete the cumulative percentage column.
(b) Draw the Pareto chart for this table.
(c) Which areas would you work on in order to reduce the number of complaints received?

5 A researcher collected the following information as to the causes of hair loss.

Cause	Frequency
Genetics	55
Cosmetic damage	25
Stress	11
Smoking	10
Vitamin deficiency	4
Medication	3
Other treatments	2
Total	**110**

(a) Add the percentage and cumulative percentage columns to this table.
(b) Draw the Pareto chart for this information.

6 A survey of 100 high school students was carried out asking what type of phone they owned. The results are contained in the following table.

	Apple	Samsung	LG	Other	Total
Boys	15	12	15	8	50
Girls	20	10	15	5	50
Total	35	22	30	13	100

(a) How many students had an Apple phone?
(b) How many students did not have a Samsung phone?
(c) What proportion of the students had a LG phone?
(d) If a student was selected at random from the group, what is the probability that their phone was not an Apple, Samsung or LG?

7 Information about the numbers from 1 to 50 inclusive is contained in the following table.

	Multiple of 3	Multiple of 7	Total
Odd	8	4	12
Even	8	3	11
Total	16	7	23

(a) How many multiples of 3 are there?
(b) Which two numbers have been included in two different groups?
(c) How many numbers from 1 to 50 inclusive are not included in this table?

8 Draw a segmented bar graph for the following data using the following steps:

1 Use the total grams per serving to determine a convenient length for the segmented bar graph.
2 Calculate the length of each component to 1 decimal place, and then draw the segmented bar graph to represent the composition of the food source.

(a) Chick peas

Total grams per serving	Protein	Total fat	Carbohydrate	Dietary fibre	Other
85	6	1.9	16.3	6.4	54.4

(i) The total grams per serving is 85, hence a convenient length for the graph is [15/16/17/18/19] cm

(ii) The length of each component, to 1 decimal place:

Protein: Dietary fibre:

Total fat: Other:

Carbohydrate:

(iii) If 68 cm is drawn to represent the total, then how many cm should protein and carbohydrate take?

(b) A can of baked beans

Total grams per serving	Protein	Total fat	Carbohydrate	Dietary fibre	Other
125	6	0.9	14.3	6.5	97.3

(i) The total grams per serving is 125, hence a convenient length for the graph is [12.5, 18, 19, 22, 29] cm

(ii) The length of each component, to 1 decimal place, is:

Protein: Dietary fibre:

Total fat: Other:

Carbohydrate:

(iii) If 50 cm is drawn to represent the total, then how many cm should protein and carbohydrate take?

(c) A jar of reduced-sugar fruit jam

Total grams per serving	Protein	Total fat	Carbohydrate	Dietary fibre	Other
15	0.1	0.1	4.8	0.3	9.7

(i) The total grams per serving is 15, hence a convenient length for the chart is [13, 15, 16, 17, 19] cm

(ii) The length of each component, to 1 decimal place, is:

Protein: Dietary fibre:

Total fat: Other:

Carbohydrate:

(iii) If 60 cm is drawn to represent the total, then how many cm should protein and carbohydrate take?

(d) Curry paste

Total grams per serving	Protein	Total fat	Carbohydrate	Other
50	1.6	8.4	3.1	36.9

(i) The total grams per serving is 50, hence a convenient length for the chart is [6, 7, 8, 9, 10] cm.

(ii) The length of each component, to 1 decimal place, is:

Protein: Dietary fibre:

Total fat: Other:

Carbohydrate:

(iii) If a 20 cm bar is drawn to represent the total, what length should be allocated to protein and what length should be allocated to carbohydrate?

9 Show the following data in a stem-and-leaf plot with a class interval of 10.

(a) A class obtained the following marks, out of 100, on a mathematics test.
36, 47, 44, 93, 82, 80, 70, 75, 66, 51, 56, 84, 92, 97, 35, 48, 91, 90, 82, 67, 54, 77, 82, 84

(b) A class obtained the following marks, out of 100, on a science test.
42, 67, 81, 37, 78, 98, 35, 42, 66, 78, 50, 72, 71, 92, 67, 56, 66, 87, 92, 93, 91, 65, 39, 32

10 Which of the following statements is incorrect?

A A histogram is used for numerical (continuous) data.
B A histogram is used for numerical (discrete) data.
C A dot plot can be used for numerical (discrete) data.
D A bar graph should never be used for numerical data.

11 The scores in the first round of a club golf championship were as follows:

68, 72, 76, 84, 82, 80, 69, 71, 70, 73, 77, 76, 81, 80, 87, 88, 79, 76, 69, 71, 74, 75, 77, 77, 78, 82, 84, 83, 82, 77, 76, 70, 88, 73, 72, 71, 74, 66, 77, 79, 64

(a) What is the range or difference between the highest and the lowest scores?
(b) Create a stem-and-leaf plot showing the data.

12 Draw dot plots to represent the following numerical (discrete) data sets.

(a) 2, 0, 1, 5, 3, 4, 2, 1, 0, 2, 3, 0, 1, 3, 1, 2, 2, 4, 3, 1, 1, 2, 4, 2
(b) Sanjay, the best bowler in the local cricket team, recorded the number of wickets he took in each innings of the season. The following values were recorded:
4, 3, 5, 2, 1, 0, 3, 4, 2, 5, 6, 3, 2, 4, 5, 3, 0, 3, 5, 4.

13 **(a)** Draw a histogram to represent the following set of continuous numerical data.

Score	Frequency
0–<10	9
10–<20	18
20–<30	13
30–<40	15
40–<50	14
50–<60	16

What are the lowest and highest scores that could have been included in the data?

(b) Draw a histogram to represent the following set of continuous numerical data.

Score	Frequency
141–<146	22
146–<151	31
151–<156	21
156–<161	26
161–<166	22
166–<171	24

What are the lowest and highest scores that could have been included in the data?

14 Show the following data in a stem-and-leaf plot with a class interval of 5.

(a) A class obtained the following marks, out of 50, on a mathematics test.
22, 24, 26, 45, 46, 15, 24, 26, 31, 37, 42, 30, 40, 44, 25, 41, 27, 22, 30, 31, 35, 45, 18, 21
(b) A class obtained the following marks, out of 50, on a science test.
32, 41, 30, 20, 17, 22, 31, 43, 29, 33, 35, 42, 41, 45, 47, 22, 16, 19, 31, 30, 29, 27, 36, 40

15 The following data represents the number of kilometres travelled each day by a solar vehicle. The distances have been rounded to the nearest whole number of kilometres.

112, 95, 122, 78, 99, 145, 133, 160, 145, 144, 78, 66, 68, 79, 93, 125, 133, 89, 67, 78, 114, 134, 80, 65, 105, 115, 127, 134, 117, 84

(a) Given the way the data has been recorded would you consider it to be discrete or continuous? Give a brief justification for your answer.
(b) Construct a frequency table for the data, using a class interval of 10 and a starting value of 61.
(c) Looking only at the frequency table, what can you say about the maximum distance travelled on any one day?

(d) Construct a stem-and-leaf plot for the data.
(e) What can this stem-and-leaf plot tell us that the frequency table cannot?
(f) What does this question reinforce about stem-and-leaf plots compared to frequency tables?

16 A survey of students asking them their favourite sport produced the information given in the following table.

	Football	Swimming	Basketball	Total
Male	32	30		70
Female		28	12	
Total	50		20	

(a) Copy the table and fill in the missing information.
(b) How many students gave swimming as their favourite sport?
(c) What proportion of the males gave basketball as their favourite sport?
(d) A student is chosen at random from the group. What is the probability that they gave football as their favourite sport?

17 A manufacturer received the following complaints from purchasers of their new ovens.

Defect	Frequency
Faulty light	17
Faulty fan	12
Faulty element	9
Damaged box	4
Scratched glass	3
Other	2
Total	**47**

(a) Add the percentage and cumulative percentage columns to this table.
(b) Draw the Pareto chart for this information.

18 A hotel chain conducted a survey of their guests and received the following responses.

Type of complaint	Frequency
Poor housekeeping	45
Slow check-in/checkout	25
Poor breakfast	15
Not value for money	9
Poor WiFi	6
Television	4
Parking	2
Total	**106**

(a) Add the percentage and cumulative percentage columns to this table.
(b) Draw the Pareto chart for this information.
(c) How could the hotel chain improve their customers' satisfaction?

11.4 MEASURES OF CENTRAL TENDENCY

When dealing with data, the three most common measures of central tendency are mean, median and mode. You should recall that a measure of central tendency is a value that can be used to represent the data set.

Measures of central tendency—mean

To calculate the mean ($\bar{x}$) of a data set, you can use the formula:

$$\bar{x} = \frac{\text{sum of data values}}{\text{number of data values}}$$

$$= \frac{\sum_{i=1}^{n} x_i}{n}$$, where Σ(capital Greek letter sigma) stands for 'the sum of'.

To calculate the mean of data in a frequency table, you can use the formula:

$$\bar{x} = \frac{\Sigma xf}{\Sigma f}$$, where f stands for 'frequency'.

To calculate the mean of grouped data, you can use the formula:

$$\bar{x} = \frac{\Sigma x_m f}{\Sigma f}$$, where the midpoint (x_m) of each class interval represents the data value.

Calculating the mean of discrete data

Example 10

Find the mean of the following data sets. Give your answer correct to 2 decimal places, if necessary.

(a) The number of cars that passed the school gate in a number of 5-minute periods:
10, 5, 12, 13, 14, 7, 12, 15, 7, 3, 2, 8

(b) The number of siblings (brothers and sisters) for the students in two Year 11 classes:

Number of siblings (x)	Frequency (f)
0	7
1	11
2	15
3	3
4	1

Solution

(a) Add all the data values and divide by the number of data values:

$$\bar{x} = \frac{10+5+12+13+14+7+12+15+7+3+2+8}{12}$$

$$= \frac{108}{12}$$

$$= 9$$

(b) Add a new column to the frequency table, labelled *xf* (*x* stands for the data value, *f* for frequency). Fill in this column by multiplying the data values (in this case, the number of siblings) by the frequency for each value. Add up the values in the *f* and *xf* columns.

Number of siblings (x)	Frequency (f)	(xf)
0	7	0
1	11	11
2	15	30
3	3	9
4	1	4
Total	$\sum f = 37$	$\sum xf = 54$

Substitute the values into the formula $\overline{x} = \dfrac{\sum xf}{\sum f}$: $\overline{x} = \dfrac{54}{37} = 1.46$

Example 11

Find the mean of the following data set. Give your answer correct to 2 decimal places, if necessary.

The following data represents the time in seconds taken by the competitors in the under-17 100 m dash at the school athletics carnival.

Time (x)	Frequency (f)
12–<12.5	3
12.5–<13	5
13–<13.5	12
13.5–<14	28
14–<14.5	10

Solution

Add two new columns to the frequency table, one labelled x_m and the other labelled $x_m f$. The values for x_m are calculated by adding together the endpoints of the class interval and dividing by 2. Add up the values in the frequency column and the $x_m f$ column.

Time (x)	Frequency (f)	x_m	$x_m f$
12–<12.5	3	12.25	36.75
12.5–<13	5	12.75	63.75
13–<13.5	12	13.25	159
13.5–<14	28	13.75	385
14–<14.5	10	14.25	142.5
	$\sum f = 58$		$\sum x_m f = 787$

Substitute the values into the formula $\overline{x} = \dfrac{\sum x_m f}{\sum f}$: $\overline{x} = \dfrac{787}{58} = 13.57$

This information could be entered into a spreadsheet. Then, by entering the appropriate formula the mean is calculated.

Measures of central tendency—median

The median is the middle data value of an ordered data list. The method for finding the median depends on whether there is an odd or even number of data values.

For odd n, the median is the $\left(\frac{n+1}{2}\right)$th value.

For even n, the median is the average of the $\left(\frac{n}{2}\right)$th and the $\left(\frac{n+2}{2}\right)$th values.

When data is grouped, you should state the class interval in which the median falls.

Example 12

Find the median of the following data set.

(a) Number of cars that passed the school gate in a number of 5 minute periods.
10, 5, 12, 13, 14, 7, 12, 15, 7, 3, 2, 8

(b) Number of brothers and sisters for the students in two Year 11 classes.

Number of siblings	Frequency
0	7
1	11
2	15
3	3
4	1

Solution

(a) Rewrite the data in numerical order from smallest to largest:
2, 3, 5, 7, 7, 8, 10, 12, 12, 13, 14, 15

There is an even number of data values, so identify the $\left(\frac{n}{2}\right)$th and $\left(\frac{n+2}{2}\right)$th values:

$\left(\frac{n}{2}\right)$th $= \frac{12}{2} =$ 6th value so count from the left until you get to the 6th value which is 8.

$\left(\frac{n+2}{2}\right)$th = 7th value which is 10.

The median is the average of these two values: $\frac{8+10}{2} = 9$
The median is 9.

(b) The frequency table has effectively put the data in order already. Find Σf to decide if you are dealing with an odd or even number of data values.

Number of siblings	Frequency
0	7
1	11
2	15
3	3
4	1
	$\Sigma f = 37$

There is an odd number of data values so the median will be the $\left(\frac{n+1}{2}\right)$th value: $\left(\frac{n+1}{2}\right)$th value $= \frac{37+1}{2} = 19$

Count down the frequency column to find the required value and state the answer: 19th value is 2.
The median number of brothers and sisters is 2.

Example 13

Find the median of the following data set, which shows the time taken by the competitors in the under-17 100 m dash at the school athletics carnival.

Time (x)	Frequency (f)
12–<12.5	3
12.5–<13	5
13–<13.5	12
13.5–<14	28
14–<14.5	10

Solution

The frequency table has ordered data. Find Σf and see whether there is an odd or even number of data values:

Time (x)	Frequency (f)
12–<12.5	3
12.5–<13	5
13–<13.5	12
13.5–<14	28
14–<14.5	10
	$\Sigma f = 58$

There is an even number of data values, so the median will be the average of the $\left(\frac{n}{2}\right)$th and $\left(\frac{n+2}{2}\right)$th values:

$\left(\frac{n}{2}\right)$th value $= \frac{58}{2} = 29$th value.

Count down the frequency column until you get to the 29th value: in the interval 13.5–<14

$\left(\frac{n+2}{2}\right)$th value $= 30$th value

Count down the frequency column until you get to the 30th value: in the interval 13.5–<14

The median is in the interval 13.5–<14.

Measures of central tendency—mode

The mode is the most frequently occurring data value. A data set with a unique mode is sometimes referred to as **unimodal**. If a data set has two results with the same equal highest frequency, then the data is **bimodal**. If more than two results share the highest frequency we usually say the data has no mode. Sometimes this is described as **multimodal**.

Example 14

Find the mode of the following data sets.

(a) The number of cars that passed the school gate in a number of 5 minute periods:

10, 5, 12, 13, 14, 7, 12, 15, 12, 3, 2, 8

(b) The number of brothers and sisters for the students in two Year 11 classes:

Number of siblings (x)	Frequency (f)
0	7
1	11
2	15
3	3
4	1

Solution

(a) Write the data in order to make it easier to identify the mode: 2, 3, 5, 7, 8, 10, 12, 12, 12, 13, 14, 15
The score of 12 has the highest frequency 3 so the mode is 12.

(b) Identify the highest frequency from the frequency table: The score 2 has a frequency of 15.
The mode is 2.

Example 15

Find the mode of the following data set, which shows the time taken by the competitors in the under-17 100 metre dash at the school athletics carnival.

Time (x)	Frequency (f)
12–<12.5	3
12.5–<13	5
13–<13.5	12
13.5–<14	28
14–<14.5	10

Solution

Identify the highest frequency from the frequency table: The modal class is 13.5–<14

Technology does not generally display the mode as one of the summary statistics. However, it is easy enough to find without the use of technology.

Quartiles

The median divides the data set into two equal 'halves'. The data set could be divided into many equal groups called quantiles. In this subject you will only consider dividing the data into four groups called quartiles.

- The median is referred to as the second quartile, Q_2.
- The median of the lower half of the data is called the first quartile Q_1 or lower quartile Q_L.
- The median of the upper half of the data is called the third quartile Q_3 or upper quartile Q_U.

Remember to order the data from smallest to largest before you find these values.

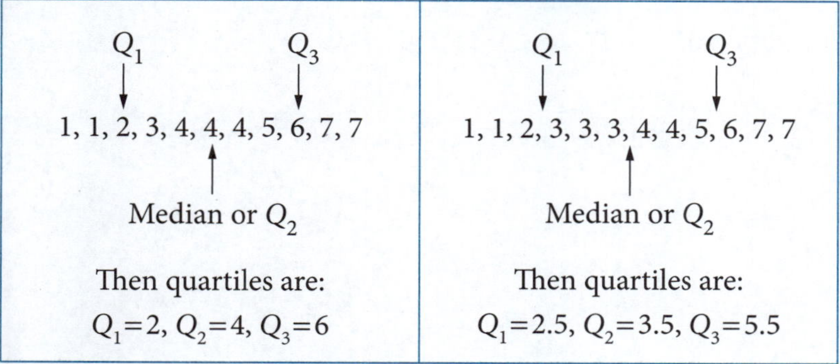

Measures of spread—range

A measure of spread gives an indication of how close or spread out the values are in a data set. The measure of spread you would have used most frequently is the **range**.

Range = maximum value – minimum value

The range is influenced by extreme values. For example, the following table shows the number of pets owned by each of the students in a class.

Number of pets (x)	Frequency (f)
0	5
1	9
2	4
3	2
4	2
5	1
6	1

Range = 6 – 0 = 6 pets

A new student joins the class. She has an aviary that contains 25 birds. The new range is now:

Range = 25 – 0 = 25 pets

This additional value has distorted the range. This could be regarded as not being a fair representation of the class as a whole. You will further consider this issue of outlier values later in this chapter.

Measures of spread—interquartile range (IQR)

The interquartile range is a measure of spread based on the quartile values. It is often referred to by the abbreviation IQR.

Interquartile range = third quartile – first quartile

$\text{IQR} = Q_3 - Q_1$

The IQR represents the middle 50% of the ordered data set. It is a more reliable measure of spread than the range because it does not use extreme values at the start or end of the ordered data list.

Example 16

Find the interquartile range for the data set: 1, 14, 20, 1, 105, 30, 27, 16, 24, 24, 29.

Solution

Order the data from smallest to largest: 1, 1, 14, 16, 20, 24, 24, 27, 29, 30, 105

Identify Q_1, Q_2 (median) and Q_3 for the data set:

Q_1 ↓ (14) Q_3 ↓ (29)

1, 1, 14, 16, 20, 24, 24, 27, 29, 30, 105

↑ Q_2 (median) (24)

$$\text{IQR} = Q_3 - Q_1 = 29 - 14$$
$$= 15$$

Five-number summary

The five-number summary for a data set consists of:

- the minimum value
- the lower quartile
- the median
- the upper quartile
- the maximum value

For the data in Example **16** this is:

Minimum = 1, $Q_1 = 14$, Median $Q_2 = 24$, $Q_3 = 29$, Maximum = 105

EXPLORE FURTHER

Five-number summary

Explore how to find the five-number summary of a data set using a spreadsheet.

Box plots

The five-number summary can be used to draw a statistical graph called a **box plot** or box-and-whisker plot.

A box plot needs to have a scale line associated with it. The 'box' is the central box structure and the 'whiskers' are the lines going out from the box to the extreme values.

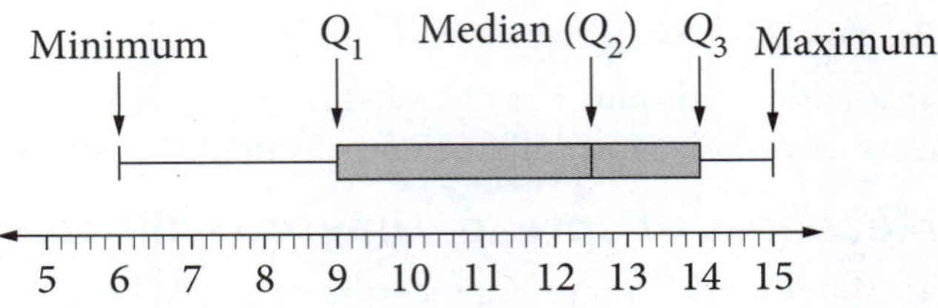

Example 17

Find the five-number summary and use it to draw a box plot for this data set: 2, 3, 5, 4, 5, 5, 12, 9, 2, 6, 7, 5.

Solution

Write the data in ascending order: 2, 2, 3, 4, 5, 5, 5, 5, 6, 7, 9, 12

Identify the location of Q_1, Q_2 and Q_3: There are 12 data values.

So, the median is between the 6th and 7th values.

Also, Q_1 is between the 3rd and 4th values and Q_3 is between the 9th and 10th values.

State the five-number summary:

Minimum = 2
$Q_1 = 3.5$
Median $Q_2 = 5$
$Q_3 = 6.5$
Maximum = 12

$Q_1 = 3.5$ ↓ $Q_3 = 6.5$ ↓

2, 2, 3, 4, 5, 5, 5, 5, 6, 7, 9, 12

↑ $Q_2 = 5$

Draw the box plot. Remember to include a scale:

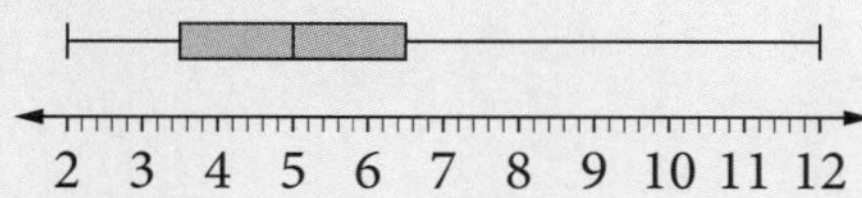

You can also find the five-number summary if the data is presented in a frequency table or a stem-and-leaf plot.

Example 18

Find the interquartile range and draw a box plot for the data in the table.

Score	Frequency
3	2
4	8
5	9
6	6
7	4
8	1

Solution

Find the sum of the frequencies, n: $2 + 8 + 9 + 6 + 4 + 1 = 30$

Identify the position of the median, and its value: The median is between the 15th and 16th values.

The 15th and 16th values are both 5, so median = 5.

Identify the position of Q_1, and its value: There are 15 values in the lower half of the data; the middle value here is the 8th.

$Q_1 = 4$

Identify the position of Q_3, and its value: There are 15 values in the upper half of the data; the middle value here is the 8th from the end:

$Q_3 = 6$

Calculate the interquartile range: $\text{IQR} = Q_3 - Q_1$
$= 6 - 4 = 2$

Write the five-number summary as a basis for drawing the box plot:

Minimum = 3
$Q_1 = 4$
Median $Q_2 = 5$
$Q_3 = 6$
Maximum = 8

Draw the box plot. (Don't forget to include a scale line.)

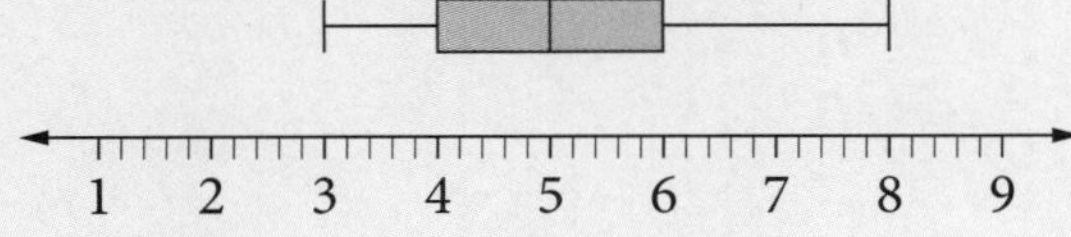

Example 19

Find the interquartile range and then draw a box plot.

Stem	Leaf
3	0 3 6
4	1 3 5 7 9
5	4 5 7 8
6	2 4 7 9 9
7	0 0 0 1 4 7 8
8	3 5 7 8
9	0 0 0 0 0 1

Key: $3|0 = 3.0$

Solution

Count the number of values, n, in the data set: $n = 34$

Identify the position of the median and its value:
The median is between the 17th and 18th values,
$\frac{6.9+7}{2} = \frac{13.9}{2} = 6.95$

Identify the position of Q_1 and its value:
There are 17 values in the lower half, so Q_1 is the 9th score.

$Q_1 = 5.4$

Identify the position of Q_3 and its value:
There are 17 values in the upper half, so Q_3 is the 26th score.

$Q_3 = 8.5$

Stem	Leaf
3	0 3 6
4	1 3 5 7 9
5	4 5 7 8
6	2 4 7 9 **9**
7	**0** 0 0 1 4 7 8
8	3 5 7 8
9	0 0 0 0 0 1

Key: $3|0 = 3.0$

Calculate the interquartile range: $\text{IQR} = Q_3 - Q_1$
$= 8.5 - 5.4 = 3.1$

Write the five-number summary:

Minimum = 3.0
$Q_1 = 5.4$
Median $Q_2 = 6.95$
$Q_3 = 8.5$
Maximum = 9.1

Draw the box plot.

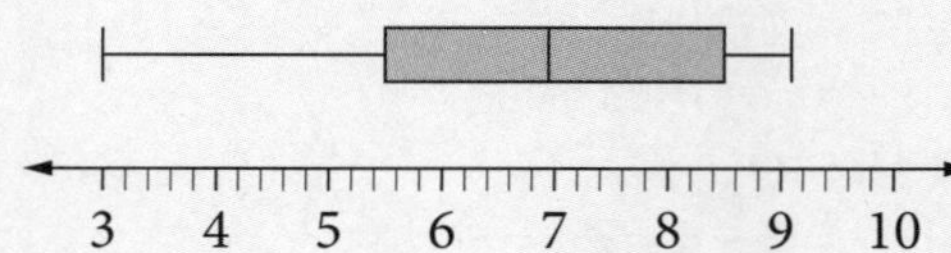

Outliers

An outlier is an extreme value that does not fit in with the rest of the data set. However, this is not a good definition as each person could have a slightly different interpretation of 'does not fit in'. There is a mathematical measure that is used to identify outliers consistently:

An outlier is a piece of data that lies outside the 'fences' set up as follows.

Lower fence $= Q_1 - 1.5 \times \text{IQR}$
Upper fence $= Q_3 + 1.5 \times \text{IQR}$

Example 20

Determine if any of the values in the following data set would be regarded as outliers.

1, 5, 7, 8, 8, 8, 9, 11, 13, 18, 23

Solution

Find Q_1: There are 11 data values and the values are already in order, so the median is the 6th value. $Q_2 = 8$.

The lower half contains 5 values: Q_1 is the 3rd value = 7

Find Q_3: The upper half contains 5 values, Q_3 is 3rd from the end = 13

Calculate the interquartile range: IQR = 13 − 7 = 6

Calculate IQR × 1.5 = 6 × 1.5 = 9

Calculate the fence values:

$$\text{Lower fence} = Q_1 - 1.5 \times \text{IQR} = 7 - 9 = -2$$

$$\text{Upper fence} = Q_3 + 1.5 \times \text{IQR} = 13 + 9 = 22$$

State which, if any, of the values are outliers: The value 23 is an outlier because it is above the upper fence.

It is possible for outliers to be present at both ends of the data list and for there to be more than one outlier at either or both ends.

When drawing a box plot, the whiskers should not extend to outliers. Any outlier is marked with a symbol such as ∗ or • and the whisker is drawn only as far as the last data value before the fence.

EXERCISE 11.4 MEASURES OF CENTRAL TENDENCY

1 Find the mean of the following data sets. Where necessary, state your answers correct to 2 decimal places.

(a) The following data represents the number of cars that passed the school gate in a number of 5-minute periods.

7, 14, 12, 11, 13, 16, 17, 9, 7, 12, 13, 15

(b) The following data represents the number of brothers and sisters for the students in two Year 11 classes.

Number of brothers and sisters (x)	Frequency (f)
0	6
1	11
2	18
3	9
4	3

(c) The following data represents the time taken by the competitors in the under-17 100 m dash at the school athletics carnival.

Time (x)	Frequency (f)
12–<12.5	4
12.5–<13	9
13–<13.5	15
13.5–<14	22
14–<14.5	17

2 Find the median of the following data sets.

(a) 1, 4, 3, 2, 6, 4, 8, 9, 4, 3, 1

(b) 2, 3, 6, 2, 9, 6, 9, 6, 3, 2, 5, 2

(c)

Score (x)	Frequency (f)
0	3
1	6
2	7
3	4
4	3
5	1

3 Find the median class interval for the following data sets.

(a)

Time (x)	Frequency (f)
12–<12.5	3
12.5–<13	5
13–<13.5	4
13.5–<14	11
14–<14.5	6

(b)

Time (x)	Frequency (f)
12–<12.5	6
12.5–<13	8
13–<13.5	4
13.5–<14	5
14–<14.5	7

4 Find the mode of the following data sets.

(a) 4, 6, 4, 2, 7, 8, 3, 4, 9, 1, 1 **(b)** 3, 5, 2, 1, 7, 4, 9, 3, 5, 6, 8 **(c)** 1, 4, 2, 1, 5, 4, 3, 1, 4, 2, 2

5 Find the mean and the median for each of the following discrete data sets. Where necessary, state your answers correct to 2 decimal places.

(a)

Score	Frequency (f)
4	15
5	23
6	14
7	23
8	17
9	19
10	20

(b)

Score	Frequency (f)
201	49
202	83
203	99
204	121
205	143
206	117

6 Find the interquartile range for the following data sets.

(a) 2, 3, 5, 8, 9, 11, 15, 16, 18, 25, 36

(b) 4, 5, 2, 6, 7, 9, 1, 9, 11, 12, 16, 18, 21, 17, 16, 17

(c) 1, 7, 2, 7, 3, 1, 9, 11, 14, 12, 16, 17, 22, 11

7 Complete the five-number summary and use it to draw the box plot for each of the following data sets.

(a) 2, 3, 5, 5, 5, 6, 8, 9, 10, 11, 12, 15

(b) 2, 5, 9, 11, 12, 9, 8, 7, 12, 14, 13, 22, 15, 18

8 Find the five-number summary, IQR and range for each of the following data sets. Then create a box plot.

(a)

Score	Frequency (f)
0	3
1	6
2	7
3	4
4	3
5	1

(b)

Score	Frequency (f)
4	15
5	23
6	14
7	23
8	17
9	19
10	20

(c)

Stem	Leaf
1	1 2 3 3 4 4 5 8
2	2 3 4 5
3	0 1 6 6 2 1
4	2 2 3 9
5	0 9
6	1 8

Key: 5|9 = 59

(d)

Stem	Leaf
15	2 5 8
16	0 1 3 4 5
17	1 8 9 9
18	9 9
19	8 9
20	0 2

Key: 18|9 = 189

9 Determine if any of the values in the following data sets would be regarded as outliers.

(a) 1, 2, 17, 18, 19, 22, 24, 26, 27, 28

(b) 5, 9, 11, 13, 17, 18, 19, 19, 29

(c) 7, 22, 23, 26, 29, 31, 34, 38, 58

(d) 3, 23, 24, 27, 28, 29, 32, 34, 48

10 Look at the following data set.

x	1	2	3	4	5	6
f	5	8	3	2	4	5

The mean and median (in that order) are:

A 2, 3 **B** 3.26, 3 **C** 3.26, 4 **D** 3, 3.26

11 Find the five-number summary and use it to create a box plot for each of the following data sets. Use technology if possible.

(a) 1, 4, 6, 3, 2, 4, 5, 7, 8, 9, 10, 2, 5, 6, 7, 4, 5, 3, 5, 1, 7, 9, 6, 9, 2, 4, 8, 10

(b) 8, 12, 15, 18, 23, 34, 32, 12, 16, 13, 18, 19, 13, 15, 21, 32, 23, 34, 23, 12, 18, 19, 17, 15

(c)

Score	Frequency (f)
8	10
9	12
10	1
11	11
12	5
13	7
14	9

(d)

Score	Frequency (f)
23	8
24	11
25	14
26	12
27	13
28	7
29	3

12 Find the five-number summary and use it to create a box plot for each of the following data sets. Also state which two statistical values are equal.

(a) 2, 3, 4, 6, 6, 6, 8 **(b)** 2, 2, 4, 5, 6, 7, 9 **(c)** 2, 3, 4, 5, 8, 9, 9 **(d)** 2, 2, 2, 2, 5, 7, 8

13 Look at the following box plot.

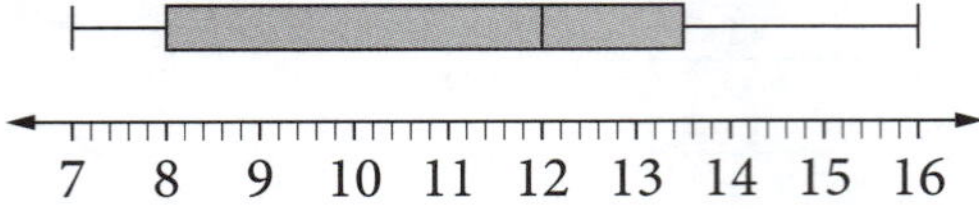

The IQR is closest to: **A** 9 **B** 1 **C** 5.5 **D** 1.5

14 Look at the following box plot. Find the range.

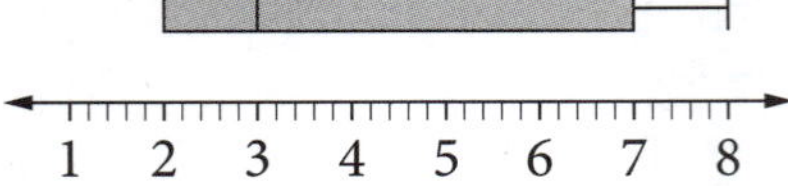

15 Find the mean, the median and the mode for each of the following data sets.

(a) 2, 3, 4, 5, 6, 6, 7, 8, 9, 10
(b) 2, 2, 2, 2, 2, 3, 4, 5, 6, 7, 8, 9, 10, 11, 12, 13
(c) 2, 3, 4, 5, 6, 7, 8, 9, 21
(d) 2, 3, 4, 5, 6, 7, 11, 11

16 Draw a box plot for each of the following data sets, and then use it to identify any outliers.

(a) 22, 24, 25, 19, 45, 73, 21, 25, 18, 16, 22, 25, 28, 19, 17, 16, 23
(b) 101, 108, 113, 121, 113, 67, 97, 135, 141, 123, 116, 119, 121, 123, 134, 132, 119, 118
(c) 38, 38, 37, 39, 41, 42, 42, 45, 43, 44, 41, 42, 46, 18, 42, 45, 47, 72, 32, 43, 41, 40, 42, 45
(d) 113, 134, 123, 143, 123, 129, 127, 137, 136, 139, 129, 130, 140, 76, 72, 112, 142, 132, 141, 132, 137

17 For each of the following data sets, calculate:

(i) the value below which outliers would occur
(ii) the value above which outliers would occur.

(a) 13, 15, 22, 27, 36, 41, 28, 17, 25, 36, 15
(b) 33, 35, 28, 27, 46, 51, 29, 36, 49, 26

(c)

x	5	6	7	8
f	3	4	2	1

(d)

x	11	12	13	14
f	4	5	3	2

18 The following table gives the maximum temperature (°C) reached on a particular day in a number of cities across the world. Answer the following questions about this data set.

City	Temp (°C)	City	Temp (°C)	City	Temp (°C)
Amsterdam	3	Copenhagen	0	New York	13
Athens	18	Helsinki	–4	Oslo	–3
Barcelona	9	Istanbul	18	Prague	0
Belgrade	3	London	5	Singapore	33
Budapest	1	Los Angeles	14	Taipei	27
Cairo	34	Madrid	6	Toronto	10
Chicago	13	Moscow	1	Warsaw	–1

(a) Find the mean of the temperatures. Give your answer correct to 2 decimal places.
(b) Calculate the median temperature.
(c) Draw a grouped data frequency table using –5 to <0 as the first class interval.
(d) Use the frequency table to find an estimate for the mean temperature. Give your answer correct to 2 decimal places.
(e) In which class interval is the median temperature?

11.5 STANDARD DEVIATION

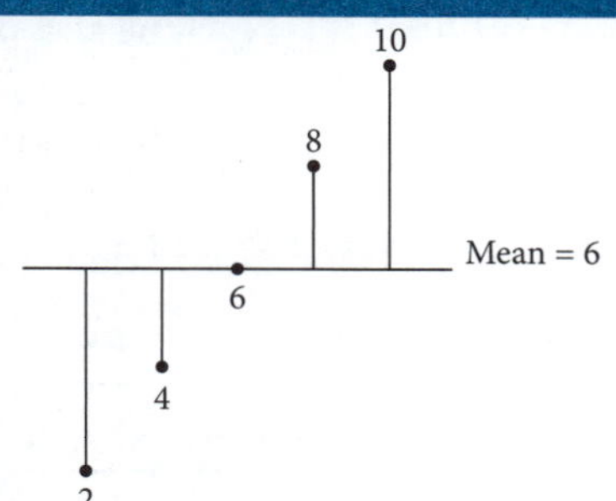

The **standard deviation** is a measure of spread. It uses all the data values, so is not influenced by outliers as much as the range. The range only uses the highest and lowest score, so is affected by outliers. The standard deviation, as a measure of spread, is often paired with the mean, a measure of central tendency. Statisticians like to use a measure of central tendency combined with a measure of spread to describe a data set.

Consider the following very simple data set: 2, 4, 6, 8, 10. Here, the mean is 6.

The differences between each data value and the mean, in order from left to right, are: −4, −2, 0, 2, 4. The sum of these differences is always 0. To avoid this you square the deviations to make them all positive: 16, 4, 0, 4, 16.

The average of these squared deviations is: $\dfrac{16+4+0+4+16}{5} = \dfrac{40}{5}$

$= 8$

Now find the square root of this value (to undo the squaring you did earlier): $\sqrt{8} = 2.83$.

So, here the standard deviation has a value of 2.83, correct to 2 decimal places.

The **population standard deviation**, represented by the Greek letter sigma σ, uses all data values in the population.

The **sample standard deviation**, represented by *s*, uses only a sample taken from the population. To find the sample standard deviation you divide by one less than the number of results. This means the sample standard deviation will always be greater than the population standard deviation. This is reasonable because taking a sample has introduced a degree of uncertainty to our calculations.

In either case, the bigger the value of the standard deviation, then the more spread out the data is considered to be.

If you are unsure whether you have a population or a sample, then calculate the sample standard deviation so that you do not draw conclusions that are unjustified by the statistical evidence.

Population standard deviation: $\sigma = \sigma_x = \sqrt{\dfrac{\Sigma(x-\mu)^2}{n}}$, where x refers to the individual scores, μ is the population mean and n is the number of values.

Sample standard deviation: $s = s_x = \sqrt{\dfrac{\Sigma(x-\bar{x})^2}{n-1}}$, where x refers to the individual scores, $\bar{x}$ is the sample mean and n is the number of values.

Calculating the standard deviation by hand is quite complicated so you should calculate it using technology.

Example 21

The results of a survey show the number of brothers and sisters for each member of the class.

1, 3, 4, 2, 1, 0, 3, 1, 2, 4, 0, 1, 0, 2, 1, 0, 1, 2, 2, 3

Calculate the standard deviation.

Solution

$n = 20$ the size of the population

$\mu = \frac{33}{20} = 1.65$ the population mean

To calculate the standard deviation, either a frequency table, scientific calculator or spreadsheet needs to be used.

Method 1—frequency table

x	f	$x-\mu$	$(x-\mu)^2$	$f(x-\mu)^2$
0	4	−1.65	2.7225	10.89
1	6	−0.65	0.4225	2.535
2	5	0.35	0.1225	0.6125
3	3	1.35	1.8225	5.4675
4	2	2.35	5.5225	11.045
	$\sum f = 20$			$\sum f(x-\mu)^2 = 30.55$

$$\sigma_x = \sqrt{\frac{\Sigma(x-\mu)^2}{n}} = \sqrt{\frac{30.55}{20}} = 1.2359 \approx 1.24$$

Method 2—scientific calculator

Set your calculator in Statistics mode.

Enter the data score by score.

Use $\bar{x}$ key to find the mean: $\bar{x} = 1.65$

Use σ_x to find the standard deviation: $\sigma_x = 1.2359 \approx 1.24$

Use the s_x if you need the sample standard deviation: $s_x \approx 1.27$

Note: Each calculator is different so make sure you learn how to use your model.

Method 3—spreadsheet

x	f	xf	$x-\mu$	$(x-\mu)^2$	$f(x-\mu)^2$
0	4	0	−1.65	2.7225	10.89
1	6	6	−0.65	0.4225	2.535
2	5	10	0.35	0.1225	0.6125
3	3	9	1.35	1.8225	5.4675
4	2	8	2.35	5.5225	11.045
Σ	20	33			30.55

$\mu = 1.65$

$s_x = 1.268$

$\sigma_x = 1.2359$

EXPLORE FURTHER

Population and sample standard deviation

Explore how to use a spreadsheet to find the standard deviation of samples drawn from a data set with a known population standard deviation.

Example 22

A sample of members at a running club is asked to keep track of the distance, to the nearest whole kilometre, covered in training over the course of a week. The frequency table summarises the data collected.

Find the mean and the standard deviation for the data set shown.

Distance covered (nearest km)	Number of runners
21–30	3
31–40	5
41–50	4
51–60	2
61–70	1
71–80	1

Solution

Add a column for the midpoint of each class interval, x_m, and enter the data into a spreadsheet.

x	f	x_m	$x_m f$	$x_m - \mu$	$(x_m - \mu)^2$	$f(x_m - \mu)^2$
21–30	3	25.5	76.5	−17.5	306.25	918.75
31–40	5	35.5	177.5	−7.5	56.25	281.25
41–50	4	45.5	182	2.5	6.25	25
51–60	2	55.5	111	12.5	156.25	312.5
61–70	1	65.5	65.5	22.5	506.25	506.25
71–80	1	75.5	75.5	32.5	1056.25	1056.25
Σ	16		688			3100

$\mu = 43$
$\sigma_x = 13.9194$
$s_x = 14.3759$

The mean is 43 and the standard deviation is 13.92, correct to 2 decimal places.

The sample standard deviation is 14.38.

Limitations of using the standard deviation

The standard deviation uses every piece of data, so it can be affected by outliers in the same way that those values can affect the mean.

Example 23

The number of goals Eloise scored in 15 games of a netball competition was recorded.

22, 27, 32, 48, 29, 36, 29, 41, 44, 29, 38, 26, 39, 42, 40

(a) Find the mean and standard deviation for the data. Assume the data is a sample of her goals for the season. Give answers rounded to 2 decimal places if necessary.

(b) In the following week Eloise is injured early in the game and scores only 4 goals. Find the new mean and standard deviation.

Solution

(a) Using a scientific calculator, $n = 15$: $\bar{x} = \dfrac{522}{15} = 34.8$

$\sigma_x = 7.3774 \approx 7.38$

$s_x = 7.6364 \approx 7.64$

Considering the data as a sample of the goals for the season, use $s_x = 7.64$ as the standard deviation. Mean is 34.8.

(b) All that has to be done is to add the new data value into your calculator and then obtain the answers:

$n = 16$

$\bar{x} = \dfrac{526}{16} = 32.875 \approx 32.88$

$\sigma_x = 10.3252 \approx 10.33$

$s_x = 10.6638 \approx 10.66$

Considering the data as a sample of the goals for the season, use $s_x = 10.66$ as the standard deviation. Mean is 32.88.

Example 24

A teacher recorded the results for a test, marked out of 50, as follows:

12, 17, 22, 23, 26, 26, 27, 29, 30, 31, 31, 32, 32, 35, 35, 37, 38, 40, 41, 42, 42, 49.

After the tests were returned, the student who had scored 12 pointed out that several pages had not been corrected. Their mark increased to 36.

Which of the following statements correctly describes what would happen to the class statistics? (Do not use technology for this question.)

- **A** The standard deviation and mean would both decrease.
- **B** The standard deviation and mean would both increase.
- **C** The standard deviation would increase and the mean would decrease.
- **D** The standard deviation would decrease and the mean would increase.
- **E** The median would not change.

Solution

The median is 31.5 and will become 32 once the 12 is removed and the 36 inserted. This means **E** is incorrect.

The lowest value has been removed and replaced by a value nearer the mean. This means the standard deviation must be smaller. So, **B** and **C** can both be eliminated.

A smaller value has been replaced by a bigger value; the mean must increase. So, **A** is incorrect.

D is the correct response.

EXERCISE 11.5 STANDARD DEVIATION

1 You conduct a survey of your class to find the number of pets for each member of the class.

Number of pets: 0, 4, 1, 1, 6, 3, 8, 4, 2, 0, 0, 0, 1, 1, 4, 6, 2, 3, 4, 1.

(a) What is the mean number of pets?

(b) When finding the standard deviation you [**can** / **cannot**] regard this as a population, so you will find the [**population** / **sample**] standard deviation.

(c) Use a spreadsheet structure on your technology to find the standard deviation, correct to 2 decimal places. What is its value?

2 The data set shown represents the length of a sample of fish, measured to the nearest centimetre, caught by a commercial fishing boat.

Length of fish (nearest cm)	Number
20–29	8
30–39	12
40–49	24
50–59	38
60–69	54
70–79	49
80–89	21
90–99	7

Use appropriate technology to find the following values, correct to 2 decimal places.

(a) the mean **(b)** the standard deviation

3 The number of goals Irene scored each week in her netball competition was recorded for part of the season.

Goals: 42, 39, 42, 43, 29, 33, 30, 45, 40, 27, 29, 33, 38, 42, 46

(a) You can reasonably assume these values refer to a [**sample / population**].

(b) Find the mean and the standard deviation for the data. Give the values rounded to 2 decimal places, if necessary.

(c) In the next round Irene is injured early in the game and scores only 3 goals. Find the new mean and standard deviation.

4 A teacher recorded the results for a test, marked out of 50.

Test results: 26, 29, 31, 32, 34, 36, 36, 39, 40, 40, 41, 42, 43, 44, 46, 46, 47

Before the tests were returned a student who had been absent sat the test. The student scored 27 on the test. Which of the following statements correctly describes what would happen to the class statistics? Do not use technology for this question.

A The standard deviation and mean would both decrease.
B The standard deviation and mean would both increase.
C The standard deviation would increase and the mean would decrease.
D The standard deviation would decrease and the mean would increase.

5 Consider the following simple data set which represents the number of people in the queue for an ATM at a bank at different times of day: 7, 3, 2, 4, 6, 7, 3, 5, 7, 3, 7, 2, 4, 6, 7, 2, 6, 4, 5.

(a) The interquartile range is: **A** 4 **B** 4.7 **C** 5 **D** 7

(b) Calculate the standard deviation. **(c)** Calculate the population standard deviation.

6 Calculate the population standard deviation for the following data sets, to 1 decimal place.

(a) 3, 3, 4, 6, 2, 1, 3, 4, 6, 7, 5, 3, 2, 1, 7, 9

(b) 25, 21, 22, 25, 26, 29, 28, 22, 24, 25, 27, 26, 30

(c) 7, 5, 2, 4, 7, 8, 11, 13, 16, 15, 11, 16, 17, 18, 13, 12

(d) 15, 19, 22, 28, 21, 23, 26, 27, 21, 20, 17, 13, 18, 17

(e) 15, 19, 16, 17, 21, 29, 33, 21, 27, 28, 26, 25, 20

(f) 56, 53, 57, 52, 59, 61, 66, 65, 60, 61, 54, 65, 67, 68, 60

(g)

Score	Frequency
44	8
45	12
46	13
47	10
48	9
49	5

(h)

Score	Frequency
101	11
102	14
103	18
104	19
105	13
106	12

(i)

Score	Frequency
10–14	6
15–19	7
20–24	9
25–29	10
30–34	4
35–39	1

(j)

Score	Frequency
80–<90	16
90–<100	15
100–<110	17
110–<120	19
120–<130	12
130–<140	11

7 A factory has two machines which make the same product. The weights, correct to the nearest gram, of samples of 10 cakes from the two machines are shown in the table below. The target weight is 60 g.

Machine A	59	56	57	62	63	63	60	59	58	59
Machine B	57	58	63	60	61	62	61	62	58	58

(a) Find the mean weight for each machine.
(b) If production is stopped if the mean weight is more than 0.5 g from the target weight of 60 g, would either machine be shut down?
(c) Find the sample standard deviation for each machine, correct to 2 decimal places.
(d) If production is stopped when the sample standard deviation is more than 2.25 g, would either machine be shut down?

8 Use the following data sets to draw conclusions about the relationship between the interquartile range and the population standard deviation.

(a) 1, 14, 15, 17, 17, 18, 19, 22, 24, 25, 26, 26, 26, 26, 29
What is the interquartile range and the population standard deviation, correct to 1 decimal place?
(b) 11, 14, 15, 17, 17, 18, 19, 22, 24, 25, 26, 26, 26, 26, 39
What is the interquartile range and the population standard deviation, correct to 1 decimal place?
(c) 1, 14, 15, 17, 17, 18, 19, 22, 24, 25, 26, 26, 26, 26, 39
What is the interquartile range and the population standard deviation, correct to 1 decimal place?
(d) What conclusions can you draw from the answers to the first three parts of the question?

11.6 ANALYSIS OF DATA

Statistical displays: symmetry and skew

Statistical displays are used to analyse data, and to determine the shape of the data distribution. A data set can be symmetrical (evenly distributed), or skewed in the negative (left) or positive (right) direction. A skewed data display can also be described as stretched.

You can use histograms and box plots to illustrate the terms.

Negative skew (stretched towards the negative, or left, side of the diagram)

Symmetrical (relatively evenly distributed)

Positive skew (stretched towards the positive, or right, side of the diagram)

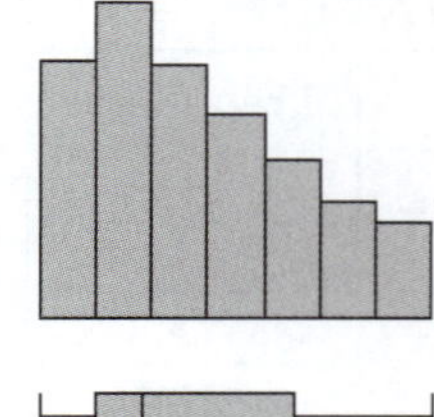

When describing the shape of a data distribution, the median and IQR are included, rather than the mean and standard deviation. This is because the mean and standard deviation can be affected by outliers.

Example 25

Describe the following data set, including a comment about its shape. The data set represents the marks obtained on a test, out of 50, by a class of 20 students.

14, 26, 49, 45, 46, 23, 24, 25, 25, 48, 49, 28, 33, 35, 37, 38, 38, 39, 41, 43

Solution

Enter the data into your technology to find the five-number summary as well as the mean and population standard deviation. Write your answers correct to 2 decimal places, when necessary:

Minimum = 14, $Q_1 = 25.5$, Median = 37.5, $Q_3 = 44$, Maximum = 49, Mean = 35.3, Standard deviation = 9.89

$$\text{IQR} = 44 - 25.5 = 18.5$$
$$1.5 \times \text{IQR} = 1.5 \times 18.5 = 27.75$$
$$Q_1 - 1.5 \times \text{IQR} = 25.5 - 27.75 = -2.250$$
$$Q_3 + 1.5 \times \text{IQR} = 44 + 27.75 = 71.75$$

All marks are in the interval from −2.25 to 71.75, so there are no outliers.

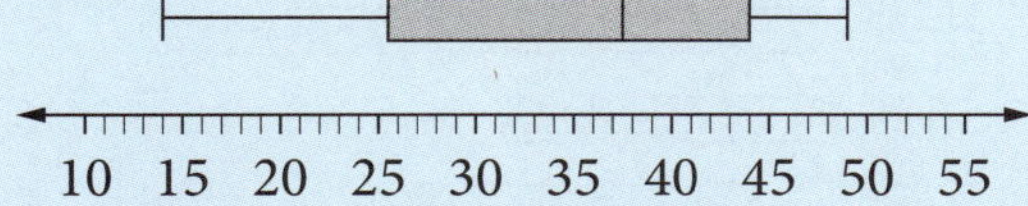

The data set contains no outliers, but is slightly negatively skewed. This is shown in the box plot where the median is closer to the upper quartile. At least 50% of the students achieved 37.5 or more, with no more than 25% obtaining less than 25.5.

Composite bar graphs

A **composite bar graph** allows two data sets to be visually compared. The example below shows a comparison of a population based on gender and age. This particular sort of graph is sometimes referred to as a population pyramid.

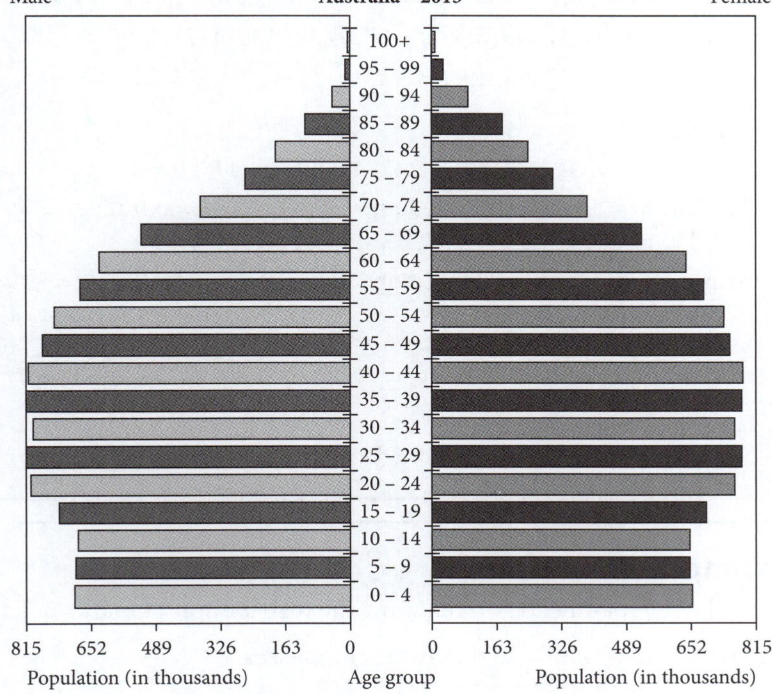

Back-to-back stem-and-leaf plots

Another way of comparing two data sets is by using a **back-to-back stem-and-leaf plot**. These have a central stem with the leaves moving away from the stem in either direction. The example below shows the height data, measured to the nearest centimetre, for two different hockey teams.

Hockey players' heights (cm)

Team A		Team B
9 8 5	15	1 4 6 7
6 5 4 4 1	16	0 8
9 8 8 6 5	17	2 4 4 6
	18	3 6
	19	0

Key: 17|9 = 179

Note: Both sides have the lowest leaf values next to the stem column. So, the left-hand data set is written right to left, but the right-hand data set is written left to right.

Parallel box plots

A **parallel box plot** draws several box plots on the same scale. The example below compares the marks for three different classes on the same test.

Distribution of marks in test for three classes

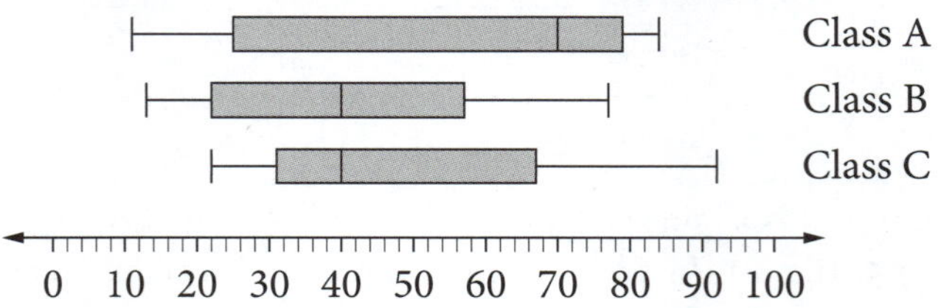

Example 26

After first obtaining the five-number summary, draw parallel box plots for these two data sets.

Class A: 22, 25, 34, 12, 49, 50, 44, 48, 27, 32, 42, 44, 28, 50

Class B: 33, 37, 30, 40, 48, 49, 50, 42, 46, 21, 17, 18, 22, 25

Solution

Class A: 12, 22, 25, 27, 28, 32, 34, 42, 44, 44, 48, 49, 50, 50.

Minimum = 12, $Q_1 = 27$, Median = 38, $Q_3 = 48$, Maximum = 50

Class B: 17, 18, 21, 22, 25, 30, 33, 37, 40, 42, 46, 48, 49, 50.

Minimum = 17, $Q_1 = 22$, Median = 35, $Q_3 = 46$, Maximum = 50

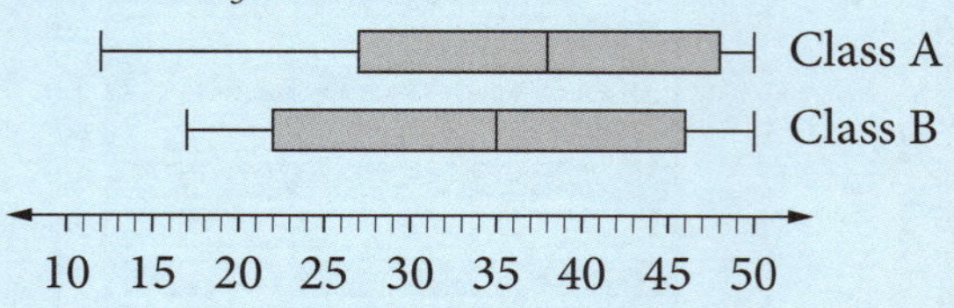

Example 27

The following parallel box plots show the distribution of marks in a test for three classes.

Distribution of marks in test for three classes

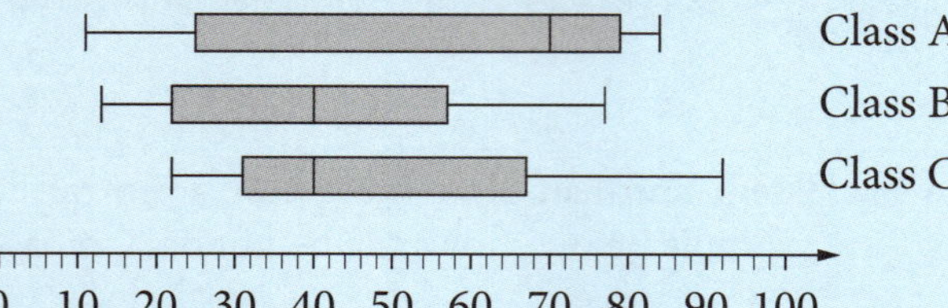

Compare the box plots.

Solution

The box plots indicate that Class A is negatively skewed. Class B is close to symmetrical and Class C is positively skewed.

The highest mark was in Class C and the lowest was in Class A.

At least 50% of Class A obtained 70 or more, less than 25% of Class B obtained 70 or more and about 25% of Class C obtained 70 or more.

Classes B and C have the same median mark which is much less than the median mark for Class A.

EXERCISE 11.6 ANALYSIS OF DATA

1 The following data set represents the marks obtained on a test, out of 66, by a class of 21 students.

60, 51, 47, 42, 53, 34, 47, 39, 56, 63, 35, 34, 50, 35, 41, 19, 48, 42, 37, 45, 29

(a) Is it reasonable to assume the data set represents a sample or a population?

(b) Find the mean, standard deviation, minimum value, Q_1, median, Q_3, and maximum value. Round to 1 decimal place, where necessary.

(c) Describe the data set by choosing the appropriate words for the following sentences:
The data set [**does / does not**] contain outliers and is [**positively skewed / negatively skewed / symmetrical**]. [**25 / 50 / 75**]% of the data set was above 42 and no more than [**25 / 50 / 75**]% was below 35. The [**lowest / median / highest**] student scored a mark of 63 out of 66.

2 The parallel box plots below compare the yield per hectare of wheat for paddocks treated in various ways.

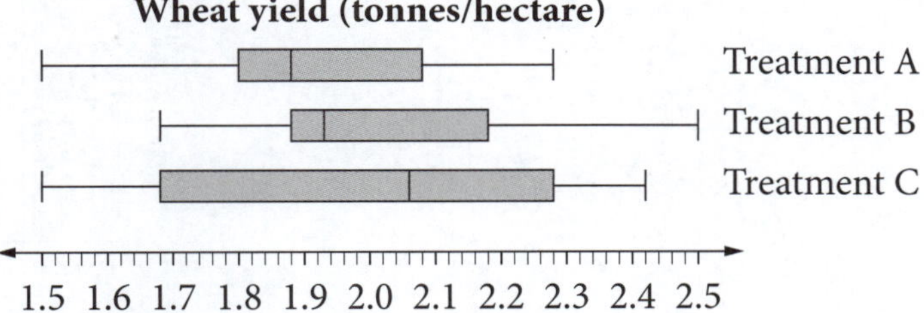

Which of the following statements is incorrect?

- **A** The interquartile range for treatment A is the smallest.
- **B** The highest yielding paddock has been given treatment B.
- **C** About 50% of the paddocks for treatment B had a yield less than 2.2 tonnes/hectare.
- **D** About 25% of the paddocks for treatment B had a yield between 1.875 and 1.925 tonnes/hectare.
- **E** Treatment A produced the most consistent results.

3 The table below gives the life expectancy for males and females in 20 of the most populated countries of the world. Draw two parallel box plots for male life expectancy and female life expectancy, and then discuss your findings.

Country	Male	Female	Country	Male	Female
Bangladesh	54	53	Germany	73	79
Brazil	62	68	India	57	58
China	68	72	Indonesia	59	63
Egypt	60	61	Iran	64	65
Ethiopia	50	53	Japan	76	82
France	74	82	Mexico	68	76

Country	Male	Female	Country	Male	Female
Pakistan	56	57	Turkey	68	72
Philippines	62	67	United Kingdom	73	79
Russia	64	74	United States	72	79
Thailand	66	71	Vietnam	63	67

4 The following table shows the average batting performances (the number of runs scored) by the Waugh twins, Steve and Mark, in cricket Test matches against other Test-playing nations. The figures cover the entire careers of both players. Draw a composite bar graph and use it to assist you in deciding who was the better batsman. Although Steve played one Test against Bangladesh, making 256 without being dismissed, Mark never played against them, so those figures are not included.

	England	India	New Zealand	Pakistan	South Africa	Sri Lanka	West Indies	Zimbabwe	Total Career
Steve	58.18	41.92	38.52	34.59	49.87	87.63	49.82	145	51.06
Mark	50.09	33.24	42.56	42.41	42.04	24.64	41.29	90	41.82

5 The bar graph below shows the number of people killed in traffic accidents in Australia in the years 2011 and 2012, by state and territory. Compare the data in terms of the two years.

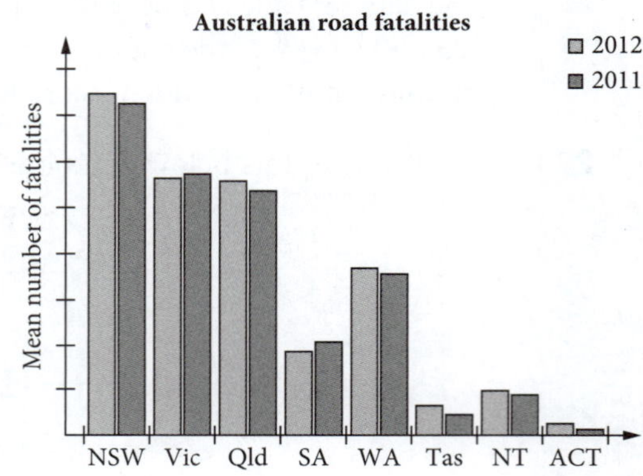

6 The bar graph compares the accumulated rainfall, per month, in Melbourne for a particular year against the average rainfall. Discuss the content of the graph.

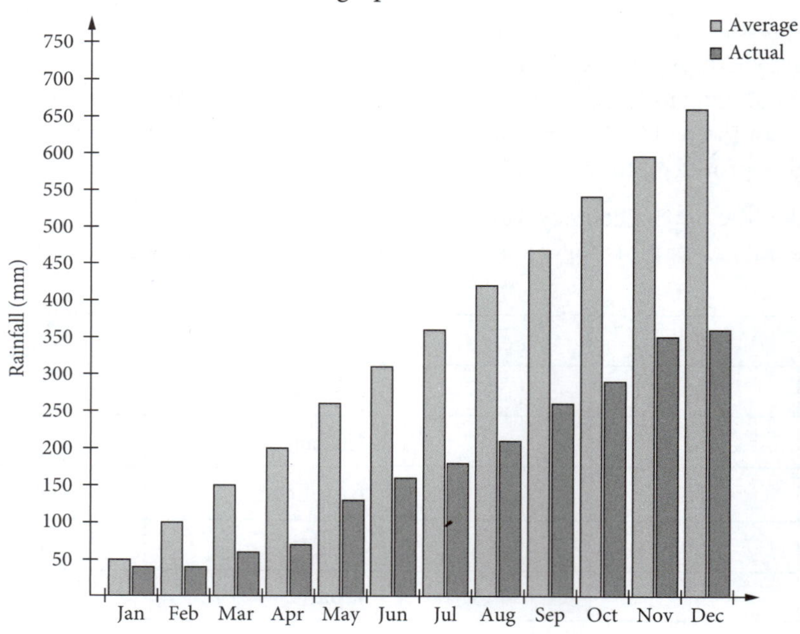

7 The data below represents the height (cm) of the senior list players for the Geelong AFL team for the years 2003 and 2013.

2003: 182, 184, 181, 197, 179, 183, 190, 186, 190, 192, 190, 196, 192, 192, 185, 181, 189, 183, 202, 193, 189, 196, 186, 192, 197, 189, 181, 194, 198, 187, 183, 189, 185, 192, 176, 208, 193, 195

2013: 196, 189, 187, 192, 180, 177, 193, 185, 183, 188, 191, 199, 197, 182, 182, 206, 203, 184, 183, 189, 190, 187, 186, 186, 192, 197, 175, 177, 186, 198, 193, 183, 186, 179, 188, 190, 190, 182, 187, 194

Draw a back-to-back stem-and-leaf plot of this data and use it to make comparisons between the two lists.

8 The table below gives data for the population of Australia in 1998 and 2013. The population figures are given in 1000s.

(a) Draw a population pyramid (composite bar graph) for 1998.
(b) Draw a population pyramid for 2013.
(c) Compare the two diagrams, discussing the changes in population distribution.
(d) Draw four parallel box plots for the data, clearly indicating any outliers. Discuss your findings.

	1998		2012	
Age	Male	Female	Male	Female
0–4	661	628	692	656
5–9	682	650	689	653
10–14	680	648	681	648
15–19	645	613	731	692
20–24	686	654	800	762
25–29	728	703	815	782
30–34	729	713	796	762
35–39	733	724	814	781
40–44	688	689	808	783
45–49	665	658	771	753
50–54	572	554	745	737
55–59	445	433	676	684
60–64	360	365	633	642
65–69	328	347	521	527
70–74	290	338	375	392
75–79	198	269	265	301
80+	188	348	348	546

CHAPTER REVIEW 11

1 Your class is conducting a statistical investigation. You are in the group that is searching the Australian Bureau of Statistics website for statistics relevant to your investigation. In which part of the investigation process are you involved?

A identification of a problem and the setting for a statistical question
B collection of the statistical data
C analysis of the data
D interpretation and communication of the results

2 The collar size of shirts worn by the pupils in a class is recorded. Select the option that shows the most precise way of describing the data collected.

A continuous B categorical C nominal D discrete

3 The height of each member of the class is recorded correct to the nearest centimetre. Select the option that shows the most precise way of describing the data collected.

A continuous B nominal C ordinal D discrete

The data set shown is to be used for questions **4**, **5** and **6**.

x	1	2	3	4	5
f	3	5	2	4	2

4 Select the option closest to the mean of the data set.

A 2.5 **B** 5 **C** 2.8 **D** 2

5 Select the option closest to the median of the data set.

A 2.5 **B** 5 **C** 2.8 **D** 2

6 Select the option closest to the mode of the data set.

A 2.5 **B** 5 **C** 2.8 **D** 2

7 What is the mean of the following data set? Select the closest answer.

x	0–9	10–19	20–29	30–39	40–49
f	6	2	7	8	3

A 25 **B** 25.5 **C** 29 **D** 24.5

The data set shown is to be used for questions **8** and **9**.

Stem	Leaf
3	0 1 4 6 7
4	3 3 3 3 4 5
5	8 9
6	1 3 4 6 6 6 7 8
7	0 6 7 7 9
8	1 2 2 2 5 6
9	2 2 2 5 7

Key: 5|8 = 58

8 Select the option that best represents the median of the data set.

A 6 **B** 43 **C** 63 **D** 66

9 Select the option that best represents the IQR of the data set.

A 38.5 **B** 43.5 **C** 38 **D** 39

10 For the following data set, select the answer that best represents the IQR.

2, 4, 5, 2, 4, 3, 1, 4, 6, 5, 4, 7, 8, 6, 8, 4, 5, 6, 2, 4, 3, 1, 3, 5, 7, 9, 1, 3, 2, 4, 6, 5, 3, 2, 1, 5, 2, 5, 6

A 2 **B** 4 **C** 4 to 5 **D** 3 to 6

11 For the following data set, select the answer that best represents the IQR.

x	8	13	15	19	21	25
f	4	8	6	3	1	2

A 2.5 **B** 4 **C** 8 **D** 13

12 A teacher recorded the results for a test, marked out of 50, as follows:

12, 17, 22, 23, 26, 26, 27, 29, 30, 31, 31, 32, 32, 35, 35, 37, 38, 40, 41, 42, 42, 49

After the tests were returned the student who had scored 38 pointed out that several pages had not been corrected. This increased the mark to 48. Which of the following statements correctly describes what would happen to the class statistics?

- **A** The standard deviation and mean would both decrease.
- **B** The standard deviation and mean would both increase.
- **C** The standard deviation would increase and the mean would decrease.
- **D** The standard deviation would decrease and the mean would increase.

13 You have recorded, as a sample, the number of cars that pass the school, in 5 minute blocks, for 2 hours:

20, 26, 24, 26, 19, 18, 10, 33, 28, 42, 5, 6, 18, 12, 16, 32, 41, 38, 25, 26, 15, 10, 29, 30

In your presentation about traffic volume outside your school, what should you report as the value for the standard deviation?

A 10.19 **B** 14 **C** 10.41 **D** 22.88

14 The height of every student in the school was recorded. The results are shown in the following table.

Height of student (cm)	Frequency
140–<145	5
145–<150	11
150–<155	18
155–<160	26
160–<165	44
165–<170	92
170–<175	64
175–<180	32
180–<185	17
185–<190	9
190–<195	4

What is the best estimate for the standard deviation of the school population?

A 9.65 **B** 9.66 **C** 10 **D** 11.27

15 The parallel box plots below compare the yield per hectare of wheat for paddocks treated in various ways.

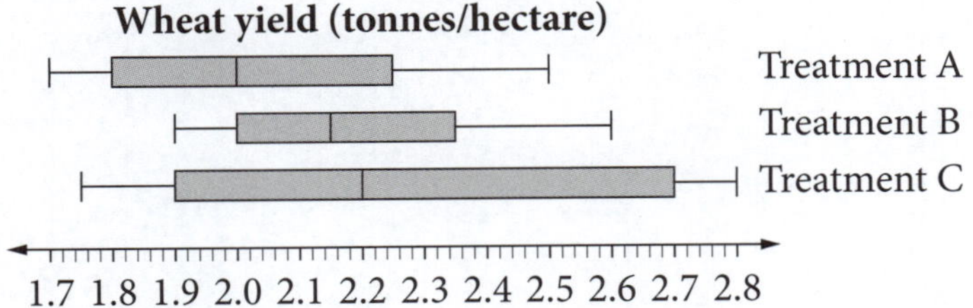

Which of the following statements is incorrect?

- **A** The interquartile range for treatment B is the smallest.
- **B** Treatment C produced the most consistent results.
- **C** About 50% of the paddocks for treatment C had a yield greater than 2.2 tonnes/hectare.
- **D** About 25% of the paddocks for treatment A had a yield between 1.7 and 1.8 tonnes/hectare.

16 The following table appeared in an article in *The Age* newspaper in 2013 under the headline 'Victoria's organ donor rate plummets in 2012' and with the sub-headline 'State still leads nation in numbers'.

Donors per million of population			
State	2011	2012	
NSW	11.0	12.4	↑
Vic	19.3	16.3	↓
Qld	15.0	17.0	↑
SA	21.4	17.5	↓
WA	14.0	13.2	↓
Tas	11.7	29.0	↑
NT	17.3	33.5	↑
ACT	13.7	20.3	↑
Australia	15.1	15.6	↑

Are the following statements true or false? If false, rewrite the statement to make it true.

(a) The data in the table does not seem to support the sub-headline because the figure of 16.3 is less than those for the majority of other states.

(b) On reading the table heading 'Donors per million of population', Victoria obviously has a higher population than most states. In 2012, for every million Victorian residents only 16.3 people were organ donors.

(c) The article indicates there were 92 Victorian organ donors in 2012. Based on this figure, calculate an estimate for the 2012 population of Victoria, to the nearest thousand.

(d) Based on the data in the table, if Australia's population in 2012 was approximately 22 692 308, estimate the number of organ donors.

(e) However, the article indicates that 1052 Australians received transplants from deceased donors in 2012. This is because people donated more than one organ.

(f) The figures presented in the table are most likely to represent estimates based on a sample.

(g) This article represents the analysis and communication part of the statistical investigation process.

17 State the correct data type for each of the following examples.

(a) The height jumped by each of the competitors in the school pole vault.

(b) The shirt size of each of the students in your class.

(c) The favourite colour of each of the students in your class.

(d) The highest level of education achieved by each of the respondents to a survey.

18 Into which interval of the following data set does the median fall?

x	Frequency
0–4	2
5–9	5
10–14	7
15–19	3
20–24	1

19 Find the five-number summary for the data set below.

Stem	Leaf
13	1 1 2 3 5 6 7
14	2 3 3 3 5 7 9 9
15	0 0 0 1 2 3 3 3 3
16	2 3 4 4 5 8 9
17	0 0 1 2 3 4
18	2 2 2 5 7 8 9
19	0 0 1

Key: 13|2 = 132

20 Use technology to answer the questions about the following data set.

1, 1, 2, 2, 2, 2, 3, 4, 4, 5, 6, 7, 8, 8, 8, 9, 9, 9, 9

(a) Find the population standard deviation, correct to 2 decimal places.

(b) Find the sample standard deviation, correct to 2 decimal places.

21 Draw a stem-and-leaf plot for the following data set.

25, 38, 65, 32, 77, 21, 79, 81, 66, 50, 47, 53, 25, 30, 42, 60, 70, 29, 51, 63, 68, 82, 40, 33, 22, 45, 37, 65, 74, 70, 35, 61, 81, 77, 65, 66

22 Draw a dot plot for the data set shown.

Number of pets: 2, 0, 1, 3, 3, 2, 5, 2, 0, 3, 6, 0, 5, 1, 3, 3, 1, 2, 2, 4, 3

23 The data below represents the weight (kg) of the senior list players for Western Bulldogs, in the AFL, for the years 2003 and 2013.

2003: 77, 98, 91, 77, 89, 81, 90, 76, 94, 80, 98, 84, 79, 72, 77, 75, 75, 98, 94, 82, 94, 96, 84, 89, 85, 72, 94, 74, 81, 88, 83, 85, 82, 83, 78, 101, 83, 85

2013: 84, 83, 84, 87, 88, 76, 89, 85, 92, 84, 80, 100, 81, 83, 77, 87, 87, 88, 97, 77, 81, 84, 100, 77, 83, 80, 107, 76, 82, 82, 92, 90, 87, 93, 90, 77, 80, 104, 99

Draw a back-to-back stem-and-leaf plot and use it to help make comparisons between the two lists.

24 The bar graph shown gives the number of fatalities on West Australian roads for 2010–2012 for a number of categories of road user. Discuss the content of the graph.

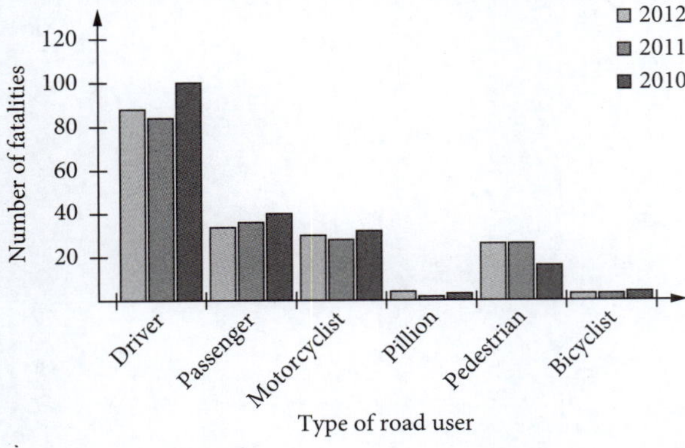

25 Two brands of rechargeable batteries were tested to compare their lifetimes, in hours, before they required recharging. Samples of 10 batteries of each brand were tested in the same equipment, producing the following times before requiring recharging:

Big Zap: 7, 22, 9, 24, 16, 22, 25, 26, 23, 26
Sparky: 19, 20, 21, 15, 17, 15, 39, 23, 14, 15

(a) Calculate the mean, median, standard deviation and interquartile range of the lifetime of each brand of battery before recharging.
(b) Represent the data as a back-to-back stem-and-leaf plot.
(c) Construct parallel box plots for the lifetimes of the two brands of batteries.
(d) If you wanted to be confident that the battery would last at least 14 hours, which battery would you buy?
(e) If you wanted to maximise your chances of the battery you purchased lasting more than 20 hours, which battery would you buy?
(f) Which battery tends to last longer? Justify your answer with reference to the summary statistics and statistical displays.

26 A refrigerator manufacturer received the following complaints from purchasers of new refrigerators.

Defect	Frequency
Dent in door	16
Door seal	12
Damaged box	8
Faulty light	6
Wiring defect	2
Refrigerant leak	1
Total	**45**

(a) Add the percentage and cumulative percentage columns to this table.
(b) Draw the Pareto chart for this information.
(c) What steps should the manufacturer take to reduce the number of complaints?

CHAPTER 12
Trigonometric functions and graphs

12.1 TRANSFORMATION OF GRAPHS OF THE TRIGONOMETRIC FUNCTIONS

In the diagram below, the graph of $y = \sin x$ is the unbroken curve and the graph of $y = \cos x$ is the dashed curve. The graphs are drawn in the domain $-2\pi \le x \le 2\pi$. For a domain of real numbers, the graphs continue to repeat in both directions, so the graphs for the domains $-2\pi \le x \le 0$, $0 \le x \le 2\pi$ and $2\pi \le \theta \le 4\pi$ are the same.

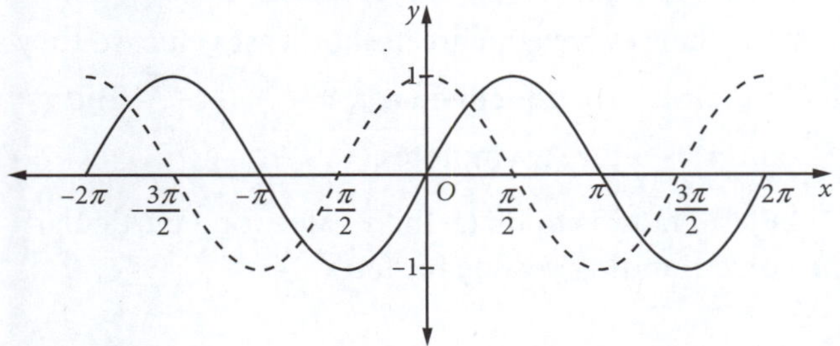

As x increases, the values of $\sin x$ and $\cos x$ repeat themselves after an interval or **period** of 2π. Sine and cosine are therefore called **periodic functions**. This means that the points $P(\theta)$, $P(2\pi + \theta)$, $P(4\pi + \theta)$ on the unit circle all coincide and hence $\sin(2\pi + \theta) = \sin\theta$, $\cos(2\pi + \theta) = \cos\theta$ and so on.

The maximum and minimum values of $\sin x$ and $\cos x$ are 1 and -1 respectively, so we say that their **amplitude** is 1.

If the graph of $y = \cos x$ is translated $\frac{\pi}{2}$ units to the right, parallel with the x-axis, it coincides with the graph of $y = \sin\theta$. You can check that this follows from $\cos x = \sin\left(\frac{\pi}{2} + x\right)$. You can also check by sketching graphs by hand or by using graphing software.

MAKING CONNECTIONS

Transforming the sine graph into the cosine graph

Move the slider to see how the translated sine graph matches the cosine graph.

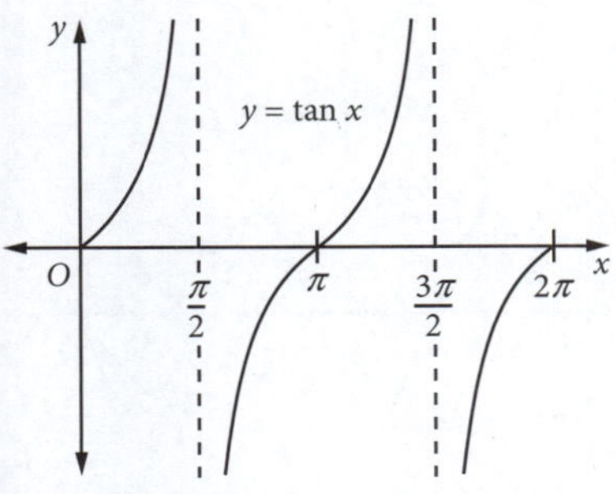

The diagram on the left shows the graph of $y = \tan x$ for $0 \le x \le 2\pi$. As x increases, the values of $\tan x$ repeat after an interval or **period** of π. Because $\tan x = \frac{\sin x}{\cos x}$, the graph of $y = \tan x$ does not exist when $\cos x = 0$, so the tan function is undefined at $x = \frac{\pi}{2}, \frac{3\pi}{2}, \ldots$

The domain of $\tan x$ is all real x except for $x = \frac{(2n-1)\pi}{2}$, where n is an integer (i.e. odd multiples of $\frac{\pi}{2}$). The range of $\tan x$ is all real y.

Graphs of $y = k\sin x$, $y = k\cos x$ and $y = k\tan x$

Consider the graphs of $y = \sin x$ and $y = 2\sin x$, drawn on the same axes for $0 \le x \le 2\pi$.

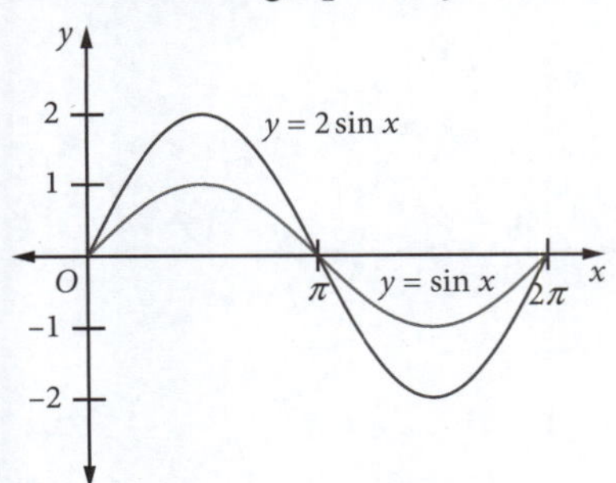

The graphs have the same shape and the same period, 2π. They both cut the x-axis at $x = 0, \pi, 2\pi$. The greatest value of $\sin x$ is 1, at $x = \frac{\pi}{2}$, while the greatest value of $2\sin x$ is 2, at $x = \frac{\pi}{2}$. The least value of $\sin x$ is -1, at $x = \frac{3\pi}{2}$, while the least value of $2\sin x$ is -2, at $x = \frac{3\pi}{2}$. The amplitude of $y = \sin x$ is 1 and the amplitude of $y = 2\sin x$ is 2.

Now consider the graphs of $y = \cos x$ and $y = 3\cos x$, drawn on the same axes for $0 \leq x \leq 2\pi$.

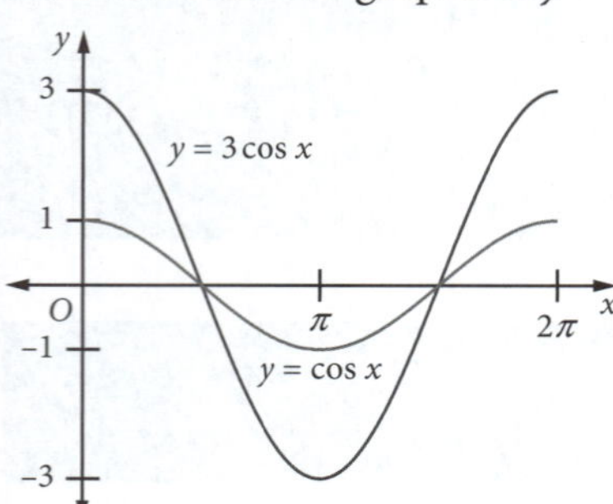

The graphs have the same shape and the same period, 2π. They both cut the x-axis at $x = \frac{\pi}{2}, \frac{3\pi}{2}$. The greatest value of $\cos x$ is 1, at $x = 0, 2\pi$, while the greatest value of $3\cos x$ is 3, at $x = 0, 2\pi$. The least value of $\cos x$ is -1, at $x = \pi$, while the least value of $3\cos x$ is -3, at $x = \pi$. The amplitude of $y = \cos x$ is 1 and the amplitude of $y = 3\cos x$ is 3.

Finally, consider the graphs of $y = \tan x$ and $y = 2\tan x$, drawn on the same axes for $-\frac{\pi}{2} \leq x \leq \frac{3\pi}{2}$.

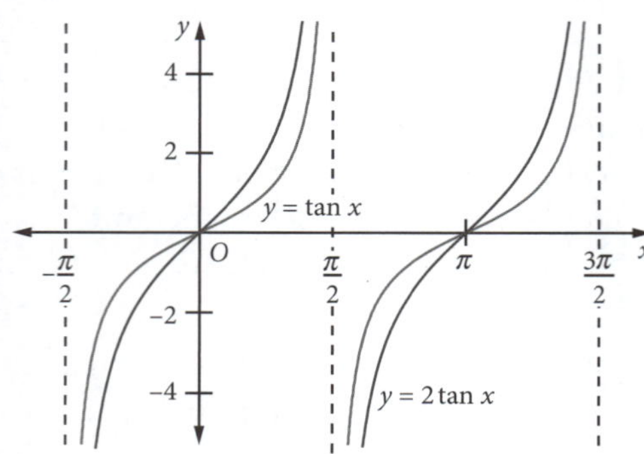

The graphs have the same shape and the same period, π. They both cut the x-axis at $x = 0, \pi$. The curves have no greatest or least value, so they have no amplitude. The curves are undefined at $x = -\frac{\pi}{2}, \frac{\pi}{2}, \frac{3\pi}{2}$. The lines $x = -\frac{\pi}{2}$, $x = \frac{\pi}{2}$, and $x = \frac{3\pi}{2}$ are asymptotes.

The effect of the '2' in $y = 2\tan x$ is to make the y value for a particular value of x twice the corresponding y value for $\tan x$.

In general, the effect of the k multiplier on a trigonometric function is to multiply the function's amplitude by k. If the function has no amplitude, as for $y = k\tan x$, then the effect is to stretch the curve vertically by a factor of k.

MAKING CONNECTIONS

Transformations of the graphs of $y=k\sin x$, $y=k\cos x$ and $y=k\tan x$

Move the sliders to explore transformations of the trigonometric graphs.

Graphs of $y=\sin ax$, $y=\cos ax$ and $y=\tan ax$

Consider the graphs of $y = \sin x$ and $y = \sin 2x$, drawn on the same axes for $0 \leq x \leq 2\pi$.

The graphs have the same shape, but in one period of $y = \sin x$ the graph of $y = \sin 2x$ occurs twice. This means that the period of $y = \sin 2x$ is half the period of $y = \sin x$. The period of $y = \sin 2x$ is π.

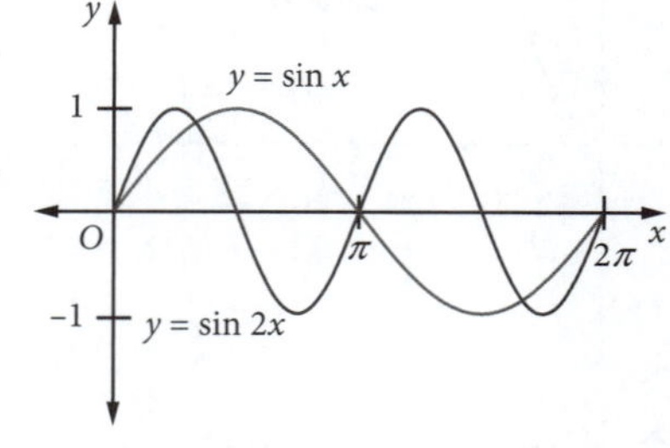

Both graphs cut the x-axis when $x = 0, \pi, 2\pi$. The greatest value of $\sin x$ is 1, at $x = \frac{\pi}{2}$, while the greatest value of $\sin 2x$ is also 1, at $x = \frac{\pi}{4}, \frac{5\pi}{4}$.
The least value of $\sin x$ is -1, at $x = \frac{3\pi}{2}$, while the least value of $\sin 2x$ is also -1, at $x = \frac{3\pi}{4}, \frac{7\pi}{4}$. The amplitude of both functions is 1.

Consider the graphs of $y = \cos x$ and $y = \cos 3x$, drawn on the same axes for $0 \leq x \leq 2\pi$.

The graphs have the same shape, but in one period of $y = \cos x$ the graph of $y = \cos 3x$ occurs three times. This means that the period of $y = \cos 3x$ is one-third the period of $\cos x$. The period of $y = \cos 3x$ is $\frac{2\pi}{3}$.

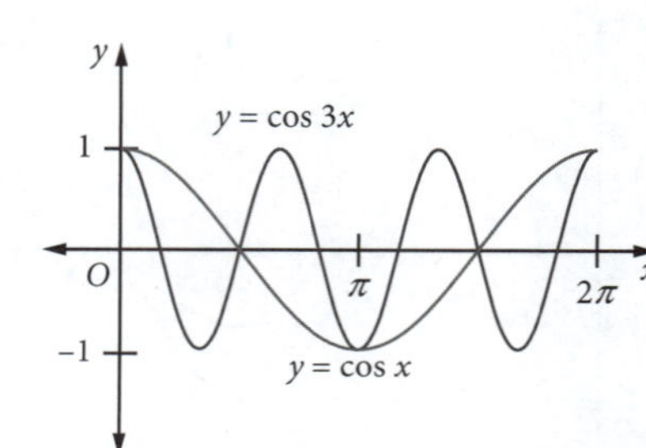

Both graphs cut the x-axis at $x = \frac{\pi}{2}, \frac{3\pi}{2}$. The greatest value of $\cos x$ is 1, at $x = 0, 2\pi$, while the greatest value of $\cos 3x$ is also 1, at $x = 0, \frac{2\pi}{3}, \frac{4\pi}{3}, 2\pi$.
The least value of $\cos x$ is -1, at $x = \pi$, while the least value of $\cos 3x$ is also -1, at $x = \frac{\pi}{3}, \pi, \frac{5\pi}{3}$. The amplitude of both functions is 1.

Consider the graphs of $y = \tan x$ and $y = \tan 2x$, drawn on the same axes for $-\frac{\pi}{2} \le x \le \frac{\pi}{2}$.
The graphs have the same shape, but in one period of $y = \tan x$ the graph of $y = \tan 2x$ occurs twice. This means that the period of $y = \tan 2x$ is half the period of $\tan x$. The period of $y = \tan 2x$ is $\frac{\pi}{2}$.
Both graphs cut the x-axis when $x = 0$. The curves have no greatest or least value, so they have no amplitude. $y = \tan x$ is undefined at $x = -\frac{\pi}{2}, \frac{\pi}{2}$.
$y = \tan 2x$ is undefined at $x = -\frac{\pi}{4}, \frac{\pi}{4}$. The lines $x = -\frac{\pi}{2}$ and $x = \frac{\pi}{2}$ are asymptotes to $y = \tan x$, while the lines $x = -\frac{\pi}{4}$ and $x = \frac{\pi}{4}$ are asymptotes to $y = \tan 2x$.

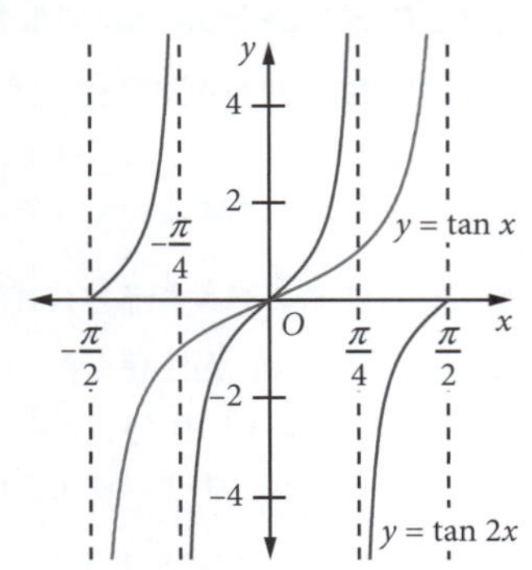

In general, the effect of the a multiplier in each equation is to divide the function's period by a. Thus the period of $y = \sin ax$ is $\frac{2\pi}{a}$, the period of $y = \cos ax$ is $\frac{2\pi}{a}$ and the period of $y = \tan ax$ is $\frac{\pi}{a}$.

MAKING CONNECTIONS

Transformations of the graphs of $y = \sin ax$, $y = \cos ax$ and $y = \tan ax$

Move the sliders to explore transformations of the trigonometric graphs.

Graphs of $y = \sin x + c$, $y = \cos x + c$ and $y = \tan x + c$

Consider the graphs of $y = \sin x$ and $y = 2 + \sin x$, drawn on the same axes for $0 \le x \le 2\pi$.
The graphs have the same shape and the same period, 2π. The greatest value of $\sin x$ is 1, at $x = \frac{\pi}{2}$, while the greatest value of $2 + \sin x$ is 3, at $x = \frac{\pi}{2}$. The least value of $\sin x$ is -1, at $x = \frac{3\pi}{2}$, while the least value of $2 + \sin x$ is 1, at $x = \frac{3\pi}{2}$.
The amplitude of both curves is 1.
The effect of adding 2 to $\sin x$ is to translate the curve vertically upwards 2 units (i.e. 'move up 2').

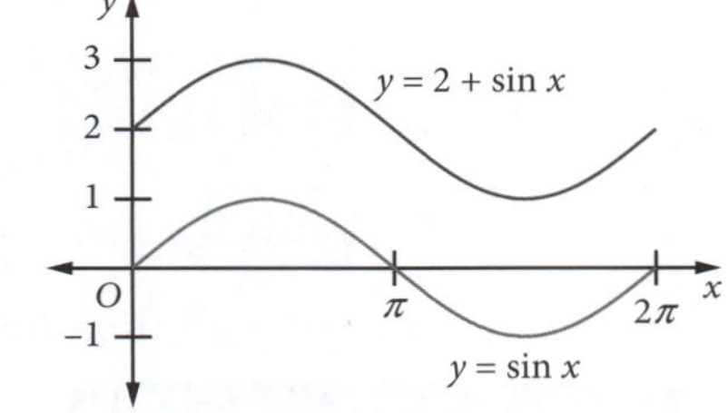

Consider the graphs of $y = \cos x$ and $y = 3 + \cos x$, drawn on the same axes for $0 \le x \le 2\pi$.
The graphs have the same shape and the same period, 2π. The greatest value of $\cos x$ is 1, at $x = 0, 2\pi$, while the greatest value of $3 + \cos x$ is 4, at $x = 0, 2\pi$. The least value of $\cos x$ is -1, at $x = \pi$, while the least value of $3 + \cos x$ is 2, at $x = \pi$.
The amplitude of both curves is 1.
The effect of adding 3 to $\cos x$ is to translate the curve vertically upwards 3 units (move up 3).

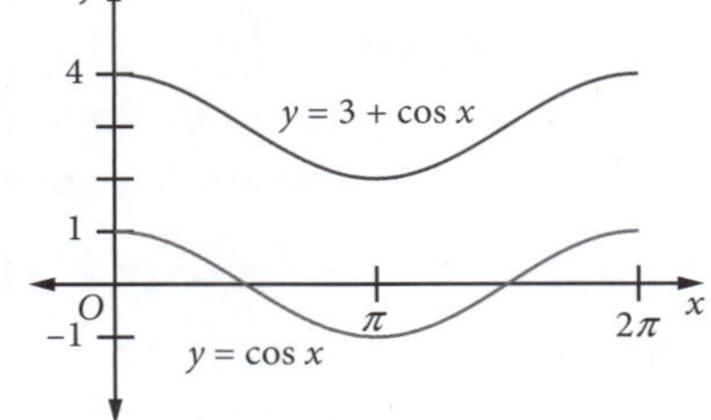

Consider the graphs of $y = \tan x$ and $y = \tan x - 1$, drawn on the same axes for $-\frac{\pi}{2} \le x \le \frac{\pi}{2}$.
The graphs have the same shape and the same period, π. The curves have no greatest or least value, so they have no amplitude.
The curves are undefined at $x = -\frac{\pi}{2}, \frac{\pi}{2}$. The lines $x = -\frac{\pi}{2}$ and $x = \frac{\pi}{2}$ are asymptotes.
The effect of the -1 in $y = \tan x - 1$ is to translate the curve vertically downwards 1 unit (move down 1).
In general, the effect of the c in each equation is to translate the curve vertically by c units.

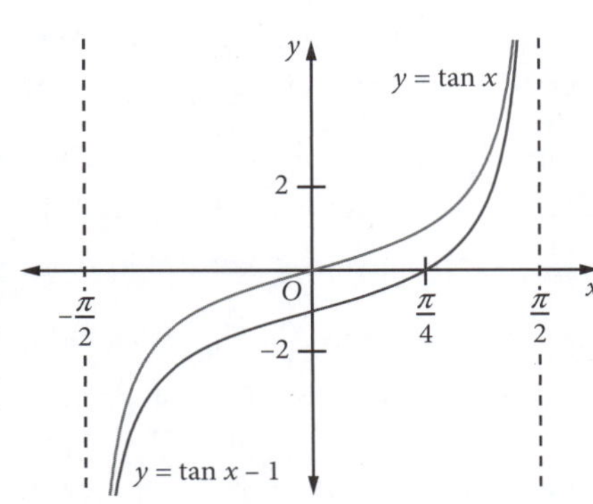

MAKING CONNECTIONS

Transformations of the graphs of $y = \sin x + c$, $y = \cos x + c$ and $y = \tan x + c$

Move the sliders to explore vertical translations of the trigonometric graphs.

Graphs of $y = k\sin a(x+b)$, $y = k\cos a(x+b)$ and $y = k\tan a(x+b)$

These graphs combine the effects seen so far. The value k is still the amplitude of the sine and cosine functions and affects the size of the y value in the tan function, and a still divides the period of the functions. The value b now changes the position along the x-axis where the functions occur. It is called the **phase** of the function.

Consider the graphs of $y = \sin 2x$ and $y = \sin 2\left(x - \frac{\pi}{4}\right)$, drawn on the same axes for $0 \le x \le 2\pi$.

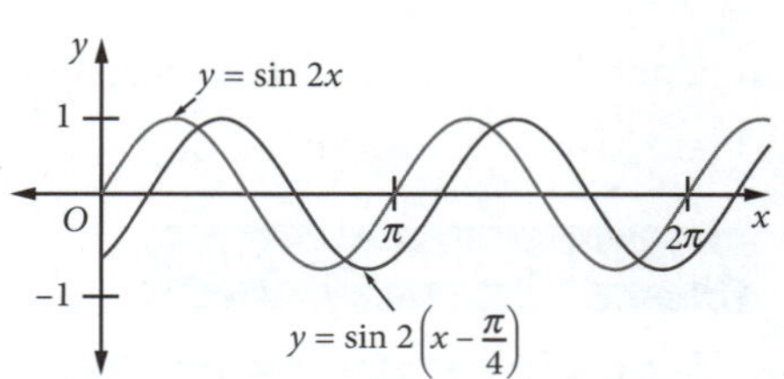

The graphs have the same shape, amplitude 1 and period π. The graph of $y = \sin 2x$ has been translated $\frac{\pi}{4}$ units to the right to obtain the graph of $y = \sin 2\left(x - \frac{\pi}{4}\right)$.

$y = \sin 2x$ cuts the x-axis when $x = 0, \frac{\pi}{2}, \pi, \frac{3\pi}{2}, 2\pi$.

$y = \sin 2\left(x - \frac{\pi}{4}\right)$ cuts the x-axis at $x = \frac{\pi}{4}, \frac{3\pi}{4}, \frac{5\pi}{4}, \frac{7\pi}{4}$. The greatest value of each function is 1 and the least value of each function is -1.

Consider the graphs of $y = 2\cos 3x$ and $y = 2\cos 3\left(x + \frac{\pi}{9}\right)$, drawn on the same axes for $0 \le x \le 2\pi$.

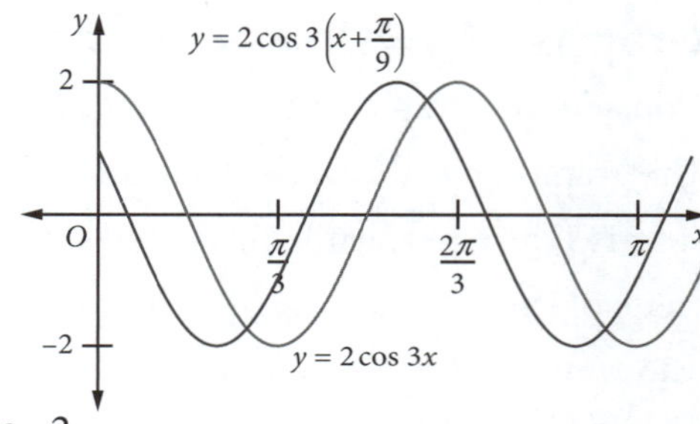

The graphs have the same shape, amplitude 2 and period $\frac{2\pi}{3}$. The graph of $y = 2\cos 3x$ has been translated $\frac{\pi}{9}$ units to the left to obtain $y = 2\cos 3\left(x + \frac{\pi}{9}\right)$.

$y = 2\cos 3x$ cuts the x-axis when $x = \frac{\pi}{6}, \frac{\pi}{2}, \frac{5\pi}{6}$.

$y = 2\cos 3\left(x + \frac{\pi}{9}\right)$ cuts the x-axis when $x = \frac{\pi}{18}, \frac{7\pi}{18}, \frac{13\pi}{18}$.

The greatest value of each function is 2 and the least value of each function is -2.

MAKING CONNECTIONS

Transformations of the graphs of $y = k\sin a(x+b)$, $y = k\cos a(x+b)$ and $y = k\tan a(x+b)$

Move the sliders to explore horizontal translations of the trigonometric graphs.

Graphs of $y = k\sin a(x+b)+c$, $y = k\cos a(x+b)+c$ and $y = k\tan a(x+b)+c$

These graphs have the same shape as $y = k\sin a(x+b)$, $y = k\cos a(x+b)$ and $y = k\tan a(x+b)$. The effect of the c is to change the y-value by c units, or graphically, to translate the graph vertically a distance of $|c|$ units, upwards if c is positive.

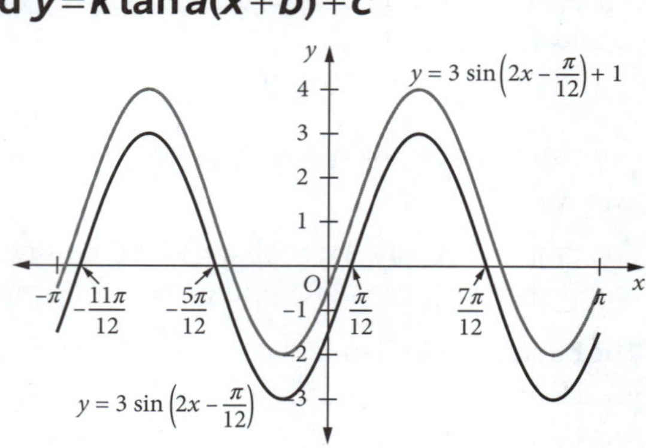

Consider the graphs of $y = 3\sin 2\left(x - \frac{\pi}{12}\right)$ and $y = 3\sin 2\left(x - \frac{\pi}{12}\right) + 1$ drawn on the same axes for $-\pi \le x \le \pi$.

The graphs have the same shape, amplitude 3 and period π. The graph of $y = 3\sin 2\left(x - \frac{\pi}{12}\right)$ has been translated one unit upwards to give the graph of $y = 3\sin 2\left(x - \frac{\pi}{12}\right) + 1$.

The curve $y = 3\sin 2\left(x - \frac{\pi}{12}\right)$ cuts the x-axis at $x = -\frac{11\pi}{12}, -\frac{5\pi}{12}, \frac{\pi}{12}, \frac{7\pi}{12}$.

The curve $y = 3\sin 2\left(x - \frac{\pi}{12}\right) + 1$ cuts the line $y = 1$ at $x = -\frac{11\pi}{12}, -\frac{5\pi}{12}, \frac{\pi}{12}, \frac{7\pi}{12}$

The amplitude, period and phase of each function is the same. The greatest value of $3\sin\left(2x - \frac{\pi}{12}\right)$ is 3, its least value is −3. The greatest value of $y = 3\sin 2\left(x - \frac{\pi}{12}\right) + 1$ is 4, its least value is −2.

MAKING CONNECTIONS

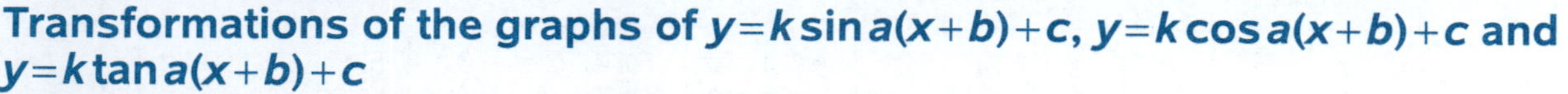

Transformations of the graphs of $y = k\sin a(x+b)+c$, $y = k\cos a(x+b)+c$ and $y = k\tan a(x+b)+c$

Move the sliders to explore transformations to the trigonometric graphs.

Summary of trigonometric functions

Function	Period	Amplitude	Domain	Range	Symmetry
$y = k\sin x$	2π	k	Real x	$-k \le y \le k$	Rotational symmetry
$y = k\cos x$	2π	k	Real x	$-k \le y \le k$	Symmetrical about y-axis
$y = k\tan x$	π	none	Real x, $x \ne \frac{(2n-1)\pi}{2}$ for integer n	Real y	Rotational symmetry
$y = k\sin ax$	$\frac{2\pi}{a}$	k	Real x	$-k \le y \le k$	Rotational symmetry
$y = k\cos ax$	$\frac{2\pi}{a}$	k	Real x	$-k \le y \le k$	Symmetrical about y-axis
$y = k\tan ax$	$\frac{\pi}{a}$	none	Real x, $x \ne \frac{(2n-1)\pi}{2a}$ for integer n	Real y	Rotational symmetry
$y = k\sin ax + c$	$\frac{2\pi}{a}$	k	Real x	$-k + c \le y \le k + c$	
$y = k\cos ax + c$	$\frac{2\pi}{a}$	k	Real x	$-k + c \le y \le k + c$	Symmetrical about y-axis
$y = k\tan ax + c$	$\frac{\pi}{a}$	none	Real x, $x \ne \frac{(2n-1)\pi}{2}$ for integer n	Real y	
$y = k\sin a(x+b)$	$\frac{2\pi}{a}$	k	Real x	$-k \le y \le k$	
$y = k\cos a(x+b)$	$\frac{2\pi}{a}$	k	Real x	$-k \le y \le k$	
$y = k\tan a(x+b)$	$\frac{\pi}{a}$	none	Real x, $x \ne \frac{(2n-1)\pi - 2c}{2a}$ for integer n	Real y	

When first sketching graphs of trigonometric functions, you should draw up a table of values and plot the points, joining the points with an appropriate curve. When you are familiar with the shapes of the standard curves, you will be able to draw them confidently using only the properties of amplitude, period and asymptotes. You might also use a standard template to obtain the shape of a curve and then mark the appropriate scale on the axes.

After you become more experienced at drawing these graphs by hand, draw them using graphing software and compare the results to check that you have not missed anything.

Example 1

Sketch the graph of $y = 3 \sin 2x$ for $0 \le x \le \pi$.

Solution

Complete a table of values.

x	0	$\frac{\pi}{12}$	$\frac{\pi}{8}$	$\frac{\pi}{6}$	$\frac{\pi}{4}$	$\frac{\pi}{3}$	$\frac{3\pi}{8}$	$\frac{5\pi}{12}$	$\frac{\pi}{2}$	$\frac{7\pi}{12}$	$\frac{5\pi}{8}$	$\frac{2\pi}{3}$	$\frac{3\pi}{2}$	$\frac{5\pi}{3}$	$\frac{7\pi}{8}$	$\frac{11\pi}{12}$	π
y	0	1.5	2.1	2.6	3	2.6	2.1	1.5	0	−1.5	−2.1	−2.6	−3	−2.6	−2.1	−1.5	0

Plot the points and sketch the graph, marking scales on each axis. Make sure that your curve is smooth, not pointed or jagged.

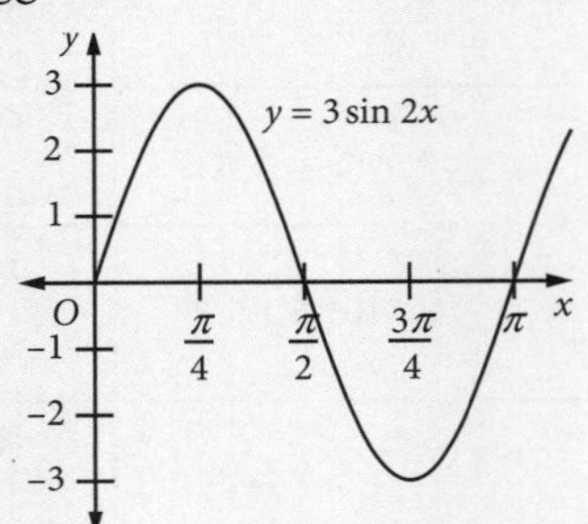

An alternative method is to first sketch $y = \sin x$ for $0 \le x \le 2\pi$. You can then draw $y = 3 \sin 2x$ by noting that it has 3 times the amplitude and half the period of $y = \sin x$.

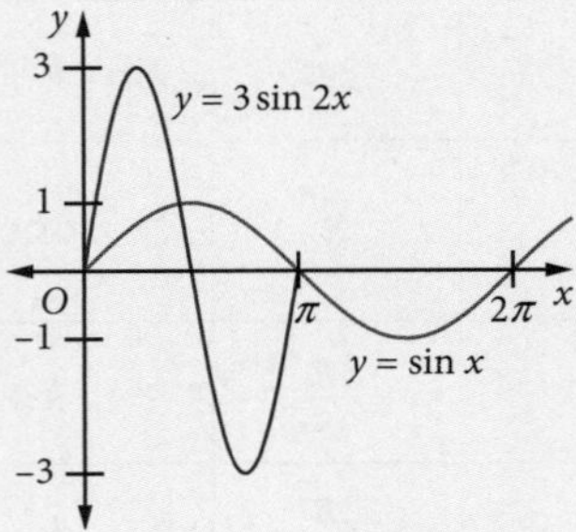

Example 2

Sketch the graph of **(a)** $y = 2 \cos 3x$ **(b)** $y = \tan \pi x$, showing one complete cycle.

Solution

(a) Amplitude = 2

Period = $\frac{2\pi}{3}$

The midpoint of the cycle is $\frac{\pi}{3}$

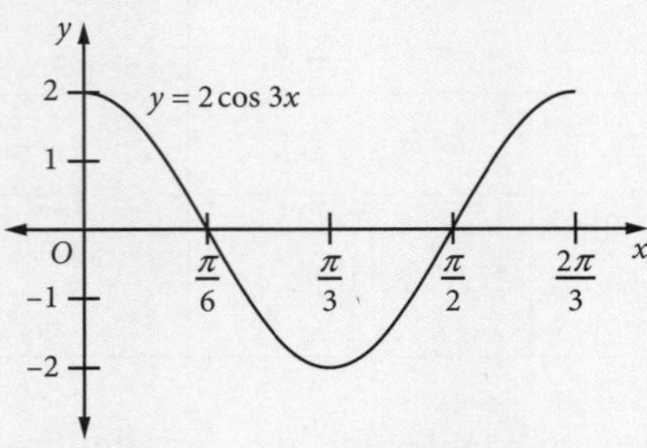

(b) No amplitude

Period = 1 (variable is πx)

Cycle is $-0.5 < x < 0.5$

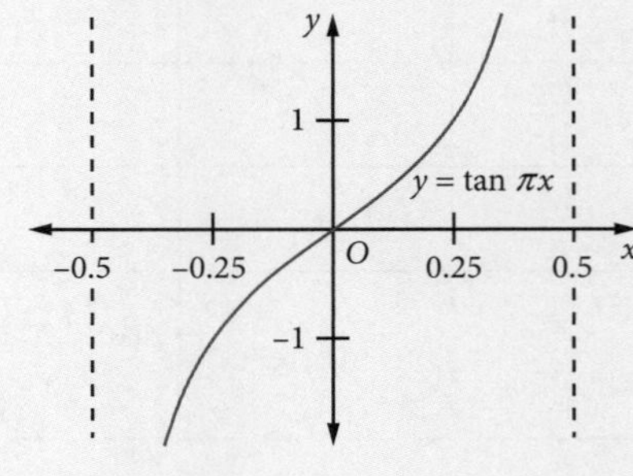

Example 3

(a) Sketch the graph of $y = \cos x$ for $0 \le x \le 2\pi$.

(b) On the same set of axes sketch $y = \cos\frac{x}{2}$.

(c) Hence sketch $y = -\cos\frac{x}{2}$.

(d) Use your answer to **(c)** to sketch $y = 1 - \cos\frac{x}{2}$.

Solution

(a) $y = \cos x$ for $0 \le x \le 2\pi$.

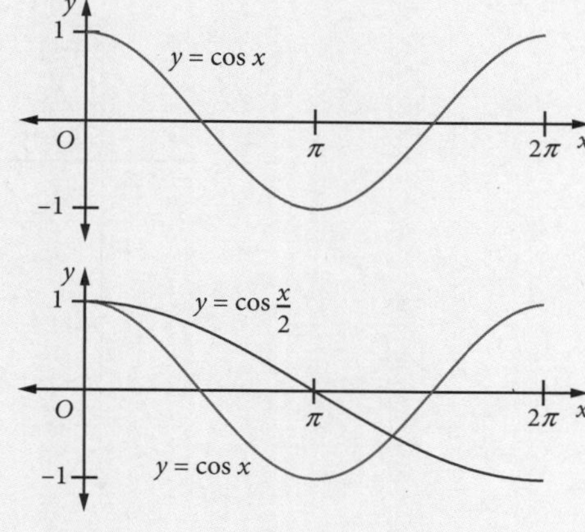

(b) The period of $y = \cos\frac{x}{2}$ is 4π, so half a cycle of the curve will fit in the domain $0 \le x \le 2\pi$.

$y = \cos x$ and $y = \cos\frac{x}{2}$ both have an amplitude of 1.

(c) $y = -\cos\frac{x}{2}$ is just $y = \cos\frac{x}{2}$ flipped over (reflected in the x-axis).

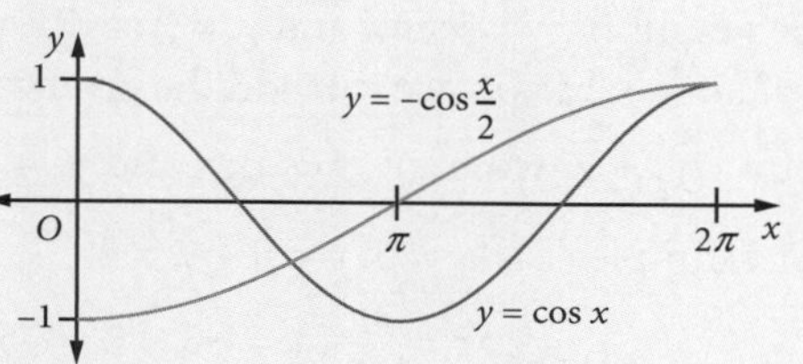

(d) Obtain $y = 1 - \cos\frac{x}{2}$ by moving $y = -\cos\frac{x}{2}$ up 1 unit.

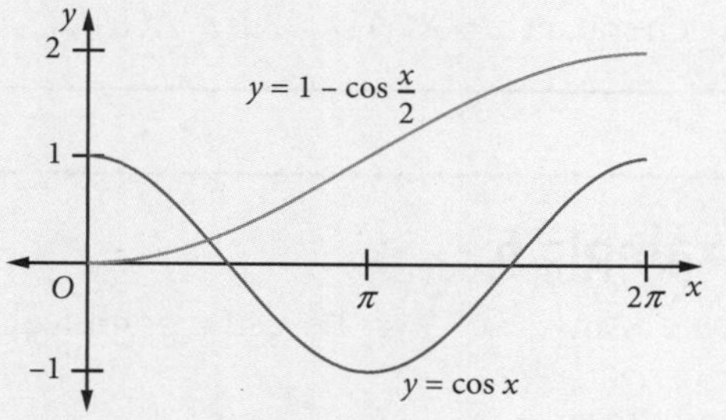

Example 4

Sketch graphs of the following, showing one complete cycle (period) of each:

(a) $y = 2\cos 3\left(x + \frac{\pi}{3}\right)$

(b) $y = 3\sin\left(2x - \frac{\pi}{2}\right)$

Solution

(a)

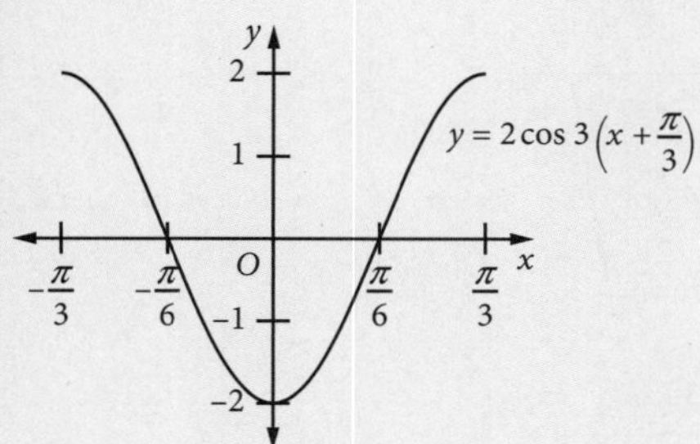

(b) Rewrite as $y = 3\sin 2\left(x - \frac{\pi}{4}\right)$:

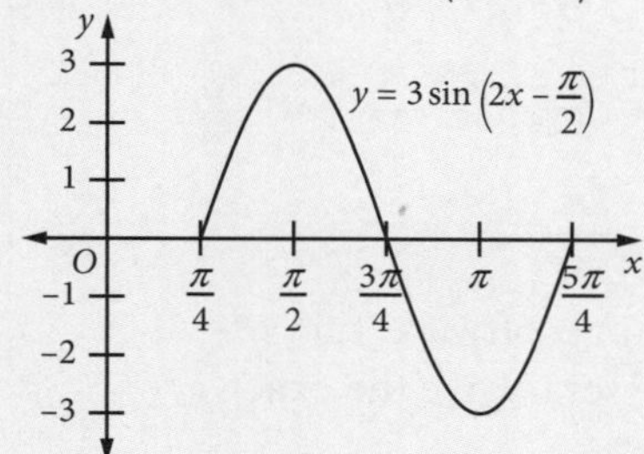

Example 5

Using the same scale and axes, sketch the graphs of $y = 3\sin x$ and $y = 2\cos 2x$. Hence sketch the graph of $y = 3\sin x + 2\cos 2x$ for $0 \le x \le 2\pi$.

Solution

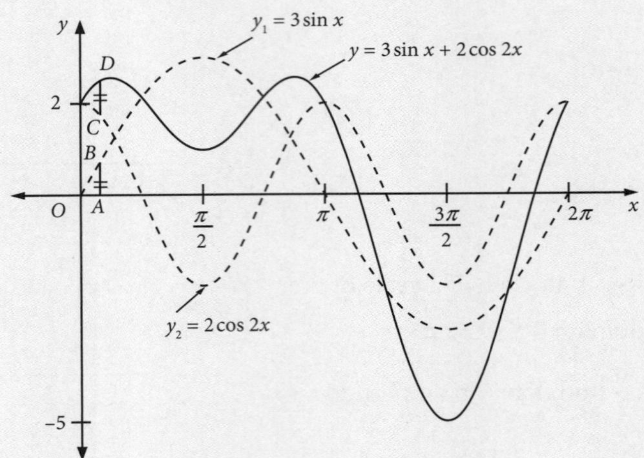

The graphs of $y_1 = 3\sin x$ and $y_2 = 2\cos 2x$ are shown here as dashed lines. To obtain points on the graph of $y = 3\sin x + 2\cos 2x$ we can add the ordinates of the component curves, remembering to take the sign into account.

Let $y = y_1 + y_2$ where $y_1 = 3\sin x$ and $y_2 = 2\cos 2x$.

For example: at $x = 0$, $y = 0 + 2 = 2$; at $x = \frac{\pi}{2}$, $y = 3 - 2 = 1$; at $x = \pi$, $y = 0 + 2 = 2$;
at $x = \frac{3\pi}{2}$, $y = -3 - 2 = -5$; at $x = 2\pi$, $y = 0 + 2 = 2$; and so on.

In general, at $x = A$, $AD = AB + AC$.

Example 6

Solve $3\sin\left(2x - \frac{\pi}{2}\right) = 1.5$ using graphical methods. Check your answer algebraically.

Solution

The solution is $x = \frac{\pi}{3}, \frac{2\pi}{3}$.

$$3\sin\left(2x - \frac{\pi}{2}\right) = 1.5$$

$$\sin\left(2x - \frac{\pi}{2}\right) = \frac{1}{2}$$

$$2x - \frac{\pi}{2} = \frac{\pi}{6}, \pi - \frac{\pi}{6}$$

$$2x = \frac{\pi}{6} + \frac{\pi}{2}, \frac{5\pi}{6} + \frac{\pi}{2}$$

$$2x = \frac{2\pi}{3}, \frac{4\pi}{3}$$

$$x = \frac{\pi}{3}, \frac{2\pi}{3}$$

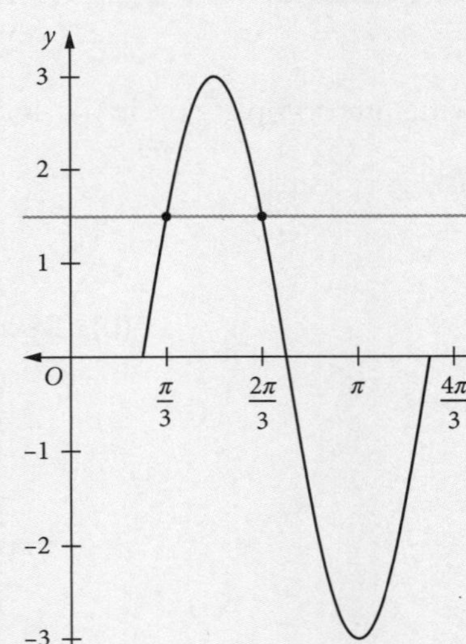

Note: Using graphing software, the scales can often be set to give the exact values in terms of π.

EXERCISE 12.1 TRANSFORMATION OF GRAPHS OF THE TRIGONOMETRIC FUNCTIONS

1 Sketch the graph of each of the following, stating the period and amplitude of the function:

(a) $y = 4\sin x,\ 0 \le x \le 2\pi$ **(b)** $y = \cos 2x,\ 0 \le x \le 2\pi$ **(c)** $y = 3\tan x,\ -\frac{\pi}{2} \le x \le \frac{\pi}{2}$

(d) $y = 4\sin 3x,\ 0 \le x \le 2\pi$ **(e)** $y = 3\cos 2x,\ 0 \le x \le 2\pi$ **(f)** $y = 3\tan 2x,\ -\frac{\pi}{2} \le x \le \frac{\pi}{2}$

(g) $y = \sin\frac{x}{2},\ 0 \le x \le 4\pi$ **(h)** $y = \cos\frac{x}{2},\ -2\pi \le x \le 2\pi$ **(i)** $y = \tan\frac{x}{2},\ -\pi \le x \le \pi$

2 Which diagram shows the graph of $y = 3\sin\frac{x}{4}$ for $0 \le x \le 4\pi$?

A

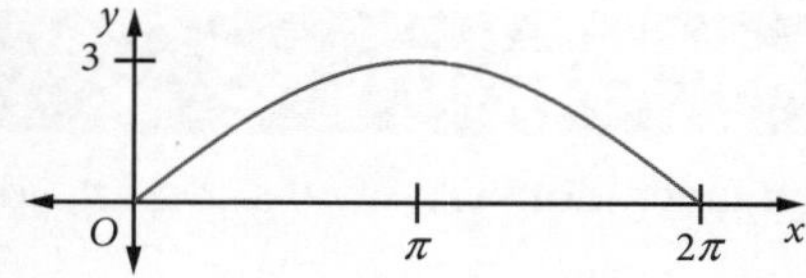

B

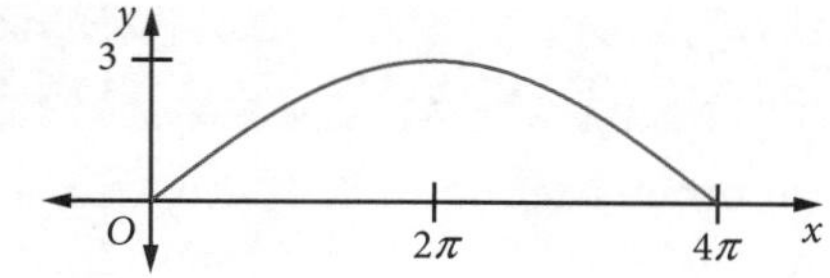

C

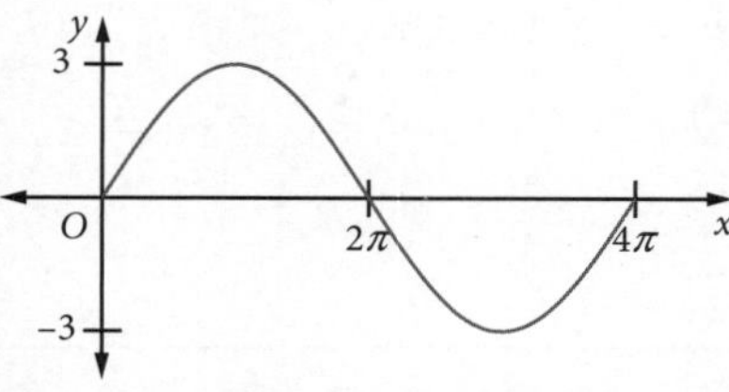

D

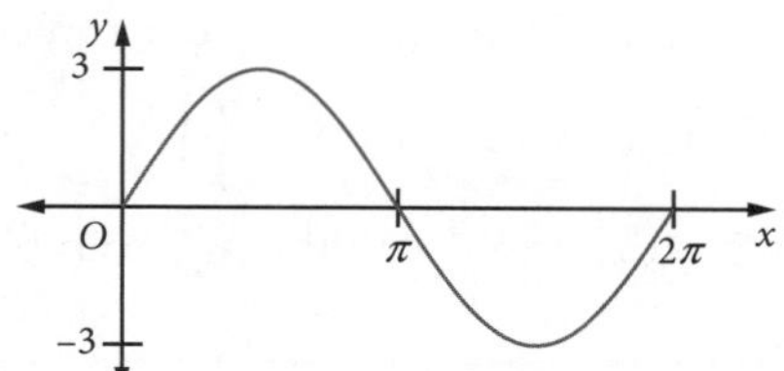

3 Sketch the graph of each of the following, stating the period and amplitude of each function:

(a) $y = 3\sin\frac{x}{2},\ -\pi \le x \le \pi$ **(b)** $y = 2\cos\frac{x}{2},\ 0 \le x \le 2\pi$ **(c)** $y = 3\tan\frac{x}{2},\ 0 \le x \le 2\pi$

4 Sketch the graph of $y = \cos x,\ 0 \le x \le \pi$. On the same axes, sketch the graph of:

(a) $y = -\cos x$ **(b)** $y = 1 - \cos x$

5 Sketch the graph of $y = \sin x,\ -\pi \le x \le \pi$. On the same axes, sketch the graph of:

(a) $y = 2\sin x$ **(b)** $y = -2\sin x$ **(c)** $y = 3 - 2\sin x$

6 Sketch the graph of each of the following, stating the period and amplitude of the function:

(a) $y = \sin \pi x,\ -1 \le x \le 1$ **(b)** $y = \cos 2\pi x,\ 0 \le x \le 2$ **(c)** $y = \tan\frac{\pi x}{2},\ 0 \le x \le 2$

7 Sketch the graph of $y = \tan x,\ -\pi \le x \le \pi$. On the same axes, sketch the graph of:

(a) $y = 2\tan x$ **(b)** $y = -2\tan x$ **(c)** $y = 2 - 2\tan x$

8 Sketch the graph of each of the following for $0 \le x \le 2\pi$.

(a) $y = \sin x - 1$ **(b)** $y = 1 + 2\sin x$ **(c)** $y = 2 - \cos x$

(d) $y = 2\tan 2x$ **(e)** $y = 2 + \sin 2x$ **(f)** $y = \cos 2x - 1$

9 Sketch the graph of each of the following.

(a) $y = 2\sin\left(\theta - \frac{\pi}{2}\right)$ **(b)** $y = 3\cos\left(\theta + \frac{\pi}{4}\right)$ **(c)** $y = 2\sin(\theta - \pi)$

(d) $y = 5\cos 3\left(\theta + \frac{\pi}{3}\right)$ **(e)** $y = \frac{1}{2}\tan 2(\theta + \pi)$ **(f)** $y = \sqrt{2}\sin\left(2\theta - \frac{\pi}{2}\right)$

10 Sketch the graph of each:

(a) $y = \sin 2\theta + 1$ **(b)** $y = 3\cos\theta - 2$ **(c)** $y = \frac{1}{2}\sin\left(\theta - \frac{\pi}{2}\right) + 3$

(d) $y = 2\cos 2\left(\theta - \frac{\pi}{4}\right) + 1$ **(e)** $y = 4\sin 3\left(\theta - \frac{\pi}{6}\right) - 2$ **(f)** $y = 3 - \sin\left(\theta - \frac{\pi}{2}\right)$

11 By adding ordinates, sketch the graphs of:

(a) $y = \sin\theta + \cos\theta$ (b) $y = 3\sin 2\theta + 4\sin\theta$ (c) $y = 2\cos 3\theta + 3\sin 2\theta$
(d) $y = \sin 2\theta - \cos\theta$ (e) $y = \frac{1}{2}\cos 2\theta - \sin\theta$ (f) $y = \sin\theta + \sin 2\theta$

12 Ocean tides can be approximated by the equation $y = k\sin ax + c$. Find a website that records tidal data, select a location and find the times and heights of high and low tides over a four-day period. Graph this information to see how close the graph's shape is to a sine curve.

13 By drawing appropriate graphs, solve each equation for $0 \le x \le 2\pi$.

(a) $\sqrt{3}\tan\left(x - \frac{\pi}{3}\right) - 1 = 0$ (b) $2\sqrt{3}\cos\left(x + \frac{\pi}{4}\right) - 3 = 0$ (c) $\sqrt{2}\sin\left(x + \frac{\pi}{6}\right) + 1 = 0$

12.2 FURTHER SOLUTION OF TRIGONOMETRIC EQUATIONS

When solving trigonometric equations of the form $k\sin(ax + b) = c$ over a given domain, say $0 \le x \le 2\pi$, you need to be aware of how the $(ax + b)$ affects the domain for your solution.

Given $0 \le x \le 2\pi$, then if solving $\sin(2x) = 0.5$ you can only consider values for x in $0 \le 2x \le 4\pi$, or x in $[0, 4\pi]$.

Given $0 \le x \le 2\pi$, then if solving $\sin\left(x + \frac{\pi}{4}\right) = \frac{\sqrt{3}}{2}$ you can only consider values for x in $0 + \frac{\pi}{4} \le x + \frac{\pi}{4} \le 2\pi + \frac{\pi}{4}$, or $x \in \left[\frac{\pi}{4}, \frac{9\pi}{4}\right]$. This is shown in the following examples.

Example 7

Solve, for $0 \le x \le 2\pi$:

(a) $2\cos\left(x + \frac{\pi}{6}\right) = 1$ (b) $2\sin\left(x - \frac{\pi}{4}\right) = -\sqrt{3}$ (c) $\sqrt{3}\tan\left(x + \frac{\pi}{3}\right) = 1$

Solution

(a) $2\cos\left(x + \frac{\pi}{6}\right) = 1$

$0 \le x \le 2\pi$

$\frac{\pi}{6} \le x + \frac{\pi}{6} \le \frac{13\pi}{6}$

Divide by 2: $\cos\left(x + \frac{\pi}{6}\right) = \frac{1}{2}$

Evaluate: $x + \frac{\pi}{6} = \frac{\pi}{3}, \frac{5\pi}{3}$

Simplify: $x = \frac{\pi}{3} - \frac{\pi}{6}, \frac{5\pi}{3} - \frac{\pi}{6}$

$x = \frac{\pi}{6}, \frac{3\pi}{2}$

(b) $2\sin\left(x - \frac{\pi}{4}\right) = -\sqrt{3}$

$0 \le x \le 2\pi$

$-\frac{\pi}{4} \le x - \frac{\pi}{4} \le \frac{7\pi}{4}$

Divide by 2: $\sin\left(x - \frac{\pi}{4}\right) = -\frac{\sqrt{3}}{2}$

Evaluate: $x - \frac{\pi}{4} = \pi + \frac{\pi}{3}, 2\pi - \frac{\pi}{3}$

$x - \frac{\pi}{4} = \frac{4\pi}{3}, \frac{5\pi}{3}$

Simplify: $x = \frac{4\pi}{3} + \frac{\pi}{4}, \frac{5\pi}{3} + \frac{\pi}{4}$

$x = \frac{19\pi}{12}, \frac{23\pi}{12}$

(c) $\sqrt{3}\tan\left(x + \frac{\pi}{3}\right) = 1$

$0 \le x \le 2\pi$

$\frac{\pi}{3} \le x + \frac{\pi}{3} \le \frac{7\pi}{3}$

Divide by $\sqrt{3}$:

$\tan\left(x + \frac{\pi}{3}\right) = \frac{1}{\sqrt{3}}$

$x + \frac{\pi}{3} = \frac{\pi}{6}, \pi + \frac{\pi}{6}, 2\pi + \frac{\pi}{6}$

$x + \frac{\pi}{3} = \frac{\pi}{6}, \frac{7\pi}{6}, \frac{13\pi}{6}$

$x = -\frac{\pi}{6}, \frac{5\pi}{6}, \frac{11\pi}{6}$

$x = -\frac{\pi}{6}$ is outside the given domain so the solution is: $x = \frac{5\pi}{6}, \frac{11\pi}{6}$

Note: In **(c)** as $0 \le x \le 2\pi$ so $\frac{\pi}{3} \le x + \frac{\pi}{3} \le 2\pi + \frac{\pi}{3}$ which is why you must consider $2\pi + \frac{\pi}{6}$ as a possible solution.

Example 8

Solve, for $-\pi \le x \le \pi$:

(a) $2\sin 3x = 1$ **(b)** $2\cos\left(2x + \frac{\pi}{3}\right) = 1$ **(c)** $2\sin\left(2x - \frac{\pi}{12}\right) = -\sqrt{3}$

Solution

(a) $2\sin 3x = 1$

Divide by 2: $\sin 3x = \frac{1}{2}$

$-\pi \le x \le \pi$ then $-3\pi \le 3x \le 3\pi$: $3x = -2\pi + \frac{\pi}{6}, -\pi - \frac{\pi}{6}, \frac{\pi}{6}, \pi + \frac{\pi}{6}, 2\pi + \frac{\pi}{6}$

$$3x = -\frac{11\pi}{6}, -\frac{7\pi}{6}, \frac{\pi}{6}, \frac{7\pi}{6}, \frac{13\pi}{6}$$

$$x = -\frac{11\pi}{18}, -\frac{7\pi}{18}, \frac{\pi}{18}, \frac{7\pi}{18}, \frac{13\pi}{18}$$

(b) $2\cos\left(2x + \frac{\pi}{3}\right) = 1$

Divide by 2: $\cos\left(2x + \frac{\pi}{3}\right) = \frac{1}{2}$

$-\pi \le x \le \pi$ then $-2\pi \le 2x \le 2\pi$

so $-\frac{5\pi}{3} \le 2x + \frac{\pi}{3} \le \frac{7\pi}{3}$: $2x + \frac{\pi}{3} = -2\pi + \frac{\pi}{3}, -\frac{\pi}{3}, \frac{\pi}{3}, 2\pi - \frac{\pi}{3}, 2\pi + \frac{\pi}{3}$

$$2x + \frac{\pi}{3} = \frac{5\pi}{3}, -\frac{\pi}{3}, \frac{\pi}{3}, \frac{5\pi}{3}, \frac{7\pi}{3}$$

$$2x = -2\pi, -\frac{2\pi}{3}, 0, \frac{4\pi}{3}, 2\pi$$

$$x = -\pi, -\frac{\pi}{3}, 0, \frac{2\pi}{3}, \pi$$

(c) $2\sin\left(2x - \frac{\pi}{12}\right) = -\sqrt{3}$

Divide by 2: $\sin\left(2x - \frac{\pi}{12}\right) = -\frac{\sqrt{3}}{2}$

If $-\pi \le x \le \pi$ then $-2\pi \le 2x \le 2\pi$

so $-\frac{25\pi}{12} \le 2x - \frac{\pi}{12} \le \frac{23\pi}{12}$ $2x - \frac{\pi}{12} = -\frac{2\pi}{3}, -\frac{\pi}{3}, \pi + \frac{\pi}{3}, 2\pi - \frac{\pi}{3}$

$$2x - \frac{\pi}{12} = -\frac{2\pi}{3}, -\frac{\pi}{3}, \frac{4\pi}{3}, \frac{5\pi}{3}$$

$$2x = -\frac{2\pi}{3} + \frac{\pi}{12}, -\frac{\pi}{3} + \frac{\pi}{12}, \frac{4\pi}{3} + \frac{\pi}{12}, \frac{5\pi}{3} + \frac{\pi}{12}$$

$$2x = -\frac{7\pi}{12}, -\frac{\pi}{4}, \frac{17\pi}{12}, \frac{21\pi}{12}$$

$$x = -\frac{7\pi}{24}, -\frac{\pi}{8}, \frac{17\pi}{24}, \frac{21\pi}{24}$$

Example 9

(a) Solve, for $0 \le x \le 2\pi$, $\sqrt{2}\sin\left(\frac{x}{2}-\frac{\pi}{3}\right)=1$.

(b) Solve, for $-3\pi \le x \le 3\pi$, $\tan\left(\frac{x}{3}+\frac{\pi}{6}\right)=1$.

(c) Solve, for $-2\pi \le x \le 2\pi$, $2\cos 2\left(\frac{x}{4}-\frac{\pi}{6}\right)=\sqrt{3}$.

Solution

(a) $\sqrt{2}\sin\left(\frac{x}{2}-\frac{\pi}{3}\right)=1$

Divide by $\sqrt{2}$: $\sin\left(\frac{x}{2}-\frac{\pi}{3}\right)=\frac{1}{\sqrt{2}}$

$0 \le x \le 2\pi$ so $0 \le \frac{x}{2} \le \pi$: $\frac{x}{2}-\frac{\pi}{3}=\frac{\pi}{4},\frac{3\pi}{4}$

$$\frac{x}{2}=\frac{\pi}{4}+\frac{\pi}{3},\frac{3\pi}{4}+\frac{\pi}{3}$$

$$x=\frac{7\pi}{12},\frac{13\pi}{12}$$

The only valid solution is: $x=\frac{7\pi}{12}$

(b) $\tan\left(\frac{x}{3}+\frac{\pi}{6}\right)=1$

$-3\pi \le x \le 3\pi$ so $-\pi \le \frac{x}{3} \le \pi$: $\frac{x}{3}+\frac{\pi}{6}=-\frac{3\pi}{4},\frac{\pi}{4}$

$$\frac{x}{3}=-\frac{3\pi}{4}-\frac{\pi}{6},\frac{\pi}{4}-\frac{\pi}{6}$$

$$\frac{x}{3}=-\frac{11\pi}{12},\frac{\pi}{12}$$

$$x=-\frac{11\pi}{4},\frac{\pi}{4}$$

(c) $2\cos 2\left(\frac{x}{4}-\frac{\pi}{6}\right)=\sqrt{3}$

$$2\cos\left(\frac{x}{2}-\frac{\pi}{3}\right)=\sqrt{3}$$

$$\cos\left(\frac{x}{2}-\frac{\pi}{3}\right)=\frac{\sqrt{3}}{2}$$

$-2\pi \le x \le 2\pi$ so $-\pi \le \frac{x}{2} \le \pi$: $\frac{x}{2}-\frac{\pi}{3}=-\frac{\pi}{6},\frac{\pi}{6}$

$$\frac{x}{2}=\frac{\pi}{6},\frac{\pi}{2}$$

$$x=\frac{\pi}{3},\pi$$

Another transformation that can be used to solve equations is to square both sides of the equation. This may introduce solutions that are not valid for the original equation, so all answers need to be checked by substituting them into the original equation.

Example 10

Solve $\sin x + 2\cos x = 1$ for $0 \le x \le 2\pi$.

Solution

$\sin x + 2\cos x = 1$

Rearrange the equation: $2\cos x = 1 - \sin x$ [1]

Square both sides of [1]: $4\cos^2 x = 1 - 2\sin x + \sin^2 x$

Use the Pythagorean identity: $4(1 - \sin^2 x) = 1 - 2\sin x + \sin^2 x$

$$5\sin^2 x - 2\sin x - 3 = 0$$

Factorise: $(5\sin x + 3)(\sin x - 1) = 0$

$$\sin x = -\frac{3}{5}, 1 \quad [2]$$

$$x = \pi + 0.6435, 2\pi - 0.6435, \frac{\pi}{2}$$

$$x = 3.784, 5.640, \frac{\pi}{2}$$

To check if all these values are valid, you can substitute them into the original equation. Alternatively, without a calculator it is easier to substitute the solutions for $\sin(x)$, as follows.

Substitute [2] (first value) into [1] to check: $2\cos x = 1 - \left(-\frac{3}{5}\right)$

$$\cos x = \frac{4}{5}$$

As this is positive, x must be in Q1 or Q4.

Hence the only valid answer is $x = 5.640$

Substitute [2] (second value) into [1] to check: $2\cos x = 1 - (-1)$

$$\cos x = 0$$

The only valid answer is $x = \frac{\pi}{2}$

The remaining answer $x = 3.785$ is not valid as $\cos(3.785) = -0.8$

Hence $x = \frac{\pi}{2}$, 5.640 are the only valid solutions.

EXERCISE 12.2 FURTHER SOLUTION OF TRIGONOMETRIC EQUATIONS

1 Solve for $0 \le x \le 2\pi$:

(a) $\sin(2x) = 0.5$ **(b)** $\cos(3x) = 1$ **(c)** $\tan(2x) = 1$

2 Solve for $-\pi \le x \le \pi$:

(a) $\tan\left(\frac{x}{2}\right) = \frac{1}{\sqrt{3}}$ **(b)** $\sin\left(\frac{x}{3}\right) = -\frac{\sqrt{3}}{2}$ **(c)** $\cos\left(\frac{x}{4}\right) = \frac{1}{\sqrt{2}}$

3 Solve, for $0 \le x \le 2\pi$:

(a) $2\cos\left(x - \frac{\pi}{6}\right) = 1$ **(b)** $\sqrt{2}\sin\left(x + \frac{\pi}{4}\right) = 1$ **(c)** $2\cos\left(x - \frac{\pi}{3}\right) = \sqrt{3}$

(d) $\sqrt{3}\tan\left(x + \frac{5\pi}{6}\right) = 1$ **(e)** $\sin\left(x - \frac{\pi}{4}\right) = \sqrt{3}\cos\left(x - \frac{\pi}{4}\right)$

4 Solve, for $-\pi \le x \le \pi$:

(a) $\sqrt{2}\cos 2x = 1$ **(b)** $\cos\left(2x - \frac{\pi}{2}\right) = 1$ **(c)** $\sin\left(2x + \frac{\pi}{6}\right) = -1$

(d) $3\tan\left(3x - \frac{\pi}{4}\right) = \sqrt{3}$ **(e)** $\sin\left(\frac{x}{2} + \frac{\pi}{18}\right) = \frac{1}{2}$

5 **(a)** Solve, for $0 \le x \le 2\pi$, $\sqrt{2}\cos\left(\frac{x}{2}+\frac{\pi}{6}\right)=1$.

(b) Solve, for $-3\pi \le x \le 3\pi$, $\tan\left(\frac{x}{3}-\frac{\pi}{6}\right)=1$.

(c) Solve, for $-2\pi \le x \le 2\pi$, $2\sin 2\left(\frac{x}{4}-\frac{\pi}{3}\right)=\sqrt{3}$.

(d) Solve, for $-2\pi \le x \le 2\pi$, $2\cos 2\left(\frac{x}{3}-\frac{2\pi}{3}\right)=\sqrt{3}$.

6 **(a)** Solve $\sin\left(2x+\frac{\pi}{6}\right)=\sqrt{3}\cos\left(2x+\frac{\pi}{6}\right)$ for $0 \le x \le 2\pi$.

(b) By drawing appropriate graphs using graphing software, solve $\sin\left(2x+\frac{\pi}{6}\right)=\sqrt{3}\cos\left(2x+\frac{\pi}{6}\right)$ for $0 \le x \le 2\pi$.

7 Solve over the given domain:

(a) $\cos x + 3\sin x = 1$ for $0 \le x \le 2\pi$

(b) $3\sin x + 4\sin x = 5$ for $-\pi \le x \le \pi$

(c) $\sec x + \tan x = 1.5$ for $0 \le x \le 180°$

(d) $\sin 2x = 1 + \cos 2x$ for $0 \le x \le \pi$

12.3 GRAPHICAL SOLUTION OF EQUATIONS

There are usually standard techniques for solving equations associated with various functions, e.g. linear equations, quadratic equations and trigonometric equations. However, there is usually no standard technique for solving equations that combine two or more different functions, for example:

- $\sin x = 1 - 2x$
- $\cos x = x^2 + 3$

Equations like this may 'transcend' (go beyond) algebraic equation-solving techniques and hence are called **transcendental equations**. You must instead use various non-algebraic methods to find approximate solutions. The most accurate method is to calculate numerical solutions using technology. Another method is to sketch the graphs of the two functions, use the graphs to find approximate x-values of any intersection points, then use a calculator to refine these approximate x values as closely as possible.

Example 11

Solve $\sin x = 1 - 2x$.

Solution

Sketch the functions.

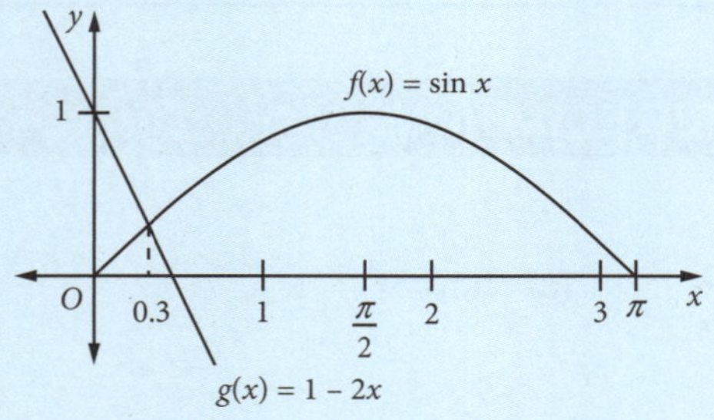

This shows that the graphs of $f(x) = \sin x$ and $g(x) = 1 - 2x$ intersect at only one point, approximately $x = 0.3$, so 0.3 is an approximate solution of $\sin x = 1 - 2x$. Testing values of x to refine this approximation:

x	0.3	0.4	0.35	0.335
$\sin x$	0.2955	0.3894	0.3429	0.3288
$1 - 2x$	0.4	0.2	0.3	0.33

This table shows the values of $\sin x$ and $1 - 2x$ for values of x between $x = 0.3$ and $x = 0.4$. $x = 0.35$ is a better solution than $x = 0.3$. Further trial and error leads to a solution of $x = 0.335$.

A 'find the intersection' function in a graphing software program would allow you to find this solution quickly and accurately. Unfortunately, under examination conditions you may have to use the sketching method.

EXPLORE FURTHER

Graphical solutions to transcendental equations

Use technology to solve transcendental equations graphically.

EXERCISE 12.3 GRAPHICAL SOLUTION OF EQUATIONS

1 Draw the graph of $y = \sin x$ for $0 \le x \le \pi$. Use this graph to find the number of solutions in $0 \le x \le \pi$ for:

(a) $\sin x = \dfrac{x}{2}$ (b) $\sin x = \dfrac{x}{2} - 1$ (c) $\sin x = \dfrac{x}{2} - x$

2 Draw the graph of $y = \cos x$ for $-\pi \le x \le \pi$ and use it to solve the following equations.

(a) $\cos x = x$ (b) $\cos x = \dfrac{x}{2}$ (c) $\cos x = 1 - x$

3 Use a graphical method to solve each equation. Check your answer using graphing software.

(a) $\sin 2x = \dfrac{x}{2}$ (b) $\cos x = x, x \ge 0$ (c) $\tan x = 1 - x, 0 \le x \le 2\pi$

4 Which graph could be used to solve the equation $\sin x = x - 1$?

A

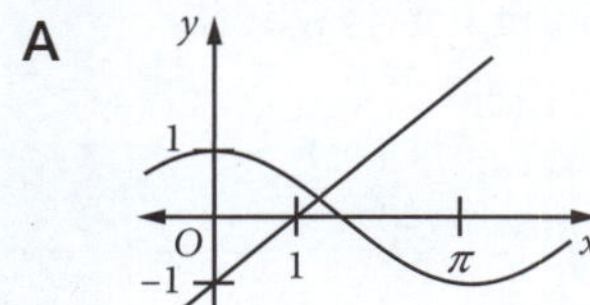

B

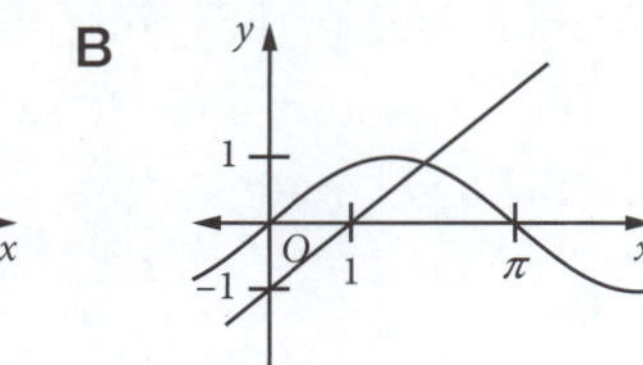

C

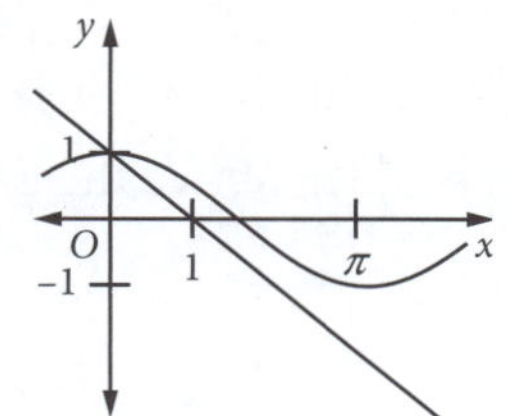

D

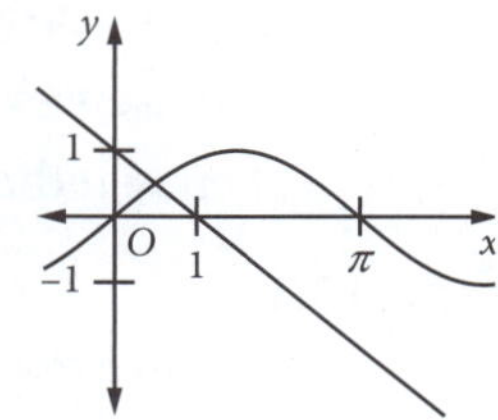

5 By drawing appropriate graphs, determine the number of solutions to the equation $\sin 2x = 1 - \dfrac{x}{4}$ in the domain $0 \le x \le 2\pi$.

6 A rectangular strip of metal 9 cm wide is bent to form a water channel. The section perpendicular to the length is a circular arc whose chord is 6 cm long.

(a) Show that if the circular arc subtends an angle of $2x$ radians at the centre of the circle then $3 \sin x = 2x$.

(b) Solve the equation in part **(a)** graphically, giving your answer correct to 1 decimal place.

(c) Find the area of the cross-section of the channel in cm^2.

7 A semicircle of radius r is divided into two parts of equal area by a chord parallel to the base (diameter).

(a) If the chord subtends an angle of x radians at the centre, prove that $x - \dfrac{\pi}{2} = \sin x$.

(b) Solve the equation in part **(a)** graphically, giving your answer correct to 1 decimal place.

8 The chord of a segment of a circle subtends an angle of $\dfrac{\pi}{2} + x$ radians at the centre. The area of the segment is one-quarter the area of the circle.

(a) Prove that $x = \cos x$.

(b) Solve the equation in part **(a)** graphically, giving your answer correct to 2 decimal places.

9 Show from a rough sketch that the equation $2 - \dfrac{x}{2} = \tan x$ has a solution that is approximately 1. Using a calculator, find this solution correct to 2 decimal places.

12.4 APPLICATIONS INVOLVING TRIGONOMETRIC FUNCTIONS AND GRAPHS

Trigonometric functions can be used to model periodic phenomena that occur in the physical world—that is, real-life situations that can be modelled by a function f where $f(x + b) = f(x)$ for all x and some non-zero real number, b. The constant b is the period of the function. Average monthly temperatures, seasonal climate variations, sound waves, variations in tides and the phases of the moon are all real phenomena that display periodic behaviour and can be modelled by the sine or cosine functions.

In these situations, the independent variable on the graph is typically time rather than an angle. Because of this, the scale on the horizontal axis will usually not be represented in units that are exact fractions or multiples of π.

It is useful to sketch a graph of the trigonometric function and it may be necessary to determine the axes intercepts.

Finding axes intercepts:

- To find the x-intercept(s), let $y = 0$ and solve for x.
- To find the y-intercept, let $x = 0$ and evaluate.

Example 12

A Ferris wheel with a radius of 30 m rotates once every 60 seconds. Passengers get into a cabin 1 m above level ground, which is the wheel's lowest point. The height $h(t)$ metres above the ground is given by $h(t) = -30\cos\dfrac{\pi t}{30} + 31$, where t seconds is the time after leaving the lowest point.

(a) Sketch the graph of $h(t)$ for a 2-minute period after a passenger has got into the Ferris wheel.

(b) What is a passenger's height above the ground, to the nearest m, after 20 seconds?

(c) When will the cabin be 50 m above the ground for the first time?

Solution

(a) Write the value of k and its effect:

$k = -30$

The amplitude is $|-30| = 30$ metres.

Write the value of a and its effect:

$a = \dfrac{\pi}{30}$

The period is $2\pi \div \dfrac{\pi}{30} = 60$ seconds

The graph will show 2 periods.

Write the value of c and its effect:

$c = 31$

The median value will be 31.

The minimum value of $h(t)$ will be $31 - 30 = 1$ metre and the maximum value of $h(t)$ will be $31 + 30 = 61$ metres.

(b) Substitute the given time into the equation to find the required height above the ground:

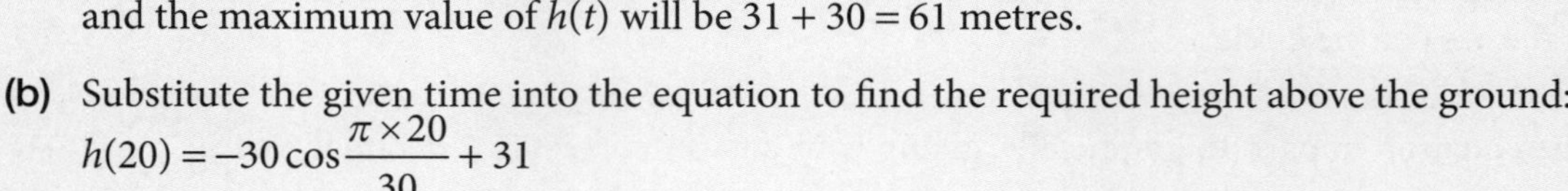

$$h(20) = -30\cos\frac{\pi \times 20}{30} + 31$$
$$\approx 46 \text{ metres}$$

(c) Solve the equation $50 = -30\cos\dfrac{\pi t}{30} + 31$ for t

$$\cos\frac{\pi t}{30} = -\frac{19}{30}$$
$$\frac{\pi t}{30} = \pi - \cos^{-1}\left(\frac{19}{30}\right)$$
$$\frac{\pi t}{30} = 2.2566$$
$$t = 21.55 \text{ seconds}$$

The cabin will be 50 m above the ground for the first time after approximately 21.55 seconds.

Example 13

The tide at a point on the NSW coast can be modelled using the equation $y = k\cos nt$. At Nobby's Beach in NSW, over two consecutive days, the average difference between high and low tides is 1.6 metres and the average time between high tide and low tide is 6.25 hours.

(a) What is the amplitude of the tide function at Nobby's Beach?

(b) How much time passes between successive high tides (i.e. the period) and what is the value of n?

(c) Use this information to obtain the Nobby's Beach tide function and draw its graph.

(d) If the depth of water at low tide is 0.2 metres, what is the depth of the water 2 hours after low tide?

Solution

(a) $2a = 1.6$ so $a = 0.8$

(b) Time between successive high tides $= 2 \times 6.25 = 12.5$ hours

Period $T = \frac{2\pi}{n}$ so $12.5 = \frac{2\pi}{n}$ hence $n = \frac{4\pi}{25}$.

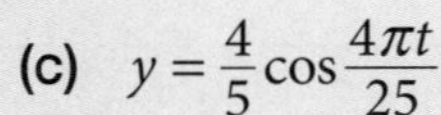

(c) $y = \frac{4}{5}\cos\frac{4\pi t}{25}$

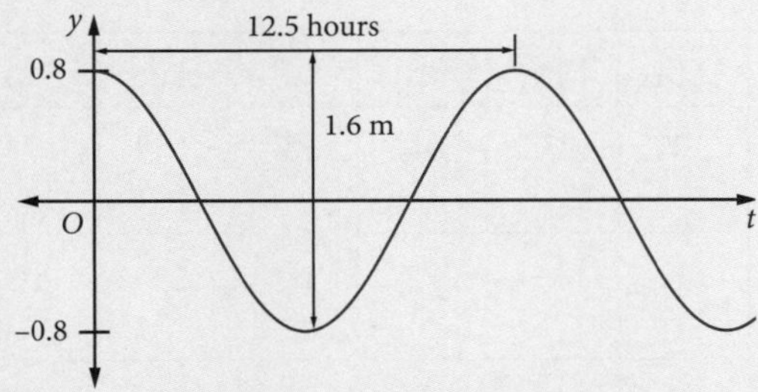

(d) From the graph, low tide occurs at $t = 6.25$ and $y = -0.8$.
2 hours after low tide means $t = 6.25 + 2 = 8.25$

$t = 8.25$: $\quad y = \frac{4}{5}\cos\frac{33\pi}{25} = -0.43$ (to 2 d.p.)

Hence the depth is $0.8 - 0.43 = 0.37$ m above low tide.
$\therefore$ depth of the water 2 hours after low tide $= 0.2 + 0.37 = 0.57$ m (to 2 d.p.)

EXERCISE 12.4 APPLICATIONS INVOLVING TRIGONOMETRIC FUNCTIONS AND GRAPHS

1 A Ferris wheel with a radius of 35 m rotates once every 90 seconds. Passengers enter a cabin at the loading point, which is 2 m above level ground. The height, $h(t)$ metres, above the ground is given by $h(t) = -35\cos\frac{\pi t}{45} + 37$, where t seconds is the time after leaving the lowest level.

(a) Sketch the graph of $h(t)$ for a 3-minute period after a cabin passes the loading point and state the coordinates of the h intercept.

(b) What is a cabin's height above the ground in metres after 30 seconds? Give your answer correct to 1 decimal place.

(c) After how many seconds will the cabin will be 40 m above the ground for the first time? Give your answer correct to two decimal places.

2 The temperature T°C inside the school classroom during a day in February can be modelled by the equation $T = 15\sin\frac{\pi t}{12} + 20$, where t is the number of hours after 9.00 am and $0 \le t \le 12$.

(a) Sketch the graph of the temperature, T°C, inside the school classroom over the 12-hour period and state the coordinates of the T intercept.

(b) What is the temperature inside the room at 9:00 am?

(c) What is the temperature in the room at midday (12:00 pm)?

(d) At what times is the temperature inside the room equal to 30°C?

(e) State the percentage of time that the temperature inside the room was at least 30°C.

3 The height of the water, h metres, in a bay varies over time and can be modelled by a cosine function of the form $h(t) = k\cos at + c$, where k, a and c are constants and t represents the number of hours after midnight. On a particular day off the coast of South Australia, a high tide of 4.5 metres occurred at 6:00 am and a low tide of 0.5 metre occurred at 12:00 pm. Write a model for the height, h metres, of the water as a function of time, t hours, since midnight.

(a) Determine the values of k, a and c.

(b) Hence write the rule of $h(t)$.

(c) At what time does the second high tide occur?

4 The voltage, V volts, supplied by an electrical outlet is described by the sine function $V = k\sin at + c$, where k, a and c are constants, t is measured in seconds and $k > 0$.

(a) If the voltage, V, oscillates between −240 volts and 240 volts, find the value of k and c.

(b) If it has a frequency of 50 cycles per second, find the value of a.

(c) Determine the voltage (V) involving a transformation of the sine function.

5 The average daily maximum and minimum temperatures (in °C) for Melbourne are given in the table. Time, t, is measured in months, with $t = 0$ representing 1 January.

The average temperatures for Melbourne are as follows:

Temperature	Jan	Feb	Mar	Apr	May	Jun	Jul	Aug	Sep	Oct	Nov	Dec
Maximum Temperature	26°	26°	24°	20°	17°	14°	13°	15°	17°	20°	22°	24°
Minimum Temperature	14°	14°	13°	11°	8°	7°	6°	7°	8°	9°	11°	13°

Write trigonometric models, involving transformation of the sine function, that give the average daily maximum and minimum temperatures for Melbourne as functions of time, t.

(a) Average daily maximum temperature.

(b) Average daily minimum temperature.

6 Scientists monitoring a population of penguins on Heard Island found that the number of penguins varies between a high of 900 at the start of the year and a low of 700 on 30 June and is approximated by the cosine function $P(t) = k\cos at + c$, where t is the number of months since the start of the year, and k, a and c are constants.

(a) Sketch a graph of the penguin population over a whole year.

(b) Find the values of k, a and c for the function.

(c) According to the population function, how many penguins will there be on Penguin Island on 31 March? 20 June? 23 October?

7 The tide at a point on the WA coast can be modelled using the equation $y = k\cos nt$. At Cable Beach in WA, over two consecutive days, the average difference between high and low tides is 9.0 metres and the average time between high tide and low tide is 6.1 hours.

(a) What is the amplitude of the tide function at Cable Beach?

(b) How much time passes between successive high tides (i.e. the period) and what is the value of n?

(c) Use this information to obtain the tide function and draw its graph.

(d) If the depth of water at low tide is 0.5 metres, what is the depth of the water 1 hour after low tide?

CHAPTER REVIEW 12

1 Solve the trigonometric equations. Round answers to the nearest minute.

(a) $5\sin 2x = 2\tan 30°, 0° < x < 180°$

(b) $3\sin x = \tan x, 0° < x < 360°$

2 Sketch each trigonometric function. Show at least one period. Assume x is in radians.

(a) $y = 5\sin 3x$

(b) $y = 2\tan\frac{\pi x}{3} + 4$

(c) $y = 4\cos 2\left(x + \frac{\pi}{3}\right)$

3 On the same set of axes, for $0 \le x \le 2\pi$, sketch:

(a) $y = \cos x$ and $y = 3 - 2\cos\frac{x}{2}$ **(b)** $y = 2\sin x$, $y = 3\cos 2x$ and $y = 2\sin x + 3\cos 2x$.

4 Draw the graph of $y = \cos 2x$ for $-\pi \le x \le \pi$. On the same set of axes, draw the graph of $y = -\frac{x}{2}$. Use your graphs to solve the equation $\cos 2x = -\frac{x}{2}$.

5 State the function represented by the graph, assuming it is a cosine function.

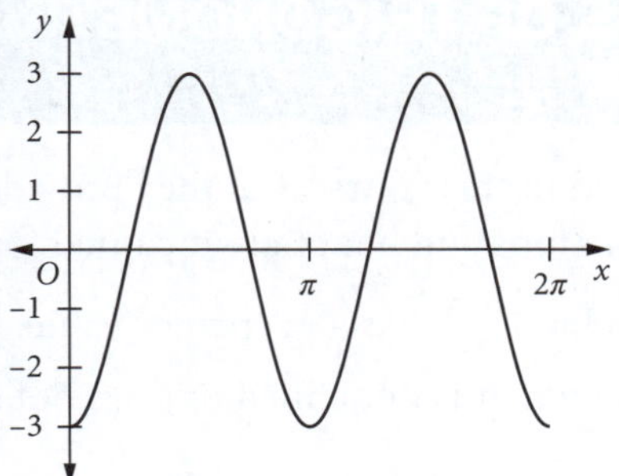

6 The temperature of a pond, in °C, is given by the function $f:[0,24] \to \mathbb{R}, f(x) = k\sin(ax) + c$ where x is time in hours after midnight. $y = f(x)$ is shown.

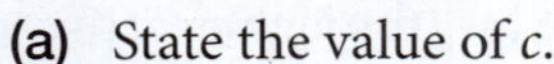

(a) State the value of c.
(b) State the value of k.
(c) State the period of f.
(d) Find the value of a.
(e) Find the minimum temperature that the pond reaches.
(f) By solving an appropriate equation, find the times at which the temperature of the pond is 0°C over the given domain.

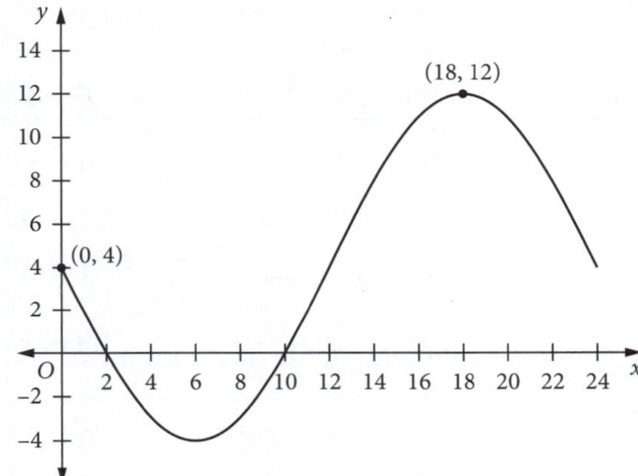

7 Tess likes to swim out to a small island and collect shells at low tide. However, for much of the time this island is submerged. To avoid the disappointment of going to collect shells and finding the island submerged, Tess consults the local marine and tide charts for next Sunday. She discovers that the island is above water when the tide is 1.5 metres above the low water mark. The height, h, in metres of the tide above the low-water mark is given by the equation $h(t) = 2\cos\left(\frac{8\pi}{45}t\right) + 2$, where t is the time in hours after midday Sunday.

(a) Draw a graph showing the height above the low water mark for the 24 hours commencing at midday Sunday.
(b) To the nearest minute, what time will the island first appear above the water on Sunday afternoon?
(c) To the nearest minute, how long does Tess have to collect shells before the island becomes submerged again?

8 On a carnival ride the height, h metres, of Oscar from the ground is modelled by the equation $h:[0,180] \to R$, $h(t) = -8\cos\left(\frac{\pi t}{10}\right) + 9$, where t is the number of seconds after the ride has started.

(a) How far from the ground is Oscar at the beginning of the ride?
(b) What is the greatest distance Oscar is from the ground during the course of the ride?
(c) How long does it take for the ride to complete one full rotation?
(d) How many rotations are completed before the ride comes to a stop?
(e) Draw a graph of Oscar's distance from the ground versus time during the first minute of the ride.
(f) When Oscar is 9 m or higher above the ground he can see the ocean in the distance. During the 3-minute ride, for how much of the time can Oscar see the ocean?

9 Solve **(a)** $\cos\left(2x - \frac{\pi}{4}\right) = -1$ for $0 \le x \le 2\pi$ **(b)** $\sin\left(2x + \frac{\pi}{3}\right) = 0.5$ for $-\pi \le x \le \pi$

CHAPTER 13
Differential calculus

13.1 APPROXIMATIONS OF TRIGONOMETRIC FUNCTIONS WHEN x IS SMALL

The limits in this section are not required in this course, but they provide the mathematically correct way to derive the derivatives of the trigonometric functions and hence are included for completeness.

It may seem strange that the expression $\lim\limits_{x\to 0}\frac{\sin x}{x}$ has a particular value. As you look at this fraction, which appears to be approaching $\frac{0}{0}$, you would assume that it is undefined or does not exist. However, consider the following table of values (where x is in radians):

x	0.1	0.01	0.001	0.0001	−0.0001	−0.001	−0.01	−0.1
$\frac{\sin x}{x}$	0.998334	0.999983	0.9999998	0.999999998	0.999999998	0.9999998	0.999983	0.998334

As $x \to 0$ from above (through positive values) it seems that $\frac{\sin x}{x} \to 1$. As $x \to 0$ from below (through negative values) it also seems that $\frac{\sin x}{x} \to 1$. You can use a spreadsheet to continue this investigation with even smaller values for x, and ultimately seem to find the result: $\lim\limits_{x\to 0}\frac{\sin x}{x} = 1$

Consider the following table of values for $\frac{\tan x}{x}$.

x	0.1	0.01	0.001	0.0001	−0.0001	−0.001	−0.01	−0.1
$\frac{\tan x}{x}$	1.003347	1.000033	1.0000003	1.000000003	1.000000003	1.0000003	1.000033	1.003347

As $x \to 0$ from above (through positive values) it seems that $\frac{\tan x}{x} \to 1$. As $x \to 0$ from below (through negative values) it seems that $\frac{\tan x}{x} \to 1$. Again, you can use a spreadsheet to continue this investigation with even smaller values for x and seem to find the result: $\lim\limits_{x\to 0}\frac{\tan x}{x} = 1$

From these tables of values it would be reasonable to say that when x is small, $\frac{\tan x}{x} > \frac{\sin x}{x}$. This leads to another useful result: when x is small, you have $\tan x > \sin x$ for small x.

Consider the following table of values for $\frac{1-\cos x}{x^2}$.

x	0.1	0.01	0.001	−0.001	−0.01	−0.1
$\frac{1-\cos x}{x^2}$	0.49958	0.4999958	0.49999999	0.49999999	0.4999958	0.49958

As $x \to 0$ from above (through positive values) it seems that $\frac{1-\cos x}{x^2} \to 0.5$. As $x \to 0$ from below (through negative values) it seems that $\frac{1-\cos x}{x^2} \to 0.5$. A spreadsheet investigation can confirm this for even smaller values of x, hence the result: $\lim\limits_{x\to 0}\frac{1-\cos x}{x^2} = 0.5$

Using a spreadsheet to check all these results with smaller and smaller values of x may seem convincing, but spreadsheet value tests like this are *not* actual proofs. However, in fact all these results can be actually proved. One possible proof is discussed below.

Derivation of trigonometric limit results

Consider a sector OAB of a circle of unit radius. The tangent at A meets OB produced at C.

Let the size of $\angle AOB$ be θ radians.

$\therefore$ Area $\Delta AOB <$ area of sector $AOB <$ area ΔAOC

But: Area $\Delta AOB = \frac{1}{2} \times 1 \times 1 \times \sin\theta = \frac{1}{2}\sin\theta$, Area of sector $AOB = \frac{1}{2} \times 1^2 \times \theta = \frac{\theta}{2}$,

Area $\Delta AOC = \frac{1}{2} \times 1 \times \tan\theta = \frac{1}{2}\tan\theta$

So: $\quad \frac{1}{2}\sin\theta < \frac{\theta}{2} < \frac{1}{2}\tan\theta \qquad \left(0 < \theta < \frac{\pi}{2}\right)$

Hence: $\quad \sin\theta < \theta < \tan\theta$

Take reciprocals: $\quad \frac{1}{\sin\theta} > \frac{1}{\theta} > \frac{1}{\tan\theta}$

or $\quad \frac{1}{\sin\theta} > \frac{1}{\theta} > \frac{\cos\theta}{\sin\theta}$

Hence: $\quad 1 > \frac{\sin\theta}{\theta} > \cos\theta$

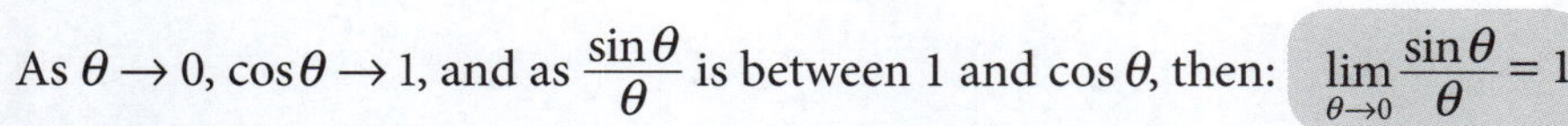

As $\theta \to 0$, $\cos\theta \to 1$, and as $\frac{\sin\theta}{\theta}$ is between 1 and $\cos\theta$, then: $\lim\limits_{\theta \to 0} \frac{\sin\theta}{\theta} = 1$

Another way of thinking about this limit is to see that when θ is small, $\sin\theta \approx \theta$ and $\cos\theta \approx 1$.

Looking again at: $\quad \frac{1}{\sin\theta} > \frac{1}{\theta} > \frac{1}{\tan\theta}$

$$\frac{\tan\theta}{\sin\theta} > \frac{\tan\theta}{\theta} > 1$$

Hence: $\quad \frac{1}{\cos\theta} > \frac{\tan\theta}{\theta} > 1$

As $\theta \to 0$, $\cos\theta \to 1$, and because $\frac{\tan\theta}{\theta}$ is between 1 and $\cos\theta$, then: $\lim\limits_{\theta \to 0} \frac{\tan\theta}{\theta} = 1$

Now consider $\frac{\sin^2\theta}{\theta^2} = \frac{1-\cos^2\theta}{\theta^2} = \frac{(1-\cos\theta)(1+\cos\theta)}{\theta^2}$.

This can be written: $\quad \frac{(1-\cos\theta)(1+\cos\theta)}{\theta^2} = \frac{\sin^2\theta}{\theta^2}$

Hence: $\quad \frac{1-\cos\theta}{\theta^2} = \frac{1}{1+\cos\theta}\left(\frac{\sin\theta}{\theta}\right)^2$

Thus as $x \to 0$: $\quad \frac{1-\cos\theta}{\theta^2} \to \frac{1}{1+1}(1)^2 = \frac{1}{2}$

$$\lim_{\theta \to 0} \frac{1-\cos\theta}{\theta^2} = \frac{1}{2}$$

Some of these results you will use later to find the derivatives of $\sin x$, $\cos x$ and $\tan x$ from first principles.

Example 1

Find: **(a)** $\lim\limits_{x \to 0} \frac{\sin 3x}{3x}$ **(b)** $\lim\limits_{x \to 0} \frac{\sin 4x}{2x}$ **(c)** $\lim\limits_{x \to 0} \frac{\tan 3x}{6x}$

Solution

(a) $\lim\limits_{x \to 0} \frac{\sin 3x}{3x}$

$= \lim\limits_{x \to 0} \frac{\sin\theta}{\theta}$ where $\theta = 3x$

$= 1$

(b) $\lim\limits_{x \to 0} \frac{\sin 4x}{2x}$

$= 2 \times \lim\limits_{x \to 0} \frac{\sin 4x}{4x}$

$= 2 \times 1 = 2$

(c) $\lim\limits_{x \to 0} \frac{\tan 3x}{6x}$

$= \frac{1}{2} \times \lim\limits_{x \to 0} \frac{\tan 3x}{3x}$

$= \frac{1}{2} \times 1 = \frac{1}{2}$

EXERCISE 13.1 APPROXIMATIONS OF TRIGONOMETRIC FUNCTIONS WHEN x IS SMALL

1 Evaluate the following limits.

(a) $\lim_{x\to 0}\frac{\sin 2x}{x}$ (b) $\lim_{x\to 0}\frac{\sin x}{3x}$ (c) $\lim_{\theta\to 0}\frac{\sin 5\theta}{2\theta}$ (d) $\lim_{x\to 0}\frac{\tan 2x}{2x}$

(e) $\lim_{x\to 0}\frac{\sin\frac{x}{2}}{x}$ (f) $\lim_{x\to 0}\frac{\sin\frac{x}{3}}{3x}$ (g) $\lim_{x\to 0}\frac{\tan 3x}{x}$ (h) $\lim_{x\to 0}\frac{3\sin 2x}{4x}$

2 Indicate whether each statement is correct or incorrect.

(a) $\sin(\pi - x) = -\sin x$ (b) $\sin(\pi - x) = \sin x$ (c) $\lim_{x\to 0}\frac{\sin(\pi - x)}{x} = -1$ (d) $\lim_{x\to 0}\frac{\sin(\pi - x)}{x} = 1$

3 Evaluate the following limits.

(a) $\lim_{x\to 0}\frac{1-\cos 2x}{x^2}$ (b) $\lim_{h\to 0}\frac{\tan 2h}{3h}$ (c) $\lim_{\theta\to 0}\frac{1-\cos\theta}{\theta^2}$ (d) $\lim_{x\to 0}\frac{\sin^2 x}{x}$

(e) $\lim_{x\to 0}\frac{\sin(\pi + x)}{x}$ (f) $\lim_{x\to 0}\frac{\cos\left(\frac{\pi}{2} - x\right)}{x}$ (g) $\lim_{x\to 0}\frac{1-\sin\left(\frac{\pi}{2} - x\right)}{x^2}$ (h) $\lim_{x\to 0}\frac{2\sin\frac{x}{2}}{x}$

13.2 DERIVATIVES OF TRIGONOMETRIC FUNCTIONS

The result in Example **2**, verified and then derived below, is needed to differentiate $\sin x$ from first principles. You do not need to memorise this result for this course, but you may be asked to use it in a question where it is given.

Example 2

Show that $\sin(x + y) = \sin x\cos y + \cos x\sin y$, for:

(a) $x = \frac{\pi}{3}, y = \frac{\pi}{6}$ (b) $x = \frac{\pi}{2}, y = \frac{\pi}{4}$ (c) $x = \frac{5\pi}{6}, y = \frac{\pi}{3}$

Solution

(a) $x = \frac{\pi}{3}, y = \frac{\pi}{6}$: LHS: $\sin\left(\frac{\pi}{3} + \frac{\pi}{6}\right) = \sin\frac{\pi}{2} = 1$

RHS: $\sin\frac{\pi}{3}\cos\frac{\pi}{6} + \cos\frac{\pi}{3}\sin\frac{\pi}{6}$

$= \frac{\sqrt{3}}{2}\times\frac{\sqrt{3}}{2} + \frac{1}{2}\times\frac{1}{2}$

$= \frac{3}{4} + \frac{1}{4} = 1$

$=$ LHS

(b) $x = \frac{\pi}{2}, y = \frac{\pi}{4}$: LHS: $\sin\left(\frac{\pi}{2} + \frac{\pi}{4}\right) = \sin\frac{3\pi}{4} = \frac{1}{\sqrt{2}}$

RHS: $\sin\frac{\pi}{2}\cos\frac{\pi}{4} + \cos\frac{\pi}{2}\sin\frac{\pi}{4}$

$= 1\times\frac{1}{\sqrt{2}} + 0\times\frac{1}{\sqrt{2}}$

$= \frac{1}{\sqrt{2}}$

$=$ LHS

(c) $x = \frac{5\pi}{6}, y = \frac{\pi}{3}$: LHS: $\sin\left(\frac{5\pi}{6} + \frac{\pi}{3}\right) = \sin\frac{7\pi}{6} = -\frac{1}{2}$

$$\begin{aligned} \text{RHS: } & \sin\frac{5\pi}{6}\cos\frac{\pi}{3} + \cos\frac{5\pi}{6}\sin\frac{\pi}{3} \\ &= \frac{1}{2}\times\frac{1}{2} + \left(\frac{-\sqrt{3}}{2}\right)\times\frac{\sqrt{3}}{2} \\ &= \frac{1}{4} - \frac{3}{4} = -\frac{1}{2} \\ &= \text{LHS} \end{aligned}$$

Important result

This verifies that $\sin(\alpha + \beta) = \sin\alpha\cos\beta + \cos\alpha\sin\beta$.

Proof of result: $\sin(\alpha + \beta) = \sin\alpha\cos\beta + \cos\alpha\sin\beta$

Consider ΔABC:

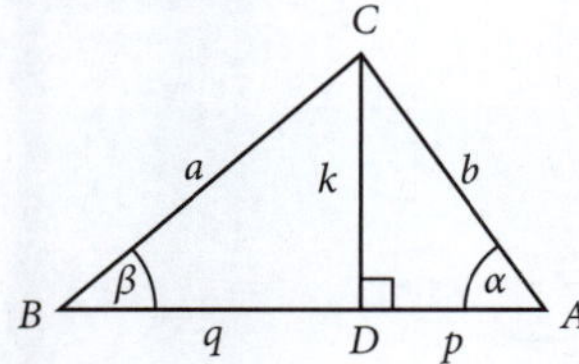

$$\begin{aligned} \text{Area } \Delta ABC &= \frac{1}{2}ab\sin C \\ &= \frac{1}{2}ab\sin(\pi - (\alpha + \beta)) \\ \sin(\pi - \theta) = \sin\theta: \quad &= \frac{1}{2}ab\sin(\alpha + \beta) \\ \text{But:} \quad & \text{Area } \Delta ABC = \frac{1}{2}kp + \frac{1}{2}kq \\ \text{Hence:} \quad & \frac{1}{2}ab\sin(\alpha+\beta) = \frac{1}{2}kp + \frac{1}{2}kq \\ & \therefore \sin(\alpha+\beta) = \frac{kp}{ab} + \frac{kq}{ab} \\ & \sin(\alpha+\beta) = \frac{k}{a}\times\frac{p}{b} + \frac{k}{b}\times\frac{q}{a} \\ & \sin(\alpha+\beta) = \sin\beta\cos\alpha + \sin\alpha\cos\beta \end{aligned}$$

Hence the general result is:

$$\sin(\alpha + \beta) = \sin\alpha\cos\beta + \cos\alpha\sin\beta$$

Note: This result is a part of the Mathematics Extension 1 course. It does not need to be memorised for this course.

Derivative of $\sin x$

First, draw the graphs of $y = \frac{\sin(x + 0.001) - \sin x}{0.001}$ and $y = \cos x$ for $0 \le x \le 2\pi$.

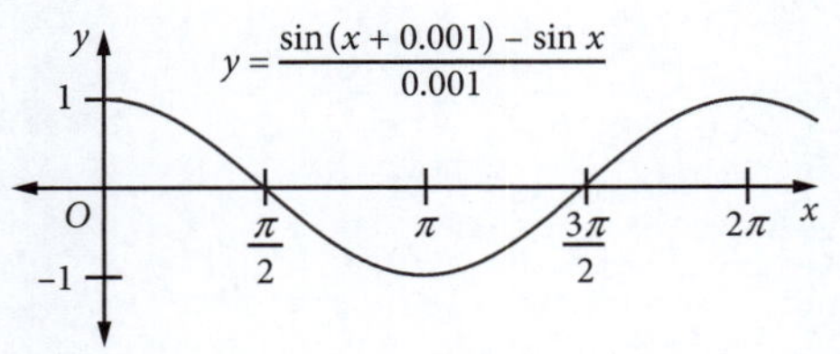

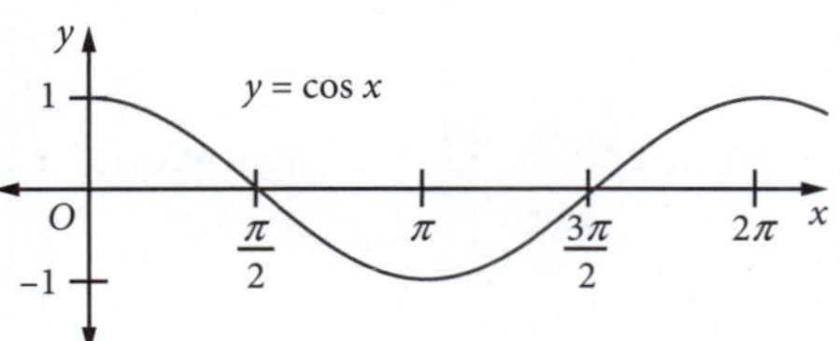

These two graphs appear the same. If you use graphing software to draw the two graphs on the same set of axes, you will be unable to distinguish between them. It seems reasonable to assume the result:

$$\frac{d}{dx}(\sin x) = \cos x$$

Proof

If $y = \sin x$, then from first principles you have $\dfrac{dy}{dx} = \lim\limits_{h \to 0} \dfrac{\sin(x+h) - \sin x}{h}$.

From Example **2** you have the result $\sin(x+h) = \sin x \cos h + \cos x \sin h$, so:

$$\begin{aligned}\frac{dy}{dx} &= \lim_{h\to 0}\frac{\sin x\cos h + \cos x\sin h - \sin x}{h}\\ &= \lim_{h\to 0}\frac{\sin x(\cos h - 1) + \cos x\sin h}{h}\\ &= \lim_{h\to 0}\frac{\sin x(\cos h - 1)}{h} + \lim_{h\to 0}\frac{\cos x\sin h}{h}\\ &= \sin x\lim_{h\to 0}\frac{h(\cos h - 1)}{h^2} + \cos x\lim_{h\to 0}\frac{\sin h}{h}\end{aligned}$$

But $\lim\limits_{x\to 0}\dfrac{\sin x}{x} = 1$ and $\lim\limits_{x\to 0}\dfrac{1-\cos x}{x^2} = \dfrac{1}{2}$, so: $\dfrac{dy}{dx} = \sin x \times 0 \times \left(-\dfrac{1}{2}\right) + \cos x \times 1$

$$\therefore \frac{d}{dx}(\sin x) = \cos x$$

Derivative of cos *x*

First, draw the graphs of $y = \dfrac{\cos(x+0.001) - \cos x}{0.001}$ and $y = -\sin x$ for $0 \le x \le 2\pi$.

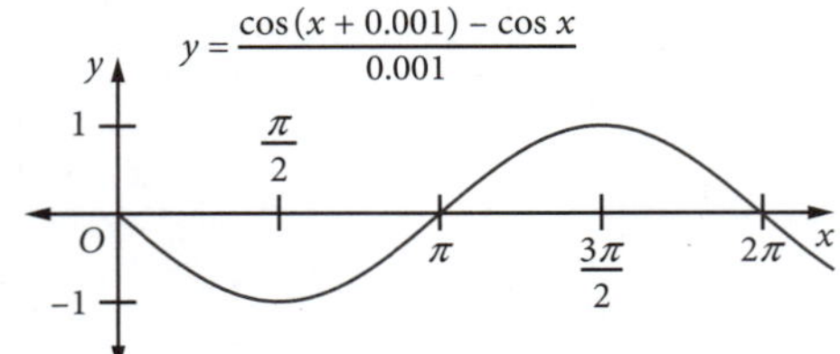

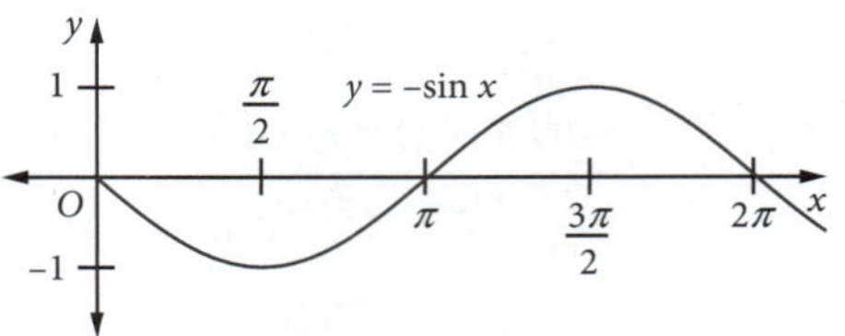

These two graphs appear the same. If you use graphing software to draw the two graphs on the same set of axes, you will be unable to distinguish between them. It seems reasonable to assume the result:

$$\frac{d}{dx}(\cos x) = -\sin x$$

Proof

If $y = \cos x$, then from first principles you have $\dfrac{dy}{dx} = \lim\limits_{h\to 0}\dfrac{\cos(x+h) - \cos x}{h}$. But this result can be derived more easily using the fact that $\cos x = \sin\left(\dfrac{\pi}{2} - x\right)$ with the chain rule.

If $y = \cos x$, then: $\quad y = \sin\left(\dfrac{\pi}{2} - x\right)$

Differentiate using chain rule: $\quad \dfrac{dy}{dx} = -\cos\left(\dfrac{\pi}{2} - x\right)$

$$\therefore \frac{dy}{dx} = -\sin x$$

Derivative of tan *x*

This result can be derived from $\tan x = \dfrac{\sin x}{\cos x}$ using the quotient rule.

If $y = \tan x$, then: $\quad y = \dfrac{\sin x}{\cos x}$

Differentiate using quotient rule:

$$\begin{aligned}\frac{dy}{dx} &= \frac{\cos x \times \cos x - \sin x \times (-\sin x)}{\cos^2 x}\\ &= \frac{\cos^2 x + \sin^2 x}{\cos^2 x}\\ &= \frac{1}{\cos^2 x}\end{aligned}$$

$$\therefore \frac{d}{dx}(\tan x) = \sec^2 x$$

EXPLORE FURTHER

Derivatives of trigonometric functions

Use technology to explore the graphs of the derivatives of trigonometric functions.

Derivatives of sin(*f*(*x*)), cos(*f*(*x*)), tan(*f*(*x*))

Use the chain rule to differentiate these functions.

$y=\sin\left(f(x)\right)$ $\qquad$ $y=\cos\left(f(x)\right)$ $\qquad$ $y=\tan\left(f(x)\right)$

$\frac{dy}{dx}=f'(x)\cos\left(f(x)\right)$ $\qquad$ $\frac{dy}{dx}=-f'(x)\sin\left(f(x)\right)$ $\qquad$ $\frac{dy}{dx}=f'(x)\sec^2\left(f(x)\right)$

When $f(x) = ax + b$, these results may be summarised as follows.

Summary—derivatives of trigonometric functions

$\frac{d}{dx}(\sin x)=\cos x$	$\frac{d}{dx}(\sin(ax+b))=a\cos(ax+b)$
$\frac{d}{dx}(\cos x)=-\sin x$	$\frac{d}{dx}(\cos(ax+b))=-a\sin(ax+b)$
$\frac{d}{dx}(\tan x)=\sec^2 x$	$\frac{d}{dx}(\tan(ax+b))=a\sec^2(ax+b)$

Differentiating the reciprocal trigonometric functions

Derivative of cosec *x*

$y=\text{cosec } x=\frac{1}{\sin x}$

Rewrite this as $y=(\sin x)^{-1}$ and use the chain rule to differentiate with respect to x.

$$\begin{aligned}\frac{dy}{dx}&=(-1)\,(\sin x)^{-2}\times\cos x\\&=-\frac{\cos x}{\sin^2 x}\\&=-\frac{1}{\sin x}\times\frac{\cos}{\sin x}\\&=-\text{cosec } x\cot x\end{aligned}$$

Derivative of sec *x*

$y=\sec x=\frac{1}{\cos x}$

Rewrite this as $y=(\cos x)^{-1}$ and use the chain rule to differentiate with respect to x.

$$\begin{aligned}\frac{dy}{dx}&=(-1)(\cos x)^{-2}\times(-\sin x)\\&=\frac{\sin x}{\cos^2 x}\\&=\frac{1}{\cos x}\times\frac{\sin x}{\cos x}\\&=\sec x\tan x\end{aligned}$$

Derivative of cot *x*

$y=\cot x=\frac{1}{\tan x}$

Rewrite this as $y=(\tan x)^{-1}$ and use the chain rule to differentiate with respect to x.

$$\begin{aligned}\frac{dy}{dx}&=(-1)(\tan x)^{-2}\times(\sec^2 x)\\&=-\frac{1}{\tan^2 x}\times\frac{1}{\cos^2 x}\\&=-\frac{\cos^2 x}{\sin^2 x}\times\frac{1}{\cos^2 x}\\&=-\frac{1}{\sin^2 x}\\&=-\text{cosec}^2 x\end{aligned}$$

Summary—derivatives of reciprocal trigonometric functions

$\frac{d}{dx}(\text{cosec } x) = -\text{cosec } x \cot x$ $\quad$ $\frac{d}{dx}(\text{cosec}(ax+b)) = -a\text{cosec}(ax+b)\cot(ax+b)$

$\frac{d}{dx}(\sec x) = \sec x \tan x$ $\quad$ $\frac{d}{dx}(\sec(ax+b)) = a\sec(ax+b)\tan(ax+b)$

$\frac{d}{dx}(\cot x) = -\text{cosec}^2 x$ $\quad$ $\frac{d}{dx}(\cot(ax+b)) = -a\text{cosec}^2(ax+b)$

Example 3

Differentiate: **(a)** $f(x) = \sin(2x+3)$ **(b)** $f(x) = 5\cos x^2$ **(c)** $f(x) = 3\tan 4x$

Solution

(a) $f(x) = \sin(2x+3)$

$f'(x) = 2\cos(2x+3)$

(b) $f(x) = 5\cos x^2$

$f'(x) = -5\sin x^2 \times 2x$

$= -10x\sin x^2$

(c) $f(x) = 3\tan 4x$

$f'(x) = 3 \times 4\sec^2(4x)$

$= 12\sec^2 4x$

Example 4

Differentiate: **(a)** $f(x) = \sin^2 x$ **(b)** $f(x) = \sin x \cos x$ **(c)** $f(x) = \cos \pi x - \tan \pi x$

Solution

(a) $f(x) = \sin^2 x$

$f(x) = (\sin x)^2$

$f'(x) = 2\sin x \cos x$

(b) $f(x) = \sin x \cos x$

$f'(x) = \cos x \cos x + \sin x(-\sin x)$

$= \cos^2 x - \sin^2 x$

(c) $f(x) = \cos \pi x - \tan \pi x$

$f'(x) = -\pi \sin \pi x - \pi \sec^2 \pi x$

$= -\pi(\sin \pi x + \sec^2 \pi x)$

Example 5

Differentiate: **(a)** $f(x) = \frac{\sin(3x-1)}{\cos x}$ **(b)** $f(x) = \frac{e^x}{\sin x}$

Solution

(a) $f(x) = \frac{\sin(3x-1)}{\cos x}$

$$f'(x) = \frac{3\cos(3x-1)\cos x - \sin(3x-1)(-\sin x)}{\cos^2 x}$$

$$= \frac{3\cos(3x-1)\cos x + \sin(3x-1)\sin x}{\cos^2 x}$$

(b) $f(x) = \frac{e^x}{\sin x}$

$$f'(x) = \frac{e^x \sin x - e^x \cos x}{\sin^2 x}$$

$$= \frac{e^x(\sin x - \cos x)}{\sin^2 x}$$

Example 6

Differentiate: **(a)** $f(x) = \sin(3x-1)\sec x$ **(b)** $e^x\text{cosec } x$ **(c)** $x^2\cot x$

Solution

(a) $f(x) = \sin(3x-1)\sec x$

$f'(x) = 3\sec x\cos(3x-1) + \sin(3x-1)\sec x\tan x$

$= \sec x(3\cos(3x-1) + \sin(3x-1)\tan x)$

or $\dfrac{3\cos(3x-1)\cos x + \sin(3x-1)\sin x}{\cos^2 x}$

(b) $f(x) = e^x\operatorname{cosec} x$

$f'(x) = e^x\operatorname{cosec} x - e^x\operatorname{cosec} x\cot x$

$= e^x\operatorname{cosec} x(1-\cot x)$

or $\dfrac{e^x(\sin x - \cos x)}{\sin^2 x}$

(c) $f(x) = x^2\cot x$

$f'(x) = 2x\cot x - x^2\operatorname{cosec}^2 x$

$= \dfrac{x(2\sin x\cos x - x)}{\sin^2 x}$

or $\dfrac{x\left(2\tan x - x\sec^2 x\right)}{\tan^2 x}$

Note: With reciprocal trigonometric function ratios there may be more than one simplest form of an expression.

EXERCISE 13.2 DERIVATIVES OF TRIGONOMETRIC FUNCTIONS

1 Differentiate with respect to x:

(a) $\sin 3x$ (b) $3\sin x$ (c) $\cos 2x$ (d) $2\cos x$
(e) $\sin x + 4\cos x$ (f) $\tan 2x$ (g) $\sin 2x - \cos 2x$ (h) $\sin\left(x + \frac{\pi}{4}\right)$

2 The derivative of $\cos^2 5t$ is:

A $-10\sin 5t\cos 5t$ B $-10\cos 5t$ C $-5\sin 5t\cos 5t$ D $-2\sin 5t\cos 5t$

3 Differentiate with respect to x:

(a) $\sin x\cos x$ (b) $x\sin x$ (c) $2x\tan x$ (d) $x^2\cos x$
(e) $\dfrac{x}{\sin x}$ (f) $\dfrac{\sin x}{x}$ (g) $x^3\sin 2x$ (h) $\sin(x^3)$
(i) $x\sec x$ (j) $\dfrac{\operatorname{cosec} x}{x}$ (k) $x^2\cot x$ (l) $\dfrac{\sec x}{\operatorname{cosec} x}$

4 Differentiate with respect to t:

(a) $\sin\frac{t}{2} + \frac{1}{2}\cos t$ (b) $\cos^3 t$ (c) $\cos(t^2+1)$ (d) $\sin\left(2t + \frac{\pi}{2}\right)$
(e) $t^2 + \tan\frac{t}{2}$ (f) $(t^2-1)\cos 3t$ (g) $\cos\left(2t + \frac{\pi}{3}\right)$ (h) $\cos(3t-2)$

5 Differentiate with respect to x:

(a) $\cos^2 2x$ (b) $\sin^2 3x$ (c) $\cos^3 x$ (d) $\cos(x^3)$
(e) $\sin 2x\cos 2x$ (f) $3x + 2\cos x - \sin\frac{x}{2}$ (g) $2\cos^2 x$ (h) $\dfrac{\sin x}{x^2}$
(i) $\tan(x^2-1)$ (j) $x\cos\frac{x}{2}$ (k) $\sin x(1+\cos x)$ (l) $\frac{1}{2}\cos 2x + x\sin x$
(m) $\operatorname{cosec}(x^2-1)$ (n) $\cot^2(2x)$ (o) $\sec x + \operatorname{cosec} x$ (p) $\sec^2 x$

6 Find $f'(x)$ for $f(x) = 3\sin\frac{x}{2} - 4\cos\frac{3x}{2} - x^3$. Indicate whether each statement below is a correct or incorrect step in finding $f'(x)$.

(a) $\dfrac{d}{dx}\left(3\sin\frac{x}{2}\right) = \frac{3}{2}\cos\frac{x}{2}$ (b) $\dfrac{d}{dx}\left(4\cos\frac{3x}{2}\right) = 6\sin\frac{3x}{2}$
(c) $f'(x) = \frac{3}{2}\cos\frac{3x}{2} - 6\sin\frac{3x}{2} - 3x^2$ (d) $f'(x) = \frac{3}{2}\cos\frac{x}{2} + 6\sin\frac{3x}{2} - 3x^2$

7 Differentiate with respect to x:

(a) $\sqrt{\sin 2x}$ (b) $(\sin x - \cos x)^2$ (c) $\sin^2 x + \cos^2 x$
(d) $\sin^2 x - \cos^2 x$ (e) $\sin(\cos x)$ (f) $\cos(\sin x)$
(g) $\tan(\pi - x)$ (h) $\sqrt{1 - \cos x}$ (i) $\tan\sqrt{x}$

8 Differentiate with respect to x:

(a) $e^x \sin x$ (b) $e^{2x}\cos\frac{x}{2}$ (c) $e^{-x}\sin 3x$ (d) $e^x \cos 4x$
(e) $(\cos x + \sin x)e^{-x}$ (f) $e^{\sin 2x}$ (g) $e^{\cos x}$ (h) $e^{\sin x + \cos x}$

13.3 DERIVATIVE OF THE LOGARITHM FUNCTION

You have previously looked at the relationship between the expressions $y = a^x$ and $x = \log_a y$. You have also defined the exponential function $y = e^x$ so that $x = \log_e y$. This logarithm to the base e is called the **natural logarithm**, or sometimes the 'Naperian' (or 'Napierian') logarithm after John Napier, the Scottish mathematician who introduced logarithms in the 1600s. The function $\log_e x$ is sometimes written as $\ln x$, and calculators usually have a natural logarithm key labelled **ln**.

The diagram shows the graphs of $g(x) = e^x$ and $f(x) = \log_e x$.

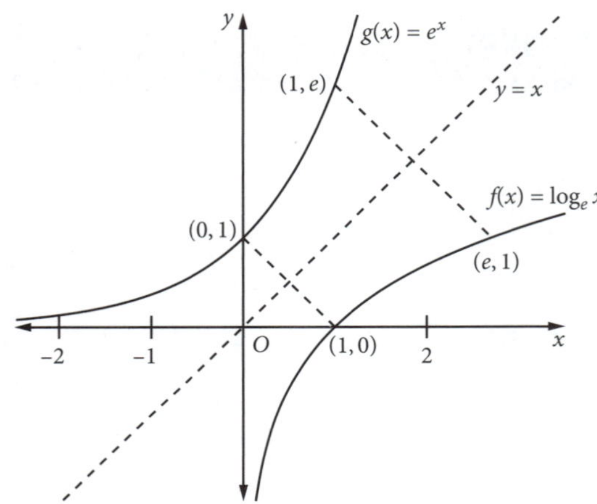

These two graphs are symmetrical about the line $y = x$. You can say that $f(x)$ is the **reflection** of $g(x)$ in the line $y = x$. To 'reflect' a curve of $y = e^x$ in the line $y = x$ means that you interchange (swap) the x and y values, so the equation of the reflection becomes $x = e^y$. Because you know how the exponential and logarithmic functions are linked, you can rewrite $x = e^y$ as $y = \log_e x$.

Because of this property, $y = e^x$ and $y = \log_e x$ are known as **inverse functions**.

On the graph of $g(x) = e^x$ the point $(1, e)$ shows that $e^1 = e$. On the graph of $f(x) = \log_e x$ the point $(e, 1)$ shows that $\log_e e = 1$.

Hence you can write that $e^{\log_e e} = e$ and in general that $e^{\log_e x} = x$ and $\log_e e^x = x$.

Summary of important results—exponentials and logarithms

- If $y = e^x$ then $\log_e y = x$ or $\ln y = x$
- If $y = \log_e x$ or $y = \ln x$ then $e^y = x$
- $e^{\log_e x} = x$ or $e^{\ln x} = x$
- $\log_e e^x = x$ or $\ln e^x = x$
- $e^0 = 1$ and $\log_e 1 = 0$
- $e^x > 0$ for all x
- Domain of $y = \log_e x$ is $x > 0$

The derivative of $\log_e x$

You have seen that $g(x) = e^x$ and $f(x) = \log_e x$ define an inverse pair of functions. The graph of one of these functions is the reflection of the graph of the other function in the line $y = x$. This means that the rule for one function is obtained by writing x in place of y in the rule for the other. (This also means that their domains and ranges interchange.) This all leads to an important property relating to their gradients at any point:

Consider the chain rule: $$\frac{dy}{dx} = \frac{dy}{du} \times \frac{du}{dx}$$

Now write x in place of y: $$\frac{d(x)}{dx} = \frac{dx}{du} \times \frac{du}{dx}$$

Hence, because $\frac{d}{dx}(x) = 1$, we have: $1 = \frac{dx}{du} \times \frac{du}{dx}$

Thus: $\frac{du}{dx} = \frac{1}{\left(\frac{dx}{du}\right)}$

This result is most frequently used in the form $\frac{dy}{dx} = \frac{1}{\left(\frac{dx}{dy}\right)}$. You can now apply this to the exponential and logarithmic functions.

Consider $y = \log_e x$ and rewrite it as $x = e^y$.

Differentiating with respect to y: $\frac{dx}{dy} = e^y = x$

Hence: $\frac{dy}{dx} = \frac{1}{\left(\frac{dx}{dy}\right)} = \frac{1}{x}$

So you have: $\frac{d}{dx}(\log_e x) = \frac{1}{x}, \quad x > 0$

Graph of $f(x) = \log_e x$, $x > 0$

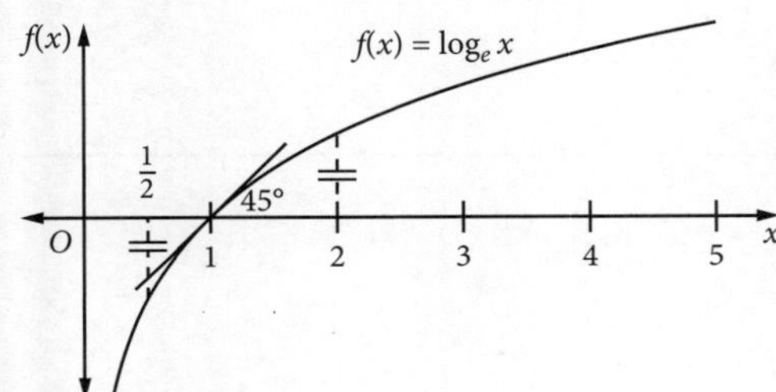

The slope of the curve is $\frac{1}{x}$, which is positive for all x in the domain ($x > 0$). The graph rises steadily with a very steep slope for small x, but this slope decreases as x increases. When x is large, the slope approaches zero.

The curve crosses the x-axis at $x = 1$ (because $\log_e 1 = 0$), at an angle of 45° because the gradient is 1.

For all $x > 1$, $f(x) > 0$

For $0 < x < 1$, $f(x) < 0$

Because $\log_e x = -\log_e\left(\frac{1}{x}\right)$, you have the interesting property that $\log_e 2 = -\log_e \frac{1}{2}$, so the lengths of the ordinates $x = 2$ and $x = \frac{1}{2}$ are equal. This is similarly true for any other value of x and its reciprocal.

Derivative of $\log_e(ax)$, $a > 0$

Let $y = \log_e(ax)$, so $y = \log_e(a) + \log_e(x)$.

$$\frac{dy}{dx} = 0 + \frac{1}{x}$$
$$= \frac{1}{x}$$

Or, let $y = \log_e u$, where $u = ax$.

$$\frac{dy}{dx} = \frac{1}{u} \times a$$
$$= \frac{a}{ax}$$
$$= \frac{1}{x}, \quad x > 0$$

Hence:

$$\frac{d}{dx}(\log_e(ax)) = \frac{1}{x}, \quad x > 0$$

The derivatives of $\log_e x$ and $\log_e ax$ are both $\frac{1}{x}$, so the only difference between the graphs of the functions is a vertical translation of $\log_e a$ a units.

If $a > 1$ then $y = \log_e(ax)$ is above $y = \log_e(x)$, but if $0 < a < 1$ then $y = \log_e(ax)$ is below $y = \log_e(x)$.

These are 'parallel curves', because they are a fixed distance apart and their gradients are equal at each corresponding point.

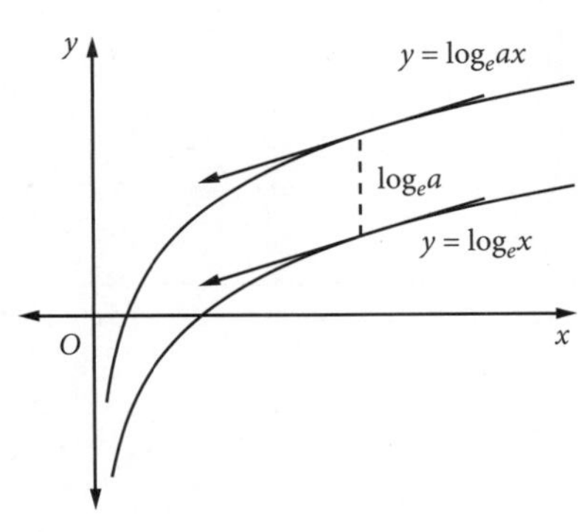

Example 7

Differentiate with respect to x:

(a) $x^2\log_e(2x)$ (b) $\dfrac{\log_e 3x}{e^x}$ (c) $f(x)=\dfrac{\log_e x}{\tan x}$

Solution

(a) Let $y=x^2\log_e(2x)=uv$, where $u=x^2$ and $v=\log_e(2x)$

$$\frac{dy}{dx}=v\frac{du}{dx}+u\frac{dv}{dx}$$

$$=2x\log_e 2x+x^2\times\frac{1}{x},\ x>0$$

$$=x(2\log_e 2x+1)$$

(b) Let $y=\dfrac{\log_e 3x}{e^x}=\dfrac{u}{v}$, where $u=\log_e(3x)$ and $v=e^x$

$$\frac{dy}{dx}=\frac{v\frac{du}{dx}-u\frac{dv}{dx}}{v^2}$$

$$\frac{dy}{dx}=\frac{e^x\times\frac{1}{x}-\log_e 3x\times e^x}{e^{2x}}$$

$$=\frac{e^x\left(1-x\log_e 3x\right)}{xe^{2x}}$$

$$=\frac{1-x\log_e 3x}{xe^x}$$

(c) $f(x)=\dfrac{\log_e x}{\tan x}$

$$f'(x)=\frac{\frac{1}{x}\times\tan x-\sec^2 x\log_e x}{\tan^2 x}$$

$$=\frac{\tan x-x\sec^2 x\log_e x}{x\tan^2 x}$$

Derivative of $\log_e(ax+b)$

Let $y=\log_e(ax+b)=\log_e u$, where $u=ax+b$.

Chain rule: $$\frac{dy}{dx}=\frac{dy}{du}\times\frac{du}{dx}$$

$$=\frac{1}{u}\times a$$

$$=\frac{a}{ax+b}$$

Example 8

Differentiate:

(a) $\log_e(x^3+1)$ (b) $\log_e(x^2+2x-1)$

Solution

(a) Let $y=\log_e(x^3+1)$

$$=\log_e u,\text{ where } u=x^3+1$$

$$\frac{dy}{dx}=\frac{dy}{du}\times\frac{du}{dx}$$

$$=\frac{1}{u}\times 3x^2$$

$$=\frac{3x^2}{x^3+1}$$

(b) Let $y=\log_e(x^2+2x-1)$

$$=\log_e u,\text{ where } u=x^2+2x-1$$

$$\frac{dy}{dx}=\frac{dy}{du}\times\frac{du}{dx}$$

$$=\frac{1}{u}\times(2x+2)$$

$$=\frac{2x+2}{x^2+2x-1}$$

Example 9

Use the logarithm laws and then find the derivative of each function.

(a) $y = \log_e\left(\frac{x^2+1}{x}\right)$ (b) $f(x) = \log_e\left(e^x\left(x^2+3\right)\right)$ (c) $g(x) = \log_e\left(\frac{x^2(e^x-1)}{e^{-x}+1}\right)$

Solution

(a) $y = \log_e\left(\frac{x^2+1}{x}\right) = \log_e\left(x^2+1\right) - \log_e x$

$$\frac{dy}{dx} = \frac{2x}{x^2+1} - \frac{1}{x}$$

If a stationary point had to be found then you would write this answer as a single fraction, otherwise leave it. It is a good idea to practice this algebraic simplification before it is needed.

$$\frac{dy}{dx} = \frac{2x^2-(x^2+1)}{x(x^2+1)} = \frac{x^2-1}{x(x^2+1)}$$

(b) $f(x) = \log_e\left(e^x\left(x^2+3\right)\right)$

$$= \log_e e^x + \log_e (x^2+3)$$

$$= x + \log_e (x^2+3)$$

$$f'(x) = 1 + \frac{2x}{x^2+3}$$

$$= \frac{x^2+3+2x}{x^2+3}$$

$$= \frac{x^2+2x+3}{x^2+3}$$

(c) $g(x) = \log_e\left(\frac{x^2(e^x-1)}{e^{-x}+1}\right)$

$$= \log_e x^2 + \log_e (e^x-1) - \log_e (e^{-x}+1)$$

$$= 2\log_e x + \log_e (e^x-1) - \log_e (e^{-x}+1)$$

$$g'(x) = \frac{2}{x} + \frac{e^x}{e^x-1} - \frac{e^{-x}}{e^{-x}+1}$$

$$= \frac{2\left(e^x-1\right)\left(e^{-x}+1\right) + xe^x\left(e^{-x}+1\right) - xe^{-x}\left(e^x-1\right)}{x\left(e^x-1\right)\left(e^{-x}+1\right)}$$

$$= \frac{2\left(1+e^x-e^{-x}-1\right) + x + xe^x - x + xe^{-x}}{x\left(e^x-1\right)\left(e^{-x}+1\right)}$$

$$= \frac{2e^x - 2e^{-x} + x + xe^x - x + xe^{-x}}{x\left(e^x-1\right)\left(e^{-x}+1\right)}$$

In Example 8, each derivative is a quotient in which the numerator is the derivative of the denominator. It can be seen that:

$$\frac{d}{dx}\left(\log_e\left(f(x)\right)\right) = \frac{f'(x)}{f(x)}$$

Summary—derivatives of logarithms

$\frac{d}{dx}\left(\log_e x\right) = \frac{1}{x}, x > 0$	$\frac{d}{dx}\left(\log_e(ax+b)\right) = \frac{a}{ax+b}$
$\frac{d}{dx}\left(\log_e(ax)\right) = \frac{1}{x}, x > 0$	$\frac{d}{dx}\left(\log_e\left(f(x)\right)\right) = \frac{f'(x)}{f(x)}$

The function $y = a^x$

When introducing the exponential function you considered the graph of $y = a^x$. From this graph you developed the function, $f(x) = e^x$, which was its own derivative. This led to the natural logarithm function, $g(x) = \log_e x$ and its derivative.

It is now time to look at $y = a^x$ in its own right.

Given $y = a^x$, take the logarithm to base e of each side: $\log_e y = x\log_e a$

Raise both sides as a power of e: $e^{\log_e y} = e^{x\log_e a}$

Simplify the left hand side: $y = e^{x\log_e a}$

So: $a^x = e^{x\log_e a}$

Use the chain rule to differentiate both sides with respect to x: $\frac{d}{dx}\left(a^x\right) = \log_e a \times e^{x\log_e a}$

Hence: $\frac{d}{dx}\left(a^x\right) = \left(\log_e a\right)a^x$

Now consider the function $y = \log_a x$:

Using the change of base rule: $y = \frac{\log_e x}{\log_e a}$

Differentiate both sides with respect to x: $\frac{dy}{dx} = \frac{1}{\log_e a} \times \frac{1}{x}$

But $y = \log_a x$: $\frac{d}{dx}(\log_a x) = \frac{1}{x\log_e a}$

Hence:

$$\frac{d}{dx}\left(a^x\right) = \left(\log_e a\right)a^x \quad \text{or} \quad \frac{d}{dx}\left(a^x\right) = a^x \ln a$$

$$\frac{d}{dx}\left(\log_a x\right) = \frac{1}{x\log_e a} \quad \text{or} \quad \frac{d}{dx}\left(\log_a x\right) = \frac{1}{x\ln a}$$

EXERCISE 13.3 DERIVATIVE OF THE LOGARITHM FUNCTION

1 Differentiate:

(a) $\log_e 2x$ **(b)** $2\log_e x$ **(c)** $\log_e x^2$

(d) $\log_e(3x-5)$ **(e)** $\log_e x + 3$ **(f)** $x^2 - \log_e(4x-1)$

2 State the largest possible domain of each of the following functions and find their derivatives.

(a) $f(x) = \log_e(3x+2)$ **(b)** $f(x) = \log_e(x^2+1)$ **(c)** $f(x) = \log_e(x^2-4x+4)$

(d) $f(x) = \log_e(4x+3)$ **(e)** $\ln\sqrt{x}$ **(f)** $\log_e\left(x+\sqrt{x}\right)$

3 The derivative of $\log_e(3x^2+1)$ is:

A $6x$ **B** $\frac{6}{x}$ **C** $\frac{6x}{3x^2+1}$ **D** $\frac{1}{x^3+x}$

4 Differentiate:

(a) $x\ln x$ **(b)** $x^3\ln x$ **(c)** $(x+2)\ln(x+2)$ **(d)** $(x^2+1)\ln 2x$

(e) $(2x-5)\ln x$ **(f)** $e^x\ln x$ **(g)** $e^{2x}\ln 2x$ **(h)** $\frac{x}{\log_e x}$

(i) $\frac{\log_e x}{x}$ **(j)** $\frac{\log_e x}{e^x}$ **(k)** $\frac{\log_e\left(x^2+1\right)}{x}$ **(l)** $e^x\log_e(e^x+1)$

5 Differentiate:

(a) $\log_e(x^2 - 2x)$ (b) $\log_e(e^x)$ (c) $\log_e(x^3)$ (d) $\log_e\left(\frac{x-3}{x+3}\right)$

(e) $\log_e(x^2 - 1)$ (f) $\log_e(x^4)$ (g) $e^{\log_e x}$ (h) $x^2\log_e(x^2)$

(i) $\log_e x\sqrt{x}$ (j) $\log_e(\log_e x)$ (k) $e^{x\log_e x}$ (l) $\log_e\left(\frac{e^x+1}{e^x-1}\right)$

6 Find the gradient of the curve $y = \log_e(x^2 + 1)$ at the point where $x = 3$.

7 If $f(x) = \log_e x$, find: (a) $f'(x)$ (b) $f''(x)$ (c) $f'(2)$ (d) $f''(2)$

8 Find the equation of the tangent and normal to the curve $y = \log_e x$ at the point where it crosses the x-axis.

9 (a) On the same axes, sketch the graphs of $y = \log_e x$ and $y = \log_e\left(\frac{x}{2}\right)$ and show that their gradients are equal for all values of x.

(b) Write the equation of the tangent to the curve $y = \log_e\left(\frac{x}{2}\right)$ at the point where it crosses the x-axis.

10 Show that $y = \log_e(e^x)$ is equivalent to $y = x$ for all values of x and hence state its gradient.

11 Sketch the graph of $y = \log_e(x - 1)$, stating its largest possible domain. Find the equation of the tangent at the point where $x = 2$.

12 Solve: (a) $e^x = 2$ (b) $e^{3x} = 5$ (c) $e^{2x+3} = 7$ (d) $e^{x^2-1} = 10$

13 Differentiate:

(a) $y = \log_e\left(\frac{x^3-1}{x}\right)$ (b) $f(x) = \log_e\left(e^x(x+2)\right)$ (c) $y = \log_e\left(\sqrt{x}(x+1)^5\right)$

(d) $f(x) = \log_e\left(\frac{e^x+x}{\sqrt{x}}\right)$ (e) $g(x) = \log_e\left(\frac{x^3(e^x+1)}{e^{-x}+1}\right)$ (f) $y = \log_e\left(\frac{\sqrt[3]{x^2}\sin x}{1-2e^x}\right)$

14 Differentiate:

(a) 2^x (b) $e^x + 3^x$ (c) $\log_2 x$ (d) $x + \log_3 x$

(e) $x^2\,4^x$ (f) $x^3\log_5 x$ (g) $\frac{2^x}{x}$ (h) $\frac{\log_a x}{x^2}$

15 Differentiate:

(a) a^{-x} (b) $a^x\log_a x$ (c) $\frac{\log_a x}{a^x}$ (d) $\sqrt{\log_a x}$ (e) $\sqrt{xa^x}$ (f) $(x\log_2 x)^2$

16 (a) Find the equation of the tangent to the curve $y = 10^x$ at the point (1, 10).

(b) Find the equation of the normal to the curve $y = 5^x$ at the point (2, 25).

(c) At what point(s) are the tangents to $y = 10^x$ and $y = 5^x$ parallel?

13.4 DERIVATIVE OF $e^{f(x)}$

In Chapter 8 you derived the following results for the exponential function.

$$\frac{d}{dx}\left(e^x\right) = e^x \qquad \frac{d}{dx}\left(e^{kx}\right) = ke^{kx} \qquad \frac{d}{dx}\left(e^{ax+b}\right) = ae^{ax+b} \qquad \frac{d}{dx}\left(e^{f(x)}\right) = f'(x)e^{f(x)}$$

In this section you will use these results with a range of other functions from the course.

Example 10

Find the derivative of each function.

(a) $e^{\sin x}$ (b) $x e^{\tan x}$ (c) $e^{x \ln x}$

Solution

(a) Use $\frac{d}{dx}\left(e^{f(x)}\right) = f'(x)e^{f(x)}$

In this case, $f(x) = \sin x$

so $f'(x) = \cos x$.

$\therefore \frac{d}{dx}\left(e^{\sin x}\right) = \cos x\, e^{\sin x}$

(b) Use $\frac{d}{dx}\left(e^{f(x)}\right) = f'(x)e^{f(x)}$

$f(x) = \tan x$

$f'(x) = \sec^2 x$

$\frac{d}{dx}\left(e^{\tan x}\right) = \sec^2 x\, e^{\tan x}$

$\frac{d}{dx}(uv) = v\frac{du}{dx} + u\frac{dv}{dx}$

$u = x,\ v = e^{\tan x}$

$\therefore \frac{d}{dx}\left(x e^{\tan x}\right) = e^{\tan x} \times 1 + x \times \sec^2 x\, e^{\tan x}$

$= e^{\tan x}\left(1 + x\sec^2 x\right)$

(c) Use $\frac{d}{dx}\left(e^{f(x)}\right) = f'(x)e^{f(x)}$

$f(x) = x \ln x$

Use the product rule to find $f'(x)$:

$\frac{d}{dx}(uv) = v\frac{du}{dx} + u\frac{dv}{dx}$

$u = x,\ v = \ln x$

$\frac{d}{dx}(x \ln x) = x \times \frac{1}{x} + \ln x \times 1$

$= \ln x + 1$

$\therefore \frac{d}{dx}\left(e^{x \ln x}\right) = (\ln x + 1)e^{x \ln x}$

In the example above, $e^{x \ln x}$ may be written as $e^{\ln x^x}$ which is x^x. This means you have also found the derivative of x^x.

EXERCISE 13.4 DERIVATIVE OF $e^{f(x)}$

1 Differentiate:

(a) $e^{x^2} + 2$ (b) $\left(e^x + x^2\right)^4$ (c) $e^x + ex$

(d) $4e^{\cos x}$ (e) $e^{\sqrt{x}+1}$ (f) $e^{x+\ln x}$

2 Differentiate:

(a) $x e^{\sin x}$ (b) $e^x \log_e x$ (c) $e^{\cos(2x+1)}$ (d) $1 + x + x^2 e^x$

3 Given $y = \dfrac{100}{1 + 15e^{-0.5t}}$, find $\dfrac{dy}{dt}$.

4 (a) Sketch the graphs of $f(x) = e^{\sin x}$ and $g(x) = e^{\cos x}$ on the same diagram for $0 \le x \le 2\pi$, using appropriate technology.

(b) Write the coordinates of their points of intersection (correct to 3 decimal places where necessary). Check your solutions algebraically.

(c) Find the gradient of the tangent to each curve at their points of intersection.

(d) Do the curves intersect at right angles at these points? Justify your answer.

5 In statistics, the normal probability density function is given by $f(x) = \frac{1}{\sqrt{2\pi}}e^{-\frac{x^2}{2}}$. Find $f'(0)$.

CHAPTER REVIEW 13

1 Differentiate:

(a) $\sin x + \tan 2x$ (b) $3\cos 4x - 5\sin 2x$ (c) $x\cos x$

(d) $\dfrac{\cos x}{\sin x}$ (e) $2e^{-x}\cos 3x$ (f) $\log_e(\sin 2x)$

2 Differentiate with respect to x:

(a) $(x^2 + 2x)e^x$ (b) $2e^{-x}\ln x$ (c) $\log_e(1 + e^x)$

(d) $\log_e(x^2 + 2x)$ (e) $(x^2 + 3x)e^{-3x}$ (f) $e^{\sqrt{x}} + \log_e \sqrt{x}$

3 The position x of a particle moving along a straight line at any time t is given by $x = 3 + 6\cos\dfrac{\pi t}{6}$.

(a) Find the position of the particle for values of $t = 0, 2, 4, 6, 8, 10, 12$.

(b) Find the velocity and acceleration of the particle when it first reaches the position $x = 0$.

4 Differentiate:

(a) $x^2 \sin x$ (b) $\dfrac{x^2}{\cos x}$ (c) $\sin x \cos x$

(d) $\sqrt{\tan x}$ (e) $\cos(x^2)$ (f) $\dfrac{\sin x}{2x + 1}$

5 Differentiate:

(a) $\log_e(x^2 + 2x + 1)$ (b) $e^x \log_e x$ (c) $\dfrac{\log_e x}{e^x}$

(d) $\log_e(\tan x)$ (e) $\dfrac{x^2 + \log_e x}{x}$ (f) $x^4 + \cos(x^2) + \log_e(\sin x)$

6 Differentiate:

(a) $\log_e(x\tan x)$ (b) $\log_e\left(\dfrac{x^3 - 6}{e^{-x} - 1}\right)$ (c) $\log_e = \left(\dfrac{\sqrt{x}\cos x}{1 - \sin^2 x}\right)$

7 Differentiate:

(a) $x^3\, 10^x$ (b) $\sin x + \log_a x$ (c) $2^x + 3^x + 4^x$ (d) $\dfrac{a^x}{\log_a x}$

CHAPTER 14
The first and second derivative

14.1 THE SIGN OF THE DERIVATIVE

Remember that the first derivative of a function (as studied earlier, see Chapters 7 and 12) is also known as the gradient function, because it allows you to calculate the gradient at any point on the curve. You have used the process of differentiation to find this gradient function or derivative of a function. This can be positive, negative, zero or undefined. Each possible value has a geometrical application.

Example 1

A sketch of the function $y = 4x - x^2$ is given in the diagram.

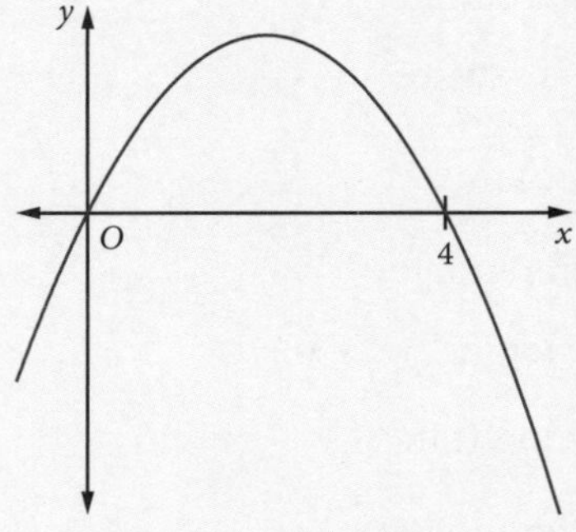

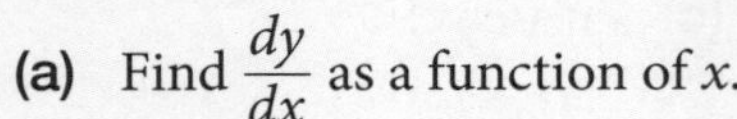

(a) Find $\dfrac{dy}{dx}$ as a function of x.

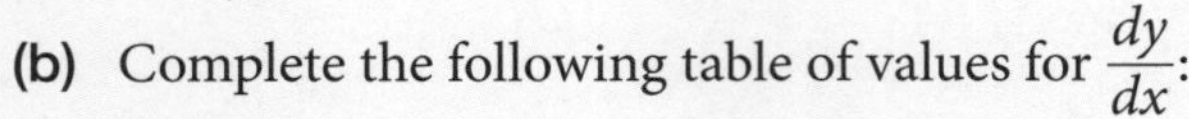

(b) Complete the following table of values for $\dfrac{dy}{dx}$:

x	0	1	2	3	4	5
$\dfrac{dy}{dx}$						

(c) Draw the graph of $\dfrac{dy}{dx}$ on the same diagram as a graph of y.

(d) For what values of x is **(i)** $\dfrac{dy}{dx} > 0$ **(ii)** $\dfrac{dy}{dx} = 0$ **(iii)** $\dfrac{dy}{dx} < 0$?

(e) Describe the function $y = 4x - x^2$ where $\dfrac{dy}{dx} > 0$.

(f) Describe the function $y = 4x - x^2$ where $\dfrac{dy}{dx} < 0$.

(g) Describe the function $y = 4x - x^2$ where $\dfrac{dy}{dx} = 0$.

Solution

(a) $\dfrac{dy}{dx} = 4 - 2x$

(b)

x	0	1	2	3	4	5
$\dfrac{dy}{dx}$	4	2	0	−2	−4	−6

(c) 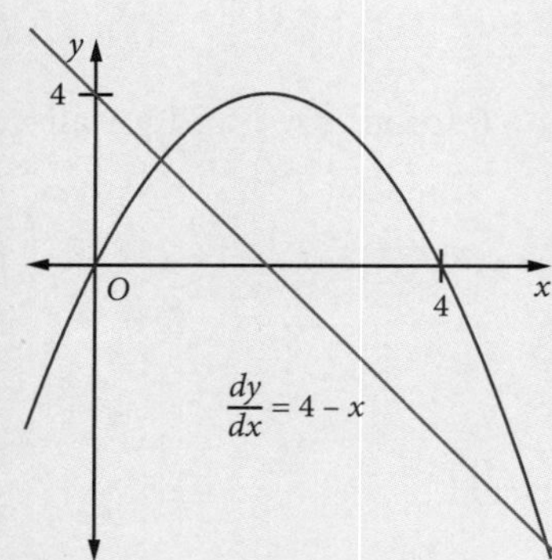

(d) (i) $x < 2$ (ii) $x = 2$ (iii) $x > 2$

(e) Where $\frac{dy}{dx} > 0$, y increases as x increases. The curve slopes up.

(f) Where $\frac{dy}{dx} < 0$, y decreases as x increases. The curve slopes down.

(g) Where $\frac{dy}{dx} = 0$, y neither increases nor decreases. It is at its highest point and the tangent at this point is horizontal.

The sign of the first derivative

- If $\frac{dy}{dx} > 0$ as x increases, the function is an **increasing** function.
- If $\frac{dy}{dx} < 0$ as x increases, the function is a **decreasing** function.
- If $\frac{dy}{dx} = 0$ at a given value of x, the function is **stationary** at that point; the point is called a **stationary point**. At this point, the tangent to the curve is parallel to the x-axis.

Each definition and its converse can be used to determine a function's characteristics:

- If $\frac{dy}{dx} > 0$, then the function is increasing; if the function is increasing, then $\frac{dy}{dx} > 0$.
- If $\frac{dy}{dx} < 0$, then the function is decreasing; if the function is decreasing, then $\frac{dy}{dx} < 0$.
- If $\frac{dy}{dx} = 0$ at a point, then it is a stationary point; at a stationary point, $\frac{dy}{dx} = 0$.

Example 2

For what values of x is the function $f(x) = 2x^3 - 9x^2 - 24x + 1$:

(a) stationary (b) increasing (c) decreasing?

Solution

Find $f'(x)$: $f'(x) = 6x^2 - 18x - 24$

Remove common factor: $f'(x) = 6(x^2 - 3x - 4)$

Factorise: $f'(x) = 6(x + 1)(x - 4)$

(a) For stationary points, $f'(x) = 0$: $6(x + 1)(x - 4) = 0$

$x = -1, 4$

Stationary points occur at $x = -1$ or $x = 4$.

(b) Increasing where $f'(x) > 0$: $(x + 1)(x - 4) > 0$ **OR** Graphically: 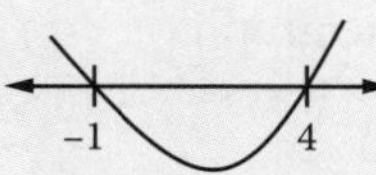

Test $x = 0$: LHS $= 1 \times (-4)$

$= -4$

< 0

Function is not increasing for $-1 < x < 4$

Hence function is increasing for $x < -1$ or $x > 4$.

Graph is above axis for $x < -1$ or $x > 4$

Function is increasing for $x < -1$ or $x > 4$

(c) Decreasing where $f'(x) < 0$, so function is decreasing for $-1 < x < 4$.

Example 3

Sketch a graph of $f(x)$ such that $f(4) = 2$, $f'(4) = 0$, $f(1) = 5$, $f'(1) = 0$, $f'(x) > 0$ for all $x < 1$ and for all $x > 4$; also $f'(x) < 0$ for $1 < x < 4$.

Solution

Mark this information on a number plane:

Use straight lines to indicate gradients.

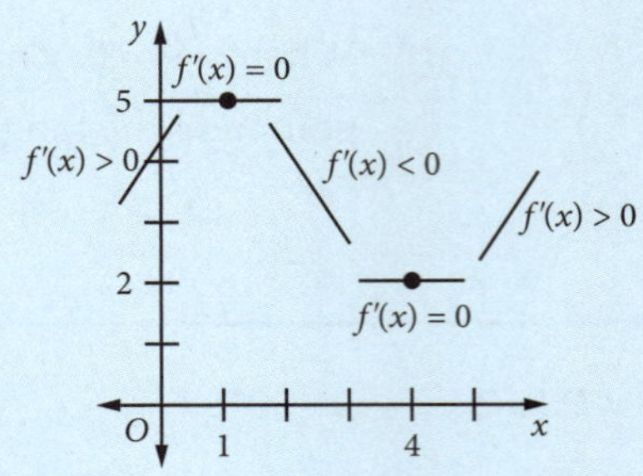

Draw a curve using this information:

The only points that you know for sure on $y = f(x)$ are $(1, 5)$ and $(4, 2)$.

From the gradients you know that the curve changes from increasing to decreasing at $(1, 5)$ and from decreasing to increasing at $(4, 2)$. Fit the curve to this information.

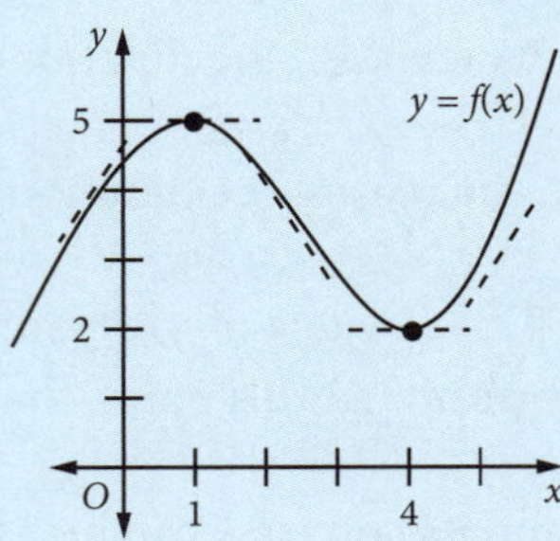

EXERCISE 14.1 THE SIGN OF THE DERIVATIVE

1 A function $f(x)$ has the following properties: $f(3) = 4$, $f'(3) = 1$. Which sketches fit the graph of $y = f(x)$ near $x = 3$?

A

B

C

D

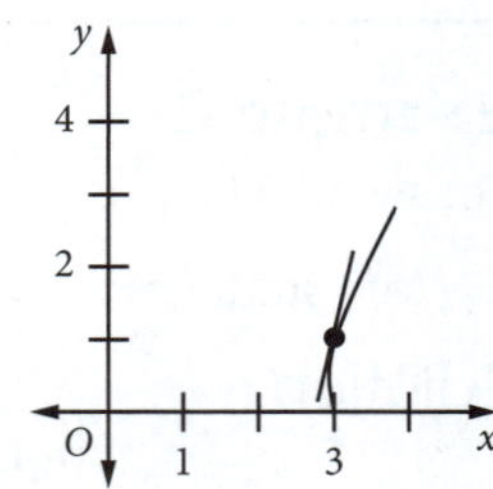

2 Sketch the graph of $y = f(x)$ with the following properties: $f(1) = 0$, $f'(x) = 2$ for all x. State the rule that defines the function.

3 Sketch the graph of a function given that $f(2) = 0$, $f'(2) = 0$, $f'(x) < 0$ for all $x < 2$, and $f'(x) > 0$ for all $x > 2$.

4 Sketch the graph of $y = f(x)$, such that $f(3) = 3$, $f'(3) = 0$, $f(1) = 5$, $f'(1) = 0$, $f'(x) > 0$ for all $x < 1$ and for all $x > 3$; also $f'(x) < 0$ for $1 < x < 3$.

5 For the function $f(x) = x^2 - 5x + 6$, sketch the graph of $f'(x)$ and hence find the values of x for which:

(a) $f'(x) < 0$ **(b)** $f'(x) = 0$ **(c)** $f'(x) > 0$.

6 Sketch the curve $y = 4x - x^2$. For what values of x is $\frac{dy}{dx} = 0$? What is the sign of the gradient to the left and right of this point?

7 For the graph of $f(x) = 6 - 3x - x^2$, find the values of x for which the function:

(a) increases when x increases **(b)** decreases when x increases

(c) changes from increasing to decreasing.

8 For the graph of $f(x) = x^3 - 6x^2 + 9x + 2$, find:

(a) $f'(x)$
(b) the values of x for which the function increases when x increases
(c) the values of x for which the function decreases when x increases
(d) the values of x for which the function changes from increasing to decreasing.

9 For the graph of $f(x) = x^3 - 1$, find the values of x for which the function:

(a) increases when x increases (b) decreases when x increases
(c) changes from increasing to decreasing.

10 For the graph of $f(x) = (x - 1)^2(x + 1)$, find the values of x for which the function is:

(a) stationary (b) increasing (c) decreasing.

14.2 THE FIRST DERIVATIVE AND TURNING POINTS

In functions you have seen so far, a stationary point is usually the point where the function changes from increasing to decreasing (or from decreasing to increasing), so that the stationary point is the highest (or lowest) point in the neighbourhood. (Here 'neighbourhood' means 'near the point on the graph'.) However, sometimes the function does not change the sign of its gradient at the stationary point. This situation will be considered later.

When a function changes from increasing to decreasing at a stationary point, the sign of $f'(x)$ changes from positive to negative.

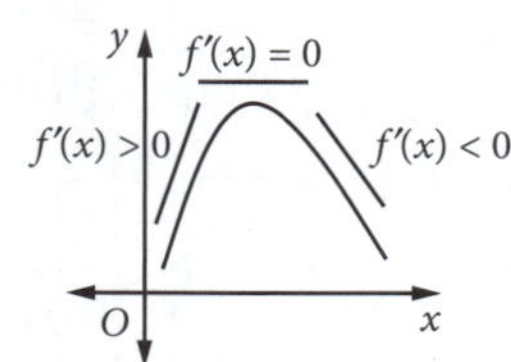

When a function changes from decreasing to increasing at a stationary point, the sign of $f'(x)$ changes from negative to positive.

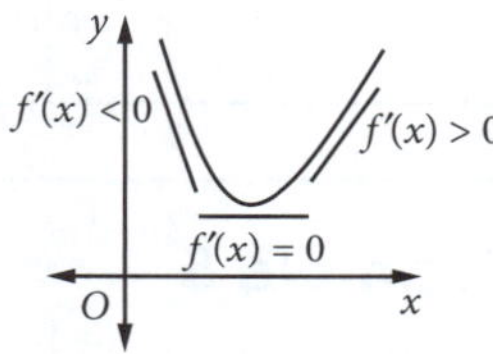

These points are called **turning points**. If the turning point is higher than the other points in its neighbourhood, it is called a **local maximum** turning point. If the point is lower than the other points in its neighbourhood, it is called a **local minimum** turning point.

A turning point of $f(x)$ is a point where the curve $y = f(x)$ is locally a maximum or a minimum.

For a differentiable function $f(x)$, all turning points are stationary points.
(However, note that not all stationary points are turning points.)

First derivative test for local maxima and minima

A turning point of the differentiable function $f(x)$ may occur when $f'(x) = 0$. The type of turning point will depend on the change in sign of $f'(x)$ as x passes through the abscissa (x-coordinate) of the stationary point, so you need to find the sign of $f'(x)$ on either side of the stationary point.

If $f'(x) = 0$ when $x = x_1$, $f'(x) > 0$ when $x < x_1$, and $f'(x) < 0$ when $x > x_1$, then the sign of $f'(x)$ changes from positive to negative as x passes through x_1. The point $(x_1, f(x_1))$ must be a **local maximum turning point**.

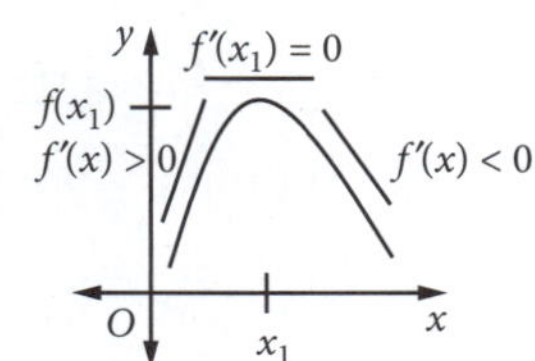

If $f'(x) = 0$ when $x = x_2$, $f'(x) < 0$ when $x < x_2$, and $f'(x) > 0$ when $x > x_2$, then the sign of $f'(x)$ changes from negative to positive as x passes through x_2. The point $(x_2, f(x_2))$ is a **local minimum turning point**.

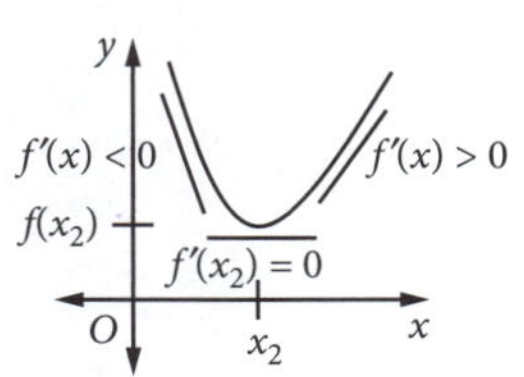

Thus there are two conditions required to find turning points of the function $y = f(x)$:

1 $f'(x) = 0$ at $x = x_1$, and

2 $f'(x)$ changes sign as x passes through x_1.

The way that $f'(x)$ changes will tell you whether the turning point is a maximum or minimum turning point.

Example 4

The diagrams show the graphs of $f(x) = x^2 - 6x + 8$ and $f'(x) = 2x - 6$ drawn with a common x scale.

(a) Find the coordinates of any stationary points on $f(x)$.

(b) Use the graph of $f'(x)$ to determine the nature of the turning points (i.e. maximum or minimum).

(c) What is the least value of $f(x)$?

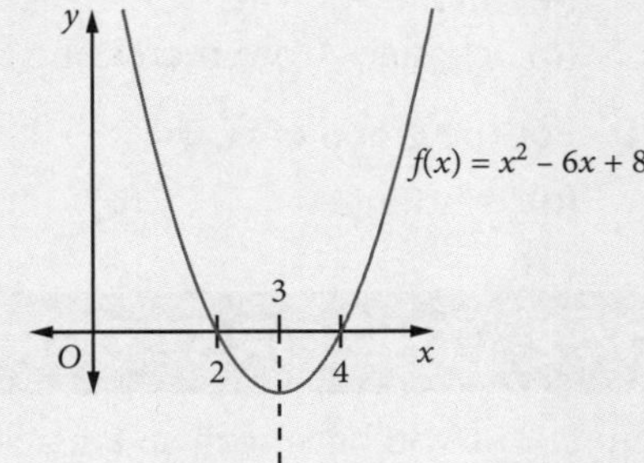

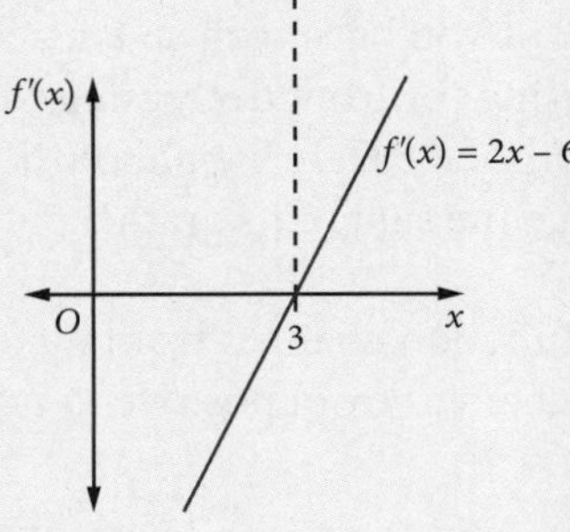

Solution

(a) The graph of $f'(x)$ gives that $f'(x) = 0$ when $x = 3$.

$f(3) = 9 - 18 + 8 = -1$

The coordinates of the stationary point are $(3, -1)$.

(b) When $x < 3$, $f'(x) < 0$; when $x > 3$, $f'(x) > 0$.

$f'(x)$ changes from negative to positive as x passes through 3.

$\therefore$ The stationary point is a relative minimum turning point (local minimum).

(c) The least value of $f(x)$ is -1.

Example 5

A function is given by $f(x) = x^3 - 12x + 16$.

(a) Find $f'(x)$.

(b) Find the coordinates of any stationary points.

(c) Determine the nature of the stationary points.

(d) Sketch $y = f(x)$.

Solution

(a) $f'(x) = 3x^2 - 12$

(b) For stationary points, $f'(x) = 0$, so:

$$3(x^2 - 4) = 0$$

$$(x + 2)(x - 2) = 0$$

$$x = -2, 2$$

$f(-2) = -8 + 24 + 16 = 32$ $\quad f(2) = 8 - 24 + 16 = 0$

Stationary points are $(-2, 32)$ and $(2, 0)$.

(c) Consider the stationary point $(-2, 32)$.

At $x = -3$, $f'(-3) = 27 - 12 = 15 > 0$

At $x = -1$, $f'(-1) = 3 - 12 = -9 < 0$

$f'(x)$ changes from +ve to −ve, so $f(x)$ has a maximum turning point at $(-2, 32)$.

Consider the stationary point $(2, 0)$.

At $x = 1$, $f'(1) = 3 - 12 = -9 < 0$

At $x = 3$, $f'(3) = 27 - 12 = 15 > 0$

$f'(x)$ changes from +ve to −ve, so $f(x)$ has a minimum turning point at $(2, 0)$.

$\therefore$ $(-2, 32)$ is a maximum turning point and $(2, 0)$ is a minimum turning point.

(d)

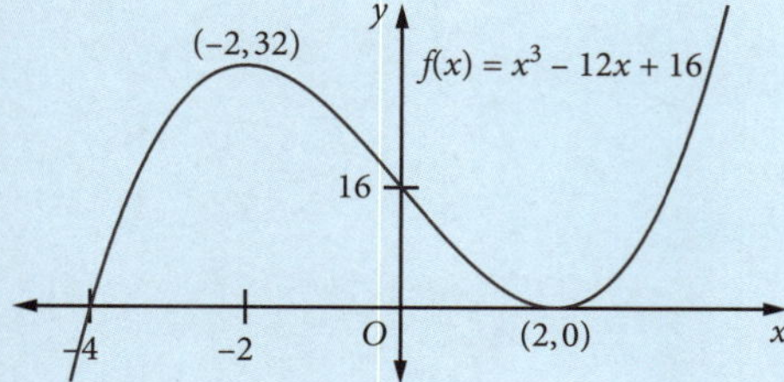

In calculating the change in gradient, a sketch of $f'(x) = 3x^2 - 12$ could have been used to investigate the change in sign of $f'(x)$:

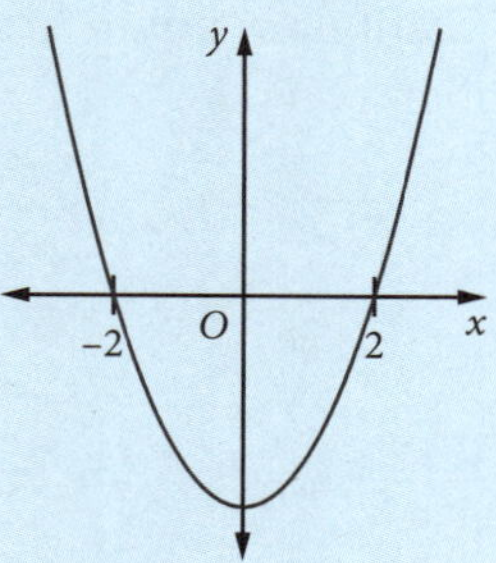

$x < -2$, $f'(x) > 0$
$x > -2$, $f'(x) < 0$: maximum turning point at $(-2, 32)$

$x < 2$, $f'(x) < 0$
$x > 2$, $f'(x) > 0$: minimum turning point at $(2, 0)$

When selecting values of x to substitute into $f'(x)$, you need to pick values near the abscissa (x-coordinate) of the stationary point. But how near is 'near'? In Example **5** you used values that were 1 unit either side of the stationary point, and in this example that was close enough. You could have used −2.1 and −1.9 and reached the same answer, but the calculations would have been more time-consuming.

Calculations for the stationary point at $(-2, 32)$ could have been summarised in tables as follows:

x-value	−3	−2	−1
$f'(x)$	15	0	−9
Sign of $f'(x)$	+	0	−
Direction of curve	↗	→	↘
Shape of curve	⌒		

x-value	−2.1	−2	−1.9
$f'(x)$	1.2	0	−1.2
Sign of $f'(x)$	+	0	−
Direction of curve	↗	→	↘
Shape of curve	⌒		

This would then allow you to say that the point $(-2, 32)$ is a local maximum.

Turning points—theoretical method

Another method to investigate turning points is to use the vanishingly small positive number ε (the Greek letter epsilon), defined so that ε is infinitesimally small but $\varepsilon > 0$.

To apply the ε method to Example **5**, at the stationary point $(-2, 32)$ consider $x = -2 - \varepsilon$ and $x = -2 + \varepsilon$ as the values of x either side of $x = -2$. Factorise the expression $f'(x) = 3x^2 - 12$ into $f'(x) = 3(x + 2)(x - 2)$ and examine the values of f':

$f'(-2 - \varepsilon) = 3(-2 - \varepsilon + 2)(-2 - \varepsilon - 2) = 3(-\varepsilon)(-4 - \varepsilon) > 0$, because $(-) \times (-) = (+)$

$f'(-2 + \varepsilon) = 3(-2 + \varepsilon + 2)(-2 + \varepsilon - 2) = 3(\varepsilon)(-4 + \varepsilon) < 0$, because $(+) \times (-) = (-)$

$f'(x)$ changes from +ve to −ve, so $f(x)$ has a maximum turning point at $(-2, 32)$.

Example 6

The function $y = 3x^2 - x^3$ is defined on the domain $-2 \le x \le 4$.

(a) Find the stationary points and determine their nature.

(b) Find the greatest and least values of y in the domain.

(c) Sketch the function for the given domain.

Solution

(a) $y = 3x^2 - x^3$

$\frac{dy}{dx} = 6x - 3x^2$

$= 3x(2 - x)$

For stationary points, $\frac{dy}{dx} = 0$

$3x(2 - x) = 0$

$\therefore x = 0$ or $x = 2$

Stationary points are $(0, 0)$ and $(2, 4)$.
For $\varepsilon > 0$:

$x = 0 - \varepsilon$, $\frac{dy}{dx} = 3(-\varepsilon)(2 - (-\varepsilon))$
$= -3\varepsilon(2 + \varepsilon)$
< 0

$x = 0 + \varepsilon$, $\frac{dy}{dx} = 3\varepsilon(2 - \varepsilon)$
> 0

$\frac{dy}{dx}$ changes from –ve to +ve, so $(0, 0)$ is a minimum turning point.

$x = 2 - \varepsilon$, $\frac{dy}{dx} = 3(2 - \varepsilon)(2 - (2 - \varepsilon))$
$= 3(2 - \varepsilon)(\varepsilon)$
> 0

$x = 2 + \varepsilon$, $\frac{dy}{dx} = 3(2 + \varepsilon)(2 - (2 + \varepsilon))$
$= 3(2 + \varepsilon)(-\varepsilon)$
< 0

$\frac{dy}{dx}$ changes from +ve to –ve, so $(2, 4)$ is a maximum turning point.

(b) The turning points give the local maximum and minimum values. The values at the endpoints of a given domain may be greater than or less than these local values. In this example you need to find the value of y at $x = -2$ and $x = 4$.

$x = -2$: $y = 12 + 8 = 20$

$x = 4$: $y = 48 - 64 = -16$

Hence the greatest value of the function in the domain is 20 and the least value is –16. These occur at the endpoints of the domain.

(c)

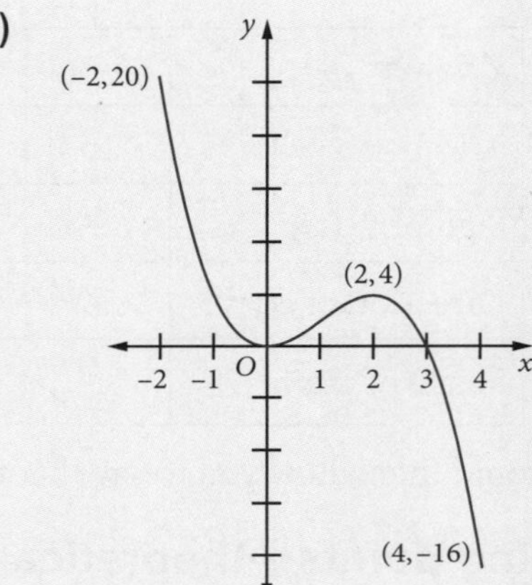

MAKING CONNECTIONS

Tangent to a curve at turning points

Explore the changing gradient of the tangent to the curve near its turning points.

EXERCISE 14.2 THE FIRST DERIVATIVE AND TURNING POINTS

1 A function is given by $f(x) = x^2 - 6x + 8$.

(a) Find $f'(x)$. **(b)** Find the coordinates of any stationary points and determine their nature.
(c) Sketch $y = f(x)$.

2 If $f'(x) = x^2 - 5x - 6$ then stationary points may occur when:

A $x = 1, -6$ **B** $x = -2, -3$ **C** $x = -1, 6$ **D** $x = -3, 2$

3 A function is given by $f(x) = x^3 - 6x^2 + 16$.

(a) Find $f'(x)$. **(b)** Find the coordinates of any stationary points and determine their nature.
(c) Sketch $y = f(x)$.

4 A function is given by $f(x) = x^3 - 9x^2 + 15x + 16$. You are asked to find the stationary points and determine their nature. Four steps in answering this question are shown below. Indicate whether each step is correct or incorrect.

(a) $f'(x) = 3x^2 - 18x + 15$ (b) Stationary points occur when $x = -1, 5$.
(c) $(1, 23)$ is a maximum turning point. (d) $(5, -9)$ is a minimum turning point.

5 A function is given by $f(x) = 2x^3 - 15x^2 + 36x$.

(a) Find $f'(x)$. (b) Find the coordinates of any stationary points and determine their nature.
(c) Sketch $y = f(x)$.

6 A function is given by $f(x) = x^3 - x^2 - x + 1$.

(a) Find $f'(x)$. (b) Find the coordinates of any stationary points and determine their nature.
(c) Sketch $y = f(x)$ over the domain $-2 \leq x \leq 2$, showing the turning points.

7 Find the maximum value of $5x - 2x^2$.

8 Find the minimum value of $x(x - 2) + 3$.

9 Sketch the curve $y = x^3 - 6x^2$ over the domain $-1 \leq x \leq 6$, showing the maximum and minimum turning points.

10 Find the turning points of $y = 2x^3 + 3x^2 - 12x + 7$. Hence sketch the curve, showing the turning points and the y-intercept.

11 Sketch the curve $y = (2 - x)(1 + x^2)$, locating the turning points and the points where it crosses the coordinate axes over the domain $-1 \leq x \leq 3$.

12 Consider the function $f(x) = 9x(x - 2)^2$, $-1 \leq x \leq 3$. Find the values of x for which:

(a) $f'(x) = 0$ (b) $f'(x) > 0$ (c) $f'(x) < 0$
(d) Sketch the graph of $f(x)$ and state its range and greatest and least values.

13 Prove that the parabola $y = ax^2 + bx + c$ has a turning point at $x = \dfrac{-b}{2a}$.

14 Show that the hyperbola $y = \dfrac{1}{x}$ has no turning points. Also show that its gradient is always negative throughout its domain.

14.3 THE SECOND DERIVATIVE AND CONCAVITY

Differentiating a function once to obtain $f'(x)$ or $\dfrac{dy}{dx}$ gives you the first derivative of the original function.

Differentiating the first derivative gives you $f''(x)$ or $\dfrac{d^2y}{dx^2}$, which is called the **second derivative** of the original function. The second derivative is the rate of change of the first derivative (i.e. of the gradient function): $\dfrac{d}{dx}\left(f'(x)\right)$. Differentiating again will give the third derivative, and so on. The differentiation process may be continued for as long as a derivative exists.

Notation

Several different notations can be used for derivatives. If $y = f(x)$, then:

- the first derivative can be written $\dfrac{dy}{dx}$, $f'(x)$, $\dfrac{d}{dx}\left(f(x)\right)$ or y'
- the second derivative can be written $\dfrac{d}{dx}\left(\dfrac{dy}{dx}\right)$, $\dfrac{d^2y}{dx^2}$, $f''(x)$, $\dfrac{d}{dx}\left(f'(x)\right)$ or y''.

Example 7

Find the second derivative of each function.

(a) $y = 4x^3 - 2x^2 + 3x + 7$ (b) $f(x) = (2x + 1)^5$ (c) $y = \dfrac{x^2}{x+1}$

Solution

(a) $y = 4x^3 - 2x^2 + 3x + 7$:

$$\frac{dy}{dx} = 12x^2 - 4x + 3$$

$$\frac{d^2y}{dx^2} = 24x - 4$$

(b) $f(x) = (2x+1)^5$:

$$f'(x) = 5(2x+1)^4 \times 2$$
$$= 10(2x+1)^4$$
$$f''(x) = 10 \times 4(2x+1)^3 \times 2$$
$$= 80(2x+1)^3$$

(c) $y = \dfrac{x^2}{x+1}$:

$$y' = \frac{2x(x+1) - x^2 \times 1}{(x+1)^2}$$
$$= \frac{2x^2 + 2x - x^2}{(x+1)^2}$$
$$= \frac{x^2 + 2x}{(x+1)^2}$$

$$y'' = \frac{(2x+2)(x+1)^2 - \left(x^2 + 2x\right) \times 2(x+1)}{(x+1)^4}$$
$$= \frac{2(x+1)^3 - 2(x+1)\left(x^2 + 2x\right)}{(x+1)^4}$$
$$= \frac{2(x+1)\left((x+1)^2 - \left(x^2 + 2x\right)\right)}{(x+1)^4}$$
$$= \frac{2\left(x^2 + 2x + 1 - x^2 - 2x\right)}{(x+1)^3}$$
$$= \frac{2}{(x+1)^3}$$

Concavity

The **concavity** of a function describes the general curvature of a graph of a non-linear function. Graphs can be 'concave upwards' and 'concave downwards':

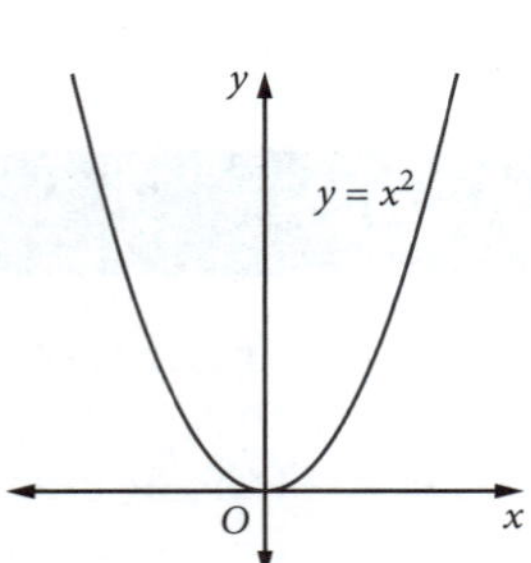

The function is concave up.
Note: This has a minimum turning point.

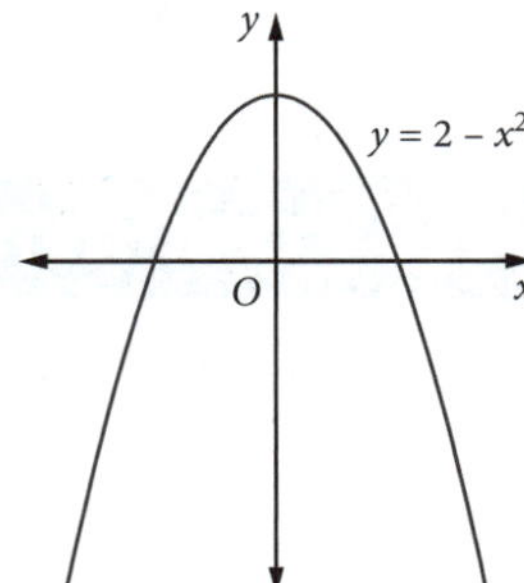

The function is concave down.
Note: This has a maximum turning point.

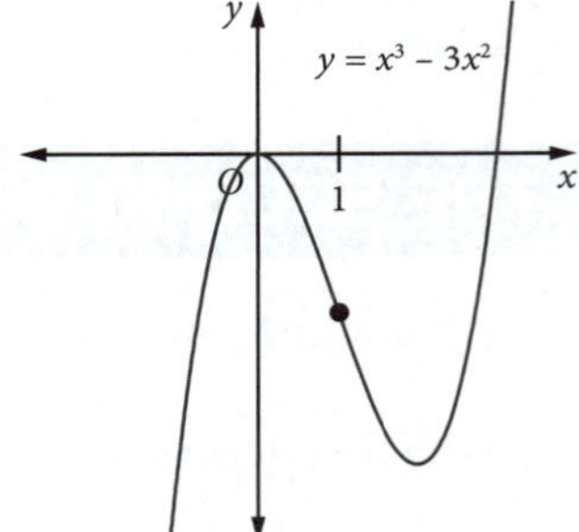

The left-hand part is concave down.
The right-hand part is concave up.
Concavity changes at $x = 1$.

The second derivative is the rate at which the first derivative is changing. This gives information about the concavity of a function.

In each diagram below, a series of tangents have been drawn.

1 The tangents have a positive gradient and the gradient is increasing from left to right.

If the gradient of the tangent $= \dfrac{dy}{dx}$, then $\dfrac{dy}{dx} > 0$.

The rate at which the gradient is increasing $= \dfrac{d}{dx}\left(\dfrac{dy}{dx}\right) = \dfrac{d^2y}{dx^2}$.

Because the gradient is increasing, $\dfrac{d^2y}{dx^2} > 0$: the curve is concave up.

2 The tangents have a negative gradient and the gradient is increasing (becoming less negative) from left to right.
If the gradient of the tangent $= \frac{dy}{dx}$, then $\frac{dy}{dx} < 0$.
Because the gradient is increasing, $\frac{d^2y}{dx^2} > 0$: the curve is concave up.

3 The tangents have a negative gradient and the gradient is decreasing (becoming more negative) from left to right.
If the gradient of the tangent $= \frac{dy}{dx}$, then $\frac{dy}{dx} < 0$.
Because the gradient is decreasing, $\frac{d^2y}{dx^2} < 0$: the curve is concave down.

4 The tangents have a positive gradient and the gradient is decreasing from left to right.
If the gradient of the tangent $= \frac{dy}{dx}$, then $\frac{dy}{dx} > 0$.
Because the gradient is decreasing, $\frac{d^2y}{dx^2} < 0$: the curve is concave down.

5 The gradient is positive.
Initially the gradient is decreasing, $\frac{d^2y}{dx^2} < 0$, but then it starts increasing, $\frac{d^2y}{dx^2} > 0$.
This means that at some point $\frac{d^2y}{dx^2} = 0$. This is also where the concavity changes from concave down to concave up, so this point is called a **point of inflection**.

6 The gradient is negative.
Initially the gradient is increasing, $\frac{d^2y}{dx^2} > 0$, but then it starts decreasing, $\frac{d^2y}{dx^2} < 0$.
This means that at some point $\frac{d^2y}{dx^2} = 0$. This is also where the concavity changes from concave down to concave up, so this is a point of inflection.

The sign of the second derivative

- If $\frac{d^2y}{dx^2} > 0$ on an interval then the curve is concave upwards on that interval.
- If $\frac{d^2y}{dx^2} < 0$ on an interval then the curve is concave downwards on that interval.
- If $\frac{d^2y}{dx^2} = 0$ at a point on the curve and the concavity changes at this point, then the point is called a **point of inflection.**

MAKING CONNECTIONS

The second derivative and points of inflection

Use graphing software to explore the changing value of the second derivative of a function.

Example 8

For what values of x is the curve given by $y = x^3 - 3x^2 + 6x + 3$:

(a) concave up **(b)** concave down?

(c) Find the coordinates of the point of inflection. **(d)** Sketch the curve.

Solution

Find $\frac{dy}{dx}$: $\frac{dy}{dx} = 3x^2 - 6x + 6$

Find $\frac{d^2y}{dx^2}$: $\frac{d^2y}{dx^2} = 6x - 6$

(a) Concave up, $\frac{d^2y}{dx^2} > 0$: $6x - 6 > 0$

$x > 1$

The curve is concave up for $x > 1$.

(b) Concave down, $\frac{d^2y}{dx^2} < 0$: $x < 1$

The curve is concave down for $x < 1$.

(c) Inflection point, $\frac{d^2y}{dx^2} = 0$: $x = 1$

$x < 1$, curve is concave down

$x > 1$, curve is concave up

$\therefore$ concavity changes at $x = 1$

$x = 1, y = 1 - 3 + 6 + 3 = 7$

$\therefore$ point of inflection is $(1, 7)$

(d)

y
(1,7)
O
x

EXERCISE 14.3 THE SECOND DERIVATIVE AND CONCAVITY

1 Find $f''(x)$ for each function.

(a) $f(x) = 3x^2 + 5x + 6$ **(b)** $f(x) = x^3 + 2x^2 + 4x + 2$ **(c)** $f(x) = 24 - x^2$

(d) $f(x) = x^5 + 2x^3 - 4x$ **(e)** $f(x) = 12 - x^4 + 2x^2$ **(f)** $f(x) = 5x - 4$

2 Given $f(x) = x^6 + 3x^3 - 4x + 2$, $f''(x) = \ldots$

A $6x^5 + 3x^2 - 4$ **B** $6x^5 + 9x^2 - 4$ **C** $30x^4 + 6x$ **D** $30x^4 + 18x$

3 Given $y = \frac{x^2 - 1}{x}$, find $\frac{d^2y}{dx^2}$. Indicate whether each statement below is a correct or incorrect step in finding $\frac{d^2y}{dx^2}$.

(a) $y = x - \frac{1}{x}$ **(b)** $\frac{dy}{dx} = \frac{x^2 - 1}{x^2}$ **(c)** $\frac{dy}{dx} = 1 + \frac{1}{x^2}$ **(d)** $\frac{d^2y}{dx^2} = \frac{-2}{x^2}$

4 Find $\frac{d^2y}{dx^2}$ given: **(a)** $y = \sqrt{x}$ **(b)** $y = \sqrt{x - 2}$ **(c)** $y = x\sqrt{x^2 + 1}$ **(d)** $y = \frac{1}{x}$

(e) $y = \frac{1}{x + 1}$ **(f)** $y = \frac{x}{x + 3}$ **(g)** $y = \frac{x^2 + 1}{\sqrt{x}}$ **(h)** $y = (x^2 - 1)\sqrt{x}$ **(i)** $y = \frac{\sqrt{x - 1}}{x + 1}$

5 For what values of x is $y = 5x^2 - 1$ concave up?

6 For what values of x is $y = 6 - 3x^2$ concave down?

7 For what values of x is the curve $y = x^3 + 6x^2 - x + 4$:

(a) concave up **(b)** concave down? **(c)** Find the coordinates of the point of inflection.

8 For what values of x is the graph of $y = \sqrt{x + 1}$ concave down? Explain why there is no point of inflection.

9 For what values of x is the curve $y = \frac{1}{x}$:

(a) concave up **(b)** concave down? **(c)** Explain why there is no point of inflection.

10 Explain why the graph of $y = \frac{1}{x^2}$ is concave up over its domain.

14.4 THE SECOND DERIVATIVE AND TURNING POINTS

Just as you compared the graphs of $f(x)$ and $f'(x)$ drawn together on the same horizontal scale, you now need to consider the graphs of $f(x)$, $f'(x)$ and $f''(x)$ together.

Consider: $y = f(x) = x^3 - x^2 - x + 1$

Differentiate: $\frac{dy}{dx} = y' = f'(x) = 3x^2 - 2x - 1$

Differentiate again: $\frac{d}{dx}\left(\frac{dy}{dx}\right) = \frac{d^2y}{dx^2} = y'' = f''(x) = 6x - 2$

For stationary points, $f'(x) = 0$: $3x^2 - 2x - 1 = 0$

Factorise: $(3x+1)(x-1) = 0$

Solve: $x = -\frac{1}{3}, 1$

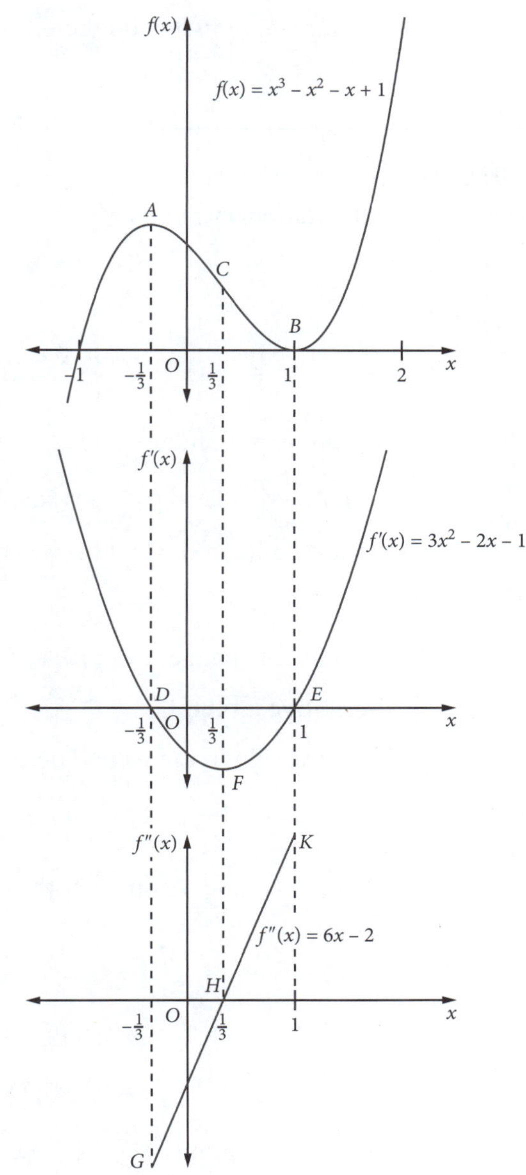

On the graph, the vertical lines ADG and BEK correspond to the lines $x = -\frac{1}{3}$ and $x = 1$ respectively.

Using the first derivative test (or the graph of $f'(x)$), you can say that A is a maximum turning point and that B is a minimum turning point.

- Consider where the line ADG cuts the three curves. The point A on $f(x)$ is a maximum turning point. The point D on $f'(x)$ is where $f'(x)$ cuts the x-axis, that is, $f'(x) = 0$. At the point G on $f''(x)$, $f''(x) < 0$. Hence the curve $y = f(x)$ is concave down at a maximum turning point.
- Consider where the line BEK cuts the three curves: at point B, $f(x)$ is a minimum turning point; at point E, $f'(x)$ cuts the x-axis, i.e. $f'(x) = 0$; at point K, $f''(x) > 0$. Hence the curve $y = f(x)$ is concave up at a minimum turning point.
- Consider where the line CFH cuts the three curves: at point C, $f(x)$ seems to have its steepest tangent; at point F, $f'(x)$ has its least value, i.e. the gradient of $f(x)$ is at its most negative (steepest); at point H, $f''(x) = 0$. The concavity changes either side of C, so C is a point of inflection on $y = f(x)$.

For $y = f(x) = x^3 - x^2 - x + 1$, you can say that the function has a maximum turning point at $\left(-\frac{1}{3}, 1\frac{5}{27}\right)$, a minimum turning point at $(1, 0)$, is concave down for $x < \frac{1}{3}$, is concave up for $x > \frac{1}{3}$ and has a point of inflection at $\left(\frac{1}{3}, \frac{16}{27}\right)$.

Second derivative test for turning points

- If $\frac{dy}{dx} = 0$ at (x_1, y_1) and $\frac{d^2y}{dx^2} < 0$ then the point (x_1, y_1) is a maximum turning point.
- If $\frac{dy}{dx} = 0$ at (x_2, y_2) and $\frac{d^2y}{dx^2} > 0$ then the point (x_2, y_2) is a minimum turning point.
- If $\frac{dy}{dx} = 0$ and $\frac{d^2y}{dx^2} = 0$ at (x_3, y_3) then the point may be a turning point **OR** it may be a horizontal point of inflection. Further investigation is needed.

Point of inflection test

- If $\frac{d^2y}{dx^2} = 0$ at (x_4, y_4) and the concavity changes either side of this point, then (x_4, y_4) is a point of inflection.
- If $\frac{dy}{dx} = 0$ at a point of inflection then it is called a horizontal point of inflection.

Example 9

Investigate the stationary points of:

(a) $y = x^4$ (b) $y = 3x^2 - 3x - x^3$

Solution

(a) $y = x^4$: $\frac{dy}{dx} = 4x^3$; $\frac{d^2y}{dx^2} = 12x^2$

For stationary points, $\frac{dy}{dx} = 0$: $x = 0$

At $x = 0$, $\frac{d^2y}{dx^2} = 0$, so you can't identify the nature of this stationary point.

Check the sign of the first derivative: For $x < 0$, e.g. $x = -1$: $\frac{dy}{dx} = -4 < 0$

For $x > 0$, e.g. $x = 1$: $\frac{dy}{dx} = 4 > 0$

The gradient changes from negative to positive through $x = 0$, so $(0, 0)$ is a minimum turning point.

Alternative method:

Check the sign of the second derivative: For $x < 0$, e.g. $x = -1$: $\frac{d^2y}{dx^2} = 12 > 0$

For $x > 0$, e.g. $x = 1$: $\frac{d^2y}{dx^2} = 12 > 0$

The curve is concave up on both sides of the stationary point, so $(0, 0)$ is a minimum turning point.

(b) $y = 3x^2 - 3x - x^3$: $\frac{dy}{dx} = 6x - 3 - 3x^2$ $\frac{d^2y}{dx^2} = 6 - 6x$

$$= -3(x^2 - 2x + 1)$$

$$= -3(x-1)^2$$

For stationary points, $\frac{dy}{dx} = 0$: $(x-1)^2 = 0$, hence $x = 1$

At $x = 1$, $\frac{d^2y}{dx^2} = 0$, so you can't identify the nature of this stationary point.

Check the sign of the second derivative: For $x < 1$, e.g. $x = 0.1$: $\frac{d^2y}{dx^2} = 6 - 0.6$

$$= 5.4 > 0$$

For $x > 1$, e.g. $x = 1.1$: $\frac{d^2y}{dx^2} = 6 - 6.6$

$$= -0.6 < 0$$

The concavity changes either side of $x = 1$, so $(1, -1)$ is a horizontal point of inflection.

Example 10

Consider the graph of $y = 2x^4 - x + 1$.

(a) Find any turning points and points of inflection. **(b)** For what values of x is the curve concave up?

(c) On the same set of axes sketch y, $\frac{dy}{dx}$ and $\frac{d^2y}{dx^2}$.

Solution

(a) $y = 2x^4 - x + 1$: $\quad \frac{dy}{dx} = 8x^3 - 1$

For stationary points, $\frac{dy}{dx} = 0$: $\quad 8x^3 - 1 = 0$, hence $x^3 = \frac{1}{8}$ and so $x = \frac{1}{2}$.

At $x = \frac{1}{2}$, $y = 2 \times \frac{1}{2^4} - \frac{1}{2} + 1$

$= 1\frac{3}{8}$ $\quad \therefore$ stationary point is at $\left(\frac{1}{2}, 1\frac{3}{8}\right)$.

Find $\frac{d^2y}{dx^2}$: $\quad \frac{d^2y}{dx^2} = 24x^2$

At $x = \frac{1}{2}$, $\frac{d^2y}{dx^2} = 6 > 0 \quad \therefore \left(\frac{1}{2}, 1\frac{3}{8}\right)$ is a minimum turning point.

$\frac{d^2y}{dx^2} = 0$ at $x = 0$, so check concavity to see if there is an inflection point at $(0, 1)$:

$x < 0$: $\frac{d^2y}{dx^2} > 0 \qquad x > 0$: $\frac{d^2y}{dx^2} > 0$

The concavity does not change either side of $x = 0$, so the curve does not have a point of inflection at $(0, 1)$.

(b) Concave up when $\frac{d^2y}{dx^2} > 0$, hence concave up for all real x.

(c)

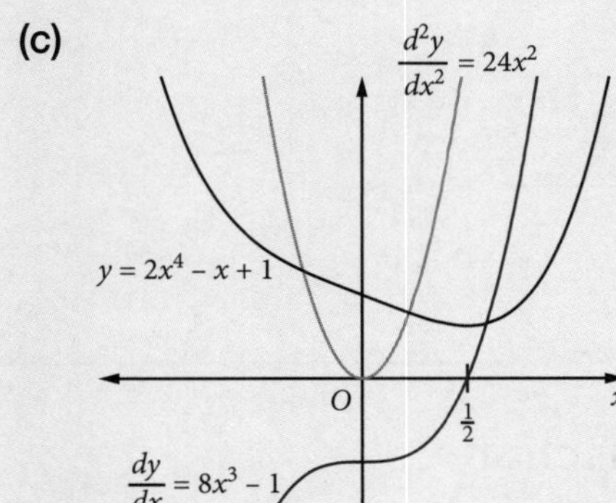

This example shows that $\frac{d^2y}{dx^2} = 0$ is not enough to confirm a point of inflection. You must also check that the concavity changes at the point.

Example 11

Sketch the graph of $y = x^3(4 - x)$, showing any turning points and points of inflection. For what values of x is the curve concave down?

Solution

$y = 4x^3 - x^4$: $\quad \dfrac{dy}{dx} = 12x^2 - 4x^3 = 4x^2(3-x)$ $\quad \dfrac{d^2y}{dx^2} = 24x - 12x^2 = 12x(2-x)$

For stationary points, $\dfrac{dy}{dx} = 0$: $\quad 4x^2(3-x) = 0$

$x = 0$ or 3

$y = 0$ or 27 $\quad \therefore$ stationary points at $(0, 0)$ and $(3, 27)$

At $x = 0$, $\dfrac{d^2y}{dx^2} = 0$. Investigate further: $\quad x = -1$: $\dfrac{d^2y}{dx^2} = -12(2+1) < 0$

$x = 1$: $\dfrac{d^2y}{dx^2} = 12(2-1) > 0$

The sign of $\dfrac{d^2y}{dx^2}$ changes, so the concavity changes. Hence $(0, 0)$ is a horizontal point of inflection.

At $x = 3$: $\quad \dfrac{d^2y}{dx^2} = 36(2-3) < 0$ $\quad$ Hence $(3, 27)$ is a relative maximum turning point.

For points of inflection, $\dfrac{d^2y}{dx^2} = 0$: $\quad 12x(2-x) = 0$

$x = 0$ or 2

$(0, 0)$ has already been identified as a horizontal point of inflection. For $(2, 16)$:

$x = 1$: $\dfrac{d^2y}{dx^2} = 12(2-1) > 0$ $\quad x = 3$: $\dfrac{d^2y}{dx^2} = 36(2-3) < 0$

The sign of $\dfrac{d^2y}{dx^2}$ changes, so the concavity changes. Hence $(2, 16)$ is a point of inflection.

Curve is concave downwards for $x < 0$ and $x > 2$.

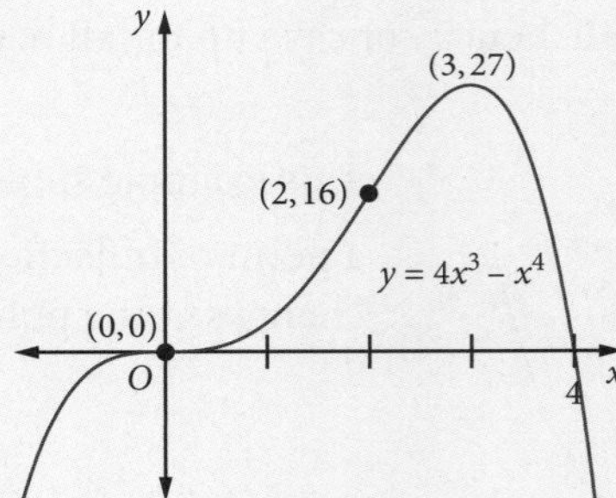

Summary of tests for turning points and points of inflection

Stationary point: $\dfrac{dy}{dx} = 0$

Using the first derivative only:

Turning point: $\dfrac{dy}{dx} = 0$, $\dfrac{dy}{dx}$ changes sign on passing through the stationary point.

Maximum turning point: $\dfrac{dy}{dx} = 0$, $\dfrac{dy}{dx}$ changes sign from positive to negative on passing through the stationary point.

Minimum turning point: $\dfrac{dy}{dx} = 0$, $\dfrac{dy}{dx}$ changes sign from negative to positive on passing through the stationary point.

Using the first and second derivatives:

Turning point: $\dfrac{dy}{dx} = 0$, $\dfrac{d^2y}{dx^2}$ does not change sign on passing through the stationary point.

Maximum turning point: $\frac{dy}{dx}=0,\ \frac{d^2y}{dx^2}<0$ at the stationary point.

Minimum turning point: $\frac{dy}{dx}=0,\ \frac{d^2y}{dx^2}>0$ at the stationary point.

Point of inflection: $\frac{d^2y}{dx^2}=0,\ \frac{d^2y}{dx^2}$ changes sign on passing through the stationary point (i.e. concavity changes).

Special situation:

$\frac{dy}{dx}=0,\ \frac{d^2y}{dx^2}=0$ at a point: may be a turning point **or** a horizontal point of inflection. You must check to see if either the gradient or the concavity changes.

Global maxima and minima

In Example **10** the minimum turning point gives the least value of the function over its domain. It is called the global minimum value of the function. The function has no greatest value.

In Example **11** the maximum turning point gives the greatest value of the function over its domain. It is called the global maximum value of the function. The function has no least value.

The greatest and least values of a function, also called the global maxima and global minima, may occur at the endpoints of the domain or at the turning points of the function.

When asked to find the global maximum or global minimum value of a function, as well as considering the values of the function at the turning points, you also need to find the value of the function at the endpoints of the given domain (or the natural domain if no restrictions are given). You then compare the value of the function at these points to find the global maximum and the global minimum.

Example 12

(a) Find the coordinates of the stationary points of $y=x^2e^x$ and determine their nature.

(b) Find the coordinates of any points of inflection.

(c) Sketch the graph of this function.

(b) Find the global maximum and global minimum values of this function.

Solution

(a)

$$y=x^2e^x$$

Differentiate: $\frac{dy}{dx}=2xe^x+x^2e^x$

Factorise: $\frac{dy}{dx}=xe^x(2+x)$

For stationary points, $\frac{dy}{dx}=0$: $x=0,-2$

$$y=0,\ 4e^{-2}$$

Find second derivative: $\frac{d^2y}{dx^2}=2e^x+2xe^x+2xe^x+x^2e^x$

Factorise: $\frac{d^2y}{dx^2}=e^x\left(2+4x+x^2\right)$

$x=0$: $\frac{d^2y}{dx^2}=2>0$ so minimum turning point at (0, 0).

$x=-2$: $\frac{d^2y}{dx^2}=-2e^{-2}<0$ so maximum turning point at $(0,\ 4e^{-2})$.

(b) Require $\dfrac{d^2y}{dx^2} = 0$: $\quad e^x\left(2 + 4x + x^2\right) = 0$

$e^x > 0$: $\quad x^2 + 4x + 2 = 0$

$$x = \frac{-4 \pm \sqrt{8}}{2}$$

$$= -2 \pm \sqrt{2}$$

$$\approx -0.586, -3.41$$

$x = -1$: $\quad \dfrac{d^2y}{dx^2} = -e^{-1} < 0$

$x = 0$: $\quad \dfrac{d^2y}{dx^2} = 2 > 0$ and concavity changes at $x = -0.586$

Hence a point of inflection when $x = -0.586$

$x = -4$: $\quad \dfrac{d^2y}{dx^2} = 2e^{-4} > 0$

$x = -3$: $\quad \dfrac{d^2y}{dx^2} = -e^{-3} < 0$ and concavity changes at $x = -3.41$

Hence a point of inflection when $x = -3.41$

The points of inflection are (−0.586, 0.191) and (−3.41, 0.384).

(c)

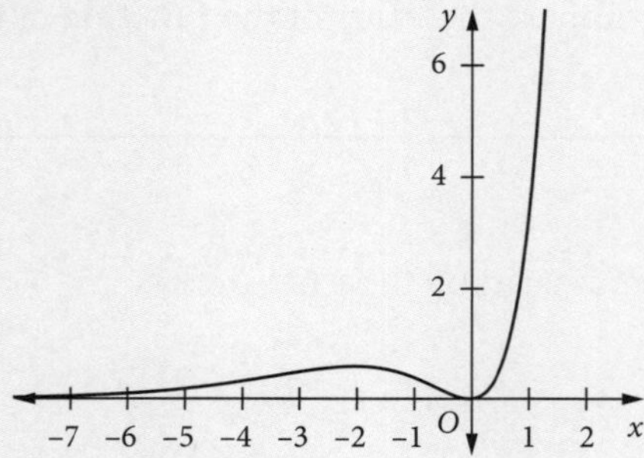

(d) From the graph the global minimum value is 0. There is no global maximum value.
Algebraically, as $x \to -\infty$, $y \to 0$ from above (asymptote).
As $x \to \infty$, y increases without bound. Hence the global minimum value occurs at the minimum turning point.

EXERCISE 14.4 THE SECOND DERIVATIVE AND TURNING POINTS

1 For $y = 2x^3 + 3x^2 - 12x + 2$, find any stationary points and determine their nature. Sketch the curve, showing the turning points and any points of inflection.

2 For what values of x is the graph of the function $f(x) = x^3 - 3x^2 + 1$ concave up?

A $x > 1$ **B** $x > -2$ **C** $x < 1$ **D** $x < -2$

3 Find the local maxima, minima and points of inflection of $f(x) = x^2(3 - x)$. Sketch the graph of f.

4 Sketch the graph of $8y = 8 + 8x^2 - x^4$, showing any turning points. What is the greatest value of the function?

5 A function $f(x)$ is defined by $y = x^3(x - 2)$.

(a) Find the coordinates of the turning points of $y = f(x)$.

(b) Find the coordinates of the points of inflection.

(c) Hence sketch the graph of $y = f(x)$, showing the turning point, the points of inflection and the points where the curve meets the x-axis.

(d) What is the minimum value of $f(x)$ for $-1 \le x \le 3$?

6 If $f(x) = x^3 - 6x^2 + 2$, find the values of x for which: **(a)** $f''(x) = 0$ **(b)** $f''(x) > 0$ **(c)** $f''(x) < 0$

7 Find the coordinates of the points of inflection of $y = x^4 - 2x^3 - 12x^2$. For what values of x is the curve concave up?

8 Let $f(x) = x^4 - x^2$.

(a) Find the coordinates of the points where the curve crosses the axes.
(b) Find the coordinates of the stationary points and determine their nature.
(c) Find the coordinates of the points of inflection.
(d) Sketch the graph of $y = f(x)$ for $-1.5 \le x \le 1.5$, indicating clearly the intercepts, stationary points and points of inflection.
(e) For what values of x is the curve concave down?

9 The graph represents a function of the form $y = ax^3 + bx^2 + cx + d$, where a, b, c and d are real numbers, $a \neq 0$. $(1, 2)$ is a point of inflection on the curve. Indicate whether each statement below is correct or incorrect for this graph.

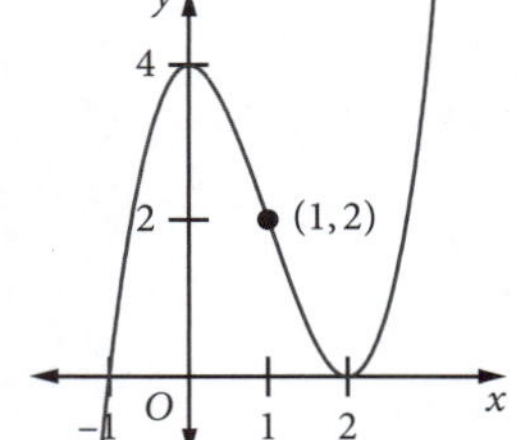

(a) The curve is concave up for $x > 1$.
(b) $\frac{dy}{dx} = mx(x-2)$ where m is a positive constant. **(c)** The least value of y is zero.
(d) The coordinates of the point of inflection can be found by taking the average of the coordinates of the maximum and minimum turning points.

10 Find the turning points and points of inflection of $y = -x^3 + 3x^2 - 3x$ and sketch its graph. Show that it crosses the x-axis at one point only. Show that $\frac{dy}{dx} < 0$ for all x except $x = 1$.

11 **(a)** Find the stationary points of $y = x^4 - 4x + 3$ and determine their nature.
(b) Show that $y = x^4 - 4x + 3$ has no points of inflection. Comment on the concavity of the curve.
(c) What is the global minimum value of this function?

12 **(a)** Find the greatest and least values of the function $y = x^3 - 6x^2 + 6x$ over the domain $[0, 6]$.
(b) Sketch the graph of the function.

13 The revenue function for a magazine is given by $R = 4500x - 500x^2$, where x is the cost per issue of the magazine. What will be the cost per issue of the magazine to achieve maximum revenue?

14 The revenue equation for a manufacturer is $R = \frac{80x - x^2}{4}$, where x is the number of units sold. How many units must be sold to achieve maximum revenue?

15 A supplier has a monopoly on sales of books. The supplier's profit function is given by $P = 396x - 2.2x^2 - 400$, where x is the number of books sold.

(a) How many books must the supplier sell to maximise the profit?
(b) What is the maximum profit?
(c) If the government imposes a new 'monopoly tax' of \$22 per book on the supplier, what is the new profit equation?
(d) Under the monopoly tax, how many books must the supplier now sell to maximise the profit? What is the new maximum profit?

16 $G = f(t)$ is the Gross Domestic Product (GDP) for Australia, where t is the number of years after 2012. The GDP growth rate $f'(t)$ for three years is given in the table.

Year	2012	2013	2014
$f'(t)$	3.9	3.2	3.1

(a) Is $f'(t)$ a decreasing or increasing function?
(b) What can you say about change in $f(t)$? **(c)** Is $f(t)$ concave up or concave down?
(d) There are two different predictions for the rate of growth in 2015: that the growth rate will be 3.0, or that the growth rate will be 3.2. Sketch $G = f(t)$ for each growth rate.
(e) For which growth rate does the concavity of $f(t)$ change? What is the name of the kind of point at which this change occurs?
(f) If the growth rate for 2015 is actually 2.9, what can you say about the GDP for Australia?

14.5 PROBLEM SOLVING WITH DERIVATIVES

A function may not always be given algebraically. Sometimes you must interpret the information given to construct the function. Remember to show clearly what each variable represents. It often helps to draw a diagram of the situation being considered.

Example 13

A piece of wire 12 cm long is bent in the shape of a rectangle. Find the maximum area of the rectangle.

Solution

Let $A(x)$ cm^2 be the area of the rectangle. Let the side lengths be x cm and y cm.

Because the wire is 12 cm long: $2x + 2y = 12 \quad \therefore y = (6 - x)$

Because the wire is 12 cm, the longest side of the rectangle cannot be longer than 6 cm.

Area of rectangle: $A(x) = x(6 - x)$ for $0 < x < 6$

$A(x) = 6x - x^2$

Differentiate: $A'(x) = 6 - 2x$

For stationary points, $A'(x) = 0$: $6 - 2x = 0$

$x = 3$

Differentiate again: $A''(x) = -2 < 0$ for all x

Hence $A(x)$ is concave down for all values of x in the domain. It has a maximum value when $x = 3$.

$A(3) = 3 \times 3 = 9\text{ cm}^2$

The maximum area of the rectangle is 9 cm^2.

Example 14

A sheet of cardboard measures 15 cm by 7 cm. Four equal squares are cut out of the corners and the sides are turned up to form an open rectangular box. Find the edge length of the squares that were cut out to give the box a maximum volume.

Solution

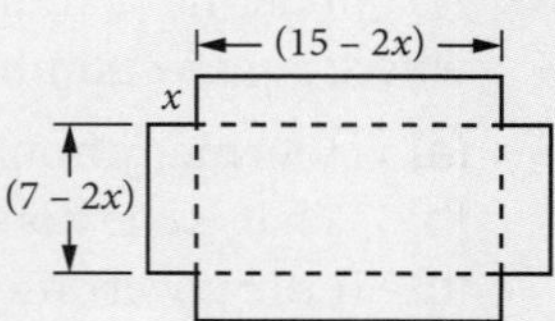

Let the edge length of the squares that were cut out be x cm.

The dimensions of the base of the box will be $(15 - 2x)$ cm and $(7 - 2x)$ cm.

The height will be x cm.

Let the volume of the box be $V(x)$ cm^3. $\quad \therefore V(x) = x(15 - 2x)(7 - 2x)$

$= 4x^3 - 44x^2 + 105x$

Differentiate: $V'(x) = 12x^2 - 88x + 105$

For stationary points, $V'(x) = 0$: $12x^2 - 88x + 105 = 0$

$(2x - 3)(6x - 35) = 0$

$\therefore x = 1\frac{1}{2}$ or $5\frac{5}{6}$

But x must be less than half the shortest side, i.e. $x < 3.5$, so we can disregard $x = 5\frac{5}{6}$. The only possible value for x is $x = 1.5$.

Use the first derivative test:

For $x < 1.5$, test $x = 1.4$: $\quad V'(1.4) = 12 \times 1.4^2 - 88 \times 1.4 + 105 = 5.32 > 0$

For $x > 1.5$, test $x = 1.6$: $\quad V'(1.6) = 12 \times 1.6^2 - 88 \times 1.6 + 105 = -5.08 < 0$

$V'(x)$ changes from positive to negative on passing through $x = 1.5$, so a maximum volume occurs when $x = 1.5$.

$V(1.5) = 1.5(15 - 3)(7 - 3) = 72\text{ cm}^3$

A graph of $V(x)$ for $0 \le x \le 7$ helps to see what is happening.

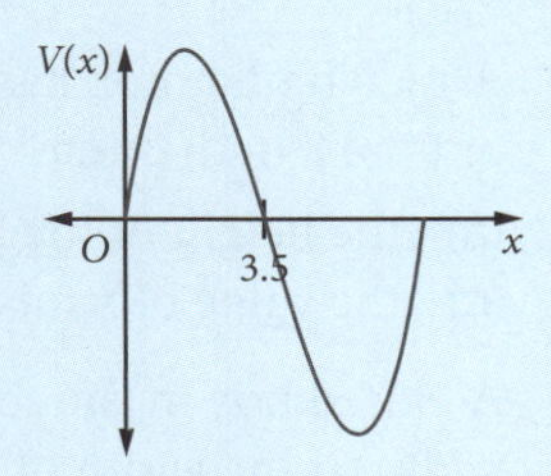

Because $V(x) > 0$, the domain of the function is $0 < x < 3.5$.

The part of the graph below the x-axis is not relevant to the problem.

EXERCISE 14.5 PROBLEM SOLVING WITH DERIVATIVES

1 A rectangular block of land is enclosed by 160 m of fencing. If the breadth of the block is x m:

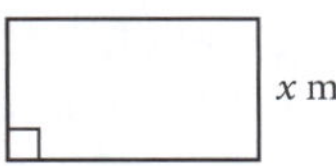

(a) express the length of the block in terms of x
(b) find the function $A(x)$ for the area of the block
(c) find the maximum area of the block that can be fenced using this fencing.

2 The sum of two numbers is 12. If one number is x, the value of x for which the product of the two numbers is a maximum is:

A 3 **B** 6 **C** 12 **D** 36

3 A rectangular block of land has one side along a river. The other three sides are to be fenced using 160 m of fencing. If the breadth of the block is x m:

(a) express the length of the block in terms of x
(b) find the function $A(x)$ for the area of the block
(c) find the maximum area of the block that can be fenced using this fencing.

x m

River

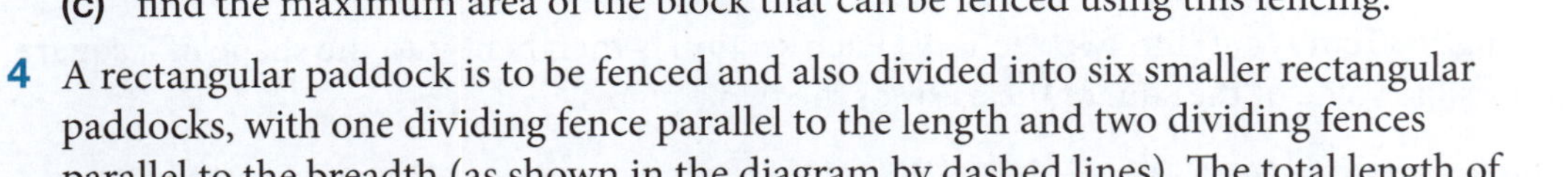

4 A rectangular paddock is to be fenced and also divided into six smaller rectangular paddocks, with one dividing fence parallel to the length and two dividing fences parallel to the breadth (as shown in the diagram by dashed lines). The total length of fencing to be used is 120 m. If the width of the paddock is x m and the breadth is y m, find:

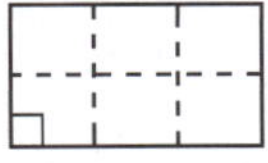

(a) an expression giving y as a function of x
(b) the function $A(x)$ for the area of the original paddock
(c) the maximum possible area of the paddock.
(d) To allow access to the paddocks, six gates are to be added to the fences. Each gate is 3 m wide. What is the new maximum area of the large paddock?

5 $ABCD$ is a square of unit length. Points E and F are on the sides AB and AD respectively so that $AE = AF = x$.

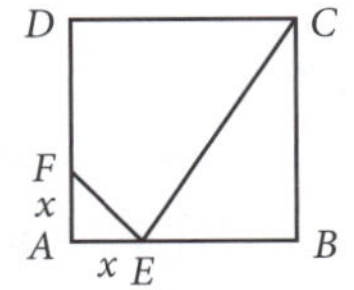

(a) Express the area of the quadrilateral $CDFE$ as a function of x.
(b) Find the greatest area that the quadrilateral can have.

6 A rectangular sheet of cardboard measures 16 cm by 6 cm. Equal squares are cut out of each corner and the sides are turned up to form an open rectangular box. What is the maximum volume of the box?

7 A block of wood in the shape of a cuboid has square ends of edge length x cm. The length of the block is y cm. The sum of length of the block and the perimeter of one end is 12 cm.

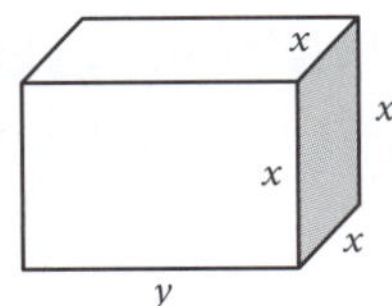

(a) Express y in terms of x.
(b) Find the volume V as a function of x.
(c) What is the largest possible volume of the block?

8 A box in the shape of a cuboid with a square base is to be made so that the sum of its dimensions $(l+b+h)$ is 20 cm. Find its maximum volume.

9 A rectangular box has a square base of edge length x cm. Its framework of 12 edges is constructed from wire of total length 36 cm. Find:

(a) the height of the box in terms of x **(b)** the volume of the box in terms of x
(c) the value of x for which the volume is a maximum.

10 A closed box in the shape of a cuboid has a total surface area of 216 cm^2 and a base length that is twice the width. If the width of the base is x cm, find:

(a) the length of the base and the height in terms of x **(b)** the volume of the box in terms of x
(c) the maximum volume of the box.

11 A rectangular field is to be fenced along three sides using 300 m of fencing. The length of each equal end is x m and the length of the other side is y m. To find the dimensions of the field if its area is as large as possible, the following statements are made. Indicate whether each statement is correct or incorrect.

(a) $y = 300 - 2x$ **(b)** $A = 300x - 2x^2$ **(c)** $x = 75$ **(d)** Maximum area $= 11\,250$ m^2

12 The diagonal of the base of a box in the shape of a cuboid has a length of 10 cm. One edge of the base has a length of x cm, as shown in the diagram.

(a) Express, in terms of x, the length of the other edge of the base.
(b) The height of the box is equal to the length of this other edge. Find the volume of the box in terms of x.
(c) Calculate the maximum volume of the box.

13 The slant edge AB of a right circular cone is 6 cm. The vertical height of the cone is x cm, as shown in the diagram.

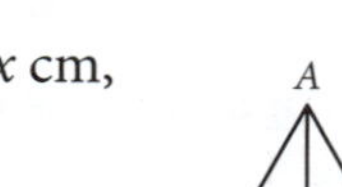

(a) Express the radius of the base in terms of x.
(b) Express the volume of the cone in terms of x.
(c) Find the vertical height of the cone when the volume is a maximum.

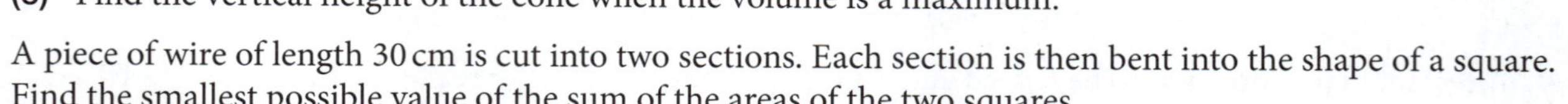

14 A piece of wire of length 30 cm is cut into two sections. Each section is then bent into the shape of a square. Find the smallest possible value of the sum of the areas of the two squares.

15 A block of metal is cast into the shape of a right cylinder with a total surface area of 20π cm^2. The radius of the base is r cm and the height is h cm. The total surface area of a cylinder is given by $A = 2\pi r^2 + 2\pi rh$.

(a) Express h in terms of r. **(b)** Express the volume V in terms of r.
(c) Find the value of r for which the volume is greatest.

16 A piece of wire of length 50 cm is cut into two sections. One section is used to construct a rectangle whose dimensions are in the ratio 3 : 1; the other section is used to construct a square. Find the dimensions of the rectangle and the square so that the total enclosed area is a minimum.

17 Whale-watching boat trips go to sea with 20 or more passengers. For 20 passengers the charge is \$380 per person. For groups of more than 20, the price per person is reduced by \$12 for each additional person over 20 passengers.

(a) Show that the revenue function for the boat trip is given by $R = 620n - 12n^2$, where R is the revenue in dollars and n is the number of people in the group, $n \geq 20$.
(b) What number of passengers will produce the greatest revenue for the company?

18 Wobbly Skateboards looked at their sales in relation to their advertising budget. They found that the relationship between sales, $f(x)$, and thousands of dollars spent on advertising, x, was given by $f(x) = \frac{x^3}{3} - \frac{45x^2}{2} + 450x$, $10 \leq x \leq 40$.

(a) What number of advertising dollars can be expected to produce a maximum number of sales?
(b) What sales can be expected for that amount of advertising?

19 A company finds that the function $f(x) = x^3 - 96x^2 + 2880x$ provides a good approximation for their profit $f(x)$ in dollars, where x is the advertising expenditure in thousands of dollars.

(a) What expenditure on advertising would produce the maximum profit?

(b) What is this maximum profit?

20 A cylinder is inscribed in a sphere of radius a, centred at O. The height of the cylinder is $2h$ and the radius of the base is r, as shown in the diagram.

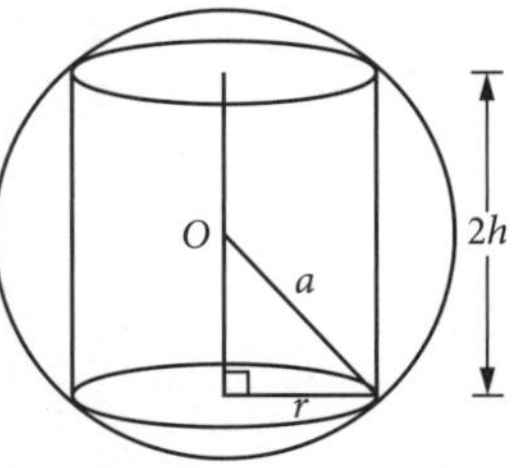

(a) Show that the volume V of the cylinder is given by $V = 2\pi r^2\sqrt{a^2 - r^2}$.

(b) Find the value of r for which the volume of the cylinder is a maximum. Explain why your value of r gives the maximum volume.

14.6 APPLICATIONS OF THE EXPONENTIAL AND LOGARITHMIC FUNCTIONS

Example 15

For the function $f(t) = 2te^{-0.5t}$, find the value of t for which $f(t)$ has a maximum and hence calculate the maximum value. Sketch the graph of $f(t)$.

Solution

$f(t) = 2te^{-0.5t}$ Let $u = t$, $v = e^{-0.5t}$

$$f'(t) = 2\left(e^{-0.5t} + t \times \left(-\frac{1}{2}\right)e^{-0.5t}\right)$$

$$= e^{-0.5t}(2 - t)$$

For stationary points, $f'(t) = 0$: $e^{-0.5t}(2 - t) = 0$

But $e^{-0.5t} > 0$ for all t, so $t = 2$ is the only solution and $f(2) = \frac{4}{e}$

For $t < 2$: $f'(t) > 0$

For $t > 2$: $f'(t) < 0$

Gradient changes from positive to negative as x increases, so $\left(2, \frac{4}{e}\right)$ is a maximum turning point.

The maximum value of the function is $\frac{4}{e} \approx 1.472$

$f(t) = 0$ at $t = 0$ as $e^{-0.5t} > 0$ for all t.

$t < 0$, $f(t) < 0$ $t < 2$, $f'(t) > 0$

$t > 0$, $f(t) > 0$ $t > 2$, $f'(t) < 0$

$t \to \infty$, $f(t) \to 0$ from above

$t = 0$ is a horizontal asymptote

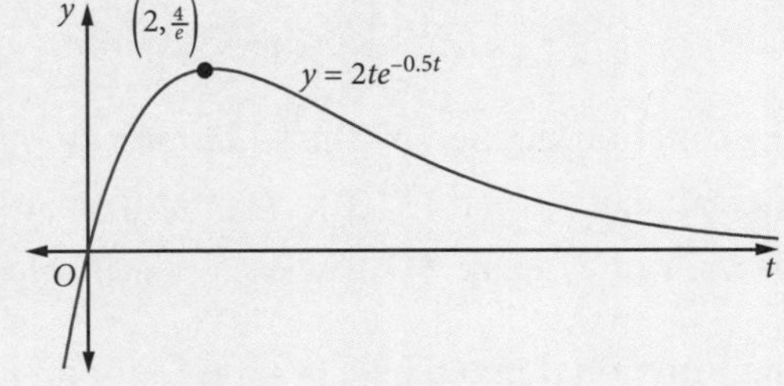

Example 16

Find the coordinates of any maximum or minimum turning points of the curve $y = \frac{\ln x}{e^x}$ given that $x \ln x = 1$ when $x = 1.76$.

Solution

$$y = \frac{\ln x}{e^x} = \ln x \times e^{-x}$$

$$\frac{dy}{dx} = \ln x \times -e^{-x} + \frac{1}{x} \times e^{-x}$$

$$= \frac{1 - x\ln x}{xe^x}$$

For stationary points, $\frac{dy}{dx} = 0$:

$$\frac{1 - x\ln x}{xe^x} = 0$$

$$1 - x\ln x = 0$$

$$x\ln x = 1$$

$$\therefore x = 1.76$$

$x = 1.7$: $\quad \frac{dy}{dx} = \frac{1 - 1.7\ln 1.7}{1.7e^{1.7}} \approx 0.01 > 0$

$x = 1.8$: $\quad \frac{dy}{dx} = \frac{1 - 1.8\ln 1.8}{1.8e^{1.8}} = -0.005 < 0$

The gradient changes from positive to negative on passing through the stationary point so the function has a maximum value when $x = 1.76$.

$x = 1.76, y = 0.097$

The coordinates of the maximum turning point are (1.76, 0.097).

EXERCISE 14.6 APPLICATIONS OF THE EXPONENTIAL AND LOGARITHMIC FUNCTIONS

1 Find the minimum value of $(x - 2)e^x$.

2 Find the coordinates of the turning point of the curve $y = xe^{-0.5x}$ and state whether it is a maximum or minimum. Find the values of x for which:

(a) $y > 0$ (b) $\frac{dy}{dx} > 0$

3 Consider the function defined by the rule $f(x) = 3 - e^{-x}, x \geq 0$.

(a) Find the value of $f(0)$ and $f'(0)$.
(b) Show that $f'(x) > 0$ for all values of x in the domain.
(c) What is the value of $\lim_{x \to \infty} f(x)$?
(d) Sketch the graph of $f(x)$.

4 Consider the function defined by $f(x) = e^{-x^2}$ for all values of x.

(a) Find $f'(x)$.
(b) Find the values of x for which: (i) $f'(x) = 0$ (ii) $f'(x) > 0$ (iii) $f'(x) < 0$.
(c) Sketch the graph of the function.

5 The concentration of a certain drug in the blood at a time t hours after taking the dose is x units, where $x = 0.3te^{-1.1t}$.

(a) Determine the maximum concentration and the time at which this is reached.
(b) Plot the graph of $x = 0.3te^{-1.1t}$ for $t = 0, 0.1, 0.5, 1, 2, 3$ using graph paper or graphing software.
(c) This drug kills germs only while its concentration is at least 0.06 units. From the graph, find the length of time during which the drug will kill germs.

6 For $y = e^t + 4e^{-t}$, find the minimum value of y. Indicate whether each of the statements below is a correct or incorrect step in solving this problem.

(a) $y' = e^t - 4e^{-t}$ (b) Stationary point when $e^t = \pm 2$

(c) $y'' = e^t + 4e^{-t}$ (d) Minimum value is 4

7 Sketch the graph of $f(t) = \dfrac{5}{2 + 3e^{-t}}$, $t \geq 0$.

(a) Show that $f'(t) > 0$ for all values of t in the domain. (b) Find $\lim\limits_{t \to \infty} f(t)$.

(c) State the range of the function.

8 The rectangle $PQRS$ has two vertices on the x-axis and two on the curve $y = e^{-x^2}$, as shown in the diagram. Find:

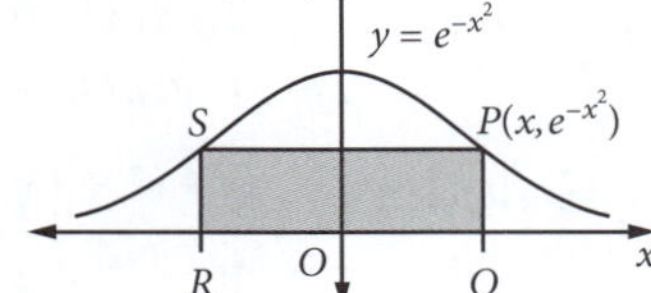

(a) the value of x for which the rectangle has a maximum area

(b) the maximum area of the rectangle.

9 Find the coordinates of any maximum or minimum turning points on the curve $y = \dfrac{\ln x}{x}$.

10 Show that $y = e^{2x} + 4e^{-2x}$ has a minimum value when $x = \dfrac{\ln 2}{2}$. What is the minimum value?

11 Find the minimum value of y if $y = x \ln x$ for $x > 0$.

12 (a) Find all values of x between 0 and 2π for which $\log_e (\sin x)$ is defined.

(b) Find the maximum value of $\log_e (\sin x)$ and when it occurs.

13 The sales revenue (in dollars) that a manufacturer receives for selling x units of a certain product can be approximated by the function $R(x) = 900 \log_e \left(1 + \dfrac{x}{300}\right)$.

Each unit costs the manufacturer \$1 to produce and the initial cost of adjusting the machinery for production is \$200, so that the total cost (in dollars) of the production of x units is $C(x) = 200 + x$.

(a) Find the profit, $P(x)$ dollars, obtained by the production and sale of x units and find the number of units which should be produced and sold for maximum profit. Calculate this maximum profit.

(b) Using technology, draw on the same diagram the graphs of $C(x)$ and $R(x)$ for $0 \leq x \leq 1500$. Show on the scale every 100 units on each axis. Use your graph to determine how many units must be produced to break even (i.e. where revenue first equals cost).

(c) If before being sold, the units are packaged in batches of ten, estimate from your graph minimum and maximum numbers of batches that can be produced and sold so that a profit is made.

(d) On the same diagram draw the graphs of $y = C(x)$, $y = R(x)$ and $y = P(x)$ and determine where $P(x) \geq 0$.

14 When a uniform chain is suspended at two fixed points, it hangs in a catenary whose equation is $y = \dfrac{1}{2a}\left(e^{ax} + e^{-ax}\right)$

(a) Sketch the curve when $a = 0.5$ and the fixed points are at the same horizontal level and 8 units apart.

(b) Find the sag at the centre.

(c) Find the angle of inclination of the chain at the supports.
(Using graphing software, set a slider for a and observe what happens as a changes.)

15 $f(x)$ is defined as $f(x) = e^{-x} \cos x$ in the domain $[0, \pi]$.

(a) Find $f(0)$, $f\left(\dfrac{\pi}{2}\right)$ and $f(\pi)$. (b) Find $f'(x)$.

(c) Evaluate $f'(0)$ and $f'\left(\dfrac{3\pi}{4}\right)$. (d) Sketch the graph of $y = f(x)$.

14.7 FURTHER APPLICATIONS OF TRIGONOMETRIC FUNCTIONS

Example 17

Given the function $y = 3\sin t$, for $t > 0$:

(a) Sketch the graph of this function for $0 \le t \le 4\pi$.

(b) Find the greatest and least values of the function and where they occur for $0 \le t \le 4\pi$.

(c) Describe the behaviour of the curve if y is the distance in metres to the right of a fixed point after a time t hours.

Solution

(a) $y = 3\sin t$, for $0 \le t \le 4\pi$:

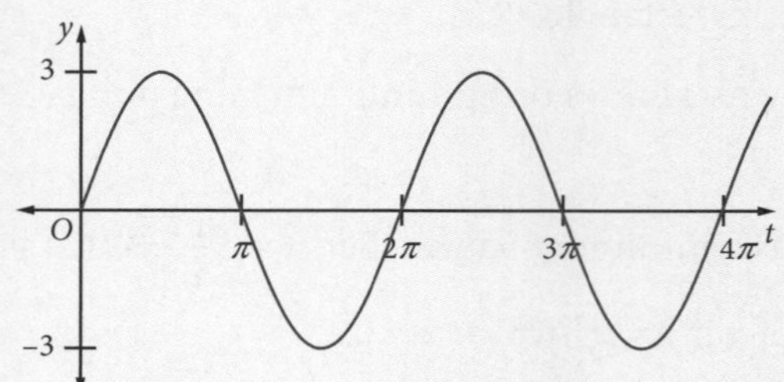

(b) $\frac{dy}{dt} = 3\cos t$. For stationary points $\frac{dy}{dt} = 0$, so $\cos t = 0$.

For $0 \le t \le 4\pi$: $t = \frac{\pi}{2}, \frac{3\pi}{2}, \frac{5\pi}{2}, \frac{7\pi}{2}$

$\frac{d^2y}{dt^2} = -3\sin t$: at $t = \frac{\pi}{2}, \frac{5\pi}{2}$ we have $\frac{d^2y}{dt^2} < 0$ $\therefore$ maximum turning point

at $t = \frac{3\pi}{2}, \frac{7\pi}{2}$ we have $\frac{d^2y}{dt^2} > 0$ $\therefore$ minimum turning point

At $t = \frac{\pi}{2}, \frac{5\pi}{2}$, $y = 3$. The greatest value of the function is 3 and occurs at $t = \frac{\pi}{2}, \frac{5\pi}{2}$.

At $t = \frac{3\pi}{2}, \frac{7\pi}{2}$, $y = -3$. The least value of the function is -3 and occurs at $t = \frac{3\pi}{2}, \frac{7\pi}{2}$.

(c) As t increases, y increases and then decreases in a repeating pattern. The function is periodic with a period of 2π hours (i.e. about 6 h 17 min). Its greatest distance from the fixed point is 3 metres in either direction. After moving 3 metres to the right, the object moves back through its starting point until it reaches a point 3 metres to the left; it then starts moving back to the right again until the pattern repeats.

The kind of wave-like repeating movement described in Example **17** is called **simple harmonic motion**. It is a type of motion that can be described using sine and cosine functions. It occurs naturally in the motion of pendulums and masses on springs, and it can also be used to model ocean tides and other wave-like movements.

EXERCISE 14.7 FURTHER APPLICATIONS OF TRIGONOMETRIC FUNCTIONS

1 Find the derivative of $\log_e(\cos x)$.

2 Find the equation of the tangent to the curve $y = \tan x$ at $x = \frac{\pi}{4}$.

3 For $f(x) = \sin x + \cos x$ over the domain $0 \le x \le 2\pi$, find:

(a) $f'(x)$ (b) $f''(x)$ (c) the coordinates of any turning points
(d) the coordinates of any points of inflection (e) the maximum value of $f(x)$.

4 For $y = e^{\sin x}$, find the equation of the normal to the curve at the point where $x = 0$.

5 Find all the points on the graph of $y = 2\sin x + \sin^2 x$, $0 \le x \le 4\pi$, at which the tangent is horizontal.

6 The tide at a point on the WA coast can be modelled using the equation $y = a\cos nt$. At Cable Beach in WA, over two consecutive days, the average difference between high and low tides is 9.0 metres and the average time between high tide and low tide is 6.1 hours.
   (a) What is the amplitude of the tide function at Cable Beach?
   (b) How much time passes between successive high tides (i.e. the period) and what is the value of n?
   (c) Use this information to obtain the tide function and draw its graph.
   (d) If the depth of water at low tide is 0.5 metres, what is the depth of the water 1 hour after low tide?

7 Consider the function $y = \sin x + \cos x$ for $0 < x < 2\pi$.
   (a) For what values of x is $\frac{dy}{dx} = 0$?
   (b) Find the greatest and least values of y and when they occur.
   (c) For what values of x is $\frac{dy}{dx} > 0$?
   (d) Find the coordinates of any points of inflection.

8 Find the equation of the normal to the curve $y = \cot x$ at the point $P\left(\frac{\pi}{4}, 1\right)$.

9 If $y = e^{\operatorname{cosec} x}$, find the equation of the tangent to the curve at $x = \frac{\pi}{2}$.

10 If $y = 3\cos 4x$, prove that $\frac{d^2y}{dx^2} + 16y = 0$.

11 The population of rock wallabies on an island is given by $P(t) = 200 + 40\cos\left(\frac{\pi}{6}t\right)$, where t is the time in months after the population was first measured.
   (a) Find all times during the first 12 months when the population is 180 rock wallabies.
   (b) Sketch the graph of $P(t)$ for $0 \le t \le 12$.
   (c) When is $\frac{dP(t)}{dt} > 0$?
   (d) What is the greatest population of rock wallabies over this twelve-month period?
   (e) Between what values does the population of rock wallabies range over this twelve-month period?

14.8 USING DERIVATIVES IN MOTION IN A STRAIGHT LINE

In Chapter 7 you met the terms displacement, velocity and acceleration. They were linked to the derivative of a function.

Displacement

Displacement is defined as the position relative to a starting point. It can be positive or negative. Displacement does not necessarily represent the total distance travelled.

Unlike displacement, distance is always a positive quantity.

Velocity

Velocity is defined as the rate of change of position (i.e. of displacement) with respect to time, or as the time rate of change of position in a given direction.

$$v(t) = f'(t) = \frac{dx}{dy} = \dot{x} = \lim_{h \to 0} \frac{f(t+h) - f(t)}{h}$$

Velocity can be positive or negative, depending on the direction of travel.

Speed is the magnitude of the velocity and is always positive.

Acceleration

Acceleration is defined as the rate of change of velocity with respect to time. Acceleration, like velocity, can be positive or negative. Positive acceleration indicates that the velocity is increasing, while negative acceleration indicates that the velocity is decreasing, which is often called deceleration or retardation.

(Note that 'increasing velocity' is not necessarily 'faster speed'; it only means acceleration in the direction of positive displacement.)

If you denote the velocity by $v(t)$, then the average acceleration over the interval from t to $(t+h)$ is $\frac{v(t+h)-v(t)}{h}$.

The instantaneous acceleration at time t is defined by $\lim_{h\to 0}\frac{v(t+h)-v(t)}{h}$. It may be denoted by

$$v'(t),\ a(t),\ f''(t),\ \frac{dv}{dt},\ \frac{d^2x}{dt^2},\text{ or }\ddot{x}: a(t)=v'(t)=\frac{d^2x}{dt^2}=\ddot{x}=\lim_{h\to 0}\frac{v(t+h)-v(t)}{h}$$

Summary of important terms

'initially': $t=0$ 'at the origin': $x=0$

'at rest': $v=0$ 'velocity is constant': $a=0$

Units and symbols

Physical quantity	Unit	Symbol
Time	s	t
Displacement	cm, m	x (or s in physics)
Velocity	cm s^{-1}, m s^{-1}	$v, \frac{dx}{dt}, \dot{x}$
Acceleration	cm s^{-2}, m s^{-2}	$a, \frac{dv}{dt}, \frac{d^2x}{dt^2}, \ddot{x}$

Note that 's' is the abbreviation for second, 'cm' for centimetre and 'm' for metre.
Constant acceleration due to gravity = $9.8\ \text{m s}^{-2}$.

Example 18

A ball is projected vertically upwards from the top of a building 30 metres high. The equation for its motion is given by $x = 30 + 25t - 5t^2$, where x is the displacement in metres above the top of the building and t is in seconds.

- **(a)** Graph the displacement function.
- **(b)** Find the velocity as a function of time.
- **(c)** What is the initial velocity of the ball?
- **(d)** The ball reaches its greatest height when $\frac{dx}{dt}=0$. When does it reach its greatest height and how high above the ground is it then?
- **(e)** How long will it take for the ball to hit the ground?
- **(f)** What is the ball's speed when it hits the ground?
- **(g)** Find the expression for the acceleration of the ball.

Solution

(a) $x = 30 + 25t - 5t^2$

The graph will be a parabola so the axis of symmetry is given by $t = \dfrac{-25}{2\times(-5)} = 2.5$

t	0	1	2	2.5	3	4	5	6
x	30	50	60	61.25	60	50	30	0

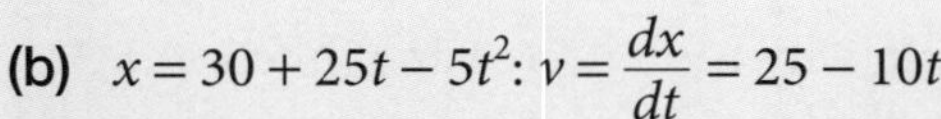

(b) $x = 30 + 25t - 5t^2$: $v = \dfrac{dx}{dt} = 25 - 10t$

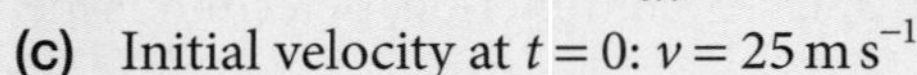

(c) Initial velocity at $t = 0$: $v = 25\ \text{m s}^{-1}$

(d) $\dfrac{dx}{dt} = 0$: $25 - 10t = 0$

$t = 2.5$ seconds (This could have been obtained from the graph).

$t = 2.5$: $x = 61.25$ m

The highest point is 61.25 metres above the ground, which the ball reaches after 2.5 seconds.

(e) The ball reaches the ground when $x = 0$: using the graph the answer is 6 seconds.

Using the displacement function:

$$\begin{aligned} 30 + 25t - 5t^2 &= 0 \\ 5(6 + 5t - t^2) &= 0 \\ (6 - t)(1 + t) &= 0 \end{aligned}$$

As $t \geq 0$, $t = 6$.

The ball hits the ground after 6 seconds.

(f) $t = 6$: $v = 25 - 60 = -35$

The ball hits the ground with a speed of $35\ \text{m s}^{-1}$.

As the initial upwards velocity is positive, the velocity when it hits the ground is negative as the ball is moving downwards.

(g) $a = \dfrac{dv}{dt} = -10\ \text{m s}^{-2}$

This means that the acceleration is acting in the opposite direction to the initial upwards velocity.

EXERCISE 14.8 USING DERIVATIVES IN MOTION IN A STRAIGHT LINE

1 A particle is moving in a straight line so that its displacement x metres is given by $x = \dfrac{t^3}{2} - 3t^2 + 5$.

(a) Find an expression for its velocity. **(b)** Find an expression for its acceleration.

(c) When is the velocity zero? **(d)** Find the displacement, velocity and acceleration after 4 seconds.

2 The displacement x metres at time t seconds, $t \geq 0$, of a particle moving in a straight line is given by $x = 4t^3 - 3t^2 + 5t - 1$. Its acceleration is given by:

A $a = 4t^3 - 3t^2 + 5t - 1$ **B** $a = 12t^2 - 6t + 5$ **C** $a = 24t - 6$ **D** $a = 24$

3 The displacement x metres at time t seconds, $t \geq 0$, of a particle moving in a straight line is given by $x = 2t^3 - 6t^2 - 30t$.

(a) Find the velocity and acceleration at any time t.

(b) Find the initial velocity and acceleration.

(c) At what time is the velocity zero? What is the acceleration at this time?

(d) During what time interval is the velocity negative?

4 A particle is projected vertically upwards from the ground. The equation for its motion is given by $x = 30t - 5t^2$, where x is the displacement in metres above the ground and t is in seconds.

(a) Graph the displacement function.
(b) Find the velocity as a function of time.
(c) What is the initial velocity of the particle?
(d) When does the particle reach its greatest height and how high above the ground is it then?
(e) How long will it take before the particle returns to the ground?
(f) What is the particle's speed when it hits the ground?
(g) Find the expression for the acceleration of the particle.

5 The velocity of a function is given by $v = 5e^{-t}$.

(a) Find the expression for a, the acceleration.
(b) Explain why v is a decreasing function.
(c) The displacement function is not given. Without finding the displacement function, determine whether it has any stationary points. Justify your answer.

6 An object moves with a velocity v given by $v = 20 + (2t - 1)e^{-0.5t}$, where t is in hours and v is in km h^{-1}. Calculate:

(a) the velocity after 1 hour
(b) the time taken to reach its maximum velocity.

CHAPTER REVIEW 14

1 For the graph of $y = 15x + 12x^2 - 4x^3$ for $-1 \le x \le 3$, find the values of x for which:

(a) y increases as x increases
(b) y decreases as x increases
(c) y is a maximum
(d) y is a minimum.

2 If $f(x) = 2x^3 + 3x^2 - 12x$, find the values of x for which: (a) $f'(x) = 0$ (b) $f'(x) > 0$ (c) $f'(x) < 0$

3 Sketch the graph of $y = f(x)$, given that:

(a) $f(3) = 5,\ f'(3) = 0,\ f'(x) > 0$ for $x < 3$ and $f'(x) < 0$ for $x > 3$
(b) $f(-1) = 8,\ f'(-1) = 0, f(2) = 3,\ f'(2) = 0,\ f'(x) < 0$ for $-1 < x < 2$, and $f'(x) > 0$ for $x < -1$ and for $x > 2$.

4 Find the coordinates of the points on the following curves where the gradient is zero. Determine whether these points are local maximum or minimum points.

(a) $y = 3x^3 - 2x^2$
(b) $y = x^3 - 3x^2 - 9x$

5 A figure $ABCED$ consists of a rectangle $ABCD$ topped by an equilateral triangle CED as shown in the diagram. If the perimeter of the figure is 45 cm, find the dimensions of the rectangle when the total area is a maximum.

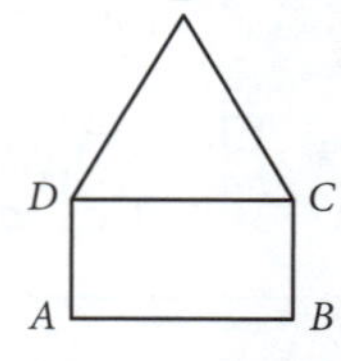

6 A piece of wire 8 m long is cut into two parts. One part is bent into the shape of a square and the other part is bent into a rectangle whose length is twice its breadth. Calculate the length of each part if the sum of the areas of the square and the rectangle is to be a minimum.

7 A figure consists of a semicircle with a rectangle constructed on its diameter, as shown in the diagram. If the perimeter of the figure is 50 cm, find the dimensions of the rectangle such that the area of the figure is as large as possible. What is this largest area?

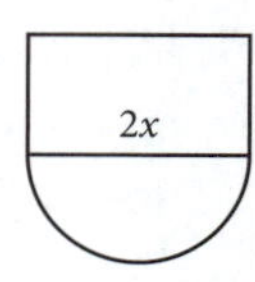

8 For the graph of $y = f(x)$ where $f(x) = \dfrac{x^3}{3} - 4x + 3$, find the following:

(a) the values of x for which $f'(x) = 0$
(b) the values of x for which $f'(x) < 0$
(c) any local maximum and minimum values of $f(x)$.
(d) Sketch the curve.

9 Sketch the graph of $y = 3x^3 - 5x^2$ for values of x in the domain $-0.5 \le x \le 1.5$, locating the turning points.

10 A rectangular sheet of metal measures 6 cm by 4 cm. Four equal squares are cut out of the corners and the sides are turned up to form an open rectangular box. Find the edge length of the squares cut so that the box has a maximum volume.

11 Prove that the curve $y = x^2(3 - x)$ has a horizontal tangent where $x = 2$ and crosses the y-axis at right angles at the origin.

12 Sketch the graph of $y = x^3(3 - x)$ in the domain $-1 \le x \le 3$, giving the coordinates of the turning points and points of inflection.

13 Jack is in the bush at point A, 3 km from the nearest point C, which is at one end of a straight 4 km path CB, as shown in the diagram. Jack wants to get to point B, the other end of the path, as quickly as possible. He can run at a speed of 20 km h^{-1} along the path CB but only at $10\sqrt{2}\text{ km h}^{-1}$ in the bush off the path. He runs in a straight line through the bush from A to a point X on the path CB, then along the path from X to B.

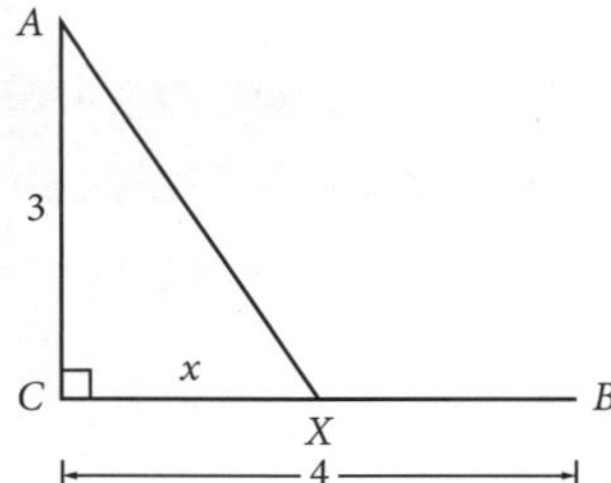

(a) Find, in terms of x, the time taken for Jack to go from:
(i) A to X **(ii)** X to B.
(b) Find, in terms of x, the total time t hours to get from A to B.
(c) Find the position of the point X for which t is a minimum. Find this minimum time.

14 A haulage company makes frequent deliveries from Sydney to Cairns and calculates that the overhead cost \$$C$ depends on the average delivery speed $v\text{ km h}^{-1}$ according to the rule $C = v + \frac{3600}{v}$. Find the average delivery speed to minimise the overhead cost.

15 **(a)** Find the maximum value of $2xe^{-1.5x}$ and the value for which this function has a maximum value.
(b) If $f(x) = 2xe^{-1.5x}$, find $f(0)$, $f(0.5)$, $f(1)$ and hence graph the function in the domain $0 \le x \le 1$.

16 If $\theta = \theta_0 e^{-kt}$, show that $\frac{d\theta}{dt} = -k\theta$.

17 A car is worth \$10 000 when new. After t years, the value of the car (in dollars) is given by the formula $V = Ae^{-0.2t}$.
(a) Find the value of A. **(b)** Find the value of the car after 6 years.
(c) Find the rate in dollars per year at which the car's value is depreciating, when:
(i) $t = 6$ **(ii)** $V = 5000$
(d) How much time will it take until the car is worth only \$1000?

18 Find the turning points of $y = 3\sec 2x$ for $-\frac{\pi}{4} < x < \frac{3\pi}{4}$ and determine their nature.

CHAPTER 15
Graphing techniques

15.1 TRANSFORMATION OF GRAPHS USING $y = f(x + b)$ AND $y = f(x) + c$

Consider the following graphs:

$y = x^2$

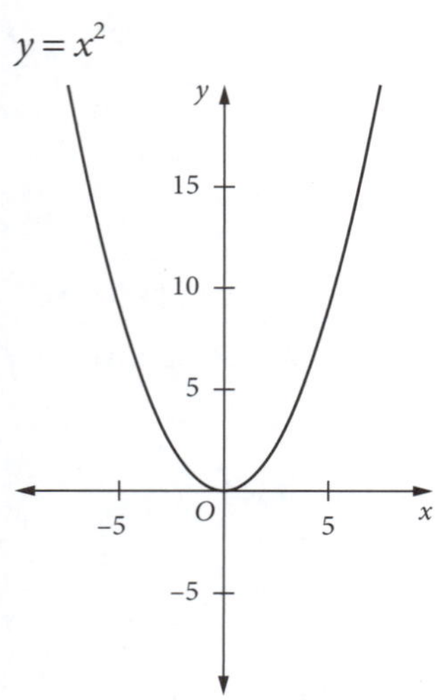

$y = (x + 2)^2$

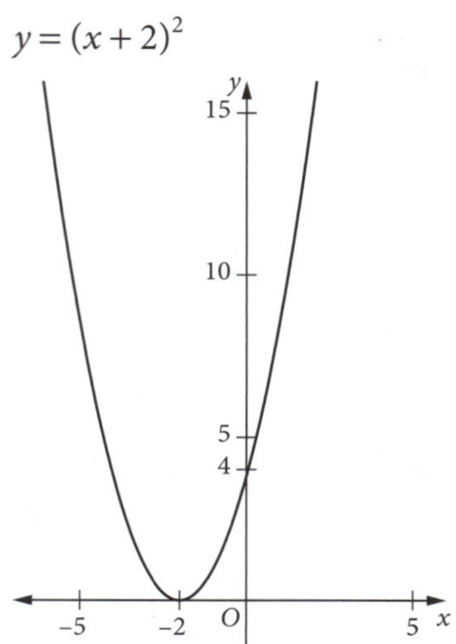

$y = x^2 + 2$

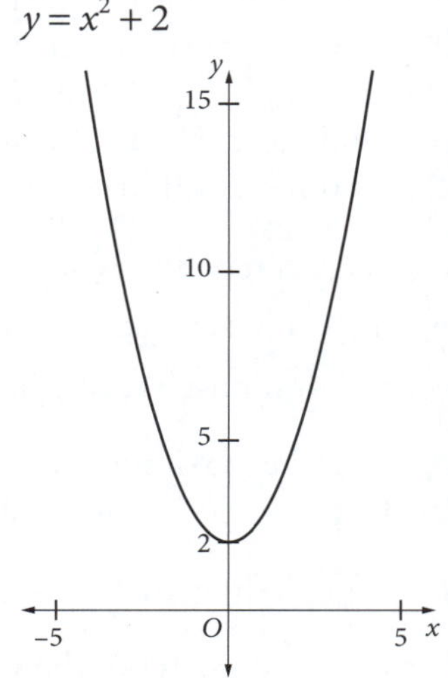

If $y = x^2$ is written as $y = f(x)$ then $y = (x + 2)^2$ becomes $y = f(x + 2)$ and $y = x^2 + 2$ becomes $y = f(x) + 2$.
In $y = f(x + 2)$ the curve for $y = f(x)$ has been moved 2 units to the left, (0, 0) moved to (−2, 0).
In $y = f(x) + 2$ the curve for $y = f(x)$ has been moved 2 units upwards, (0, 0) moved to (0, 2).

Now consider the following similar graphs:

$y = x^3$

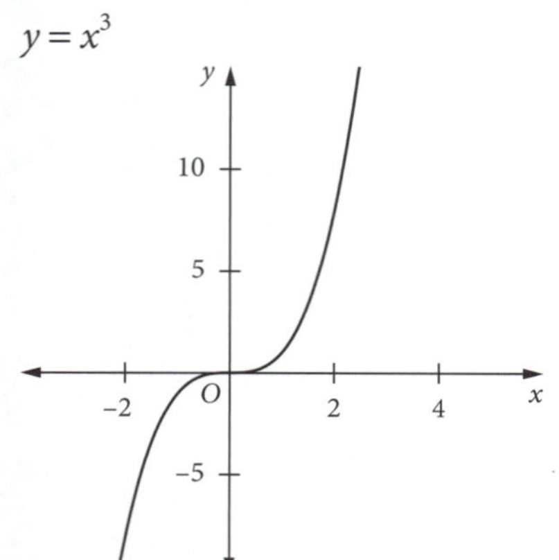

$y = (x - 2)^3$

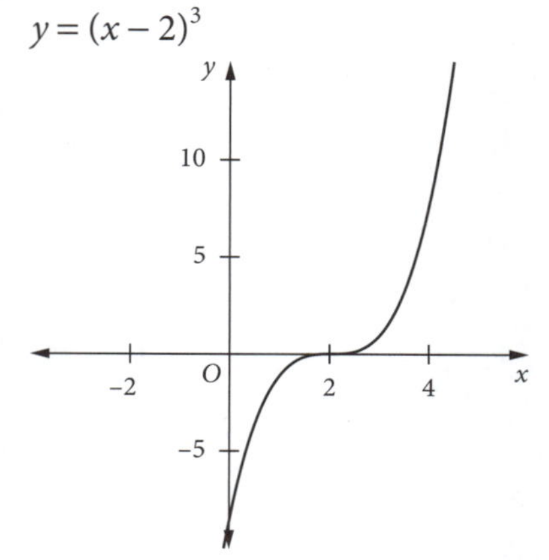

$y = x^3 - 2$

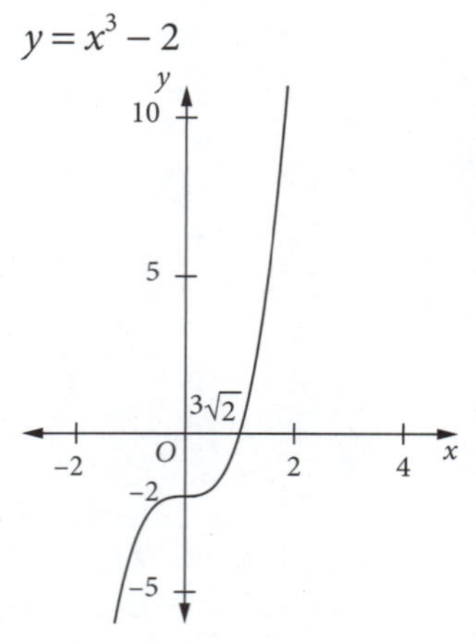

If $y = x^3$ is written as $y = f(x)$ then $y = (x - 2)^3$ becomes $y = f(x - 2)$ and $y = x^3 - 2$ becomes $y = f(x) - 2$.
In $y = f(x - 2)$ the curve for $y = f(x)$ has been moved 2 units to the right, (0, 0) moved to (2, 0).
In $y = f(x) - 2$ the curve for $y = f(x)$ has been moved 2 units downwards, (0, 0) moved to (0, −2).

In the cases above, the same function is translated horizontally and vertically by changing the function. The same changes may also be applied to functions involving square roots, for example:

$y = \sqrt{x}$

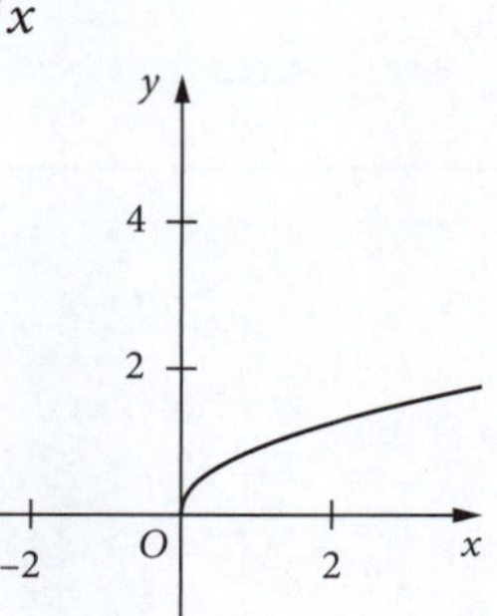

$y = \sqrt{x+2}$

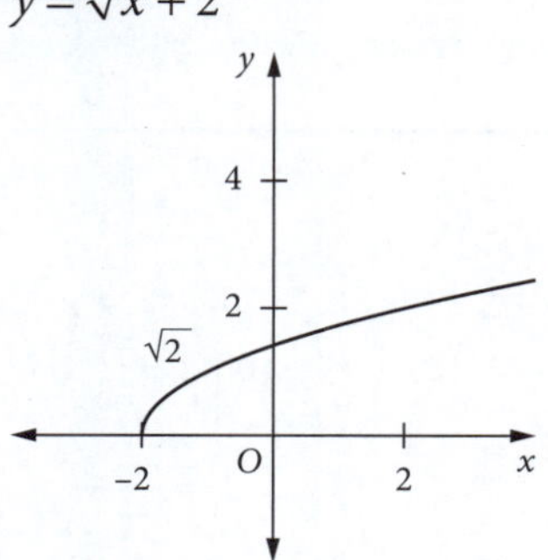

$y = \sqrt{x} + 2$

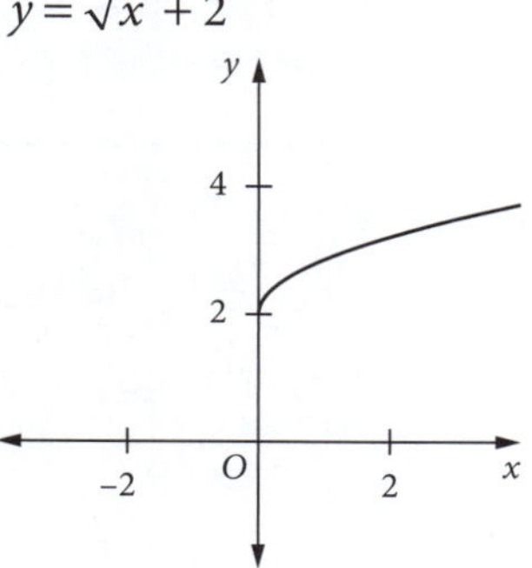

If $y = \sqrt{x}$ is written as $y = f(x)$ then $y = \sqrt{x+2}$ becomes $y = f(x+2)$ and $y = \sqrt{x} + 2$ becomes $y = f(x) + 2$.
In $y = f(x+2)$ the curve for $y = f(x)$ has been moved 2 units to the left, (0, 0) moved to (−2, 0).
In $y = f(x) + 2$ the curve for $y = f(x)$ has been moved 2 units upwards, (0, 0) moved to (0, 2).

Similarly, consider the following reciprocal functions:

$y = \dfrac{1}{x}$

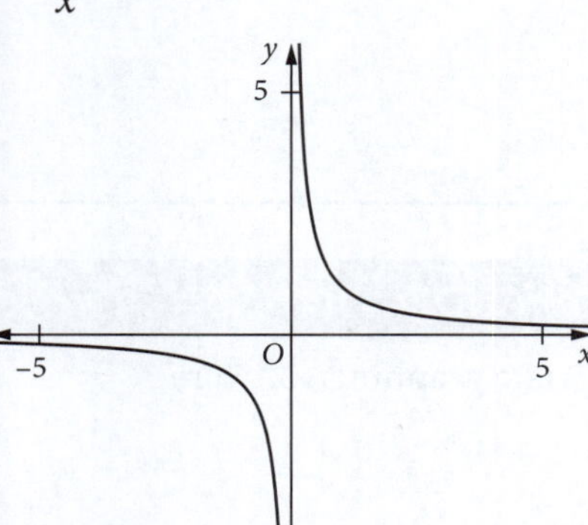

$y = \dfrac{1}{x-2}$

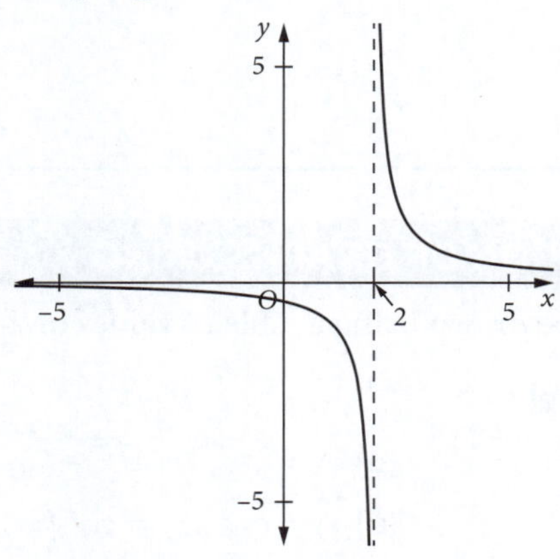

$y = \dfrac{1}{x} - 2$

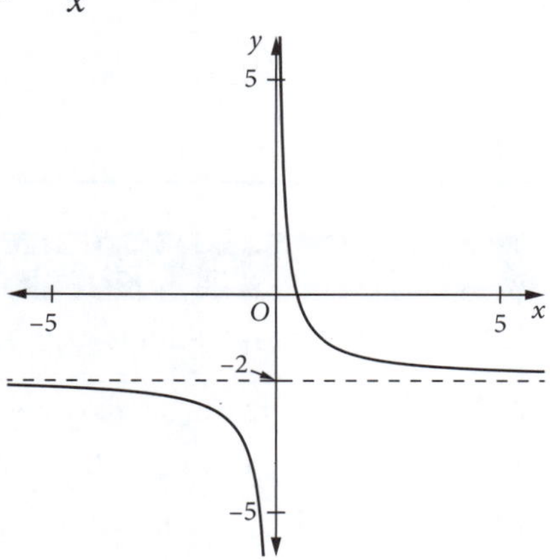

If $y = \dfrac{1}{x}$ is written as $y = f(x)$ then $y = \dfrac{1}{x-2}$ becomes $y = f(x-2)$ and $y = \dfrac{1}{x} - 2$ becomes $y = f(x) - 2$.
In $y = f(x-2)$ the curve for $y = f(x)$ has been moved 2 units to the right.
In $y = f(x) - 2$ the curve for $y = f(x)$ has been moved 2 units downwards.

Hence the graph of $y = f(x+b)$ is just the graph of $y = f(x)$ moved b units to the left.

Also the graph of $y = f(x) + c$ is just the graph of $y = f(x)$ moved c units upwards.

MAKING CONNECTIONS

Horizontal translation of the graph of a function

Move the slider to explore the effect of changing b on the graph of a function $f(x+b)$.

MAKING CONNECTIONS

Vertical translation of the graph of a function

Move the slider to explore the effect of changing c on the graph of a function $f(x) + c$.

Example 1

On the same set of axes, for $0 \le x \le 2\pi$, draw the graphs of:

(a) $y = \sin x$ (b) $y = \sin\left(x + \frac{\pi}{6}\right)$ (c) $y = \sin\left(x - \frac{\pi}{3}\right)$

Solution

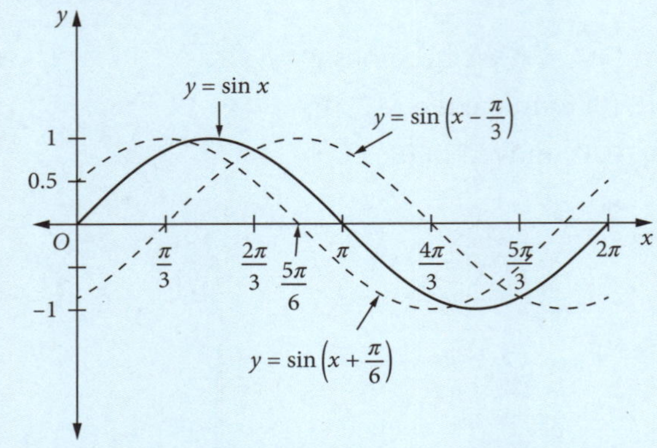

EXERCISE 15.1 TRANSFORMATION OF GRAPHS USING $y = f(x + b)$ AND $y = f(x) + c$

For the following questions, graphs may be drawn using a table of values or appropriate graphing software.

1 On the same diagram, draw the graphs of:

(a) $y = x^2, y = (x - 2)^2, y = x^2 - 2$

(b) $y = \sqrt{x}, y = \sqrt{x - 4}, y = \sqrt{x} - 4$

(c) $y = \frac{1}{x}, y = \frac{1}{x + 3}, y = \frac{1}{x} + 3$

(d) $y = \cos x, y = \cos\left(x + \frac{\pi}{6}\right), y = \cos\left(x - \frac{\pi}{3}\right)$ for $0 \le x \le 2\pi$.

2 On the same diagram, draw the graphs of:

(a) $y = e^x, y = e^{x+2}, y = e^x + 2$

(b) $y = \ln x, y = \ln(x - e), y = \ln x - e$

(c) $y = e^{-x}, y = e^{-x-1}, y = e^{-x} - 1$

(d) $y = \ln(-x), y = \ln(1 - x), y = \ln(-x) + 1$

3 On the same diagram, draw the following graphs for $0 \le x \le 2\pi$:

(a) $y = \cos x, y = \cos\left(x - \frac{\pi}{6}\right), y = \cos\left(x + \frac{\pi}{3}\right)$

(b) $y = \tan x, y = \tan\left(x + \frac{\pi}{4}\right), y = \tan\left(x - \frac{\pi}{4}\right)$

(c) $y = \operatorname{cosec} x, y = \operatorname{cosec}\left(x + \frac{\pi}{4}\right), y = \operatorname{cosec}\left(x - \frac{\pi}{2}\right)$

(d) $y = \cos x, y = \cos x + 2, y = \cos x - 1$

4 On the same diagram, draw the graphs of:

(a) $y = \frac{1}{x^2}, y = \frac{1}{(x - 1)^2}, y = \frac{1}{x^2} - 1$

(b) $y = x^3 + x, y = (x - 1)^3 + (x - 1), y = x^3 + x - 1$

(c) $y = \sqrt{x^2}, y = \sqrt{(x + 2)^2}, y = \sqrt{x^2} + 2$

5 The diagram below shows the graph of $y = f(x)$. Which of the options shows the graph of $y = f(x - 1)$?

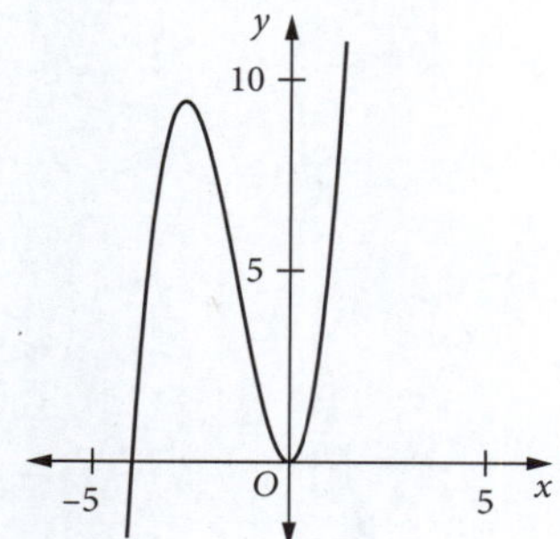

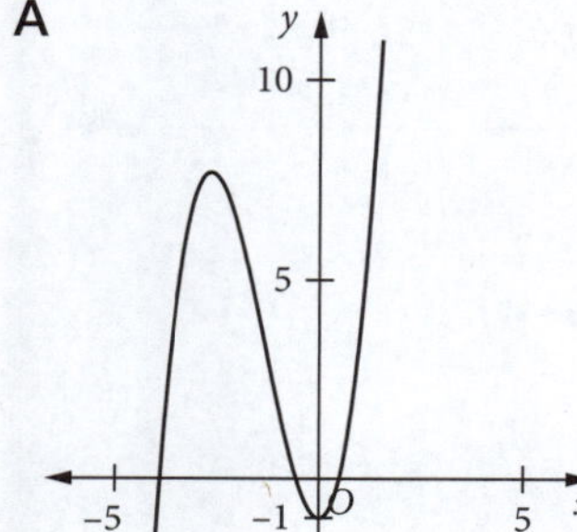

B

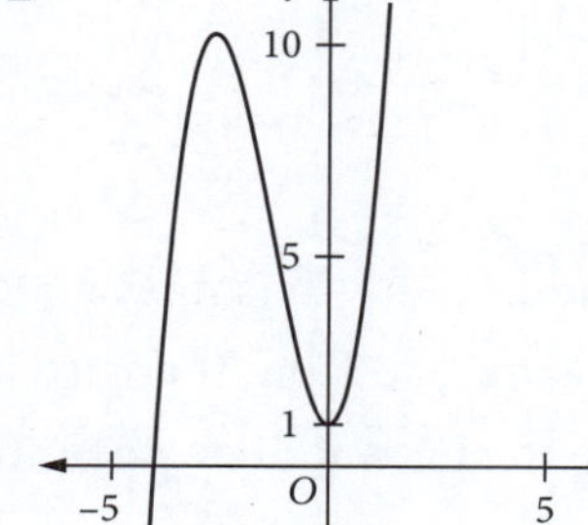

C

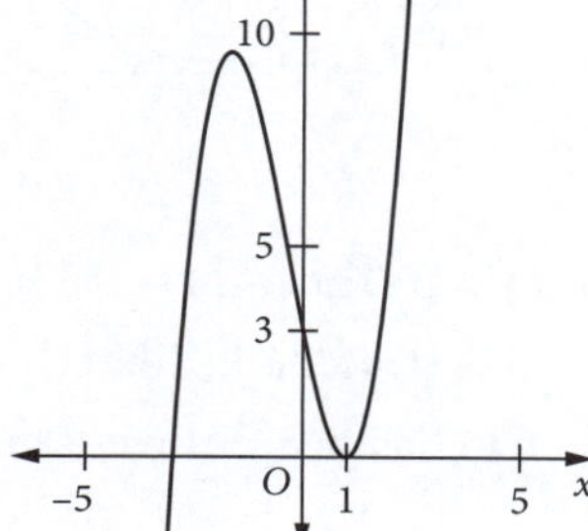

D

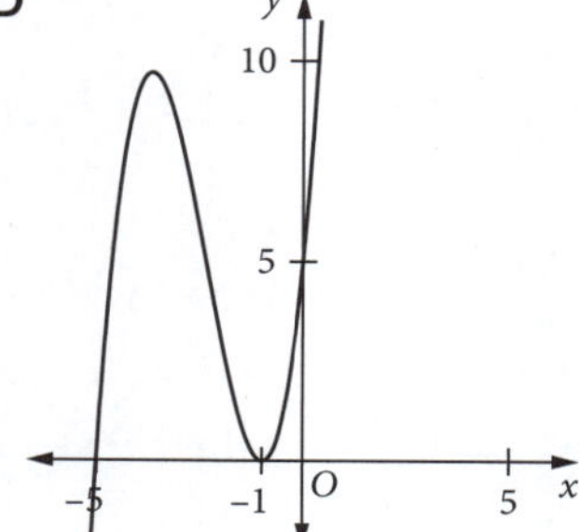

15.2 TRANSFORMATION OF GRAPHS USING $y = kf(x)$ AND $y = kf(x + b)$

Consider the following graphs:

$y = x^2$

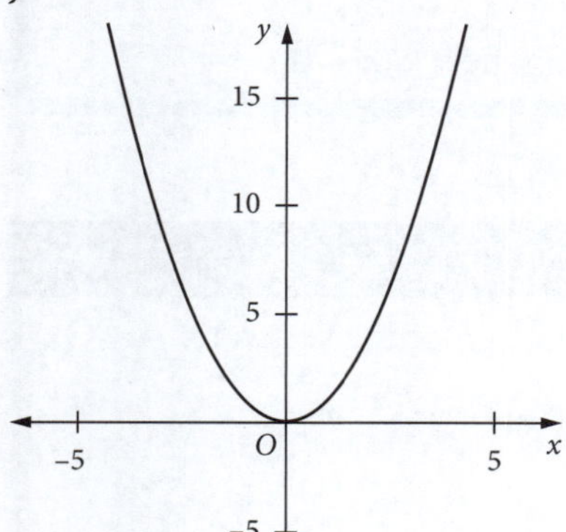

$y = 2x^2$

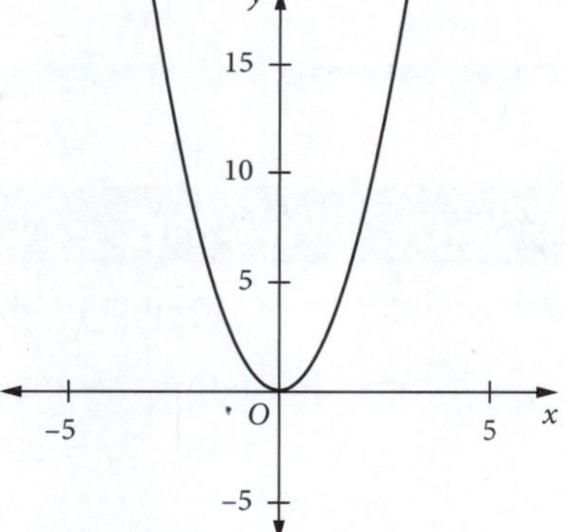

$y = 2(x - 1)^2$

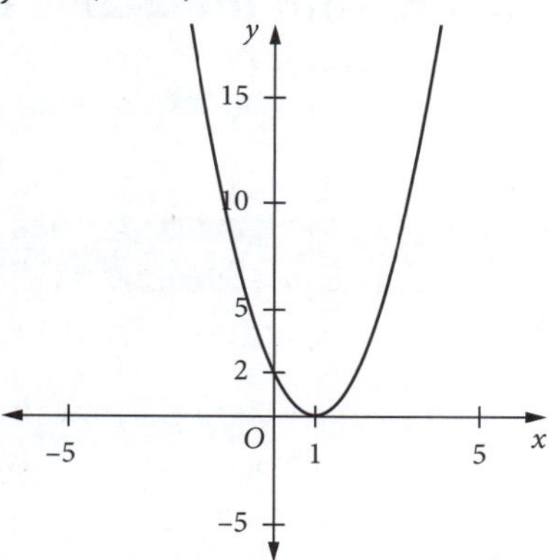

If $y = x^2$ is written as $y = f(x)$ then $y = 2x^2$ becomes $y = 2f(x)$ and $y = 2(x - 1)^2$ becomes $y = 2f(x - 1)$.

In $y = 2f(x)$ the curve for $y = f(x)$ has stretched (dilated) by a factor of 2 from the x-axis.

In $y = 2f(x - 1)$ the curve for $y = f(x)$ has been moved 1 unit to the right and then stretched by a factor of 2 from the x-axis.

Now consider the following exponential graphs:

$y = e^x$

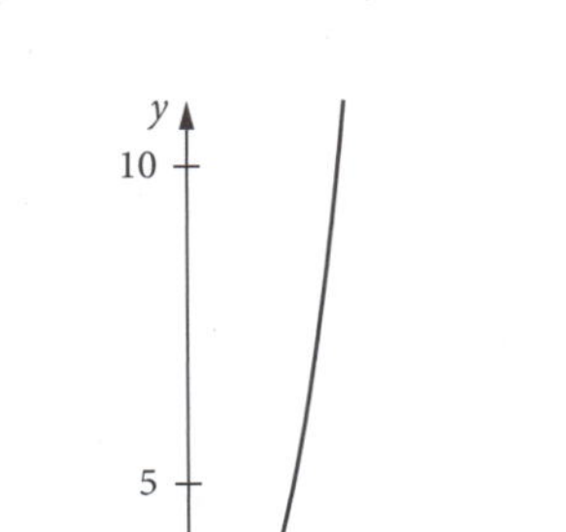

$y = 3e^x$

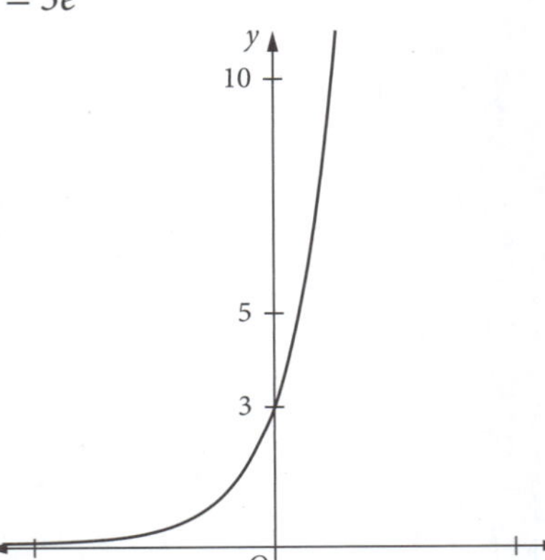

$y = 3e^{x+1}$

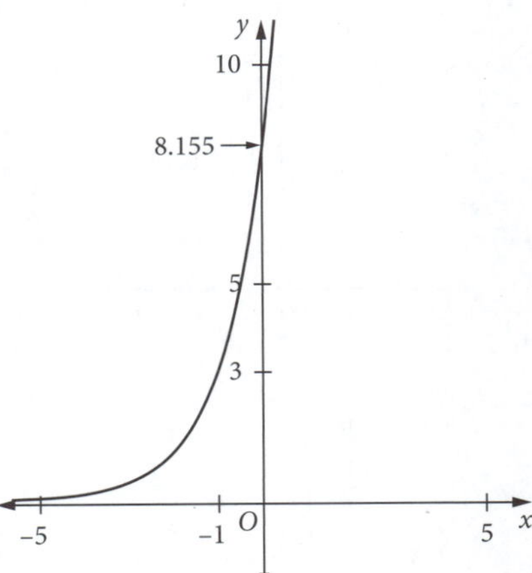

If $y = e^x$ is written as $y = f(x)$ then $y = 3e^x$ becomes $y = 3f(x)$ and $y = 3e^{x+1}$ becomes $y = 3f(x + 1)$.

In $y = 3f(x)$ the curve for $y = f(x)$ has stretched (dilated) by a factor of 3 from the x-axis.

In $y = 3f(x + 1)$ the curve for $y = f(x)$ has been moved 1 unit to the left and then been stretched by a factor of 3 from the x-axis.

In all these cases, the graph of $y = kf(x)$ is just the graph of $y = f(x)$ stretched (dilated) by a factor of k.

As for the cases in the previous section, the graph of $y = kf(x + b)$ is just the graph of $y = f(x)$ stretched (dilated) by a factor of k and also translated horizontally b units to the left.

MAKING CONNECTIONS

Dilation from the x-axis of the graph of a function

Move the sliders to explore the effects of changing k and b on the graph of a function $kf(x + b)$.

EXERCISE 15.2 TRANSFORMATION OF GRAPHS USING $y = kf(x)$ AND $y = kf(x + b)$

For the following questions, graphs may be drawn using a table of values or appropriate graphing software.

1 On the same diagram, draw the graph of each equation, stating the dilation factor.

(a) $y = x^2, y = 3x^2, y = 3(x + 2)^2$

(b) $y = x^3, y = 2x^3, y = 2(x - 1)^3$

(c) $y = \sqrt{x}, y = 5\sqrt{x}, y = 5\sqrt{x - 4}$

(d) $y = x^2, y = \frac{x^2}{4}, y = \frac{(x - 2)^2}{4}$

2 On the same diagram, draw the graph of each equation, stating the dilation factor.

(a) $y = e^x, y = \frac{e^x}{2}, y = \frac{e^{x-1}}{2}$

(b) $y = \sin x, y = 2\sin x, y = 2\sin\left(x - \frac{\pi}{2}\right)$ for $-\pi \le x \le \pi$

(c) $y = \sec x, y = \frac{\sec x}{2}, y = \frac{\sec\left(x - \frac{\pi}{6}\right)}{2}$ for $-\pi \le x \le \pi$

3 On the same diagram, draw the graph of each equation, stating the dilation factor.

(a) $y = x^2 + x, y = 2(x^2 + x), y = 2[(x - 1)^2 + (x - 1)]$

(b) $y = -x^2, y = -3x^2, y = -3(x + 2)^2$

(c) $y = x - x^3, y = 2(x - x^3), y = 2[(x + 1) - (x + 1)^3]$

4 The diagram to the right shows the graph of $y = f(x)$.
Which diagram below shows the graph of $y = 3f(x)$?

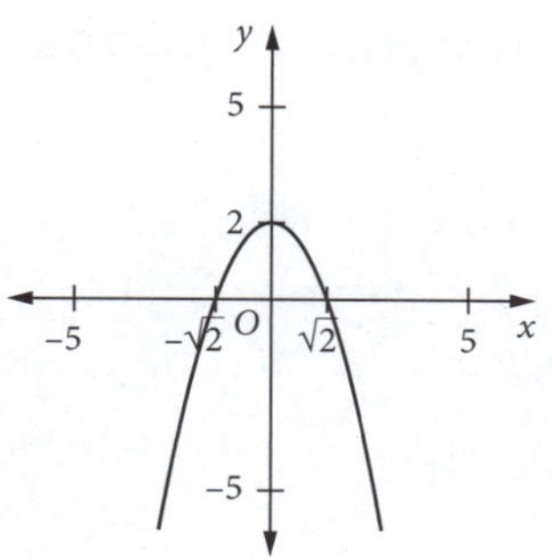

A

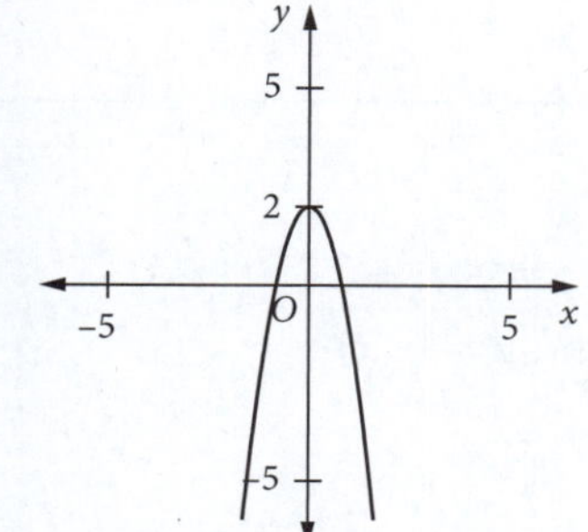

B

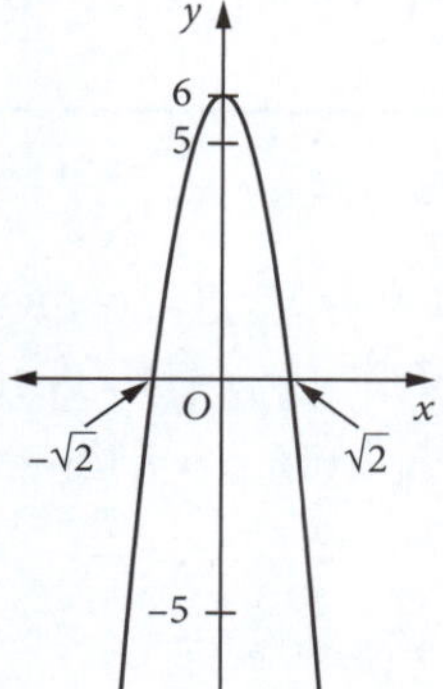

C

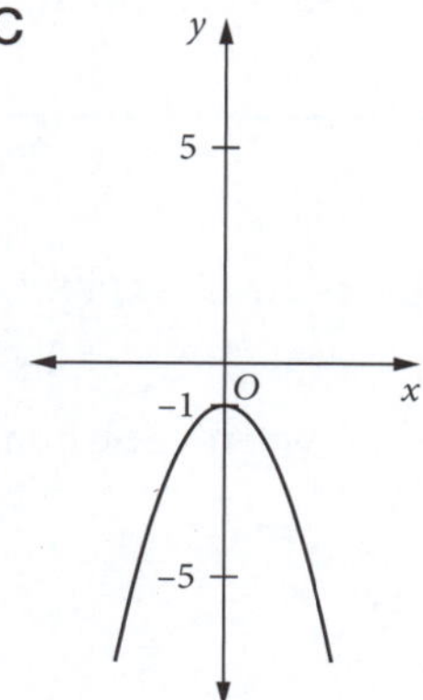

D

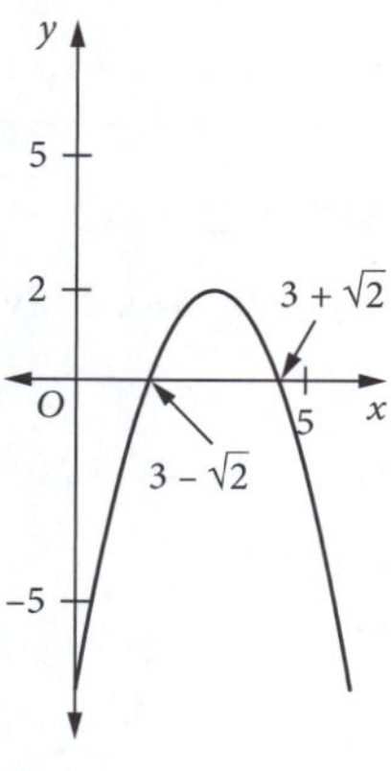

15.3 TRANSFORMATION OF GRAPHS USING $y = f(ax)$ AND $y = f(a(x + b))$

Consider the following graphs.

$y = x^2$

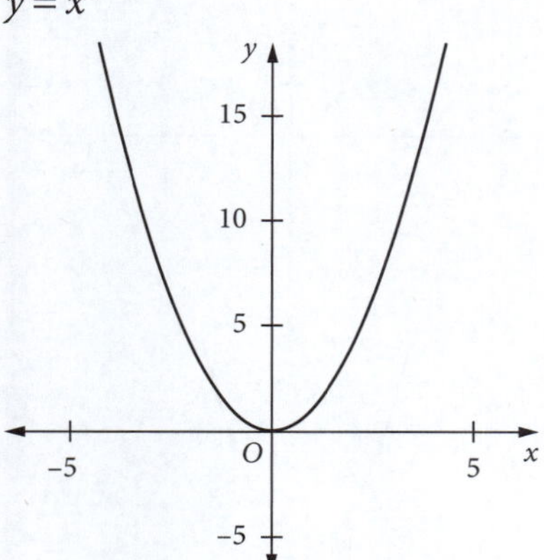

$y = (2x)^2$

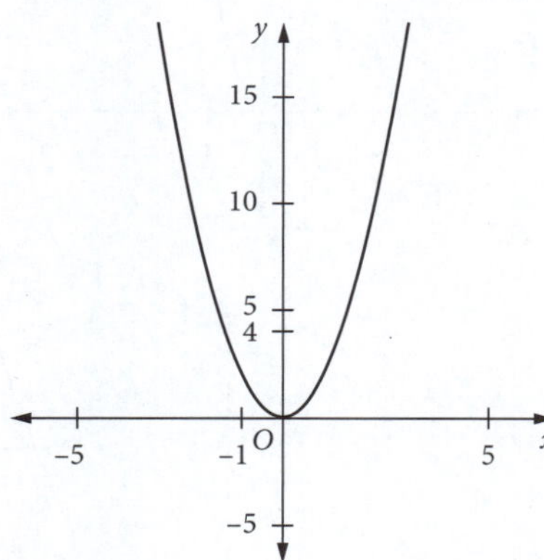

$y = (2(x + 1))^2$

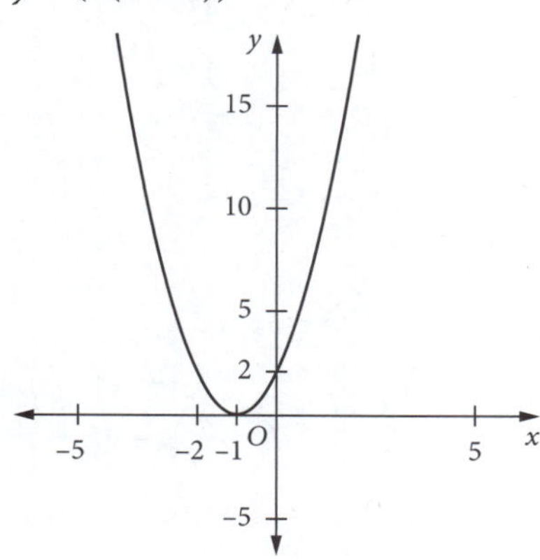

If $y = x^2$ is written as $y = f(x)$ then $y = (2x)^2$ becomes $y = f(2x)$ and $y = (2(x + 1))^2$ becomes $y = f(2(x + 1))$.

In $y = f(2x)$ the curve for $y = f(x)$ has stretched by a factor of $\frac{1}{2}$ unit from the y-axis.

In $y = f(2(x + 1))$ the curve for $y = f(x)$ has been moved 1 unit to the left and then been stretched by a factor of $\frac{1}{2}$ unit from the y-axis.

MAKING CONNECTIONS

Dilation from the *y*-axis of the graph of a function

Move the sliders to explore the effects of changing *a* and *b* on the graph of a function $f(a(x + b))$.

Summary—transformations of graphs

Given $y = f(x)$, then:

- $y = f(x) + c$ translates the curve c units up
- $y = f(x + b)$ translates the curve b units to the left
- $y = kf(x + b)$ stretches (dilates) the curve by a factor of k from the x-axis
- $y = f(a(x + b))$ stretches (dilates) the curve by a factor $\frac{1}{a}$ from the y-axis.

The order in which transformations are performed is important. Applying transformations in a different order may change the shape of the graph.

Example 2

(a) On successive diagrams, draw the graphs of $f(x) = x^2$, $g(x) = 3f(x)$ and $y = g(x) + 5$.

(b) On successive diagrams, draw the graphs of $f(x) = x^2$, $g(x) = f(x) + 5$ and $y = 3g(x)$.

(c) Discuss the differences between your final graphs in parts **(a)** and **(b)**.

Solution

(a)

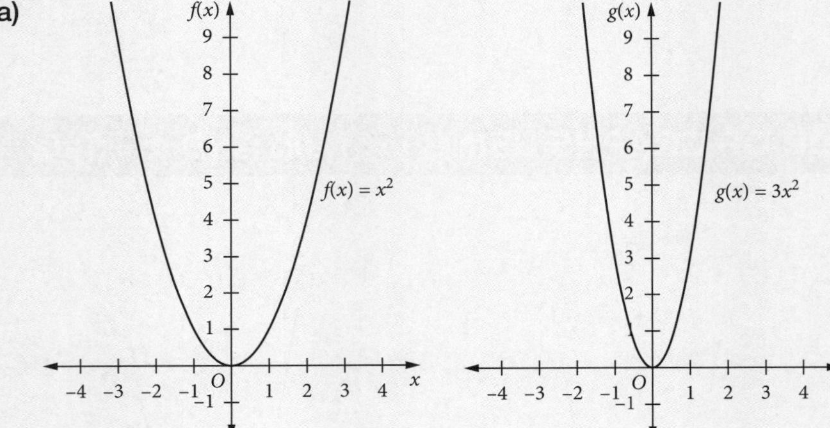

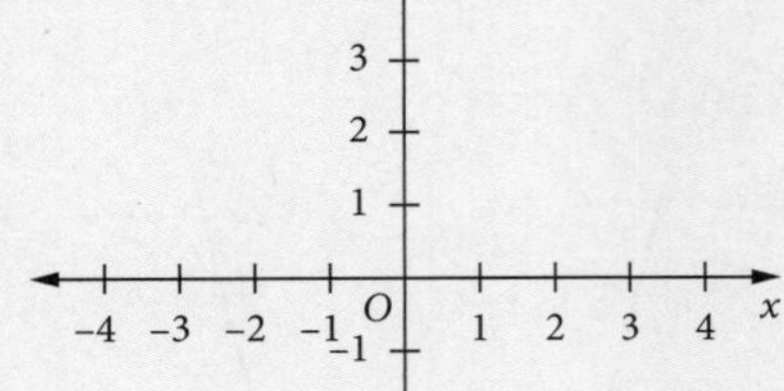

(b)

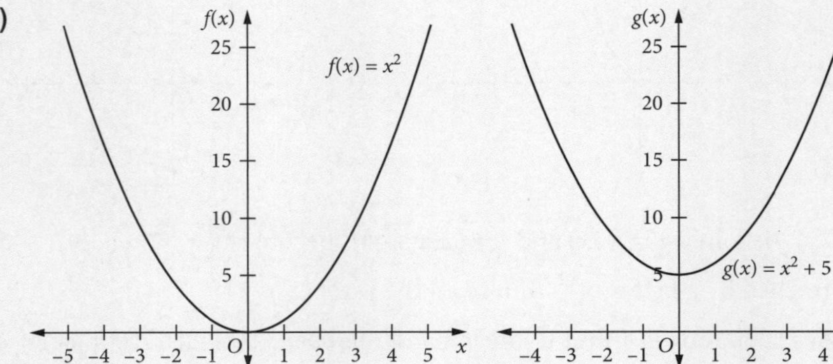

$y = 3x^2 + 15$

(c) In part **(b)** the vertical translation was 15 instead of 5. The dilation was the same in each case.

The order in which a series of transformations is applied to a function is important.

Reversing or changing the order can change the final function.

EXPLORE FURTHER

The order of transformations

Use technology to explore how the order of transformations changes the graph of the function.

EXERCISE 15.3 TRANSFORMATION OF GRAPHS USING $y = f(ax)$ AND $y = f(a(x + b))$

For the following questions, graphs may be drawn using a table of values or appropriate graphing software.

1 On the same diagram, draw the graph of each equation, stating the dilation and describe any changes to the position of the original graph:

(a) $y=\sqrt{x}, y=\sqrt{2x}, y=\sqrt{2(x-3)}$

(b) $y=\tan x, y=\tan 2x, y=\tan 2\left(x+\frac{\pi}{6}\right)$ for $-\frac{\pi}{2} \le x \le \frac{\pi}{2}$

(c) $y=e^{-x}, y=e^{-3x}, y=e^{-3(x-1)}$

(d) $y=\cos x, y=\cos\frac{x}{2}, y=\cos\left(\frac{1}{2}\left(x-\frac{\pi}{4}\right)\right)$ for $0 \le x \le 2\pi$

2 If $f(x)=x^3$, write down the new equation obtained by applying the condition given in each part. Simplify your answer where possible.

(a) $f(2x)$ (b) $f(x-1)$ (c) $f(x)+3$ (d) $2f(x)+1$ (e) $3f(2(x+2))-4$

3 If $f(x)=\cos\frac{x}{2}$, write down the new equation obtained by applying the condition given in each part.

(a) $f(2x)$ (b) $f\left(x-\frac{\pi}{3}\right)$ (c) $2f(x)$ (d) $f(x)-1$ (e) $2f\left(x+\frac{\pi}{6}\right)+1$

4 Using graphing software, draw the graphs of the new equation for each part of question **2**.

5 The diagram to the right shows the graph of $y=f(x)$. Which diagram below shows the graph of $y=f\left(\frac{x}{2}\right)$?

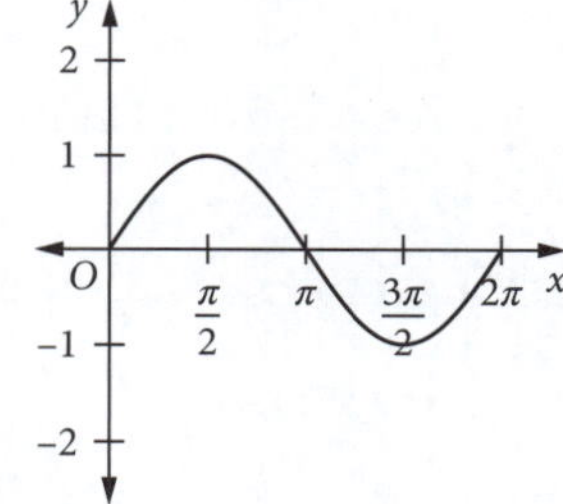

A

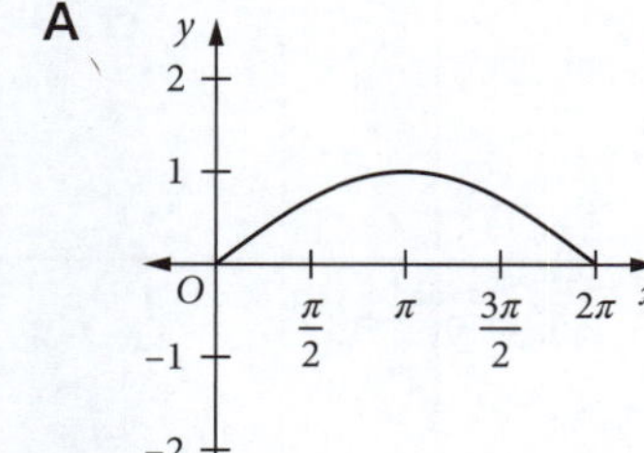

B

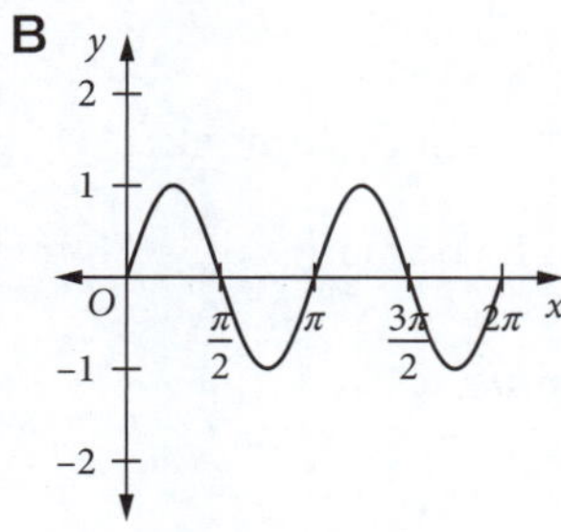

C

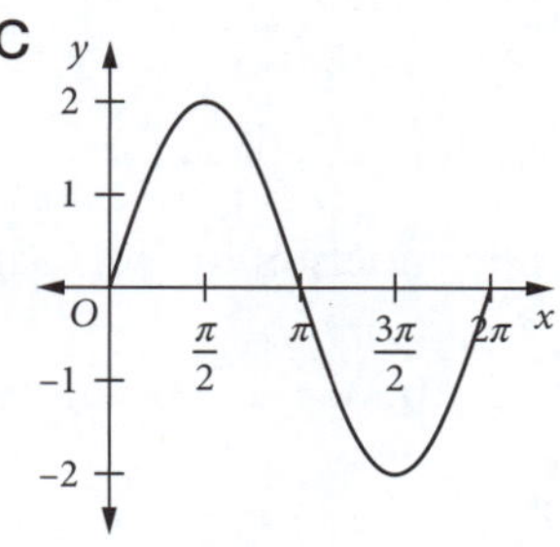

D

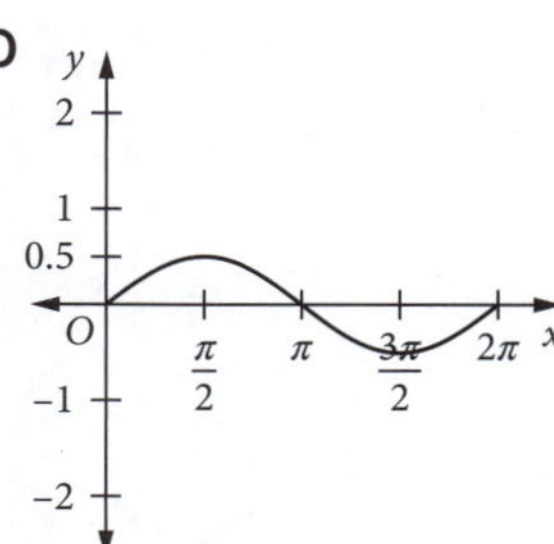

6 (a) On successive diagrams, draw the graphs of $f(x)=x^2$, $g(x)=2f(x)$ and $y=g(x)-3$.

(b) On successive diagrams, draw the graphs of $f(x)=x^2$, $g(x)=f(x)-3$ and $y=2g(x)$.

(c) Discuss the differences between your final graphs in parts **(a)** and **(b)**.

7 (a) On successive diagrams, draw the graphs of $f(x)=x+1$, $g(x)=[f(x)]^2$ and $y=g(x)+2$.

(b) On successive diagrams, draw the graphs of $f(x)=x+1$, $g(x)=f(x)+2$ and $y=[g(x)]^2$.

(c) Discuss the differences between your final graphs in parts **(a)** and **(b)**.

8 (a) On successive diagrams, draw the graphs of $f(x) = x^3$, $g(x) = f(2x)$ and $y = g(x) + 1$.
(b) On successive diagrams, draw the graphs of $f(x) = x^3$, $g(x) = f(x) + 1$ and $y = g(2x)$.
(c) Discuss the differences between your final graphs in parts **(a)** and **(b)**.

9 (a) On successive diagrams, draw the graphs of $f(x) = \sin x$, $g(x) = f(2x)$ and $y = g(x) - 1$ for $0 \le x \le 2\pi$.
(b) On successive diagrams, draw the graphs of $f(x) = \sin x$, $g(x) = f(x) - 1$ and $y = g(2x)$.
(c) Discuss the differences between your final graphs in parts **(a)** and **(b)**.

10 (a) On successive diagrams, draw the graphs of $f(x) = e^x$, $g(x) = f(x - 1)$ and $y = 2g(x)$.
(b) On successive diagrams, draw the graphs of $f(x) = e^x$, $g(x) = 2f(x)$ and $y = g(x - 1)$.
(c) Discuss the differences between your final graphs in parts **(a)** and **(b)**.

15.4 GRAPHING RATIONAL ALGEBRAIC FUNCTIONS

Functions with the independent variable in the denominator generate curves that are not continuous and may have asymptotes. They may not have any turning points. You need to consider what happens to the function for very large positive and negative values of the variable.

Example 3

Sketch the graph of $y = \dfrac{1}{x-2}$.

Solution

Because $x - 2 \neq 0$, the function is not defined for $x = 2$, so at $x = 2$ there is a vertical asymptote.

For $x > 2$, $x - 2 > 0$, so $y > 0$. As $x \to 2$ from above, $x - 2$ is a very small positive number and so $y \to \infty$. This can be written as: $x \to 2^+$, $y \to \infty$.

For $x < 2$, $x - 2 < 0$, so $y < 0$. As $x \to 2$ from below, $x - 2$ is a very small negative number and so $y \to -\infty$. This can be written as: $x \to 2^-$, $y \to -\infty$.

The numerator of $y = \dfrac{1}{x-2}$ is never zero, so the curve does not cut the x-axis.

For $x = 0$, $y = -0.5$, so the y-intercept is -0.5.

As $x \to \infty$, $y \to 0$ from above; as $x \to -\infty$, $y \to 0$ from below. Thus $y = 0$ is a horizontal asymptote.

For stationary points, find $\dfrac{dy}{dx}$: $\quad \dfrac{dy}{dx} = \dfrac{-1}{(x-2)^2}$, $x \neq 2$

Hence $\dfrac{dy}{dx} < 0$ for all x in the domain, because $(x - 2)^2 > 0$ in the domain.

Thus $y = \dfrac{1}{x-2}$ is a decreasing function in each part of its domain.

Also $\dfrac{dy}{dx} \neq 0$ in the domain, so there are no turning points.

For $x > 2$, $y > 0$; for $x < 2$, $y < 0$.

The curve is concave down for $x < 2$ and concave up for $x > 2$.

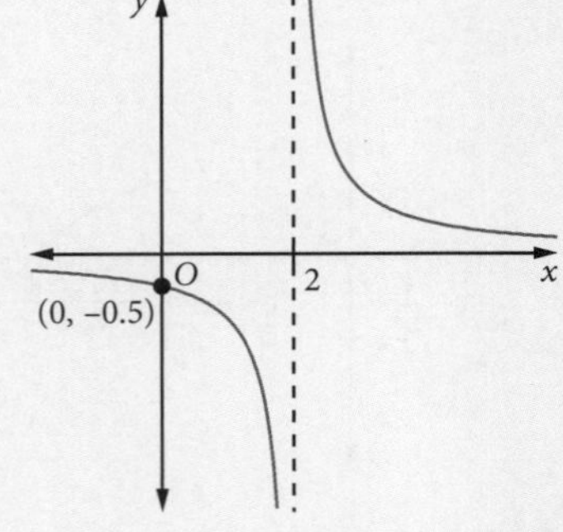

Example 4

Sketch the graph of $y = x + \dfrac{1}{x}$. For what values of x is the curve concave up? What is the range of the function?

Solution

$x \neq 0$: does not cut y-axis and $x = 0$ is a vertical asymptote.

$y = 0$: $x + \frac{1}{x} = 0$ or $\frac{x^2+1}{x} = 0$

Because $x^2 + 1 \neq 0$ for real x: does not cut x-axis.

As $x \to \infty$, $y \to x + [\text{very small amount}] \to x + 0$, so $y \to x$ from above.

As $x \to -\infty$, $y \to x - [\text{very small amount}] \to x - 0$, so $y \to x$ from below.

$\therefore$ $y = x$ is a sloping asymptote.

For stationary points, find $\frac{dy}{dx}$: $\frac{dy}{dx} = 1 - \frac{1}{x^2}, x \neq 0$

Hence for $\frac{dy}{dx} = 0$: $\frac{x^2-1}{x^2} = 0$, so $x = -1, 1$ (for which $y = -2, 2$)

$\therefore$ stationary points at $(-1,-2)$ and $(1,2)$.

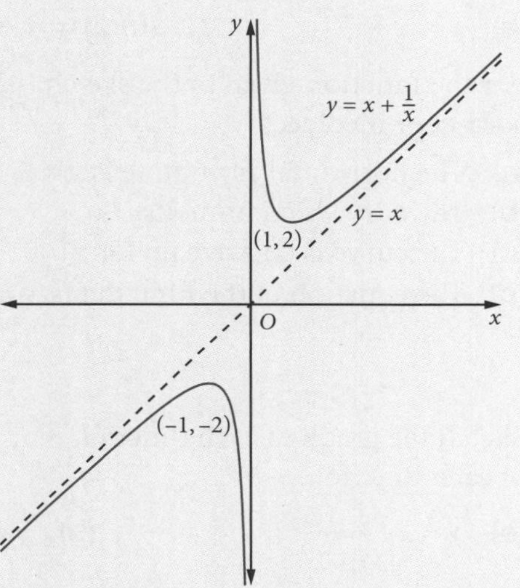

Second derivative: $\frac{d^2y}{dx^2} = 0 - \frac{-2}{x^3} = \frac{2}{x^3}$

At $(-1,-2)$: $\frac{d^2y}{dx^2} = -2 < 0$

$\therefore$ $(-1,-2)$ is a local maximum turning point.

At $(1,2)$: $\frac{d^2y}{dx^2} = 2 > 0$

$\therefore$ $(1,2)$ is a local minimum turning point.

Because $\frac{d^2y}{dx^2} = \frac{2}{x^3}$ and $x \neq 0$, there are no points of inflection.

The curve is concave up for $x > 0$.

The range of the function is real $y, |y| \geq 2$.

Consider: why is the curve between the lines $x = 0$ and $y = x$?

Summary—rational algebraic function graphs

When sketching rational algebraic functions:

- identify any restrictions on the domain and the range
- find the intercepts on the coordinate axes where possible (it is usually easy to find the y-intercept, if it exists, but it is not always possible to find the x-intercept)
- use the symmetry properties of odd and even functions whenever possible
- find stationary points and determine their nature
- find asymptotes and use them to guide the shape of the curve. Asymptotes may be horizontal, vertical, sloping or occasionally another curve.

Remember that the shape of your graph sketches can be verified using graphing software.

EXERCISE 15.4 GRAPHING RATIONAL ALGEBRAIC FUNCTIONS

1 The asymptotes of $y = \dfrac{1}{x+2}$ are:

A $y = 0$ and $x = -2$ **B** $y = 0$ and $x = 2$ **C** $x = 0$ and $y = -2$ **D** $x = 0$ and $y = 2$

2 Sketch the graph of each function. For what values of x is the curve concave down? State the range of each function.

(a) $y = \dfrac{1}{x+2}$ **(b)** $y = \dfrac{1}{x-1}$ **(c)** $y = \dfrac{1}{2-x}$

3 **(a)** Show that the function $y = \dfrac{x-1}{x-2}$ can be written as $y = 1 + \dfrac{1}{x-2}$.

(b) Hence sketch the graph of $y = \dfrac{x-1}{x-2}$, showing all the asymptotes.

4 Sketch the graph of each function, showing all turning points and points of inflection. For what values of x is the curve concave up? State the range of each function.

(a) $y = x + \frac{4}{x}$ **(b)** $y = x - \frac{1}{x}$ **(c)** $y = 2x + \frac{8}{x}$

5 For the function given in the sketch, state whether each statement below is correct or incorrect.

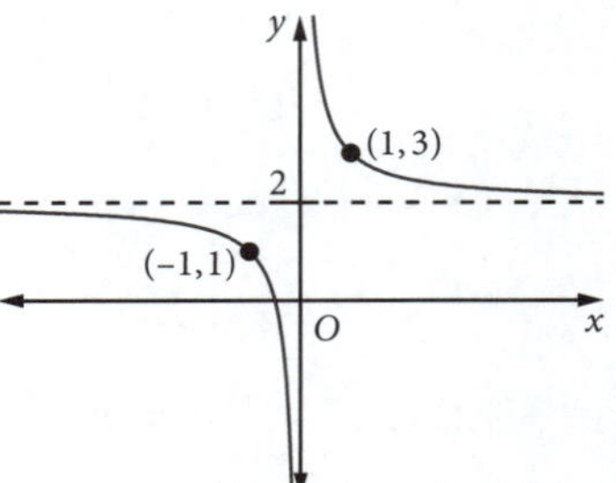

(a) The horizontal asymptote is $y = 2$.
(b) The curve is continuous.
(c) The curve is concave up for $x > 0$.
(d) The equation of the function is $y = 2 + \frac{1}{x}$.

6 Sketch the graph of each function. For what values of x is the curve concave down? State the range of each function.

(a) $y = 1 + \dfrac{1}{x+2}$ **(b)** $y = \dfrac{x-1+1}{x-1} = 1 + \dfrac{1}{x-1}$ **(c)** $y = \dfrac{x-2}{x-3}$

7 Sketch the graph of each function, showing all turning points and points of inflection. State the range of each function.

(a) $y = x + \dfrac{4}{x-1}$ **(b)** $y = x + 2 + \dfrac{4}{x-1}$ **(c)** $y = x + 3 + \dfrac{1}{x-7}$

(d) $y = \dfrac{1}{2x-3}$ **(e)** $y = \frac{x}{4} + \frac{1}{x}$ **(f)** $y = |x| + \frac{1}{x}$

15.5 APPLICATIONS INVOLVING GRAPHING FUNCTIONS

Graphs of functions with algebraic denominators may have asymptotes. You also have information about turning points and points of inflection obtained by using calculus to differentiate the function. You can use all these skills to sketch a variety of functions.

Example 5

(a) Sketch the graph of $y = \dfrac{1}{4x-1}$.
(b) Find the equation of the tangent to the curve at the point where $x = 1$.
(c) Find the equation of the normal to the curve at point where $x = -1$.
(d) Find the coordinates of the point of intersection of the tangent and normal found in parts **(b)** and **(c)**.

Solution

(a)

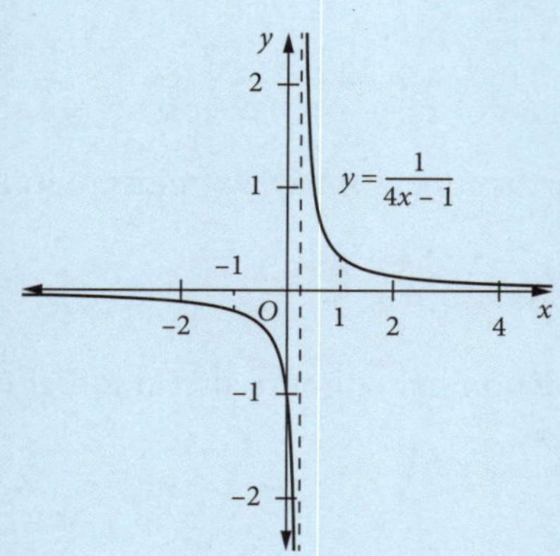

(b) $y = \frac{1}{4x-1}$: $\frac{dy}{dx} = -\frac{4}{(4x-1)^2}$

$x = 1$: Gradient of tangent $= -\frac{4}{9}$

$x = 1, y = \frac{1}{3}$

Equation of tangent: $y - \frac{1}{3} = -\frac{4}{9}(x-1)$

$9y - 3 = -4x + 4$

$4x + 9y - 7 = 0$

(c) $x = -1$: Gradient of tangent $= -\frac{4}{25}$

Gradient of normal $= \frac{25}{4}$

$x = -1, y = -\frac{1}{5}$

Equation of normal: $y + \frac{1}{5} = \frac{25}{4}(x+1)$

$20y + 4 = 125x + 125$

$125x - 20y + 121 = 0$

(d) $4x + 9y - 7 = 0$ [1]

$125x - 20y + 121 = 0$ [2]

[1] × 20: $80x + 180y - 140 = 0$ [3]

[2] × 9: $1125x - 180y + 1089 = 0$ [4]

[3] + [4]: $1205x + 949 = 0$

$x = -\frac{949}{1205}$

Substitute into [1]: $-\frac{3796}{1205} + 9y - 7 = 0$

$9y = \frac{12231}{1205}$

$y = \frac{1359}{1205}$

The point of intersection is $\left(-\frac{949}{1205}, \frac{1359}{1205}\right)$

EXERCISE 15.5 APPLICATIONS INVOLVING GRAPHING FUNCTIONS

1 **(a)** Sketch the graph of $y = \frac{1}{2x-1}$.

(b) Find the equation of the tangent to the curve at the point where $x = 1$.

(c) Find the equation of the normal to the curve at point where $x = -1$.

(d) Find the coordinates of the point of intersection of the tangent and normal found in parts **(b)** and **(c)**.

2 **(a)** Sketch the curve $y = x + \frac{1}{x}$, showing its asymptotes.

(b) Find the coordinates of the turning points of $y = x + \frac{1}{x}$ and determine their nature.

(c) What is the least value of $x + \frac{1}{x}$ over the domain $x > 0$?

3 **(a)** Sketch the graph of $f(x) = e^x + 4e^{-x}$.

(b) For what values of x is $f'(x) > 0$?

(c) What is the minimum value of $f(x)$ and when does it occur?

4 **(a)** Sketch the graph of $f(t) = \frac{5}{2+3e^{-t}}$, $t \geq 0$.

(b) Show that $f'(t) > 0$ for all values of t in the domain.

(c) Find $\lim_{t \to \infty} f(t)$.

(d) What is the range of the function?

5 $f(x)$ is defined by the rule $f(x) = e^{-x}\cos x$ over the domain $-\frac{\pi}{2} \leq x \leq \frac{3\pi}{2}$.

(a) Find the values of $f(0)$, $f\left(\frac{\pi}{2}\right)$, $f(\pi)$.

(b) Find $f'(x)$.

(c) Show that $f'(0) = -1$, $f'\left(\frac{3\pi}{4}\right) = 0$ and $f'\left(-\frac{\pi}{4}\right) = 0$.

(d) Sketch the graph of $y = f(x)$.

(e) Find the maximum value of $f(x)$ over the domain and the value of x for which it occurs.

6 **(a)** For what values of x is $f(x) = \log_e(\sin x)$ defined over the domain $0 \le x \le 2\pi$?
(b) Find $f'(x)$ and state its domain.
(c) Find the maximum value of $f(x)$ over its domain.
(d) Sketch the graph of $y = f(x)$ over its domain.

7 **(a)** If $y = \log_e(1 + \sin x)$, find $\frac{dy}{dx}$ and $\frac{d^2y}{dx^2}$.
(b) For what values of x is $\frac{dy}{dx} = 0$ over $0 \le x \le 4\pi$?
(c) When is the function a maximum?
(d) Explain why the function has no points of inflection.
(e) Sketch $y = \log_e(1 + \sin x)$ for $0 \le x \le 4\pi$.

8 The concentration C of insects per square metre in a forest is given by
$C(t) = 1000\left[\cos\left(\frac{\pi(t-8)}{2}\right) + 2\right]^2 - 1000$, for $8 \le t \le 16$, where t is the number of hours after midnight.
(a) What is the minimum concentration of insects and when does it occur?
(b) Sketch the function for the given domain.

9 The diagram consists of a rectangle surmounted by an isosceles triangle with dimensions as shown.

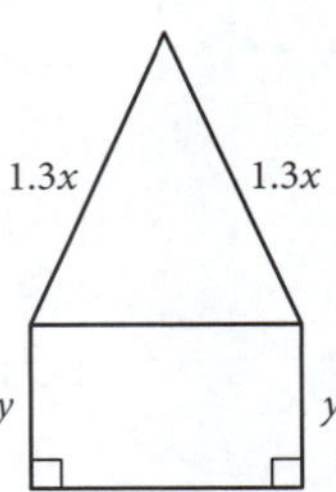

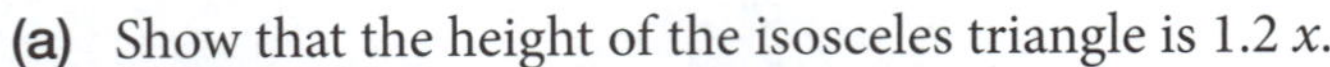
(a) Show that the height of the isosceles triangle is $1.2\,x$.
(b) Show that the total area of the figure is given by $A = xy + 0.6x^2$.
(c) If the perimeter of the figure is 48 metres, express y in terms of x.
(d) Find the expression for $A(x)$ as a function of x only.
(e) Sketch the graph of $y = A(x)$.
(f) Find the dimensions of the diagram that give a maximum area and state that area.

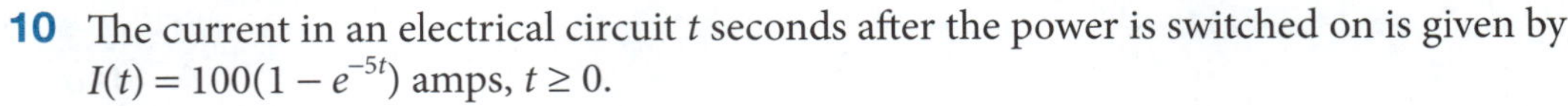
10 The current in an electrical circuit t seconds after the power is switched on is given by $I(t) = 100(1 - e^{-5t})$ amps, $t \ge 0$.

(a) Find the current when $t = 0$, 0.2 and 1 second.
(b) As t increases, describe what happens to the current.
(c) Find $I'(t)$ and draw the graph of $y = I'(t)$.
(d) Draw the graph of $y = I(t)$.
(e) What does the graph in part **(c)** represent on the graph in part **(d)**?

15.6 GRAPHICAL SOLUTION OF EQUATIONS

There are usually standard techniques for solving equations associated with various functions, e.g. linear equations, quadratic equations and trigonometric equations. However, there is usually no standard technique for solving equations that combine two or more different functions, for example:

- $\sin x = e^x$
- $\cos x = \log x$

Equations like this 'transcend' (go beyond) algebraic equation-solving techniques and hence are called **transcendental equations**. You must instead use various non-algebraic methods to find approximate solutions. The most accurate method is to calculate numerical solutions using graphing software. Another method is to sketch the graphs of the two functions, use the graphs to find approximate x-values of any intersection points, then use a calculator to refine these approximate x values as closely as possible. This section extends the scope of the equations covered in Section 12.3.

EXPLORE FURTHER

Graphical solutions of equations

Use technology to solve transcendental equations graphically.

EXERCISE 15.6 GRAPHICAL SOLUTION OF EQUATIONS

1 By drawing graphs of the given functions, determine how many solutions exist for the given equation.

(a) $y = x^2, y = 2x - 1$
Equation: $x^2 - 2x + 1 = 0$

(b) $y = x^2, y = 3x + 1$
Equation: $x^2 - 3x - 1 = 0$

(c) $y = x^2, y = x - 4$
Equation: $x^2 - x + 4 = 0$

(d) $y = x^3, y = 2x$
Equation: $x^3 - 2x = 0$

(e) $y = x^3 - x, y = x^2$
Equation: $x^3 - x^2 - x = 0$

(f) $y = e^x, y = x + 2$
Equation: $e^x - x - 2 = 0$

2 By drawing graphs of the given functions for the given domain, determine how many solutions exist for the given equation.

(a) $y = \sin x, y = \frac{x}{2}$
Domain: $-2\pi \le x \le 2\pi$
Equation: $\sin x - \frac{x}{2} = 0$

(b) $y = \log_e x, y = x - 2$
Domain: $0 \le x \le 2\pi$
Equation: $\log_e x - x + 2 = 0$

(c) $y = 2\cos x, y = \log_e x.$
Domain: $0 \le x \le 2\pi$
Equation: $2\cos x - \log_e x = 0$

(d) $y = e^x, y = \sin x$
Domain: $-2\pi < x < 2\pi$
Equation: $e^x - \sin x = 0$

(e) $y = e^{-x}, y = \sin x$
Domain: $-2\pi \le x \le 2\pi$
Equation: $e^{-x} - \sin x = 0$

(f) $y = e^{-x}, y = \tan x$
Domain: $-\frac{\pi}{2} < x < \frac{3\pi}{2}$
Equation: $e^{-x} - \tan x = 0.$

3 Show graphically that the equation $8\log_{10}(0.1x + 0.5) = 2 - x$ has a solution between $x = 2$ and $x = 4$. Find this solution correct to 2 decimal places.

15.7 REGIONS AND INEQUALITIES

A straight line represents a function (unless it is a vertical line). If the equation of a straight line is changed to an inequality, then the function becomes a relation. It is no longer a straight line, but instead it can be represented graphically by a region in the number plane.

To graph a region on the number plane:

- Graph the equation of the region's boundary.
- Select a point not on the boundary.
- Substitute the coordinates of this point into the equation of the boundary.
- If the point makes the inequality true, then all points on that side of the inequality will also make it true. Shade that side of the boundary to indicate the region. (If the point does not make the inequality true, then the points on the other side of the inequality must, so shade that side instead.)

If the inequality includes '... or equal to', then the boundary is part of the region. If the inequality does not include '... or equal to', then the boundary is not part of the region, so it should be dashed to show this.

Example 6

Graph the region in the number plane represented by each inequality.

(a) $x + y \ge 1$ **(b)** $x + y < 1$ **(c)** $x + y \le 1$ **(d)** $x + y > 1$

Solution

In each case, first draw the line $x + y = 1$. In parts **(b)** and **(d)**, dash the line to show that the boundary is not part of the region. Substitute the non-boundary point $(1, 1)$ into each inequality to find the region side.

(a) $x + y \ge 1$
Solid line
$2 > 1$
LHS $= 1 + 1 = 2 > 1$
Result true:
shade above the line

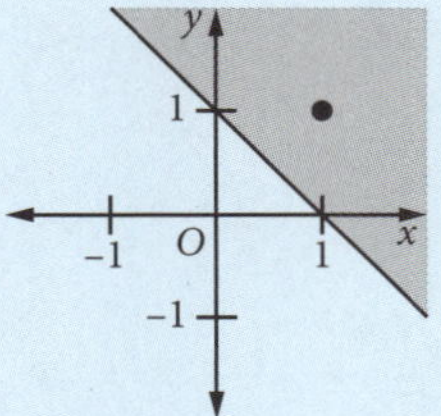

(b) $x + y < 1$
Dashed line
$2 > 1$
LHS $= 1 + 1 = 2 > 1$
Result not true:
shade below the line

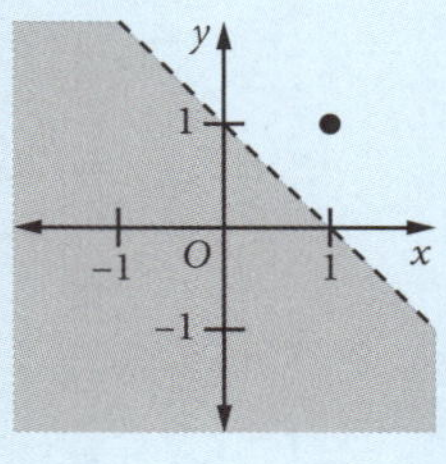

(c) $x + y \leq 1$

Solid line

$2 > 1$

LHS $= 1 + 1 = 2 > 1$

Result not true:
shade below the line

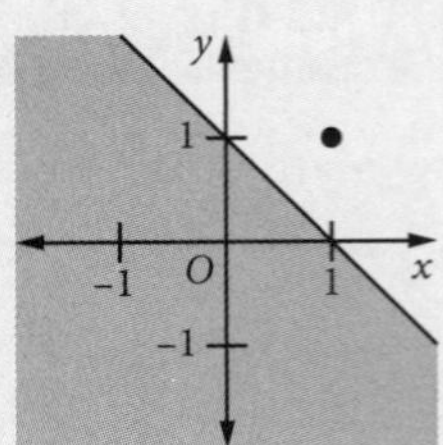

(d) $x + y > 1$

Dashed line

$2 > 1$

LHS $= 1 + 1 = 2 > 1$

Result true:
shade above the line

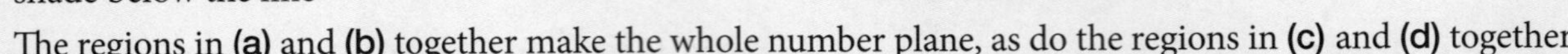

The regions in **(a)** and **(b)** together make the whole number plane, as do the regions in **(c)** and **(d)** together.

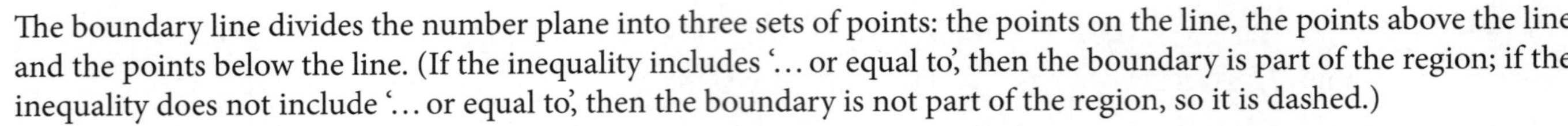

The boundary line divides the number plane into three sets of points: the points on the line, the points above the line and the points below the line. (If the inequality includes '… or equal to', then the boundary is part of the region; if the inequality does not include '… or equal to', then the boundary is not part of the region, so it is dashed.)

Thus for the example above, the region in part **(a)** could be described as 'the set of points on or above the line with equation $x + y = 1$'. The region in part **(b)** could be described as 'the set of points below the line with equation $x + y = 1$'.

Example 7

Graph the region in the number plane represented by each inequality.

(a) $y \leq 2$ **(b)** $-1 < y \leq 2$ **(c)** $x > 1$ **(d)** $x > 1$ or $x \leq -1$

Solution

These inequalities have only one variable, so their boundaries are either horizontal lines (as in **(a)** and **(b)**) or vertical lines (as in **(c)** and **(d)**).

(a) $y \leq 2$

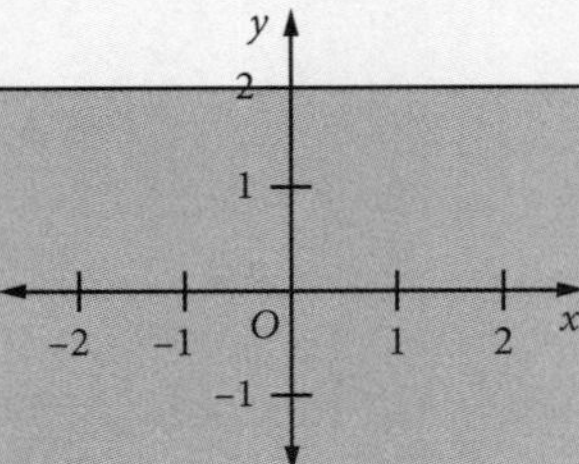

Region is below the line

(b) $-1 < y \leq 2$

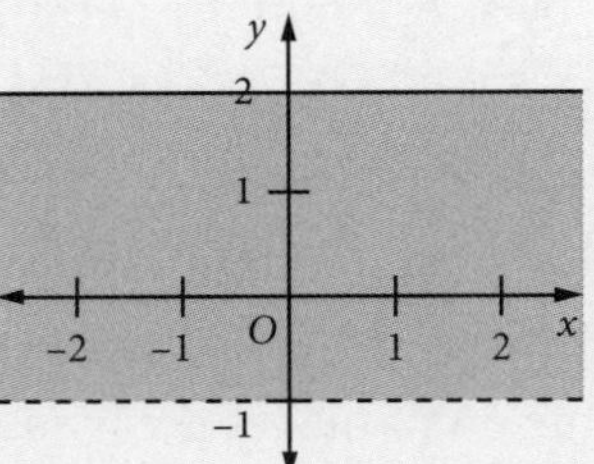

Region is between the lines

(c) $x > 1$

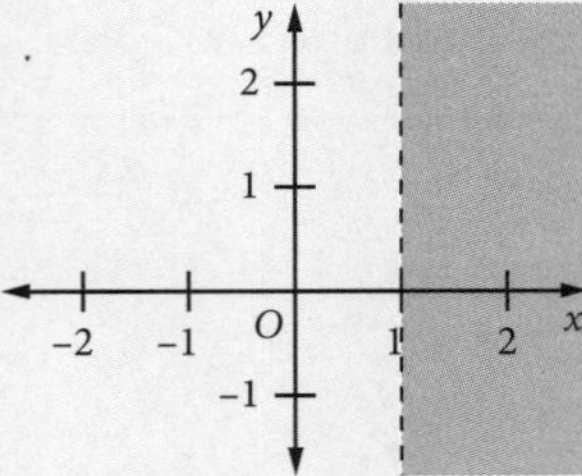

Region is to the right of the line

(d) $x > 1$ or $x \leq -1$

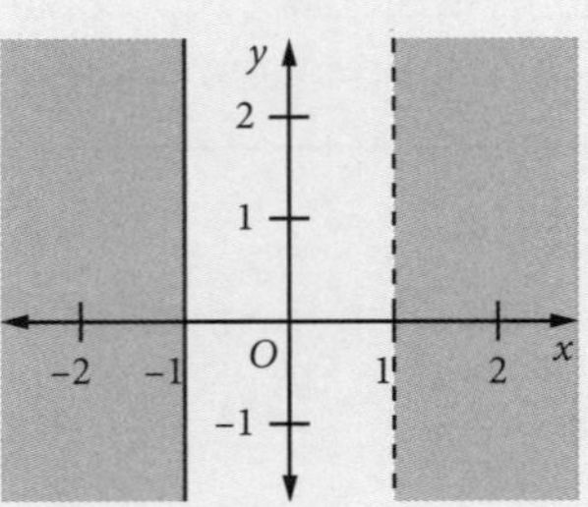

Region is outside the lines

Example 8

(a) Sketch the region defined by the intersection $y \geq x^2 - 1$ and $y \leq 3x + 3$.

(b) Hence write the solution to $x^2 - 3x - 4 \leq 0$.

(c) Solve $x^2 - 3x - 4 \leq 0$ algebraically to check your solution to **(b)**.

(d) What would be different in this process if you were solving $x^2 - 3x - 4 < 0$?

Solution

(a)

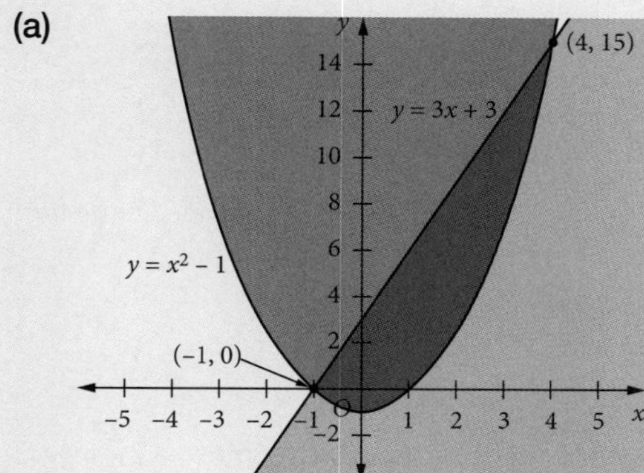

(b)

	$x^2 - 3x - 4 \leq 0$
Rewrite in terms of given equations:	$x^2 - 1 \leq 3x + 3$
Find the points of intersection of the graphs by using simultaneous equations, setting each rule equal to the other:	(−1, 0) and (4, 15)
Roots of $x^2 - 1 = 3x + 3$ are:	$x = -1, 4$
Required region is between the curves:	Solution is $-1 \leq x \leq 4$

(c)

	$x^2 - 3x - 4 \leq 0$
Factorise:	$(x + 1)(x - 4) \leq 0$
The roots of $x^2 - 3x - 4 = 0$ are:	$x = -1, 4$

Pick a value of x between −1 and 4, e.g. $x = 0$ and substitute into the quadratic expression.

	To test $x^2 - 3x - 4 \leq 0$ use $x = 0$.
$x = 0$:	$0^2 - 3(0) - 4 \leq 0$
	$-4 \leq 0$

Since this value makes the inequality true, it must lie in the region defined where $-1 \leq x \leq 4$.

(d) Graphically, the boundary would not be included so the parabola and the lines would be dashed. The solution would not include equality, it would be $-1 < x < 4$.

EXERCISE 15.7 REGIONS AND INEQUALITIES

Shade the region represented by each inequality.

1 $y > 2x$ **2** $y < x + 1$ **3** $x \leq 3$ **4** $y \geq 1$

5 $y \leq x + 2$ **6** $2x + 3y \geq 6$ **7** $2x + y < 1$ **8** $0 \leq x \leq 2$

9 $-2 < y \leq 3$ **10** $3x - 4y \leq 6$ **11** $x - y < -2$ **12** $2y - 5x < 10$

13 $x \leq 1.5$ **14** $\frac{x}{2} + \frac{y}{3} < 1$ **15** $-3 < x + y < 2$ **16** $0 \leq x - y < 3$

17 Which inequality defines the shaded region?

A $3x - 2y + 6 \geq 0$
B $3x - 2y + 6 \leq 0$
C $3x + 2y - 6 \geq 0$
D $3x + 2y - 6 \leq 0$

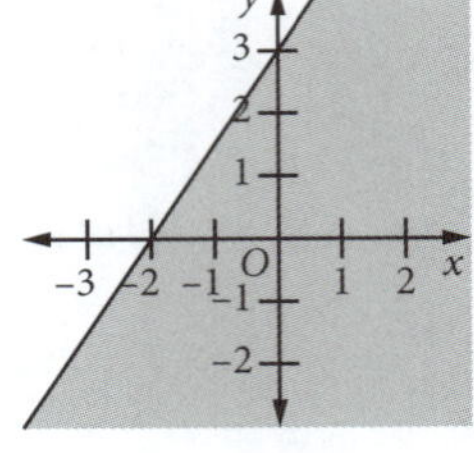

18 Shade the region in the number plane that:

(a) is above the line $y = 3x + 2$
(b) is on or below the line $x - 2y + 4 = 0$
(c) is above the line $y = -1$
(d) is on or to the left of the line $x = 3$
(e) is between the lines $2x + 3y + 6 = 0$ and $2x + 3y - 6 = 0$
(f) is on or above the x-axis.

19 Describe in words the regions defined by the following inequalities:

(a) $y < x + 2$ (b) $y \geq x$ (c) $x > 3$ (d) $y \leq 4$ (e) $x + 3y \leq 9$ (f) $-2 \leq x < 3$

20 For this graph, indicate whether each statement is correct or incorrect.

(a) The equation of the boundary is $x + 2y - 2 = 0$.
(b) The gradient of the boundary line is $\frac{1}{2}$.
(c) The inequality for the region is $x + 2y - 2 > 0$.
(d) The inequality for the region is $x + 2y - 2 < 0$.

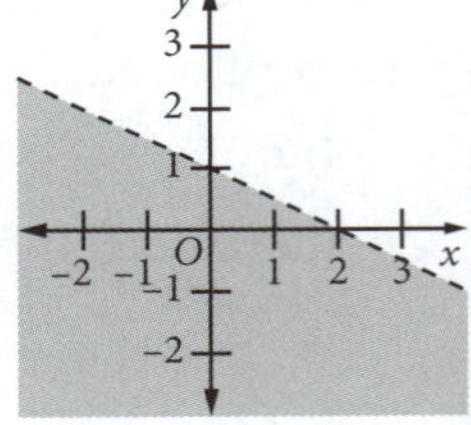

21 (a) Sketch the region defined by the intersection $y \geq x^2 - 1$ and $y \leq 3 - 3x$.
(b) Hence write the solution to $x^2 + 3x - 4 \leq 0$.

22 (a) Sketch the region defined by the intersection $y \leq 3 - x^2$ and $y \geq 2x$.
(b) Hence write the solution to $x^2 + 2x - 3 \leq 0$.

23 (a) Sketch the region defined by the intersection $y \geq x^2 + x$ and $y \leq 2x + 2$.
(b) Hence write the solution to $x^2 - x - 2 \leq 0$.

24 (a) Sketch the region defined by the intersection $y < 2x - x^2$ and $y \geq 3x - 2$.
(b) Hence write the solution to $x^2 + x - 2 < 0$.
(c) Solve $x^2 + x - 2 > 0$ algebraically to check your solution to **(b)**.

25 (a) Sketch the region defined by the intersection $y \geq x^2$ and $y \leq x + 3$.
(b) Hence write the solution to $x^2 - x - 3 \leq 0$.
(c) Solve $x^2 - x - 3 \leq 0$ algebraically to find the exact solution to **(b)**.

15.8 SIMULTANEOUS LINEAR INEQUALITIES

Two linear equations may intersect at a point. The intersection of two linear inequalities is the region common to the two inequalities. This is the region where both inequalities hold simultaneously.

Example 9

(a) Sketch the region given by $y \geq x$. **(b)** Sketch the region given by $x + y < 2$.

(c) Sketch the region common to $y \geq x$ and $x + y < 2$.

Solution

(a) $y \geq x$

Test the point $(0, 1)$

Test: LHS ≥ RHS using (0,1)

$y > -x$, sub in $y = 1$ and $x = 0$

$1 >= 0$, which is true.

Shade region above line

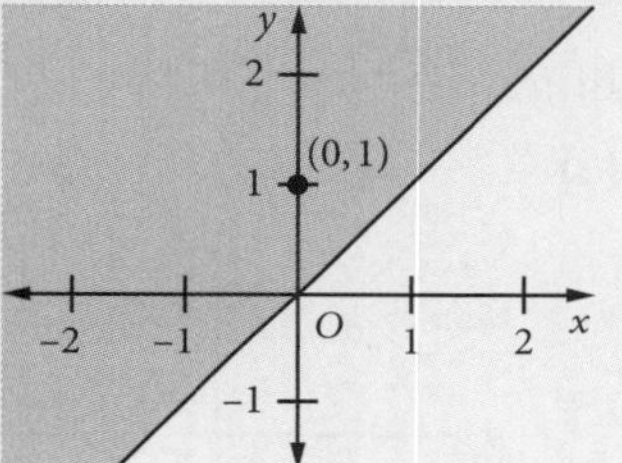

(b) $x + y < 2$

Test the point $(0, 0)$

Test LHS < RHS using the point (0,0)

$x + y < 2$, where $x = 0$ and $y = 0$

$0 + 0 < 2$, which is true.

Shade region below dashed line

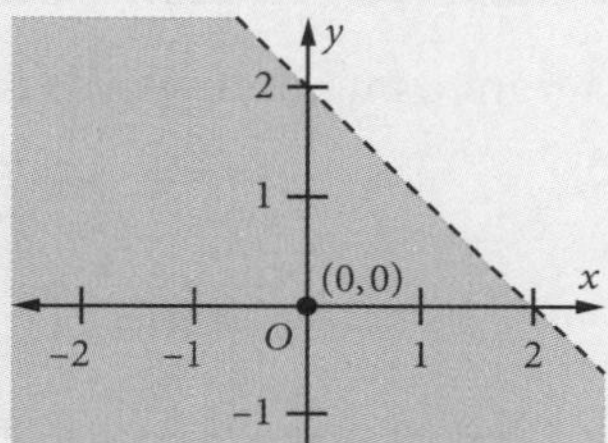

(c) $y \geq x$ and $x + y < 2$

Identify common region

Shade clearly, using a darker shading for the common region

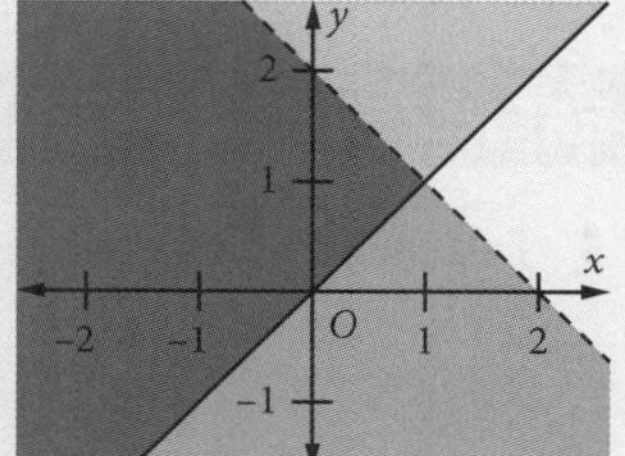

Example 10

Sketch the region defined by each pair of inequalities. Describe the region in words.

(a) $x \leq 2, y > -1$ **(b)** $y \leq x + 1, x \leq 1$

Solution

Draw each boundary and shade the two regions, then shade the common region differently.

(a) $x \leq 2, y > -1$

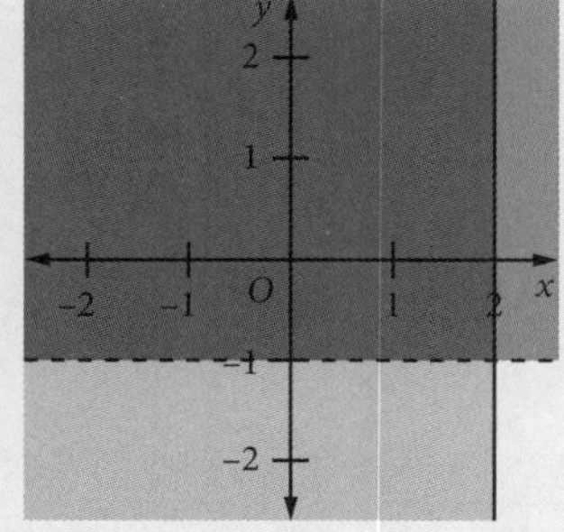

The region on and to the left of the line $x = 2$ that is also above the line $y = -1$.

(b) $y \leq x + 1, x \leq 1$

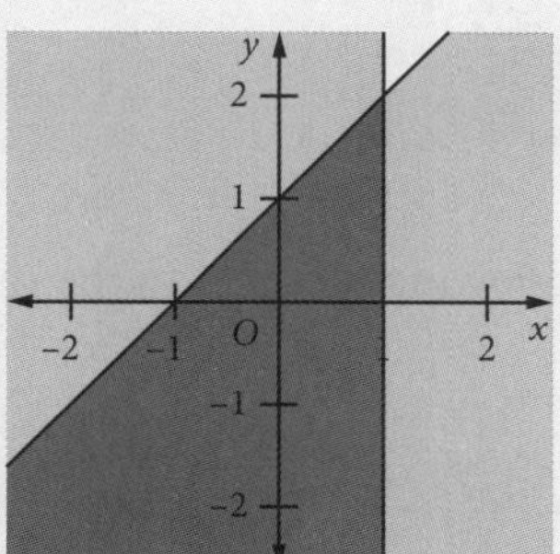

The region on and below the line $y = x + 1$ that is also on and to the left of the line $x = 1$.

If the shading of different regions becomes difficult to show, you should just lightly shade the original regions before darkening the final answer, as in part **(b)**.

Example 11

Sketch the region defined by the three inequalities $x - y \geq -1$, $x + 3y \geq -1$, $5x + 3y \leq 19$. Show the points of intersection of the lines. Describe the region in words.

Solution

To find the points of intersection of the lines, solve pairs of equations simultaneously:

- The lines $x - y = -1$ and $5x + 3y = 19$ intersect at $A(2, 3)$
- The lines $x - y = -1$ and $x + 3y = -1$ intersect at $B(-1, 0)$
- The lines $5x + 3y = 19$ and $x + 3y = -1$ intersect at $C(5, -2)$

The shaded region is the interior of the triangle bounded by the lines $x - y = -1$, $x + 3y = -1$ and $5x + 3y = 19$. The vertices of the region are the intersection points A, B and C.

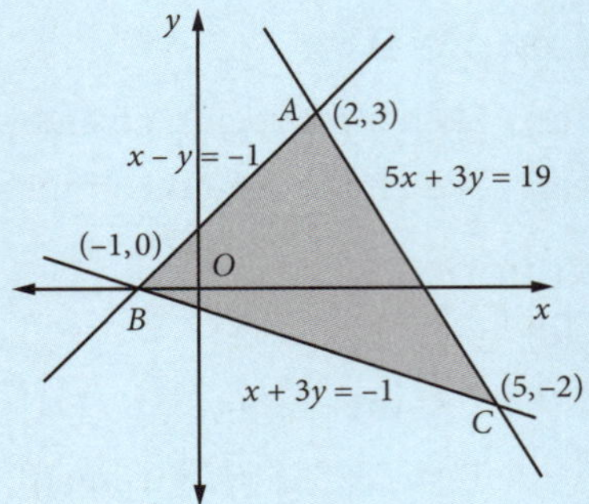

EXERCISE 15.8 SIMULTANEOUS LINEAR INEQUALITIES

1 Describe the shaded region in each diagram using both words and inequalities.

(a)

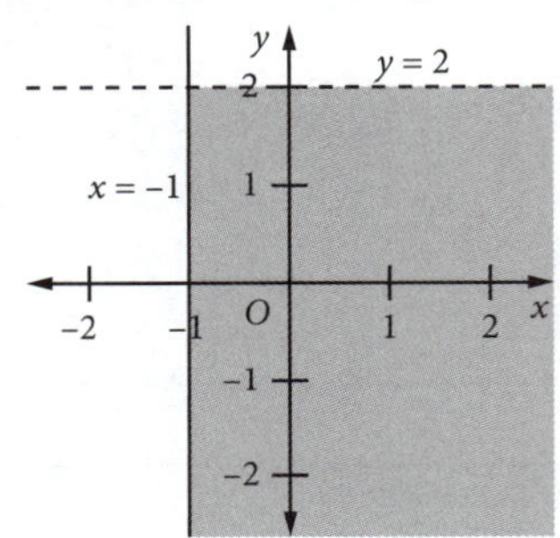

(b)

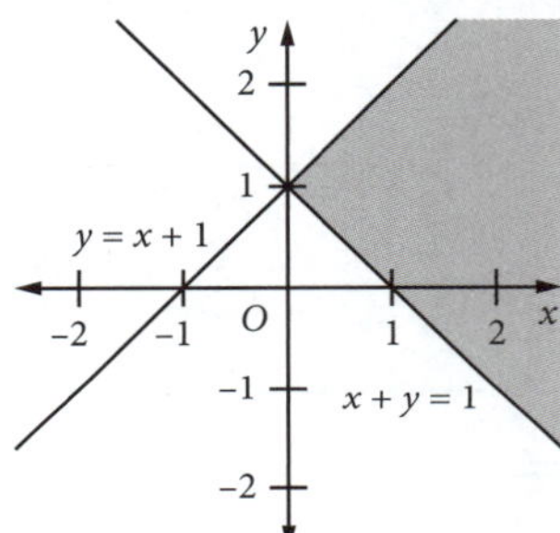

(c)

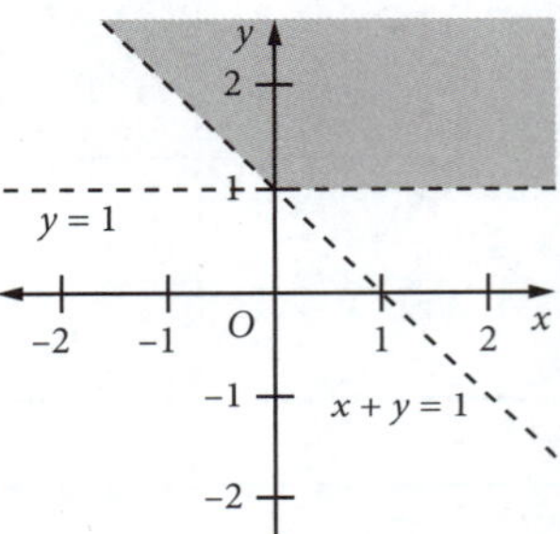

(d)

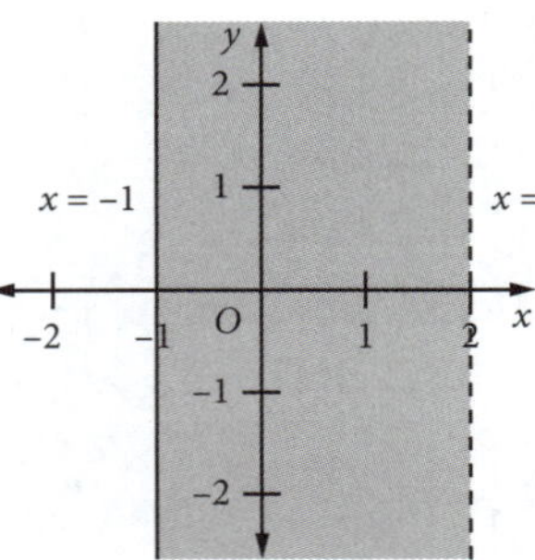

(e)

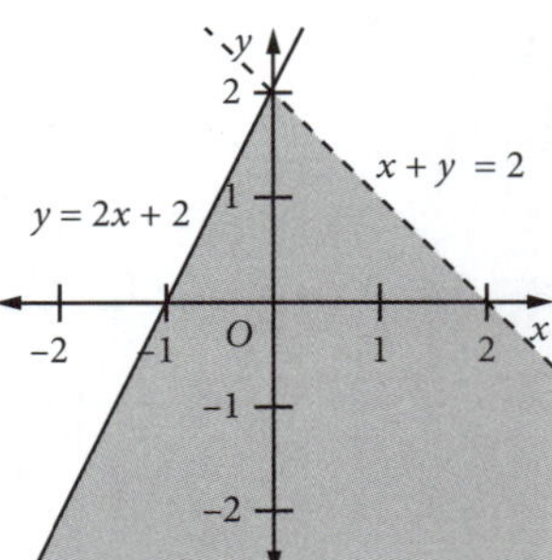

(f)

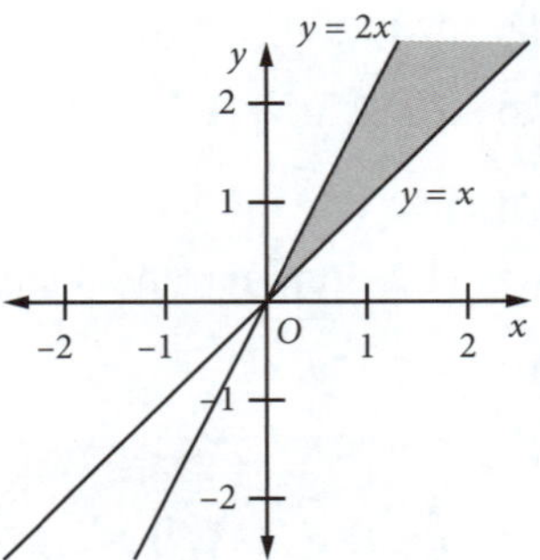

2 Which of the following points is in the shaded region?

A $(1, 3)$ **B** $(1, 1)$
C $(3, 1)$ **D** $(-1, 3)$

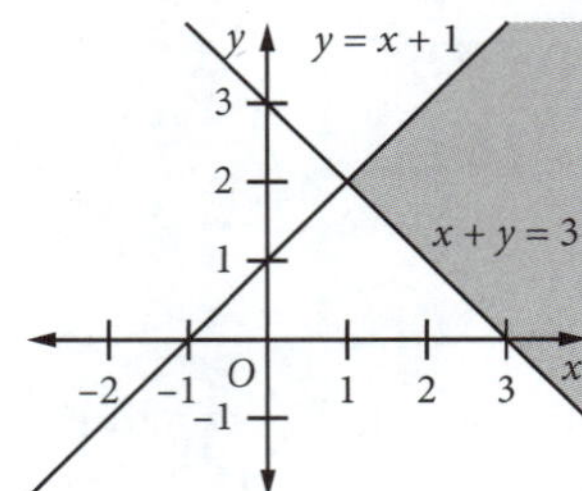

3 What are the inequalities that define the region graphed in question **2**?

4 Graph the regions defined by each set of inequalities. State whether each of the given points is in the region.

(a) $x+y \le 3, y \le x$
$(0,0), (2,3), (-1,-2)$

(b) $2y > x+2, x+y > -1$
$(0,0), (0,1), (2,5)$

(c) $x+2y \ge 8, y < 7$
$(0,4), (-1,1), (9,2)$

(d) $3y \le 2x+6, x+y > 2$
$(2,0), (3,3), (4,-1)$

(e) $4x+y \le 4, x \ge -2$
$(0,0), (-3,1), (1,0)$

(f) $y > 3x+3, x+y < 3$
$(0,3), (2,7), (-1,4)$

5 The two lines divide the number plane into four regions labelled on the diagram A, B, C and D. Indicate whether the inequalities given for each region are correct or incorrect.

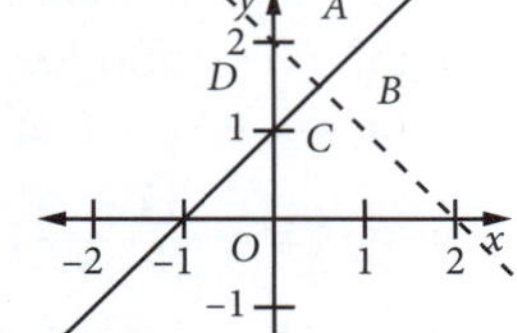

(a) $A: x+y > 2, x-y \le -1$
(b) $B: x+y > 2, x-y \le -1$
(c) $C: x+y < 2, x-y \ge -1$
(d) $D: x+y < 2, x-y \le -1$

6 Shade the region defined by the inequalities. By solving the inequalities simultaneously in pairs, find the coordinates of the vertices of the figure.

(a) $y \le x+2, 2x+y \le 4, x+y \ge 2$
(b) $2y-x \le 4, y > 3x-6, 3x+y > -6$
(c) $y-2x < 4, y+2x < 6, y \ge x+2$
(d) $y-3x < 3, 3x+4y < 12, x-2y < 4$
(e) $y \ge x-1, x+y \le 2, x \ge 0, y \ge 0$
(f) $x+y \le 2, x+y \ge -2, x-y \le 2, x-y \ge -2$

7 Describe (in words and using inequalities) the regions that are shaded in each part. Find the coordinates of the points A, B and C in each case by solving simultaneously.

(a)

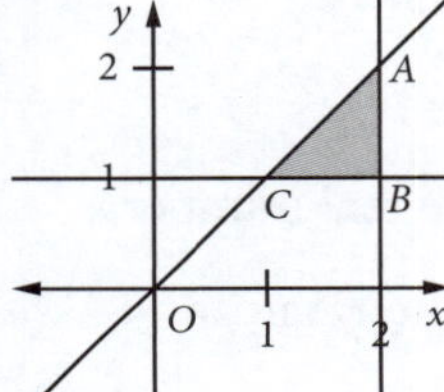

(b)

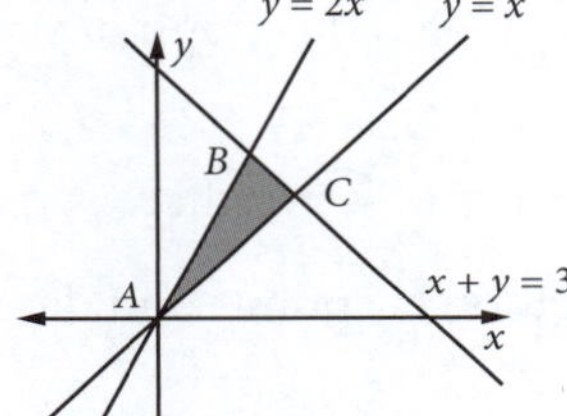

(c)

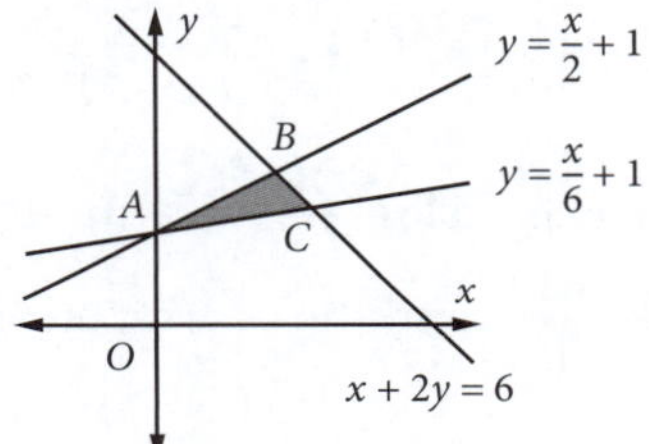

CHAPTER REVIEW 15

1 On the same diagram, draw the graphs of $y = x^3$, $y = (x+3)^3$, $y = x^3 + 3$.

2 On the same diagram, draw the graph of each equation, stating the dilation and other transformations from the first graph: $y = \ln x$, $y = 4 \ln x$, $y = 4 \ln(x+2)$.

3 On the same diagram, draw the graphs for $0 \le x \le 2\pi$:

(a) $y = \sin x, y = \sin\left(x - \frac{\pi}{6}\right), y = \sin\left(x + \frac{\pi}{3}\right)$
(b) $y = \sec x, y = \sec x - 2, y = \sec x + 1$

4 If $f(x) = e^x$, write down the new equation obtained by applying the condition given in each part.

(a) $f(2x)$ **(b)** $f(x-3)$ **(c)** $f(x)+1$ **(d)** $2f(x)+4$ **(e)** $3f(2(x+2)) - 1$

5 Haulage Company makes frequent deliveries from Sydney to Melbourne and calculates that the overhead cost \$$C$ depends on the average delivery speed v km h^{-1} according to the rule $C = v + \frac{6400}{v}$.

(a) What is the domain of this function?
(b) Sketch the function for this domain.
(c) Find the average delivery speed to minimise the overhead cost.

6 Figure $AEBFCD$ consists of the rectangle $ABCD$ topped by an equilateral triangle AEB on the side AB, with another equilateral triangle BFC on the side BC, as shown in the diagram. The perimeter of the figure is 54 cm.

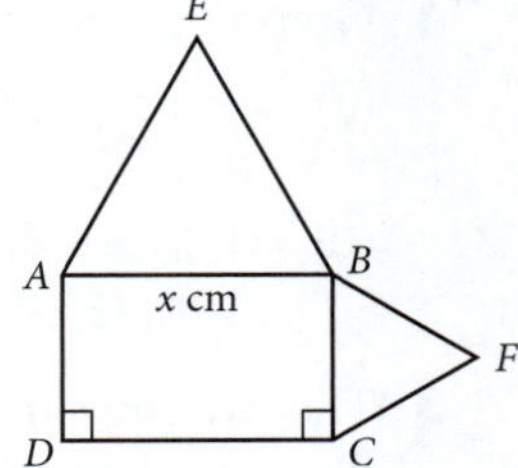

(a) Obtain a formula for the area of the figure, $A(x)$, in terms of x, the length of the side AB.
(b) What is the domain of this function?
(c) Sketch the function for this domain.
(d) Find the dimensions of this rectangle when the total area is a maximum.

7 (a) Given $f(x) = 4 + \frac{3x-1}{x^2}$, find:
(i) the values of x for which $f(x) = 0$ **(ii)** $\lim_{x\to\infty} f(x)$
(iii) the value of $f(x)$ as $x \to 0$ **(iv)** the equation of any asymptotes
(v) the coordinates of the stationary points and determine their nature
(vi) the coordinates of any points of inflection.
(b) Sketch the graph of $y = f(x)$, showing the information obtained from part **(a)**.

8 (a) If $y = \frac{4x}{(x-1)^2}$, find:
(i) the coordinates of the stationary points and determine their nature
(ii) the coordinates of any points of inflection **(iii)** the equations of any asymptotes.
(b) Sketch the graph of $f(x) = \frac{4x}{(x-1)^2}$. **(c)** Sketch the graph of $y = f'(x)$.

9 (a) Write the expression $\frac{x(x+1)}{x-1}$ in the form $px + q + \frac{r}{x-1}$, where p, q and r are real numbers.
(b) If $y = \frac{x(x+1)}{x-1}$, find the coordinates of the stationary points and determine their nature.
(c) Sketch the graph of $y = \frac{x(x+1)}{x-1}$. **(d)** On your diagram, draw the lines $y = x$ and $y = x + 3$.
(e) For what value of c will the line $y = x + c$ not intersect the graph?

10 Draw the graph of $y = \sin 2x$ for $-\frac{\pi}{2} \le x \le \frac{\pi}{2}$ and use it to solve: **(a)** $\sin 2x = \frac{x}{3}$ **(b)** $\sin 2x = 1 - x$.

11 (a) On successive diagrams, draw the graphs of $f(x) = x^2$, $g(x) = f(3x)$ and $y = g(x) - 2$.
(b) On successive diagrams, draw the graphs of $f(x) = x^2$, $g(x) = f(x) - 2$ and $y = g(3x)$.
(c) Discuss the differences between your final graphs in parts **(a)** and **(b)**.

12 (a) On successive diagrams, draw the graphs of $f(x) = \cos x$, $g(x) = [f(x)]^2$ and $y = g(x) + 1$ for $0 \le x \le 2\pi$.
(b) On successive diagrams, draw the graphs of $f(x) = \cos x$, $g(x) = f(x) + 1$ and $y = [g(x)]^2$ for $0 \le x \le 2\pi$.
(c) Discuss the differences between your final graphs in parts **(a)** and **(b)**.

13 (a) Sketch the region defined by the intersection $y \ge x^2 + 4$ and $y \le 2x + 4$.
(b) Hence write the solution to $x^2 - 2x \le 0$.

CHAPTER 16
The anti-derivative

16.1 PRIMITIVE FUNCTIONS

Previously you have practised finding the derivative of a function. But can the process be reversed? Given a derivative, can you find the original function that was differentiated? The short answer is: sometimes!

The process of finding the function from the derivative is called **anti-differentiation** or **finding the primitive** of the function.

If $F(x)$ is a function such that $F'(x) = f(x)$, then $F(x)$ is called the **anti-derivative** or **primitive** of $f(x)$. Similarly, $f(x)$ is the primitive of $f'(x)$. For example:

- The derivative of x^2 is $2x$, so x^2 is a primitive of $2x$.
- The derivative of $x^2 + 3$ is $2x$, so $x^2 + 3$ is a primitive of $2x$.
- For any real number C, the derivative of $x^2 + C$ is $2x$, so $x^2 + C$ is a primitive of $2x$.

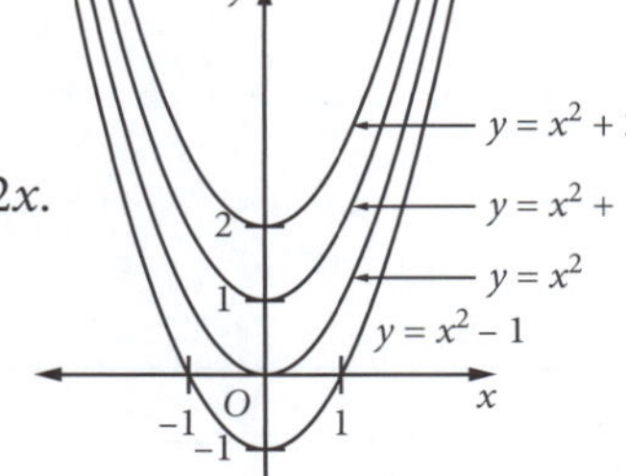

These primitives all vary only by a constant. Graphically, they are the same curve translated vertically by a fixed amount. The equations are all of the form $y = x^2 + C$ where C is any real number.

If the tangent at $x = 1$ is drawn for these curves then the tangents form a system of parallel lines, because each curve has the same gradient function.

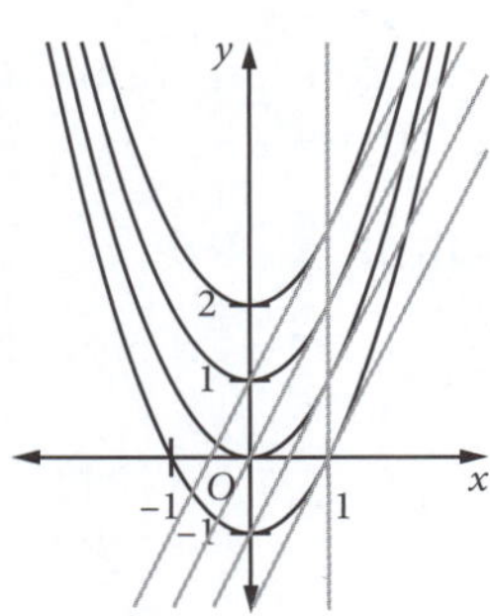

Example 1

Find the equation of the curve defined by $f'(x) = 2x$ that passes through the point $(2, 8)$.

Solution

$f'(x) = 2x$, find the primitive: $\quad f(x) = x^2 + C$

Passes through $(2, 8)$: $\quad 8 = 2^2 + C$

$C = 4$

Hence $f(x) = x^2 + 4$ is the equation of the curve.

From your experience differentiating powers of x, you can say:

- the primitive of x^2 is $\dfrac{x^3}{3} + C$
- the primitive of x^3 is $\dfrac{x^4}{4} + C$
- the primitive of $x^{\frac{1}{2}}$ is $\dfrac{x^{\frac{3}{2}}}{\left(\frac{3}{2}\right)} + C$ or $\dfrac{2}{3}x^{\frac{3}{2}} + C$

- the primitive of x^n is $\dfrac{1}{n+1}x^{n+1}+C$

This result is easily verified by differentiating $\dfrac{1}{n+1}x^{n+1}+C$ to obtain x^n.

In words: 'to find the primitive of a power of x, increase the power of x by one and divide the expression by the new power'.

Example 2

Find the primitive of each expression.

(a) $5x^4+3x^2+4$ (b) $16-8x+x^3$ (c) $x^{10}+4$

Solution

(a) $f'(x)=5x^4+3x^2+4$:

$$f(x)=x^5+x^3+4x+C$$

(b) $f'(x)=16-8x+x^3$:

$$f(x)=16x-\frac{8x^2}{2}+\frac{x^4}{4}+C$$

$$f(x)=16x-4x^2+\frac{x^4}{4}+C$$

(c) $f'(x)=x^{10}+4$:

$$f(x)=\frac{x^{11}}{11}+4x+C$$

Example 3

The gradient function of a curve is $3x^2-2x$ and the curve passes through the point $(2, 1)$. Find the equation of the curve.

Solution

$\dfrac{dy}{dx}=3x^2-2x$, find the primitive: $y=\dfrac{3x^3}{3}-\dfrac{2x^2}{2}+C$

Simplify: $y=x^3-x^2+C$

At point $x=2, y=1$: $1=8-4+C$

$C=-3$

The equation of the curve is $y=x^3-x^2-3$.

Example 4

Find $f(x)$ given $f'(x)$.

(a) $f'(x)=x+\dfrac{1}{x^2}$ (b) $f'(x)=x\sqrt{x}+1$ (c) $f'(x)=\dfrac{x^3+2x^2+1}{x^2}$

Solution

(a) $f'(x)=x+\dfrac{1}{x^2}$, rewrite using index notation: $f'(x)=x+x^{-2}$

Find primitive: $f(x)=\dfrac{x^2}{2}+\dfrac{x^{-1}}{-1}+C$

$$f(x)=\frac{x^2}{2}-\frac{1}{x}+C$$

(b) $f'(x) = x\sqrt{x} + 1$, rewrite using index notation: $f'(x) = x^{\frac{3}{2}} + 1$

Find primitive: $f(x) = \dfrac{x^{\frac{5}{2}}}{\frac{5}{2}} + x + C$

$f(x) = \dfrac{2}{5}x^{\frac{5}{2}} + x + C$

(c) If there is a single term in the denominator, divide the numerator by the denominator before finding the primitive.

$f'(x) = \dfrac{x^3 + 2x^2 + 1}{x^2}$: $\quad f'(x) = x + 2 + \dfrac{1}{x^2}$

Rewrite using indices: $\quad f'(x) = x + 2 + x^{-2}$

Find primitive: $\quad f(x) = \dfrac{x^2}{2} + 2x + \dfrac{x^{-1}}{-1} + C$

$f(x) = \dfrac{x^2}{2} + 2x - \dfrac{1}{x} + C$

Example 5

A colony of bats has a growth rate $\dfrac{dN}{dt} = 80t$, where t is the number of years since 2018 and $N(t)$ is the population. If there were 3000 bats in the year 2018, how many bats are in the colony in 2048?

Solution

$t = 0, N = 3000, \dfrac{dN}{dt} = 80t$

Find the primitive: $\quad N(t) = 40t^2 + C$

But $N(0) = 3000$: $\quad 3000 = C$

$N(t) = 40t^2 + 3000$

In 2048, $t = 30$: $\quad N(30) = 40 \times 900 + 3000 = 39\,000$

There would be 39 000 bats in the colony in 2048.

EXERCISE 16.1 PRIMITIVE FUNCTIONS

1 Find the primitive of:

(a) $6x^2 - 4x + 5$ (b) $3 + 5x + x^2 - 3x^3$ (c) $x^2 - 1$
(d) $x^5 - 4x^3 - 2x^2 + 1$ (e) $3x - 5$ (f) $2x^2 - 7x + 5$

2 If $f'(x) = (x - 1)(x - 2)$, indicate whether each statement below is correct or incorrect.

(a) $f'(x) = x^2 - 3x + 2$ (b) $f(x) = \dfrac{x^3}{3} - \dfrac{3x^2}{2} + 2x + C$

(c) $f(x) = (x - 1)^2(x - 2)^2 + C$ (d) $f(x) = \dfrac{(x-1)^2(x-2)^2}{4} + C$

3 Find an expression for $f(x)$ given:

(a) $f'(x) = (2x + 1)^2$ (b) $f'(x) = 5$ (c) $f'(x) = x^2 + 3x$
(d) $f'(x) = (2x - 1)(x + 2)$ (e) $f'(x) = 4x^3 - 6x^2 + x$ (f) $f'(x) = (x + 3)^2$

4 Show that $\dfrac{(2x+1)^3}{6} = \dfrac{4x^3}{3} + 2x^2 + x + C$ where C is an arbitrary constant.

5 Express y in terms of x, given that:

(a) $\frac{dy}{dx}=3+2x-3x^2$ (b) $\frac{dy}{dx}=x^3+2x^2$ (c) $\frac{dy}{dx}=x^4-x^3$

(d) $\frac{dy}{dx}=(x-3)(x+4)$ (e) $\frac{dy}{dx}=x^4-\frac{1}{3}x^3+\frac{1}{2}x^2$ (f) $\frac{dy}{dx}=\sqrt{x}$

6 For each of the following find $F(x)$, given that $F'(x)=f(x)$.

(a) $f(x)=x^3-3x^2$ (b) $f(x)=\frac{x^2}{5}-\frac{x^3}{4}$ (c) $f(x)=x^2(1-3x)$

(d) $f(x)=(x^2-1)(x^2+1)$ (e) $f(x)=x^{-\frac{3}{2}}+x^{-\frac{5}{2}}$ (f) $f(x)=\sqrt{x}+\sqrt[3]{x}$

7 Find $f(x)$ given $f'(x)=2x-2$ and $f(1)=4$.

8 If $f'(x)=4x^2-3x$ and $f(-1)=3$ then $f(x)=\ldots$

A $f(x)=\frac{4x^3}{3}-\frac{3x^2}{2}+C$ **B** $f(x)=8x-3$ **C** $f(x)=\frac{4x^3}{3}-\frac{3x^2}{2}+\frac{35}{6}$ **D** $f(x)=\frac{x^3}{12}-\frac{x^2}{6}+\frac{13}{4}$

9 At all points on a certain curve, $\frac{dy}{dx}=4x-6$. The point $(2,4)$ is on the curve. Find the equation of the curve.

10 Find the equation of a curve that passes through the point $(3,3)$ and for which the gradient function at any point $P(x,y)$ is $3x^2-2x+3$.

11 For the function $f(x)=(x+2)^2$, find the rule that defines the function F where $F'(x)=f(x)$ and $F(1)=4$.

12 Find the equation of a curve given that $\frac{dy}{dx}=2x+b$ at any point P and that when $x=3$, $\frac{dy}{dx}=2$ and $y=-3$.

13 Find the rule that defines a function f for which $f'(x)=x^2-2x+b$ for all x, $f'(0)=1$ and $f(0)=2$.

14 A curve contains the point $(0,4)$ and its gradient is $(x-1)(x+2)$ at any point on the curve. Find the equation of the curve.

15 Find the anti-derivative of:

(a) $3+\frac{1}{x^2}-\frac{2}{x^3}$ (b) $\frac{x^2+2}{x^2}$ (c) $\frac{3x^3-2x^2+x^{-1}}{x^2}$ (d) $\frac{1}{x^3}$ (e) $\frac{5}{x^2}+2x^{\frac{1}{2}}+3$ (f) $\frac{1}{\sqrt{x}}+\frac{1}{x\sqrt{x}}$

16 For $\frac{ds}{dt}=12t^2-6t+1$, find s in terms of t given that $s=4$ when $t=1$.

17 A new school started in 2006 with 50 students and grew at 60 students per year for the next six years. If $N(t)$ is the number of students in the school and t is the number of years after 2006, then the growth rate is given by $\frac{dN}{dt}=60$. Use calculus to find the population of the school in 2012.

18 If velocity v is the rate of change of distance d as a function of time t, find the distance function if $v=3t^2+4$ and $d=0$ when $t=0$.

16.2 INDEFINITE INTEGRALS

The process of finding a function's primitive has its own notation and is called the **indefinite integral**, $\int f(x)dx$. This uses the integral sign $\int$ as a way of saying 'find the primitive of $f(x)$'. The indefinite integral gives a function to represent all possible values of the primitive by adding C, the **constant of integration**:

> If $F(x)$ is a primitive of $f(x)$, then $\int f(x)dx=F(x)+C$, where C is the constant of integration.

This is called the indefinite integral of $f(x)$ and represents all the primitives of $f(x)$. If further information is given, then a value for C may be calculated.

Note that the indefinite integral is a function of x.

You can now rewrite the formula for finding the primitive of x^n as

$\int x^n\,dx = \frac{1}{n+1}x^{n+1} + C$ where n is an integer and $n \neq -1$.

The constant of integration can be written as C or c.

An extension of the formula is $\int f'(x)\left[f(x)\right]^n dx = \frac{1}{n+1}\left[f(x)\right]^{n+1} + C$, where n is an integer and $n \neq -1$.

This result is really using the chain rule in reverse. Consider $y = [f(x)]^{n+1}$.

Using the chain rule gives $\frac{dy}{dx} = (n+1)\,[f(x)]^n \times f'(x)$

Rearranging this gives: $f'(x) \times [f(x)]^n = \frac{1}{n+1} \times \frac{dy}{dx}$

Find the primitive of each side: $\int f'(x)\left[f(x)\right]^n dx = \frac{1}{n+1}\int \frac{dy}{dx} \times dx + C$

So: $\int f'(x)\left[f(x)\right]^n dx = \frac{1}{n+1} \times y + C$

Or: $\int f'(x)\left[f(x)\right]^n dx = \frac{1}{n+1}\left[f(x)\right]^{n+1} + C$, where n is an integer and $n \neq -1$.

Example 6

Find: **(a)** $\int \left(x^3 + 3x^2 - 2x + 1\right)dx$ **(b)** $\int \left(\sqrt{x} - 2\right)dx$

Solution

(a) $\int \left(x^3 + 3x^2 - 2x + 1\right)dx = \frac{x^4}{4} + x^3 - x^2 + x + C$

(b) $\int \left(\sqrt{x} - 2\right)dx = \int \left(x^{\frac{1}{2}} - 2\right)dx = \frac{2}{3}x^{\frac{3}{2}} - 2x + C$ or $\frac{2x\sqrt{x}}{3} - 2x + C$

Example 7

If $\frac{dy}{dx} = 1 + x - x^2$, find the equation of the curve y that passes through the point $(6, -40)$.

Solution

$\frac{dy}{dx} = 1 + x - x^2$: $\quad y = \int \left(1 + x - x^2\right)dx$

$$y = x + \frac{x^2}{2} - \frac{x^3}{3} + C$$

Point $(6, -40)$ is on the curve: $\quad -40 = 6 + 18 - 72 + C$

$$C = 8$$

The equation of the curve is $y = x + \frac{x^2}{2} - \frac{x^3}{3} + 8$.

EXERCISE 16.2 INDEFINITE INTEGRALS

1 Find: (a) $\int x\,dx$ (b) $\int (x^2+x+1)\,dx$ (c) $\int (3-x^2)\,dx$

(d) $\int (6x^5-4x^3+2x)\,dx$ (e) $\int dx$ (f) $\int x^n\,dx$

2 Find: (a) $\int \sqrt{x}\,dx$ (b) $\int \frac{1}{x^2}\,dx$ (c) $\int (1+\sqrt{x}+x)\,dx$

(d) $\int \left(x+\frac{1}{x^2}\right)dx$ (e) $\int \left(x+\frac{1}{x}\right)^2 dx$ (f) $\int (1-\sqrt{x})^2\,dx$

3 $\int (1+2x+3x^2)\,dx$ is equal to:

A $x+x^2+\frac{x^3}{3}+C$ **B** $x+\frac{x^2}{2}+\frac{x^3}{3}+C$ **C** $x+x^2+x^3+C$ **D** $2+6x+C$

4 If $\frac{dy}{dx}=1+x+3x^2$, find the equation of the curve that passes through the point $(2,6)$.

5 If $\frac{dy}{dx}=1+\sqrt{x}$, find the equation of the curve that passes through the point $(4,10)$.

16.3 PRIMITIVES OF TRIGONOMETRIC FUNCTIONS

Because $\frac{d}{dx}(\sin x)=\cos x$, $\int \cos x\,dx=\sin x+C$.

Because $\frac{d}{dx}(\cos x)=-\sin x$, $\int \sin x\,dx=-\cos x+C$.

Because $\frac{d}{dx}(\tan x)=\sec^2 x$, $\int \sec^2 x\,dx=\tan x+C$.

Thus, provided $a\neq 0$:

$\frac{d}{dx}(\sin(ax+b))=a\cos(ax+b)$, so $\int \cos(ax+b)\,dx=\frac{1}{a}\sin(ax+b)+C$

$\frac{d}{dx}(\cos(ax+b))=-a\sin(ax+b)$, so $\int \sin(ax+b)\,dx=-\frac{1}{a}\cos(ax+b)+C$

$\frac{d}{dx}(\tan(ax+b))=a\sec^2(ax+b)$, so $\int \sec^2(ax+b)\,dx=\frac{1}{a}\tan(ax+b)+C$

Summary—Trigonometric integrals

$\int \sin x\,dx=-\cos x+C$	$\int \sin ax\,dx=-\frac{1}{a}\cos ax+C$	$\int \sin(ax+b)\,dx=-\frac{1}{a}\cos(ax+b)+C$
$\int \cos x\,dx=\sin x+C$	$\int \cos ax\,dx=\frac{1}{a}\sin ax+C$	$\int \cos(ax+b)\,dx=\frac{1}{a}\sin(ax+b)+C$
$\int \sec^2 x\,dx=\tan x+C$	$\int \sec^2 ax\,dx=\frac{1}{a}\tan ax+C$	$\int \sec^2(ax+b)\,dx=\frac{1}{a}\tan(ax+b)+C$

Example 8

Find: (a) $\int (\sin x+2\cos x)\,dx$ (b) $\int (\cos 2x+\sin 3x)\,dx$ (c) $\int \sin\frac{x}{2}\,dx$ (d) $\int \left(2\sin x+3\cos\frac{x}{2}\right)dx$

Solution

(a) $\int(\sin x + 2\cos x)\,dx$
$= -\cos x + 2\sin x + C$

(b) $\int(\cos 2x + \sin 3x)\,dx$
$= \frac{1}{2}\sin 2x - \frac{1}{3}\cos 3x + C$

(c) $\int \sin\frac{x}{2}\,dx$
$= -2\cos\frac{x}{2} + C$

(d) $\int\left(2\sin x + 3\cos\frac{x}{2}\right)dx$
$= -2\cos x + 6\sin\frac{x}{2} + C$

EXERCISE 16.3 PRIMITIVES OF TRIGONOMETRIC FUNCTIONS

1 Write the primitive function of:

(a) $\sin 2x$ (b) $\cos 3x$ (c) $\sec^2 x$ (d) $\sin x + \cos x$

(e) $2\sin x - 3\cos x$ (f) $\sin\left(x + \frac{\pi}{4}\right)$ (g) $\cos\frac{x}{2}$ (h) $2\sin 2x$

2 The primitive of $3\cos\frac{x}{3}$ is:

A $-\sin\frac{x}{3}$ B $-9\sin\frac{x}{3}$ C $\sin\frac{x}{3}$ D $9\sin\frac{x}{3}$

3 Find:

(a) $\int\left(\sin\frac{\pi}{4} + \cos\frac{\pi}{4}\right)dx$ (b) $\int(\sin x - \cos 2x)\,dx$ (c) $\int \sin\left(2x + \frac{\pi}{2}\right)dx$

(d) $\int \cos\left(2x - \frac{\pi}{4}\right)dx$ (e) $\int \sec^2 3x\,dx$ (f) $\int\left(\frac{1}{2}\sin 2x - \cos x\right)dx$

(g) $\int\left(\cos\frac{x}{2} - \frac{1}{2}\sin 2x\right)dx$ (h) $\int\left(x^2 + \sin 2x\right)dx$ (i) $\int\left(\sqrt{x} - \frac{1}{2}\cos x\right)dx$

4 Find: (a) $\int \frac{\sin x}{1 - \cos x}\,dx$ (b) $\int \frac{\cos x}{\sin x}\,dx$

5 The gradient of a curve is given by $\frac{dy}{dx} = 2\sin 3x$. If the curve passes through the point $\left(\frac{\pi}{3}, 3\right)$, find the equation of the curve.

16.4 INTEGRATING THE EXPONENTIAL FUNCTION

Indefinite integral of e^x

Because the exponential function e^x is its own derivative, it is also its own integral: $\int e^x\,dx = e^x + C$.

The integrals of related exponential functions can similarly be determined from their derivatives.

Indefinite integral of e^{kx}, k constant

Because $\frac{d}{dx}\left(e^{kx}\right) = ke^x$, it follows that $\int e^{kx}\,dx = \frac{1}{k}e^{kx} + C$.

Consider some values for k, for example: $k = 2$: $\int e^{2x}\,dx = \frac{1}{2}e^{2x} + C$

$k = \frac{2}{5}$: $\int e^{\frac{2x}{5}}\,dx = \frac{5}{2}e^{\frac{2x}{5}} + C$

$k = -3$: $\int e^{-3x}\,dx = -\frac{1}{3}e^{-3x} + C$

Indefinite integral of e^{ax+b}, *a* and *b* constant

Because $\frac{d}{dx}\left(e^{ax+b}\right) = ae^{ax+b}$, it follows that $\int e^{ax+b}\,dx = \frac{1}{a}e^{ax+b} + C$.

Indefinite integral of $f'(x)e^{f(x)}$

Because $\frac{d}{dx}\left(e^{f(x)}\right) = f'(x)e^{f(x)}$, it follows that $\int f'(x)e^{f(x)}\,dx = e^{f(x)} + C$.

You can only find $\int e^{f(x)}\,dx$ when $f'(x)$ is a constant, i.e. if $f(x) = ax + b$.

Summary—Integrals involving e^x

$$\int e^x\,dx = e^x + C \qquad \int e^{ax+b}\,dx = \frac{1}{a}e^{ax+b} + C$$

$$\int e^{kx}\,dx = \frac{1}{k}e^{kx} + C \qquad \int f'(x)e^{f(x)}\,dx = e^{f(x)} + C$$

Example 9

Find: (a) $\int e^{4x-1}\,dx$ (b) $\int 3x^2e^{x^3+1}\,dx$

Solution

(a) $\frac{d}{dx}(4x-1) = 4$: $\int e^{4x-1}\,dx = \frac{1}{4}e^{4x-1} + C$

(b) $\frac{d}{dx}\left(x^3+1\right) = 3x^2$:

The integral must be of the form $\int f'(x)e^{f(x)}\,dx = e^{f(x)} + C$ where $f(x) = x^3 + 1$.

$\int 3x^2e^{x^3+1}\,dx = e^{x^3+1} + C$

EXERCISE 16.4 INTEGRATING THE EXPONENTIAL FUNCTION

1 Write a primitive function for each of the following:

(a) e^{2x} (b) e^{5x} (c) $e^{-0.4x}$ (d) $5e^{2.5x}$ (e) $e^x + e^{-3x}$ (f) $e^{-2x} - e^{-x}$

2 Find: (a) $\int e^{-x}\,dx$ (b) $\int e^{\frac{x}{2}}\,dx$ (c) $\int e^{-3x}\,dx$

(d) $\int \left(e^{-t} - 1\right)dt$ (e) $\int \left(e^{2u} + u^2\right)du$ (f) $\int \left(e^{-2.5x} + e^{0.4x}\right)dx$

16.5 INTEGRALS RESULTING IN LOGARITHMIC FUNCTIONS

In Chapter 13 you differentiated the logarithm function. You will now look at finding primitive functions that involve the logarithm function. Recall that $\log_e(x)$ can be written as $\ln(x)$.

1 $\frac{d}{dx}(\log_e x) = \frac{1}{x}$

Hence: $\int \frac{1}{x}\,dx = \log_e |x| + C$

This fills the gap in the earlier result for integrating powers of x, $\int x^n\,dx = \frac{x^{n+1}}{n+1} + c$, $n \neq -1$, by providing the result for $n = -1$.

2 $\frac{d}{dx}(\log_e(ax+b)) = \frac{a}{ax+b}$

Hence: $\int \frac{1}{ax+b}\,dx = \frac{1}{a}\log_e|ax+b| + C$

3 $\frac{d}{dx}\left(\log_e|f(x)|\right) = \frac{f'(x)}{f(x)}$

Hence: $\int \frac{f'(x)}{f(x)}\,dx = \log_e|f(x)| + C$

4 $\frac{d}{dx}\left(a\log_e[f(x)]\right) = \frac{af'(x)}{f(x)}$

Hence: $\int \frac{af'(x)}{f(x)}\,dx = a\log_e|f(x)| + C$

Anti-derivative of a^x

Let $y = a^x$. In Chapter 13 it was shown that:

$$\frac{d}{dx}\left(a^x\right) = a^x \log_e a$$

Hence: $a^x = \frac{1}{\log_e a}\frac{d}{dx}\left(a^x\right)$

So $\int a^x\,dx = \frac{a^x}{\log_e a} + C$

You can see why the abbreviation 'ln' is often used instead of '$\log_e$' when working with natural logarithms and calculus. $y = e^{x\ln a}$ is easier to write than $e^{x\log_e a}$.

Summary of results

$$\int \frac{1}{x}\,dx = \log_e|x| + C$$

$$\int \frac{f'(x)}{f(x)}\,dx = \log_e|f(x)| + C$$

$$\int \frac{1}{ax+b}\,dx = \frac{1}{a}\log_e|ax+b| + C$$

$$\int \frac{af'(x)}{f(x)}\,dx = a\log_e|f(x)| + C$$

$$\int a^x\,dx = \frac{a^x}{\log_e a} + C$$

Example 10

Find the indefinite integral of the following:

(a) $\frac{2}{2x-3}$ (b) $\frac{x}{x^2+4}$ (c) $\frac{4x-6}{x^2-3x}$ (d) $\frac{e^x}{1+e^x}$

Solution

(a) $\int \frac{2}{2x-3}\,dx = \int \frac{f'(x)}{f(x)}\,dx$ where $f(x) = 2x - 3$ and $f'(x) = 2$

$\int \frac{2}{2x-3}\,dx = \log_e|2x-3| + C$

(b) $\int \frac{x}{x^2+4}\,dx = \frac{1}{2}\int \frac{2x}{x^2+4}\,dx$

$= \frac{1}{2}\int \frac{f'(x)}{f(x)}\,dx$ where $f(x) = x^2 + 4$ and $f'(x) = 2x$

$= \frac{1}{2}\log_e|x^2+4| + C$

(c) $\int \frac{4x-6}{x^2-3x}\,dx = 2\int \frac{2x-3}{x^2-3x}\,dx$

$= 2\int \frac{f'(x)}{f(x)}\,dx$ where $f(x) = x^2 - 3x$ and $f'(x) = 2x - 3$

$= 2\log_e|x^2-3x| + C$

(d) $\int \frac{e^x}{1+e^x}\,dx = \int \frac{f'(x)}{f(x)}\,dx$ where $f(x) = 1 + e^x$ and $f'(x) = e^x$

$= \log_e|1+e^x| + C$

Example 11

Given $\frac{dy}{dx} = \frac{1}{x}$ and $y = 0$ when $x = 0.5$, express y in terms of x.

Solution

$\frac{dy}{dx} = \frac{1}{x}$

$y = \int \frac{1}{x}\,dx$

$y = \log_e|x| + C$

Where $x = 0.5$, $y = 0$, so: $0 = \log_e \frac{1}{2} + C$

$0 = -\log_e 2 + C$

$C = \log_e 2$

$\therefore y = \log_e|x| + \log_e 2$ or $y = \log_e|2x|$

Example 12

The gradient of a curve at any point is $\frac{4x}{x^2+1}$ and the curve passes through the point $(0,0)$. Find the equation of the curve.

Solution

$f'(x) = \frac{4x}{x^2+1}$

$\therefore f(x) = \int \frac{4x}{x^2+1}\,dx$

$= 2\int \frac{2x}{x^2+1}\,dx$

$= 2\log_e(x^2+1) + C$

As $f(0) = 0$: $0 = 2\log_e 1 + C$

$C = 0$

$\therefore f(x) = 2\log_e(x^2+1)$

EXERCISE 16.5 INTEGRALS RESULTING IN LOGARITHMIC FUNCTIONS

1 Find the primitive of the following:

(a) $\frac{2}{x}$ (b) $\frac{1}{x+1}$ (c) $\frac{2}{2x+1}$ (d) $\frac{x}{x^2-4}$

(e) $\frac{1}{2x-1}$ (f) $\frac{e^x}{4+e^x}$ (g) $\frac{x^3}{x^4+1}$ (h) $\frac{e^{2x}}{4-e^{2x}}$

2 If $\frac{dy}{dx}=\frac{1}{x}$ and $y=0$ where $x=2$, then the correct expression for y in terms of x is:

A $y=\log_e x-2$ B $y=\frac{1}{2}\log_e x$ C $y=\log_e\left(\frac{x}{2}\right)$ D $y=2\log_e x$

3 The gradient of a curve at any point is $\frac{2}{2x+1}$ and the curve passes through the point $(1,\log_e 3)$. Find the equation of the curve.

4 Find the rule that defines $f(x)$ given that $f'(x)=\frac{x}{x^2+9}$ and $f(0)=\log_e 3$. Indicate whether each statement below is a correct or incorrect step in the solution of this problem.

(a) $f(x)=\int\frac{x}{x^2+9}dx$ (b) $f(x)=\frac{1}{2}\ln\left(x^2+9\right)+C$ (c) $C=2\ln 3$ (d) $f(x)=\frac{1}{2}\ln\left(x^2+9\right)$

5 Write the derivative of $\log_e(\cos x)$ and hence find the primitive function of $\tan x$.

6 Differentiate: (a) 2^x (b) $x+10^x$ (c) e^x+5^x (d) 5^{x^2} (e) $a^{\sqrt{x}}$

7 Find: (a) $\int 3^x\,dx$ (b) $\int\left(x+10^x\right)dx$ (c) $\int\left(\frac{1}{x}+1+e^x+a^x\right)dx$

CHAPTER REVIEW 16

1 Find the primitive of the following:

(a) $x+9$ (b) $3x^2-2x+4$ (c) x^4+x^3-2 (d) $(x-2)(x+3)$ (e) $(x+2)^2$ (f) 7

2 Express y in terms of x, given the following:

(a) $\frac{dy}{dx}=5x+4$ (b) $\frac{dy}{dx}=5-4x+3x^2+x^3$ (c) $\frac{dy}{dx}=2x+\sqrt{x}+3$

3 Find $f(x)$ in terms of x, given the following:

(a) $f'(x)=x^2+x^3+1, f(0)=2$ (b) $f'(x)=3-x+6x^3, f(1)=3$ (c) $f'(x)=1-\frac{1}{x^2},\ f(2)=\frac{1}{2}$

4 During a storm, water flows into a 5000-litre tank at the rate of $\frac{dV}{dt}$ litres per minute, where $\frac{dV}{dt}=140+13t-t^2$ and t is the time in minutes since the storm began.

(a) Find the volume of water that has flowed into the tank since the start of the storm as a function of t.

(b) How much water has flowed into the tank after 12 minutes?

5 (a) Show that $\frac{d}{dx}\left(xe^x\right)=e^x+xe^x$. (b) Hence find $\int xe^x\,dx$.

6 Find: (a) $\int 3\sin\frac{x}{2}dx$ (b) $\int\left(x+\sec^2 2x\right)dx$ (c) $\int\frac{\cos t}{\sin t}dt$

7 Find: (a) $\int\frac{5}{x}dx$ (b) $\int\frac{3}{x+4}dx$ (c) $\int\frac{4x}{x^2+1}dx$ (d) $\int\frac{e^x}{e^x+2}dx$

CHAPTER 17
Integral calculus

You should already be familiar with the formulae used to calculate the areas of plane figures such as squares, rectangles, trapezia, triangles and circles. In this chapter you will develop a way to calculate areas bounded by curves and straight lines, using calculus.

17.1 AREA UNDER A CURVE

In the diagram below, the shaded area *ABCD* is bounded by the continuous increasing function $y = f(x)$, the x-axis and the lines $x = a$ and $x = b$. You will call this the area 'under the curve' between $x = a$ and $x = b$. By convention this means that the other boundary is the x-axis. The lines $x = a$ and $x = b$ represent the ordinates at a and b respectively.

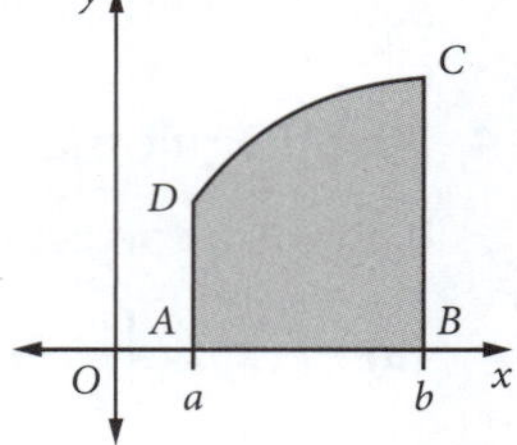

There are several ways to find an approximation to this area *ABCD*. For example:

Draw the rectangles *ABED* and *ABCF*.

By comparing areas: Area *ABED* < Area *ABCD* < Area *ABCF*

Or: $(b - a)\,f(a) <$ Area *ABCD* $< (b - a)\,f(b)$

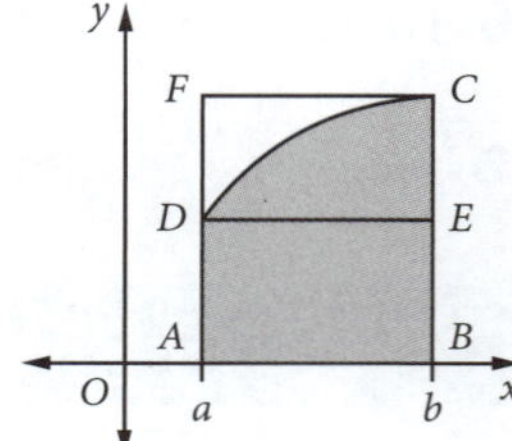

Now divide the interval *AB* into 4 equal portions of length h, where $h = \dfrac{b-a}{4}$, as shown in the diagram. Draw each of the interior and exterior rectangles.

Total area of inside rectangles < Area *ABCD* < Total area of outside rectangles

The sides of the rectangles are at $x = a$, $x = a + h$, $x = a + 2h$, $x = a + 3h$, $x = b$.

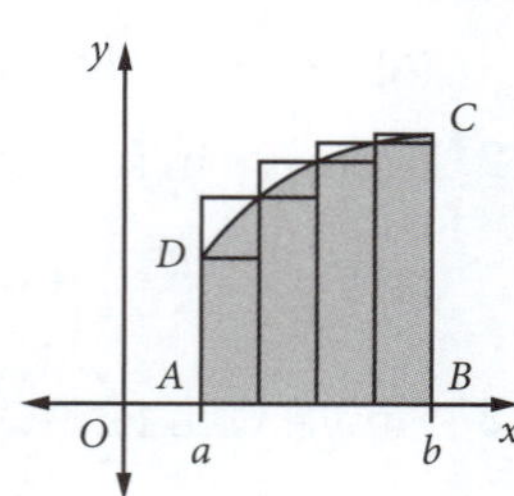

So: $h\,f(a) + h\,f(a+h) + h\,f(a+2h) + h\,f(a+3h) <$ Area $ABCD < h\,f(a+h) + h\,f(a+2h) + h\,f(a+3h) + h\,f(b)$

$h(f(a) + f(a+h) + f(a+2h) + f(a+3h)) <$ Area $ABCD < h(f(a+h) + f(a+2h) + f(a+3h) + f(b))$

This is a better estimate of the area than you obtained from the first diagram. The more parts that you divide *AB* into, the smaller the rectangles, and the better the approximation obtained for the area under the curve.

Using trapezia

A better approximation of the area can be obtained by using trapezia instead of rectangles. Joining *DC* with a straight line gives a shape whose area is closer to the shaded area than using a rectangle.

Remember: Area of a trapezium = half the sum of the parallel sides × the distance between them.

By comparing areas: Area of trapezium *ABCD* < Shaded area *ABCD*

Or: $(b - a)\dfrac{f(a) + f(b)}{2} <$ Area *ABCD*

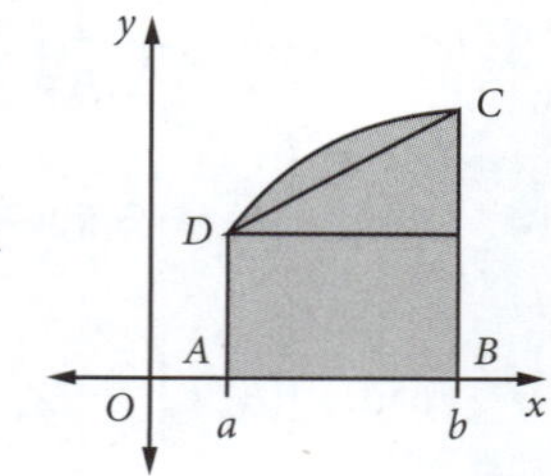

Now divide the interval AB into 4 equal portions of length h, where $h = \frac{b-a}{4}$, as shown in the diagram.

Join the tops of each column with a straight line to create 4 trapezia.

Total area of 4 trapezia < Area $ABCD$

The more trapezia used, the closer the area becomes to the area under the curve.

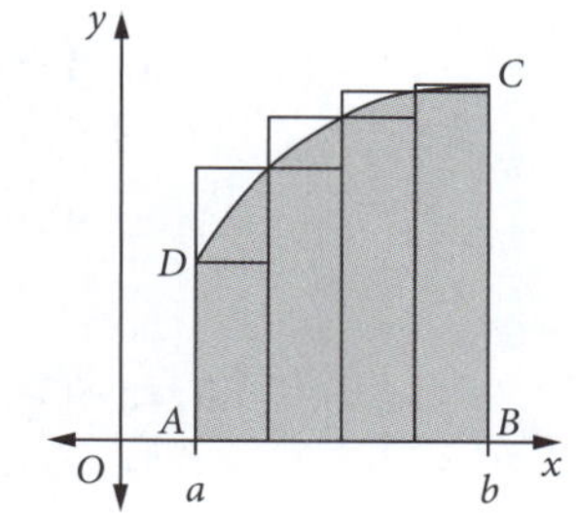

Now consider the continuous function $y = f(x)$, where $f(x) > 0$ for values of x in $a \le x \le b$. Over this domain, $f(x)$ will have a greatest value and a least value.

There is an area enclosed by the curve $y = f(x)$, $x = a$, $x = b$ and the x-axis.
Let A be the size of this area and let h and H be the minimum and maximum values respectively of $f(x)$ in $a \le x \le b$. You then have $h(b-a) \le A \le H(b-a)$.

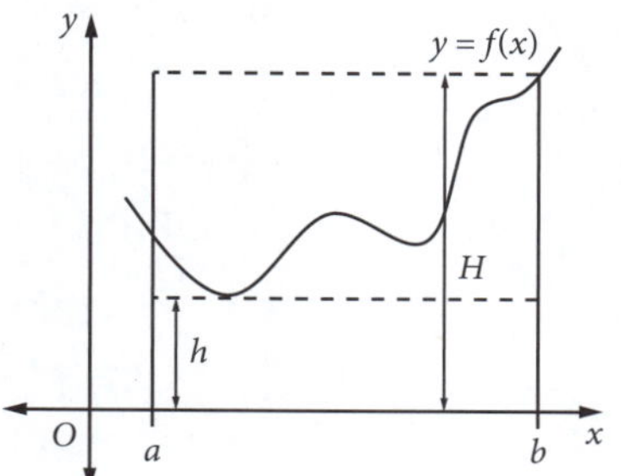

Now split $a \le x \le b$ into two subintervals, $a \le x \le \frac{a+b}{2}$ and $\frac{a+b}{2} \le x \le b$.

Take minimum and maximum values h_1, H_1 in the first subinterval and h_2, H_2 in the second subinterval as shown in the diagram.

Hence: $h(b-a) \le (h_1 + h_2) \times \frac{b-a}{2} \le A \le (H_1 + H_2) \times \frac{b-a}{2} \le H(b-a)$

The more equal subintervals you make, the closer the corresponding area sums will approach the value for A.

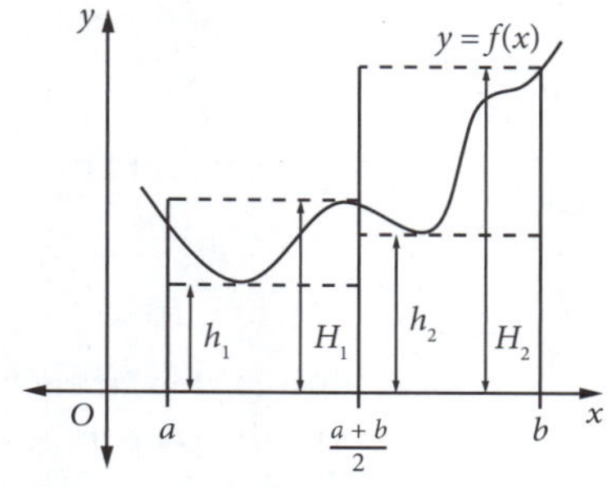

Example 1

Find an approximation for the area bounded by the line $y = 2x + 1$, the x-axis and the ordinates at $x = 0$ and $x = 4$, using:

(a) rectangles above and below the line, taking:

(i) one subinterval **(ii)** two subintervals **(iii)** four subintervals.

(b) trapezia with the line as one side, taking:

(i) one subinterval **(ii)** two subintervals **(iii)** four subintervals.

Solution

(a) **(i)** Draw a diagram showing the region.
For $0 \le x \le 4$ and one subinterval:

Area of inner rectangle $= 4 \times 1 = 4$

Area of outer rectangle $= 4 \times 9 = 36$

Hence $4 < A < 36$

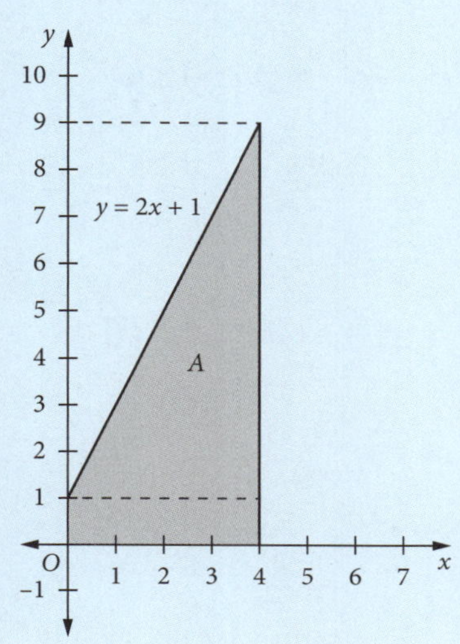

(ii) Draw a diagram showing the region.
For $0 \le x \le 4$ and two subintervals:

Area of inner rectangles $= 2 \times 1 + 2 \times 5$

$= 12$

Area of outer rectangles $= 2 \times 5 + 2 \times 9$

$= 28$

Hence $12 < A < 28$

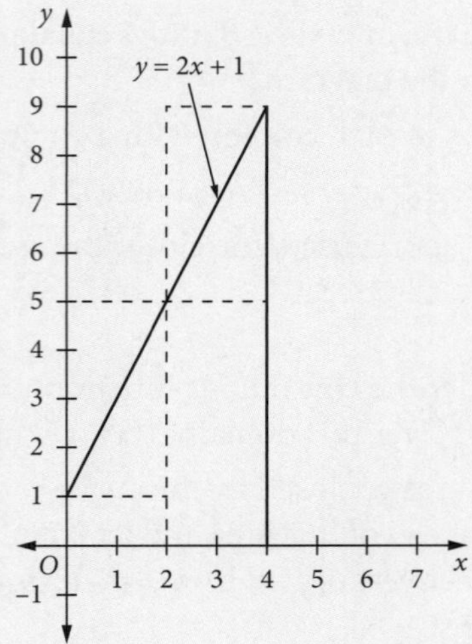

(iii) Draw a diagram showing the region.
For $0 \le x \le 4$ and four subintervals:

Area of inner rectangles $= 1 \times 1 + 1 \times 3 + 1 \times 5 + 1 \times 7$

$= 16$

Area of outer rectangles $= 1 \times 3 + 1 \times 5 + 1 \times 7 + 1 \times 9$

$= 24$

Hence $16 < A < 24$

Note that in each of these three parts, the average of the two values in the inequality is 20.

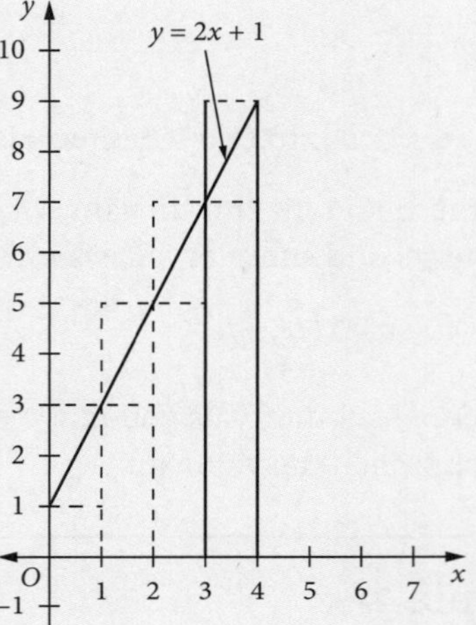

(b) (i) Draw a diagram showing the region.
The area of a trapezium is given by:
Area = half the sum of the parallel sides × the distance between them.

With one subinterval:

$A = \frac{1}{2}(f(0) + f(4)) \times (4 - 0)$

$= \frac{1}{2} \times (1 + 9) \times 4$

$= 20u^2$

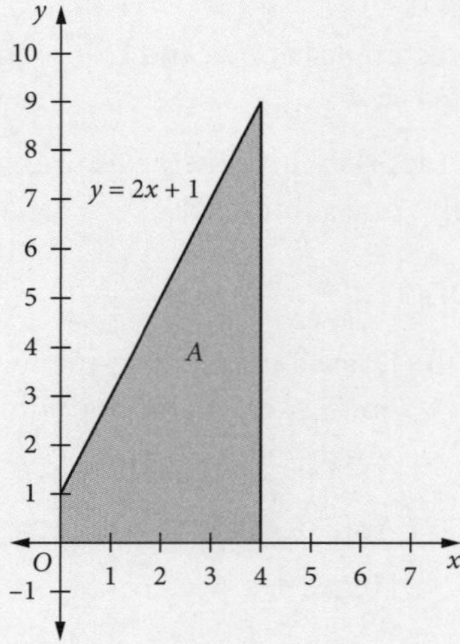

(ii) Draw a diagram showing the region.
With two subintervals:

$$A = \frac{1}{2}(f(0)+f(2))\times(2-0)+\frac{1}{2}(f(2)+f(4))\times(4-2)$$
$$= \frac{1}{2}\times(1+5)\times 2+\frac{1}{2}\times(5+9)\times 2$$
$$= 20u^2$$

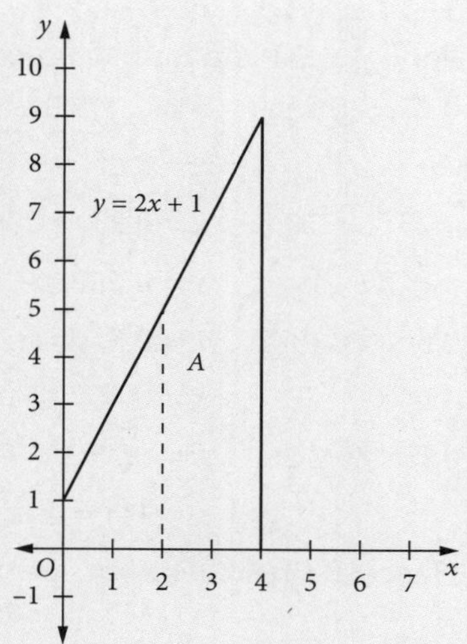

(iii) With four subintervals the area will again be $20u^2$.
For a straight line, using trapezia to find the area under the curves gives the exact value.

Example 2

Find an approximation of the area of the region bounded by the curve $y = x^2$, the x-axis and the ordinates at $x = 0$ and $x = 1$, using rectangles with:

(a) one subinterval **(b)** two subintervals **(c)** four subintervals.

Solution

(a) Draw a diagram showing the region:
For $0 \le x \le 1$ and one subinterval,
$x = 0, y = 0$, is the smallest value;
$x = 1, y = 1$, is the largest value.
Hence $0 < A < 1$

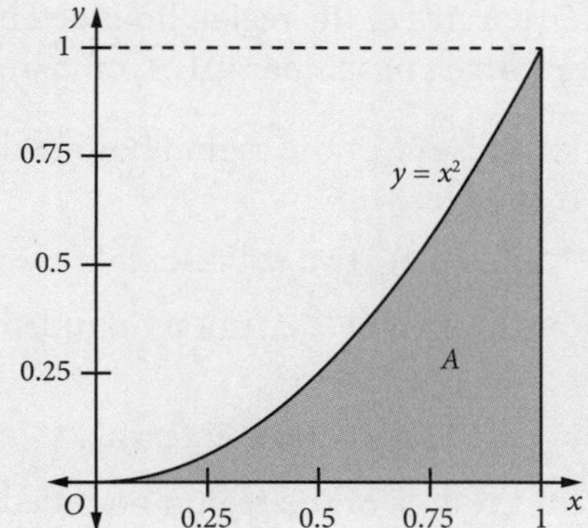

(b) For $0 \le x \le 1$ and two subintervals:
For $0 \le x \le 0.5$:
$x = 0, y = 0$ is smallest value; $x = 0.5, y = 0.25$ is largest value.
For $0.5 \le x \le 1$:
$x = 0.5, y = 0.25$ is smallest value; $x = 1, y = 1$ is largest value.
Hence: $(0 + 0.25) \times 0.5 \le A \le (0.25 + 1) \times 0.5$
$0.125 \le A \le 0.625$

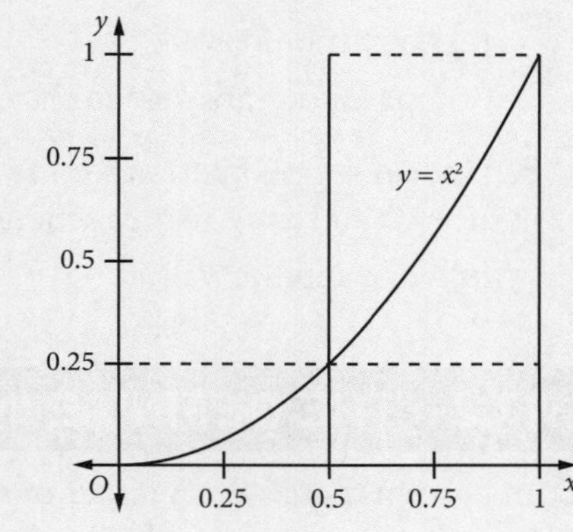

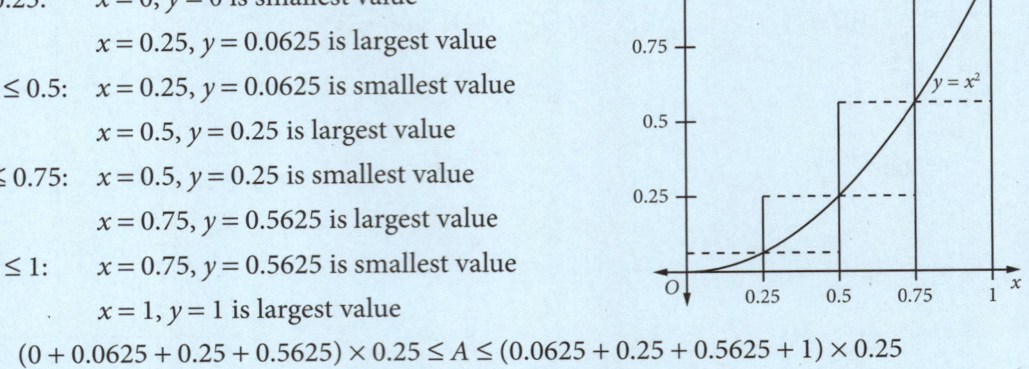

(c) For $0 \le x \le 1$ and four subintervals:

$0 \le x \le 0.25$: $x = 0, y = 0$ is smallest value

$x = 0.25, y = 0.0625$ is largest value

$0.25 \le x \le 0.5$: $x = 0.25, y = 0.0625$ is smallest value

$x = 0.5, y = 0.25$ is largest value

$0.5 \le x \le 0.75$: $x = 0.5, y = 0.25$ is smallest value

$x = 0.75, y = 0.5625$ is largest value

$0.75 \le x \le 1$: $x = 0.75, y = 0.5625$ is smallest value

$x = 1, y = 1$ is largest value

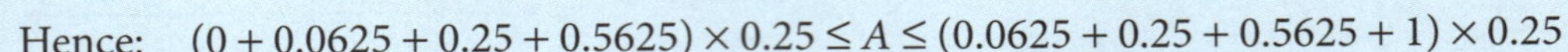

Hence: $(0 + 0.0625 + 0.25 + 0.5625) \times 0.25 \le A \le (0.0625 + 0.25 + 0.5625 + 1) \times 0.25$

$$0.218\,75 \le A \le 0.468\,75$$

The area is between 0.218 75 and 0.468 75. Taking the average of these two values gives an approximate area of 0.343 75.

MAKING CONNECTIONS

Approximating the area beneath a curve

Move the slider to increase the number of subintervals and approximate the area under the curve of $y = x^2$.

EXERCISE 17.1 AREA UNDER A CURVE

1 Find an approximation for the area of the region bounded by the curve $y = x^2$, the x-axis and the ordinates at $x = 0$ and $x = 1$ using rectangles with eight subintervals. Compare your answer to the answers obtained in Example **1**.

2 Find an approximation for the area of the region bounded by the curve $y = x^2$, the x-axis and the ordinates at $x = 1$ and $x = 3$ using rectangles with:

(a) one subinterval **(b)** two subintervals **(c)** four subintervals.

3 Find an approximation for the area of the region bounded by the curve $y = x^2 + 1$, the x-axis and the ordinates at $x = 0$ and $x = 1$ using rectangles with:

(a) one subinterval **(b)** two subintervals **(c)** four subintervals.

4 Find an approximation for the area of the region bounded by the curve $y = x + 1$, the x-axis and the ordinates at $x = 1$ and $x = 3$ using rectangles with:

(a) one subinterval **(b)** two subintervals **(c)** four subintervals.

Compare your answers to the exact area of the region.

5 Find an approximation for the area of the region bounded by the curve $y = x^3$, the x-axis and the ordinates at $x = 0$ and $x = 2$ using rectangles with:

(a) one subinterval **(b)** two subintervals **(c)** four subintervals.

17.2 THE DEFINITE INTEGRAL AND THE AREA UNDER A CURVE

It can be seen that as the number of subdivisions increases, the sums of the areas of rectangles become closer to the area under the curve bounded by the two ordinates. Integral calculus is based on the infinitesimal limit of these rectangular subdivisions. The process that you are about to develop was discovered independently by Leibniz and Newton in the seventeenth century, when they were the first to make the link between the area under a curve and calculus by way of the primitive function.

Let f be a continuous and increasing function in the interval $a \le x \le b$, $f(x) \ge 0$. Divide this interval into n subintervals, where n is large. Each rectangle will have a small base of length dx (where dx is a single symbol—dx is **not** 'd multiplied by x') and a height that is close to $f(x)$ for any value of x in the base. Because f is continuous, all values of $f(x)$ are close together if the values of x are close together. Thus the area of a typical rectangle is $f(x)\,dx$ and the sum of these areas can be written as $\sum f(x)dx$. The large Greek letter Σ (capital sigma) denotes 'the sum of'.

As n increases (and hence dx decreases), the limiting value of this sum is denoted by $\int_a^b f(x)dx$. The integral symbol $\int$ represents a large elongated S, which stands for 'sum', while the bounds a and b at either end of the $\int$ indicate the interval from $x = a$ to $x = b$ over which the sum is taken. This $\int_a^b f(x)dx$ is called the **definite integral** of the function $f(x)$ between $x = a$ and $x = b$. The value A of this definite integral is the size of the area under the curve $y = f(x)$ between $x = a$ and $x = b$.

For a definite integral, you write: $A = \int_a^b f(x)dx$ **OR** $A = \int_a^b y\,dx$

Example 3

Write the definite integral for the area of the region under the line $y = x + 1$ between the ordinates $x = 1$ and $x = 3$. By using appropriate area formulae, find the value of this area.

Solution

Sketch the region: bounds are $f(x) = x + 1$, $x = 1$, $x = 3$, x-axis

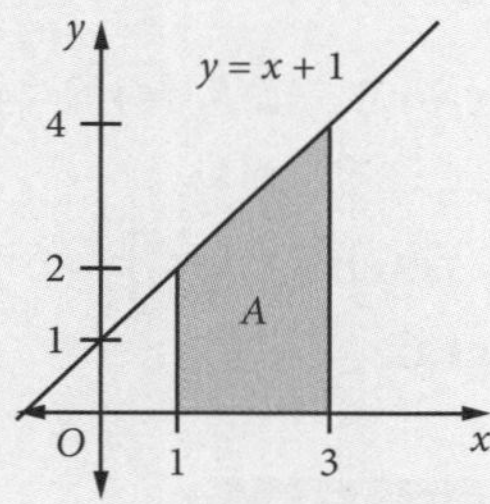

$$A = \int_a^b f(x)dx$$
$$= \int_1^3 (x+1)dx$$

The required area is a trapezium. $A = \frac{(2+4)}{2} \times (3-1)$

$$= 6 \text{ units}^2$$

Example 4

Write the definite integral for the area of the region bounded by the lines $y = 2x$, $x = t$ and the x-axis. By using appropriate area formulae, find the value of this area as a function of t.

Solution

Sketch the region: bounds are $f(x) = 2x$, $x = 0$, $x = t$, x-axis

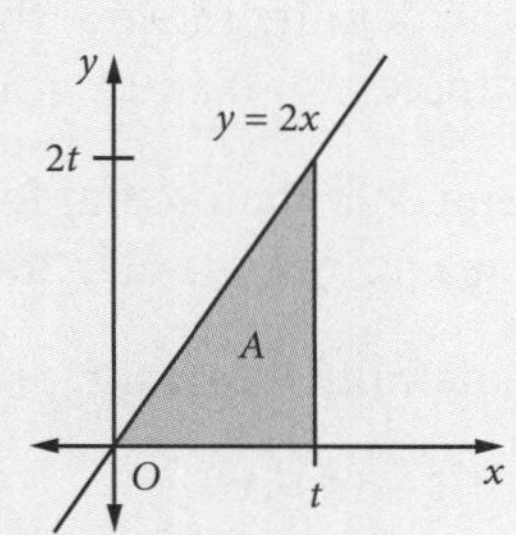

$$A = \int_a^b f(x)dx$$
$$= \int_0^t 2x\,dx$$

Area of triangle $= \frac{1}{2} \times t \times 2t$

$$= t^2$$

It is worth noting here that x^2 is a primitive of $2x$.

Definite integral—formal development

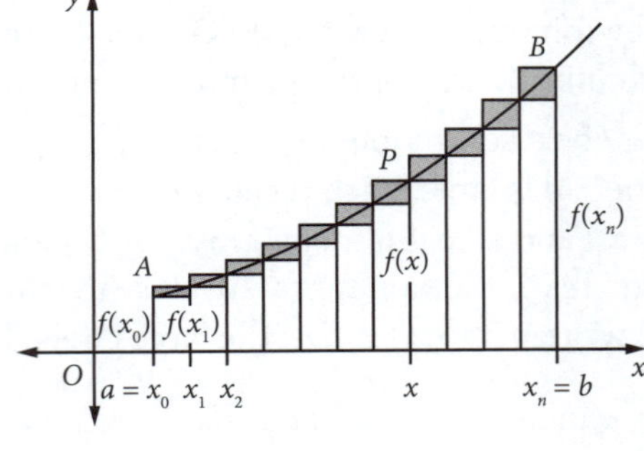

Let f be a continuous and increasing function in the interval $a \le x \le b$. Divide this interval into n parts, not necessarily equal, using the points $x_1, x_2, x_3, \ldots x_n$ such that $a = x_0 < x_1 < x_2 < x_3 \ldots < x_n = b$.

Using the subintervals $(x_1 - x_0), (x_2 - x_1), \ldots (x_n - x_{n-1})$ as bases, construct upper and lower rectangles as shown in the diagram.

If you denote the sum of the areas of the lower rectangles by S_L, then:

$$S_L = f(x_0)(x_1 - x_0) + f(x_1)(x_2 - x_1) + \ldots + f(x_{n-1})(x_n - x_{n-1}).$$

If you denote the sum of the areas of the upper rectangles by S_U, then:

$$S_U = f(x_1)(x_1 - x_0) + f(x_2)(x_2 - x_1) + \ldots + f(x_n)(x_n - x_{n-1}).$$

It is clear from the diagram that $S_U > S_L$ and that $S_U - S_L$ equals the sum of the areas of the shaded rectangles. If you denote the area bounded by the curve, the x-axis and the ordinates $x = a$ and $x = b$ by A, then $S_L < A < S_U$.

If you divide the interval $a \le x \le b$ into a very large number of parts, i.e. $n \to \infty$, then the sum of the areas of the shaded rectangles becomes very small: $S_U - S_L \to 0$ and S_U and S_L approach the same limit A.

Consider now a typical rectangle of width δx at any point x in the interval $a \le x \le b$. The area of a typical lower rectangle is $f(x)\,\delta x$ and the area of a typical upper rectangle is $f(x + \delta x)\,\delta x$.

Hence: $S_L = \sum_a^b f(x)\delta x$ and $S_U = \sum_a^b f(x + \delta x)\delta x$

The Σ notation here denotes 'the sum of' the areas of the n lower rectangles (S_L) or upper rectangles (S_U) in the interval $a \le x \le b$.

Thus: $\sum_a^b f(x)\delta x < A < \sum_a^b f(x + \delta x)\delta x$

As $n \to \infty$, $\delta x \to 0$ and $S_U \to S_L$, so: $A = \lim\limits_{\delta x \to 0} \sum_a^b f(x)\delta x$

This limit is written as $A = \int_a^b f(x)\,dx$, and is read as 'the integral of f from a to b'. This is known as the **definite integral**.

EXPLORE FURTHER

Calculating the definite integral

Use technology to calculate the definite integral for a curve.

EXERCISE 17.2 THE DEFINITE INTEGRAL AND THE AREA UNDER A CURVE

1 Write the definite integral that you would use to find the area of the region under the line $y = x + 2$ between the ordinates $x = 0$ and $x = 3$. By using appropriate area formulae, find the value of this area.

2 Write the definite integral for the area of the region under the line $y = 2x + 1$ between $x = 1$ and $x = 4$. By using appropriate area formulae, find the value of this area.

3 Which definite integral represents the area bounded by the curve $y = 4 - x^2$ and the x-axis?

A $\int_0^2 (4 - x^2)\,dx$ **B** $\int_{-2}^0 (4 - x^2)\,dx$ **C** $\int_{-2}^2 (4 - x^2)\,dx$ **D** $\int_{-\sqrt{2}}^{\sqrt{2}} (4 - x^2)\,dx$

4 Write the definite integral for the area of the region bounded by the lines $y = 3x$, $x = t$ and the x-axis. By using appropriate area formulae, find the value of this area.

5 Write the definite integral for the area of the region under the curve $y = x^2$ between $x = 0$ and $x = 2$. By drawing this graph on a 5 mm grid, count squares to find an approximation for this area.

6 Write the definite integral for the area of the region under the line $y = 2 - x$ between $x = 0$ and where it cuts the x-axis. By using appropriate area formulae, find the value of this area.

7 Write the definite integral for the area of the region under the curve $y = \sqrt{9 - x^2}$. By using appropriate area formulae, find the value of this area.

8 Write the definite integral for the area of the region under the curve $y = \sqrt{9 - (x - 2)^2}$. By using appropriate area formulae, find the value of this area.

17.3 THE DEFINITE INTEGRAL AND THE PRIMITIVE FUNCTION

In Example 3, you found that
$$\begin{aligned} A &= \int_a^b f(x)\,dx \\ &= \int_0^t 2x\,dx \\ &= t^2. \end{aligned}$$

This means that the **primitive** of $2x$ is $x^2 + C$.

If $F'(x) = f(x)$, then $F(x)$ is a primitive of $f(x)$.
Here $f(x) = 2x$ so $F(x) = x^2$ and $F(t) = t^2$, $F(0) = 0$;
you can see that $F(t) - F(0) = t^2 - 0 = t^2$.
This leads you to a possible way to evaluate the definite integral:

$$A = \int_a^b f(x)\,dx = F(b) - F(a) \quad \text{where } F(x) \text{ is a primitive of } f(x)$$

In Example 2 you had
$$\begin{aligned} A &= \int_a^b f(x)\,dx \\ &= \int_1^3 (x+1)\,dx. \end{aligned}$$

By finding the area of the triangle, you found that $A = 6$ units^2.

Following the definitions above, for $f(x) = x + 1$ you have $F(x) = \frac{x^2}{2} + x$.

Hence $F(3) = \frac{9}{2} + 3 = 7.5$ and $F(1) = \frac{1}{2} + 1 = 1.5$:

$$\begin{aligned} F(3) - F(1) &= 7.5 - 1.5 \\ &= 6 \end{aligned}$$

This verifies that $\int_1^3 (x+1)\,dx = F(3) - F(1) = 6$ where $F(x)$ is a primitive of $x + 1$. You now need to derive this result to be sure that it holds true for all primitives.

Fundamental theorem of calculus

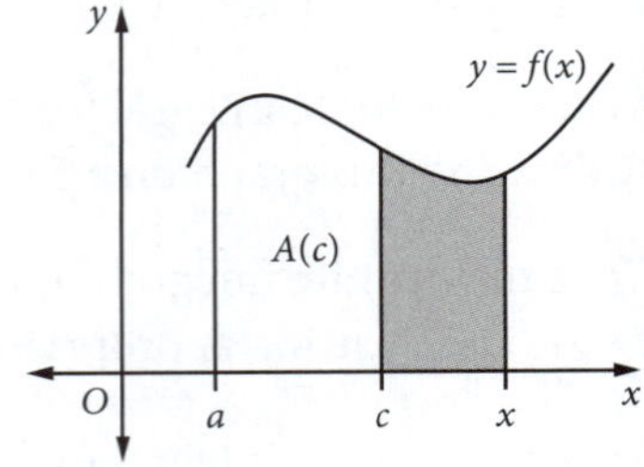

Let $y = f(x)$ be a continuous positive curve defined for values of x including $x \geq a$.

For $c > a$, let $A(c)$ denote the area under the curve between a and c, and $A(x)$ denote the area under the curve between a and x. The shaded area of the diagram is $A(x) - A(c)$. It can be approximated by a rectangle of height $f(x)$ and base $(x - c)$:

$$A(x) - A(c) \approx (x - c)\,f(x) \quad \text{or} \quad f(x) \approx \frac{A(x) - A(c)}{x - c}$$

If the maximum and minimum values of $f(t)$ for $c \leq t \leq x$ are M and m respectively, then:

$$m \leq \frac{A(x) - A(c)}{x - c} \leq M$$

These inequalities also hold when $x < c$.

As x approaches c, both m and M approach $f(c)$.

Hence: $\displaystyle\lim_{x \to c} \frac{A(x) - A(c)}{x - c} = f(c)$

But this is the same as the earlier definition of the derivative of a function $A(x)$.

Thus you have: $A'(c) = \displaystyle\lim_{x \to c} \frac{A(x) - A(c)}{x - c}$. Hence $A'(c) = f(c)$, $c > a$.

Because this statement is true for all $c > a$, the two functions $y = f(x)$ and $y = A'(x)$ are equal for all $x > a$. Thus $A(x)$ is a primitive function of $f(x)$ for $x > a$.

If you already know a primitive function $F(x)$ of $f(x)$, then you must have $A(x) = F(x) + C$ for some constant C.

As $x \to a$, $A(x) \to 0$, $F(x) \to F(a)$, so: $0 = F(a) + C$ or $C = -F(a)$

Thus $A(x) = F(x) - F(a)$ for $x > a$.

If $b > a$, then substituting b for x you have: $A(b) = \displaystyle\int_a^b f(x)\,dx = F(b) - F(a)$

Hence the area and the definite integral can be found using the known primitive function $F(x)$.

Fundamental theorem of calculus:

$$\int_a^b f(x)\,dx = F(b) - F(a)$$ where $F(x)$ is a primitive function of $f(x)$.

In deriving the above result $f(x)$ was considered as an increasing function in the interval $a \leq x \leq b$. However, the theorem is still true for a decreasing function or for a function that is sometimes increasing and sometimes decreasing in the interval, as long as the function is continuous.

When evaluating an integral, it is convenient to set it out as follows:

$$\int_a^b f(x)\,dx = \big[F(x)\big]_a^b = F(b) - F(a)$$

If you can find the primitive function, you can now evaluate any definite integral. Unfortunately, not all primitive functions are easy to find. At this stage you will be limited to polynomial functions and rational powers of the independent variable. Other functions will be looked at later. Later you will also use the trapezoidal rule to find numerical approximations for definite integrals.

Example 5

Evaluate:

(a) $\displaystyle\int_1^3 \left(x^2 - x\right)dx$

(b) $\displaystyle\int_{-1}^2 \left(x^3 - 2x^2 + 3x - 4\right)dx$

Solution

(a) $\int_1^3 (x^2 - x)\,dx = \left[\frac{x^3}{3} - \frac{x^2}{2}\right]_1^3$

$= \left(9 - \frac{9}{2}\right) - \left(\frac{1}{3} - \frac{1}{2}\right)$

$= 4\frac{2}{3}$

(b) $\int_{-1}^2 (x^3 - 2x^2 + 3x - 4)\,dx = \left[\frac{x^4}{4} - \frac{2x^3}{3} + \frac{3x^2}{2} - 4x\right]_{-1}^2$

$= \left(4 - \frac{16}{3} + 6 - 8\right) - \left(\frac{1}{4} + \frac{2}{3} + \frac{3}{2} + 4\right)$

$= -9\frac{3}{4}$

In Example **5** you are evaluating a definite integral using the given process. The significance of the negative answer in **(b)** will be explained later. For now, draw the graph of $f(x)$ in each part to see what might be happening if you were finding areas. What is different about the graph in part **(b)**?

Example 6

Calculate the area of the region bounded by the graph of the straight line $y = 2x + 3$, the x-axis and the ordinates $x = 1$ and $x = 5$.

Solution

Sketch the region:

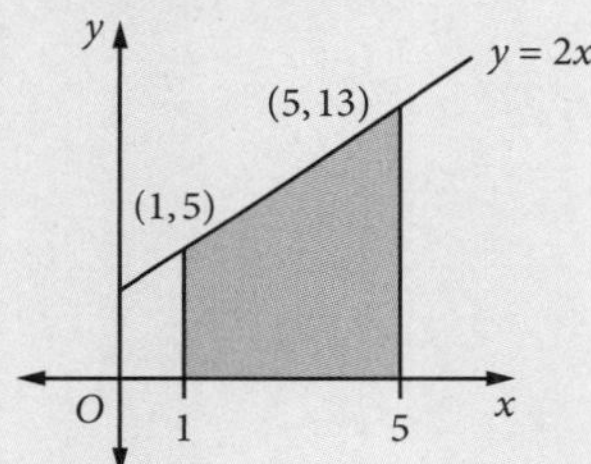

$$\text{Area} = \int_1^5 (2x+3)\,dx$$

$$= \left[x^2 + 3x\right]_1^5$$

$$= (25+15) - (1+3)$$

$$= 36 \text{ units}^2$$

Example 7

Use integration to find the area of the region enclosed by the parabola $y = x^2$, the x-axis and the line $x = 5$.

Solution

Sketch the region:

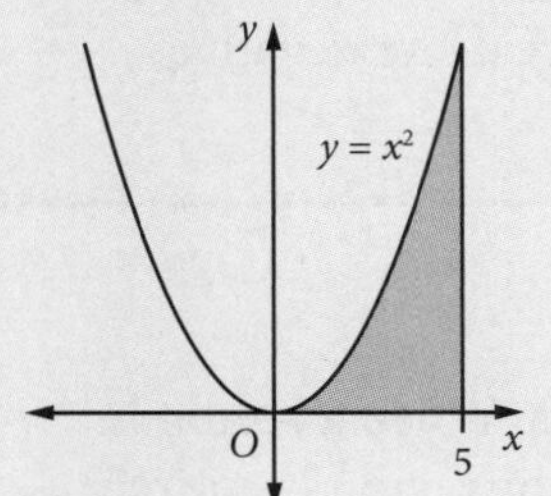

$$\text{Area} = \int_0^5 x^2\,dx$$

$$= \left[\frac{x^3}{3}\right]_0^5$$

$$= \frac{5^3}{3} - 0$$

$$= 41\frac{2}{3} \text{ units}^2$$

Example 8

Calculate the area of the region bounded by:

(a) the graph of the parabola $y = 7x - x^2 - 10$ and the x-axis

(b) the graph of the parabola $y = x^2 - 7x + 10$ and the x-axis.

(c) Compare the answers.

Solution

(a) Sketch the region:

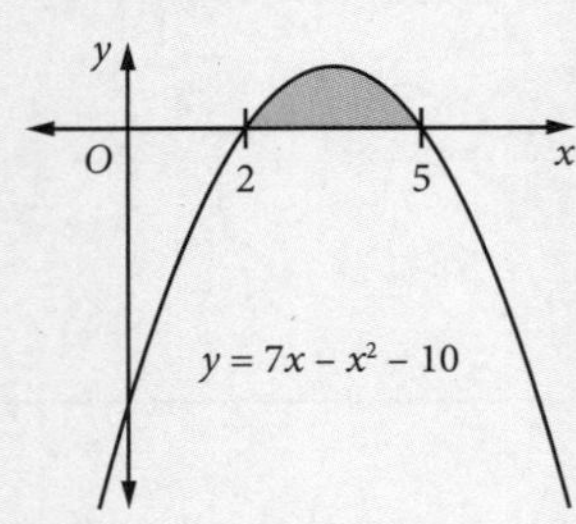

Cuts x-axis when $7x - x^2 - 10 = 0$:

$$x^2 - 7x + 10 = 0$$
$$(x-2)(x-5) = 0$$
$$x = 2, 5$$

$$\text{Area} = \int_2^5 \left(-x^2 + 7x - 10\right) dx$$
$$= \left[-\frac{x^3}{3} + \frac{7x^2}{2} - 10x\right]_2^5$$
$$= \left(-\tfrac{125}{3} + \tfrac{175}{2} - 50\right) - \left(-\tfrac{8}{3} + 14 - 20\right)$$
$$= 4.5$$

(b) Sketch the region:

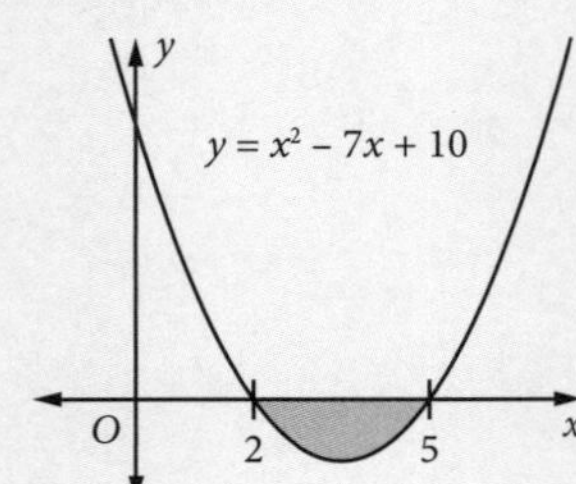

Cuts x-axis when $x^2 - 7x + 10 = 0$:

$$(x-2)(x-5) = 0$$
$$x = 2, 5$$

$$\text{Area} = \int_2^5 \left(x^2 - 7x + 10\right) dx$$
$$= \left[\frac{x^3}{3} - \frac{7x^2}{2} + 10x\right]_2^5$$
$$= \left(\tfrac{125}{3} - \tfrac{175}{2} + 50\right) - \left(\tfrac{8}{3} - 14 + 20\right)$$
$$= -4.5$$

Negative area does not make sense, but you can overcome this problem by taking the absolute value of the answer. The working needs to be written as:

$$\text{Area} = \left|\int_2^5 \left(x^2 - 7x + 10\right) dx\right| = \left|-4.5\right| = 4.5$$

(c) The two integrals have the same size but a different sign. The second curve and region are just the first curve and region reflected in the x-axis, so the regions must have the same area. The negative sign indicates that the region is below the x-axis.

Evaluating a definite integral and finding an area

The value of an integral and the area of a region can have different signs, as Example **8** shows. When a region is entirely above the x-axis, the value of the integral is positive and is equal to the area. When the region is entirely below the x-axis, the value of the integral is negative. To find the area, you take the absolute value:

$$\text{Area} = \left|\int_a^b f(x)\,dx\right| = \left|\left[F(x)\right]_a^b\right| = \left|F(b) - F(a)\right|$$

What happens when the curve defining a region is both above and below the x-axis? In a situation like that you must find where the curve cuts the x-axis and then consider the regions above and below the x-axis separately.

Example 9

The curve $y = x^3 - 6x^2 + 8x$ cuts the x-axis at 0, 2, and 4 as shown in the diagram. By first finding the areas A_1 and A_2, find the total area enclosed by the curve and the x-axis.

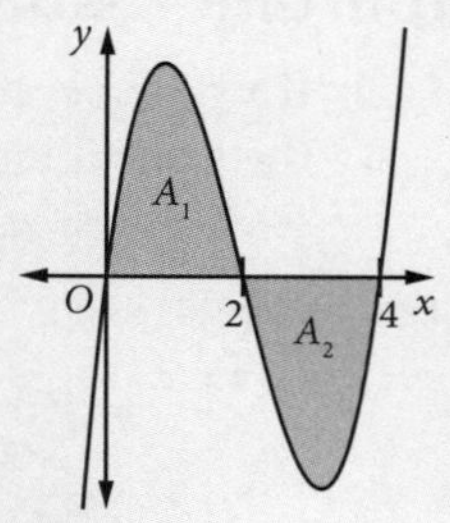

Solution

$$\begin{aligned} A_1 &= \left| \int_0^2 \left(x^3 - 6x^2 + 8x\right) dx \right| \\ &= \left| \left[\frac{x^4}{4} - 2x^3 + 4x^2 \right]_0^2 \right| \\ &= \left| (4 - 16 + 16) - (0) \right| \\ &= 4 \end{aligned}$$

$$\begin{aligned} A_2 &= \left| \int_2^4 \left(x^3 - 6x^2 + 8x\right) dx \right| \\ &= \left| \left[\frac{x^4}{4} - 2x^3 + 4x^2 \right]_2^4 \right| \\ &= \left| (64 - 128 + 64) - (4 - 16 + 16) \right| \\ &= |-4| \\ &= 4 \end{aligned}$$

$$\begin{aligned} \text{Total area} &= A_1 + A_2 \\ &= 8 \text{ units}^2 \end{aligned}$$

Example 10

Find the value of the definite integral $\int_0^4 \left(x^3 - 6x^2 + 8x\right) dx$.

Solution

$$\begin{aligned} \int_0^4 \left(x^3 - 6x^2 + 8x\right) dx &= \left[\frac{x^4}{4} - 2x^3 + 4x^2 \right]_0^4 \\ &= (64 - 128 + 64) - (0) \\ &= 0 \end{aligned}$$

If you used this integral to find the area of the region in Example 9, you would obtain an area of zero, which is clearly wrong. You should be aware that finding the value of a definite integral is not always the same as finding the area of a region. This is why it is important for you to sketch a region before setting up an integral to calculate its area. You must calculate the areas above and below the x-axis separately. Areas above and below the x-axis have opposite signs.

MAKING CONNECTIONS

Definite integrals with signed and unsigned areas

Move the sliders to calculate the signed and unsigned areas between the curve and the x-axis.

Important results

1 If $y = f(x)$ is an increasing function over the interval $a \le x \le b$, then $y = -f(x)$ is a decreasing function over the same interval. $y = -f(x)$ is the reflection of $y = f(x)$ in the x-axis. Thus the area of the region bounded by $y = f(x)$, $x = a$, $x = b$ and the x-axis is equal to the area of the region bounded by $y = -f(x)$, $x = a$, $x = b$ and the x-axis.

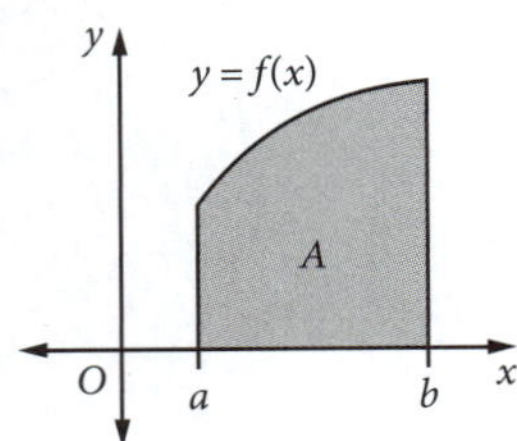

$$A = \left| \int_a^b f(x)dx \right|$$

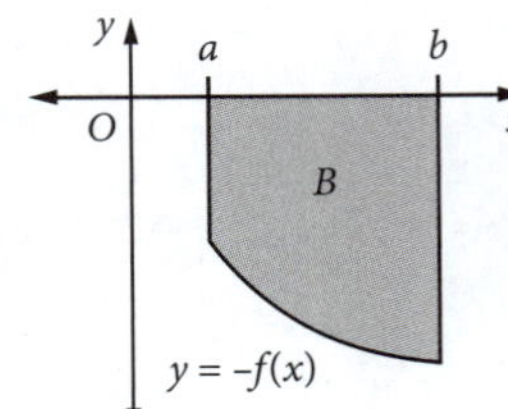

$$B = \left| \int_a^b (-f(x))dx \right|$$

$$B = \left| -\int_a^b f(x)dx \right| = \left| \int_a^b f(x)dx \right|$$

2 If $y = f(x)$ is a continuous function over the interval $a \le x \le b$, and $a \le c \le b$, then:

$$\int_a^b f(x)dx = \int_a^c f(x)dx + \int_c^b f(x)dx$$

3 If F, G are primitives of f, g respectively, then $F + G$ is a primitive of $f + g$. Thus:

$$\begin{aligned}\int_a^b (f(x)+g(x))dx &= (F(b)+G(b))-(F(a)+G(a)) \\ &= (F(b)-F(a))+(G(b)-G(a)) \\ &= \int_a^b f(x)dx + \int_a^b g(x)dx\end{aligned}$$

4 If you reverse the limits of integration, you change the sign of the integral: $\int_a^b f(x)dx = -\int_b^a f(x)dx$

5 $\int_a^b f(x)dx \pm \int_a^b g(x)dx = \int_a^b (f(x) \pm g(x))dx$

6 $\int_a^b cf(x)dx = c\int_a^b f(x)dx$ where c is a constant.

Example 11

Calculate the area between $y = \sqrt{x}$, the x-axis and $x = 4$.

Solution

Sketch the region:

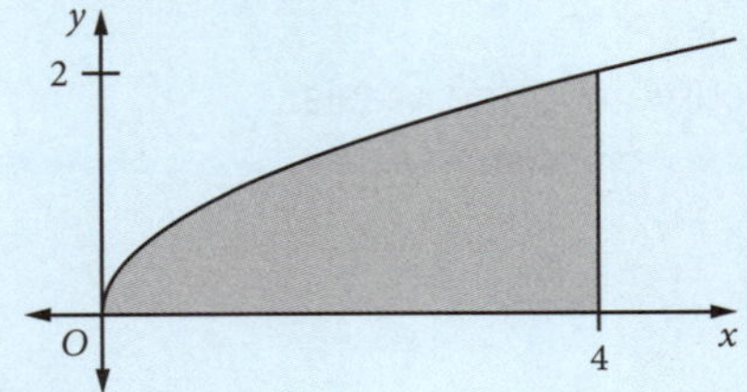

It is clear from the diagram that the region is above the x-axis.

Hence: Area $= \displaystyle\int_0^4 \sqrt{x}\,dx$

$$= \left[\frac{x^{\frac{3}{2}}}{\frac{3}{2}} \right]_0^4$$

$$= \left[\frac{2}{3} x^{\frac{3}{2}} \right]_0^4$$

$$= \frac{2}{3}(8-0)$$

$$= 5\frac{1}{3} \text{ units}^2$$

Example 12

Find the area of the region bounded by the curve $f(x) = x(x-2)(x+1)$ and the x-axis.

Solution

Sketch the region:

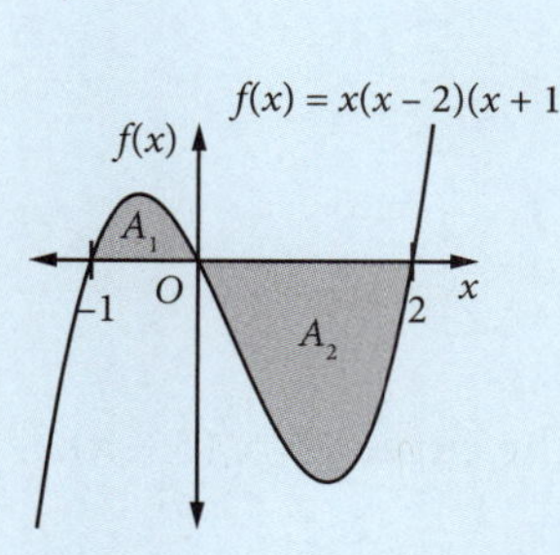

$$f(x) = x(x-2)(x+1) = x^3 - x^2 - 2x$$

$$A_1 = \int_{-1}^{0} \left(x^3 - x^2 - 2x\right) dx = \left[\frac{x^4}{4} - \frac{x^3}{3} - x^2\right]_{-1}^{0} = 0 - \left(\frac{1}{4} + \frac{1}{3} - 1\right) = \frac{5}{12}$$

$$A_2 = \left|\int_{0}^{2} \left(x^3 - x^2 - 2x\right) dx\right| = \left|\left[\frac{x^4}{4} - \frac{x^3}{3} - x^2\right]_{0}^{2}\right| = \left|\left(\frac{16}{4} - \frac{8}{3} - 4\right) - 0\right| = 2\frac{2}{3}$$

$$\text{Total area} = \frac{5}{12} + 2\frac{2}{3} = 3\frac{1}{12} \text{ units}^2$$

You can show by evaluating the integral that $\int_{-1}^{2} \left(x^3 - x^2 - 2x\right) dx \neq 3\frac{1}{12}$.

EXERCISE 17.3 THE DEFINITE INTEGRAL AND THE PRIMITIVE FUNCTION

1 Evaluate $\int_0^3 x^2\,dx$, $\int_3^0 x^2\,dx$ and show that $\int_0^3 x^2\,dx = -\int_3^0 x^2\,dx$.

2 Evaluate $\int_1^2 x^2\,dx$, $\int_2^3 x^2\,dx$, $\int_1^3 x^2\,dx$ and show that $\int_1^3 x^2\,dx = \int_1^2 x^2\,dx + \int_2^3 x^2\,dx$.

3 Evaluate $\int_1^3 x^2\,dx$, $\int_1^3 2x\,dx$, $\int_1^3 \left(x^2 + 2x\right) dx$ and verify that $\int_1^3 \left(x^2 + 2x\right) dx = \int_1^3 x^2\,dx + \int_1^3 2x\,dx$.

4 Evaluate $\int_0^3 \left(4x^2 + 8x\right) dx$, $4\int_0^3 \left(x^2 + 2x\right) dx$ and verify that $\int_0^3 \left(4x^2 + 8x\right) dx = 4\int_0^3 \left(x^2 + 2x\right) dx$.

5 Evaluate: **(a)** $\int_0^1 (2x-5)\,dx$ **(b)** $\int_{-1}^4 (3x+1)\,dx$ **(c)** $\int_{-2}^2 \left(x^2 - 4\right) dx$

(d) $\int_1^3 (2t-4)\,dt$ **(e)** $\int_{-2}^3 \left(u^2 - 2u\right) du$ **(f)** $\int_{-2}^3 (2x+1)^2\,dx$

6 The value of $\int_0^5 \left(6 + 8x - 3x^2\right) dx$ is: **A** 29 **B** 5 **C** −5 **D** −29

7 Evaluate: **(a)** $\int_{-1}^1 x\left(1 - x^2\right) dx$ **(b)** $\int_{-1}^3 \left(2x^2 - 8x + 8\right) dx$ **(c)** $\int_0^4 (x-2)^3\,dx$

(d) $\int_{-1}^1 x^2(1-x)\,dx$ **(e)** $\int_0^4 x^{\frac{1}{2}}\,dx$ **(f)** $\int_{-1}^{-\frac{1}{2}} \left(-x^3\right) dx$

(g) $\int_{-8}^8 x^{\frac{2}{3}}\,dx$ **(h)** $\int_0^1 \left(x^{\frac{1}{5}} - x^{\frac{1}{3}}\right) dx$ **(i)** $\int_0^1 \left(\sqrt{x} - \sqrt[3]{x}\right) dx$

8 Find an expression for: **(a)** $\int_b^a 3x^2\,dx$ **(b)** $\int_0^2 \left(ax^2 + bx + c\right) dx$ **(c)** $\int_0^a \left(x^2 - 2x\right) dx$

9 The following expressions are steps in the evaluation of $\int_2^4 \left(x+\frac{1}{x}\right)^2 dx$. Indicate whether each step is correct or incorrect.

(a) $\int_2^4 \left(x^2+2+\frac{1}{x^2}\right)dx$ **(b)** $\left[\frac{x^3}{3}+2x-\frac{2}{x^3}\right]_2^4$ **(c)** $\left[\frac{x^3}{3}+2x-\frac{1}{x}\right]_2^4$ **(d)** $22\frac{11}{12}$

10 Find the area of the region bounded by the x-axis and the graph of $y = 2 - x - x^2$.

11 Find the area of the region bounded by the x-axis and the graph of $y = x^2(1 - x)$.

12 Find the area of the region bounded by the graph of $y = 3 + 5x - 2x^2$ and the x-axis.

13 If $\int_0^a (4-2x)\,dx = 4$, find the value of a.

14 If $\int_{-1}^a x\,dx = 0$, find the value of a.

15 If $c\int_{-2}^2 (x-5)\,dx = 1$, find the value of c.

16 Calculate the area bounded by the curve $y = x^2(3 - x)$.

17.4 MORE AREAS

Example 13

Calculate the area of the region bounded by the graph of the parabola $f(x) = x^2 - 7x + 10$, the x-axis, and the ordinates $x = 3$ and $x = 4$.

Solution

Sketch the region:

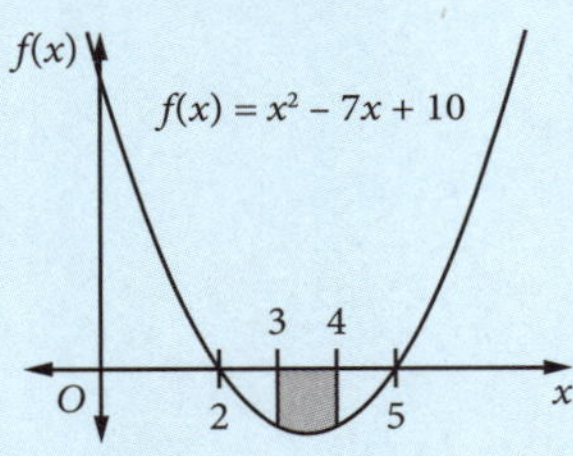

$$\text{Area} = \left|\int_3^4 \left(x^2-7x+10\right)dx\right|$$

$$= \left|\left[\frac{x^3}{3}-\frac{7x^2}{2}+10x\right]_3^4\right|$$

$$= \left|\left(\frac{64}{3}-56+40\right)-\left(9-\frac{63}{2}+30\right)\right|$$

$$= 2\frac{1}{6}$$

Odd and even functions

For an even function, the area bounded by the curve $y = f(x)$, the x-axis and the ordinates $x = a$ and $x = -a$ is:

$$\text{Area} = \left|\int_{-a}^a f(x)\,dx\right| = 2\left|\int_0^a f(x)\,dx\right|$$

For an odd function, the area bounded by the curve $y = f(x)$, the x-axis and the ordinates $x = a$ and $x = -a$ is:

$$\text{Area} = \left|\int_{-a}^0 f(x)\,dx\right| + \left|\int_0^a f(x)\,dx\right| = 2\left|\int_0^a f(x)\,dx\right|$$

Example 14

Find the area of the region bounded by the curve $y = f(x)$, the x-axis and the ordinates $x = -2$ and $x = 2$, for:

(a) $f(x) = x^2 - 4$ **(b)** $f(x) = x^3$

Solution

(a) Sketch the region:

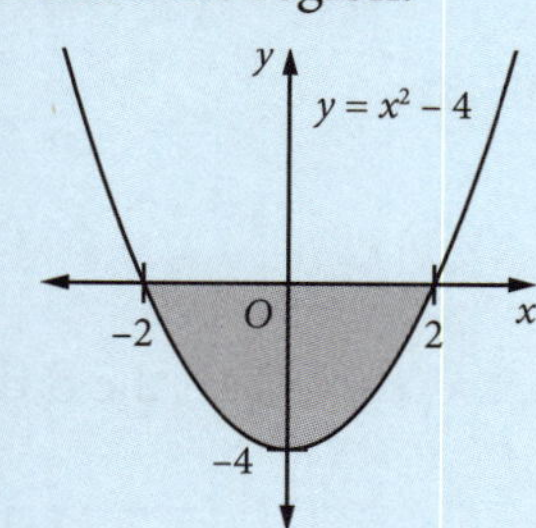

$f(x) = x^2 - 4$ is an even function.

$$\int_{-2}^{0}\left(x^2 - 4\right)dx = \int_{0}^{2}\left(x^2 - 4\right)dx$$

$$\therefore \text{ Area} = 2\left|\int_{0}^{2}\left(x^2 - 4\right)dx\right|$$

$$= 2\left|\left[\frac{x^3}{3} - 4x\right]_0^2\right|$$

$$= 2\left|\frac{8}{3} - 8\right| = 2\left|-\frac{16}{3}\right|$$

$$= 10\frac{2}{3} \text{ units}^2$$

(b) Sketch the region:

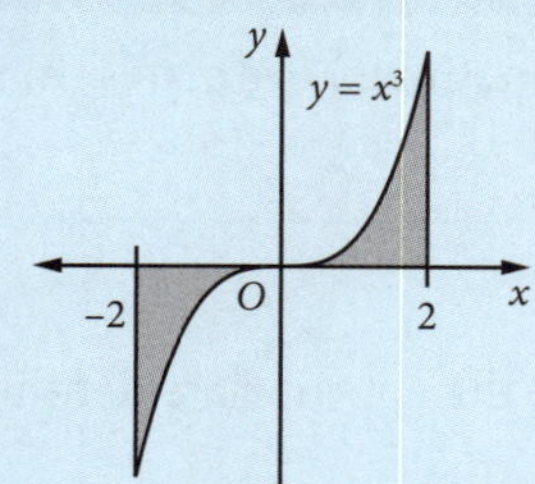

$f(x) = x^3$ is an odd function.

$$\int_{-2}^{0} x^3\,dx = -\int_{0}^{2} x^3\,dx \quad \text{and} \quad \int_{-2}^{2} x^3\,dx = 0$$

$$\therefore \text{ Area} = \left|\int_{-2}^{0} x^3\,dx\right| + \left|\int_{0}^{2} x^3\,dx\right| = 2\left|\int_{0}^{2} x^3\,dx\right|$$

$$= 2\left|\left[\frac{x^4}{4}\right]_0^2\right|$$

$$= 2|4 - 0|$$

$$= 8 \text{ units}^2$$

EXERCISE 17.4 MORE AREAS

1 Calculate the area of the region bounded by the line $y = 3x + 5$, the x-axis and the lines $x = 3$ and $x = 6$.

2 Find the area of the region bounded by the parabola $y = x^2 + 2$, the x-axis and the lines $x = -1$ and $x = 2$.

3 Calculate the area of the region bounded by the graph of $f(x) = x^2 - 4x + 4$, the x-axis and the lines $x = 1$ and $x = 4$.

4 Calculate the area of the region bounded by the curve $y = 4 - x^2$ and the x-axis.

5 Which of these integrals will give the area of the region bounded by the curve $y = 16 - x^4$ and the x-axis? Indicate whether each answer is correct or incorrect.

(a) $\int_{-2}^{2}\left(16 - x^4\right)dx$ **(b)** $\int_{-4}^{4}\left(16 - x^4\right)dx$ **(c)** $2\int_{0}^{2}\left(16 - x^4\right)dx$ **(d)** $\left|\int_{-2}^{2}\left(16 - x^4\right)dx\right|$

6 Calculate the area of the region bounded by the curve $y = -x^3$, the x-axis and the ordinates $x = -3$ and $x = 3$.

7 The value of the definite integral $\int_{-4}^{4} x^3\,dx$ is: **A** -128 **B** 0 **C** 64 **D** 128

8 Calculate the area of the region bounded by the graph of $f(x) = (x - 2)^3$, the x-axis, $x = 2$ and $x = 3$.

9 Find a positive number k such that the area of the region bounded by the graph of $f(x) = kx(2 - x)^2$ and the x-axis is equal to 1 unit^2.

10 For the graph of $f(x) = (x + 1)(x - 1)^2$, calculate:

(a) the area bounded by the curve, the x-axis, $x = 0$ and $x = 0.5$

(b) the area bounded by the curve and the x-axis

(c) the area to the right of the origin bounded by the curve and the coordinate axes.

11 Find the area bounded by the parabola $y = 8x - x^2$ and the x-axis.

12 Find the area of the region bounded by the curve $y = x(x - 2)^2$ and the x-axis.

13 Calculate the area of the region bounded by the curve $y = (x + 1)(x - 1)(x - 3)$, the x-axis and the ordinates at $x = 0$ and $x = 2$.

14 Show by integration that the area of a unit square (see diagram) is:

(a) bisected by the line $y = x$

(b) trisected by the curves $y = x^2$ and $y = \sqrt{x}$

(c) divided into four equal parts by the curves $y = x^3$, $y = x$ and $y = \sqrt[3]{x}$.

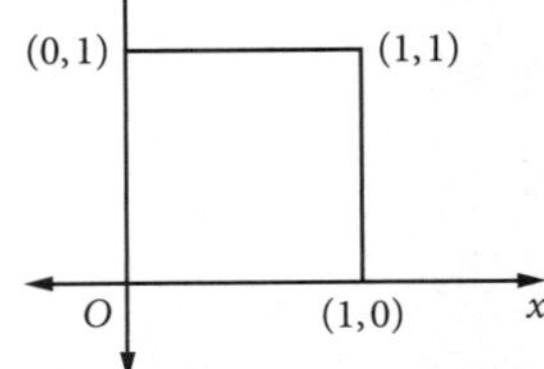

17.5 AREA BETWEEN TWO CURVES

So far you have used integration to find areas bounded by a curve, an axis and (sometimes) ordinates. In this section you will use integration to calculate the area between two curves.

Example 15

Sketch the region bounded by the curves $y = 4 - x^2$ and $y = x^2 - 4$. By evaluating the appropriate definite integrals, calculate the area of this region.

Solution

Sketch the region:

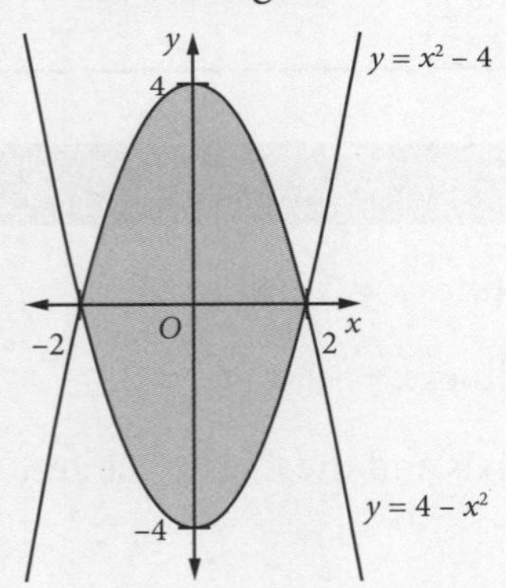

Both curves cut the x-axis at $x = \pm 2$.

Call the area above the x-axis A_1 and the area below the x-axis A_2.

A_1 is bounded by $y = 4 - x^2$ and the x-axis, hence:

$$A_1 = \int_{-2}^{2} \left(4 - x^2\right) dx$$

You can see from the sketch that this area is positive, so the absolute value bars can be left out.

A_2 is bounded by $y = x^2 - 4$ and the x-axis, hence:

$$A_2 = \left| \int_{-2}^{2} \left(x^2 - 4\right) dx \right|$$

You can see from the sketch that the integral for this area will be negative, so you must include absolute value bars.

Evaluate the integrals:

$$\begin{aligned} A_1 &= \int_{-2}^{2} \left(4 - x^2\right) dx \\ &= \left[4x - \frac{x^3}{3}\right]_{-2}^{2} \\ &= \left(8 - \frac{8}{3}\right) - \left(-8 + \frac{8}{3}\right) \\ &= \frac{32}{3} \\ &= 10\frac{2}{3} \end{aligned}$$

$$\begin{aligned} A_2 &= \left| \int_{-2}^{2} \left(x^2 - 4\right) dx \right| \\ &= \left| \left[\frac{x^3}{3} - 4x\right]_{-2}^{2} \right| \\ &= \left| \left(\frac{8}{3} - 8\right) - \left(-\frac{8}{3} + 8\right) \right| \\ &= \left| -\frac{32}{3} \right| \\ &= 10\frac{2}{3} \end{aligned}$$

Area of shaded region $= A_1 + A_2$

$= 10\frac{2}{3} + 10\frac{2}{3} = 21\frac{1}{3}$ units^2

Alternate method:

$$A_1 + A_2 = \int_{-2}^{2}(4 - x^2)dx + \left|\int_{-2}^{2}(x^2 - 4)dx\right|$$

Now $\left|\int_{-2}^{2}(x^2 - 4)dx\right| = -\left(\int_{-2}^{2}(x^2 - 4)dx\right)$, so:

$$\begin{aligned} A_1 + A_2 &= \int_{-2}^{2}(4 - x^2)dx - \int_{-2}^{2}(x^2 - 4)dx \\ &= \int_{-2}^{2}(4 - x^2)dx + \int_{-2}^{2}(4 - x^2)dx \\ &= 2\int_{-2}^{2}(4 - x^2)dx \\ &= 2\left[4x - \frac{x^3}{3}\right]_{-2}^{2} = 21\tfrac{1}{3} \end{aligned}$$

$$\begin{aligned} A_1 + A_2 &= \int_{-2}^{2} f(x)dx + \left|\int_{-2}^{2} g(x)dx\right| \\ &= \int_{-2}^{2} f(x)dx - \int_{-2}^{2} g(x)dx \\ &= \int_{-2}^{2}(f(x) - g(x))dx \end{aligned}$$

The process here becomes more clear if you write $f(x) = 4 - x^2$ and $g(x) = x^2 - 4$:

$$\begin{aligned} &= \int_{-2}^{2}\left((4 - x^2) - (x^2 - 4)\right)dx \\ &= \int_{-2}^{2}(8 - 2x^2)dx \\ &= 2\int_{-2}^{2}(4 - x^2)dx = 2\left[4x - \frac{x^3}{3}\right]_{-2}^{2} = 21\tfrac{1}{3} \end{aligned}$$

Area between two curves—general result

If f and g are two continuous functions whose graphs do not intersect in the interval $a \le x \le b$, and $f(x) > g(x)$ over this interval, then the area of the region bounded by the two curves and the ordinates $x = a$ and $x = b$ is given by:

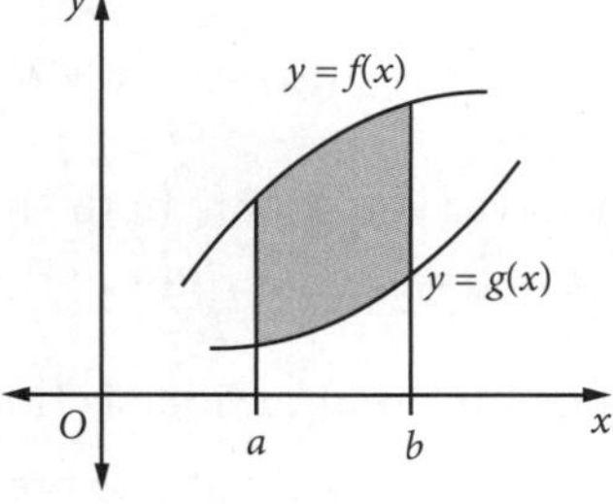

$$\begin{aligned} \text{Area} &= \int_a^b f(x)dx - \int_a^b g(x)dx \\ &= \int_a^b (f(x) - g(x))dx \end{aligned}$$

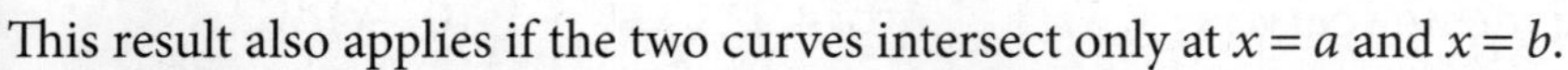

This result also applies if the two curves intersect only at $x = a$ and $x = b$.

Example 16

Calculate the area of the region enclosed by the graphs of $f(x) = x + 1$ and $g(x) = x^2 - x - 2$.

Solution

Find the x-values of the points of intersection of the two curves by solving $f(x)$ and $g(x)$ simultaneously:

$$\begin{aligned} x^2 - x - 2 &= x + 1 \\ x^2 - 2x - 3 &= 0 \\ (x + 1)(x - 3) &= 0 \\ x &= -1, 3 \\ y &= 0, 4 \end{aligned}$$

Sketch the region:

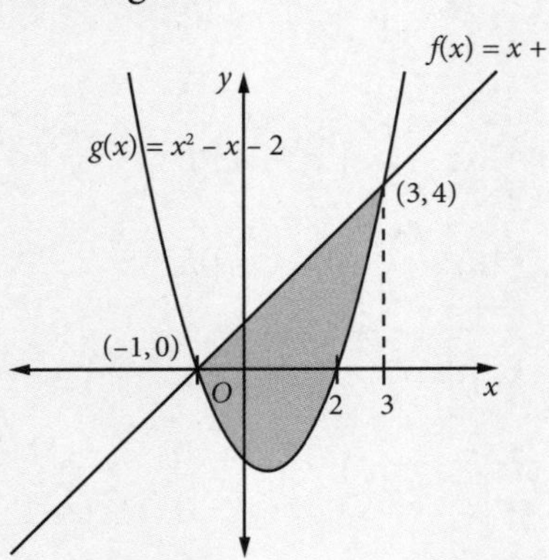

$f(x) \geq g(x)$ for $-1 \leq x \leq 3$. Area of the region A is given by:

$$A = \int_{-1}^{3} f(x)\,dx - \int_{-1}^{3} g(x)\,dx$$

or $$A = \int_{-1}^{3} \left(f(x) - g(x)\right)dx$$

Hence: $$A = \int_{-1}^{3} \left((x+1) - \left(x^2 - x - 2\right)\right)dx$$

$$= \int_{-1}^{3} \left(-x^2 + 2x + 3\right)dx$$

$$= \left[-\frac{x^3}{3} + x^2 + 3x\right]_{-1}^{3}$$

$$= (-9+9+9) - \left(\frac{1}{3} + 1 - 3\right) = 10\frac{2}{3}$$

The area between the curves is $10\frac{2}{3}$ units^2.

Example 17

Calculate the area of the region bounded by the x-axis and the curves with equations $y = \sqrt{x}$ and $y = 6 - x$.

Solution

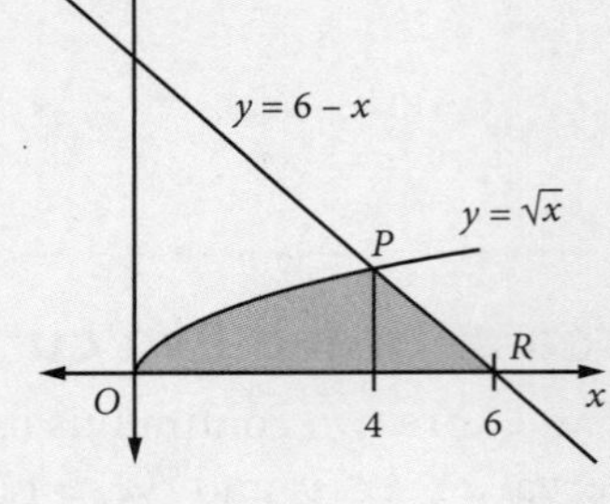

Find the point of intersection: $\sqrt{x} = 6 - x$

Square both sides: $(\sqrt{x})^2 = (6 - x)^2$

$$x = 36 - 12x + x^2$$

$$x^2 - 13x + 36 = 0$$

$$(x-4)(x-9) = 0$$

$$x = 4, 9$$

Only $x = 4$ is a root of $\sqrt{x} = 6 - x$. Check by substituting $x = 9$ into the equation.

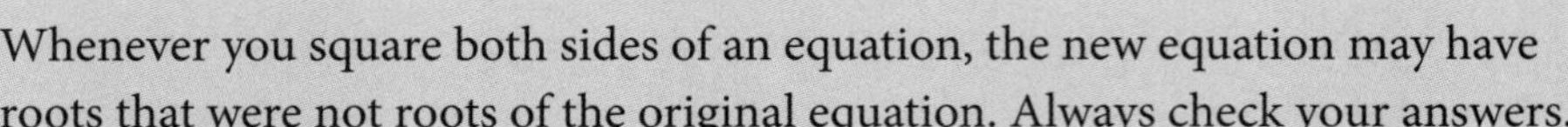

Whenever you square both sides of an equation, the new equation may have roots that were not roots of the original equation. Always check your answers.

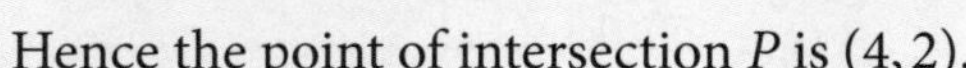

Hence the point of intersection P is $(4, 2)$.

The region consists of two areas: the area under the curve $y = \sqrt{x}$ between O and P (i.e. for $0 \leq x \leq 4$) and the area under the curve $y = 6 - x$ between P and R (i.e. for $4 \leq x \leq 6$).

$$\text{Area} = \int_{0}^{4} \sqrt{x}\,dx + \int_{4}^{6} (6 - x)\,dx$$

$$= \left[\frac{2}{3}x^{\frac{3}{2}}\right]_{0}^{4} + \left[6x - \frac{x^2}{2}\right]_{4}^{6}$$

$$= \left(\frac{2}{3} \times 8 - 0\right) + (36 - 18 - (24 - 8))$$

$$= 5\frac{1}{3} + 2$$

$$= 7\frac{1}{3}$$

The area of the region is $7\frac{1}{3}$ units^2.

EXPLORE FURTHER

Area between two curves

Use technology to graph and calculate the area between two curves.

EXERCISE 17.5 AREA BETWEEN TWO CURVES

1 Calculate the area of the region bounded by the line $y = 2x$ and the parabola $y = x^2$.

2 Calculate the area of the region bounded by the line $y = x + 1$ and the parabola $y = \frac{x^2}{4} - 2$.

3 The area of the region bounded by the line $y = x + 2$ and the parabola $y = x^2 - 4$ is given by:

A $\int_{-2}^{3}(6+x-x^2)dx$ **B** $\int_{-3}^{2}(6+x-x^2)dx$ **C** $\int_{-2}^{3}(x^2-x-6)dx$ **D** $\int_{-3}^{2}(x^2-x-6)dx$

4 Calculate the area bounded by $f(x) = x^2$, $g(x) = \frac{1}{x^2}$, $x > 0$, the x-axis and the line $x = 3$.

5 Calculate the area bounded by $f(x) = x^2$, $g(x) = \frac{1}{x^2}$, $x > 0$, and the line $x = 3$.

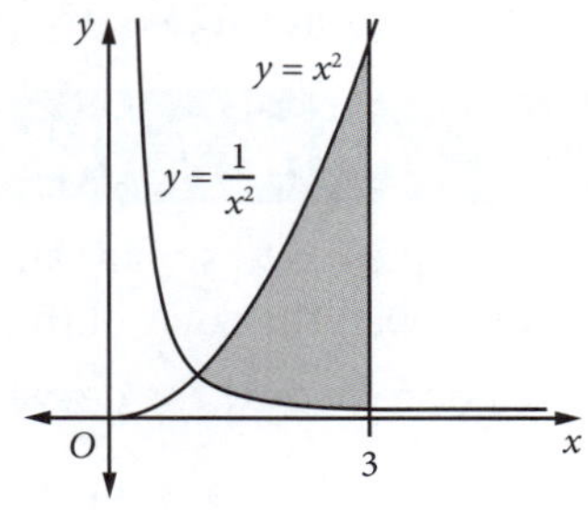

6 Find the area enclosed by the line $y = 2x + 3$ and the parabola $y = x^2$.

7 Find the area enclosed by the line $y = 2x$, the parabola $y = -x^2$ and the line $x = 2$.

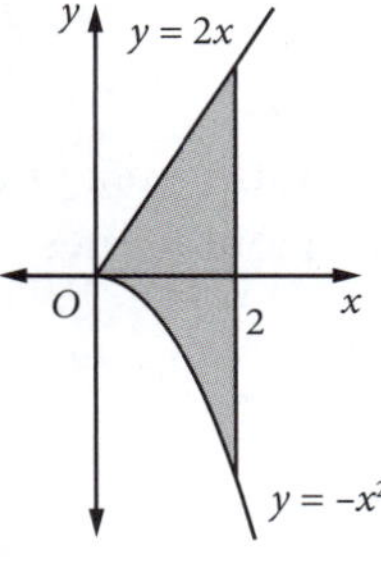

8 Calculate the area of the region enclosed by the graphs of the parabola $y = x^2 + 4$ and the line $y = 5$.

9 Calculate the area of the region enclosed by the graphs of the parabola $y = 2x^2 - 5x - 3$ and the line $y = 3x - 3$. Indicate whether each statement below is a correct or incorrect step in calculating this area.

(a) Intersection points: $(0,-3)$ and $(4,9)$ **(b)** Area $= \int_{0}^{4}(8x-2x^2)dx$

(c) Area $= \int_{-3}^{9}(8x-2x^2)dx$ **(d)** Area $= 21\frac{1}{3}$ units^2

10 Calculate the area of the region enclosed by the graphs of the parabolas $f(x) = 6x^2 - 5x$ and $g(x) = 5x - 4x^2$.

11 Calculate the area of the region defined by the inequalities $y \geq x$ and $y \leq 3x - x^2$.

12 Calculate the area of the region defined by the inequalities $y \leq 6 - x$ and $y \geq x^2 - 5x + 6$.

13 A straight line through the origin cuts the parabola $y = 4x - x^2$ at the point where $x = 3$.

(a) Find the equation of this line.
(b) Calculate the area of the region bounded by the parabola and the straight line.
(c) Calculate the area of the region bounded by the parabola, the straight line and the x-axis.

14 **(a)** Show that the area bounded by the curves $y = \sqrt{x}$, $y = \frac{1}{x}$ and $x = 3$ is given by $\int_{1}^{3}\left(\sqrt{x} - \frac{1}{x}\right)dx$.

(b) Use the trapezoidal rule with two subintervals to find an approximation for this area correct to 2 decimal places.

15 Part of a vertical section of an ore deposit takes the form shown in the diagram (not drawn to scale). The distances shown on the diagram have been indicated by vertical drilling at P and Q and by geological interpretation of the local rock formations. Taking the x-axis and y-axis as indicated, find:

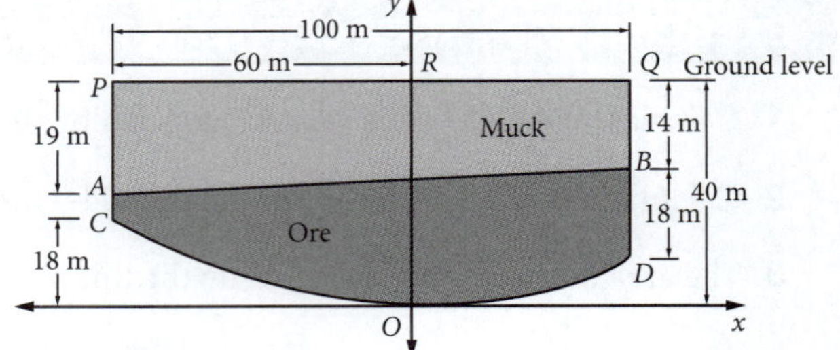

(a) the equation of the straight line AB

(b) the depth of muck at R

(c) the equation of the curve COD, assuming it is of the form $y = ax^n$

(d) the total area of the section $ABDC$ of the ore deposit.

17.6 AREA BOUNDED BY THE *y*-AXIS

The techniques used so far can also be used to find areas where one of the boundaries is the y-axis. You can effectively swap the roles of the x and y variables.

The area bounded by the curve $y = f(x)$, the x-axis and the ordinates at $x = a$ and $x = b$ is given by:

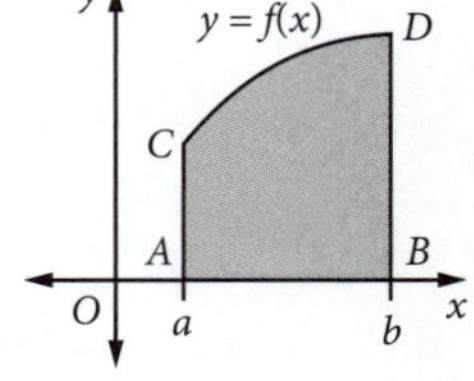

$$\text{Area} = \int_a^b f(x)dx \qquad \text{or} \qquad \text{Area} = \int_a^b y\,dx$$

Similarly, the area bounded by the curve $x = g(y)$, the y-axis and the abscissae (i.e. horizontal lines) at $y = c$ and $y = d$ is given by:

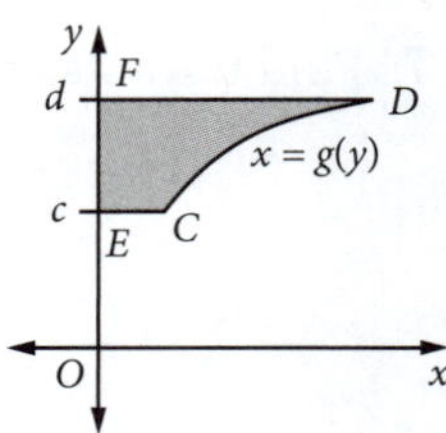

$$\text{Area} = \int_c^d g(y)dy \qquad \text{or} \qquad \text{Area} = \int_c^d x\,dy$$

For two curves $x = f(y)$ and $x = g(y)$ that intersect at $y = c$ and $y = d$, where $f(y) \ge g(y)$ over the interval $c \le y \le d$, the area bounded by the curves is given by:

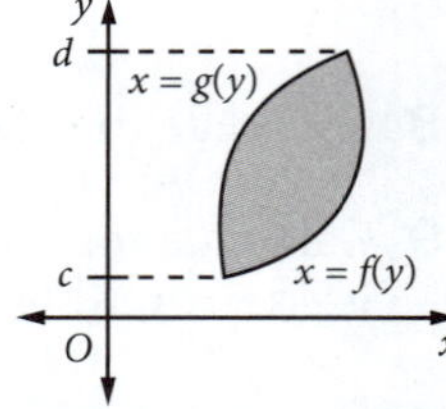

$$\text{Area} = \int_c^d \big(f(y) - g(y)\big)dy$$

Example 18

Calculate the area of the region bounded by the curve $y = \sqrt{x}$, the y-axis and the line $y = 2$.

Solution

Sketch the region:

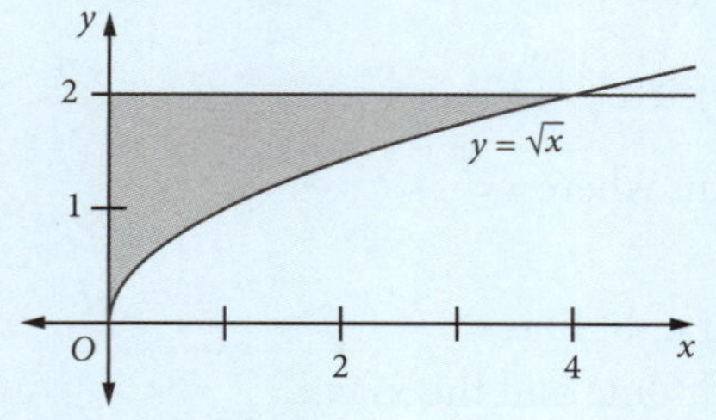

Write the equation as a function of y (i.e. make x the subject):

$$y = \sqrt{x} \to \sqrt{x} = y$$

$$x = y^2$$

$$\text{Area} = \int_0^2 x\,dy = \int_0^2 y^2\,dy$$

$$= \left[\frac{y^3}{3}\right]_0^2$$

$$= \frac{8}{3} - 0$$

$$= 2\frac{2}{3}$$

Area of the region is $2\frac{2}{3}$ units^2.

Example 19

Calculate the area of the region bounded by the curve $y=\sqrt{x}$, the line $y=6-x$ and the y-axis.

Solution

Find the point of intersection: $\sqrt{x}=6-x$

Square both sides: $(\sqrt{x})^2=(6-x)^2$

$$x=36-12x+x^2$$
$$x^2-13x+36=0$$
$$(x-4)(x-9)=0$$
$$x=4, 9$$

Only $x=4$ is a root of $\sqrt{x}=6-x$: intersection point is $(4,2)$.

Sketch the region:

Write the equations as functions of y:

$$y=\sqrt{x} \rightarrow y^2=x$$
$$y=6-x \rightarrow x=6-y$$
$$\text{Area}=\int_0^2 y^2\,dy+\int_2^6 (6-y)\,dy$$
$$=\left[\frac{y^3}{3}\right]_0^2+\left[6y-\frac{y^2}{2}\right]_2^6$$
$$=\left(\frac{8}{3}-0\right)+(36-18)-(12-2)$$
$$=10\frac{2}{3}$$

Area of the region is $10\frac{2}{3}$ units^2.

EXERCISE 17.6 AREA BOUNDED BY THE y-AXIS

1 Calculate the area of the region bounded by the line $y=x+1$, the y-axis and the line $y=4$.

2 Calculate the area of the region bounded by the curve $y=\sqrt{x}$, the y-axis and the line $y=3$.

3 The area of the region bounded by the curve $y=\sqrt[3]{x}$ and the line $y=2$ is given by:

A $\int_0^2 y\,dy$ **B** $\int_0^8 y\,dy$ **C** $\int_0^2 y^3\,dy$ **D** $\int_0^8 y^3\,dy$

4 Calculate the area of the region bounded by the curve $y=\frac{1}{x^2}$, the y-axis and the lines $y=1$ and $y=9$.

5 Calculate the area of the region bounded by the curve $y=x^2$ and the line $y=4$.

6 **(a)** Show that the equation of the tangent to the parabola $y=x^2+1$ at the point where $x=2$ is $y=4x-3$.

(b) Hence find the area enclosed by the parabola, the tangent and the y-axis.

(c) Find the area enclosed by the parabola, the tangent and the coordinate axes.

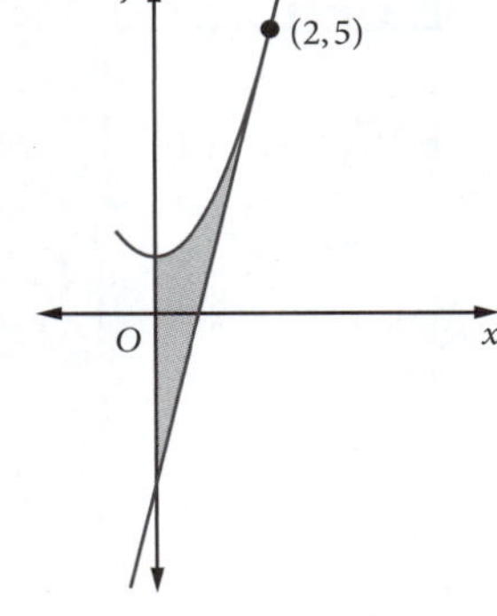

7 Calculate the area of the region bounded by the curve $x=\frac{1}{y}$, the y-axis and the lines $y=1$ and $y=4$ using the trapezoidal rule with three subintervals.

8 **(a)** Calculate the area of the region bounded by the parabolas $y = x^2$ and $y = 4 - x^2$.
(b) Calculate the area of the region bounded by the x-axis and the parabolas $y = x^2$ and $y = 4 - x^2$.

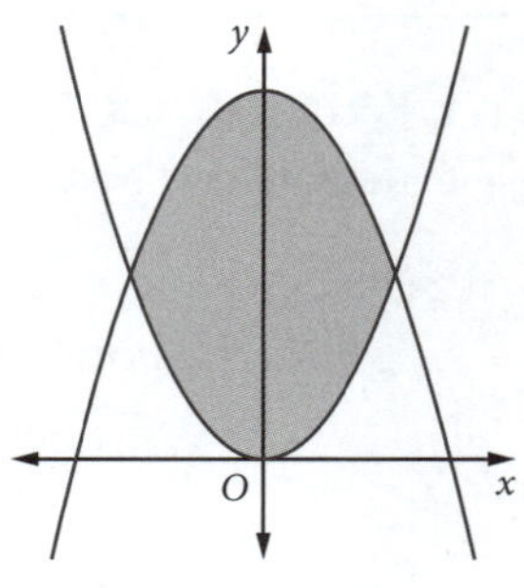

9 The area between the curves $x = \sqrt{y}$ and $y = 2x$ is calculated. Indicate whether each statement below is a correct or incorrect step in the calculation.

(a) The curves intersect at $(0, 0)$ and $(2, 4)$

(b) Area $= \int_0^4 \left(\sqrt{y} - \frac{y}{2}\right) dy$

(c) Area $= \int_0^2 \left(2x - x^2\right) dx$

(d) Area $= 1\frac{1}{3}$ units^2

10 Calculate the area of the region bounded by the lines $y = 4 - x$, $y = x$ and the y-axis.

11 Calculate the area of the region bounded by the curve $x = \sqrt{y + 4}$, the y-axis and the line $y = 2$.

12 **(a)** Sketch the region defined by $y \le 3x$, $y \ge x^2 - 4$ and $x \ge 0$.
(b) Calculate the area of this region.

13 **(a)** Sketch the region defined by $y \le 3x$, $y \ge x^2 - 4$.
(b) Show that the area of this region is given by $\int_{-3}^{12} \left(\sqrt{y + 4} - \frac{y}{3}\right) dy + 2\int_{-4}^{-3} \sqrt{y + 4}\, dy$.
(c) Calculate the area of this region.
(d) What other integral could have been used to find this area? Use it to verify your answer for **(c)**.

17.7 DEFINITE INTEGRALS INVOLVING TRIGONOMETRIC FUNCTIONS

Summary—trigonometric integrals

$\int \sin x\, dx = -\cos x + C$ | $\int \sin ax\, dx = -\frac{1}{a}\cos ax + C$ | $\int \sin(ax + b)\, dx = -\frac{1}{a}\cos(ax + b) + C$

$\int \cos x\, dx = \sin x + C$ | $\int \cos ax\, dx = \frac{1}{a}\sin ax + C$ | $\int \cos(ax + b)\, dx = \frac{1}{a}\sin(ax + b) + C$

$\int \sec^2 x\, dx = \tan x + C$ | $\int \sec^2 ax\, dx = \frac{1}{a}\tan ax + C$ | $\int \sec^2(ax + b)\, dx = \frac{1}{a}\tan(ax + b) + C$

Example 20

Evaluate: **(a)** $\int_{\frac{\pi}{3}}^{\frac{\pi}{2}} \cos x\, dx$ **(b)** $\int_0^{\pi} \sin 2x\, dx$ **(c)** $\int_0^{\frac{\pi}{4}} \sec^2 x\, dx$

(d) $\int_0^{\frac{\pi}{2}} \cos\left(2x - \frac{\pi}{2}\right) dx$ **(e)** $\int_{\frac{\pi}{6}}^{\frac{\pi}{3}} (\cos 2x - 3\sin x)\, dx$

Solution

(a) $\int_{\frac{\pi}{3}}^{\frac{\pi}{2}} \cos x\, dx = [\sin x]_{\frac{\pi}{3}}^{\frac{\pi}{2}}$

$= \sin\frac{\pi}{2} - \sin\frac{\pi}{3}$

$= 1 - \frac{\sqrt{3}}{2}$

(b) $\int_0^{\pi} \sin 2x\, dx = \left[-\frac{1}{2}\cos 2x\right]_0^{\pi}$

$= -\frac{1}{2}\cos 2\pi + \frac{1}{2}\cos 0$

$= -\frac{1}{2} + \frac{1}{2}$

$= 0$

(c) $\int_0^{\frac{\pi}{4}} \sec^2 x\, dx = [\tan x]_0^{\frac{\pi}{4}}$

$= \tan\frac{\pi}{4} - \tan 0$

$= 1 - 0$

$= 1$

(d) $\int_0^{\frac{\pi}{2}} \cos\left(2x - \frac{\pi}{2}\right)dx = \left[\frac{1}{2}\sin\left(2x - \frac{\pi}{2}\right)\right]_0^{\frac{\pi}{2}}$

$= \frac{1}{2}\sin\frac{\pi}{2} - \frac{1}{2}\sin\left(-\frac{\pi}{2}\right)$

$= \frac{1}{2} + \frac{1}{2}$

$= 1$

(e) $\int_{\frac{\pi}{6}}^{\frac{\pi}{3}} (\cos 2x - 3\sin x)\, dx = \left[\frac{1}{2}\sin 2x + 3\cos x\right]_{\frac{\pi}{6}}^{\frac{\pi}{3}}$

$= \left(\frac{1}{2}\sin\frac{2\pi}{3} + 3\cos\frac{\pi}{3}\right) - \left(\frac{1}{2}\sin\frac{\pi}{3} + 3\cos\frac{\pi}{6}\right)$

$= \frac{\sqrt{3}}{4} + \frac{3}{2} - \left(\frac{\sqrt{3}}{4} + \frac{3\sqrt{3}}{2}\right)$

$= \frac{3 - 3\sqrt{3}}{2}$

Example 21

Calculate the area bounded by the curve $y = \sin x$, the x-axis and the ordinates $x = \frac{\pi}{3}$ and $x = \frac{\pi}{2}$.

Solution

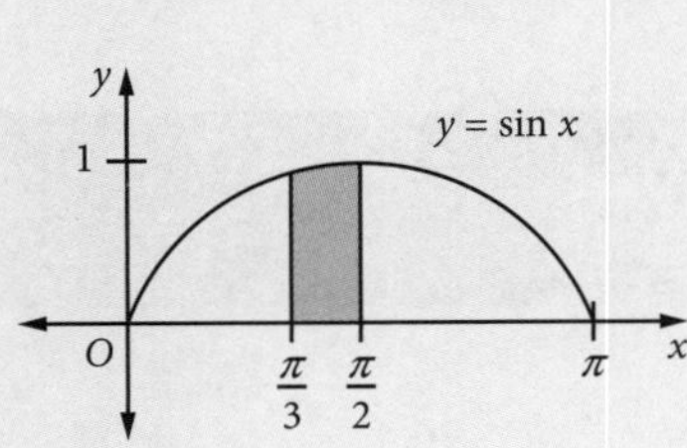

$\text{Area} = \int_{\frac{\pi}{3}}^{\frac{\pi}{2}} \sin x\, dx$

$= [-\cos x]_{\frac{\pi}{3}}^{\frac{\pi}{2}}$

$= \left(-\cos\frac{\pi}{2}\right) - \left(-\cos\frac{\pi}{3}\right)$

$= 0 + \frac{1}{2}$

$= \frac{1}{2}$

$\therefore \text{Area} = \frac{1}{2}\text{ unit}^2$

EXERCISE 17.7 DEFINITE INTEGRALS INVOLVING TRIGONOMETRIC FUNCTIONS

1 Sketch the graph of $f(x) = 1 - \cos x$, $0 \le x \le \pi$. Evaluate $\int_0^{\pi} f(x)\, dx$ and indicate on the sketch the area represented by this integral. What is the exact value of this area?

2 Evaluate:

(a) $\int_0^{\pi} \sin x\,dx$ (b) $\int_0^{\frac{\pi}{3}} \sec^2 x\,dx$ (c) $\int_{\frac{\pi}{3}}^{\pi} \cos\frac{x}{2}\,dx$ (d) $\int_0^{\frac{\pi}{2}} \cos x\,dx$ (e) $\int_0^{\pi} (\sin x + \cos x)\,dx$

(f) $\int_{-\pi}^{\pi} \sin x\,dx$ (g) $\int_{\frac{\pi}{6}}^{\frac{\pi}{2}} (\cos x - \sin 2x)\,dx$ (h) $\int_{\frac{\pi}{8}}^{\frac{\pi}{4}} (\sin 2x - \cos 2x)\,dx$ (i) $\int_{-\frac{\pi}{4}}^{\frac{\pi}{4}} (2\sin x + \sin 2x)\,dx$

(j) $\int_0^{\pi} \left(\sin\frac{x}{4} + \cos\frac{x}{4}\right)dx$ (k) $\int_0^{\frac{\pi}{3}} \left(3\cos 3x - \frac{\sin 2x}{2}\right)dx$ (l) $\int_{-\pi}^{\pi} \left(\frac{\sin x}{2} - \cos x\right)dx$

3 Evaluate $I = \int_{-\frac{\pi}{2}}^{\frac{\pi}{2}} (3\sin 2x + 4\cos 2x)\,dx$. Indicate whether each statement below is a correct or incorrect step in evaluating this integral.

(a) $I = 2\int_0^{\frac{\pi}{2}} (3\sin 2x + 4\cos 2x)\,dx$ (b) $\left[\frac{-3\cos 2x}{2} + 2\sin 2x\right]_{-\frac{\pi}{2}}^{\frac{\pi}{2}}$

(c) $\frac{-3\cos\pi}{2} + 2\sin\pi + \frac{3\cos(-\pi)}{2} - 2\sin(-\pi)$ (d) 0

4 Sketch the graph of $f(x) = \sqrt{1-\cos^2 x}$, $0 \le x \le 2\pi$, and evaluate $\int_{\pi}^{\frac{3\pi}{2}} f(x)\,dx$. Illustrate your result on your sketch.

5 By integration, find the area bounded by the curve $y = \cos x$ and the x-axis between $x = -\frac{\pi}{2}$ and $x = \frac{\pi}{2}$.

6 Evaluate: (a) $\int_0^{\frac{\pi}{6}} \frac{\cos x}{1+\sin x}\,dx$ (b) $\int_0^{\frac{\pi}{2}} \frac{\sin x}{2-\cos x}\,dx$

7 Evaluate $\int_{\frac{\pi}{3}}^{\frac{5\pi}{3}} \left(\frac{\cos 2x}{4} + \cos x\right)dx$.

8 Calculate the area of the region enclosed by the graph of $y = \cos 2x$, the x-axis and the ordinates $x = 0$ and $x = \frac{\pi}{4}$.

9 Calculate the area bounded by the given curves, the x-axis and the given ordinates:

(a) $y = 1 + \cos x$, $x = 0$, $x = \pi$ (b) $y = 1 + \frac{\sin 2x}{2}$, $x = 0$, $x = \frac{\pi}{2}$

(c) $y = \frac{1+\cos x}{2}$, $x = -\frac{\pi}{4}$, $x = \frac{\pi}{4}$ (d) $y = a\sin\frac{\pi x}{b}$, $x = 0$, $x = \frac{b}{2}$

10 Find the points of intersection of the curves $y = \sin\theta$ and $y = \cos\theta$ for $0 \le \theta \le 2\pi$ and calculate the area between the two curves.

17.8 DEFINITE INTEGRALS INVOLVING EXPONENTIAL AND LOGARITHMIC FUNCTIONS

Review of integrals involving e^x

$\int e^x\,dx = e^x + C$ $\int e^{ax+b}\,dx = \frac{1}{a}e^{ax+b} + C$

$\int e^{kx}\,dx = \frac{1}{k}e^{kx} + C$ $\int f'(x)e^{f(x)}\,dx = e^{f(x)} + C$

Example 22

Evaluate: (a) $\int_0^2 e^{2x}\,dx$ (b) $\int_{-0.5}^{1.5} (e^x - e^{-x})\,dx$

Solution

(a) $\int_0^2 e^{2x}\,dx = \left[\frac{e^{2x}}{2}\right]_0^2$

$= \frac{1}{2}\left(e^4 - e^0\right)$

$= \frac{e^4 - 1}{2}$ is the exact value.

$= 26.80$ using a calculator and writing the answer correct to 2 d.p.

(b) $\int_{-0.5}^{1.5}\left(e^x - e^{-x}\right)dx = \left[e^x + e^{-x}\right]_{-0.5}^{1.5}$

$= \left(e^{1.5} + e^{-1.5}\right) - \left(e^{-0.5} + e^{0.5}\right)$

$= 2.4496$ correct to 4 d.p.

It is an interesting exercise to show that the exact value of this integral is $\frac{(e+1)(e-1)^2}{e^{1.5}}$.

Important note about definite integrals

Before now, when you have evaluated a definite integral at a limit of 0 the result has usually been zero. But do not assume that an integral at 0 is always zero! As you have now seen, substituting a value of 0 into an exponential integral will often produce a value of 1 or some other non-zero constant. When evaluating definite integrals you must always take care to substitute both limit values of the integral, even when one value is zero.

Example 23

Calculate the area bounded by the curve $y = e^{1.5x}$, the coordinate axes and the line $x = 2$.

Solution

$y = e^{1.5x}$, $y = 0$, $x = 2$

Area $= \int_0^2 e^{1.5x}\,dx$

$= \left[\frac{2}{3}e^{1.5x}\right]_0^2$

$= \frac{2}{3}\left(e^3 - e^0\right)$

$= \frac{2\left(e^3 - 1\right)}{3} \approx 12.72 \text{ units}^2$

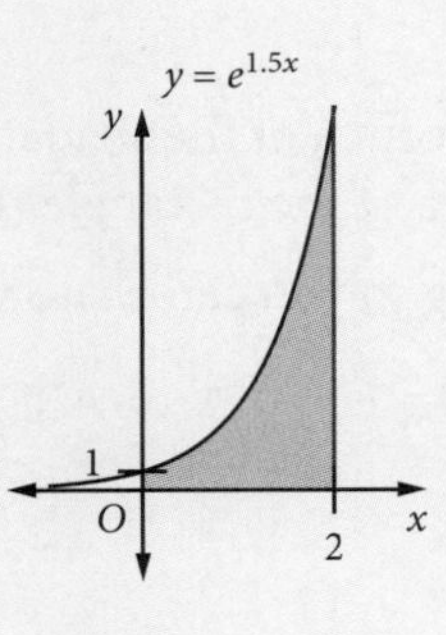

Review of integrals involving logarithms

$\int \frac{1}{x}\,dx = \log_e|x| + C$

$\int \frac{1}{ax+b}\,dx = \frac{1}{a}\log_e|(ax+b)| + C$

$\int \frac{f'(x)}{f(x)}\,dx = \log_e|[f(x)]| + C$

$\int \frac{af'(x)}{f(x)}\,dx = a\log_e|[f(x)]| + C$

$\int a^x\,dx = \frac{a^x}{\ln a} + C$

Example 24

Evaluate: (a) $\int_1^2 \frac{3}{x+1}\,dx$ (b) $\int_3^4 \frac{2x-1}{x^2-x-2}\,dx$ (c) $\int_2^4 \frac{x^2-1}{x}\,dx$ (d) $\int_1^3 2^x\,dx$

Solution

(a) $\int_1^2 \frac{3}{x+1}\,dx = 3\int_1^2 \frac{1}{x+1}\,dx$

$= 3\left[\log_e(x+1)\right]_1^2$

$= 3(\log_e 3 - \log_e 2)$

$= 3\log_e 1.5$

≈ 1.216

(b) $\int_3^4 \frac{2x-1}{x^2-x-2}\,dx = \left[\log_e(x^2-x-2)\right]_3^4$

$= \log_e 10 - \log_e 4$

$= \log_e 2.5$

≈ 0.916

(c) $\frac{x^2-1}{x} = \frac{x^2}{x} - \frac{1}{x} = x - \frac{1}{x}$, so: $\int_2^4 \frac{x^2-1}{x}\,dx = \int_2^4 \left(x - \frac{1}{x}\right)dx$

$= \left[\frac{x^2}{2} - \ln x\right]_2^4$

$= 8 - \ln 4 - (2 - \ln 2)$

$= 6 - \ln 2$

≈ 5.307

(d) $\int_1^3 2^x\,dx = \left[\frac{2^x}{\ln 2}\right]_1^3$

$= \frac{1}{\ln 2}(2^3 - 2^1)$

$= \frac{7}{\ln 2} \approx 10.10$ (2 d.p.)

If question **(c)** had asked for an exact answer, the answer would be written as $6 - \ln 2$.

Example 25

Find the area bounded by the curve $y = \log_e x$, the x-axis and the ordinate $x = 2$.

Solution

Area $= \int_1^2 \log_e x\,dx$

You don't learn how to evaluate this integral in this course. Instead, draw a diagram to see whether there may be another way to calculate the area.

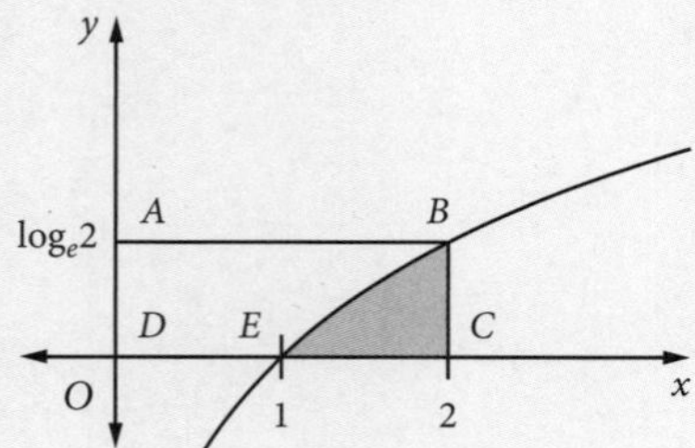

You require the area of the shaded region BCE. It can be obtained by finding the area of the rectangle $ABCD$ and subtracting the area $ABED$.

Because $y = \log_e x$, you can write $x = e^y$.

At $x = 2$, $y = \log_e 2$: Area $ABED = \int_0^{\log_e 2} e^y\,dy$

$= \left[e^y\right]_0^{\log_e 2}$

$= e^{\log_e 2} - e^0$

$= 2 - 1$

$= 1$

Area $ABCD = 2\log_e 2$

$\therefore$ Area $BCE = 2\log_e 2 - 1$

≈ 0.386 units^2

Example 26

Differentiate $x\log_e x$. Hence find the primitive of $\log_e x$ and so evaluate $\int_1^2 \log_e x\,dx$.

Solution

Let $y = x\log_e x = uv$, where $u = x$ and $v = \log_e x$.

Product rule, $\frac{dy}{dx} = v\frac{du}{dx} + u\frac{dv}{dx}$: $\frac{d}{dx}(x\log_e x) = 1 \times \log_e x + x \times \frac{1}{x}$

$$\therefore \frac{d}{dx}(x\log_e x) = \log_e x + 1$$

Rearrange: $\log_e x = \frac{d}{dx}(x\log_e x) - 1$

The primitive of the derivative of a function is the function itself, so:

$$\int \frac{d}{dx}(x\log_e x)dx = x\log_e x + C$$

Hence: $\int \log_e x\,dx = \int \frac{d}{dx}(x\log_e x)dx - \int 1\,dx$

$$= x\log_e x - x + C$$

$$\therefore \int_1^2 \log_e x\,dx = \left[x\log_e x - x\right]_1^2$$

$$= (2\log_e 2 - 2) - (\log_e 1 - 1)$$

$$= 2\log_e 2 - 1$$

$$\approx 0.386$$

This is the same answer as in Example **25**, obtained by a different method.

EXERCISE 17.8 DEFINITE INTEGRALS INVOLVING EXPONENTIAL AND LOGARITHMIC FUNCTIONS

1 Find the value of:

(a) $\int_{-1}^{1} e^x\,dx$ **(b)** $\int_0^2 e^{2x}\,dx$ **(c)** $\int_{-1}^{3} e^{-\frac{x}{2}}\,dx$ **(d)** $\int_0^1 e^{1.5t}\,dt$ **(e)** $\int_{-0.5}^{0.5} e^{-2t}\,dt$

(f) $\int_{-1}^{0} e^{-3u}\,du$ **(g)** $\int_0^1 \left(e^{2x} - e^{-2x}\right)dx$ **(h)** $\int_{-3}^{3} \left(e^{\frac{t}{2}} - e^{\frac{-t}{2}}\right)dt$ **(i)** $\int_{0.5}^{1.5} \left(e^{\theta} + e^{-3\theta}\right)d\theta$

2 Indicate whether each statement below is a correct or incorrect step in the evaluation of $I = \int_{-1}^{1} \left(e^x - e^{-x}\right)^2 dx$.

(a) $I = \int_{-1}^{1} \left(e^{2x} - 2 + e^{-2x}\right)dx$ **(b)** $I = \left[\frac{e^{2x}}{2} - 2x - \frac{e^{-2x}}{2}\right]_{-1}^{1}$

(c) $I = \left[e^{2x} - 4x - e^{-2x}\right]_0^1$ **(d)** $I = \frac{e^4 - 4e^2 - 1}{4e^2}$

3 Calculate the area enclosed between the curve $y = e^{2x} + e^{-2x}$, the x-axis and the lines $x = 1$ and $x = -1$.

4 **(a)** Calculate the area bounded by the curve $y = e^x$, the coordinate axes and the line $x = 2$.
(b) Write the equation of the tangent to $y = e^x$ at the point where $x = 2$.
(c) Calculate the area bounded by $y = e^x$, the coordinate axes and the tangent at $x = 2$.
(d) Calculate the area bounded by $y = e^x$, the y-axis and the line $y = e^2$.

5 Calculate the area bounded by the curves $y = e^x$, $y = e^{-x}$ and the ordinate $x = 2$.

6 Calculate the area bounded by the curve $y = e^{0.5x} - e^{-0.5x}$, the x-axis and the line $x = 1$.

7 Evaluate:

(a) $\int_2^3 \frac{1}{x-1}dx$ (b) $\int_0^3 \frac{2}{x+3}dx$ (c) $\int_{-2}^0 \frac{dx}{5+2x}$ (d) $\int_2^4 \frac{2}{2t-3}dt$

(e) $\int_3^5 \frac{2x}{x^2-1}dx$ (f) $\int_3^6 \frac{3}{x-1}dx$ (g) $\int_5^7 \frac{x-1}{x^2-2x}dx$ (h) $\int_0^4 \frac{6}{2x+3}dx$

8 Evaluate:

(a) $\int_1^3 \left(x^2+\frac{1}{x}\right)dx$ (b) $\int_{-3}^3 \frac{x}{x^2+1}dx$ (c) $\int_2^4 \frac{x^2-1}{x}dx$ (d) $\int_2^4 \left(\frac{1}{x}+\frac{1}{x^2}\right)dx$

(e) $\int_2^3 \left(x+\frac{1}{x-1}\right)dx$ (f) $\int_1^2 \left(x-\frac{1}{x^2}\right)^2 dx$ (g) $\int_1^3 \left(e^x+\frac{1}{x}\right)dx$ (h) $\int_1^4 \left(\sqrt{x}-\frac{1}{x}\right)dx$

9 Sketch the graph of $f(x)=\frac{1}{2x-1}$ and find the area enclosed by the curve, the x-axis and the lines $x=1$ and $x=4$.

10 Find the area of the region enclosed by the curve $y=\frac{x}{x^2+1}$, the x-axis and the ordinates $x=2$ and $x=4$.

11 For the curve whose equation is $y=x+\frac{4}{x+1}$, find the area enclosed by the curve, the x-axis and the lines $x=0$ and $x=2$. Indicate whether each statement below is a correct or incorrect step in the solution of this problem.

(a) Area $=\int_0^2 \left(x+\frac{4}{x+1}\right)dx$ (b) $\left[\frac{x^2}{2}+4\ln(x+1)\right]_0^2$ (c) $\left[1-\frac{4}{(x+1)^2}\right]_0^2$ (d) $2+4\ln 3$

12 Sketch the curve $y=x+\frac{1}{x}$, $x>0$, showing its asymptotes and the coordinates of the turning points. Also find the area enclosed by the curve, the x-axis and the lines $x=1$ and $x=2$.

13 The value of $\int_0^3 \frac{2x\,dx}{x^2+9}=\ldots$

A $\frac{1}{2}\ln 2$ B $\ln 2$ C $2\ln 2$ D $\ln 18$

14 $\int_0^1 \frac{e^x}{1+e^x}dx=\log_e c$. Find the value of c.

15 Find the area of the region enclosed by the curve $y=\frac{x}{x^2+1}$, the x-axis and the line $x=1$.

16 Given $a>1$, sketch the curve $y=\log_e x$ for $1\le x\le a$. Find the area enclosed by the curve and the lines $y=0$ and $x=a$.

17 (a) Find $\frac{d}{dx}(\log_e(\cos x))$.

(b) Find the area enclosed by the curve $y=\tan x$, the x-axis and the ordinate $x=\frac{\pi}{3}$.

17.9 APPLICATIONS INVOLVING INTEGRALS

Example 27

A large cube of ice has an edge length of 10 cm. It melts so that its volume decreases at a constant rate of 25 cm^3 per hour. Find:

(a) the volume V at time t (b) the time required to completely melt the ice.

Solution

(a) Because the volume decreases at a constant rate of 25 cm^3 per hour, $\frac{dV}{dt} = -25$.

$$\therefore V = -\int 25\,dt$$
$$= -25t + C$$

When $t = 0$, $V = 10^3 = 1000$: $\quad 1000 = C$

$$\therefore V = 1000 - 25t$$

(b) When $V = 0$: $\quad 0 = 1000 - 25t$

$$t = 40$$

The volume at any time t is given by $V = 1000 - 25t$, $0 \le t \le 40$.

Note that the domain of the volume function is restricted to values of t from 0 to 40. The volume function is a linear function and its graph has a constant gradient of -25.

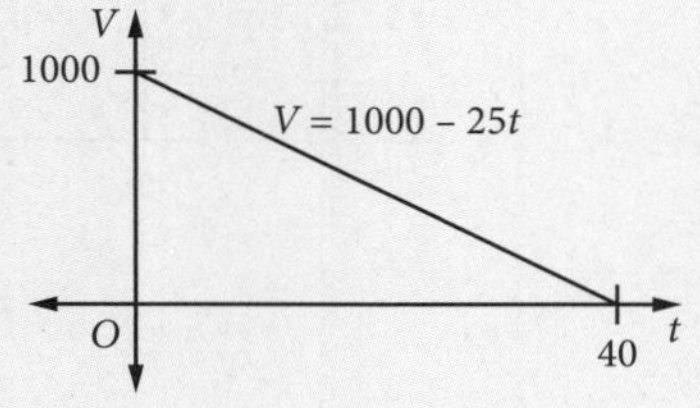

Example 28

A chemical solution is being filtered into an empty beaker. The rate at which the level of the solution is rising in the beaker is given by $\frac{dh}{dt} = \frac{1}{10}\left(1 - \frac{t}{100}\right)$, where h is the depth of the solution in cm and t is the time in seconds after the solution starts flowing through the filter.

(a) What is the depth of the solution in the beaker after 40 seconds?

(b) What will be the depth of the solution in the beaker when the solution has stopped flowing through the filter?

Solution 1

(a) When $t = 0$, $h = 0$: $\quad \frac{dh}{dt} = \frac{1}{10}\left(1 - \frac{t}{100}\right)$

$$h = \int \frac{1}{10}\left(1 - \frac{t}{100}\right) dt$$
$$= \frac{1}{10}\left(t - \frac{t^2}{200}\right) + C$$

When $t = 0$, $h = 0$: $\quad \therefore C = 0$

Hence: $\quad h = \frac{1}{10}\left(t - \frac{t^2}{200}\right)$

When $t = 40$: $\quad h = \frac{1}{10}\left(40 - \frac{1600}{200}\right)$

$$h = 3.2$$

The depth of the solution after 40 seconds is 3.2 cm.

(b) The solution will have stopped flowing when $\frac{dh}{dt} = 0$, as at this time the depth of the solution has stopped changing.

When $\frac{dh}{dt} = 0$: $\quad \frac{1}{10}\left(1 - \frac{t}{100}\right) = 0$

$$t = 100$$

Stops flowing after 100 seconds.

When $t = 100$: $\quad h = \frac{1}{10}\left(100 - \frac{100^2}{200}\right) = 5$

The depth when it stops flowing is 5 cm.

Solution 2

(a) $h=\int dh=\int \frac{dh}{dt}dt$, so the area under $\frac{dh}{dt}$ can be used to find the value of h between appropriate limits.

Sketch $\frac{dh}{dt}=\frac{1}{10}\left(1-\frac{t}{100}\right)$:

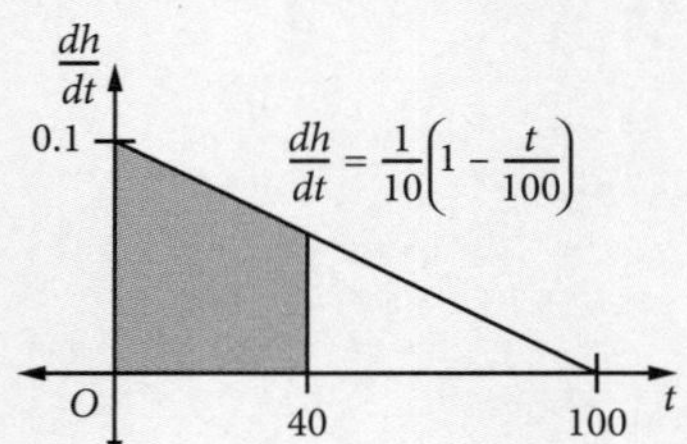

The shaded region is bounded by $\frac{dh}{dt}=\frac{1}{10}\left(1-\frac{t}{100}\right)$, $\frac{dh}{dt}=0$, $t=0$, $t=40$, because the depth is after 40 seconds.

$$\text{Shaded area}=\int_0^{40}\frac{1}{10}\left(t-\frac{t}{100}\right)dt$$

$$=\frac{1}{10}\left[t-\frac{t^2}{200}\right]_0^{40}$$

$$=\frac{1}{10}\left(40-\frac{1600}{200}-0\right)$$

$$=3.2$$

The shaded area represents the depth of the solution after 40 seconds, so:

$$h=\int_0^{40}\frac{1}{10}\left(t-\frac{t}{100}\right)dt$$

$$=3.2\text{ cm}$$

(b) The graph shows that $\frac{dh}{dt}=0$ when $t=100$, so this is when the water has stopped flowing.

$$h=\int_0^{100}\frac{1}{10}\left(t-\frac{t}{100}\right)dt$$

$$=\frac{1}{10}\left[t-\frac{t^2}{200}\right]_0^{100}$$

$$=\frac{1}{10}\left(100-\frac{10\,000}{200}-0\right)$$

$$=5\text{ cm}$$

Example 29

A particle starts from rest 5 m from a fixed point O and moves in a straight line with an acceleration a m s^{-1}, where $a=3t-4$. Find the velocity and position of the particle at any time t.

Solution

When $t=0$, $v=0$, $x=5$: $\quad a=3t-4$

Integrate for v: $\quad v=\int(3t-4)\,dt$

$$v=\frac{3t^2}{2}-4t+C_1$$

When $t=0$, $v=0$: $\quad 0=0-0+C_1$

$$C_1=0$$

Hence: $\quad v=\frac{3t^2}{2}-4t$

Integrate for x: $\quad x=\int\left(\frac{3t^2}{2}-4t\right)dt$

$$x=\frac{t^3}{2}-2t^2+C_2$$

When $t = 0, x = 5$: $\quad 5 = 0 - 0 + C_2$

$$C_2 = 5$$

Hence $\quad x = \frac{t^3}{2} - 2t^2 + 5$

Units and symbols

Physical quantity	Unit	Symbol
Time	s	t
Displacement	cm, m	x (or s in Physics)
Velocity	cm s^{-1}, m s^{-1}	$v, \frac{dx}{dt}, \dot{x}$
Acceleration	cm s^{-2}, m s^{-2}	$a, \frac{dv}{dt}, \frac{d^2x}{dt^2}, \ddot{x}$

Note that 's' is the abbreviation for second, 'cm' for centimetre and 'm' for metre.

Constant acceleration due to gravity = $9.8\ \text{m s}^{-2}$ ($\approx 10\ \text{m s}^{-2}$)

Example 30

A ball is projected vertically upwards, with a velocity of $25\ \text{m s}^{-1}$, from the top of a building 30 m high. If the acceleration due to gravity is taken to be $10\ \text{m s}^{-2}$, find:

(a) the time taken for the ball to reach its highest point, and the height of this highest point

(b) how much time the ball will take to reach the ground

(c) the speed with which the ball hits the ground

(d) when the ball is 60 m above the ground.

Solution

Take the upward direction as positive. Thus when $t = 0$, $a = -10$, $v = 25$, $x = 30$.

$$a = \ddot{x} = -10$$

Integrate for v: $\quad v = \dot{x} = -\int 10\, dt$

$$v = \dot{x} = -10t + C_1$$

When $t = 0, v = 25$: $\quad C_1 = 25$

$$\dot{x} = 25 - 10t$$

Integrate for x: $\quad x = \int (25 - 10t)\, dt$

$$x = 25t - 5t^2 + C_2$$

When $t = 0, x = 30$: $\quad C_2 = 30$

$$x = 30 + 25t - 5t^2$$

or $\quad x = 5(6 + 5t - t^2)$

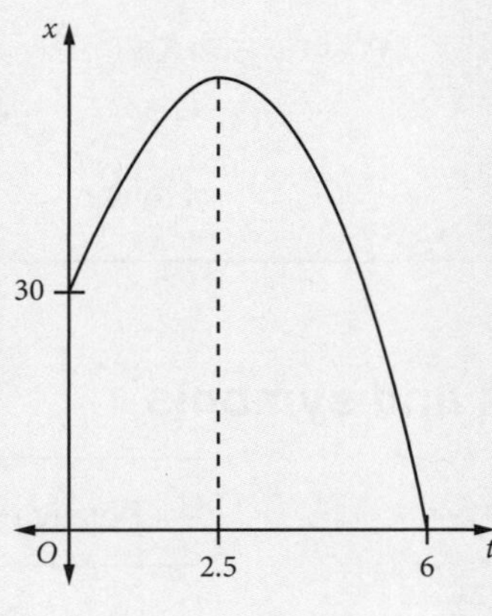

(a) At the highest points, $\dot{x} = 0$: $0 = 25 - 10t$

$t = 2.5$

It takes 2.5 seconds to reach the highest point.

When $t = 2.5$: $x = 5(6 + 12.5 - 6.25)$

$x = 61.25$

Highest point is 61.25 m above the ground.

(b) Reaches the ground when $x = 0$: $5(6 + 5t - t^2) = 0$

$$t^2 - 5t - 6 = 0$$

$$(t - 6)(t + 1) = 0$$

Because $t \geq 0$, the ball reaches the ground after 6 seconds.

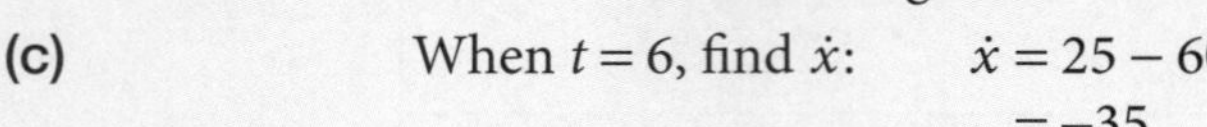

(c) When $t = 6$, find $\dot{x}$: $\dot{x} = 25 - 60$

$= -35$

The ball is falling when it hits the ground, hence the negative velocity, and it strikes the ground with a speed of 35 m s^{-1}.

(d) When $x = 60$: $60 = 5(6 + 5t - t^2)$

$$12 = 6 + 5t - t^2$$

$$t^2 - 5t + 6 = 0$$

$$(t - 2)(t - 3) = 0$$

$$t = 2 \text{ or } 3$$

The ball is 60 m above the ground at 2 seconds (on the way up) and again at 3 seconds (on the way down).

MAKING CONNECTIONS

Motion of a particle in a straight line

Move the slider to see the relationship between the velocity and displacement graphs for the motion of a particle in a straight line.

Example 31

A particle moves in a straight line so that at any time t seconds, its velocity $v\text{ m s}^{-1}$ is given by $v = e^{-t}$. If initially the particle is 2 metres from a fixed point O in the line, find its position x at any time t. Sketch the graph of x as a function of t.

Solution

$$v = \frac{dx}{dt} = e^{-t}$$

$$\therefore x = \int e^{-t}\, dt$$

$$x = -e^{-t} + C$$

When $t = 0$, $x = 2$: $2 = -1 + C$

$$C = 3$$

$$x = 3 - e^{-t}$$

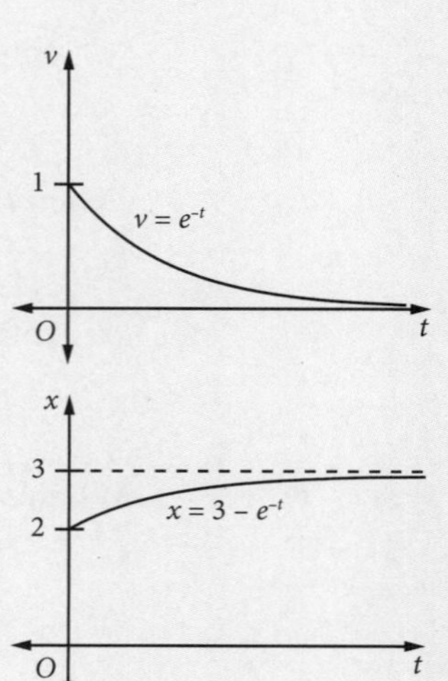

The velocity–time graph for $v = e^{-t}$ shows that initially the particle moves with a velocity of 1 m s^{-1}, but as time goes by this velocity gets smaller and smaller so that as $t \to \infty$, $v \to 0$ from above.

The displacement–time graph shows that the particle starts at $x = 2$ and as time goes by it moves closer and closer to $x = 3$, but it never reaches it. As $t \to \infty$, $x \to 3$ from below.

Displacement, velocity, acceleration—important links

Given the displacement function, you can find the velocity and acceleration functions by differentiating with respect to time.

Given the velocity function, you can find the displacement function by integrating with respect to time, and you can find the acceleration function by differentiating with respect to time.

EXERCISE 17.9 APPLICATIONS INVOLVING INTEGRALS

1 A cube of ice has an edge length of 10 cm. It melts so that its volume decreases at a constant rate and the block remains a cube. If the edge length measures 5 cm after 70 minutes, find:

(a) the rate at which the volume decreases **(b)** the volume at any time t.

2 A machine manufactures items at a variable rate given by $\frac{dQ}{dt} = 2t + 1$, $t \geq 0$, where Q is the number of items manufactured in a time t minutes.

(a) At what rate is the machine working: **(i)** initially **(ii)** after 10 minutes?
(b) What is the total number of items manufactured in the first 10 minutes?

3 The sluice gates of a dam are operated by an automatic program that controls the flow of water out of the dam. The program is set so that t hours after 7 am the flow of water will be given by
$\frac{dV}{dt} = 500 - 15t^2 + t^3$ megalitres (ML) per hour.

(a) If no water flows from the dam before 7 am, calculate:
(i) the flow of the water at 9 am
(ii) the total volume of water released between 7 am and 9 am

(b) **(i)** Sketch $\frac{dV}{dt} = 500 - 15t^2 + t^3$ for $0 \leq t \leq 10$.
(ii) When does the flow of water stop?
(iii) If the sluice gates close at the moment when $\frac{dV}{dt} = 0$, how much water has been released altogether?

4 The rate at which carbon dioxide is produced by the action of yeast in a dough is given by
$\frac{dV}{dt} = \frac{1}{10000}\left(200t - t^2\right)$, where V cm^3 is the volume of carbon dioxide produced after t seconds. What is the volume of carbon dioxide produced in the first 3 minutes after the yeast starts to work? Indicate whether each statement below is a correct or incorrect step in the solution of this problem.

(a) $V = \frac{1}{10000}\int_0^4 \left(200t - t^2\right)dt$ **(b)** $V = \frac{1}{10000}\int_0^{180} \left(200t - t^2\right)dt$

(c) $V = \left[100t^2 - \frac{t^3}{3}\right]_0^{180}$ **(d)** $V = 129.6\,\text{cm}^3$

5 A body starts from O and moves in a straight line. At any time t its velocity is given by $\dot{x} = 6t - 4$. Indicate whether each statement below is correct or incorrect.

(a) $x = 3t^2 - 4t + C$ **(b)** $x = 3t^2 - 4t$ **(c)** $\ddot{x} = 3t^2 - 4t$ **(d)** $\ddot{x} = 6$

6 A body starts from O and moves in a straight line. At any time t, its velocity is $t^2 - 4t^3$. Find, in terms of t:

(a) the displacement x **(b)** the acceleration.

7 The velocity v m s^{-1} at time t seconds ($t \geq 0$) of a body moving in a straight line is given by $v = 6t^2 + 6t - 12$. Its initial displacement is 7 m from O. Find:

(a) the displacement and acceleration at any time t
(b) the acceleration when the velocity is zero **(c)** the initial velocity and acceleration.

8 The acceleration of a body moving in a straight line is $10 - 2t$ m s^{-2} at any time t seconds. The body is initially at zero displacement with a velocity of 11 m s^{-1}. Find:

(a) the velocity and displacement at any time t
(b) when the body has zero velocity, and its displacement at this time.

9 A body is projected vertically upwards with an initial velocity of 30 m s^{-1}. It rises with a deceleration of 10 m s^{-2}. Find:

(a) its velocity at any time t
(b) its height h m above the point of projection at any time t
(c) the greatest height reached
(d) the time taken to return to the point of projection.

10 A particle is projected vertically upwards from a point O with a velocity of 25 m s^{-1} and a downward acceleration of 10 m s^{-2}.

(a) Find its velocity and height above O at any time t.
(b) What maximum height does the particle reach?
(c) At what time has its velocity been reduced to half the velocity of projection?

11 The velocity v m s^{-1} of a body moving in a straight line is given by $v = 3t^2 - 2t - 1$. The body initially has a displacement 1 m from O. Find:

(a) the displacement and acceleration at any time t
(b) when the body has zero displacement, and its velocity and acceleration at this time
(c) the distance travelled in the first two seconds.

12 The velocity $v(t)$ of a particle moving in a straight line at any time $t \geq 0$ is $v(t) = 12t^2 - 6t + 1$. Find its position $s(t)$ given that $s(1) = 4$.

13 The acceleration of a particle moving in a straight line is $10 - 2t$ m s^{-2} at any time $t \geq 0$. The particle starts from O with a velocity of 24 m s^{-1}. At what time is its velocity zero, and what is its displacement at this time?

14 Two cars A and B travel along a straight road in the same direction. Their respective distances x km from a fixed point O at any time t hours are given by the following rules:

A: $x = 50t - 20t^2$ $\qquad$ B: $x = 80t^2 + 20t$

(a) Calculate each car's speed at the point O.
(b) At what time are the cars travelling at the same speed?
(c) Both cars reach a point Q at the same time. Calculate the distance from O to Q.
(d) A third car, travelling at uniform speed, is 2 km ahead of A and B when they pass the point O. If this car arrives at Q at the same time as A and B, find a rule connecting x and t for it.

15 A particle moves in a straight line so that its displacement $x(t)$ in metres to the right from a fixed point in the line at time $t \geq 0$ is given by $x(t) = 3 + 4t - 5\sqrt{t^2 + 4}$. Find the particle's displacement when it comes to rest.

16 A particle is moving so that, for $0 < t < 1$, its velocity is positive and its acceleration is negative. Which graph could represent the displacement function of this particle?

A
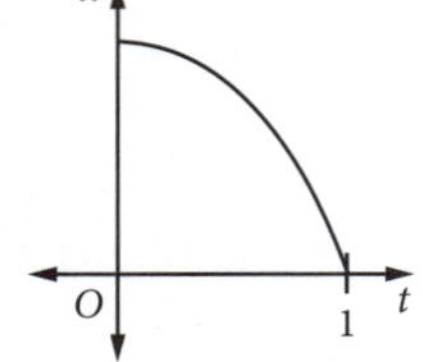

B
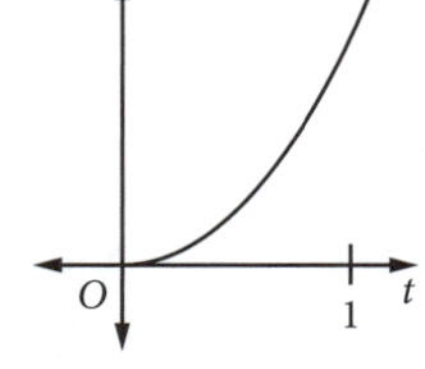

C
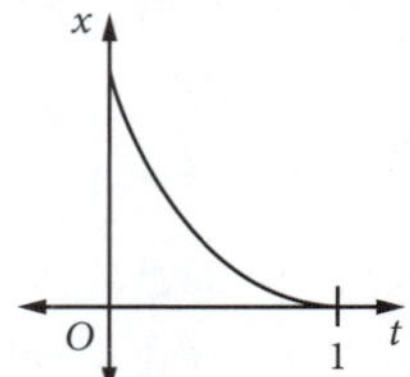

D
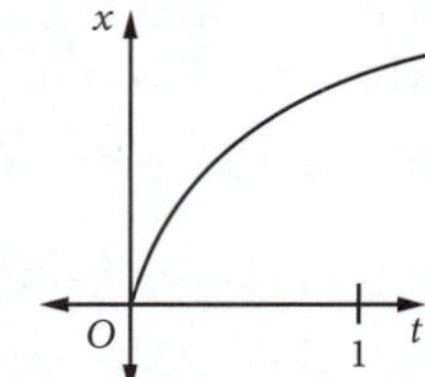

17 A particle moves in a straight line so that at time $t \geq 0$ its velocity v in metres per second is given by $v = \frac{1}{1+t}$.

(a) If its distance from a fixed point O in the line is x, show that $x = \log_e(1 + t)$, given that $x = 0$ when $t = 0$.
(b) Show that as t increases, the particle is moving away from O.
(c) Find the acceleration at $t = 0$.

18 A driver takes 3 hours to travel the distance between two points A and B on a country road. At time t hours after passing A, her speed v km h^{-1} is given by $v = 60 + 40e^{-t}$.

(a) Calculate the speeds when she passes points A and B.
(b) Write the acceleration in terms of: **(i)** t **(ii)** v
(c) Sketch the velocity–time curve and comment on the motion for large t.
(d) Calculate the distance from A to B.

19 A particle moves in a straight line. At time t its displacement from a fixed origin in the line is x and its velocity is v, where $v = e^{-t}$.

(a) Express x in terms of t, given that $x = 1$ when $t = 0$. Draw the graph of the function.
(b) Show that $\frac{d^2x}{dt^2} = x - 2$.

20 A particle moves in a straight line so that at time t its displacement from a fixed origin is x and its velocity is v. If its acceleration is $2\sin t$, and $v = 1$ and $x = 1$ when $t = 0$, find x as a function of t.

21 A particle moves in a straight line so that at time t its displacement from a fixed origin is x and its velocity is v.

(a) If its acceleration is $2\cos t$, and $v = 1$ and $x = 0$ when $t = 0$, find x in terms of t.
(b) If its acceleration is $-3e^{-t}$ and $v = 0$ when $t = 0$, find the time at which $v = -2$.

22 A particle moves in a straight line. At time t its distance x from a fixed point O in the line is given by $x = 4\log_e(1 + t) - 2t$, $t \geq 0$.

(a) Find x when $t = 0$. **(b)** Find the initial velocity, i.e. for $t = 0$.
(c) Find the acceleration at any time t and show that it is always negative.
(d) At what time is the velocity zero?

23 The acceleration of a particle moving in a straight line is given by $\frac{d^2x}{dt^2} = 12\cos 2t$. Initially $v = 0$ and $x = 6$. Find its velocity v and displacement x at any time t seconds and sketch the graph of the displacement. How many times does the particle change direction in the first 10 seconds?

24 A body falls from rest. Its velocity v at any time t is given by $v = 49(1 - e^{-0.5t})$. Find:

(a) the acceleration at any time t
(b) the initial velocity
(c) the terminal velocity, i.e. its velocity as $t \to \infty$
(d) the distance fallen in the first 5 seconds.

17.10 APPROXIMATE METHODS OF INTEGRATION—TRAPEZOIDAL RULE

For many functions you cannot easily find the primitive. In those cases, you can use numerical methods of integration to find the approximate value of a definite integral. There are many different methods of numerical approximation; computer software that evaluates definite integrals usually uses sophisticated numerical methods.

Approximating areas under curves

Example 32

(a) Sketch the graph of $y = \ln x$ for $0 < x \leq 5$.
(b) By drawing inner and outer rectangles and averaging the result, find an approximation for $\int_1^4 \ln x\, dx$, using three subintervals.
(c) Find an approximation for $\int_1^4 \ln x\, dx$ using trapezia and three subintervals.

Solution

(a)

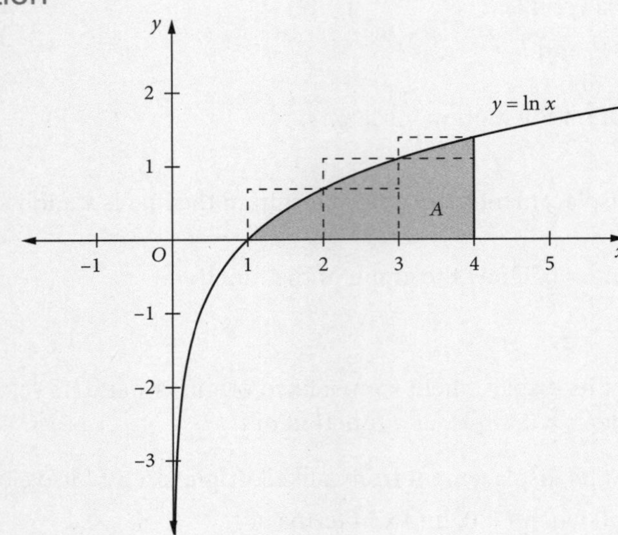

(b) For $1 \le x \le 4$ and three subintervals:
Area of inner rectangles
$= 1 \times 0 + 1 \times \ln 2 + 1 \times \ln 3 = \ln 6$
Area of outer rectangles
$= 1 \times \ln 2 + 1 \times \ln 3 + 1 \times \ln 4 = \ln 24$
Hence $\ln 6 < A < \ln 24$
Thus $\int_1^4 \ln x\,dx \approx \dfrac{\ln 6 + \ln 24}{2} = \dfrac{\ln 144}{2} = \ln 12$

(c) One side of each trapezium is very close to the graph of $y = \ln x$.

The first trapezium is really a right-angled triangle.

$$\int_1^4 \ln x\,dx \approx \frac{1}{2}\ln 2 + \frac{1}{2}(\ln 2 + \ln 3) \times 1 + \frac{1}{2}(\ln 3 + \ln 4) \times 1$$
$$= \frac{1}{2}(2\ln 2 + 2\ln 3 + \ln 4)$$
$$= 2\ln 2 + \ln 3$$
$$= \ln 12$$

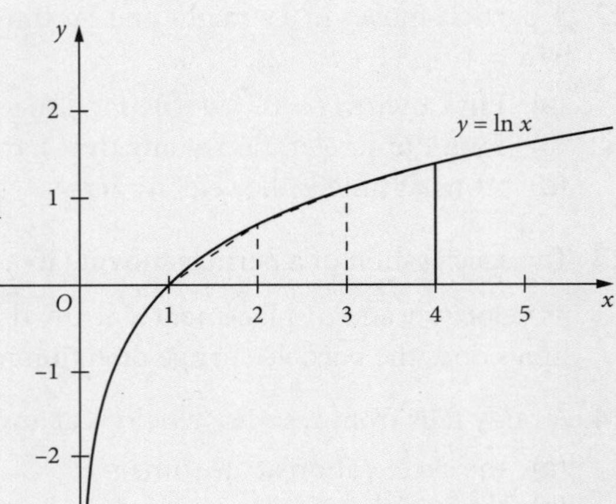

The **trapezoidal rule** is a method where trapezia are used to approximate the area under the curve.

In Example **1** the area under the curve $y = x^2$ was approximated using rectangles drawn above and below the curve. The more rectangles used, the better the approximation obtained for the area. In Example **33** you will redo this problem using trapezia.

Recall the area of a trapezium: Area $= \dfrac{\text{sum of lengths of parallel sides}}{2} \times$ distance between them

Example 33

Calculate the area of the region bounded by the curve $y = x^2$, the x-axis and the ordinates at $x = 0$ and $x = 1$, using trapezia with:

(a) one subinterval (b) two subintervals (c) four subintervals.

Solution

(a) Draw a diagram showing the region.

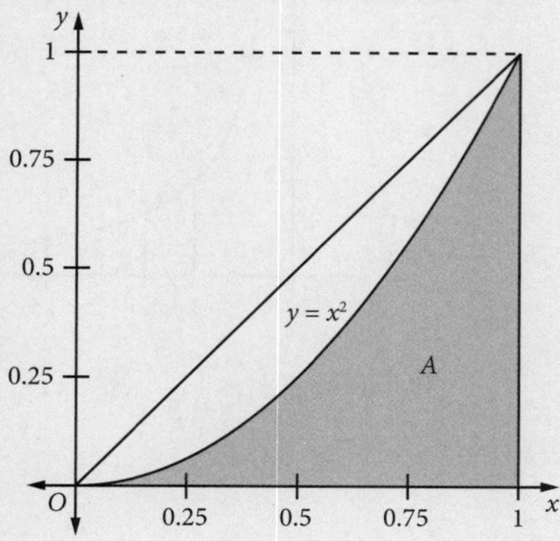

For $0 \le x \le 1$ and one subinterval, the trapezium is a triangle, but for consistency use the trapezium area formula.

$f(x) = x^2$: $f(0) = 0, f(1) = 1$, base $= 1 - 0 = 1$

$$\text{Area} = \frac{f(0) + f(1)}{2} \times (1 - 0)$$

$$= \frac{0+1}{2} \times 1 = 0.5 \text{ units}^2$$

(b) For $0 \le x \le 1$ and two subintervals:

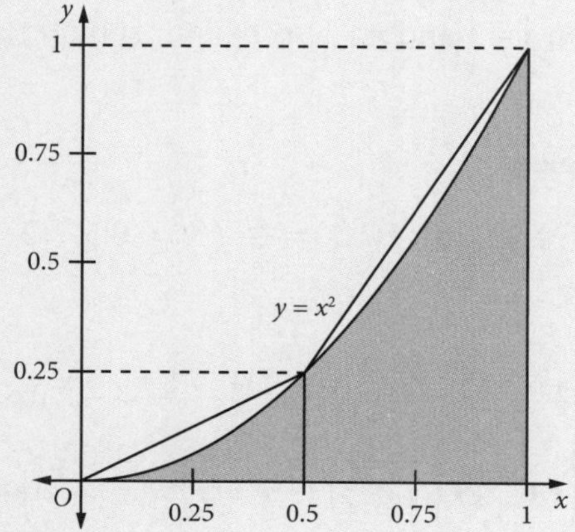

$f(x) = x^2$: $f(0) = 0, f(0.5) = 0.25, f(1) = 1$,
base $= 0.5 - 0 = 1 - 0.5 = 0.5$

$$\text{Area} = \frac{f(0) + f(0.5)}{2} \times 0.5 + \frac{f(0.5) + f(1)}{2} \times 0.5$$

$$= \frac{(0 + 0.25)}{2} \times 0.5 + \frac{(0.25 + 1)}{2} \times 0.5$$

$$= 0.0625 + 0.3125 = 0.375 \text{ units}^2$$

(c) For $0 \le x \le 1$ and four subintervals:

$f(x) = x^2$: $\quad f(0) = 0, f(0.25) = 0.0625, f(0.5) = 0.25, f(0.75) = 0.5625, f(1) = 1$, base $= 0.25 - 0 = 0.25$

$$\text{Area} = \frac{f(0) + f(0.25)}{2} \times 0.25 + \frac{f(0.25) + f(0.5)}{2} \times 0.25 + \frac{f(0.5) + f(0.75)}{2} \times 0.25 + \frac{f(0.75) + f(1)}{2} \times 0.25$$

$$= \frac{0 + 0.0625}{2} \times 0.25 + \frac{0.0625 + 0.25}{2} \times 0.25 + \frac{0.25 + 0.5625}{2} \times 0.25 + \frac{0.5625 + 1}{2} \times 0.25$$

$$= 0.34375 \text{ units}^2$$

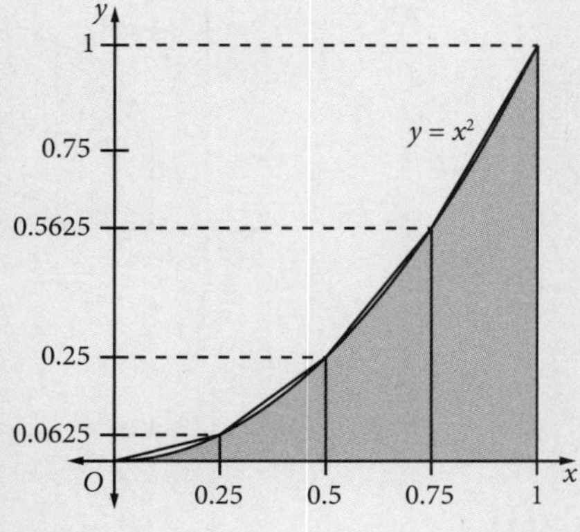

By evaluating $\int_0^1 x^2\,dx$ you know that the exact area is $\frac{1}{3}$ units^2. You can see that increasing the number of subintervals (and hence trapezia) gives a better approximation to the area.

Example 34

Use trapezia to find an approximation for the area under the curve $y = \frac{1}{x}$ between $x = 1$ and $x = 3$ using four subintervals.

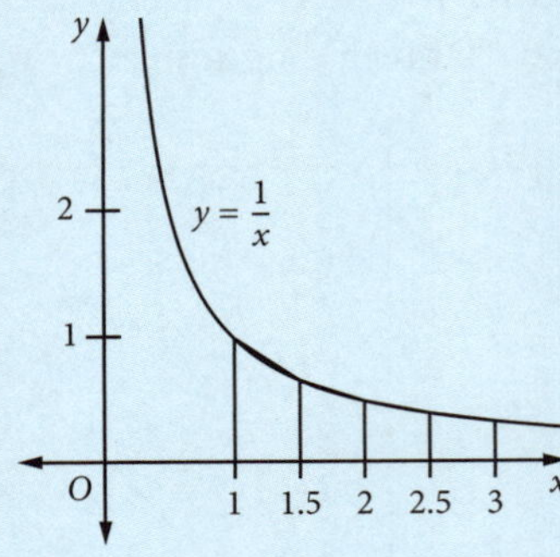

Solution

Sketch the region:

$f(x) = \frac{1}{x}$: $f(1) = 1$, $f(1.5) = \frac{2}{3}$, $f(2) = 0.5$, $f(2.5) = 0.4$, $f(3) = \frac{1}{3}$,
base $= 1.5 - 1 = 0.5$

$$\begin{aligned}\text{Area} &= \frac{f(1)+f(1.5)}{2}\times 0.5 + \frac{f(1.5)+f(2)}{2}\times 0.5 + \frac{f(2)+f(2.5)}{2}\times 0.5 + \frac{f(2.5)+f(3)}{2}\times 0.5\\ &= \left(\left(1+\frac{2}{3}\right)+\left(\frac{2}{3}+\frac{1}{2}\right)+\left(\frac{1}{2}+\frac{2}{5}\right)+\left(\frac{2}{5}+\frac{1}{3}\right)\right)\times\frac{0.5}{2}\\ &= \frac{1}{4}\left(1+2\times\frac{2}{3}+2\times\frac{1}{2}+2\times\frac{2}{5}+\frac{1}{3}\right)\\ &= 1.11\dot{6}\ \text{units}^2\end{aligned}$$

In this example you have found an approximation for the definite integral, $\int_1^3 \frac{dx}{x}$. The exact value of this integral is ln 3 so using trapezia gives a good approximation for this area.

The trapezoidal rule

If $f(x)$ is a continuous function and $f(x) \geq 0$ on the interval $a \leq x \leq b$, then you can find an approximation to the definite integral $\int_a^b f(x)\,dx$ by dividing the area into a number of trapezia of equal width, where the parallel sides of the trapezia are ordinates.

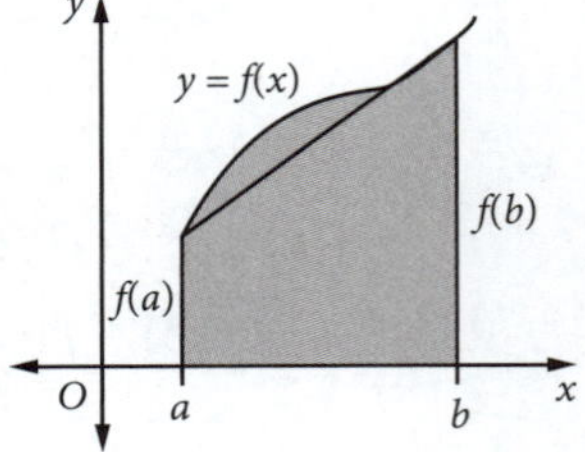

- One trapezium: $$\int_a^b f(x)\,dx \approx (b-a)\left(\frac{f(a)+f(b)}{2}\right) = \frac{(b-a)}{2}\left(f(a)+f(b)\right)$$

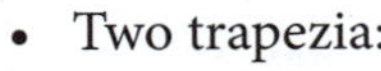

- Two trapezia:

$$\int_a^b f(x)\,dx \approx \frac{(b-a)}{2}\left(\frac{f(a)+f\left(\frac{a+b}{2}\right)}{2}\right)+\frac{(b-a)}{2}\left(\frac{f\left(\frac{a+b}{2}\right)+f(b)}{2}\right) = \frac{(b-a)}{4}\left(f(a)+2f\left(\frac{a+b}{2}\right)+f(b)\right)$$

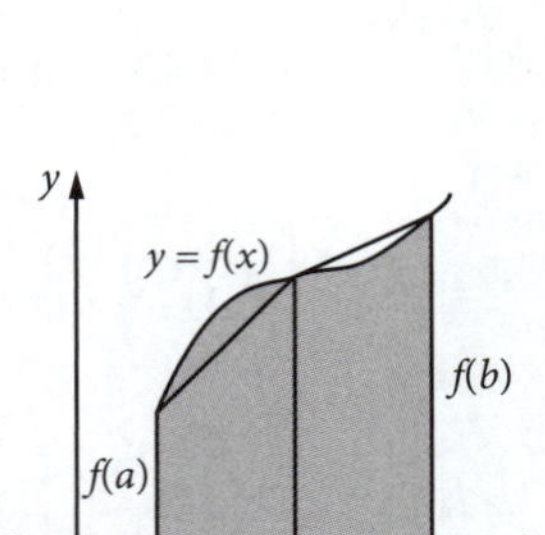

- n trapezia: the base of each trapezium is $\frac{b-a}{n} = h$. If $a = x_0$, $b = x_n$ and the points between are $x_1, x_2, \ldots x_{n-1}$, then the trapezoidal rule becomes:

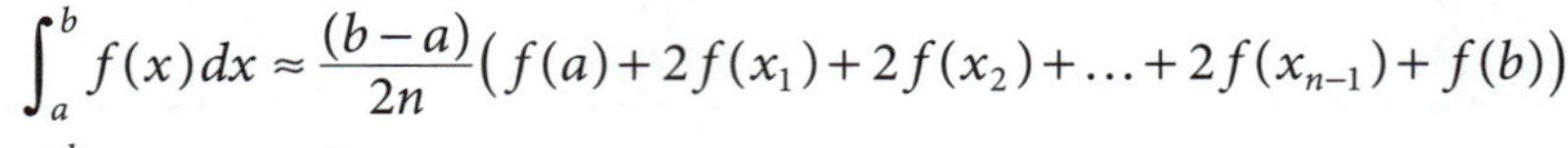

$$\int_a^b f(x)\,dx \approx \frac{(b-a)}{2n}\left(f(a)+2f(x_1)+2f(x_2)+\ldots+2f(x_{n-1})+f(b)\right)$$

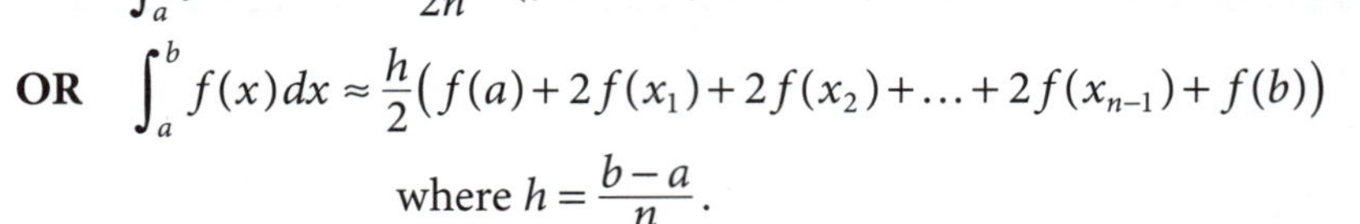

OR $$\int_a^b f(x)\,dx \approx \frac{h}{2}\left(f(a)+2f(x_1)+2f(x_2)+\ldots+2f(x_{n-1})+f(b)\right)$$

where $h = \frac{b-a}{n}$.

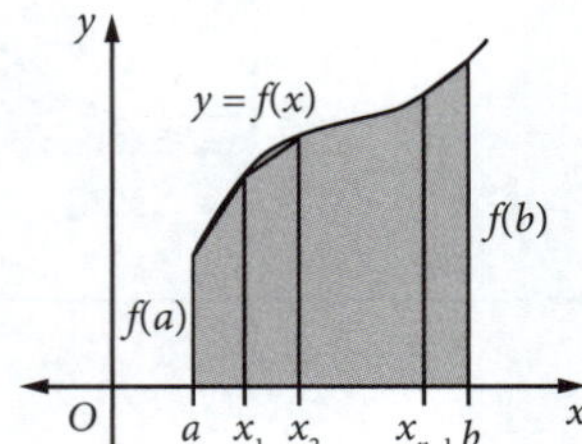

For $h=\frac{b-a}{n}$, therefore:

$n=1, h=b-a$: $\int_a^b f(x)dx \approx \frac{(b-a)}{2}(f(a)+f(b))=\frac{h}{2}(f(a)+f(b))$

$n=2, h=\frac{b-a}{2}$: $\int_a^b f(x)dx \approx \frac{h}{2}\left(f(a)+2f\left(\frac{a+b}{2}\right)+f(b)\right)$

$n=n, h=\frac{b-a}{n}$: $\int_a^b f(x)dx \approx \frac{h}{2}(f(a)+2f(x_1)+2f(x_2)+\ldots+2f(x_{n-1})+f(b))$ where $a=x_0, b=x_n$.

Example 35

Evaluate $\int_0^2 \sqrt{4-x^2}\, dx$ using the trapezoidal rule with:

(a) two subintervals **(b)** four subintervals. **(c)** What is the exact value of this integral?

Solution

(a) Use a table of values to sketch $f(x)=\sqrt{4-x^2}$ with two subintervals:

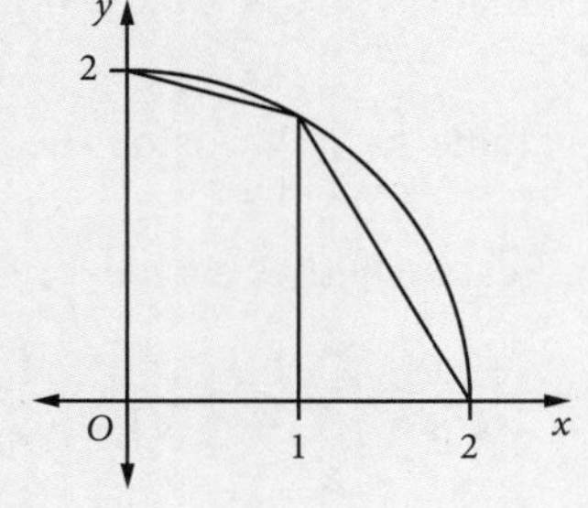

x	0	1	2
$f(x)$	2	$\sqrt{3}$	0

$$h=\frac{2-0}{2}=1: \quad \int_0^2 \sqrt{4-x^2}\, dx \approx \frac{1}{2}(f(0)+2f(1)+f(2))$$
$$=\frac{1}{2}\left(2+2\sqrt{3}+0\right)$$
$$=2.732$$

(b) Use a table of values to sketch $f(x)=\sqrt{4-x^2}$ with four subintervals:

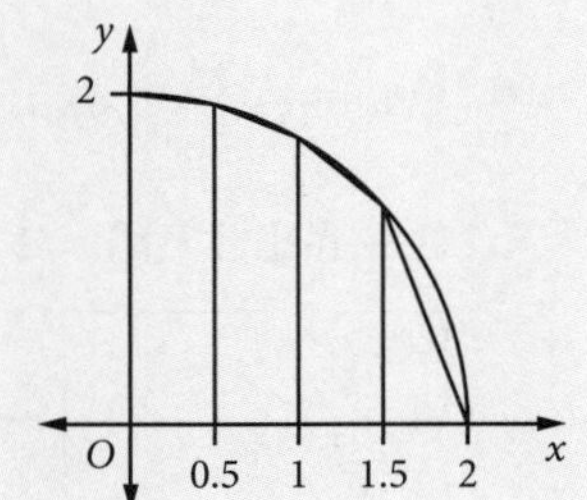

x	0	0.5	1	1.5	2
$f(x)$	2	$\sqrt{3.75}$	$\sqrt{3}$	$\sqrt{1.75}$	0

$$h=\frac{2-0}{4}=0.5: \quad \int_0^2 \sqrt{4-x^2}\, dx \approx \frac{0.5}{2}\left(f(0)+2\left(f(0.5)+f(1)+f(1.5)\right)+f(2)\right)$$
$$=\frac{1}{4}\left(2+2\left(\sqrt{3.75}+\sqrt{3}+\sqrt{1.75}\right)+0\right)$$
$$=2.996$$

(c) The region is a quarter of a circle of radius 2, so the exact value of $\int_0^2 \sqrt{4-x^2}\, dx$ is π.

Example 36

To measure the cross-sectional area of a river, a boat travels directly across the river and measures the river's depth as shown in the following table. The depth of the river at the banks is zero metres. Use the trapezoidal rule and all the values in the table to calculate the cross-sectional area of the river.

Distance from riverbank (in metres)	0	5	10	15	20	25
Depth (in metres)	0	1.2	5	4.8	1.3	0

Solution

The distance from the bank is the independent variable and the depth of the river is the dependent variable.

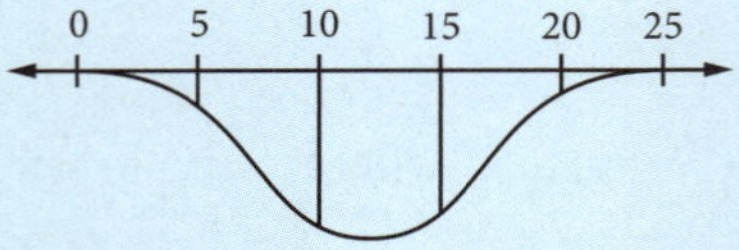

$$\text{Area of cross-section} = \int_0^{25} f(x)dx \approx \frac{5}{2}\big(f(0)+2\big(f(5)+f(10)+f(15)+f(20)\big)+f(25)\big)$$
$$= \frac{5}{2}(0+2(1.2+5.0+4.8+1.3)+0)$$
$$= 5\times 12.3$$
$$= 61.5$$

The area of the cross-section is approximately 61.5 m^2.

If an average flow rate of the river was known, then the amount of water flowing past in a given time could be calculated.

EXERCISE 17.10 APPROXIMATE METHODS OF INTEGRATION—TRAPEZOIDAL RULE

1 Evaluate $\int_0^4 \sqrt{x^2+9}\,dx$ using the trapezoidal rule and two subintervals.

2 Evaluate $\int_0^1 \frac{dx}{x^2+1}$ using the trapezoidal rule with: **(a)** two subintervals **(b)** four subintervals.

3 Indicate whether each expression is a correct or incorrect example of the use of the trapezoidal rule.

(a) $\int_1^3 \frac{dx}{\sqrt{x}} \approx \frac{3-1}{2}\left(\frac{1}{\sqrt{1}}+\frac{2}{\sqrt{2}}+\frac{1}{\sqrt{3}}\right)$

(b) $\int_1^3 \frac{dx}{\sqrt{x}} \approx \frac{2-1}{2}\left(\frac{1}{\sqrt{1}}+\frac{2}{\sqrt{2}}+\frac{1}{\sqrt{3}}\right)$

(c) $\int_1^3 \frac{dx}{\sqrt{x}} \approx \frac{2-1}{2}\left(\frac{1}{\sqrt{1}}+\frac{1}{\sqrt{3}}\right)$

(d) $\int_1^3 \frac{dx}{\sqrt{x}} \approx \frac{0.5}{2}\left(\frac{1}{\sqrt{1}}+2\left(\frac{\sqrt{2}}{\sqrt{3}}+\frac{1}{\sqrt{2}}+\frac{\sqrt{2}}{\sqrt{5}}\right)+\frac{1}{\sqrt{3}}\right)$

4 Evaluate $\int_1^2 x^3\,dx$ by: **(a)** direct integration **(b)** using the trapezoidal rule with two subintervals.

5 Given that $f(x) = 2^x$, complete the following table:

x	0	1	2	3	4
2^x					

Using these five function values and the trapezoidal rule, estimate $\int_0^4 2^x\,dx$.

6 Evaluate $\int_{-1}^1 4^x\,dx$ using the trapezoidal rule with four subintervals.

7 Evaluate $\int_0^1 \sqrt{x}\,dx$ by: **(a)** direct integration **(b)** using the trapezoidal rule with three subintervals.

8 Evaluate $\int_0^2 \sqrt{1+x^3}\,dx$ using the trapezoidal rule with two subintervals.

9 Evaluate $\int_0^4 \frac{x}{1+x}\,dx$ using the trapezoidal rule with four subintervals.

10 Evaluate $\int_0^4 \sqrt{1+x^4}\,dx$ using the trapezoidal rule with four subintervals.

11 A mining engineer is trying to estimate the volume of earth above a coal seam prior to open-cut mining. The area to be mined is flat and rectangular. In a straight 500-metre line across the middle of the rectangle, the engineer drills down to record the depth from the ground surface to the coal seam at 50-metre intervals. The information is recorded in the following table:

Distance from edge (m)	0	50	100	150	200	250	300	350	400	450	500
Depth (m)	200	150	100	150	100	50	50	100	150	200	250

(a) The engineer needs to calculate the vertical area beneath this line down to the coal seam. Use the trapezoidal rule to calculate this area.

(b) The rectangular area to be mined is 2.5 km long. Estimate the volume of soil to be removed (in cubic metres) to expose the coal seam. (Assume that the cross-sectional area is the same for the whole length of the rectangle.)

12 Use the trapezoidal rule with two subintervals to estimate the value of:

(a) $\int_{\frac{\pi}{6}}^{\frac{5\pi}{6}} \sin^2 x\,dx$ **(b)** $\int_{-\frac{\pi}{3}}^{\frac{\pi}{3}} \cos^2 x\,dx$

13 Use the trapezoidal rule with two subintervals to estimate $\int_0^{\pi} \sqrt{\sin x}\,dx$ correct to 2 decimal places.

14 Given $\int_0^1 \frac{4}{1+x^2}\,dx = \pi$, find an approximate value of π using the trapezoidal rule with two subintervals.

15 Using the trapezoidal rule with two subintervals, find an approximate value for $\int_{-1}^{1} x^2 e^x\,dx$.

16 The length of one arch of the curve $y = \sin x$ is given by $\int_0^{\pi} \sqrt{1+\cos^2 x}\,dx$. Find this length using the trapezoidal rule with four subintervals.

17 Use the trapezoidal rule with four subintervals to evaluate $\int_0^{0.8} xe^{-x}\,dx$.

18 Using the trapezoidal rule with four subintervals, find an approximate value for $\int_1^5 xe^{0.4x}\,dx$.

19 **(a)** Show that the area between the curves $y = \frac{1}{x^2}$ and $y = \frac{-1}{x}$ for $1 \le x \le 4$ is given by $\int_1^4 \left(\frac{1}{x^2} + \frac{1}{x}\right)dx$.

(b) Use the trapezoidal rule with three subintervals to find an approximation for this area correct to 2 decimal places.

17.11 AVERAGE VALUE OF A FUNCTION—AN APPLICATION OF INTEGRATION

(**Note**: This section is not part of this course; it is given as a useful application of integration.)

You are used to finding the average of a set of numbers by adding up all the numbers and dividing by how many there are. Hence the average of the n numbers $y_1, y_2, \ldots y_n$ is given by $y_{\text{ave}} = \frac{y_1 + y_2 + \ldots + y_n}{n}$.

Similarly, to find the average value of a function $f(x)$ over the interval $a \le x \le b$, you can divide the interval $a \le x \le b$ into n equal subintervals of length $\Delta x = \frac{b-a}{n}$. Take the points $x_1, x_2, \ldots x_n$ in successive intervals and calculate the average of the numbers $f(x_1), f(x_2), \ldots, f(x_n)$: Average $= \frac{f(x_1) + f(x_2) + \ldots + f(x_n)}{n}$.

But $\Delta x = \frac{b-a}{n}$ so $n = \frac{b-a}{\Delta x}$, so you have:

$$\text{Average} = \frac{f(x_1)+f(x_2)+\ldots+f(x_n)}{n} = \frac{f(x_1)+f(x_2)+\ldots+f(x_n)}{\frac{b-a}{\Delta x}}$$

$$= \frac{f(x_1)\Delta x + f(x_2)\Delta x + \ldots + f(x_n)\Delta x}{b-a}$$

$$= \frac{1}{b-a}\sum_{i=1}^{n} f(x_i)\Delta x$$

As n increases, Δx decreases and so this expression can be written as a definite integral:

$$\lim_{n\to\infty}\frac{1}{b-a}\sum_{i=1}^{n} f(x_i)\Delta x = \frac{1}{b-a}\int_a^b f(x)\,dx$$

Thus you can define the **average value** of f on the interval $a \le x \le b$ as: $f_{\text{ave}} = \frac{1}{b-a}\int_a^b f(x)\,dx$

Example 37

Find the average value of each function over the interval $1 \le x \le 3$:

(a) $f(x) = x + 1$ **(b)** $g(x) = x^2 + 1$ **(c)** $h(x) = x^3 + 1$

Solution

For the integrals $a = 1$, $b = 3$:

(a) $f(x) = x + 1$

$$f_{\text{ave}} = \frac{1}{3-1}\int_1^3 (x+1)\,dx = \frac{1}{2}\left[\frac{x^2}{2}+x\right]_1^3 = \frac{1}{2}\left(\frac{9}{2}+3-\left(\frac{1}{2}+1\right)\right) = 3$$

(b) $g(x) = x^2 + 1$

$$g_{\text{ave}} = \frac{1}{3-1}\int_1^3 (x^2+1)\,dx = \frac{1}{2}\left[\frac{x^3}{3}+x\right]_1^3 = \frac{1}{2}\left(\frac{27}{3}+3-\left(\frac{1}{3}+1\right)\right) = 5\frac{1}{3}$$

(c) $h(x) = x^3 + 1$

$$h_{\text{ave}} = \frac{1}{3-1}\int_1^3 (x^3+1)\,dx = \frac{1}{2}\left[\frac{x^4}{4}+x\right]_1^3 = \frac{1}{2}\left(\frac{81}{4}+3-\left(\frac{1}{4}+1\right)\right) = 11$$

EXERCISE 17.11 AVERAGE VALUE OF A FUNCTION—AN APPLICATION OF INTEGRATION

1 Find the average value of f on the interval $0 \le x \le 2$ for $f(x) = x^2 - x$.

2 Find the average value of f on the interval $-2 \le x \le 2$ for $f(x) = x^4$.

3 Find the average value of f on the interval $-2 \le x \le 0$ for $f(x) = x^3$.

4 For $f(x) = x^2 - 2x + 1$, the average value of f on the interval $-1 \le x \le 3$ is:

A $-\frac{1}{3}$ **B** $\frac{1}{3}$ **C** $\frac{2}{3}$ **D** $\frac{4}{3}$

5 Find the average value of f on the interval $0 \le x \le 9$ for $f(x) = \sqrt{x}$.

6 The linear density of a rod 7 m long is given by $d(x) = \frac{12}{\sqrt{x+2}}$ kg m^{-1}, where x is measured in metres from one end of the rod. Find the average density of the rod.

7 Find the values of b for which the average value of $f(x) = 1 - 6x + 3x^2$ on the interval $1 \le x \le b$ is equal to 2.

8 The average value of f on the interval $a \le x \le b$ is defined $f_{\text{ave}} = \frac{1}{b-a}\int_a^b f(x)dx$. Find the average value of each function on the interval given.

(a) $f(x) = e^x, 0 \le x \le 2$
(b) $f(x) = e^{-x}, 0 \le x \le 2$
(c) $g(x) = e^{2x}, 1 \le x \le 5$
(d) $f(x) = e^x + e^{-x}, -2 \le x \le 2$

9 We define the average value of f on the interval $a \le x \le b$ as $f_{\text{ave}} = \frac{1}{b-a}\int_a^b f(x)dx$. Find the average value of each function on the interval given.

(a) $f(x) = \frac{1}{x}, 1 \le x \le 3$
(b) $f(x) = \frac{2x}{x^2+1}, 0 \le x \le 2$
(c) $g(x) = \frac{1}{x+1}, 0 \le x \le 2$
(d) $f(x) = \frac{x}{x^2+4}, -2 \le x \le 2$

10 The average value of a function $f(x)$ between $x = a$ and $x = b$ is defined as $\frac{1}{b-a}\int_a^b f(x)dx$. Find the average value of $f(x) = 2\cos x + 3\sin x$ between $x = -\frac{\pi}{2}$ and $x = \frac{\pi}{2}$.

11 The average value of a function $f(x)$ in the interval $[a, b]$ is defined as $\frac{1}{b-a}\int_a^b f(x)dx$. Find the average value of $f(x) = 2\cos 2x + \sin 2x$ in the interval $\frac{\pi}{6} \le x \le \frac{\pi}{3}$.

CHAPTER REVIEW 17

1 Evaluate the following:

(a) $\int_{-1}^{3}\left(x^3 - 3x^2 + 2x - 5\right)dx$
(b) $\int_{-2}^{2}\left(x^2 - 2x\right)^2 dx$
(c) $\int_{-3}^{1}(5-2x)^3\,dx$

2 Find the area of the region bounded by the parabola $y = 4x - x^2$ and the x-axis.

3 Evaluate the following:

(a) $\int_{-1}^{2} 3x(2-x)dx$
(b) $4\int_{-3}^{-1} x(x+1)^2\,dx$
(c) $\int_{-2}^{4}\left(x^3 - 2\right)dx$

4 Find the equation of the tangent to the parabola $y = x^2$ at the point where $x = 2$. By integration, find the area of the region bounded by the parabola, the tangent and the x-axis.

5 Find the area of the region defined by the inequalities $y \ge -5$ and $y \le 4x - x^2$.

6 Calculate the area of the region bounded by the curve $y = 2x^2(4 - x)$ and the x-axis.

7 Evaluate: (a) $\int_{\frac{\pi}{8}}^{\frac{\pi}{4}}(\sin 2x - \cos 2x)dx$ (b) $\int_0^{\frac{\pi}{2}}(\cos 2x - x)dx$ (c) $\int_{\frac{\pi}{6}}^{\frac{\pi}{3}}\frac{\sec^2 x}{\tan x}dx$

8 Find the function f, defined for positive x only, for which $f'(x) = \left(\frac{1}{2x} + x\right)^2$ and $f(2) = 6$.

9 Calculate:

(a) the area between the curve $y = x^2(3 - x)$ and the x-axis
(b) the area bounded by the curve $y = x^2(3 - x)$ and the lines $y = 3 - x$, $x = 1.5$ and $x = 2$.

10 Find the average value of the function $f(x) = 1 + \sqrt{x}$ on the interval $1 \le x \le 9$.

11 Evaluate $\int_{\frac{\pi}{4}}^{\frac{\pi}{3}}\cos(x + \pi)dx$.

12 (a) Find the values of x for which $4x - x^2 > 0$.
(b) Find the area under the graph of $y = 4x - x^2$ between $x = 1$ and $x = 2$.
(c) Find the angle between the x-axis and the tangent to the curve $y = 4x - x^2$ at $x = 1$.

13 **(a)** Sketch the graph of $y = 3x^2 - x^3$.
(b) The tangent to the curve $y = 3x^2 - x^3$ at the point A, where $x = 2$, cuts the graph at another point B. Show that the coordinates of B are $(-1, 4)$.
(c) Calculate the area enclosed by the graph and the line AB.

14 Use the trapezoidal rule with four subintervals to find an approximation for $\int_{-2}^{2} \frac{dx}{1+x^2}$.

15 A speleologist is in a cave with a flat, circular floor. She walks across the diameter of the floor and measures the height of the cave ceiling every four metres, as recorded in the following table:

Distance from edge, x m	0	4	8	12	16	20	24
Height, y m	0	1	3	5	4	2.5	0

(a) Use the trapezoidal rule and all the values in the table to find an approximation for the area of the cave's vertical cross-section to the nearest m^2.
(b) Considering the diameter of the cave to be the x-axis (for $-12 \le x \le 12$), and considering the vertical at 12 m from the edge to be the y-axis, discuss how you might use the trapezoidal rule to estimate the volume of the cave.

16 The average value of a function $f(x)$ in the interval $[a, b]$ is defined as $\frac{1}{b-a}\int_a^b f(x)dx$. Find the average value of $f(x) = 2\cos 2x + \sin 2x$ between $x = \frac{\pi}{4}$ and $x = \frac{\pi}{2}$.

17 Evaluate: **(a)** $\int_{-2}^{2}\left(e^x - e^{-x}\right)dx$ **(b)** $\int_{-1}^{2}\left(e^x - e^{-x}\right)^2 dx$ **(c)** $\int_{1}^{3}\left(e^x + \frac{1}{x}\right)dx$

18 Sketch the graphs of $y = e^{-x}$, $y = x + 1$ and the line $x = 2$. By integration, find the area of the region bounded by them.

19 **(a)** Find the maximum value of $2xe^{-1.5x}$ and the value for which this function has a maximum value.
(b) If $f(x) = 2xe^{-1.5x}$, find $f(0)$, $f(0.5)$, $f(1)$ and hence graph the function in the domain $0 \le x \le 1$.
(c) Use the trapezoidal rule with two subintervals to find the value of $2\int_0^1 xe^{-1.5x}\, dx$ correct to 3 decimal places.

20 Sketch the graph of $y = e^{-\frac{x}{2}}$, $x \ge 0$, and indicate the region whose area is represented by the integral $\int_0^1 e^{-\frac{x}{2}}\, dx$. Evaluate this integral.

21 The diagram shows the graphs of $y = \sqrt{3}\sin x$ and $y = \cos x$. The first two points of intersection to the right of the y-axis are labelled P and Q.
(a) Solve the equation $\sqrt{3}\sin x = \cos x$ to find the abscissae of P and Q.
(b) Find the area of the shaded region in the diagram.

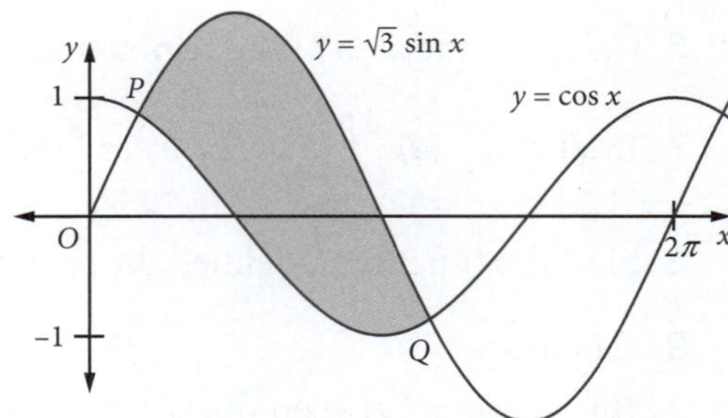

22 **(a)** Show that $\frac{d}{dx}(\log_e(\sin x)) = \cot x$.
(b) The shaded region in the diagram is bounded by the curve $y = \cot x$ and the lines $y = \frac{\pi}{2} - x$ and $x = \frac{\pi}{4}$. Using the result of part **(a)**, or otherwise, find the exact area of the shaded region.

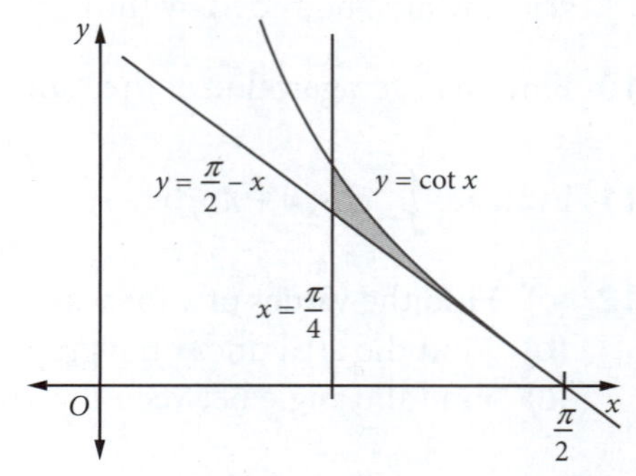

23 The shaded region in the diagram is bounded by the curves $y = \sin\frac{\pi x}{2}$, $y = x^2$ and the x-axis.

(a) Show that the two curves meet at $x = 1$.

(b) Calculate the exact area of the shaded region.

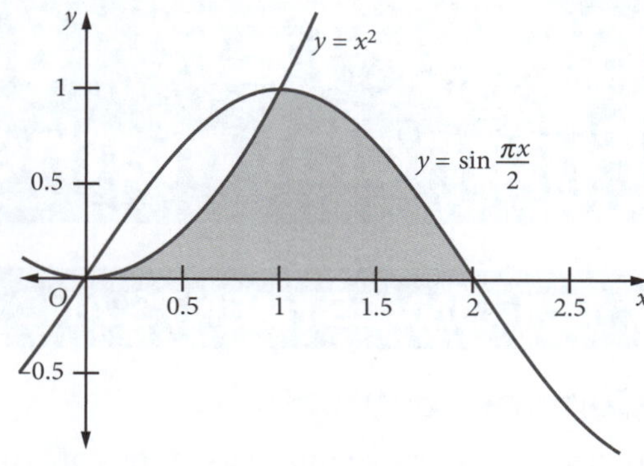

24 The acceleration of a particle moving in a straight line is $6 - 8t\ \text{m s}^{-2}$ at any time $t \geq 0$. The particle starts from O with a velocity of $10\ \text{m s}^{-1}$. When is its velocity zero, and what is its displacement at that time?

25 A body starts from rest and moves in a straight line so that its velocity $v\ \text{m s}^{-1}$ after t seconds is given by $v = 2t + 6t^2$. Calculate:

(a) its acceleration at the end of the first second

(b) its displacement after 5 seconds, given that the body is initially at zero displacement.

26 The velocity $v\ \text{m s}^{-1}$ of a particle moving in a straight line is given by $v = 6t^2 - 4t + 1\ (t \geq 0)$. The particle initially has a displacement -10 m from O. Find:

(a) the displacement and acceleration at any time t

(b) the acceleration when the velocity is $3\ \text{m s}^{-1}$

(c) the velocity when the acceleration is $20\ \text{m s}^{-2}$.

27 A full water tank holds 4000 litres. When the tap is turned on, water flows out from the tank at a rate of $\frac{dV}{dt} = 110 + 17t - t^2$ litres per minute, where t is the time in minutes since the tap was turned on.

(a) At what time is the tank emptying at a rate of 50 litres per minute?

(b) Find the volume of water that has flowed out of the since the tap was turned on as a function of t.

(c) How much water has flowed out of the tank 12 minutes after the tap was turned on?

(d) When does the water stop flowing out of the tank?

(e) How much water is left in the tank (to the nearest litre) when the water stops flowing out of the tank?

28 A particle is initially at rest at the origin. Its acceleration as a function of time t is given by $\ddot{x} = 9\sin 3t$.

(a) Show that the velocity of the particle is given by $\dot{x} = 3 - 3\cos 3t$.

(b) Sketch the graph of the velocity for $0 \leq t \leq \frac{4\pi}{3}$ and determine the time at which the particle first comes to rest after $t = 0$.

(c) Find the distance travelled by the particle between $t = 0$ and the time found in **(b)**.

(d) Find an expression for x as a function of t.

29 **(a)** Sketch the graphs of $y = 2^x$ and $y = 3^{-x}$ over the domain $-1 \leq x \leq 2$.

(b) Calculate the area of the region bounded by the curves $y = 2^x$, $y = 3^{-x}$ and the ordinates $x = -1$ and $x = 2$.

30 **(a)** Use graphing software to draw the graphs of $y = 2^x$ and $y = 2 - x^2$ over the domain $-2 \leq x \leq 2$. From your graph, write down the coordinates of their points of intersection, correct to 2 decimal places.

(b) Hence calculate the area enclosed between the two curves.

(c) Calculate the area enclosed by the curves and the x-axis.

CHAPTER 18
Financial mathematics

18.1 INVESTMENTS AND LOANS

Compound interest

In previous years you have used the compound interest formula in calculations:

$A = P(1 + r)^n$

where P is the principal (or starting amount)

r is the interest rate for the period, expressed as a decimal

n is the number of compounding periods

A is the amount to which P has grown or reduced (final amount)

This formula for compound interest may also be written as a recurrence relation.

If $P = A_0$

then $A_1 = A_0 \times R$

and $A_2 = A_1 \times R$

and eventually $A_n = A_{n-1} \times R$

where A_n is the value of the initial amount after n time periods and $R = 1 + \frac{r}{100}$, the interest rate being r% per time period.

When money is invested or borrowed, the compound interest formula is used to calculate what the investment or loan becomes over a period of time.

If payments are made to the investment or loan, the final amount changes, either increasing more rapidly in the case of an investment, or decreasing in the case of a loan.

Example 1

Anh deposits $1000 in an investment account that is paying a monthly interest rate of 0.25%, with the interest compounded monthly. Calculate the value of the investment after 12 months.

Solution

$P = 1000$, $r = 0.25\% = 0.0025$, $n = 12$

$A = P(1 + r)^n$: $A = 1000 \times 1.0025^{12}$

$= 1030.42$

After 12 months, there is $1030.42 in Anh's investment account.

Example 2

Caleb knows that he will need \$10 000 in five years time to pay for a study tour. He wishes to deposit enough in an investment account to achieve this goal. The account interest will be fixed at 3.6% per annum, compounded monthly. How much does he need to deposit into the account to achieve this goal? Round your answer to the next dollar.

Solution

$P = ?$, $n = 5 \times 12 = 60$ months, $A = 10\,000$, $r = \frac{3.6}{12} = 0.3\%$ per month $= 0.003$

Use $A = P(1 + r)^n$: $10\,000 = P \times 1.003^{60}$

$$P = \frac{10\,000}{1.003^{60}}$$

$$P = 8354.95$$

Caleb should deposit \$8355.

Example 3

Maria deposits \$1000 in an investment account that is paying a monthly interest rate of 0.25%, with the interest compounded monthly. She adds a further \$1000 to the account at the start of each subsequent month. How much is her investment worth at the end of six months?

Solution

$P = 1000$, $r = 0.25\% = 0.0025$, $n = 6, 5, 4, 3, 2, 1$

The first \$1000 is invested for 6 months: $A = 1000 \times 1.0025^6 = 1015.09$

The second \$1000 is invested for 5 months: $A = 1000 \times 1.0025^5 = 1012.56$

The third \$1000 is invested for 4 months: $A = 1000 \times 1.0025^4 = 1010.04$

The fourth \$1000 is invested for 3 months: $A = 1000 \times 1.0025^3 = 1007.52$

The fifth \$1000 is invested for 2 months: $A = 1000 \times 1.0025^2 = 1005.01$

The sixth \$1000 is invested for 1 month: $A = 1000 \times 1.0025^1 = 1002.50$

The value of Maria's investment after 6 months = \$1015.09 + \$1012.56 + \$1010.04 + \$1007.52 + \$1005.01 + \$1002.50 = \$6052.72

This could have been found using the recurrence relation in a spreadsheet.

$A_0 = 1000$, $r = 0.0025$, $R = 1.0025$, $n = 6$

	A	B	C	D
1	A0	1000.00		
2	r =	0.0025		
3	R =	1.0025		
4				
5	n	An	Deposit	Total
6	0	1000.00		1000.00
7	1	1002.50	1000.00	2002.50
8	2	2007.51	1000.00	3007.51
9	3	3015.03	1000.00	4015.03
10	4	4025.06	1000.00	5025.06
11	5	5037.63	1000.00	6037.63
12	6	6052.72	0.00	6052.72

Later in this chapter you will use the sum of a geometric series to find the value of an investment like Maria's.

Future value

In Example 1, the value of the investment, **Future value** (FV), could have been found from the **Present value** (PV), using the formula $FV = PV(1 + r)^n$, where r is the interest rate per period, expressed as a decimal and n is the number of compounding periods.

For Anh, $PV = 1000$, $r = 0.0025$ and $n = 12$: $FV = 1000 \times 1.0025^{12} = \$1030.42$.

This formula can be rearranged so that if you know the future value that you require then you can calculate the present value. The formula is $PV = \frac{FV}{(1+r)^n}$. The present value is the single amount, which if invested at the same rate for the same period, would give that future value.

If Anh wished to have \$1500 after 6 months at an interest rate of 0.25% per month, then this formula will give the amount that she has to invest.

$FV = 1500$, $r = 0.0025$, $n = 6$, find the PV: $PV = \frac{1500}{1.0025^6} = 1477.70$

Anh would need to make an initial deposit of \$1477.70.

PV = present value, FV = future value

r = interest rate per period as a decimal

n is the number of compounding periods

$FV = PV(1 + r)^n$

$PV = \frac{FV}{(1+r)^n}$

Annuities

An annuity is a compound interest investment from which equal payments are received on a regular basis (at equal periods of time) for a fixed period of time.

The payment is usually made at the end of the time period so that no interest is received until the end of the second time period.

Future value

The future value of an investment or annuity is the total value of the investment at the end of the term of investment, including all contributions and interest earned.

Present value

The present value of an investment or annuity is the single sum of money (or principal) that could be initially invested to produce a given future value over a given period of time.

The information needed to answer questions about annuities will be given in a table. The entries in this table have been calculated using the following formula.

$$FVA = a\left\{\frac{(1+r)^n - 1}{r}\right\} \text{ and } PVA = a\left\{\frac{(1+r)^n - 1}{r(1+r)^n}\right\}$$

where

FVA is the Future Value of an Annuity

PVA is the Present Value of an Annuity

a is the contribution per period paid at the end of the period

r is the interest rate per compounding period as a decimal

n is the number of periods.

You can set up a spreadsheet using these formulae to see if your own calculations agree with the following for $a = 1$.

This table gives the future value of an annuity of $1 at the given interest rate for the given period.

Future value interest factors (FVA)					
$1	**Interest rate per period**				
N	**1%**	**2%**	**3%**	**4%**	**5%**
1	1.0000	1.0000	1.0000	1.0000	1.0000
2	2.0100	2.0200	2.0300	2.0400	2.0500
3	3.0301	3.0604	3.0909	3.1216	3.1525
4	4.0604	4.1216	4.1836	4.2465	4.3101
5	5.0101	5.2040	5.3091	5.4163	5.5256
6	6.1520	6.3081	6.4684	6.6330	6.8019

This table gives the present value of an annuity of $1 at the given interest rate for the given period.

Present value interest factors (PVA)					
$1	**Interest rate per period**				
N	**1%**	**2%**	**3%**	**4%**	**5%**
1	0.9901	0.9804	0.9709	0.9615	0.9524
2	1.9704	1.9416	1.9135	1.8861	1.8594
3	2.9410	2.8839	2.8286	2.7751	2.7232
4	3.9020	3.8077	3.7171	3.6299	3.5460
5	4.8534	4.7135	4.5797	4.4518	4.3295
6	5.7955	5.6014	5.4172	5.2421	5.0757

The following examples show how to use these tables.

Example 4

Use the table of future value interest factors to find:

(a) the future value of an annuity of $1000 per year for 4 years at 3% per annum

(b) the future value of an annuity of $1500 per year for 6 years at 4% per annum.

Solution

(a) 4 years at 3% gives a multiplication factor of 4.1836 from the table.
FVA = $1000 × 4.1836 = $4183.60

(b) 6 years at 4% gives a multiplication factor of 6.6330 from the table.
FVA = $1000 × 6.6330 = $6633.00

Example 5

Use the table of present value interest factors for an annuity of $1 per period.

(a) Jake plans to invest $5000 per year for 6 years in an annuity. His investment will earn interest at the rate of 4% per annum. Calculate the present value of this annuity.

(b) Ari takes out a personal loan of $9000 to be repaid over 4 years at an interest rate of 5% per year. Use the PVA table to find his yearly repayments.

Solution

(a) 6 years at 4% gives a multiplication factor of 5.2421.
PVA = $5000 × 5.2421 = $26 210.50

(b) Let M be the yearly repayments.
Using the PVA table with $r = 5\%$ and the period 4 years, the present value of $1 is 3.5460.
The formula used to create the table was PVA = $a \times$ value in table, where $a = \$1$.
In this application, PVA = value of the loan and $a = M$, the repayments.
Hence $3.5460 \times M = 9000$

$$M = \frac{9000}{3.5460}$$
$$= \$2538.07$$

Ari will have to repay $2538.05 every year for 4 years.

Summary of present and future value interest factors

1. The table of future value interest factors is used to find the future value of an annuity for a given interest rate and a set time period. The table gives the future value of each dollar of the annuity. Multiply this value by the value of the annuity.
2. The table of present value interest factors is used to find the present (or current) value of an annuity for a given interest rate and a set time period. The table gives the present value of each dollar of the annuity. Multiply this value by the value of the annuity to obtain its present value.
3. The table of present value interest factors can be used to find the repayments required to pay off a loan of a given amount at a set rate of compound interest over a fixed time period. The repayments are obtained by dividing the value of the loan by the value of the cell in the table corresponding to the interest rate and time period.

EXPLORE FURTHER

Present value and future value interest factors

Use spreadsheet software to create tables of present value and future value interest factors.

Example 6

The table gives the present value interest factors for an annuity of $1 for various interest rates, *r*, and number of periods, *N*.

Present value interest factors					
$1	**Interest rate per period (as a decimal) (r)**				
N	**0.0025**	**0.005**	**0.0075**	**0.008**	**0.009**
71	64.98140	59.64121	54.89293	54.00754	52.29657
72	65.81686	60.33951	55.47685	54.57097	52.82118
73	66.65023	61.03434	56.05643	55.12993	53.34111
74	67.48153	61.72571	56.63169	55.68446	53.85641
75	68.31075	62.41365	57.20267	56.23458	54.36710
76	69.13791	63.09815	57.76940	56.78034	54.87324

(a) Kumba plans to invest $300 each month for 76 months. Her investment will earn interest at a rate of 0.0025 (as a decimal) per month. Use the information in the table to calculate the present value of this annuity.

(b) Zephan uses the same table to calculate the loan repayments for his car loan. His loan is for $22 000 and will be repaid in equal monthly repayments over 6 years. The interest rate on his loan is 9.6% per annum. Calculate the amount of each monthly repayment, rounded to the next dollar.

Solution

(a) 76 months at 0.0025 gives a multiplication factor of 69.13791 from the table.

Present value of the annuity $= \$300 \times 69.13791$

$= \$20\,741.37$

(b) 6 years $= 6 \times 12 = 72$ months, $N = 72$

9.6% p.a. $= 0.8\%$ per month so $r = 0.008$ as a decimal

From the table, the value to be used is 54.57097.

Hence $22\,000 = M \times 54.57097$

$$M = \frac{22\,000}{54.57097}$$

$$= 403.14$$

The repayments would be $404 per month, rounded to the next dollar.

Effective annual rate of interest

The **effective annual rate** of interest is a way of comparing interest rates on a common basis even when they are quoted as per annum, per quarter, per month, per day etc. Without a common basis, it is very hard to compare the overall effects of different interest rates that are applied at different time periods.

By converting each interest rate to its effective annual rate, you can compare the costs of loans or the returns on investments so as to decide which one is best for you.

$$\text{Effective annual interest rate} = \left(1 + \frac{r}{n}\right)^n - 1$$

Example 7

What is the effective annual interest rate for a loan advertised as:

(a) 6% p.a. compounded monthly
(b) 6% p.a. compounded quarterly
(c) 6% p.a. compounded daily (use 365 days in a year).

Solution

(a) $r = 0.06$, $n = 12$

Effective annual interest rate $= \left(1 + \frac{0.06}{12}\right)^{12} - 1 = 6.168\%$

(b) $r = 0.06$, $n = 4$

Effective annual interest rate $= \left(1 + \frac{0.06}{4}\right)^{4} - 1 = 6.136\%$

(c) $r = 0.06$, $n = 4365$

Effective annual interest rate $= \left(1 + \frac{0.06}{365}\right)^{365} - 1 = 6.183\%$

Example 8

A home loan from Bank B is advertised with an interest rate of 5% p.a. compounded monthly. Credit Union C advertises their home loan with an interest rate of 4.9% p.a. compounded daily. If there are no other fees involved, which financial institution offers the better deal?

Solution

Bank B: $r = 0.05$, $n = 12$

Effective annual interest rate $= \left(1 + \frac{0.05}{12}\right)^{12} - 1 = 5.116\%$

Credit Union C: $r = 0.049$

Effective annual interest rate $= \left(1 + \frac{0.049}{365}\right)^{365} - 1 = 5.022\%$

Credit Union C offers the better deal.

EXERCISE 18.1 INVESTMENTS AND LOANS

1 (a) Julian deposits \$1000 in an investment account that is paying a monthly interest rate of 0.3%, with the interest compounded monthly. Calculate the value of the investment after 12 months.
(b) Minh deposits \$2000 in an investment account that is paying a monthly interest rate of 0.5%, with the interest compounded monthly. Calculate the value of the investment after 10 months.
(c) Ben deposits \$3400 in an investment account that is paying a monthly interest rate of 0.35%, with the interest compounded monthly. Calculate the value of the investment after 20 months.

2 The table shows the future value of an investment of $1000, compounding yearly, at varying interest rates for different periods of time.

Future values of an investment of $1000					
Number of years	**Interest rate per annum**				
	1%	**2%**	**3%**	**4%**	**5%**
1	1010.00	1020.00	1030.00	1040.00	1050.00
2	1020.10	1040.40	1060.90	1081.60	1102.50
3	1030.30	1061.21	1092.73	1124.86	1157.63
4	1040.60	1082.43	1125.51	1169.86	1215.51
5	1051.01	1104.08	1159.27	1216.65	1276.28
6	1061.52	1126.16	1194.05	1265.32	1340.09

(a) Based on the information provided, what is the future value of an investment of $3000 over 4 years at 3% p.a.?
A $3374.58 **B** $3376.53 **C** $3278.19 **D** $4502.04

(b) Based on the information provided, what is the future value of an investment of $4500 over 6 years at 5% p.a.?
A $5053.28 **B** $5684.94 **C** $5360.36 **D** $6030.41

(c) Based on the information provided, what is the future value of an investment of $500 over 5 years at 4% p.a.?
A $608.33 **B** $607.76 **C** $2431.02 **D** $2433.30

(d) Based on the information provided, what is the future value of an investment of $700 over 3 years at 2% p.a.?
A $7426.30 **B** $7428.47 **C** $742.85 **D** $742.63

3 Brian and Faye plan to have $25 000 in an investment account in 10 years time to pay for a cruise. The interest rate for the account will be fixed at 4.2% per annum, compounded monthly. How much do they need to deposit into the account to achieve this goal? Round your answer to the next dollar.

4 What amount must be invested now at 6% per annum, compounded monthly, so that in 4 years it will have grown to $50 000?

5 The table shows the compounded values of $1.

Future values of the compounded values of $1						
Period	**Interest rate per period**					
	1%	**2%**	**3%**	**4%**	**5%**	**6%**
1	1.010	1.020	1.030	1.040	1.050	1.060
2	1.020	1.040	1.061	1.082	1.103	1.124
3	1.030	1.061	1.093	1.125	1.158	1.191
4	1.041	1.082	1.126	1.170	1.216	1.262
5	1.051	1.104	1.159	1.217	1.276	1.338
6	1.062	1.126	1.194	1.265	1.340	1.419
7	1.072	1.149	1.230	1.316	1.407	1.504
8	1.083	1.172	1.267	1.269	1.477	1.594

Use this table to calculate the value of the following investments.

(a) Stuart invests $2500 for 4 years at an interest rate of 3% per half year compounded half yearly.

(b) Karina invests $5600 for 2 years at an interest rate of 1% per quarter, compounded quarterly.

(c) Aaliyah invests $3000 for 5 years at an interest rate of 5% per annum compounded yearly.

(d) Achim makes two investments of $4000 each. His first investment is for 3 years at an interest rate of 6% per annum compounded yearly. His second investment is for 3 years at an interest rate of 3% per half year compounded half yearly. Which investment earns the most interest and by how much?

6 Following is a table of future value interest factors for a $1 contribution to an annuity.

Future value interest factors					
Period	**Interest rate per period**				
	1%	**2%**	**3%**	**4%**	**5%**
1	1.0000	1.0000	1.0000	1.0000	1.0000
2	2.0100	2.0200	2.0300	2.0400	2.0500
3	3.0301	3.0604	3.0909	3.1216	3.1525
4	4.0604	4.1216	4.1836	4.2465	4.3101
5	5.0101	5.2040	5.3091	5.4163	5.5256
6	6.1520	6.3081	6.4684	6.6330	6.8019

Answer the following questions using this table.

(a) A certain annuity involves making equal contributions of $1000 into an account every 4 months for 2 years at an interest rate of 6% per annum. The future value of this annuity is:
A $6308.10 B $6468.40 C $6633.00 D $4121.60

(b) A certain annuity involves making equal contributions of $5000 into an account every 6 months for 2 years at an interest rate of 4% per annum. The future value of this annuity is:
A $10 100.00 B $10 200.00 C $21 232.50 D $20 608.00

(c) A certain annuity involves making equal contributions of $500 into an account every year for 6 years at an interest rate of 5% per annum. The future value of this annuity is:
A $3316.50 B $6633.00 C $3400.95 D $6801.90

7 Following is a table of future value interest factors for a $1 contribution to an annuity.

Future value interest factors						
Period	**Interest rate per period**					
	0.5%	**1%**	**2%**	**3%**	**4%**	**5%**
6	6.0755	6.1520	6.3081	6.4684	6.6330	6.8019
12	12.3356	12.6825	13.4121	14.1920	15.0258	15.9171
18	18.7858	19.6147	21.4123	23.4144	25.6454	28.1324
24	25.4320	26.9735	30.4219	34.4265	39.0826	44.5020
36	39.3361	43.0769	51.9944	63.2759	77.5983	95.8363
42	46.6065	51.8790	64.8622	82.0232	104.8196	135.2318
48	54.0978	61.2226	79.3535	104.4084	139.2632	188.0254

Use the table of future value interest factors to find:

(a) the future value of an annuity of $1000 per year for 6 years at 3% per annum
(b) the future value of an annuity of $1500 per year for 18 years at 4% per annum
(c) the future value of an annuity of $200 per month for 48 months at 6% per annum
(d) the future value of an annuity of $100 per quarter for 9 years at 8% per annum
(e) the future value of an annuity of $400 per half year for 21 years at 6% per annum.

8 The table gives the present value interest factors for an annuity of $1 per period, for various interest rates *r* and number of periods *N*.

Present value interest factors					
N	**Interest rate per period (as a decimal) (*r*)**				
	0.0025	**0.005**	**0.0075**	**0.008**	**0.009**
71	64.9814	59.6412	54.8929	54.0075	52.2966
72	65.8169	60.3395	55.4769	54.5710	52.8212
73	66.6502	61.0343	56.0564	55.1299	53.3411
74	67.4815	61.7257	56.6317	55.6845	53.8564
75	68.3108	62.4137	57.2027	56.2346	54.3671
76	69.1379	63.0982	57.7694	56.7803	54.8732

Use the table to answer the following questions.

(a) What is the present value of an annuity of $200 per month for 71 months if the interest rate is 0.75% per month?

A $11 095.38 B $10 978.58 C $11 928.24 D $10 801.50

(b) What is the present value of an annuity of $150 per month for 74 months if the interest rate is 0.8% per month?

A $8078.46 B $8268.49 C $8352.68 D $8435.19

(c) A loan of $20 000 is to be repaid in equal monthly instalments over 6 years. The interest rate is 10.8% per annum. What are the monthly repayments, rounded to the next dollar?

A $379 B $367 C $361 D $375

(d) A loan of $8000 is to be repaid in equal monthly instalments over 6 years. The interest rate is 9% per annum. What are the monthly repayments, rounded to the next dollar?

A $146 B $143 C $140 D $145

9 The table gives the contribution per period for an annuity with a future value of $1 at different interest rates and different periods of time.

Contributions per period for an annuity with a future value of $1						
Number of periods	**Interest rate (% per period)**					
	0.25%	**0.5%**	**0.75%**	**1%**	**1.25%**	**1.5%**
1	1.0000	1.0000	1.0000	1.0000	1.0000	1.0000
3	0.3325	0.3317	0.3308	0.3300	0.3292	0.3284
6	0.1656	0.1646	0.1636	0.1625	0.1615	0.1605
9	0.1100	0.1089	0.1078	0.1067	0.1057	0.1046
12	0.0822	0.0811	0.0800	0.0788	0.0778	0.0767
15	0.0655	0.0644	0.0632	0.0621	0.0610	0.0599
18	0.0544	0.0532	0.0521	0.0510	0.0499	0.0488
21	0.0464	0.0453	0.0441	0.0430	0.0419	0.0409
24	0.0405	0.0393	0.0382	0.0371	0.0360	0.0349

Use the table to answer the following questions.

(a) Barbie and Ken need to save \$100 000 over 3 years for a deposit on a new house. They make regular quarterly contributions into an investment account which pays interest at 4% p.a. How much do they need to contribute each quarter to reach this savings goal?

(b) Beatrice is saving for a deposit to buy a car. She needs to save \$3000 in a year. How much must she pay into an investment account each month, if the interest is 3% p.a., calculated monthly, to reach this goal?

(c) Danh is starting a superannuation fund. He wishes to have \$1 000 000 in the account after 12 years. Interest is 3% p.a. If he makes a deposit to this account every six months, how much should he deposit, rounded to the next dollar, to be sure that he reaches his goal?

Use the table of values given in question **5** to answer question **10**.

10 Fabio makes a one-off deposit of \$3000 into an investment account that pays interest at the rate of 3% p.a., compounded yearly.

(a) How much money is in the account after 3 years?

(b) As an incentive, the bank said that they would double the interest rate after 3 years. If Fabio leaves the money in this account for another 3 years, what is the value of the investment at the end of this time?

(c) Another bank offered Fabio an interest rate of 4% p.a., compounded six monthly for the first 3 years and then an interest rate of 5% p.a. compounded yearly for the next 3 years. What is the value of the investment at the end of 6 years with this bank?

(d) Which investment strategy gives the best result?

11 What is the effective annual interest rate for a loan advertised as:

(a) 4% p.a. compounded quarterly (b) 4% p.a. compounded monthly

(c) 4% p.a. compounded daily (use 365 days in a year)?

12 A special investment account from Bank B is advertised with an interest rate of 5.1% compounded monthly. Credit Union C advertises their investment account with an interest rate of 5% compounded daily. Which financial institution offers the better return?

13 What is the effective annual interest rate for each term deposit at:

(a) 2.6% p.a. compounded quarterly (b) 2.5% p.a. compounded monthly

(c) 2.4% p.a. compounded daily (use 365 days in a year)?

14 The home loan from Building Society N is advertised with an interest rate of 5.4% compounded monthly. Credit Union T advertises their home loan with an interest rate of 5.35% compounded daily. If there are no other fees involved, which financial institution offers the better deal?

18.2 ARITHMETIC SEQUENCES

Sequences

The word **sequence** is often used in everyday language. You say 'a sequence of events'; you complete tasks in a certain sequence; the order of chapters in this book is the sequence in which the Mathematics syllabus is treated.

Whenever the word 'sequence' is used, you are considering a set of objects, ideas, steps or events in some definite order. This order can be associated with the set of positive integers $\{1, 2, 3, \ldots\}$, as is indicated when you use the terms first, second, third etc. to describe a position in the sequence of events.

An **arithmetic sequence** is a sequence of terms in which each term after **the first term** is formed by adding a constant number, called the **common difference**, to the preceding term.

For arithmetic sequences: a is the first term, d is the common difference and n is the number of terms.

The sequence $a, a + d, a + 2d, \ldots, a + (n - 1)d$ is an arithmetic sequence with n terms, where a and d are real numbers and n is a positive integer. The value of each term can be labelled as T_n and defined as follows:

$$T_n = T_{n-1} + d$$

This definition is called a **recursive definition** because it defines each term of the sequence as a function of the preceding term.

Rewriting this definition gives $d = T_n - T_{n-1}$ and this result can be used to identify whether a sequence is arithmetic. If there is a constant difference between all successive terms then the sequence is arithmetic.

An arithmetic sequence may be defined by its nth term, given by:

$T_n = a + (n - 1)d$

Graphically, T_n as a function of n is linear, consisting of a series of points for integer values of n, $n > 0$, which follow the straight line given by $T_n = a - d + nd$. The independent variable is n and the dependent variable is T_n.

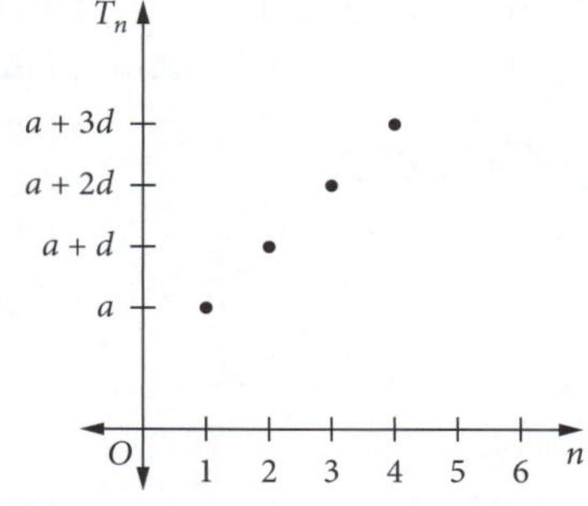

Some examples of arithmetic sequences are:

$2, 5, 8, 11, \ldots$	$a = 2, d = 3$
$12, 5, -2, -9, \ldots$	$a = 12, d = -7$
$\pi + 1, 2\pi + 4, 3\pi + 7, 4\pi + 10, \ldots$	$a = \pi + 1, d = \pi + 3$
$2 + \sqrt{3}, 2 + 2\sqrt{3}, 2 + 3\sqrt{3}, 2 + 4\sqrt{3}, \ldots$	$a = 2 + \sqrt{3}, d = \sqrt{3}$

Example 9

By finding the difference between successive terms, identify the sequences that are arithmetic in the following:

(a) $11, 15, 19, 23, 27, \ldots$ **(b)** $10, 12, 15, 19, 24, \ldots$ **(c)** $22, 16, 10, 4, -2, \ldots$

(d) $8, 6, 8, 6, 8, \ldots$ **(e)** $1\frac{1}{2}, 2\frac{1}{4}, 3, 3\frac{3}{4}, 4\frac{1}{2}, \ldots$

Solution

(a) $15 - 11 = 19 - 15 = 23 - 19 = 27 - 23 = 4$, a constant. The sequence is arithmetic.

(b) $12 - 10 = 2$, $15 - 12 = 3$, $19 - 15 = 4$, $24 - 19 = 5$. The difference between successive terms is not constant. The sequence is not arithmetic, even though it follows a pattern.

(c) $16 - 22 = 10 - 16 = 4 - 10 = -2 - 4 = -6$, a constant. The sequence is arithmetic.

(d) $6 - 8 = -2$, $8 - 6 = 2$. The difference between successive terms is not constant. The sequence is not arithmetic, even though it follows a pattern.

(e) $2\frac{1}{4} - 1\frac{1}{2} = 3 - 2\frac{1}{4} = 3\frac{3}{4} - 3 = 4\frac{1}{2} - 3\frac{3}{4} = \frac{3}{4}$, a constant. The sequence is arithmetic.

You can use a spreadsheet to check for an arithmetic sequence. First, enter the terms into a column. In the next column, starting next to the second term, enter a formula to represent $T_n - T_{n-1}$ e.g. '=A2–A1'. Fill down to copy this formula down the column, to the end of the list of terms. As this formula finds the difference between each pair of terms, it should be obvious whether or not there is a common difference.

Generating an arithmetic sequence

Use spreadsheet software to generate consecutive terms of an arithmetic sequence.

Example 10

For the arithmetic sequence 4, 9, 14, 19, … find:

(a) the value of a **(b)** the value of d **(c)** the expression for T_n

(d) the 13th term **(e)** the value of k if $T_k = 99$.

Solution

(a) $a = 4$ (b) $d = 9 - 4 = 5$ (c) $T_n = a + (n-1)d$
$4 + 5(n-1) = 5n - 1$
$T_n = 5n - 1$
(d) $T_{13} = 4 + 12 \times 5 = 64$ (e) $99 = 5k - 1$
$100 = 5k$
$k = 20$

Example 11

If $T_5 = 7$ and $T_9 = -5$ are two terms of an arithmetic sequence, find the values of a and d and use them to write the first three terms of the sequence.

Solution

$T_5 = 7$:	$a + 4d = 7$	[1]
$T_9 = -5$:	$a + 8d = -5$	[2]
[1] – [2]:	$-4d = 12$	
	$d = -3$	
Substitute into [1]:	$a - 12 = 7$	
	$a = 19$	

The sequence begins 19, 16, 13.

EXERCISE 18.2 ARITHMETIC SEQUENCES

1 Which of the following are arithmetic sequences?

(a) 7, 17, 27, 37, … (b) 5, 2, −1, −4, … (c) $\frac{1}{2}, \frac{1}{3}, \frac{1}{4}, \frac{1}{5}, \ldots$

(d) $\frac{5}{8}, 1, 1\frac{3}{8}, 1\frac{3}{4}, \ldots$ (e) $\sqrt{2} - 1, \sqrt{2} + 1, \sqrt{2} + 3, \sqrt{2} + 5, \ldots$ (f) $\pi, \pi^2 + 1, \pi^3 + 2, \pi^4 + 3, \ldots$

2 For the arithmetic sequence 5, 8, 11, 14, … find:

(a) the value of a (b) the value of d (c) the expression for T_n
(d) the 13th term (e) the value of k if $T_k = 98$.

3 For the arithmetic sequence 7, 3, −1, −5, … find:

(a) the value of a (b) the value of d (c) the expression for T_n
(d) the 13th term (e) the value of k if $T_k = -101$.

4 Find the eighth and fourteenth terms of the arithmetic sequence 8, 14, 20, 26, … .

5 For the arithmetic sequence 17.2, 16.6, 16, 15.4, … , find T_6 and T_{11}.

6 The first and second terms of an arithmetic sequence are p and q respectively. Write an expression for the tenth term.

7 Find T_{n+2} for the sequence 14, 11, 8, 5, … .

8 Find the arithmetic sequence in which $T_5 = 17$ and $T_{12} = 52$.

9 Find T_6 of the arithmetic sequence in which $T_3 = 5.6$ and $T_{12} = -7$.

10 Write an expression for the seventh term of an arithmetic sequence in which the fifth term is m and the eleventh is n.

11 Find the value of p so that $p + 5$, $4p + 3$, $8p - 2$ form the first three terms of an arithmetic sequence.

12 How many terms are there in the arithmetic sequence 9, 12, 15, 18, …, $(6p + 15)$?

13 The first term of an arithmetic sequence is −8 and the seventh term is 22. Find the missing five terms of the sequence.

14 In an arithmetic sequence the first term is 6 and the fifth term is twice the fourth term. Find the common difference.

15 Given 36, 31, 26, are the first three terms of an arithmetic sequence, find the value of n if the nth term is −4.

16 Cans of juice in a supermarket display are stacked so that there are 3 cans in the top row, 5 in the next row, 7 in the next and so on. If there are 12 rows in the display, how many cans of juice are in the bottom row of the display?

17 The lengths of the sides of a right-angled triangle form the terms of an arithmetic sequence. If the hypotenuse is 25 cm long, find the lengths of the other two sides.

18 What is the sixth term of the arithmetic sequence $5\sqrt{2}-\sqrt{3}, 3\sqrt{2}, \sqrt{3}+\sqrt{2}$?

A $2\sqrt{3}-3\sqrt{2}$ **B** $3\sqrt{3}-4\sqrt{2}$ **C** $4\sqrt{3}-5\sqrt{2}$ **D** $5\sqrt{3}-6\sqrt{2}$

19 Alexandra starts a new job with an initial salary of $28 000 per annum. She is promised an increase of $300 per quarter for the first 4 years of her employment.

(a) What will her salary be in 3 months time?
(b) Write an expression for her salary, S, in dollars after n quarters.
(c) What is the domain of the function $S(n)$?
(d) What is the maximum salary that she can expect to receive?

20 $1000 is borrowed for two years at a simple interest rate of 0.5% per month, with the interest added monthly.

(a) How much interest is added each month?
(b) How much is owed after 1 month?
(c) How much is owed after one year?
(d) How much is to be paid back at the end of two years?

18.3 SERIES AND SIGMA NOTATION (Σ)

Series

When you write a set of numbers in order according to some rule, the addition of all numbers is called a **series** and each number in the set is called a **term** of the series. In other words, a series is the sum of the terms of its corresponding sequence.

For example, $1+4+9+16+\ldots+n^2$ is a series in which the first term is 1, the second term is 4, the third term is 9 and the nth term is n^2.

The symbols T_n or u_n are commonly used to represent the nth term of a series. Thus in the series $1+4+9+16+\ldots+n^2$, $T_1=1$, $T_2=4$, $T_3=9$ and $T_n=n^2$. T_n is also called the **general term** of the series, because it allows you to find any other term of the series according to its rule.

$S_n=T_1+T_2+T_3+\ldots+T_n$ is the sum of the series.

A useful result is that $S_{n-1}=T_1+T_2+T_3+\ldots+T_{n-1}$

so that $S_n=S_{n-1}+T_n$

and hence $T_n=S_n-S_{n-1}$

Example 12

What are the first three terms of the series whose general term is given by $T_n=n^2+2$?

Solution

$T_1=1^2+2=3$ $\quad T_2=2^2+2=6$ $\quad T_3=3^2+2=11$

This series can be written as $S_n=3+6+11+\ldots+(n^2+2)$.

Example 13

Find the nth term of the series $4 + 7 + 10 + 13 + \ldots$

Solution

Because the terms increase by 3 each time, you can rewrite each term as a number plus an increasing multiple of 3:

$$\begin{aligned} T_1 &= 4 & T_2 &= 7 & T_3 &= 10 \\ &= 1 + 3 & &= 1 + 6 & &= 1 + 9 \\ &= 1 + 3 \times 1 & &= 1 + 3 \times 2 & &= 1 + 3 \times 3 \end{aligned}$$

$\therefore T_n = 1 + 3n = 3n + 1$

Sigma notation (Σ)

The symbol Σ (the Greek capital letter sigma) is used in mathematics to mean 'the sum of'. When you sum the terms of a sequence you have a series, so sigma notation provides a shorter way to write the series.

You can rewrite the series discussed above using sigma notation:

$$\sum_{k=1}^{n} k^2 = 1 + 4 + 9 + 16 + \ldots + n^2$$

$$\sum_{k=1}^{n} \left(k^2 + 2\right) = 3 + 6 + 11 + \ldots + (n^2 + 2)$$

$$\sum_{k=1}^{n} (3k+1) = 4 + 7 + 10 + \ldots + (3n + 1)$$

Each sigma means 'the sum of these terms, starting with the value underneath the Σ and continuing to the value above the Σ'. For example:

- $\sum_{n=1}^{5} x^n$ denotes the sum of terms of the form x^n, where n has the values 1, 2, 3, 4, 5

  Thus: $\sum_{n=1}^{5} x^n = x^1 + x^2 + x^3 + x^4 + x^5$

- $\sum_{k=0}^{n} (2k+1)$ denotes the sum of terms of the form $2k + 1$, where $k = 0, 1, 2, 3, \ldots n$

  (so k is an integer from 0 to n inclusive): $\sum_{k=0}^{n} (2k+1) = 1 + 3 + 5 + \ldots + (2n + 1)$

  The right-hand side is the expansion of the series defined by $\sum_{k=0}^{n} (2k+1)$.

- $\sum_{r=1}^{10} rx^{r-1} = 1 + 2x + 3x^2 + \ldots + 10x^9$

It is also useful to use sigma notation to obtain a shorter expression for the sum of a number of terms.

- $1^2 + 2^2 + 3^2 + \ldots + n^2 = \sum_{k=1}^{n} k^2$ The left-hand expression has n terms.
- $1 \times 2 + 2 \times 3 + 3 \times 4 + \ldots + 10 \times 11 = \sum_{n=1}^{10} n(n+1)$ The left-hand expression has 10 terms.
- $2 + 2x + 2x^2 + \ldots + 2x^8 = \sum_{k=0}^{8} 2x^k$ The left-hand expression has 9 terms.

Example 14

Evaluate: (a) $\sum_{n=2}^{5} n^2$ (b) $\sum_{n=1}^{5} (2n-1)$

Solution

Write the series in full, then add the terms.

(a) $\sum_{n=2}^{5} n^2 = 2^2 + 3^2 + 4^2 + 5^2$
$= 4 + 9 + 16 + 25$
$= 54$

(b) $\sum_{n=1}^{5} (2n-1) = 1 + 3 + 5 + 7 + 9$
$= 25$

EXERCISE 18.3 SERIES AND SIGMA NOTATION (Σ)

1 Write expansions of the series defined by the following:

(a) $\sum_{k=1}^{4} k^2$ (b) $\sum_{r=0}^{5} 3^r$ (c) $\sum_{k=1}^{p} (2k-1)$ (d) $\sum_{k=1}^{5} k(k+1)$

(e) $\sum_{r=1}^{8} rx^r$ (f) $\sum_{r=1}^{k} \frac{1}{x^r}$ (g) $\sum_{k=1}^{p} (2k+1)^2$ (h) $\sum_{k=1}^{n+1} (3k-2)$

2 Indicate whether each statement below is a correct or an incorrect expression for $1 + x + x^2 + \ldots + x^{10}$.

(a) $\sum_{n=0}^{10} x^n$ (b) $\sum_{n=1}^{10} x^n$ (c) $\sum_{n=1}^{11} x^{n-1}$ (d) $\sum_{n=1}^{10} x^{n-1}$

3 Use sigma notation to represent each of the following:

(a) $1^2 + 2^2 + 3^2 + \ldots + 9^2$ (b) $1 \times 3 + 2 \times 4 + 3 \times 5 + \ldots + 10 \times 12$ (c) $1 + 6 + 11 + \ldots + (5p-4)$

(d) $\frac{1}{2 \times 3} + \frac{1}{3 \times 4} + \frac{1}{4 \times 5} + \ldots + \frac{1}{p(p+1)}$ (e) $2x^2 + 3x^3 + 4x^4 + \ldots + 12x^{12}$ (f) $a + ar + ar^2 + \ldots + ar^{n-1}$

4 Evaluate: (a) $\sum_{n=1}^{4} n^2$ (b) $\sum_{n=1}^{6} (2n+1)$ (c) $\sum_{k=1}^{4} (3k-2)$ (d) $\sum_{r=1}^{4} 2^r$

(e) $\sum_{n=0}^{5} n^2$ (f) $\sum_{n=0}^{5} (2n-1)$ (g) $\sum_{n=1}^{4} \left(n^2 + n\right)$ (h) $\sum_{n=1}^{6} (12-3n)$ (i) $\sum_{r=1}^{4} r^r$

18.4 ARITHMETIC SERIES

An **arithmetic series** is the sum of the terms in an arithmetic sequence. The series starts with the **first term**, which is usually denoted by a and is the first term of the sequence. The constant amount added each time to the term is called the **common difference** and is usually denoted by d.

Thus $S_n = a + (a+d) + (a+2d) + \ldots + (a+(n-1)d)$ is a series with a first term a and a common difference d.

- The nth term is $T_n = a + (n-1)d$.
- The common difference is given by $d = T_2 - T_1 = T_3 - T_2 = \ldots = T_n - T_{n-1}$.
- Also note that $S_n = T_1 + T_2 + T_3 + \ldots + T_n$.

We call S_n the 'sum to n terms' of the series. This is a finite series.

Using sigma notation: $S_n = \sum_{k=1}^{n} (a + (k-1)d)$

Sum of an arithmetic sequence

You can obtain a formula for S_n so that you don't always have to add the terms to find the sum. First, note that the nth term of a series is sometimes called the 'last' term and denoted by l, so $T_1 = a$ and $T_n = l = a + (n-1)d$.

Write: $S_n = a + (a+d) + (a+2d) + \ldots + (a+(n-3)d) + (a+(n-2)d) + (a+(n-1)d)$ [1]

Let $l = a + (n-1)d$, so: $S_n = a + (a+d) + (a+2d) + \ldots + (l-2d) + (l-d) + l$

Write this series in reverse: $S_n = l + (l-d) + (l-2d) + \ldots + (a+2d) + (a+d) + a$ [2]

Add the two expressions [1] + [2]: $2S_n = (a+l) + (a+l) + (a+l) + \ldots + (a+l) + (a+l) + (a+l)$

There are n factors of $(a+l)$: $2S_n = n(a+l)$

$$S_n = \frac{n}{2}(a+l)$$

But $l = a + (n-1)d$, so: $S_n = \frac{n}{2}(a+a+(n-1)d)$

$$\therefore S_n = \frac{n}{2}(2a+(n-1)d)$$

It is also important to realise that $S_n = S_{n-1} + T_n$, so $T_n = S_n - S_{n-1}$, for $n > 1$.

Summary of formulae

- $a + (a+d) + (a+2d) + \ldots + (a+(n-1)d)$ is an arithmetic series
- d is the common difference
- $T_1 = a$
- $T_n = a + (n-1)d$
- $S_n = \frac{n}{2}\{2a+(n-1)d\}$
- $S_n = \frac{n}{2}(a+l)$ where $l = a + (n-1)d$
- $T_n = S_n - S_{n-1}$, $n > 1$

Example 15

Find the sum of the first twenty terms of the series $3 + 5 + 7 + \ldots$

Solution

$5 - 3 = 7 - 5 = 2$, so the series is arithmetic with $a = 3$, $d = 2$

$n = 20$:
$$\begin{aligned} S_n &= \frac{n}{2}(2a+(n-1)d) \\ &= \frac{20}{2}(6+19\times 2) \\ &= 440 \end{aligned}$$

Example 16

How many terms of the series $4 + 7 + 10 + \ldots$ must be taken to give a sum of 531?

Solution

$7 - 4 = 10 - 7 = 3$, so the series is arithmetic with $a = 4$, $d = 3$

$S_n = 531$:
$$\begin{aligned} S_n &= \frac{n}{2}(2a+(n-1)d) \\ 531 &= \frac{n}{2}(8+(n-1)3) \\ 1062 &= n(3n+5) \\ 3n^2 + 5n - 1062 &= 0 \\ (3n+59)(n-18) &= 0 \\ n &= -19\tfrac{2}{3}, 18 \end{aligned}$$

n must be a positive integer, so $n = 18$ and 18 terms must be taken.

Example 17

Find the sum of the first twenty terms of an arithmetic series, given that the tenth term is 39 and the sum of the first ten terms is 165.

Solution

$T_{10} = 39$: $a + 9d = 39$ [1]

$S_{10} = 165$: $\frac{10}{2}(2a + 9d) = 165$

So: $2a + 9d = 33$ [2]

[2] − [1]: $a = -6$

Substitute into [1]: $-6 + 9d = 39$

$d = 5$

$S_n = \frac{n}{2}(2a + (n-1)d)$: $S_{20} = \frac{20}{2}(-12 + 19 \times 5) = 10(-12 + 95) = 830$

Example 18

Find the first three terms of the arithmetic series defined by $S_n = 2n^2 + n$.

Solution

$T_n = S_n - S_{n-1}$

$S_n = 2n^2 + n$

and $S_{n-1} = 2(n-1)^2 + (n-1)$

$S_{n-1} = 2n^2 - 3n + 1$

$\therefore T_n = 2n^2 + n - (2n^2 - 3n + 1)$

$T_n = 4n - 1$

$\therefore T_1 = 3, T_2 = 7, T_3 = 11$

Hence the series is $3 + 7 + 11 + \ldots$

Alternatively:

$S_n = 2n^2 + n$

$S_1 = 3 = T_1$

$S_2 = 10$

$T_2 = S_2 - S_1 = 10 - 3 = 7$

$S_3 = 21$

$T_3 = S_3 - S_2 = 21 - 10 = 11$

Hence the series is $3 + 7 + 11 + \ldots$

EXERCISE 18.4 ARITHMETIC SERIES

1 Find the sum of the first 16 terms of the arithmetic series $3 + 4\frac{1}{4} + 5\frac{1}{2} + \ldots$

2 Find the sum of the first 12 terms of the arithmetic series in which the first term is 8 and the twelfth term is 41.

3 Find the sum of all integers between 20 and 50 that are divisible by 3.

4 The first three terms of an arithmetic series are $-2 + 3 + 8 + \ldots$

(a) Find the 60th term. **(b)** Hence, or otherwise, find the sum of the first 60 terms of the series.

5 The sum of the first 40 terms of the arithmetic series $40 + 36 + 32 + \ldots$ is:

A 3920 **B** −1520 **C** −2320 **D** 4720

6 An object falling freely from a height travels 4.9 metres in the first second, 14.7 metres in the second second and 24.5 metres in the third second. How far has it fallen:

(a) after six seconds **(b)** between the fifth and the sixth second?

7 Find the sum of the first 50 terms of an arithmetic series, given that the fifteenth term is 34 and the sum of the first eight terms is 20.

8 Find the sum of the first 20 terms of an arithmetic series whose eighth term is 6 and whose twelfth term is 9.

9 Show that the sum of the first n odd positive integers is a perfect square.

10 Evaluate: **(a)** $\sum_{k=1}^{10}(3k-7)$ **(b)** $\sum_{k=1}^{8}(4k+1)$ **(c)** $\sum_{k=1}^{n}(4k-1)$

(d) $\sum_{k=1}^{n}(6k+2)$ **(e)** $\sum_{k=1}^{n}(2k-1)$ **(f)** $\sum_{k=1}^{10}(3k+2)$

11 The first term of an arithmetic series is 7, the common difference is 2 and the sum of the first n terms is 247. Find the value of n.

12 In an arithmetic series, $T_3 = -2$ and $T_9 = 28$. How many terms of this series are required to give a sum of 1092?

13 The sum of the first three terms of the arithmetic series $a + (a + d) + (a + 2d) + \ldots$ is 15 and the product of the three terms is 105. Indicate whether each statement below is a correct or incorrect step in finding the terms of the series.

(a) $a + d = 5$ **(b)** $a(5 + d) = 21$ **(c)** $a^2 - 10a + 21 = 0$ **(d)** The series is $3 + 5 + 7 + \ldots$

14 Find the sum of the integers between 0 and 101 that are:

(a) divisible by 2 **(b)** divisible by 5 **(c)** divisible by 2 and 5 **(d)** divisible by 2 or 5 but not both.

15 The sum of the first six terms of an arithmetic series is −12 and the sum of the first fourteen terms is 196. Find:

(a) the sum of n terms **(b)** the smallest value of n for the sum of the series to exceed 250.

16 The sum of the magnitudes of the angles of an irregular pentagon (five-sided polygon) is 540°. The magnitudes of the angles form the terms of an arithmetic series. If the largest angle has a magnitude of 136°, find the magnitude of each of the other four angles.

17 The sum of the first four terms of an arithmetic series is 34 and the sum of the next four terms is 146. Find the sum of the ninth and tenth terms.

18 The rungs of a ladder decrease in length uniformly from 40 cm at the bottom rung to 30 cm at the top rung. How many rungs are there if the total length of all the rungs is 5.25 m?

19 How many terms of the series $7 + 9 + 11 + \ldots$ must be added to give 352?

20 The first three terms of an arithmetic series are 45, 41, 37. If the nth term is 1, find the value of n and the sum of these n terms.

21 The track of a vinyl record is in the shape of a spiral curve, but it may be considered as a number of concentric circles of minimum and maximum radius 5.25 cm and 10.5 cm respectively. The record rotates at $33\frac{1}{3}$ revolutions per minute and takes 18 minutes to play from start to finish. Find an approximation to the length of the track.

22 Given $S_n = 3n^2 - 11n$, find T_n and hence show that the series is arithmetic.

23 A series is such that the sum of its first n terms is $n(3n + 2)$, where n is a positive integer. Prove that the series is arithmetic and find the eighth term.

24 The sum of the first nine terms of an arithmetic series is 81. If the sum of the first and third terms is zero, find the first term and the common difference.

25 Find the sum of all the integers between 200 and 400 that are divisible by 6.

26 The first and second terms of an arithmetic series are a and b respectively. If the nth term is c, express n in terms of a, b and c, and hence find the sum of these n terms.

27 Find the first three terms of an arithmetic series in which the fifth term is three times the second term, and the sum of the first six terms is 36.

28 Logs of wood are stacked in a pile so that there are 15 logs on the top row, 16 on the next row, 17 on the next, and so on. If there are 246 logs altogether:
(a) how many rows are there (b) how many logs are on the bottom row?

29 The lengths of the sides of a right-angled triangle form the terms of an arithmetic series. If the hypotenuse is 15 cm, what is the length of each of the other two sides?

30 How many terms of the series $6 + 10 + 14 + \ldots$ must be taken to give a sum of 880?

31 The sum of the first n terms of the series $30 + 26 + 22 + \ldots$ is 120. Find two possible values of n.

32 Find the sum of: (a) the first n odd positive integers (b) the first n even positive integers
(c) the first n positive integers, and find this value of n if the sum is 210.

33 For a potato race, a straight line is marked on the ground from a point A and points $B, C, D, \ldots$ are marked on the line so that $AB = BC = CD = \ldots = 2$ metres. A potato is placed at each of the points $B, C, D, \ldots$ A runner has to start from A and bring each potato back to a basket at A, carrying only one potato at a time. Find the number of potatoes so that the total distance run during the race will be 480 metres.

34 Cans of fruit in a supermarket display are stacked so that there are 3 cans in the top row, 5 in the next row, 7 in the next row and so on. If there are 10 rows in the display, find:
(a) the number of cans in the bottom row (b) the total number of cans in the display.

35 The first term of an arithmetic series is 5. The ratio of the sum of the first four terms to the sum of the first ten terms is 8 : 35. Find the common difference.

36 The angles of an irregular hexagon form the terms of an arithmetic series. If the smallest angle is 95°, find the size of each of the other angles. (Hint: you need to know the angle sum of a hexagon.)

18.5 GEOMETRIC SEQUENCES

A **geometric sequence** is a sequence of terms in which each term after the first term is formed by multiplying the preceding term by a constant number called the **common ratio**.

a is the first term, r is the common ratio and n is the number of terms.

The sequence $a, ar, ar^2, ar^3, \ldots, ar^{n-1}$ is a geometric sequence with n terms, where a and r are real numbers and n is a positive integer.

$$T_n = rT_{n-1}$$

This definition is called a recursive definition because it defines a term of the sequence in terms of the preceding term. Rewriting this definition gives $r = \frac{T_n}{T_{n-1}}$ and this result is used to identify whether a sequence is geometric. If there is a constant ratio between successive terms then the sequence is geometric.
A geometric sequence may be defined by its nth term, given by the following equation.

$$T_n = ar^{n-1}$$

Graphically, T_n as a function of n is exponential in shape, consisting of a series of points for integer values of n, $n > 0$, which follow the exponential function given by $T_n = ar^{n-1}$, where the independent variable is n and the dependent variable is T_n.

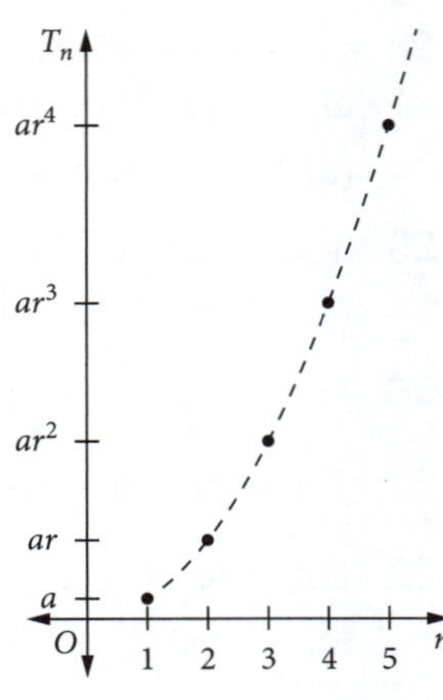

Some examples of geometric sequences are:

$2, 6, 18, 54, \ldots$	$a = 2, r = 3$
$3, -15, 75, -375, \ldots$	$a = 3, r = -5$
$18, 9, 4\frac{1}{2}, 2\frac{1}{4}, \ldots$	$a = 18, r = \frac{1}{2}$
$3, 3\sqrt{2}, 6, 6\sqrt{2}, \ldots$	$a = 3, r = \sqrt{2}$
$x, x^2, x^3, x^4, \ldots$	$a = x, r = x$

EXPLORE FURTHER

Generating a geometric sequence

Use spreadsheet software to generate consecutive terms of a geometric sequence.

Example 19

By finding the ratio between successive terms, identify the sequences that are geometric.

(a) $2, 4, 8, 16, \ldots$ **(b)** $3, 6, 24, 192, \ldots$ **(c)** $48, 24, 12, 6, \ldots$

(d) $x, x^2 + 1, x^3 + 2, x^4 + 3, \ldots$ **(e)** $\frac{1}{\sqrt{2} - 1}, 3 + 2\sqrt{2}, 7 + 5\sqrt{2}, 17 + 12\sqrt{2}, \ldots$

Solution

(a) $\frac{4}{2} = \frac{8}{4} = \frac{16}{8} = 2$

$r = 2$, a constant ratio. The sequence is geometric.

(b) $\frac{6}{3} = 2, \frac{24}{6} = 4$

The ratio is not constant so the sequence is not geometric.

(c) $\frac{24}{48} = \frac{12}{24} = \frac{6}{12} = \frac{1}{2}$

$r = \frac{1}{2}$, a constant ratio. The sequence is geometric.

(d) $\frac{x^2 + 1}{x} \neq \frac{x^2 + 2}{x^2 + 1}$

The ratio is not constant so the sequence is not geometric.

(e) $\frac{3 + 2\sqrt{2}}{\left(\frac{1}{\sqrt{2} - 1}\right)} = (3 + \sqrt{2})(\sqrt{2} - 1) = \sqrt{2} + 1$

$\frac{7 + 5\sqrt{2}}{3 + 2\sqrt{2}} = \frac{(7 + 5\sqrt{2})(3 - 2\sqrt{2})}{9 - 8} = 21 - 14\sqrt{2} + 15\sqrt{2} - 20 = \sqrt{2} + 1$

$\frac{17 + 12\sqrt{2}}{7 + 5\sqrt{2}} = \frac{(17 + 12\sqrt{2})(7 - 5\sqrt{2})}{49 - 50} = -(119 - 85\sqrt{2} + 84\sqrt{2} - 120) = \sqrt{2} + 1$

$r = \sqrt{2} + 1$, a constant ratio. The sequence is geometric.

Example 20

For the geometric sequence 2, 6, 18, 54, … find:

(a) the value of a (b) the value of r (c) the expression for T_n
(d) the 10th term (e) the value of k if $T_k = 4374$.

Solution

(a) $a = 2$ (b) $r = \frac{6}{2} = 3$ (c) $T_n = ar^{n-1} = 2 \times 3^{n-1}$

(d) $n = 10$
$T_{10} = 2 \times 3^9$
$= 39\,366$

(e) $2 \times 3^{k-1} = 4374$
$3^{k-1} = 2187 = 3^7$
$k - 1 = 7$
$k = 8$

Example 21

If $T_3 = 12$ and $T_6 = 96$ are two terms of a geometric sequence, find the values of a and r and use them to write down the first three terms of the sequence.

Solution

$T_3 = 12 = ar^2$ [1]

$T_6 = 96 = ar^5$ [2]

[2] ÷ [1]: $\frac{ar^5}{ar^2} = \frac{96}{12}$

$r^3 = 8$

$r = 2$

$a \times 2^2 = 12$

$a = 3$

The geometric sequence begins: 3, 6, 12.

EXERCISE 18.5 GEOMETRIC SEQUENCES

1 Which of the following are geometric sequences?

(a) 3, 6, 12, 24, … (b) $8, -2, \frac{1}{2}, -\frac{1}{8}, \ldots$ (c) 2, 5, 11, 23, … (d) $\frac{1}{9}, \frac{1}{3}, 1, 3, \ldots$ (e) $\frac{1}{2}, \frac{1}{4}, \frac{1}{6}, \frac{1}{8}, \ldots$

2 For the geometric sequence 1, 3, 9, 27, … find:

(a) the value of a (b) the value of r (c) the expression for T_n
(d) the 10th term (e) the value of k if $T_k = 6561$.

3 For the geometric sequence 144, 48, 16, $5\frac{1}{3}$, … find:

(a) the value of a (b) the value of r (c) the expression for T_n
(d) the 10th term (e) the value of k if $T_k = \frac{16}{81}$.

4 Find the sixth term of the sequence 4, 6, 9, 13.5, … .

5 Find the first term and the common ratio of the geometric sequence in which $T_3 = 25$ and $T_5 = 156.25$.

6 Find the first term and the common ratio of the geometric sequence in which $T_6 = \frac{32}{9}$ and $T_7 = \frac{64}{27}$.

7 Find the first three terms of a geometric sequence in which:

(a) the sixth term is 160 and the seventh term is 320

(b) the fifth term is 4 and the eighth term is $-\frac{1}{2}$.

8 If $2p + 1$, $5p$ and $12p - 4$ are successive terms of a geometric sequence, find the value of p.

9 If $2k + 3$, $5k + 1$ and $8k - 1$ are three successive terms of a geometric sequence, then the value of k is:

A $\frac{3}{2}$ B $-\frac{3}{2}$ C $-\frac{2}{3}$ D $\frac{2}{3}$

10 For x and y it is given that 1, x and y form an arithmetic sequence while 1, y and x form a geometric sequence. Find the value of x and y.

11 The first two terms of a geometric sequence are $\sqrt{7} + \sqrt{5}$ and $\sqrt{7} - \sqrt{5}$. Find the common ratio, expressing in its simplest surd form with a rational denominator.

12 The population of a certain town is 10 000. If the population decreases each year by 10% of the population in the preceding year, find the population in 5 years' time.

13 The value of a car when new is $40 000. It depreciates at the rate of 15% of its value at the beginning of each year. Find its value after 6 years.

14 Given that $a, ar, ar^2, ar^3, \ldots$ is a geometric sequence, show that the sequence $\log a, \log ar, \log ar^2, \log ar^3, \ldots$ is arithmetic.

15 Find a number which, when added to each of 2, 6, 13 gives three numbers in geometric sequence.

16 In a homogeneous nutrient medium the number of bacteria present doubles every hour. If there are initially 200 bacteria present, how many will be present after:

(a) 4 hours (b) 10 hours?

(c) If there are N bacteria present after t hours, write down a formula that gives the number of bacteria present after t hours.

17 The isotope carbon-11 is used in the treatment of cancers. It has a half-life of 20 minutes. This means that after 20 minutes only half the initial amount of carbon-11 remains.

(a) What fraction of the initial dose of carbon-11 remains after 1 hour?

(b) When will there only be 0.025% of the initial dose left in a patient's body?

18 The population of a colony of wading birds is decreasing by 15% each year. The initial population is 30 000 birds.

(a) Find how many birds remain after 5 years (to the nearest thousand).

(b) Find when the population is equal to 9000.

(c) Sketch the graph of the population as a function of time.

18.6 FINITE GEOMETRIC SERIES

A **geometric series** is a set of terms in which each term is formed by multiplying the preceding term by a constant number. The series starts with the **first term**, which is usually denoted by a. The constant multiplier is called the **common ratio** and is usually denoted by r.

The 'sum to n terms' of a geometric series is given by $S_n = a + ar + ar^2 + \ldots + ar^{n-1}$.

- The nth term of this series is $T_n = ar^{n-1}$
- The common ratio is given by $r = \frac{T_2}{T_1} = \frac{T_3}{T_2} = \frac{T_4}{T_3} = \ldots - \frac{T_n}{T_{n-1}}$
- Also note that $S_n = T_1 + T_2 + T_3 + \ldots + T_n$

You call S_n the 'sum to n terms' of the series. This is a finite series.

$S_\infty = T_1 + T_2 + T_3 + \ldots + T_n + \ldots$ is an infinite geometric series.

Using sigma notation, we have: $S_n = \sum_{k=1}^{n} ar^{k-1}$

Thus $T_1 = a$ and $T_n = ar^{n-1}$.

Sum of a geometric series

You can obtain a formula for S_n so that you don't always have to add the terms to find the sum.

Write: $S_n = a + ar + ar^2 + \ldots + ar^{n-1}$ [1]

Multiply both sides of [1] by r: $rS_n = ar + ar^2 + \ldots + ar^{n-1} + ar^n$ [2]

[1] – [2]: $S_n - rS_n = a - ar^n$

$S_n(1 - r) = a(1 - r^n)$

Hence: $S_n = \dfrac{a(1-r^n)}{1-r}$ for $r < 1$

Alternatively, finding [2] – [1] gives: $S_n = \dfrac{a(r^n-1)}{r-1}$ for $r > 1$

Note that when $r = 1$, the series becomes $a + a + a + \ldots$ to n terms and $S_n = na$.

It is also important to realise that $S_n = S_{n-1} + T_n$, so $T_n = S_n - S_{n-1}$, $n > 1$.

Summary of formulae

- $a + ar + ar^2 + \ldots + ar^{n-1}$ is a geometric series
- r is the common ratio
- $T_1 = a$
- $T_n = ar^{n-1}$
- $S_n = \dfrac{a(1-r^n)}{1-r}$ for $r < 1$
- $S_n = \dfrac{a(r^n-1)}{r-1}$ for $r > 1$
- $T_n = S_n - S_{n-1}$, $n > 1$

Example 22

Find the sum of the first eight terms of the geometric series $3 + 6 + 12 + \ldots$

Solution

$a = 3, r = 2, n = 8$: $S_n = \dfrac{a(r^n-1)}{r-1}$

$S_8 = \dfrac{3(2^8-1)}{2-1}$

$S_8 = 3(2^8 - 1) = 765$

Example 23

For the series $4 + 2 + 1 + \ldots$, find S_6 and S_{10}.

Solution

$\dfrac{T_2}{T_1} = \dfrac{2}{4} = \dfrac{1}{2}$ and $\dfrac{T_3}{T_2} = \dfrac{1}{2}$, so the series is geometric with $a = 4$ and $r = \dfrac{1}{2}$.

Now: $S_n = \dfrac{a(1-r^n)}{1-r}$

For $n = 6$: $S_6 = \dfrac{4\left(1-\frac{1}{2^6}\right)}{1-\frac{1}{2}}$

$S_6 = 8\left(1-\frac{1}{64}\right) = 8\times\left(\frac{64-1}{64}\right)$

$= 8\times\frac{63}{64} = \frac{63}{8} = 7\frac{7}{8}$

For $n = 10$: $S_{10} = \dfrac{4\left(1-\frac{1}{2^{10}}\right)}{1-\frac{1}{2}}$

$S_{10} = 8\left(1-\frac{1}{1024}\right) = 8\times\left(\frac{1024-1}{1024}\right)$

$= 8\times\frac{1023}{1024} = \frac{1023}{128} = 7\frac{127}{128}$

These values of S_n for $n = 6$ and $n = 10$ appear to be getting closer to 8.

The graph of $S_n = 8\left(1-\frac{1}{2^n}\right)$ for $n = 1, 2, 3, \ldots 10$ is shown here.

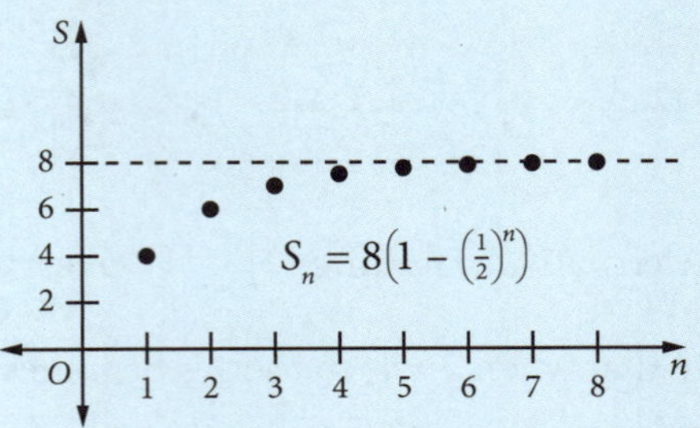

EXERCISE 18.6 FINITE GEOMETRIC SERIES

1 Find the sum of the first six terms of the series $4 + 6 + 9 + \ldots$

2 Find the sum of the first ten terms of the series $8 - 4 + 2 - \ldots$

3 The sum of the first ten terms of the series $2 + 2\sqrt{3} + 6 + \ldots$ is:

A $242(\sqrt{3}+1)$ B $162\sqrt{3}$ C $244(\sqrt{3}+1)$ D $121(\sqrt{3}+1)$

4 Evaluate the series $16 - 8 + 4 - 2 + \ldots + \frac{1}{16}$.

5 Find the sum of nine terms of the series $3 + 3^{\frac{4}{3}} + 3^{\frac{2}{3}} + \ldots$

6 The first four terms of a series are $x + \frac{2x^2}{3} + \frac{4x^3}{9} + \frac{8x^4}{27}$. Find the sum of n terms and evaluate the sum when $n=6$ and $x = 3$.

7 How many terms of the series $6 + 3 + \frac{3}{2} + \ldots$ must be taken to give a sum of $11\frac{13}{16}$? Indicate whether each statement below is a correct or incorrect step in solving this problem.

(a) $a = 6, r = \frac{1}{2}$ (b) $\frac{189}{16} = 12\left(1-\frac{1}{2^n}\right)$ (c) $\frac{1}{2^n} = 1 + \frac{63}{64}$ (d) $n = 6$

8 Find the sum of the first ten terms of the series $\log_{10} 3 + \log_{10} 6 + \log_{10} 12 + \ldots$

9 The sum of the first eight terms of a geometric series is seventeen times the sum of its first four terms. Find the common ratio.

10 An A0-sized sheet of paper is folded along its long side and then cut to create two sheets of A1-sized paper. Each sheet of A1-sized paper is folded along its long side and then cut to create two sheets of A2-sized paper. This process is repeated many times.

(a) How many sheets of A3-sized paper are created?

(b) You have eleven sheets of A0-sized paper. You cut each sheet into one of the eleven different sizes A0, A1, A2, … A10, creating as many sheets of each size as possible from each sheet of A0.

(i) How many sheets of A10 are created?

(ii) You stack all the sheets of paper on top of each other. with the A0 sheet on the bottom and the A10 sheets on top. How many sheets of paper are in the pile?

(iii) If a pack of 500 sheets of A4 paper is 55 mm thick, approximately how high is the stack of sheets in part **(ii)**?

11 Find the sum of the first seven terms of the series $\log_{10} 27 + \log_{10} 9 + \log_{10} 3 + \ldots$

18.7 INFINITE GEOMETRIC SERIES

In Example **23** you observed that for $a = 4, r = \frac{1}{2}$:

$$S_1 = 4, S_2 = 7, S_3 = 7, S_4 = 7\frac{1}{2}, S_5 = 7\frac{3}{4}, S_6 = 7\frac{7}{8}, S_7 = 7\frac{15}{16}, S_{10} = 7\frac{127}{128}$$

It appears that as n increases, $S_n = 7 +$ a fraction less than 1, so that as $n \to \infty$, $S_n \to 8$. Because the series approaches a limiting value, it is said that it **converges**.

You can define S_∞, the limiting sum of S_n, as $S_\infty = \lim_{n\to\infty} S_n$. Hence in this case $S_\infty = 8$.

Consider a piece of string 8 cm long. It is trimmed so that the first piece cut off is 4 cm long. The remainder is then cut in half so that the next piece is 2 cm long. The remainder is then cut in half so that the next piece is 1 cm long, and so on. You can imagine that there will always be a piece left over to be cut in half again, however small; but as the number of pieces cut off becomes increasingly large, their total length will get closer and closer to 8. That is:

$$S_\infty = 4 + 2 + 1 + \frac{1}{2} + \frac{1}{4} + \ldots = 8$$

You have seen that $S_n = \dfrac{a(1-r^n)}{1-r}$, which can be written as $S_n = \dfrac{a}{1-r} - \dfrac{ar^n}{1-r}$.

If $|r| < 1$, i.e. r is between -1 and 1, then $r^n \to 0$ as $n \to \infty$. In this example you are looking at $r = \frac{1}{2}$, so $r^n = \left(\frac{1}{2}\right)^n = \frac{1}{2^n}$:

n	1	4	10	$\to \infty$
$\frac{1}{2^n}$	$\frac{1}{2}$	$\frac{1}{16}$	$\frac{1}{1024}$	$\to 0$

As $n \to \infty$, $\frac{1}{2^n} \to 0$. Thus $\lim_{n\to\infty} r^n = 0$ and so $\lim_{n\to\infty} \frac{ar^n}{1-r} = 0$, which means that the geometric series converges for $|r| < 1$:

$$S_\infty = \frac{a}{1-r} \quad \text{for } |r| < 1$$

If $|r| > 1$ then $r^n \to \infty$ and there is no limiting sum, so the geometric series does **not** converge for $|r| > 1$.

Example 24

A rubber ball is dropped from a height of 20 m. Each time it strikes the ground it rebounds to $\frac{3}{4}$ of the height of the previous fall. Find the total number of metres it travels.

Solution

For the downward motion: $a = 20, r = \frac{3}{4}, n \to \infty, S_\infty = \frac{a}{1-r}$

$$S_\infty = \frac{20}{1-\frac{3}{4}} = 80$$

For the upward motion: $a = 15, r = \frac{3}{4}, n \to \infty, S_\infty = \frac{a}{1-r}$

$$S_\infty = \frac{15}{1-\frac{3}{4}} = 60$$

Total distance = 80 m + 60 m = 140 m

Example 25

Find the values of x for which the series $1 + (x-3)^2 + (x-3)^4 + \ldots$ converges.

Solution

For the series to be geometric, $a = 1$ and $r = (x-3)^2$.

Series converges for $|r| < 1$:

$$\begin{aligned}(x-3)^2 &< 1\\ (x-3)^2 - 1 &< 0\\ (x-3-1)(x-3+1) &< 0\\ (x-4)(x-2) &< 0\\ \therefore\ 2 < x &< 4\end{aligned}$$

Example 26

Express as a rational number: **(a)** $0.\dot{2}\dot{3}$ **(b)** $0.5\dot{7}$

Solution

(a) $0.\dot{2}\dot{3} = 0.23 + 0.0023 + 0.000\,023 + \ldots$

Geometric series with $a = 0.23$, $r = \frac{1}{100}$, $S_\infty = \frac{a}{1-r}$

$$0.\dot{2}\dot{3} = \frac{0.23}{1 - \frac{1}{100}} = \frac{23}{99}$$

(b) $0.5\dot{7} = 0.5 + 0.07 + 0.007 + 0.0007 + \ldots$

Geometric series with $a = 0.07$, $r = 0.1$, $S_\infty = \frac{a}{1-r}$

$$0.5\dot{7} = 0.5 + \frac{0.07}{1-0.1} = \frac{1}{2} + \frac{7}{90} = \frac{52}{90} = \frac{26}{45}$$

EXERCISE 18.7 INFINITE GEOMETRIC SERIES

1 Evaluate the following: **(a)** $8 - 4 + 2 - \ldots$ **(b)** $4 + 3 + 2\frac{1}{4} + \ldots$

(c) $25 - 10 + 4 + \ldots$ **(d)** $\left(\sqrt{3}+1\right) + 1 + \frac{\left(\sqrt{3}-1\right)}{2} + \ldots$

2 If $1 + 2x + 4x^2 + \ldots = \frac{3}{4}$, find the value of x.

3 Find the first three terms of a geometric series given that the sum of the first four terms is $21\frac{2}{3}$ and the sum to infinity is 27.

4 The limiting sum of the series $6 + 3 + \frac{3}{2} + \ldots$ is: **A** 10.5 **B** 11.25 **C** 11.625 **D** 12

5 The sum of the first four terms of a geometric series is 30 and the sum of the infinite series is 32. Find the first three terms.

6 Find the sum of the series $12 + 8 + 5\frac{1}{3} + \ldots$

7 Find the sum of the series $1 + \frac{1}{a+1} + \frac{1}{(a+1)^2} + \ldots$ For what values of a does this infinite series have a sum?

8 For the geometric series $\left(\sqrt{5}+\sqrt{3}\right)+\left(\sqrt{5}-\sqrt{3}\right)+\dfrac{\left(\sqrt{5}-\sqrt{3}\right)^2}{\sqrt{5}+\sqrt{3}}+\ldots$, find the common ratio and the sum of the infinite series.

9 Find the fractional equivalent of: **(a)** $2.3\dot{8}$ **(b)** $4.\dot{6}\dot{2}$ **(c)** $0.417\,17\ldots$

10 $0.323\,232\ldots = \dfrac{p}{q}$, where p and q are integers with no common factor. Find the value of p and q.

11 Show that 1.2888... is a rational number by expressing it in the form $\frac{m}{n}$, where m and n are integers with no common factor.

12 Evaluate $\dfrac{1+2+3+\ldots+10}{1+\frac{1}{2}+\frac{1}{4}+\ldots+\frac{1}{512}}$.

13 A ball dropped from a height of 12 metres rebounds from the floor, rising to a height of 8 metres. At subsequent rebounds the ball rises to a height equal to two-thirds of the previous height.

(a) To what height does the ball rise on the third rebound?

(b) By considering two infinite geometric series, find the total number of metres travelled by the ball before it comes to rest.

14 The Sierpinski triangle can be constructed by starting with an equilateral triangle, dividing it into four congruent equilateral triangles and then removing the central one. The process is then repeated in each of the remaining smaller triangles as shown in the diagram.

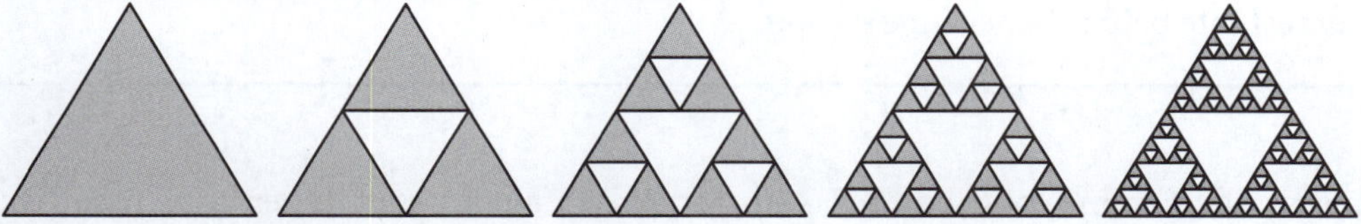

Let the initial triangle have an area of 1 square unit.

(a) How much of the original triangle remains shaded in the third diagram?

(b) How much of the original triangle remains shaded in the tenth diagram?

(c) Discuss what happens to the remaining area as the number of triangles increases.

MAKING CONNECTIONS

The Sierpinski triangle

Use technology to view successive iterations constructing a Sierpinski triangle.

18.8 COMPOUND INTEREST APPLICATIONS

A practical example of a geometric sequence is the growth of money invested at compound interest. If \$100 is invested at the start of a year at compound interest of 8% p.a. (where 'p.a.' = 'per annum' = per year), its value is:

$\$100 \times 1.08$ at the end of the first year,

$\$100 \times (1.08)^2$ at the end of the second year,

$\$100 \times (1.08)^3$ at the end of the third year,

$\$100 \times (1.08)^n$ at the end of n years.

In general, if $\$P$ is invested at compound interest of r% p.a., it grows to:

PR at the end of the first year, where $R = 1 + \dfrac{r}{100}$,

PR^2 at the end of the second year,

PR^3 at the end of the third year,

PR^n at the end of n years.

These amounts form a geometric sequence whose common ratio is R. Thus you have the compound interest formula:

$$A_n = PR^n \text{ or } A_n = P\left(1+\frac{r}{100}\right)^n$$

Here P is the initial amount (or 'principal') and A_n is the amount that P grows to after n periods of time, where interest is applied at r% per period. The period is not necessarily one year:

18% p.a. = 9% per 6 months = 4.5% per 3 months = 1.5% per month, etc.

(For compound interest, this division of interest into shorter time periods is not precisely correct, but for simplicity you will calculate it as shown above.)

A_n is the nth term of the sequence. The corresponding series is investigated in the following examples.

Recurrence relation

This result for compound interest may also be written as a recurrence relation.

If $P = A_0$

then $A_1 = A_0 \times R$

and $A_2 = A_1 \times R$

and eventually $A_n = A_{n-1} \times R$

where A_n is the value of the initial amount after n time periods and

$R = 1 + \frac{r}{100}$, the interest rate being r% per time period.

Example 27

At the beginning of each year, Clementine contributes \$500 to her investment account. If her investment account earns compound interest paid at 8% p.a.,

(a) Calculate the value of the first contribution (\$500) at the end of 10 years.

(b) Draw the graph of $A_n = 500 \times 1.08^n$ for $0 \le n \le 10$.

(c) Calculate the accumulated value of her total investment (all contributions and interest) at the end of 10 years.

Solution

(a) Value of first contribution at end of first year $= 500 \times 1.08$

Value of first contribution at end of second year $= 500 \times 1.08^2$

Value of first contribution at end of tenth year $= 500 \times 1.08^{10}$

$= 1079.46$

The value of Clementine's first contribution at the end of 10 years is \$1079.46.

(b)

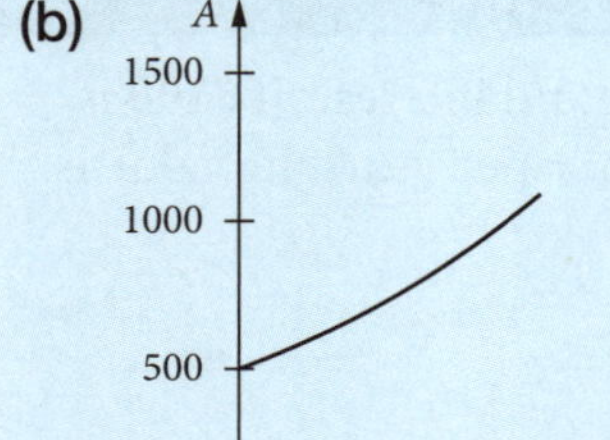

This graph shows that the amount is growing exponentially.

(c) Value of last contribution at end of tenth year $= 500 \times 1.08$ (1 year's interest)

Value of second-last contribution at end of tenth year $= 500 \times 1.08^2$ (2 years' interest)

Value of first contribution at end of tenth year $= 500 \times 1.08^{10}$ (10 years' interest)

The sum of all ten contributions $= 500 \times 1.08 + 500 \times 1.08^2 + 500 \times 1.08^3 + \ldots + 500 \times 1.08^{10}$

This is a finite geometric series of ten terms, with first term 500×1.08 and common ratio 1.08.

$$
\begin{aligned}
S_{10} &= \frac{a\left(R^{10}-1\right)}{R-1} \\
&= \frac{500 \times 1.08\left(1.08^{10}-1\right)}{1.08-1} \\
&= \frac{500 \times 1.08\left(1.08^{10}-1\right)}{0.08} \\
&= 7822.74
\end{aligned}
$$

The total investment is worth $7822.74 after 10 years.

In general, if someone invests $\$P$ at the beginning of each period of time at $r\%$ per period compound interest, the sum V of all investments at the end of n periods of time is given by:

$$
\begin{aligned}
V &= PR + PR^2 + PR^3 + \ldots + PR^n \\
&= PR(1 + R + R^2 + \ldots + R^{n-1}) \\
&= \frac{PR\left(R^n - 1\right)}{R-1} \qquad \text{where } R = 1 + \frac{r}{100}
\end{aligned}
$$

Note that the *process* of obtaining this result is the important thing to remember, not the result.

Example 28

Your employer offers you two different salary packages: Package 1 offers a raise of $1300 each year for 5 years, while Package 2 offers a raise of $100 per month for 5 years.

(a) Which package will give you the largest salary after 5 years?

(b) Package 3 offers an annual increase of 4%. What further information will you need to determine the best of all three packages?

Solution

(a) Package 1: Starting salary $= \$P$, increase $1300 p.a. for 5 years

Salary after 5 years $= P + 5 \times 1300 = \$(P + 6500)$

Package 2: Starting salary $= \$P$, increase $100 p.m. for 60 months

Salary after 5 years $= P + 60 \times 100 = \$(P + 6000)$

Package 1 offers the largest salary after 5 years.

(b) To compare the first two salary packages to a percentage increase, you will need to know the starting salary.

Example 29

Amira takes out an interest-only loan for $400 000 at an interest rate of 6% p.a. compounded monthly. She pays back the loan amount plus interest after 6 months.

(a) Set up the recurrence relation for this loan where A_0 is the amount borrowed and A_n is the amount owing after n months (i.e. n lots of interest added).

(b) How much does she pay back after 6 months?

(c) How much interest does she pay after 6 months?

Solution

(a) $A_0 = 400\,000$, $r = 6\%\text{ p.a.} = \frac{6}{12}\%\text{ p.m.}$
$= 0.5\%\text{ p.m.}$

$R = 1 + \frac{0.5}{100} = 1.005$

$A_1 = 400\,000 \times 1.005 = A_0 \times 1.005$

$A_2 = A_1 \times 1.005$

so $A_n = A_{n-1} \times 1.005$

(b) $n = 6$:

$A_6 = A_5 \times 1.005 = A_4 \times 1.005 \times 1.005 + A_4 \times 1.005^2$ etc.

This is obviously not an efficient way to find the amount.

Use $A_6 = A_0 \times 1.005^6$

$= 400\,000 \times 1.005^6$

$= 412\,151.00$

She pays back $412 151 after 6 months.

(c) Interest $= A_6 - A_0$

$= \$412\,151 - \$400\,000$

$= \$12\,151$

Example 30

Georgia borrows $5000 and agrees to make repayments of $100 at the end of each month, calculated from the date of the loan. Interest is charged on the unpaid debt at 1.5% per month.

(a) How much does Georgia still owe after the eighth repayment?

(b) How much time will it take to pay off the loan?

(c) If Georgia doubles the repayments, how much time will it take to pay off the loan?

Solution

(a) $P = 5000$, $r = 0.015$, $R = 1.015$

Amount owing after first repayment $= 5000 \times 1.015 - 100$

Amount owing after second repayment $= (5000 \times 1.015 - 100) \times 1.015 - 100$

$= 5000 \times 1.015^2 - 100(1 + 1.015)$

Amount owing after third repayment $= (5000 \times 1.015^2 - 100(1 + 1.015)) \times 1.015 - 100$

$= 5000 \times 1.015^3 - 100(1 + 1.015 + 1.015^2)$

Amount owing after eighth repayment $= 5000 \times 1.015^8 - 100(1 + 1.015 + 1.015^2 + \ldots + 1.015^7)$

Now $1 + 1.015 + 1.015^2 + \ldots + 1.015^7$ is a geometric series with $a = 1$, $r = 1.015$, $n = 8$.

$\therefore$ Amount owing after eighth repayment $= 5000 \times 1.015^8 - \dfrac{100\left(1.015^8 - 1\right)}{1.015 - 1}$

$= 5000 \times 1.015^8 - \dfrac{100\left(1.015^8 - 1\right)}{0.015}$

$= 4789.18$

Thus Georgia still owes $4789.18 after the eighth repayment.

(b) From part **(a)** you can see that after n repayments, the amount owing $= 5000 \times 1.015^n - \dfrac{100(1.015^n - 1)}{0.015}$

When the loan is paid off, the amount owing is 0, so you must solve $5000 \times 1.015^n - \dfrac{100(1.015^n - 1)}{0.015} = 0$

Solve:

$$0.015 \times 5000 \times 1.015^n - 100 \times 1.015^n + 100 = 0$$
$$1.015^n(100 - 75) = 100$$
$$1.015^n = 4$$
$$n\log_{10} 1.015 = \log_{10} 4$$
$$n = \frac{\log_{10} 4}{\log_{10} 1.015} = 93.1$$

Georgia would take 94 months (7 years 10 months) to pay off the loan. The final payment is less than \$100.

(c) As for part **(b)**, but replacing the repayment value 100 with 200:

Solve:

$$0.015 \times 5000 \times 1.015^n - 200 \times 1.015^n + 200 = 0$$
$$1.015^n(200 - 75) = 200$$
$$1.015^n = \frac{200}{125}$$
$$1.015^n = 1.6$$
$$n = \frac{\log_{10} 1.6}{\log_{10} 1.015} = 31.6$$

Georgia would take 32 months (2 years 8 months) to pay off the loan. Doubling the repayments reduces the time to pay off the loan by almost two-thirds.

The terms **Future Value** and **Present Value** are frequently used in financial circles.

The Future Value of an investment is what it grows to, that is A_n that has been calculated in previous examples.

The Present Value of an investment is the amount that you need to start with to obtain a particular amount in the future. This is A_0 used in previous examples.

These terms were used when you looked at annuities at the beginning of the chapter.

EXERCISE 18.8 COMPOUND INTEREST APPLICATIONS

1 At the beginning of 2013, Amélie invests \$1000. If compound interest is paid at 6% p.a., the value of the investment after 8 years is:

A \$60 **B** \$480 **C** \$1593.85 **D** \$10 491.32

2 Write a recursion relation for each of the following.

(a) An investment of \$1000 at an interest of 3.1% with the interest compounded annually.
(b) A loan of \$5000 at an interest rate of 6.2% with the interest compounded six-monthly.
(c) An investment of \$10 000 at an interest of 2.7% with the interest monthly.

3 What is the annual compound interest rate for the information given in each recurrence relation?

(a) $A_0 = 5000$, $A_n = 1.027 \times A_{n-1}$ if the interest is charged annually.
(b) $A_0 = 10\,000$, $A_n = 1.015 \times A_{n-1}$ if the interest is charged six-monthly.
(c) $A_0 = 4000$, $A_n = 1.005 \times A_{n-1}$ if the interest is charged monthly.

4 In 2006, 5000 students entered for a particular examination. The number increased each year by 20% of the number who entered the previous year. Calculate:

(a) the number who entered in 2011
(b) the total number who entered between 2006 and 2011 inclusive.

5 If the amount of wheat harvested in a certain area in 2012 is 4.2 thousand tonnes and it is anticipated that this will increase annually by 2.5%, estimate:

(a) the amount harvested in 2020
(b) the total amount harvested over the 9 years from 2012 to 2020 inclusive.

6 At the beginning of 2011, a mining town had a population of 15 000. It was estimated that this would increase each year by 8% of its population at the beginning of the year. What should the population be at the beginning of 2019?

7 The value of a new car is $35 000. Its value depreciates (falls) each year by 15% of its value at the beginning of that year. After how many years will its value be $15 000?

8 Megan is on a fixed salary of $83 000 per year and is offered a performance bonus of 5% of this salary each year. Paul is on a salary of $83 000 per year with an annual increase of 3% p.a.

(a) How much has each person earned after working 4 years, if Megan receives her bonus in the last pay for each year?
(b) When will Paul's annual earnings first exceed Megan's annual earnings, if Megan receives her bonus every year?

9 Kris contributes towards a pension for his retirement by depositing into a superannuation fund an amount of $5000 on each of his 44 birthdays from his 21st to his 64th inclusive.

(a) If the money is invested at 7% p.a. compound interest, how much money is in Kris's superannuation fund on his 65th birthday?
(b) As Kris is about to make the deposit on his 43rd birthday, he is told that he can only expect to earn 3% p.a. compound interest for the remainder of the time. How much less will be in the superannuation fund on his 65th birthday?

10 Two employees start work on a salary of $40 000 per year. Eleanor decides to take an increase of 4% p.a. while Henry accepts an annual increase of $2000 per year.

(a) What are their salaries at the start of the fourth year of employment?
(b) What is the total earned by each person after they have been working for 10 years?
(c) When does Eleanor's salary first exceed Henry's?

11 Tranh borrowed $20 000 on 1 January 2008. He agreed that on 1 January in each succeeding year he would pay back $3000 and add to the debt 6% p.a. interest on the amount owing during the year just completed. Find:

(a) the amount still owing immediately after 1 January 2013
(b) the number of years necessary to pay off the debt.

12 Kathryn borrows $200 000, to be repaid in equal monthly instalments. The interest rate is 8.4% p.a., calculated monthly.

(a) Show that the interest for the first month is $1400.
(b) Why should Kathryn's payments be more than $1400 per month?
(c) Kathryn decides to pay $3000 per month off the loan. Show that the amount owing after two repayments, $\$A_2$, is given by $A_2 = 200\,000 \times 1.007^2 - 3000(1 + 1.007)$.
(d) Hence find an expression for $\$A_n$, the amount owing after the nth repayment.
(e) How long will it take Kathryn to pay off the loan?

13 Phuong decides to set up a trust fund to provide for the education of her grandchildren. She deposits $60 at the beginning of each month into an account that earns 7.2% p.a. compounded monthly. The trust fund will mature at the end of the month of her final investment, 20 years after her first investment, so Phuong will need to make 240 monthly deposits.

(a) After 20 years, what is the value of the first $60 deposited?
(b) Write a geometric series for the value of all the deposits and calculate the final value of the trust fund.

18.9 FURTHER APPLICATIONS OF SERIES

Annuities

An **annuity** (from the same Latin word that gives us 'per annum' and 'annual') is a series of payments made at equal intervals of time: payments are traditionally once per year (hence 'annuity'), but may also be half-yearly, quarterly or at more frequent intervals.

In this course, an annuity is defined as a form of investment in which periodical equal contributions are made to or taken from the investment, with interest compounding at the conclusion of each period.

For example, a person may deposit the same amount of money each time period (e.g. year) into a superannuation fund or an investment account, with the interest compounding at the end of the time period. At some point in the future, the funds may be taken out of the account as a lump sum or as a regular payment or pension.

Example 31

Ian deposits \$50 000 in an account which will earn interest at a rate of 3% p.a., paid monthly. Ian wishes to withdraw \$200 per month.

(a) Set up a recurrence relation that models this account and gives the balance after n payments.

(b) What is the balance after two withdrawals?

(c) Explain what will happen to the balance if the interest was 5% per annum from the start.

(d) What monthly payment would keep the balance at \$50 000? Calculate the answer for rates of 5% and 3%.

Solution

(a) Amount deposited $= A_0 = \$50\,000$.

Interest rate per month $= r = \dfrac{3}{1200} = 0.0025$ so $R = 1 + r = 1.0025$

Time period $= n$ months so that A_1 is the balance at the end of the first month after interest has been added and the first payment made.

Monthly payment: $M = \$200$

After 1 month: $A_1 = 50\,000 \times 1.0025 - 200$

This can be written as: $A_1 = A_0 \times R - M$

$= A_0 \times 1.0025 - 200$

After 2 months: $A_2 = A_1 \times 1.0025 - 200$

After n months (and n payments) this becomes: $A_n = A_{n-1} \times 1.0025 - 200$

(b) $A_1 = 50\,000 \times 1.0025 - 200$

$= 49\,925$

$A_2 = 49\,925 \times 1.0025 - 200$

$= 49\,849.81$

After two withdrawals, the balance remaining is \$49 849.81.

If the payment made each month is greater than the interest earned, then the value of A_n decreases as n increases and eventually $A_n \to 0$, with the last payment most likely being less than \$200.

(c) $r = 5\%$ p.a., $r = \dfrac{5}{1200}$ p.m. $= \dfrac{1}{240}$ p.m.

$R = 1 + \dfrac{1}{240} = \dfrac{241}{240}$

$A_1 = 50\,000 \times \dfrac{241}{240} - 200$

$= 50\,008.33$

After the first payment, the balance remaining is larger than the initial deposit, so the amount in the account will continue to grow. At an interest rate of 5% p.a., the account earns more interest than the amount withdrawn each month.

(d) $\$50\,000 \times \frac{5}{100} \times \frac{1}{12} \approx \$208.33$

A monthly payment of \$208.33 will leave \$50 000 in the account if the annual interest rate remained at 5%.

$\$50\,000 \times \frac{3}{100} \times \frac{1}{12} = \$125$

If the interest rate is only 3%, then a monthly payment of \$125 will leave \$50 000 in the account. This means that for the account balance to remain unchanged, the payment taken each time period must be the same as the interest earned during that time period.

Annuities are usually designed to reduce in value over time. It is possible to calculate the payments to be made on a regular basis to meet individual needs. Of course, if the interest rate falls then the payments will cease earlier and if the interest rate rises then the payments will last longer.

Example 32

Sam lends \$5000 on the condition that she is repaid the money, plus interest, in 10 equal quarterly instalments. The first amount is to be repaid 3 months from the loan date. If this loan earns interest at a rate of 4% per quarter, what is the amount of each instalment?

Solution

$P = 5000$, $r = 0.04$, $R = 1.04$, repayment amount (instalment) is Q per quarter

Amount owing after first repayment: $= 5000 \times 1.04 - Q$

Amount owing after second repayment: $= (5000 \times 1.04 - Q) \times 1.04 - Q$

$= 5000 \times 1.04^2 - Q(1 + 1.04)$

Amount owing after third repayment: $= (5000 \times 1.04^2 - Q(1 + 1.04)) \times 1.04 - Q$

$= 5000 \times 1.04^3 - Q(1 + 1.04 + 1.04^2)$

Amount owing after tenth repayment: $= 5000 \times 1.04^{10} - Q(1 + 1.04 + 1.04^2 + \ldots + 1.04^9)$

Now $1 + 1.04 + 1.04^2 + \ldots + 1.04^9$ is a geometric series with $a = 1$, $r = 1.04$, $n = 10$.

After the tenth repayment, the amount owing is zero, so $A_{10} = 0$.

$\therefore$ Amount owing after tenth repayment: $5000 \times 1.04^{10} - \frac{Q(1.04^{10} - 1)}{1.04 - 1} = 0$

Hence: $\frac{Q(1.04^{10} - 1)}{1.04 - 1} = 5000 \times 1.04^{10}$

$$Q = \frac{5000 \times 1.04^{10} \times 0.04}{1.04^{10} - 1} = 616.45$$

Thus Sam receives \$616.45 per quarter.

This result can be generalised into a formula if you let $P = 5000$, $R = 1.04$. with n repayments:

Amount owing after n repayments $= P \times R^n - \frac{Q(R^n - 1)}{R - 1} = 0$

So repayments are: $Q = \frac{PR^n(R-1)}{R^n - 1}$

Although this formula could have been used from the start, it is important to understand how it is obtained.

Example 33

When Peggy started work she began paying \$120 at the beginning of each month into a superannuation fund. These contributions are compounded monthly at an interest rate of 6% p.a. Peggy intends to retire after having worked for 40 years.

(a) Let $\$P$ be the final value of Peggy's superannuation when she retires after 40 years (480 months). Show that $\$P = \$240\,174$ to the next dollar.

(b) After working for 20 years, Peggy decides that she needs to have \$750 000 in her fund before she can retire. At this stage the fund has only \$55 722. Peggy decides to increase the amount that she pays into the fund for the next 20 years to $\$M$ at the beginning of each month. The contributions will continue to attract the same interest rate of 6% p.a. compounded monthly. At the end of n months after starting the new contributions, the amount in the fund is $\$A_n$.

(i) Show that $A_2 = 55\,722 \times 1.005^2 + M(1.005 + 1.005^2)$

(ii) Find the value of M so that Peggy will have \$750 000 in her fund after the remaining 20 years (240 months).

Solution

(a) Monthly investment = \$120, $r = 0.005\%$ p.m., $n = 480$

Value of last investment at end of 40th year: $= 120 \times 1.005$ (1 month's interest)

Value of second-last investment at end of 40th year: $= 120 \times 1.005^2$ (2 months' interest)

Value of first investment at end of 40th year: $= 120 \times 1.005^{480}$ (480 months' interest)

Sum of the investments $= 120 \times 1.005 + 1200 \times 1.005^2 + 120 \times 1.005^3 + \ldots + 120 \times 1.005^{480}$

This is a finite geometric series with first term 120×1.005, common ratio 1.005, $n = 480$.

$$P = S_{480} = \frac{a\left(R^{480} - 1\right)}{R - 1}$$

$$= \frac{120 \times 1.005\left(1.05^{480} - 1\right)}{1.005 - 1} = \frac{120 \times 1.005\left(1.005^{480} - 1\right)}{0.005} = 240173.78$$

$\$P = \$240\,174$ (to the next dollar)

(b) (i) Monthly investment $= \$M$, $r = 0.005\%$ p.m., $n = 2$, initial balance = \$55 722

Value of balance at end of 2 months: $= 55\,722 \times 1.005^2$

Value of last investment at end of 1 month: $= M \times 1.005$

Value of second-last investment at end of 2 months: $= M \times 1.005^2$

$\therefore A_2 = 55\,722 \times 1.005^2 + M(1.005 + 1.005^2)$

(ii) $A_{240} = 55\,722 \times 1.005^{240} + M(1.005 + 1.005^2 + 1.005^3 + \ldots + 1.005^{240})$

$$A_{240} = 55\,722 \times 1.005^{240} + \frac{M \times 1.005\left(1.005^{240} - 1\right)}{1.005 - 1}$$

We require $A_{240} = 750\,000$:

$$750000 = 55722 \times 1.005^{240} + \frac{M \times 1.005\left(1.005^{240} - 1\right)}{0.005}$$

$$750000 - 55722 \times 1.005^{240} = \frac{M \times 1.005\left(1.005^{240} - 1\right)}{0.005}$$

$$M = \frac{0.005\left(750000 - 55722 \times 1.005^{240}\right)}{1.005\left(1.005^{240} - 1\right)} = 1217.93$$

The monthly payment would need to be \$1218 per month (to the next dollar).

Example 34

An investment of \$20 000 earns interest at a rate of 2.7% p.a. compounded annually.

(a) What is the future value of this investment after 8 years?

(b) Use a recurrence relation to set up a spreadsheet to model this investment over 10 years.

Solution

(a) $P = 20\,000$, $r = 2.7\% = 0.027$. $R = 1.027$

$A_8 = PR^8$

$= 20\,000 \times 1.027^8$

$= 24\,751.05$

The future value of this investment is \$24 751.05.

(b) Set up the recurrence relation: $A_n = A_{n-1} \times 1.027$

	A	B	C	D	E
1	*P* =	20000	*R* =	1.027	*A* = *PR*^*n*
2	Amount 1 =	\$20,540.00			
3	Amount 2 =	\$21,094.58			
4	Amount 3 =	\$21,664.13			
5	Amount 4 =	\$22,249.07			
6	Amount 5 =	\$22,849.79			
7	Amount 6 =	\$23,466.73			
8	Amount 7 =	\$24,100.34			
9	Amount 8 =	\$24,751.05			
10	Amount 9 =	\$25,419.32			
11	Amount 10 =	\$26,105.65			

Calculating this with a spreadsheet shows that $A_8 = \$24\,751.05$ (which matches the figure calculated in (a)) and that $A_{10} = \$26\,105.65$.

EXPLORE FURTHER

Future values for an annuity

Use spreadsheet software to set up a recurrence relation to model future values of an annuity.

Verifying the table of future value of an annuity

Annuity

An annuity is a compound interest investment from which equal payments are made or received on a regular basis (at equal periods of time) for a fixed period of time.

The payment is usually made at the end of the time period so that no interest is received until the end of the second time period.

Future value, FVA

The future value of an investment or annuity is the total value of the investment at the end of the term of investment, including all contributions and the interest earned.

At the start of the chapter, the formula used to obtain the FVA was given as $\text{FVA} = a\left\{\frac{(1+r)^n - 1}{r}\right\}$, where

FVA is the Future value of an annuity

a is the contribution per period paid at the end of the period

r is the interest rate per compounding period, as a decimal

n is the number of periods.

The formula was used to obtain the values in the table.

Consider an annuity in which 6 annual payments each of \$500 are made into an account where the compound interest rate is 4% per annum. The payments are made at the end of each year.

This information is shown in the following table.

Year	Balance at beginning of year (\$)	Interest on balance (\$)	Annual payment (\$)	Total at end of year (\$)
1	0.00	0.00	500	500.00
2	500.00	20.00	500	1020.00
3	1020.00	40.80	500	1560.80
4	1560.80	62.43	500	2123.23
5	2123.23	84.93	500	2708.16
6	2708.16	108.33	500	3316.49

$$\begin{aligned}\text{FVA} &= 500 + 500 \times 1.04 + 500 \times 1.04^2 + 500 \times 1.04^3 + 500 \times 1.04^4 + 500 \times 1.04^5\\ &= 500(1 + 1.04 + 1.04^2 + 1.04^3 + 1.04^4 + 1.04^5)\\ &= \frac{500\left(1.04^6 - 1\right)}{1.04 - 1}, \text{ by summing a geometric series with 6 terms}\\ &= \$3316.49\end{aligned}$$

If there had been n annual payments of $\$a$ made at the end of each year with a compound interest rate of r% per year, then if you would have

$$\begin{aligned}\text{FVA} &= a(1 + (1 + r) + (1 + r)^2 + \ldots + (1 + r)^{n-1})\\ &= a\left\{\frac{(1+r)^n - 1}{r}\right\}\end{aligned}$$

This verifies the formula that was used to obtain the FVA table used at the beginning of the chapter.

EXERCISE 18.9 FURTHER APPLICATIONS OF SERIES

1 Noor invests \$10 000 on the condition that the money is repaid in 12 equal quarterly instalments. If the investment earns interest at the rate of 1% per quarter, what is the amount of each instalment?

2 Ava invests \$20 000 on the condition that she is repaid the money in 16 equal quarterly instalments. If the investment earns interest at the rate of 0.75% per quarter, what is the amount of each instalment?

3 **(a)** When Nazeera started a new job, \$400 was deposited into her superannuation fund at the beginning of each month. The money was invested at 0.4% per month, compounded monthly.
Let $\$Y$ be the value of the investment after 360 months, when Nazeera retires. Show that $Y = 322\,142.43$.

(b) After retirement, Nazeera withdraws \$3000 from her superannuation fund at the end of each month without making any further deposits. The account continues to earn interest at 0.4% per month.
Let $\$A_n$ be the amount of money left in the account n months after Nazeera's retirement.

(i) Show that $A_n = (Y - 750\,000) \times 1.004^n + 750\,000$.

(ii) For how many months after retirement will there be money left in the account?

4 One year ago William and Kate borrowed \$400 000 to buy an apartment. The interest rate was 6% p.a., compounded monthly. They agreed to repay the loan over 25 years with equal monthly repayments of \$2578.

(a) Calculate how much money William and Kate owed after their first monthly repayment.

(b) After making their twelfth monthly repayment, William and Kate owe \$392 870. The interest rate now increases to 9% p.a., compounded monthly. The amount $\$A_n$ owing on the loan after the nth monthly repayment is now calculated using the formula $A_n = 392\,870 \times 1.0075^n - 1.0075^{n-1}M - \ldots - 1.005M - M$, where $\$M$ is the monthly repayment and $n = 1, 2, \ldots 288$. (Do **not** prove this formula.)
Calculate the monthly repayment if the loan is to be repaid over the remaining 24 years (288 months).

(c) William and Kate now decide to increase their monthly repayments to \$3500. How long will it take them to repay \$392 870?

(d) How much money will William and Kate save over the term of the loan by making these higher monthly repayments, compared to keeping the repayments at the amount calculated in **(b)**?

5 Which will give the better financial result after 20 years: a lump sum of \$100 000 invested at 5% p.a. compounded annually, or a monthly payment of \$600 with interest at 5% p.a. compounded monthly?

6 A lottery offers a prize of \$100 000 immediately, or \$10 000 now plus \$10 000 per year for the next 11 years. You take the prize of \$100 000, keep \$10 000 to spend over the next year and invest the remaining \$90 000 as an annuity at 5% p.a. You plan to withdraw \$10 000 at the end of each year for the next 11 years.

(a) Is this a realistic plan?

(b) How much money is left in your annuity after the eleventh payment?

(c) How much money is left in your annuity after the twelfth payment?

7 Prior to retirement, Cassie deposits \$100 000 in an account which will earn interest at a rate of 2.7% p.a., paid monthly. Cassie wishes to withdraw \$300 per month.

(a) Set up a recurrence relation that models this account and gives the balance after n payments.

(b) What is the balance after two withdrawals?

(c) Explain what will happen to the balance if the interest was 5% per annum from the start.

(d) What monthly payment would keep the balance at \$100 000? Calculate the answer for rates of 5% and 2.7%.

Another practical use for series—the rule of 72

(**Note**: the following information is not part of the syllabus.)

The '**rule of 72**' is a rule that can be used to find roughly how much time it will take for an investment to double in value if invested at a particular interest rate. It should technically be the rule of $100 \ln 2$ (or 69.3), but 72 is used instead because it has more factors, which makes it easier to divide without a calculator than 69.3. (A good rule for approximations is that they must be easy to use.)

The rule of 72 says that if you divide 72 by the interest rate, the answer will be a good approximation of how much time it will take to double your investment. For example:

- At 12% p.a. it takes $\approx 72 \div 12 = 6$ years to double your money.
- At 6% p.a. it takes ≈ 12 years to double your money.

The rule also works for depreciation by half. For example, if the depreciation rate of a car is 15% then it will take $\approx 72 \div 15 = 4.8$ years for the car to halve in value.

Why does this rule work? Consider:

$$A = P(1+r)^n \quad \text{where } r \text{ is a decimal}$$

To double in value to $2P$:

$$2P = P(1+r)^n$$

$$2 = (1+r)^n$$

Take $\log_e$ of both sides:

$$\log_e 2 = n \log_e (1+r)$$

$$n = \frac{\log_e 2}{\log_e (1+r)}$$

When r is small, you have $\log_e (1+r) \approx r$, so that $n \approx \frac{\log_e 2}{r} = \frac{0.693}{r}$. If you consider $r = \frac{\alpha}{100}$, meaning that the interest rate is $\alpha\%$, then you have:

$$n \approx \frac{0.693}{\alpha} \times 100 = \frac{69.3}{\alpha} \approx \frac{72}{\text{interest rate}}$$

Therefore 72 divided by the interest rate is approximately equal to the length of time n for the investment to double.

As this table of values shows, for various values of r and n we have:

r	0.01	0.03	0.06	0.10	0.20
n	69.7	23.4	11.9	7.3	3.8
$r \times n$	0.697	0.702	0.714	0.730	0.760
$r\% \times n$	69.7	70.2	71.4	73.0	76.0

In practice, 72 is a good approximation for rates up to about 12%.

Knowing the rule of 72 can help people to avoid bad investment decisions. It is a good backup for that other important rule: 'if it seems too good to be true, then it probably is'.

Consider the case of the 'Wattle' scheme in the late 1990s, which was one of Australia's worst investment disasters. More than 3000 Australians invested almost $200 million in an illegal investment scheme that promised a return of 50% per year, paid each month.

Using the rule of 72, $72 \div 50 \approx 1.5$. This means that an investor in the scheme would double their money in less than 1.5 years. It sounds too good to be true, and it was. The scheme was supposed to make its profits by lending funds to new businesses, but this never happened; instead, the initial investors were paid using the funds collected from later investors. (This type of scam is so common, it has a name: it is called a 'Ponzi' scheme.) The initial investors, impressed with their profits, put even more money into the scheme and encouraged their friends to do so too. Of course, the scheme soon turned into a disaster for investors and its promoter was sentenced to jail for 10 years.

CHAPTER REVIEW 18

1 Are the following either arithmetic or geometric sequences? Explain.

(a) $2, -6, 10, -14, \ldots$ **(b)** $4+\sqrt{3},\ 1+3\sqrt{3},\ -2+5\sqrt{3},\ -5+7\sqrt{3},\ \ldots$ **(c)** $1.6, 2.4, 3.6, 5.4, \ldots$

2 For the arithmetic sequence 22, 15, 8, 1, … find:

(a) the 10th term **(b)** the value of k if $T_k = -90$.

3 For the geometric sequence 36, 126, 441, 1543.5, … find:

(a) the 7th term **(b)** the smallest value of k for which $T_k > 1\,000\,000$.

4 If $T_4 = 600$ and $T_{10} = 75$ are two terms of a sequence. Find the first three terms if:

(a) the sequence is arithmetic **(b)** the sequence is geometric.

5 Find the sum of all the numbers between 20 and 200 that are divisible by 9.

6 Evaluate $6 + 3 + 1.5 + \ldots$

7 A ball is dropped from a height of 20 m and rebounds to a height of 18 m. If every time it rebounds it rises to nine-tenths of its previous height, calculate the total number of metres it could travel.

8 Three numbers whose sum is 15 are successive terms of an arithmetic series. If 1, 1 and 4 are added to these three numbers respectively, the resulting numbers are successive terms of a geometric series. Find the numbers.

9 If $\sqrt{3} - 1$ and $2 - \sqrt{3}$ are the first two terms of a geometric series, write the next two terms and the sum to infinity in simplest surd form.

10 Express 0.2333… in the form $\frac{m}{n}$, where m and n are integers.

11 The first, third and ninth terms of an arithmetic series are also the first three terms of a geometric series. Find the common ratio of the geometric series.

12 For the function defined by $S_n = n^2 - 3n$ for $n = 1, 2, 3, \ldots$ find T_n and hence show that the series is arithmetic.

13 For a geometric series, the second term is 6 and the fifth term is 48. Find the sum of the first five terms.

14 The population of a certain town in the year 2015 was 24 000. Every year its population increases by 25% of its population during the previous year. What will the town's population be in 2035?

15 Find the sum of the series $1 - \frac{1}{4} + \frac{1}{16} - \frac{1}{64} + \ldots$

16 Find the sum of the first n terms of a series given $T_r = 2^r + 2r - 1$.

17 A snail moves 1 m west, turns 90° to the left, then moves half the previous distance before again turning 90° to the left. If the snail continues in this way until it is hardly moving at all, where will it end up? Write your answer using a distance west and a distance south of the starting position.

18 This table gives the future value of an annuity of $1 at the given interest rate for the given period.

Future value interest factors					
$1	**Interest rate per period**				
N	**1%**	**2%**	**3%**	**4%**	**5%**
1	1.0000	1.0000	1.0000	1.0000	1.0000
2	2.0100	2.0200	2.0300	2.0400	2.0500
3	3.0301	3.0604	3.0909	3.1216	3.1525
4	4.0604	4.1216	4.1836	4.2465	4.3101
5	5.0101	5.2040	5.3091	5.4163	5.5256
6	6.1520	6.3081	6.4684	6.6330	6.8019

Use the table of future value interest factors to find:

(a) the future value of an annuity of $4500 per year for 6 years at 4% per annum
(b) the future value of an annuity of $700 per year for 5 years at 2% per annum.

19 This table gives the present value of an annuity of $1 at the given interest rate for the given period.

Present value interest factors					
$1	**Interest rate per period**				
N	**1%**	**2%**	**3%**	**4%**	**5%**
1	0.9901	0.9804	0.9709	0.9615	0.9524
2	1.9704	1.9416	1.9135	1.8861	1.8594
3	2.9410	2.8839	2.8286	2.7751	2.7232
4	3.9020	3.8077	3.7171	3.6299	3.5460
5	4.8534	4.7135	4.5797	4.4518	4.3295
6	5.7955	5.6014	5.4172	5.2421	5.0757

Use the table of present value interest factors for an annuity of $1 per period.

(a) Jake plans to invest $6000 per year for 5 years in an annuity. His investment will earn interest at the rate of 3% per annum. Calculate the present value of this annuity.
(b) Ari takes out a personal loan of $10 000 for 6 years at an interest rate of 5% per year. Use the table to find his yearly repayments.

20 The table gives the present value interest factors for an annuity of $1 per period, for various interest rates, r, and number of periods, N.

Present value interest factors					
r	**Interest rate per period (as a decimal)**				
N	**0.0025**	**0.005**	**0.0075**	**0.008**	**0.009**
71	64.98140	59.64121	54.89293	54.00754	52.29657
72	65.81686	60.33951	55.47685	54.57097	52.82118
73	66.65023	61.03434	56.05643	55.12993	53.34111
74	67.48153	61.72571	56.63169	55.68446	53.85641
75	68.31075	62.41365	57.20267	56.23458	54.36710
76	69.13791	63.09815	57.76940	56.78034	54.87324

(a) Randall plans to invest $200 each month for 74 months. Her investment will earn interest at a rate of 0.005 (as a decimal) per month. Use the information in the table to calculate the present value of this annuity.

(b) Rosa uses the same table to calculate the loan repayments for her car loan. Her loan is for $19000 and will be repaid in equal monthly repayments over 5 years and 11 months. The interest rate on her loan is 9% per annum. Calculate the amount of each monthly repayment, rounded to the next dollar.

21 Malcolm deposits $3000 in an investment account that is paying a monthly interest rate of 0.4%, with the interest compounded monthly. Calculate the value of the investment after 24 months.

22 Tanya and Alan plan to have $20 000 in an investment account in 5 years time to pay for a holiday. The interest rate for the account will be fixed at 3.6% per annum, compounded monthly. How much do they need to deposit into the account to achieve this goal? Round your answer to the next dollar.

23 Sanjay borrows $5000 at an interest rate of 1.5% per month and pays it off in equal monthly instalments. What should the instalment be in order to pay off the loan at the end of 3 years?

24 (a) At the beginning of each year, $100 is placed in an investment fund. Calculate the accumulated value at the end of 12 years if the interest is 6% p.a. compounded yearly.

(b) If the $100 due at the beginning of the fifth year was not placed in the fund, what would be the accumulated value at the end of 12 years?

25 A family borrows $300000 from a bank. Interest is charged at 6% p.a.

(a) How much must be repaid each year, rounded to the next dollar, if the loan is to be repaid over 30 years?

(b) How much of the loan will remain after the 18th payment?

(c) If repayments are made at $40000 per year, how long will it take to repay the loan? What will be the last repayment?

26 (a) Find the limiting sum of the geometric series $3+\frac{3}{\sqrt{3}+1}+\frac{3}{\left(\sqrt{3}+1\right)^2}+\ldots$

(b) Explain why the geometric series $3+\frac{3}{\sqrt{3}-1}+\frac{3}{\left(\sqrt{3}-1\right)^2}+\ldots$ does not have a limiting sum.

27 Merv retires with a lump sum of $200000. The money is invested in a fund that pays interest each month at a rate of 6% p.a., and Merv receives a fixed monthly payment of $$M$ from the fund. Thus the amount left in the fund after the first monthly payment is $$(201000 - M)$.

(a) Find a formula for the amount $$A_n$ left in the fund after n monthly payments.

(b) Merv chooses the value of M so that there will be nothing left in the fund at the end of the 15th year (after 180 payments). Find the value of M.

CHAPTER 19
Bivariate data analysis

19.1 TYPES OF DATA AND THEIR INTERPRETATION

Comparing sets of data includes looking for patterns or associations between the sets. The method you choose to compare data sets depends on the type of data.

Data can be divided into two basic categories, numerical and categorical, both of which can be further divided into sub-categories.

Numerical data

Discrete data: Each piece of data is one of a set you can write down so that the number of possibilities can be counted, e.g. the number of children in a family.

Continuous data: The possible outcomes cannot be written down as all decimal places are possible. You usually have to measure the outcomes to a set number of decimal places, e.g. the time for the 100 m sprint at the Olympics is measured to the nearest hundredth of a second.

Categorical data (non-numerical)

Ordinal: Ordered in some way, e.g. rating a statement as strongly disagree, disagree, neutral, agree or strongly agree.

Nominal: There is no order, e.g. recording people's favourite search engine.

Comparing numerical variables

A visual display of this type of data can be done using a parallel box plot or a back-to-back stem-and-leaf plot.

For example, the same information concerning the age of Year 12 parents attending a Year 12 information evening is represented by both these displays.

Back-to-back stem-and-leaf plot:

Fathers	Stem	Mothers
9 7	3	8 8 9
8 7 5 3 2	4	2 3 5 6 6 7
8 8 4 3 1	5	2 4
0	6	

Parallel box plots:

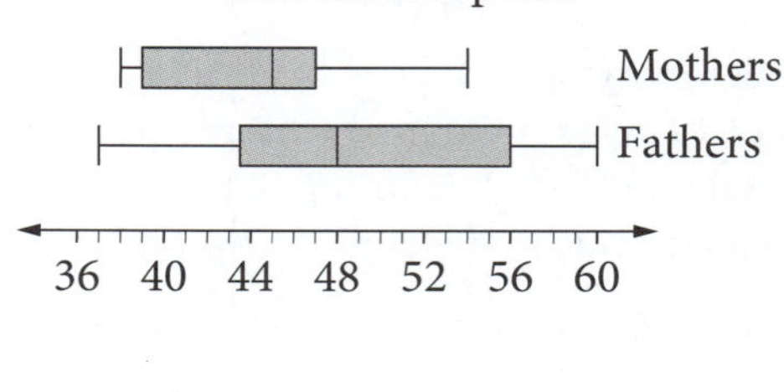

Parallel box plots can be used to compare more than two data sets. The back-to-back stem-and-leaf plot can only be used when there are two sets of data.

Comparing two categorical variables

The simplest way to compare two sets of nominal categorical data is to draw a back-to-back frequency table such as the following.

Students	Favourite film genre	Teachers
5	Horror	18
16	Comedy	20
22	Action	4
16	Science fiction	16
2	Romantic comedy	48
0	Animation	30
18	Thriller	32

This information can also be shown in a clustered column graph or a stacked column graph.

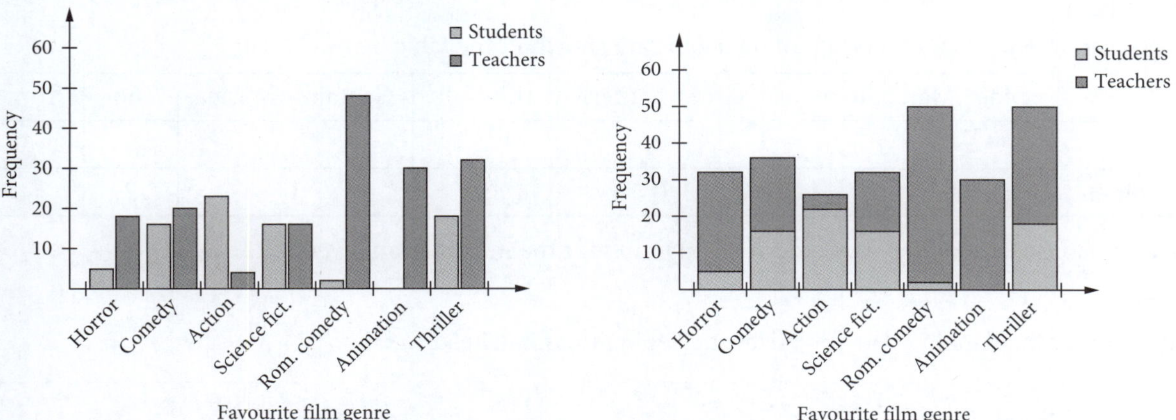

The unequal number of students and teachers in the survey (the total number of entries in each student and teacher column) makes it difficult to see any difference or similarity that may exist in the answers for the two groups.

In this case, converting the data to percentages may show any differences or similarities in the groups.

The table then becomes:

% Students	Favourite film genre	% Teachers
6.3	Horror	10.7
20.3	Comedy	11.9
27.8	Action	2.4
20.3	Science fiction	9.5
2.5	Romantic comedy	28.6
0	Animation	17.9
22.8	Thriller	19.0

And the column graph becomes:

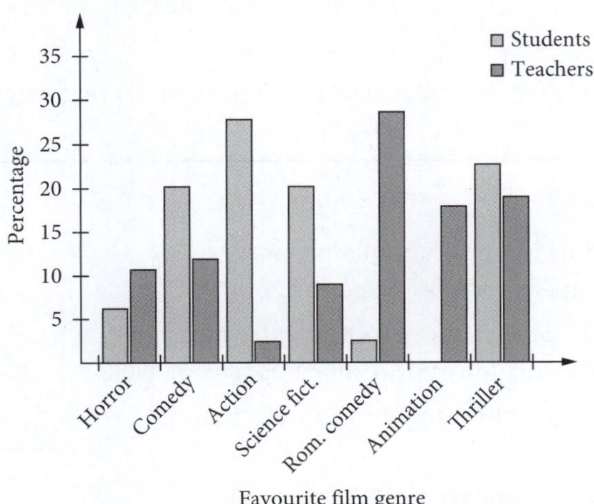

The differences between the percentages are more clearly seen here. The students surveyed preferred 'Action' and 'Science fiction', whereas the teachers surveyed preferred 'Romantic comedy' and 'Animation'.

These graphs may also be drawn using a spreadsheet.

Example 1

The favourite superheroes of a group of 62 people are shown in the table below.

	Spider-Man	Batman	Green Lantern	The Hulk	Captain America	Thor	Iron Man
Men	1	5	2	4	2	5	12
Women	6	14	0	3	1	2	5

Using a spreadsheet, draw the clustered column graph for the information given in the table.

Solution

Step 1: Enter the data into the spreadsheet, keeping the same headings.

	A	B	C	D	E	F	G	H
1	**Superheroes**	**Spider-Man**	**Batman**	**Green Lantern**	**The Hulk**	**Captain America**	**Thor**	**Iron Man**
2	Males	1	5	2	4	2	5	12
3	Females	6	14	0	3	1	2	5

Step 2: Select Insert > Column > 2D cluster. (The exact selection may vary depending on spreadsheet software, but it should be available as a kind of column graph.)

Step 3: A graph like the following should be produced.

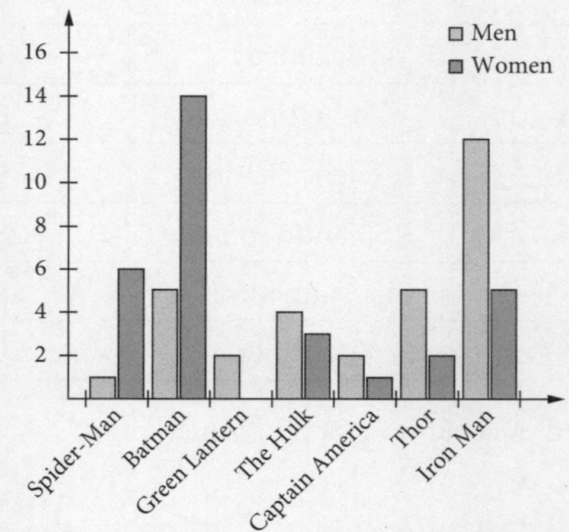

Note: There are other useful graphs (such as the segmented bar graph) that can be drawn by selecting a different graph output.

Two-way frequency tables

Consider the results of surveying 200 shoppers about a proposal to change the arrangements for childcare at a shopping centre.

	Male	Female	Total
Agree	25	88	113
Disagree	45	42	87
Total	70	130	200

This is called a two-way frequency table as it shows not just the 'agree' and 'disagree' totals but breaks down the figures into subcategories as well.

From the table it can be seen that of the 200 shoppers, 70 were male and 130 were female.

In total, 113 shoppers agreed with the proposal and 87 disagreed.

While it can be seen that more females (88) than males (25) agree with the proposal, the numbers disagreeing appear approximately the same.

In this case, the gender of the person being interviewed may influence their response but a person's response will not influence their gender. The variable influencing a response is called an independent variable. The independent variable is given in the vertical columns of the two-way table. The variable reacting to the independent variable is called the response variable, or dependent variable.

Since the number of males and females interviewed is different, the trend will become clearer if you convert the figures to percentages. Remove the last column (since this refers to the dependent variable), and convert each of the figures into a percentage of the total in the last row of the column.

	Number of men	Percentage	Number of women	Percentage
Agree	25	$\frac{25}{70} \times 100 = 35.7\%$	88	$\frac{88}{130} \times 100 = 67.7\%$
Disagree	45	64.3%	42	32.3%
Total	70	100%	130	100%

	Male %	Female %
Agree	35.7	67.7
Disagree	64.3	32.3
Total	100	100

Now the data shows more clearly that a woman is almost twice as likely as a man to agree with the proposal.

Example 2

The following table represents the results obtained on a particular test (marked out of 40), taken by two groups: Group A (students 16 years of age) and Group B (students 18 years of age). The pass mark for the test is a score of 25.

Group A (16 years of age): 37, 26, 31, 23, 34, 38, 29, 17, 33, 26
Group B (18 years of age): 24, 28, 29, 34, 18, 19, 32, 29, 37, 28

Present the data as a percentage-based two-way frequency diagram using class intervals of 5.

Solution

By looking at the range of values for both sets of data decide upon the class boundaries:

Count the number of figures in each interval and add the frequency to the relevant column.

	15–<20	20–<25	25–<30	30–<35	35–<40	Total
Group A	1	1	3	3	2	10
Group B	2	1	4	2	1	10
Total	3	2	7	5	3	20

Convert each frequency to a percentage of the total in each column:

$\frac{1}{10} \times 100\% = 10\%$ $\frac{3}{10} \times 100\% = 30\%$ $\frac{2}{10} \times 100\% = 20\%$ $\frac{4}{10} \times 100\% = 40\%$

Redraw the table to show the percentages:

	15–<20	20–<25	25–<30	30–<35	35–<40	Total
Group A	10%	10%	30%	30%	20%	100%
Group B	20%	10%	40%	20%	10%	100%

80% of Group A passed compared to 70% of Group B (pass mark 25).

50% of Group A achieved 75% or more compared with 30% of Group B.

EXERCISE 19.1 TYPES OF DATA AND THEIR INTERPRETATION

Give answers correct to 2 decimal places, unless otherwise stated.

1 A group of students and teachers were asked the following question: "Which of the following movie villains do you like the most?" Their responses are summarised in the table below.

Draw a clustered column graph to represent the following data and state the most popular villain for each gender. You may choose to use a spreadsheet to do so.

Villain	Lex Luthor	Goldfinger	The Penguin	The Joker	KAOS
Students	46	10	80	40	12
Teachers	84	22	60	55	28

2 The following data was collected about the travel time for classes of Year 7 and Year 12 students.

Number of Year 7 students	Travel time (minutes)	Number of Year 12 students
25	0–<10	2
40	10–<20	8
24	20–<30	25
12	30–<40	55
4	40–<50	24
2	50–<60	6

Convert the table to a percentage-based two-way frequency diagram using the class intervals shown. Give your answers correct to 1 decimal place.

3 Consider the following raw data as an example of a percentage of people of various ages with high blood pressure.

Note: For this survey, high blood pressure is defined as greater than 140/90 mmHg and the sample size for each age group is 100 males and 100 females.

Male	Age group	Female
7	18–24	5
13	25–34	4
19	35–44	11
29	45–54	22
33	55–64	27
37	65–74	41
42	>74	52

(a) Complete the following two-way frequency table.

	Age group						
	18–24	25–34	35–44	45–54	55–64	65–74	>74
Male							
Female							

(b) Consider the two-way frequency table you just completed. In which age group does the percentage of women with high blood pressure start to be higher than the percentage of men with high blood pressure?

(c) Explain that it is not correct to say that 12% of people aged 18–24 years old have high blood pressure.

4 Vitamin C is thought by some people to help avoid colds, or at least shorten the length of a cold. In order to test this theory, a Year 12 Biology class conducted a test with all 250 Year 11 and 12 students at their school. A controlled experiment was conducted by giving half of the students vitamin C tablets to take for the winter and the other half sugar tablets that looked the same as the vitamin C tablets. Students were not told which type of tablet they were given.

The results of the study are shown in the following two-way table.

(a) Convert this table to the appropriate percentages. Give answers to the nearest whole number.

(b) Does this small sample say anything about the effectiveness of vitamin C in reducing colds? Explain.

	Sugar tablet	Vitamin C tablet	Total
Cold	29	24	53
No cold	96	101	197
Total	125	125	250

5 The data shown is adapted from data available on the Australian Bureau of Statistics website. The men and women surveyed have first been categorised as being either overweight to obese, or normal to underweight. They were then put into subgroups of being diabetic, pre-diabetic (in danger of developing diabetes in the near future) or within the healthy range for diabetes.

Data was collected for 800 males and females aged between 24 and 34.

	Overweight to obese		Normal to underweight		Total
	males	females	males	females	
Diabetic	14	8	20	10	52
Pre-diabetic	9	7	13	5	34
Healthy range	77	105	267	265	714
Total	100	120	300	280	800

(a) Complete the two-way frequency table below for the 'Overweight to obese' data. Give your answers correct to 1 decimal place.

	Overweight to obese					
	males		females		Total	
	Number	%	Number	%	Number	%
Diabetic	14		8		22	
Pre-diabetic	9		7		16	
Healthy range	77		105		182	
Total	100	100	120	100	220	100

(b) Complete the two-way frequency tables below for the 'Normal to underweight' data.

	Normal to underweight					
	Males		Females		Total	
	Number	%	Number	%	Number	%
Diabetic	20		10		30	
Pre-diabetic	13		5		18	
Healthy range	267		265		532	
Total	300		280		580	100

(c) What conclusions can you make about the risk of diabetes for the two groups of people?

6 A group of 100 tennis players were observed serving a tennis ball. The table below shows the number of unreturnable serves (aces) per set for tennis players with a fast serve, compared to those with a slow serve. However, some of the data was misplaced.

	0 aces	1 ace	2 aces	3 aces	4 aces	Total
Slow serve	7		8	4	0	27
Fast serve	1	8				
Total			22	34		100

(a) Complete the table using the data provided to find the missing values.
(b) How many players served three aces?
(c) Calculate the percentage of fast and slow serves that resulted in more than 2 aces. Give your answer correct to 1 decimal place.
(d) What conclusion can you draw from these survey results?

7 Information for the average weekly salary of men and women measured every six months from May 2010 to May 2013 is shown in the graphs below.

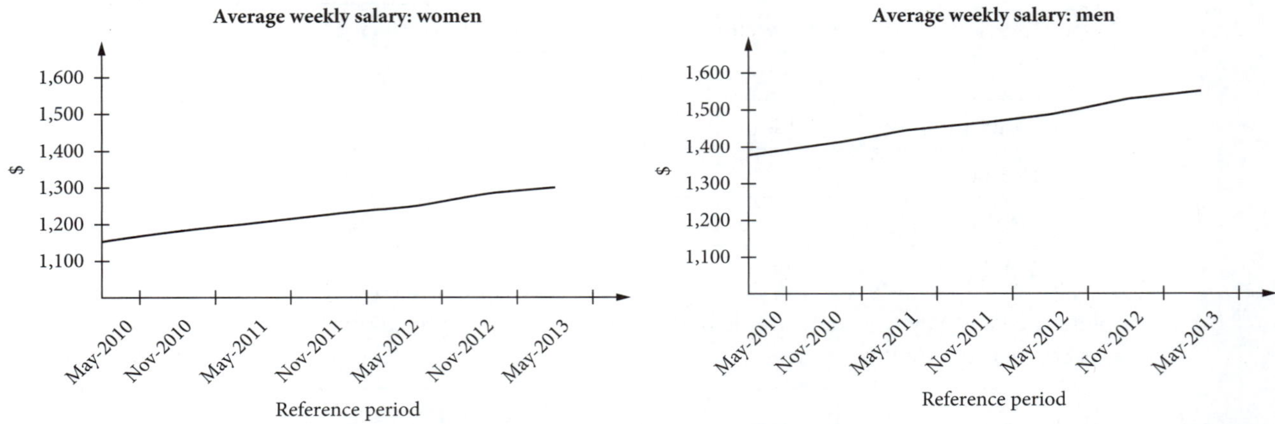

(a) Complete the following two-way frequency diagram by estimating values from the graphs to the nearest 50.

	Weekly salary May '10	Weekly salary Nov. '10	Weekly salary May '11	Weekly salary Nov. '11	Weekly salary May '12	Weekly salary Nov. '12	Weekly salary May '13
Men							
Women							
Difference							

(b) What conclusion can you draw about the difference in the average weekly salary for men and women in Australia?

8 The lengths of rivers at least 100 km long in the North and South Islands of New Zealand are shown below.

North Island (km) 172, 290, 425, 193, 105, 175, 241, 161, 158, 154, 182, 137, 143, 132, 119, 137

South Island (km) 209, 169, 288, 209, 138, 121, 177, 161, 322, 145, 121, 108

(a) Construct a two-way frequency table using class size of 50 km.
(b) What is the total number of rivers in New Zealand in the group 150–<200 km?
(c) How many rivers in New Zealand are 200 km or longer?
(d) How many rivers are at least 100 km in length on each island?
(e) The approximate area of the North and South Islands of New Zealand are 114 000 km^2 and 150 000 km^2 respectively. Does the size have any influence on the number of rivers over 100 km?

19.2 SCATTERPLOTS AND ASSOCIATION

It is common for two data sets to have an association where one set of variables may influence the other. For example if you were to measure a random group of people's height and weight you would probably find a positive association, i.e. taller people are more likely to be heavier.

Drawing a scatterplot provides a visual representation of any trend or underlying pattern in the data. A scatterplot is drawn by treating the bivariate data as a series of coordinate pairs and plotting the pairs on a suitable set of axes.

The data used often comes from measurements of real-life situations, so the graph is quite often limited to the positive part of the scale.

To draw a scatterplot:

Step 1 Draw a suitable set of axes by noting the range of each set of figures. If it is sensible, start with zero and have between 8 and 12 convenient, evenly spaced points on each scale.

Step 2 Treat each data pair as a coordinate point and place a mark on the grid. Do not join the points.

Example 3

Draw a scatterplot of the following bivariate data set and describe any trend you see.

Person	A	B	C	D	E	F	G	H
Height (cm)	140	160	180	90	100	50	60	120
Weight (kg)	60	75	95	40	50	20	35	65

Solution

Draw a suitable set of axes by noting the range of each set of figures:

In this case Height is from 0 (cm) to 180 (cm) so start at zero and use 9 divisions of 20 (cm).

Weight varies from 0 (kg) to 95 (kg) so start at zero and use 10 divisions of 10 (kg) to cover the required range.

It appears that, in general, the taller the person is, the heavier they are.

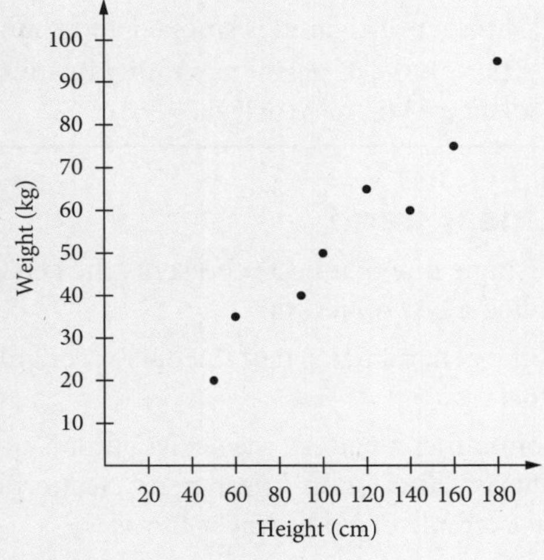

Often one of the variables can explain the association and this is known as the explanatory variable, or independent variable. The other variable responds to a change in the independent variable and is known as the response variable, or dependent variable. In the case of height versus weight, a person's weight is more likely to be explained by their height than the other way around.

EXPLORE FURTHER

Scatterplots using technology

Use spreadsheet software to construct a scatterplot displaying a bivariate data set.

The horizontal axis is used for the independent variable and the dependent variable is on the vertical axis.

Examples:

- Temperature of the body responds to, or depends upon, the time spent in front of a heater.
- The number of ice-creams sold will respond to or depend on the temperature of the day.

Warning

Not all bivariate data sets have clear independent and dependent variables. They may both be dependent on a third variable. For example, a student's study scores in English and Music may appear related but both depend on other variables such as effort, intelligence, hours of practice and study.

Example 4

Assuming an association exists, identify the independent variable in each of the following pairs:

- age and wealth
- age and the number of offspring
- temperature and the number of people at the beach
- number of cigarettes smoked and chance of cancer
- the volume of petrol remaining in your tank and the distance you have driven.

Solution

Ask yourself which of the variables could cause or explain a change in the other:

Age and wealth: The independent variable is age.

Age and the number of offspring: The independent variable is age.

Temperature and the number of people at the beach: The independent variable is temperature.

Number of cigarettes smoked and chance of cancer: The independent variable is number of cigarettes smoked.

The volume of petrol remaining in your tank and the distance you have driven: The independent variable is distance you have driven.

Linear trend

A single line that best represents the general pattern of the data is called a line of best fit.

If the general pattern of the data is roughly a straight line, a linear association exists.

Sometimes a pattern may exist but it is not a straight line. This is called a non-linear trend. Scatterplot **(a)** to the right is an example of a non-linear trend.

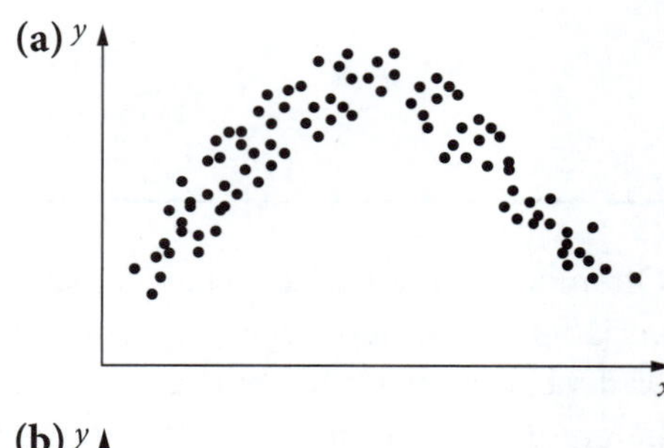

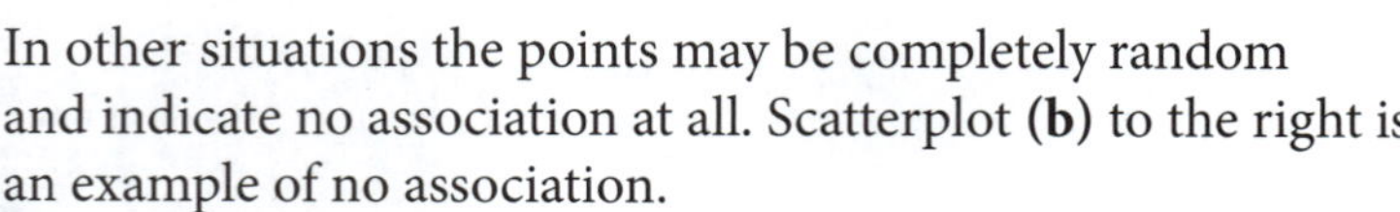

In other situations the points may be completely random and indicate no association at all. Scatterplot **(b)** to the right is an example of no association.

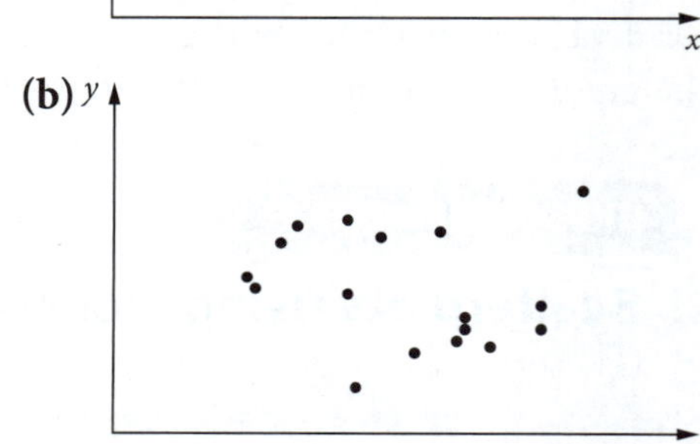

The pattern may show either a positive, increasing trend (going up as you move to the right) or a negative, decreasing trend (going down as you move to the right).

When analysing the scatterplot, look for the following characteristics:

- Is there is an observable pattern?
- Does the pattern show a linear or non-linear association?
- Is the slope positive or negative?
- Identify any outliers, that is, single points that seem to be well outside the general pattern of the rest of the data. If the outliers are excluded from the data, the pattern will be easier to see.
- How well does the pattern fit represent the data? (Is it strong, moderate or weak?) A good fit represents a strong trend and a poor fit represents a weak trend.

Examples are as follows:

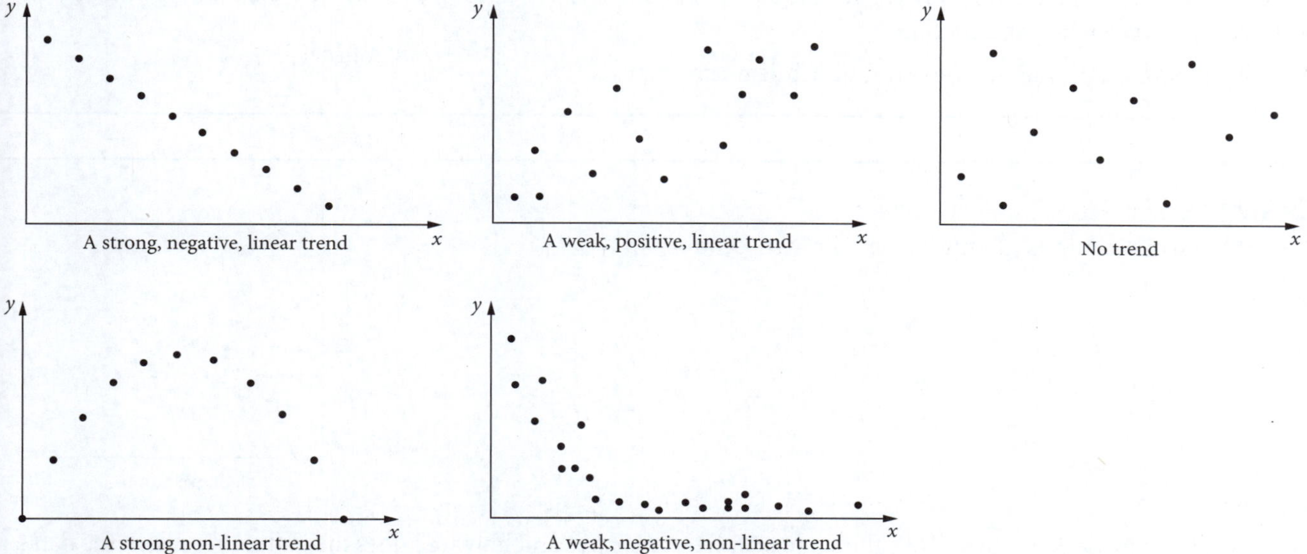

Example 5

Describe the associations between the variables represented in the following scatterplot.

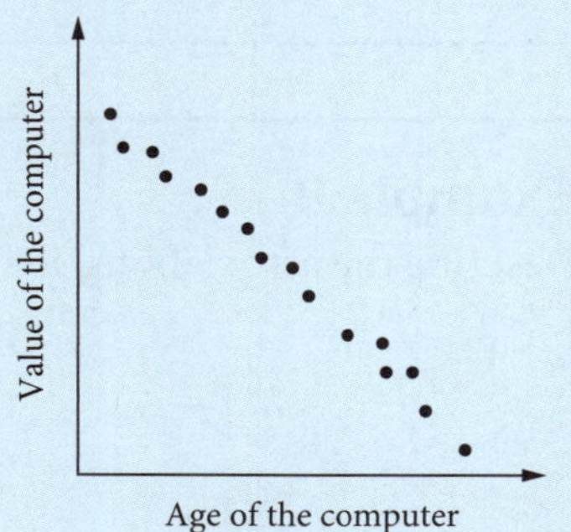

Solution

Is there a pattern? Check for linearity. Is the slope positive or negative and how well does the pattern represent the data?

There is a pattern, with a straight line and the slope is negative. The scatterplot shows a strong negative linear relationship between the age of the computer and the value of the computer.

Therefore, there is a strong, negative, linear association.

Example 6

Describe the associations, if any, in the following scatterplot.

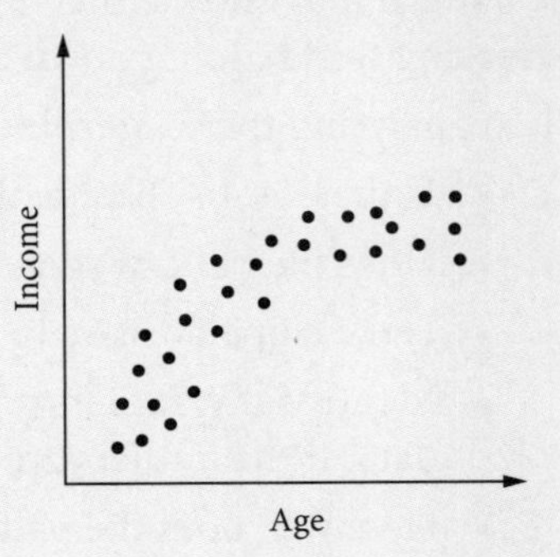

Solution

Is there a pattern? Check for linearity. Is the slope positive or negative and how well does the pattern represent the data?

There is a pattern, which is non-linear and the slope is positive. The scatter plot shows a moderate positive linear relationship between age and income.

Therefore, there is a moderate, positive, non-linear association.

Example 7

Describe the associations, if any, in the following scatterplot.

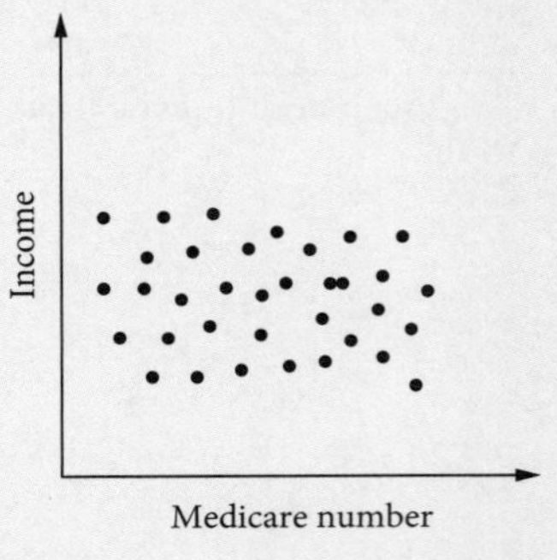

Solution

Is there a pattern? Check for linearity. Is the slope positive or negative and how well does the pattern represent the data?

There is no pattern. The scatter plot shows no relationship between Medicare number and income.

Therefore, there is no association between the variables.

Example 8

Describe the associations, if any, in the following scatterplot.

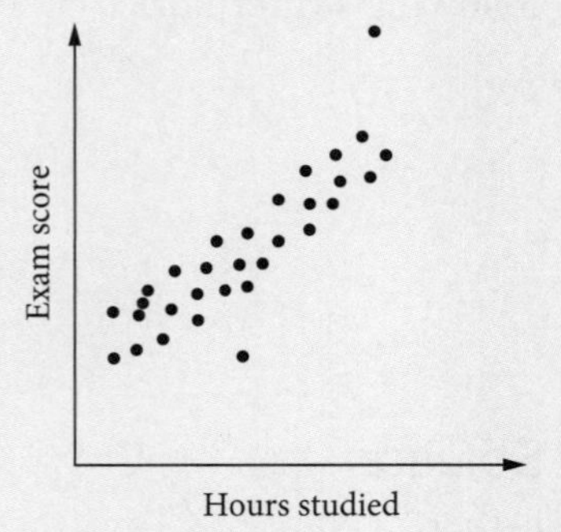

Solution

Is there a pattern? Check for linearity. Is the slope positive or negative and how well does the pattern represent the data?

(Note the presence of possible outliers in this graph.)

There is a pattern, which is linear and the slope is positive. The scatterplot shows a strong positive linear relationship between hours studied and exam score.

Therefore, there is strong, positive, linear association.

EXERCISE 19.2 SCATTERPLOTS AND ASSOCIATION

1 Five friends compete with each other to do the best on their upcoming Mathematics Advanced Examination. In the final month before the exam they record how much time is spent studying, and their final results are given in the table below.

Name	Alex	Bel	Cam	Darmi	Echo
Hours of study	30	15	5	10	35
Mathematics Advanced result (%)	70	45	60	80	95

Draw a scatterplot of hours of study versus maths results and describe any pattern you see.

2 Assuming an association exists, choose the independent variable in each of the following pairs.

(a) Choose the independent variable.
A years of employment **B** value of superannuation

(b) Choose the independent variable.
A rainfall **B** size of plants

(c) Choose the independent variable.
A temperature **B** number of ice-creams sold

(d) Choose the independent variable.
A waist measurement **B** cans of soft drink consumed

3 Describe the associations, if any, in the following scatterplots.

(a)
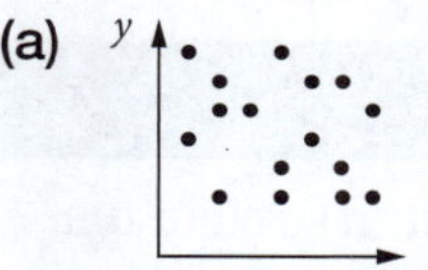

(b)
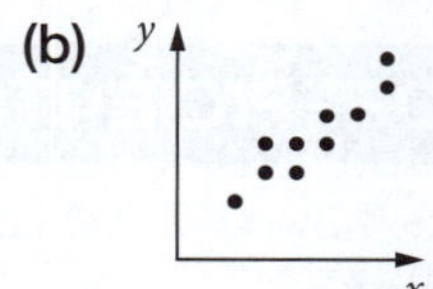

(c)
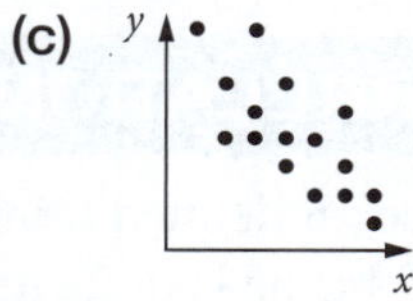

(d)
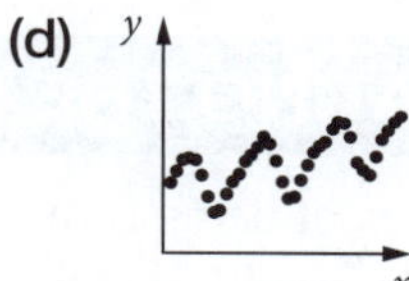

4 The following data set shows the final percentage results for 10 students in the three subjects they have in common.

Maths	40	60	75	60	80	30	45	75	90	55
English	50	70	70	80	90	50	40	65	75	60
Drama	60	80	50	90	75	40	55	80	80	35

In each case put the first named variable on the horizontal axis.

(a) Draw a scatterplot to show any potential association between their Maths and English scores.
(b) Draw a scatterplot to show any potential association between their Maths and Drama scores.
(c) Draw a scatterplot to show any potential association between their Drama and English scores.
(d) Which association appears to be the strongest?

5 The following experimental data has been collected by scientists doing research on climate change. They measured the volume of carbon dioxide CO_2 produced by vehicles using either petrol or LPG (liquid petroleum gas).

Volume of petrol burnt (L)	10	20	30	40	50	60	70	80
Volume of carbon dioxide produced (L)	13.5	25.6	38	53.5	67	74.2	25	108

Volume of LPG burnt (L)	10	20	30	40	50	60	70	80
Volume of carbon dioxide produced (L)	11.1	22	32.5	46	57.5	62.8	80.2	90

(a) Draw a scatterplot for each set of data.
(b) Which set of data contains an outlier? Give the coordinates of the outlier.
(c) Describe the trend for each scatterplot.
(d) Which fuel produces the least amount of carbon dioxide per litre of fuel burnt? Explain your answer.

6 The census results published by the Australian Bureau of Statistics (ABS) in 2013 contained the following information about the percentage of 15 to 24 year olds who were engaged in Higher Education. The results comparing Indigenous and non-Indigenous students are shown on the graph below.

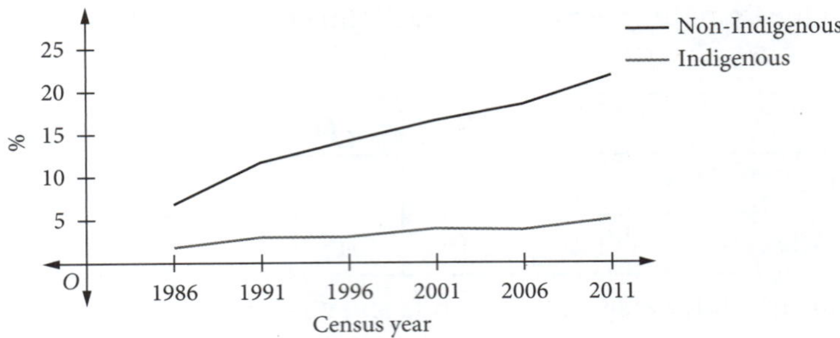

(a) Complete the table below by estimating the percentage values from the graph.

Year	1986	1991	1996	2001	2006	2011
Non-Indigenous	7					22
Indigenous	2					5

(b) Draw a scatterplot of the Indigenous and non-Indigenous data.

(c) Describe the trend in the scatterplot.

(d) By examining the figures for 1986 and 2011, determine the rate of growth in engagement for each population group and comment upon which population group has increased its engagement the most.

19.3 CALCULATING THE CORRELATION COEFFICIENT

As mathematicians, the idea of using rather vague, undefined terms like 'how well the data fits a straight line' and having broad categories like weak, moderate and strong is quite unsettling.

Mathematicians, including young developing ones such as yourself, always prefer to have a number that can be used to describe the strength of the association.

Correlation is the numerical measure to express how closely the individual coordinate pairs fit the line of best fit.

The correlation coefficient *r*

The correlation coefficient r is a number between -1 and 1. The value of r is known as the product moment correlation coefficient; it is also sometimes called the Pearson correlation coefficient, after its developer Karl Pearson.

Correlation coefficient	Description
$0.75 \leq r \leq 1$	This correlates to a strong, positive, linear association with all points on the line of best fit. 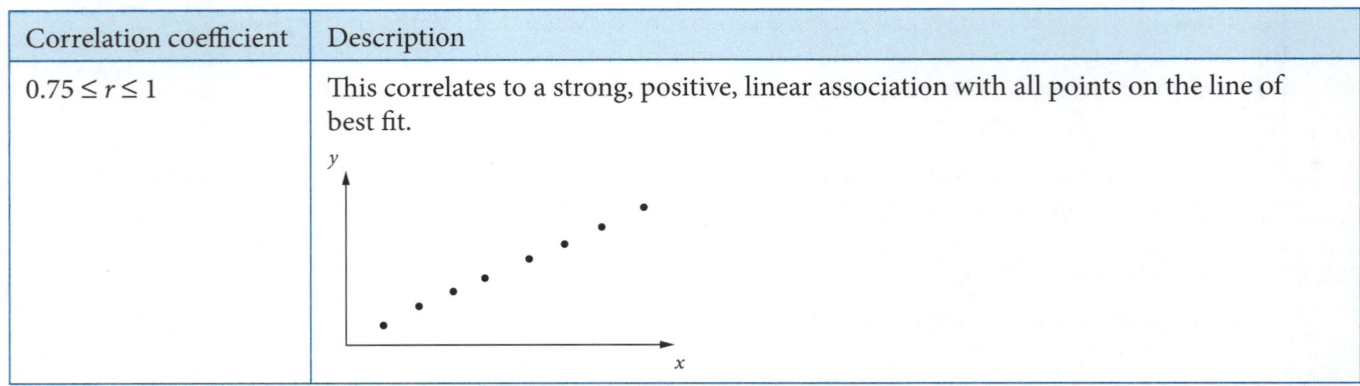

$0.5 \le r \le 0.75$	This correlates to a moderate, positive, linear association with the points spaced on either side of the line of best fit. 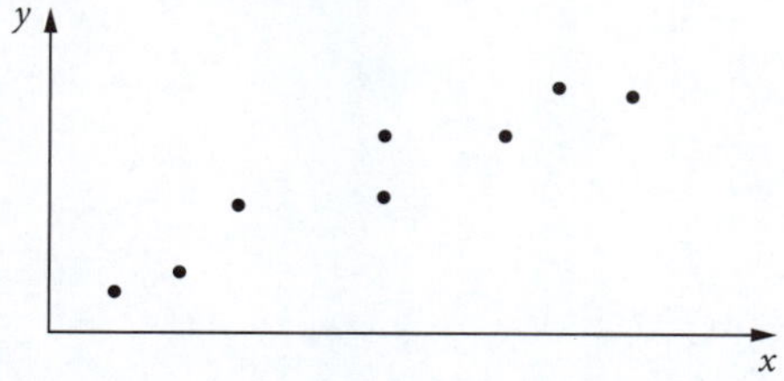
$0.25 \le r \le 0.5$	This correlates to a weak, positive, linear association with the points spaced on either side of the line of best fit. 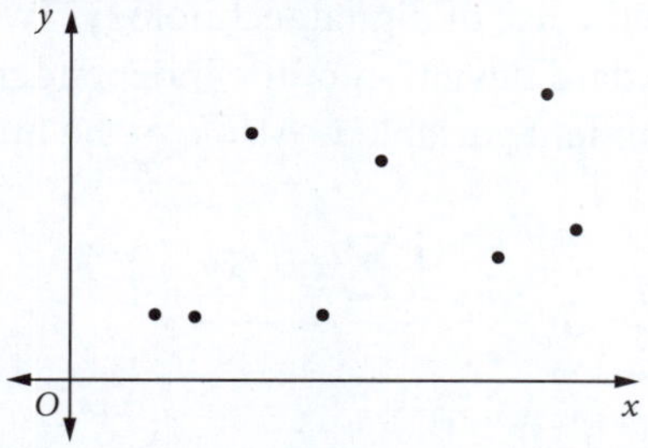
$-0.25 \le r \le 0.25$	There is no association. 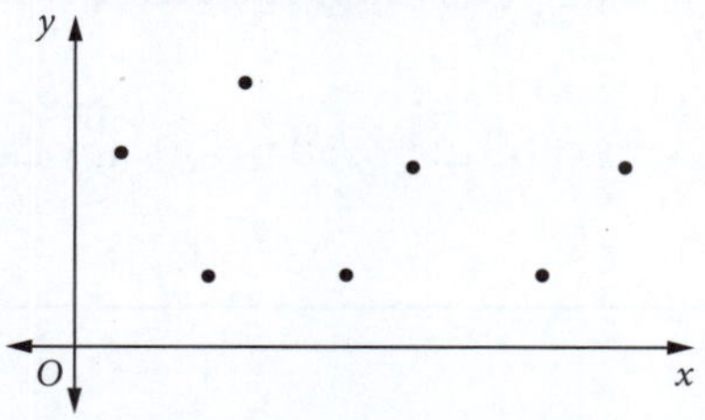
$-0.5 \le r \le -0.25$	This correlates to a weak, negative, linear association with the points spaced on either side of the line of best fit. 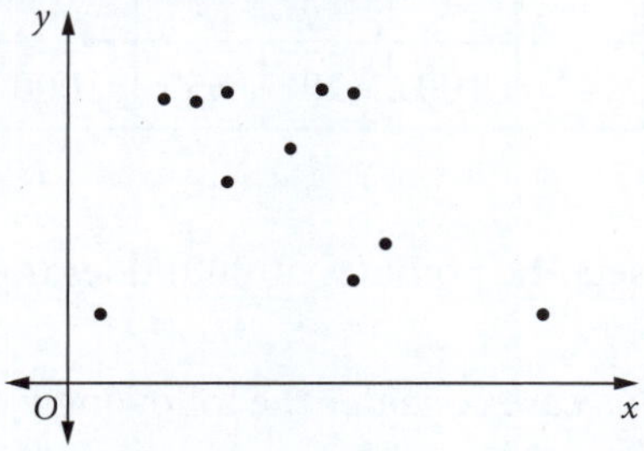
$-0.75 \le r \le -0.5$	This correlates to a moderate, negative, linear association with the points spaced on either side of the line of best fit. 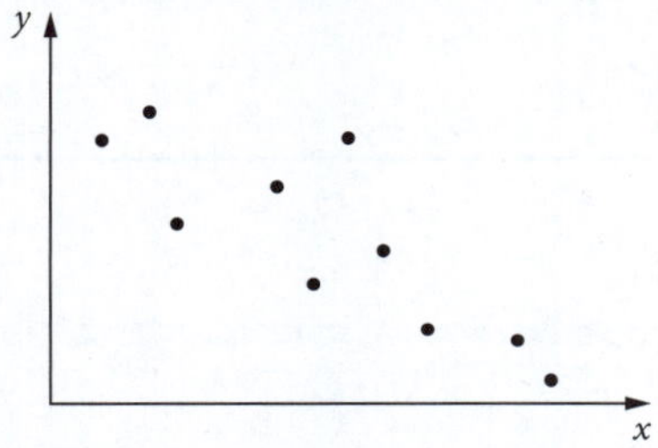

$-1 \le r \le -0.75$	This correlates to a strong, negative, linear association with all points on the line of best fit.

The formula for r is quite complex and often will require the use of digital technology. Two forms of the formula are given below. Here $\bar{x}$ is the mean and s_x is the sample standard deviation of the independent variable x, while $\bar{y}$ is the mean and s_y is the sample standard deviation of the dependent variable y. Also n is the number of points used.

$$r = \frac{\sum(x-\bar{x})(y-\bar{y})}{ns_x s_y} \quad \text{or} \quad r = \frac{\frac{1}{n}\sum xy - \overline{x}\overline{y}}{s_x s_y}$$

You can set up a spreadsheet to calculate the value of the correlation coefficient r. On scientific calculators, the correlation coefficient may be calculated after entering the data pairs into the statistics mode.

Outliers have a large impact on the calculation of r and must be removed before any calculation is done.

Example 9

Find the value of r, correct to 2 decimal places, for the following bivariate data set.

Study time (hours)	1	6	4	12	3	2	9
Number of Facebook friends	900	40	100	20	650	1000	40 000

Solution

Identify any outliers and remove them from the data set: The point (9, 40 000) does not fit the other six points so it should be removed.

Identify the independent variable, if there is one: In this case consider the following two options.

- Having more friends could mean the person becomes more distracted and less study will be done.
- Alternatively, increasing your study time is unlikely to increase the number of Facebook friends you have.

Independent variable is the number of Facebook friends.

Use digital technology to calculate the value of r: $r = -0.77$

EXPLORE FURTHER

The correlation coefficient r

Use spreadsheet software to find the correlation coefficient r for a bivariate data set.

Association and causation—not the same thing

Consider the following associations of pairs of variables.

Variable 1	Variable 2	r	r^2
Hours of structured practice	Piano-playing ability	0.8	0.64
Hours of training	Tiredness after a 5 km run	0.75	0.56
Hours of productive study	Exam result	0.9	0.81
Maths exam result	English exam result	0.95	0.90
Number of newsagencies	Number of hospitals	0.9	0.81

Each of these associations has a strong positive correlation, but can you say that the change in one variable can be caused by a change in the other?

Simply knowing that two variables are associated, no matter how strongly, is not sufficient evidence to conclude that the two variables are causally related.

A causal relationship is demonstrated when a change in the independent variable causes a change in the dependent variable.

Other reasons for an association may be:

- Both variables may be responding to a third variable. This is known as a common response as both variables share a common response to a third variable.
- There may be causation, but the change may also be caused by one or more uncontrolled variables whose effects cannot be disentangled from the effect of the independent variable. This is known as **confounding**.
- There also remains the possibility that what is considered to be the dependent variable is actually the independent variable and is causing the change.
- It may also simply be a coincidence.

Statisticians calculate a **coefficient of determination**, r^2, to give the level of association, but you should not assume that this implies the level of causation.

For the first three associations in the table above, there seems to be a clear cause-and-effect situation. A change in variable 1 (the independent variable) causes a responsive change in variable 2, the dependent variable.

In fact, the value of r^2 can be used to be even more precise about this:

- 64% of the change in piano-playing ability is due to a change in the hours of structured practice. 36% of the change is due to other factors.
- 56% of the change in tiredness after a 5 km run is due to the change in the hours of training. 44% of the change is due to other factors.
- 81% of the change in exam results is due to the change in the hours of productive study. 19% of the change is due to other factors.

The strong correlation in Maths and English results suggest that it is most likely that both of these results are due to other uncontrolled variables such as intelligence, completion of work and hours of study. This would be an example of confounding causation.

The association between the number of hospitals and newsagencies is a good example of a common response. Both of the variables are responding to the population of the town.

Example 10

Determine if there is a causal relationship between each pair of variables:

(a) hours of coaching versus tennis ability: $r = 0.8$ and $r^2 = 0.64$.

(b) number of hours of sunlight versus temperature: $r = 0.5$ and $r^2 = 0.25$

Where causation is observed, complete the following statement.

__% of the change in variable 2 is due to the change in variable 1. __% of the change is due to other factors.

Where causation cannot be verified discuss the reason.

Solution

(a) Decide whether one variable directly influences the other: More hours of coaching would increase a person's tennis ability so a causal relationship is established.
64% of the change in tennis ability is due to the change in hours of coaching. 36% of the change is due to other factors.

(b) Decide whether one variable directly influences the other: These factors are more likely to respond to a common response in this case such as season or location.
Hours of sunlight and temperature both respond to the season and location so they both respond to a common response.

Note: If, for example, you studied the link between the number of phone calls made and the amount of donations received, and found r^2 to be 0.36, the causation is still there. The amount of donations received has many other factors like the persuasiveness of the caller, the financial situation at the time, the areas the calls are made to and if there is a current high profile emergency situation.

Causation is not an exact science and whilst there are fairly obvious examples of each type, in the real world the situation is often somewhere in between and open to some interpretation or further research. In the following exercise only more obvious examples have been chosen.

EXERCISE 19.3 CALCULATING THE CORRELATION COEFFICIENT

1 Calculate the correlation coefficient r to 2 decimal places for each of the following bivariate data sets.

(a)

Length of trip (km)	0.5	10	40	60	80	100
Number of times parents are asked 'Are we there yet?'	2	5	18	20	25	40

(b)

Computer time (h)	1	6	4	12	3	2
Number of 'Likes' added	24	40	50	120	50	30

(c)

IQ	90	120	85	105	88	103	145	94
Distance of home from the post office (km)	4	2	14	6	48	50	25	32

(d)

Minutes played	30	40	50	60	70	80
Points scored	25	15	30	50	45	65

2 Consider the following scatterplot.

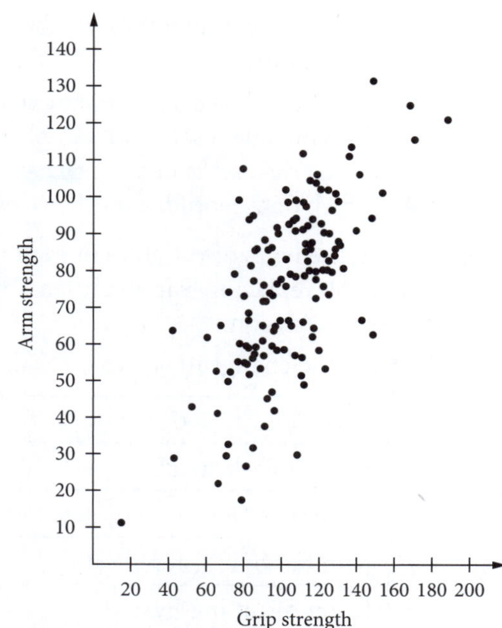

(a) Which pattern describes the scatterplot?

A strong, positive linear

B moderate, negative linear

C no association

D moderate, positive linear

(b) Which value would r be closest to?

A 0.5

B 0.2

C 0

D −0.2

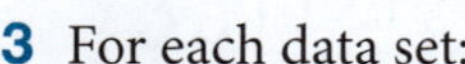

3 For each data set:

(i) use your technology to draw a scatterplot and calculate r

(ii) comment on the association (if any).

(a)

x	2	5	10	18	22
y	18	12	7	4	2

(b)

Length (m)	1	2	4	7	9	15
Cost ($)	2.80	5.60	11.20	19.60	25.20	42.00

(c)

Swimmers in the pool	10	25	45	80	100	120	180
Bacteria level (ppm)	0.2	0.5	0.8	0.6	2.3	3.0	2.8

4 The following experimental data was collected.

Rate of reaction (mol/min)	4.2	6.8	10.3	15.6	182	19.5
Temperature (K)	300	350	400	450	500	550

(a) Without removing any outliers in the data, calculate the value of the correlation coefficient r to 2 decimal places.

(b) Identify the outlier in the data.

(c) An outlier exerts a large influence on the data. Which statement is true?

A Outliers are generally excluded from the data set, so calculations without it are an accurate representation of the study.

B Outliers must be included because they form a significant part of the data set.

(d) After excluding the outlier, use technology to draw a scatterplot of the data.

(e) How is it evident in this scenario that excluding the outlier will mean that the points line up with a stronger linear relationship?

(f) What error may have occurred when recording the data?

5 Three bivariate data sets and their scatterplots are shown below.

Set A

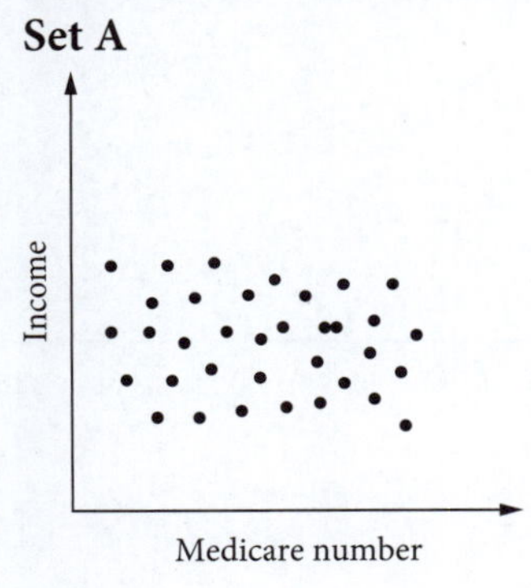

Set B

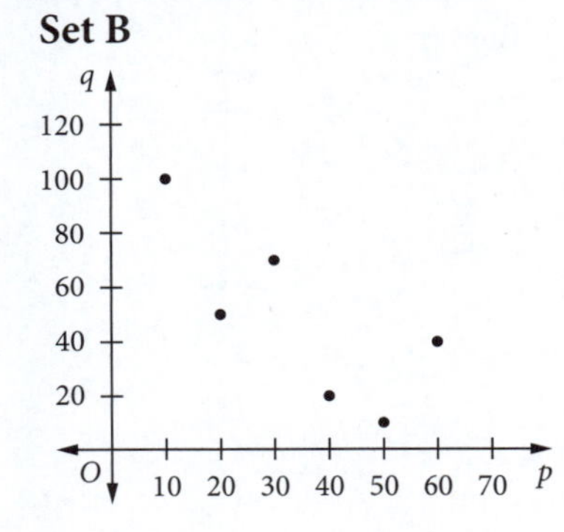

Set C

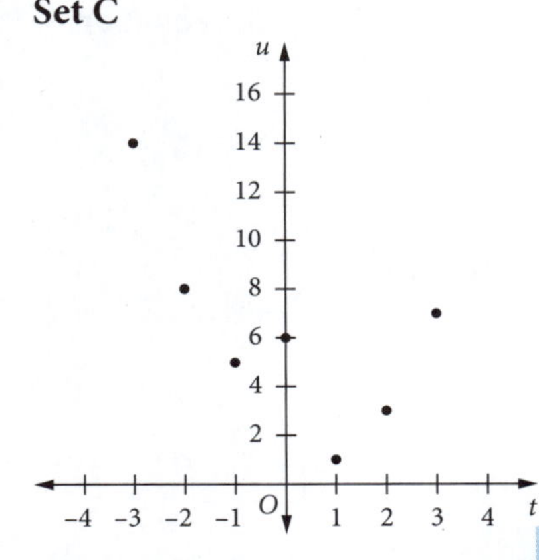

Answer the following questions about the bivariate associations. Note the same response may be true for more than one question.

(a) For which data set is the sample correlation coefficient r closest to -1?
(b) For which data set is the sample correlation coefficient r closest to 1?
(c) Which data set is non-linear?
(d) Which data set indicates the strongest correlation between its two variables?

6 Look at the pairs of variables in each row of the table. Where causation is observed, complete the following statement, replacing 'variable 1' and 'variable 2' with the actual names. Where causation cannot be verified, discuss the reason.

'__% of the change in variable 2 is due to the change in variable 1. __% of the change is due to other factors.'

	Variable 1	Variable 2	r	r^2
(a)	Hours of rehearsal	Quality of performance	0.8	0.64
(b)	Number of fast food outlets in a town	Number of hospitals	0.95	0.90
(c)	Number of registered cars	Number of registered motorcycles	0.6	0.36
(d)	Exam result in Physics	Exam result in Maths	0.9	0.81
(e)	Episode number of a reality TV series	Number of contestants remaining	0.98	0.96

19.4 MODELLING BY FINDING THE EQUATION OF THE LINE OF BEST FIT

There are many different ways to approximate a **line of best fit** for a series of data points. One of the most useful standard procedures in statistics is the method called **least squares regression** analysis. Whenever you are asked to draw a line of best fit using software, you should usually assume that you are finding the least squares regression line.

Drawing a line of best fit by eye

When placing a straight line on a coordinate grid to represent a trend, it is good to aim to have an equal number of points on both sides of the line, with the total distance of points from the line on each side being the same.
Note: The line may not necessarily pass through any of the actual points.

Once you have positioned a line of best fit, you can find the gradient, the y-intercept and consequently the equation for this line. A line of best fit drawn by eye will be placed in slightly different places by different people, so an answer found in this way can only be considered to be an estimate.

Example 11

Use the given scatterplot to answer the questions.

(a) Place a line of best fit by eye.
(b) Calculate the gradient of the line.
(c) Find the y-intercept.
(d) Write the equation of your line of best fit.

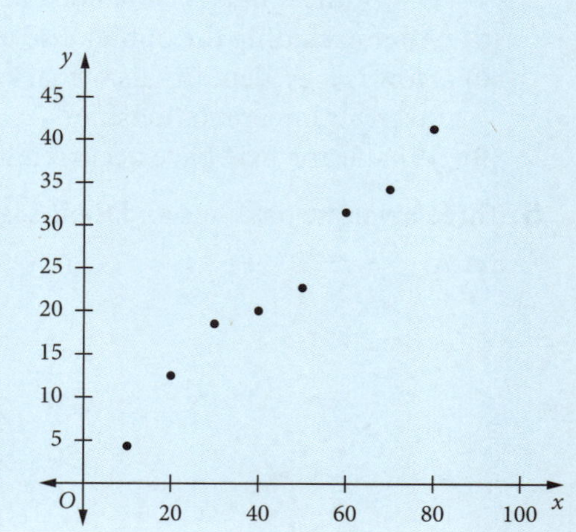

Solution

(a) Use a straight edge to position a line so that there are approximately the same number of points on either side of the line, and so the total distance of the points from the line on both sides is roughly the same.

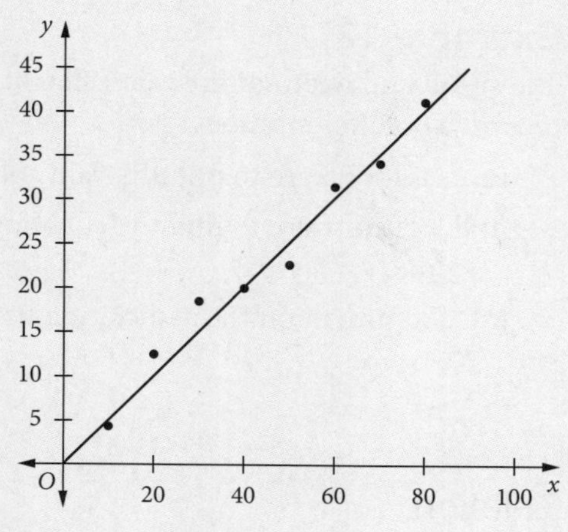

(b) Choose two convenient points from the line and calculate the gradient: Gradient $= \dfrac{\text{rise}}{\text{run}}$

For this line choose the points (20, 11) and (60, 31).

$$\begin{aligned}\text{Gradient} &= \frac{\text{rise}}{\text{run}}\\ &= \frac{31-11}{60-20}\\ &= \frac{20}{40}\\ &= 0.5\end{aligned}$$

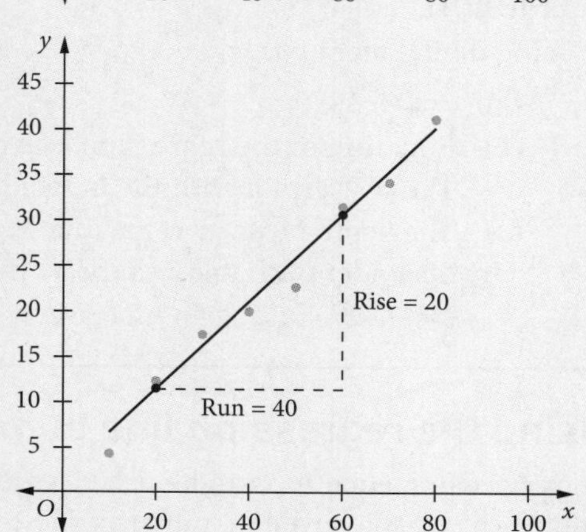

(c) Remember the general equation for a linear graph is $y = mx + c$.
Use $m = 0.5$ from part **(b)** and substitute either of the points used in the calculation for m to calculate c.
(20, 11) is used here but (60, 31) would produce the same result.

$$y = mx + c$$

Substitute $m = 0.5$, $x = 20$ and $y = 11$:

$$\begin{aligned}11 &= 0.5 \times 20 + c\\ 11 &= 10 + c\\ c &= 1\end{aligned}$$

If possible, check this result by reading the value of the y-intercept directly from the graph:
Extending the line of best fit through to intersect with the y-axis, the coordinates of the y-intercept are (0, 1).

(d) Write the equation by substituting in the values for m and c that have been calculated: $y = mx + c$ becomes $y = 0.5x + 1$.

Using technology to find the regression equation

You can use technology to find the equation to a line of best fit directly, which will typically use least squares regression analysis. This course does not expect you to understand how a least squares regression line of best fit is calculated, but you should know how to use digital technology to get the equation. Refer to the 'Explore further' activity below for a worked example using digital technology. The least squares regression line is the line that minimises the sum of the squares of the residuals (i.e. the differences between the data points and the line).

EXPLORE FURTHER

Line of best fit

Use spreadsheet software to find a line of best fit using least squares regression.

Example 12

The speed and depth of the water flowing at a particular point in a river is recorded at different times.

(a) Use software to find the value of r.

(b) Use software to find the equation of the least squares regression line, $y = a + bx$.

(c) Comment on the association using the statistics found.

Depth (m)	Speed (m s^{-1})
0.22	0.47
0.6	0.33
0.78	0.43
0.95	0.42
1.40	0.44
1.75	0.34

Solution

Using digital technology:

(a) $r = -0.3929$

(b) Equation of the regression line is $y = 0.4434 - 0.0404x$.
The scatterplot with the line of best fit follows.

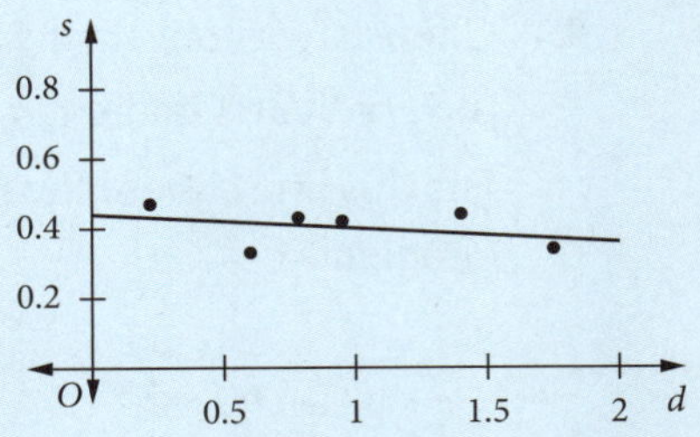

(c) The depth of water is the independent variable and the speed is the dependent variable. It appears that there is a weak, negative association, $r = -0.39$, linking an increase in water depth to a slowing of the speed.

Using the regression line to make predictions

The regression equation can be used to model the situation described by the data. This allows you to predict an unknown variable given a known variable.

If the value used for prediction comes from within the boundaries of the original data, then this is called **interpolation** and the prediction should be relatively reliable.

If the regression equation is used to predict a value outside the data boundaries, then this is called **extrapolation** and can sometimes be misleading, as data relationships often do not continue without changing unpredictably at some point.

Generally, the further the regression line prediction is from the original data, the less reliable the result.

To demonstrate this, consider bivariate data collected concerning the shoe size of people under 18 years of age. For this data you can assume a strong, positive linear association: as people get older, their feet get larger.

Using the regression equation to predict the shoe size of a 15-year-old would be interpolation as 15 lies within the age range of the original data, 0–18. Even allowing for individual variation , you would expect this to provide a reasonable level of accuracy.

However, if you were to extrapolate beyond 18 years of age, then the regression line assumes that feet continue to grow larger and larger as people get older, forever—even though, feet generally stop growing before 25 years of age. It would also make no sense to extrapolate the shoe size of a person 500 years old, or to extrapolate anything that involved negative age or negative shoe size, even though the regression line will extend to all these values.

Generally, the further away from the original data the less reliable the prediction will be.

Example 13

It is given that water flow speed = −0.04 × (depth of water) + 0.44 or $v = -0.04 \times d + 0.44$ where $0.22 \leq d \leq 1.75$. Use this equation to complete the following table.

Speed (m s^{-1})	Depth (m)	Interpolation/Extrapolation
	1	
0.38		
	5	

Solution

Find the speed if the depth is 1 m: $s = -0.04 \times d + 0.44$

$$s = -0.04 \times 1 + 0.44$$

$$s = 0.40\ \text{ms}^{-1}$$

Depth in the original data ranges from 0.22 m to 1.75 m. As 1 m is within these boundaries, this is interpolation.

Find the depth if the speed is 0.38 m s^{-1}: $s = -0.04 \times d + 0.44$

$$0.38 = -0.04 \times d + 0.44$$

$$0.38 - 0.44 = -0.04 \times d$$

$$-0.06 = -0.04 \times d$$

$$\frac{-0.06}{-0.04} = d$$

$$d = 1.5\ \text{m}$$

Speed in the original data ranges from 0.33 m s^{-1} to 0.47 m s^{-1}. As 0.38 m s^{-1} is within these boundaries, this is interpolation.

Find the depth if the speed is 5 m s^{-1}: $s = -0.04 \times d + 0.44$

$$= -0.04 \times 5 + 0.44$$

$$= 0.24\ \text{m}$$

Depth in the original data ranges from 0.22 m to 1.75 m. As 5 m is outside these boundaries, this is extrapolation.

Speed (m s^{-1})	Depth (m)	Interpolation/Extrapolation
0.40	1	Interpolation
0.38	1.5	Interpolation
0.24	5	Extrapolation

Understanding the gradient and intercept values

In the general linear equation $y = mx + c$, the value of m is the gradient or slope of the line and c represents the y-intercept. In modelling situations, these values can often be linked to a practical understanding of the data.

Consider this situation. The cost of phone call with a particular company is a 17 cent connection fee and then an additional fee of 98 cents per minute.

This can be shown graphically as:

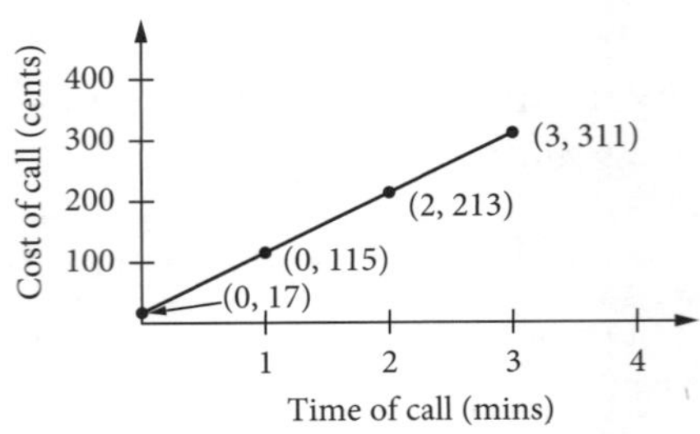

(This assumes that the cost for partial minutes is a proportion of the cost per minute, so that the cost increases steadily as time increases. The graph would be different if the charges were applied in blocks of a minute or 30 seconds, in which case it would look like a step graph.)

The equation of the graph is: Cost of call (cents) $= 98 \times$ time of call (minutes) $+ 17$

Note that the y-intercept (17) represents the fixed costs and the gradient (98) represents the cost per minute of the call.

Example 14

Jayde is in charge of arranging a round robin tournament for her local tennis club. Her fixed costs are the cost of hiring the courts and buying the trophies. The extra costs, which depend upon the number of entrants, are for lunch and tennis balls. Due to court space, the maximum number of people who can enter is 48. The graph below shows the total cost of running the event depending upon the number of entrants.

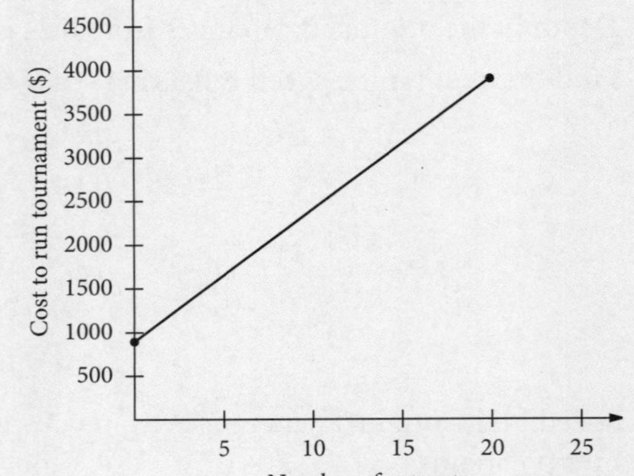

(a) Find the cost of hiring the courts and buying the trophies.

(b) Find the extra cost for every entrant.

(c) Write the equation and find the cost for 25 entrants.

Solution

(a) Read the y-intercept from the graph: 900
The fixed costs are \$900.

(b) Find the gradient: Gradient $= \dfrac{\text{rise}}{\text{run}}$

$$\text{Gradient} = \frac{2400 - 900}{10 - 0} = \frac{1500}{10} = 150$$

The extra cost per entrant is \$150.

(c) Write a linear equation for the situation: $y = mx + c$

$$\text{cost}(\$) = 150 \times (\text{number of entrants}) + \$900$$
$$c = 150n + 900$$

Substitute into the given equation and find the value of the missing variable: $n = 25$

$$c = 150n + 900$$
$$= 150 \times 25 + 900$$
$$= \$4650$$

If there are 25 entrants then Jayde's total cost is \$4650.

EXERCISE 19.4 MODELLING BY FINDING THE EQUATION OF THE LINE OF BEST FIT

1 Consider the following scatterplot.

(a) Copy the scatterplot and on it sketch a line of best fit.
(b) Find the y-intercept and the gradient of this line of best fit.
(c) What is the equation of this line? Round the gradient and y-intercept to the nearest whole number.

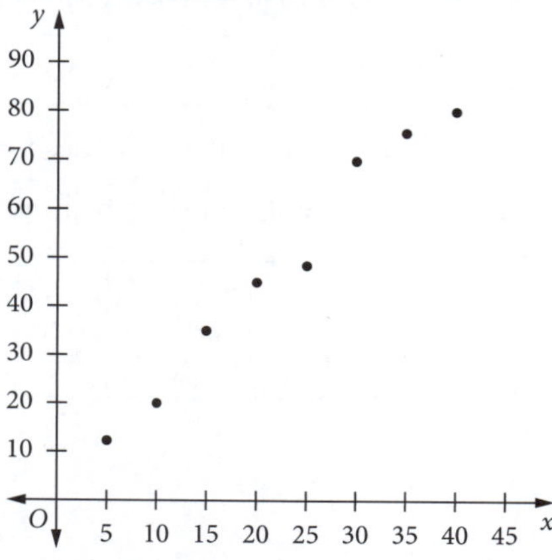

2 The data from a weather balloon measuring the air temperature every kilometre as it rises through the atmosphere are shown below.

Altitude (km)	0	1	2	3	4	5	6	7
Temperature (°C)	15.0	8.5	2.0	−4.5	−11.0	−17.5	−23.9	−30.5

(a) Use technology to find r and the least squares regression equation for the line of best fit.
(b) Comment on the association.

3 The following experimental equation links the length and diameter of octopus tentacles. The original data set for this equation included lengths between 0 and 4 m: length (m) = 0.2 + 0.7 × diameter (mm)

Complete the following table:

Length (m)	Diameter (mm)	Interpolation/Extrapolation
	1	
2.3		
10		

4 Sonya is arranging a round robin tournament for the local Netball Association. Her fixed costs are the cost of hiring the courts and buying the trophies. The extra costs, which depend upon the number of teams that enter, are for umpires and catering. The maximum number of teams who can be accepted is 24 due to the availability of courts. The graph below shows the total cost of running the event depending upon the number of teams entered.

(a) Find the cost of hiring the courts and buying the trophies.
(b) Find the extra cost for every team that enters.

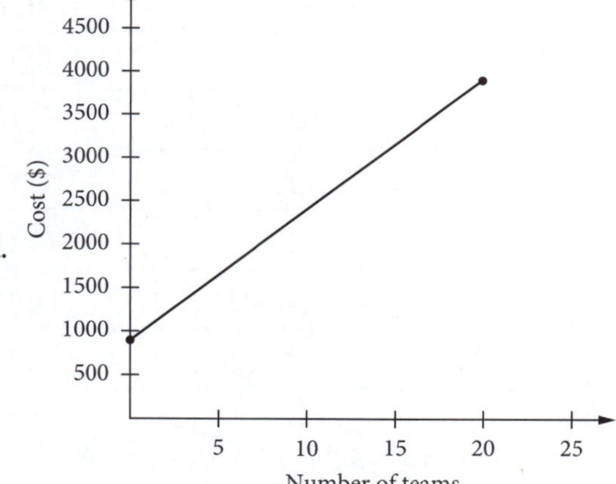

5 For the following bivariate data sets, calculate to 2 decimal places the correlation coefficient and the linear regression equation.

(a) Set A

Husband's age	25	56	80	45	23	32	62	40
Wife's age	23	56	75	46	22	36	57	38

(b) Set B

Husband's age	26	40	70	37	23	32	61	70
Wife's age	22	46	66	41	19	30	57	32

(c) Explain why one set has a much higher correlation than the other.

6 The scatterplot and data shown represent the experimental data collected linking the variables p and s.

p	5	10	15	20	25	30	35
s	3	10	12	15	30	28	30

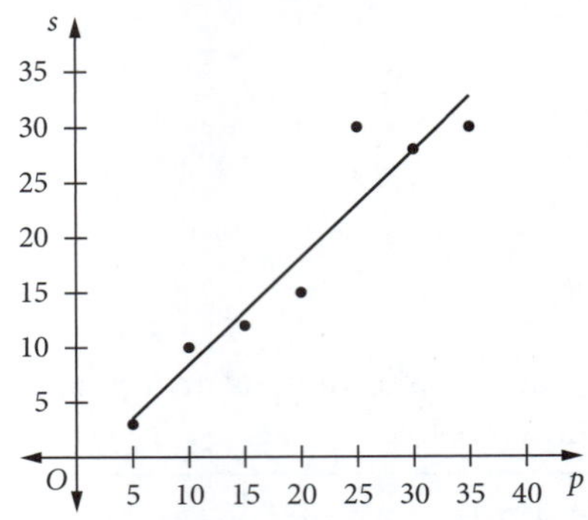

(a) Without calculation, which of the following is the most likely linear regression equation linking the variables?
A $s = 0.9p - 2$ **B** $s = -0.9p - 2$ **C** $s = -p - 1$ **D** $s = p - 1$

(b) Without calculation, if the point (35, 30) should have been recorded as (30, 35), what effect would this have on the gradient?
A the same **B** steeper **C** more shallow **D** negative

7 Some believe that you can predict a person's height by measuring the length of their arm span. The results for 20 students were recorded as shown in the table below.

Person	Height, h (cm)	Arm span, a (cm)
1	128	132
2	138	134
3	141	136
4	142	138
5	145	138
6	144	140
7	152	141
8	145	142
9	148	143
10	144	144

Person	Height, h (cm)	Arm span, a (cm)
11	148	144
12	149	144
13	152	145
14	150	145
15	150	146
16	147	148
17	152	148
18	154	155
19	172	164
20	168	170

(a) Use the table of data to choose the correct options given for the following sentences:
  (i) This is [**bivariate / univariate**] data, since there are [**one / two**] measurements for each person.
  (ii) The data is ranked according to increasing [**arm span / height**]
  (iii) The other variable follows [**a decreasing / an increasing**] linear trend.
  (iv) As both sets of data are increasing together, this would suggest a [**negative / positive**] linear trend.

(b) Choose the correct options to complete the statements below.

Statement
To show any association you could draw a
As this is recorded data you need to check for
To find the equation for the line of best fit you find the
To check on the strength of the association you calculate the
The independent variable is most likely to be
The dependent variable is most likely to be
The independent variable should be plotted on the
The dependent variable should be plotted on the

Options
scatterplot
outliers
least squares regression line
correlation coefficient r
height
arm span
horizontal axis
vertical axis

(c) Use the table of data to complete the following:
- **(i)** The range of heights measured go from [ ] cm to [ ] cm.
- **(ii)** The range of arm spans measured go from [ ] cm to [ ] cm.

(d) Which is a suitable scale to use, so that there are between 5 and 10 even divisions on both the vertical and horizontal axes?

A From 0 to 180 with divisions every 10 cm **B** From 0 to 180 with divisions every 50 cm
C From 125 to 175 with divisions of 2 cm **D** From 125 to 175 with divisions of 5 cm

(e) Draw the scatterplot using the scale from part **(d)** and describe the association.
(f) Calculate the value of r, correct to 2 decimal places. **(g)** Find the regression equation.
(h) From this limited data set, do you think the arm span is a good way of predicting height? Explain.

8 Two of the many variables that influence temperature and rainfall in Australia are the altitude (the height above sea level) and latitude (a measure of the distance from the equator, where a larger number represents a greater distance from the equator).

The data below records the average maximum daily temperature and average yearly rainfall recorded against latitude, all measured at the same altitude.

Latitude (°S)	10	15	20	25	30	35	40	45
Average temperature (°C)	31	28	24	21	16	14	15	10
Average rainfall (mm)	630	550	600	450	250	654	1168	800

(a) Firstly, consider the association between latitude and temperature. Identify the independent variable and draw a scatterplot using technology or in your workbook on a suitably scaled set of axes.
(b) Comment on the association between latitude and average maximum temperature.
(c) Calculate the Pearson correlation coefficient.
(d) Find the regression equation, giving values to 1 decimal place.

Now, consider the association between latitude and rainfall.

(e) Identify the independent variable and draw a scatterplot.
(f) Comment on the association between latitude and average rainfall.
(g) Calculate the Pearson correlation coefficient.
(h) Find the regression equation, giving values to 1 decimal place.

Now, compare the assocations between latitude and temperature, and latitude and rainfall.

(i) Which of the two associations is stronger? Explain your answer.
(j) For what range of values could interpolation be used to predict the temperature, given the latitude?
(k) Find the value of r between temperature and rainfall.

9 The information below about each Australian state from June 2013 is taken from the Australian Bureau of Statistics website.

	Area (km^2)	Population (millions)	Motorcar thefts	Average weekly earnings of full-time workers
New South Wales	800 642	7407.7	23 100	$1480
Victoria	227 416	5737.6	12 100	$1400
Queensland	1 730 648	4658.6	9800	$1560
South Australia	983 482	1670.8	3800	$1320
Western Australia	2 529 875	2517.2	7900	$1800
Tasmania	68 401	513.0	3000	$1280
Northern Territory	1 349 129	239.5	900	$1580
Australian Capital Territory	2358	383.4	400	$1820

(a) Complete the following table without doing any calculations.

Association	Independent variable	Strength and direction of association	Correlation coefficient estimate
Area and population			
Area and motorcar thefts			
Area and average weekly earnings			
Population and motorcar thefts			
Population and average weekly earnings			
Motorcar thefts and average weekly earnings			

(b) Calculate the value of r, to 2 decimal places, for each of the six associations.
(c) For which associations would it be appropriate to calculate the regression equation?
(d) Calculate the regression equation for the associations found in part **(c)**.

19.5 THE STATISTICAL PROCESS

Real-world investigation

Statistics can easily be misused or misrepresented, sometimes deliberately. For example, headlines such as 'Children with bigger feet read better' and 'Head injuries have soared since bike helmets were made compulsory' could lead you to conclude that stretching your feet will improve your reading or that not wearing a bike helmet will prevent head injury.

Both statements in the headlines may be partly true but neither represents the whole truth.

The reality is that older children have bigger feet and also better reading skills, simply because they are older. The introduction of compulsory helmets has meant that bike accidents which previously would cause death, now instead cause only injuries, leading to a higher rate of injuries overall but a lower rate of deaths. Removing the helmet would of course increase the death rate again.

Statistics are created by humans. The process of collection, analysis and interpretation of data must be carefully done to ensure the statistics are as accurate and 'fair' as possible. This may seem straightforward, but there are issues to be addressed at each stage if you want the result to be fair and unbiased.

One way of describing the statistical investigation process is as follows.

Step 1: Clarify the problem and formulate one or more questions that can be answered with data.

Write research questions in such a way that answers can be acted upon.

For example, 'Should the canteen sell sushi?' rather than, 'Do we need more options on the canteen menu?'

Step 2: Design and implement a plan to collect or obtain appropriate data.

The questions asked must be clear and easily understood.

The questions asked must be closed, i.e. have only a small number of alternative answers. Open-ended questions will usually produce data that is hard to categorise; it is best to design your questions so the categories are already there.

The questions must be neutral and not lead the person completing the questionnaire to a particular conclusion. For example, both of the following are 'leading questions':

- Street violence is on the increase, are you in favour of teaching people to use guns?
- Unemployment is an issue, are you in favour of compulsory military training for all?

In statistics, a population is defined as every possible member of the relevant group. If you are unable to interview or survey the population then a sample must be selected without bias.

For example:

Your school is interested in finding out if all students support the idea of the Year 12 students having a special lounge to themselves. If you want a 'yes' from the survey then only ask the Year 12 students, but a more representative sample would be to take 10 students from each year level. This is called taking a stratified sample.

Phone surveys taken during the day may be biased towards the opinions of those who are not working; surveys taken before 11 am on weekends are unlikely to include the opinions of many teenagers, etc.

Step 3: Select and apply appropriate graphical or numerical techniques to analyse the data.

Step 4: Interpret the results of this analysis and relate the interpretation to the original question; communicate findings in a systematic and concise manner.

It is important to remember that in the following exercise you are only looking at bivariate data. Any investigations considered here are looking at the presence or absence of an association between two variables.

Using the statistical method: a case study

Joseph is establishing a new business across different countries and wants to set salary levels that will attract the best people, while keeping within his budget. Two of the positions he needs to fill are the positions of financial manager and human resources manager. Help Joseph do the research and investigation into the relative salaries he should offer for both positions.

Step 1: Clarify the problem and formulate one or more questions that can be answered with data.

Data needs to be found to answer the question 'What is the relationship between the salary of the financial manager and the human resources manager in a company?'

Step 2: Design and implement a plan to collect or obtain appropriate data.

The following data shows the average salary of the same two positions in cities across the world.

City	Financial manager ($AUD)	Human resources manager ($AUD)
Adelaide	88 070	76 790
Albury–Wodonga	81 570	59 570
Alor Setar	28 670	74 020
Amsterdam	87 150	57 760

City	Financial manager ($AUD)	Human resources manager ($AUD)
Auckland	66 680	53 340
Bangkok	37 180	27 440
Beijing	61 480	54 730
Belgrade	61 580	57 910

City	Financial manager ($AUD)	Human resources manager ($AUD)
Berlin	108 530	81 240
Berwick	71 320	60 220
Brisbane	85 430	82 910
Canberra	85 720	82 560
Christchurch	64 610	52 280
Damascus	59 780	61 420
Darwin	74 960	56 230
Geelong	85 630	81 540
Glasgow	85 210	69 540
Hobart	84 800	79 770
Istanbul	57 810	53 530
Kabul	43 880	37 200
Kuala Lumpur	29 950	18 400
Las Vegas	120 360	86 240
London	80 800	70 770
Luang Prabang	40 280	33 240
Madrid	86 940	67 760
Manchester	80 180	55 960
Melbourne	100 660	82 890

City	Financial manager ($AUD)	Human resources manager ($AUD)
Moscow	63 980	50 490
Newcastle	83 720	81 420
New Dehli	42 130	29 020
New York	92 710	78 300
Paris	95 480	74 430
Perth	84 380	76 620
Phakse	21 820	15 850
Santiago	35 760	28 070
Singapore	89 560	70 960
Sydney	97 370	92 920
Tenterfield	65 070	49 260
Tokyo	99 570	77 710
Tripoli	64 950	58 000
Vienna	97 480	77 840
Vientiane	23 580	20 550
Wagga Wagga	79 790	64 690
Warrnambool	72 940	61 550
Wellington	60 470	53 910
Yogyakarta	43 390	31 310

There are 50 cities within this table and using all 50 would be quite time-consuming, so a random sample of 15 cities is selected using a random selection process.

There are many ways to randomise the sample selection, such as using random numbers, drawing the names of cities out of a hat or using every fifth city. The method used here is rolling a die to find the first city and then rolling the die again repeatedly to move down the list until the bottom is reached.

The list of randomly chosen cities is shown below.

City	Financial manager ($AUD)	Human resources manager ($AUD)
Alor Setar	28 670	74 020
Amsterdam	87 150	57 760
Berwick	71 320	60 220
Christchurch	64 610	52 280
Glasgow	85 210	69 540
Kuala Lumpur	29 950	18 400
Luang Prabang	40 280	33 240

City	Financial ($AUD)	Human resources manager ($AUD)
New Dehli	42 130	29 020
Phakse	21 820	15 850
Singapore	89 560	70 960
Tripoli	64 950	58 000
Vienna	97 480	77 840
Wagga Wagga	79 790	64 690
Wellington	60 470	53 910

Step 3: Select and apply appropriate graphical or numerical techniques to analyse the data.

Check to see if there is an independent variable. In this case there is not an independent variable, as both variables seem to vary according to the relative wealth of the country.

Draw a scatterplot to visually show any trend.

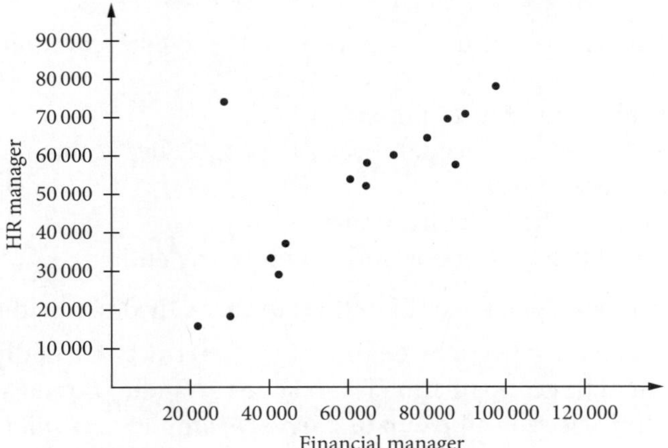

The scatterplot picks up an outlier at Alor Setar (28 670, 74 020). This is removed from the data set before placing the line of best fit and calculating the regression analysis.

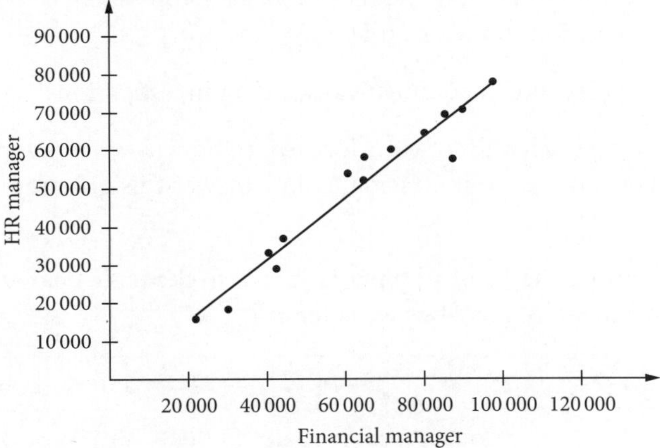

Line of best fit can be obtained using your calculator or approximating the gradient and the y-intercept from the graph.

From a calculator, the line of best fit is: $y = 0.81x - 632.7$

The correlation coefficient is: $r = 0.97$

Step 4: Interpret the results of this analysis and relate the interpretation.

The linear regression equation can be written as:

(Human Resources Manager Salary) $= 0.81 \times$ (Financial Manager Salary) $- 632.7$

There is a strong, positive, linear correlation, although in this case there is no obvious independent variable and the data is exhibiting a common response to the local situation.

EXERCISE 19.5 THE STATISTICAL PROCESS

In these exercises, express answers correct to 2 decimal places unless otherwise stated.

1 From the statements below, find and order the correct four steps that describe the statistical process.

A Interpret the results of this analysis and relate the interpretation to the original question; communicate findings in a systematic and concise manner.

B If there is a large population, design a method for collecting the results that will lead to the conclusion you are after.

C Design and implement a plan to collect or obtain appropriate data.
D Clarify the problem and formulate one or more questions that can be answered with data.
E Design open-ended questions to give the responders the maximum opportunity to express their feelings.
F Select and apply appropriate graphical or numerical techniques to analyse the data.

2 State whether each of the following statements is true or false.

(a) To best understand Australians' attitude to sport, you should take a random survey of people attending a football match.
(b) Statistics can be easily misused or misrepresented.
(c) In statistics, if a population is referenced it means all people alive at that moment in time.
(d) All surveys will produce useful results.
(e) It is important to avoid bias when creating a survey.
(f) When using experimental data it is important to remove any outliers.

3 Select the correct words from each given pair of options to correctly describe the statistical process.

When researching an issue you must begin by defining very [**narrowly / broadly**] the question you wish to answer. If the potential target population is too large to handle you must design a method of [**randomly / conveniently**] selecting a [**group of friends / sample**] that will provide sufficient data, even if you need to eliminate any [**poorly written surveys / outliers**]. Identify, if they exist, the [**independent / discrete**] and [**continuous / dependent**] variables. Use a combination of appropriate statistical tools including: [**scatterplots / Pythagoras' theorem**], back-to-back stem-and-leaf plots, [**frequency diagrams / trigonometry**], line graphs and [**regression analysis / Pythagoras' theorem**]. Summarise your findings in relation to the original question.

The following questions contain suggestions for some bivariate data investigations.

4 Is there an association between the length of your forearm (elbow to wrist) and the length of your right foot? Collect data from 10 or more of your classmates and answer this question using statistical calculations and graphs.

5 The frequency of medical appointments of all types is shown in the table below. The data is found in figures available on the Australian Bureau of Statistics website.

Age group	15–24	25–34	35–44	45–54	55–64	65–74	75 and over
Consultations: Males ('000s)							
GP	967.7	1093.7	1127.6	1162.4	1086.6	805.5	543.4
Specialist	291.7	343.5	382.2	453.5	493.4	438.9	321.1
Dentist	705.8	614.7	618.1	708.4	625.9	440.6	259.6
Hospital admission	126.3	105.9	115.6	143.8	179.0	158.5	137.4
Visited emergency	252.2	214.2	175.9	183.9	137.8	120.2	102.3
Consultations: Females ('000s)							
GP	1158.1	1382.7	1331.7	1292.8	1167.9	830.5	669.5
Specialist	363.1	526.0	556.7	522.4	555.6	440.3	351.6
Dentist	812.4	721.1	832.3	868.2	797.1	479.4	285.3
Hospital admission	163.2	293.4	222.3	169.6	176.2	170.3	142.8
Visited emergency	247.7	251.6	193.6	184.2	162.6	126.2	114.3

Consider the following research questions. Use the data provided to statistically answer some or all of the questions posed or you might prefer to develop your own questions.

(a) What, if any, is the relationship between the number of GP visits made by males and the number of specialist visits made by males?

(b) What, if any, is the relationship between the number of GP visits made by females and the number of specialist visits made by females?

(c) What, if any, is the relationship between the number of GP visits made by males and the number of GP visits made by females?

(d) What, if any, is the relationship between the total number of emergency department visits and the total number of hospital admissions?

(e) What, if any, is the relationship between the total number of GP visits and the total number of dental visits?

6 The table below (continuing over the page) shows the age, sex, height and lung capacity of young smokers and non-smokers. Use this data and the statistical process to develop an investigation into the effect of smoking on the height and lung capacity of young people. Your report should include the following:

The exact question you wish to answer, your method of randomly selecting the sample, the identification of the independent and dependent variables (if possible), appropriate scatterplots and residual graphs, regression analysis and a summary of your findings.

Non-Smokers				Smokers			
Age (years)	Height (cm)	Lung capacity (L)	Sex	Age (years)	Height (cm)	Lung capacity (L)	Sex
9	145	1.708	female	9	147	1.953	male
11	175	2.884	male	14	168	2.236	female
10	163	2.328	male	14	163	3.428	female
14	160	3.381	male	13	155	3.208	female
11	147	2.17	female	11	152	1.694	male
11	169	3.47	male	14	183	3.957	male
12	154	3.058	female	13	175	4.789	male
10	145	1.811	male	12	161	2.384	female
11	163	2.524	male	14	165	3.074	female
17	165	2.742	female	10	168	2.387	female
14	174	3.741	male	12	177	3.835	female
13	177	4.336	male	13	159	2.599	female
14	183	4.842	male	13	173	4.756	male
12	180	4.55	male	13	171	3.086	female
12	160	2.841	female	14	175	4.309	male
18	165	3.566	female	10	168	3.413	female
13	161	3.816	female	10	160	2.975	female
13	173	3.549	male	11	159	3.169	female
13	163	3.147	female	12	173	3.343	male
14	174	4.683	male	12	183	3.751	male
13	170	3.994	male	13	173	2.216	female
12	174	4.393	male	13	168	3.078	female
13	173	3.745	female	12	170	3.186	female
11	170	3.774	female	13	165	3.297	female
17	174	3.155	male	12	169	2.304	male

Non-Smokers			
Age (years)	Height (cm)	Lung capacity (L)	Sex
11	178	2.988	male
11	152	2.498	male
14	163	3.169	female
11	159	2.887	male
13	155	2.704	female
11	163	3.515	female
11	166	3.425	male
15	155	2.287	female
13	166	2.434	female
12	157	2.868	female
16	156	2.813	female
12	168	3.255	female
11	175	4.593	male
14	180	4.111	male
12	154	1.916	male
10	147	1.858	male
16	175	3.88	male
18	192	4.126	male
12	163	2.241	male
13	188	4.225	male
13	171	3.089	male
11	152	2.465	male
12	163	2.913	male
13	185	4.877	male
12	179	3.279	male
10	168	2.581	male
12	156	2.347	female
11	159	2.827	female
14	180	2.538	female
12	174	4.073	male
13	175	4.448	male
13	180	3.984	male
10	147	2.25	female
12	161	2.752	female
14	170	3.68	male
15	160	3.982	male
15	165	4.011	male
18	172	3.702	female
19	176	3.84	female
17	185	4.125	male

Smokers			
Age (years)	Height (cm)	Lung capacity (L)	Sex
11	163	3.102	female
13	170	2.677	female
13	165	3.297	female
10	173	3.498	male
12	156	2.759	female
11	170	2.953	female
13	160	3.785	female
14	168	2.276	male
11	183	4.637	male
10	165	3.038	female
11	155	3.12	female
11	174	3.339	male
13	157	3.152	female
11	171	3.104	female
13	175	4.045	male
14	173	4.763	male
11	165	3.069	female
15	180	4.506	male
19	168	3.519	female
16	173	3.688	male
15	168	2.679	female
15	157	2.198	female
19	166	3.345	female
17	170	3.082	male
16	160	2.903	female
15	163	3.004	female
17	175	3.406	male
15	163	3.122	female
15	174	3.33	female
16	157	2.608	female
15	169	3.799	male
18	170	4.086	male
16	177	4.07	male
15	160	2.264	female
18	179	4.404	male
15	152	2.278	female
16	183	4.872	male
16	170	4.27	male
15	173	3.727	male
16	160	2.795	female

CHAPTER REVIEW 19

1 Which statement characterises the data shown on the scatterplot?

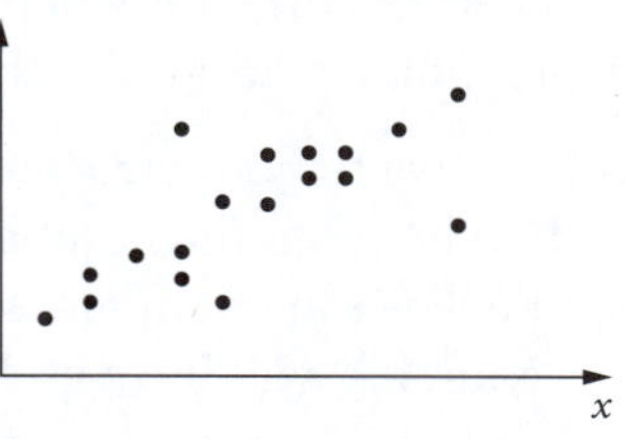

A Weak, positive, linear trend
B Moderate, positive, linear trend
C Moderate, negative, linear trend
D Strong, negative, linear trend

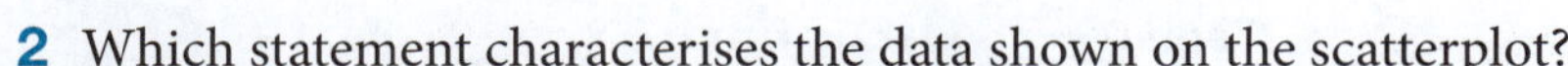

2 Which statement characterises the data shown on the scatterplot?

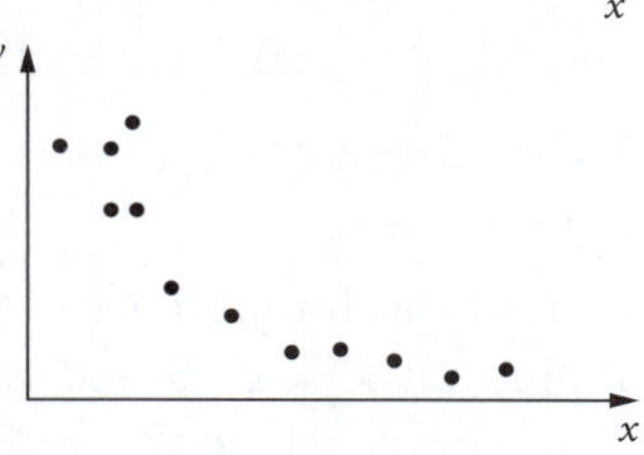

A Weak, positive, linear trend
B Non-linear trend
C Moderate, negative, linear trend
D Strong, negative, linear trend

3 For the scatterplot shown, what is the value of r?

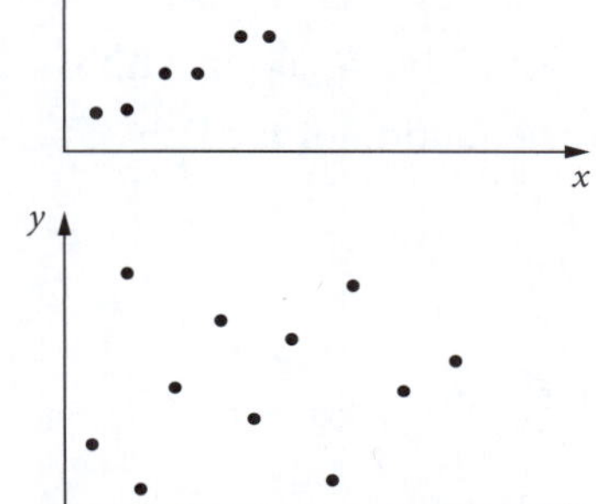

A $-1 < r < -0.7$
B $-0.5 < r < -0.3$
C $0.3 < r < 0.5$
D $0.7 < r < 1$

4 For the scatterplot shown, what is the value of r?

A $-1 < r < -0.7$
B $-0.5 < r < -0.3$
C $-0.2 < r < 0.2$
D $0.3 < r < 0.5$

5 For the scatterplot shown, what is the value of r?

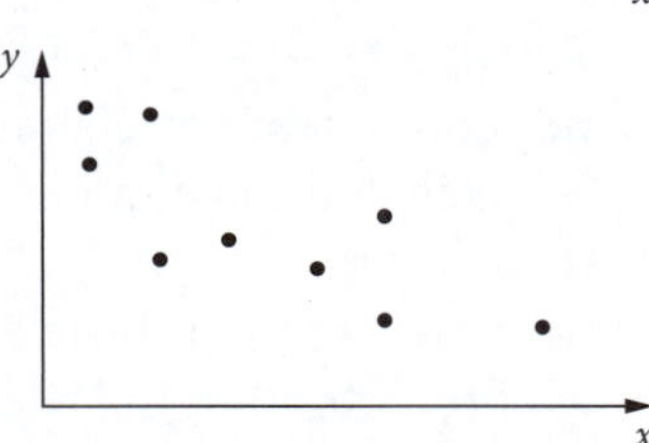

A $-1 < r < -0.7$
B $-0.5 < r < -0.3$
C $0.3 < r < 0.5$
D $0.7 < r < 1$

6 Experimental results indicate the correlation between the mass of food remaining, m, and the number of ants eating the food, n, is -0.4. Which of the following statements is true?

A 40% of the food has been eaten by the ants.
B Increasing the amount of food increases the number of ants.
C Increasing the number of ants decreases the amount of food.
D 60% of the change in number of ants is due to random factors.

7 Consider this table of data.

Name	Abhu	Kiet	Carli	Danni	Ek	Liang	Ramone
Age	1	3	6	9	10	11	12
Shoe size	1	4	7	6	9	9	12

(a) When comparing age and shoe size, r is closest to which value?
A 0.81 **B** 0.88 **C** 0.94 **D** 0.85

(b) After performing a linear regression analysis, what is the equation of the line of best fit?
A shoe size = −0.81 × Age + 0.85 **B** shoe size = 0.81 × Age − 0.85
C shoe size = 0.81 × Age + 0.85 **D** shoe size = 0.85 × Age + 0.81

8 If the coefficient of determination between the variables x and y is 0.64, which of the following statements must be true?

A There is a weak correlation between the variables. **B** The value of r is either 0.8 or −0.8.
C The regression equation is in the form $y = ax + b$. **D** An increase in x will result in an increase in y by 64%.

9 Jezza and Joffa like to note their knowledge of Rugby Union by creating their own statistics. Jezza investigates the association between the number of tries and the number of games played, while Joffa looks at the association between the number of games played and the number of letters in the players' surnames. What respective values of r will Jezza and Joffa likely find?

A 0.7 and 0.8 **B** 0.2 and 0 **C** 0.8 and 0 **D** 0 and 0.8

10 The product moment correlation coefficient for the scatterplot shown was found to be −0.6.

The point (4, 9) was found to be recorded incorrectly and should have been plotted as (4, 1). Based on this change, what is the correct product moment correlation?

A Positive but closer to 0 **B** Positive but closer to 1
C Negative but closer to 0 **D** Negative but closer to −1

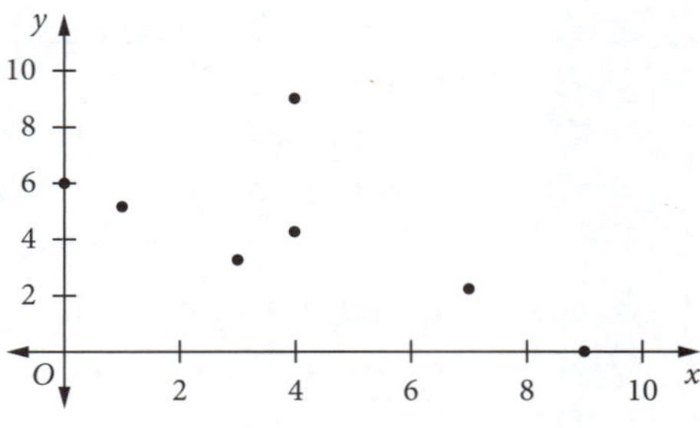

11 A scatterplot is shown along with its line of best fit.
What is the equation of the line of best fit?

A $K = 4B - 3$
B $K = \frac{4}{3}B + 1$
C $K = \frac{4}{3}B - 1$
D $K = \frac{3}{4}B + 1$

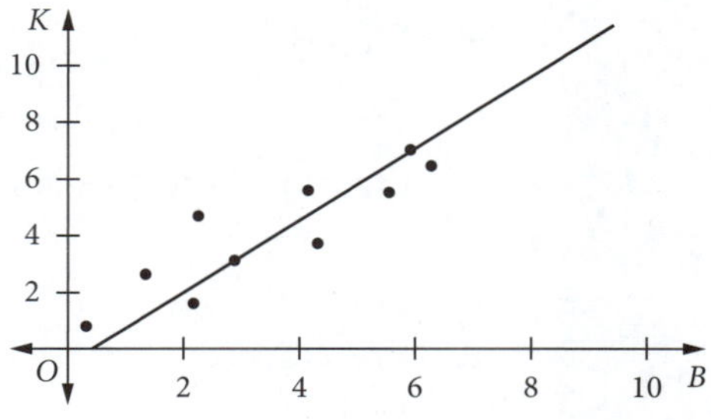

12 Interpret the following information.

(a) The product moment correlation coefficient for the number of points won by Helena on serve during a tennis match and the number of her first serves in play is 0.6. Select the best interpretation of this information.
A 36% of the variation in Helena's success is dependent on the racquet used.
B 36% of the variation in Helena's success is dependent on her ability.
C 36% of the variation in Helena's success is dependent on the number of games of tennis played.
D 36% of the variation in Helena's success is dependent on her first serve.

(b) The product moment correlation coefficient for the number of hospital beds and number of fast food outlets in a town is 0.8. Select the best interpretation of this information.
A 64% of the variation in the number of hospital beds is dependent on the population of a town.
B 64% of the variation in the number of hospital beds is dependent on the number of hospitals in a town.
C 64% of the variation in the number of hospital beds is dependent on the average income of the population.
D 64% of the variation in the number of hospital beds is dependent on the number of fast food outlets.

13 A survey of 150 people of various ages was conducted to learn their opinion about hip-hop dance. The data collected is shown in the two-way table below.

Age bracket	Enjoy	No opinion	Dislike	Total
10–20	4	18		24
20–30	5		5	16
30–40	3			50
>40			40	
Total	16	60	74	

(a) Complete the table by calculating the missing figures.
(b) Which of the variables, age or opinion, is the independent variable?
(c) Convert the figures from part **(a)** into percentages. Give your answers correct to 1 decimal place.
(d) Discuss any trends shown in the data.

14 The following table shows the height (cm) and weight (kg) of players on the roster for the NBL club Perth Wildcats for 2013.

(a) Which of the variables, height or weight, is the dependent variable?
(b) Use technology to help you construct a scatterplot and comment on the association.
(c) Find the value of r.
(d) Find the regression equation. Use h to represent height.

Name	Height (cm)	Weight (kg)
Jermaine Beal	191	92
Greg Hire	201	99
James Ennis	200	95
Matthew Knight	204	109
Tom Jervis	211	105
Drake U'u	192	97
Mathiang Muo	196	95
Jesse Wagstaff	203	100
Erik Burdon	188	82
Shawn Redhage	202	103
Damian Martin	186	92

15 The English Premier League table for 20/1/2014 is shown below.

P—games played
W—wins
L—losses
D—draws
F—goals for
A—goals against
Pts—points

The number of wins is the best predictor of ladder position, but is it more important to score goals or stop the other team from scoring goals?

Find out by comparing the association between goals for and points with the association between goals against and points. You will need to calculate r for both data sets and draw a conclusion.

Team	P	W	L	D	F	A	Pts
1. Arsenal	22	16	3	3	43	19	51
2. Manchester City	22	16	4	2	63	25	50
3. Chelsea	22	15	3	4	43	20	49
4. Liverpool	22	13	5	4	53	27	43
5. Tottenham	22	13	5	4	29	26	43
6. Everton	22	11	2	9	35	20	42
7. Manchester United	22	11	7	4	36	27	37
8. Newcastle United	22	11	8	3	32	28	36
9. Southampton	22	8	7	7	29	25	31
10. Aston Villa	22	6	10	6	22	29	24
11. Hull City	22	6	11	5	22	28	23
12. Norwich City	22	6	11	5	18	35	23
13. West Bromwich Albion	22	4	8	10	24	29	22
14. Stoke City	22	5	10	7	21	36	22
15. Swansea City	22	5	11	6	27	33	21
16. Crystal Palace	22	6	14	2	14	31	20
17. Fulham	22	6	15	1	22	48	19
18. West Ham United	22	4	12	6	22	33	18
19. Sunderland	22	4	12	6	21	36	18
20. Cardiff City	22	4	12	6	17	38	18

16 A group of friends challenge each other to do their best on their upcoming exams. For the month before the exams they record the time they spend studying. The hours of study and exam results for Maths and English are shown in the table below.

Name	Anika	Brae	Chao	Damon	Ed	Fai
Total hours of study	50	32	10	30	70	85
Maths	70	50	70	80	95	90
English	85	75	60	45	90	96

Consider the association between the hours of study and their Maths result.

(a) Which variable is the independent variable? **(b)** Draw a scatterplot of the data.
(c) Describe the association. **(d)** Estimate a value for r.
(e) Draw in a line of best fit by eye and find its equation. Give your answer correct to 1 decimal place.
(f) The correlation coefficient is 0.66 and the regression equation for this association is $y = 0.4x + 58.0$. Explain why it is unlikely everyone in the class would get exactly these results with the method used here.
(g) Determine the relationship between total hours of study and the English result.
  (i) Use technology to find the regression equation. **(ii)** Find the value of r.
(h) Use your regression equation to predict the English scores with the following hours of study (correct to 1 decimal place): 40 hours; 60 hours; 120 hours.
(i) Which of the predictions in the previous question is least likely to be accurate? Why?
(j) Draw a conclusion about the impact of extra study on the end result for Maths and English.
(k) This is a very small data set and a much larger study would be needed to draw any firm conclusions. However, based on these figures, do you consider there is a clear association between hours of study and overall result in English and Maths, or is there a common response or confounding data? Explain.

CHAPTER 20
Random variables

20.1 CONTINUOUS PROBABILITY DISTRIBUTIONS

When a variable can take any value in a particular interval, for example when you measure it rather than count it, you say you have a **continuous random variable**. Quantities which can be modelled using continuous random variables include height, weight, time and mass.

A continuous random variable is defined by its **probability density function** ('pdf') which is usually represented by $f(x)$. A continuous random variable is defined over the interval $(-\infty, \infty)$ although, in practice, it is common that the probability associated with much of this interval is 0. As a result of this, $f(x)$ is usually defined as a hybrid or 'piece-wise', function.

To help understand continuous random variables, consider the following example.

Example 1

Dirk travels to work by train every day. He is interested in mathematics and decides to collect some data about the lateness of his train service. Over a ten-week period (50 work days) he found the following, where 0–<2 means from 0 to less than 2 minutes late.

Minutes late	0–<2	2–<4	4–<6	6–<8	8–<10
Frequency	5	15	17	10	3

(Dirk's train service runs every 10 minutes so a train cannot be more than 10 minutes late.)

If Dirk wants to calculate the probability of his train being less than 4 minutes late, he can add together the 5 and 15 and say the probability is $\frac{20}{50} = \frac{2}{5}$.

This is written as: $P(\text{train is less than 4 minutes late}) = \frac{2}{5}$

This is a simple calculation. But how can Dirk calculate the probability of his train being less than 3 minutes late? The grouped data table is not useful for calculations within individual data bins.

Dirk thinks a histogram might help, but decides to use the relative probability density values for the vertical axis. The relative probability values are found by dividing the probabilities by the bin width, which in this case is 2.

Minutes late	0–<2	2–<4	4–<6	6–<8	8–<10
Probability	0.1	0.3	0.34	0.2	0.06
Relative probability density	0.05	0.15	0.17	0.1	0.03

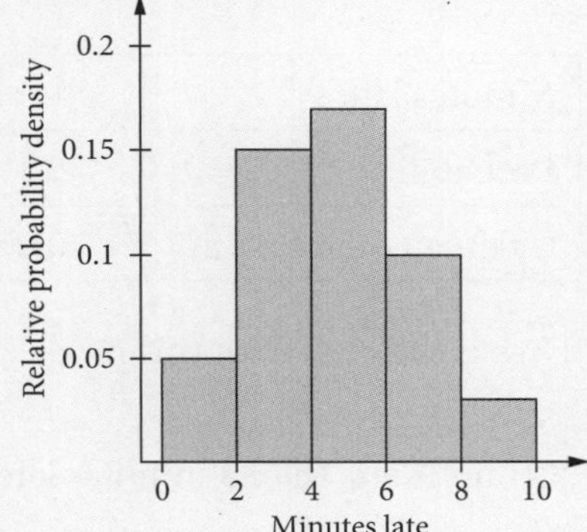

If you calculate the area of each of the rectangles in this histogram showing relative frequencies you will find that their sum is 1. This is because the total area represents the total probability of all possibilities.

Does this help Dirk find the probability of his train being less than 3 minutes late? No, but it gives Dirk another idea. He decides to model the data using a parabola, as he can imagine a negative parabola (concave down) running over the histogram. Dirk decides that the turning point for his parabola will be (5, 0.17) as this represents the median of the time values and corresponds to the highest point in the histogram. This gives an equation in the form $y = -k(x-5)^2 + 0.17$.

Dirk now has to find the value of k. The area under the curve must be 1, so calculus can be used to express the area as an integral.

This can now be solved to find the value of k: $\int_0^{10} \left(-k(x-5)^2 + 0.17\right) dx = 1$

Evaluate the integral: $\left[\frac{-k(x-5)^3}{3} + 0.17x\right]_0^{10} = 1$

$$\frac{-k \times 5^3}{3} + 1.7 - \left(\frac{-k \times (-5)^3}{3} + 0\right) = 1$$

$$\frac{-125k}{3} + 0.7 - \frac{125k}{3} = 0$$

$$250k = 2.1$$

$$k = \frac{2.1}{250} = 0.0084$$

This gives the probability density function (that is, the parabola that models the data) as:

$$f(x) = \begin{cases} -0.0084(x-5)^2 + 0.17 & 0 \le x \le 10 \\ 0 & \text{otherwise} \end{cases}$$

To check the accuracy of this Dirk can calculate the probability for each of the intervals in the original grouped data table.

$$\int_0^{2} \left(-0.0084(x-5)^2 + 0.17\right) dx = \left[\frac{-0.0084(x-5)^3}{3} + 0.17x\right]_0^{2} = 0.0656$$

$$\int_2^{4} \left(-0.0084(x-5)^2 + 0.17\right) dx = \left[\frac{-0.0084(x-5)^3}{3} + 0.17x\right]_2^{4} = 0.2672$$

$$\int_4^{6} \left(-0.0084(x-5)^2 + 0.17\right) dx = \left[\frac{-0.0084(x-5)^3}{3} + 0.17x\right]_4^{6} = 0.3344$$

$$\int_6^{8} \left(-0.0084(x-5)^2 + 0.17\right) dx = \left[\frac{-0.0084(x-5)^3}{3} + 0.17x\right]_6^{8} = 0.2672$$

$$\int_8^{10} \left(-0.0084(x-5)^2 + 0.17\right) dx = \left[\frac{-0.0084(x-5)^3}{3} + 0.17x\right]_8^{10} = 0.0656$$

Minutes late	0–<2	2–<4	4–<6	6–<8	8–<10
Probability	0.1	0.3	0.34	0.2	0.06
Calculated probability	0.07	0.27	0.33	0.27	0.07

Dirk is satisfied with this model as the values are close, so he uses it to answer his original question:

$$P(\text{train is less than 3 minutes late}) = \int_0^{3} \left(-0.0084(x-5)^2 + 0.17\right) dx = \left[\frac{-0.0084(x-5)^3}{3} + 0.17x\right]_0^{3} = 0.1824$$

Correct to 2 decimal places, $P(\text{train is less than 3 minutes late}) = 0.18$.

The model in the example above cannot be used to calculate the probability that, for example, the train will be *exactly* 3 minutes late, as the upper and lower limits in the integral would be the same and hence its value would be zero. Models like this must deal with intervals, even if the interval is very small.

If X is a continuous random variable then $P(X = x) = 0$, for all possible values of x.
The probability is represented by area, so zero width results in zero area.

The function f is the probability density function ('pdf') of the variable X. It is important to note that $f(x)$ is not the probability. You need to integrate between two limit values to obtain a probability.

As $P(X = x) = 0$, all of the following expressions have the same value:

$P(a \le X \le b) = P(a \le X < b) = P(a < X \le b) = P(a < x < b)$

As seen above, you can find the probabilities associated with continuous probability density functions by integrating the function between the values that specify an interval.

A quadratic model will not suit every data set. It may be that a linear model will work best, or a model based on some other mathematical function, perhaps even a piece-wise (hybrid) function.

An example of this type of function would be one that follows a 'triangular' distribution. Consider the following piece-wise function and a sketch of its graph:

$$f(x) = \begin{cases} 4(x-4.5), & 4.5 < x \le 5.0 \\ 4(5.5-x), & 5.0 < x \le 5.5 \\ 0, & \text{otherwise} \end{cases}$$

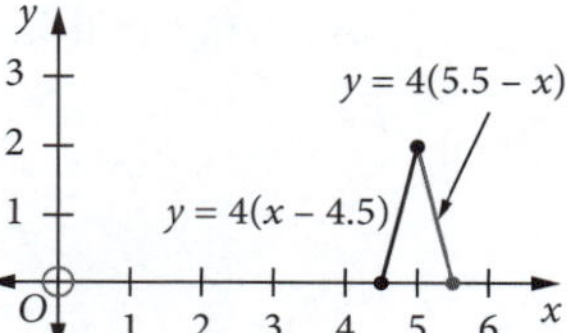

An inspection of the graph will reveal that the area under the graph is 1, as required for the function to represent a probability density function.

For a function $f(x)$ to be a probability density function:

- $f(x) \ge 0$ for all values of x
- $\int_{-\infty}^{\infty} f(x)dx = 1$, i.e. the area enclosed by the graph $y = f(x)$ and the x-axis is equal to 1.

Note that a probability density function can take on values greater than 1. You must remember that it is the *area* bounded by the curve and the x-axis that must be 1.

Calculating a value associated with a continuous density function

Example 2

A particular continuous random variable has the following probability density function:

$$f(x) = \begin{cases} -3(x-1)^2 + 2, & 0 \le x \le 1 \\ 0, & \text{otherwise} \end{cases}$$

Find $P(X \le 0.7)$.

Solution

Identify the bounds of integration to use: the lower bound is 0 and the upper bound is 0.7.

Expand the quadratic expression: $-3(x-1)^2+2 = -3(x^2-2x+1)+2$

$$= -3x^2+6x-3+2$$

$$= -3x^2+6x-1$$

Find the required integral: $\int_0^{0.7}\left(-3x^2+6x-1\right)dx$

$$= \left[-x^3+3x^2-x\right]_0^{0.7}$$

$$= (-(0.7)^3+3\times(0.7)^2-0.7)-0$$

$$= 0.427$$

Calculating a value associated with a continuous density function for a piece-wise function

Example 3

A particular continuous random variable has the following probability density function:

$$f(x)=\begin{cases}4(x-4.5), & 4.5<x\le 5.0\\ 4(5.5-x), & 5.0<x\le 5.5\\ 0, & \text{otherwise}\end{cases}$$

(a) Find $P(X\le 4.7)$.

(b) Find $P(X\le 5.2)$.

Solution

Define the piece-wise function: $f(x)=\begin{cases}4(x-4.5), & 4.5<x\le 5.0\\ 4(5.5-x), & 5.0<x\le 5.5\\ 0, & \text{otherwise}\end{cases}$

(a) Find the definite integral for $P(X\le 4.7)$: $\int_{4.5}^{4.7}4(x-4.5)\,dx$

Evaluate this integral:

$$= [2(x^2-9x)]_{4.5}^{4.7}$$

$$= 2[4.7^2-9\times 4.7-(4.5^2-9\times 4.5)]$$

$$= 0.08$$

(b) Find the definite integral for $P(X\le 5.2)$: $\int_{4.5}^{5.0}4(x-4.5)\,dx+\int_{5.0}^{5.2}4(5.5-x)\,dx$

$$= [2(x^2-9x)]_{4.5}^{5}+[2(11x-x^2)]_5^{5.2}$$

$$= 2[25-45-(4.5^2-40.5)+57.2-5.2^2-(55-25)]$$

$$= 0.82$$

If any of your integrals evaluate to more than 1 for any probability density function, then you can be sure your answer is incorrect. The entire area under the curve must be 1, so part of it cannot be greater than this.

Similarly, if any of your integrals give a negative value, then you can be sure this is incorrect.

The value of the integral represents probability, and probability cannot be less than 0 or greater than 1.

As for discrete probability distributions, common calculations with continuous probability density functions involve finding the mean, variance and standard.

You should recall that for any discrete probability distribution, the expected value (mean) is found by adding the sum of the products of the values and individual probability values ($\Sigma x_i\, p_i$) and that the variance can be found from $E(X^2) - [E(X)]^2$. The integral is the continuous distribution equivalent of summation, Σ, so the following formulas should not surprise you.

The mean of a continuous probability density function is found using the following formula:

$$\mu = \int_{-\infty}^{\infty} x f(x)\,dx$$

The variance of the continuous probability density function is found using the following formula:

$$\begin{aligned}\sigma^2 &= E\left(X^2\right) - [E(X)]^2 \\ &= \left(\int_{-\infty}^{\infty} x^2 f(x)\,dx\right) - \mu^2\end{aligned}$$

As usual, the standard deviation is the square root of the variance.

Although the formulae say to find the integrals from $-\infty$ to ∞, in practice you calculate the integrals over the interval where the function is non-zero.

Example 4

For a particular Infant Welfare Centre the probability density function of the age of the children(x years) brought to the centre is given by:

$$f(x) = \begin{cases} \frac{3}{4}x(2-x), & 0 \le x \le 2 \\ 0, & \text{otherwise} \end{cases}$$

Find the following correct to three decimal places.

(a) mean

(b) standard deviation

Solution

(a) Find the expression for $xf(x)$: $xf(x) = x \times \frac{3}{4}x(2-x)$

$$= \frac{3}{4}\left(2x^2 - x^3\right)$$

Find the mean μ by calculating the integral $\int_{-\infty}^{\infty} x f(x)\,dx$:

(Remember, you actually calculate the integral over the interval for which the function is non-zero.)

$$\mu = \frac{3}{4}\int_0^2 (2x^2 - x^3)\,dx$$

$$= \frac{3}{4}\left[\frac{2x^3}{3} - \frac{x^4}{4}\right]_0^2$$

$$= \frac{3}{4}\left(\frac{16}{3} - 4 - 0\right)$$

$$= 1$$

(b) Find the value of the integral $\int_{-\infty}^{\infty} x^2 f(x)\,dx$:

(Remember, you actually calculate the integral over the interval for which the function is non-zero.)

$$\int_{-\infty}^{\infty} x^2 f(x)\,dx = \frac{3}{4}\int_0^2 x^2(2x - x^2)\,dx$$

$$= \frac{3}{4}\int_0^2 (2x^3 - x^4)\,dx$$

$$= \frac{3}{4}\left[\frac{x^4}{2} - \frac{x^5}{5}\right]_0^2$$

$$= \frac{3}{4}\left(8 - \frac{32}{5}\right)$$

$$= \frac{6}{5}$$

From (a), μ^2=1.

Find the value of μ^2 (you know the value of μ from **(a)**): $\mu^2 = \frac{256}{225}$

Calculate the variance using $\sigma^2 = \left(\int_{-\infty}^{\infty} x^2 f(x)\,dx\right) - \mu^2$: $\sigma^2 = \frac{6}{5} - 1$

$$= 0.2$$

$$\sigma = \sqrt{0.2}$$

$$= 0.447$$

You should check that your answers are plausible. For example, is the mean value within the interval for which the probability is non-negative? Also, as a general rule, the range of the distribution should be roughly five times the standard deviation. Knowing this will assist you when trying to sketch some graphs. For the example above, the mean is near the middle of the non-negative interval and five standard deviations is $5 \times 0.377 = 1.885$, compared to the actual range of 2.

You can also find the median of a continuous pdf. As should be expected, the median m is the value such that $P(X < m) = 0.5$. This means you need to set up and solve an equation of the form:

$\int_0^a f(x)dx = 0.5$, assuming the non-zero part of the piece-wise function starts at 0.

Example 5

A particular continuous random variable X has the following probability density function:

$$f(x) = \begin{cases} \frac{x}{16}, & 0 \le x \le 4\sqrt{2} \\ 0, & \text{otherwise} \end{cases}$$

Find the median.

Solution

Write an equation that states what you are required to solve: $\int_0^a f(x)dx = 0.5$

Write an expression for $\int f(x)dx$: $\int f(x)dx = \int \frac{x}{16}dx$

Call the upper boundary a, and write the definite integral in the equation: $\int_0^a \frac{x}{16}dx = 0.5$

Find the primitive: $\left[\frac{x^2}{32}\right]_0^a = \frac{1}{2}$

Evaluate and solve the equation: $\frac{a^2}{32} = \frac{1}{2}$

$a^2 = 16$

$a = 4$, taking the positive square root as $a > 0$.

The median is 4.

Cumulative distribution function

Thinking back to Dirk and his investigation into the lateness of his train (Example 1 above), to find P(train less than three minutes late) he needed to calculate $\int_0^3 f(x)dx$ and to find P(train less than five minutes late) the calculation would be $\int_0^5 f(x)dx$.

To save recalculating $\int f(x)dx$ each time a new function, the **cumulative distribution function** ('cdf'), can be defined. The cdf is usually designated as $F(x)$. For Dirk's train investigation the following would apply:

$f(x) = -0.0084(x-5)^2 + 0.17$, or after anti-differentiating the function:

$$F(x) = \begin{cases} -0.0028x^3 + 0.042x^2 - 0.04x, & 0 \le x \le 10 \\ 0, & \text{elsewhere} \end{cases}$$

Note that $0 \le F(x) \le 1$, as the probability must be between 0 and 1.

The graphs of $F(x)$ and $y = 0.5$ can be graphed on the same set of axes as an alternative way of finding the median.

For the probability density function in the previous example the corresponding cumulative density function would be:

$$F(x) = \begin{cases} \dfrac{x^2}{32}, & 0 \le x \le 4\sqrt{2} \\ 0, & \text{otherwise} \end{cases}$$

The value of the median is the x-coordinate of the point of intersection, which is the same value as found earlier.

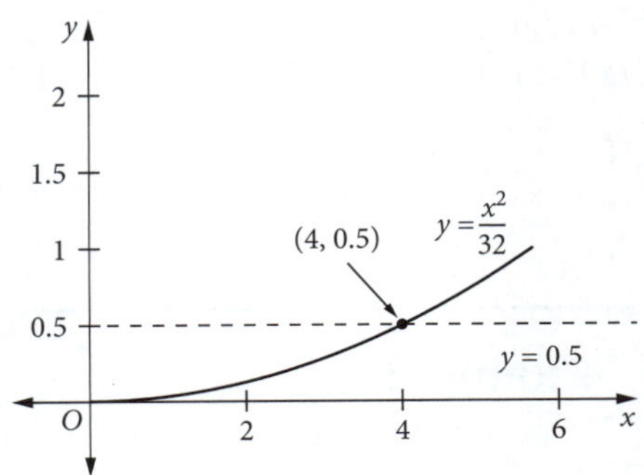

EXERCISE 20.1 CONTINUOUS PROBABILITY DISTRIBUTIONS

1 For the continuous random variable with probability density function:

$$f(x) = \begin{cases} \dfrac{3}{13}\left(x^2 + 4\right), & 0 \le x \le 1 \\ 0, & \text{otherwise} \end{cases}$$

find, correct to four decimal places:

(a) $P(X \le 0.75)$ **(b)** $P(X \le 0.9)$ **(c)** $P(X \ge 0.25)$ **(d)** $P(X \ge 0.65)$

2 For the continuous random variable with probability density function:

$$f(x) = \begin{cases} \dfrac{15000(x-50)}{x^4}, & x \ge 50 \\ 0, & \text{otherwise} \end{cases}$$

find the exact values for the following:

(a) $P(X \le 60)$ **(b)** $P(X \le 75)$ **(c)** $P(X \ge 65)$ **(d)** $P(X \ge 99)$

3 For the continuous random variable with probability density function:

$$f(x) = \begin{cases} \dfrac{3}{13}\left(x^2 + 4\right), & 0 \le x \le 1 \\ 0, & \text{otherwise} \end{cases}$$

find, correct to four decimal places, the following values:

(a) the mean

(b) the standard deviation.

4 For the continuous random variable with probability density function:

$$f(x) = \begin{cases} \dfrac{15000(x-50)}{x^4}, & x \ge 50 \\ 0, & \text{otherwise} \end{cases}$$

find the value of the mean.

5 For the continuous random variable with probability density function:

$$f(x) = \begin{cases} \dfrac{3}{13}\left(x^2 + 4\right), & 0 \le x \le 1 \\ 0, & \text{otherwise} \end{cases}$$

find the median, correct to four decimal places.

6 For the continuous random variable with probability density function:

$$f(x) = \begin{cases} \dfrac{15000(x-50)}{x^4}, & x \geq 50 \\ 0, & \text{otherwise} \end{cases}$$

find the median, correct to the nearest whole number.

7 If $f(x) = \begin{cases} \dfrac{x}{45} + k, & 0 \leq x \leq 6 \\ 0, & \text{otherwise} \end{cases}$

defines a probability density function, then the value of k is:

A 0 **B** $-\dfrac{1}{10}$ **C** $\dfrac{1}{10}$ **D** $\dfrac{1}{2}$

8 For the continuous random variable with probability density function:

$$f(x) = \begin{cases} \dfrac{2x}{25}, & 0 \leq x \leq 5 \\ 0, & \text{otherwise} \end{cases}$$

find, correct to four decimal places:

(a) $P(1.5 \leq X \leq 2.5)$

(b) $P(2 \leq X \leq 4.5)$

(c) $P(1.75 \leq X \leq 3.15)$

9 If $f(x) = \begin{cases} \dfrac{\pi}{12}\sin\left(\dfrac{\pi x}{6}\right), & 0 \leq x \leq 6 \\ 0, & \text{otherwise} \end{cases}$

then $P(X \leq 3)$ is equal to:

A 0.45 **B** 0.5 **C** 0.55 **D** 0.6

10 For the uniform continuous random variable with probability density function:

$$f(x) = \begin{cases} k, & 0 \leq x \leq 6 \\ 0, & \text{otherwise} \end{cases}$$

find the following values:

(a) k **(b)** $P(X \leq 3)$ **(c)** $P(X \leq 5)$

11 For the continuous random variable with probability density function:

$$f(x) = \begin{cases} \dfrac{\sin(x)}{2}, & 0 \leq x \leq \pi \\ 0, & \text{otherwise} \end{cases}$$

find the exact value of the median.

12 In the javelin competition at a primary schools athletics carnival, it is found that the distance s metres that the javelin is thrown is a continuous random variable with probability density function:

$$f(s) = \begin{cases} \dfrac{1}{486}\left(81 - s^2\right), & 0 \leq x \leq 9 \\ 0, & \text{otherwise} \end{cases}$$

Find:

(a) the mean distance the javelin is thrown

(b) the standard deviation for the distance the javelin is thrown

(c) the median distance the javelin is thrown.

13 Let the duration, X minutes of a telephone conversation be represented by a continuous random variable with probability density function:

$$f(x) = \begin{cases} Ce^{-\frac{x}{10}}, & 0 \le x \le \infty \\ 0, & \text{otherwise} \end{cases}$$

(a) Find the exact value of C.
(b) Find the following values, correct to four decimal places:
(i) $P(X \le 2.5)$ **(ii)** $P(X \ge 6.4)$ **(iii)** $P(2 \le X \le 7)$
(c) Find $E(X)$.
(d) Find $\text{Var}(X)$.
(e) What is $P(X \le E(X))$, correct to 4 decimal places?

14 A physical therapist uses a particular practical test to determine the reaction time of her patients. From experience, she has determined that the reaction time, t seconds, can be modelled by a continuous random variable with probability density function:

$$f(t) = \begin{cases} k\left(t - \frac{1}{4}\right)\left(\frac{5}{4} - t\right)^2, & \frac{1}{4} \le x \le \frac{5}{4} \\ 0, & \text{otherwise} \end{cases}$$

Unless otherwise stated, find the following answers correct to 3 decimal places.
(a) Find the exact value of k. **(b)** Find the mean reaction time.
(c) Find the median reaction time.
(d) Find the proportion of patients who react in less than one second.
(e) Find the proportion of patients who react in a time between $\frac{2}{5}$ and $\frac{4}{5}$ of a second.
(f) Find the proportion of patients who take longer than $\frac{3}{4}$ of a second to react.

15 The local council needs to decide what to do about collecting recyclable materials. Each household has a 'recyclable materials' bin that is collected fortnightly. Each truck that collects these bins will contain an amount X of materials that are *not* recyclable. (X is measured in units of 100 kg.) It is known that the random variable X has a probability density function:

$$f(x) = \begin{cases} kx(4 - x)^2, & 0 \le x \le 4 \\ 0, & \text{otherwise} \end{cases}$$

(a) Show that $k = \frac{3}{64}$. **(b)** Find the mean and standard deviation of X.
(c) Find the probability that a bin chosen at random has more than 3 units of non-recyclable material in it.

The council has two choices as to how to proceed with the collection of recyclable material. The first is as follows.

If the truck contains less than 3 units of non-recyclable material they will be able to sell the materials for \$500, but if the quantity is more than this they will only receive \$300.
(d) Find the expected value of X for a truck full of material.

The second choice for the council is to sort the material before selling it. By doing this they will ensure that there are virtually no non-recyclable materials being sold. As a result, they will get \$650 per load. However, this means they cannot process as many trucks, so they only collect the material from households every 3 weeks.
(e) Is it financially viable for the council to follow this second approach?

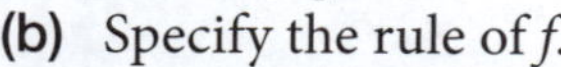

16 Consider the function f whose graph is shown.

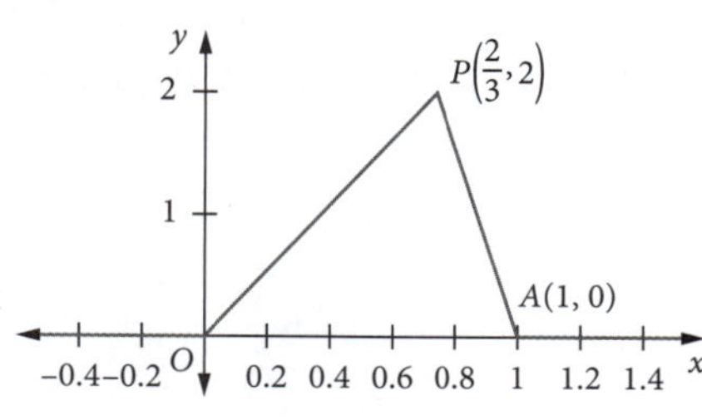

(a) Show that the function could be used to represent a probability density function. Explain your answer using mathematical reasoning.

(b) Specify the rule of f.

(c) Consider a continuous random variable, X, with probability density function f. Calculate $P\left(X < \frac{2}{3}\right)$ and $P\left(X > \frac{2}{3}\right)$ using the areas of the triangles.

(d) Consider a continuous random variable X with probability density function f. Calculate $P\left(X < \frac{2}{3}\right)$ and $P\left(X > \frac{2}{3}\right)$ using integration. Comment on the results from parts **(c)** and **(d)**.

(e) Calculate $P\left(\frac{1}{2} < X < \frac{5}{6}\right)$, using integration.

20.2 THE NORMAL DISTRIBUTION

The following histogram shows the height of 200 randomly selected Australian women, measured in cm accurate to one decimal place.

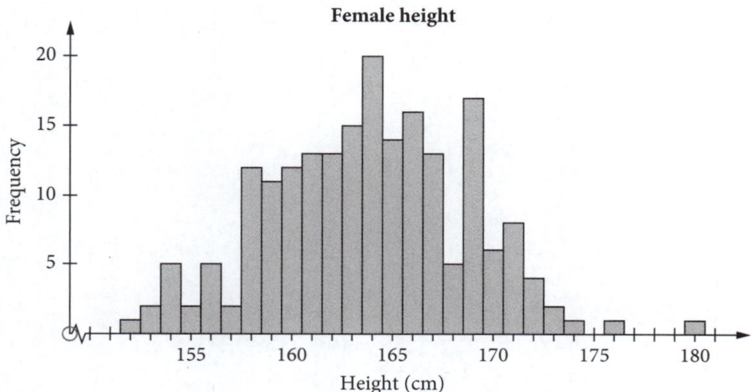

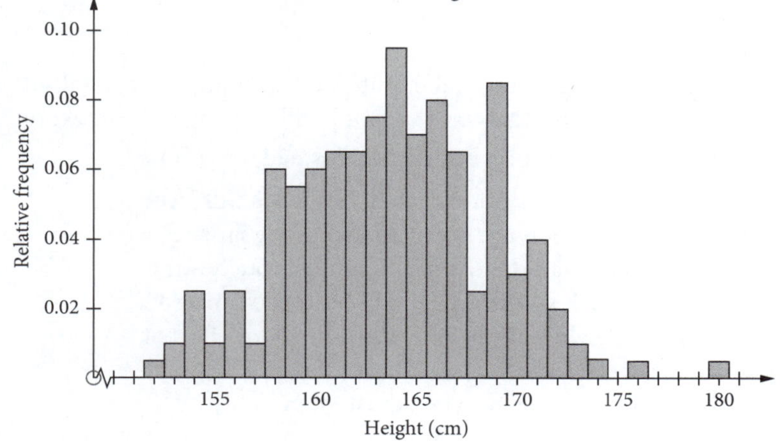

Can you find a function that will model the data, just like Dirk did with the train data in the previous module?

The total area of the rectangles in the histogram is 200, the size of the sample, so the first thing that needs to be done is to make the sum of the areas of the rectangles equal to 1.

To do this, divide the vertical scale values by 200. As the bin width (the width of the columns in the histogram) is set to 1, this means that the area of each bin represents the probability of a woman having a height in that 1 cm range. More importantly, the total area of the bins is now 1, so the situation is like a probability density function.

From the raw data, the mean and standard deviation can be calculated:

Mean = 163.8085

Standard deviation = 4.858 855

What sort of function might be suitable as a model for this data? The shape appears similar to a parabola, but the ends seem to drop away too quickly for that to be a successful model. Perhaps a shape such as the one shown below might be suitable.

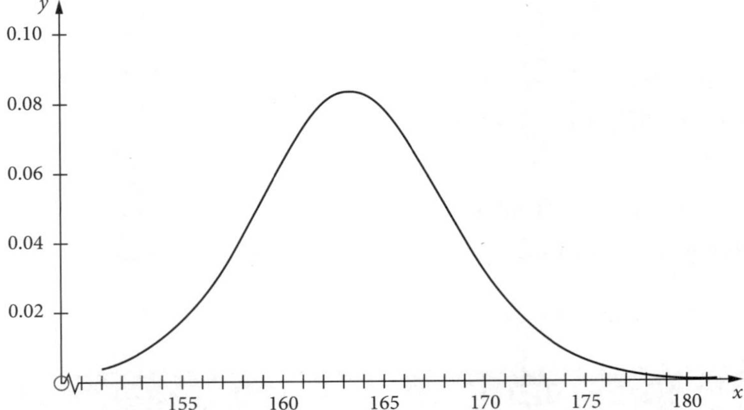

If the two images are put together it can be seen that this function certainly has some potential as a model for the distribution of heights.

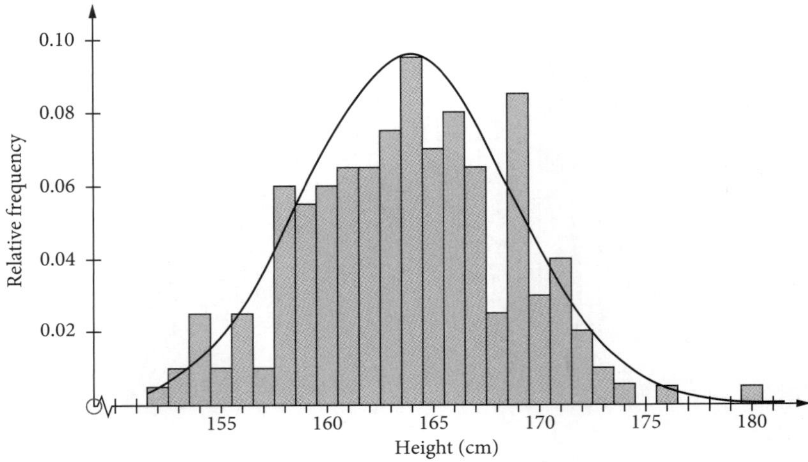

This matches the general shape quite well, although as with all models, it is not perfect. What function might produce this graph? The way the graph falls off suggests an exponential function of some kind, but unlike a typical exponential it falls off on both sides. Instead, try $f(x) = 2^{-x^2}$.

This shape looks like a good match, but is it a pdf? The values of the function are non-negative, so it satisfies that condition for a pdf. The graph almost touches the x-axis at -3 and 3, so the area could be approximated using a triangle with a base of 6 and a height of 1. That gives an area of 3, which is too large a value. The graph of the function will need to be adjusted, possibly by using a vertical dilation, for example $f(x) = k \times 2^{-x^2}$ for some suitable value of k. This can be determined by using a definite integral to obtain a more accurate approximation.

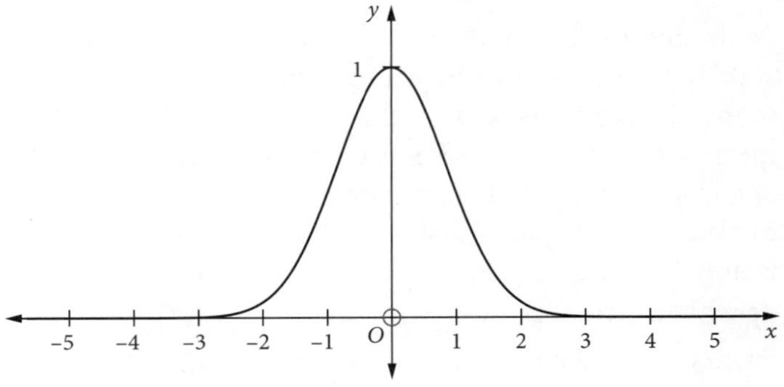

Note that while the domain of the function is $\mathbb{R}$ (real numbers), the value of the function is increasingly close to 0 to the left and right of the graph, for example, $x < -3$ and $x > 3$.

Using a suitable definite integral to find an approximation for the area beneath the graph, therefore find the value of $\int_{-3}^{3} 2^{-x^2}\,dx$. The integration of this function is not covered in this course, but if you have access to graphing software you can use that to evaluate this integral. The integral could also be approximated using the trapezoidal rule.

MAKING CONNECTIONS

Definite integral of an approximate distribution curve

Use technology to find the value of the definite integral $\int_{-3}^{3} 2^{-x^2}\,dx$

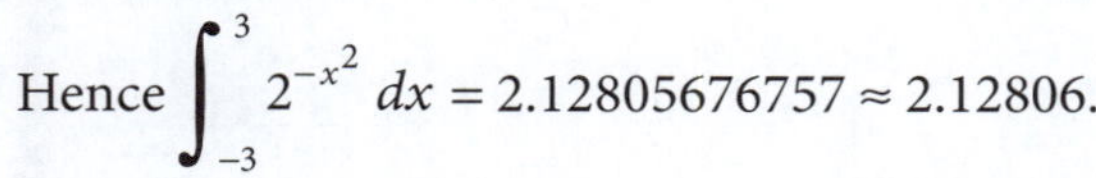

Hence $\int_{-3}^{3} 2^{-x^2}\,dx = 2.12805676757 \approx 2.12806$.

This shows that using $k = 2.12806$, and hence $f(x) = \frac{1}{2.12806}2^{-x^2}$, will produce a pdf with the desired shape. However, it will also need to be translated and dilated to match the data presented at the beginning. Mathematicians have found that a base of e rather than 2 works well, and in this case instead of $\frac{1}{2.12806}$ the value $\frac{1}{\sqrt{2\pi}}$ should be used. This gives the function with domain $\mathbb{R}$ and rule $f(x) = \frac{1}{\sqrt{2\pi}}e^{\frac{-x^2}{2}}$, defined using constants you are already familiar with. This graph is shown at right and has the same general shape.

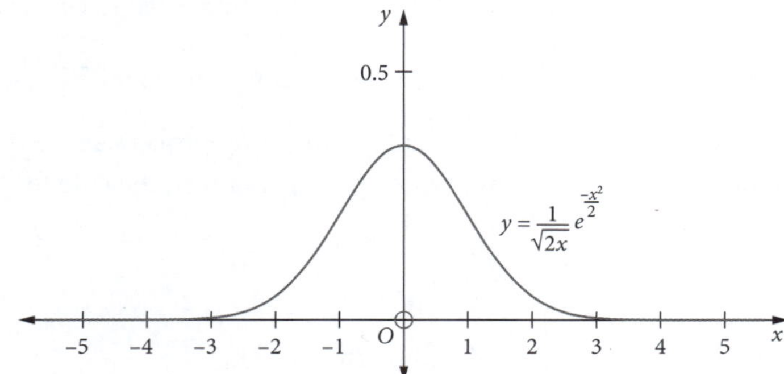

Although it looks like the graph meets the x-axis, it is asymptotic to the x-axis in both directions. That is, as $x \to \infty, f(x) \to 0$ and as $x \to -\infty, f(x) \to 0$ where $f(x) > 0$ for all $x \in \mathbb{R}$.

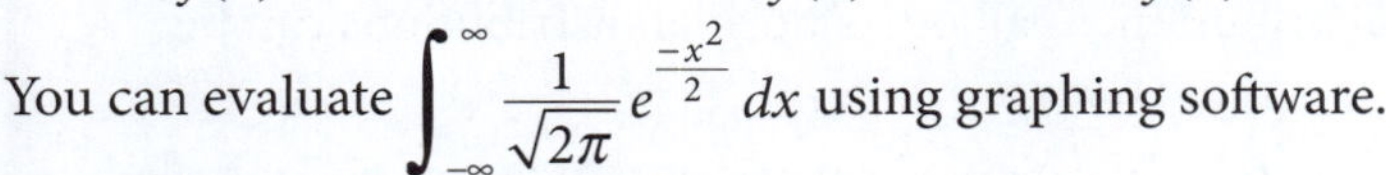

You can evaluate $\int_{-\infty}^{\infty} \frac{1}{\sqrt{2\pi}}e^{\frac{-x^2}{2}}\,dx$ using graphing software.

MAKING CONNECTIONS

Definite integral of a normal distribution curve

Use technology to find the value of the definite integral $\int_{-\infty}^{\infty} \frac{1}{\sqrt{2\pi}}e^{\frac{-x^2}{2}}\,dx$

$$\int_{-\infty}^{\infty} \frac{1}{\sqrt{2\pi}}e^{\frac{-x^2}{2}}\,dx = \int_{-100}^{100} \frac{1}{\sqrt{2\pi}}e^{\frac{-x^2}{2}}\,dx = 1$$

(The limits on the integral have been changed as ∞ cannot be used by all software, but $[-100,100]$ is a close enough approximation to infinity; −50 to 50 also gives a value of 1.)

You now have a pdf, but you still need to translate and dilate this to match your data. From your previous work of transformations of graphs of functions, you should recall that a horizontal translation by h units will transform the graph of $y = f(x)$ onto the graph of $y = f(x - h)$. You should be able to see that the required horizontal translation is equal to the mean; however, a horizontal dilation will also be required. What value might be related to the horizontal dilation? The effect of the horizontal dilation is to spread out the values. The standard deviation is a measure of the spread of the data, so that is a value worth trying.

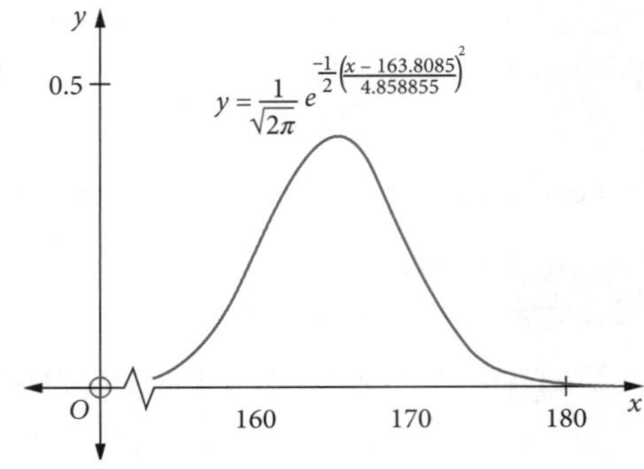

In combination, a horizontal translation by h units combined with a horizontal dilation by b units will transform the graph of $y = f(x)$ onto the graph of $y = f\left(\frac{x-h}{b}\right)$. In this case, the value of h is given by the mean of your data, 163.8085, and the value of b is given by the standard deviation, 4.858855. The function obtained is $f(x) = \frac{1}{\sqrt{2\pi}} e^{\frac{-1}{2}\left(\frac{x-163.8085}{4.858855}\right)^2}$

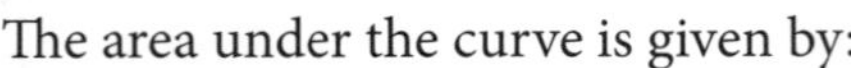

The area under the curve is given by:

$$\int_{-\infty}^{\infty} \frac{1}{\sqrt{2\pi}} e^{\frac{-1}{2}\left(\frac{x-163.8085}{4.858855}\right)^2} dx = \int_{133}^{193} \frac{1}{\sqrt{2\pi}} e^{\frac{-1}{2}\left(\frac{x-163.8085}{4.858855}\right)^2} dx$$

$$= 4.858855$$

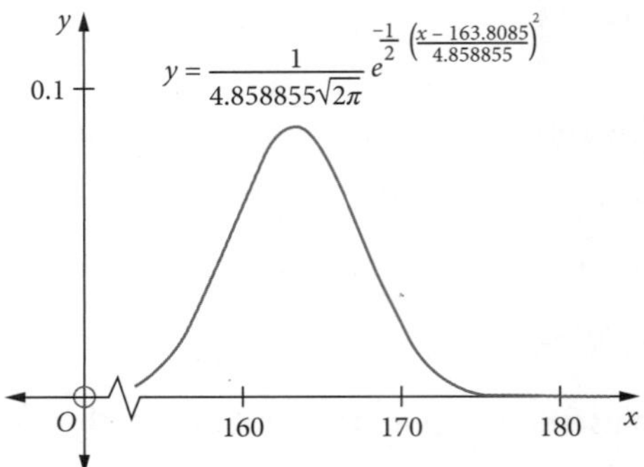

The area should be related to the total probability (1), but it is too big by a factor of the standard deviation. Divide by this to obtain a vertical dilation which reduces the area to one, so that

$f(x) = \frac{1}{4.858855\sqrt{2\pi}} e^{\frac{-1}{2}\left(\frac{x-163.8085}{4.858855}\right)^2}$ is the final result. This is a pdf with a mean of 163.8085 and a standard deviation of 4.858855.

You would not want to go through this long process every time you have a data set to analyse. However, you can generalise the rule for this function to the form: $f(x) = \frac{1}{\sigma\sqrt{2\pi}} e^{\frac{-1}{2}\left(\frac{x-\mu}{\sigma}\right)^2}$

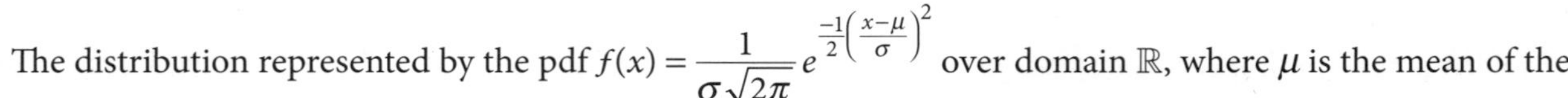

The distribution represented by the pdf $f(x) = \frac{1}{\sigma\sqrt{2\pi}} e^{\frac{-1}{2}\left(\frac{x-\mu}{\sigma}\right)^2}$ over domain $\mathbb{R}$, where μ is the mean of the distribution and σ is the standard deviation of the distribution, is known as the **normal distribution**. The special case where $\mu = 0$ and $\mu = 1$ is called the standard normal distribution, as all other normal distributions can be obtained from it by translation and dilation.

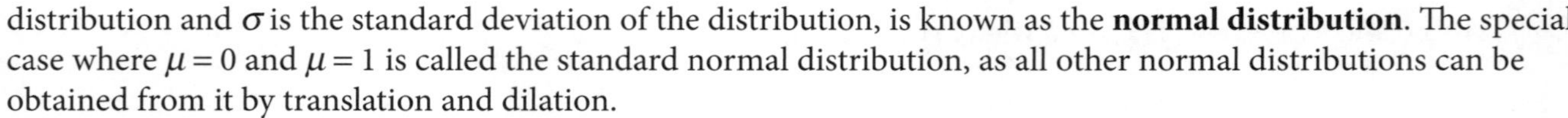

Note that this distribution is defined over $\mathbb{R}$, that is from $-\infty$, to ∞. In practical situations which can be modelled by a normal distribution, there will usually be an interval $[a, b]$ for which the value of the pdf can 'reasonably' be interpreted in that situation.

The normal distribution is the most useful of all the probability distributions for continuous random variables. As has been seen, its graph is characterised by a symmetrical 'bell' shape. This shape can be used as a model for the data collected from many naturally occurring variables, such as the height of a population, the intelligence quotient (IQ) of a population, or the distribution of errors in a production process. It is also frequently used as an approximation for the sums and averages of samples taken from fixed populations and can be essential when making inferences about populations from samples.

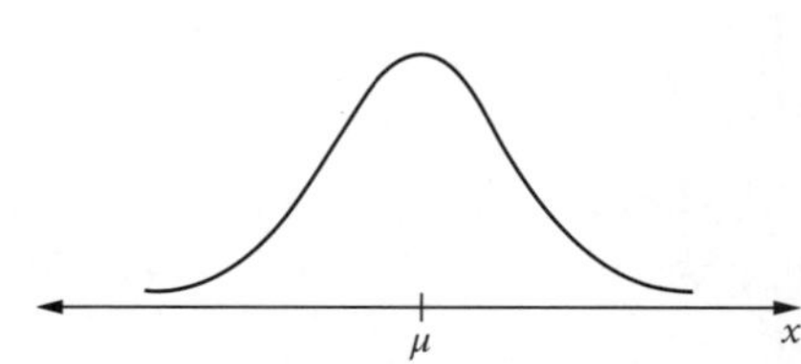

The symmetrical nature of the graph means that the mean and median coincide for the normal distribution and this value becomes the axis of symmetry. It should also be noted that, this graph extends infinitely (in theory) in both positive and negative directions, but it is always located above the x-axis. Thus, the normally distributed variable can assume any value.

Looking at the graph above it can be seen that the maximum value occurs when $x = \mu$. If you substitute this into the equation you get $f(\mu) = \frac{1}{\sigma\sqrt{2\pi}} e^0$. This simplifies to $\frac{1}{\sigma\sqrt{2\pi}}$, as shown in the diagram on the next page.

(This maximum value is useful when you are setting the window size for your graphing software.)

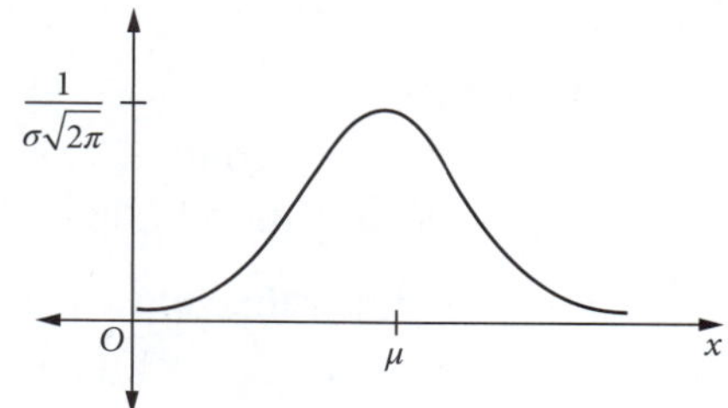

You need to check that the conditions for a pdf have been met.

By inspection, $f(x) \geq 0$ for all values of x and $\int_{-\infty}^{\infty} \frac{1}{\sigma\sqrt{2\pi}} e^{-\frac{1}{2}\left(\frac{x-\mu}{\sigma}\right)^2} dx = 1$.

However, the proof of this is beyond the scope of this course.

MAKING CONNECTIONS

Transformations of the normal distribution

Move the sliders to explore the effects of changing μ and σ on the graph of a normal distribution.

These investigations show that:

- increasing μ shifts the graph to the right, i.e. μ affects the location of the curve
- increasing σ makes the curve flatter and wider, but does not alter its axis of symmetry. If the size of the standard deviation is increased then the range $\mu \pm 3\sigma$ also increases, thus demonstrating the increased width required for the graph.

The common feature for all of the graphs is that the area under the curve is always 1.

Remember that the total area under the curve is 1 (a condition of being a pdf). Probability is related to the proportion of the total area which is being considered in a particular example.

One feature of the normal distribution that is frequently used, especially in situations where technology cannot be used, is related to the percentage of observations that you would expect to be within a certain number of standard deviations of the mean.

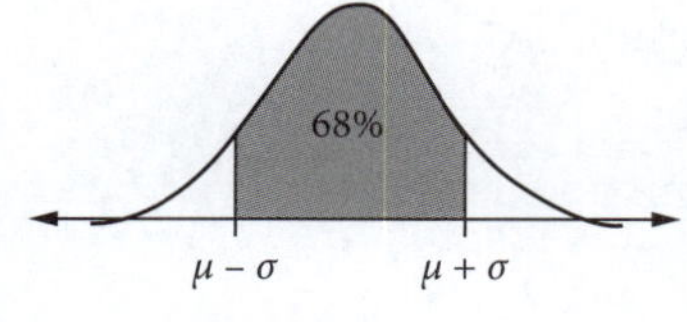

Approximately 68% of the observations will lie within one standard deviation of the mean.

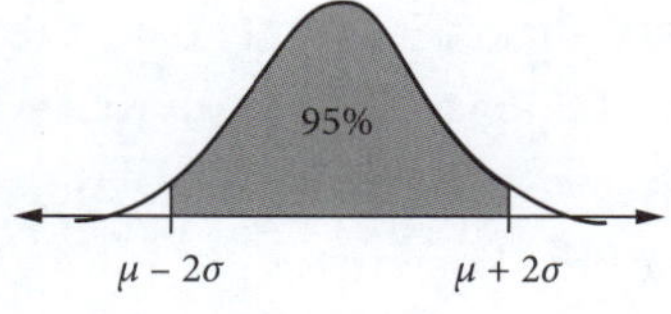

Approximately 95% of the observations will lie within two standard deviations of the mean. It can be said that it is *very likely* that any particular observation will lie within this range.

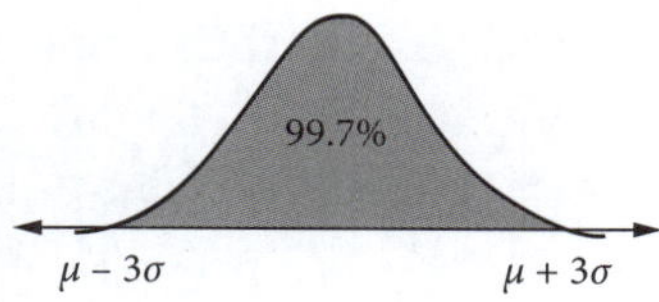

Approximately 99.7% of the observations will lie within three standard deviations of the mean. It can be said that it is *almost certain* that any particular observation will lie within this range.

For instance, if you know that the height of a particular population is normally distributed with a mean of 170 cm and a standard deviation of 6 cm, you would expect observations roughly as follows:

- 68% in the range $\mu \pm \sigma$, i.e. 170 ± 6, which gives the range 164 to 176 cm
- 95% in the range $\mu \pm 2\sigma$, i.e. 170 ± 12, which gives the range 158 to 182 cm
- 99.7% in the range $\mu \pm 3\sigma$, i.e. 170 ± 18, which gives the range 152 to 188 cm.

You can use the 68%, 95% and 99.7% rules to help estimate and check values in many situations.

There is a short way of writing that a random variable has a normal distribution:

A continuous random variable that has a normal distribution with mean μ and variance σ^2 is written as: $X \sim N(\mu, \sigma^2)$.

Example 6

The time taken in seconds, X, for all competitors to finish a 100 m race at the school athletics carnival was found to follow a normal distribution where $X \sim N(15, 4)$. Find the following values.

(a) The time range in which you would expect to find the middle 68%.

(b) The percentage of students you would expect to take more than 19 seconds.

Solution

(a) Write the mean μ and the standard deviation σ: $\mu = 15$, $\sigma^2 = 4$, $\sigma = \sqrt{4} = 2$

The middle 68% describes the values within one standard deviation of the mean, i.e. $\mu \pm \sigma$:

15 ± 2 gives the range 13 to 17 seconds.

You would expect the middle 68% of participants to take between 13 and 17 seconds to complete the race.

(b) Identify the value in terms of the mean and the standard deviation: $19 = \mu + 2\sigma$

(You may need to use a guess-and-check process.)

Use the symmetry of the curve to answer the question.

(In this case you are using only one tail, so you need to halve the percentage.)

You would expect 5% outside the range $\mu \pm 2\sigma$, so you would expect 2.5% of competitors to take more than 19 seconds.

EXPLORE FURTHER

Calculating probabilities using the normal distribution

Use technology to explore normal distribution probabilities within 1, 2 and 3 standard deviations from the mean.

Beware:

The formal description of the normal distribution $N(\mu, \sigma^2)$ uses the variance σ^2 as a parameter. However, worded questions often give the standard deviation σ. Make sure you read the questions carefully. When using technology you will also usually need to use the standard deviation and not the variance.

EXERCISE 20.2 THE NORMAL DISTRIBUTION

1 The time taken for all competitors to finish the 50 m freestyle at the school swimming carnival, X seconds, was found to follow a normal distribution where $X \sim N(45, 9)$. Find the following values.

- **(a)** The time range in which you would expect to find the middle 95% of results.
- **(b)** The percentage of students you would expect to take more than 48 seconds.
- **(c)** The percentage of students you would expect to take less than 36 seconds.

2 $X \sim N(10, 4)$.

- **(a)** What is the range of x values in which you would expect to find the middle 68%?
- **(b)** What is the range of x values in which you would expect to find the middle 95%?
- **(c)** What is the range of x values in which you would expect to find the middle 99.7%?

3 The height X cm of a population is known to be distributed as $X \sim N(170, 81)$.

- **(a)** What is the percentage of the population expected to be found in the range 152–188 cm?
- **(b)** What is the percentage of the population expected to be taller than 197 cm?
- **(c)** What is the percentage of the population expected to be shorter that 170 cm?
- **(d)** What is the percentage of the population expected to be found in the range 161–197 cm?

4 The graph represents a continuous random variable which has a normal distribution.

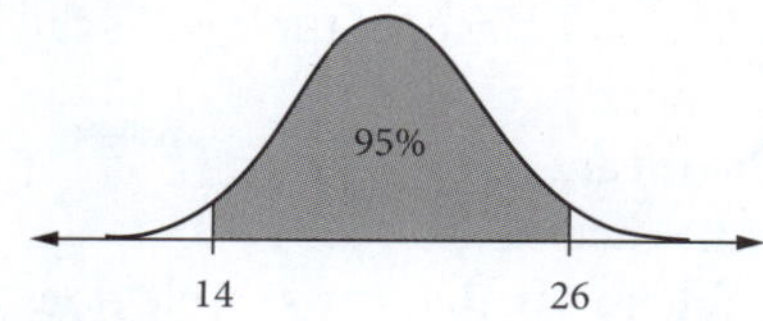

The distribution is best represented by:

A $N(14, 26)$ **B** $N(26, 12)$ **C** $N(20, 9)$ **D** $N(20, 3)$

5 Given X is normally distributed with a mean of 100 and a variance of 16, which of the following graphs correctly represents the distribution?

A

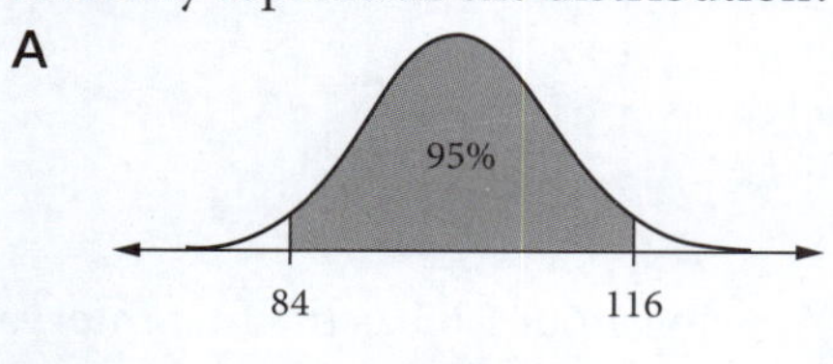

C

68%
84
116

B

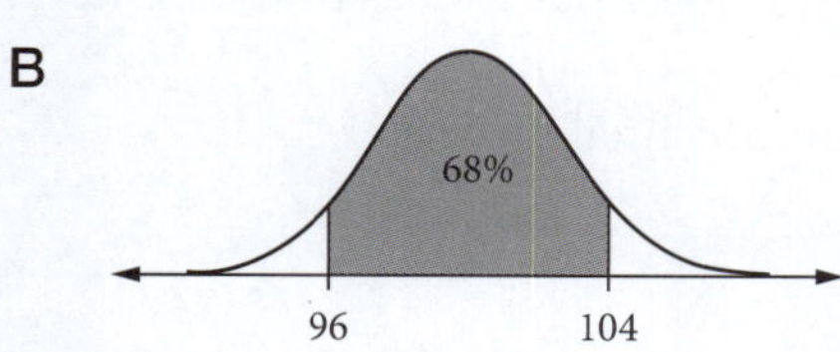

D

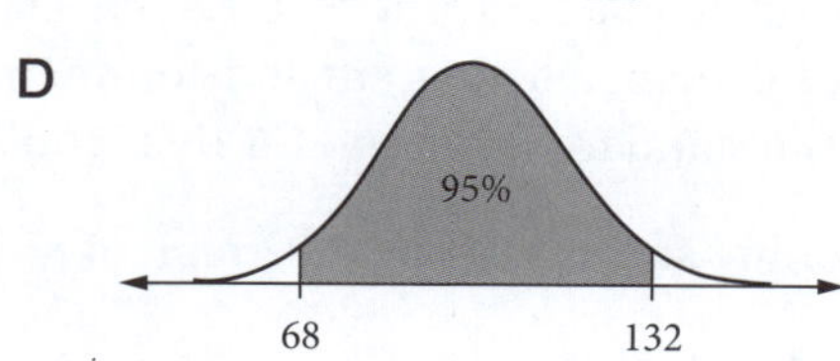

6 The marks X obtained by students in an examination were normally distributed with a mean of 85 and a standard deviation of 4. If the top 2.5% of students received a prize, find the minimum whole number score possible to receive a prize.

7 The mass M grams of a batch of commemorative coins is such that $M \sim N(50, 9)$. Each coin is weighed before packaging and will be rejected if its mass is less than 47 g. What is the percentage of coins expected to be rejected?

8 Packets of 'Greatstart' breakfast cereal are labelled as having a mass of 500 g. However, the machine that fills the packets actually follows a normal distribution with a mean of 510 g and a standard deviation of 5 g.
What percentage of packets, correct to two decimal places, will have a mass less than 500 g?

9 'Statzone' potato chips are packed by two different machines. Machine A fills the packets following a normal distribution with a mean of 100 g and a standard deviation of 3 g. Machine B fills the packets following the normal distribution $N(104, 16)$.

(a) **(i)** Between what values would you expect to find the middle 68% of packets for Machine A?
(ii) Between what values would you expect to find the middle 68% of packets for Machine B?

(b) **(i)** Between what values would you expect to find the middle 95% of packets for Machine A?
(ii) Between what values would you expect to find the middle 95% of packets for Machine B?

(c) **(i)** Between what values would you expect to find the middle 99.7% of packets for Machine A?
(ii) Between what values would you expect to find the middle 99.7% of packets for Machine B?

(d) If you bought a packet of the chips, which machine would you prefer to have packed it? Explain your answer.

10 Seventy students enrolled in a university course had their height measured in metres, correct to 2 decimal places where necessary. Their heights were as follows:

1.54	1.57	1.63	1.65	1.68	1.69	1.7	1.7	1.7	1.7
1.7	1.7	1.7	1.7	1.7	1.7	1.72	1.72	1.73	1.73
1.73	1.73	1.73	1.73	1.75	1.75	1.75	1.75	1.75	1.75
1.76	1.76	1.76	1.76	1.77	1.78	1.78	1.78	1.78	1.78
1.78	1.78	1.78	1.78	1.78	1.78	1.8	1.8	1.8	1.8
1.8	1.8	1.8	1.82	1.83	1.83	1.83	1.83	1.83	1.84
1.85	1.85	1.85	1.85	1.85	1.85	1.85	1.88	1.93	1.95

(a) Group the results using 1.5–<1.55, 1.55–<1.6, … as the intervals and draw a histogram to represent the data.

(b) Comment on the shape of the histogram and whether it would be reasonable to suggest that the sample follows a normal distribution.

(c) Find the mean and standard deviation, correct to three decimal places, of the original ungrouped data.
(d) What percentage, correct to one decimal place, of the results fall within:
  (i) one standard deviation of the mean
  (ii) two standard deviations of the mean
  (iii) three standard deviations of the mean?
(e) Do you think it would be reasonable to say that the sample does follow a normal distribution?

11 Ivana sells freshly cut Christmas trees. She has 50 trees on her property ready to be cut and delivered. Prices depend on the height of the tree. She has recorded the heights of the trees, in metres, as shown.

1.61	1.66	1.68	1.68	1.71	1.71	1.71	1.72	1.73	1.73
1.73	1.76	1.76	1.77	1.77	1.77	1.77	1.79	1.79	1.79
1.81	1.81	1.81	1.81	1.82	1.83	1.83	1.83	1.83	1.84
1.86	1.86	1.86	1.86	1.86	1.86	1.87	1.87	1.88	1.91
1.91	1.91	1.93	1.93	1.94	1.94	1.96	1.96	1.96	2.01

(a) Calculate the average height of the Christmas trees.
(b) Display this data in a frequency table using class intervals of 0.05, using 1.60–1.65 as the first interval.
(c) Display this data on a histogram and comment on its shape.
(d) Calculate the standard deviation for this set of data, correct to four decimal places.

12 (a) Using graphing software draw the graph of $f(x) = e^{-x^2}$.
(b) Use software to evaluate $\int_{-4}^{4} e^{-x^2}\, dx$.

13 (a) Using graphing software, draw the graph of $f(x) = \frac{1}{\sigma\sqrt{2\pi}} e^{-\frac{(x-\mu)^2}{2\sigma^2}}$, for:
  (i) $\mu = 10, \sigma = 3$
  (ii) $\mu = 0, \sigma = 1$
(b) Use the integration tool in the software to evaluate $\int_{\mu-3\sigma}^{\mu+3\sigma} \frac{1}{\sigma\sqrt{2\pi}} e^{-\frac{(x-\mu)^2}{2\sigma^2}}\, dx$ in each case in part **(a)**.

14 Use the trapezoidal rule with six sub-intervals to find the approximate value of the following:

(a) $\int_{1}^{19} \frac{1}{3\sqrt{2\pi}} e^{-\frac{(x-10)^2}{18}}\, dx$ (b) $\int_{-3}^{3} \frac{1}{\sqrt{2\pi}} e^{\frac{-x^2}{2}}\, dx$

Hint: Consider the symmetry of the function when constructing your table.
(c) Compare your answers to those obtained in question **13(b)**.

20.3 THE STANDARD NORMAL DISTRIBUTION

Although it is usually easy to use technology to find values associated with any normal distribution, it is often useful, especially when comparing distributions, to use what is called the **standard normal distribution**. This is a normal distribution that has a mean of 0, a variance of 1 and probability density function $f(x) = \frac{1}{\sqrt{2\pi}} e^{\frac{-x^2}{2}}$ with domain $\mathbb{R}$. Due to its importance this has a special letter, Z, reserved for the random variable of the standard normal distribution.

The standard normal distribution is such that $Z \sim N(0, 1)$, i.e. it has a mean of 0 and a variance of 1.

In the previous section you saw that the probability density function for $X \sim N(\mu, \sigma^2)$ can be obtained from that of $Z \sim N(0, 1)$ by the transformation $x = \sigma z + \mu$. Thus, to transform a normal distribution into the standard normal distribution the transformation $z = \frac{x-\mu}{\sigma}$ is applied. That is, the mean μ is subtracted from the observed value, X, and the result is divided by the standard deviation σ. As stated above, the resulting standard normal distribution is usually referred to by the letter Z.

If $X \sim N(\mu, \sigma^2)$ then $\frac{x-\mu}{\sigma} = z$, $Z \sim N(0, 1)$.

Before digital technology was easily available, tables of values were used to calculate probabilities associated with normal distributions. This was one of the reasons it was so important to be able to change a distribution to the standard normal: it was the only distribution for which tables of values were easily available. In subjects such as this it is now used mainly for comparisons.

However, these standard z values (also called z-scores) have some important applications in science and elsewhere, such as in the field of paediatric health. Children grow at different rates, so it is difficult to use the more standard statistical tests when assessing particular children for serious health defects or conditions. For these applications to be useful, large statistical samples are needed for children of various ages, heights and weights. Then, comparisons can be made for a particular child against other children of the same age/height/weight, or some combination of these factors. For example, it may be suspected that a child has an inappropriately large dilation of a ventricle in their heart. If the dilation increases over time, this may not be a problem, because the dilation would always be expected to increase as the child grows older. However, if the z value increases over time, this is evidence that there may be a problem.

Calculating a *z* value

Example 7

X is a random variable following a normal distribution with mean 10 and variance 9 (i.e. $X \sim N(10, 9)$).

Find the z value that would be used to represent an x value of 14.

Solution

Identify σ and μ: $\mu = 10$, $\sigma = \sqrt{9} = 3$

Use the formula $z = \frac{x-\mu}{\sigma}$ to convert the x value to the equivalent z value: $z = \frac{14-10}{3}$

$$= \frac{4}{3}$$

$$\approx 1.33$$

This means that the observed value of X, 14, is 1.33 standard deviations greater than the mean.

Comparing *z* values

Example 8

Juan has been applying for scholarships. On one particular test he obtained 45 on a test that followed the distribution $X \sim N(40, 4)$ and on another he obtained 85 on a test that followed the distribution $Y \sim N(75, 25)$. On which test did Juan do better?

Solution

Find the z value for the first test: $\frac{X-\mu}{\sigma} = \frac{45-40}{2}$

$$= \frac{5}{2} = 2.5$$

Find the z value for the second test: $\frac{Y-\mu}{\sigma} = \frac{85-75}{5}$

$$= \frac{10}{5}$$

$$= 2$$

Juan did better on the first test, as his z value was further to the right.

The symmetry of the normal distribution helps with some calculations. This is most easily seen using the standard normal curve, where $\mu = 0$ and $\sigma^2 = 1$ (and therefore $\sigma = 1$).

In the diagram the two shaded areas are equal. For this to be true, the distance between a and 0 (μ) must be the same as the distance between 0 and b. So $a = -b$.

This result is useful as it shows that $P(X < a) = P(X > b)$. Also worth noting from this diagram is $P(X > b) = 1 - P(X < b)$ as the total probability is 1.

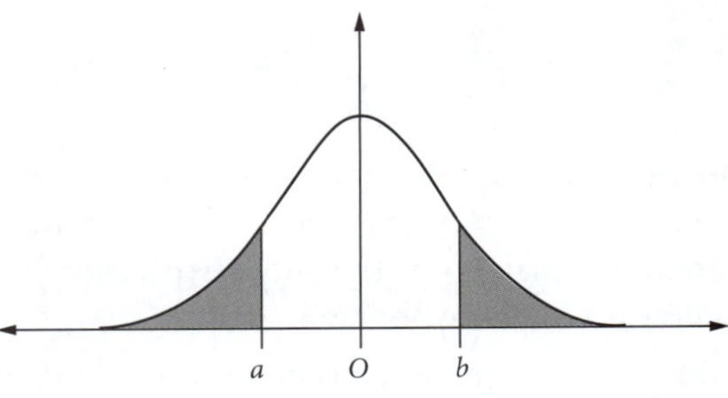

$P(Z < -z) = P(Z > z) = 1 - P(Z < z)$

A negative z value indicates that the observed value is less than the mean.

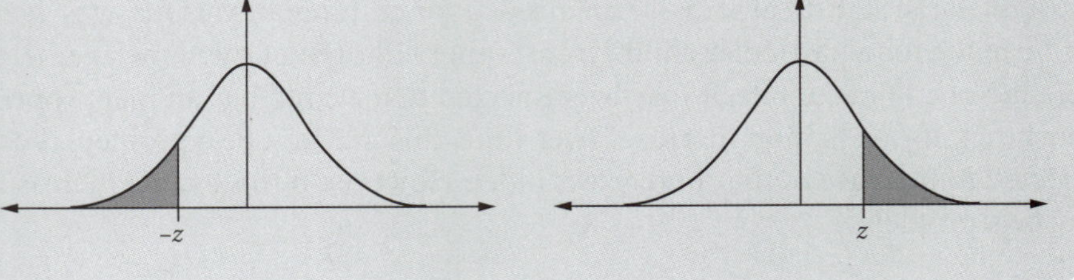

This result can be extended to any variable X that follows a normal distribution.

Assume $\mu = 20$. The value 5 units to the left of the mean is 15 and the value 5 units to the right of the mean is 25. This means, in this case, that $P(X < 15) = P(X > 25)$. This can also be expressed as $P(X < 15) = 1 - P(X < 25)$.

Expressing this result in general terms, for a variable X that follows a normal distribution with a mean of μ:

$P(X < \mu - k) = P(X > \mu + k)$ which leads to $P(X < \mu - k) = 1 - P(X < \mu + k)$, where k is the distance from the mean.

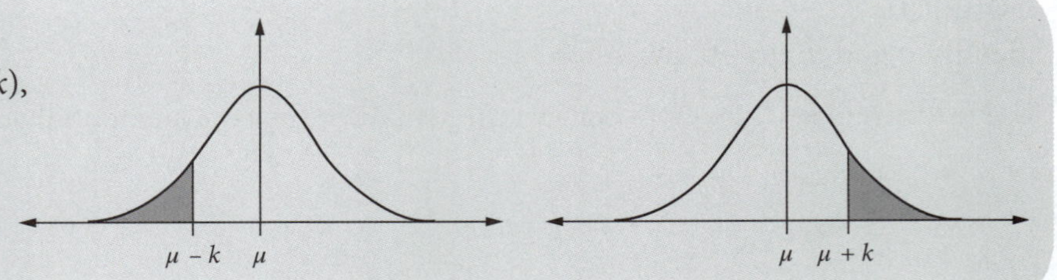

EXPLORE FURTHER

Symmetry of the normal distribution

Use technology to explore the symmetry of the normal distribution graph

This result is useful when dealing with technology-free problems where the symmetry of the distribution is the only real information available to you.

For example, if you needed to find $P(10 < X < 15)$ where $\mu = 12$, you could find $P(X < 10)$ and subtract this from $P(X < 15)$.

Using general terms, to find $P(a < X < b)$:

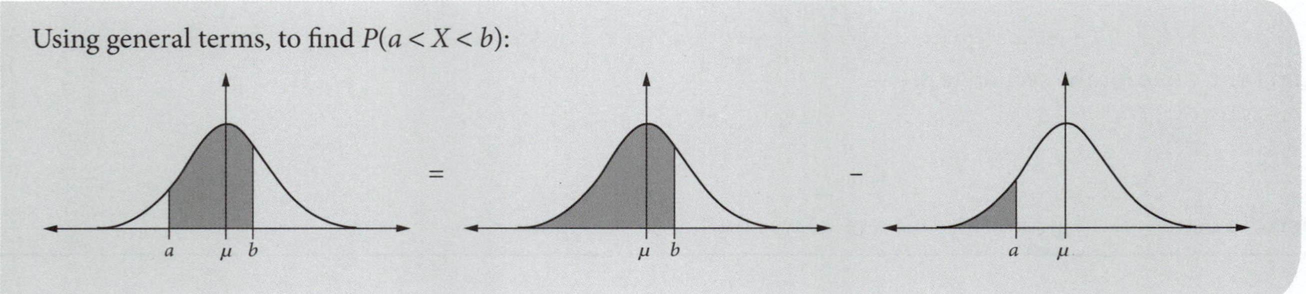

Using the symmetry of the normal distribution

Example 9

A normal distribution graph is shown.

If $P(X > 14) = 0.22$ then $P(X > 6)$ is equal to:

A 0.22 **B** 0.44

C 0.56 **D** 0.78

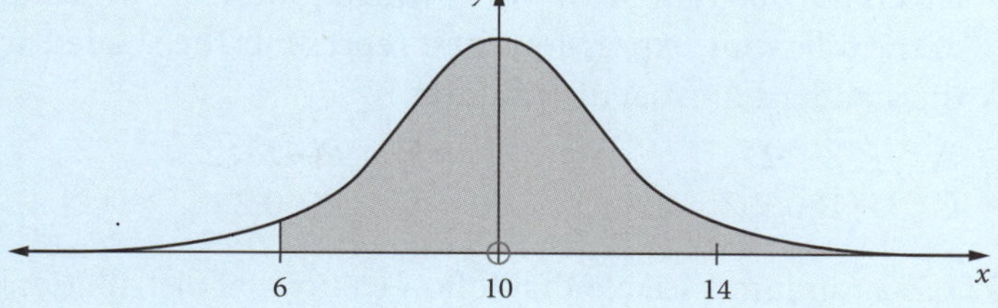

Solution

Find $P(X > 6)$.

Identify the relevant symmetry aspects of the diagram in terms of probabilities: $P(X > 14) = P(X < 6)$

Write an expression using the required probability: $P(X > 6) = 1 - P(X < 6)$

Substitute known values and calculate the numerical answer: $P(X > 6) = 1 - 0.22 = 0.78$

Hence the answer is D.

Empirical rule

The results obtained earlier apply for the standard normal distribution:

For normally distributed random variables,

- approximately 68% of data will have z-scores between −1 and 1
- approximately 95% of data will have z-scores between −2 and 2
- approximately 99.7% of data will have z-scores between −3 and 3.

Sections 20.4 and 20.5 are available as online resources only.

EXERCISE 20.3 THE STANDARD NORMAL DISTRIBUTION

1 If $X \sim N(15, 9)$, find the exact z values corresponding to the following x values:

(a) 18 **(b)** 21 **(c)** 22 **(d)** 16

2 Felipe has been applying for scholarships. On one particular test that followed the distribution $X \sim N(40, 4)$, he obtained a 42; and on another test that followed the distribution $\sim N(75, 25)$, he obtained an 82.

(a) Find the z value for the first test.
(b) Find the z value for the second test.
(c) On which test did Felipe do better?

3 A normal distribution graph is shown.

If $P(X > 14) = 0.35$ then $P(X > 6)$ is equal to:

A 0.35 **B** 0.55
C 0.65 **D** 0.7

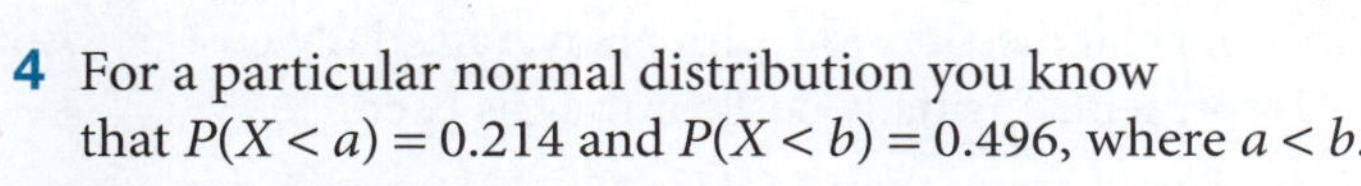

4 For a particular normal distribution you know that $P(X < a) = 0.214$ and $P(X < b) = 0.496$, where $a < b$.

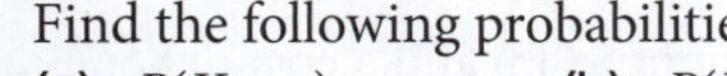

Find the following probabilities.

(a) $P(X > a)$ **(b)** $P(X > b)$ **(c)** $P(a < X < b)$

5 For $Z \sim N(0, 1)$ it is known that $P(Z < 0.85) = 0.8023$. Find:

(a) $P(Z > 0.85)$ **(b)** $P(Z < -0.85)$ **(c)** $P(Z > -0.85)$ **(d)** $P(-0.85 < Z < 0.85)$

6 For $Z \sim N(0, 1)$ it is known that $P(-1.5 < Z < 1.5) = 0.8864$. Find:

(a) $P(Z < -1.5)$ **(b)** $P(Z > -1.5)$ **(c)** $P(Z < 1.5)$

7 Given the diagram below for X which follows a normal distribution, which of the following expressions best represents the shaded area? (Z represents the standard normal distribution.)

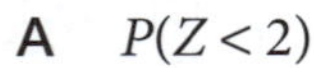

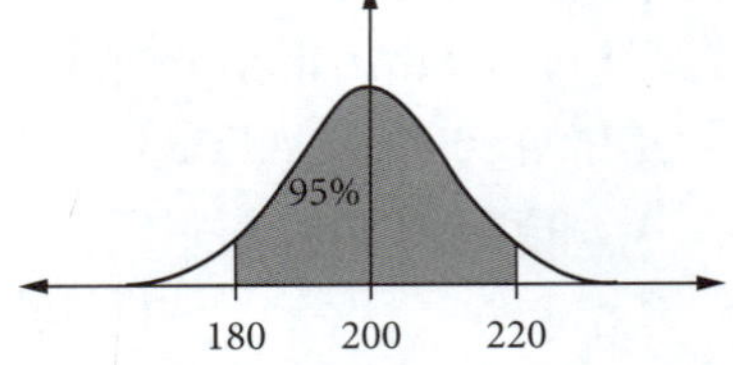

A $P(Z < 2)$ **B** $P(-2 < Z < 2)$
C $P(180 < Z < 220)$ **D** $P(-0.2 < Z < 0.2)$

8 X is a random variable that follows a normal distribution with a mean of 35 and a standard deviation of 7. The values of a and b are such that $P(a < X < b) = 0.95$ where this represents the middle 95% of values. a and b are best represented by:

A $a = 23.49, b = 46.51$ **B** $a = 21.28, b = 48.72$
C $a = 29.11, b = 40.89$ **D** $a = 21, b = 49$

9 The distribution of weights for all 80.5 cm-tall girls in a population is such that the mean weight is 10.3 kg with a standard deviation of 0.8 kg.

(a) Adah is 80.5 cm tall and weighs 8.3 kg. What is Adah's z value for weight?

These z values are used as a definition for some forms of malnutrition. Moderate acute protein-energy malnutrition is defined as having a z value in the range $[-3.0, -2.0)$ and severe acute protein-energy malnutrition is defined as having a z value less than -3.0.

(b) Based on these definitions, what sort of acute protein-energy malnutrition would Adah be diagnosed with?

(c) Jamilah, who is also 80.5 cm tall, has been diagnosed with severe acute protein-energy malnutrition. What is Jamilah's weight, correct to one decimal place, less than?

(d) Another girl who is 80.5 cm tall, Xhosa, is not diagnosed with either form of acute protein-energy malnutrition. What is Xhosa's minimum weight?

10 The percentages obtained by a group of students in a Mathematics examination are represented by a random variable M and are normally distributed with a mean of 72 and a variance of 121. All percentages are rounded to the nearest whole percentage.

(a) Calculate the probability that a student obtained a mark of at least 50% (when rounded to the nearest whole percentage) in this examination, correct to four decimal places, and the number of standard deviations that this mark is below the mean.

(b) Determine the z-score of a student who obtained a mark of 45%. What is the expected mark of a student whose z-score has the same size but opposite sign from the student who scored 45%?

To obtain an A^{++} mark, a student has to be in the top 2.5% of the group of students who have undertaken this examination.

(c) Calculate the minimum mark a student should obtain in this examination to be awarded an A^{++} by first finding the corresponding z value.

The marks in the previous year's Mathematics examination were normally distributed with a mean of 70 and a variance of 144.

(d) Would a student who obtained a mark of 94% have been awarded an A^{++} grade? Use appropriate calculations in your explanation.

11 Anita's daily charges for gas usage in her home form a normal distribution with an average daily cost of \$7.65 and a variance of 1.44, where the random variable C represents the daily cost for the gas used.

(a) What is the probability that in any one day Anita's cost is more than \$6.45?

(b) Determine the number of standard deviations from the mean for a cost of \$8.05 and a cost of \$6.65.

(c) Plot the two z values from part **(b)** on the normal distribution curve $N(7.65, 1.2^2)$.

CHAPTER REVIEW 20

1 For the uniform continuous variable with probability density function $f(x) = \begin{cases} k, & 0 \le x \le 10 \\ 0, & \text{otherwise} \end{cases}$, find the following values.

(a) k (b) $P(X \le 3)$ (c) $P(X \le 5)$ (d) $P(2 \le X \le 9)$

2 Does the hybrid function $f(x) = \begin{cases} x^2 + 4x - \frac{10}{3}, & 0 \le x \le 3 \\ 0, & \text{otherwise} \end{cases}$ represent a probability density function?

3 For the continuous variable with probability density function $f(x) = \begin{cases} \frac{x^2}{12}, & 0 \le x \le \sqrt[3]{36} \\ 0, & \text{otherwise} \end{cases}$, find the exact value of the median.

4 Honey is packed in jars with a labelled mass of 500 g. The actual amount X g of honey in the jar is such that $X \sim N(500, 25)$.

(a) Find the probability that a jar chosen at random contains less than 495 g.

(b) Jars containing less than 490 g cannot be sold. In a batch of 1000 how many jars would you expect to be rejected?

5 If $X \sim N(12, 16)$, find the exact z values corresponding to the following x values.

(a) $x = 8$ (b) $x = 20$

6 If $X \sim N(28, 16)$, find the exact z values corresponding to the following x values.

(a) $x = 24$ (b) $x = 40$

7 If $f(x) = \begin{cases} \frac{2x}{15 + k}, & 0 \le x \le 3 \\ 0, & \text{otherwise} \end{cases}$ defines a probability density function, then the value of k is:

A $\frac{2}{15}$ B -6 C $-\frac{2}{15}$ D $\frac{1}{5}$

8 $X \sim N(10, 9)$. You would expect about 95% of observations to be in the range:

A -8 to 28 B 0 to 28 C 10 to 19 D 4 to 16

9 Consider the following graph.

The graph is best described by:

A $N(111, 129)$ B $N(120, 81)$
C $N(120, 3)$ D $N(120, 9)$

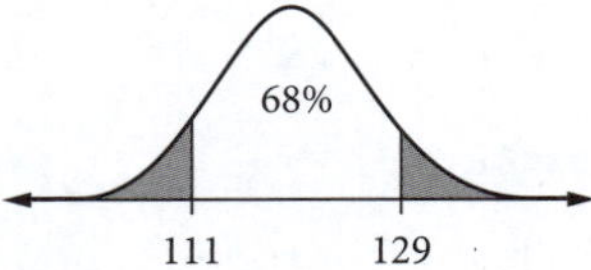

10 The graph shows two normal distributions drawn on the same set of axes. Graph 1 has a marker on it.

Compared to Graph 1, Graph 2 has:

A the same mean but a larger variation
B the same mean but a smaller variation
C the same variation but a larger mean
D the same variation but a smaller mean

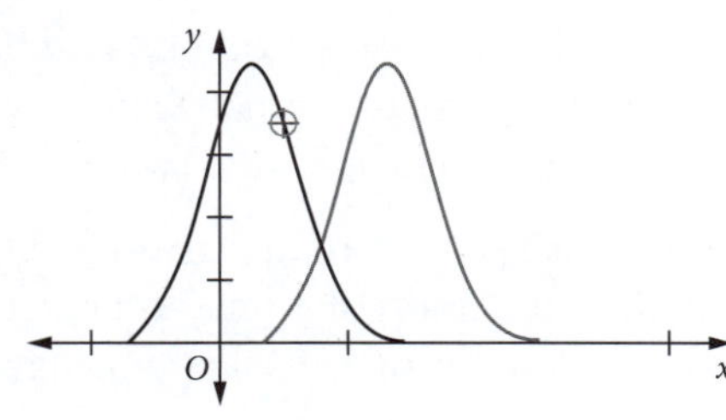

11 The graph shows two normal distributions drawn on the same set of axes. Graph 1 has a marker on it.

Compared to Graph 1, Graph 2 has:

A the same mean but a larger variation
B the same mean but a smaller variation
C the same variation but a larger mean
D the same variation but a smaller mean

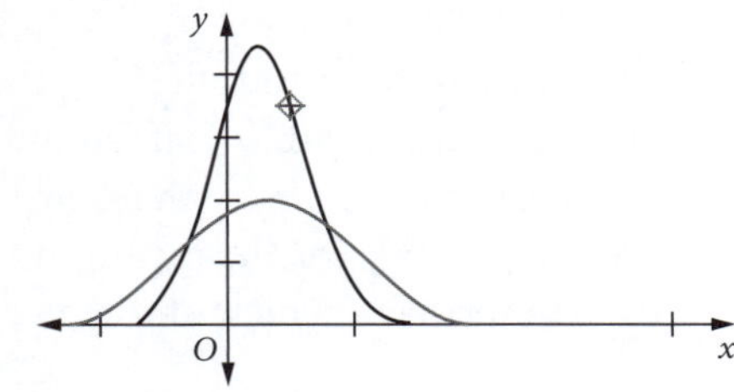

In any of the following questions that involve evaluating an integral, you may need to use software to evaluate the integral if the integrand is outside the scope of this course, or use the trapezoidal rule.

12 An arrow is fired at a target. The distance X m from the centre of the target that the arrow hits pierces follows the function $f(x) = \begin{cases} \dfrac{3}{\pi\left(1+x^2\right)}, & 0 \le x \le \sqrt{3} \\ 0, & \text{otherwise} \end{cases}$. You are given that $\displaystyle\int_0^{\sqrt{3}} \frac{dx}{1+x^2} = \frac{\pi}{3}$.

(a) Explain why it is reasonable to say that this represents a probability density function.

(b) Find:

(i) $P(X < 1)$ **(ii)** $P(X > 1.25)$ **(iii)** $P(0.25 < X < 1.1)$

(c) Find $E(X)$.

(d) Find $\text{Var}(X)$.

13 Andrea is waiting for a taxi at a popular venue at a busy time of the day. The continuous random variable T, in hours, that represents the time spent by people waiting for a taxi at this venue, is given by the probability density function: $f(t) = \begin{cases} \dfrac{5}{2}t\sqrt{t}, & 0 \le t \le 1 \\ 0, & \text{otherwise} \end{cases}$

(a) Calculate the average time Andrea should expect to wait to catch a taxi. Give your answer correct to the nearest minute.

(b) Calculate the probability that Andrea will have to wait more than 15 minutes to catch a taxi, correct to two decimal places.

(c) What is the probability that Andrea's wait for the taxi is between 10 and 40 minutes?

(d) If 10 people are waiting for a taxi, how many would be expected to wait more than 30 minutes?

14 For their Mathematics assignment, Xandra and Zack have decided to record the times they spend completing the daily homework. Xandra's times are represented by the continuous random variable, X, in hours, with the probability density function:

$$f(x) = \begin{cases} -\frac{1}{9}\left(x^2 - 4x\right), & 0 \le x \le 3 \\ 0, & \text{otherwise} \end{cases}$$

Zack's times are represented by the continuous random variable, Z, in hours, with the probability density function:

$$f(z) = \begin{cases} \frac{2}{57}\left(z^2 - 9z + 20\right), & 0 \le z \le 3 \\ 0, & \text{otherwise} \end{cases}$$

(a) Calculate the average time each student spends completing homework each day and compare the results.

Xandra has a school concert to attend and has to complete her set homework within 1.5 hours.

(b) Calculate the probability that Xandra will complete her homework in the required time.

(c) Calculate the variance for the two variables.

15 When fishing, fish shorter than a given length have to be thrown back into the water. Red snapper have a minimum allowed length of 30 cm and barramundi have a minimum allowed length of 55 cm. Let the variable R represent the lengths of red snapper caught, in cm, and the variable B represent the lengths of barramundi caught, in cm.

Observations of caught red snapper show that R is normally distributed with mean 36 and a variance of 9. Observations of caught barramundi show that B is normally distributed with a variance of 16.

(a) Calculate the mean length of barramundi caught, correct to the nearest whole number, if 2.5% of the barramundi caught are less than 54 cm.

(b) Calculate the z-scores for the minimum allowed lengths for both variables, R and B, and interpret what this means in terms of which of the two species are more likely to be too small to keep.

SUMMARY

1 ALGEBRAIC TECHNIQUES

Simplifying algebraic expressions

When adding and subtracting algebraic expressions, you can only combine like terms. You may first have to expand brackets by multiplying terms inside the brackets by the common factor outside the brackets.

Substitution in formulae

Replace the pronumerals with their values. Sometimes this will produce an equation which should be solved.

Common terms

- A **monomial** is an expression that contains only **one** term, e.g. $5x$, x^2, $2ab$, $5a^2b^3$.
- A **binomial** is an expression that contains **two** terms added or subtracted, e.g. $x + y$, $3a - 2b$, $x^2 + 1$, $3y - 4$.
- A **trinomial** is an expression that contains **three** terms added or subtracted, e.g. $x^2 - 5x + 6$, $x + y - 4$, $4x^2 - 2xy + y^2$, $m + n - p$.
- A **quadratic trinomial** is a trinomial of the form $ax^2 + bx + c$ $(a \neq 0, b \neq 0, c \neq 0)$. a is the coefficient of x^2, b is the coefficient of x, and c is the constant term.

Standard results for basic polynomials

$$(x + m)(x + n) = x^2 + (m + n)x + mn$$
$$(a + b)^2 = a^2 + 2ab + b^2$$
$$(a - b)^2 = a^2 - 2ab + b^2$$
$$(a - b)(a + b) = a^2 - b^2$$

Factorising by grouping in pairs

When there are four terms, factorise two pairs of terms, then use an expression for the common factor.

$$\begin{aligned} ax + bx + ay + by &= x(a + b) + y(a + b) \\ &= (a + b)(x + y) \end{aligned}$$

Factorising using the difference of two squares

$$a^2 - b^2 = (a - b)(a + b)$$

Factorising quadratic trinomials

To factorise quadratic trinomials, you must remember how to expand binomial products and then work backwards. We know the following:

- $(x + m)(x + n) = x^2 + (m + n)x + mn$
  $= x^2 + (\text{sum of } m \text{ and } n)x + (\text{the product of } m \text{ and } n)$
- $(x - m)(x - n) = x^2 - (m + n)x + mn$
  $= x^2 + (\text{sum of } -m \text{ and } -n)x + (\text{the product of } -m \text{ and } -n)$
- $(x + m)(x - n) = x^2 + (m - n)x - mn$
  $= x^2 + (\text{sum of } m \text{ and } -n)x + (\text{the product of } m \text{ and } -n)$

To factorise $x^2 + 5x + 6$ you must write it in the form $(x + m)(x + n)$, where $m + n = 5$ and $mn = 6$. This means you must find two numbers whose sum is 5 and whose product is 6.

Non-monic trinomials

The coefficient of x^2 is not 1, so you must consider factors of the coefficient of x^2 as well as of the constant term.

Use trial and error or the cross method.

Mixed factorisations

To factorise quadratic expressions, you may need to use **one or more** of the following techniques:

- remove a common factor
- group in pairs and remove common factors
- difference of two squares
- quadratic trinomials.

Sum and difference of two cubes

$$a^3 + b^3 = (a + b)(a^2 - ab + b^2)$$
$$a^3 - b^3 = (a - b)(a^2 + ab + b^2)$$

Algebraic fractions

To **simplify** algebraic fractions:

1. Factorise the numerator and denominator.
2. Cancel any common factors.

Adding and subtracting algebraic fractions

- When adding and subtracting fractions, first rewrite each fraction with the same (common) denominator, then add or subtract the numerators.
- More complex algebraic fractions require you to factorise the denominator before you find the common denominator. Write each fraction with the common denominator before you add or subtract the numerators.

Rational and irrational numbers

- Rational numbers can be expressed in the form $\frac{a}{b}$ where a and b are integers and $b \neq 0$.
  They include positive and negative integers, fractions, terminating decimals and recurring decimals.
- Irrational numbers are numbers that are not rational, like $\sqrt{2}$, $\sqrt{15}$, $2\sqrt{7}$ and π. They cannot be represented as a fraction. Irrational numbers such as $\sqrt{2}$, $\sqrt{15}$, $2\sqrt{7}$ (i.e. any square root or higher root) are called surds; π is called a transcendental number (i.e. it cannot be solved algebraically).

Basic rules for surds

If a and b are positive numbers then:

(a) $\sqrt{ab} = \sqrt{a} \times \sqrt{b}$

(b) $\sqrt{\frac{a}{b}} = \frac{\sqrt{a}}{\sqrt{b}}$

(c) $\sqrt{a^2} = a$

(d) $\sqrt{a} \times \sqrt{b} = \sqrt{ab}$

(e) $\frac{\sqrt{a}}{\sqrt{b}} = \sqrt{\frac{a}{b}}$

(f) $\sqrt{a} \times \sqrt{a} = a$

(g) $\sqrt{a} \div \sqrt{b} = \frac{\sqrt{a}}{\sqrt{b}} = \frac{\sqrt{a} \times \sqrt{b}}{\sqrt{b} \times \sqrt{b}} = \frac{\sqrt{ab}}{b}$ to write the expression with a rational denominator.

Adding and subtracting surds

- Surds of the same kind can be added and subtracted just like pronumerals, using the distributive law to collect like terms:

  $ab + ac = a(b + c)$

- You can only add and subtract like algebraic terms, such as $2a + 3a - a = 4a$. Similarly, you can only add and subtract like surds, such as $2\sqrt{2} + 3\sqrt{2} = 5\sqrt{2}$. It may be necessary to first simplify the surd terms by removing perfect square factors.

Rationalising surds

- The expressions $\sqrt{a} \times \sqrt{a}$ and $\left(\sqrt{a} - \sqrt{b}\right)\left(\sqrt{a} + \sqrt{b}\right)$ both have rational answers (that is, answers that do not involve surds): $\sqrt{a} \times \sqrt{a} = a$ and $\left(\sqrt{a} - \sqrt{b}\right)\left(\sqrt{a} + \sqrt{b}\right) = \left(\sqrt{a}\right)^2 - \left(\sqrt{b}\right)^2 = a - b$.
- The expansion $\left(\sqrt{a} - \sqrt{b}\right)\left(\sqrt{a} + \sqrt{b}\right) = a - b$ is known as the 'difference of two squares'.
- When you have a surd expression in the denominator of a fraction, it is normal to make the denominator into a rational number. This process is called **rationalising the denominator**.
- The expressions $\sqrt{a} - \sqrt{b}$ and $\sqrt{a} + \sqrt{b}$ are called conjugate surds, with each expression being the conjugate of the other. To convert any surd into a rational number, multiply the surd by its conjugate.

2 TRIGONOMETRY

Trigonometric ratio definitions

These trigonometric ratios describe the relationship between the angles and sides of a right-angled triangle:

$$\sin A = \frac{\text{opposite side}}{\text{hypotenuse}} = \frac{a}{c}$$

$$\cos A = \frac{\text{adjacent side}}{\text{hypotenuse}} = \frac{b}{c}$$

$$\tan A = \frac{\text{opposite side}}{\text{adjacent side}} = \frac{a}{b}$$

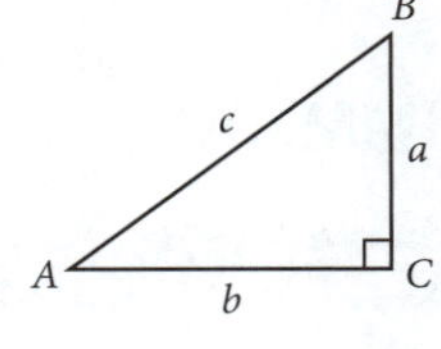

To remember these ratios you can use the mnemonic **SOHCAHTOA**, which stands for **S**in is **O**pposite over **H**ypotenuse, **C**os is **A**djacent over **H**ypotenuse, **T**an is **O**pposite over **A**djacent.

Pythagoras' theorem

The square of the hypotenuse of a right-angled triangle is equal to the sum of the squares of the other two sides $c^2 = a^2 + b^2$.

The unit circle

A unit circle is a circle of unit radius whose centre is at the origin. The equation of the circle is $x^2 + y^2 = r^2$.

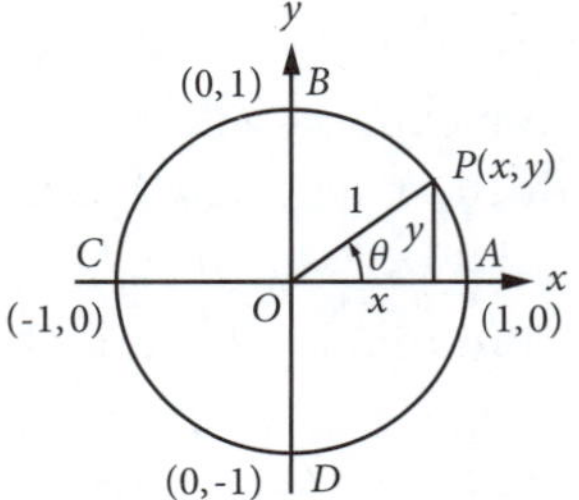

Take any point P on the circumference of this unit circle whose coordinates are (x, y). Consider the point P as starting at A and rotating in an anticlockwise direction, taking various positions around the circumference as shown in the diagrams above. In each position $\angle AOP = \theta$. Then we can define:

- $\cos\theta = x = x$-coordinate of P (the abscissa)
- $\sin\theta = y = y$-coordinate of P (the ordinate)
- $\tan\theta = \frac{y}{x} = \frac{\sin\theta}{\cos\theta}$, $\cos\theta \neq 0$
- $\cot\theta = \frac{x}{y} = \frac{\cos\theta}{\sin\theta} = \frac{1}{\tan\theta}$, $\sin\theta \neq 0$
- $\sec\theta = \frac{1}{x} = \frac{1}{\cos\theta}$, $\cos\theta \neq 0$
- $\operatorname{cosec}\theta = \frac{1}{y} = \frac{1}{\sin\theta}$, $\sin\theta \neq 0$

Symmetry properties of trigonometric ratios

Second quadrant $90° < \theta < 180°$

$\cos(180° - \theta) = -a = -\cos\theta$

$\sin(180° - \theta) = b = \sin\theta$

$\tan(180° - \theta) = \frac{b}{-a} = -\tan\theta$

Third quadrant $180° < \theta < 270°$

$\cos(180° + \theta) = -a = -\cos\theta$

$\sin(180° + \theta) = -b = -\sin\theta$

$\tan(180° + \theta) = \dfrac{-b}{-a} = \tan\theta$

Fourth quadrant $270° < \theta < 360°$

$\cos(360° - \theta) = a = \cos\theta$

$\sin(360° - \theta) = -b = -\sin\theta$

$\tan(360° - \theta) = \dfrac{-b}{a} = -\tan\theta$

Sign of the trigonometric ratios

The sign of cos, sin and tan for the first four quadrants can be summarised as follows:

- First quadrant: All are positive (A)
- Second quadrant: sin only is positive (S)
- Third quadrant: tan only is positive (T)
- Fourth quadrant: cos only is positive (C)

The positive signs can be remembered with the mnemonic **ASTC**: **A**ll **S**tations **T**o **C**entral.

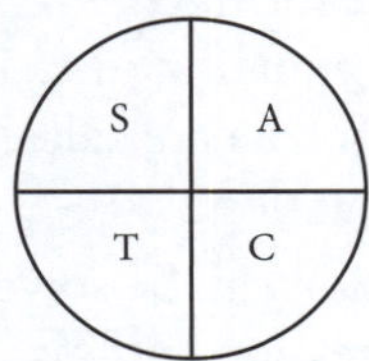

Complementary angles: θ and $(90° - \theta)$

$\sin(90° - \theta) = \cos\theta$ $\quad$ $\cos(90° - \theta) = \sin\theta$

$\tan(90° - \theta) = \cot\theta$ $\quad$ $\cot(90° - \theta) = \tan\theta$

$\sec(90° - \theta) = \operatorname{cosec}\theta$ $\quad$ $\operatorname{cosec}(90° - \theta) = \sec\theta$

Negative angles: $\theta < 0°$

- $\cos(-\theta) = \cos\theta$
- $\sin(-\theta) = -\sin\theta$
- $\tan(-\theta) = -\tan\theta$

Graphs of trigonometric functions

- $y = \sin\theta$ and $y = \cos\theta$

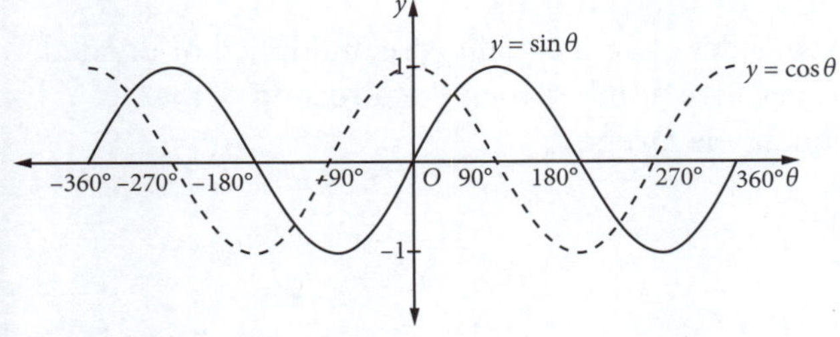

Continuous curves
Period = 360° $\quad$ Amplitude = 1

- $y = \tan\theta$

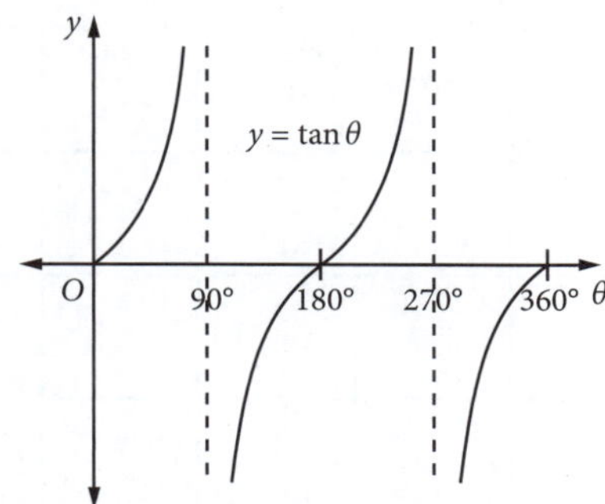

Discontinuous, asymptotes at discontinuities
Period = 180° $\quad$ No amplitude

- $y = \cot\theta$

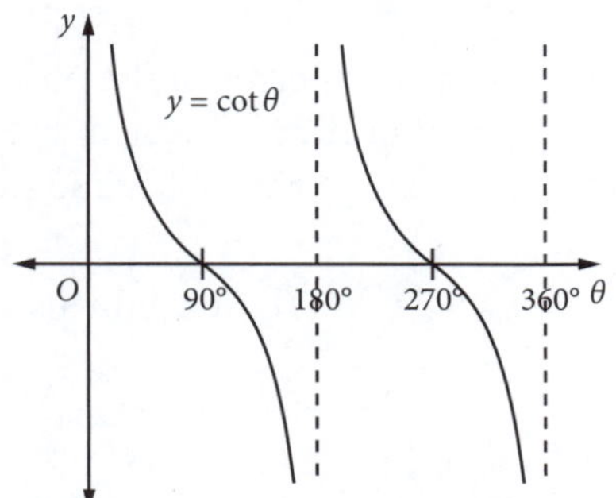

Discontinuous, asymptotes at discontinuities
Period = 180° $\quad$ No amplitude

- $y = \operatorname{cosec}\theta$

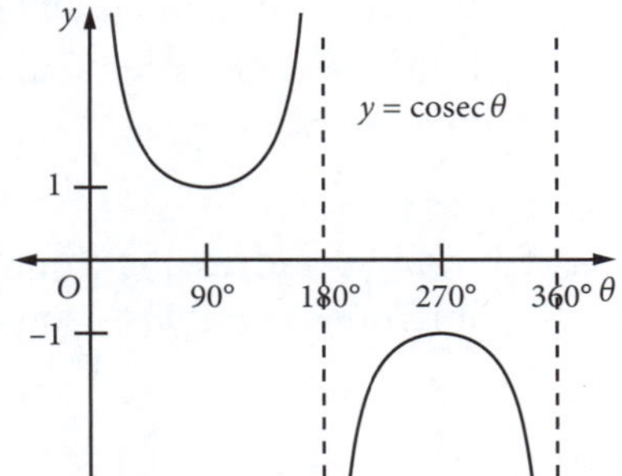

Discontinuous, asymptotes at discontinuities
Period = 360° $\quad$ No amplitude

- $y = \sec\theta$

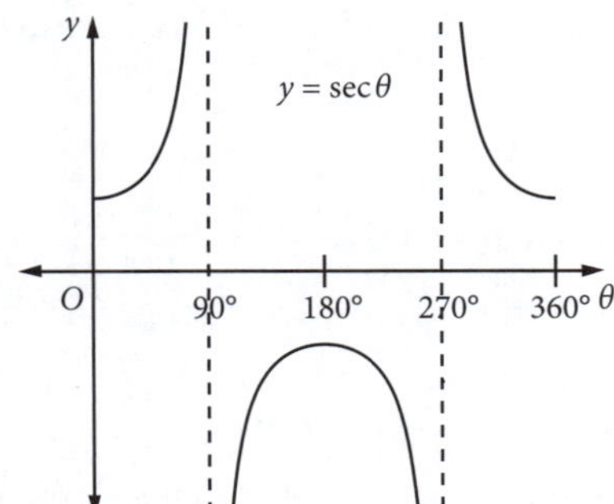

Discontinuous, asymptotes at discontinuities
Period = 360° $\quad$ No amplitude

Exact values of trigonometric ratios

θ	0°	30°	45°	60°	90°
$\sin\theta$	0	$\frac{1}{2}$	$\frac{1}{\sqrt{2}}$	$\frac{\sqrt{3}}{2}$	1
$\cos\theta$	1	$\frac{\sqrt{3}}{2}$	$\frac{1}{\sqrt{2}}$	$\frac{1}{2}$	0
$\tan\theta$	0	$\frac{1}{\sqrt{3}}$	1	$\sqrt{3}$	undefined

Direction and bearing

A bearing is a direction angle that indicates the direction of one point relative to another point.

Directions are indicated in terms of the number of degrees east or west of north or south, or are measured clockwise from north and written in standard three-figure notation.

Examples: N 70° E, S 54° W. These could also be written as 070° and 234°.

The sine rule

The sine rule: $\frac{a}{\sin A}=\frac{b}{\sin B}=\frac{c}{\sin C}$

The sine rule can be used with a triangle when you are given:

(a) the size of two angles and the length of one side, **OR**

(b) the lengths of two sides and the size of an angle opposite one of these sides.

The ambiguous case

From two sides of a triangle and an angle opposite one of these sides, it is sometimes possible to construct two different triangles.

The cosine rule

For ΔABC, the cosine rule states $a^2=b^2+c^2-2bc\cos A$. In words:

The square of one side of a triangle is equal to the sum of the squares of the other two sides minus twice the product of those two sides and the cosine of the included angle.

- To find a side, use: $a^2=b^2+c^2-2bc\cos A$
- To find an angle, use: $\cos A=\frac{b^2+c^2-a^2}{2bc}$

The cosine rule can be used with a triangle when you are given:

(a) the lengths of three sides, **OR**

(b) the lengths of two sides and the size of the included angle.

Area of a triangle

Area of $\Delta ABC=\frac{bc\sin A}{2}$

= half the product of two sides and the sine of the angle between them.

3 FURTHER ALGEBRAIC TECHNIQUES

Linear equations in one variable

- To **solve** a linear equation in one variable is to find the value of the variable that will make the equation true.
- For equations involving **fractions**, write each term with the common denominator and then solve the equation given by the numerators.

  e.g. $\frac{x}{3}=\frac{4}{5}$ becomes $\frac{5x}{15}=\frac{12}{15}$ and you can solve $5x=12$.

  If the equation has only one term on each side of the equals sign, you can multiply each numerator by the other term's denominator to achieve the same result.

Quadratic equations

- An equation of the form $ax^2+bx+c=0,\ a\neq 0$ is called a quadratic equation in x. The values of x that make this equation true are called the solutions or the roots of the equation.
- A quadratic equation can be solved by factorising the quadratic expression and making the factored expression equal to zero: if $AB=0$, then $A=0$ or $B=0$ or $A=B=0$.
- A quadratic equation which cannot be factorised can be solved by completing the square. This is done by adding a constant term equal to the square of half the coefficient of the x term, squared. The final step is to take the square root of both sides.
- When completing the square, if the coefficient of x^2 is not 1, first divide by the coefficient of x^2.
- The equation $ax^2+bx+c=0,\ a\neq 0$ is the general quadratic equation. Its roots are: $x=\frac{-b\pm\sqrt{b^2-4ac}}{2a}$

Rules for inequalities

If both sides of an inequality are multiplied or divided by a negative number, then the direction of the inequality is reversed.

If $a > b$, then:		If $a < b$, then:	
$a + c > b + c$		$a + c < b + c$	
$a - c > b - c$		$a - c < b - c$	
$ac > bc$	if $c > 0$	$ac < bc$	if $c > 0$
$ac < bc$	if $c < 0$	$ac > bc$	if $c < 0$
$\frac{a}{c} > \frac{b}{c}$	if $c > 0$	$\frac{a}{c} < \frac{b}{c}$	if $c > 0$
$\frac{a}{c} < \frac{b}{c}$	if $c < 0$	$\frac{a}{c} > \frac{b}{c}$	if $c < 0$

On number lines:

- $a > b$ means that a is to the right of b.
- $a < b$ means that a is to the left of b.
- $x \geq 2$ is shown by a solid circle over 2 and an arrow to the right
- $x > 2$ is shown by an empty circle over 2 and the arrow to the right.

4 FUNCTIONS

Square roots

- If $a \geq 0$, then $\sqrt{a}$ is a non-negative number such that $\left(\sqrt{a}\right)^2 = a$. $\sqrt{a}$ is called the positive square root of a. The value of $\sqrt{a}$ will always be either positive or zero (non-negative).
- An equation like $x^2 = 9$ always has two solutions: the positive and negative square roots of 9. $x = \sqrt{9} = 3$ and $x = -\sqrt{9} = -3$.
- Thus we have:

$$\begin{aligned} \sqrt{x^2} &= x && \text{if } x > 0 \\ &= -x && \text{if } x < 0 \\ &= 0 && \text{if } x = 0 \end{aligned}$$

Here $-x$ means the opposite sign of x, hence $-x > 0$.

Absolute value

- The absolute value (also called 'modulus') of a real number x is written $|x|$. It is the non-negative number that defines the magnitude of the given number.
- Thus we have:

$$\begin{aligned} |x| &= x && \text{if } x > 0 \\ &= -x && \text{if } x < 0 \\ &= 0 && \text{if } x = 0 \end{aligned}$$

This is identical to $\sqrt{x^2}$, so it leads to another definition: $|x| = \sqrt{x^2}$

- Because x is a real number, it can be represented by a point on the number line, and $|x|$ is the distance of the point x from the origin. Distance is always positive, so $|x| > 0$ for all $x \neq 0$.
- For example: the expression $|x| = 2$ describes a distance of 2 from the origin. The two points that are 2 units distant from 0 are $x = 0 + 2 = 2$ and $x = 0 - 2 = -2$.
- In general:

$$\begin{aligned} |x - y| &= x - y && \text{if } x > y \\ &= y - x && \text{if } x < y \\ &= 0 && \text{if } x = y \end{aligned}$$

Thus e.g. $|5 - 3| = |3 - 5| = 2$.

Absolute value—important results

1 $|xy| = |x| \times |y|$

2 $|x + y| \leq |x| + |y|$ (the **triangle inequality**) and $|x + y| = |x| + |y|$ if and only if x and y are either zero or have the same sign.

Absolute value functions

- $f(x) = |x|$ is the absolute value function (or 'numerical value function').
- $f(x) = |x| = \begin{cases} x & \text{for } x \geq 0 \\ -x & \text{for } x < 0 \end{cases}$ or equivalently

  $f(x) = |x| = \sqrt{x^2}$ for all real x.
- The domain of $f(x) = |x|$ is all real x. The range of $f(x) = |x|$ is non-negative real numbers (i.e. $f(x)$ is zero or a positive real number).

When graphing absolute value functions, consider the sign of the function within the absolute value.

Functions

- A function is a type of mathematical object that precisely describes a relationship between variables.
- A real function f of a real variable x assigns to each element x of a given set of real numbers **exactly one** real number y, which is the value of the function f at x. The notation $f(x)$ means the value of f at x: $y = f(x)$.
- The set of real numbers x on which f is defined is called the **domain** of f, while the set of values $f(x)$ obtained from x over the domain of f is called the **range** or **image** of f.
- The variable x is the **independent variable**, because it can be chosen freely from the domain of f, while y is the **dependent variable**, because its value depends on x.
- A **function** can also be defined as a set of ordered pairs with the special property that no two pairs have the same first element (x value).

Vertical line test

- If you can draw a vertical line to cut the graph more than once, then the graph does not represent a function.
- The x values for which a vertical line cannot cut the graph are not in the domain.

Relations

- A relation is like a multi-valued function: it is a set of ordered pairs, usually defined by some rule, but each first member of the ordered pairs can have more than one second member.
- The language of functions also applies to relations, so the terms independent variable, dependent variable, domain and range have the same meanings.

Function rules

- When a function rule f is given and a domain is not specified, it is assumed that the domain of the function is the set of real numbers for which $f(x)$ defines a real number range. To find the domain, the solution of an inequality may be needed.
- When the domain of f is all values of x over an interval, the graph of $y = f(x)$ is called the **curve** $y = f(x)$ and a part of the curve between two points is called an **arc**.

Odd and even functions

- An **odd function** has the property that $f(-x) = -f(x)$. The graph of f for $x \leq 0$ can be obtained by rotating the graph for $x \geq 0$ through an angle of 180° about the origin.
- An **even function** has the property that $f(-x) = f(x)$. The graph of an even function is symmetrical about the y-axis. The graph for $x \leq 0$ can be obtained by reflecting the portion for $x \geq 0$ in the y-axis.
- Note that the statement $f(-a) = f(a)$ implies that the function is defined at both $x = a$ and $x = -a$.

The circle

- A circle can be defined as the set of all points P in a plane at a given distance from a fixed point in the plane. The fixed point is the centre of the circle and the given distance is the radius.
- The equation of a circle is given by $(x-h)^2 + (y-k)^2 = r^2$, with the values for x and y restricted so that:

  $h - r \leq x \leq h + r$

  $k - r \leq y \leq k + r$
- If the centre of the circle is at the origin, then $h = 0$, $k = 0$ and the circle equation is $x^2 + y^2 = r^2$.

The circle $x^2 + y^2 = r^2$ can be represented as two functions, $y = \sqrt{r^2 - x^2}$ and $y = -\sqrt{r^2 - x^2}$. The circle $(x-h)^2 + (y-k)^2 = r^2$ can be represented as two functions, $y = k + \sqrt{r^2 - x^2}$ and $y = k - \sqrt{r^2 - x^2}$.

5 EQUATIONS AND FUNCTIONS

Gradient of a straight line, *m*

- The **gradient** (or slope) m of a straight line joining the points (x_1, y_1) and (x_2, y_2) is given by $m = \dfrac{y_2 - y_1}{x_2 - x_1}$, if $x_1 \neq x_2$.
- If $x_1 = x_2$, the gradient is undefined and the line is vertical (that is, parallel to the y-axis).
- If $y_1 = y_2$, the gradient is zero and the line is horizontal (that is, parallel to the x-axis).

Angle of inclination, θ

- The angle that a line makes with the positive direction of the x-axis is called the **angle of inclination** and usually denoted by θ, where $0° \leq \theta < 180°$.
- The gradient m is also equal to $\tan\theta$, hence $\tan\theta = \dfrac{y_2 - y_1}{x_2 - x_1}$, if $x_1 \neq x_2$.

Straight line—gradient-intercept form

- $y = mx + c$ is the equation of a straight line with gradient m that intersects the y-axis at $y = c$.

Straight line—point-gradient form

- $y - y_1 = m(x - x_1)$ is the equation of a line with gradient m that passes through the point (x_1, y_1).

Straight line—two-point form

- $\dfrac{y - y_1}{y_2 - y_1} = \dfrac{x - x_1}{x_2 - x_1}$ or $y - y_1 = \dfrac{y_2 - y_1}{x_2 - x_1}(x - x_1)$ is the equation of a line that passes through the points (x_1, y_1) and (x_2, y_2).

Straight line—general form

- $ax + by + c = 0$ is called the **general form** of the equation of a straight line. Here a is the coefficient of x, b is the coefficient of y and c is the constant term
- From the general form: gradient $= -\dfrac{a}{b}$ and y-intercept $= -\dfrac{c}{b}$
- If $a = 0$, the line is parallel to the x-axis.
- If $b = 0$, the line is parallel to the y-axis.
- If $c = 0$, the line passes through the origin.

Parallel lines

- Two lines are parallel if their gradients are equal, so $m_1 = m_2$.
- The equation of a line parallel to $ax + by + c = 0$ is $ax + by + d = 0$, where $c \neq d$.

Perpendicular lines

- Two lines are perpendicular if the product of their gradients is -1, so $m_1 m_2 = -1$.

  This can also be written as $m_2 = \dfrac{-1}{m_1}$. Each gradient is the negative reciprocal of the other.
- The equation of a line perpendicular to $ax + by + c = 0$ is $bx - ay + d = 0$.

Intersection of two lines

Two straight lines must either:

(a) intersect at a point, **OR**

(b) be parallel, **OR**

(c) coincide.

When you solve a pair of simultaneous linear equations, you will always get either:

(a) a unique solution (as the lines intersect), **OR**

(b) no solution (as the lines are parallel), **OR**

(c) an infinite number of solutions (as the lines coincide).

You can determine which situation is the case without actually solving the equations.

For $a_1x + b_1y + c_1 = 0$ and $a_2x + b_2y + c_2 = 0$, the lines are:

(a) intersecting lines if $\frac{a_1}{a_2} \neq \frac{b_1}{b_2}$ (the gradients are not equal)

(b) parallel lines if $\frac{a_1}{a_2} = \frac{b_1}{b_2} \neq \frac{c_1}{c_2}$ (the gradients are equal, but the lines are different)

(c) coincident lines if $\frac{a_1}{a_2} = \frac{b_1}{b_2} = \frac{c_1}{c_2}$ (the equations are equivalent).

Simultaneous equations

The most common algebraic methods of solving a simultaneous pair of linear equations in two variables are (a) elimination and (b) substitution.

(a) Solving simultaneous equations by elimination:
- If the coefficients have the same size but opposite sign, eliminate by addition.
- If the coefficients are equal, eliminate by subtraction.

(b) Solving simultaneous equations by substitution:
- Rewrite one of the equations to make it equal to one of the variables, then substitute this into the other equation.

Simultaneous equations may be used to solve worded problems, including break-even analysis. The break-even point occurs when the revenue is equal to the cost.

Solving simultaneous equations where one is not linear

- Equations like $y = x^2 - 5x + 6$ (a parabola when graphed), $x^2 + y^2 = 4$ (a circle when graphed) and $xy = 6$ (a rectangular hyperbola when graphed) may be intersected by a straight line. To find the intersection points, solve the pair of simultaneous equations.
- To solve non-linear simultaneous equations, the substitution method is usually best. You might also find the solutions graphically (by hand or by using graphing software).

Quadratic functions

- $ax^2 + bx + c$, a polynomial of the second degree, is called a quadratic polynomial or a quadratic expression.
- A function defined by the rule $y = ax^2 + bx + c$, where a, b and c are constants, $a \neq 0$, is a **quadratic function**. The domain of this function is the set of real numbers (unless otherwise stated or implied).
- When $y = 0$ the quadratic function becomes the quadratic equation $ax^2 + bx + c = 0$. The values of x that satisfy this equation are called the roots, solutions or zeros of the equation.
- The graph of a quadratic function has the characteristic shape of a **parabola**. The parabola has a turning point at its vertex, where the function has a minimum value if $a > 0$ or a maximum value if $a < 0$.

Maximum or minimum value of a quadratic function

All quadratic polynomials can be expressed in the form $a(x + B)^2 + C$, where a, B and C are constants, $a \neq 0$. Because $(x + B)^2$ is a perfect square, it is non-negative for all real x, i.e. $(x + B)^2 \geq 0$. Thus the smallest possible value of $(x + B)^2$ is zero, when $x = -B$, and so:

(a) If $a > 0$, the **minimum** value of $a(x + B)^2 + C$ is C.

(b) If $a < 0$, the **maximum** value of $a(x + B)^2 + C$ is C.

Turning point of a parabola

The turning point (minimum or maximum value) of the quadratic function $y = ax^2 + bx + c$ occurs at $x = -\frac{b}{2a}$, so it has the coordinates $\left(-\frac{b}{2a}, f\left(-\frac{b}{2a}\right)\right)$.

(a) If $a < 0$ this will be a **maximum** turning point, so the **maximum value** of the function is $f\left(-\frac{b}{2a}\right)$.

(b) If $a > 0$ this will be a **minimum** turning point, so the **minimum value** of the function is $f\left(-\frac{b}{2a}\right)$.

Quadratic equations

- The roots of the general quadratic equation $ax^2 + bx + c = 0, a \neq 0$ are $x = \dfrac{-b \pm \sqrt{b^2 - 4ac}}{2a}$.

Discriminants

- The **discriminant** of a quadratic equation is $\Delta = b^2 - 4ac$, so the roots of the quadratic equation can be written $x = \dfrac{-b + \sqrt{\Delta}}{2a}$ and $x = \dfrac{-b - \sqrt{\Delta}}{2a}$.
- If $\Delta > 0$, $\sqrt{\Delta}$ is a real number and so the roots are two real numbers. We say that the equation has two unequal roots, or two different roots.
- If $\Delta = 0$, $\sqrt{\Delta} = 0$ and so the roots consist of one real number: $-\dfrac{b}{2a}$. We say that the equation has only one root, or two equal roots.

- If Δ is a perfect square, $\sqrt{\Delta}$ is a rational number and so the roots are two rational numbers. We say that the equation has two unequal, rational roots.
- If $\Delta < 0$, $\sqrt{\Delta}$ does not exist in the field of real numbers and so there are no real roots. We say that the equation has no real roots.

Intersection of the parabola with the *x*-axis

- The roots of $ax^2 + bx + c = 0$ are the x values of the intersection points of the graph of $y = ax^2 + bx + c = 0$ with the x-axis. Hence at the points where the parabola cuts the x-axis, $y = 0$. Thus, the parabola:
  (a) cuts the x-axis at two distinct points if $\Delta > 0$
  (b) touches the x-axis at one point only (two coincident points) if $\Delta = 0$
  (c) does not touch the x-axis if $\Delta < 0$.

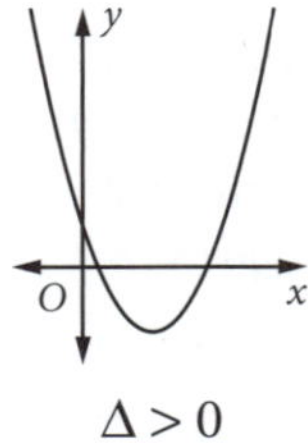

$\Delta > 0$

$a > 0$

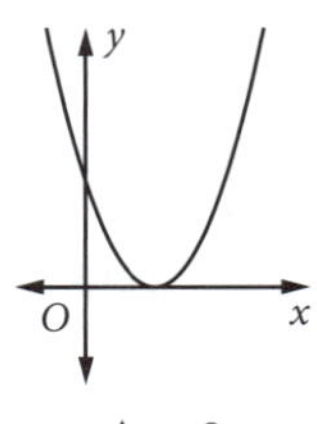

$\Delta = 0$

$a > 0$

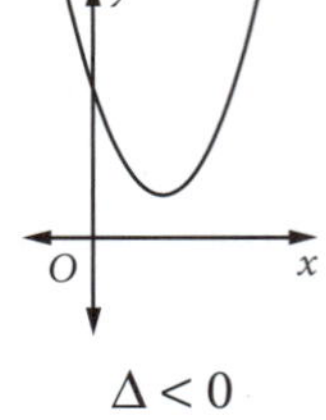

$\Delta < 0$

$a > 0$

When the coefficient of x^2 is positive, the vertex is at the bottom and the curve opens upwards.

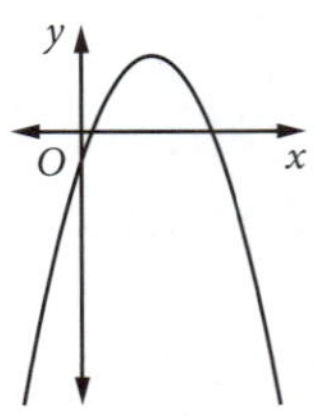

$\Delta > 0$

$a < 0$

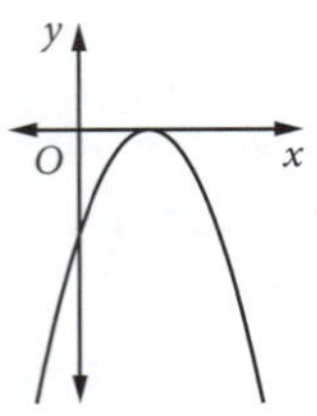

$\Delta = 0$

$a < 0$

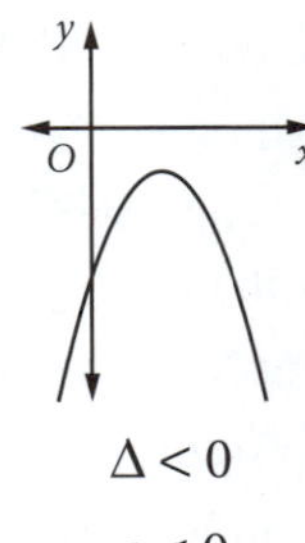

$\Delta < 0$

$a < 0$

When the coefficient of x^2 is negative, the vertex is at the top and the curve opens downwards.

- When $a > 0$ and $\Delta < 0$, the graph of $f(x) = ax^2 + bx + c = 0$ is always above the x-axis, i.e. $f(x) > 0$ for all values of x. We say f is a **positive definite** function.
- When $a < 0$ and $\Delta < 0$, the graph of $f(x) = ax^2 + bx + c = 0$ is always below the x-axis, i.e. $f(x) < 0$ for all values of x. We say f is a **negative definite** function.
- If a quadratic function is neither positive definite nor negative definite, it is **indefinite**.

6 FURTHER TRIGONOMETRY

Radian measure of an angle

1 radian is the angle subtended by an arc of 1 unit in a circle of radius 1 unit, so $\pi^c = 180°$:

$$1^c = \left(\frac{180}{\pi}\right)^{\circ} \approx 57°\,18' \quad \text{and} \quad 1° = \left(\frac{\pi}{180}\right)^{c} \approx 0.017\,45$$

$$\text{So:} \quad \theta^c = \left(\frac{180\theta}{\pi}\right)^{\circ} \quad \text{and} \quad \theta° = \left(\frac{\pi\theta}{180}\right)^{c}$$

Common conversions

Degrees	0°	30°	45°	60°	90°	120°	135°	150°	180°	210°	225°	240°	270°	300°	315°	330°	360°
Radians	0	$\frac{\pi}{6}$	$\frac{\pi}{4}$	$\frac{\pi}{3}$	$\frac{\pi}{2}$	$\frac{2\pi}{3}$	$\frac{3\pi}{4}$	$\frac{5\pi}{6}$	π	$\frac{7\pi}{6}$	$\frac{5\pi}{4}$	$\frac{4\pi}{3}$	$\frac{3\pi}{2}$	$\frac{5\pi}{3}$	$\frac{7\pi}{4}$	$\frac{11\pi}{6}$	2π

Adding 360° to an angle in degrees is the same as adding 2π to an angle in radians. Both 360° and 2π each represent a full turn.

Arc length

- Arc length $\ell = r\theta$
  The angle θ (in radians) is subtended at the centre of a circle of radius r units.

Area of a sector

- Area of a sector $A = \frac{1}{2}r^2\theta$
  The angle θ (in radians) is subtended at the centre of a circle of radius r units.
- Because $\ell = r\theta$, $A = \frac{1}{2}r^2\theta = \frac{r\ell}{2}$.

Area of a segment of a circle

- Area of segment $= \frac{1}{2}r^2(\theta - \sin\theta)$
  The angle θ (in radians) is subtended at the centre of a circle of radius r units.

Symmetry properties of trigonometric functions

Degrees	Radians
$\sin(90° - \theta) = \cos\theta$	$\sin\left(\frac{\pi}{2} - \theta\right) = \cos\theta$
$\cos(90° - \theta) = \sin\theta$	$\cos\left(\frac{\pi}{2} - \theta\right) = \sin\theta$
$\tan(90° - \theta) = \cot\theta$	$\tan\left(\frac{\pi}{2} - \theta\right) = \cot\theta$
$\sin(180° - \theta) = \sin\theta$	$\sin(\pi - \theta) = \sin\theta$
$\cos(180° - \theta) = -\cos\theta$	$\cos(\pi - \theta) = -\cos\theta$
$\tan(180° - \theta) = -\tan\theta$	$\tan(\pi - \theta) = -\tan\theta$
$\sin(180° + \theta) = -\sin\theta$	$\sin(\pi + \theta) = -\sin\theta$
$\cos(180° + \theta) = -\cos\theta$	$\cos(\pi + \theta) = -\cos\theta$

Degrees	Radians
$\tan(180° + \theta) = \tan\theta$	$\tan(\pi + \theta) = \tan\theta$
$\sin(360° - \theta) = -\sin\theta$	$\sin(2\pi - \theta) = -\sin\theta$
$\cos(360° - \theta) = \cos\theta$	$\cos(2\pi - \theta) = \cos\theta$
$\tan(360° - \theta) = -\tan\theta$	$\tan(2\pi - \theta) = -\tan\theta$
$\sin(360° + \theta) = \sin\theta$	$\sin(2\pi + \theta) = \sin\theta$
$\cos(360° + \theta) = \cos\theta$	$\cos(2\pi + \theta) = \cos\theta$
$\tan(360° + \theta) = \tan\theta$	$\tan(2\pi + \theta) = \tan\theta$

Sign of the trigonometric ratios

For radians as well as for degrees, the sign of cos, sin and tan for the first four quadrants can be remembered with the mnemonic **ASTC**: **A**ll **S**tations **T**o **C**entral.

Exact values in radians

θ	0	$\frac{\pi}{6}$	$\frac{\pi}{4}$	$\frac{\pi}{3}$	$\frac{\pi}{2}$
$\sin\theta$	0	$\frac{1}{2}$	$\frac{1}{\sqrt{2}}$	$\frac{\sqrt{3}}{2}$	1
$\cos\theta$	1	$\frac{\sqrt{3}}{2}$	$\frac{1}{\sqrt{2}}$	$\frac{1}{2}$	0
$\tan\theta$	0	$\frac{1}{\sqrt{3}}$	1	$\sqrt{3}$	undefined
$\text{cosec}\,\theta$	undefined	2	$\sqrt{2}$	$\frac{2}{\sqrt{3}}$	1
$\sec\theta$	1	$\frac{2}{\sqrt{3}}$	$\sqrt{2}$	2	undefined
$\cot\theta$	undefined	$\sqrt{3}$	1	$\frac{1}{\sqrt{3}}$	0

Trigonometric equations

Solve to find the trigonometric ratio, and then find the angles within the given parameters. You should know the exact values and may need to use the Pythagorean identities.

Trigonometric identities

- $\cos^2\theta + \sin^2\theta = 1$
- $\cos^2\theta = 1 - \sin^2\theta$
- $\sin^2\theta = 1 - \cos^2\theta$
- $1 + \tan^2\theta = \sec^2\theta$
- $\cot^2\theta + 1 = \text{cosec}^2\theta$

7 INTRODUCTION TO DIFFERENTIATION

Continuity

- Functions such as $f(x) = x, f(x) = x^3, f(x) = \sin x$ and $f(x) = |x|$ are **continuous**: their graphs do not have any gaps or jumps in them.
- Functions such as $f(x)=\frac{1}{x}, f(x)=\tan x$, $f(x)=\frac{1}{x^2}$ and $f(x)=\frac{|x|}{x}$ are **not continuous**, because their graphs all have gaps or jumps.
- A simple way to describe a continuous function is that it can be drawn smoothly without lifting your pencil from the paper. Continuous functions like $f(x) = x, f(x) = x^3, f(x) = \sin x$ are smooth graphs, but $f(x) = |x|$ is not a smooth graph because the slope of its curve changes suddenly at $x = 0$ to create a sharp corner.

Limit theorems

1 For the constant function f, where $x = c$:
$$\lim_{x\to a} f(x)=c$$

2 If $\lim_{x\to a} f(x)=L$ and $\lim_{x\to a} g(x)=M$, then:
$$\begin{aligned}\lim_{x\to a}\big(f(x)\pm g(x)\big) &= \lim_{x\to a} f(x)\pm \lim_{x\to a} g(x)\\ &= L\pm M\end{aligned}$$
The limit of a sum = the sum of the limits.
The limit of a difference = the difference of the limits.

3 If $\lim_{x\to a} f(x)=L$ and $\lim_{x\to a} g(x)=M$, then:
$$\begin{aligned}\lim_{x\to a}\big(f(x)\times g(x)\big) &= \lim_{x\to a} f(x)\times \lim_{x\to a} g(x)\\ &= L\times M\end{aligned}$$
The limit of a product = the product of the limits.

4 If $\lim_{x\to a} f(x)=L$ and $\lim_{x\to a} g(x)=M$, then:
$$\begin{aligned}\lim_{x\to a}\left(\frac{f(x)}{g(x)}\right) &= \frac{\lim_{x\to a} f(x)}{\lim_{x\to a} g(x)}\\ &= \frac{L}{M} \quad \text{for } M\neq 0\end{aligned}$$
The limit of a quotient = the quotient of the limits.

Continuity at a point—formal definition

- A function f that is defined in some neighbourhood of $x = c$ is continuous at c if:
  (a) the function has a definite value $f(c)$ at c, **AND**
  (b) as $x \to c$, $f(x)\to f(c)$ as a limit, i.e. $\lim_{x\to c} f(x)=f(c)$.

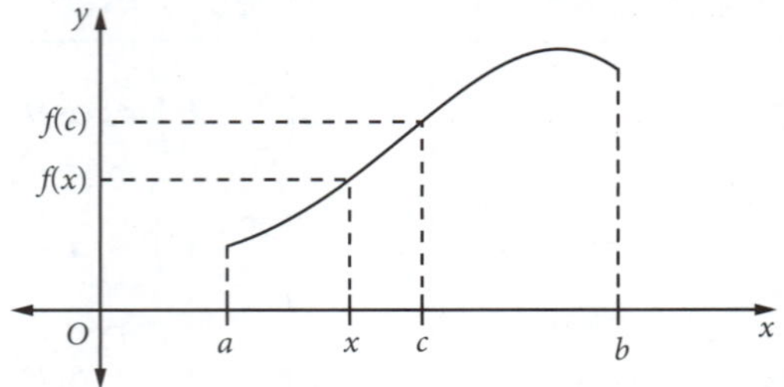

- A function f is continuous in an (open or closed) interval if it is continuous at all points of that interval. For a closed interval $[a, b]$, continuity at a and b implies that $\lim_{x\to a^+} f(x)=f(a)$ and $\lim_{x\to b^-} f(x)=f(b)$.
- In simple language, a function is continuous in the interval $[a, b]$ if its graph can be drawn from $x = a$ to $x = b$ without lifting your pencil off the paper.
- All polynomial functions are continuous.

Gradient and derivative notation

The gradient at any point (x, y) on the curve $y = f(x)$ can be represented by any of the following notations: $f'(x)$, $\frac{dy}{dx}$, y', $\frac{d}{dx}\big(f(x)\big)$.

Finding the derivative from first principles

- The derivative at any point (x, y) on the curve $y = f(x)$ is given by $f'(x)=\lim_{h\to 0}\frac{f(x+h)-f(x)}{h}$.
- If $f(x)$ possesses a derivative $f'(x)$ for each x belonging to the domain of f, then $f(x)$ is called a **differentiable function**.
- The statement '$f(x)$ is differentiable' means '$f(x)$ has a derivative at each point of its domain'.

Differentiation results

- **Binomial factorisation:** $a^n - b^n = (a - b)(a^{n-1} + a^{n-2}b + a^{n-3}b^2 + \ldots + ab^{n-2} + b^{n-1})$

1 If $y = x^n$, then $\frac{dy}{dx}=nx^{n-1}$ where n is any real number

2 If $y = cu$, where c is a constant and $u = g(x)$, then $\frac{dy}{dx}=c\frac{du}{dx}$

3 If $y = u \pm v$ where $u = f(x)$ and $v = g(x)$, then $\frac{dy}{dx}=\frac{du}{dx}\pm\frac{dv}{dx}$

4 If $y=\sqrt{x}$ then $\frac{dy}{dx}=\frac{1}{2\sqrt{x}}$

5 If $y=\frac{1}{x}$ then $\frac{dy}{dx}=-\frac{1}{x^2}$

6 From results **1** and **3**, it follows that if
$y = x^n + x^{n-1} + x^{n-2} + \ldots + x$, then
$\frac{dy}{dx} = nx^{n-1} + (n-1)x^{n-2} + (n-2)x^{n-3} + \ldots + 1$

7 **Product rule:** If u and v are functions of x, and $y = uv$, then $\frac{d}{dx}(uv) = v\frac{du}{dx} + u\frac{dv}{dx}$

8 **Chain rule:** $\frac{dy}{dx} = \frac{dy}{du} \times \frac{du}{dx}$ **OR** $\frac{d}{dx}F(u) = F'(u) \times \frac{du}{dx}$ for any differentiable function F

9 **Quotient rule:** If u and v are functions of x, and $y = \frac{u}{v}$, $v \neq 0$, then $\frac{dy}{dx} = \frac{v\frac{du}{dx} - u\frac{dv}{dx}}{v^2}$

Tangents and normals to a curve

- Differentiation gives the gradient function of a curve. This function can be evaluated at a point on the curve to obtain the gradient of the curve's **tangent** at that point. The point-gradient form of the straight line can then be used to find the equation of that tangent.
- A **normal** to a point on a curve is the line that is perpendicular to the tangent at that point. Because the lines are perpendicular, the gradient of the normal can be obtained from the gradient of the tangent.

Properties of the derivative

1 If $f(x)$ is a polynomial then the degree of $f'(x)$ will be one less than the degree of $f(x)$.

2 $f'(x)$ is also a function.

3 Functions that only differ by a constant will have the same derivative.

4 If the derivative is a constant then the original function increases or decreases at a constant rate.

5 If the derivative is itself a function of the variable, then the original function will increase and/or decrease at a variable rate

Variables in proportion

- **Direct proportion** means that the absolute values of the variables change in the same direction: as the magnitude of the independent variable increases, the magnitude of the dependent variable increases proportionally (and vice versa). Mathematically, the two variables are linear functions of each other (e.g. $y = kx$).
- **Inverse proportion** means that the absolute values of the variables change oppositely (inversely) to each other: as the magnitude of the independent variable increases, the magnitude of the dependent variable decreases proportionally (and vice versa). Mathematically, the two variables are reciprocal functions of each other (e.g. $y = \frac{k}{x}$).

Motion of a particle in a straight line

A **particle** is a body that behaves such that all forces acting on it can be regarded as acting through a single point. This means we can represent the body as a single point, regardless of its actual size and shape. Quite large bodies, e.g. trains, can still be classified as 'particles' provided this condition applies.

Displacement

- For a particle represented by a point P, moving in a straight line $X'OX$:

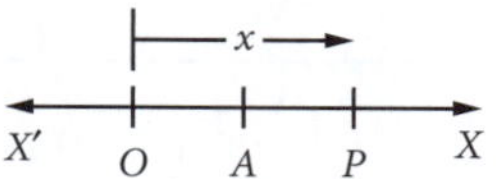

The **displacement** x is the particle's position relative to the fixed point O. It may be a positive or negative number, according to whether P is to the right or left of O, so it does not necessarily represent the total **distance** travelled. Unlike displacement, distance is always a positive quantity.

- Note that the origin of the motion (when $t = 0$) is not necessarily at O.

Velocity

- **Velocity** is defined as the rate of change of position (i.e. of displacement) with respect to time, or as the time rate of change of position in a given direction:

$$v(t) = f'(t) = \frac{dx}{dt} = \dot{x} = \lim_{h \to 0} \frac{f(t+h) - f(t)}{h}$$

- Velocity can be positive or negative, depending on the direction of travel. Speed is the magnitude of the velocity and is always positive.

Acceleration

- **Acceleration** is defined as the rate of change of velocity with respect to time:

$$a(t) = v'(t) = \frac{d^2x}{dt^2} = \frac{dv}{dt} = \ddot{x} = \lim_{h \to 0} \frac{v(t+h) - v(t)}{h}$$

- Acceleration can be positive or negative. Positive acceleration means the velocity is increasing (i.e. in the direction of positive displacement), while negative acceleration means the velocity is decreasing, which is often called deceleration or retardation.

Important motion terms

- 'initially': $t = 0$
- 'at the origin': $x = 0$
- 'at rest': $v = 0$
- 'velocity is constant': $a = 0$

Units and symbols

Physical quantity	Unit	Symbol
Time	s	t
Displacement	cm, m	x (or s in Physics)
Velocity	cm s^{-1}, m s^{-1}	$v, \frac{dx}{dt}, \dot{x}$
Acceleration	cm s^{-2}, m s^{-2}	$a, \frac{dv}{dt}, \frac{d^2x}{dt^2}, \ddot{x}$

- Note that 's' is the abbreviation for second, 'cm' for centimetre and 'm' for metre.
- Constant acceleration due to gravity $= 9.8\,\text{m s}^{-2}$ ($\approx 10\,\text{m s}^{-2}$)

Displacement, velocity, acceleration—important links

- Given the displacement function, you can find velocity and acceleration functions by differentiating with respect to time.
- Given the velocity function, you can find the acceleration function by differentiating with respect to time.
- If motion information is given as a graph, you must remember that the definite integral is linked to the area under the graph, while the derivative is linked to the gradient of the curve.
- For example: given the graph of velocity $\frac{dx}{dt}$ against time, then the value of the definite integral gives the displacement and the slope of the curve gives the acceleration.

8 EXPONENTIAL AND LOGARITHMIC FUNCTIONS

Index laws

Let $a^x = m$ and $a^y = n$

1 $a^x \times a^y = a^{x+y} \rightarrow mn = a^{x+y}$

2 $\frac{a^x}{a^y} = a^{x-y} \rightarrow \frac{m}{n} = a^{x-y}$

3 $\frac{1}{a^x} = a^{-x} \rightarrow \frac{1}{m} = a^{-x}$

4 $\left(a^x\right)^p = a^{xp} \rightarrow m^p = a^{xp}$

5 $a^1 = a$

6 $a^0 = 1$

7 $a^{\frac{1}{n}} = \sqrt[n]{a}$

8 $a^{\frac{m}{n}} = \left(\sqrt[n]{a}\right)^m = \sqrt[n]{a^m}$

9 $(ab)^x = a^x \times b^x$

Logarithm laws

Let $\log_a m = x$ and $\log_a n = y$

1 $\log_a(mn) = x + y = \log_a m + \log_a n$

$\log_a 15 = \log_a 3 + \log_a 5$

2 $\log_a\left(\frac{m}{n}\right) = x - y = \log_a m - \log_a n$

$\log_a\left(\frac{17}{5}\right) = \log_a 17 - \log_a 5$

3 $\log_a\left(\frac{1}{m}\right) = -x = -\log_a m$

$\log_a\left(\frac{1}{2}\right) = -\log_a 2$

4 $\log_a m^p = px = p\log_a m$

$\log_a 81 = \log_a(3^4) = 4\log_a 3$

5 $\log_a a = 1$

$\log_{10} 10 = 1$

6 $\log_a 1 = 0$

Note that these laws are all for $a > 0$, and so we also have $m > 0$ and $n > 0$.

- For $m > 0$, $\log_a m > 0$; for $m = 1$, $\log_a 1 = 0$; for $0 < m < 1$, $\log_a m < 0$.
  Because $a^x \geq 0$, m is never negative, so you cannot find the logarithm of a negative number.
- Most calculators have two logarithm keys: **log** (which is logarithm to base 10, also called 'common logarithm') and **ln** (which is logarithm to base e, usually called 'natural logarithm').

Change of base rule

- $\log_a n = \frac{\log_b n}{\log_b a}$

Exponential function (Euler's number e)

- $e = \lim_{n\to\infty}\left(1 + \frac{1}{n}\right)^n$
- $e = 2.718\,28$ is an approximation correct to 5 decimal places.

Important properties of e^x and $\log_e x$

- If $y = e^x$ then $\log_e y = \ln y = x$
- If $y = \log_e x = \ln x$ then $e^y = x$
- $e^{\log_e x} = e^{\ln x} = x$
- $\log_e e^x = \ln e^x = x$
- $e^0 = 1$ and $\log_e 1 = 0$
- $e^x > 0$ for all x
- Domain of $y = \log_e x$ is $x > 0$

Graphs of exponential and logarithmic functions

There are many transformations of exponential and logarithmic functions of the types $f(x) = ke^{ax+b} + c$ and $g(x) = k\log_e(ax+b) + c$.

In each case, adding c provides a vertical translation. Multiplying by k extends the range of the function (a dilation) by a factor of $|k|$. Reflection in the x-axis occurs if k is negative.

In general, you can write $a^{\log_a x} = x$ for all real x, and $\log_a a^x = x$ for all $x > 0$.

Logarithms in science

Decibels:

The difference in loudness $= 10\log_{10}\left(\frac{P_2}{P_1}\right)$

Loudness $L = 10\log_{10}\left(\frac{P}{P_0}\right)$

The Richter scale:

$M_L = \log_{10}\left(\frac{A}{A_0}\right)$

$\text{pH} = -\log_{10}\left[\text{H}^+\right]$

Number of octaves higher $= \log_2\left(\frac{f_2}{f_1}\right)$

Reciprocal property of $\frac{dy}{dx}$

- $\frac{dy}{dx} = \frac{1}{\frac{dx}{dy}}$
- $\frac{d}{dx}(\log_a x) = \frac{1}{x\log_e x}$
- $\frac{d}{dx}(a^x) = (\log_e a)a^x$

9 PROBABILITY

- The probability $P(A)$ of a particular result A is:
  $P(A) = \frac{\text{number of favourable outcomes}}{\text{number of possible outcomes}}, 0 \le P(A) \le 1$
- If $P(A) = 0$, then A is impossible.
- If $P(A) = 1$, then A is certain.

Complementary events

- The **complement** of event A is the event 'A **does not occur**', and is denoted by $\bar{A}$. A and $\bar{A}$ are complementary events.
- $P(A) + P(\bar{A}) = 1$
- $P(\bar{A}) = 1 - P(A)$

Mutually exclusive events

- If two or more events cannot occur simultaneously then the events are called **mutually exclusive** or **disjoint**.
- If the events A and B have no points in common and thus are mutually exclusive, then $P(A \text{ or } B) = P(A) + P(B)$.
- If the events A and B have points in common, then $P(A \text{ or } B) = P(A) + P(B) - P(A \text{ and } B)$.

Multiplication principle

- Tree diagrams can be used to calculate probabilities. To obtain the probability of a final outcome, you can multiply the probabilities of the successive intermediate outcomes along the branches.
- Number of ways a two-part outcome can occur $= m \times n$ (where m = number of ways the first part can occur, n = number of ways the second part can occur).

Independent events

- Two events are **independent** if the outcome of each event is not affected by the outcome of the other event.
- If events A and B are independent, then: $P(A \text{ and } B) = P(A) \times P(B) = P(AB)$.

Dependent events

- Two events are **dependent** if the outcome of one event is affected by the outcome of the other event.
- $P(B \mid A)$ is the probability of B occurring given that A has already occurred.
- $P(AB) = P(A) \times P(B \mid A)$
- $P(B \mid A) = \frac{P(AB)}{P(A)}$

Venn diagrams

- The set made by combining sets D and F is the **union**, written $D \cup F$. It consists of all the elements that are in D or in F or in both D and F.
- The set of the elements common to sets D and F is the **intersection**, written $D \cap F$.
- $n(D \cup F) = n(D) + n(F) - n(D \cap F)$ where n is the number of elements in a set D or F.
- If sets are **mutually exclusive** or **disjoint**, then $n(D \cap F) = 0$ and so $n(D \cup F) = n(D) + n(F)$.

10 DISCRETE PROBABILITY DISTRIBUTIONS

Discrete random variables

- For a discrete random variable, the number of outcomes is countable.
- In a discrete probability distribution, $0 \leq P(X = x) \leq 1$ for all values of x.
- $\Sigma P(X = x) = 1$
- The expected value of the variable X is $E(X) = \mu = \Sigma\, x_i\, p_i$.
- A game is fair if the cost to play is equal to the expected return.
- $E\,[f(X)] = \Sigma f(x_i)\, p_i$
- $E(aX + b) = aE(X) + b$
- $E(X + Y) = E(X) + E(Y)$
- $\text{Var} = \sigma^2 = E(X - \mu)^2 = \Sigma(x - \mu)^2\, p_i = E(X^2) - [E(X)]^2$
- $\sigma = \sqrt{\sigma^2} = \sqrt{\text{Var}(X)}$
- $\sigma(ax + b) = |a|\sigma(x)$
- For many distributions, about 95% of results lie within two standard deviations of the mean. This interval can be represented as $\mu \pm 2\sigma$.

The uniform distribution

A discrete probability distribution is said to be uniform if all values of the random variable are equally likely.

For a discrete uniform probability distribution, where $X = 1, 2, \ldots, n$,

- $P(X = x) = \dfrac{1}{n}$ for $x \in \{1, 2, 3, \ldots, n\}$ and zero otherwise.
- $E(X) = \dfrac{n+1}{2}$
- $\text{Var}(X) = \dfrac{n^2 - 1}{12}$

11 DESCRIPTIVE STATISTICS

Statistical investigation process

There are four parts to a statistical investigation:

- identification of a problem and the setting of a statistical question
- collection of the statistical data
- analysis of the data
- interpretation and communication of the results.

Data types

There are two types of categorical data:

- nominal (such as type of fruit)
- ordinal (such as strongly agree, agree, disagree).

There are two types of numerical data:

- discrete (such as number of brothers and sisters)
- continuous (such as height of players in a team).

Simple data displays

- A frequency table is used to collect numerical data. It can be interpreted on its own or used as an aid in drawing a statistical graph.

x	f
1	3
2	5
3	2
4	4
5	2

- A bar chart is used to represent categorical (nominal) data.

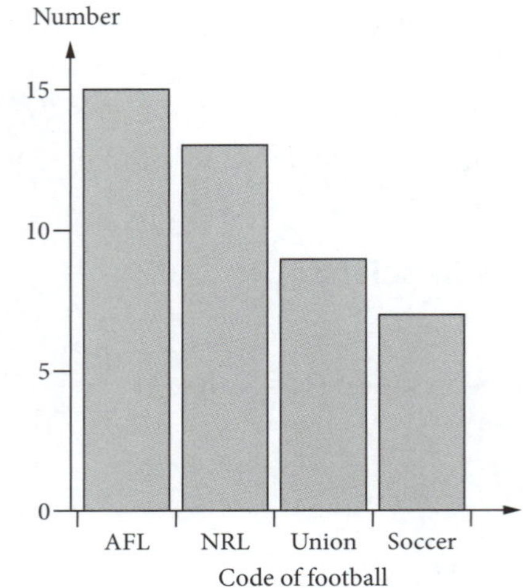

- A segmented bar chart is used to display information for which there is some sense of whole.

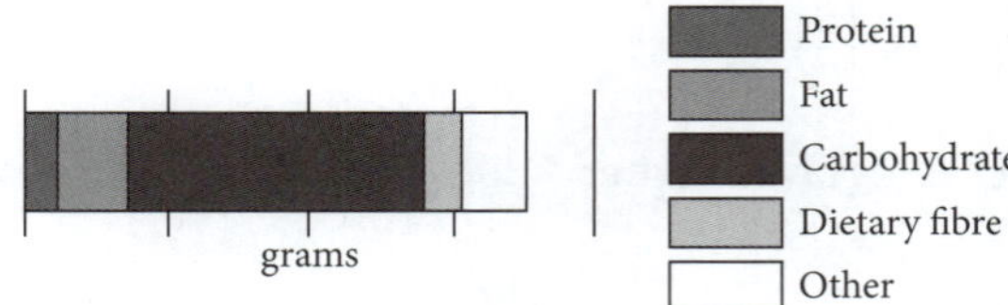

- A dot plot can be used with either numerical or categorical data.

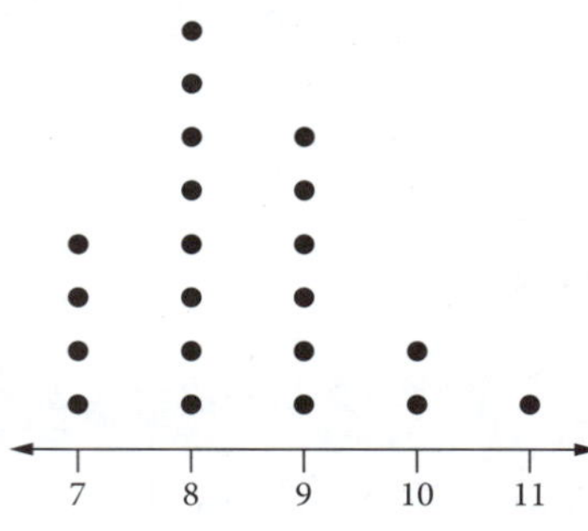

- A histogram is used to display continuous numerical data.

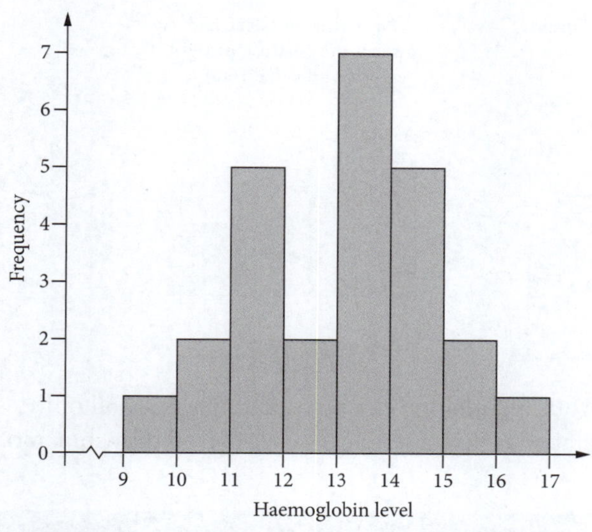

A stem plot is used to display discrete numerical data. A stem plot should include a key to enable the values to be interpreted. A stem plot does not lose any of the original information.

Stem	Leaf
4	2 6 9
5	0 5 6 7 7 8
6	0 6
7	2 8 8 8
8	1 2 3 3 4 5
9	2 3 9

Key: 5|0 = 50

Values of central tendency and spread

- There are three measures of central tendency: mean, median and mode.
- For ungrouped data the mean, $\overline{x}$, is found using:

$$\overline{x} = \frac{\text{sum of the data values}}{\text{number of data values}}$$
$$= \frac{\Sigma x}{n}$$
$$= \frac{\Sigma x}{\Sigma f}$$

- For ungrouped data presented in a frequency table use: $\overline{x} = \frac{\Sigma xf}{\Sigma f}$
- For grouped data, find the midpoint of the class interval x_m and multiply this by f to find the approximate total for each interval. $\overline{x} = \frac{\Sigma x_m f}{\Sigma f}$
- The median is the physical middle of the data set. For odd values of n, the median is the $\left(\frac{n}{2}+0.5\right)$th value; for even values of n, the median is the mean of the $\left(\frac{n}{2}\right)$th and $\left(\frac{n}{2}+1\right)$th values.
- The mode is the most frequently occurring data value.
- A bimodal data set has two values with the same highest frequency. If there are more than two values with the highest frequency, then the set has no mode or is multimodal.
- Grouped data is difficult to deal with as far as the median is concerned. You can, at best, estimate the value or just record the class interval that contains the value.
- There are three measures of spread: range, interquartile range (IQR) and standard deviation.
- Range is the highest value less the lowest value.
- To find the IQR, first divide the data set into four pieces, called quartiles. The dividing lines are called Q_1, median (Q_2) and Q_3.

  $\text{IQR} = Q_3 - Q_1$
- The IQR, unlike the range, is not affected by extreme values.
- An outlier lies below $Q_1 - 1.5 \times \text{IQR}$ or above $Q_3 + 1.5 \times \text{IQR}$.
- Standard deviation is a measure of spread which uses all of the data values. It can be affected by extreme values so some caution is required in its use.
- Population standard deviation is calculated using $\sigma = \sqrt{\frac{\Sigma(x-\mu)^2}{n}}$, where x refers to the individual scores, μ is the population mean and n is the number of values.
- Sample standard deviation is calculated using: $s = \sqrt{\frac{\Sigma(x-\overline{x})^2}{n-1}}$, where x refers to the individual scores, $\overline{x}$ is the sample mean and n is the number of values.

Data analysis methods and tools

- The five-figure summary arises from dividing the data set into quartiles. It consists of the lowest value, Q_1, median, Q_3 and the maximum value.
- The statistical graph that uses the five-figure summary is the box plot.

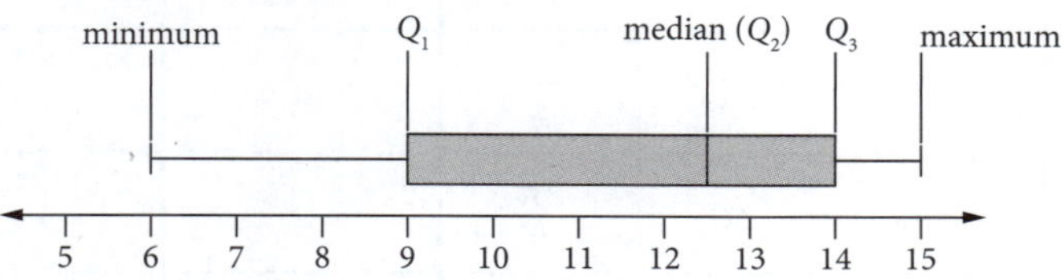

- The shape of a data set can be described using terms such as negative skew, symmetrical and positive skew.

Negative skew (stretched towards the negative, or left, side of the diagram)

Symmetrical (relatively evenly distributed)

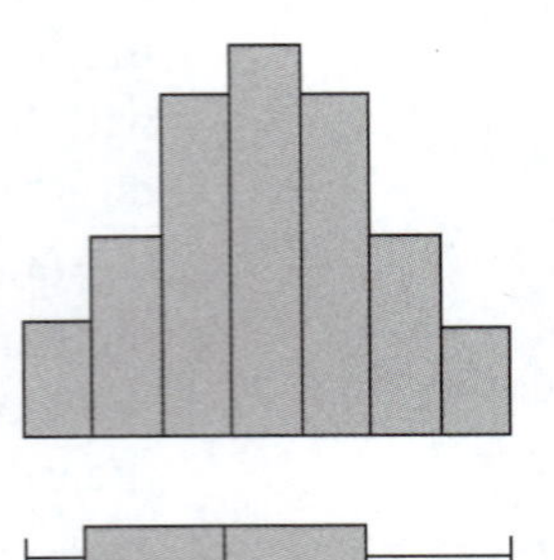

Positive skew (stretched towards the positive, or right, side of the diagram)

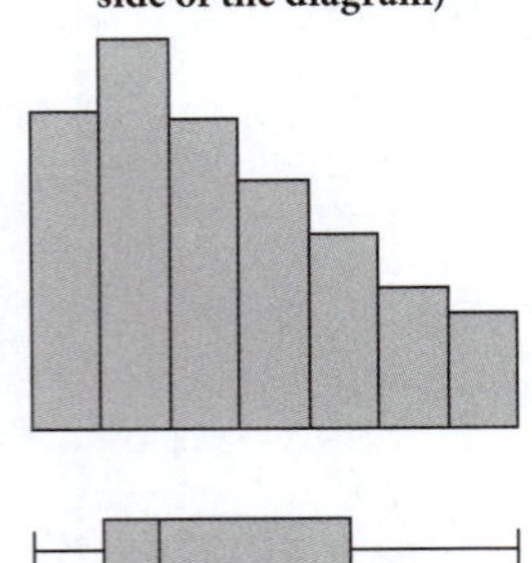

- Composite bar charts, back-to-back stem plots and parallel dot plots are all used to assist in the analysis of data sets.
- Your technology can be used to produce all of the statistics discussed here, as well as to draw box plots.

12 TRIGONOMETRIC FUNCTIONS AND GRAPHS

Summary of trigonometric functions

Function	Period	Amplitude	Domain	Range	Symmetry
$y = a\sin x$	2π	a	Real x	$-a \le y \le a$	Rotational symmetry
$y = a\cos x$	2π	a	Real x	$-a \le y \le a$	Symmetrical about y-axis
$y = a\tan x$	π	none	Real x, $x \ne \frac{(2n-1)\pi}{2}$ for integer n	Real y	Rotational symmetry
$y = a\sin bx$	$\frac{2\pi}{b}$	a	Real x	$-a \le y \le a$	Rotational symmetry
$y = a\cos bx$	$\frac{2\pi}{b}$	a	Real x	$-a \le y \le a$	Symmetrical about y-axis
$y = a\tan bx$	$\frac{\pi}{b}$	none	Real x, $x \ne \frac{(2n-1)\pi}{2b}$ for integer n	Real y	Rotational symmetry
$y = a\sin bx + c$	$\frac{2\pi}{b}$	a	Real x	$-a + c \le y \le a + c$	
$y = a\cos bx + c$	$\frac{2\pi}{b}$	a	Real x	$-a + c \le y \le a + c$	Symmetrical about y-axis
$y = a\tan bx + c$	$\frac{\pi}{b}$	none	Real x, $x \ne \frac{(2n-1)\pi}{2}$ for integer n	Real y	
$y = a\sin(bx + \alpha)$	$\frac{2\pi}{b}$	a	Real x	Real y	
$y = a\cos(bx + \alpha)$	$\frac{2\pi}{b}$	a	Real x	Real y	
$y = a\tan(bx + \alpha)$	$\frac{\pi}{b}$	none	Real x, $x \ne \frac{(2n-1)\pi - 2\alpha}{2b}$ for integer n	Real y	

Further solution of trigonometric equations

When solving trigonometric equations of the form sin $(ax + b)$ over a given domain, say $0 \le x \le 2\pi$, you need to be aware of how $(ax + b)$ affects the domain for your solution.

For example, if you are solving sin $(2x) = 0.5$, and are given $0 \le x \le 2\pi$, then you can only consider values for x in $0 \le 2x \le 4\pi$, or $[0, 4\pi]$.

13 DIFFERENTIAL CALCULUS

Derivatives involving e^x

- $\frac{d}{dx}\left(e^x\right) = e^x$
- $\frac{d}{dx}\left(e^{kx}\right) = ke^x$
- $\frac{d}{dx}\left(e^{ax+b}\right) = ae^{ax+b}$
- $\frac{d}{dx}\left(e^{f(x)}\right) = f'(x)e^{f(x)}$

Approximations when x is small

- $\lim\limits_{x \to 0} \frac{\sin x}{x} = 1$ • $\lim\limits_{x \to 0} \frac{\tan x}{x} = 1$
- Thus: $\tan x > \sin x$ for small x
- $\lim\limits_{x \to 0} \frac{1 - \cos x}{x^2} = 0.5$

Derivatives of trigonometric functions

- $\frac{d}{dx}(\sin x) = \cos x$
- $\frac{d}{dx}(\cos x) = -\sin x$
- $\frac{d}{dx}(\tan x) = \sec^2 x$
- $\frac{d}{dx}\left(\sin\left(f(x)\right)\right) = f'(x)\cos\left(f(x)\right)$
- $\frac{d}{dx}\left(\cos\left(f(x)\right)\right) = -f'(x)\sin\left(f(x)\right)$
- $\frac{d}{dx}\left(\tan\left(f(x)\right)\right) = -f'(x)\sec^2\left(f(x)\right)$
- $\frac{d}{dx}(\operatorname{cosec} x) = -\operatorname{cosec} x \cot x$
- $\frac{d}{dx}(\sec x) = \sec x \tan x$
- $\frac{d}{dx}(\cot x) = -\operatorname{cosec}^2 x$
- $\frac{d}{dx}(\sin(ax+b)) = a\cos(ax+b)$
- $\frac{d}{dx}(\cos(ax+b)) = -a\sin(ax+b)$
- $\frac{d}{dx}(\tan(ax+b)) = a\sec^2(ax+b)$
- $\frac{d}{dx} = \left(\operatorname{cosec}(ax+b)\right) = -a \operatorname{cosec}(ax+b)\cot(ax+b)$
- $\frac{d}{dx} = \left(\sec(ax+b)\right) = -a\sec(ax+b)\tan(ax+b)$
- $\frac{d}{dx} = \left(\cot(ax+b)\right) = -a \operatorname{cosec}^2(ax+b)$

Derivatives involving $\log_e x$

- $\frac{d}{dx}(\log_e x) = \frac{1}{x}, x > 0$
- $\frac{d}{dx}(\log_e(ax)) = \frac{1}{x}, x > 0$
- $\frac{d}{dx}(\log_e(ax+b)) = \frac{a}{ax+b}$
- $\frac{d}{dx}(\log_e[f(x)]) = \frac{f'(x)}{f(x)}$
- $\frac{d}{dx}(\log_a x) = \frac{1}{x\log_e x}$
- $\frac{d}{dx}\left(a^x\right) = (\log_e a)a^x$

14 THE FIRST AND SECOND DERIVATIVE

The sign of the derivative

- If $\frac{dy}{dx} > 0$ as x increases, the function is an **increasing** function.
- If $\frac{dy}{dx} < 0$ as x increases, the function is a **decreasing** function.
- If $\frac{dy}{dx} = 0$ at a given value of x, the function is **stationary** at that point. At this **stationary point**, the tangent to the curve is parallel to the x-axis.

These results and their converses can determine a function's characteristics:

- If $\frac{dy}{dx} > 0$, then the function is increasing; if the function is increasing, then $\frac{dy}{dx} > 0$.
- If $\frac{dy}{dx} < 0$, then the function is decreasing; if the function is decreasing, then $\frac{dy}{dx} < 0$.
- If $\frac{dy}{dx} = 0$, then the point is a stationary point; at a stationary point, $\frac{dy}{dx} = 0$.

Turning points

- A turning point of $f(x)$ is a point where the curve $y = f(x)$ is locally a maximum or a minimum.
- For a differentiable function $f(x)$, all turning points are stationary points.
(**However**, not all stationary points are turning points.)

The second derivative

- Differentiating a function once to obtain $f'(x)$ or $\frac{dy}{dx}$ gives you the first derivative of the original function.
- Differentiating the first derivative gives you $f''(x)$ or $\frac{d^2y}{dx^2}$, which is the **second derivative** of the original function. This is the rate of change of the first derivative: $\frac{d}{dx}\left(f'(x)\right)$.
- If $y = f(x)$, then the first derivative can be written $\frac{dy}{dx}$, $f'(x)$, $\frac{d}{dx}\left(f(x)\right)$ or y'.
- If $y = f(x)$, then the second derivative can be written $\frac{d}{dx}\left(\frac{dy}{dx}\right)$, $\frac{d^2y}{dx^2}$, $f''(x)$, $\frac{d}{dx}\left(f'(x)\right)$ or y''.

Concavity

The **concavity** of a function describes how the shape of the function's graph 'opens'. Graphs can be 'concave up' and 'concave down'.

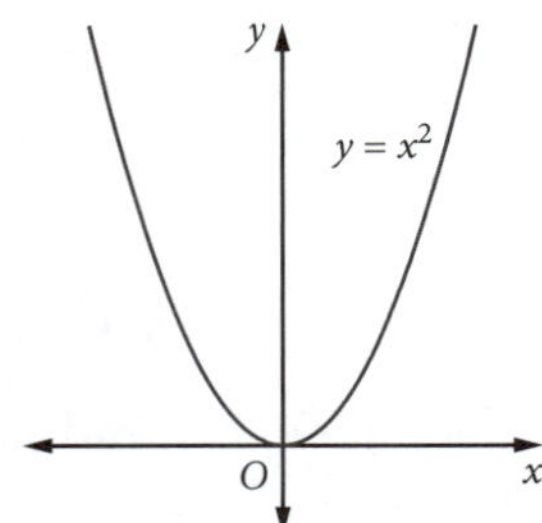

The function is concave up.
Note: has a minimum turning point.

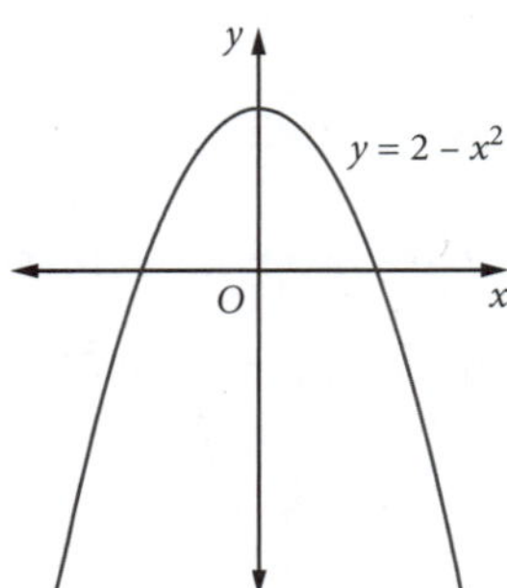

The function is concave down.
Note: has a maximum turning point.

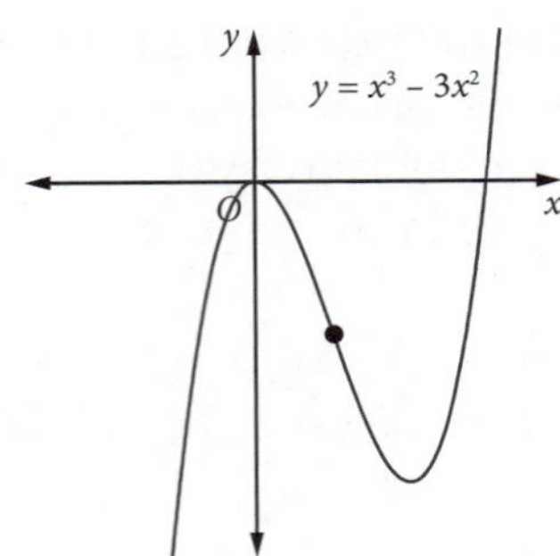

The left-hand part is concave down. The right-hand part is concave up. Concavity changes at $x = 1$.

The sign of the second derivative

- If $\frac{d^2y}{dx^2} > 0$ on an interval then the curve is concave up on that interval.
- If $\frac{d^2y}{dx^2} < 0$ on an interval then the curve is concave down on that interval.
- If $\frac{d^2y}{dx^2} = 0$ at a point on the curve **and** the concavity changes at this point, then the point is called a **point of inflection**.

Tests for stationary points—summary

- For stationary points: $\frac{dy}{dx} = 0$

Using only the first derivative:

- Turning points: $\frac{dy}{dx} = 0$ and $\frac{dy}{dx}$ changes sign on passing through the point.
- Maximum turning points: $\frac{dy}{dx} = 0$ and $\frac{dy}{dx}$ changes from positive to negative through the point.
- Minimum turning points: $\frac{dy}{dx} = 0$ and $\frac{dy}{dx}$ changes from negative to positive through the point.

Using the first and second derivatives:

- Turning points: $\frac{dy}{dx} = 0$ and $\frac{d^2y}{dx^2}$ does not changes sign on passing through the point.
- Maximum turning points: $\frac{dy}{dx} = 0$ and $\frac{d^2y}{dx^2} < 0$ at the point.
- Minimum turning points: $\frac{dy}{dx} = 0$ and $\frac{d^2y}{dx^2} > 0$ at the point.
- Point of inflection: if $\frac{dy}{dx} = 0$ and $\frac{d^2y}{dx^2} = 0$ at the point, it may be a turning point **OR** a horizontal point of inflection. If $\frac{d^2y}{dx^2}$ also changes sign on passing through the point, then it is a point of inflection.

Global maxima and minima

The global maxima and global minima are the greatest and least values of a function. If the function has a restricted domain, these may occur at the endpoints of the domain or at the turning points of the function. You also need to find the value of the function at the endpoints of the given domain, or the natural domain if no restrictions are given.

Motion in a straight line

Displacement is defined as the position relative to a starting point.

Velocity is defined as the rate of change of position (i.e. of displacement) with respect to time.

$$v(t) = f'(t) = \frac{dx}{dt} = \dot{x} = \lim_{h\to 0}\frac{f(t+h)-f(t)}{h}$$

Speed is the magnitude of the velocity and is always positive.

Acceleration is defined as the rate of change of velocity with respect to time.

$$a(t) = v'(t) = \frac{d^2x}{dt^2} = \ddot{x} = \lim_{h\to 0}\frac{v(t+h)-v(t)}{h}$$

Summary of important terms:

'initially': $t = 0$ 'at the origin': $x = 0$

'at rest': $v = 0$ 'velocity is constant': $a = 0$

15 GRAPHING TECHNIQUES

Summary of the transformations

Given $y = f(x)$, then:

$y = f(x) + k$ translates the curve k units upwards

$y = f(x + c)$ translates the curve c units to the left

$y = af(x + c)$ stretches the curve by a factor of a from the x-axis

$y = f(b(x + c))$ stretches the curve by a factor $\frac{1}{b}$ from the y-axis.

The order in which transformations are performed is important. Using an incorrect order may change the shape of the graph.

Graphing rational algebraic functions

Functions with the independent variable in the denominator generate curves that are not continuous and may have asymptotes. They may not have any turning points. You need to consider what happens to the function for very large positive and negative values of the variable.

When sketching rational algebraic functions:

- identify any restrictions on the domain and the range
- find the intercepts on the coordinate axes where possible (it is usually easy to find the y-intercept, if it exists, but it is not always possible to find the x-intercept)
- use the symmetry properties of odd and even functions whenever possible
- find stationary points and determine their nature
- find asymptotes and use them to guide the shape of the curve. Asymptotes can be horizontal, vertical, sloping and occasionally another curve.

Graphs of functions with variable denominators may have asymptotes. You also have information about turning points and points of inflection obtained by using calculus on the function.

Graphical solution of equations

When it is not possible to use standard techniques for solving equations associated with various functions, draw graphs and find the x values of any intersection points. This can be done with graphing software. Alternatively, carefully sketch each graph, then use a claculator to refine these approximate x values as accurately as possible.

Regions and inequalities

- A straight line represents a function. If the equation is changed to an inequality, then the function becomes a relation and is represented by a region in the number plane.
- A boundary line divides the number plane into three sets of points: points on the line, points above the line and points below the line.
- An inequality with $\geq$ or $\leq$ includes the boundary line as part of the region; an inequality with $>$ or $<$ does not include the boundary line as part of the region, so the line is dashed.
- To identify which region should be shaded, substitute the coordinates of a point (not on the boundary) into the inequality. If the point makes the inequality true, then all points on that side of the inequality will also make it true. Shade that side of the boundary to indicate the region. If the point does not make the inequality true, shade the other side instead.

Simultaneous linear inequalities

Two linear equations may intersect at a point. The intersection of two linear inequalities is the region common to the two inequalities. This is the region where the inequalities hold simultaneously. Shade the area which is common to both regions.

Simultaneous parabolic and linear inequalities

The parabolic inequality will be either above or below the curve. Use a point to test which it is as before.

First find the intersection of the equalities, then shade the region common to both inequalities.

The solution can be used to solve quadratic inequalities.

16 THE ANTI-DERIVATIVE

Primitive functions

- The process of finding the function from the derivative is called **antidifferentiation, finding the primitive** of the function or **finding the indefinite integral**.
- If $F(x)$ is a function such that $F'(x)=f(x)$, then $F(x)$ is called the **antiderivative** or **primitive** of $f(x)$:

  $$F(x)=\int f(x)dx$$

  Similarly, $f(x)$ is the primitive of $f'(x)$.
- The primitive of x^n is $\frac{1}{n+1}x^{n+1}+C, n\neq 1$.

17 INTEGRAL CALCULUS

The definite integral and the area under a curve

- $\int_a^b f(x)dx$ is called the **definite integral** of the function $f(x)$ between $x=a$ and $x=b$.
- The value A of this definite integral is the size of the area under the curve $y=f(x)$ between $x=a$ and $x=b$.
- For a definite integral, we write:

  $$A=\int_a^b f(x)dx \quad \textbf{OR} \quad A=\int_a^b y\,dx$$

Fundamental theorem of calculus

- $\int_a^b f(x)dx=F(b)-F(a)$ where $F(x)$ is a primitive function of $f(x)$.
- When evaluating an integral, it is convenient to set it out as:

  $$\int_a^b f(x)dx=[F(x)]_a^b$$
  $$=F(b)-F(a)$$

Integrals—important results

- If $y=f(x)$ is an increasing function over the interval $a\leq x\leq b$, then $y=-f(x)$ is a decreasing function over the same interval.
- $y=-f(x)$ is the reflection of $y=f(x)$ in the x-axis. Thus the area of the region bounded by $y=f(x)$, $x=a$, $x=b$ and the x-axis is equal to the area of the region bounded by $y=-f(x)$, $x=a$, $x=b$ and the x-axis:

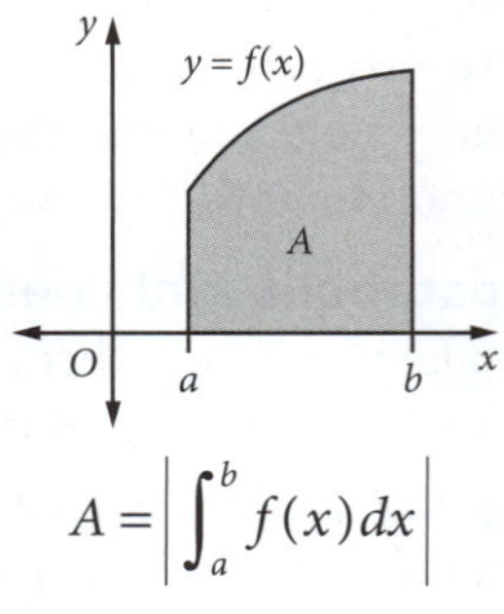

$$A=\left|\int_a^b f(x)dx\right|$$

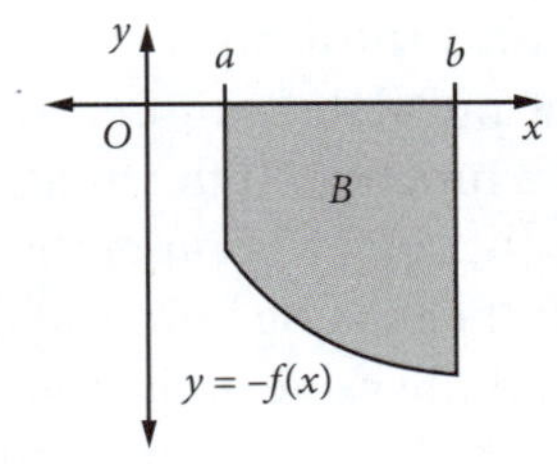

$$B=\left|\int_a^b(-f(x))dx\right|$$

$$B=\left|-\int_a^b f(x)dx\right|=\left|\int_a^b f(x)dx\right|$$

- If $y=f(x)$ is a continuous function over the interval $a\leq x\leq b$, and $a\leq c\leq b$, then:

  $$\int_a^b f(x)dx=\int_a^c f(x)dx+\int_c^b f(x)dx$$
- If F, G are primitives of f, g respectively, then $F+G$ is a primitive of $f+g$. Thus:

  $$\int_a^b(f(x)+g(x))dx=(F(b)+G(b))-(F(a)+G(a))$$
  $$=(F(b)-F(a))-(G(b)-G(a))$$
  $$=\int_a^b f(x)dx+\int_a^b g(x)dx$$
- If you reverse the limits of integration, you change the sign of the integral:

  $$\int_a^b f(x)dx=-\int_b^a f(x)dx$$
- $\int_a^b f(x)dx\pm\int_a^b g(x)dx=\int_a^b(f(x)\pm g(x))dx$
- $\int_a^b cf(x)dx=c\int_a^b f(x)dx$ where c is a constant.

Odd and even functions

- For an even function, the area bounded by the curve $y=f(x)$, the x-axis, $x=a$ and $x=-a$ is:

  $$\text{Area}=\left|\int_{-a}^a f(x)dx\right|=2\left|\int_0^a f(x)dx\right|$$
- For an odd function, the area bounded by the curve $y=f(x)$, the x-axis, $x=a$ and $x=-a$ is:

  $$\text{Area}=\left|\int_{-a}^0 f(x)dx\right|+\left|\int_0^a f(x)dx\right|=2\left|\int_0^a f(x)dx\right|$$

Trapezoidal rule

If $f(x)$ is a continuous function and $f(x)\geq 0$ on the interval $a\leq x\leq b$, then you can find an approximation to the definite integral $\int_a^b f(x)dx$ by dividing the area into a number of trapezia of equal width, where the parallel sides of the trapezia are ordinates.

- One trapezium:

  $$\int_a^b f(x)dx\approx\frac{(b-a)}{2}(f(a)+f(b))$$

- Two trapezia:

$$\int_a^b f(x)dx \approx \frac{(b-a)}{2}\left(\frac{f(a)+f\left(\frac{a+b}{2}\right)}{2}\right)+\frac{(b-a)}{2}\left(\frac{f\left(\frac{a+b}{2}\right)+f(b)}{2}\right)$$

$$=\frac{(b-a)}{4}\left(f(a)+2f\left(\frac{a+b}{2}\right)+f(b)\right)$$

- n trapezia: the base of each trapezium is $\frac{b-a}{n}=h$. For $a=x_0$, $b=x_n$ and other points are $x_1, x_2, \dots x_{n-1}$, so the trapezoidal rule is:

$$\int_a^b f(x)dx \approx \frac{h}{2}\big(f(a)+2f(x_1)+2f(x_2)+\dots+2f(x_{n-1})+f(b)\big)$$

where $h=\frac{b-a}{n}$

Area between two curves

If f and g are two continuous functions whose graphs do not intersect in the interval $a \le x \le b$, and $f(x) > g(x)$ over this interval, then the area of the region bounded by the two curves and the ordinates $x=a$ and $x=b$ is given by:

$$\text{Area} = \int_a^b f(x)dx - \int_a^b g(x)dx$$

$$= \int_a^b \big(f(x)-g(x)\big)dx$$

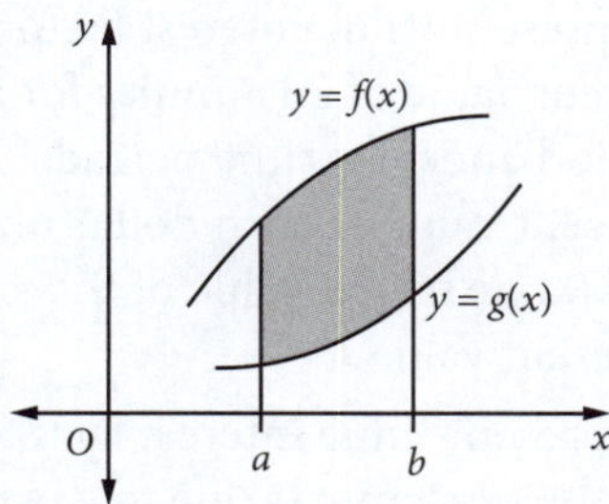

This result also applies if the two curves intersect only at $x=a$ and $x=b$.

Area bounded by the y-axis

- To find areas where one of the boundaries is the y-axis, you can effectively swap the roles of the x and y variables.
- The area bounded by the curve $y=f(x)$, the x-axis and the ordinates at $x=a$ and $x=b$ is given by:

$$\text{Area} = \int_a^b f(x)dx = \int_a^b y\,dx$$

- Similarly, the area bounded by the curve $x=g(y)$, the y-axis and the abscissae at $y=c$ and $y=d$ is given by:

$$\text{Area} = \int_c^d g(y)dy = \int_c^d x\,dy$$

- For two curves $x=f(y)$ and $x=g(y)$ that intersect at $y=c$ and $y=d$, where $f(y) \ge g(y)$ over the interval $c \le y \le d$, the area bounded by the curves is given by:

$$\text{Area} = \int_c^d \big(f(y)-g(y)\big)dy$$

The indefinite integral

- The notation $\int f(x)dx$ represents the **indefinite integral**. This uses the integral sign $\int$ to say 'find the primitive of $f(x)$'. It gives a function to represent all possible values of the primitive by adding C, the **constant of integration**.
- If $F(x)$ is a primitive of $f(x)$, then:

$$\int f(x)dx = F(x)+C$$

- The indefinite integral is a function of x, whereas the definite integral has a (definite) value.

Integrals involving e^x

- $\int e^x\,dx = e^x + C$
- $\int e^{kx}\,dx = \frac{1}{k}e^{kx} + C$
- $\int e^{ax+b}\,dx = \frac{1}{a}e^{ax+b} + C$
- $\int f'(x)e^{f(x)}\,dx = e^{f(x)} + C$

Caution: do not assume an integral at 0 is always zero; exponential integrals at 0 often produce a non-zero constant. Always evaluate both limit values of a definite integral.

Integrals involving $\log_e x$

- $\int \frac{1}{x}dx = \log_e x + C$
- $\int \frac{1}{ax+b}dx = \frac{1}{a}\log_e|ax+b| + C$
- $\int \frac{f'(x)}{f(x)}dx = \log_e|f(x)| + C$
- $\int \frac{af'(x)}{f(x)}dx = a\log_e|f(x)| + C$
- $\int a^x\,dx = \frac{a^x}{\log_e a} + C$

Integrals of trigonometric functions

- $\int \sin x\,dx = -\cos x + C$
- $\int \sin(ax)\,dx = -\frac{1}{a}\cos(ax) + C$
- $\int \sin(ax+b)\,dx = -\frac{1}{a}\cos(ax+b) + C$
- $\int \cos x\,dx = \sin x + C$
- $\int \cos ax\,dx = \frac{1}{a}\sin ax + C$
- $\int \cos(ax+b)\,dx = \frac{1}{a}\sin(ax+b) + C$
- $\int \sec^2 x\,dx = \tan x + C$
- $\int \sec^2 ax\,dx = \frac{1}{a}\tan ax + C$
- $\int \sec^2(ax+b)\,dx = \frac{1}{a}\tan(ax+b) + C$

18 FINANCIAL MATHEMATICS

Compound interest

The compound interest formula

$A = P(1+r)^n$

P is the principal (or starting amount)

r is the interest rate for the period expressed as a decimal

n is the number of compounding periods

A is the amount to which P has grown or reduced (final amount)

Interest is $A - P$

Compound interest as a recurrence relation.

$P = A_0$, $A_1 = A_0 \times R$, $A_2 = A_1 \times R$, … $A_n = A_{n-1} \times R$.

A_n is the value of the initial amount after n time periods and $R = 1 + \frac{r}{100}$. In this case r is expressed as a percentage rate, r%, per period. If r is expressed as a decimal, then $R = 1 + r$.

Future value and present value

The starting amount is also called the present value (PV), and the amount to which it grows is called the future value (FV). The compound interest formula becomes $FV = PV(1+r)^n$.

If you know the future value you require, you can express this formula with PV as the subject.

$$PV = \frac{FV}{(1+r)^n}$$

Annuities

An annuity is a compound interest investment from which equal payments are made or received on a regular basis (at equal periods of time) for a fixed period of time.

$$\text{FVA} = a\left\{\frac{(1+r)^n - 1}{r}\right\} \text{ and } \text{PVA} = a\left\{\frac{(1+r)^n - 1}{r(1+r)^n}\right\}$$

FVA is the Future Value of an Annuity

PVA is the Present Value of an Annuity

a is the contribution per period paid at the end of the period

r is the interest rate per compounding period as a decimal

n is the number of periods

Annuities can also be expressed as recurrence relations. This enables spreadsheets to be used to predict their value over time.

Summary of present and future value interest factors

You can use tables to calculate annuities.

1. The table of future value interest factors is used to find the value of an annuity for a given rate per time period and a set time period. The table gives the value of each dollar of the annuity. Multiply this value by the value of the annuity.
2. The table of present value interest factors is used to find the current value of an annuity for a given rate per time period and a set time period. The table gives the present value of each dollar of the annuity. Multiply this value by the value of the annuity to obtain its present value.
3. The table of present value interest factors can be used to find the repayments due to pay off a loan of a given amount at a set rate of compound interest over a fixed time period. The repayments are obtained by dividing the value of the loan by the value of the cell in the table corresponding to the interest rate and time period.

Effective annual rate of interest

Effective annual interest rate $= \left(1 + \frac{r}{n}\right)^n - 1$ where n is the number of times the principal is compounded per year and r is the per annum interest rate as a decimal.

Arithmetic sequences

A sequence is a set of objects, in this case numbers, in some definite order.

An **arithmetic sequence** is a sequence of terms in which each term after **the first term** is formed by adding a constant number, called the **common difference**, to the preceding term.

a is the first term, d is the common difference and n is the number of terms.

The sequence $a, a + d, a + 2d, \ldots, a + (n - 1)d$ is an arithmetic sequence with n terms, where a and d are real numbers and n is a positive integer.

Arithmetic series—summary of formulae

- $a + (a + d) + (a + 2d) + \ldots + (a + (n - 1)d)$ is an arithmetic series
- d is the common difference
- $T_1 = a$ is the first term
- $T_n = a + (n - 1)d$
- $S_n = \frac{n}{2}(a + l)$ where $l = a + (n - 1)d$
- $S_n = \frac{n}{2}[2a + (n-1)d]$ or $n\left(a + \frac{n-1}{2}d\right)$
- $S_n = \sum_{k=1}^{n}[a + (k-1)d]$
- $T_n = S_n - S_{n-1}, n > 1$

Geometric sequences

A geometric sequence is a sequence of terms in which each term after the first term is formed by multiplying the preceding term by a constant number called the common ratio.

a is the first term, r is the common ratio and n is the number of terms.

The sequence $a, ar, ar^2, ar^3, \ldots, ar^{n-1}$ is a geometric sequence with n terms, where a and r are real numbers and n is a positive integer.

Recursive definition of geometric sequences

$T_n = rT_{n-1}$

This leads to $r = \frac{T_n}{T_{n-1}}$

Geometric series—summary of formulae

- $a + ar + ar^2 + \ldots + ar^{n-1}$ is a geometric series
- r is the common ratio
- $T_1 = a$ is the first term
- $T_n = ar^{n-1}$
- $S_n = \frac{a(1-r^n)}{1-r}$ for $r < 1$
- $S_n = \frac{a(r^n - 1)}{r-1}$ for $r > 1$
- $S_n = \sum_{k=1}^{n} ar^{k-1}$
- $T_n = S_n - S_{n-1}, n > 1$

Infinite geometric series

- If $\lim_{n\to\infty} r^n = 0$ then $\lim_{n\to\infty} \frac{ar^n}{1-r} = 0$, so the geometric series converges for $|r| < 1$:

$$S_\infty = \frac{a}{1-r} \qquad \text{for } |r| < 1$$

- If $|r| > 1$ then $r^n \to \infty$ and there is no limiting sum, so the geometric series does **not** converge for $|r| > 1$.

Compound interest

- The compound interest formula: $A_n = PR^n$

If someone invests \$$P$ at the beginning of each period of time at r% per period compound interest, the sum V of all investments at the end of n periods of time is:

$$V = PR + PR^2 + PR^3 + \ldots + PR^n = \frac{PR(R^n - 1)}{R - 1}$$

where $R = 1 + \frac{r}{100}$

Note that the **process** of obtaining this result is the important thing to remember, not the result.

19 BIVARIATE DATA ANALYSIS

Types of data

Data can be numerical or categorical.

There are two types of numerical data:

- Discrete data can only be measured in certain finite (discrete) units. It is often, but not always, a whole number.
- Continuous data can be measured by any number, which is usually rounded because it is limited by the accuracy of the measuring device. An example might be the time for the 100 m sprint at the Olympics.

There are two types of categorical data:

- Ordinal data can be put in an order in some way.
- Nominal ('named') data has no order.

Comparing data

Comparing two numerical variables

Use a parallel box plot or a back-to-back stem-and-leaf plot.

A back-to-back frequency table can be used for any kind of data, numerical or categorical.

Fathers	Stem	Mothers
9 7	3	8 8 9
8 7 5 3 2	4	2 3 5 6 6 7
8 8 4 3 1	5	2 4
0	6	

Mothers
Fathers
36 40 44 48 52 56 60

Comparing two categorical variables

Use a back-to-back frequency table, or draw a clustered column graph or a stacked column graph.

Changing the data to percentages can help identify associations as it removes any complication arising from unequal numbers selected for each variable.

Two-way tables can be used to compare responses from two categories. If the numbers from each category are different, change the results to percentages.

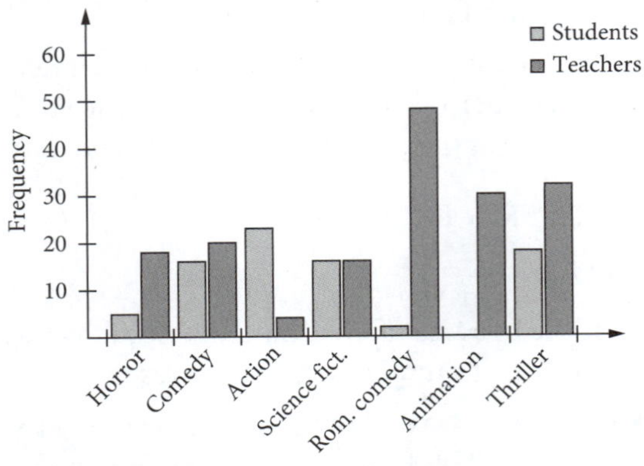

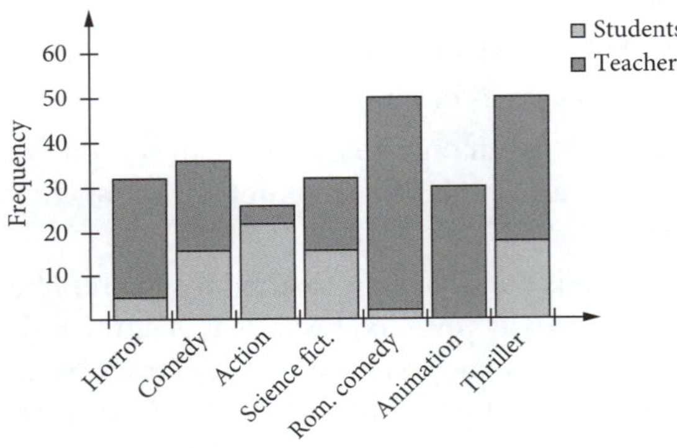

Scatterplots and association

Bivariate data is data from two sets of variables. One set of variables may have an effect on the other. This can be seen by drawing a scatterplot. The bivariate data is represented as a sequence of ordered pairs, which are plotted as points on a graph. The points are not joined. Observing the pattern of the points can indicate whether the two sets of data are connected in any way, which is described as an association.

If one variable affects, or can be used to explain, the values of the other variable, it is called the explanatory variable, or the independent variable. The other variable is known as the response variable, or the dependent variable, as its values are dependent on the independent variable. The independent or explanatory variable is represented by the horizontal axis, and the dependent or response variable by the vertical axis.

Note: Association is not the same as causality. The variables may both be caused by a third variable. For example, hot weather may increase the sales of both drinks and ice creams, but an increase in sales of drinks does not cause an increase in sales of ice creams, nor vice versa. Where an association exists, the explanatory variable is the one that could cause a change in the response variable.

Summary for scatterplot results

- A single line that best represents the general pattern of the data is called a line of best fit.
- If the pattern is roughly a straight line, a linear association exists.
- Sometimes a pattern may exist but it is not a straight line. This is called a non-linear trend.
- In other situations the points may be completely random and indicate no association at all.
- The slope can be positive (both variables increasing together) or negative (when one variable increases as the other decreases).
- The trend can be strong, moderate or weak, depending on how well the pattern fits the data.
- If outliers are ignored, the pattern can be seen more clearly.

The correlation coefficient

Correlation is a measure of the strength of the association between the variables. Where a linear trend exists you can calculate the Product moment or Pearson correlation coefficient, r.

The value of r is between -1 and $+1$. A value of -1 represents a perfect straight line with a negative slope. A value of $+1$ is a perfect straight line with a positive slope. A value of zero represents a random collection of points with no association.

The formula for r is quite complex and often will require the use of digital technology. Two forms of the formula are given below. Any outliers should be removed from the data before calculating r.

$$r = \frac{\sum(x-\bar{x})(y-\bar{y})}{ns_x s_y} \quad \text{or} \quad r = \frac{\frac{1}{n}\sum xy - \overline{xy}}{s_x s_y}$$

Where $\bar{x}$ is the mean and s_x is the sample standard deviation of the independent variable x, while $\bar{y}$ is the mean and s_y is the sample standard deviation of the dependent variable y. Also n is the number of points used.

The coefficient of determination

The coefficient of determination, r^2, can be used to determine the level of association between the two variables. It would be incorrect to assume that this implies the level of causation. The data may have a strong correlation but this does not necessarily mean there is a causal relationship. Both variables may have a common response to a third value. Causation may exist, but the change may also be caused by one or more uncontrolled variables whose effects cannot be disentangled from the effect of the explanatory variable. This is known as confounding.

If there is a clear cause and effect situation, the value of r^2 can be used to state the proportion of the change which is due to the independent (explanatory) variable. For example, if $r^2 = 0.75$, you could say three quarters, or 75% of the change in the response variable is due to the explanatory variable.

Regression analysis and the line of best fit

The line of best fit can be drawn by eye. You can then find its gradient (using rise and run), y-intercept, and hence its equation.

You can use technology to calculate the equation of the regression line, a line of best fit, as well as r. Finding the equation to the line of best fit is known as regression analysis.

Using the regression line to make predictions

You can use the line of best fit to predict the value of one variable if you know the value of the other variable. Use the relationship between gradient, y-intercept and the equation $y = mx + c$.

The statistical process

Statistics do not occur naturally and are created by mankind. To limit bias and unsubstantiated claims, follow these steps.

Step 1 Clarify the problem and formulate one or more questions that can be answered with data.

Step 2 Design and implement a plan to collect or obtain appropriate data.

Step 3 Select and apply appropriate graphical or numerical techniques to analyse the data.

Step 4 Interpret the results of this analysis and relate the interpretation.

20 RANDOM VARIABLES

Continuous probability distributions

Consider a continuous random variable X.

- For a function $f(x)$ to be a probability density function, $f(x) \geq 0$ for all values of x, and $\int_{-\infty}^{\infty} f(x)\,dx = 1$, i.e. the area enclosed by $y = f(x)$ and the x-axis is equal to 1
- $P(X = x) = 0$ for all values of x
- $P(a \leq X \leq b) = P(a \leq X < b) = P(a < X \leq b) = P(a < X < b) = \int_a^b f(x)\,dx$
- The median, m, is such that $P(X < m) = \int_{-\infty}^{m} f(x)\,dx = 0.5$
- $\mu = \int_a^b x f(x)\,dx$
- $\text{Var}(X) = \sigma^2 = \int_{-\infty}^{\infty} x^2 f(x)\,dx - \mu^2$

The cumulative distribution function (cdf)

The cumulative distribution function is usually designated as $F(x)$.

$F(a) = P(X \leq a) = \int_{-\infty}^{a} f(x)\,dx$.

$F(x)$ is the antiderivative of $f(x)$.

For all values of x, $0 \leq F(x) \leq 1$.

Since $f(x) \geq 0$ for all x, $F(x)$ will be an increasing function, although there may be values of x where $f(x)$ is zero and for these values $F(x)$ will be constant.

The normal distribution

- A continuous random variable that has a normal distribution with mean μ and variance σ^2 is written as $X \sim N(\mu, \sigma^2)$.
- The pdf of the normal distribution is $f(x) = \dfrac{1}{\sigma\sqrt{2\pi}} e^{-\frac{1}{2}\left[\frac{(x-\mu)}{\sigma}\right]^2}$ where μ is the mean of the distribution and σ is the standard deviation of the distribution.
- For the normal distribution:
  - about 68% of results lie within one standard deviation of the mean
  - about 95% of results lie within two standard deviations of the mean; we say the probability a result will lie within this range is very likely
  - about 99.7% of results lie within three standard deviations of the mean. We say the probability a result will lie within this range is almost certain.
- The graph of the pdf of the normal distribution is a symmetrical 'bell shaped' curve, with both the maximum value and the axis of symmetry at $X = \mu$:
  - increasing μ shifts the graph to the right
  - increasing σ makes the curve flatter and wider, but does not alter its axis of symmetry.
- A variable, X, which follows a normal distribution can be written as $X \sim N(\mu, \sigma 2)$, where μ is the mean and $\sigma 2$ is the variance.
- A variable, X, which has a standard normal distribution is such that $Z \sim N(0, 1)$.

The standard normal distribution

- The standard normal distribution is such that $Z \sim N(0, 1)$, i.e. it has a mean of 0 and a variance of 1.
- If $X \sim N(\mu, \sigma^2)$ then if $z = \dfrac{X - \mu}{\sigma}$, $Z \sim N(0, 1)$.

- The standardised normal score (or z-score) is calculated using the following formula:
$z = \dfrac{x - \mu}{\sigma} = \dfrac{\text{observed value} - \text{mean}}{\text{standard deviation}}$
- The symmetry of the distribution can be useful in calculating probabilities, combined with the total probability being 1. For example, $P(Z < -z) = P(Z > z) = 1 - P(Z < z)$ and $P(Z > -z) = P(Z < z)$.

The empirical rule

For normally distributed random variables,

- approximately 68% of data will have z scores between -1 and 1
- approximately 95% of data will have z scores between -2 and 2
- approximately 99.7% of data will have z scores between -3 and 3.

SOME BASIC CURVES

Parabola, $y = x^2$

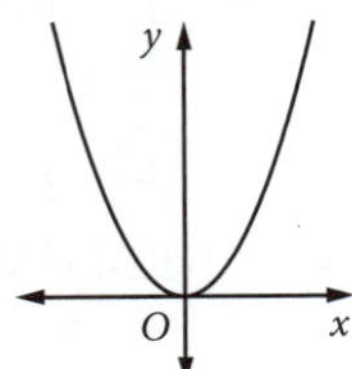

Absolute value, $y = |x|$

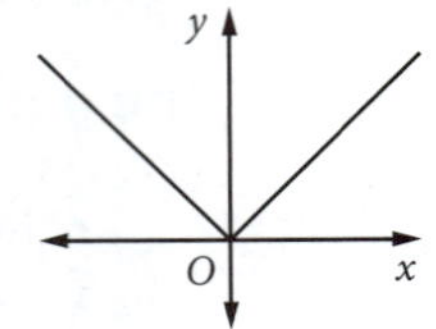

Pair of lines, $x^2 - y^2 = 0$

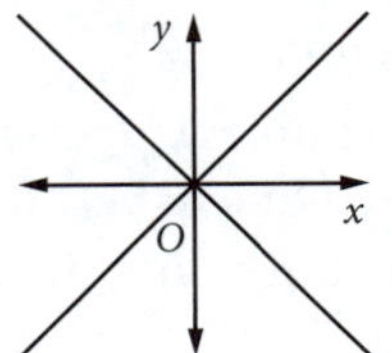

Circle, $x^2 + y^2 = r^2$

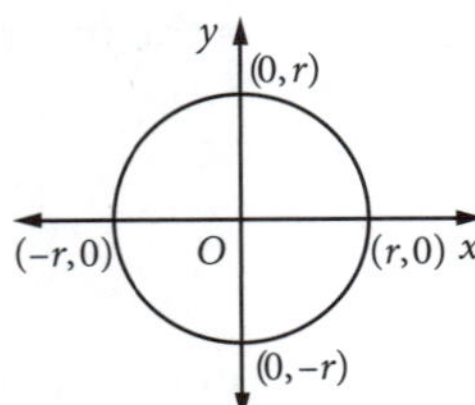

Semicircle, $y = \sqrt{r^2 - x^2}$

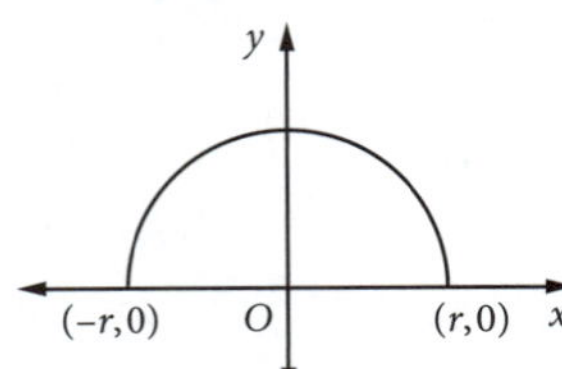

Semicircle, $y = -\sqrt{r^2 - x^2}$

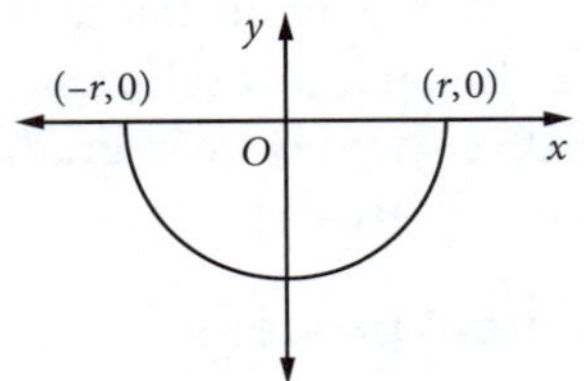

Positive square root, $y = \sqrt{x}, x \geq 0$

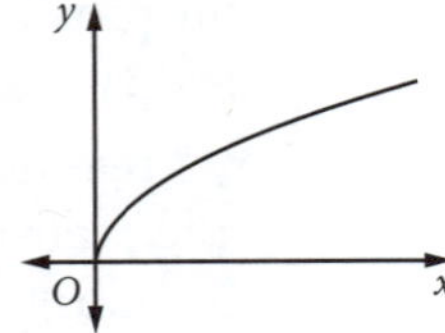

Rectangular hyperbola, $y = \frac{1}{x}, x \neq 0$

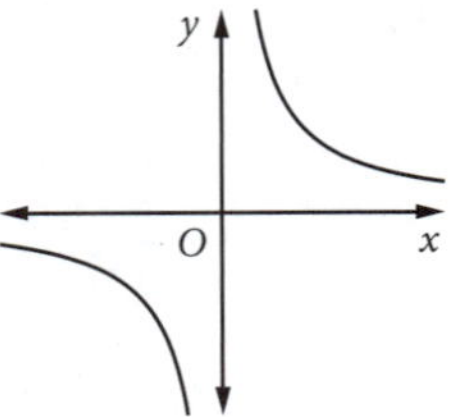

$y = x + \frac{1}{x}, x \neq 0$

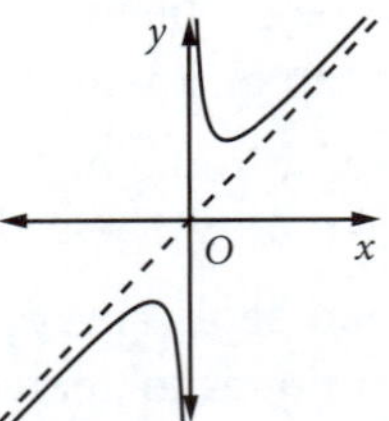

Logarithm, $y = \log_e x, x > 0$

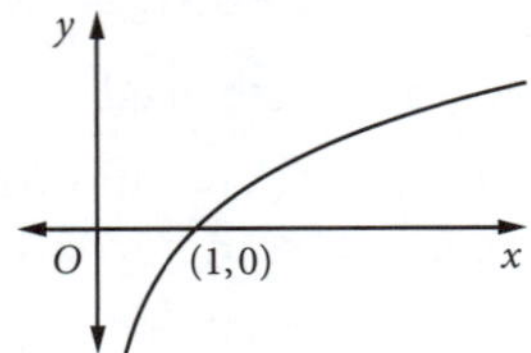

Exponential, $y = e^x$

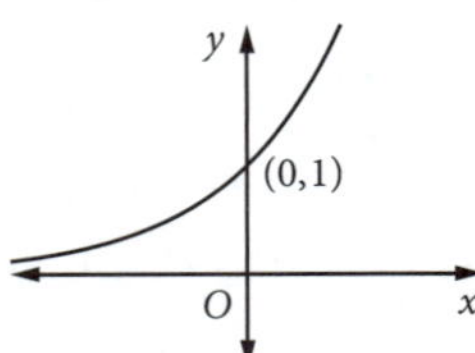

Exponential, $y = e^{-x}$

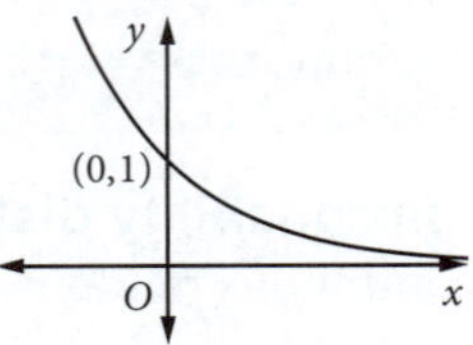

Sine, $y = \sin x, 0 \leq x \leq 2\pi$

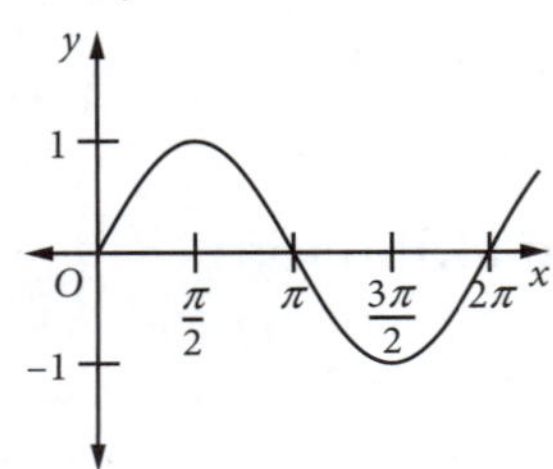

Cosine, $y = \cos x, 0 \leq x \leq 2\pi$

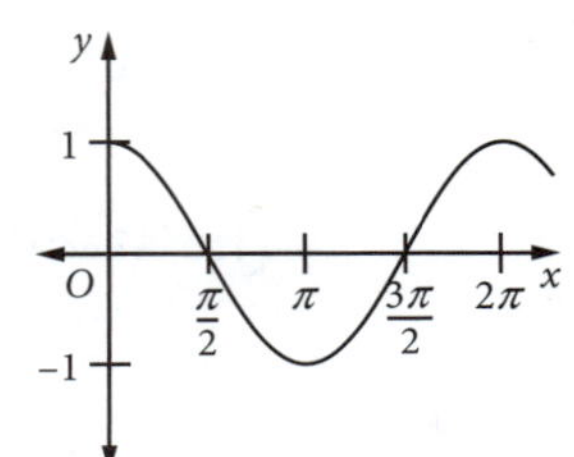

Tangent, $y = \tan x, 0 \leq x \leq 2\pi$, $x \neq \frac{\pi}{2}, \frac{3\pi}{2}$

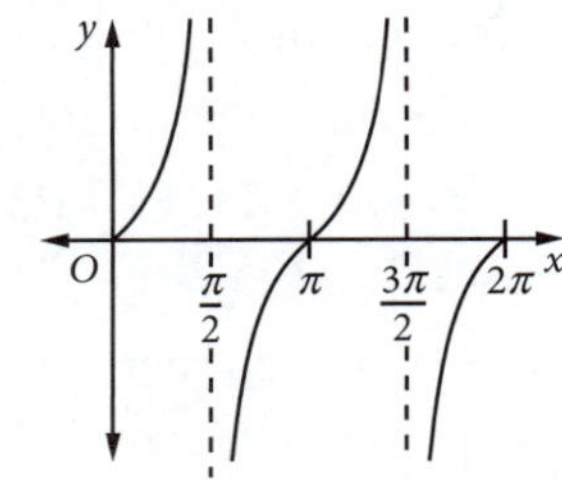

Mathematics Advanced Course Outcomes

Objective	
Students: • develop knowledge, skills and understanding about efficient strategies for pattern recognition, generalisation and modelling techniques	
Year 11 outcomes A student:	**Year 12 outcomes** A student:
MA11-1 uses algebraic and graphical techniques to solve, and where appropriate, compare alternative solutions to problems	**MA12-1** uses detailed algebraic and graphical techniques to critically construct, model and evaluate arguments in a range of familiar and unfamiliar contexts **MA12-2** models and solves problems and makes informed decisions about financial situations using mathematical reasoning and techniques **MA12-3** applies calculus techniques to model and solve problems

Objective	
Students: • develop the ability to use mathematical concepts and skills and apply complex techniques to the modelling and solution of problems in algebra and functions, measurement, financial mathematics, calculus, data and statistics and probability	
Year 11 outcomes A student:	**Year 12 outcomes** A student:
MA11-2 uses the concepts of functions and relations to model, analyse and solve practical problems	**MA12-4** applies the concepts and techniques of arithmetic and geometric sequences and series in the solution of problems
MA11-3 uses the concepts and techniques of trigonometry in the solution of equations and problems involving geometric shapes	**MA12-5** applies the concepts and techniques of periodic functions in the solution of problems involving trigonometric graphs
MA11-4 uses the concepts and techniques of periodic functions in the solutions of trigonometric equations or proof of trigonometric identities	
MA11-5 interprets the meaning of the derivative, determines the derivative of functions and applies these to solve simple practical problems	**MA12-6** applies appropriate differentiation methods to solve problems

Mathematics Advanced Course Outcomes

MA11-6 manipulates and solves expressions using the logarithmic and index laws, and uses logarithms and exponential functions to solve practical problems	**MA12-7** applies the concepts and techniques of indefinite and definite integrals in the solution of problems
MA11-7 uses concepts and techniques from probability to present and interpret data and solve problems in a variety of contexts, including the use of probability distributions	**MA12-8** solves problems using appropriate statistical processes

Objective

Students:

- develop the ability to use advanced mathematical models and techniques, aided by appropriate technology, to organise information, investigate, model and solve problems and interpret a variety of practical situations

Year 11 outcomes A student:	**Year 12 outcomes** A student:
MA11-8 uses appropriate technology to investigate, organise, model and interpret information in a range of contexts	**MA12-9** chooses and uses appropriate technology effectively in a range of contexts, models and applies critical thinking to recognise appropriate times for such use

Objective

Students:

- develop the ability to communicate and interpret mathematics logically and concisely in a variety of forms

Year 11 outcomes A student:	**Year 12 outcomes** A student:
MA11-9 provides reasoning to support conclusions which are appropriate to the context	**MA12-10** constructs arguments to prove and justify results and provides reasoning to support conclusions which are appropriate to the context

ANSWERS

CHAPTER 1

EXERCISE 1.1

1 $10x+15$ **2** $10x-5$ **3** $3a-3b$
4 $9ab+9a$ **5** $4xy$ **6** $5a^2b-3ab^2$
7 $x^2y+6x^2y^2$ **8** $6abc$ **9** $6mn+2m$
10 $5x^2-x$ **11** $5y^2-2x^2$ **12** $2a-3b$
13 $2x-4y$ **14** $15n-2m$ **15** $10x+11y$
16 $5x-20$ **17** $15a+15b$ **18** $2a^2+7a$
19 $11x^2-13xy$ **20** $a+4b$ **21** $3y-2x$
22 $9x^2+4x$ **23** $27x-42$ **24** $3x^2-10x+4$
25 $x^2+14x-1$ **26** $4x+9y-12$
27 C **28** B

EXERCISE 1.2

1 64 **2** 96 **3** 10 **4** (a) 140 (b) 5
5 38.5 **6** 554.2 **7** (a) 25.6 (b) 3 **8** 38
9 (a) 29.3 (b) 8 **10** 15 **11** 7.3 **12** 55.0
13 (a) 7.0 (b) 9.6 **14** 24 **15** (a) 7.6 (b) 5
16 $6\frac{2}{3}\approx 6.7$ **17** 2 **18** 65 **19** 92.4
20 3.6 **21** 33.0 **22** (a) 4 (b) 39.8
23 1.0 **24** (a) $11\frac{1}{9}\approx 11.1$ (b) 100 **25** 1210

EXERCISE 1.3

1 x^2+6x+5 **2** x^2-5x+6 **3** a^2+a-12
4 x^2-4x+4 **5** $y^2+14y+49$ **6** $2x^2+13x+15$
7 $3x^2-10x+8$ **8** $6m^2+11m-7$ **9** $9x^2+12x+4$
10 $4p^2-81$ **11** $6x^2+13x+6$ **12** $16p^2-40p+25$
13 $9x^2+24x+16$ **14** $2x^3-3x^2-8x-3$
15 $6x^3-22x^2+24x-8$ **16** x^3-4x **17** x^3-4x^2+5x-2
18 $2x^3-12x^2+22x-12$ **19** $x^4-2x^3+2x^2-10x-15$
20 x^3+2x^2-4x-8 **21** $x^6-3x^4y^2+3x^2y^4-y^6$ **22** C
23 (a) incorrect (b) correct (c) incorrect (d) correct

EXERCISE 1.4

1 $(a+b)(x+2)$ **2** $(3a-m)(2b-3c)$ **3** $(p+q-r)(a+b)$
4 $(x^2+4)(2x-1)$ **5** $(a+b)(x+4)$ **6** $(x+z)(x-y)$
7 $(2x+1)(y+z)$ **8** $(a-c)(a-b)$ **9** $(2x+5y)(2-5y)$
10 $(a^2+b)(a+3b)$ **11** $(c-2d)(a-2b)$ **12** $(3y+7)(x-2)$
13 $(x-z)(x-2y)$ **14** $(a^2-b)(a-b)$ **15** $(2m+pn)(n+p)$
16 $(x^2+4)(x+3)$ **17** $(pq+5)(p-q)$ **18** $(m^2+n)(p+1)$
19 $(x^2+1)(y+1)$ **20** $(a-4)(b-3)$ **21** $(2-y)(x-3y)$
22 A **23** (a) incorrect (b) correct (c) correct (d) incorrect

EXERCISE 1.5

1 $(m-1)(m+1)$ **2** $(x-4)(x+4)$ **3** $(8-m)(8+m)$
4 $(3a-5)(3a+5)$ **5** $(x-0.6)(x+0.6)$ **6** $(ab-c)(ab+c)$
7 $(3x-2y)(3x+2y)$ **8** $(x-2)(x+4)$ **9** $(x-yz)(x+yz)$
10 $\left(\frac{a}{5}-1\right)\left(\frac{a}{5}+1\right)$ **11** $\left(p-\frac{1}{2}\right)\left(p+\frac{1}{2}\right)$ **12** $\left(\frac{x}{2}-\frac{1}{3}\right)\left(\frac{x}{2}+\frac{1}{3}\right)$
13 $a(a+4)$ **14** $(x-y-z)(x+y+z)$
15 $98\times 100=9800$ **16** $46\times 1000=46\,000$
17 $ab(a-b)(a+b)$ **18** $3a(2a-b)(2a+b)$
19 $3y(x-3)(x+3)$ **20** $(x+y-2)(x+y+2)$
21 $b(2a-b)$ **22** $(x-3)(x+3)(x-y)$
23 $(x-2)(x+2)(x+3)$ **24** $(p-4)(p+4)(q-1)$
25 $x(a-1)(a+1)$ **26** $3(4a-5b)(4a+5b)$
27 $h(2+h)$ **28** $\left(\frac{x}{5}-y\right)\left(\frac{x}{5}+y\right)$ **29** D
30 (a) correct (b) correct (c) incorrect (d) correct
31 $(y-5)(y^2+5y+25)$ **32** $(z+1)(z^2-z+1)$
33 $(2p+3)(4p^2-6p+9)$ **34** $(6-a)(36+6a+a^2)$
35 $(2x+3)(x^2+3x+39)$ **36** $(x+7)(7x^2-x+13)$
37 $(b-a)(b+a)(b^4+b^2a^2+a^4)$
$=(b-a)(b+a)(b^2+ba+a^2)(b^2-ba+a^2)$
38 $8(2a+b)(4a^2-2ab+b^2)$
39 $\frac{4\pi}{3}(R-r)\left(R^2+Rr+r^2\right)$ **40** $p^4x^4(p-x)(p^2+px+x^2)$
41 $(x^2+y^2)(x^4-x^2y^2+y^4)$ **42** $\left(\frac{2}{a}-\frac{3}{b}\right)\left(\frac{4}{a^2}+\frac{6}{ab}+\frac{9}{b^2}\right)$
43 $(a-b)(m+n)(a^2+ab+b^2)(m^2-mn+n^2)$
44 $(2x-3)(2x+3)(x-1)(x^2+x+1)$
45 $h(3x^2+3xh+h^2)$ **46** $(2a-b)(a^2-ab+b^2)$
47 $2b(3a^2+b^2)$ **48** $2(12x^2+1)$
49 $x(12-6x+x^2)$ **50** $a^2b(ab-1)(a^2b^2+ab+1)$
51 $2(x-y+3)(x^2-2xy+y^2-3x+3y+9)$ **52** B **53** D

EXERCISE 1.6

1 $(x+1)(x+3)$ **2** $(x+3)(x+7)$ **3** $(x+3)(x+8)$
4 $(a+4)(a+8)$ **5** $(m+5)(m+4)$ **6** $(x+1)(x+12)$
7 $(x+6)(x+2)$ **8** $(x-3)(x-4)$ **9** $(x-12)(x-1)$
10 $(x-2)(x-6)$ **11** $(p+5)(p-3)$ **12** $(p+15)(p-1)$
13 $(p-5)(p+3)$ **14** $(p-15)(p+1)$ **15** $(x-7)(x+5)$
16 $(x-5)(x+2)$ **17** $(x+9)(x+8)$ **18** $(a-6)(a+2)$
19 $(x-6)(x-1)$ **20** $(x-9)(x+8)$ **21** $(x+12)(x-6)$
22 $(x-24)(x+3)$ **23** $(a+10)(a+3)$ **24** $(x-7)(x+6)$
25 $(x-21)(x+2)$ **26** $(x+21)(x-2)$ **27** D
28 (a) incorrect (b) correct (c) incorrect (d) correct

EXERCISE 1.7

1 $(2x+1)(x+1)$ **2** $(3x-1)(x+4)$ **3** $(2x+3)(x+2)$
4 $(4a+1)(a+3)$ **5** $(3a-2)(a-1)$ **6** $(4x-1)(2x-3)$
7 $(13c+6)(c-1)$ **8** $(4x+5)(2x+1)$ **9** $(3x-2)(x-5)$
10 $(3a+7)(2a-9)$ **11** $(3x+1)(x-4)$ **12** $(5x-8)(2x+1)$
13 $(2x-1)(x+2)$ **14** $(2x-3)^2$ **15** $(3x-2)^2$
16 $(x-2)(2x-5)$ **17** $(6x-1)(x-14)$ **18** $(y-3)(y+1)$
19 $2(3y-1)(2y+3)$ **20** $(3x-2)(2x-7)$ **21** $(3x-4)(2x-7)$
22 $(6x-7)(x-2)$ **23** $2(3x-7)(x-1)$ **24** $(4x+3)(2x-1)$
25 $(p+3)(6p+7)$ **26** $(5a+2)(2a-3)$ **27** $(6y-1)(2y+5)$
28 $(8x-9)(3x-4)$ **29** $(5x-3)(3x-2)$ **30** $(3x+1)(x-1)$
31 $(3x+5)(3x-2)$ **32** $(2x-1)(x-4)$ **33** C
34 (a) correct (b) correct (c) correct (d) incorrect

EXERCISE 1.8

1 $x(x-3)$ **2** $2a(a-2)(a+2)$ **3** $(x-3)(x+3)$
4 $(x-9)(x+1)$ **5** $3y(x-2y)(x+2y)$ **6** $5xy(x-2y)(x+2y)$
7 $(1-b-c)(1+b+c)$ **8** $(10x-1)(x+1)$
9 $a(a+2b)$ **10** $6(x-2)(x+2)$ **11** $(a-7)(a+6)$
12 $(a-b)(m+n)$ **13** $2x(x+8)(x-1)$ **14** $3a(a+7)(a+1)$
15 $(x+2y-2)(x+2y+2)$ **16** $ab(b+c+d)$
17 $(x-6y)(x+6y)$ **18** $(x+y)(y-z)$
19 $4(x-15)(x+8)$ **20** $b(x-7y)^2$
21 $3y(2y-1)(y+1)$ **22** $2y(3y+1)(y+4)$
23 $15(a-2)(a+2)$ **24** $mn(3-5mn)(3+5mn)$
25 $5x(a-5)(a+5)$ **26** $4xy$ **27** $5t(t+9)(t-8)$
28 $(m+6)(m-n)$ **29** $(x-2)(x+2)(x+3)$
30 $(x-l)(mx-y)$ **31** D **32** $4r^2-\pi r^2=(4-\pi)r^2$
33 $(\pi-2)r^2$ **34** $\pi(R-r)(R+r)$ **35** $(8-\pi)r^2$
36 $\pi a(a-b)$ **37** $\pi R^2-4r^2=\pi(R-2r)(R+2r)$
38 $16r-4\pi r^2=4(4-\pi)r^2$ **39** $a(a+b+c)$
40 (a) correct (b) correct (c) incorrect (d) correct

EXERCISE 1.9

1 $2a-b$ 2 $\frac{3x+2y}{3}$ 3 7 4 $\frac{2x-y}{2y}$ 5 $\frac{2(2a-b)}{3a}$

6 2 7 $\frac{1}{a}$ 8 $1+m$ 9 $m-n$ 10 $p-q$

11 $\frac{x+y}{2}$ 12 $\frac{2(s-6)}{r+s}$ 13 $\frac{x-y}{x+y}$ 14 k 15 $\frac{x-3}{x}$

16 $\frac{5a}{b}$ 17 $\frac{4x}{x+y}$ 18 $\frac{(x-6)(x-1)}{(x-6)(x+6)}=\frac{x-1}{x+6}$ 19 $\frac{a}{b}$

20 $\frac{a-b}{a}$ 21 $\frac{x+1}{x-4}$ 22 $\frac{(x-2)(x-4)}{(x-2)(x+1)}=\frac{x-4}{x+1}$ 23 $\frac{x+1}{x-2}$

24 $\frac{x-2}{x+4}$ 25 $\frac{x+2}{x-5}$ 26 $\frac{4xy(x-2)(x+2)}{(x+4)(x-2)}=\frac{4xy(x+2)}{x+4}$

27 1 28 $\frac{x^2}{y}$ 29 C 30 $\frac{a(2a-3b)}{b(a-b)}\times\frac{2a(a-b)}{2(2a-3b)}=\frac{a^2}{b}$

31 $\frac{5x(3x-y)}{10xy}\times\frac{2y}{3x-y}=1$ 32 $\frac{4x(3x-1)}{x(3x-1)}\times\frac{5x^2y^2}{10x^2y}=2y$ 33 1

34 $\frac{a-b}{b}$ 35 (a) incorrect (b) incorrect (c) incorrect (d) correct

36 $\frac{x^2-xy+y^2}{x-y}$ 37 $\frac{2}{2x-1}$ 38 $\frac{2}{x^2+xy+y^2}$

39 $3x^2+3xh+h^2$ 40 $\frac{3x^2-3xy+y^2}{2x-y}$

EXERCISE 1.10

1 $\frac{x}{30}$ 2 $\frac{7x}{8}$ 3 $\frac{10a+24a-5a}{30}=\frac{29a}{30}$ 4 $\frac{11y}{12}$

5 $\frac{3(a+2)-5(a-1)}{15}=\frac{11-2a}{15}$ 6 $\frac{3x+y}{6}$ 7 $\frac{3x+1}{12}$

8 $\frac{7m-3n}{10}$ 9 $\frac{6x+3y-4(x+y)}{12}=\frac{2x-y}{12}$ 10 $\frac{-(a+8b)}{18}$

11 $\frac{7a+11b}{12}$ 12 $\frac{3}{3x}-\frac{2}{3x}=\frac{1}{3x}$ 13 $\frac{3a+1}{a^2}$ 14 $\frac{1-2a}{ab}$

15 $\frac{m^2-n^2}{mn}$ 16 $\frac{4z+3x}{xyz}$ 17 $\frac{5b-2a}{a^2b^2}$ 18 $\frac{4a-11}{6a}$

19 $\frac{3+2x+2}{3(x+1)}=\frac{2x+5}{3(x+1)}$ 20 $\frac{3}{x}-\frac{1}{x^2}=\frac{3x-1}{x^2}$ 21 B

22 $(x-3)(x+3)$ 23 $x(x-2)$ 24 $6(x-2)$

25 $x^2-4x=x(x-4)$ 26 $x(x-4)(x+4)$ 27 $x^2+4x+4=(x+2)^2$

28 $(x-y)(x+y)$ 29 $xy(x-y)(x+y)$ 30 A

31 $\frac{2a}{(a-b)(a+b)}$ 32 $\frac{3(x+y)-2(x-y)}{(x-y)(x+y)}=\frac{x+5y}{(x-y)(x+y)}$

33 $\frac{x+y}{x-y}$ 34 $\frac{x^2+2xy-y^2}{(x-y)(x+y)}$ 35 $\frac{3a-b+a+b}{(a-b)(a+b)}=\frac{4a}{(a-b)(a+b)}$

36 $\frac{3-x}{(x-2)(x+2)}$ 37 $\frac{1}{x+y}$ 38 $\frac{3+2(x-2)}{(x-2)^2}=\frac{2x-1}{(x-2)^2}$

39 $\frac{4}{(x-1)(x-3)(x+1)}$ 40 $\frac{4x+7}{(x-2)(x+3)}$ 41 $\frac{-2y}{(x+y)(x-y)}$

42 $\frac{7a-15}{2(a-3)(a+3)}$ 43 $\frac{x+20}{(x-5)(x+5)}$ 44 $\frac{22}{(3x-2)(4x+1)}$

45 $\frac{-a(a+1)}{(3a-4)(2a-3)}$ 46 $\frac{y+x-y}{x(x-y)}=\frac{1}{x-y}$ 47 $\frac{2(x-2)}{(x-4)(x+4)}$

48 $\frac{2}{x-2}$ 49 $\frac{a+2+a+4}{(a+3)(a+2)}=\frac{2}{a+2}$ 50 $\frac{x-7}{(x-2)(x+2)(x-1)}$

51 $\frac{(x+1)^2-(x-1)^2}{(x-1)(x+1)}=\frac{4x}{(x-1)(x+1)}$ 52 $\frac{5x+13}{(x-2)(x+2)}$

53 $\frac{3x-2(x-4)}{(x-4)(x+4)}=\frac{x+8}{(x-4)(x+4)}$ 54 $\frac{5(x-5)-6}{3x(x-5)}=\frac{5x-31}{3x(x-5)}$

55 (a) incorrect (b) correct (c) incorrect (d) incorrect

EXERCISE 1.11

1 $2\sqrt{2}$ 2 $2\sqrt{5}$ 3 $3\sqrt{3}$ 4 $4\sqrt{2}$

5 C 6 $3\sqrt{5}$ 7 $6\sqrt{2}$ 8 $2\sqrt{21}$

9 $7\sqrt{2}$ 10 $6\sqrt{3}$ 11 $5\sqrt{5}$ 12 $9\sqrt{2}$

13 $10\sqrt{2}$ 14 $5\sqrt{64\times 2}=40\sqrt{2}$ 15 $80\sqrt{2}$

16 $10\sqrt{6}$ 17 $6\sqrt{13}$ 18 D 19 $\frac{\sqrt{64\times 5}}{2}=4\sqrt{5}$

20 $\sqrt{7}$ 21 $12\sqrt{3}$ 22 $\frac{\sqrt{10}}{3}$ 23 B

24 (a) correct (b) incorrect (c) correct (d) incorrect

25 $\sqrt{15}$ 26 4 27 $2\sqrt{3}$ 28 $2\sqrt{15}$

29 $8\sqrt{6}$ 30 8 31 $10\sqrt{10}$ 32 $24\sqrt{2}$

33 $6\sqrt{35}$ 34 $12\sqrt{6}$ 35 $8\sqrt{6}$ 36 40

37 $\frac{\sqrt{6}}{2}$ 38 $\frac{2\sqrt{3}}{3}$ 39 2 40 $\sqrt{7}$

41 4 42 $\sqrt{3}$ 43 1 44 3

EXERCISE 1.12

1 $7\sqrt{3}$ 2 $7\sqrt{7}$ 3 $6\sqrt{5}$ 4 $3\sqrt{2}+3\sqrt{3}$

5 $4\sqrt{5}-5\sqrt{2}$ 6 $\sqrt{2}$ 7 A 8 $3\sqrt{5}$

9 $6\sqrt{2}$ 10 $7\sqrt{3}$ 11 $7\sqrt{3}$ 12 $\sqrt{2}$ 13 0

14 (a) incorrect (b) incorrect (c) correct (d) correct

15 $4\sqrt{5}+4\sqrt{7}$ 16 $8\sqrt{3}-3\sqrt{5}$ 17 $3\sqrt{5}+3\sqrt{2}$

18 $5\sqrt{15}-2\sqrt{10}$ 19 $5\sqrt{7}$ 20 $\sqrt{2}+\sqrt{3}$

21 $12\sqrt{3}$ 22 $5\sqrt{6}-10\sqrt{2}$ 23 $6\sqrt{6}$

24 $\sqrt{7}+3\sqrt{5}$ 25 $15\sqrt{5}-8\sqrt{2}$ 26 $7\sqrt{2}-4\sqrt{5}-2\sqrt{3}$

27 $5\sqrt{5}$ 28 $9\sqrt{3}$ 29 $-\sqrt{6}$ 30 0

EXERCISE 1.13

1 $\sqrt{10}+\sqrt{15}$ 2 $5+\sqrt{10}$ 3 6 4 $\sqrt{6}-3\sqrt{2}$

5 $3\sqrt{2}-2\sqrt{6}$ 6 $14\sqrt{5}-7$ 7 B 8 $12\sqrt{3}-3\sqrt{10}$

9 $a+\sqrt{ab}$ 10 $x-\sqrt{xy}$ 11 $\sqrt{35}-\sqrt{10}+\sqrt{21}-\sqrt{6}$

12 $\sqrt{6}+4+\sqrt{21}+2\sqrt{14}$ 13 $\sqrt{6}+3\sqrt{3}-\sqrt{2}-3$

14 $16+7\sqrt{5}$ 15 $-3-4\sqrt{3}$ 16 $8-3\sqrt{6}$

17 $20+6\sqrt{5}-2\sqrt{10}-3\sqrt{2}$ 18 $5\sqrt{6}-8\sqrt{2}$ 19 $4+2\sqrt{3}$

20 $7-2\sqrt{10}$ 21 $27+12\sqrt{2}$ 22 7 23 21

24 4 25 3 26 D 27 $8-2\sqrt{15}$ 28 $18+2\sqrt{77}$

29 $4+3\sqrt{3}$ 30 1 31 1 32 $11+4\sqrt{6}$

33 37 34 $\sqrt{30}-\sqrt{5}+4\sqrt{3}-2\sqrt{2}$ 35 $3+15\sqrt{2}$

36 (a) incorrect (b) correct (c) correct (d) incorrect

37 34 38 $82+12\sqrt{42}$ 39 $15-25\sqrt{3}$

40 $35+12\sqrt{6}$ 41 11 42 $13-4\sqrt{3}$

EXERCISE 1.14

1 $\frac{2\sqrt{3}}{3}$ 2 $\frac{\sqrt{15}}{3}$ 3 $\sqrt{3}$ 4 $\sqrt{3}+\sqrt{2}$

5 $\frac{2\sqrt{7}-\sqrt{6}}{28-6}=\frac{2\sqrt{7}-\sqrt{6}}{22}$ 6 $\sqrt{5}-2$ 7 $\frac{2\sqrt{5}+3\sqrt{2}}{2}$

8 $\frac{3\left(\sqrt{10}+\sqrt{6}\right)}{2}$ 9 $\sqrt{30}+3\sqrt{3}$ 10 $\frac{6\sqrt{55}+8\sqrt{10}}{67}$

11 $\frac{\left(\sqrt{3}+2\right)^2}{3-4}=-\left(7+4\sqrt{3}\right)$

12 $\frac{(4\sqrt{2}+3\sqrt{5})(2\sqrt{5}+\sqrt{2})}{20-2}=\frac{38+11\sqrt{10}}{18}$

13 $\frac{3\sqrt{35}+2\sqrt{14}-30-4\sqrt{10}}{37}$ **14** $5+2\sqrt{6}$

15 $\frac{25\sqrt{15}+45+75+9\sqrt{15}}{125-27}=\frac{17\sqrt{15}+60}{49}$ **16** $\frac{18+4\sqrt{3}}{23}$

17 $\frac{12\sqrt{3}-9\sqrt{2}}{5}$ **18** $\frac{9+\sqrt{11}}{5}$ **19** $-\frac{2+\sqrt{6}}{2}$ **20** $\frac{18-3\sqrt{6}}{10}$

21 $\frac{\sqrt{3}}{2\sqrt{6}-4\sqrt{3}}=\frac{1}{2(\sqrt{2}-2)}=-\frac{\sqrt{2}+2}{4}$ **22** $3-2\sqrt{2}$

23 $\frac{11-2\sqrt{10}}{9}$ **24** $\frac{6+5\sqrt{2}}{7}$ **25** $\frac{11-2\sqrt{14}}{13}$

26 $\frac{10\sqrt{2}+\sqrt{30}+2\sqrt{30}+3\sqrt{2}}{40-6}=\frac{13\sqrt{2}+3\sqrt{30}}{34}$

27 $\frac{12\sqrt{3}+2\sqrt{30}-45\sqrt{2}-15\sqrt{5}}{39}$ **28** A

29 $(\sqrt{3}+1)^2-\frac{1}{(\sqrt{3}+1)^2}=4+2\sqrt{3}-\frac{1}{4+2\sqrt{3}}=\frac{6+5\sqrt{3}}{2}$

30 $\frac{17\sqrt{2}}{6}$ **31** $\frac{7\sqrt{5}-15}{4}$

32 $\text{LHS}=(2\sqrt{2}-3)^2+6(2\sqrt{2}-3)+1$
$=8-12\sqrt{2}+9+12\sqrt{2}-18+1=0=\text{RHS}$

33 (a) incorrect (b) correct (c) incorrect (d) correct

34 $\text{LHS}=(\sqrt{5}-1)^3+3(\sqrt{5}-1)^2-2(\sqrt{5}-1)-4$
$=(8\sqrt{5}-16)+3(6-2\sqrt{5})-2\sqrt{5}+2-4$
$=8\sqrt{5}-16+18-6\sqrt{5}-2\sqrt{5}-2$
$=0=\text{RHS}$

35 $\text{LHS}=2\left(\frac{1}{\sqrt{3}-1}\right)^2-2\left(\frac{1}{\sqrt{3}-1}\right)-1=\frac{2(\sqrt{3}+1)^2}{(3-1)^2}-\frac{2(\sqrt{3}+1)}{3-1}-1$
$=\frac{4+2\sqrt{3}}{2}-\sqrt{3}-1-1$
$=2+\sqrt{3}-\sqrt{3}-2=0=\text{RHS}$

36 $\frac{2\sqrt{3}+1}{11}+\frac{3(\sqrt{3}-1)}{2}=\frac{4\sqrt{3}+2+33\sqrt{3}-33}{22}=\frac{37\sqrt{3}-31}{22}$

37 $\frac{4\sqrt{10}}{3}$ **38** $\frac{3\sqrt{6}-7}{3\sqrt{6}+7}=\frac{103-42\sqrt{6}}{5}$

39 $\frac{5\sqrt{10}-14}{9}$ **40** $\frac{21+4\sqrt{5}}{19}-\frac{7+\sqrt{5}}{11}=\frac{98+25\sqrt{5}}{209}$

41 $4\sqrt{2}-1$ **42** $\frac{51\sqrt{3}+3\sqrt{15}-98}{17}$

43 $\frac{2}{\sqrt{2}-\sqrt{3}}-\frac{3}{\sqrt{5}-\sqrt{2}}=-(3\sqrt{2}+2\sqrt{3}+\sqrt{5})$ **44** $\frac{\sqrt{3}-1}{2}$

CHAPTER REVIEW 1

1 (a) $4x^2-20x+25$ (b) $x^2-4x-21$ (c) $6y^2+11y+4$
(d) $25x^2-16$ (e) $2x^3-3x^2y+3xy^2-y^3$

2 (a) $(9-2a)(9+2a)$ (b) $5y(2x-1)$ (c) $5xy(x-2y-1)$
(d) $(a-4)(a-14)$ (e) $(4x-1)(4x+1)$ (f) $(3x+7)(x-1)$
(g) $(a-b)(a+b)+2(a-b)=(a-b)(a+b+2)$
(h) $x(3+7x-6x^2)=x(3-2x)(1+3x)$
(i) $(2a-3)(4a^2+6a+9)$
(j) $(2-x-h)(4+2(x+h)+(x+h)^2)$
(k) $(4y+3)(2y-3)$ (l) $x(x-2)(x^2+2x+4)$

3 $\sqrt{25}=5$

4 (a) $\frac{3x^3}{4a^2}\times\frac{a(y-1)}{xy^2}\times\frac{4ay^2}{3(y-1)}=x^2$
(b) $\frac{20x-6x-9+2x}{12}=\frac{16x-9}{12}$ (c) $2n$
(d) $\frac{4-2x}{x^2}$ (e) $\frac{3}{x+1}+\frac{1}{(x+1)^2}=\frac{3x+4}{(x+1)^2}$
(f) $\frac{x(2x-y)}{y(x-y)}\times\frac{2x(x-y)}{2(2x-3y)}=\frac{x^2}{y}$ (g) $\frac{a^2-ab+b^2}{a-b}$
(h) $\frac{m-4}{(m-2)(m+2)(m-1)}$ (i) $\frac{2}{(x-4)(x-6)}$

5 (a) $a-b$ (b) $a-\frac{2\sqrt{a}}{\sqrt{b}}+\frac{1}{b}$ **6** $\frac{3}{8}$ **7** (a) 416 (b) 25

8 (a) $9\sqrt{3}-9\sqrt{6}$ (b) $\frac{14\sqrt{5}}{5}$

9 (a) $\frac{7-2\sqrt{10}}{3}$ (b) 15

10 $57-12\sqrt{15}$ **11** D **12** $\frac{3-\sqrt{2}}{7}$ **13** 289

14 $\text{LHS}=\left(\frac{\sqrt{5}-1}{2}\right)^3+3\left(\frac{\sqrt{5}-1}{2}\right)^2+\left(\frac{\sqrt{5}-1}{2}\right)-2$
$=\frac{8\sqrt{5}-16}{8}+\frac{18-6\sqrt{5}}{4}+\frac{\sqrt{5}-1}{2}-2$
$=\sqrt{5}-2+\frac{9-3\sqrt{5}}{2}+\frac{\sqrt{5}-1}{2}-2$
$=\sqrt{5}-4+4-\sqrt{5}=0=\text{RHS}$

15 (a) $\frac{1}{5}$ (b) $\frac{7}{8}$ (c) $\frac{4H\pi R^2}{27}$ (d) $\frac{x^2-2x-3}{(x-1)^2}$
(e) $\frac{1225}{2556}$ (f) t (g) $\frac{20}{3}$ (h) $(x-1)^2+y^2=4^2$

16 (a) $2(x-2)(x+2)$ (b) $6(x-6)(x+6)$

17 (a) 68 499.5 (b) $M=2997.75$
(c) $A_{20}=269\,903.63$ (d) \$44 404

18 $y=\frac{H(R-x)}{R}$

19 9

CHAPTER 2

EXERCISE 2.1

1 C
2 (a) $x=4.36$ (b) $x=4.50$ (c) $x=3.28$ (d) $x=4.73$
3 (a) $\theta=30°$ (b) $\theta=56°\,19'$ (c) $\theta=53°\,8'$ (d) $\theta=22°\,37'$
4 $x=20.0$ m, $y=45.4$ m

EXERCISE 2.2

1 (a) 1st, 2nd (b) 2nd, 4th (c) 2nd, 3rd (d) 4th
(e) 2nd (f) 3rd (g) 1st

2 (a) 1st (b) 2nd (c) 2nd (d) 3rd (e) 4th (f) 3rd
(g) 1st (h) 3rd (i) 4th (j) 2nd

3 (a) $\sin A$ (b) $\sin A$ (c) $-\tan A$ (d) $-\cos A$
(e) $-\sin A$ (f) $\tan A$

4 (a) (i) 0.8912, –0.5736, –1.4281, 1.2208, –1.7434, –0.7002
(ii) 0.4695, –0.8829, –0.5317, 2.1301, –1.1326, –1.8807
(iii) 0.8910, –0.4540, –1.9626, 1.1223, –2.2027, –0.5095
(b) (i) –0.4226, –0.9063, 0.4663, –2.3662, –1.1034, 2.1445
(ii) –0.6018, –0.7986, 0.7536, –1.6616, –1.2521, 1.3270
(iii) –0.9455, –0.3256, 2.9042, –1.0576, –3.0716, 0.3443
(c) (i) –0.9781, 0.2079, –4.7046, –1.0223, 4.8097, –0.2126

(ii) −0.8572, 0.5150, −1.6643, −1.1666, 1.9416, −0.6009
(iii) −0.3090, 0.9511, −0.3249, −3.2361, 1.0515, −3.0777
(d) (i) −0.4226, 0.9063, −0.4663, −2.3662, 1.1034, −2.1445
(ii) −0.8480, −0.5299, 1.6003, −1.1792, −1.8871, 0.6249
(iii) 0.5736, −0.8192, −0.7002, 1.7434, −1.2208, −1.4281

5 (a) 0.2 (b) −0.2 (c) −0.2 (d) 0.2 (e) −0.2 (f) 5

6 (a) $\frac{1}{t}$ (b) t (c) $-t$ (d) $-t$ (e) $-\frac{1}{t}$ (f) t

7 (a) $\frac{1}{c}$ (b) c (c) $-c$ (d) c (e) $\frac{1}{c}$ (f) $-c$

8 (a) −1.4281 (b) −0.5299 (c) −0.7660 (d) −0.3420 (e) 1.1106 (f) −0.1736

9 (a) positive (b) positive (c) negative (d) negative (e) positive (f) negative

10 (a)

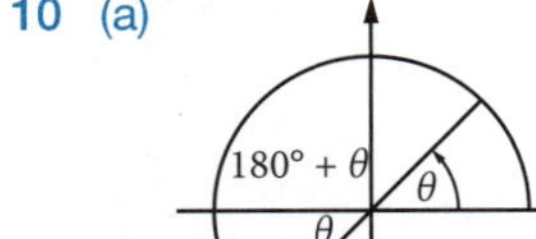
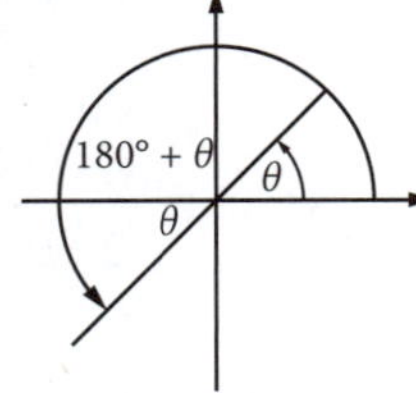

$\cos(180° + \theta) = -\cos\theta$

(b)

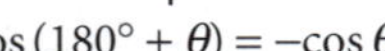

$\sin(360° - \theta) = -\sin\theta$

11 D

12 (a) correct (b) incorrect (c) correct (d) correct

EXERCISE 2.3

1 (a)

θ	0°	30°	90°	150°	180°	210°	270°	330°	360°
$\sin\theta$	0	0.5	1	0.5	0	−0.5	−1	−0.5	0

(b)

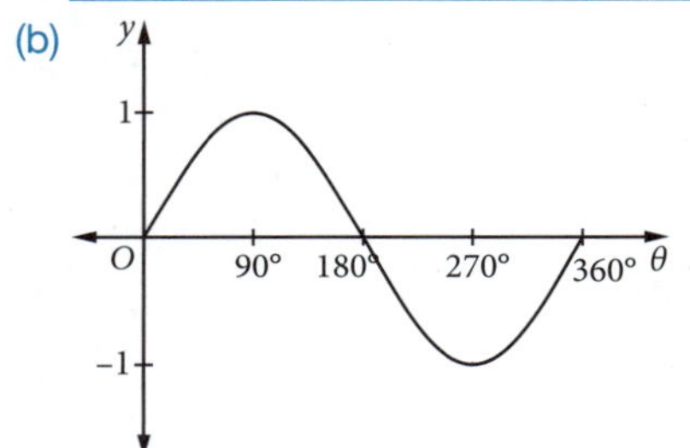

2 (a)

θ	0°	60°	90°	120°	180°	240°	270°	300°	360°
$\cos\theta$	1	0.5	0	−0.5	−1	−0.5	0	0.5	1

(b)

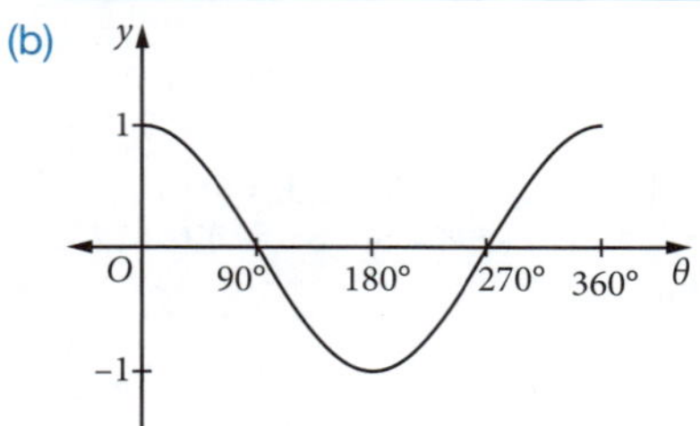

3

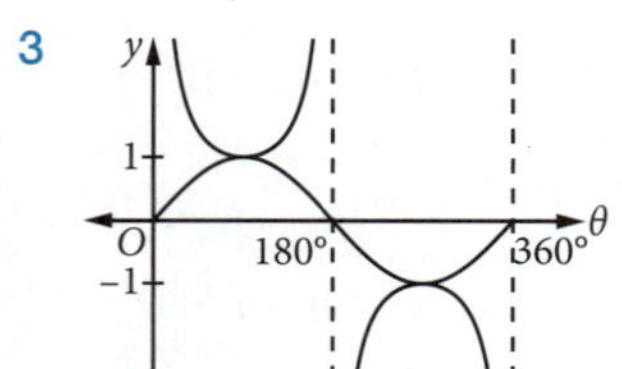

4

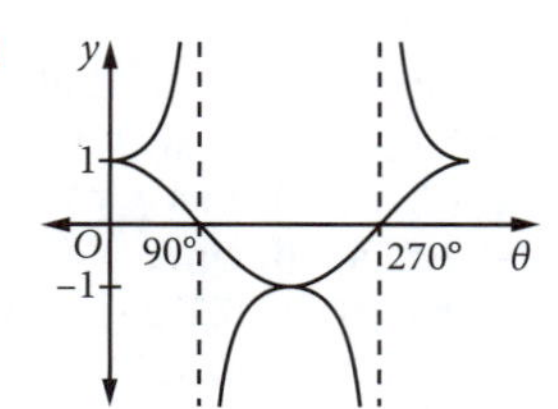

5 (a)

θ	0°	30°	90°	150°	180°	210°	270°	330°	360°
$\tan\theta$	0	0.577	undefined	−0.577	0	0.577	undefined	−0.577	0

(b)

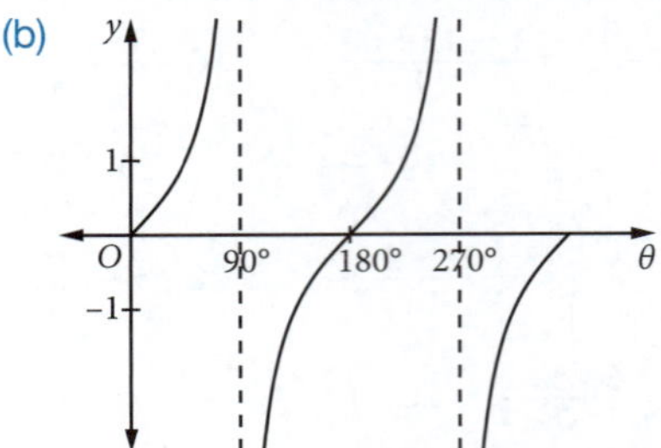

6

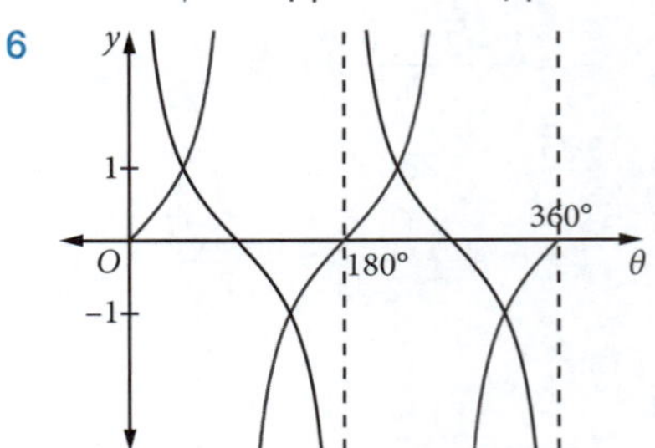

7 (a)

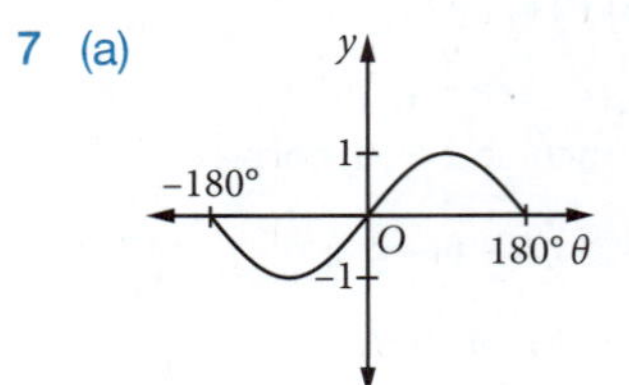

(b)

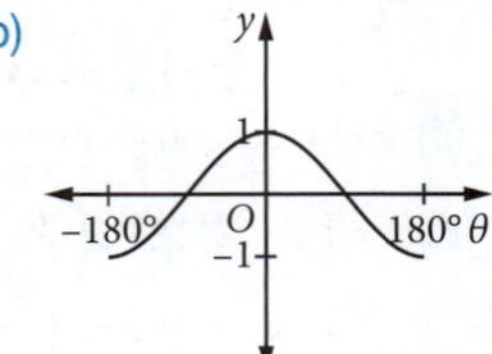

(c)

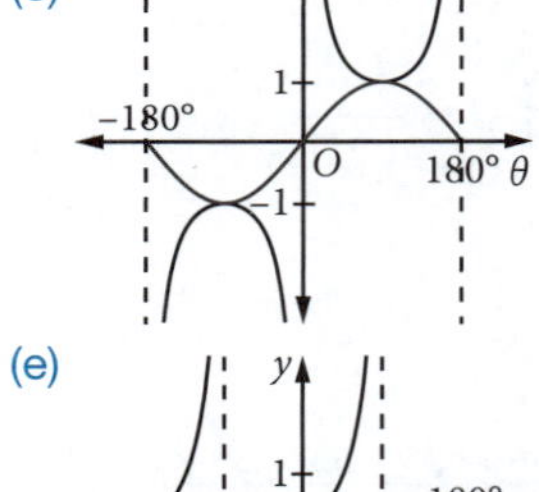

(d)

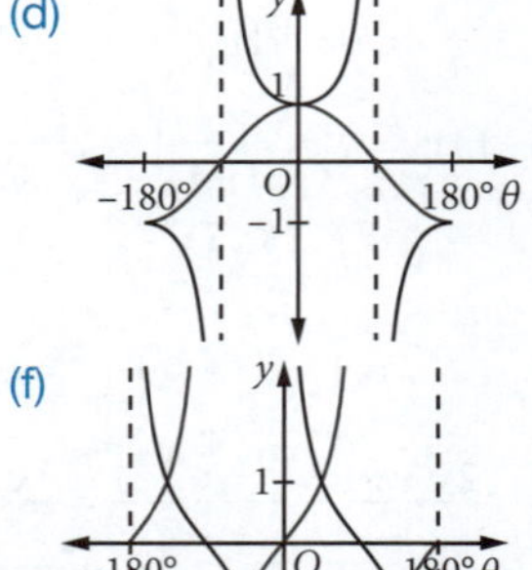

(e)

(f)

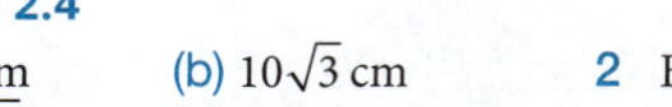

EXERCISE 2.4

1 (a) 10 cm (b) $10\sqrt{3}$ cm

2 B

3 (a) $15\sqrt{3}$ m (b) 30°

4 (a) $5\sqrt{3}$ m (b) 5 m

5 (a) 4 cm (b) $4\sqrt{3}$ cm (c) $4\sqrt{7}$ cm

6 (a) 10 cm (b) $10\sqrt{3}$ cm (c) $5\sqrt{2}$ cm (d) $5\sqrt{2}$ cm

7 (a) $6\sqrt{3}$ cm (b) 6 cm (c) $\left(6\sqrt{3}-6\right)$cm (d) $6\sqrt{6}$ cm

8 $\left(40+8\sqrt{2}+8\sqrt{3}\right)$cm

9 $\sqrt{3}$ m

10 $30\sqrt{3}$ cm, 30 cm

11 (a) correct (b) correct (c) incorrect (d) correct

EXERCISE 2.5

1 (a) 1 (b) $-\frac{1}{2}$ (c) $-\frac{1}{\sqrt{3}}$ (d) -1 (e) $\frac{\sqrt{3}}{2}$

2 (a) 0 (b) $-\frac{\sqrt{3}}{2}$ (c) 1 (d) $-\frac{1}{2}$ (e) 0

3 (a) -1 (b) $-\sqrt{3}$ (c) -1 (d) $\frac{\sqrt{3}}{2}$ (e) $-\frac{\sqrt{3}}{2}$

4 (a) 0 (b) $\frac{\sqrt{3}}{2}$ (c) 1 (d) 0 (e) $\frac{\sqrt{3}}{2}$

5 A 6 (a) correct (b) incorrect (c) correct (d) correct

7 240°, 300° 8 135°, 315° 9 180° 10 45°, 225°

11 180° 12 120°, 240° 13 60°, 120° 14 120°, 300°

15 15°, 75°, 195°, 255° 16 120°

17 22.5°, 112.5°, 202.5°, 292.5° 18 no solution

19 (a) (i) 1.2208, −1.7434, −0.7002 (ii) 2.1301, −1.1326, −1.8807
(iii) 1.1223, −2.2027, −0.5095
(b) (i) −2.3662, −1.1034, 2.1445 (ii) −1.6616, −1.2521, 1.3270
(iii) −1.0576, −3.0716, 0.3443
(c) (i) −1.0223, 4.8097, −0.2126 (ii) −1.1666, 1.9416, −0.6009
(iii) −3.2361, 1.0515, −3.0777
(d) (i) −2.3662, 1.1034, −2.1445 (ii) −1.1792, −1.8871, 0.6249
(iii) 1.7434, −1.2208, −1.4281

20 (a) $\frac{1}{c}$ (b) $-\frac{1}{c}$ (c) $\frac{1}{c}$ (d) $-\frac{1}{c}$

21 (a) −1.2208 (b) 1.6003 (c) −1.3054 (d) 1.1106

22 D

23 (a) correct (b) incorrect (c) incorrect (d) correct

24 (a) 1 (b) $-\frac{2}{\sqrt{3}}$ (c) $\frac{1}{\sqrt{3}}$ (d) undefined (e) 2 (f) −1
(g) $\frac{2}{\sqrt{3}}$ (h) 2 (i) $\sqrt{3}$ (j) 0

25 C

26 (a) $\theta = 240°, 300°$ (b) $\theta = 30°, 210°$ (c) $\theta = 180°$
(d) $\theta = 135°, 315°$ (e) no solution (f) $\theta = 60°, 300°$
(g) $\theta = 30°, 120°, 210°, 300°$ (h) $\theta = 60°$

EXERCISE 2.6

1 C 2 (a) SW (b) N 60° W or 300° (c) S

3 051° 20′

4 13 km, 112° 37′

5 50 m, N 30° W

6 13 km, 210° 23′

7 (a) 25 km (b) 24 min (c) 306° 52′

8 16.4 km, 11.5 km

9 39.74 km, 2.54 h

10 20 km, 233° 8′

11 (a) 41.45 km (b) 17.41 km (c) 202° 47′

EXERCISE 2.7

1 (a) incorrect (b) correct (c) correct (d) incorrect

2 12.4 m 3 (a) 8395 m (b) 504 km h^{-1}

4 (a) 45 m (b) 75 m (c) 67° 23′

5 614.4 m

6 (a) 137.4 m (b) 96.2 m (c) $\angle DBE = 18° 35'$

7 (a) 23.32 m (b) 18° 26′

8 (a) 13 m (b) 13 m

9 7 m

10 (a) 21.16 m (b) 46° 37′

11 (a) 129 m (b) 37 m

12 (a) $25\sqrt{3}$ m, $75\sqrt{3}$ m (b) $50\sqrt{3}$ m

13 $\frac{80\sqrt{3}}{3}$ m

EXERCISE 2.8

1 A 2 $A = 15°, a = 3.66, c = 12.2$

3 18.8 cm 4 $\frac{5\sqrt{6}}{3}$ 5 $\frac{8}{15}$

6 $\frac{3}{\sin A} = \frac{5}{\sin 2A}$ 7 14.7 cm

8 (a) incorrect (b) correct (c) incorrect (d) correct

9 (a) 0.3 (b) 4.8

10 3.26 m from A

11 $C = 61° 58'$ or $118° 2'$, $A = 81° 2'$ or $24° 58'$

12 $\angle RPQ = 16° 32'$, $PR = 40.41$ cm;
$\angle RPQ = 163° 28'$, $PR = 2.07$ cm

13 20.2 cm

14 (a) 0.5 (b) 30°, 150°

15 (a) 2.15 m (b) 1.76 m

16 (a) 8.6 cm (b) 1.9 cm

17 (a) 14.3 cm (b) 90°; triangle is right-angled, sine rule still works

18 9.90 m

19 (a) 561 km (b) 116° 9′

20 74 km

21 $CA = 1578$ m, $CB = 1149$ m

22 93 m

23 3.88 km

24 $\sin\angle BAD = \frac{BD \sin\angle B}{AD}$, $\sin\angle DAC = \frac{DC \sin\angle ADC}{AC}$,

$\therefore \frac{BD \sin\angle B}{AD} = \frac{DC \sin\angle ADC}{AC}$

$\sin\angle ADC = \sin\angle ADB$, so $\frac{\sin\angle B}{\sin\angle ADB} = \frac{DC \times AD}{BD \times AC}$,

$\therefore \frac{\cancel{AD}}{AB} = \frac{DC \times \cancel{AD}}{BD \times AC}$, $\therefore \frac{AB}{AC} = \frac{BD}{DC}$

25 (a) b lies between about 8.8 cm and 11 cm.
(b) $b = 11 \sin 53° \approx 8.8$ cm or $b > 11$ cm
(c) $b < 11 \sin 53° \approx 8.8$ cm

26 if B is given

27 $c > 9$

28 $B = 57°12'$ or $122°48'$

29 $C = 65° 35'$ or $114° 25'$

30 $B = 88°14'$ or $B = 91°46'$

31 $A = 56° 28'$, $B = 61° 2'$, $b = 8.2$ cm

EXERCISE 2.9

1 D 2 0.7308

3 (a) 6.1 (b) 43° 31′

4 11.7 cm, 24.5 cm

5 13 cm

6 (a) incorrect (b) incorrect (c) correct (d) correct

7 (a) 24.2 (b) 23° 22′

8 109° 28′

9 (a) 73° 24′ (b) 7 cm (c) 25° 13′

10 (a) 101° 32′, 78° 28′ (b) 11.5 cm

11 (a) 14.4 (b) 52° 25′

12 (a) incorrect (b) correct (c) correct (d) correct

13 10.6 km 14 44.9 km

15 (a) 030° (b) 068° 13′

16 (a) 8.62 km (b) 034° 53′

17 (a) 40 m (b) angle of elevation = 0° 57′

18 212.3 m 19 2.654 km 20 418 km

EXERCISE 2.10

1 (a) $a = 6.1$ (b) $\angle B = 43° 31'$ (c) 21 cm^2

2 8 cm^2 3 (a) 24° 37′ (b) 3.33 cm (c) 18.3 cm^2

4 B

5 (a) 10 cm (b) 29.4 cm^2

6 (a) $\angle APB = 101° 32'$, $\angle APC = 127° 10'$, $\angle BPC = 131° 18'$, $BC^2 = 244 - 240 \cos 131° 18'$, $BC = 20$ cm

(b) 122.5 cm^2

7 (a) $47° 09'$ (b) 3.7 cm (c) 9.2 cm^2

8 (a) $AC = 11.4$ cm, $BD = 7$ cm (b) $20\sqrt{3}$ cm^2

9 (a) correct (b) correct (c) incorrect (d) incorrect

10 3420 m^2

EXERCISE 2.11

1 20.06 cm, 122.5 cm^2

2 39.97 m, elevation $0° 56'$

3 212.3 m

4 B, C

5 (a) 30 m, 40 m (b) $10\sqrt{7}$ m

6 $\frac{h}{y} = \tan\beta$ so $y = h\cot\beta$

$\frac{h}{x+y} = \tan\alpha$ so $h = (x+y)\tan\alpha$

$h = x\tan\alpha + h\tan\alpha\cot\beta$

$x = h(\cot\alpha - \cot\beta)$

7 $h = y\tan 62°$, $h + 5 = y\tan 68°$

$h + 5 = \frac{h\tan 68°}{\tan 62°}$

$h = \frac{5\tan 62°}{\tan 68° - \tan 62°}$

8 Area of triangular cross-section $= \frac{1}{2}a \times a\cot\theta = \frac{1}{2}a^2\cot\theta$

Depth of water $= \frac{\text{Area}}{\text{Base}} = \frac{a^2\cot\theta}{2b}$

9 $AC = OC\tan\alpha$, $CG = \frac{1}{3}OC$

$\tan\angle AGC = \frac{AC}{CG} = 3\tan\alpha$

$\angle AGC = \alpha + \beta$

$\tan(\alpha+\beta) = \frac{\tan\alpha + \tan\beta}{1 - \tan\alpha\tan\beta} = 3\tan\alpha$

$\tan\alpha + \tan\beta = 3\tan\alpha - 3\tan^2\alpha\tan\beta$

$\tan\beta + 3\tan^2\alpha\tan\beta = 2\tan\alpha$

$\tan\beta = \frac{2\tan\alpha}{1 + 3\tan^2\alpha}$

10 A

11 0.9067 km

12 $70° 32'$

13 16 m

14 71 $km h^{-1}$

15 $x = \frac{h}{\tan 47°}$, $40 - x = \frac{h}{\tan 28°}$

$40 - \frac{h}{\tan 47°} = \frac{h}{\tan 28°}$

$h = \frac{40\tan 47°\tan 28°}{\tan 47° + \tan 28°}$

16 $\frac{h-2}{2\tan 75°} = \tan 30°$

$h = 2(1 + \tan 30°\tan 75°)$ metres

17 131 m

18 C is the foot of the tower.

$AC = h\tan 70°$, $BC = h\tan 72°$

$AC^2 + BC^2 = 100^2$

$h^2(\tan^2 70° + \tan^2 72°) = 100^2$

$h = \frac{100}{\sqrt{\tan^2 70° + \tan^2 72°}}$

19 26.55 cm

20 $AC^2 = 8^2 + 7^2 - 2\times 8\times 7\cos\theta$

$AC^2 = 5^2 + 6^2 - 2\times 5\times 6\cos(180° - \theta)$

$113 - 112\cos\theta = 61 + 60\cos\theta$

$\cos\theta = \frac{52}{172} = \frac{13}{43}$

21 6488 m

22 147.4 m

23 3.881 km

24 $\frac{x}{\sin(\beta-\alpha)°} = \frac{h}{\sin\beta°} \times \frac{1}{\sin\alpha°}$

$h = \frac{x\sin\alpha°\sin\beta°}{\sin(\beta-\alpha)°}$

25 $\frac{x}{\sin 30°} = \frac{AC}{\sin 45°}$, $AC = \sqrt{2}x$

$\frac{AC}{\sin 60°} = \frac{y}{\sin 45°}$, $y = \frac{2AC}{\sqrt{6}}$

$y = \frac{2}{\sqrt{6}} \times \sqrt{2}x$, $\frac{x}{y} = \frac{\sqrt{3}}{2}$

Area $\Delta ABC = \frac{1}{2}x \times AC\sin 105° = \frac{1}{2}x \times AC\sin 75°$

Area $\Delta ADC = \frac{1}{2}y \times AC\sin 75°$

$\frac{\text{Area } \Delta ABC}{\text{Area } \Delta ADC} = \frac{x}{y} = \frac{\sqrt{3}}{2}$

26 (a) 34.2 m (b) $18° 32'$ (c) $73°$

27 (a) $\angle NBQ = 50°$, $\angle AQB = 140°$ (exterior angle of triangle)

(b) $\frac{AQ}{h} = \tan 77°$, $AQ = h\tan 77°$

(c) $\frac{BQ}{h} = \tan 80°$, $BQ = h\tan 80°$

(d) $h = 106$ m

28 $BC = 1264$ m

29 (a) Use the angle between directions is 78° (b) $x = h\tan 79°$

(c) $y = h\tan 82°$ (d) $h = 381.4$ m

CHAPTER REVIEW 2

1 (a) $\sin\theta$ (b) $-\cos\theta$ (c) $-\sin\theta$ (d) $\tan\theta$ (e) $-\tan\theta$ (f) $-\sin\theta$

2 (a) -1 (b) $-\frac{1}{\sqrt{2}}$ (c) -1 (d) 0 (e) $\frac{\sqrt{3}}{2}$ (f) $-\frac{\sqrt{3}}{2}$

3 (a) 1 (b) 0

4 (a) $-\frac{3}{\sqrt{34}}$ (b) $-\frac{5}{\sqrt{34}}$

5 (a) 0.6 (b) 0.6 (c) -0.8 (d) $\frac{3}{4}$ (e) -0.75 (f) -0.6

6 (a) $\frac{1}{t}$ (b) t (c) $-\frac{1}{t}$ (d) $-t$ (e) $-t$ (f) $-\frac{1}{t}$

7 $\frac{5}{7}$

8 $\theta = 92° 52'$

9 (a) 9.8 cm (b) 29.2 cm^2

10 $4(\sqrt{3} - \sqrt{2})$

11 (a) 25.17 m (b) 39.16 m (c) $\theta = 22° 45'$

12 (a) 64.3 m (b) 54 m (c) $20° 28'$

13 (a) 26.46 km (b) $359° 6'$

14 (a) $r = 100 - 130t$

(b) $r = 87$

15 312 m

CHAPTER 3

EXERCISE 3.1

1 4.5

2 4.6

3 $3x + 9 = 27 - 9x$, $12x = 18$, $x = 1.5$

4 5

5 -8.75

6 1

7 -1.25

8 -1

9 -1

10 $12a + 8 - 18 + 6a = 8$, $18a = 18$, $a = 1$

11 2.5

12 1.5

13 2.25

14 6.5

15 $4x + 20 - x + 1 = 9$, $3x = -12$, $x = -4$

16 4

17 -1

18 -14.75

19 $12x - 18 = 8 - 6x - 2$, $18x = 24$, $x = 1\frac{1}{3}$

20 -1.5

EXERCISE 3.2

1 24

2 40

3 30

4 $14x + 35 = 3x + 6 + 84$, $x = 5$

5 $2\frac{1}{3}$

6 $\frac{1}{6}$

7 -0.75

8 $10\frac{2}{3}$

9 $4\frac{6}{7}$

10 3 11 4.6 12 $4(x+4)-2(3-4x)=5-x,\ x=-\frac{5}{13}$
13 −2 14 −3.5 15 3.25 16 −2
17 −0.25 18 13 19 −0.5 20 −2.5
21 $-\frac{1}{7}$ 22 C 23 15 24 −4.2
25 $11\frac{2}{3}$ 26 $3(x+2)-2(x-2)=1,\ x\neq\pm 2,\ x=-9$
27 −7 28 $\frac{2}{3}$ 29 1
30 $x+2+x+1=1,\ x\neq -1,\ x\neq -2$, no solution 31 0.5
32 (a) incorrect (b) incorrect (c) incorrect (d) incorrect

EXERCISE 3.3

1 $x\geq 3$ 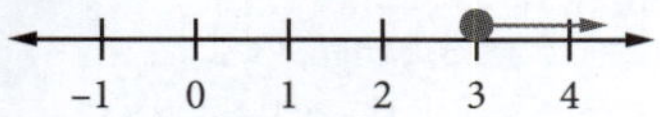
2 $x<1$
3 $x<-5$
4 $x\geq -2$ 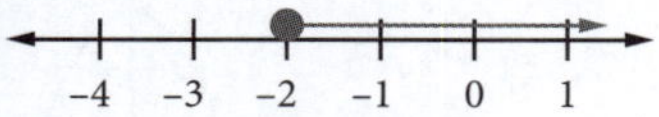
5 $x>12$ 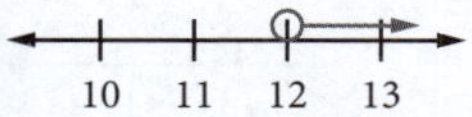
6 $x\geq 2$ 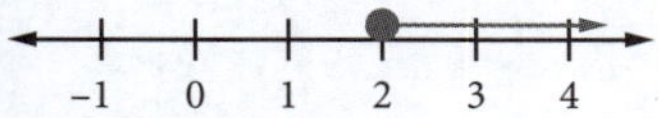
7 $x<3$
8 $x\geq 10$ 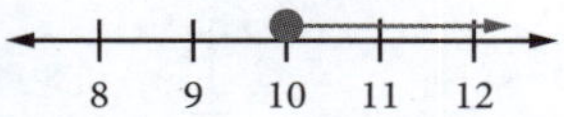
9 $4x-3x>12,\ x>12$ 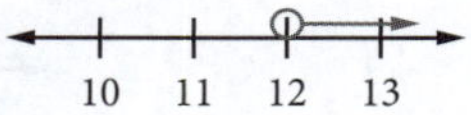
10 $x<1\frac{2}{7}$
11 $x\geq 1.5$ 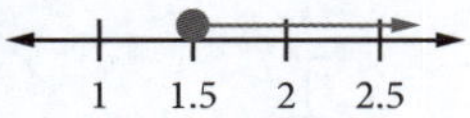
12 $x>-3$ 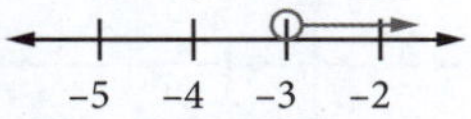
13 $x>-2$ 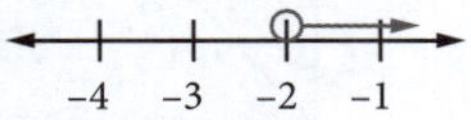
14 $x<32$
15 $x>6$ so $x=7, 8, 9\ldots$ 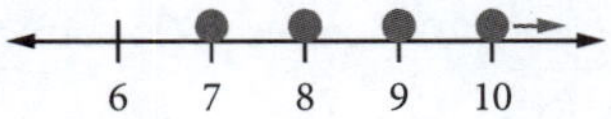
16 D
17 $-x>-30,\ x<30$
18 $x<3$
19 $x<-6$
20 $x<2\frac{1}{3}$
21 $-9<x<-3$
22 $5\leq x\leq 7$ 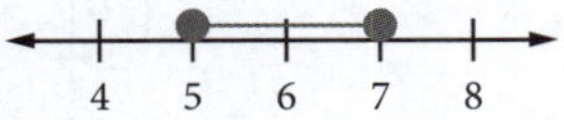
23 $-4\leq x<5$

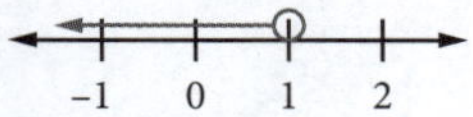

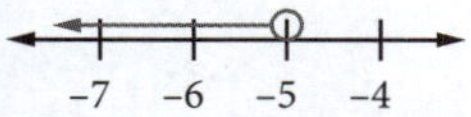

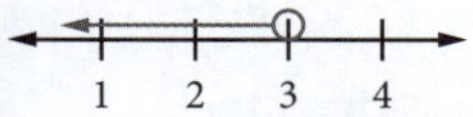

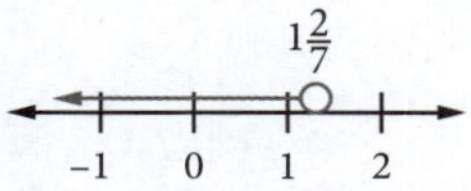

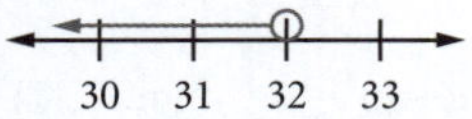

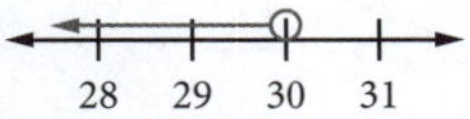

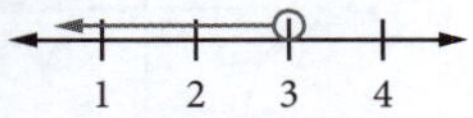

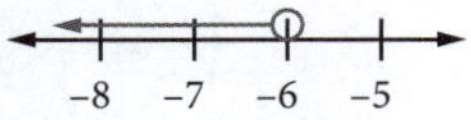

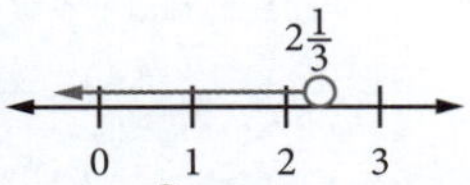

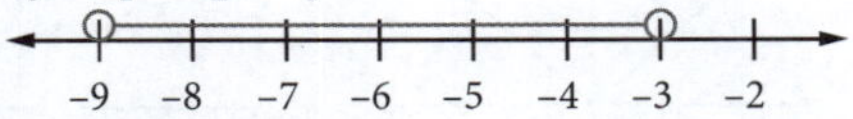

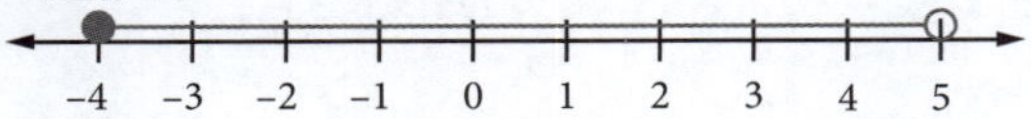

24 (a) incorrect (b) correct (c) incorrect (d) incorrect
25 $5<x-5<12,\ x=11, 12, 13, 14, 15, 16$
26 $8<x<16$
27 $x+(x+1)\leq 35,\ x=1, 2, 3, \ldots 17$
28 5, 6, 7, 8, 9
29 $\frac{x+5}{5}\leq\frac{x+13}{9},\ x\leq 5,\ x=5$
30 base length = x cm, equal sides = y cm
$2y=x+4,\ 2y+x<80$
$0<x<38$ where x is an integer
31 $7<x+(x+1)+(x+2)<25,\ x=2, 3, 4, 5, 6, 7$
32 $2<x<18$

EXERCISE 3.4

1 0, 5 2 2, 3 3 0, −0.5 4 −2.5, 7
5 0, 4.5 6 0, −1 7 a, b 8 $3a, -2b$
9 2, −2 10 5.5, −5.5 11 1 12 −1.5

EXERCISE 3.5

1 1, −1 2 5, −5 3 7, −7
4 4, −4 5 C 6 1.5, −1.5
7 2.5, −2.5 8 2.5, −2.5 9 1.5, −1.5
10 1, −1 11 $\frac{1}{4}, -\frac{1}{4}$ 12 4, −4
13 1.4, −1.4
14 (a) correct (b) correct (c) incorrect (d) correct
15 6, −2 16 7, −7 17 3, −3
18 1, −0.6 19 −1, −5 20 $\sqrt{5}, -\sqrt{5}$
21 $\sqrt{2}, -\sqrt{2}$ 22 $-1+2\sqrt{2}, -1-2\sqrt{2}$

EXERCISE 3.6

1 0, 6 2 0, 5 3 0, −5 4 0, −10
5 D 6 0, 2.5 7 0, 7 8 0, 7
9 0, −10 10 0, 4 11 0, 0.5 12 0, −0.2
13 0, −3 14 $0, \frac{5}{12}$ 15 0, 15

EXERCISE 3.7

1 1, 2 2 1, 5 3 4, −2 4 1, 3
5 3 6 1, 4 7 −1, −8 8 $-1, \frac{5}{9}$
9 B 10 2, −6 11 2, 0.2 12 3.5, −0.5
13 2.5, −2 14 −5 15 −1, −4 16 3.5, −1.5
17 $1, 8\frac{1}{3}$ 18 4 19 −4, −1.2 20 $12, 1\frac{2}{3}$
21 3, −10 22 $5x^2-8x+3=0,\ (5x-3)(x-1)=0,\ x=1, 0.6$
23 $2x^2-11x-6=0,\ (2x+1)(x-6)=0,\ x=6, -0.5$
24 1, −6 25 $-9, 2\frac{2}{3}$ 26 3, 5 27 3, −2.5
28 2, −6 29 0.5, −1.5
30 $x^2+2x+1=4x,\ x^2-2x+1=0,\ (x-1)^2=0,\ x=1$
31 $(x+6)(x+6-1)=0,\ x=-5, -6$ 32 $\frac{2}{3}, -2\frac{1}{2}$ 33 $4, \frac{6}{7}$

EXERCISE 3.8

1 4 2 9 3 49 4 1 5 A
6 $\frac{1}{4}$ 7 $6\frac{1}{4}$ 8 $2\frac{1}{4}$ 9 $12\frac{1}{4}$ 10 $\frac{1}{4}$
11 a^2 12 b^2 13 $\frac{c^2}{4}$
14 (a) incorrect (b) correct (c) correct (d) incorrect

EXERCISE 3.9

1 1, 5 2 4, −2 3 1, −5 4 2, −6 5 D
6 −3, 7 7 1, 25 8 1, 2 9 3, −4
10 1, 4 11 3, −10 12 −1, 12 13 −2, 5
14 2, 5 15 8, −9 16 −1, 11 17 0, 10
18 (a) correct (b) incorrect (c) correct (d) correct

EXERCISE 3.10

1 $3.24, -1.24$ 2 $-2 \pm 2\sqrt{2}$ 3 $2.79, -1.79$ 4 $3 \pm \sqrt{7}$

5 $4.79, 0.21$ 6 $-1 \pm \sqrt{3}$ 7 $5.24, 0.76$ 8 $\frac{-1 \pm \sqrt{5}}{2}$

9 $6.74, -0.74$ 10 $-2 \pm \sqrt{5}$ 11 $3.45, -1.45$ 12 $\frac{-3 \pm \sqrt{33}}{2}$

EXERCISE 3.11

1 $1.85, -1.35$ 2 $\frac{-3 \pm \sqrt{19}}{2}$ 3 $0.78, -1.28$ 4 $\frac{-3 \pm \sqrt{17}}{4}$

5 $1.85, -0.18$ 6 $\frac{-2 \pm \sqrt{19}}{3}$ 7 $1.54, -0.87$ 8 $\frac{3 \pm \sqrt{7}}{2}$

9 $2.70, -0.37$ 10 $\frac{-1 \pm \sqrt{6}}{2}$ 11 $3.71, -1.21$ 12 $\frac{1 \pm \sqrt{7}}{3}$

13 $1.19, -1.69$ 14 $\frac{4 \pm \sqrt{7}}{3}$ 15 $1.27, 0.39$

EXERCISE 3.12

1 $-1, -5$ 2 $2, -4$ 3 $7, -1$ 4 $2, 5$

5 $0.41, -2.41$ 6 $3 \pm \sqrt{5}$ 7 $3.45, -1.45$ 8 $\frac{-5 \pm \sqrt{29}}{2}$

9 $4.16, -2.16$ 10 $-2 \pm \sqrt{2}$ 11 $7, 8$ 12 $3, -5$

13 $-0.22, -2.28$ 14 $\frac{4 \pm \sqrt{10}}{2}$ 15 $-\frac{1}{2}, -1$ 16 $0, 1.5$

17 $-1.22, 0.55$ 18 $1, -2.5$ 19 $-0.17, -5.83$ 20 4

21 $1.5, -1$ 22 $\frac{7 \pm \sqrt{105}}{14}$ 23 $1.64, 0.61$ 24 $-\frac{1}{3}, 4$

25 $-1.5, 1$ 26 $\frac{-3 \pm \sqrt{17}}{2}$ 27 $-2.82, -0.18$ 28 $\frac{3 \pm \sqrt{15}}{2}$

29 $2.73, -0.73$ 30 $3 \pm \sqrt{11}$ 31 $2.35, -0.85$ 32 $\frac{-5 \pm \sqrt{15}}{2}$

33 $3, -20$ 34 $\frac{-9 \pm \sqrt{21}}{6}$ 35 ± 2.24 36 $\frac{-1 \pm \sqrt{5}}{2}$

EXERCISE 3.13

1 (a) 6 (b) 4 (c) 3 (d) 8 (e) 2 2 D
3 (a) 5 (b) 9 (c) 7 4 3 5 8, 11 or −8, −11
6 8, 9 or −8, −9 7 (a) 2 s, 6 s (b) 4 s 8 6 9 4 m
10 1 m 11 4 cm 12 4 cm, 12 cm 13 12 cm, 16 cm, 20 cm
14 (a) $(20 - x)$ cm (b) $20x - x^2$ (c) $x^2 - 20x + 84 = 0$, 14 cm, 6 cm

CHAPTER REVIEW 3

1 (a) -6 (b) 1 (c) $-\frac{5}{3}$ 2 (a) $\frac{3}{4}$ (b) $\frac{4}{11}$ (c) 9.5

3 (a) $x > 4$

2 3 4 5 6

(b) $-2 \le x < 6$

−2 −1 0 1 2 3 4 5 6

(c) $x < 0, x > 2$

−1 0 1 2 3

4 (a) $x = \pm 2$ (b) $x = 0, 4$ (c) $x = 2$ (d) $x^2 - 3x = \pm 4, x = -1, 4$
(e) $(x-5)(x+2)(x-4)(x+1) = 0, x = -2, -1, 4, 5$
(f) $x = -1.5, \frac{1}{3}$

5 $(x+1)^2 = x^2 + (x-7)^2, x = 12$. Sides 5 cm, 12 cm, 13 cm

6 (a) $x = \frac{1 \pm \sqrt{41}}{4}$ (b) $x = -1.35, 1.85$

7 $x = \frac{-4 \pm \sqrt{2}}{2}$

8 $2x^3 - 3x^2y + 3xy^2 - y^3$ 9 $\frac{m-4}{(m-2)(m+2)(m-1)}$

10 $n = 7$

11 (a) $9 - x$ (b) $9x - x^2$ (c) 5 cm, 4 cm

12 $n = 13$ 13 $x = -2, 0, 1$ 14 $x = \frac{ab}{a+b}$

15 $x = 23.6 \approx 24$

16 $x = 0, 1, 5$

CHAPTER 4

EXERCISE 4.1

1 (a) relation, domain {1, 2, 3}, range {1, 2, 5, 7}
(b) function, domain {−3, 3, 8, 9, 11}, range {−2, 1, 6, 7}
(c) function, domain: real x, range {5}
(d) relation, domain {2}, range: real y
(e) relation, domain $-3 \le x \le 3$, range $-3 \le y \le 3$
(f) function, domain $-4 \le x \le 4$, range $0 \le y \le 4$
(g) function, domain: real x, range: real y
(h) function, domain: real x, range $y \le 2$

2 C

3 (a)

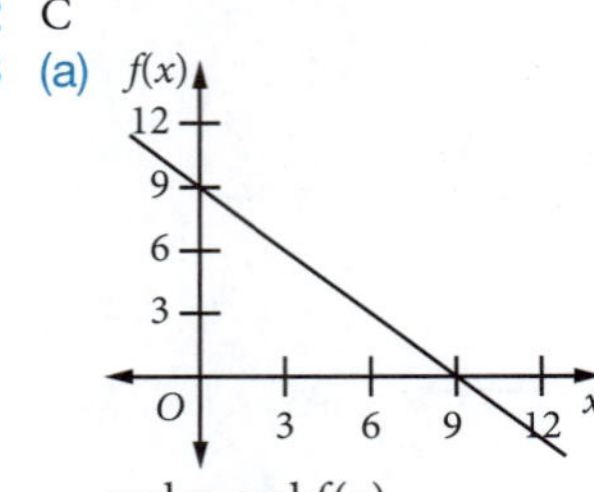

real x, real $f(x)$

(b)

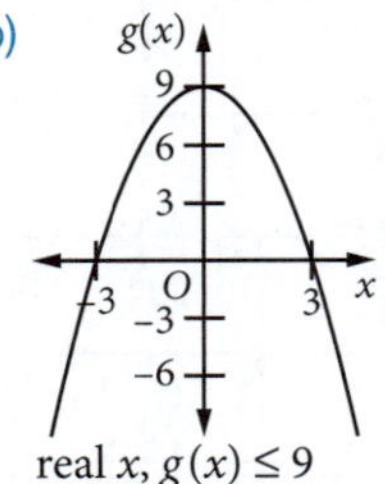

real x, $g(x) \le 9$

(c)

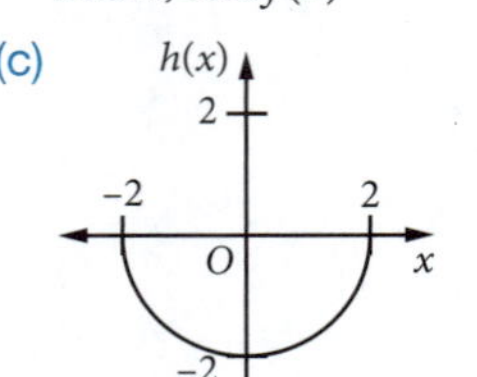

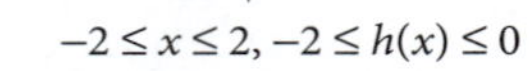

$-2 \le x \le 2, -2 \le h(x) \le 0$

(d)

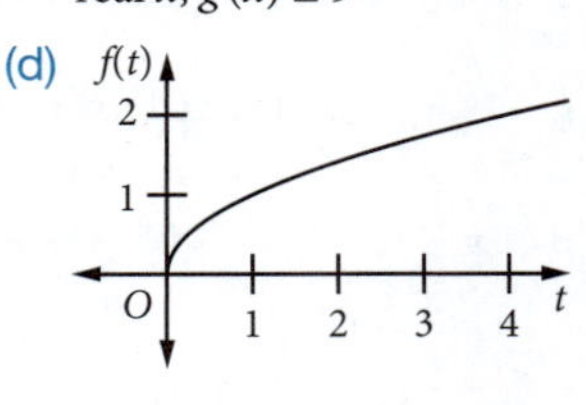

$t \ge 0, f(t) \ge 0$

4 (a) $-3, -12, 3a - 6$
(b) 3
(c) $x > 3$

(d)

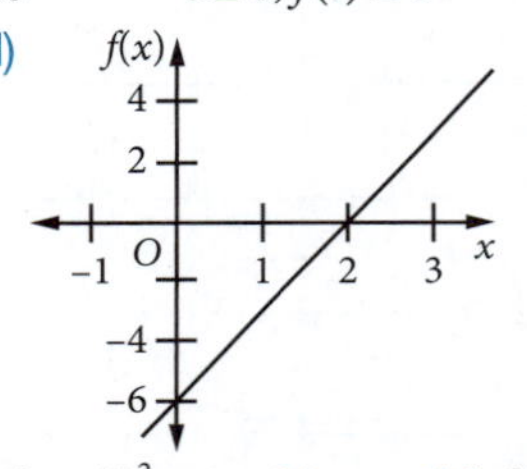

5 (a) 8, 8 (b) $a^2 - 1, b^2 - 1, (a+b)^2 - 1$ (c) no (d) $f(x) \ge -1$

6 (a) correct (b) incorrect (c) correct (d) correct

7 (a)

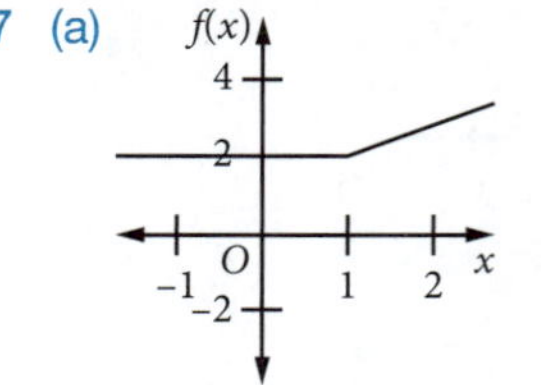

real x, $y \ge 2$

(b)

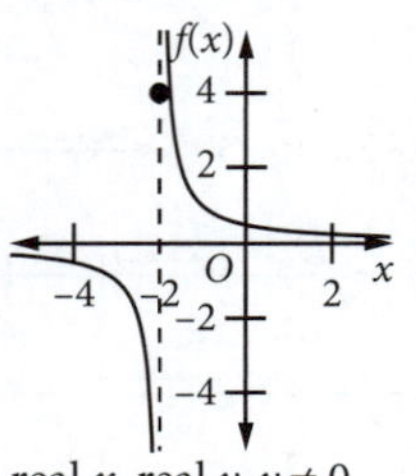

real x, real y, $y \ne 0$

(c)

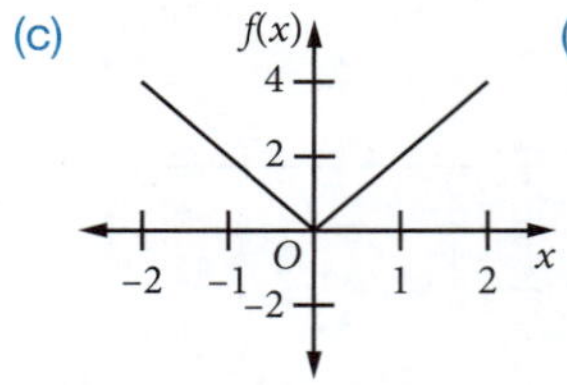

real x, $y \ge 0$

(d)

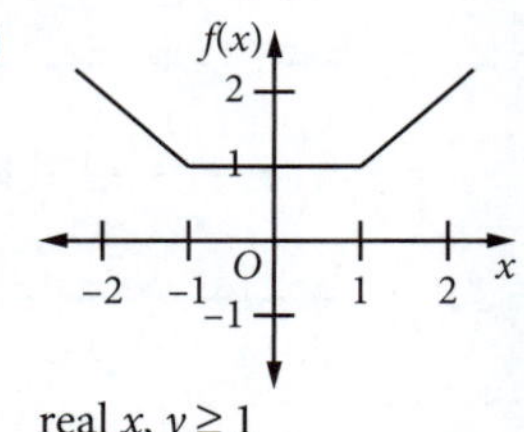

real x, $y \ge 1$

8 (a) $x \geq 2, f(x) \geq 0$ (b) $x \leq 3, f(x) \geq 0$ (c) $|x| \geq 3, f(x) \geq 0$
(d) real x, $x \neq 0$; real y, $g(x) \neq 0$ (e) real t, real $h(t)$
(f) real k, $g(k) \leq 5$
9 (a) 2 (b) does not exist (c) -1 (d) 6
10 (a) 0 (b) 2 (c) -0.5 (d) a^2
11 (a) 0 (b) -2 (c) -8 (d) 2

EXERCISE 4.2

1 (a) $m = 3; -\frac{1}{3}; 1$

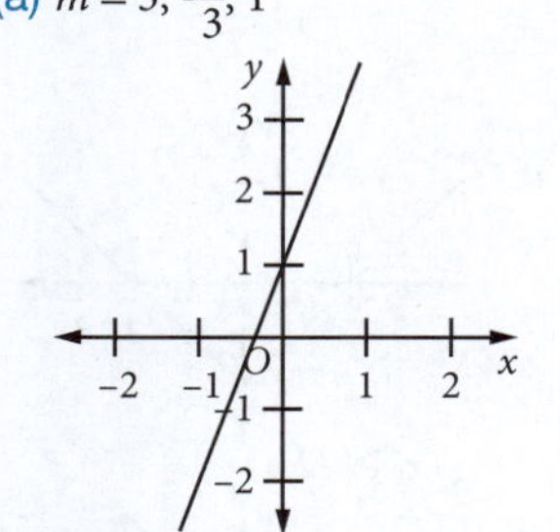

(b) $m = -1.5; 2; 3$

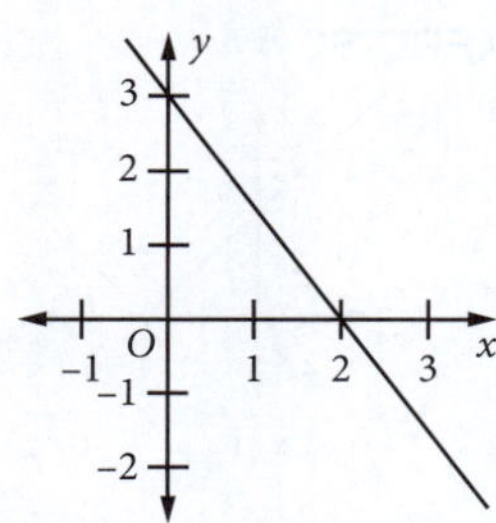

(c) $m = -2; 2; 4$

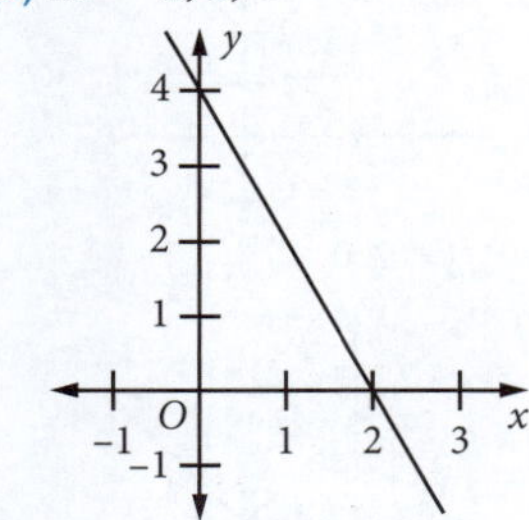

(d) $m = 1; 1; -1$

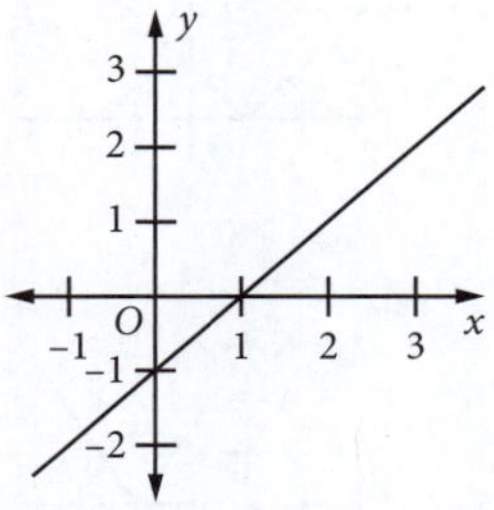

(e) $m = 4; 2; -8$

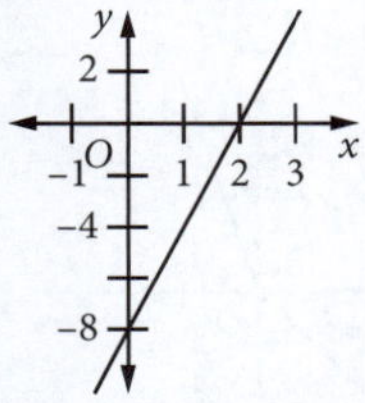

(f) $m = -1; 0; 0$

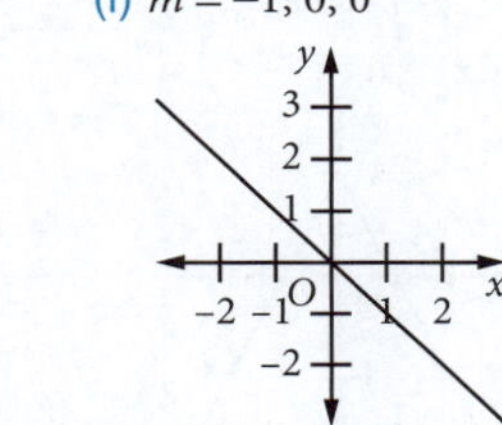

(g) $m = 0$; none; 3

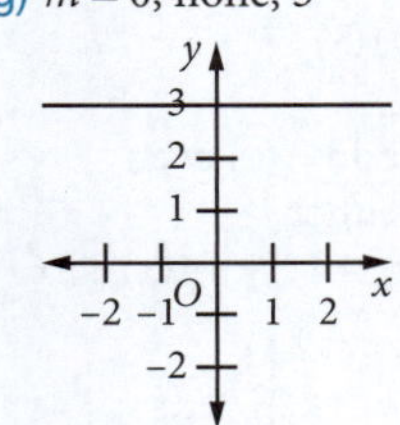

(h) $m =$ undefined; 4; none

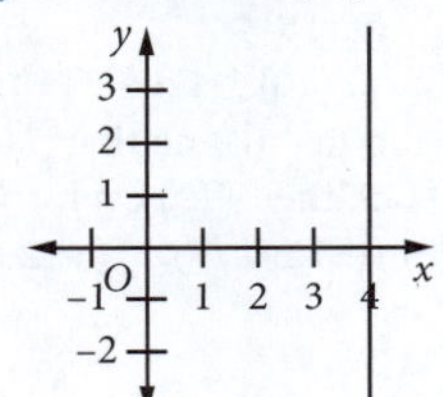

(i) $m = -0.5; -5; -2.5$

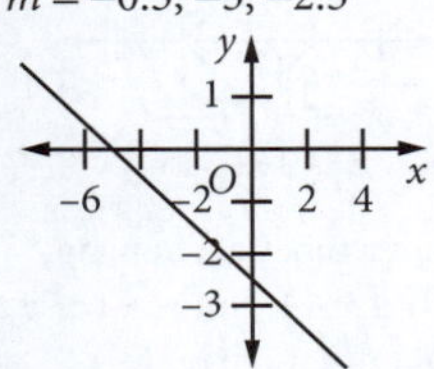

2 (a) increasing, gradient positive
(b) decreasing, gradient negative
(c) decreasing, gradient negative
(d) increasing, gradient positive
(e) increasing, gradient positive
(f) decreasing, gradient negative
(g) neither, gradient zero
(h) neither, gradient undefined
(i) decreasing, gradient negative

3 (a)

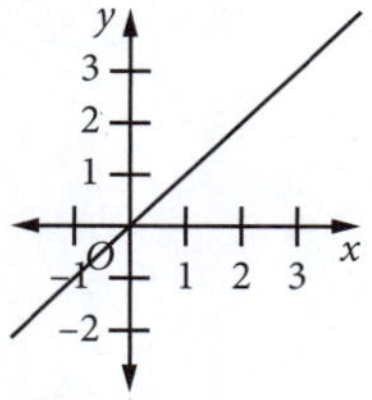

all x

(b)

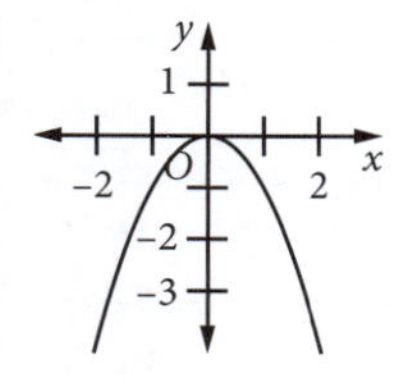

$x < 0$

(c)

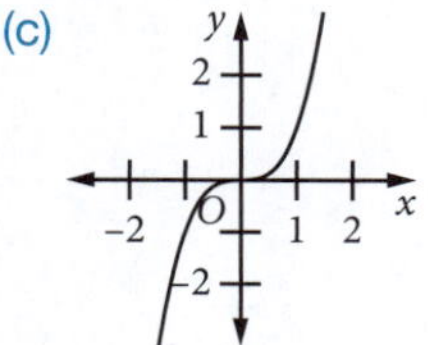

all x

(d)

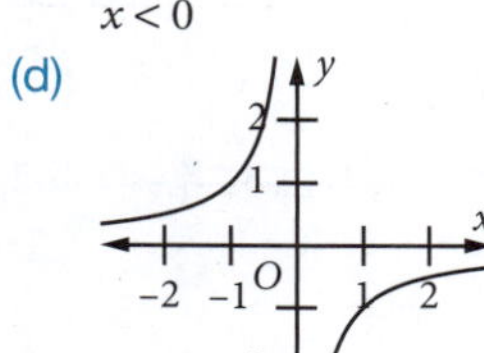

$x \neq 0$

(e)

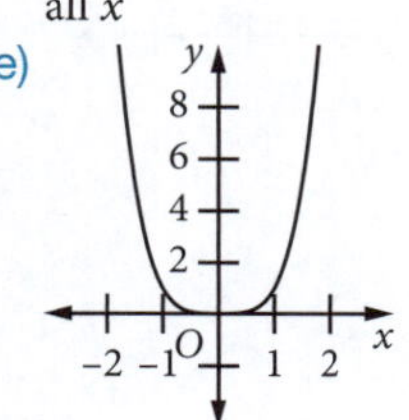

$x > 0$

(f)

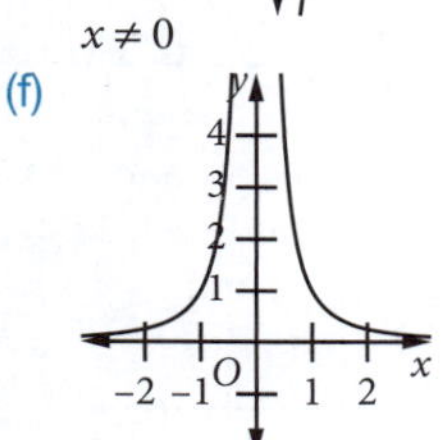

$x < 0$

4 (a) odd (b) even (c) odd (d) odd (e) even (f) even
5 B **6** (a) correct (b) incorrect (c) correct (d) incorrect

EXERCISE 4.3

1 9 **2** 2.5 **3** $6x - 4$ if $x > \frac{2}{3}$; $4 - 6x$ if $x < \frac{2}{3}$; 0 if $x = \frac{2}{3}$
4 $2 - \sqrt{3}$ **5** $(x + y)$ if $x + y > 0$; $-(x + y)$ if $x + y < 0$; 0 if $x + y = 0$
6 $(x + y)$ if $x, y > 0$; $(x - y)$ if $x > 0, y < 0$; $(y - x)$ if $x < 0, y > 0$; $-(x + y)$ if $x, y < 0$; 0 if $x = y = 0$
7 $2x$ if $x \geq 0$; 0 if $x < 0$ **8** $2x$ if $x \geq 5$; 10 if $-5 < x < 5$; $-2x$ if $x \leq -5$
9 8 **10** $2x + 3$ if $x \geq -1.5$; $-(2x + 3)$ if $x < -1.5$ **11** $x - 3$
12 (a) $x - 2 = 3$ or $-(x - 2) = 3$, $x = 5, -1$ (b) $x = -10, 4$
(c) $x = -1, 9$ (d) $x = -9, -5$ (e) $x = 6$ (f) $x = 4, 6$
(g) $x = -1$ (h) $x = -13, -7$ (i) $0.5, -1.5$ (j) 4, 1 (k) $0.6, -1$
(l) $x = \frac{-1}{3}, 3$ (m) $x = \frac{-1}{3}$ (n) $x = \frac{-4}{3}, 1$ (o) $x = \frac{1}{4}$ (p) $x = -2, 11$
13 C
14 (a) $x - 1 < 3$ or $x - 1 > -3$; $-2 < x < 4$ (b) $y < -6$ or $y > 2$
(c) $4 \leq t \leq 8$ (d) $x \leq -6$ or $x \geq -2$ (e) real m
(f) $-5 \leq 3 - x \leq 5$; $-8 \leq -x \leq 2$; $-2 \leq x \leq 8$ (g) $y = -1$
(h) $-10 < x < -4$ (i) $3 < 2x + 1 < -3$; $x < -2$ or $x > 1$
(j) $\frac{4}{3} < z < 2$ (k) $x \leq -2$ or $x \geq \frac{1}{2}$ (l) $-1 < t < \frac{7}{3}$ (m) no solution
(n) $x < -2.6$ or $x > 1$ (o) real x, $x \neq \frac{1}{2}$ (p) $x \leq -2$ or $x \geq 9$

15

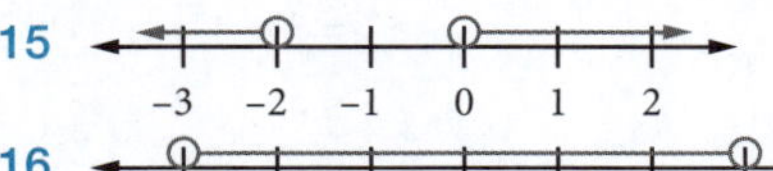

16

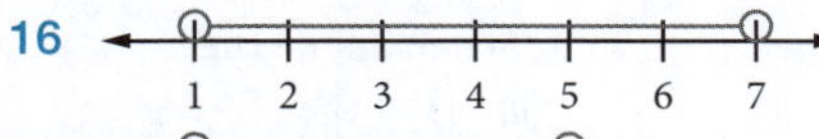

17

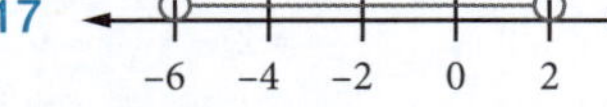

18

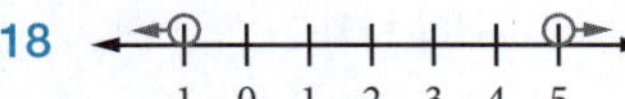

19

20

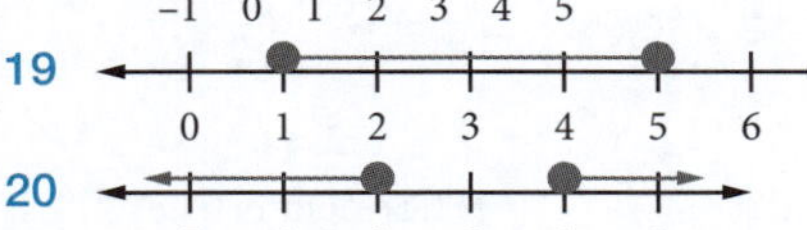

21

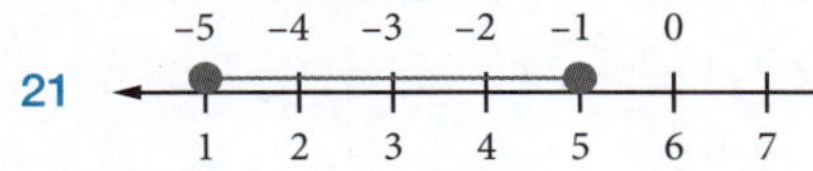

22

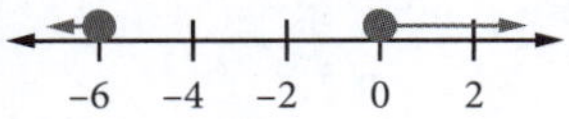

23 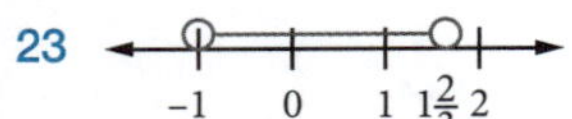

24 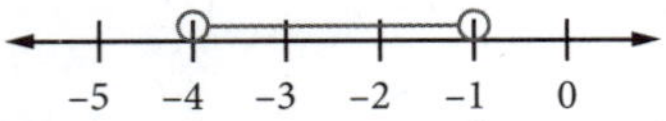 −5 −4 −3 −2 −1 0

25 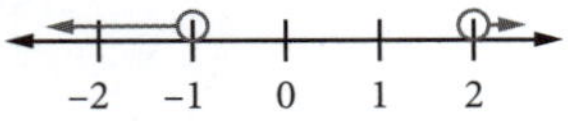 −2 −1 0 1 2

26 −3 −2 −1 0 1 2

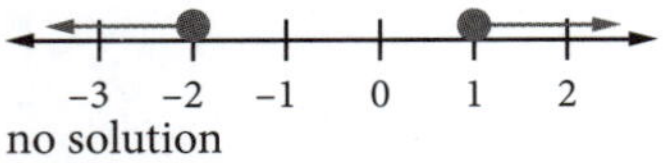

27 no solution

28 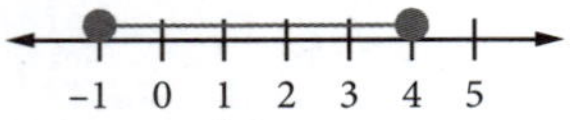 −1 0 1 2 3 4 5

29 $-1\frac{1}{3}$

−3 −2 −1 0 1 2

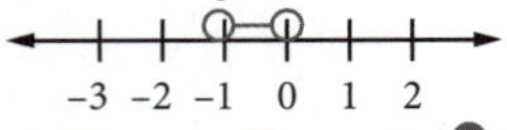

30 $-\sqrt{5} \le x \le \sqrt{5}$ $-\sqrt{5}$ −1 0 1 $\sqrt{5}$

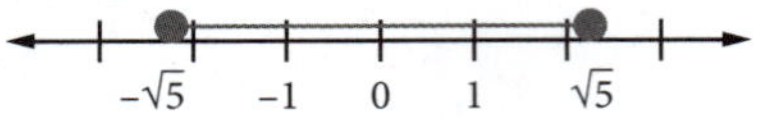

31 $x+3>1, x+3<-1$ and $-3<2x+5<3$

$x>-2, x<-4$ and $-8<2x<-2$

$x>-2$ or $x<-4$ and $-4<x<-1$

Hence $-2<x<-1$:

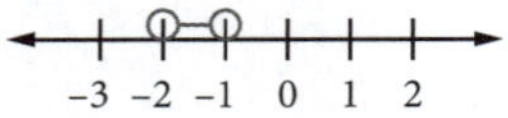

−3 −2 −1 0 1 2

32 $-3<2x+5<3$ or $2+4x\ge 6$, $2+4x\le -6$

$-8<2x<-2$ or $4x\ge 4$, $4x\le -8$

$-4<x<-1$ or $x\ge 1$, $x\le -2$

Hence $x<-1$ or $x\ge 1$

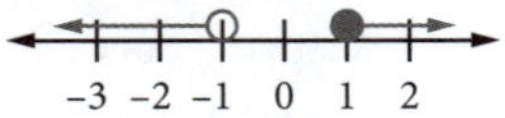

−3 −2 −1 0 1 2

33 (a) correct (b) incorrect (c) correct (d) correct

34 1 for all $x\ne 0$ 35 1 if $x>0$; −1 if $x<0$

36 $\frac{x-4}{x}$ if $x<0$; $\frac{4-x}{x}$ if $0<x<4$; $\frac{x-4}{x}$ if $x\ge 4$

37 1 if $x<1$; −1 if $x>1$ 38 $x-5$ if $x\ge 5$; $5-x$ if $x<5$

39 $x+1$ if $x<-1$ or $x>1$; $-(x+1)$ if $-1\le x<1$

40 (a) (i) $|xy| = |5\times 2|$
$= 10$
$|x|\times|y| = |5|\times|2|$
$= 10$
Statement is true

(ii) $|x+y| = |5+2|$
$= 7$
$|x|+|y| = |5|+|2|$
$= 7$
Statement is true

(b) (i) $|xy| = |3\times -2|$
$= |-6|$
$= 6$
$|x|\times|y| = |3|\times|-2|$
$= 3\times 2$
$= 6$
Statement is true

(ii) $|x+y| = |3-2|$
$= 1$
$|x|+|y| = |3|+|-2|$
$= 3+2$
$= 5$
$1<5$
Statement is true

(c) (i) $|xy| = |-6\times 8|$
$= |-48|$
$= 48$
$|x|\times|y| = |-6|\times|8|$
$= 6\times 8$
$= 48$
Statement is true

(ii) $|x+y| = |-6+8|$
$= 2$
$|x|+|y| = |-6|+|8|$
$= 6+8$
$= 14$
$2<14$
Statement is true

(d) (i) $|xy| = |-4\times -3|$
$= 12$
$|x|\times|y| = |-4|\times|-3|$
$= 4\times 3$
$= 12$
Statement is true

(ii) $|x+y| = |-4-3|$
$= |-7|$
$= 7$
$|x|+|y| = |-4|+|-3|$
$= 4+3$
$= 7$
Statement is true

EXERCISE 4.4

1 (a)

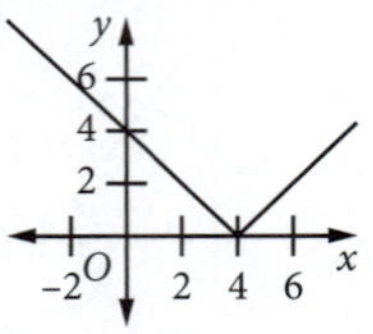

$f(x)\ge 0$

(b) 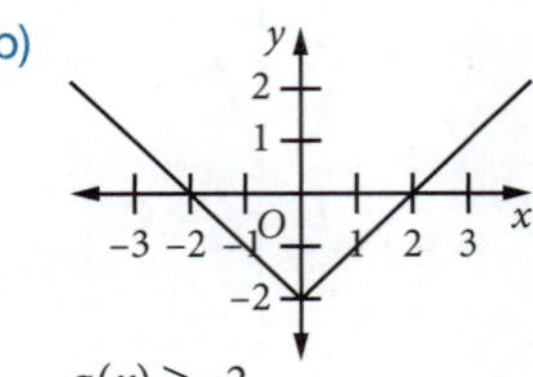

$g(x)\ge -2$

(c) 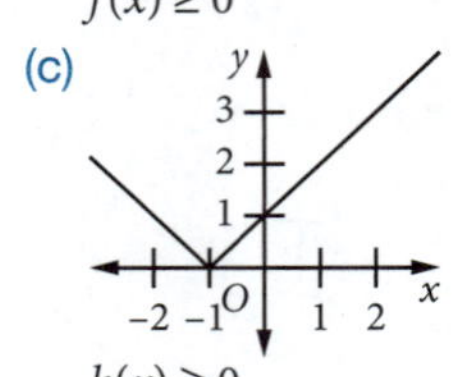

$h(x)\ge 0$

(d) 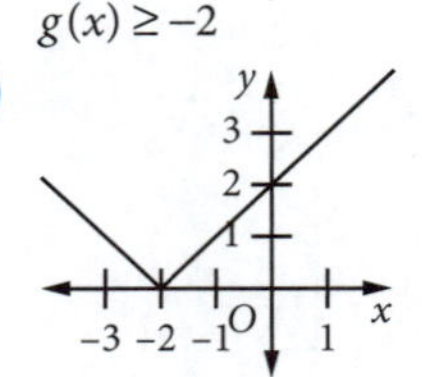

$f(x)\ge 0$

(e) 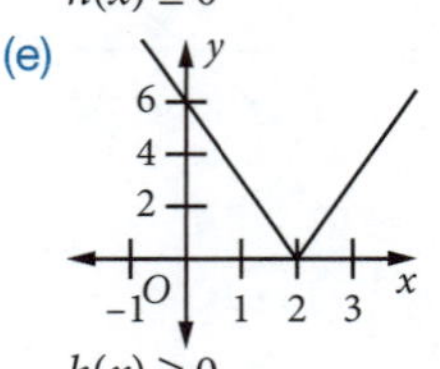

$h(x)\ge 0$

(f) 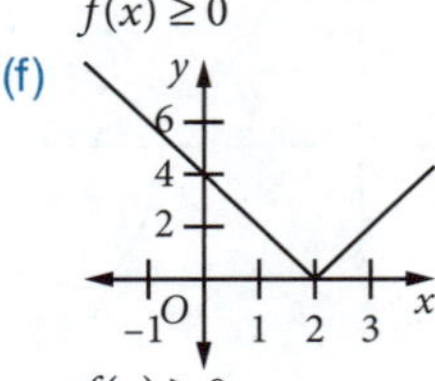

$f(x)\ge 0$

(g) 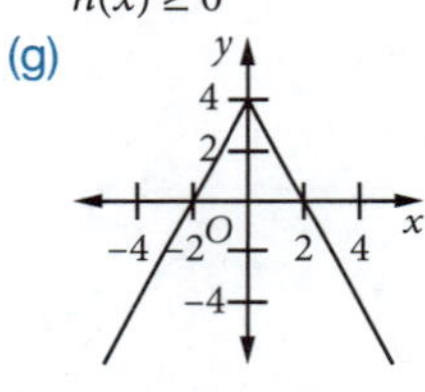

$g(x)\le 4$

(h) 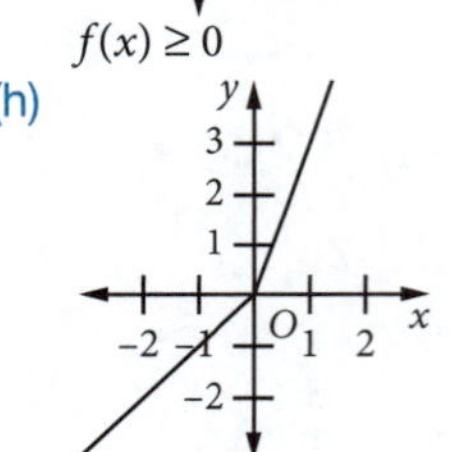

real $f(x)$

2 C 3 (a) $2\le x\le 3$ (b) real x, $x\ne 0$

4 (a) odd (b) neither (c) even (d) odd (e) even (f) neither (g) even (h) odd (i) neither

5 (a) $f(x)\ge 0$ (b) $f(x)\ge 0$ (c) $0\le f(x)\le 4$ (d) $f(x)\le 16$

6 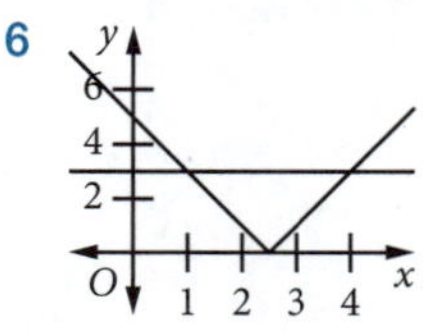

$f(x)=3$ for $x=1$ or 4

7 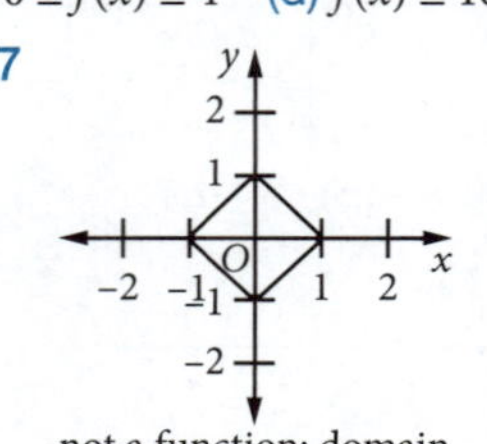

not a function; domain $-1\le x\le 1$, range $-1\le y\le 1$

8 (a) correct (b) correct (c) correct (d) correct

9 (a) correct (b) incorrect (c) correct (d) correct

EXERCISE 4.5

1 (a) $(x-3)^2+(y-2)^2=16$ (b) $(x+1)^2+(y+4)^2=9$

(c) $(x-3)^2+(y+3)^2=5$ (d) $(x+2)^2+\left(y-\frac{5}{2}\right)^2=\frac{49}{4}$

(e) $x^2+\left(y+\frac{3}{2}\right)^2=16$ (f) $(x-4)^2+y^2=9$

2 B

3 (a) $(x-3)^2+(y-2)^2=53$ (b) $(x+1)^2+(y-4)^2=17$

(c) $x^2+y^2=25$

4 (a) $(3,-2), r=4$ (b) $(-2,-1), r=3$ (c) $(3,0), r=\sqrt{3}$
(d) $(-a,b), r=2\sqrt{2}$ (e) $(2.5,-1.5), r=\frac{\sqrt{38}}{2}$
(f) $(-2,-1), r=\sqrt{10}$ (g) $\left(2,-1\frac{1}{4}\right), r=\frac{\sqrt{65}}{4}$
(h) $\left(-1\frac{1}{2},\frac{2}{3}\right), r=\frac{\sqrt{385}}{6}$

5 (a) $(x-6)^2+(y+1)^2=34$ (b) $\left(x-\frac{5}{2}\right)^2+\left(y+\frac{3}{2}\right)^2=\frac{17}{2}$
(c) $\left(x-\frac{3}{2}\right)^2+\left(y-\frac{11}{2}\right)^2=\frac{37}{2}$

6 (a) incorrect (b) incorrect (c) incorrect (d) correct

7 (a) $(-3,4), r=5$ (b) LHS = 0 = RHS (c) $4x+3y=0$

8 (a) $y=\sqrt{4-x^2}$, $y\in[0,2]$

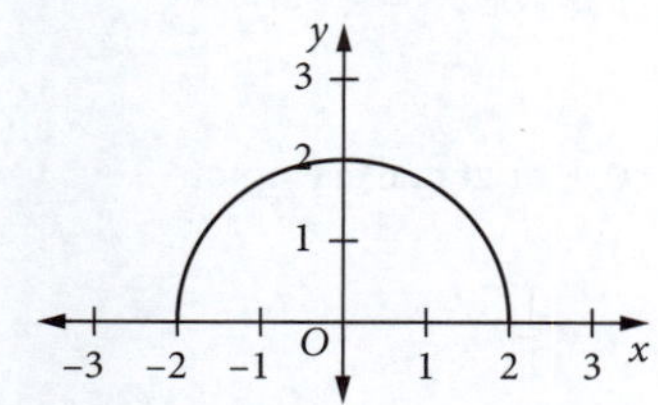

(b) $y=\sqrt{36-x^2}$, $y\in[0,6]$

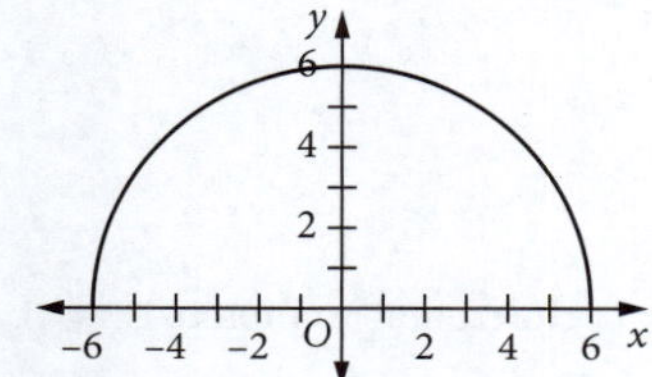

(c) $y=\sqrt{5-x^2}$, $y\in\left[0,\sqrt{5}\right]$

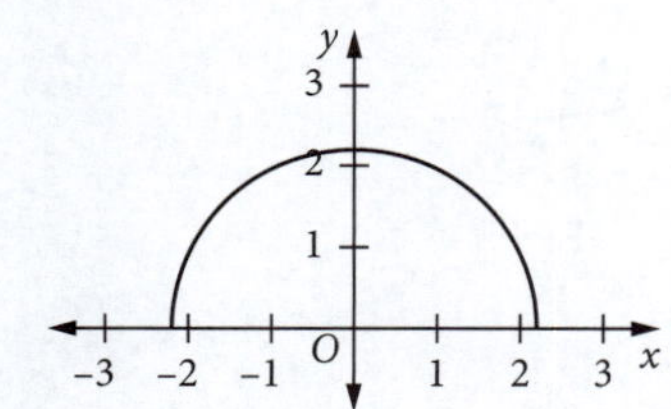

(d) $y=\sqrt{\frac{25}{9}-x^2}$, $y\in\left[0,\frac{5}{3}\right]$

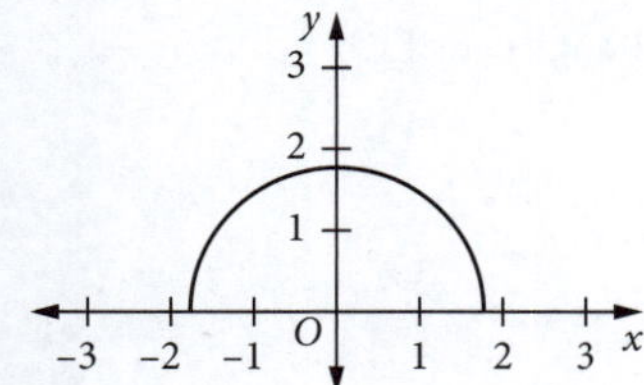

9 (a) $y=-\sqrt{4-x^2}$, $y\in[-2,0]$

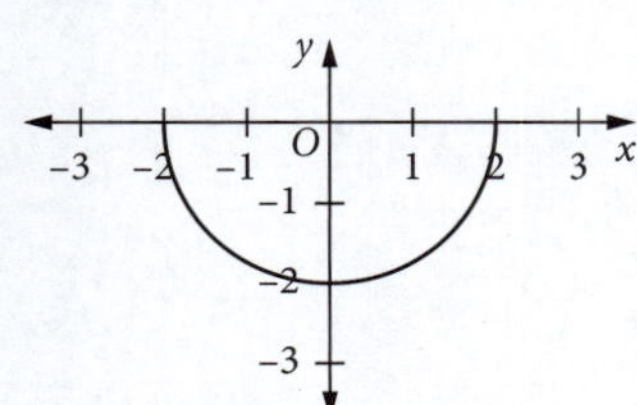

(b) $y=-\sqrt{36-x^2}$, $y\in[-6,0]$

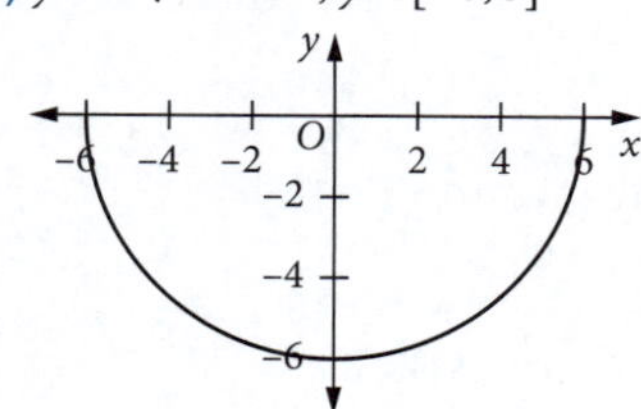

(c) $y=-\sqrt{5-x^2}$, $y\in\left[-\sqrt{5},0\right]$

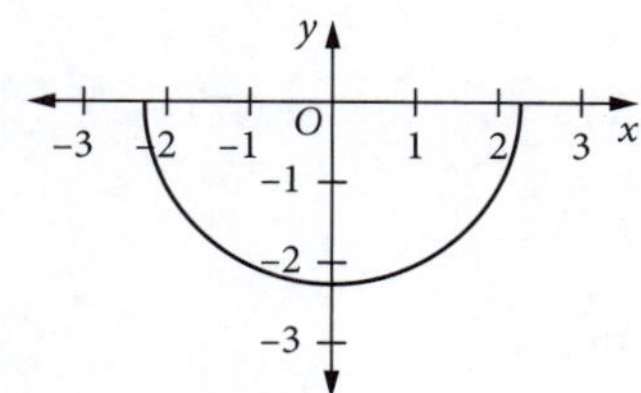

(d) $y=-\sqrt{\frac{25}{9}-x^2}$, $y\in\left[-\frac{5}{3},0\right]$

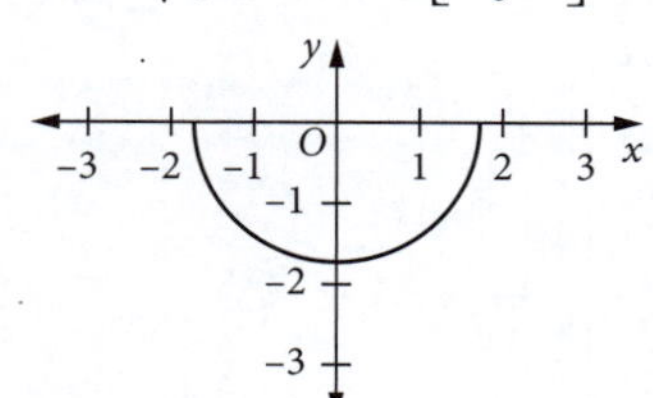

10 (a) $(-1,3)$
(b) $y=\sqrt{9-(x+1)^2}+3$
(c)
(d) Domain: $-4\le x\le 2$
Range: $3\le y\le 6$

11 (a) $(0,4)$
(b) $y=-\sqrt{16-x^2}+4$
(c)
(d) Domain: $-4\le x\le 4$
Range: $0\le y\le 4$

12 $(x-4)^2+(y-4)^2=16$ **13** inside **14** outside

15 (a) $x^2+y^2+2x-4y-20=0$ (b) $2\sqrt{21}$ units (c) 4 units

16 (a) 5 units (b) $2\sqrt{21}$ units

17 (a) $(2,1)$, $\sqrt{41}$ units (b) $2\sqrt{6}$ units

18 (a) $(2,5)$ (b) $x^2+y^2-4x-10y+16=0$ (c) $(0,2)$, $(0,8)$

19 (a) $(2,4)$, 5 units (b) $\sqrt{5}$ units, $4\sqrt{5}$ units

20 (a) (i) $r^2=(6-2)^2+(1-1)^2=16$
Equation is: $(x-2)^2+(y-1)^2=16$
(ii) $x^2-4x+4+y^2-2y+1=16$
$x^2+y^2-4x-2y-11=0$
(b) (i) $r^2=(-5+2)^2+(3+1)^2=3^2+4^2=25$
Equation is: $(x+2)^2+(y+1)^2=25$

(ii) $x^2 + 4x + 4 + y^2 + 2y + 1 = 25$
$x^2 + y^2 + 4x + 2y - 20 = 0$

(c) (i) $r^2 = (-3 - 0)^2 + (2 - 1)^2 = 3^2 + 1^2 = 10$
Equation is: $(x + 3)^2 + (y - 2)^2 = 10$

(ii) $x^2 + 6x + 9 + y^2 - 4y + 4 = 10$
$x^2 + y^2 + 6x - 4y + 3 = 0$

(d) (i) $r^2 = (3 - 1)^2 + (-4 + 2)^2 = 2^2 + 2^2 = 18$
Equation is: $(x - 3)^2 + (y + 4)^2 = 8$

(ii) $x^2 - 6x + 9 + y^2 + 8y + 16 = 8$
$x^2 + y^2 - 6x + 8y + 17 = 0$

EXERCISE 4.6

1 (a) $x = -2$ (b) $x = -1$ (c) $x = 0$ (d) $x = -0.74 \approx -0.7$

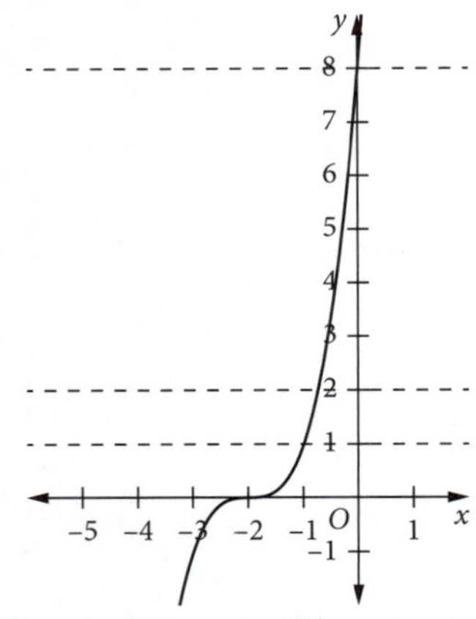

2 (a) $x + 2 = 0, x = -2$ (b) $x + 2 = 1, x = -1$
(c) $x + 2 = \sqrt[3]{8}, x + 2 = 2, x = 0$ (d) $x + 2 = \sqrt[3]{2}, x = \sqrt[3]{2} - 2$
(e) $x - 1 = \sqrt[3]{4}, x = 1 + \sqrt[3]{4}$
(f) $(x - 4)^3 = \frac{5}{3}, x - 4 = \sqrt[3]{\frac{5}{3}}, x = 4 + \frac{\sqrt[3]{45}}{3}$

3 (a) $x = -2, -1, 1$ (b) $x = -2.4, 0, 0.4$ (c) $x = 1.5$

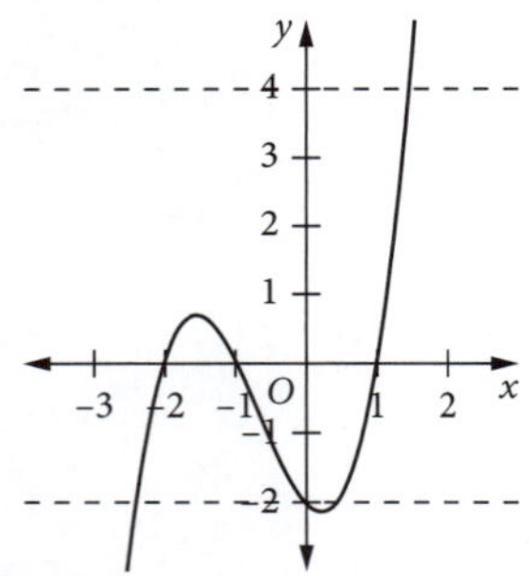

4 (a) $x = -2, -1, 1$ (b) $x^3 + 2x^2 - x - 2 = -2, x = 0, -\sqrt{2} - 1, \sqrt{2} - 1$

5 $-2.1 < c < 0.6$

6 (a) $x = 1, 2$ (b) $x = 0$ (c) $x = 3$ (d) $-0.1 < c < 0$ (e) -1

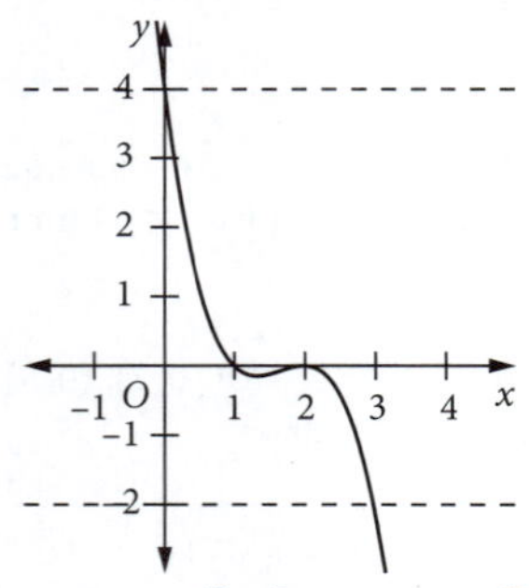

7 D **8** B **9** A

10 $y = a(x + 1)(x - 2)(x - 3), a = 1. y = (x + 1)(x - 2)(x - 3)$

11 $(1, 0): 0 = 1 \times (1 - a)(1 + b) + 4$
$-4 = 1 - a + b - ab$
$5 = a - b + ab$ [1]
$(-2, 12): 12 = (-2)(-2 - a)(-2 + b) + 4$
$8 = 2(a + 2)(b - 2)$
$4 = ab - 2a + 2b - 4$
$8 = -2a + 2b + ab$ [2]
[1] – [2]: $-3 = 3a - 3b$
$a = b - 1$ [3]
Substitute into [1]: $5 = -1 + b(b - 1)$
$b^2 - b - 6 = 0$
$(b - 3)(b + 2) = 0$
$b = 3, -2 \rightarrow a = 2, -3$
Solutions: $a = 2, b = 3$ or $a = -3, b = -2$

EXERCISE 4.7

1 Vertical asymptote is $x = 0$. Horizontal asymptote is $y = 0$.

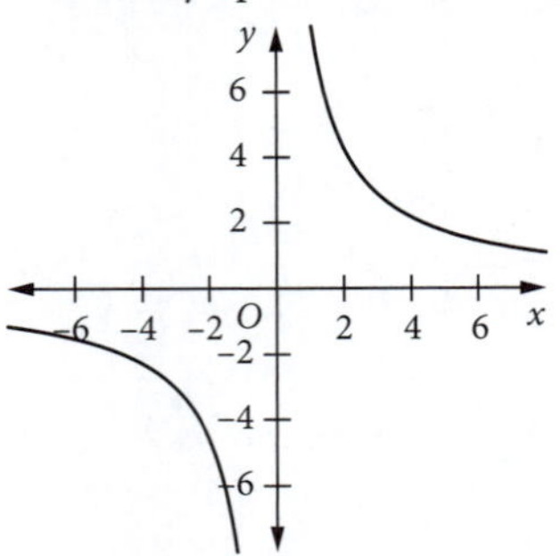

2 Vertical asymptote is $x = 0$. Horizontal asymptote is $y = 0$.

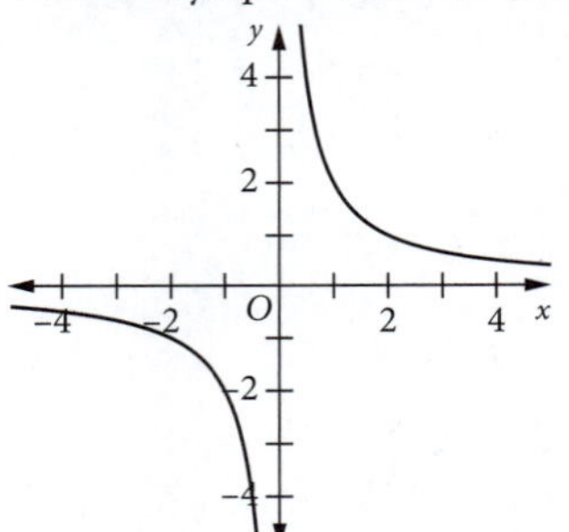

3 $y = 1 + \frac{2}{x}$. Vertical asymptote is $x = 0$. Horizontal asymptote is $y = 1$.
Domain is real $x, x \neq 0$. Range is real $y, y \neq 1$.

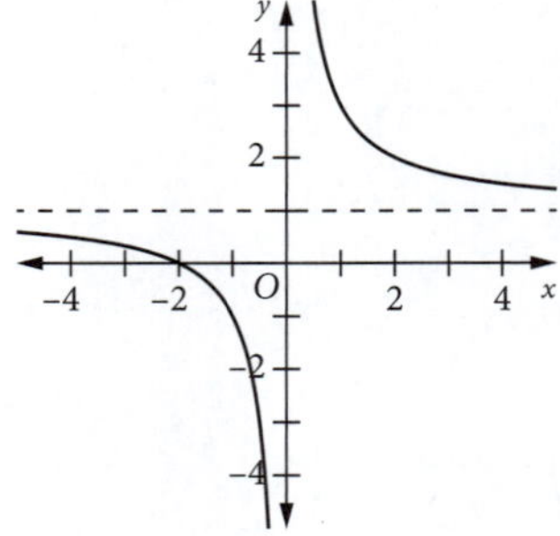

4 $y = 1 + \frac{2}{x-2}$. Vertical asymptote is $x = 2$. Horizontal asymptote is $y = 1$.
Domain is real x, $x \neq 2$. Range is real y, $y \neq 1$.

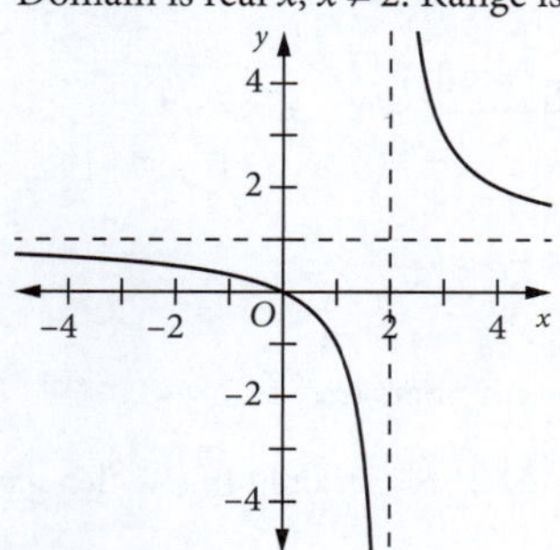

5 (a) $y = 2 + \frac{1}{x}$ solid curve. $y = 2 - \frac{1}{x}$ dotted curve.

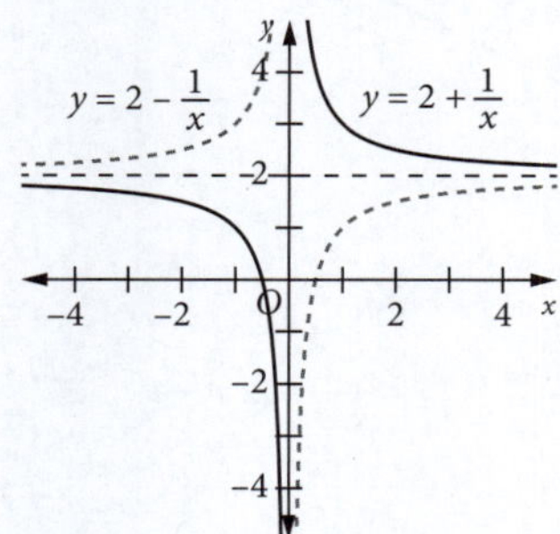

(b) No
(c) They have the same asymptotes but approach them from opposite sides.

6 (a) $V = 2$, $P = 15.28$, $PV = k$: $k = 2 \times 15.28 = 30.56$
(b) $V = 4$, $P = ?$: $P = \frac{30.56}{4} = 7.64$ atmospheres
(c) $P = 90$, $V = ?$: $V = \frac{30.56}{90} = 0.34$ litres

EXERCISE 4.8

1 B **2** D **3** A

4 (a) $f(x) + g(x) = x + 4 + x^2 - 6 = x^2 + x - 2$
Domain: Real x
Range is $[-2.25, \infty)$
(b) $f(x) - g(x) = x + 4 - (x^2 - 6) = -x^2 + x + 10$
Domain: Real x
Range is $(-\infty, 9.75]$

5 (a) $f(x) + g(x) = x^2 + 4 + x^3 - 3x^2 + 2x + 6 = x^3 - 2x^2 + 2x + 10$
Domain: Real x
Since $f(x) + g(x)$ is a cubic polynomial, range is the set of real numbers.
(b) $f(x) - g(x) = x^2 + 4 - (x^3 - 3x^2 + 2x + 6) = -x^3 + 4x^2 - 2x - 2$
Domain: Real x
Since $f(x) - g(x)$ is a cubic polynomial, range is the set of real numbers.

6 (a) $f(x) \cdot g(x) = x(x + 4) = x^2 + 4x$
Domain: Real x
Since $f(x) \cdot g(x)$ is a quadratic polynomial, the axis of symmetry: $x = -\frac{4}{2} = -2$.
$x = -2$: $f(x) \cdot g(x) = 4 - 8 = -4$. Range is $[-4, \infty)$

(b) $\frac{g(x)}{f(x)} = \frac{x+4}{x} = 1 + \frac{4}{x}$
Domain: Real x, $x \neq 0$.
As x gets larger, $x \to \infty$, $1 + \frac{4}{x}$ gets smaller and approaches 1 from above
As x gets smaller, $x \to -\infty$, $1 + \frac{4}{x}$ gets larger and approaches 1 from below
Range is the set of real numbers except 1.
Graphically this is:

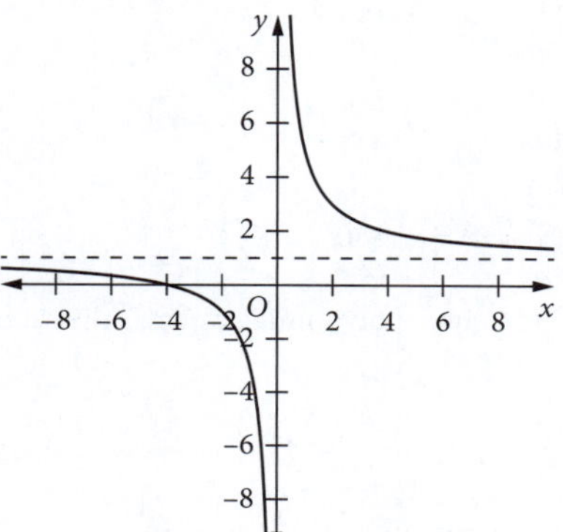

(c) $\frac{f(x)}{g(x)} = \frac{x}{x+4}$
Denominator is zero when $x = -4$. Domain: Real x, $x \neq -4$
$\frac{f(x)}{g(x)} = \frac{x}{x+4} = \frac{x+4-4}{x+4} = 1 - \frac{4}{x+4}$
This now looks like the reverse of (b):
As x gets larger, $x \to \infty$, $1 - \frac{4}{x+4}$ gets smaller and approaches 1 from below.

As x gets smaller, $x \to -\infty$, $1 - \frac{4}{x+4}$ gets smaller and approaches 1 from above.
Range is the set of real numbers except 1.
Graphically this is:

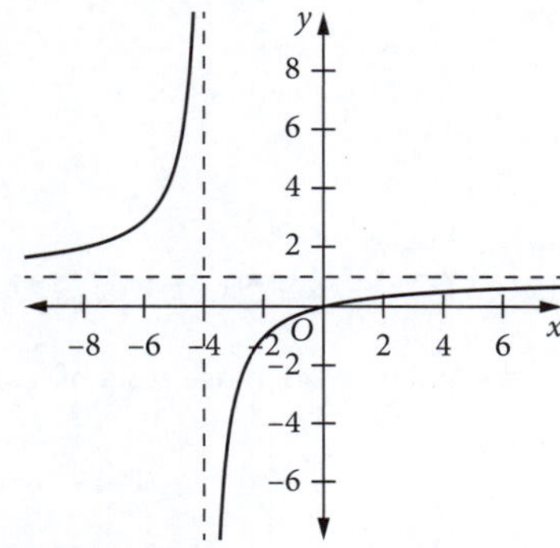

7 (a) $f(x) \cdot g(x) = (x - 1)(x + 1) = x^2 - 1$
Domain: Real x.
The least value of $f(x) \cdot g(x) = -1$ so the range is $[-1, \infty)$
(b) $\frac{f(x)}{g(x)} = \frac{x-1}{x+1} = \frac{x+1-2}{x+1} = 1 - \frac{2}{x+1}$
Denominator is zero when $x = -1$. Domain: Real x, $x \neq -1$

If $x > -1$, as x gets larger, $x \to \infty$, $1 - \frac{2}{x+1}$ gets larger and approaches 1 from below

If $x < -1$, as x gets smaller, $x \to -\infty$, $1 - \frac{2}{x+1}$ gets smaller and approaches 1 from above

Range is the set of real numbers except 1.

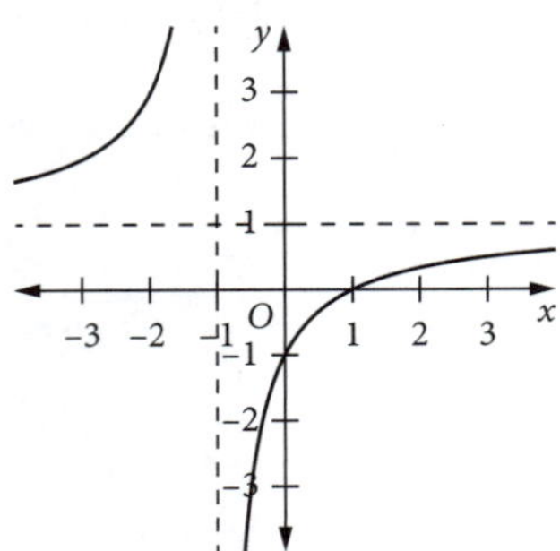

8 (a) $f(x) \cdot g(x) = x(x^2 + 4) = x^3 + 4x$

Domain: Real x

Since $f(x) \cdot g(x)$ is a cubic polynomial, range is the set of real numbers.

(b) $\frac{f(x)}{g(x)} = \frac{x}{x^2 + 4}$

$g(x) > 0$ for all x so domain of $\frac{f(x)}{g(x)}$ is real x.

Range is $[-0.25, 0.25]$

(c) $\frac{g(x)}{f(x)} = \frac{x^2 + 4}{x}$

$f(x) = 0$ for $x = 0$. Domain: Real x, $x \neq 0$.

$g(x) \geq 4$ for all x. $f(x) > 0$ for $x > 0$, $f(x) < 0$ for $x < 0$.

There is still not enough information to determine the range.

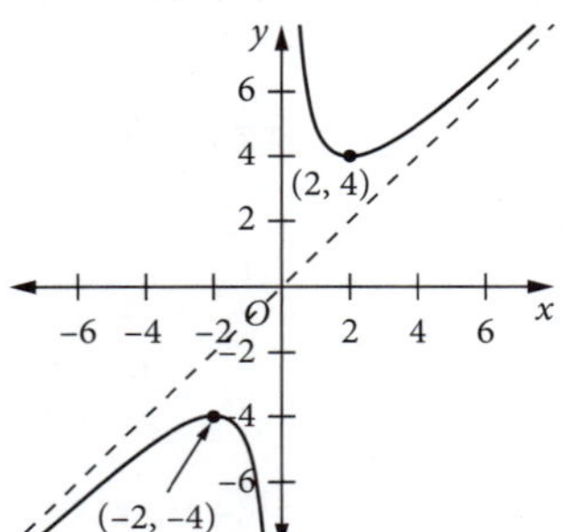

The range is $(-\infty, -4]$ and $[4, \infty)$

9 (a) $f(x) \cdot g(x) = (x + 6)(x^2 - 9) = x^3 + 6x^2 - 9x - 54$

Domain: Real x

Since $f(x) \cdot g(x)$ is a cubic polynomial, range is set of real numbers.

(b) $\frac{g(x)}{f(x)} = \frac{x^2 - 9}{x + 6}$

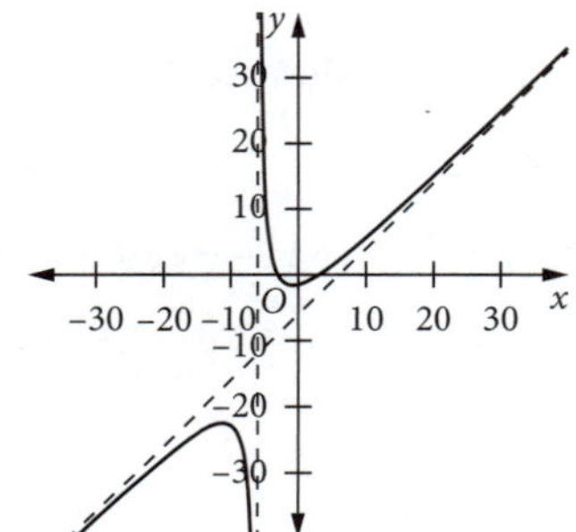

10 (a) $f(x) \cdot g(x) = (x + 4)(x^2 - 16) = x^3 + 4x^2 - 16x - 64$

Domain: Real x

Since $f(x) \cdot g(x)$ is a cubic polynomial, range is the set of real numbers.

(b) $\frac{g(x)}{f(x)} = \frac{x^2 - 16}{x + 4} = \frac{(x + 4)(x - 4)}{x + 4} = x - 4, x \neq -4.$

Domain: Real x, $x \neq -4$

$\frac{g(x)}{f(x)}$ is not defined at $x = -4$ so

it cannot take on the value $-4 - 4 = -8$.

The range is the set of real numbers except for -8.

Graphically, the function $\frac{g(x)}{f(x)}$ is a straight line with a gap at $(-4, -8)$.

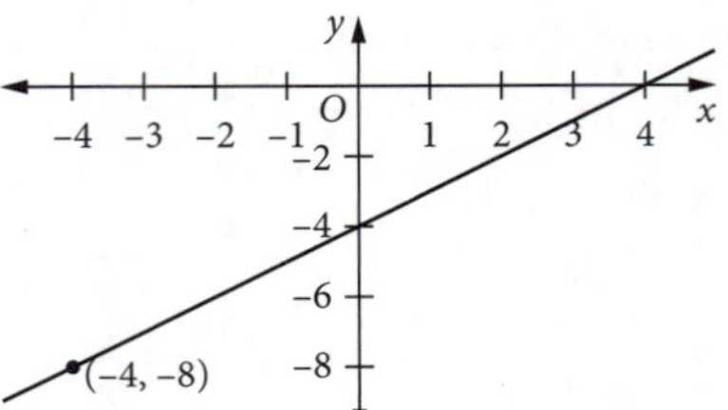

(c) $\frac{f(x)}{g(x)} = \frac{x + 4}{x^2 - 16} = \frac{x + 4}{(x + 4)(x - 4)} = \frac{1}{x - 4}, x \neq -4.$

Domain: Real x, $x \neq \pm 4$

When $x = -4$, $\frac{1}{x - 4} = \frac{-1}{8}$

$\frac{f(x)}{g(x)}$ cannot be zero or $\frac{-1}{8}$ so the range is the set of real numbers, $x \neq 0, \frac{-1}{8}$.

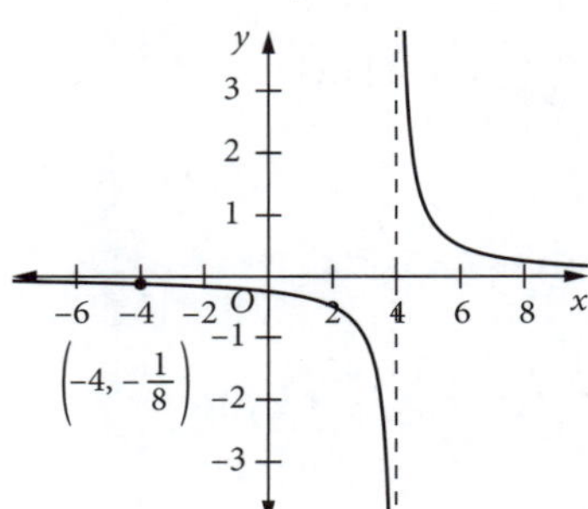

11 (a) $f(x) + g(x) = x + 3 + x^2 - 5 = x^2 + x - 2$

Domain: Real x

Range: axis of symmetry is $x = -\frac{1}{2}$, least value is $-2\frac{1}{4}$, range is real numbers $\geq -2\frac{1}{4}$

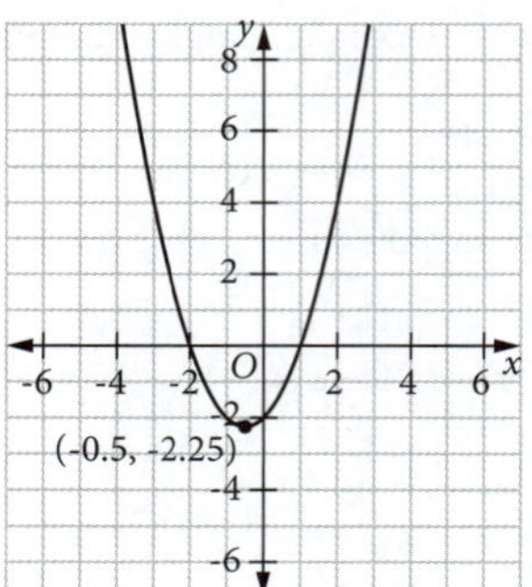

(b) $f(x) - g(x) = x + 3 - (x^2 - 5) = -x^2 + x + 8$

Domain: Real x

Range: axis of symmetry is $x = \frac{1}{2}$, greatest value is $8\frac{1}{4}$, range is real numbers $\le 8\frac{1}{4}$

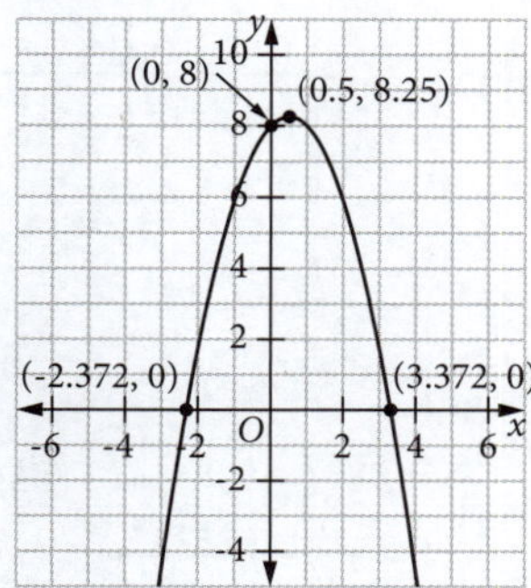

(c) $f(x) + h(x) = 3 + x + \sqrt{x-4}$

$\sqrt{x-4} \ge 0$ for $x \ge 4$. Domain is $[4, \infty)$

Where $x = 4$, $f(x) + h(x) = 7$

Where $x = 5$, $f(x) + h(x) = 9$

Where $x = 6$, $f(x) + h(x) = 9 + \sqrt{2} > 7$

It appears that the range is the set of real numbers greater than 7

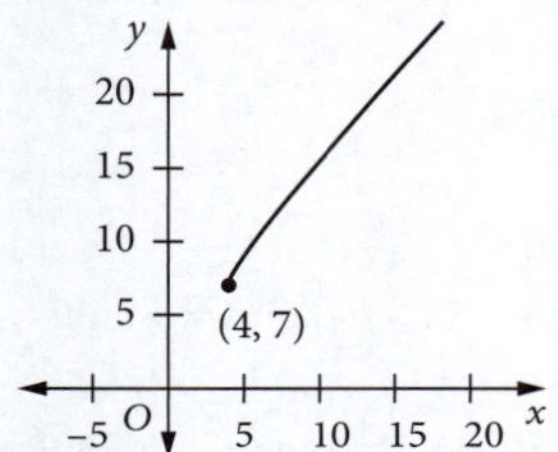

From the graph the range is $(-\infty, -0.75]$

(d) $k(x) - g(x) = x^3 + 2x^2 - 6 - (x^2 - 5) = x^3 + x^2 - 1$

Domain: Real x

Range: Real numbers

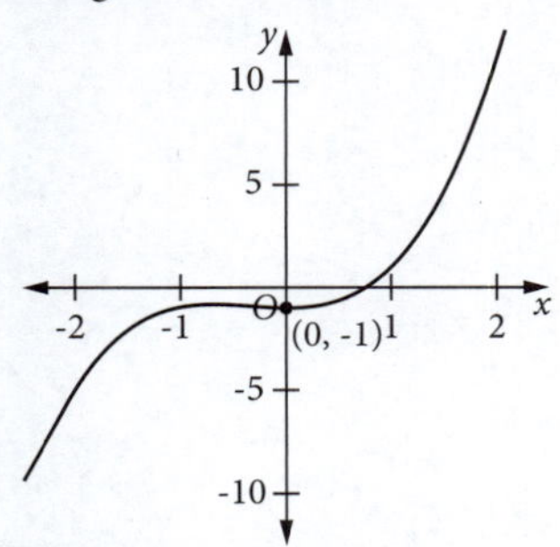

(e) $g(x) - h(x) = x^2 - 5 - \sqrt{x-4}$

$\sqrt{x-4} \ge 0$ for $x \ge 4$. Domain is $[4, \infty)$

$x^2 - 5 \ge -5$, $\sqrt{x-4} \ge 0$

$x = 4$, $g(x) - h(x) = 11$

$x = 5$, $g(x) - h(x) = 24$

It would appear that the range is $[11, \infty)$

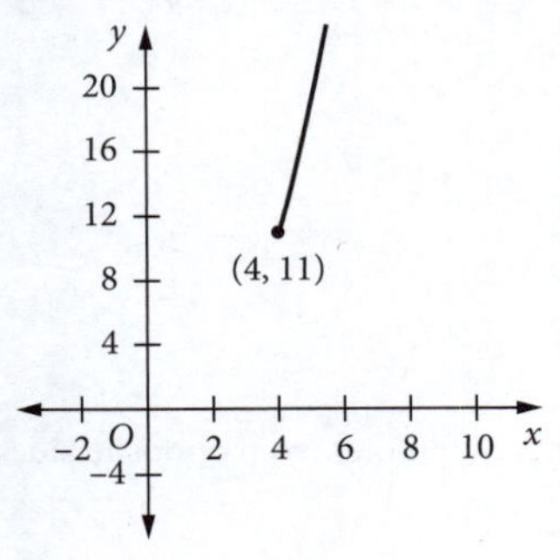

12 (a) $f(x) \cdot g(x) = (x + 3)(x^2 - 5)$

Both $f(x)$ and $g(x)$ are defined for all real x so the domain of $f(x) \cdot g(x)$ is all real x.

$f(x) \cdot g(x) = x^3 + 3x^2 - 5x - 15$ is a cubic polynomial so the range is the set of real numbers.

(b) $g(x) \cdot h(x) = (x^2 - 5)\sqrt{x-4}$

$\sqrt{x-4}$ is only defined for $x \ge 4$ so the domain of $g(x) \cdot h(x)$ is real $x \ge 4$.

Since for $x \ge 4$, $x^2 - 5 \ge 11$ and $\sqrt{x-4} \ge 0$, the product $(x^2 + 3)\sqrt{x+4} \ge 0$

The range is the set of non–negative real numbers, or $[0, \infty)$

(c) $\frac{f(x)}{g(x)} = \frac{x+3}{x^2-5}$

$x^2 - 5 = 0$ when $x = \pm\sqrt{5}$, the denominator is zero for these values. The domain of $\frac{f(x)}{g(x)}$ is real x, $x \ne \pm\sqrt{5}$.

For $-\sqrt{5} < x < \sqrt{5}$, $f(x) > 0$, $g(x) < 0$ so $\frac{f(x)}{g(x)} < 0$.

For $x > \sqrt{5}$, $f(x) > 0$, $g(x) > 0$ so $\frac{f(x)}{g(x)} > 0$.

For $-3 < x < -\sqrt{5}$, $f(x) > 0$, $g(x) > 0$ so $\frac{f(x)}{g(x)} > 0$.

$x = -3$, $f(x) = 0$ so $\frac{f(x)}{g(x)} = 0$.

For $x < -3$, $f(x) < 0$, $g(x) > 0$ so $\frac{f(x)}{g(x)} < 0$, but as $x \to -\infty$, $\frac{f(x)}{g(x)} \to 0$ from below.

It is safe to say that $\frac{f(x)}{g(x)} \ge 0$.

$x = 0$, $\frac{f(x)}{g(x)} = -0.6$ and it is safe to say that $\frac{f(x)}{g(x)} \le -0.6$.

To determine what happens to the range between -0.6 and 0 needs a graph.

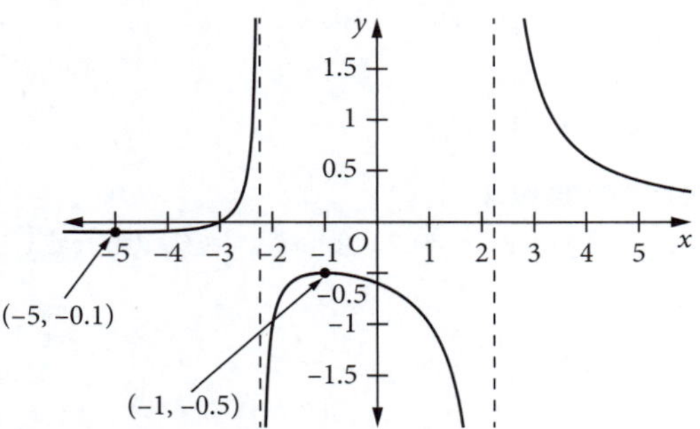

From the graph, the range is $\{-0.1, \infty)$ and $[-0.5, -\infty)$

(d) $\frac{k(x)}{f(x)} = \frac{x^3 + 2x^2 - 6}{x+3}$

$f(x) = 0$ when $x = -3$ so $\frac{k(x)}{f(x)}$ is undefined at $x = -3$.

The numerator exists for all values of x so the domain is real x, $x \ne -3$.

For the range, it looks as though the answer is the set of real numbers as when $x < -3$, $\frac{k(x)}{f(x)} > 0$, and when $x > -3$ then $\frac{k(x)}{f(x)}$ can be positive or negative $k(x) = 0$ for some value of x.

Once again, drawing a graph of the function can confirm these answers.

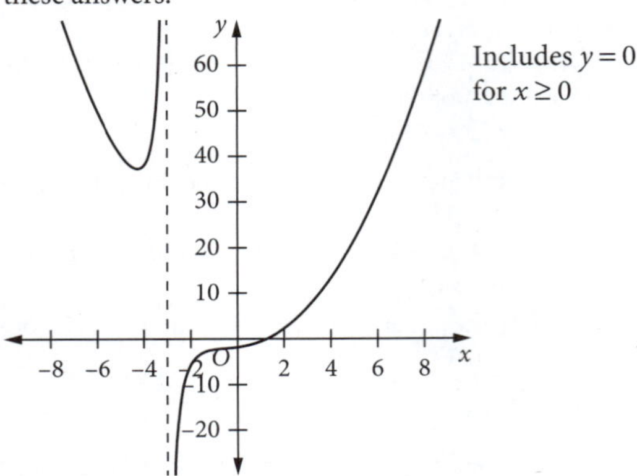

This verifies that the range is $(-\infty, \infty)$

(e) $\frac{h(x)}{f(x)} = \frac{\sqrt{x-4}}{x+3}$

$f(x) = 0$ when $x = -3$ so $\frac{h(x)}{f(x)}$ is undefined at $x = -3$.

The numerator is defined for $x \geq 4$.

The domain of $\frac{h(x)}{f(x)}$ is $x \geq 4$.

The numerator is always positive or zero, the denominator is always positive for $x \geq 4$. The denominator increases faster than the numerator as x increases so $\frac{h(x)}{f(x)}$ starts at zero, increases to some maximum value, then decreases to zero from above.
Range of the function is the set of real numbers [0, ?)

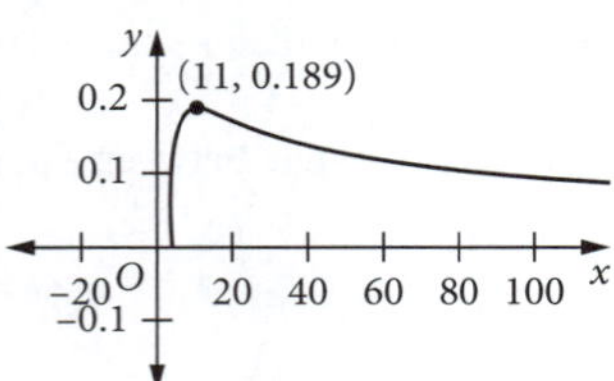

Range is [0, 0.189]

CHAPTER REVIEW 4

1 (a) $x \geq 1$ (b) real x, $x \neq \pm 2$ (c) $-5 \leq x \leq 5$ (d) real x

2 (a)

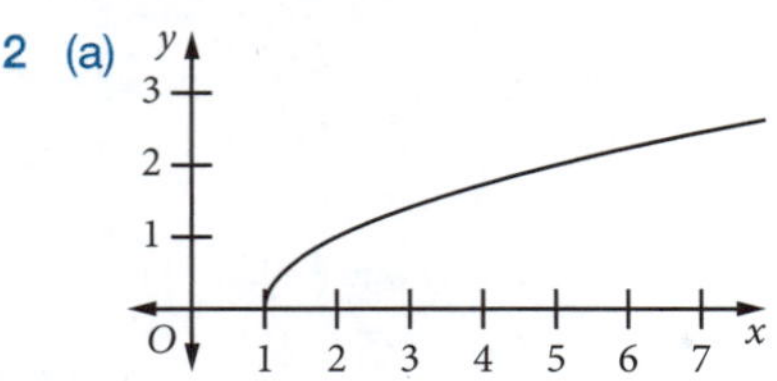

(b)

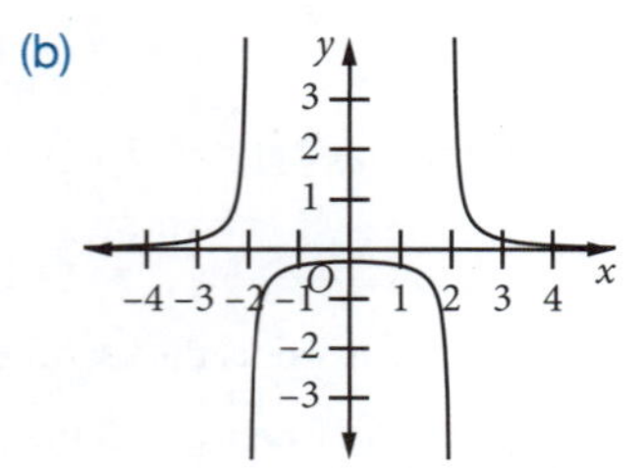

(c)

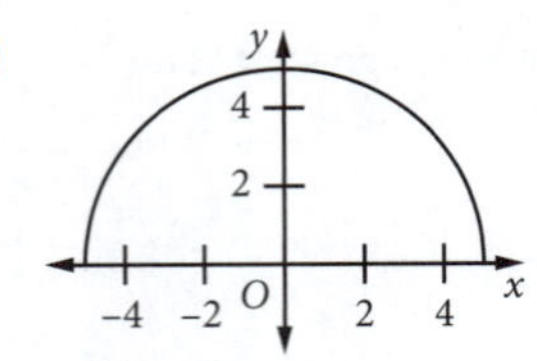

(d)

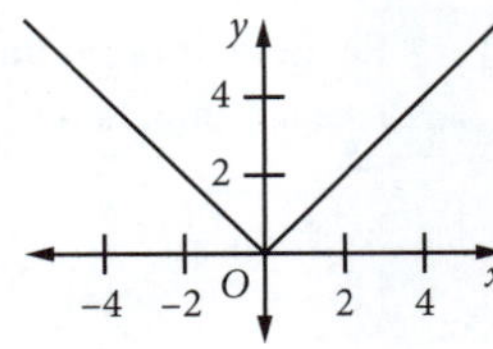

3 $g(x) = x^4 - x^2 + 1$
$g(-x) = (-x)^4 - (-x)^2 + 1$
$= x^4 - x^2 + 1$
$= g(x)$
Hence $g(x)$ is an even function.

4

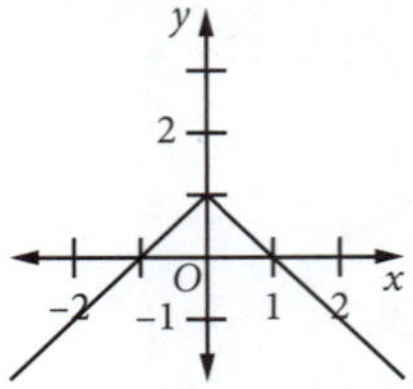

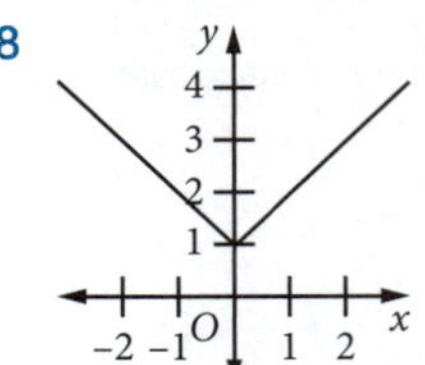

5

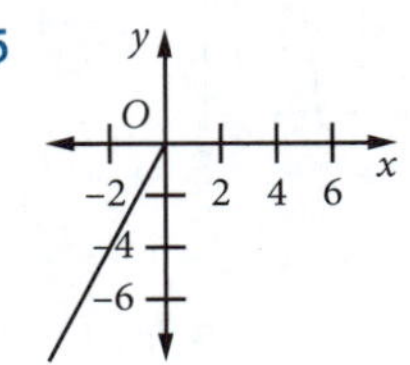

All real x

6 (a) $-3 \leq x \leq 2$ (b) real x, $x \neq 1$ 7 neither

8

9 (a) $(x-1)^2 + (y-1)^2 = 25$; centre (1, 1), radius = 5 units
(b) $\sqrt{45} = 3\sqrt{5}$ units
(c) $\sqrt{45} \approx 6.7 > 5$; distance from point to centre > radius, so point is outside circle
(d) $2\sqrt{5}$ units

10 $x^2 + (x-4)^2 = 8$, $2x^2 - 8x + 8 = 0$, $(x-2)^2 = 0$, $x = 2$; line touches circle at $(2, -2)$

11 (a) $x = -18, 4$ (b) $x \leq -\frac{1}{3}, x \geq 3$

12 B

13 (a) $x - 3 = -2$, $x = 1$
(b) $x + 5 = \sqrt[3]{4}$, $x = -5 + \sqrt[3]{4}$
(c) $x - 2 = \sqrt[3]{81}$, $x = 2 + 3\sqrt[3]{3}$

14 $y = a(x+3)(x+2)(x-1)$
$x = 0, y = -12$: $-12 = a \times 3 \times 2 \times (-1)$
$a = 2$
$y = 2a(x+3)(x+2)(x-1)$

15 C

16 (a)

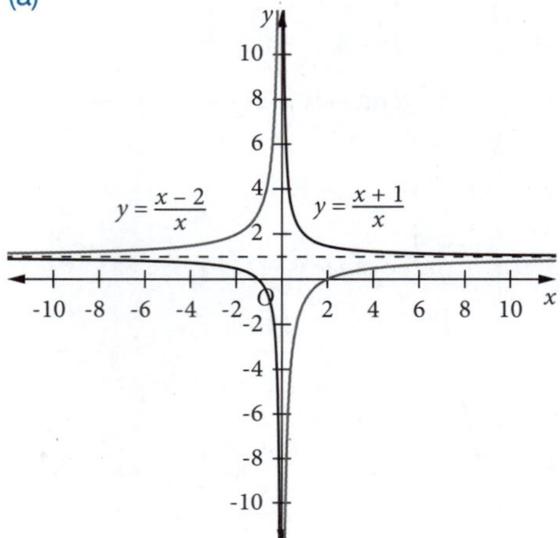

(b) Yes, $x = 0$ and $y = 1$
(c) No. They approach the asymptotes from opposite sides.

CHAPTER 5

EXERCISE 5.1

1 (a) −1 (b) −3.5 (c) 0 (d) −1 (e) −1 (f) $\frac{5c}{b}$ (g) $\frac{-2}{3}$ (h) 1.5

2 (a) $\tan\theta = 1$, $\theta = 45°$ (b) 26° 34′ (c) 159° 27′ (d) 56° 19′ (e) 153° 26′ (f) 135°

3 A 4–6 Answers will vary.

7 (a) incorrect (b) correct (c) incorrect (d) correct

8–10 Answers will vary.

EXERCISE 5.2

1 (a) $3x - 4y + 38 = 0$ (b) $x + 2y + 2 = 0$

2 (a) $8x - 7y - 3 = 0$ (b) $2x - y - 12 = 0$

3 (a) $x - y = 0$ (b) $m = \frac{4}{3}$, $4x - 3y + 17 = 0$ (c) $x + y + 7 = 0$ (d) $3x + 4y + 5 = 0$

4 $y = 2$ 5 $x = -2$ 6 B 7 $2x + 3y + 6 = 0$

8 (a) $y = \frac{-2}{3}x + \frac{4}{3}, \frac{-2}{3}$ (b) $y = \frac{3}{2}x - \frac{7}{2}$, 1.5 (c) $y = \frac{-3}{2}x + 3$, −1.5 (d) $y = \frac{2}{5}x + \frac{8}{5}$, 0.4

9 (a) correct (b) correct (c) incorrect (d) correct

10 (a) Answers will vary. (b) 3

11 (a) $3x + 2y = 0$ (b) $2x - 3y - 13 = 0$

12 (a) $4x - 5y = 0$ (b) $5x + 4y = 0$

13 (a) $2x - 3y - 6 = 0$ (b) 33° 41′ (c) 2 (d) $3x + 2y - 22 = 0$

14 Answers will vary.

15 (a) $2x - 5y + 31 = 0$ (b) $x + 2y - 7 = 0$ (c) $(-3, 5)$

16 (a) $x + 2y - 9 = 0$ (b) $2x - y + 2 = 0$ (c) −2 (d) $2x - y - 8 = 0$ (e) $x + 2y + 6 = 0$ (f) $(2, -4)$

17 (a) $x + 5y = 0$ (b) $2x + y = 0$ (c) $x - 4y + 9 = 0$

EXERCISE 5.3

1 C 2 $4x - y - 5 = 0$

3 (a) $m_{BC} = 6$; $y - 3 = \frac{-1}{6}(x + 1)$, $x + 6y - 17 = 0$ (b) $x + 6y - 32 = 0$

4 (a) $m_1 = -1 = m_2$; $m_3 = \frac{2}{3} = m_4$; first two lines are parallel, last two lines are parallel, no lines are coincident, so lines are sides of a parallelogram. (b) $\left(\frac{-4}{5}, \frac{-1}{5}\right)$, $\left(-1\frac{4}{5}, \frac{4}{5}\right)$, $(1, 1)$, $(0, 2)$ (c) $x - 14y + 13 = 0$, $11x - 4y + 8 = 0$

5 (a) $4x - 3y + 1 = 0$ (b) $2x - y - 1 = 0$ (c) $x = 2$

6 (a) correct (b) correct (c) incorrect (d) correct

7 (a) $(-9, -7)$ (b) $(-2, 0)$

8 (a) Answers will vary. (b) $(4, 3)$

9 (a) coincide (b) intersect (c) intersect (d) parallel

EXERCISE 5.4

1 $x = -2, y = 1$ 2 $x = 14, y = 4$ 3 $x = 5, y = -2$
4 $x = 0.5, y = -3.5$ 5 $m = -2, n = 0$ 6 $x = 5, y = 2$
7 $x = 2, y = -3$ 8 $x = 1, y = 2$ 9 C
10 $x = 0.5, y = 3$ 11 $x = 5.5, y = 1$ 12 $m = 4, n = 1\frac{1}{3}$
13 $x = 1, y = 2$ 14 $x = 8, y = -1$ 15 $x = 9, y = 3$
16 $x = 2, y = -4$ 17 $a = 8, b = 3$ 18 $x = 7, y = 2.5$
19 $x = -8, y = 3$ 20 $a = 5, b = 3$ 21 $x = 7, y = 4$

EXERCISE 5.5

1 250, 200 2 6, 2 3 C 4 6, 30
5 600, 400 6 2, 4 7 $\frac{7}{12}$ 8 6, 12 9 \$880, \$360

10 (a) $y = 2x - 7$ (b) $y = -4x + 4$ (c) $y = \frac{3}{4}x - 3$ (d) $y = -\frac{x}{2} + 5$

11 (a) (i) $5x + 200 = 15x$, $10x = 200$, $x = 20$ (ii) $\$R = \$300$ (iii) $P = 15x - (5x + 200) = 10x - 200$
(b) (i) $0.5x + 100 = 1.5x$, $x = 100$ (ii) $\$R = \$150$ (iii) $P = 1.5x - (0.5x + 100) = x - 100$
(c) (i) $15x + 3000 = 45x$, $30x = 3000$, $x = 100$ (ii) $\$R = \$4500$ (iii) $P = 45x - (15x + 3000) = 30x - 3000$
(d) (i) $0.3x + 5000 = 1.1x$, $0.8x = 5000$, $x = 6250$ (ii) $\$R = \$6875$ (iii) $P = 1.1x - (0.3x + 5000) = 0.8x - 5000$

12 (a) $C = 2x + 80$, $R = 4x$ (b) $2x + 80 = 4x$, $2x = 80$, $x = 40$ cups per day (c) $P = 4x - (2x + 80) = 2x - 80$ (d) $P = 200 - 80 = \$120$

13 (a) $C = 4x + 90$, $R = 10x$ (b) $4x + 90 = 10x$, $6x = 90$, $x = 15$ (c) $P = R - C = 10x - (4x + 90) = 6x - 90$; $x = 40$, $P = 240 - 90 = \$150$
(d) Saturday, $x = 30$. $P = 180 - 90 = \$90$
Sunday, $x = 10$. $P = 60 - 90 = -\$30$
Profit for the weekend = \$90 − \$30 = \$60

EXERCISE 5.6

1 $x = -1, y = 1; x = 6, y = 36$ 2 $x = 1, y = 1; x = 2, y = 4$
3 $x = -1, y = 4; x = 5, y = 10$ 4 $x = -2, y = 17; x = 7, y = 8$
5 $x = -2, y = -5; x = 5, y = 2$ 6 $x = 1, y = 3; x = -1.8, y = -2.6$
7 $x = 2, y = 3; x = 3, y = 2$ 8 $x = 0, y = -2; x = 3, y = 4$
9 $x = 2, y = 1; x = -1, y = -2$ 10 $x = 3, y = 1$
11 $x = 2, y = -2; x = 4, y = 2$ 12 $x = 3, y = 2; x = -18, y = 23$
13 $x = 1, y = 1; x = 0.4, y = -0.2$ 14 $x = -2, y = 0$
15 $x = 2, y = 5; x = -5, y = -2$ 16 $x = 3, y = 2; x = -1.4, y = 15.2$
17 $x = 1, y = 1$ 18 $x = 1, y = 1; x = -0.2, y = 0.4$
19 $x = -4, y = -2$ 20 $x = 5, y = 14; x = -3, y = 6$

EXERCISE 5.7

1 $x = 4, y = 1; x = -1, y = -1$ 2 $x = 3, y = 4; x = -3, y = -4$
3 $x = 1, y = 1; x = 5, y = -1$ 4 $x = 4, y = 5; x = -2\frac{6}{7}, y = -5\frac{2}{7}$
5 $x = 3, y = -4; x = -3\frac{2}{3}, y = 6$ 6 $x = 5, y = 3; x = -7, y = -5$

EXERCISE 5.8

1 C

2 (a) $2(x - 1)^2 - 2$; min = −2; range $y \geq -2$
(b) $-2(x - 2)^2 + 5$; max = 5; range $y \leq 5$
(c) $-4(x - 2)^2 + 23$; max = 23; range $y \leq 23$
(d) $4(x + 1)^2 - 11$; min = −11; range $y \geq -11$
(e) $-2x^2 + 8$; max = 8; range $y \leq 8$
(f) $-(x + 1)^2 + 8$; max = 8; range $y \leq 8$
(g) $2(x - 1.5)^2 - 4.5$; min = −4.5; range $y \geq -4.5$
(h) $-5(x + 1)^2 + 11$; max = 11; range $y \leq 11$

3 D 4 (a) $0 \leq t \leq 4$ (b) 20 m 5 12.5 m, 50 m

6 10, 10; 100 7 225 cm^2 8 50 m^2; 5 m, 10 m

9 $2\frac{4}{7}$ m, $3\frac{3}{7}$ m 10 30 000 m^2 11 $y = 10x - x^2$; $x = 5, y = 25$

12 area $= 1 - \frac{x^2}{2} - \frac{1 - x}{2} = \frac{1 + x - x^2}{2}$; $y = -\frac{1}{2}\left(x - \frac{1}{2}\right)^2 + \frac{5}{8}$; 0.625 units^2

EXERCISE 5.9

1 D

2 (a) 28, two (b) −23, none (c) 0, one (d) −8, none (e) 65, two

3 A

4 (a) yes (b) no (c) touches (d) yes (e) no (f) no

5 (a) 64; 3, −5 (b) 101; $\frac{9 \pm \sqrt{101}}{2}$ (c) 49; $\frac{3}{4}, 1\frac{1}{3}$ (d) 0; 1.5 (e) 1764; ±3 (f) 36; 0, 6 (g) 0; 1 (h) −368; none (i) 201; $\frac{1 \pm \sqrt{201}}{4}$ (j) 28; $\frac{1 \pm \sqrt{7}}{3}$ (k) 85; $\frac{7 \pm \sqrt{85}}{2}$ (l) 1; −5, −6

(m) 17; $\frac{9 \pm \sqrt{17}}{8}$ (n) 0; $-1\frac{1}{3}$ (o) 76; $\frac{-2 \pm \sqrt{19}}{3}$
(p) 33; $\frac{-1 \pm \sqrt{33}}{4}$ (q) 1; 7, 8 (r) 9; 0, 1.5 (s) 17; $\frac{-5 \pm \sqrt{17}}{4}$
(t) 9; 1, 0.4 (u) −39; none

EXERCISE 5.10

1 (a) (i) $k = 2.25$ (ii) $k < 2.25$ (b) (i) $k = \pm 2\sqrt{3}$ (ii) $k > 2\sqrt{3}$ or $k < -2\sqrt{3}$ (c) (i) $k = -1, -5$ (ii) $k < -5$ or $k > -1$
(d) (i) $k = \frac{1}{3}, 3$ (ii) $\frac{1}{3} < k < 3$
2 A 3 $m = 1$ or -3
4 (a) $-\sqrt{2} \le m \le \sqrt{2}$ (b) $\frac{1}{3} \le m \le 3$ (c) $m \le -4$ or $m \ge 6$
5 (a) $m = \pm\sqrt{3}$ (b) $m < -\sqrt{3}, m > \sqrt{3}$
6 $\Delta_1 = 16(m-1)^2 - 4 \times 4(m+1) \times (-3) = 16(m^2 + m + 4)$
For $m^2 + m + 4$: $\Delta_2 = 1 - 16 < 0$
$\therefore m^2 + m + 4 > 0$ for all real m.
Hence $\Delta_1 > 0$ for all real m and the original equation has real roots for all real m.
7 $\Delta = 9m^4 - 4(3m-5)(5m^2) = m^2(3m-10)^2 > 0$ for real m, $m \ne \frac{5}{3}$
Because Δ is a perfect square, the roots of the equation are rational if m is rational, $m \ne \frac{5}{3}$.
8 $\Delta = (2a+b)^2 - 4ab = 4a^2 + b^2 \ge 0$ for all real values of a and b, $\therefore$ equation has real roots.
9 (a) $p = 2$ (b) $p < 2$
10 (a) $m = 3$ or 5 (b) $m > 5$ or $m < 3$
11 $\Delta = (a+b)^2 - 4ab = a^2 - 2ab + b^2 = (a-b)^2 \ge 0$; roots real.
Because Δ is a perfect square, the roots of the equation are rational for all rational values of a and b.
12 $-2 < m < -\frac{2}{3}$

EXERCISE 5.11

1 (a) (2, 6), (−2, 18) (b) (0, 2), (2, −12) (c) (1, 2), (−5, 8)
(d) (2, 1), (−1.5, −2.5)
2 (a) (0, −3), (3, 0) (b) (2, 3), (3, 5) (c) (1, 1)
3 A 4 $c = \pm 2\sqrt{2}$ at $(\pm\sqrt{2}, \mp\sqrt{2})$ 5 0, $-1\frac{1}{3}$ 6 2, −18
7 (a) 4 or −8 (b) $m > 4$ or $m < -8$ (c) $-8 < m < 4$
8 (a) 1 or 9 (b) $m > 9$ or $m < 1$ (c) $1 < m < 9$
9 (a) −5 or 3 (b) $m > 3$ or $m < -5$ (c) $-5 < m < 3$
10 $a > 0$ or $a < -\frac{8}{9}$ 11 $-3 < a < 0$
12 $y = 2x + 1$; $y = -2x + 5$
13 $2x^2 - 6x + 7 = x^2 - 2x + 3$; $x^2 - 4x + 4 = 0$, $(x-2)^2 = 0$;
$\Delta = 0$, touches $x = 2$; point is (2, 3)

CHAPTER REVIEW 5

1 (a) $-1, \frac{2}{3}, -3.5, 0.25, -6, -1$ (b) $AB \parallel CD$ (c) $\sqrt{2}, 5\sqrt{2}, \sqrt{53}$
(d) (2, 1.5), (0, 3.5), (2.5, 1)
2 (a) $M(4, 1)$, $N(-1, -1)$
(b) gradient of $MN = 0.4$, gradient of $QR = 0.4$, $\therefore MN \parallel QR$
(c) $QR = 2\sqrt{29}$ (d) $MN = \sqrt{29}$, $\therefore MN = \frac{QR}{2}$
3 3.1 units 4 midpoint (4, 3), line $4x - 3y - 7 = 0$
5 $A(2, -3)$, $B(5, 2)$, $C(-3, 0)$; gradient of $CA = -0.6$, gradient of $BC = 0.25$, gradient of $AB = \frac{5}{3}$;
Grad $CA \times$ Grad $AB = \frac{-3}{5} \times \frac{5}{3} = -1$
$\therefore AB \perp CA$, so triangle is right-angled at A
6 $x = 7$, $y = 2.5$
7 (a) 9 (b) −13.5
8 $x = -6, y = 15$; $x = 1, y = 8$

9 (a) −1, 2, $\frac{1 \pm \sqrt{13}}{2}$ (b) ±3 (c) 1, 2
10 (a) two (b) two (c) one (d) two
11 $\Delta = (m+n)^2 - 4mn = m^2 + 2mn + n^2 - 4mn = m^2 - 2mn + n^2 = (m-n)^2$; a perfect square, so roots are rational
12 (a) $k < 7\frac{1}{4}$ (b) $k = 7\frac{1}{4}$ (c) $k > 7\frac{1}{4}$
13 $a = 2, b = 3$ 14 (2, 1), (4, 5)
15 (a) $m = 1$ or -11 (b) $m < -11$ or $m > 1$ (c) $-11 < m < 1$
16 (a) $R = 20x$
(b) $C = 10x + 10\,000$
(c) $20x = 10x + 10\,000$, $10x = 10\,000$, $x = 1000$. Break-even point is 1000 units.
(d) $P = R - C = 10x - 10\,000$
$x = 3000$, $P = \$20\,000$
17 $x + y = 24$, $y = 24 - x$. Product $= xy = x(24 - x)$
Greatest product occurs when $x = 12$. Greatest product = 144.
18 Let the dimensions of each enclosure be x m by y m as shown in the diagram.

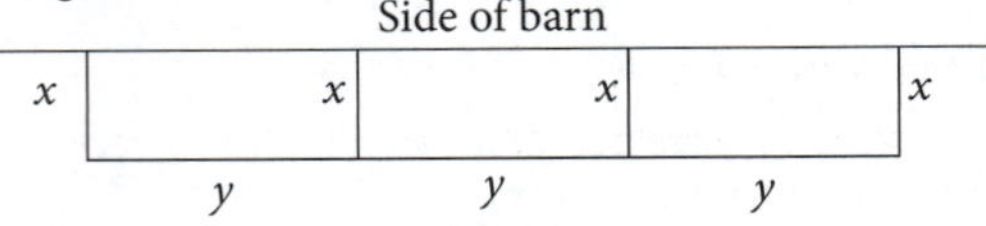

Fencing: $4x + 3y = 24$, $y = \frac{24 - 4x}{3}$
Total area $= 3xy = x(24 - 4x) = 4x(6 - x)$
Greatest area when $x = 3$, $y = 4$
Greatest area = 36 m^2.

CHAPTER 6

EXERCISE 6.1

1 (a) $\frac{\pi}{6}$ (b) $\frac{5\pi}{4}$ (c) $\frac{2\pi}{5}$ (d) $\frac{7\pi}{6}$ (e) $\frac{7\pi}{4}$ (f) $\frac{5\pi}{8}$
(g) $\frac{11\pi}{6}$ (h) $\frac{4\pi}{5}$
2 (a) 135° (b) 157.5° (c) 216° (d) 270° (e) 100° (f) 165°
(g) 324° (h) 247.5°
3 (a) 28° 39′ (b) 104° 17′ (c) 171° 53′ (d) 244° 5′
(e) 155° 51′ (f) 196° 18′ (g) 300° 14′ (h) 273° 59′
4 (a) 0.7330 (b) 1.2915 (c) 1.8326 (d) 2.8623 (e) 0.8404
(f) 3.8397 (g) 2.4120 (h) 1.2590
5 (a) incorrect (b) correct (c) correct (d) incorrect
6 (a) $\frac{2\pi}{3}$ (b) $\frac{7\pi}{4}$ (c) $\frac{7\pi}{6}$ (d) $\frac{3\pi}{4}$ (e) $\frac{13\pi}{6}$ (f) $\frac{9\pi}{4}$
(g) $\frac{8\pi}{3}$ (h) 4π (i) $-\frac{\pi}{6}$ (j) $-\frac{3\pi}{4}$ (k) $\frac{5\pi}{2}$ (l) $-\pi$ (m) $\frac{\pi}{12}$
(n) $\frac{\pi}{8}$ (o) $\frac{23\pi}{12}$ (p) $-\frac{3\pi}{8}$
7 (a) 150° (b) 300° (c) 225° (d) 540° (e) 405° (f) 630°
(g) 600° (h) −180° (i) −270° (j) −210° (k) −135° (l) −360°
(m) 22.5° (n) 75° (o) 202.5° (p) −105°

EXERCISE 6.2

1 107° 26′ 2 $15 \times 70\pi \div 180 = 18.3$ cm 3 C
4 (a) $\angle AOB = \frac{2 \times 240}{400}$ rad $= 1.2$ rad (b) 24 cm
(c) $AB^2 = 800 - 800\cos 1.2$, $AB = 22.6$ cm
5 30.6 cm
6 $\cos\angle POA = \frac{5}{8}$, $\angle AOB = 1.79$ rad
major arc $= 10\pi - 5 \times 1.79 = 22.5$ cm
7 (a) $\cos\angle AOB = -0.28$, $\angle AOB = 106°\,16'$ (b) 18.5 cm (c) 44.7 cm^2
8 16 min = 96° (a) 33.5 cm (b) $\frac{20 \sin 96°}{\sin 42°} = 29.7$ cm
9 (a) correct (b) incorrect (c) correct (d) correct

10 area $= \frac{1}{2} \times 9 \times \frac{11\pi}{6} = \frac{33\pi}{4}$ cm^2

11 arc $= 16 \times \frac{140\pi}{180} = 39.1$, chord $= \frac{16 \sin 140°}{\sin 20°} = 30.1$, difference $= 9$ cm

12 (a) $\frac{125\pi}{8}$ cm^2 (b) $\left(10 + \frac{25\pi}{4}\right)$cm

13 6πcm^2

14 small sector $= \frac{\pi r^2}{6}$, large sector $= \frac{5\pi r^2}{6}$, triangle $= \frac{\sqrt{3}r^2}{4}$; ratio is $(2\pi - 3\sqrt{3}):(10\pi + 3\sqrt{3})$

15 $\cos\theta = 0.5$, $\theta = \frac{\pi}{3}$

major sector $= \frac{1}{2} \times 20^2 \times \frac{4\pi}{3} = \frac{800\pi}{3}$ m^2,

triangle $= \frac{1}{2} \times 20^2 \times \sin\frac{2\pi}{3} = 100\sqrt{3}$ m^2

area grazed $= \frac{800\pi}{3} + 100\sqrt{3} = 1011$ m^2

16 circumference $= \frac{80\pi}{9} = 27.9$ cm

17 (a) 63.1 m (b) 163.6 m^2

18 (a) $\frac{25}{180} \times \pi \times 40 = 17.5$ cm (b) $\frac{40 \sin 25°}{\sin 77.5°} = 17.3$ cm

19 area $= 2 \times \frac{1}{2} \times 100(1.287 - \sin 1.287) = 32.7$ cm^2

20 triangle area $= \frac{1}{2} \times 100 \times \frac{\sqrt{3}}{2} = 25\sqrt{3}$,

sector area $= \frac{1}{2} \times 25 \times \frac{\pi}{3} = \frac{25\pi}{2}$

area $= 25\left(\sqrt{3} - \frac{\pi}{2}\right) = 4.0$ cm^2

21 (a) $\cos\angle LKO = \frac{3r^2 + r^2 - r^2}{2\sqrt{3}r^2} = \frac{\sqrt{3}}{2}$: $\angle LKO = 30°$ (b) $\frac{3\sqrt{3}r^2}{4}$

(c) $LM = r\sqrt{3}$ (d) $\frac{4\pi r^2 - 3\sqrt{3}r^2}{12}$

(e) area $= \frac{4\pi r^2 - 3\sqrt{3}r^2}{12} - \left(\frac{\pi r^2}{2} - \frac{3\sqrt{3}r^2}{4}\right) = \frac{3\sqrt{3}r^2 - \pi r^2}{6}$

EXERCISE 6.3

1 (a) $\sin x$ (b) $\sin x$ (c) $-\tan x$ (d) $-\cos x$ (e) $-\sin x$ (f) $\tan x$

2 A

3 (a) incorrect (b) correct (c) incorrect (d) correct

4 (a) 0.2 (b) −0.2 (c) −0.2 (d) 0.2 (e) −0.2 (f) 5

5 (a) $\frac{1}{t}$ (b) t (c) $-t$ (d) $-t$ (e) $-\frac{1}{t}$ (f) t

6 (a) $\frac{1}{c}$ (b) c (c) $-c$ (d) c (e) $\frac{1}{c}$ (f) $-c$

7 (a) 1 (b) $-\frac{1}{2}$ (c) $-\frac{1}{\sqrt{3}}$ (d) −1 (e) $-\sqrt{2}$ (f) $-\sqrt{3}$ (g) 1 (h) $\frac{\sqrt{3}}{2}$

8 (a) 0 (b) $-\frac{\sqrt{3}}{2}$ (c) 1 (d) $\frac{1}{\sqrt{3}}$ (e) $-\frac{1}{2}$ (f) $-\sqrt{2}$ (g) 0 (h) $-\frac{1}{2}$

9 (a) −1 (b) $-\sqrt{3}$ (c) −2 (d) −1 (e) −1 (f) $\frac{\sqrt{3}}{2}$ (g) $-\frac{\sqrt{3}}{2}$ (h) $-\frac{2}{\sqrt{3}}$

10 (a) 0 (b) $\frac{1}{2}$ (c) 1 (d) $\frac{1}{\sqrt{3}}$ (e) $\frac{2}{\sqrt{3}}$ (f) 0 (g) $\frac{\sqrt{3}}{2}$ (h) −1

11 (a) $\frac{\pi}{2} < x < \pi$ (b) $\frac{\pi}{2} < x < \pi$

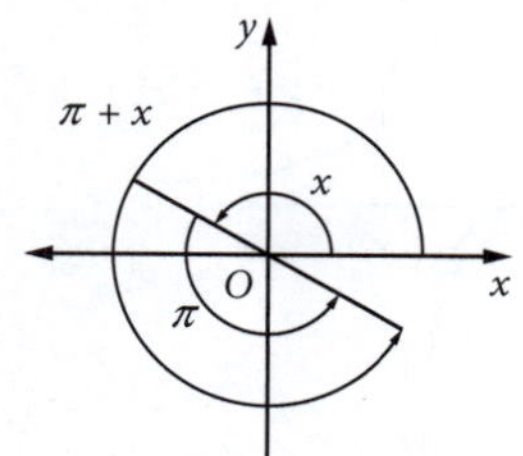

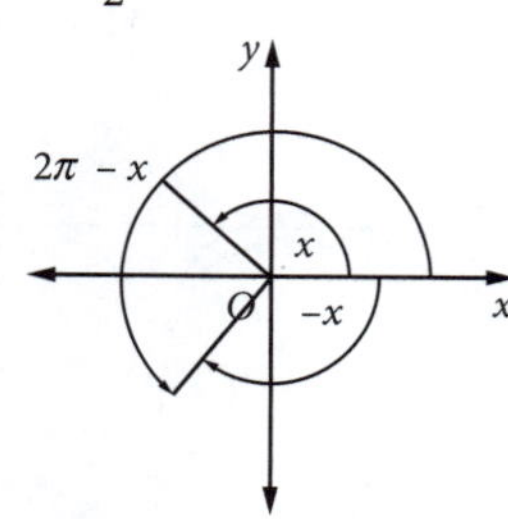

12 (a) positive (b) negative (c) negative (d) negative (e) positive (f) negative

13 (a) $\frac{4\pi}{3}, \frac{5\pi}{3}$ (b) $\frac{3\pi}{4}, \frac{7\pi}{4}$ (c) π (d) $\frac{\pi}{6}, \frac{7\pi}{6}$ (e) $\frac{3\pi}{4}, \frac{5\pi}{4}$ (f) $\frac{\pi}{4}, \frac{5\pi}{4}$ (g) $0, \pi, 2\pi$ (h) $\frac{2\pi}{3}, \frac{4\pi}{3}$ (i) $\frac{\pi}{3}, \frac{2\pi}{3}$ (j) $\frac{2\pi}{3}, \frac{5\pi}{3}$ (k) $\frac{\pi}{4}, \frac{5\pi}{4}$

14 (a) $\sin x$ (b) $\cot x$ (c) $-\cos x$

EXERCISE 6.4

1

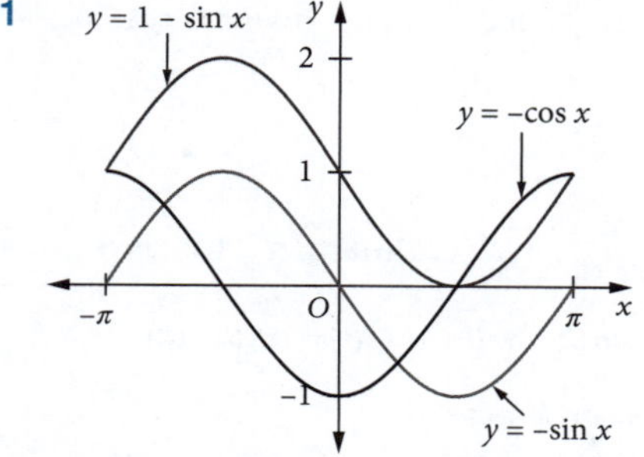

2 (a) period 2, amplitude 1 (b) period 1, amplitude 1

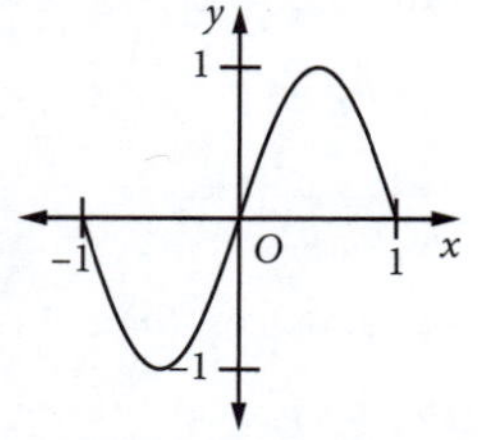

(c) period 2, no amplitude

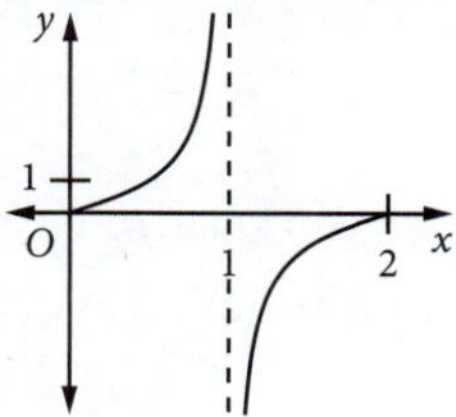

3 (a)

(b)–(c)

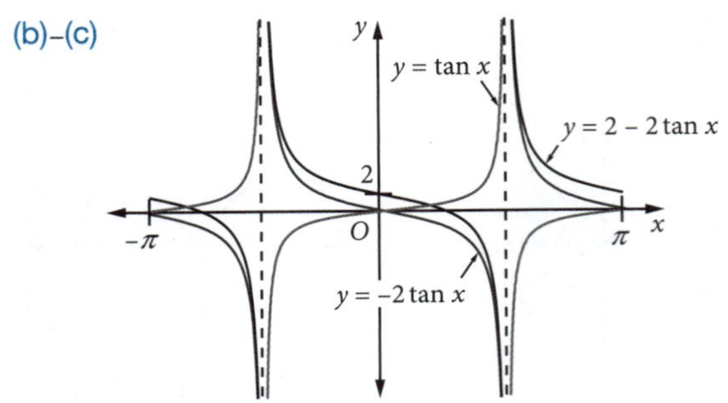

EXERCISE 6.5

1 (a) $\sec^2\theta = 1+\left(\frac{7}{24}\right)^2$, $\sec^2\theta = \frac{625}{576}$, 3rd quadrant so $\cos\theta = -\frac{24}{25}$, $\sin\theta = -\frac{7}{25}$

(b) $\cos\theta = -\frac{24}{25}$

2 (a) correct (b) incorrect (c) correct (d) incorrect

3 (triangle with sides 5, 4, 3 and angle θ)

(a) $\cot\theta = -\frac{3}{4}$

(b) 2nd quadrant, $\sin\theta > 0$, $\sin\theta = \frac{4}{5}$

(c) 2nd quadrant, $\cos\theta < 0$, $\cos\theta = -\frac{3}{5}$

4 A

5 (a) $\cot^2\theta$ (b) $\tan^2\theta$

(c) $\frac{2\cot\alpha}{\operatorname{cosec}^2\alpha} = \frac{2\cos\alpha}{\sin\alpha}\times\sin^2\alpha = 2\sin\alpha\cos\alpha$ (d) $\tan\theta$

(e) $\frac{1}{\sin A\cos A}$ (f) $\sin\theta\left(\sin^2\theta+\cos^2\theta\right) = \sin\theta$ (g) 1

(h) $1-(-\sin\theta)^2 = 1-\sin^2\theta = \cos^2\theta$

6 (a) $\frac{a^2\sec^2\theta}{\sqrt{a^2\sec^2\theta - a^2}} = \frac{a^2\sec^2\theta}{a\sqrt{\tan^2\theta}} = \frac{a\sec^2\theta}{\tan\theta} = a\sec\theta\operatorname{cosec}\theta$

(b) $a\sin\theta$

7 (a) $-\frac{8}{17}$ (b) $\frac{15}{17}$ (c) $-\frac{15}{8}$

8 3rd quadrant, $\cos^2\theta = 1-p^2$, $\cos\theta = -\sqrt{1-p^2}$

9 $-\frac{\sqrt{34}}{5}$ 10 $\sin\theta = x$, $\cos^2\theta = 1-x^2$; $\sin^2\theta\cos^2\theta = x^2(1-x^2)$

11 $a\sin^2\theta + b(1-\sin^2\theta) = c$, $(a-b)\sin^2\theta = c-b$,

$\sin\theta = \pm\sqrt{\frac{c-b}{a-b}}$, $\cos\theta = \pm\sqrt{\frac{c-a}{b-a}}$

12 Use $\sin^2\theta+\cos^2\theta = 1$, $\frac{x^2}{a^2}+\frac{y^2}{b^2} = 1$

13 $\sec^2\theta - 1 + 2\sec^2\theta = 5$, $3\sec^2\theta = 6$, $\cos^2\theta = 0.5$, $\sin^2\theta = \frac{1}{2}$

14 (a) 1 (b) $2\sec^2 V$ (c) $2\operatorname{cosec}\theta$ (d) $\cot^2\theta$ (e) 1

(f) $\sin^2 A$ (g) $\frac{1}{2}$ (h) $\cos^2\theta$

15 LHS $= 1-2\tan x+\tan^2 x+1+2\tan x+\tan^2 x = 2+2\tan^2 x$
$= 2\sec^2 x =$ RHS

16 LHS $= \left(\frac{\cos t}{\sin t}+\frac{1}{\sin t}\right)^2 = \frac{(1+\cos t)^2}{\sin^2 t} = \frac{(1+\cos t)^2}{1-\cos^2 t}$

$= \frac{(1+\cos t)^2}{(1-\cos t)(1+\cos t)} = \frac{1+\cos t}{1-\cos t} =$ RHS

17 LHS $= \sin^2\alpha(1-\sin^2\beta) - (1-\sin^2\alpha)\sin^2\beta$
$= \sin^2\alpha - \sin^2\alpha\sin^2\beta - \sin^2\beta + \sin^2\alpha\sin^2\beta =$ RHS

18 LHS $= \frac{1}{\cos\theta}+\frac{\sin\theta}{\cos\theta} = \frac{1+\sin\theta}{\cos\theta} =$ RHS

19 LHS $= \frac{\sin^3\theta}{\cos\theta}+\frac{\cos^3\theta}{\sin\theta}+2\sin\theta\cos\theta = \frac{\left(\sin^2\theta+\cos^2\theta\right)^2}{\sin\theta\cos\theta}$

$= \frac{1}{\sin\theta\cos\theta} = \frac{\sin^2\theta+\cos^2\theta}{\sin\theta\cos\theta} =$ RHS

20 LHS $= \frac{\sin\theta(\cos\theta+\sin\theta)}{\cos\theta}+\frac{\cos\theta(\sin\theta+\cos\theta)}{\sin\theta}$

$= \frac{\sin^2\theta(\cos\theta+\sin\theta)+\cos^2\theta(\sin\theta+\cos\theta)}{\sin\theta\cos\theta}$

$= \frac{\sin\theta+\cos\theta}{\sin\theta\cos\theta} =$ RHS

21 LHS $= \frac{\tan\theta(\tan^2\theta-1)}{\tan^2\theta}+\frac{1}{\tan\theta}(1-\tan^2\theta)$

$= \frac{\tan^2\theta-1+1-\tan^2\theta}{\tan\theta} = 0 =$ RHS

22 LHS $= \frac{\cos\theta}{\sin\theta}\times\cos\theta\times\frac{\sin\theta}{\cos\theta+\sin\theta\cos\theta} = \frac{\cos\theta}{1+\sin\theta} =$ RHS

23 LHS $= \sin\theta+\cos\theta-\frac{\sin\theta}{\sin^2\theta+\cos^2\theta} = \sin\theta+\cos\theta-\sin\theta =$ RHS

24 LHS $= \left(\frac{\cos^2 t}{\sin t}-\frac{\sin^2 t}{\cos t}\right)\times\frac{\sin t\cos t}{\cos t-\sin t}$

$= \frac{\cos^3 t-\sin^3 t}{\sin t\cos t}\times\frac{\sin t\cos t}{\cos t-\sin t}$

$= \sin^2 t+\sin t\cos t+\cos^2 t = 1+\sin t\cos t =$ RHS

EXERCISE 6.6

1 (a) $x = 0°$ (b) $x = 71°34'$ (c) $x = 12°50', 77°10'$
(d) $x = 45°$ (e) $x = 18°26'$ (f) $x = 15°, 75°$

2 D

3 (a) $x = 45°, 135°, 225°, 315°$ (b) $x = 90°, 270°$
(c) $x = 60°, 120°, 240°, 300°$ (d) $x = 36°52', 143°8'$
(e) $x = 0°, 45°, 180°, 225°, 360°$
(f) $x = 22.5°, 112.5°, 202.5°, 292.5°$

4 (a)

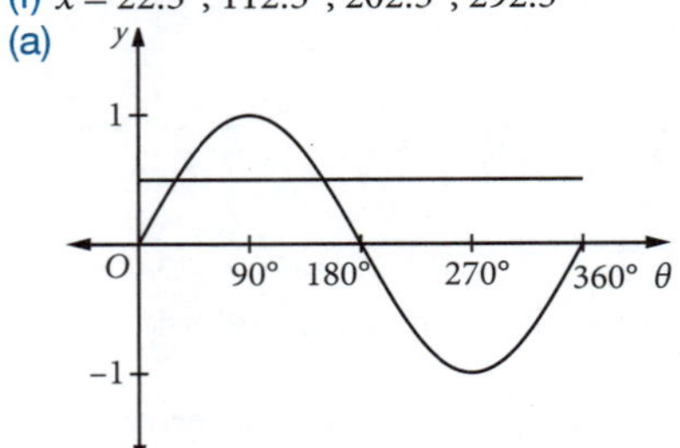

$x = 30°, 150°$

(b) Draw $y = -0.5$. $x = 210°, 330°$

5 (a) $\frac{5\pi}{4}, \frac{7\pi}{4}$ (b) $\frac{\pi}{6}, \frac{11\pi}{6}$ (c) $\frac{\pi}{4}, \frac{5\pi}{4}$

(d) $1-2\cos\theta = 0$; $\cos\theta = 0.5$; $\theta = \frac{\pi}{3}, \frac{5\pi}{3}$

(e) $\cos\theta(1-\cos\theta) = 0$; $\theta = 0, \frac{\pi}{2}, \frac{3\pi}{2}, 2\pi$

(f) $\frac{\pi}{4}, \frac{5\pi}{4}$ (g) $\frac{\pi}{3}, \frac{2\pi}{3}, \frac{4\pi}{3}, \frac{5\pi}{3}$ (h) $\frac{\pi}{4}, \frac{5\pi}{4}$ (i) $0, 2\pi$

(j) $\frac{\pi}{3}, \frac{2\pi}{3}, \frac{4\pi}{3}, \frac{5\pi}{3}$ (k) $\frac{3\pi}{4}, \frac{7\pi}{4}$

(l) $\cos\theta(2\cos\theta-1) = 0$; $\theta = 0, \frac{\pi}{3}, \frac{\pi}{2}, \frac{3\pi}{2}, \frac{5\pi}{3}$

6 A

7 (a) $\sin\theta(1+\sin\theta) = 0$; $\theta = -\pi, -\frac{\pi}{2}, 0, \pi$

(b) $2\sin^2\theta - \sin\theta - 1 = 0$; $\theta = -\frac{5\pi}{6}, -\frac{\pi}{6}, \frac{\pi}{2}$

(c) $4\cos^2\theta = 1$; $\theta = -\frac{2\pi}{3}, -\frac{\pi}{3}, \frac{\pi}{3}, \frac{2\pi}{3}$

(d) $\cos\theta(3+2\cos\theta) = 0$; $\theta = -\frac{\pi}{2}, \frac{\pi}{2}$

(e) $\theta = -\frac{5\pi}{6}, -\frac{\pi}{6}, \frac{\pi}{6}, \frac{5\pi}{6}$

(f) $(2\tan\theta - 1)(\tan\theta - 2) = 0$; $\theta = -2.678, -2.034, 0.464, 1.107$

8 (a) $\sin 2x = 2\sin x\cos x$; $\sin x(2\cos x - 1) = 0$, $\sin x = 0$, $\cos x = 0.5$; $x = 0, \frac{\pi}{3}, \pi, \frac{5\pi}{3}, 2\pi$

(b) $(5\sin x + 3)(\sin x - 1) = 0$; $x = \frac{\pi}{2}, 3.785, 5.640$

(c) $\tan x = \frac{-1}{3}, 2$; $x = 1.107, 2.820, 4.249, 5.961$

(d) $6\sin^2 x = 5\cos x$; $x = 0.841, 5.442$

(e) $\sin x = 0$, $\tan x = 7$; $x = 0, 1.429, \pi, 4.570, 2\pi$

(f) $\tan x = -3, 1$; $x = \frac{\pi}{4}, 1.893, \frac{5\pi}{4}, 5.034$

9 $x = \frac{\pi}{4}, \frac{3\pi}{4}$ (a) correct (b) incorrect (c) correct (d) incorrect

10 (a) $\frac{x}{2} = \frac{\pi}{3}, \frac{2\pi}{3}$; $x = \frac{2\pi}{3}, \frac{4\pi}{3}$ (b) $x = \frac{5\pi}{3}$ (c) $x = 0$

11 (a) $2x = \frac{\pi}{2}, \frac{3\pi}{2}$; $x = \frac{\pi}{4}, \frac{3\pi}{4}$

(b) $x = \frac{\pi}{9}, \frac{4\pi}{9}, \frac{7\pi}{9}$

(c) $x = \frac{\pi}{16}, \frac{3\pi}{16}, \frac{9\pi}{16}, \frac{11\pi}{16}$

12 (a) $x = 1.235$ (b) $x = 0.782, 1.312, 2.876$ (c) no solution

13 A

14 (a) $x = \frac{\pi}{2}, 0.644, 2.498$

(b) $x = 0.740, 2.402, 3.881, 5.543$

15 (a) $\operatorname{cosec}^2\theta$ (b) $\sin^2\theta$ (c) 1 (d) $-\frac{1}{4}$ (e) $\cos^2\theta$

16 (a) $x = 0, \frac{3\pi}{4}, \pi, \frac{7\pi}{4}, 2\pi$ (b) $x = 0, \frac{\pi}{4}, \pi, \frac{5\pi}{4}, 2\pi$

(c) $\theta = 0, \frac{2\pi}{3}, \frac{4\pi}{3}, 2\pi$ (d) $\theta = 0.340, 2.802, \frac{3\pi}{2}$

(e) $x = \frac{\pi}{2}, \frac{2\pi}{3}, \frac{4\pi}{3}$ (f) $x = 0.540, \frac{3\pi}{4}, 3.682, \frac{7\pi}{4}$

17 A

18 (a) $\theta = \frac{\pi}{4}, \frac{5\pi}{4}, \frac{\pi}{6}, \frac{5\pi}{6}, \frac{7\pi}{6}, \frac{11\pi}{6}$ (b) $\theta = 0, \frac{\pi}{2}, \frac{3\pi}{2}, 2\pi$

CHAPTER REVIEW 6

1 $1.6\pi = 5.03$ m

2 (a) 6.1 cm (b) $122\sin 28° = 5.73$ cm

3 4

4 $x = -1.231, -\frac{\pi}{3}, \frac{\pi}{3}, 1.231$

5 (a) 0.75 m (b) $\frac{3\pi}{32} = 0.3\text{ m}^2$ (c) 0.75 m

6 (a) $\frac{7\pi}{6}, \frac{11\pi}{6}$ (b) $\frac{\pi}{2}, \frac{3\pi}{2}$ (c) $\frac{3\pi}{4}, \frac{7\pi}{4}$ (d) $\frac{\pi}{6}, \frac{11\pi}{6}$

(e) $\frac{\pi}{6}, \frac{7\pi}{6}$ (f) $\frac{\pi}{4}, \frac{3\pi}{4}$

7

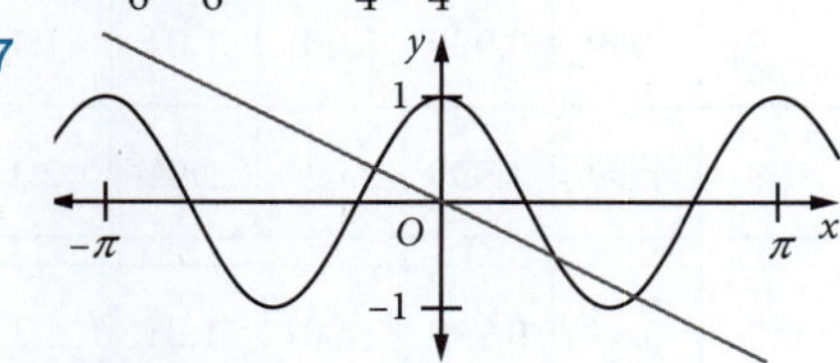

$x = -0.63, 1.07, 1.8$

8 (a) $\frac{20}{x} = \frac{7.5}{\sin\frac{x}{2}}$, $8\sin\frac{x}{2} = 3x$ (b)

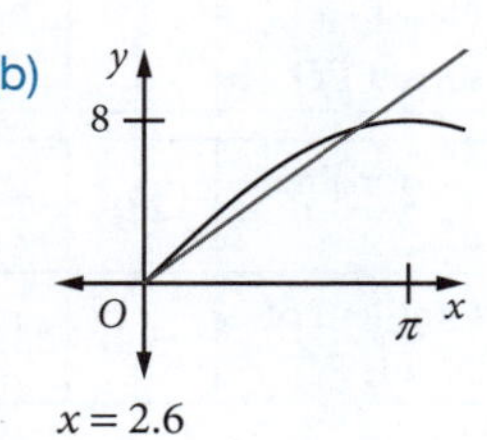

$x = 2.6$

9 (a) $\frac{\pi}{4}$ (b) $\frac{4\pi}{3}$ (c) $\frac{8\pi}{9}$ (d) $-\frac{7\pi}{6}$

10 (a) 1.1345 (b) 4.9044 (c) −1.7453 (d) −5.6898

11 (a) 144° (b) 210° (c) 345° (d) −270°

12 (a) 180°58′ (b) −80°13′ (c) 19°32′ (d) −171°53′

13 (a) $-\sin x$ (b) $\cos x$ (c) $-\tan x$

14 (a) $\cos\pi = -1$ (b) $\tan\frac{7\pi}{6} = \frac{\sqrt{3}}{3}$ (c) $\sin\frac{3\pi}{4} = \frac{\sqrt{2}}{2}$

(d) $\cos\frac{5\pi}{3} = \frac{1}{2}$

15 (a) $x = \frac{7\pi}{6}$ or $\frac{11\pi}{6}$ (b) $x = \frac{\pi}{3}$ or $\frac{4\pi}{3}$ (c) $x = \frac{3\pi}{4}$ or $\frac{5\pi}{4}$

16 (a) $\cos x$ (b) $-\sin x$ (c) $-\frac{1}{\tan x}$ or $-\cot x$

CHAPTER 7

EXERCISE 7.1

1 (a) continuous (b) discontinuous (c) continuous (d) continuous (e) discontinuous (f) continuous (g) continuous (h) continuous

2 (a), (c), (f), (h)

3 A

4

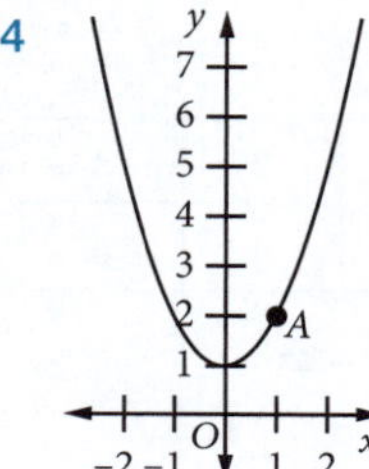

(a) $m_{AB} = 3.5$, $y = 3.5x - 1.5$
$m_{AB_1} = 3$, $y = 3x - 1$
$m_{AB_2} = 2.5$, $y = 2.5x - 0.5$
$m_{AB_3} = 2.1$, $y = 2.1x - 0.1$

(b)

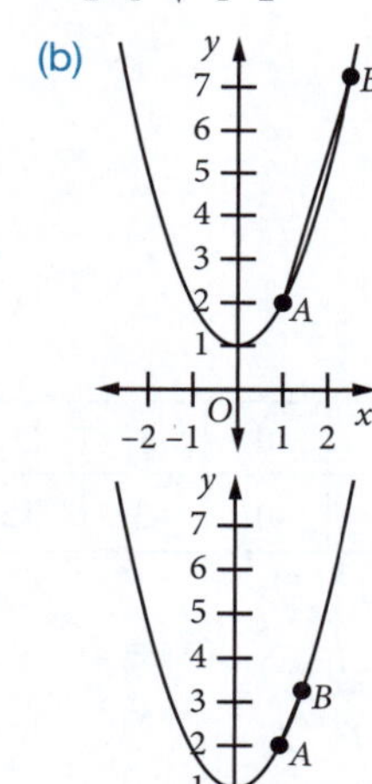

(c) 2 (d) $m_{AB_4} = 1.9$

5

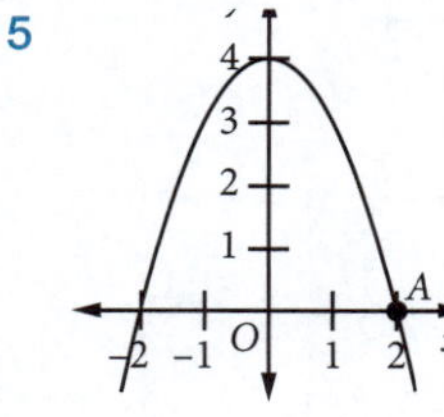

(a) $m_{AB} = -2.5$, $y = -2.5x + 5$
$m_{AB_1} = -3$, $y = -3x + 6$
$m_{AB_2} = -3.5$, $y = -3.5x + 7$
$m_{AB_3} = -3.9$, $y = -3.9x + 7.8$

(b)

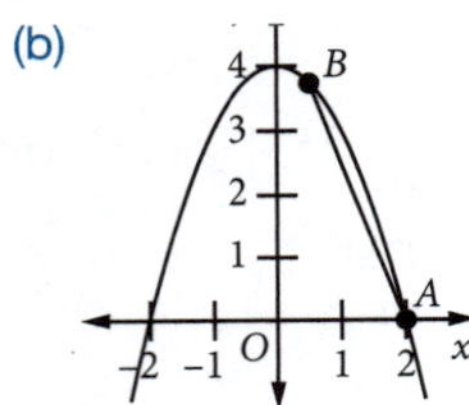

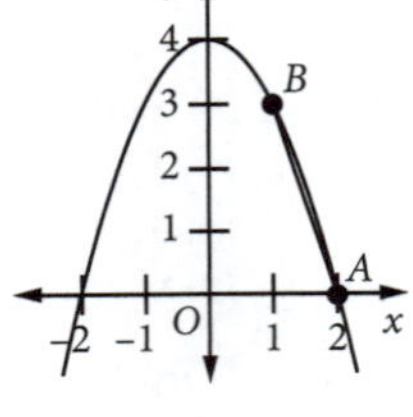

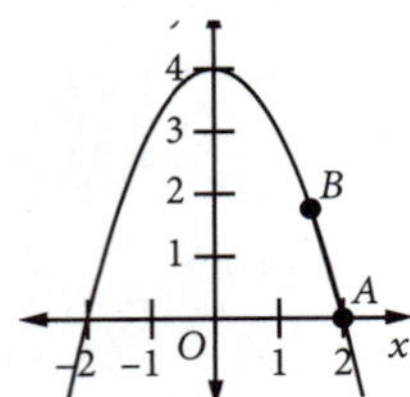

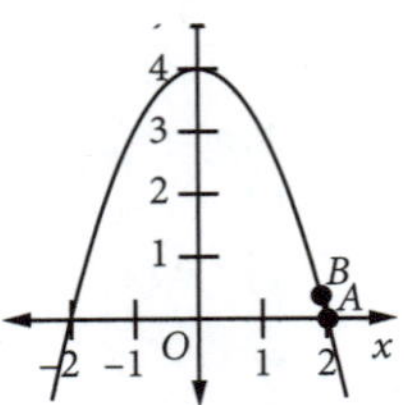

(c) −4 (d) $m_{AB_4} = -4.1$

6

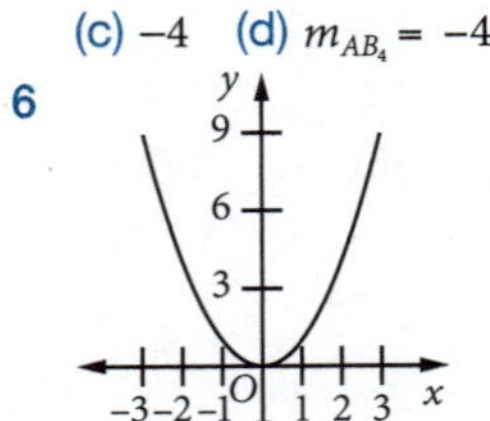

(a)

x	−2	−1	0	1	2	3
m	−4	−2	0	2	4	6

(b)

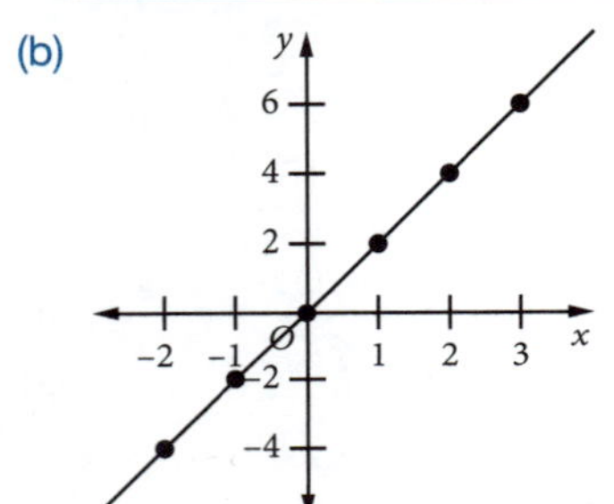

(c) $y = 2x$

7

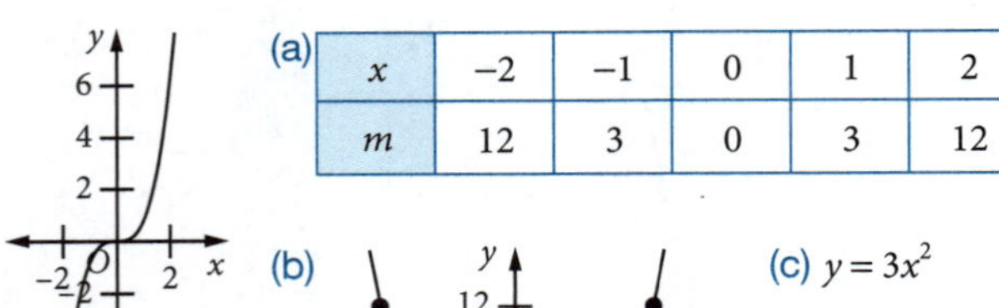

(a)

x	−2	−1	0	1	2
m	12	3	0	3	12

(b)

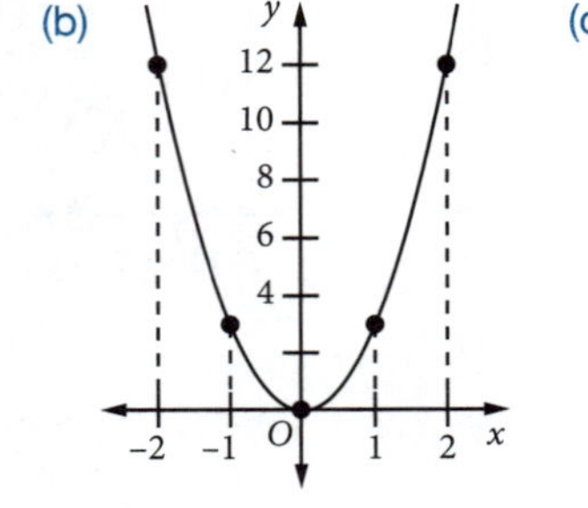

(c) $y = 3x^2$

8

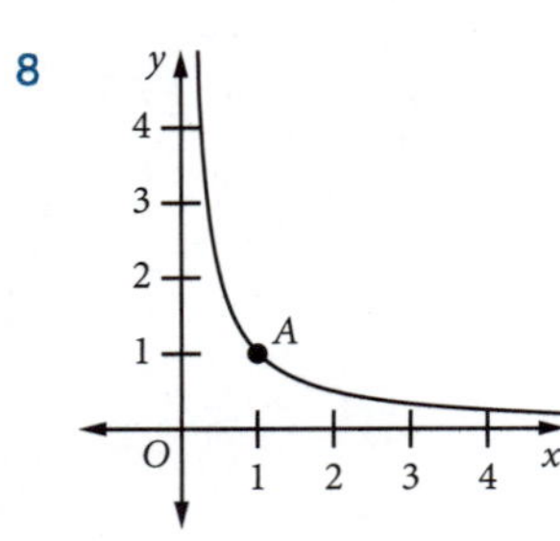

(a) $m_{AB} = -\frac{1}{3}$, $3y = 4 - x$

$m_{AB_1} = -\frac{1}{2}$, $2y = 3 - x$

$m_{AB_2} = -\frac{2}{3}$, $3y = 5 - 2x$

$m_{AB_3} = -\frac{10}{11}$, $11y = 21 - 10x$

(b)

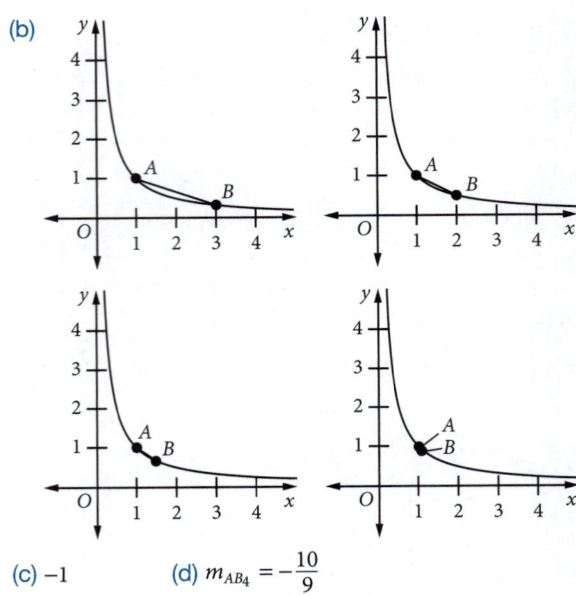

(c) −1 (d) $m_{AB_4} = -\frac{10}{9}$

EXERCISE 7.2

1 (a) 9 (b) −3 (c) 0 (d) 9 (e) −32 (f) 0 (g) −10 (h) 0.8

2 C

3 (a) 1 (b) 5 (c) 12 (d) 1 (e) 1.5 (f) $\frac{1}{11}$ (g) $\frac{1}{3}$ (h) $\frac{1}{6}$ (i) 3

4 (a) 1 (b) 2

5 (a) $2x$ (b) $4x - 3$ (c) $3x^2$ (d) $6 - 2x$

6 (a) $\frac{1}{x}$ is undefined at $x = 0$, therefore $\lim\limits_{x\to 0}\frac{1}{x}$ does not exist.

(b) $\lim\limits_{x\to 0^+} f(x) = 1$, $\lim\limits_{x\to 0^-} f(x) = 0$, $\lim\limits_{x\to 0^+} f(x) \neq \lim\limits_{x\to 0^-} f(x)$ therefore $\lim\limits_{x\to 0} f(x)$ does not exist.

(c) $\lim\limits_{x\to 0^+} f(x) = 0$, $\lim\limits_{x\to 0^-} f(x) = 0 + 1 = 1$, $\lim\limits_{x\to 0^+} f(x) \neq \lim\limits_{x\to 0^-} f(x)$ therefore $\lim\limits_{x\to 0} f(x)$ does not exist.

(d) $\lim\limits_{x\to 0^+} f(x) = 0^2 + 1 = 1$, $\lim\limits_{x\to 0^-} f(x) = 2$, $\lim\limits_{x\to 0^+} f(x) \neq \lim\limits_{x\to 0^-} f(x)$ therefore $\lim\limits_{x\to 0} f(x)$ does not exist.

7 A

EXERCISE 7.3

Values shown in these tables may be approximate.

1 2

x-coordinate of Q	0.9	0.99	0.999	1.001	1.01	1.1
Gradient of secant PQ	1.9	1.99	1.999	2.001	2.01	2.1

2 2

x-coordinate of Q	0.9	0.99	0.999	1.001	1.01	1.1
Gradient of secant PQ	2	2	2	2	2	2

3 6

x-coordinate of Q	2.9	2.99	2.999	3.001	3.01	3.1
Gradient of secant PQ	5.9	5.99	5.999	6.001	6.01	6.1

4 D

5 3

x-coordinate of Q	0.9	0.99	0.999	1.001	1.01	1.1
Gradient of secant PQ	2.71	2.97	2.997	3.003	3.03	3.31

6 −0.25

x-coordinate of Q	1.9	1.99	1.999	2.001	2.01	2.1
Gradient of secant PQ	−0.263	−0.251	−0.250	−0.250	−0.249	−0.238

7 −4

x-coordinate of Q	−2.1	−2.01	−2.001	−1.999	−1.99	−1.9
Gradient of secant PQ	−4.1	−4.01	−4.001	−3.999	−3.99	−3.9

8 (a) $\frac{2(x+h)^2-2x^2}{h}=\frac{2(x^2+2xh+h^2)-2x^2}{h}$

$=\frac{4xh+2h^2}{h}$

$=4x+2h$

(b) $\frac{2(x+h)^2+(x+h)-(2x^2+x)}{h}$

$=\frac{2(x^2+2xh+h^2)+x+h-2x^2-x}{h}$

$=\frac{4xh+h+2h^2}{h}=4x+1+2h$

(c) $\frac{4(x+h)-(x+h)^2-(4x-x^2)}{h}$

$=\frac{4x+4h-(x^2+2xh+h^2)-4x+x^2}{h}$

$=\frac{4h-2xh-h^2}{h}=4-2x-h$

(d) $\frac{(x+h)^3-x^3}{h}=\frac{(x+h-x)((x+h)^2+x(x+h)+x^2)}{h}$

$=\frac{h(x^2+2hx+h^2+x^2+hx+x^2)}{h}$

$=3x^2+3hx+h^2$

9 (a) $\lim_{h\to 0}\frac{5(x+h)+1-(5x+1)}{h}=\lim_{h\to 0}\frac{5x+5h+1-5x-1}{h}$

$=\lim_{h\to 0}\frac{5h}{h}=5$

(b) $\lim_{h\to 0}\frac{(x+h)^2-x^2}{h}=\lim_{h\to 0}\frac{(x+h-x)(x+h+x)}{h}$

$=\lim_{h\to 0}\frac{h(2x+h)}{h}$

$=\lim_{h\to 0}(2x+h)=2x$

(c) $\lim_{h\to 0}\frac{3(x+h)^2+7(x+h)-(3x^2+7x)}{h}$

$=\lim_{h\to 0}\frac{3x^2+6hx+3h^2+7x+7h-3x^2-7x}{h}$

$=\lim_{h\to 0}\frac{6hx+3h^2+7h}{h}=\lim_{h\to 0}(6x+7+3h)=6x+7$

(d) $\lim_{h\to 0}\frac{(x+h)^3+2(x+h)-(x^3+2x)}{h}$

$=\lim_{h\to 0}\frac{(x+h-x)((x+h)^2+x(x+h)+x^2)+2x+2h-2x}{h}$

$=\lim_{h\to 0}\frac{h(x^2+2hx+h^2+x^2+xh+x^2)+2h}{h}$

$=\lim_{h\to 0}(3x^2+2+3xh+h^2)=3x^2+2$

EXERCISE 7.4

1 (a) $2-2h$ (b) 2
2 (a) −4 (b) 3
3 (a) correct (b) incorrect (c) correct (d) correct
4 D
5 (a) 8, gradient of tangent at $(3,6)$
(b) $4x-4$, gradient of tangent at $(x, f(x))$
6 (a) $8x$ (b) $x-2$ (c) $3x^2-4x$

EXERCISE 7.5

1 (a) 1

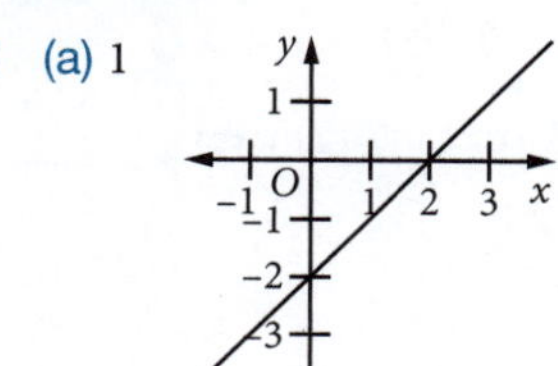

(b) 4

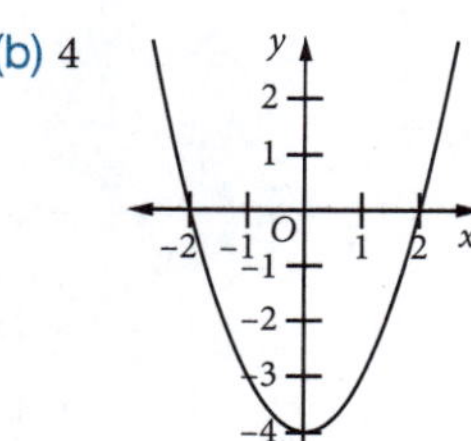

(c) 1

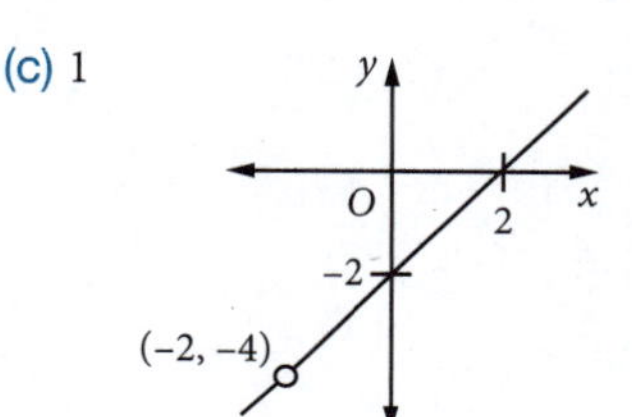

(d)

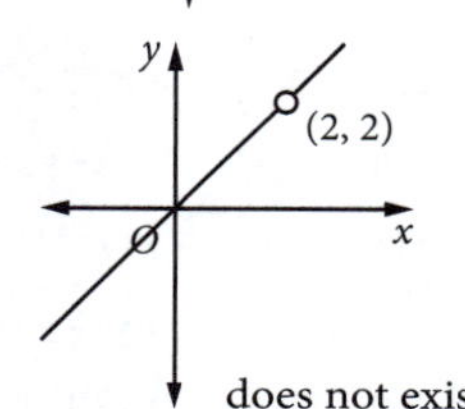

does not exist

2 $f'(3)=2$

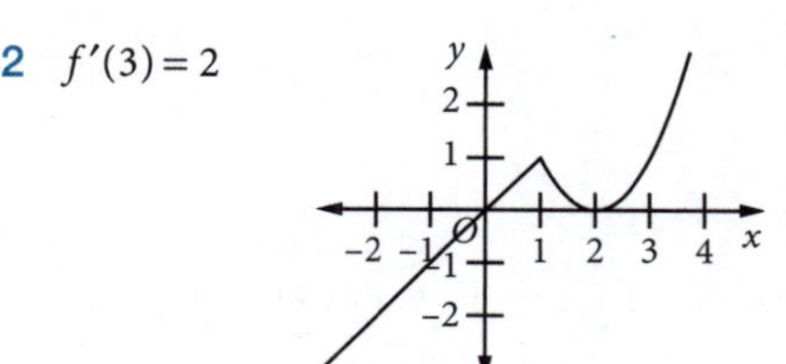

3 $f'(4)=1$

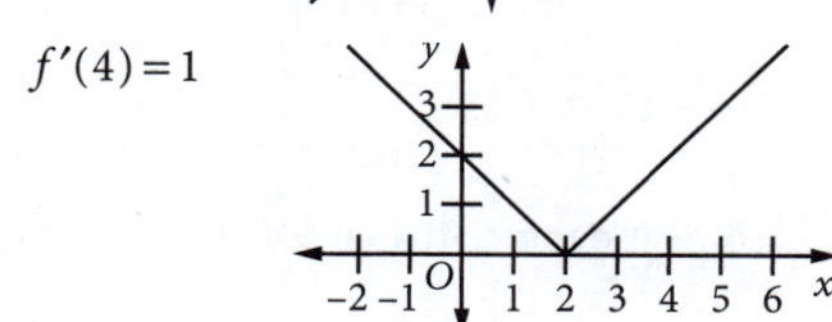

4 $f'(-1)=-2$

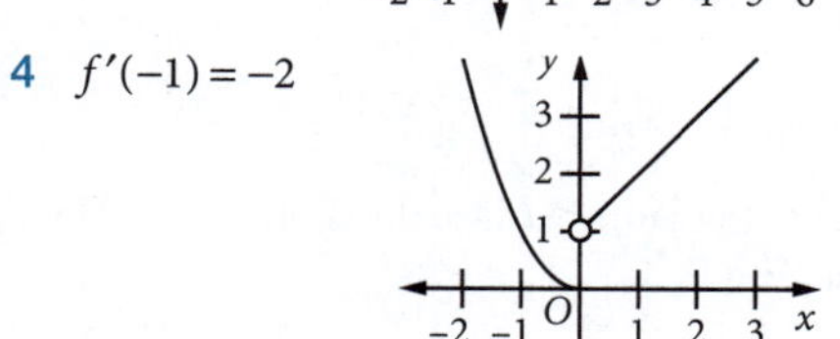

5 $f'(3)=1$

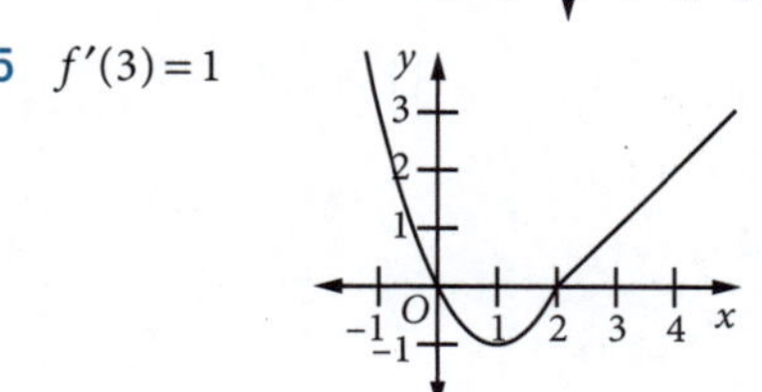

6 not differentiable at $x=3, x=-3$

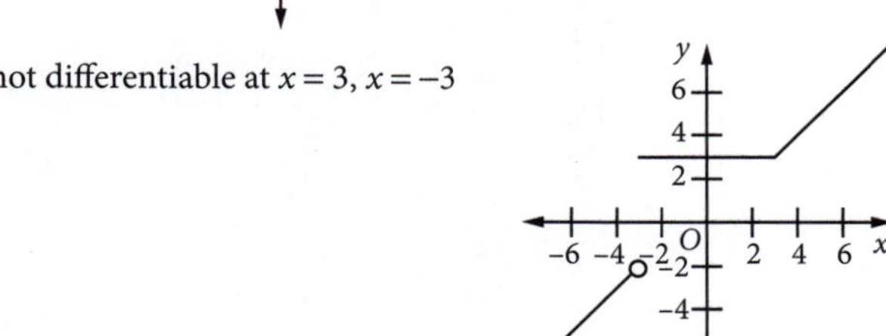

7 continuous at a and c, discontinuous at b; no derivative at a, b, c

EXERCISE 7.6

1 (a) $6x+2$ (b) $4-6x$ (c) $7-8x$ (d) $4x^3+2x$ (e) $5x^4-3x^2+1$ (f) $3t^2+8t-2$

2 (a) $\frac{3\sqrt{x}}{2}$ (b) $-\frac{2}{x^2}$ (c) $\frac{1}{\sqrt{x}}$ (d) $\frac{2}{3\sqrt[3]{t}}$ (e) $-\frac{3}{m^4}$ (f) $\frac{-1}{2x\sqrt{x}}$

3 D

4 (a) $2x+1$ (b) $9x^2-6$ (c) $4(2x-3)$ (d) $2x$ (e) $24x^2-72x+54$ (f) $9x^2-4x-7$

5 (a) $6x-6x^2$ (b) $4-16x^3$ (c) $9t^2-4t+5$ (d) $3ax^2+2bx+c$ (e) 3 (f) $6x-2$

6 (a) $1+\frac{1}{2\sqrt{x}}$ (b) $2x-\frac{1}{x^2}$ (c) $2x+1-\frac{1}{x^2}-\frac{2}{x^3}$ (d) $\frac{2}{3\sqrt[3]{x}}-\frac{1}{3x\sqrt[3]{x}}$ (e) $2x-\frac{2}{x^3}$ (f) $\frac{3\sqrt{x}}{2}$

7 (a) correct (b) incorrect (c) correct (d) correct

8 (a) $x=0$ (b) $x=\pm 1$ (c) $x=0, 2\frac{2}{3}$

9 $\frac{dy}{dx}=\pm 5$ 10 $x=-\frac{1}{3}, 1$

11 $x^2+4x-12=0$ for $x=-6, 2$; $\frac{dy}{dx}=2x+4$

$x=-6, \frac{dy}{dx}=-8$; $x=2, \frac{dy}{dx}=8$

12 (a) $x=1$ (b) $x=1$ (c) $x=0.5$

13 (a) $(3,0)$ (b) $(1,2)$ (c) $\frac{3}{2}, \frac{3}{4}$

14 (a) $x=1, 2$ (b) $x=0, 3$ (c) $x=-1, 4$

15 (a) $(1,-9)$ (b) $(0,-8)$

16 $c=0.6x^2+4x+650$. $\frac{dc}{dx}=1.2x+4$.

$x=40$: $\frac{dc}{dx}=1.2\times 40+4=52$.

$52 for each additional item.

17 (a) $6x-\frac{x^2}{2}-10=0$, $(x-2)(x-10)=0$. Breaks even point occurs when 2 items are produced in a day.

(b) $\frac{dP}{dx}=6-x$ (c) $6-x>0, x<6$

18 $\frac{dv}{dt}=-4000$. The machine is depreciating at $4000 per year for any value of t.

19 (a) (i) $\frac{dy}{dx}=2x+3$ (ii) $\frac{dy}{dx}=2x+3$ (iii) $\frac{dy}{dx}=2x+3$

(b) The derivatives are the same because the derivative of any constant term is zero.

20 (a) $f(x)=x^2+3$: $f'(x)=2x$

(b)

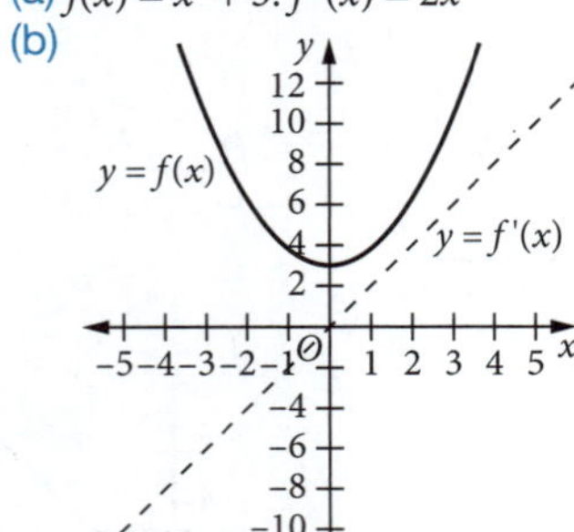

21 (a) $f(x)=x^3-1$: $f'(x)=3x^2$

(b)

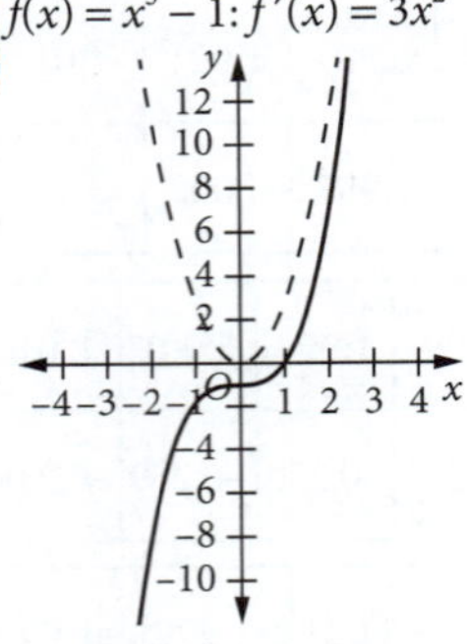

22 (a) $f(x)=x^3+x^2+x$: $f'(x)=3x^2+2x+1$

(b)

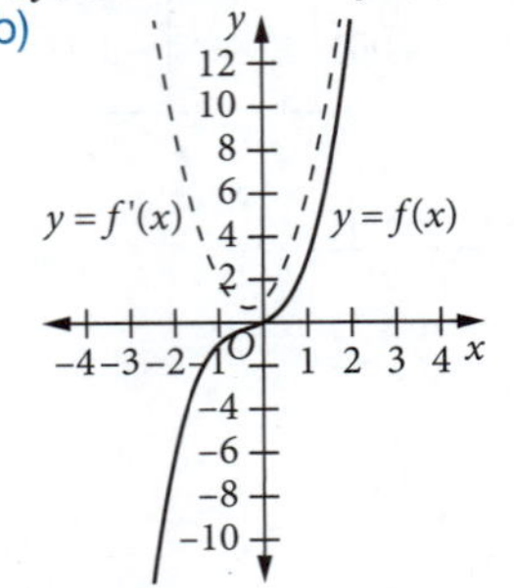

23 (a) If $g'(x)=3x^2$: $g(x)=x^3, x^3+2, x^3-4$.

(b)

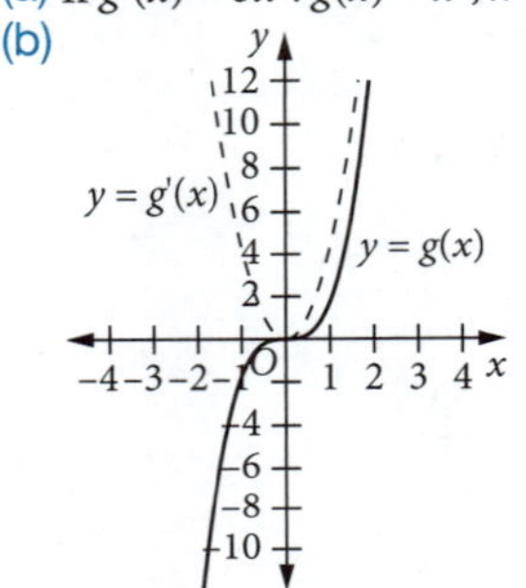

EXERCISE 7.7

1 (a) $12x-5$ (b) $4x+7$ (c) $9x^2-4x-8$ (d) $3x^2-8x+3$ (e) $6x^2-18x+10$ (f) $4x^3-12x^2+6x-12$ (g) $6x+2$ (h) $6x^2-14x-15$ (i) $60x^2-10x-28$

2 A 3 (a) incorrect (b) correct (c) correct (d) correct

4 (a) $5x^4+24x^3+15x^2+2x+5$ (b) 51 (c) -51

5 (a) $\sqrt{x}+\frac{x-1}{2\sqrt{x}}=\frac{3\sqrt{x}}{2}-\frac{1}{2\sqrt{x}}$ (b) $\frac{3\sqrt{x}}{2}-1$ (c) 1

(d) $\frac{15x\sqrt{x}}{2}+2x+\frac{6}{\sqrt{x}}$ (e) $2x+\frac{2}{x^3}$

(f) $\frac{5x\sqrt{x}}{2}+2x-\frac{1}{x\sqrt{x}}-\frac{2}{x^2}$

EXERCISE 7.8

1 (a) $10x(x^2-4)^4$ (b) $\frac{5}{2\sqrt{5x-1}}$ (c) $12(x^2-1)(x^3-3x)^3$

(d) $\frac{-2}{(2x+5)^2}$ (e) $\frac{2t+3}{3\sqrt[3]{(t^2+3t)^2}}$ (f) $4(4x+5)(2x^2+5x-4)^3$

(g) $\frac{x-1}{\sqrt{x^2-2x}}$

(h) $\frac{-4t}{(t^2+4)^3}$ (i) $\frac{-x}{\sqrt{25-x^2}}$

2 B

3 (a) $\frac{x}{\sqrt{x^2-4}}$ (b) $\frac{x}{\sqrt{x^2+1}}$ (c) $\frac{-2}{(1+2x)^2}$ (d) $4m(m^2+25)$

(e) $10(2x-1)^4$ (f) $\frac{1}{3\sqrt[3]{(t+1)^2}}$ (g) $8(3x-1)(3x^2-2x-1)^3$

(h) $\frac{-4t}{(t^2+4)^3}$ (i) $4\left(1+\frac{1}{x^2}\right)\left(x-\frac{1}{x}\right)^3$

4 (a) incorrect (b) incorrect (c) correct (d) correct

5 (a) $(21x-50)(3x+4)^5$

(b) $2x\sqrt{1-x^2}+\frac{x^2\times(-2x)}{2\sqrt{1-x^2}}=\frac{x(2-3x^2)}{\sqrt{1-x^2}}$

(c) $3t^2-4(4-t)^3$ (d) $\frac{-1}{2\sqrt{1-x}}+\frac{1}{2\sqrt{1+x}}=\frac{\sqrt{1-x}-\sqrt{1+x}}{2\sqrt{1-x^2}}$

(e) $2t-\frac{1}{(1+t)^2}$ (f) $1-\frac{1}{x^2}$

EXERCISE 7.9

1 (a) $\frac{2}{(x+1)^2}$ (b) $\frac{43}{(4x+5)^2}$ (c) $\frac{-1}{(t+2)^2}$

(d) $\frac{5-2x}{(x^2-5x+6)^2}$ (e) $\frac{2n^2-2n-11}{(2n-1)^2}$ (f) $\frac{-3}{(x-3)^2}$

(g) $\frac{8x(x+5)}{(2x+5)^2}$ (h) $\frac{44x}{(x^2+5)^2}$ (i) $\frac{-(2x^3+3x^2+1)}{(x^3-1)^2}$

2 A

3 (a) $\frac{-(x+2)}{2x^2\sqrt{x+1}}$ (b) $1-\frac{1}{x^2}$

(c) $\frac{1-x}{(x+1)^3}$ (d) $\frac{2(2x+1)^2(x^2+2x+9)}{(3-x^2)^3}$

(e) $\frac{x+2}{2(x+1)^{\frac{3}{2}}}$ (f) $\frac{-2(x+1)}{x^3}$

4 (a) correct (b) correct (c) correct (d) incorrect

5 (a) $3x^2+4x-4$ (b) $10x^{\frac{3}{2}}-3\sqrt{x}+\frac{3}{\sqrt{x}}$ (c) $\frac{5}{2\sqrt{5x-1}}$

(d) $\frac{2}{3\sqrt[3]{t}}+6(2t-1)^2$ (e) $\frac{3}{(x+5)^2}$ (f) $\frac{x}{\sqrt{x^2+2}}$

(g) $2x-5+\frac{1}{2\sqrt{x-2}}$ (h) $\frac{-9}{(2z-1)^2}$ (i) $\frac{2-2x}{(x^2-2x+2)^2}$

(j) $3x^2-12x-7$ (k) $\frac{5x\sqrt{x}}{2}+\frac{1}{2\sqrt{x}}=\frac{5x^2+1}{2\sqrt{x}}$

(l) $\frac{1-2x-x^2}{2\sqrt{x+1}(x^2+1)^{\frac{3}{2}}}$

6 $\frac{dy}{dx}=\frac{1-x^2}{(x^2+1)^2}$; $\frac{dy}{dx}=0$ at $x=-1, 1$

EXERCISE 7.10

1 $4x-y-4=0, x+4y-18=0$

2 $y=3x, x+3y=0$ 3 A

4 $3x-y-4=0, x+3y+12=0$

5 $y=4x-7, x+4y-6=0$

6 (a) incorrect (b) correct (c) correct (d) correct

7 $x+y-3=0, y=x-3$

8 $y=10x-20, x+10y-2=0$

9 $P(-2,0)$, $Q(4,6)$;
tangent at P: $2x+y+4=0$
tangent at Q: $y=4x-10$
intersection at $(1,-6)$

10 $y=4x, 12x+3y-16=0$

11 $y=-1, x=1$

12 $2x^2-6x+5=x^2-2x+1, (x-2)^2=0,$
$x=2, y=1$: intersect at $(2,1)$
$y_1'=4x-6$: at $(2,1)$ $y_1'=2$
$y_2'=2x-2$: at $(2,1)$ $y_2'=2$
curves have the same gradient at their point of intersection, so they touch at that point; equation of tangent: $y=2x-3$

13 $104°\,2'$, $80°\,32'$

14 (a) $P(-1,0), Q(3,4)$ (b) $3x+y+3=0, y=5x-11$ (c) $(1,-6)$

15 (a) $y=3$ (b) $y=2x$

16 $y=3x-7, x+3y+1=0$

17 (a) $x-2y+5=0$ (b) $\left(-\frac{1}{2}, 2\frac{1}{4}\right)$ (c) $45°$

18 $116°\,34'$, $82°\,52'$, $33°\,42'$

19 $y=6x+6, y=2-2x$
$y=3x-6$

20 (a) $(-1,-5), (2,4)$ (b) $(0,0), (1,-1)$

EXERCISE 7.11

1 (a) $0\le t\le 200$ (b) $6\,\text{m}^3\,\text{min}^{-1}$ (c) no (d) $10\,\text{m}^3\,\text{min}^{-1}$
(e) 50 min

2 D

3 (a) $12.5\,\text{cm}^3\,\text{min}^{-1}$ (b) $V=1000-12.5t$

4 (a) $2000(20-t)\,\text{L}\,\text{min}^{-1}$ (b) 20 min (c) 10 min
(d) $35\,000\,\text{L}\,\text{min}^{-1}$

5 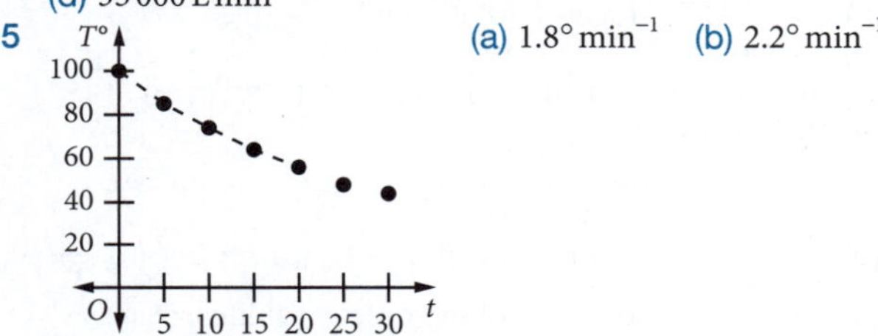

(a) $1.8°\,\text{min}^{-1}$ (b) $2.2°\,\text{min}^{-1}$

6 (a) 1 item/min (b) 21 items/min

7 $A=\pi r^2, \frac{dA}{dr}=2\pi r \propto r$; $4\pi\text{cm}^2/\text{cm}$

8 (a) πr^2 (b) $2\pi rh$ 9 -5

10 (a) -5000; depreciating by \$5000 per year (b) \$25 000

11 (a) 14.6 (b) 14 (c) 0

EXERCISE 7.12

1 (a) $x=6\,\text{m}$ (b) $v=-5\,\text{m}\,\text{s}^{-1}$ (c) $t=2\,\text{s}, v=-1\,\text{m}\,\text{s}^{-1}$
(d) $t=3\,\text{s}, v=1\,\text{m}\,\text{s}^{-1}$ (e) $t=2.5\,\text{s}, x=-0.25\,\text{m}$

2 C

3 (a) $t=2.5\,\text{s}$ (b) $a=2\,\text{m}\,\text{s}^{-2}$ (c) $4+2\times 2.25=8.5\,\text{m}$
(d) $t=6.5\,\text{s}$

4 (a) $\dot{x}=6t^2-30t+36, \ddot{x}=12t-30$ (b) $36\,\text{m}\,\text{s}^{-1}, -30\,\text{m}\,\text{s}^{-2}$
(c) $t=2, 3\,\text{s}$ (d) $t=2.5\,\text{s}, v=-1.5\,\text{m}\,\text{s}^{-1}, x=27.5\,\text{m}$ (e) $2<t<3$

5 (a) $t=1.5\,\text{s}, v=-1.5\,\text{m}\,\text{s}^{-1}$ (b) $t=1\,\text{s}, a=-6\,\text{m}\,\text{s}^{-2}$; $t=2\,\text{s}, a=6\,\text{m}\,\text{s}^{-2}$

6 (a) $t=2\,\text{s}$ (b) $t=0.5\,\text{s}$

7 (a) $t=1, a=18\,\text{m}\,\text{s}^{-2}$ (b) $v=-12\,\text{m}\,\text{s}^{-1}, a=6\,\text{m}\,\text{s}^{-2}$

8 (a) $50\,\text{km}\,\text{h}^{-1}$, $20\,\text{km}\,\text{h}^{-1}$ (b) 9 min (c) 13.2 km
(d) $x=\frac{112t}{3}+2$

9 $\dot{x}=4-\frac{5t}{\sqrt{t^2+4}}, t=2\frac{2}{3}\,\text{s}, x=-3$ 10 D

CHAPTER REVIEW 7

1 (a) 2 (b) 27 2 (a) $2x^2+3$ (b) 4 (c) 3

3 $4x-3+2h$ 4 (a) 24 (b) 10 (c) $2c+6$ (d) $c=-4$

5 $\frac{1}{\sqrt{2x-1}}$ 6 $(36x^3-14x^2-118x+40)(3x^2-2x+1)^4$

7 (a) $3(x-2)^2$ (b) $5x^9(2+3x)(1+x)^4$ (c) $\frac{-t}{\sqrt{25-t^2}}$

(d) $\frac{2(x^4-1)}{x^3}$ (e) $\frac{-2}{(x+4)^3}$ (f) $\frac{2}{x^2}$

(g) $(7x+11)(x-1)^5$ (h) $2x+3+\frac{1}{2\sqrt{x-2}}$ (i) $\frac{-2x}{(x^2-2)^2}$

(j) $\frac{20}{(2m+3)^2}$ (k) $1-\frac{2}{x^3}$ (l) $\frac{3\sqrt{t}-3}{2}$

8 (a) $4x^3$ (b) $2x-1$ (c) $3x^2-2$

9 $2x+y-9=0$ 10 $y=128-32x$; $y=0$ touches axis

11 (a) $14\,\text{m s}^{-2}$ (b) 275 m

12 (a) 0 (b) $a=\frac{18-2t^2}{(9+t^2)^2}$ (c) $t=3$ (d) $t=3,\ v=\frac{1}{3}\,\text{m s}^{-1}$

CHAPTER 8

EXERCISE 8.1

1 (a) 2^{13} (b) $a^{11}b^6$ (c) $m^{11}n^6p$ (d) 2^{6n} (e) x^5y^9 (f) $\frac{1}{2n^3}$ (g) x^9y^7 (h) a^2b^3

2 D

3 (a) $\frac{1}{mnp^8}$ (b) $\frac{q^4}{pr^4}$ (c) $\frac{2x^5}{y^2z}$ (d) 2^{n+3}

4 (a) 2^{-2} (b) 2^{-4} (c) 2^{-5} (d) 2^{-3} (e) 2^{-6} (f) 2^{-7} (g) 2^{-2} (h) 2^{-9}

5 (a) 10^2 (b) 10^1 (c) 10^0 (d) 10^{-1} (e) 10^{-2} (f) 10^{-3} (g) 10^{-3} (h) 10^{-5}

6 (a) $3^{n+1}\times5^{3n-1}$ (b) $\frac{1}{x^2}-\frac{1}{y^2}$ (c) 2^8 (d) $\frac{1}{x^4}+\frac{2}{x^3}+\frac{1}{x^2}$ (e) $x-3$ (f) $x-2$ (g) 3×2^7 (h) $\frac{5^{2n+3}}{3^{n-3}}$

7 (a) correct (b) correct (c) incorrect (d) incorrect

8 (a) $\frac{1}{a^2}-b^2$ (b) $2+xy+\frac{1}{xy}$ (c) $\frac{1}{x^4}-\frac{1}{y^4}$ (d) $3-a^2b-\frac{2}{a^2b}$ (e) $\frac{1}{ab}$ (f) $\frac{1}{y}$

EXERCISE 8.2

1 (a) 16 (b) $\frac{1}{7}$ (c) 27 (d) 9 (e) 2 (f) 64 (g) 6 (h) 5

2 C

3 (a) $6^{\frac{1}{2}}$ (b) $2^{\frac{5}{8}}$ (c) $2^{\frac{4}{3}}$ (d) $3^{\frac{11}{6}}$

4 (a) $x^{\frac{13}{6}}$ (b) $\frac{b^3}{a^2}$ (c) $x-\frac{1}{x}$ (d) $\frac{1}{y}$ (e) $x-y$ (f) $x^{\frac{2}{3}}y$ (g) $6^{\frac{1}{2}}$ (h) 1

5 (a) incorrect (b) correct (c) correct (d) correct

EXERCISE 8.3

1 (a) $x=3$ (b) $x=4$ (c) $x=-5$ (d) $x=\pm\frac{1}{9}$ (e) $x=1\frac{2}{3}$ (f) $x=2.5$ (g) $x=1.5$ (h) $x=2$ (i) $x=1$ (j) $x=-2$ (k) $x=6$ (l) $x=-3$

2 C

3 (a) $x=-3$ (b) $x=3$ (c) $x=\frac{-1}{2}$ (d) $x=\frac{-1}{2}$ (e) $x=0$ (f) $x=-3$ (g) $x=1\frac{3}{4}$ (h) $x=3$ (i) $x=1\frac{2}{3}$ (j) $x=-2, 0$ (k) $x=0, 2$ (l) $x=-2$

EXERCISE 8.4

1 (a) 2 (b) $\frac{1}{2}$ (c) 7 (d) 1 (e) −3 (f) −1 (g) 10 (h) $\frac{1}{2}$

2 A

3 (a) 4 (b) $2\frac{1}{2}$ (c) 3 (d) −4

4 (a) 7 (b) 1 (c) 8 (d) 1 (e) 3 (f) 0 (g) 1 (h) 12

5 (a) correct (b) incorrect (c) correct (d) correct

6 (a) 2 (b) x (c) 3 (d) 2 (e) $6\log_{10}5$ (f) 3 (g) $\log_{10}(2+\sqrt{5})$ (h) 2

7 (a) $x-y$ (b) $2x-2y$ (c) $y-x+1$ (d) $x+3y$ (e) $x+3y-1$ (f) $y-2x+2$ (g) $y-x+2$ (h) $3y-2$ (i) $2y-x$ (j) $x+y-1$ (k) $4y$ (l) $x+2y-1$

8 (a) 2.322 (b) 2.262 (c) 1.861 (d) 0.793 (e) 2.524 (f) 0.774 (g) 0.683 (h) 3.322

EXERCISE 8.5

1 (a) 27 (b) 9 (c) 216 (d) 7 (e) $\frac{1}{125}$ (f) 4 (g) 4 (h) $\sqrt{3}$ (i) 3.5 (j) $49\sqrt{7}$ (k) 16 (l) $16\sqrt{2}$

2 (a) 8 (b) 2 (c) 100 (d) 0.5 (e) 10 (f) ±10

3 (a) $y=\frac{x}{x-1}$ (b) $y=10x^{\frac{3}{2}}$ (c) $y=3x^{\frac{3}{2}}$ (d) $y=\frac{\sqrt{2}}{x}$ (e) $y=25x$ (f) $y=5x^3$ (g) $x^2y^3=1$ (h) $y=\frac{10^x-1}{10^x+1}$

4 (a) 30 (b) 10 (c) 10 (d) $\sqrt[3]{16}$ (e) $1\frac{1}{9}$ (f) $2\frac{1}{4}$

5 (a) incorrect (b) correct (c) correct (d) correct

6 (a) 2.807 (b) 2.631 (c) 0.431 (d) −0.756 (e) 1.699 (f) 2.096 (g) 1.723 (h) 1.161

7 (a) $x>0.43$ (b) $x\geq-1.47$ (c) $x<-1.46$ (d) $x<-3$ (e) $x\geq2.32$ (f) $x<0.29$ (g) $x>-4.92$ (h) $x\leq3.38$

8 B 9 $A=P\times10^{bt}$ 10 $y=ax^n$

11 $x=2^y$ 12 $\log x=2\log a+\frac{3\log b}{2}+\frac{\log c}{2}$ 13 1.1

14 $t=\frac{1}{4}\log_e\frac{y}{a}$ 15 (a) $2p$ (b) $\frac{1}{2p}$

16 $\log_a 2=\log_b 16$, $\log_a 2=\frac{\log_a 16}{\log_a b}$, $\log_a 2=\frac{4\log_a 2}{\log_a b}$, $\log_a b=4$, $b=a^4$

17 (a) 10.24 years (b) 20.49 years (c) 26.48 years

18 (a) 139 months (b) 278 months (c) 359.2 months

19 216 months

20 (a) $y_n=5\times(1+0.05)^n$ (b) $z_n=4\times(1+0.07)^n$ (c) 12 years

EXERCISE 8.6

1 (a) $4e^{4x}$ (b) $e^{\frac{x}{2}}$ (c) $4e^{4x}-3e^{3x}$ (d) $6e^{3x}-e^{-x}$ (e) $12e^{3x}+2e^{-2x}$ (f) $3.2e^{3.2x}-1.6e^{1.6x}$ (g) $3e^x+2e^{-x}$ (h) $8e^{2x}-e^{-2x}$ (i) $3e^{3x}-e^x$

2 B

3 (a) $xe^{3x}(2+3x)$ (b) $(1-2x)e^{-x}$ (c) $(2x^2+4x+3)e^{2x}$ (d) $(1-2x)e^{-2x}$ (e) $x^2e^{-x}(3-x)$ (f) $3x^2-e^{2x}-2xe^{2x}$ (g) $2x-3x^2e^{2x}-2x^3e^{2x}$ (h) $\frac{(x-2)e^x}{x^3}$ (i) $\frac{(3x-1)e^{3x}}{x^2}$ (j) $\frac{x^2(3-x)}{e^x}$ (k) $\frac{(4x-5)e^{4x}}{(x-1)^2}$ (l) $\frac{(2x-1)e^x}{2x\sqrt{x}}$

4 (a) $2e^{2x+3}$ (b) $2(x-1)e^{x^2-2x}$ (c) $-9x^2e^{-x^3}$ (d) $6e^{3x-1}$ (e) $3e^{3x-1}+4e^{4x+2}$ (f) $\frac{1}{2\sqrt{x}}e^{-x}-\sqrt{x}e^{-x}=\frac{(1-2x)e^{-x}}{2\sqrt{x}}$ (g) $12xe^{2x^2}$ (h) $6e^{2x-1}$ (i) $(1+2x^2)e^{x^2}$

5 (a) correct (b) incorrect (c) correct (d) correct

6 $\frac{dx}{dt}=(6+5t)e^{5t}$, $\frac{d^2x}{dt^2}=(35+25t)e^{5t}$;

$(35+25t)e^{5t}-10((6+5t)e^{5t})+25(1+t)e^{5t}=0$

7 $y=x+1$ 8 $y=1-x$

9 $y=2e^2x-e^2$ (a) $(0.5,0)$ (b) $(0,-e^2)$

10 $y = 3 - x, y = x + 3$ **11** $\frac{dV}{dn} = p \times re^{rn} = r \times Pe^{rn} = rV$

12 $100e^{-0.2t}$

EXERCISE 8.7

1 $-e$

2 $(2, \frac{2}{e})$, maximum (a) $x > 0$ (b) $x < 2$

3 (a) $f(0) = 2, f'(0) = 1$ (b) $f'(x) = e^{-x} > 0$ for all x (c) 3

(d)

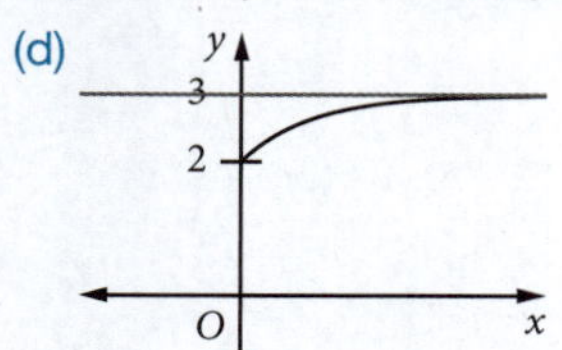

4 (a) $-2xe^{-x^2}$ (b) (i) $x = 0$ (ii) $x < 0$ (iii) $x > 0$

(c)

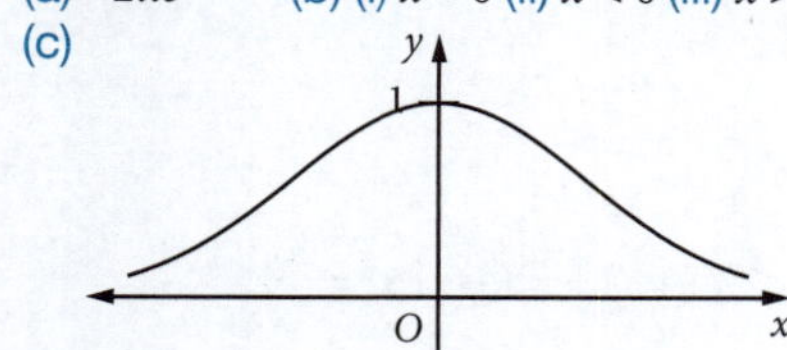

5 (a) 0.1 units after 0.91 hours

(b)

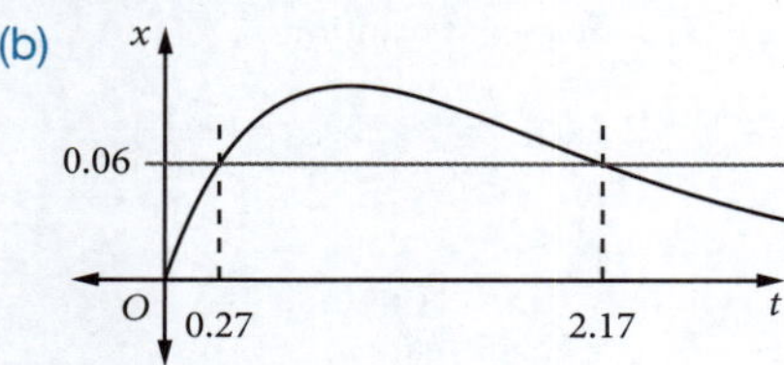

(c) $0.27 < t < 2.17$

6 (a) correct (b) incorrect (c) correct (d) correct

7

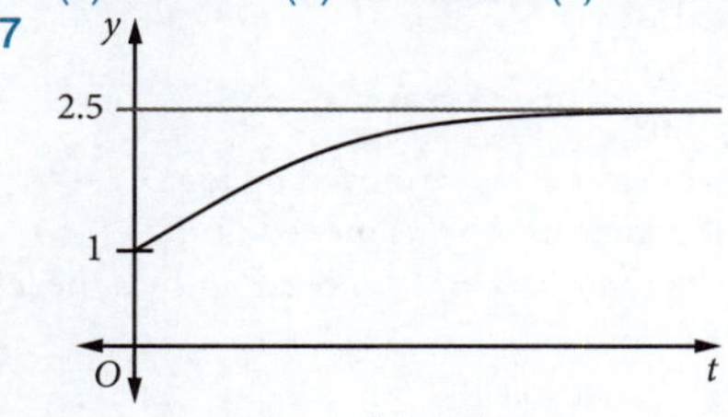

(a) $f'(t) = f\frac{-5 \times \left(-3e^{-t}\right)}{\left(2+3e^{-t}\right)^2} = \frac{15e^{-t}}{\left(2+3e^{-t}\right)^2} > 0$ because $e^{-t} > 0$ and $\left(2+3e^{-t}\right)^2 > 0$ for all t

(b) 2.5 (c) $1 \le f(t) < 2.5$

8 (a) $x = \frac{1}{\sqrt{2}}$ (b) 0.86 units^2

EXERCISE 8.8

1 (a) $2x + 3 = 5, 2x = 2, x = 1$ (b) $4x + 7 = 5x - 3, x = 10$
(c) $x = \log_e 5$

2 (a) $e^{3x+5} = e^2, 3x + 5 = 2, x = -1$ (b) $\frac{x}{4} = \log_e 3, x = 4\log_e 3$

(c) $e^4x = 1.6, 4x = \log_e 1.6, x = \frac{1}{4}\log_e 1.6$

3 (a) $x + 5 = 3, x = -2$

(b) $\log_e(x^2 - 1) = \log_e 8, x^2 - 1 = 8, x^2 = 9, x = \pm 3$. $x = 3$ is the only valid solution.

(c) $\log_e\frac{x-3}{x+1} = \log_e 9, \frac{x-3}{x+1} = 9, x - 3 = 9x + 9, 8x = -12, x = -1\frac{1}{2}$

This solution is not valid since logs are defined for positive values only.

4 (a) $x + 2 = e^3, x = e^3 - 2$

(b) $2x - 2 = e^4, 2x = e^4 + 2, x = \frac{e^4}{2} + 1$

(c) $\log_e\frac{x+2}{x-2} = 1, \frac{x+2}{x-2} = e, x + 2 = e^x - 2e, x(1 - e) = -2(1 + e)$,
$x = \frac{2(e+1)}{e-1}$

5 (a) $\log_e(x + 2e)(x - 2e) = 2, x^2 - 4e^2 = e^2, x^2 = 5e^2, x = \pm e\sqrt{5}$. The negative solution is not valid, so $x = e\sqrt{5}$.

(b) $\log_e x(x - 2e^2) = 2, x(x - 2e^2) = e^2, x^2 - 2e^2x - e^2 = 0$,
$x = \frac{2e^2 \pm \sqrt{4e^4 + 4e^2}}{2} = e^2 \pm e\sqrt{e^2+1}$. The only valid solution is $x = e^2 + e\sqrt{e^2+1}$.

6 (a) $\log_e x(x + 5) = \log_e 6(x + 2), x(x + 5) = 6(x + 2)$,
$x^2 + 5x - 6x - 12 = 0, x^2 - x - 12 = 0, (x - 4)(x + 3) = 0$,
$x = 4$ or -3. $x = -3$ is not valid so $x = 4$ is the only solution.

(b) $\log_e\frac{x}{x+5} = \log_e\frac{x-4}{x+2}, \frac{x}{x+5} = \frac{x-4}{x+2}, x(x+2) = (x-4)(x+5)$,
$x^2 + 2x = x^2 + x - 20, x = -20$. The solution is not valid. The equation has no solutions.

7 D is correct: $\log_e e^3 + \log_e x = \log_e(e^3x)$

EXERCISE 8.9

1

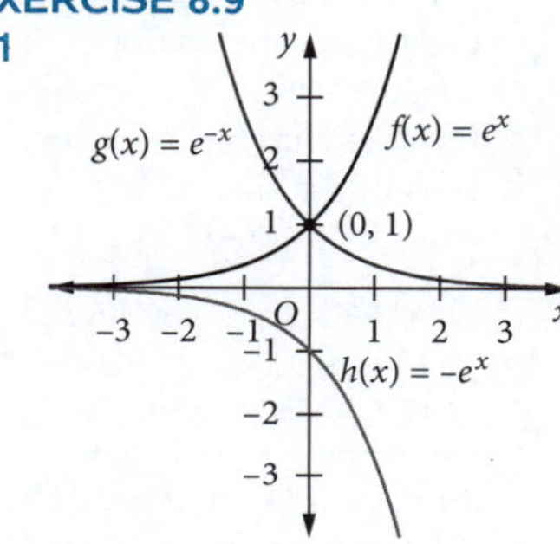

$f(x)$ passes through (0, 1) and $\to 0$ from above as $x \to -\infty$. $g(x)$ passes through (0, 1) and $\to 0$ from above as $x \to \infty$. $h(x)$ passes through (0, –1) and $\to 0$ from below as $x \to -\infty$.

2

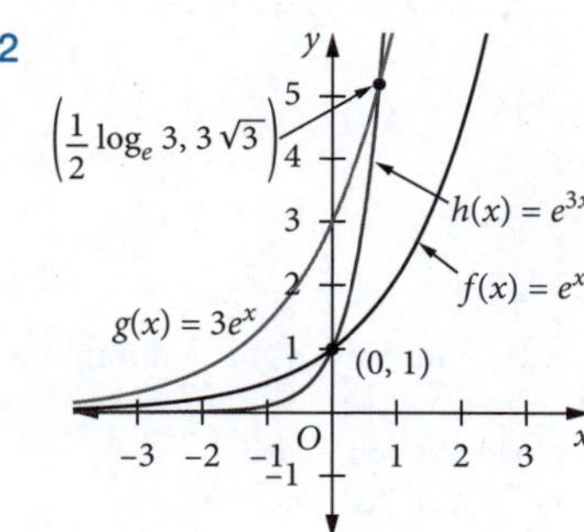

$f(x)$ and $h(x)$ intersect at (0, 1) and both $\to 0$ from above as $x \to -\infty$. When x is negative, $h(x)$ is below $f(x)$, when x is positive, $h(x)$ is above $f(x)$. The values of $g(x)$ are three times the values of $f(x)$ for a given value of x. $g(x)$ is always above $f(x)$.

3

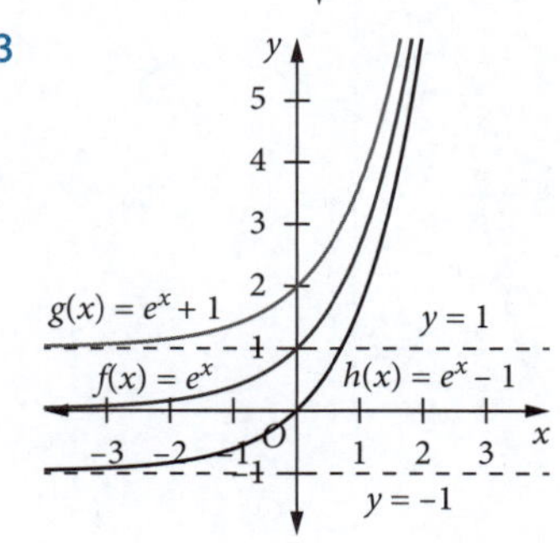

$g(x)$ is $f(x)$ moved up 1 unit, $h(x)$ is $f(x)$ moved down 1 unit. As $x \to -\infty$, $g(x) \to 1$ from above and $h(x) \to -1$ from above. Only $h(x)$ cuts the x-axis. The curves do not intersect.

4

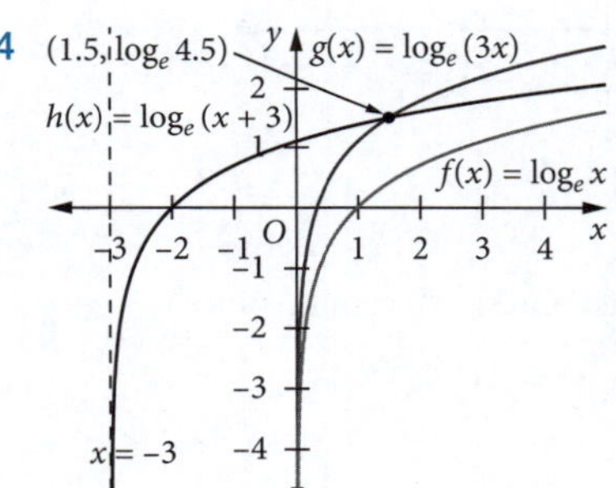

$f(x)$ cuts the x-axis at (1, 0). It $\to -\infty$ as $x \to 0$. $g(x)$ is just $f(x)$ moved up $\log_e 3$ units. It cuts the x-axis at $\left(\frac{1}{3}, 0\right)$ and $\to -\infty$ as $x \to 0$. $h(x)$ is just $f(x)$ moved 3 units to the left. It $\to -\infty$ as $x \to -3$.

5 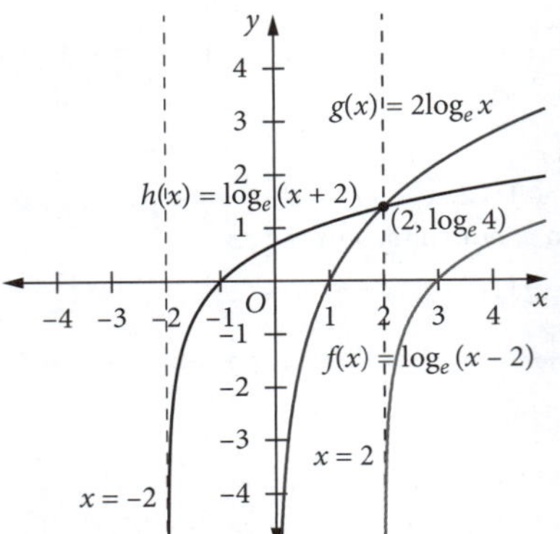

$f(x)$ is the graph of $y = \log_e x$ moved 2 units to the right. It cuts the x-axis at (3, 0) and $\rightarrow -\infty$ as $x \rightarrow 2$. $h(x)$ is the graph of $y = \log_e x$ moved 2 units to the left. It cuts the x-axis at (–1, 0) and $\rightarrow -\infty$ as $x \rightarrow -2$.

6 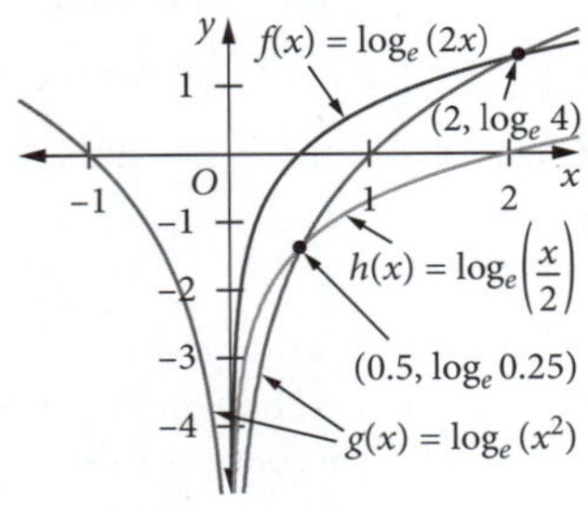

$f(x)$ cuts the x-axis at (0.5, 0) and it is $y = \log_e x$ moved up by $\log_e 2$. $h(x)$ cuts the x-axis at (2, 0) and it is $y = \log_e x$ moved down by $\log_e 2$. $g(x)$ cuts the x-axis at (1, 0) and (–1, 0). The x^2 means that the logarithm can exist for negative values of x as well so the graph has two branches. It is really $y = 2\log_e x$ for $x > 0$ and $y = 2\log_e (-x)$ for $x < 0$.

7 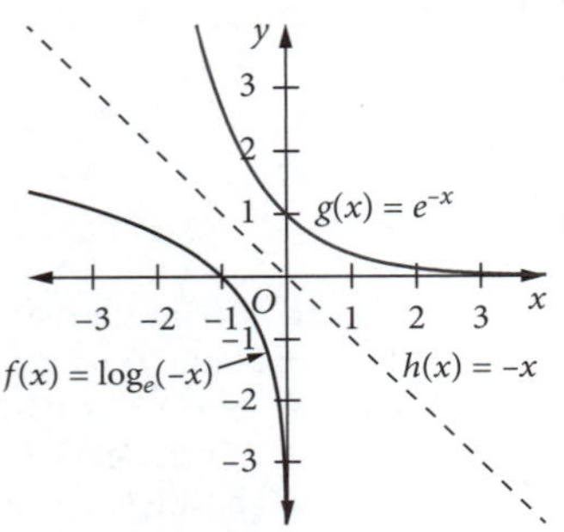

$g(x)$ cuts the y-axis at (0,1) and $\rightarrow 0$ as $x \rightarrow \infty$. $f(x)$ cuts the x-axis at (–1,0) and $\rightarrow -\infty$ as $x \rightarrow 0$. The two graphs are symmetrical about the line $y = -x$.

8 $f(x)$ and $g(x)$ intersect at (0, 1). $f(x) \rightarrow 0$ from above as $x \rightarrow -\infty$. $g(x) \rightarrow 0$ from above as x infinity. $h(x)$ has a minimum value (0, 2) and is contained entirely between the branches of $f(x)$ and $g(x)$ above their point of intersection. $h(x) \rightarrow f(x)$ as x increases, $x > 0$, and it $\rightarrow g(x)$ as x decreases, $x < 0$.

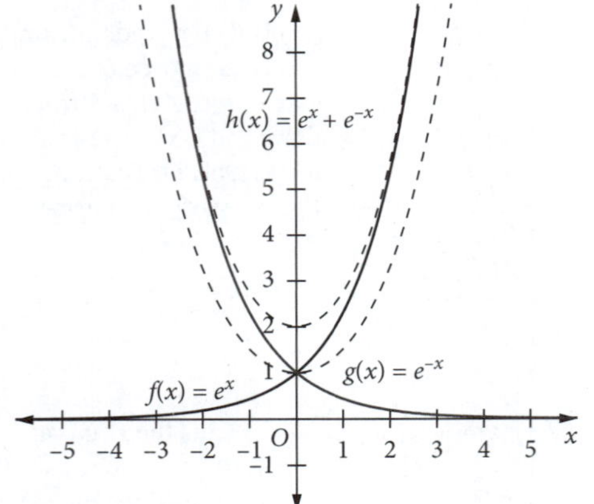

EXERCISE 8.10

1 (a) 0 (b) 50

2 (a) 100 times (b) 10 000 times (c) about 5 times

3 about 3 dB

4 about 160 times stronger

5 about 2200 times more energy

6 4.6, acidic

7 (a) 3 octaves (b) A

8 A sound of zero intensity on the decibel scale would be $10 \log_{10} 0$, which is undefined.

CHAPTER REVIEW 8

1 (a) 1 (b) $x < 5.5$ (c) $x \geq 2.26$

2 (a) 3 (b) $3\log_a y$ (c) $\log_{10} 3$ (d) 0

3 (a) $y = a^x$ (b) $y = \dfrac{100}{x}$

4 (a) $(x^2 + 4x + 2)e^x$ (b) $e^{-3x}(3 - 7x - 3x^2)$ (c) $\dfrac{1}{2\sqrt{x}}e^{\sqrt{x}}$

5 (a) $\dfrac{2^{14} \times 5^3}{3^2}$ (b) 3^2 (c) a^2b (d) $\dfrac{1}{\sqrt[4]{x^5}}$

(e) $2^{-\frac{3}{2}} \times 3^{-\frac{5}{12}} = \dfrac{1}{\sqrt{2^3} \times \sqrt[12]{3^5}}$

6 (a) $3^x = 1, x = 0.\ 2^x = 2^{-3}, x = -3.$
(b) $2^{3x+1} = 2^{-5}, 3x + 1 = -5, x = -2$
(c) $e^{4x+1} = e^{-3}, 4x + 1 = -3, x = -1$

7 (a) $\dfrac{\log_e 5}{\log_e 10}$ (b) $\log_3 27 + \log_3 6 = \log_3 162 = \dfrac{\log_e 162}{\log_e 3}$

8 (a) $x + 2 = 2x, x = 2$
(b) $x^2 - 2x - 3 = 0, (x - 3)(x + 1) = 0, x = -1, 3$
(c) $x^2 = \dfrac{x}{3}, x = \dfrac{1}{3}$ ($x = 0$ is not a valid solution.)

9 (a) $2x = \log_e 2, x = \dfrac{1}{2}\log_e 2$ (b) $x = \log_e 2e, x = 1 + \log_e 2$
(c) $4x = 4 + \log_e 2, x = 1 + \dfrac{1}{4}\log_e 2$

10 (a) $2 = 1.03^n, n = \dfrac{\log_e 2}{\log_e 1.03} \approx 23.45$ years
(b) $2.5 = 1.03^n, n = \dfrac{\log_e 2.5}{\log_e 1.03} \approx 31$ years
(c) $20 = 1.03^n, n = \dfrac{\log_e 20}{\log_e 1.03} \approx 101.35$ years

11 $f(x) = \log_e (3x)$ is the same as $y = \log_e x$ moved up $\log_e 3$,
$g(x) = \log_e (3x) + 1$ is the same as $y = f(x)$ moved up 1,
$h(x) = \log_e (3x + 1)$ is the same as $y = f(x)$ moved up 1, to the left by $\dfrac{1}{3}$.

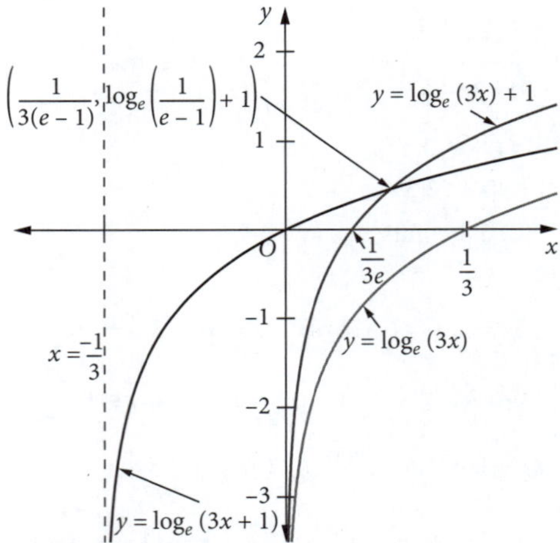

12 $\dfrac{d}{dt}(\theta_0 e^{-kt}) = -k\theta_0 e^{-kt} = -k\theta$

13 (a) \$10 000 (b) \$3012
(c) (i) \$602.39 per year (ii) \$1000 per year
(d) 11.5 years

14 (a) 192.46 (b) 79

CHAPTER 9

EXERCISE 9.1

1 no, no   2 Answers will vary.

3 (a) states have different populations
(b) there are fewer cars in Hobart

4 C   5 7, 35, 70, 105, 140   6 5, 10, 15, 20, 25

7 (a) correct (b) incorrect (c) incorrect (d) correct

8 (a) $\frac{1}{4}$ (b) $\frac{3}{4}$

9 (a) $\frac{1}{5}$ (b) $\frac{2}{5}$ (c) $\frac{3}{5}$ (d) $\frac{3}{5}$ (e) $\frac{1}{5}$ (f) $\frac{3}{5}$

10 (a) $\frac{3}{10}$ (b) $\frac{1}{5}$ (c) $\frac{9}{20}$ (d) $\frac{1}{20}$

11 (a) $\frac{2}{3}$ (b) $\frac{1}{6}$ (c) $\frac{1}{2}$ (d) $\frac{1}{6}$

12 (a) $\frac{1}{2}$ (b) $\frac{1}{2}$ (c) 1 (d) $\frac{1}{2}$ (e) $\frac{2}{5}$ (f) $\frac{1}{5}$ (g) $\frac{1}{5}$ (h) $\frac{1}{2}$

13 (a) $\frac{1}{3}$ (b) $\frac{2}{3}$ (c) $\frac{1}{3}$

14 (a) $\frac{1}{4}$ (b) $\frac{1}{13}$ (c) $\frac{4}{13}$ (d) $\frac{1}{52}$ (e) $\frac{1}{52}$

15 (a) $\frac{5}{9}$ (b) $\frac{1}{9}$ (c) $\frac{4}{9}$ (d) $\frac{4}{9}$ (e) $\frac{1}{3}$ (f) $\frac{1}{9}$ (g) $\frac{2}{3}$ (h) $\frac{5}{9}$

16 (a) $\frac{1}{2}$ (b) $\frac{3}{10}$ (c) $\frac{4}{5}$ (d) $\frac{7}{10}$ (e) $\frac{1}{5}$

17 (a) 5 (b) (i) $\frac{2}{5}$ (ii) $\frac{2}{5}$ (iii) $\frac{4}{5}$ (iv) $\frac{4}{5}$

18 (a) $\frac{3}{5}$ (b) $\frac{7}{10}$ (c) $\frac{2}{5}$ (d) $\frac{3}{10}$ (e) $\frac{2}{5}$

19 (a) $\frac{6}{11}$ (b) $\frac{8}{11}$ (c) $\frac{2}{11}$

20 (a) 0.96 (b) 0.04 (c) 0.9 (d) 0.48

21 (a) $\frac{1}{4}$ (b) $\frac{1}{8}$ (c) $\frac{3}{8}$ (d) $\frac{3}{4}$ (e) $\frac{3}{4}$

22 (a) 0.4 (b) 0.6 (c) 0.079

23 (a) $\frac{18}{37}$ (b) $\frac{18}{37}$ (c) $\frac{13}{37}$ (d) $\frac{12}{37}$ (e) $\frac{1}{37}$

EXERCISE 9.2

1 B

2 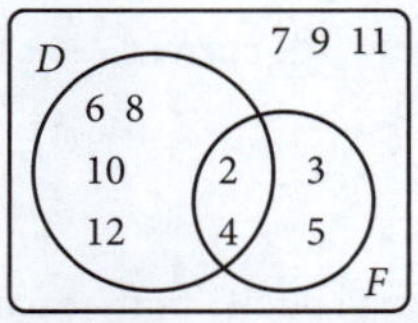

(a) {2, 3, 4, 5, 6, 8, 10, 12}
(b) {2, 4}

3 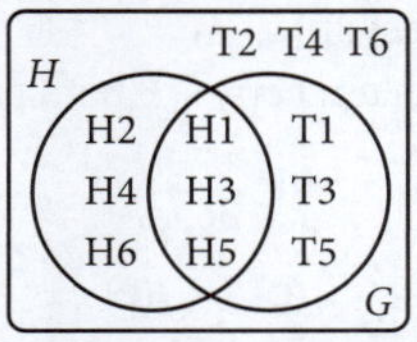

(a) {H1, H2, H3, H4, H5, H6, T1, T3, T5}
(b) 3

4 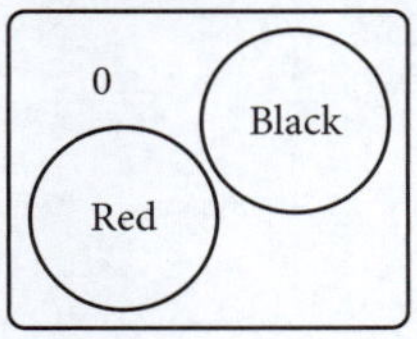

(a) {1, 2, 3, …36} (b) {} (c) 36
(d) 0 (e) 37

5 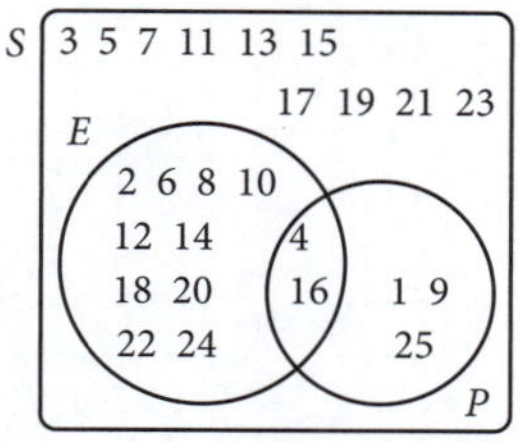

(a) 25 (b) 12 (c) 5
(d) 2 (e) 15

6 (a) $n(s) = 25$; $n(F \cup G) = 25 - 5 = 20$; $20 = 18 + 12 - n(F \cap G)$;
$n(F \cap G) = 10$; French only $= 18 - 10 = 8$
(b) German only $= 12 - 10 = 2$ (c) 10

7 (a) $\{3, 6, 9, 12, 15, 18\} = K$ (b) $\{6, 12, 18\} = L$ (c) 3

8 (a) {} (b) {1, 2, 3, 4, 5, 6} (c) {5} (d) 1

9 (a) 0 (b) 6 (c) 36 (d) 24

10 (a) incorrect (b) correct (c) correct (d) correct

EXERCISE 9.3

1 (a) $\frac{1}{4}$ (b) $\frac{1}{52}$ (c) $\frac{1}{13}$ (d) $\frac{7}{26}$ (e) $\frac{4}{13}$ (f) $\frac{1}{13}$ (g) A and B

2 (a) $\frac{5}{17}$ (b) $\frac{2}{17}$ (c) $\frac{3}{17}$ (d) $\frac{12}{17}$ (e) $\frac{15}{17}$ (f) $\frac{7}{17}$ (g) $\frac{7}{17}$
(h) $\frac{12}{17}$ (i) $\frac{15}{17}$ (j) 1 (k) A and B, B and C

3 D

4 (a) incorrect (b) incorrect (c) correct (d) correct

5 (a) $\frac{1}{2}$ (b) $\frac{1}{8}$ (c) $\frac{1}{2}$ (d) $\frac{1}{2}$ (e) 1 (f) $\frac{5}{8}$ (g) 1
(h) A and C, B and C

6 (a) $\frac{4}{25}$ (b) $\frac{1}{10}$ (c) $\frac{3}{50}$ (d) $\frac{1}{50}$ (e) $\frac{33}{50}$ (f) $\frac{3}{5}$ (g) $\frac{23}{50}$ (h) $\frac{18}{25}$

7 (a) $\frac{7}{36}$ (b) $\frac{5}{36}$ (c) $\frac{1}{18}$ (d) $\frac{1}{36}$ (e) 1 (f) $\frac{17}{18}$ (g) $\frac{7}{18}$ (h) 1

8 (a) $\frac{13}{18}$ (b) $\frac{7}{12}$ (c) $\frac{11}{36}$ (d) 1 (e) no

9 no

10 $\frac{11}{16}$ 11 (a) $\frac{1}{5}$ (b) $\frac{1}{10}$ (c) $\frac{3}{10}$

12 (a) $\frac{3}{5}$ (b) $\frac{3}{10}$ (c) $\frac{2}{3}$

EXERCISE 9.4

1 (a) $\frac{1}{4}$ (b) $\frac{1}{6}$

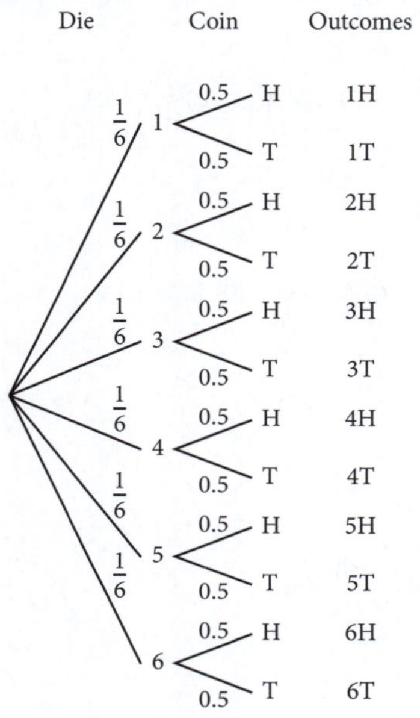

2 Cube 1 Cube 2 Outcomes (a) $\frac{1}{3}$ (b) $\frac{2}{9}$

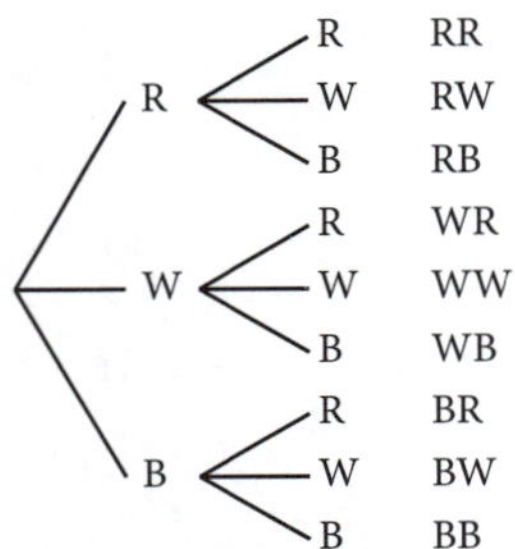

3 $\frac{1}{12}$ **4** (a) 27 (b) $\frac{1}{3}\times\frac{1}{3}\times\frac{2}{3}=\frac{2}{27}$ (c) $\frac{1}{3}\times\frac{2}{3}\times\frac{1}{3}=\frac{2}{27}$

5 C **6** (a) $\frac{1}{8}$ (b) $\frac{3}{8}$ (c) $\frac{3}{8}$ (d) $\frac{1}{8}$

7 (a) incorrect (b) correct (c) incorrect (d) correct

8 (a) Cube 1 Cube 2 Cube 3 Outcomes

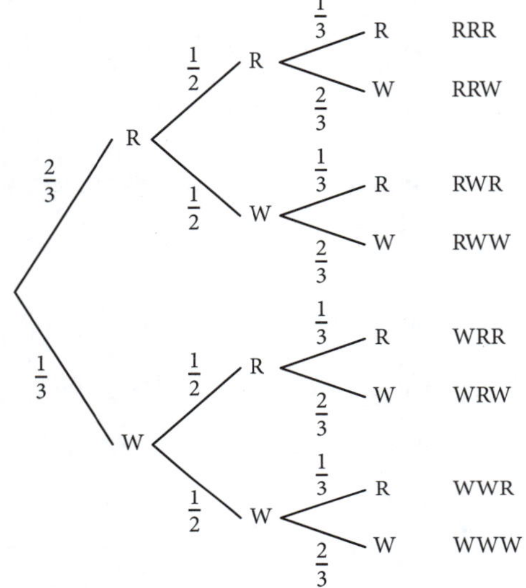

(b) $\frac{2}{3}\times\frac{1}{2}\times\frac{1}{3}=\frac{1}{9}$ (c) $\frac{2}{3}\times\frac{1}{2}\times\frac{2}{3}+\frac{2}{3}\times\frac{1}{2}\times\frac{1}{3}+\frac{1}{3}\times\frac{1}{2}\times\frac{1}{3}=\frac{7}{18}$

(d) $\frac{1}{9}+\frac{7}{18}=\frac{1}{2}$

9 (a) $\frac{1}{8}$ (b) $\frac{1}{8}$ (c) $\frac{1}{2}$ **10** (a) $\frac{1}{16}$ (b) $\frac{9}{16}$ (c) $\frac{1}{4}$

EXERCISE 9.5

1 (a) $\frac{1}{2}$ (b) $\frac{1}{3}$ (c) $\frac{1}{6}$

2 (a) correct (b) correct (c) correct (d) incorrect

3 D

4 (a) $\frac{1}{6}$ (b) $\frac{1}{6}$ (c) $\frac{1}{36}$ (d) $\frac{1}{4}$ (e) 0

5 (a) yes (b) no (c) no

6 (a) $\frac{2}{5}$ (b) $\frac{1}{2}$ (c) $\frac{1}{10}$ (d) $\frac{1}{15}$

7 (a) $\frac{9}{100}$ (b) $\frac{49}{100}$ (c) $\frac{51}{100}$ (d) $\frac{9}{100}$

8 (a) $\frac{1}{10}$ (b) $\frac{9}{20}$ (c) $\frac{3}{10}$ (d) $\frac{3}{20}$

9 (a) $\frac{64}{125}$ (b) $\frac{1}{125}$ (c) $\frac{124}{125}$

10 (a) $\frac{1}{16}$ (b) $\frac{9}{16}$ (c) $\frac{3}{8}$ (d) $\frac{3}{16}$ (e) $\frac{3}{16}$ (f) $\frac{7}{16}$

11 (a) $\frac{1}{9}$ (b) $\frac{1}{9}$ (c) $\frac{2}{9}$ (d) $\frac{2}{9}$ (e) $\frac{8}{9}$

12 (a) $\frac{1}{24}$ (b) $\frac{1}{4}$ (c) $\frac{1}{4}$ (d) $\frac{1}{12}$ (e) $\frac{1}{8}$

13 (a) $\frac{1}{8}$ (b) $\frac{63}{1000}$ (c) $\frac{3}{100}$ (d) $\frac{3}{40}$ (e) $\frac{7}{8}$ (f) $\frac{64}{125}$

14 (a) $\frac{1}{16}$ (b) $\frac{1}{16}$ (c) $\frac{1}{16}$ (d) $\frac{1}{16}$ (e) $\frac{15}{16}$

15 (a) $\frac{9}{200}$ (b) $\frac{9}{50}$ (c) $\frac{27}{50}$ (d) $\frac{1}{200}$

16 (a) $\frac{8}{27}$ (b) $\frac{1}{27}$ (c) $\frac{4}{27}$

17 $\frac{11}{21}$

18 (a) $\frac{2}{5}$ (b) $\frac{7}{15}$ (c) $\frac{13}{15}$

19 (a) (b) $\frac{37}{60}$

First branch		Second branch		Probability
A	$\frac{1}{3}$	black	$\frac{3}{5}$	$\frac{1}{5}$
		white	$\frac{2}{5}$	$\frac{2}{15}$
B	$\frac{1}{3}$	black	$\frac{3}{4}$	$\frac{1}{4}$
		white	$\frac{1}{4}$	$\frac{1}{12}$
C	$\frac{1}{3}$	black	$\frac{1}{2}$	$\frac{1}{6}$
		white	$\frac{1}{2}$	$\frac{1}{6}$

20 (a) $\frac{39}{40}$ (b) $\left(\frac{39}{40}\right)^2$ (c) $\left(\frac{39}{40}\right)^5$ (d) $1-\left(\frac{39}{40}\right)^5$

21 (a) $\frac{1}{1000}$ (b) $\frac{999}{1000}$

22 (a) $\frac{1}{208}$ (b) $\frac{1}{78}$ (c) $\frac{1}{120}$ (d) $\frac{1}{9360}$

23 (a) 0.27 (b) 0.97 (c) 0.03 (d) 0.32 (e) 0.162

EXERCISE 9.6

1 (a) $\frac{5}{14}$ (b) $\frac{13}{28}$ (c) $\frac{15}{28}$ **2** B **3** $\frac{47}{91}$ **4** $\frac{1}{220}$

5 (a) $\frac{2}{3}$ (b) $\frac{5}{12}$ (c) $\frac{1}{2}$

6 (a) correct (b) incorrect (c) correct (d) correct

7 (a) $\frac{1}{22}$ (b) $\frac{1}{11}$ (c) $\frac{5}{33}$

8 (a) $\frac{7}{22}$ (b) $\frac{5}{33}$ (c) $\frac{35}{66}$

(d) The possibilities 'one teacher and one pupil', 'both teachers' and 'both pupils' are the only possibilities, so $P(\text{one teacher one pupil}) = 1 - P(\text{both teachers}) - P(\text{both pupils})$.

9 (a) $\frac{1}{35}$ (b) $\frac{4}{35}$ (c) $\frac{4}{35}$ **10** $\frac{1}{90}$ **11** $\frac{17}{38}$

12 (a) $\frac{3}{7}$ (b) $\frac{1}{3}$ (c) $\frac{8}{15}$ **13** (a) $\frac{2}{15}$ (b) $\frac{1}{3}$ (c) $\frac{8}{15}$

14 $\frac{3}{26}$ **15** $\frac{2}{3}$ **16** $\frac{1}{504}$

17 (a) $\frac{1}{5525}$ (b) $\frac{1}{2197}$

18 (a) $\frac{36}{125}$ (b) $\frac{44}{125}$ (c) $\frac{117}{125}$

19 (a) 0.001 (b) 0.729 (c) 0.028

20 $\frac{3}{52}$ **21** (a) $\frac{20}{29}$ (b) $\frac{2}{87}$ (c) $\frac{9}{29}$

22 (a) $\frac{2}{9}$ (b) $\frac{5}{9}$ (c) $\frac{4}{45}$

23 (a) 0.343 (b) 0.973 **24** (a) 0.729 (b) 0.972 (c) 0.028

25 (a) $\frac{4}{49}$ (b) $\frac{24}{49}$ **26** (a) $\frac{13}{25}$ (b) $\frac{12}{25}$

27 (a) $\frac{1}{4}$ (b) $\frac{3}{4}$

28 (a) $\frac{1}{3}$ (b) $\frac{2}{3}$

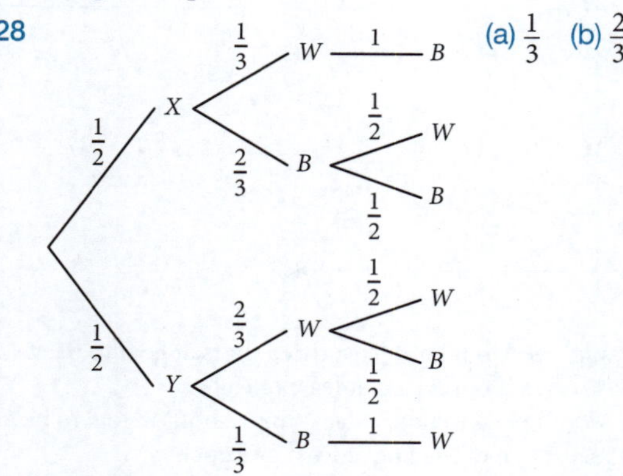

29 0.648

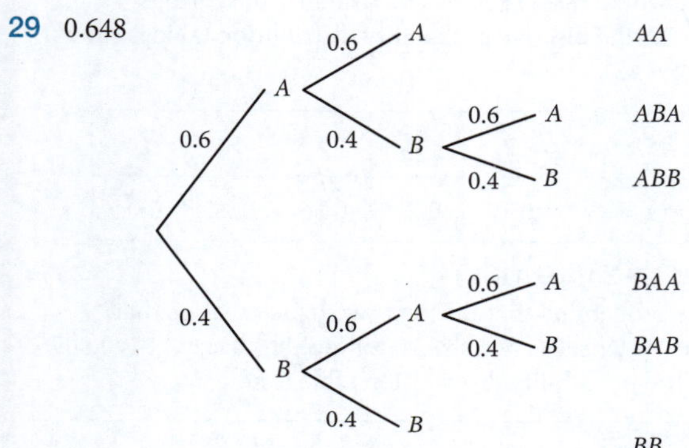

30 0.305

CHAPTER REVIEW 9

1 0.001

2 (a) $\frac{1}{6}$ (b) $\frac{5}{6}$ (c) $\frac{3}{8}$ (d) $\frac{1}{3}$ **3** (a) $\frac{5}{18}$ (b) $\frac{13}{18}$

4 (a) $\frac{54}{125}$ (b) $\frac{81}{125}$ **5** (a) $\frac{1}{6}$ (b) $\frac{1}{3}$

6 (a) $\frac{8}{25}$ (b) $\frac{2}{25}$ (c) $\frac{4}{5}$ (d) $\frac{2}{5}$

7 $\frac{7}{12}$ **8** (a) $\frac{2}{25}$ (b) $\frac{6}{25}$ (c) $\frac{4}{25}$ (d) $\frac{12}{25}$

9 (a) $\frac{1}{216}$ (b) $\frac{125}{216}$ (c) $\frac{1}{8}$ (d) $\frac{1}{8}$ (e) $\frac{5}{216}$ (f) $\frac{5}{216}$

10 (a) $\frac{8}{27}$ (b) $\frac{1}{27}$ (c) $\frac{4}{27}$ (d) $\frac{4}{27}$

11 0.89 Monday Tuesday Wednesday

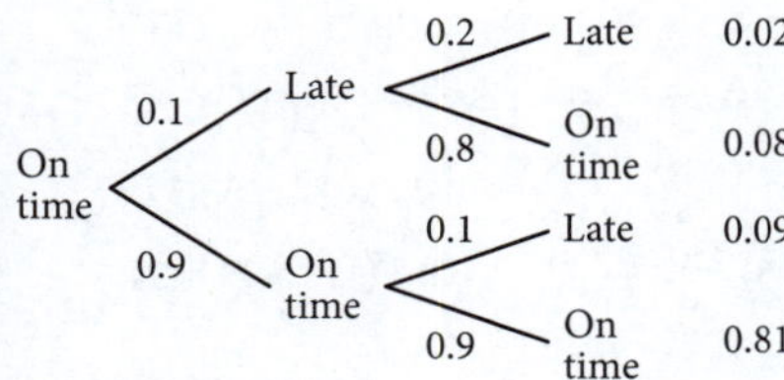

12 (a) 0.0125 (b) 0.336

13 $\frac{6}{11}$ **14** (a) $\frac{3}{10}$ (b) $\frac{1}{15}$ (c) $\frac{7}{15}$ (d) $\frac{7}{15}$

15 $\frac{22}{425}$ **16** (a) $\frac{1}{5}$ (b) $\frac{7}{25}$

17 (a) $\frac{11}{850}$ (b) $\frac{1}{5525}$ (c) $\frac{997}{1700}$ **18** (a) $\frac{124}{125}$ (b) $\frac{13}{125}$

19 (a) $\frac{135}{512}$ (b) $\frac{485}{512}$ **20** (a) 0.343 (b) 0.784 (c) 0.973

21 (a) $\frac{7}{15}$ (b) $\frac{8}{15}$ **22** (a) $\frac{2}{9}$ (b) $\frac{7}{9}$

CHAPTER 10

EXERCISE 10.1

1 (a) $\frac{3}{8}$ (b) $\frac{7}{8}$ (c) $\frac{1}{4}$ (d) $\frac{1}{2}$ (e) $\frac{1}{6}$

2 Wayne would need to take at least four shots at goal to be at least 95% certain of scoring at least one goal.

3 (a) Yes, the sum is 1 and all values are between zero and one, inclusive.
(b) No, the sum is greater than 1.
(c) Yes, the sum is 1 and all values are between zero and one, inclusive.
(d) No, the sum is 1 but not all values are between zero and one, inclusive.
(e) Yes, the sum is 1 and all values are between zero and one, inclusive.
(f) Yes, the sum is 1 and all values are between zero and one, inclusive.

4 (a) $k=\frac{2}{9}$ (b) $k=\frac{1}{6}$ (c) $k=\frac{2}{9}$ (d) $k=\frac{1}{12}$

5 (a) The number of students in each of your classes at school is associated with a discrete scale.
(b) The height of teachers at your school is associated with a continuous scale.
(c) The sizes of the shirts worn by each of the students in your mathematics class is associated with a discrete scale.
(d) The neck circumference of each of the members of the Australian netball team is associated with a continuous scale.
(e) The number of tosses of a fair coin before a head is observed is associated with a discrete scale.
(f) The time it takes for a process worker to complete 50 items is associated with a continuous scale.
(g) The distance jumped in the long jump by the competitors at the sports carnival is associated with a continuous scale.
(h) The number of red lights you stop at, per day, on the way to school over the course of a month is associated with a discrete scale.
(i) The number of whole lessons missed, per student, this year by members of your English class is associated with a discrete scale.
(j) The retail price in cents per litre charged for petrol over the course of a year is associated with a discrete scale.
(k) The number of people in the queue when you enter the bank each Friday for a year is associated with a discrete scale.
(l) The number of passionfruit collected from each vine in your orchard this season is associated with a discrete scale.

6 (a) C
(b) The probability distribution of X is shown in the table.

x	0	1	2
$P(X=x)$	0.49	0.42	0.09

(c) The probability of drawing exactly one red card from the pack is 0.32.
(d) The probability of drawing at least one red card from the pack is 0.96.

7 D

8 B

9 (a) $\frac{3}{8}$ (b) 0.24 (c) 0.06 (d) 8 spins

10 (a) $P(X=2)=0.3$, $P(X=3)=0.6$, $P(X=4)=0.1$
(b) Bar graph

11 (a) 2; 12
(b) The probability distribution of X is shown in the table.

x	2	3	4	5	6	7	8	9	10	11	12
$P(X=x)$	$\frac{1}{36}$	$\frac{1}{18}$	$\frac{1}{12}$	$\frac{1}{9}$	$\frac{5}{36}$	$\frac{1}{6}$	$\frac{5}{36}$	$\frac{1}{9}$	$\frac{1}{12}$	$\frac{1}{18}$	$\frac{1}{36}$

(c) $P(X \geq 6)=\frac{13}{18}$ (d) $P(X<10)=\frac{5}{6}$ (e) $P(4 \leq X \leq 10)=\frac{5}{6}$

12 (a) The probability distribution of T is shown in the table.

t	0	1	2
$P(T=t)$	0.25	0.5	0.25

(b) If you drew a graph of this distribution you could draw a dot graph.

13 (a)

x	1	8	27	64	125	216
$P(X=x)$	$\frac{1}{6}$	$\frac{1}{6}$	$\frac{1}{6}$	$\frac{1}{6}$	$\frac{1}{6}$	$\frac{1}{6}$

(b) $P(X<100)=\frac{2}{3}$

14 (a) The probability distribution of X is shown in the table.

x	1	2	3	4	5	6	9	16	25	36
$P(X=x)$	$\frac{1}{6}$	$\frac{1}{12}$	$\frac{1}{12}$	$\frac{1}{6}$	$\frac{1}{12}$	$\frac{1}{12}$	$\frac{1}{12}$	$\frac{1}{12}$	$\frac{1}{12}$	$\frac{1}{12}$

(b) $P(X \leq 12)=\frac{3}{4}$ (c) $P(X \geq 16)=\frac{1}{4}$ (d) $P(X \leq 9 \mid \text{H})=\frac{1}{2}$

15 (a) The probability distribution of X is shown in the table.

x	0	1	2	3
$P(X=x)$	$\frac{8}{125}$	$\frac{36}{125}$	$\frac{54}{125}$	$\frac{27}{125}$

(b) On a particular day, the probability that Enzo stops at:
(i) exactly two sets of traffic lights is $\frac{54}{125}$
(ii) no more than two sets of traffic lights is $\frac{98}{125}$
(iii) at least two sets of traffic lights is $\frac{81}{125}$
(iv) fewer than two sets of traffic lights is $\frac{44}{125}$.
(c) The answers to (b) (iii) and (iv) add to 1 because 'at least two' and 'fewer than two' are complementary events and complementary events add to 1.

16 (a) The probability distribution of Y is shown in the table.

y	2	3	4	5	6	7	8	9	10	11	12
$P(Y=y)$	$\frac{1}{81}$	$\frac{4}{81}$	$\frac{6}{81}$	$\frac{8}{81}$	$\frac{11}{81}$	$\frac{12}{81}$	$\frac{14}{81}$	$\frac{8}{81}$	$\frac{9}{81}$	$\frac{4}{81}$	$\frac{4}{81}$

(b) $P(Y \geq 4)=\frac{76}{81}$ (c) $P(Y<7)=\frac{10}{27}$
(d) $P(3 \leq Y \leq 9)=\frac{7}{9}$ (e) $P(Y \geq 6 \mid Y \leq 10)=\frac{54}{73}$

17 (a) $k=\frac{20}{49}$
(b) The probability distribution table is shown below.

X	1	2	3	4	5	6
$P(X=x)$	$\frac{20}{49}$	$\frac{10}{49}$	$\frac{20}{147}$	$\frac{5}{49}$	$\frac{4}{49}$	$\frac{10}{147}$

(c)

X	1	2	3	4	5	6
$P(X=x)$	0.41	0.20	0.14	0.10	0.08	0.07

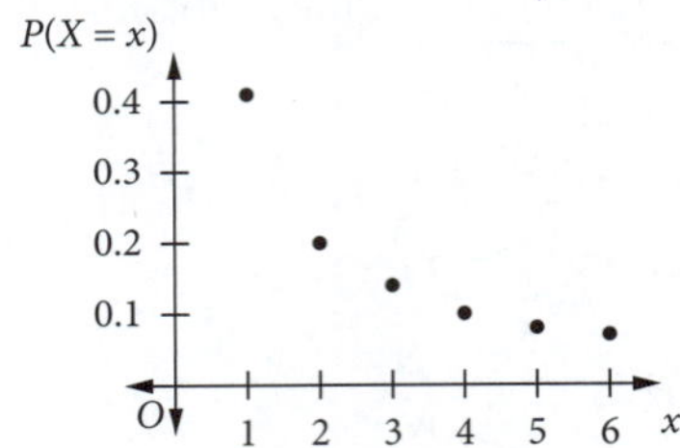

18 (a) Lou would need to take at least three shots at goal to be at least 95% certain of scoring at least one goal.
(b) Buddy would need to take at least three shots at goal to be at least 95% certain of scoring at least one goal.
(c) You would need to have a 63% chance of scoring.

19 (a) $a=48$; the discrete probability distribution table is shown below.

X	7	8	9	10	11
$P(X=x)$	0.12	0.24	0.36	0.24	0.04

(b) $P(7<X<10)=0.6$
(c) The probability that the next two pairs of shoes that Alexander sells are a size 10 followed by a size 11 is 0.0096.
(d) (i) The probability distribution table is now:

X	7	8	9	10	11
$P(X=x)$	0.125	0.24	0.350	0.24	0.045

(ii) The probability that the next two pairs of shoes that Alexander sells are a size 10 followed by a size 11 is 0.0108.

EXERCISE 10.2

1 (a) $E(X)=4.3$ (b) $E(X)=0.5$ (c) $E(X)=4\frac{17}{18}$ (d) $E(X)=1\frac{1}{4}$

2 (a) $a=0.1$ and $b=0.3$ (b) $i=0.1$ and $j=0.25$
(c) $i=\frac{1}{20}$ and $j=\frac{1}{5}$ (d) $a=\frac{1}{5}$ and $b=\frac{1}{3}$

3 (a) $E(X)=1.55$ (b) $E(2X-3)=0.1$
(c) $E(X^2-5)=-0.95$ (d) $E(X^2+X-4)=1.6$

4 (a) The expected value is 3.3 and the variance is 2.61.
(b) The expected value is 6.9 and the variance is 1.79.

5 Standard deviation of $X=1.4663 \approx 1.47$.

6 (a) The standard deviation of $Y=1.024 \approx 1.02$.
(b) $P(\mu-2\sigma \leq Y \leq \mu+2\sigma)=0.98$

7 (a) $k=\frac{1}{24}$, $E(X)=3\frac{5}{6}$ (b) $k=\frac{1}{20}$, $E(X)=-1\frac{1}{10}$

8 (a) $E(W)=4\frac{7}{8}$ (b) $E(3W-4)=10\frac{5}{8}$
(c) $E(2W+5)=14\frac{3}{4}$ (d) $E(W^2-7)=21\frac{3}{4}$

9 D

10 C

11 Eric would be better off taking the first salary package.

12 (a) If you rolled this die twenty times, the expected (mean) value of the data, stated in mixed number form, would be $3\frac{1}{2}$.
(b) The range is 0.73 to 6.27.
(c) As the range of $\mu \pm 2\sigma$ does cover all of the possible results when rolling the die, none of the values should be considered unusual.

13 (a) $w = 10$ (b) $\text{Var}(T) = 1.6475 \approx 1.65$
(c) $\sigma = 1.28$ (d) $\text{Var}(2T - 6) = 6.59$
(e) $\text{Var}(5 - 3T) = 14.8275 \approx 14.83$

14 (a) $k = \frac{1}{20}$ (b) $E(G) = 1\frac{1}{2}$
(c) $\text{Var}(G) = 1\frac{3}{4}$ (d) $\sigma = 1.3229 \approx 1.32$

15 $E(X) = 1 \times 0.2 + 2 \times 0.2 + 3 \times 0.1 + 4 \times 0.5$
$= 0.2 + 0.4 + 0.3 + 0.2$
$= 2.9$
$E(Y) = 2 \times 0.2 + 3 \times 0.3 + 4 \times 0.4 + 5 \times 0.1$
$= 0.4 + 0.9 + 1.6 + 0.5$
$= 3.4$
$E(X) + E(Y) = 2.9 + 3.4$
$= 6.3$

Y↓ X→	1	2	3	4
2	3 : 0.04	4 : 0.04	5 : 0.02	6 : 0.1
3	4 : 0.06	5 : 0.06	6 : 0.03	7 : 0.15
4	5 : 0.08	6 : 0.08	7 : 0.04	8 : 0.2
5	6 : 0.02	7 : 0.02	8 : 0.01	9 : 0.05

z	3	4	5	6	7	8	9
$P(X + Y = z)$	0.04	0.1	0.16	0.23	0.21	0.21	0.05

$E(X + Y) = 3 \times 0.04 + 4 \times 0.1 + 5 \times 0.16 + 6 \times 0.23 + 7 \times 0.21 + 8 \times 0.21 + 9 \times 0.05$
$= 0.12 + 0.4 + 0.8 + 1.38 + 1.47 + 1.68 + 0.45$
$= 6.3$
So, $E(X + Y) = E(X) + E(Y)$

16 (a) (i) If the player selects a black card, Enrico loses \$1.
(ii) If the player selects a diamond, Enrico loses \$5.
(b) Enrico keeps the game fee if a heart is drawn.
(c) Let X = amount won by Enrico (\$).

x	−5	−1	p
$P(X = x)$	$\frac{1}{4}$	$\frac{1}{2}$	$\frac{1}{4}$

(d) When $p = 7$, the expected value for the distribution is 0.
(e) Enrico must charge more than \$7 if he wishes to make a profit. Hence, the minimum whole dollar amount he should charge to play the game would be \$8.

17 (a)

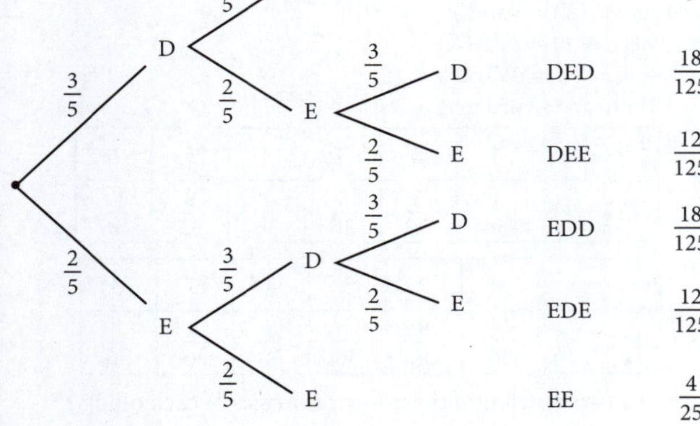

(b)

x	2	3
$P(X = x)$	$\frac{13}{25}$	$\frac{12}{25}$

(c) The expected number of sets they play in a match is $2\frac{12}{25}$.
(d) The expected result can never occur as tennis is played in whole sets. However, since the expected result $2\frac{12}{25}$ is closer to 2 than 3, this means they are more likely to play 2 sets than 3.

18 (a)

x	Y	B	R
$P(X = x)$	$\frac{5}{9}$	$\frac{1}{3}$	$\frac{1}{9}$
Payment	\$1	\$3	\$5

(b) $E(X) = 2\frac{1}{9}$
(c) If the game is to be fair, you should pay \$2.11 to play.
(d) It is unlikely that this would be the cost of any game in real life, so players would likely be charged \$2.50 or \$3.00 to play the game, in which case the operator will make a profit.

19 (a) (i) The probability that the player receives \$1 is $\frac{25}{216}$.
(ii) The probability that the player receives \$2 is $\frac{25}{216}$.
(iii) The probability that the player receives \$3 is $\frac{5}{216}$.
(iv) The probability that the player receives \$5 is $\frac{25}{216}$.
(v) The probability that the player receives \$6 is $\frac{5}{216}$.
(vi) The probability that the player receives \$7 is $\frac{5}{216}$.
(vii) The probability that the player receives \$8 is $\frac{1}{216}$.
(viii) The probability that the player receives \$0 is $\frac{125}{216}$.
(b) The expected return on this game is $\$1\frac{1}{3}$ which is approximately equal to \$1.35.
(c) If the game is fair, the operator should charge \$1.35 to play.

20 (a) $E(Y) = 3$ (b) $E(X) = 1$
(c) (i) The new probability distribution table is:

m	1	2	3	4	5	6	7	8
$P(M = m)$	0.06	0.13	0.21	0.3	0.1	0.09	0.07	0.04

(ii) $E(M) = 4$
(d) $E(M) = E(X) + E(Y)$

EXERCISE 10.3

1 (a) $E(R) = 8\frac{1}{2}$ (b) $\text{Var}(R) = 21\frac{1}{4}$

2 (a) The expected number of attempts = 5
(b) $\text{Var}(X) = 6\frac{2}{3}$

3 (a) $E(F) = 2\frac{1}{2}$ (b) $\text{Var}(F) = 1\frac{1}{4}$

4 (a) C (b) D

5 (a) (i) $E(X) = 3$ (ii) $\text{Var}(X) = 2$
(b) (i) $E(Y) = 4$ (ii) $\text{Var}(Y) = 2$
(c) $Y = X + 1$
$E(Y) = E(X) + 1$
(d) $Y = X + 1$
$\text{Var}(Y) = \text{Var}(X)$
(e) (i) $E(Z) = 8$ (ii) $\text{Var}(Z) = 2$

6 (a) The minutes possible are: 00, 01, 02, …, 58, 59.
So, there are 60 numbers that could be represented as: 00, 01, 02, …, 58, 59.
For 60 numbers, the median is between the 30th and 31st values.
Here, the 30th value is 29 and the 31st value is 30.
Therefore the median is 29.5.
This is a uniform distribution. The median value coincides with the mean, or expected value.
So, $E(T) = 29.5$.
(b) $\text{Var}(T) = 299.917$
$\sigma(T) = 17.318$
(c) If the numbers shown were 1 to 60, instead of 0 to 59, then the values would be the same.

7 (a) The probability distribution for the sum on a fair, six-sided die numbered 1 to 6 that is rolled twice is:

s	2	3	4	5	6	7	8	9	10	11	12
$P(S = s)$	$\frac{1}{36}$	$\frac{1}{18}$	$\frac{1}{12}$	$\frac{1}{9}$	$\frac{5}{36}$	$\frac{1}{6}$	$\frac{5}{36}$	$\frac{1}{9}$	$\frac{1}{12}$	$\frac{1}{18}$	$\frac{1}{36}$

(b)

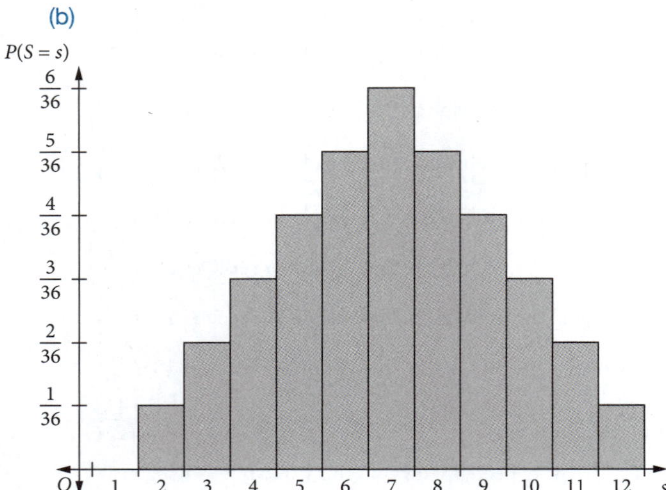

(c) The graph is symmetrical in shape around the value $s = 7$ and is triangular.
(d) (i) $E(S) = 7$ (ii) $\text{Var}(S) = 5\frac{5}{6}$
(e) The expected value and the variance for the double roll are each double the respective single roll values.
(f) (i) $E(V) = 2\frac{1}{2}$ (ii) $\text{Var}(V) = 1\frac{1}{4}$
(g) (i) $E(T) = 5$ (ii) $\text{Var}(T) = 2\frac{1}{2}$
(h) The expected value and the variance for the double spin are each double the respective single spin values.
(i) When two identical uniform discrete variables are added the expected value and variance are both double the single values.

EXERCISE 10.4

1 (a)

x	1	2	3	4
$P(X = x)$	$\frac{1}{2}$	$\frac{5}{12}$	$\frac{17}{216}$	$\frac{1}{216}$

(b) $E(X) = 1\frac{127}{216}$
(c) $\text{Var}(X) = \frac{19\,943}{46\,656}$

2 (a)

z	1	−2	−3	4	−5	6
$P(Z = z)$	$\frac{1}{6}$	$\frac{1}{6}$	$\frac{1}{6}$	$\frac{1}{6}$	$\frac{1}{6}$	$\frac{1}{6}$

(b) This is best described as a uniform distribution.
(c) $E(Z) = \frac{1}{6}$
(d) The game is not fair as it is biased in favour of the player.

3 (a) X can take three values. (b) $P(X = 2) = \frac{1}{6}$

4 (a) $k = 0.05$
(b) The probability distribution is shown in the table.

x	1	2	3	4	5	6	7	8
$P(X = x)$	0.05	0.1	0.15	0.2	0.2	0.15	0.1	0.05

(c) $E(X) = 4.5$ (d) $\text{Var}(X) = 3.25$

5 (a) All possible combinations of boys and/or girls for a family with three children is:
Sample space: {BBB, BBG, BGB, BGG, GBB, GBG, GGB, GGG}.
(b) The probability that a family with three children will have at least one boy is $\frac{8261}{9261} = 0.8920$.
(c) The proportion of families in this town with exactly four children that will have at least three girls is $\frac{54\,000}{194\,481} = 0.2777$.
(d) The proportion of families with exactly four children will have two girls and two boys is $\frac{24\,200}{64\,827} = 0.3733$.

6

x	1	2	3	4	5
$P(X = x)$	$\frac{3}{5}$	$\frac{2}{5}$	$\frac{1}{5}$	0	$-\frac{1}{5}$

Tomino's probability distribution is incorrect because there is a negative probability for $x = 5$ and probabilities cannot be negative.

7 (a) $E(X) = 2$ (b) $\text{Var}(X) = 0.5$
(c) The probability distribution of $X + Y$ is shown in the table.

z	2	3	4	5	6
$P(X + Y = z)$	$\frac{1}{16}$	$\frac{1}{4}$	$\frac{3}{8}$	$\frac{1}{4}$	$\frac{1}{16}$

(d) $E(X + Y) = 4$ (e) $\text{Var}(X + Y) = 1$
(f) $E(X) = E(Y)$:
$E(X + Y) = E(2X)$
$= 2E(X)$
$= 2E(Y)$
(g) As $\text{Var}(X) = \text{Var}(Y)$:
$\text{Var}(X + Y) = 2\text{Var}(X)$
$= 2\text{Var}(Y)$

8 (a) There are 36 ordered pairs.
(b)

z	1	2	3	4	5	6
$P(Z = z)$	$\frac{1}{36}$	$\frac{3}{36}$	$\frac{5}{36}$	$\frac{7}{36}$	$\frac{9}{36}$	$\frac{11}{36}$

(c)

y	1	2	3	4	5	6
$P(Y = y)$	$\frac{11}{36}$	$\frac{9}{36}$	$\frac{7}{36}$	$\frac{5}{36}$	$\frac{3}{36}$	$\frac{1}{36}$

(d) The two distributions are mirror images of each other.
(e) (i) $E(Z) = 4\frac{17}{36}$ (ii) $E(Y) = 2\frac{19}{36}$
(f) The difference between the greatest value of Z and $E(Z) = 1\frac{19}{36}$.

(g) The difference between the least value of Y and $E(Z)$ is $1\frac{19}{36}$.

(h) The difference between the least value of Y and $E(Y) = 1\frac{19}{36}$, which is equal to the difference between the greatest value of Z and $E(Z)$.

(i) $\text{Var}(Y) = \text{Var}(Z) = 1\frac{926}{1296}$

9 (a) The expected value for this random variable is $E(X) = 3.008$.

(b) The expected value for this random variable is $E(Y) = 3.894$

(c) (i) The probability they have the same value is $P(X = Y) = 0.207$.

(ii) The probability they have different values is $P(X \neq Y) = 0.793$.

(iii) The probability that their sum is greater than 8 is $P(X + Y > 8) = 0.151$

10 (a) $E(X) = 2.6$

(b) $E(2X) = 2 \times 0.25 + 4 \times 0.1 + 6 \times 0.45 + 8 \times 0.2$
$= 5.2$
$= 2 \times 2.6$
$= 2E(X)$

(c) $\text{Var}(X) = 1.14$

(d) (i) $\text{Var}(2X) = 4.56$ (ii) $\text{Var}(3X) = 10.26$

(e) From part (d)(i):
$\text{Var}(2X) = 4\text{Var}(X)$
From part (d)(ii):
$\text{Var}(3X) = 9\text{Var}(X)$
Conclusion:
$\text{Var}(kX) = k^2\text{Var}(X)$

11 (a) $E(X) = \sum x\, p(x)$
$= 0.80n + (n+1) \times 0.12 + (n+2) \times 0.05 + (n+3) \times 0.03$
$= 0.8n + 0.12n + 0.12 + 0.05n + 0.1 + 0.03n + 0.09$
$= n + 0.31$

(b) The probability of choosing two consecutive values is $P(X \text{ and } X+1) = 0.207$.

(c) The probability that the sum of the two values is even is $P(\text{even sum}) = 0.745$.

(d) The probability that four independent values of X are one of each type is $P(\text{one of each}) = 0.0035$.

CHAPTER REVIEW 10

1 (a) $k = \frac{5}{16}$ (b) $k = \frac{1}{18}$ (c) $k = \frac{1}{15}$

2 (a)

x	6	7	8	9	10	11	12	13	14	15	16
$P(X = x)$	$\frac{1}{36}$	$\frac{1}{18}$	$\frac{1}{12}$	$\frac{1}{9}$	$\frac{5}{36}$	$\frac{1}{6}$	$\frac{5}{36}$	$\frac{1}{9}$	$\frac{1}{12}$	$\frac{1}{18}$	$\frac{1}{36}$

(b) $P(X \geq 11) = \frac{7}{12}$ (c) $P(X < 15) = \frac{11}{12}$ (d) $P(7 \leq X \leq 11) = \frac{5}{9}$

3 (a)

x	2	3	4	5	6	7	8
$P(X = x)$	$\frac{1}{16}$	$\frac{1}{8}$	$\frac{3}{16}$	$\frac{1}{4}$	$\frac{3}{16}$	$\frac{1}{8}$	$\frac{1}{16}$

(b) (i) $E(X) = 5$ (ii) $\text{Var}(X) = 2\frac{1}{2}$

4 (a) $E(X) = 0.25$ (b) $\text{Var}(X) = 1.8875$
(c) $E(3X - 2) = -1.25$ (d) $E(X^2 - 2X) = 1.45$
(e) $\text{Var}(3X - 2) = 16.9875$

5 For a fair game the cost should be \$0.50.

6 (a) $E(R) = 10\frac{1}{2}$ (b) $\text{Var}(R) = 33\frac{1}{4}$

7 $n = 15$

8 $E(X) = 25$

9 (a) 5 points

(b) If Yehudi won 10 of the first 15 points I would *be very* surprised as the number is *well above* the expected number.

10 (a)

y	0	1	2	3
$P(Y = y)$	$\frac{7}{24}$	$\frac{21}{40}$	$\frac{7}{40}$	$\frac{1}{120}$

(b) $E(Y) = \frac{9}{10}$

11 D 12 C 13 B

14 (a) B (b) D (c) A

15 (a) D (b) B (c) A

16 C 17 A

18 (a) C (b) D (c) A

19 (a) The probability distribution of the number of times Audrey hits *Play*, given the variable X:

x	1	2	3
$P(X = x)$	$\frac{2}{31}$	$\frac{23}{31}$	$\frac{6}{31}$

(b) $E(X) = 1\frac{25}{36}$ (c) $\text{Var}(X) = \frac{347}{1296} \approx 0.268$

20 (a) $a = \frac{1}{4}$; $b = \frac{1}{12}$

(b)

x	1	2	3	4	5	6
$P(X = x)$	$\frac{1}{4}$	$\frac{1}{12}$	$\frac{1}{4}$	$\frac{1}{12}$	$\frac{1}{4}$	$\frac{1}{12}$

(c) $E(X) = 1 \times \frac{1}{4} + 2 \times \frac{1}{12} + 3 \times \frac{1}{4} + 4 \times \frac{1}{12} + 5 \times \frac{1}{4} + 6 \times \frac{1}{12}$
$= \frac{13}{4}$
$= 3\frac{1}{4}$

(d) $E(X^2) = 1 \times \frac{1}{4} + 4 \times \frac{1}{12} + 9 \times \frac{1}{4} + 16 \times \frac{1}{12} + 25 \times \frac{1}{4} + 36 \times \frac{1}{12}$
$= \frac{161}{12}$
$= 13\frac{5}{12}$

$\text{Var}(X) = E(X^2) - [E(X)]^2$
$= \frac{161}{12} - \left(\frac{13}{4}\right)^2$
$= \frac{137}{48}$
$= 2\frac{41}{48}$

(e) The probability that the sum of two independent observations from this distribution is greater than 9 is $P(Z > 9) = \frac{1}{8}$.

21 (a)

z	1	2	3	5	6	7	8	9	10
$P(Z = z)$	$\frac{1}{6}$	$\frac{1}{6}$	$\frac{1}{6}$	$\frac{7}{36}$	$\frac{7}{36}$	$\frac{1}{36}$	$\frac{1}{36}$	$\frac{1}{36}$	$\frac{1}{36}$

(b) The expected value is $E(Z) = 4\frac{1}{12}$.

(c) $P(Z > 6) = \frac{1}{9}$ (d) $P(Z < 7) = \frac{8}{9}$

(e) $P(Z > 4 \mid Z \leq 8) = \frac{8}{17}$

22 Calculate $\sum_{x=3}^{5} P(x)$ and $\sum_{x=1}^{5} P(x)$

$\sum_{x=3}^{5} P(x) = \frac{3}{15} + \frac{5}{15} + \frac{7}{15}$
$= \frac{15}{15}$
$= 1$

$\sum_{x=1}^{5} P(x) = \frac{-1}{15} + \frac{1}{15} + \frac{3}{15} + \frac{5}{15} + \frac{7}{15}$
$= \frac{15}{15}$
$= 1$

Check that all probabilities satisfy the condition $0 \le P(X = x) \le 1$ For the second set the probability when $X = 1$ is negative so this value is not valid.

23 (a) If the technology produced truly random numbers, you would expect the results to follow a uniform discrete distribution with consecutive data values.

(b) The probability distribution table is:

x	1	2	3	4	5	6
$P(X = z)$	$\frac{17}{96}$	$\frac{16}{96}$	$\frac{17}{96}$	$\frac{12}{96}$	$\frac{14}{96}$	$\frac{20}{96}$

(c) The sample is not big enough to make a definitive statement. However, the results do tend to indicate that a uniform distribution is involved as the difference between the most frequently occurring number and the least frequently appearing number is not great.

(d) The $E(X)$ for the sample is very close to the theoretical value. The Var(X) for the sample is about 9% higher than the theoretical value.

(e) Answers will vary. It is likely the results will be different to those already calculated.

(f) There are likely to be slight differences between each set of values.

(g) The bigger the sample, the closer you would expect the values to be to the theoretical values.

24 (a) The number of rolls and the associated probabilities is shown in the table.

x	1	2	3	4	5
$P(X = x)$	$\frac{1}{2}$	$\frac{1}{4}$	$\frac{1}{8}$	$\frac{1}{16}$	$\frac{1}{16}$

(b) The expected number of rolls is $E(X) = 1.9$.

(c) This is just the same as the die question as the relevant probabilities are equal. So the expected number of flips is 1.9.

25 (a) The value is $k = \frac{1}{125}$.

(b)

X	1	2	3	4	5	6
$P(X = x)$	$\frac{6}{25}$	$\frac{29}{125}$	$\frac{26}{125}$	$\frac{21}{125}$	$\frac{14}{125}$	$\frac{1}{25}$

(c)

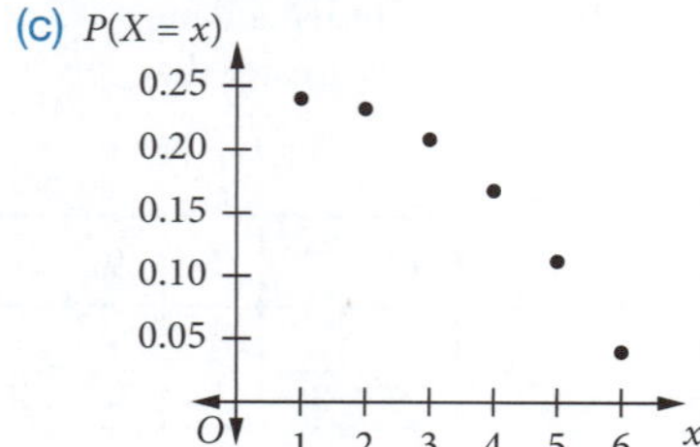

26 (a) Karina and Achim should expect to print and sell three packs of business cards per hour.

(b) Karina and Achim will make an average profit of $45.20 per hour.

27 (a) Plot 3 and plot 4 represent probability distribution functions while plot 1 and plot 2 do not represent probability distribution functions.

(b) The mean and variance of the plots that represent probability distribution functions are:
Plot 3: $E(X) = 7$ and $\text{Var}(X) = 8.2$
Plot 4: $E(X) = 23$ and $\text{Var}(X) = 421$

CHAPTER 11

EXERCISE 11.1

1 (a) The statistics were compiled in 2010.
(b) (i) 6.2%
(ii) 30.2%
(iii) 8.3%

2 (a) This graph has a vertical axis on both sides with a different scale on each because there are two different types of information shown on the graph.
(b) February
(c) February
(d) False. The average daily rainfall in January must be *less* than in February because the same number of rainy days produce *less* rainfall in January than in February.
(e) True.
(f) False. The production of this graph most likely falls into the *analysis* phase of the statistical investigation process.

3 B

4 D

5 (a) False. If asked to interpret the numbers in the graph you *would not* need to read any article that may have appeared with the graph.
(b) 19.3% (c) 224 612
(d) Some of the adoptions did not have the gender reported.
(e) 62.9%

6 (a) 2.5% (b) 32
(c) B. (C and E are not realistic suggestions.)
(d) 210

7 (a) Responses will vary but could include reference to the relative danger of the different types of exercise equipment.
(b) Missing information may include: time period of the data collection, age group and gender of people, type of medical care required.
(c) This may alter your reaction in the previous parts.
(d) Responses will vary, but could include that for children under four, accidental injuries were likely caused by unsupervised and/or incorrect use of the equipment.

8 (a)–(e) Answers may vary.
(f) Answers may vary, but percentages unlikely to be the same.

EXERCISE 11.2

1 (a) numerical (b) categorical (c) categorical
(d) numerical (e) numerical (f) numerical

2 (a) nominal (b) ordinal (c) nominal
(d) ordinal (e) nominal (f) nominal

3 (a) continuous (b) discrete (c) discrete
(d) continuous (e) continuous (f) continuous

4 (a) ordinal (b) continuous (c) discrete

5 (a) (i) Answers may vary. Sample answer: What is your favourite brand of soft drink? You may want to limit responses, so you might provide a list for the surveyed group to choose from.
(ii) Data is categorical as you are asking for 'brand' of drink. Order does not matter, so data is nominal.
(b) (i) Answers may vary. Sample answer: What is your height in cm?
(ii) Data collected will be numerical and continuous.
(c) (i) Answers may vary. Sample answer: From the list below, circle your favourite singer.
If they are ranking all the singers the question might be: Rank the singers from 1 to 10, with 1 being most popular to 10 being your least popular.
(ii) Data collected will be categorical and ordinal.
(d) (i) Answers may vary.
(ii) The collected data will be categorical and nominal.

(e) (i) Answers may vary. Sample answer: What type of dog food do you buy: dry, wet or other?
(ii) The collected data will be categorical and nominal.
(f) (i) Answers may vary. Sample answer: What was your age at your last birthday?
(ii) Collected data will be numerical and discrete.

6 Continuous data should have an infinite number of possibilities within a particular range, but measuring devices can only be accurate to hundredths or thousandths of a second. This data could then be counted, hence appears discrete.

7 The numbers actually represent a category. When completing a survey it is often easier to circle a number rather than words such as 'strongly agree'. A circled number represents a category, so the data is categorical.

EXERCISE 11.3

1 (a)

Colour	Frequency
Red	8
Blue	8
Green	7
Yellow	7

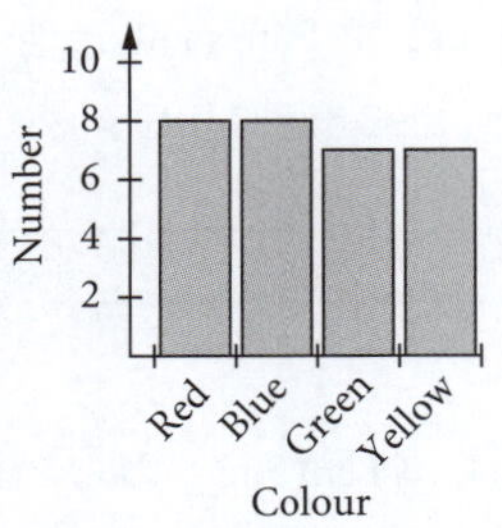

(b) Green and yellow each have a frequency of 7.

2 (a)

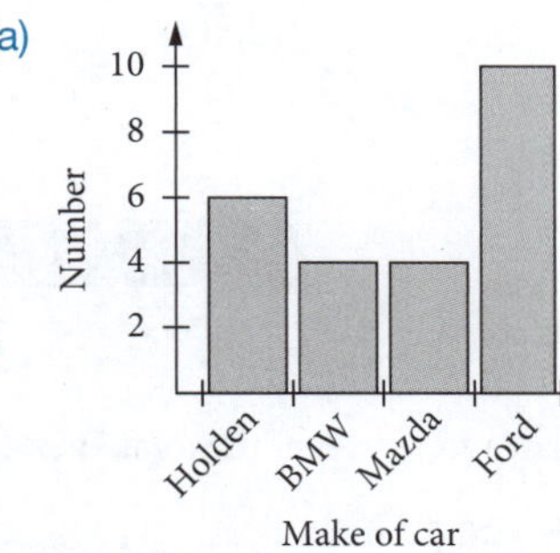

(b) The car make Holden has a frequency of 6 and the car make Ford has a frequency of 10.

3 (a)

Colour	Frequency
Blonde	9
Brown	10
Red	6

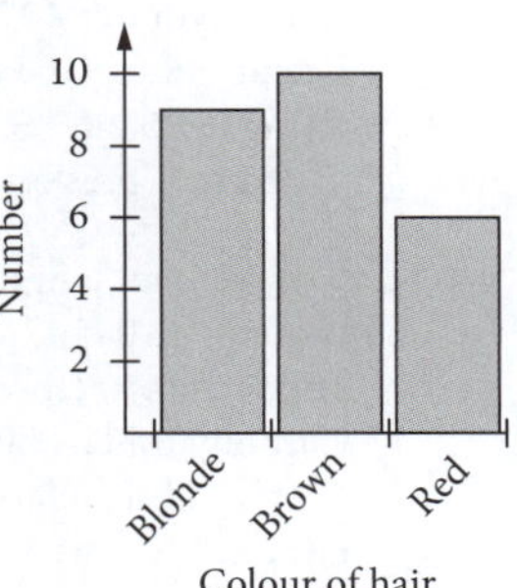

(b) The least frequently observed hair colour is red.

4 (a)

Type of complaint	Number of complaints	Percentage	Cumulative percentage
Punctuality	120	44	44
Lack of seats	80	30	74
Cleanliness	30	11	85
Slow trip	20	7	92
Cost of trip	15	6	98
Attitude of the driver	5	2	100
	270	**100**	

(b)

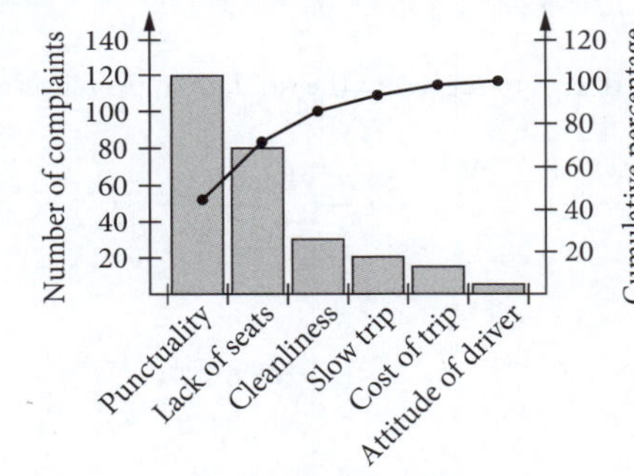

(c) The Company needs to have the buses run on time, have more buses and clean them better.

5 (a)

Cause	Frequency	Percentage	Cumulative percentage
Genetics	55	50	50
Cosmetic damage	25	23	73
Stress	11	10	83
Smoking	10	9	92
Vitamin deficiency	4	4	95
Medica-tion	3	3	98
Other treatments	2	2	100
	110	**100**	

(b)

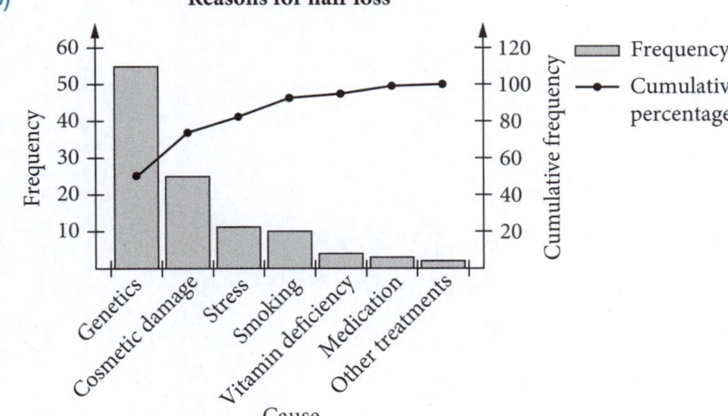

6 (a) 35 (b) 78 (c) 30% (d) 0.13

7 (a) 16 (b) 21 and 42 (c) 29

8 (a) (i) 17 cm

(ii) Protein: 1.20 cm
Total fat: 0.38 cm
Carbohydrate: 3.26 cm
Dietary fibre: 1.28 cm
Other: 10.88 cm

Grams

Protein
Total fat
Carbohydrate
Dietary fibre
Other

(iii) If a 68 cm bar is drawn to represent the total, then protein should take 4.8 cm and carbohydrate should take 13.04 cm.

(b) (i) A convenient length is 12.5 cm.

(ii) Protein: 0.60 cm
Total fat: 0.09 cm
Carbohydrate: 1.43 cm
Dietary fibre: 0.65 cm
Other: 9.73 cm

Grams

Protein
Total fat
Carbohydrate
Dietary fibre
Others

(iii) If 50 cm represented the total, then protein should take 2.4 cm and carbohydrate should take 5.72 cm.

(c) (i) A convenient length is 15 cm.

(ii) Protein: 0.10 cm
Total fat: 0.10 cm
Carbohydrate: 4.80 cm
Dietary fibre: 0.30 cm
Other: 9.70 cm

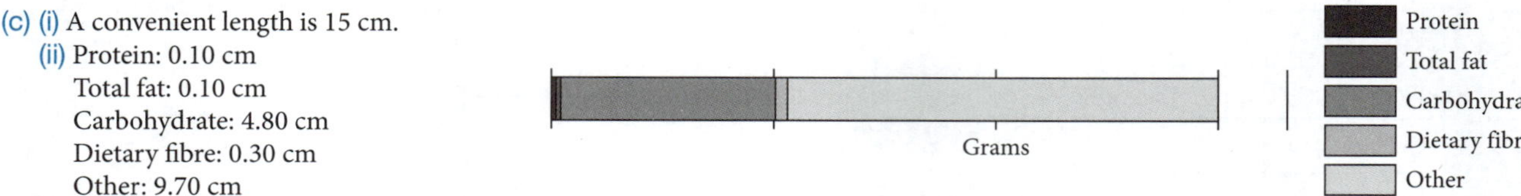

(iii) If a 60 cm bar is drawn to represent the total, then protein should take 0.4 cm and carbohydrate should take 19.2 cm.

(d) (i) A convenient length is 10 cm.

(ii) Protein: 0.32 cm
Total fat: 1.68 cm
Carbohydrate: 0.62 cm
Other: 7.38 cm

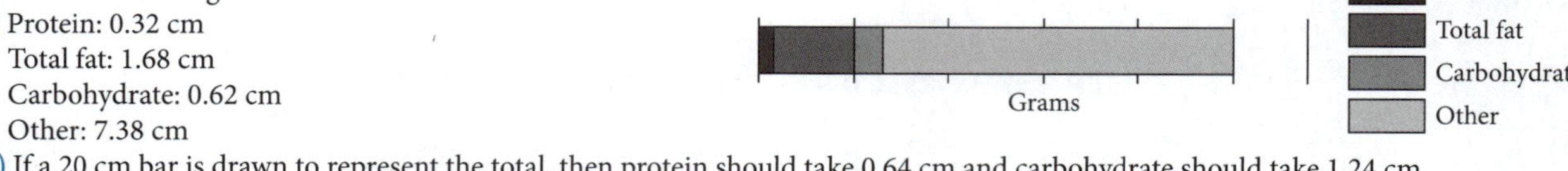

(iii) If a 20 cm bar is drawn to represent the total, then protein should take 0.64 cm and carbohydrate should take 1.24 cm.

9 (a)

Stem	Leaf
3	5 6
4	4 7 8
5	1 4 6
6	6 7
7	0 5 7
8	0 2 2 2 4 4
9	0 1 2 3 7

Key: 7|0 = 70

(b)

Stem	Leaf
3	2 5 7 9
4	2 2
5	0 6
6	5 6 6 7 7
7	1 2 8 8
8	1 7
9	1 2 2 3 8

Key: 7|0 = 70

10 B

11 (a) 88 − 64 = 24

(b)

Stem	Leaf
6_L	4
6_H	6 8 9 9
7_L	0 0 1 1 1 2 2 3 3 4 4
7_H	5 6 6 6 6 7 7 7 7 7 8 9 9
8_L	0 0 1 2 2 2 3 4 4
8_H	7 8 8

Key: 8|0 = 80

12 (a) (b)

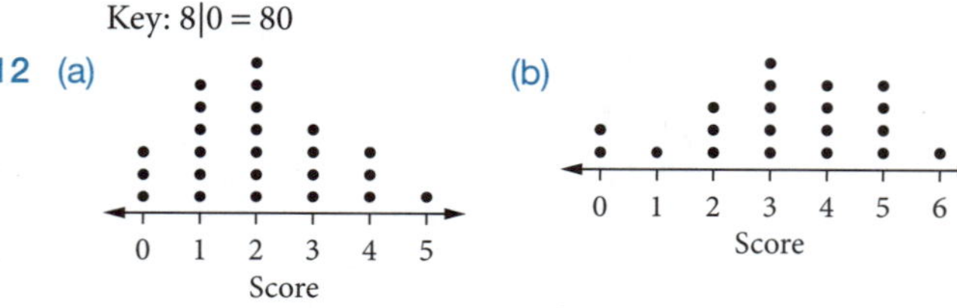

13 (a)

Frequency
18
16
14
12
10
8
6
4
2
10 20 30 40 50 60
Score

The lowest score that could have been included in the data is 0; the highest score that could have been included in the data is 59.9.

(b)

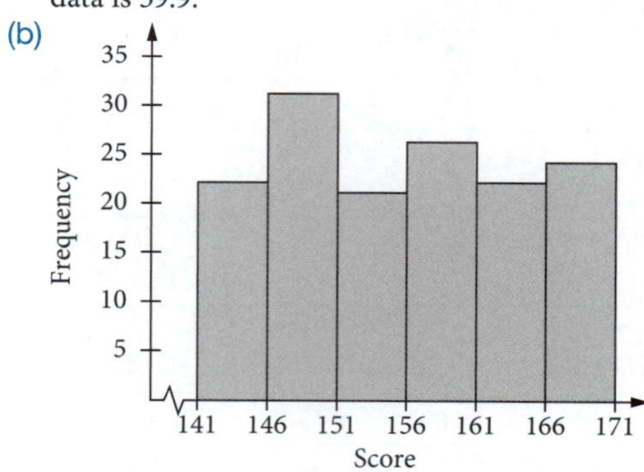

The lowest score that could have been included in the data is 141; the highest score that could have been included in the data is 170.9.

14 (a)

Stem	Leaf
1_H	5 8
2_L	1 2 2 4 4
2_H	5 6 6 7
3_L	0 0 1 1
3_H	5 7
4_L	0 1 2 4
4_H	5 5 6

Key: $3|1 = 31$

(b)

Stem	Leaf
1_H	6 7 9
2_L	0 2 2
2_H	7 9 9
3_L	0 0 1 1 2 3
3_H	5 6
4_L	0 1 1 2 3
4_H	5 7

Key: $3|1 = 31$

15 (a) Data is continuous because it has been measured.

(b)

Kilometres travelled	Frequency
61–<71	4
71–<81	5
81–<91	2
91–<1019	3
101–<111	1
111–<121	4
121–<131	3
131–<141	4
141–<151	3
151–<161	1

(c) Maximum distance travelled is between 151 and 161 km.

(d)

Stem	Leaf
6	5 6 7 8
7	8 8 8 9
8	9 0 4
9	3 5 9
10	5
11	2 4 5 7
12	2 5 7
13	3 3 4 4
14	4 5 5
15	
16	0

Key: $6|5 = 65$

(e) The stem plot gives the actual data values. The frequency table with grouped data does not show individual data values.

(f) Stem plots show actual discrete data values, whereas frequency tables do not.

16 (a)

	Football	Swimming	Basketball	Total
Male	32	30	8	70
Female	18	28	12	58
Total	50	58	20	128

(b) 58 (c) $\frac{4}{35}$ (d) $\frac{25}{64}$

17 (a)

Defect	Frequency	Percentage	Cumulative percentage
Faulty light	17	36	36
Faulty fan	12	26	62
Faulty element	9	19	81
Damaged box	4	9	90
Scratched glass	3	6	96
Other	2	4	100
	47	**100**	

(b)

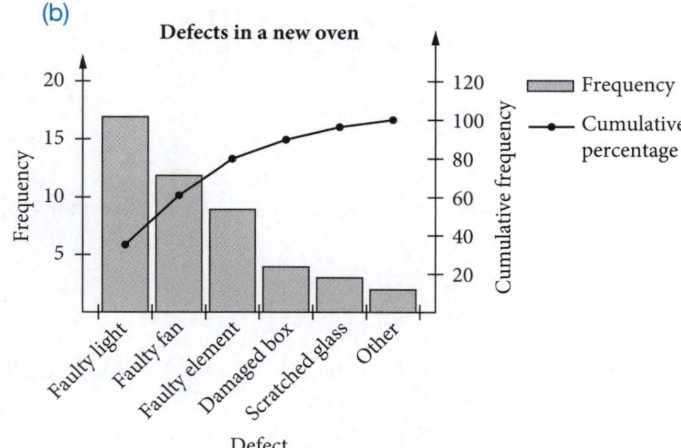

18 (a)

Type of complaint	Frequency	Percentage	Cumulative percentage
Poor housekeeping	45	42	42
Slow check-in/checkout	25	24	66
Poor breakfast	15	14	80
Not value for money	9	8	88
Poor WiFi	6	6	94
Television	4	4	98
Parking	2	2	100
	106	**100**	

(b)

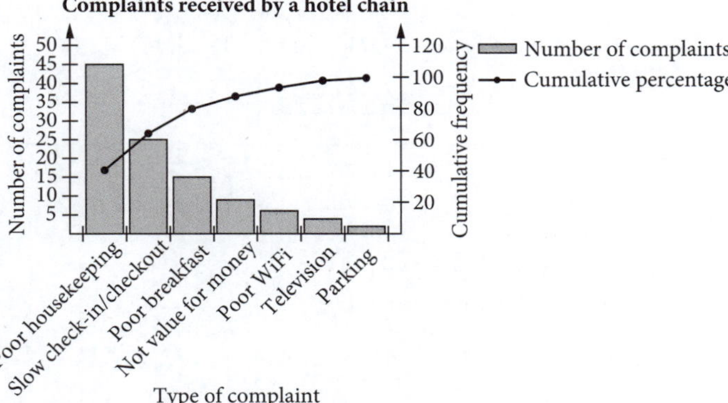

(c) Improve the housekeeping, check-in/checkout procedure and the quality of the breakfast.

EXERCISE 11.4

1 (a) 12.17 (b) 1.83 (c) 13.54

2 (a) 4 (b) 4 (c) 2

3 (a) The median is the interval 13.5–<14
(b) The median is the interval 13–<13.5

4 (a) 4 (b) 3 and 5 (c) No mode

5 (a) Mean is 7.08, median is 7
(b) Mean is 203.94, median is 204

6 (a) 13 (b) 11 (c) 11

7 (a) Minimum = 2, $Q_1 = 5$, Median $Q_2 = 7$, $Q_3 = 10.5$, Maximum = 15

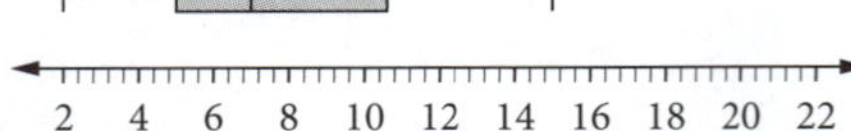

(b) Minimum = 2, $Q_1 = 8$, Median $Q_2 = 11.5$, $Q_3 = 14$, Maximum = 22

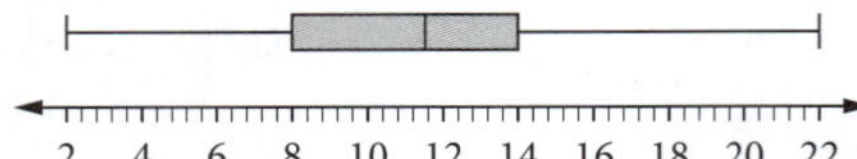

8 (a) Minimum = 0, $Q_1 = 1$, Median $Q_2 = 2$, $Q_3 = 3$, Maximum = 5, IQR = 2, Range = 5

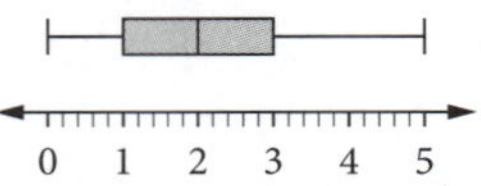

(b) Minimum = 4, $Q_1 = 5$, Median $Q_2 = 7$, $Q_3 = 9$, Maximum = 10, IQR = 4, Range = 6

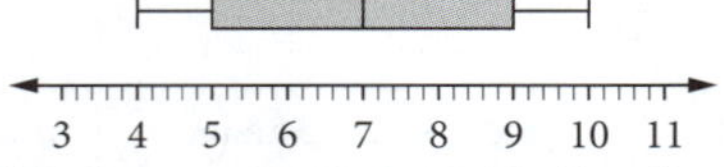

(c) Minimum = 11, $Q_1 = 15$, Median $Q_2 = 30.5$, $Q_3 = 42$, Maximum = 68, IQR = 27, Range = 57

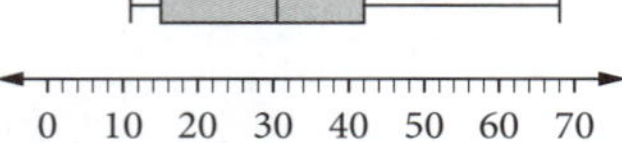

(d) Minimum = 152, $Q_1 = 161$, Median $Q_2 = 174.5$, $Q_3 = 189$, Maximum = 202, IQR = 28, Range = 50

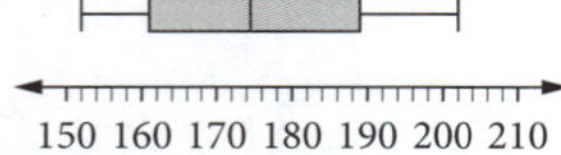

9 (a) 1 and 2 (b) No outliers (c) 58 (d) 3 and 48

10 B Mean = 3.26, Median = 3

11 (a) Minimum = 1, $Q_1 = 3.5$, Median $Q_2 = 5$, $Q_3 = 7.5$, Maximum = 10

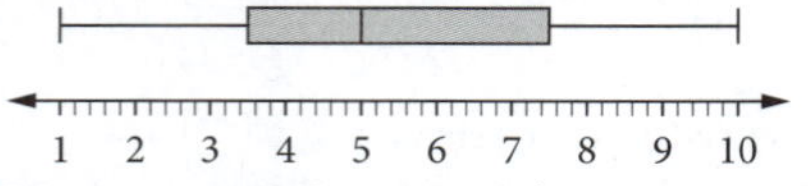

(b) Minimum = 8, $Q_1 = 14$, Median $Q_2 = 18$, $Q_3 = 23$, Maximum = 34

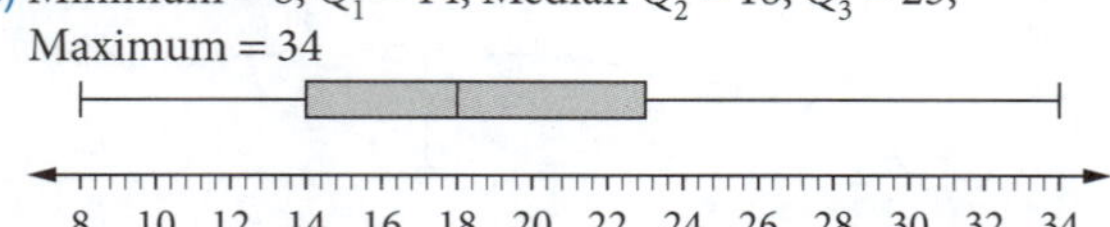

(c) Minimum = 8, $Q_1 = 9$, Median $Q_2 = 11$, $Q_3 = 13$, Maximum = 14

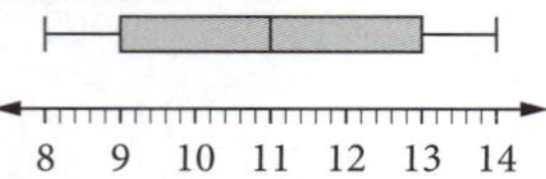

(d) Minimum = 23, $Q_1 = 24$, Median $Q_2 = 26$, $Q_3 = 27$, Maximum = 29

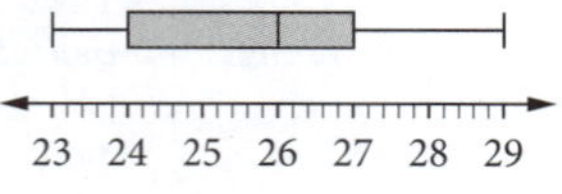

12 (a) Minimum = 2, $Q_1 = 3$, Median $Q_2 = 6$, $Q_3 = 6$, Maximum = 8
The median is equal to the upper quartile.

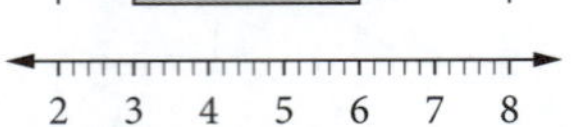

(b) Minimum = 2, $Q_1 = 2$, Median $Q_2 = 5$, $Q_3 = 7$, Maximum = 9
The minimum is equal to the lower quartile.

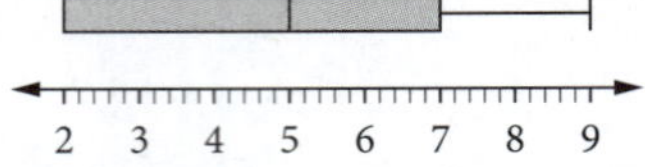

(c) Minimum = 2, $Q_1 = 3$, Median $Q_2 = 5$, $Q_3 = 9$, Maximum = 9
The upper quartile is equal to the maximum.

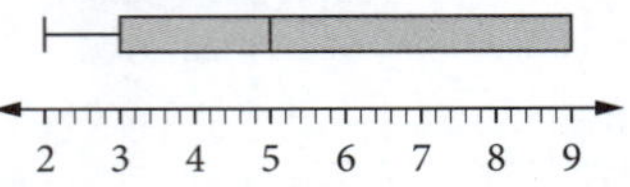

(d) Minimum = 2, $Q_1 = 2$, Median $Q_2 = 2$, $Q_3 = 7$, Maximum = 8
The minimum is equal to the lower quartile and the median.

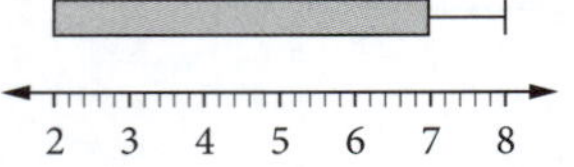

13 C

14 6

15 (a) Mean = 6, Median = 6, Mode = 6
(b) Mean = 6.125, Median = 5.5, Mode = 2
(c) Mean = 7.2, Median = 6, No Mode
(d) Mean = 6.125, Median = 5.5, Mode = 11

16 (a) From box plot, outliers are 45 and 73.

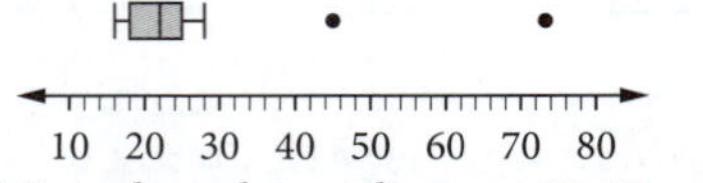

(b) From box plot, outliers are 67, 97 and 141.

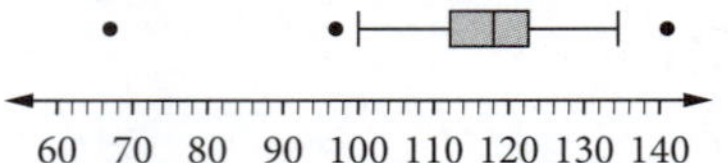

(c) From box plot, outliers are 18 and 72.

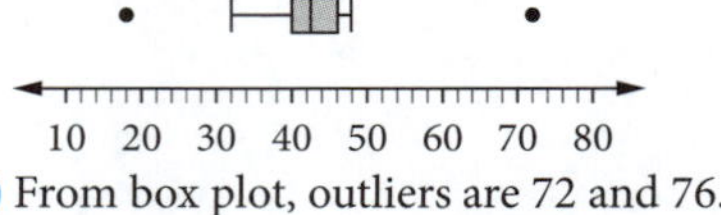

(d) From box plot, outliers are 72 and 76.

17 (a) Values below –16.5 or above 67.5 would be considered outliers.
(b) Values below 1 or above 73 would be considered outliers.
(c) Values below 2 or above 10 would be considered outliers.
(d) Values below 8 or above 16 would be considered outliers.

18 (a) 9.52°C (b) 6°C

(c)

Temperature range	Frequency
–5–<0	3
0–<5	6
5–<10	3
10–<15	4
15–<20	2
20–<25	0
25–<30	1
30–<35	2

(d) 9.88°C (e) 5–<10

EXERCISE 11.5

1 (a) Mean: 2.55
(b) When finding the standard deviation you can regard this as a population, so you will find the population standard deviation.
(c) 2.25

2 (a) 62.20 (b) $sx = 16.25$

3 (a) You can reasonably assume that these values refer to a sample.
(b) Mean: 37.2
Standard deviation 6.43
(c) Mean: 35.06
Standard deviation: 10.57

4 C

5 (a) A (b) 1.88 (c) 1.83

6 (a) 2.3 (b) 2.6 (c) 4.8 (d) 4.3 (e) 5.3
(f) 5.0 (g) 1.5 (h) 1.6 (i) 6.7 (j) 16.2

7 (a) Machine A: 59.6 g
Machine A: 60 g
(b) Neither machine would be shut down.
(c) Machine A: $sx = 2.41$
Machine B: = 2.11
(d) Machine A would be shut down.

8 (a) IQR = 9, $\sigma^x = 5.3$
(b) IQR = 9, $\sigma^x = 6.7$
(c) IQR = 9, $\sigma^x = 8.1$
(d) Answers may vary. Sample answer points may include:
- Order of data is unaltered.
- Quartiles are unchanged.
- IQR is constant.
- σx in the second and third parts alter compared with the first part, as changed data values are further away from mean, so deviations increase. The larger the change the greater the increase.

EXERCISE 11.6

1 (a) It is reasonable to assume the data set represents a population.
(b) Mean: 43.2, Standard deviation: 10.4, Minimum: 19
Q_1: 35, Median: 42, Q_3: 50.5, Maximum: 63
(c) The data set does not contain outliers and is symmetrical. 50% of the data set was above 42 and no more than 25% was below 35. The highest student scored a mark of 63 out of 66.

2 C

3

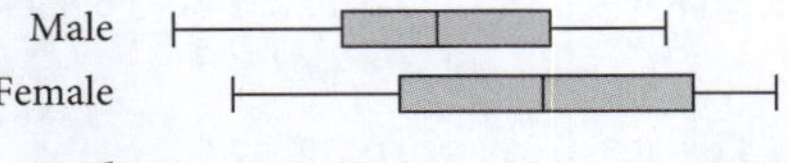

Sample answer: From the box plots we can see the shape of each set of data is approximately symmetrical.
The median for women (69.5) is greater than for men (64).

75% of women have the life expectancy of all men, 25% of women live longer than all men.

Women (max. 77.5) have higher life expectancy compared to men (max. 70).

In summary, women live longer than men.

4

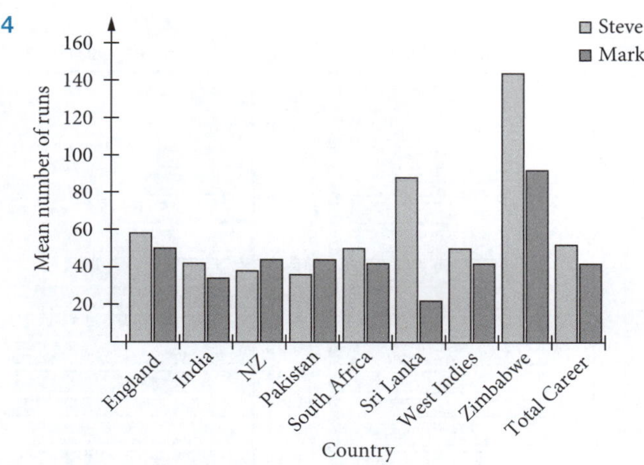

Sample answer: From the graph it can be seen that the batting averages for the brothers was similar for several countries. However, Steve had much better averages against Sri Lanka and Zimbabwe (87.63 and 145 respectively) compared with Mark's (24.64 and 90). This helped Steve maintain a higher overall career average.

5 Sample answer: From the graph we can see that road fatalities were higher in 6 out of 8 states and territories. The states with larger populations have a higher percentage of fatalities, but increases appear to be the same, which could possibly be due to an increase in population.

6 Sample answer: The average rainfall increases steadily over the course of the year. The actual rainfall increases over the year, slowly from January to April, then increases steadily to November, where it reduces again. The graph shows the rainfall for this particular year, the rainfall was well below average, nearly half the annual rainfall.

7

2003	Stem	2013
9 6	17_H	5 7 7 9
4 3 3 3 2 1 1 1	18_L	0 2 2 2 3 3 3 4
9 9 9 9 7 6 6 5 5	18_H	5 6 6 6 6 7 7 7 8 8 9 9
4 3 3 2 2 2 2 2 0 0 0	19_L	0 0 0 1 2 2 3 3 4
8 7 7 6 6 5	19_H	6 7 7 8 9
2	20_L	3
8	20_H	6

Key: 17|9 = 179

Sample answer: From the plot it can be seen that both graphs are approximately symmetrical. There are more players in the 185–189 cm group in 2013 compared with 2003 and there are fewer players in the 190–194 cm group in 2013 compared with 2003.
The minimum and maximum for each year are very similar.
The playing group in 2013 appear to be slightly shorter than the group of 2003.

8 (a)

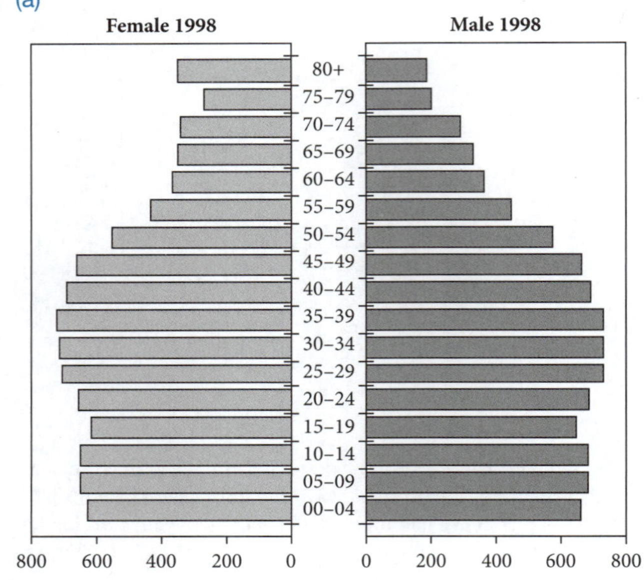

(b)

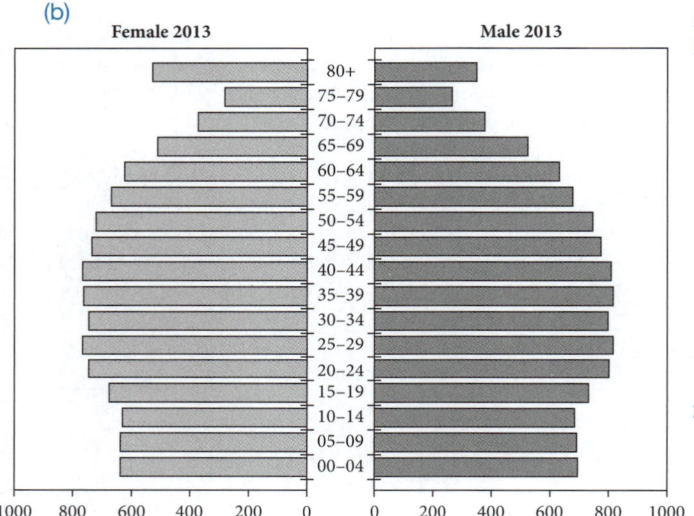

(c) Sample answer: From the graphs, the male and female populations for 1998 are very similar, except for the age groups over 70, where there are a greater number of females. A similar trend is seen in the 2013 data.
There are differences between the two years.
From 0 to 20 years the distribution of populations was similar.
From 20 to 50 years the population is greater for 2013 (approximately 800 000) than for 1998 (approximately 700 000).
For 50 to 70 years the population decreases, but is still greater for 2013 (from 750 000 to 500 000) than for 1998 (from 600 000 to 300 000).

(d)

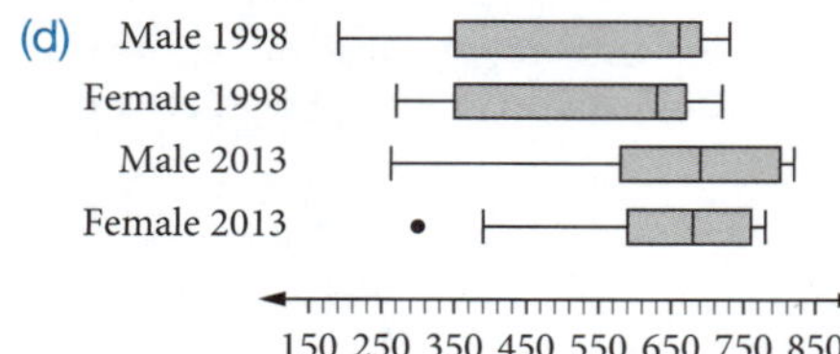

Sample answer:
There is an outlier in females for 2013.
The median population age is greater in 2013 than 1998.
The maximum is also greater for 2013.
1998 has the least of any age group, with 188 for the 80+ age group.
The middle 50% of ages are more varied in 1998 than for 2013.

CHAPTER REVIEW 11

1 B	2 D	3 D	4 C	5 A
6 D	7 D	8 D	9 A	10 B
11 B	12 B	13 C	14 A	15 B

16 (a) True (b) True (c) 5 644 000 (d) 354 (e) True
(f) False. The figures presented in the table are most likely to represent estimates based on a *census*.
(g) False. This article represents the *interpretation and communication* part of the statistical investigation process.

17 (a) continuous (b) discrete (c) nominal (d) ordinal

18 10–14

19 Minimum = 131, $Q_1 = 145$, Median $Q_2 = 153$, $Q_3 = 173$, Maximum = 191

20 (a) 2.98 (b) 3.07

21

Stem	Leaf
2	1 2 5 5 9
3	0 2 3 5 7 8
4	0 2 5 7
5	0 1 3
6	0 1 3 5 5 5 6 6 8
7	0 0 4 7 7 9
8	1 1 2

Key: 7|0 = 70

22

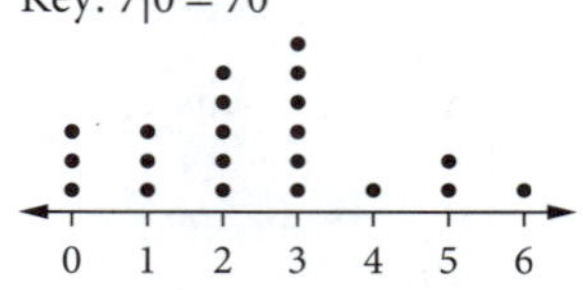

23

2003		2013
2 2 4 5 5 6 7 7 7 8 9	7	6 6 7 7 7 7
0 1 1 2 2 3 3 3 4 4 5 5 5 8 9 9	8	0 0 0 1 1 2 2 3 3 3 4 4 4 4 5 7 7 7 7 8 8 9
0 1 4 4 4 4 6 8 8 8	9	0 0 2 2 3 7 9
1	10	0 0 4 7

Key: 7|0 = 70

Sample answer:
From the plot it can be seen that both graphs are approximately positively skewed. There are more players in the 80–84 kg and 85–89 kg groups in 2013 compared to 2003 and there are fewer players in the 70–74 and 75–79 groups in 2013 compared to 2003. The minimum and maximums for 2013 are larger than in 2003. The playing group in 2013 appears to be heavier than the group of 2003. This may be due to better training and more weight training in 2013 compared to 2003.

24

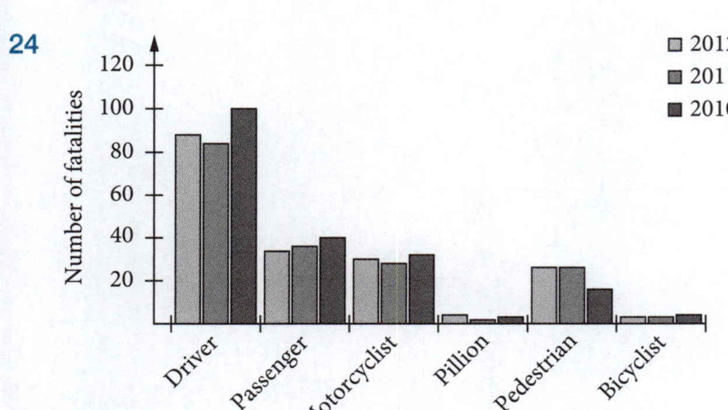

Sample response: Pedestrian fatalities have increased from 2010 to 2012; they have about doubled. The number of drivers killed decreased between 2010 and 2011 but increased a little in 2012, though not back to the 2010 level. Passenger deaths have fallen each year. There has not been much movement between 2010 to 2012 in the other figures.

25 (a) **Big Zap**
Mean: 20
Median: 22.5
Standard deviation: 6.96
IQR: 9

Sparky
Mean: 19.8
Median: 18
Standard deviation: 7.39
IQR: 6

(b) (Lifetime before recharging hours)

Big Zap		Sparky
9 7	0_U	
	1_L	4
6	1_U	5 5 5 7 9
4 3 2 2	2_L	0 1 3
6 6 5	2_U	
	3_L	
	3_U	9

Key: $1|9 = 19$

(c) Sparky
Big Zap

4 8 12 16 20 24 28 32 36 40

(d) 100% of Sparky's batteries have lives over 14 hours.
Approximately 75% of Big Zap's batteries have lives over 14 hours.
Sparky's batteries would be the better choice.

(e) More than 50% of Big Zap's batteries have lives over 20 hours. Less than 50% of Sparky's batteries has lives over 20 hours.
Big Zap's batteries would be the better choice.

(f) Various answers.
Sparky's hours are positively skewed whereas Big Zap's hours are negatively skewed.
The median for Big Zap (22.5 hours) is greater than for Sparky (18 hours).
Big Zap has the lowest life of 7 hours compared to Sparky's with 14 hours.
Sparky has the battery with the longest lifetime of 39 compared to Big Zap's 26 hours.
The IQR for Sparky batteries is 9 compared to Big Zap's IQR of 6. This tells us that the Big Zap batteries' lives are more consistent.
As Big Zap batteries have a more consistent life and a higher median, they generally tend to last longer.

26 (a)

Defect	Frequency	Percentage	Cumulative percentage
Dent in door	16	36	36
Door seal	12	27	63
Damaged box	8	18	81
Faulty light	6	13	94
Wiring defect	2	4	98
Refrigerant leak	1	2	100
	45	**100**	

(b)

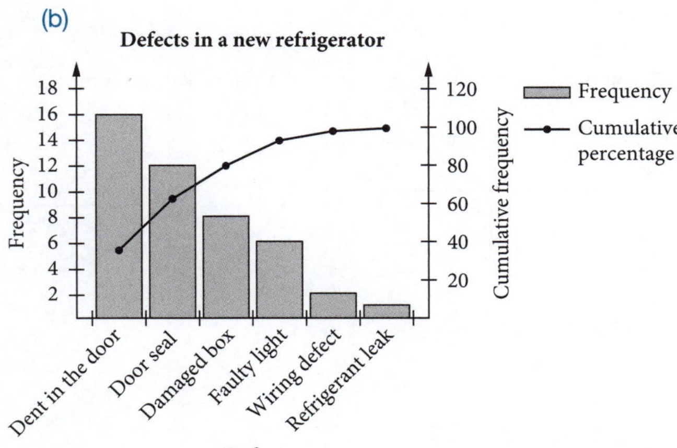

(c) Have the people handling the completed refrigerators take more care. Make sure that the door seals are fitted correctly.

CHAPTER 12

EXERCISE 12.1

1 (a) period 2π, amplitude 4

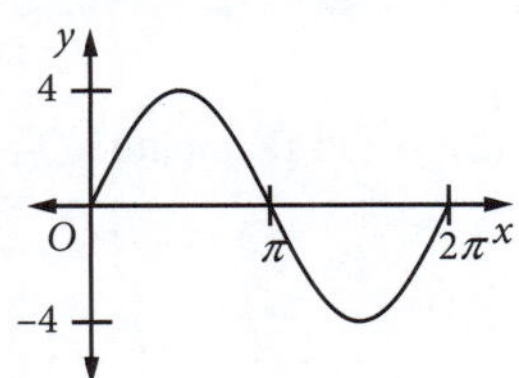

(b) period π, amplitude 1

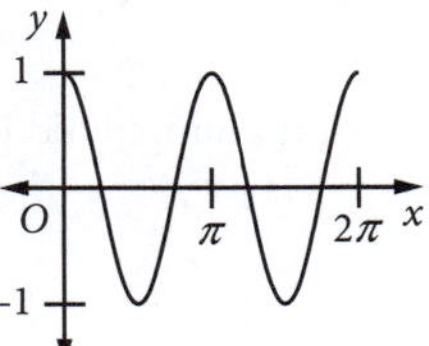

(c) period π, no amplitude

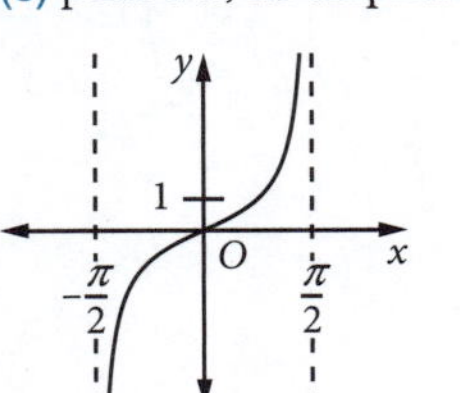

(d) period $\frac{2\pi}{3}$, amplitude 4

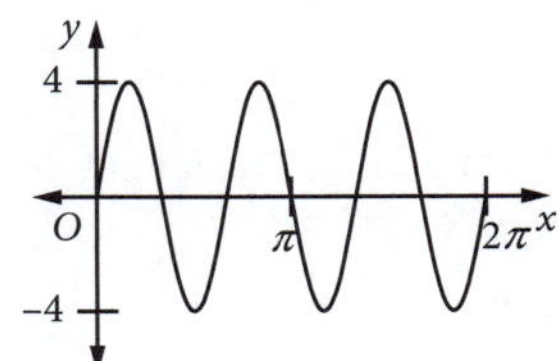

(e) period π, amplitude 3

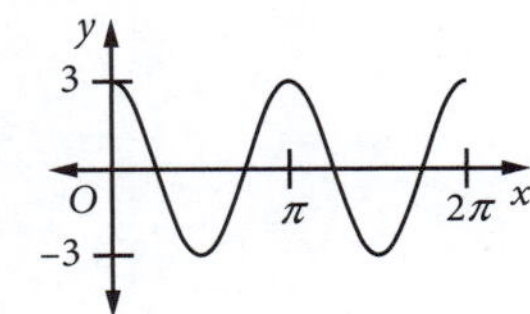

(f) period $\frac{\pi}{2}$, no amplitude

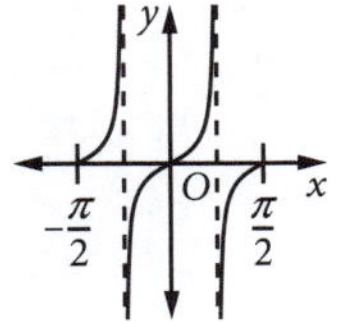

(g) period 4π, amplitude 1

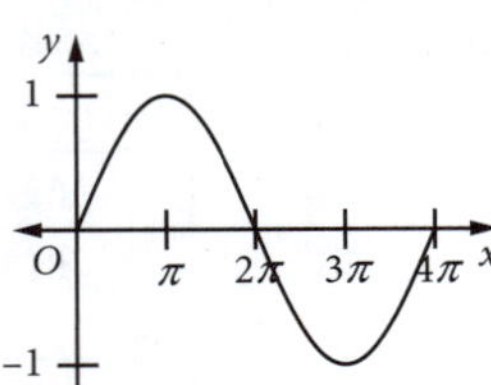

(h) period 4π, amplitude 1

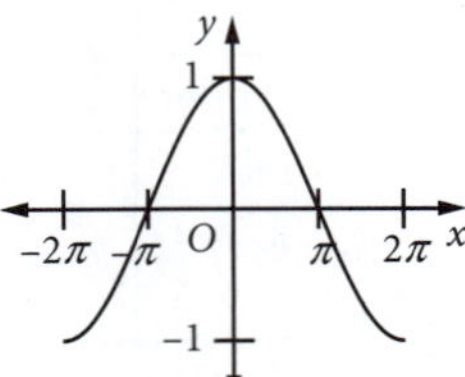

(i) period 2π, no amplitude

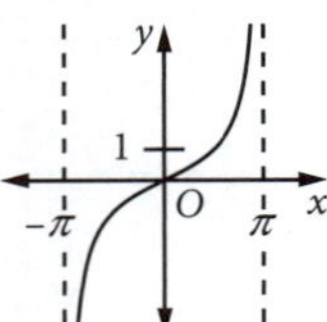

2 B

3 (a) period 4π, amplitude 3

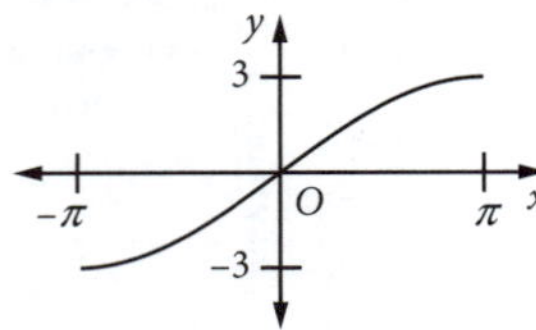

(b) period 4π, amplitude 2

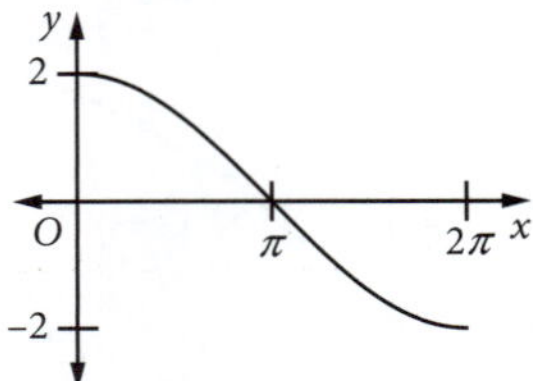

(c) period 2π, no amplitude

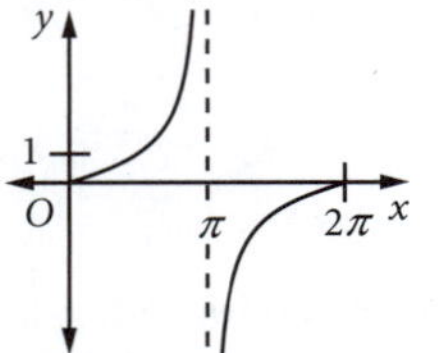

4

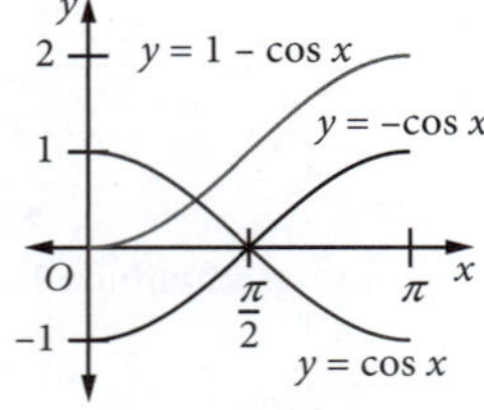

5

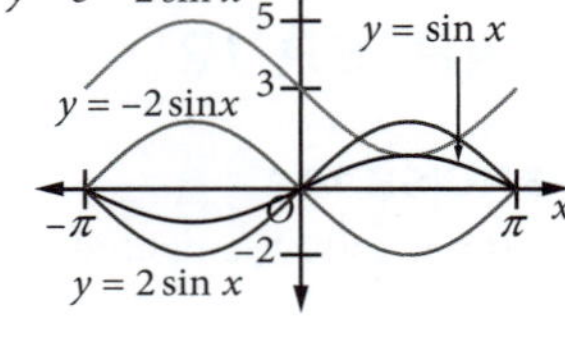

6 (a) period 2, amplitude 1

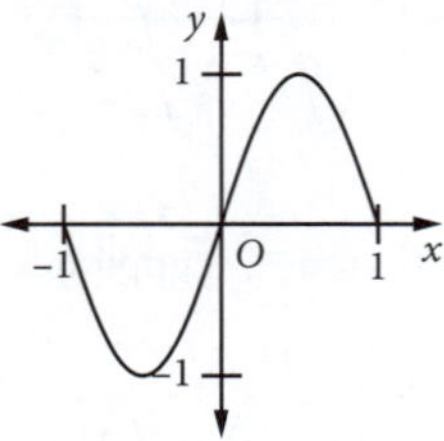

(b) period 1, amplitude 1

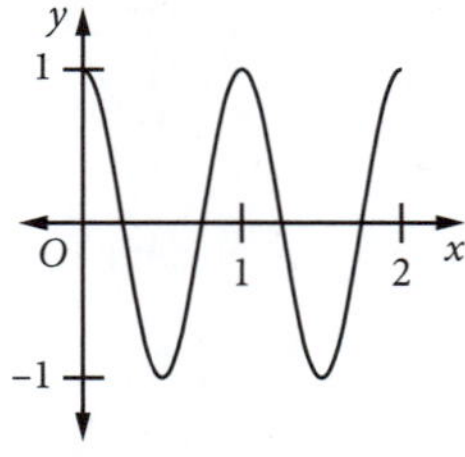

(c) period 2, no amplitude

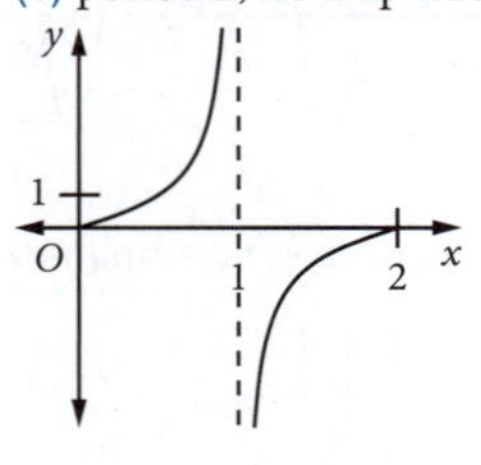

7 (a)

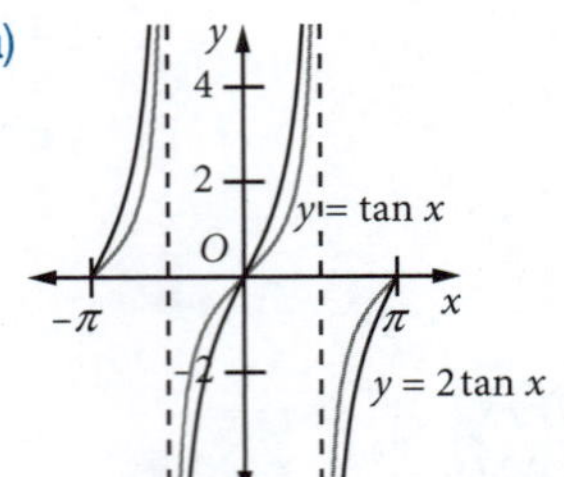

(b)–(c)

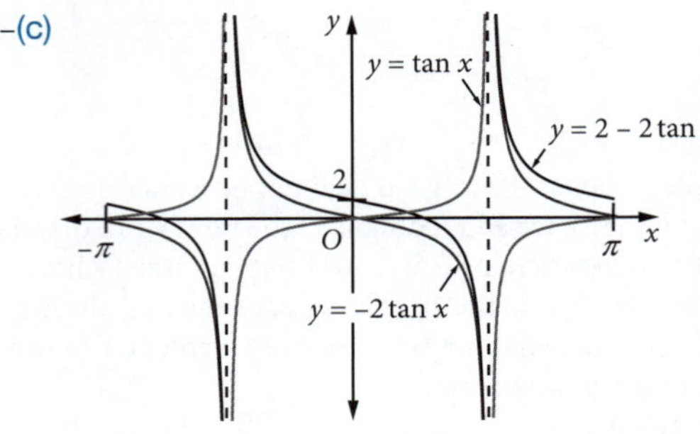

8 (a)

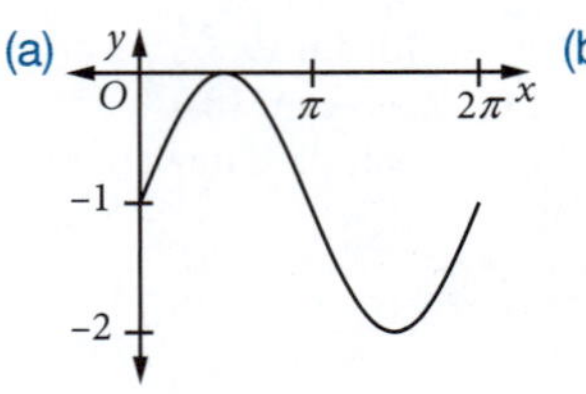

(b)

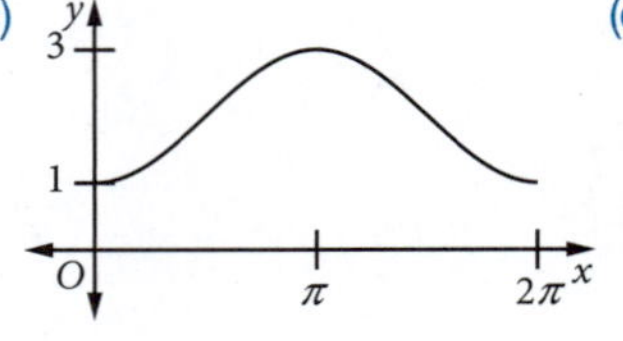

(c)

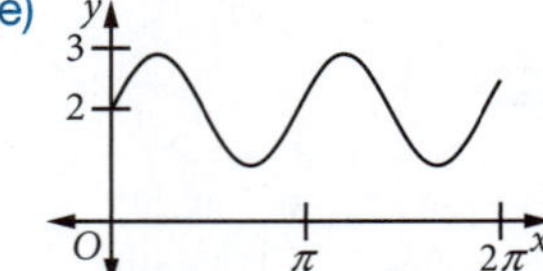

(d)

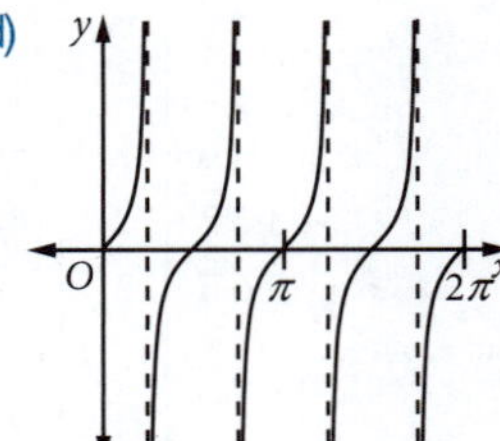

(e)

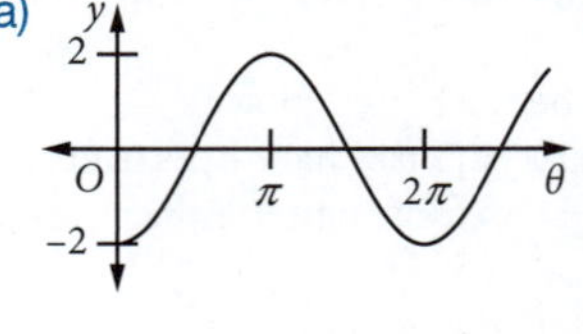

(f)

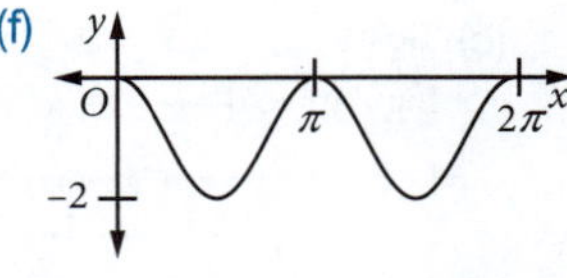

9 (a)

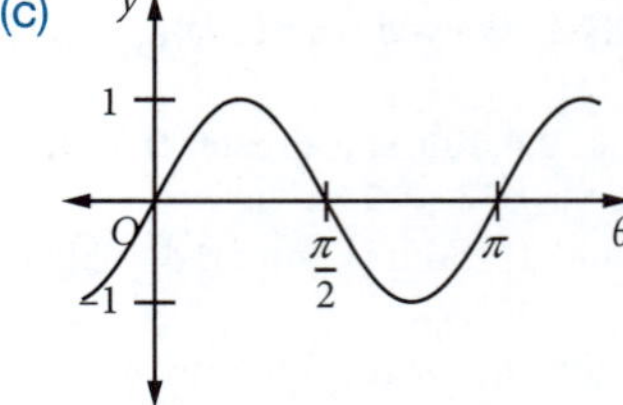

(b)

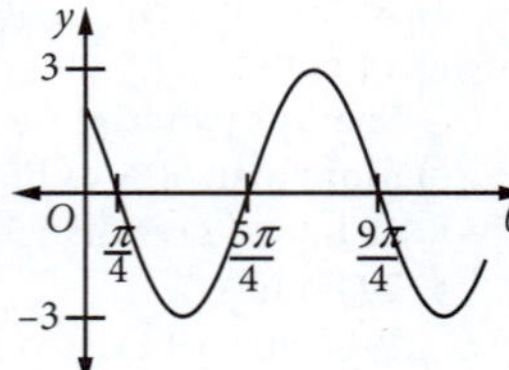

(c)

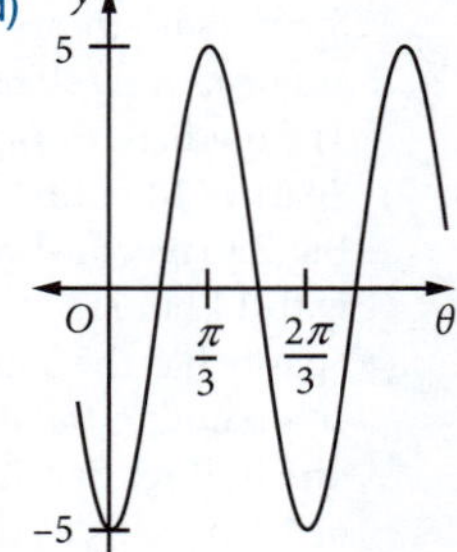

(d)

(e)

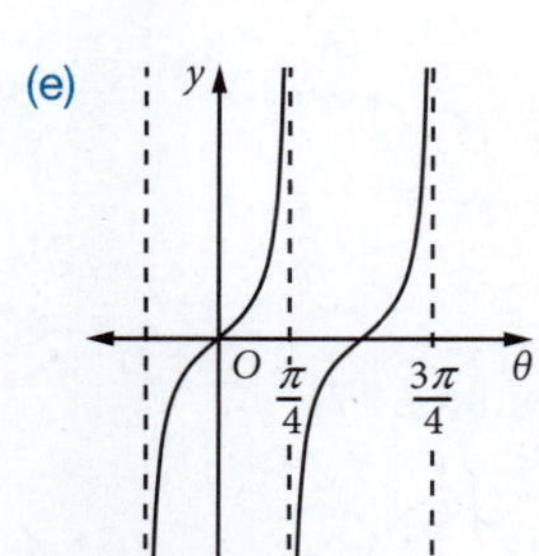

(f)

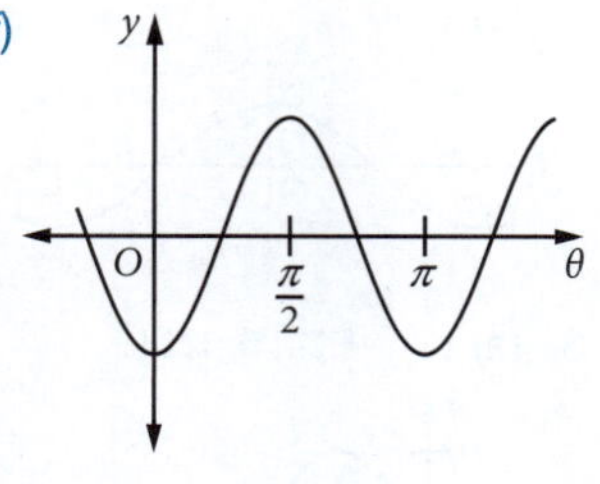

10 (a)

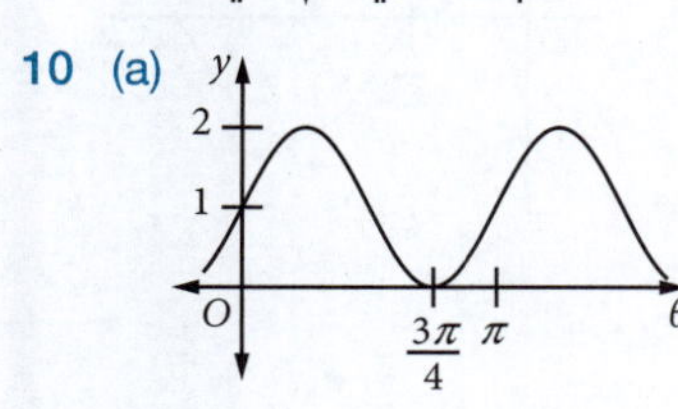

(b)

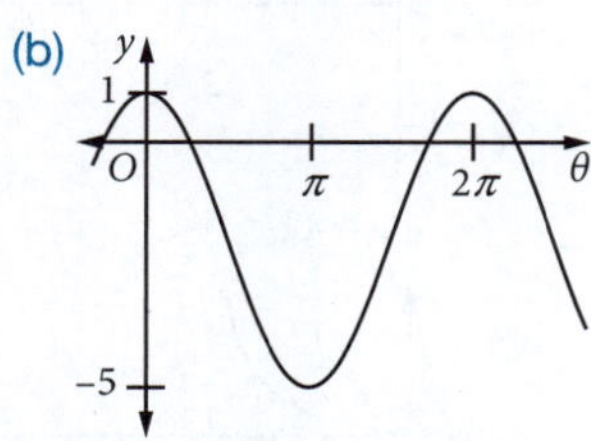

(c)

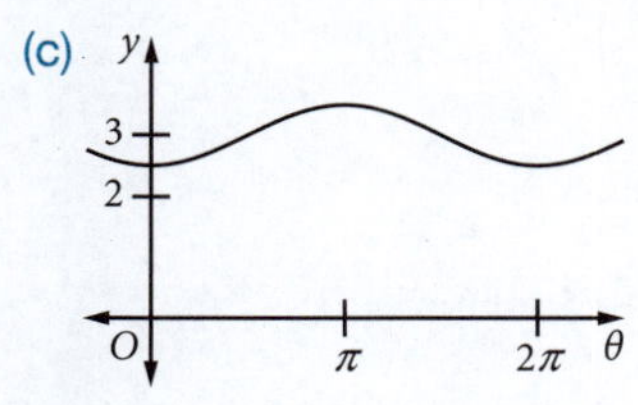

(d)

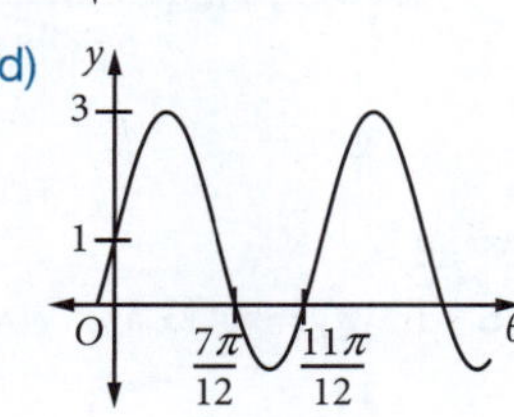

(e)

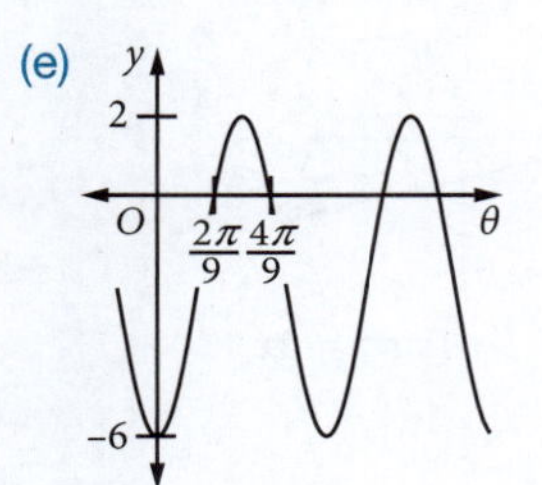

(f)

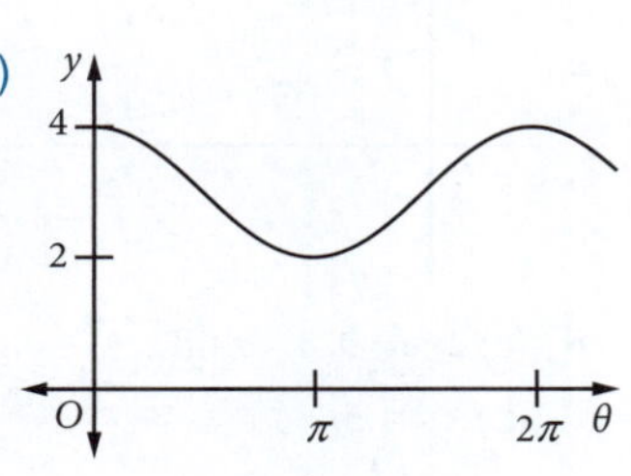

11 (a)

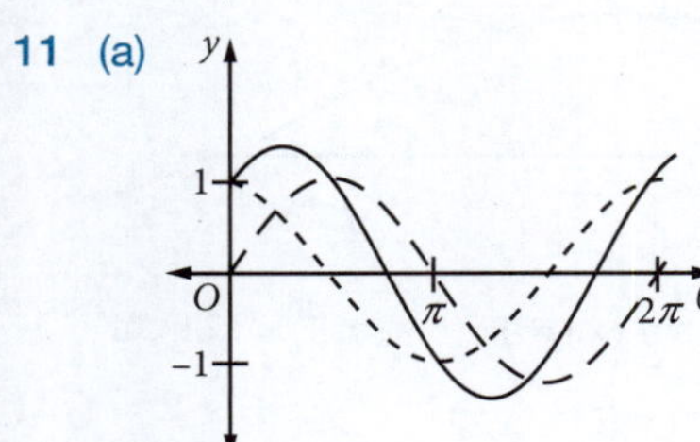

(b)

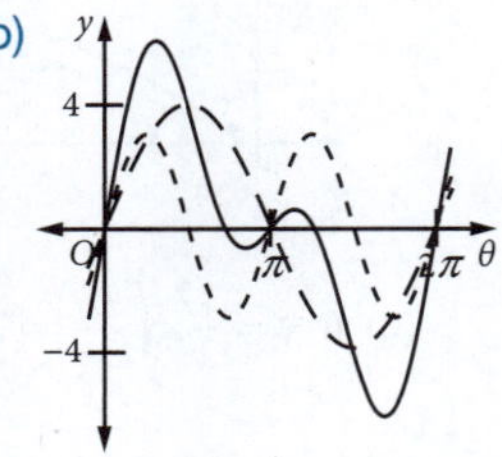

(c)

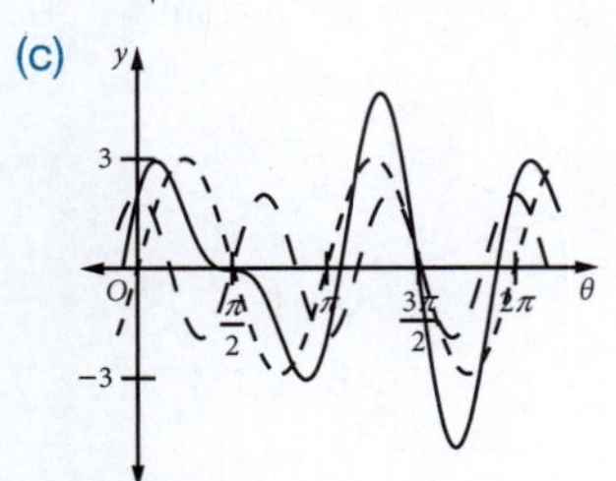

(d)

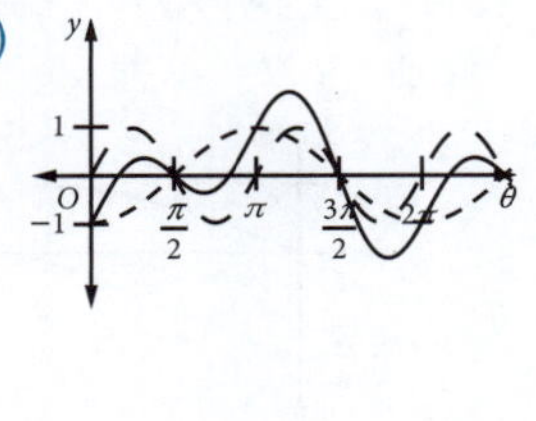

(e)

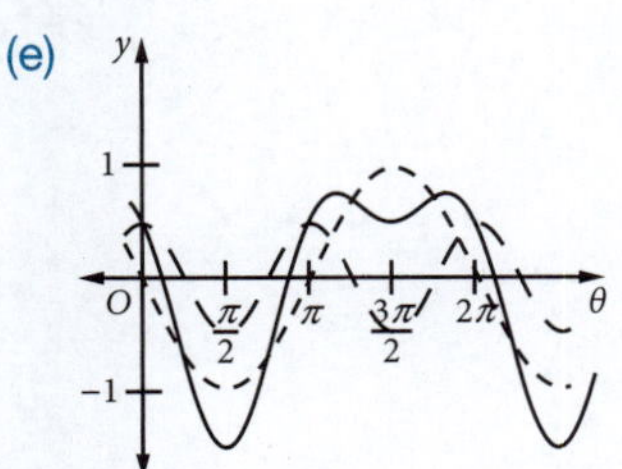

(f)

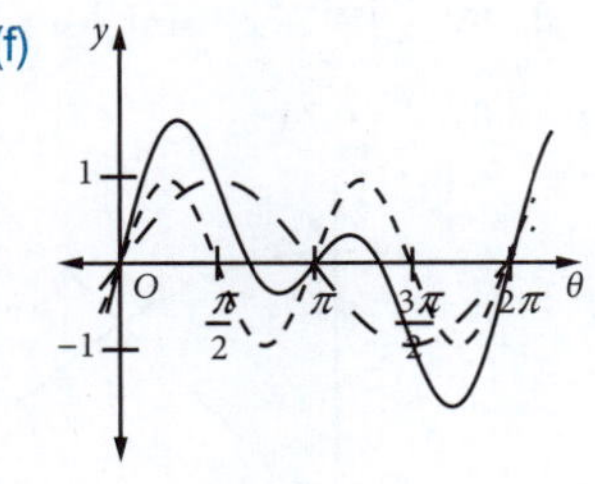

12 Answers will vary.

13 (a) $\sqrt{3}\tan\left(x - \frac{\pi}{3}\right) - 1 = 0$

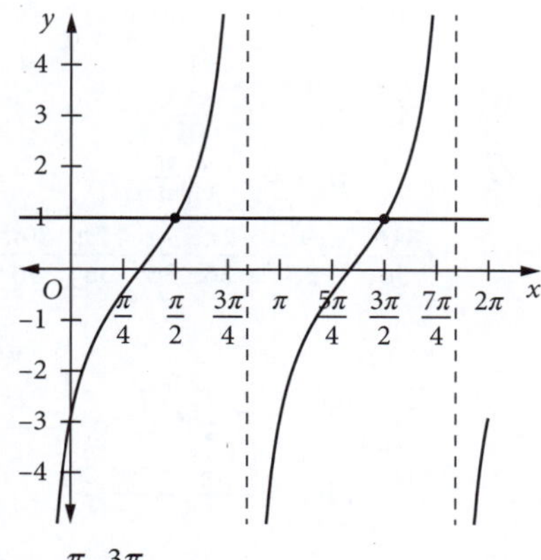

$x = \frac{\pi}{2}, \frac{3\pi}{2}$

(b) $2\sqrt{3}\cos x + \frac{\pi}{4} - 3 = 0$

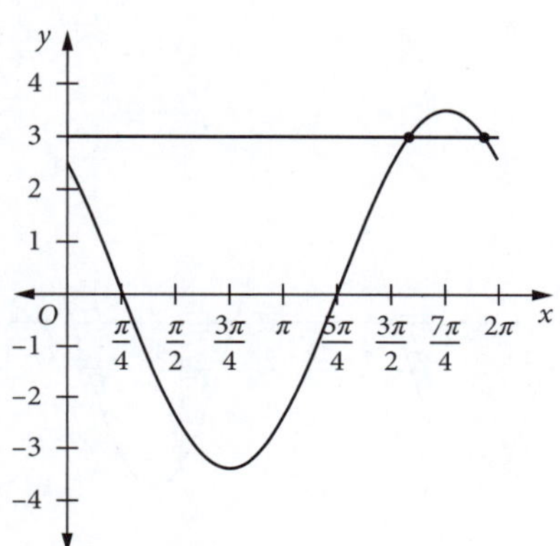

$x = \frac{19\pi}{12}, \frac{23\pi}{12}$

(c) $\sqrt{2}\sin\left(x + \frac{\pi}{6}\right) + 1 = 0$

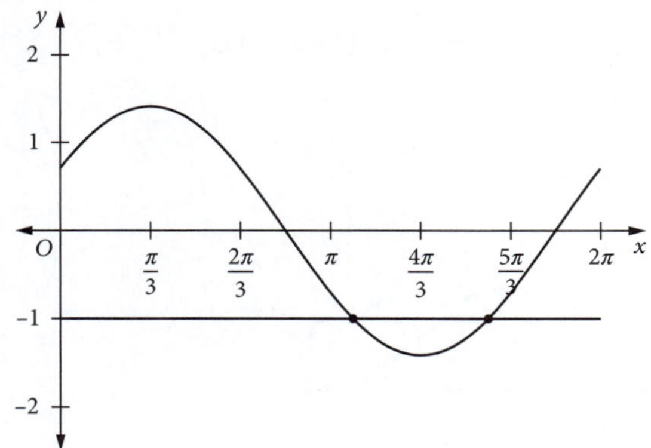

$x = \frac{13\pi}{12}, \frac{19\pi}{12}$

EXERCISE 12.2

1 (a) $x = \frac{\pi}{12}, \frac{5\pi}{12}, \frac{13\pi}{12}, \frac{17\pi}{12}$ (b) $x = 0, \frac{2\pi}{3}, \frac{4\pi}{3}, 2\pi$

(c) $x = \frac{\pi}{8}, \frac{5\pi}{8}, \frac{9\pi}{8}, \frac{13\pi}{8}$

2 (a) $x = \frac{\pi}{3}$ (b) $x = -\pi$ (c) $x = -\pi, \pi$

3 (a) $x = \frac{\pi}{2}, \frac{11\pi}{6}$ (b) $x = 0, \frac{\pi}{2}, 2\pi$ (c) $x = \frac{\pi}{6}, \frac{\pi}{2}$

(d) $x = \frac{\pi}{3}, \frac{4\pi}{3}$ (e) $x = \frac{7\pi}{12}, \frac{19\pi}{12}$

4 (a) $x = -\frac{7\pi}{8}, -\frac{\pi}{8}, \frac{\pi}{8}, \frac{7\pi}{8}$ (b) $x = -\frac{3\pi}{4}, \frac{\pi}{4}$

(c) $x = -\frac{\pi}{3}, \frac{2\pi}{3}$ (d) $x = -\frac{31\pi}{36}, -\frac{19\pi}{36}, -\frac{7\pi}{36}, \frac{5\pi}{36}, \frac{17\pi}{36}, \frac{29\pi}{36}$

(e) $x = \frac{2\pi}{9}$

5 (a) $x = \frac{\pi}{6}$ (b) $x = -\frac{7\pi}{4}, \frac{5\pi}{4}$

(c) $x = -2\pi, -\frac{4\pi}{3}, 2\pi$ (d) $x = -\frac{5\pi}{4}, -\frac{3\pi}{4}, \frac{7\pi}{4}$

6 (a) $x = \frac{\pi}{12}, \frac{7\pi}{12}, \frac{13\pi}{12}, \frac{19\pi}{12}$

(b)

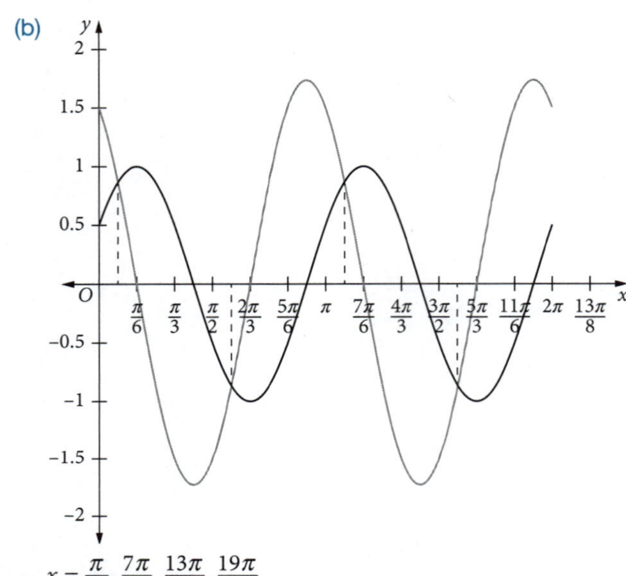

$x = \frac{\pi}{12}, \frac{7\pi}{12}, \frac{13\pi}{12}, \frac{19\pi}{12}$

7 (a) $x = 0, 2.498, 2\pi$ (b) $x = 0.6435$ (c) $x = 22° 37'$

(d) $x = \frac{\pi}{4}, \frac{\pi}{2}$

EXERCISE 12.3

1

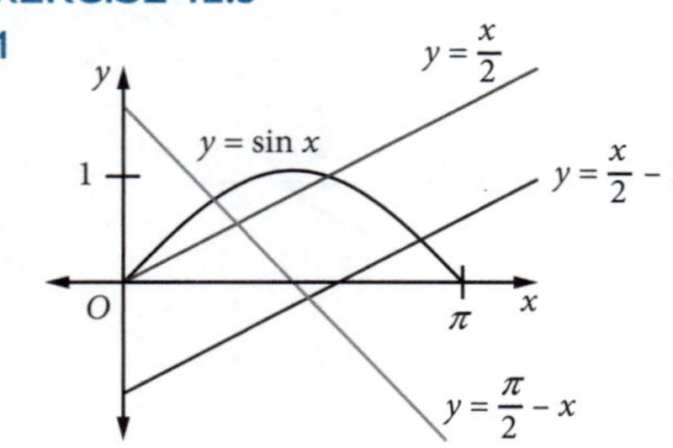

(a) 2 (b) 1 (c) 1

2 (a) $x \approx 0.74$ (b) $x \approx 1.03$

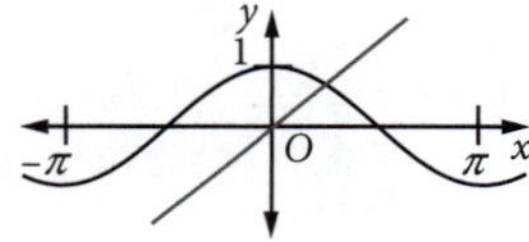

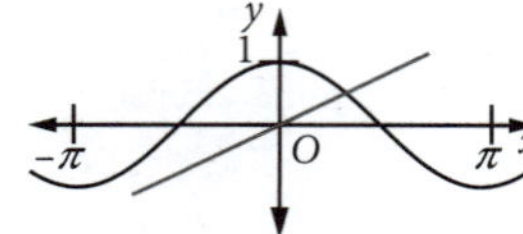

(c) $x = 0$

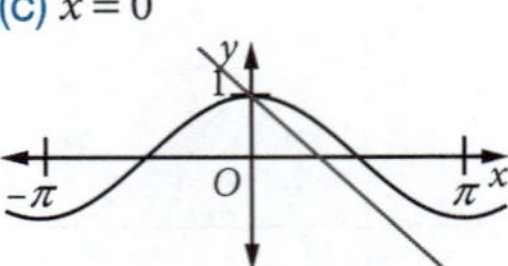

3 (a) $x \approx -1.24, 0, 1.24$ (b) $x \approx 0.74$

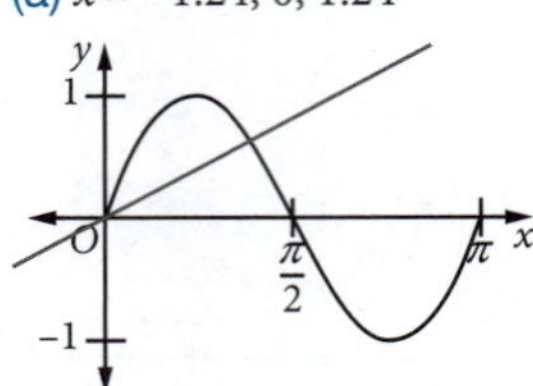

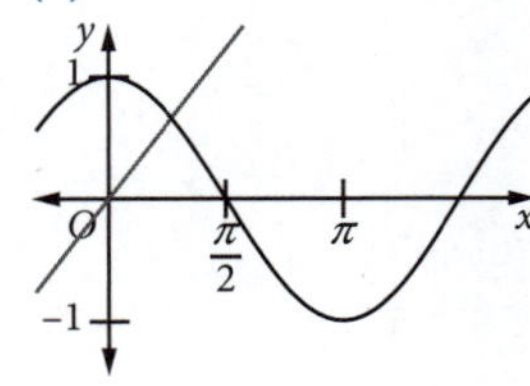

(c) $x \approx 0.48, 2.25$

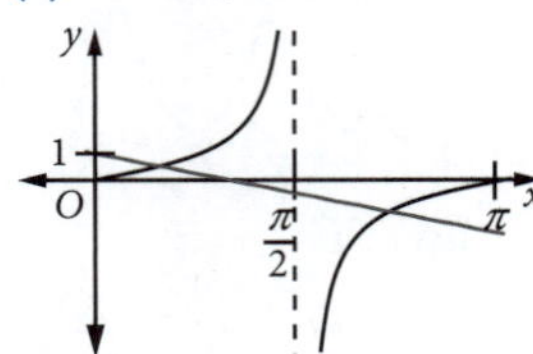

4 B

5 draw $y = \sin 2x$ and $y = 1 - \frac{x}{4}$: 5 solutions

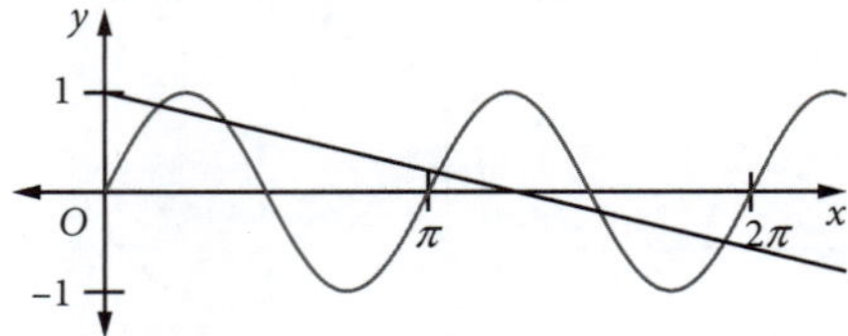

6 (a) $2rx = 9, 3 = r\sin x; \frac{9}{2x} = \frac{3}{\sin x}$ so $3\sin x = 2x$

(b) $x \approx 1.5$

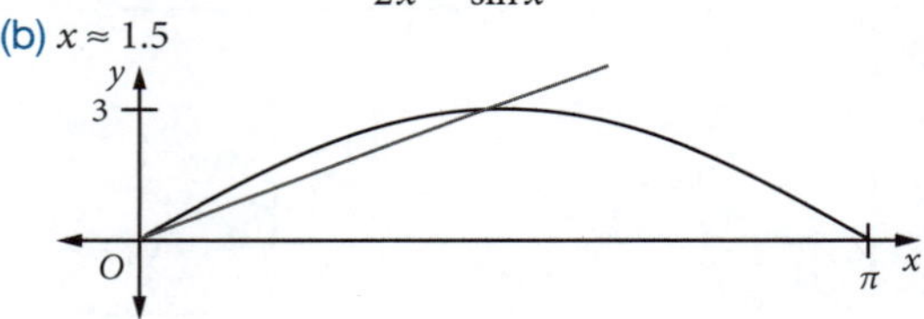

(c) area $= \frac{1}{2} \times \left(\frac{9}{2x}\right)^2 (2x - \sin 2x) = 4.5(3 - \sin 3) = 12.9\,\text{cm}^2$

7 (a) $\frac{r^2}{2}(x - \sin x) = \frac{\pi r^2}{4}, x - \sin x = \frac{\pi}{2}$ so $x - \frac{\pi}{2} = \sin x$

(b) $x \approx 2.3$

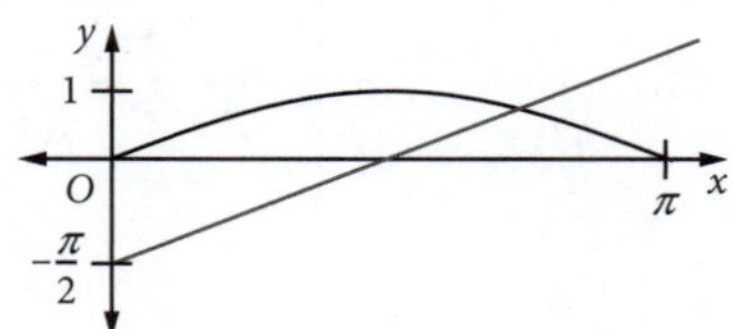

8 (a) $\frac{r^2}{2}\left(\frac{\pi}{2} + x - \sin\left(\frac{\pi}{2} + x\right)\right) = \frac{\pi r^2}{4}, \frac{\pi}{2} + x - \cos x = \frac{\pi}{2}, x = \cos x$

(b) $x \approx 0.74$

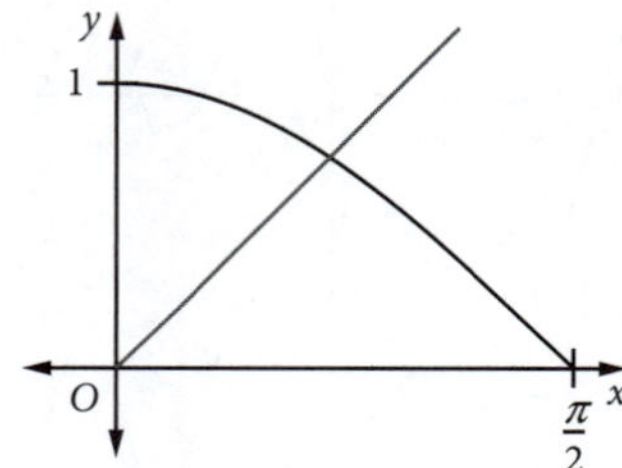

9 0.99

EXERCISE 12.4

1 (a) h intercept is at (0, 2)

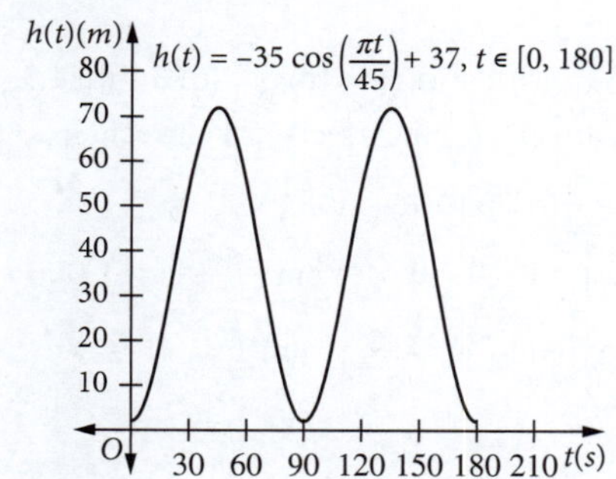

(b) 54.5 m
(c) 23.73 seconds

2 (a) T intercept is at (0, 20)

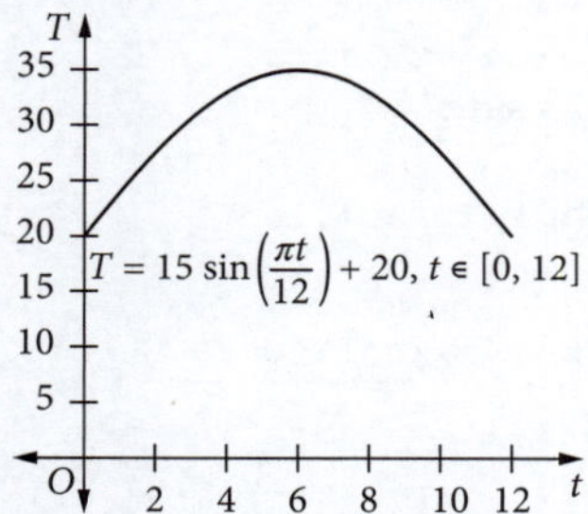

(b) The temperature inside the room at 9:00 am is 20°C.
(c) The temperature in the room is 30.6°C at midday.
(d) The times at which the temperature is 30°C inside the room, correct to one decimal place, are t = 2.8 hours or to the nearest minute 11:47 am and t = 9.2 hours or 6:13 pm.
(e) The percentage of time that the temperature inside the room was at least 30°C, correct to one decimal place, is 53.3%.

3 (a) $k = 2, a = \frac{\pi}{6}, c = 2.5$ (b) $h(t) = 2\cos\left(\frac{\pi t}{6}\right) + 2.5$
(c) The second high tide occurs at 6:00 pm.

4 (a) $k = 240$ and $c = 0$ (b) $a = 100\pi$
(c) $V = \mp 240\sin(100\pi t)$

5 (a) $6.5\sin\frac{1}{2}(t+2) + 19.5$ (b) $4\sin\frac{1}{2}(t+2) + 10$

6 (a)

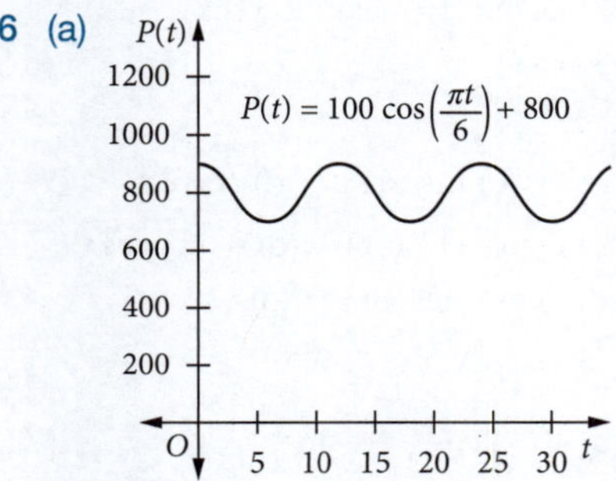

(b) $k = 100, a = \frac{\pi}{6}, c = 800$
(c) 31 March, $P = 800$
20 June, $P = 702$
23 October, $P = 838$

7 (a) 4.5 m (b) 12.2 hours; $n = \frac{10\pi}{61}$
(c) $y = 4.5\cos\frac{10\pi t}{61}$

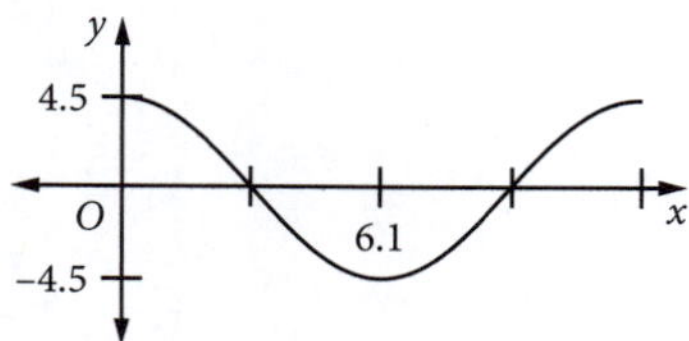

(d) low tide at $t = 6.1$; at $t = 7.1$, $y = -3.92$:
depth = 0.5 + (4.5 − 3.92) = 1.08 m

CHAPTER REVIEW 12

1 (a) $x = 6°41'$ or $83°19'$
(b) $x = 0°, 70°32', 180°, 289°28'$ or $360°$

2 (a)

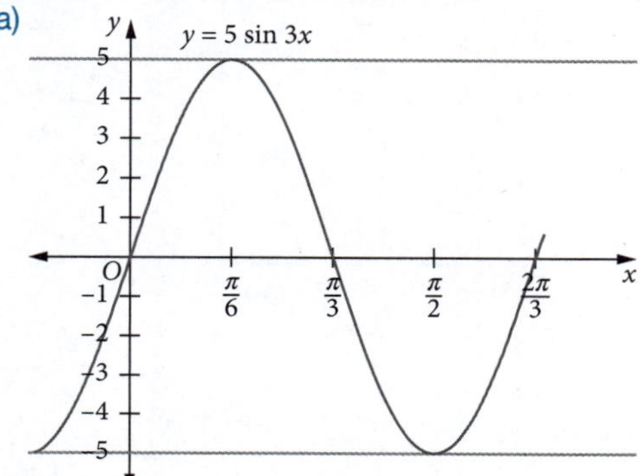

(b)

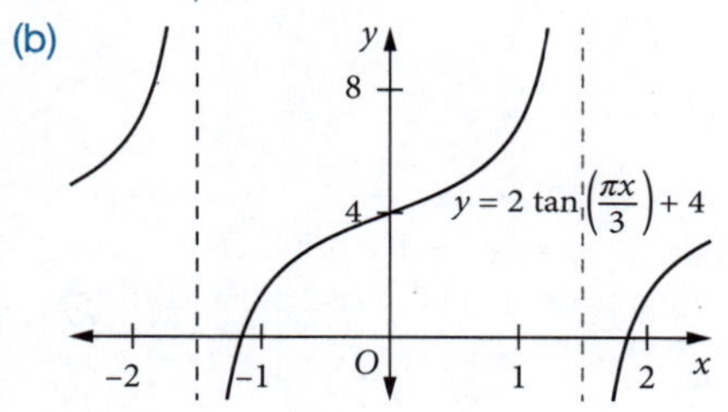

(c)

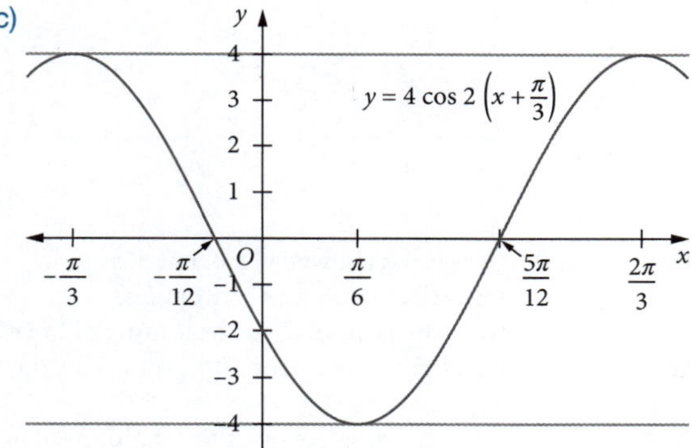

3 (a)

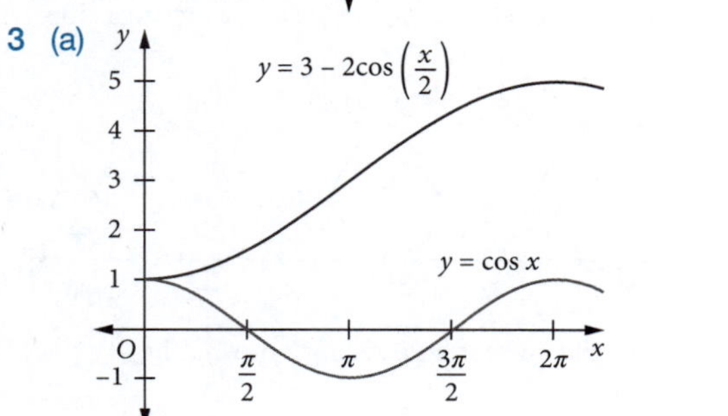

(b)

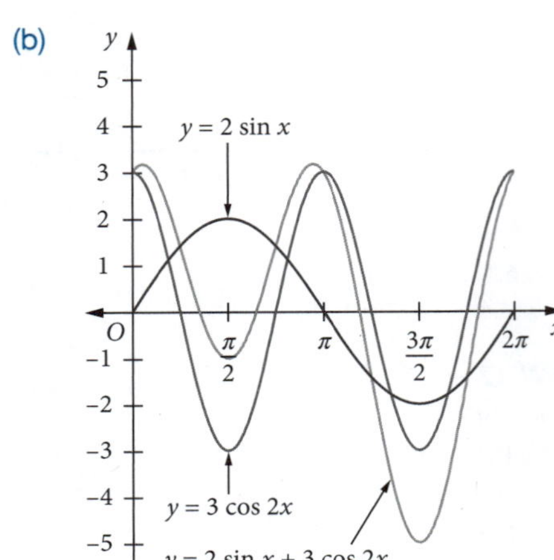

4

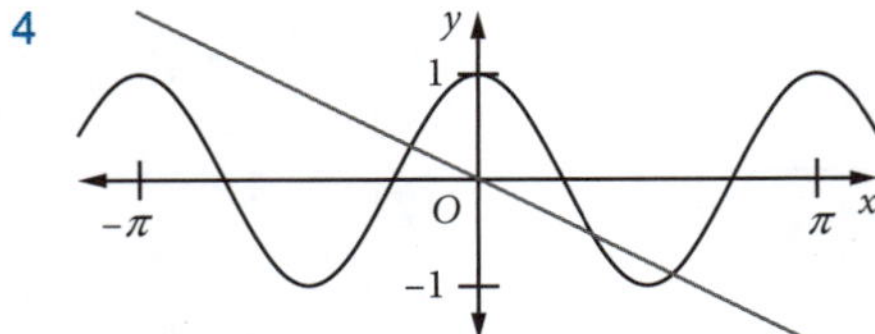

$x = -0.63, 1.07, 1.8$

5 $y = -3\cos(2x)$

6 (a) $c = 4$
(b) $k = -8$
(c) The period of f is 24.
(d) $a = \frac{\pi}{12}$
(e) The minimum temperature is −4°C.
(f) The pond has a temperature of 0°C at 2 am and 10 am.

7 (a)

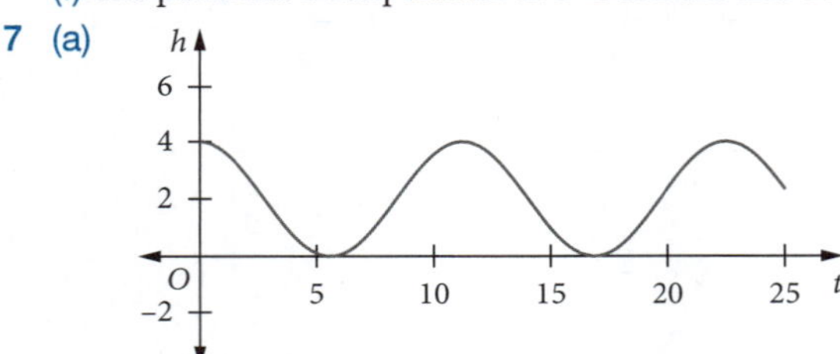

(b) The island first appears above the water at 3:16 pm.
(c) Tess can collect shells for 4 hours 43 minutes.

8 (a) Oscar is 1 m from the ground at the beginning of the ride.
(b) The greatest distance Oscar is from the ground during the ride is 17 m.
(c) The ride takes 20 s to complete one full rotation.
(d) 9 rotations are completed before the ride comes to a stop.
(e)

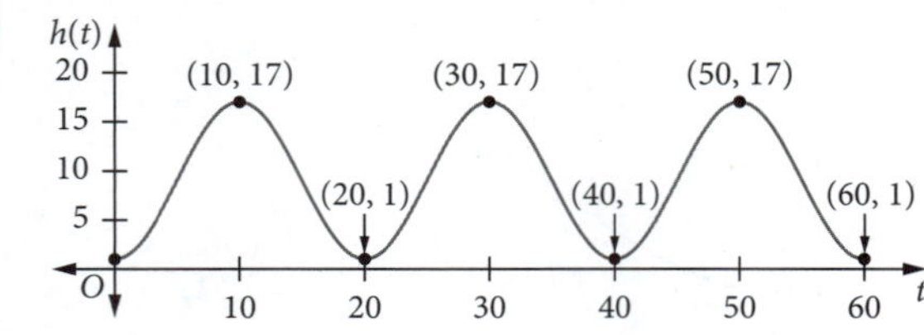

(f) Oscar can see the ocean for 90 seconds during the 3-minute ride.

9 (a) $x = \frac{5\pi}{8}, \frac{13\pi}{8}$ (b) $x = -\frac{3\pi}{4}, -\frac{\pi}{12}, \frac{\pi}{4}, \frac{11\pi}{12}$

CHAPTER 13

EXERCISE 13.1

1 (a) 2 (b) $\frac{1}{3}$ (c) 2.5 (d) 1 (e) 0.5 (f) $\frac{1}{9}$ (g) 3 (h) 1.5

2 (a) incorrect (b) correct (c) incorrect (d) correct

3 (a) $4\lim_{x\to 0}\frac{1-\cos 2x}{(2x)^2} = \frac{4}{2} = 2$ (b) $\frac{2}{3}$ (c) $\frac{1}{2}$
(d) $\lim_{x\to 0}\left(\frac{\sin x}{x}\times \sin x\right) = 1\times 0 = 0$ (e) $\lim_{x\to 0}\frac{-\sin x}{x} = -1$
(f) $\lim_{x\to 0}\frac{\sin x}{x} = 1$ (g) $\lim_{x\to 0}\frac{1-\cos x}{x^2} = \frac{1}{2}$ (h) 1

EXERCISE 13.2

1 (a) $3\cos 3x$ (b) $3\cos x$ (c) $-2\sin 2x$ (d) $-2\sin x$
(e) $\cos x - 4\sin x$ (f) $2\sec^2 2x$
(g) $2\cos 2x + 2\sin 2x$ (h) $\cos\left(x+\frac{\pi}{4}\right)$

2 A

3 (a) $\cos^2 x - \sin^2 x$ (b) $\sin x + x\cos x$
(c) $2\tan x + 2x\sec^2 x$ (d) $2x\cos x - x^2\sin x$
(e) $\frac{\sin x - x\cos x}{\sin^2 x}$ (f) $\frac{x\cos x - \sin x}{x^2}$
(g) $3x^2\sin 2x + 2x^3\cos 2x$ (h) $3x^2\cos(x^3)$
(i) $\sec x + x\sec x\tan x$
(j) $\frac{x(-\operatorname{cosec} x\cot x) - \operatorname{cosec} x}{x^2} = \frac{-\operatorname{cosec} x(x\cot x + 1)}{x^2}$
(k) $2x\cot x - x^2\operatorname{cosec}^2 x$
(l) $\frac{d}{dx}\left(\frac{\sin x}{\cos x}\right) = \frac{d}{dx}(\tan x) = \sec^2 x$

4 (a) $\frac{1}{2}\cos\frac{t}{2} - \frac{1}{2}\sin t$ (b) $-3\cos^2 t\sin t$ (c) $-2t\sin(t^2+1)$
(d) $2\cos\left(2t+\frac{\pi}{2}\right) = -2\sin 2t$ (e) $2t + \frac{1}{2}\sec^2\frac{t}{2}$
(f) $2t\cos 3t - 3(t^2-1)\sin 3t$ (g) $-2\sin\left(2t+\frac{\pi}{3}\right)$ (h) $-3\sin(3t-2)$

5 (a) $-4\sin 2x\cos 2x$ (b) $6\sin 3x\cos 3x$ (c) $-3\sin x\cos^2 x$
(d) $-3x^2\sin(x^3)$ (e) $2\cos^2 2x - 2\sin^2 2x$ (f) $3 - 2\sin x - \frac{1}{2}\cos\frac{x}{2}$
(g) $-4\sin x\cos x$ (h) $\frac{x^2\cos x - 2x\sin x}{x^4} = \frac{x\cos x - 2\sin x}{x^3}$
(i) $2x\sec^2(x^2-1)$ (j) $\cos\frac{x}{2} - \frac{x}{2}\sin\frac{x}{2}$
(k) $\cos x(1+\cos x) + \sin x(-\sin x) = \cos x + \cos^2 x - \sin^2 x$
(l) $-\sin 2x + \sin x + x\cos x$
(m) $-2x\operatorname{cosec}(x^2-1)\cot(x^2-1)$
(n) $2\cot(2x)\times(-\operatorname{cosec}^2(2x))\times 2 = -4\cot(2x)\operatorname{cosec}^2(2x)$
(o) $\sec x\tan x - \operatorname{cosec} x\cot x$
(p) $\frac{d}{dx}(1+\tan^2 x) = 2\tan x\sec^2 x$

6 (a) correct (b) incorrect (c) incorrect (d) correct

7 (a) $\frac{\cos 2x}{\sqrt{\sin 2x}}$ (b) $2(\sin^2 x - \cos^2 x)$ (c) 0 (d) $4\sin x\cos x$
(e) $-\sin x\cos(\cos x)$ (f) $-\cos x\sin(\sin x)$ (g) $-\sec^2 x$
(h) $\frac{\sin x}{2\sqrt{1-\cos x}}$ (i) $\frac{1}{2\sqrt{x}}\sec^2\sqrt{x}$

8 (a) $e^x(\sin x + \cos x)$ (b) $e^{2x}\left(2\cos\frac{x}{2} - \frac{1}{2}\sin\frac{x}{2}\right)$
(c) $e^{-x}(3\cos 3x - \sin 3x)$ (d) $e^x(\cos 4x - 4\sin 4x)$
(e) $-2e^{-x}(\sin x)$ (f) $2e^{\sin 2x}\cos 2x$ (g) $-e^{\cos x}\sin x$
(h) $(\cos x - \sin x)e^{\sin x + \cos x}$

EXERCISE 13.3

1 (a) $\frac{1}{x}$ (b) $\frac{2}{x}$ (c) $\frac{2}{x}$ (d) $\frac{3}{3x-5}$ (e) $\frac{1}{x}$ (f) $2x - \frac{4}{4x-1}$

2 (a) $x > -\frac{2}{3}, \frac{3}{3x+2}$ (b) all real x, $\frac{2x}{x^2+1}$ (c) $x \neq 2, \frac{2}{x-2}$

(d) $x > -\frac{3}{4}$, $\frac{4}{4x+3}$ (e) $x > 0$, $\frac{1}{2x}$ (f) $x > 0$, $\frac{2\sqrt{x+1}}{2x\left(\sqrt{x}+1\right)}$

3 C

4 (a) $\ln x + 1$ (b) $x^2(3\ln x + 1)$ (c) $\ln(x+2)+1$

(d) $2x\ln 2x + x + \frac{1}{x}$ (e) $2\ln x + 2 - \frac{5}{x}$ (f) $\frac{e^x}{x}(x\ln x + 1)$

(g) $\frac{e^{2x}}{x}(2x\ln 2x + 1)$ (h) $\frac{\log_e x - 1}{(\log_e x)^2}$ (i) $\frac{1-\log_e x}{x^2}$

(j) $\frac{1 - x\log_e x}{xe^x}$

(k) $\frac{x \times \frac{2x}{x^2+1} - \log_e\left(x^2+1\right)}{x^2} = \frac{2x^2 - \left(x^2+1\right)\log_e\left(x^2+1\right)}{x^2\left(x^2+1\right)}$

(l) $e^x \log_e\left(e^x+1\right) + \frac{e^{2x}}{e^x+1}$

5 (a) $\frac{2x-2}{x^2-2x}$ (b) 1 (c) $\frac{3}{x}$ (d) $\frac{6}{x^2-9}$ (e) $\frac{2x}{x^2-1}$

(f) $\frac{4}{x}$ (g) 1 (h) $2x(2\log_e x + 1)$ (i) $\frac{3}{2x}$ (j) $\frac{1}{x\log_e x}$

(k) $(\log_e x + 1)e^{x\log_e x}$ (l) $\frac{e^x}{e^x+1} - \frac{e^x}{e^x-1} = \frac{2e^x}{1-e^{2x}}$

6 $\frac{dy}{dx} = \frac{2x}{x^2+1}$, $m = 0.6$

7 (a) $\frac{1}{x}$ (b) $\frac{-1}{x^2}$ (c) 0.5 (d) −0.25

8 $y = x - 1$, $y = 1 - x$

9 (a)

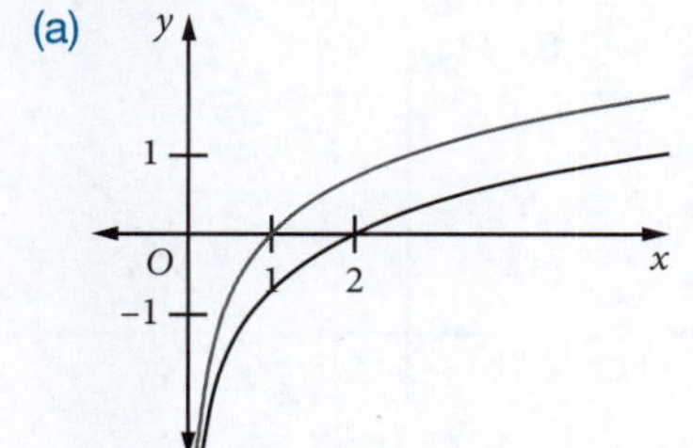

$\frac{d}{dx}(\log_e x) = \frac{1}{x}$

$\frac{d}{dx}\left(\log_e \frac{x}{2}\right) = \frac{1}{\frac{x}{2}} \times \frac{1}{2} = \frac{1}{x}$

The gradients are equal.

(b) $x - 2y - 2 = 0$

10 $\log_e(e^x) = x\log_e e = x$, so $y = \log_e(e^x)$ is the same as $y = x$; gradient $= 1$

11

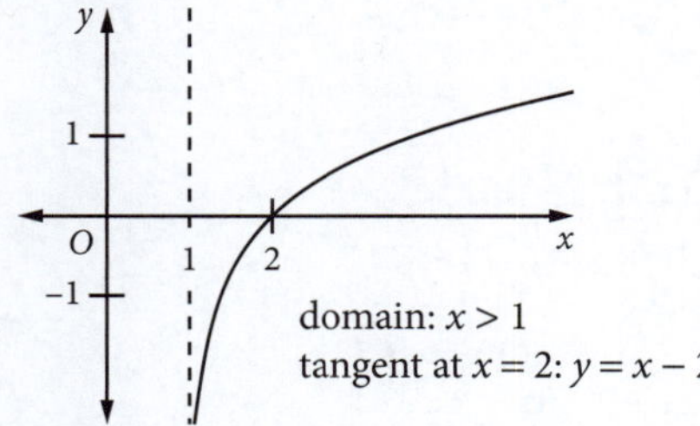

12 (a) $x = \ln 2 \approx 0.693$ (b) $x = \frac{\log_e 5}{3} \approx 0.536$

(c) $x = \frac{\log_e 7 - 3}{2} \approx -0.527$ (d) $x = \pm\sqrt{1 + \ln 10} \approx \pm 1.817$

13 (a) $y = \log_e(x^3 - 1) - \log_e x$

$\frac{dy}{dx} = \frac{3x^2}{x^3-1} - \frac{1}{x} \left(= \frac{2x^3+1}{x(x^3-1)}\right)$

(b) $f(x) = \log_e(e^x(x+2)) = x + \log_e(x+2)$

$f'(x) = 1 + \frac{1}{x+2}$

(c) $y = \log_e(\sqrt{x(x+1)^5}) = \frac{1}{2}\log_e x + 5\log_e(x+2)$

$\frac{dy}{dx} = \frac{1}{2x} + \frac{5}{x+1} \left(= \frac{11x+1}{2x(x+1)}\right)$

(d) $f(x) = \log_e\left(\frac{e^x + x}{\sqrt{x}}\right) = \log_e(e^x + x) - \frac{1}{2}\log_e x$

$f'(x) = \frac{e^x+1}{e^x+x} - \frac{1}{2x} \left(= \frac{2xe^x + x - e^x}{2x(e^x+x)}\right)$

(e) $g(x) = \log_e\left(\frac{x^3(e^x+1)}{e^{-x}+1}\right)$

$= 3\log_e x + \log_e(e^x+1) - \log_e(e^{-x}+1)$

$g'(x) = \frac{3}{x} + \frac{e^x}{e^x+1} + \frac{e^{-x}}{e^{-x}+1}$

(f) $y = \log_e \frac{\sqrt[3]{x^2 \sin x}}{1 = 2e^x} = \frac{2}{3}\log_e x + \log_e \sin x - \log_e(1 - 2e^x)$

$\frac{dy}{dx} = \frac{2}{3x} + \frac{\cos x}{\sin x} + \frac{2e^x}{1-2e^x} = \frac{2}{3x} + \cot x + \frac{2e^x}{1-2e^x}$

14 (a) $2^x \ln 2$ (b) $e^x + 3^x \ln 3$ (c) $\frac{1}{x\ln 2}$ (d) $1 + \frac{1}{x\ln 3}$

(e) $4^x(2x + x^2 \ln 4)$ (f) $x^2\left(3\log_5 x + \frac{1}{\ln 5}\right)$

(g) $\frac{2^x(x\ln 2 - 1)}{x^2}$ (h) $\frac{1 - 2x\ln a \log_a x}{x^3}$

15 (a) $-a^{-x}\ln a$ (b) $= a^x\left(\ln a \log_a x + \frac{1}{x\ln a}\right)$

(c) $\frac{1 - x\log_a x \times (\ln a)^2}{xa^x \ln a}$ (d) $\frac{1}{2x\ln a\sqrt{\log_a x}}$

(e) $\frac{\sqrt{a^x}(1 + x\ln a)}{2\sqrt{x}}$ (f) $2x\log_2 x\left(\log_2 x + x \times \frac{1}{x\ln 2}\right)$

$= \frac{(2x\log_2 x)}{(\ln 2)((\ln 2)\log_2 x + 1)}$

16 (a) $(10\ln 10)x - y + 10(1 - \ln 10) = 0$

(b) $x + 25y\ln 5 - 625\ln 5 - 2 = 0$ (c) Tangents parallel at (0, 1)

EXERCISE 13.4

1 (a) $2xe^{x^2}$ (b) $4\left(e^x + 2x\right)\left(e^x + x^2\right)^3$ (c) $e^x + e$

(d) $-4\sin x\, e^{\cos x}$ (e) $\frac{1}{2\sqrt{x}}e^{\sqrt{x}+1}$

(f) $\left(1 + \frac{1}{x}\right)e^{x+\ln x} = \left(\frac{x^2+1}{x}\right)e^{x+\ln x}$

2 (a) $e^{\sin x} + x\cos x\, e^{\sin x}$ (b) $e^x \log_e x + \frac{e^x}{x}$

(c) $-2\sin(2x+1)e^{\cos(2x+1)}$ (d) $1 + 2x\,e^x + x^2\,e^x$

3 $\frac{dy}{dx} = \frac{750e^{-0.5t}}{\left(1 + 15e^{-0.5t}\right)^2}$

4 (a)

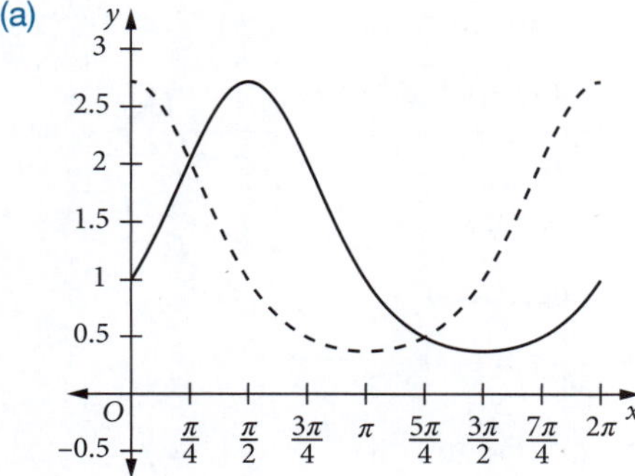

(b) $\left(\frac{\pi}{4}, 2.028\right), \left(\frac{5\pi}{4}, 0.493\right)$

$e^{\sin x} = e^{\cos x}$ $\sin x = \cos x$, $\tan x = 1$,

$x = \frac{\pi}{4}, \frac{5\pi}{4}$, $f\left(\frac{\pi}{4}\right) = e^{\frac{1}{\sqrt{2}}} = 2.028$, $f\left(\frac{5\pi}{4}\right) = e^{\frac{-1}{\sqrt{2}}} = 0.493$

(c) $f'(x) = \cos x\, e^{\sin x}$, $g'(x) = -\sin x\, e^{\cos x}$

$f'\left(\frac{\pi}{4}\right) = \frac{1}{\sqrt{2}} e^{\frac{1}{\sqrt{2}}}$, $g'\left(\frac{\pi}{4}\right) = -\frac{1}{\sqrt{2}} e^{\frac{1}{\sqrt{2}}}$

$f'\left(\frac{5\pi}{4}\right) = -\frac{1}{\sqrt{2}} e^{\frac{-1}{\sqrt{2}}}$, $g'\left(\frac{5\pi}{4}\right) = \frac{1}{\sqrt{2}} e^{\frac{-1}{\sqrt{2}}}$

(d) No

$g'\left(\frac{\pi}{4}\right) = -\frac{1}{\sqrt{2}} e^{\frac{1}{\sqrt{2}}} = -f'\left(\frac{\pi}{4}\right)$

$g'\left(\frac{5\pi}{4}\right) = \frac{1}{\sqrt{2}} e^{\frac{-1}{\sqrt{2}}} = -f'\left(\frac{5\pi}{4}\right)$

5 0

CHAPTER REVIEW 13

1 (a) $\cos x + 2\sec^2 2x$ (b) $-12\sin 4x - 10\cos 2x$

(c) $\cos x - x\sin x$ (d) $\frac{-1}{\sin^2 x} = -\operatorname{cosec}^2 x$

(e) $-2e^{-x}(\cos 3x + 3\sin 3x)$ (f) $\frac{2\cos 2x}{\sin 2x} = 2\cot 2x$

2 (a) $(x^2 + 4x + 2)e^x$ (b) $\frac{2e^{-x}(1 - x\ln x)}{x}$ (c) $\frac{e^x}{1 + e^x}$

(d) $\frac{2x+2}{x^2+2x}$ (e) $e^{-3x}(3 - 7x - 3x^2)$ (f) $\frac{1}{2\sqrt{x}} e^{\sqrt{x}} + \frac{1}{2x}$

3 (a) 9, 6, 0, −3, 0, 6, 9 (b) $t = 4$, $v = \frac{-\pi\sqrt{3}}{2}$, $a = \frac{\pi^2}{12}$

4 (a) $2x\sin x + x^2\cos x$ (b) $\frac{2x\cos x + x^2\sin x}{\cos^2 x}$

(c) $\cos x\cos x + \sin x(-\sin x) = \cos^2 x - \sin^2 x$

(d) $\frac{\sec^2 x}{2\sqrt{\tan x}}$

(e) $2x \times (-\sin(x^2)) = -2x\sin(x^2)$

(f) $\frac{(2x+1)\cos x - 2\sin x}{(2x+1)^2}$

5 (a) $\frac{2x+2}{x^2+2x+1} = \frac{2(x+1)}{(x+1)^2} = \frac{2}{x+1}$

(b) $e^x\log_e x + \frac{e^x}{x}$ (c) $\frac{\frac{1}{x} \times e^x - \log_e x \times e^x}{e^{2x}} = \frac{1 - x\log_e x}{xe^x}$

(d) $\frac{1}{\tan x} \times \sec^2 x = \frac{1}{\sin x\cos x}$

(e) $\frac{\left(2x + \frac{1}{x}\right) \times x - \left(x^2 + \log_e x\right) \times 1}{x^2} = \frac{x^2 + 1 - \log_e x}{x^2}$

(f) $4x^3 - 2x\sin(x^2) + \cot x$

6 (a) $f(x) = \log_e x + \log_e(\tan x)$

$f'(x) = \frac{1}{x} + \frac{\sec^2 x}{\tan x} = \frac{1}{x} + \frac{1}{\sin x\cos x}$

(b) $y = \log_e(x^3 - 6) - \log_e(e^{-x} - 1)$

$\frac{dy}{dx} = \frac{3x^2}{x^3 - 6} + \frac{e^{-x}}{e^{-x} - 1}$

(c) $f(x) = \frac{1}{2}\log_e x + \log_e(\cos x) - \log_e(1 - \sin^2 x)$

$f'(x) = \frac{1}{2x} - \frac{\sin x}{\cos x} - \frac{-2\sin x\cos x}{1 - \sin^2 x} = \frac{1}{2x} + \tan x$

OR

$f(x) = \log_e \frac{\sqrt{x}\cos x}{1 - \sin^2 x} = \log_e \frac{\sqrt{x}\cos x}{\cos^2 x} = \log_e x - \log_e(\cos x)$

$f'(x) = \frac{1}{2x} - \frac{-\sin x}{\cos x} = \frac{1}{2x} + \tan x$

7 (a) $x^2\, 10^x(3 + x\ln 10)$ (b) $\cos x + \frac{1}{x\ln a}$

(c) $2^x\ln 2 + 3^x\ln 3 + 4^x\ln 4$ (d) $\frac{a^x\left(x(\ln a)^2\log_a x - 1\right)}{x\ln a\left(\log_a x\right)^2}$

CHAPTER 14

EXERCISE 14.1

1 A, C

2 $f(x) = 2x - 2$

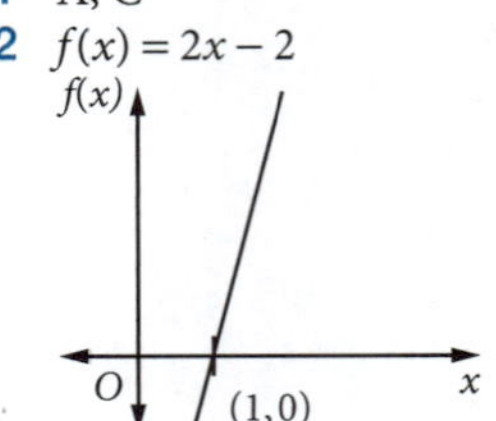

3

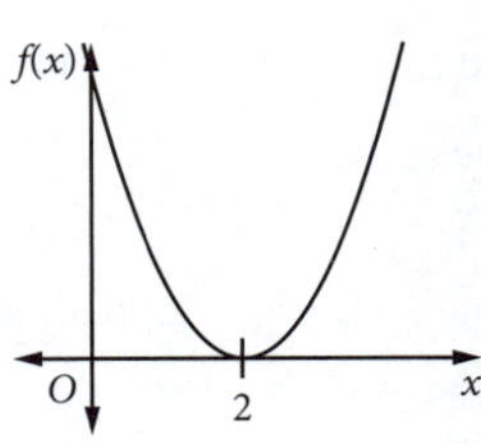

4

5

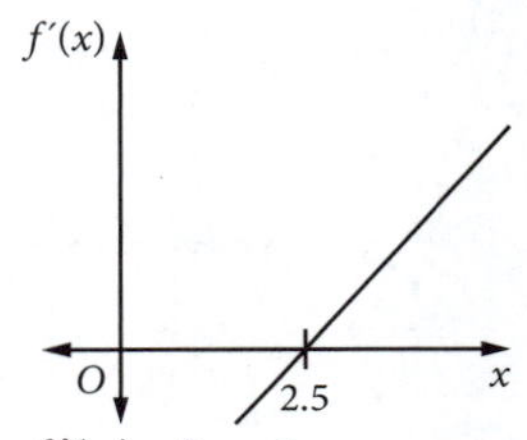

$f'(x) = 2x - 5$

(a) $x < 2.5$ (b) $x = 2.5$

(c) $x > 2.5$

6

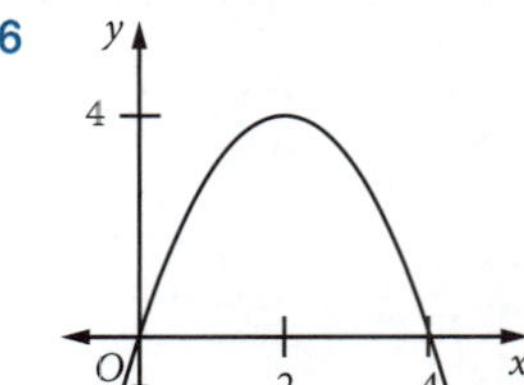

$x = 2$; positive, negative

7 (a) $x < -1.5$ (b) $x > -1.5$ (c) $x = -1.5$

8 (a) $f'(x) = 3x^2 - 12x + 9$ (b) $x < 1, x > 3$

(c) $1 < x < 3$ (d) $x = 1$

9 (a) real x (b) none (c) never

10 (a) $x = -\frac{1}{3}, 1$ (b) $x < -\frac{1}{3}, x > 1$ (c) $-\frac{1}{3} < x < 1$

EXERCISE 14.2

1 (a) $f'(x) = 2x - 6$

(b) (3, −1) minimum turning point

(c)

2 C

3 (a) $f'(x) = 3x^2 - 12x$

(b) $(0, 16)$ local maximum; $(4, -16)$ local minimum

(c)

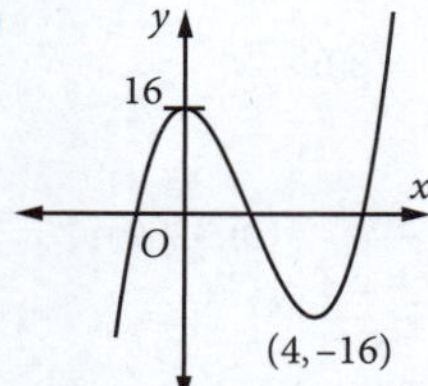

4 (a) correct (b) incorrect (c) correct (d) correct

5 (a) $f'(x) = 6x^2 - 30x + 36$

(b) $(2, 28)$ local maximum; $(3, 27)$ local minimum

(c)

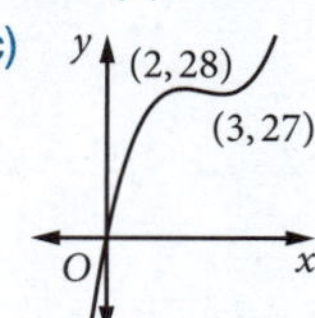

6 (a) $f'(x) = 3x^2 - 2x - 1$

(b) $-\frac{1}{3}, 1\frac{5}{27}$ local maximum; $(1, 0)$ local minimum

(c)

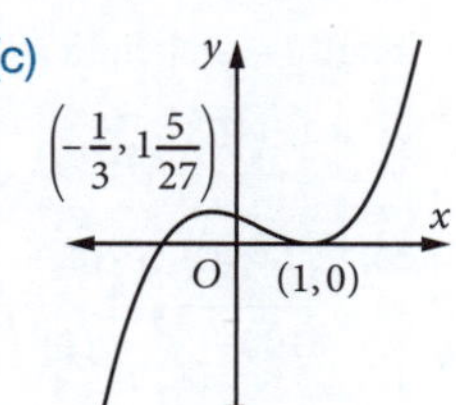

7 $3\frac{1}{8}$

8 2

9

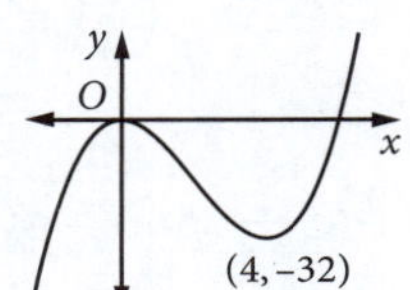

10 $(-2, 27)$ maximum; $(1, 0)$ minimum

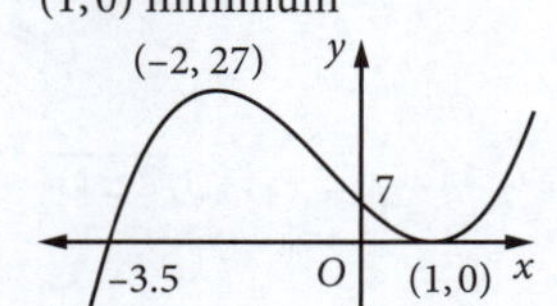

11

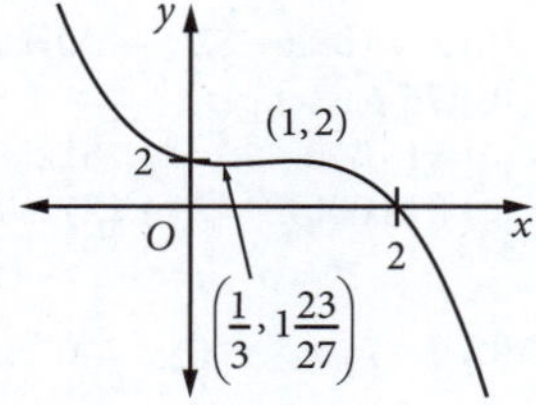

$y' = -(3x^2 - 4x + 1)$:

$\frac{1}{3}, 1\frac{23}{27}$ minimum,

$(1, 2)$ maximum

12 (a) $x = \frac{2}{3}, 2$ (b) $x < \frac{2}{3}, x > 2$ (c) $\frac{2}{3} < x < 2$

(d)

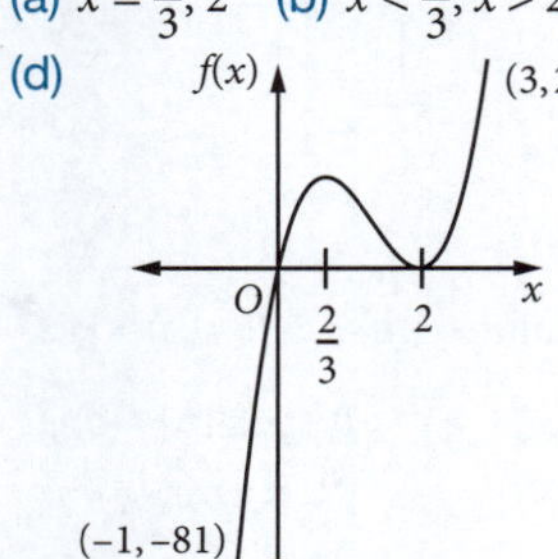

range = 108

greatest value = 27

least value = –81

13 $y' = 2ax + b$: $y' = 0$ where $2ax + b = 0$, $x = \frac{-b}{2a}$;

$y'' = 2a$ so the turning point is minimum if $a > 0$, maximum if $a < 0$

14 $\frac{dy}{dx} = -\frac{1}{x^2}$: $\frac{dy}{dx}$ is never zero, hence no stationary points, hence no turning points; $\frac{dy}{dx} < 0$ for all x in the domain

EXERCISE 14.3

1 (a) 6 (b) $6x + 4$ (c) -2 (d) $20x^3 + 12x$ (e) $-12x^2 + 4$ (f) 0

2 D

3 (a) correct (b) incorrect (c) correct (d) incorrect

4 (a) $\frac{-1}{4x\sqrt{x}}$ (b) $\frac{-1}{4(x-2)\sqrt{x-2}}$ (c) $\frac{x(2x^2+3)}{(x^2+1)^{\frac{3}{2}}}$

(d) $\frac{2}{x^3}$ (e) $\frac{2}{(x+1)^3}$ (f) $\frac{-6}{(x+3)^3}$

(g) $\frac{3(x^2+1)}{4x^2\sqrt{x}}$ (h) $\frac{15x^2+1}{4x\sqrt{x}}$ (i) $\frac{3x^2-18x+11}{4(x-1)^{\frac{3}{2}}(x+1)^3}$

5 real x

6 real x

7 (a) $x > -2$ (b) $x < -2$ (c) $(-2, 22)$

8 $x > -1$

$\frac{d^2y}{dx^2} \neq 0$ for x in the domain $\therefore$ no point of inflection

9 (a) $x > 0$ (b) $x < 0$

(c) $\frac{d^2y}{dx^2} \neq 0$ for x in the domain $\therefore$ no point of inflection

10 $\frac{d^2y}{dx^2} = \frac{6}{x^4}$; $\frac{d^2y}{dx^2} > 0$ for x in the domain $\therefore$ concave up

EXERCISE 14.4

1 $(-2, 22)$ maximum; $(1, -5)$ minimum;

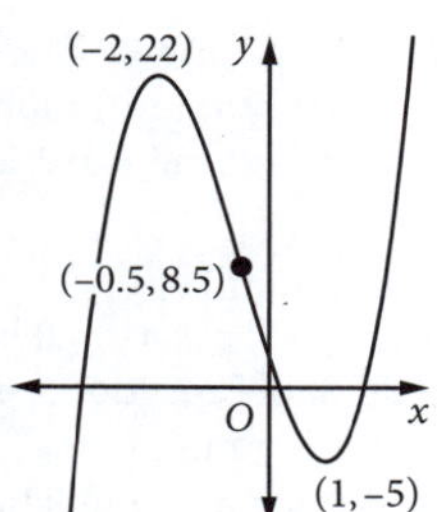

2 A

3 $(0, 0)$ minimum; $(2, 4)$ maximum; $(1, 2)$ inflection

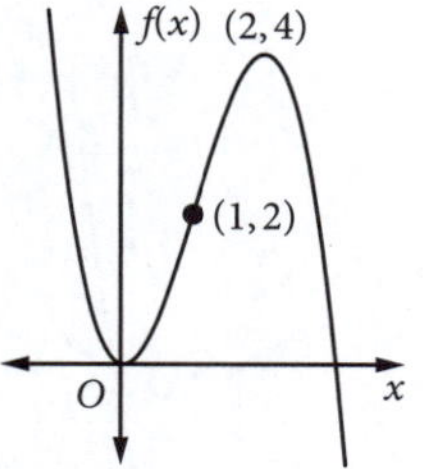

4 $(0, 1)$ min; $(2, 3)$ max; $(-2, 3)$ max; greatest value of function is 3

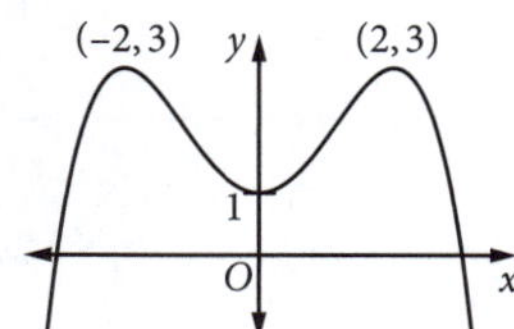

5 (a) $\left(1\frac{1}{2}, -1\frac{11}{16}\right)$ minimum

(b) $(0, 0)$ horizontal inflection, $(1, -1)$ inflection

(c)

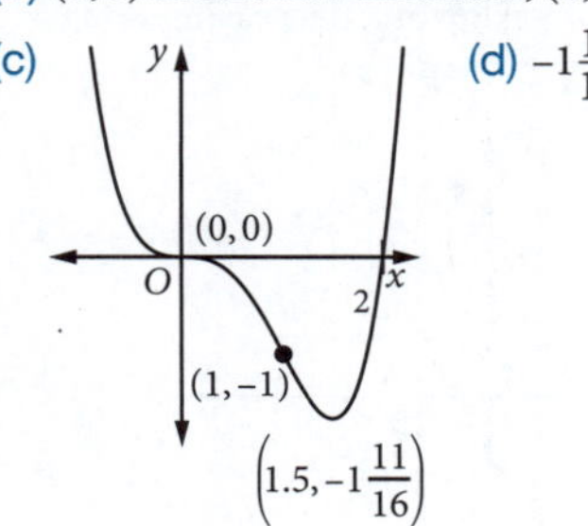

(d) $-1\frac{11}{16}$

6 (a) $x=2$ (b) $x>2$ (c) $x<2$

7 $(-1,-9)$, $(2,-48)$; $x<-1$, $x>2$

8 (a) $(-1,0)$, $(0,0)$, $(1,0)$

(b) $(0,0)$ max; $\left(\pm\frac{1}{\sqrt{2}},-\frac{1}{4}\right)$ min (c) $\left(\pm\frac{1}{\sqrt{6}},-\frac{5}{36}\right)$

(d)

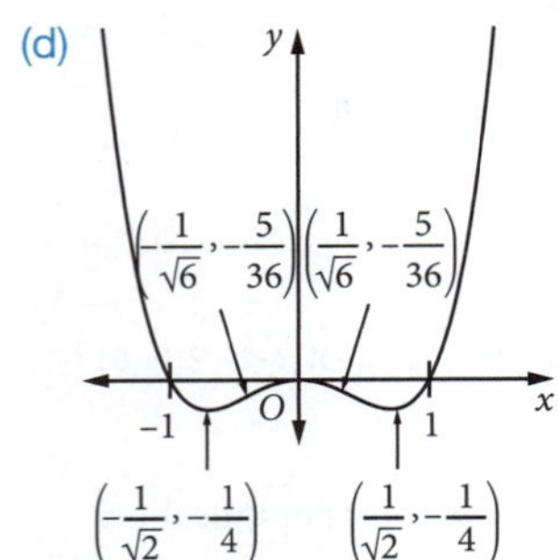

(e) $-\frac{\sqrt{6}}{6}<x<\frac{\sqrt{6}}{6}$

9 (a) correct (b) correct (c) incorrect (d) correct

10 $\frac{dy}{dx}=-3(x-1)^2<0$ for all $x\neq 1$

$\frac{d^2y}{dx^2}=-6(x-1)=0$ where $x=1$;
concavity changes, so $(1,-1)$ is a point of inflection.
$y=-x(x^2-3x+3)$, $\Delta=9-12<0$,
$x^2-3x+3=0$ has no real roots;
the curve only cuts the x-axis at $(1,-1)$

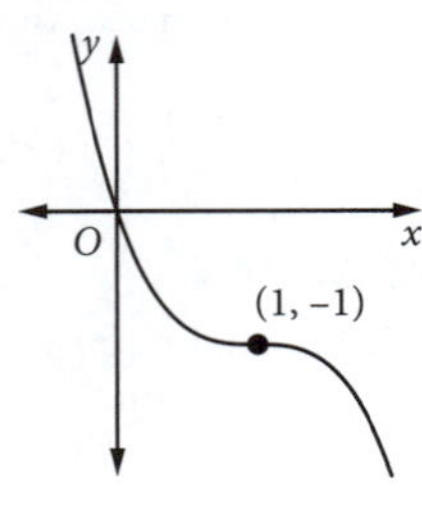

11 (a) $(1,0)$ minimum

(b) $y''=12x^2$: $y''=0$ at $x=0$ but concavity does not change
$\therefore$ no point of inflection, curve is always concave up

(c) Global minimum is 0

12 (a) Greatest value of the function is 36 when $x=6$.
Least value of the function is $-4\left(1+\sqrt{2}\right)$ when $x=2+\sqrt{2}$.

(b)

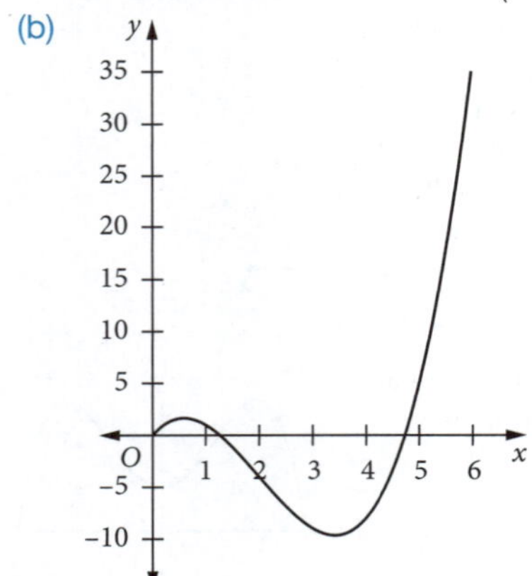

13 $4.50

14 40 units

15 (a) 90 (b) $17 420 (c) $P=374x-2.2x^2-400$ (d) 85, $15 495

16 (a) decreasing (b) it is increasing at a decreasing rate

(c) concave down

(d)

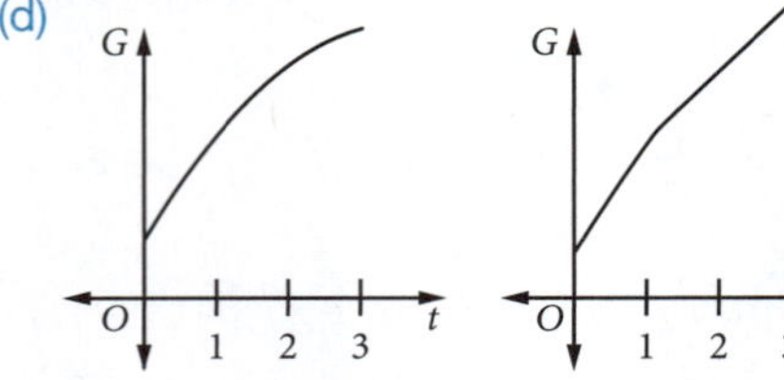

(e) $f'(t)=3.2$, point of inflection

(f) it is increasing at a decreasing rate

EXERCISE 14.5

1 (a) $l=80-x$ (b) $A(x)=x(80-x)=80x-x^2$

(c) $1600\,\text{m}^2$

2 B

3 (a) $l=160-2x$ (b) $A(x)=x(160-2x)$ (c) $3200\,\text{m}^2$

4 (a) $y=\frac{120-4x}{3}$ (b) $A(x)=\frac{x(120-4x)}{3}$

(c) $x=15$, $y=20$, $A=300\,\text{m}^2$ (d) $396.75\,\text{m}^2$

5 (a) $A=\frac{1+x-x^2}{2}$ (b) $\frac{5}{8}$ units squared

6 $\frac{1600}{27}\,\text{cm}^3$

7 (a) $y=12-4x$, $0<x<3$ (b) $V=12x^2-4x^3$ (c) $16\,\text{cm}^2$

8 $\frac{8000}{27}\,\text{cm}^3$

9 (a) $h=9-2x$, $0<x<4.5$ (b) $V=x^2(9-2x)$ (c) $x=3$

10 (a) length $=2x$, height $=\frac{108-2x^2}{3x}$

(b) $V=\frac{2x\left(108-2x^2\right)}{3}$ (c) $144\sqrt{2}\,\text{cm}^3$

11 (a) correct (b) correct (c) correct (d) correct

12 (a) $y=\sqrt{100-x^2}$ (b) $V=100x-x^3$ (c) $\frac{2000\sqrt{3}}{9}\,\text{cm}^3$

13 (a) $r=\sqrt{36-x^2}$ (b) $V=\frac{\pi x\left(36-x^2\right)}{3}$ (c) $2\sqrt{3}$ cm

14 $28\frac{1}{8}\,\text{cm}^2$

15 (a) $h=\frac{10-r^2}{r}$ (b) $V=\pi(10r-r^3)$ (c) $r=\frac{\sqrt{30}}{3}$

16 rectangle $3\frac{4}{7}$ cm $\times$ $10\frac{5}{7}$ cm; square sides $5\frac{5}{14}$ cm

17 (a) $R=(380-12(n-20))\times n=620n-12n^2$

(b) 26 passengers

18 (a) $15 000 (b) 2812 or 2813

19 (a) $24 000 (b) $27 648

20 (a) $h^2+r^2=a^2$, $h=\sqrt{a^2-r^2}$, $V=\pi r^2\times 2h=2\pi r^2\sqrt{a^2-r^2}$

(b) $V'(r)=2\pi r\left(2a^2-3r^2\right)$;

$3r^2=2a^2$, $r=\frac{a\sqrt{6}}{3}$

$V''(r)=2\pi(2a^2-9r^2)$

$V''\left(\frac{a\sqrt{6}}{3}\right)=2\pi\left(2a^2-\frac{9\times 6a^2}{9}\right)<0$;

$r=\frac{a\sqrt{6}}{3}$ gives max volume

EXERCISE 14.6

1 $-e$

2 $\frac{dy}{dx}=\frac{(2-x)e^{-0.5x}}{2}$: $\left(2,\frac{2}{e}\right)$, maximum (a) $x>0$ (b) $x<2$

3 (a) $f(0)=2$, $f'(0)=1$ (b) $f'(x)=e^{-x}>0$ for all x (c) 3

(d)

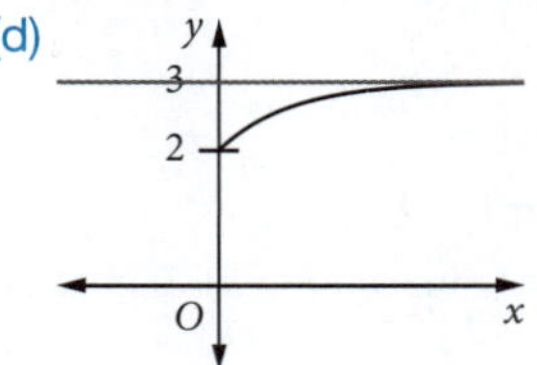

4 (a) $f'(x) = -2xe^{-x^2}$ (b) (i) $x = 0$ (ii) $x < 0$ (iii) $x > 0$

(c)

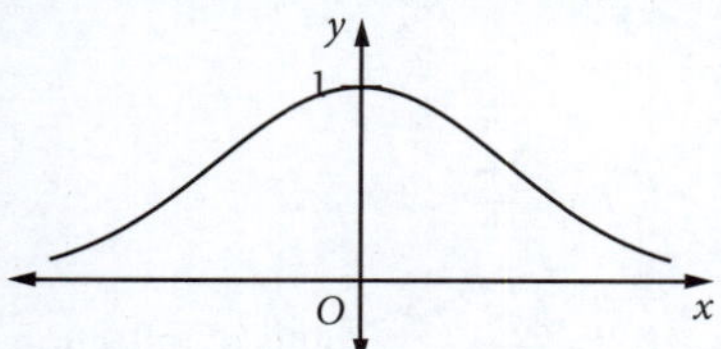

5 (a) 0.1 units after 0.91 hours

(b)

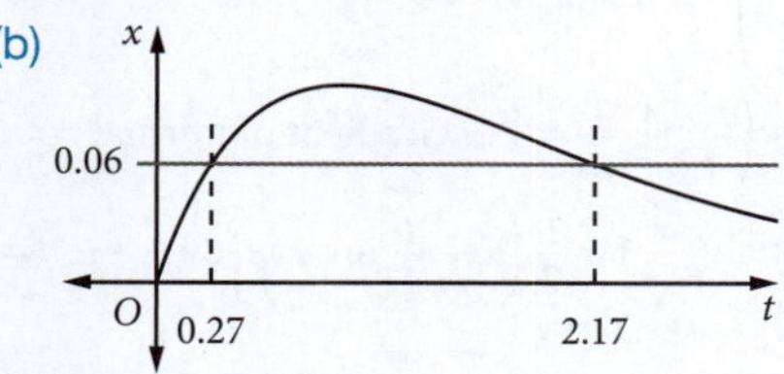

(c) 2.00 hours

6 (a) correct (b) incorrect (c) correct (d) correct

7

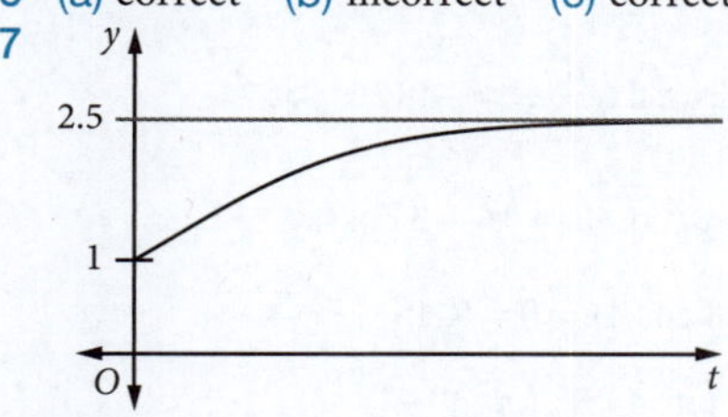

(a) $f'(t) = \dfrac{-5\times\left(-3e^{-t}\right)}{\left(2+3e^{-t}\right)^2} = \dfrac{15e^{-t}}{\left(2+3e^{-t}\right)^2} > 0$ because $e^{-t} > 0$ and $\left(2+3e^{-t}\right)^2 > 0$ for all t

(b) 2.5 (c) $1 \le f(t) < 2.5$

8 (a) $x = \dfrac{1}{\sqrt{2}}$ (b) $\sqrt{2}e^{-0.5} \approx 0.86$ units^2

9 $\dfrac{dy}{dx} = \dfrac{1-\ln x}{x^2}$. Stationary points: $x = e, y = \dfrac{1}{e}$.

$\dfrac{d^2y}{dx^2} = \dfrac{2\ln x - 3}{x^3} < 0$ when $x = e$. Maximum at $\left(e, \dfrac{1}{e}\right)$.

10 $\dfrac{dy}{dx} = 2e^{2x} - 8e^{-2x}$. Stationary points: $x = \dfrac{\ln 2}{2}, y = 4$.

$\dfrac{d^2y}{dx^2} = 16 > 0$. Minimum value is 4 when $x = \dfrac{\ln 2}{2}$.

11 $\dfrac{dy}{dx} = 1 + \ln x$. Stationary points: $x = \dfrac{1}{e}, y = -\dfrac{1}{e}$. $\dfrac{d^2y}{dx^2} = e > 0$.

Minimum when $x = \dfrac{1}{e}$ is $-\dfrac{1}{e}$.

12 (a) Require $\sin x > 0$: $0 < x < \pi$.

(b) $\dfrac{dy}{dx} = \cot x$. Stationary points: $x = \dfrac{\pi}{2}, y = 0$.

$\dfrac{d^2y}{dx^2} = -\text{cosec}^2\, x = -1 < 0$. Maximum when $x = \dfrac{\pi}{2}$ is 0.

13 (a) Maximum profit is \$188.75

(b)

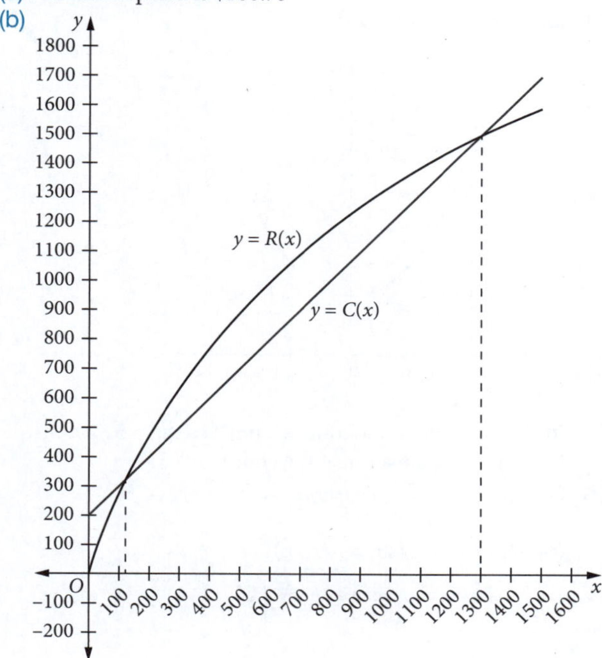

First breaks even when $x \approx 130$

(c) A profit is made when the revenue > cost. This is between about 130 and 1300, or 13 batches of ten and 130 batches of ten.

(d)

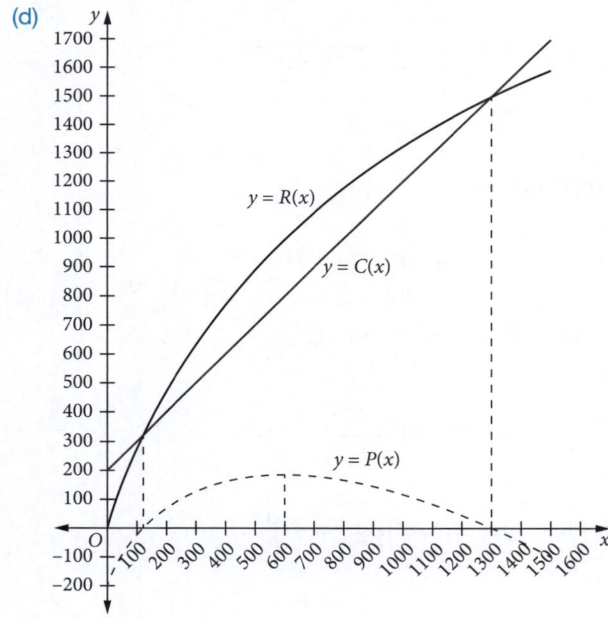

$P(x) \ge 0$ for $130 \le x \le 1300$

14 (a) $a = 0.5,\ y = e^{\frac{x}{2}} + e^{-\frac{x}{2}}$

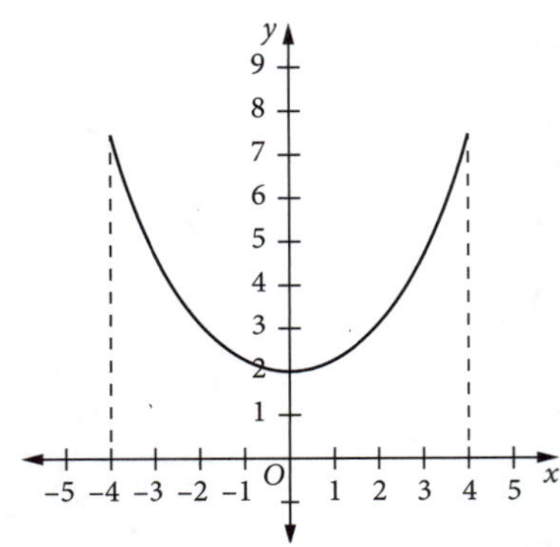

(b) Least height of function is 2 units so sag = 5.52 units.
(c) Angle of inclination at the ends is 74° 35′.

15 (a) $f(0) = 1,\ f\left(\frac{\pi}{2}\right) = 0, f(\pi) = -e^{-\pi}$

(b) $f'(x) = -e^{-x}(\sin x + \cos x)$

(c) $f'(0) = -1,\ f'\left(\frac{3\pi}{4}\right) = -e^{-\frac{3\pi}{4}}\left(\frac{1}{\sqrt{2}} - \frac{1}{\sqrt{2}}\right) = 0$

(d)

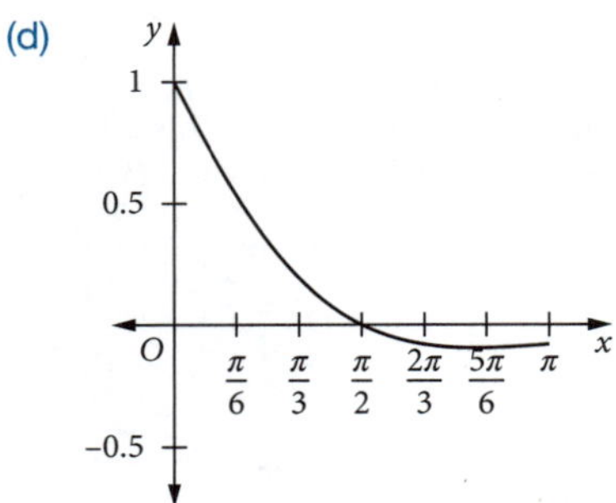

EXERCISE 14.7

1 $-\tan x$

2 $4x - 2y + 2 - \pi = 0$

3 (a) $\cos x - \sin x$ (b) $-\sin x - \cos x$ (c) $\left(\frac{\pi}{4}, \sqrt{2}\right), \left(\frac{5\pi}{4}, -\sqrt{2}\right)$
(d) $\left(\frac{3\pi}{4}, 0\right), \left(\frac{7\pi}{4}, 0\right)$ (e) $\sqrt{2}$

4 $y = 1 - x$

5 $\frac{dy}{dx} = 2\cos x + 2\sin x \cos x = 2\cos x(1 + \sin x)$; $\left(\frac{\pi}{2}, 3\right), \left(\frac{3\pi}{2}, -1\right)$,
$\left(\frac{5\pi}{2}, 3\right), \left(\frac{7\pi}{2}, -1\right)$

6 (a) 4.5 m (b) 12.2 hours; $n = \frac{10\pi}{61}$

(c) $y = 4.5\cos\frac{10\pi t}{61}$

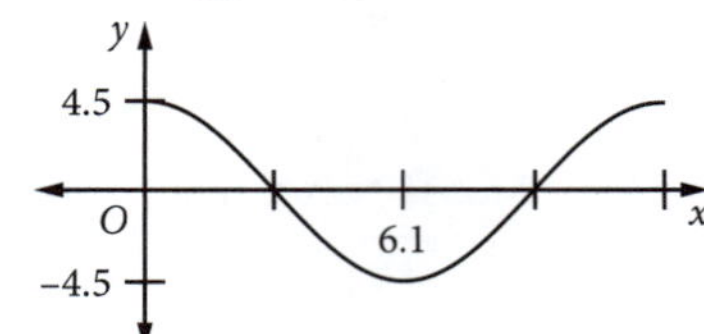

(d) 1.08 m

7 (a) $\frac{dy}{dx} = \cos x - \sin x$. $\frac{dy}{dx} = 0$ when $\tan x = 1$. $x = \frac{\pi}{4}, \frac{5\pi}{4}$.

(b) $x = \frac{\pi}{4}, y = \sqrt{2}$. $x = \frac{5\pi}{4}, y = -\sqrt{2}$. $\frac{d^2y}{dx^2} = -y$.

$x = \frac{\pi}{4}$: $\frac{d^2y}{dx^2} < 0$. Maximum at $\left(\frac{\pi}{4}, \sqrt{2}\right)$.

$x = \frac{5\pi}{4}$: $\frac{d^2y}{dx^2} > 0$. Minimum at $\left(\frac{5\pi}{4}, -\sqrt{2}\right)$.

Greatest value of y is $\sqrt{2}$ at $x = \frac{\pi}{4}$ and the least value is $-\sqrt{2}$ at $x = \frac{5\pi}{4}$.

(c) $\frac{dy}{dx} > 0$ for $0 < x < \frac{\pi}{4}$ and $\frac{5\pi}{4} < x < 2\pi$.

(d) $\frac{d^2y}{dx^2} = 0$: $\tan x = -1$, $x = \frac{3\pi}{4}, \frac{7\pi}{4}$. Points of inflection at $\left(\frac{3\pi}{4}, 0\right), \left(\frac{7\pi}{4}, 0\right)$.

8 $\frac{dy}{dx} = -\operatorname{cosec}^2 x$. $P\left(\frac{\pi}{4}, 1\right)$: $\frac{dy}{dx} = -2$. Gradient of normal $= \frac{1}{2}$.

Equation of normal: $y - 1 = \frac{1}{2}\left(x - \frac{\pi}{4}\right)$ or $4x - 8y + 8 - \pi = 0$

9 $\frac{dy}{dx} = e^{\operatorname{cosec} x}(-\operatorname{cosec} x \cot x)$. $x = \frac{\pi}{2}$: $\frac{dy}{dx} = 0, y = e$.

The equation of the tangent is $y = e$.

10 $y = 3\cos 4x$. $\frac{dy}{dx} = -12\sin 4x$. $\frac{d^2y}{dx^2} = -48\cos 4x$.

$\text{LHS} = \frac{d^2y}{dx^2} + 16y = -48\cos 4x + 16 \times 3\cos 4x$

$= -48\cos 4x + 48\cos 4x = 0 = \text{RHS}$

11 (a) $180 = 200 + 40\cos\left(\frac{\pi}{6}t\right)$. $\cos\left(\frac{\pi}{6}t\right) = -\frac{1}{2}$. $t = 4, 8$.

After 4 months and 8 months.

(b)

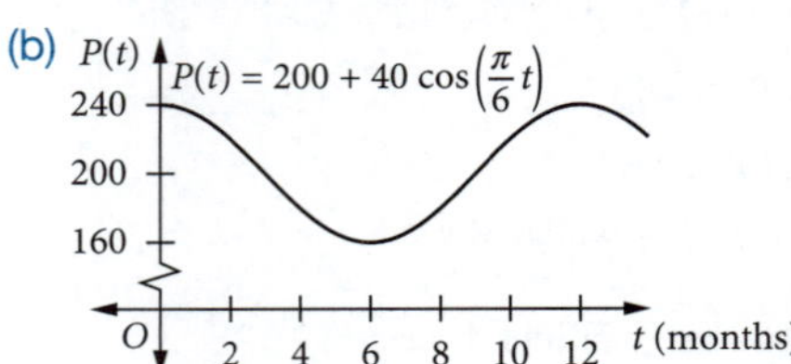

(c) $\frac{dP(t)}{dt} = -\frac{20\pi}{3}\sin\left(\frac{\pi}{6}t\right)$.

$\frac{dP(t)}{dt} > 0$: $\sin\left(\frac{\pi}{6}t\right) < 0, 6 < t < 12$.

From the graph: between 6 and 12 months.

(d) The greatest population is 240 wallabies.

(e) From 160 to 240 wallabies.

EXERCISE 14.8

1 $x = \frac{t^3}{2} - 3t^2 + 5$

(a) $v = \frac{3t^2}{2} - 6t$ (b) $a = 3t - 6$

(c) $\frac{3t^2}{2} - 3t = 0, \frac{3t}{2}(t - 4) = 0, t = 0, 4$

(d) $t = 4$: $x = -11, v = 0, a = 6$

2 C

3 $x = 2t^3 - 6t^2 - 30t$.

(a) $v = 6t^2 - 12t - 30, a = 12t - 12$

(b) $t = 0$: $v = -30, a = -12$

(c) $v = 0$: $6t^2 - 12t - 30 = 0$, $6(t^2 - 2t - 5) = 0$,

$t = \dfrac{2 \pm \sqrt{4+20}}{2} = 1 \pm \sqrt{6}$,

$t = 1 + \sqrt{6}$

$a = 12(1 + \sqrt{6}) - 12 = 12\sqrt{6}$

(d) $t^2 - 2t - 5 < 0$, $0 \le t < 1 + \sqrt{6}$

4 (a)

(b) $v = 30 - 10t$. (c) $v = 30\ \text{m s}^{-1}$

(d) $v = 0$: $t = 3\ \text{s}$, $x = 90 - 45 = 45\ \text{m}$ (e) 6 seconds

(f) $v = -30$, speed $= 30\ \text{m s}^{-1}$ (g) $a = -10\ \text{m s}^{-2}$

5 (a) $a = -5e^{-t}$. (b) $e^{-t} > 0$ for all t so $a < 0$ for all $t \ge 0$

(c) $v = 5e^{-t} > 0$ for all $t \ge 0$ as $e^{-t} > 0$ for all t. Hence $\dfrac{dx}{dt}$ is never zero so x cannot have any stationary points.

6 (a) $20.61\ \text{km h}^{-1}$

(b) Maximum velocity after 2.5 hours

CHAPTER REVIEW 14

1 (a) $-0.5 < x < 2.5$ (b) $x < -0.5, x > 2.5$ (c) $x = 2.5$ (d) $x = -0.5$

2 (a) $x = -2, 1$ (b) $x < -2, x > 1$ (c) $-2 < x < 1$

3 (a) (b)

4 (a) $(0, 0)$ max; $\left(\dfrac{4}{9}, \dfrac{-32}{243}\right)$ min (b) $(-1, 5)$ max; $(3, -27)$ min

5 $x = \dfrac{15(6+\sqrt{3})}{11}$, $y = \dfrac{45(5-\sqrt{3})}{22}$

6 $3\frac{13}{17}$ m, $4\frac{4}{17}$ m

7 height $= y$ cm; $2x + 2y + \pi x = 50$, $y = \dfrac{50 - (\pi + 2)x}{2}$;

$\dfrac{100}{\pi + 4}$ by $\dfrac{50}{\pi + 4}$, $\dfrac{1250}{\pi + 4}\ \text{cm}^2$

8 (a) $x = \pm 2$ (b) $-2 < x < 2$

(c) $8\frac{1}{3}$, local max; $-2\frac{1}{3}$, local min

(d)

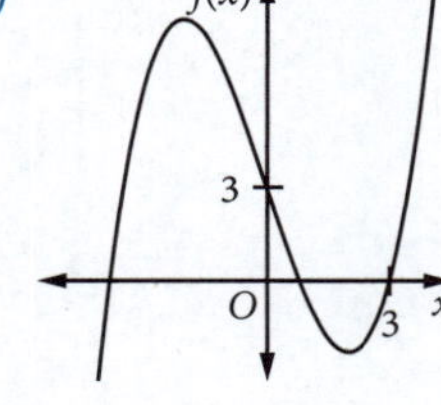

9

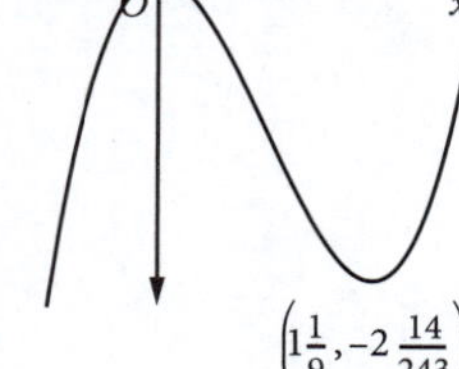

10 $\dfrac{5-\sqrt{7}}{3}$ cm

11 $y' = 6x - 3x^2 = 3x(2 - x)$

At $x = 0$, $y = 0$, $y' = 0$, so the curve is horizontal where it cuts the y-axis at the origin and therefore crosses the y-axis at right angles.

12

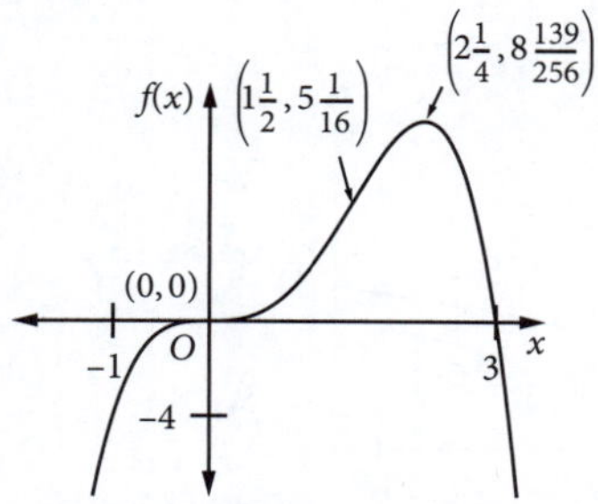

$\left(2\frac{1}{4}, 8\frac{139}{256}\right)$ maximum turning point;

$(0, 0)$ horizontal inflection, $\left(1\frac{1}{2}, 5\frac{1}{16}\right)$ inflection

13 (a) (i) $\dfrac{\sqrt{9+x^2}}{10\sqrt{2}}$ hours (ii) $\dfrac{4-x}{20}$ hours

(b) $\dfrac{\sqrt{9+x^2}}{10\sqrt{2}} + \dfrac{4-x}{20}$

(c) $x = 3\ \text{km}$, $t = 0.35\ \text{h} = 21\ \text{min}$

14 $60\ \text{km h}^{-1}$

15 (a) 0.4905 at $x = \dfrac{2}{3}$

(b) $f(0) = 0$, $f(0.5) = 0.472$, $f(1) = 0.446$

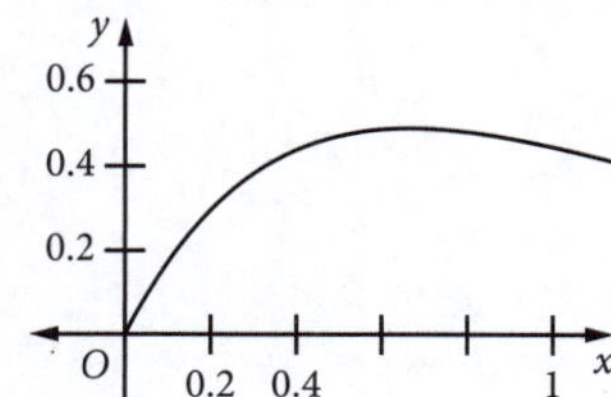

16 $\dfrac{d}{dt}(\theta_0 e^{-kt}) = -k\theta_0 e^{-kt} = -k\theta$

17 (a) \$10 000 (b) \$3012

(c) (i) \$602.39 per year (ii) \$1000 per year

(d) 11.5 years

18 $\dfrac{dy}{dx} = 6 \sec 2x \tan 2x$. Stationary points: $x = 0, \dfrac{\pi}{2}$

$\dfrac{d^2y}{dx^2} = 12 \sec 2x(\tan^2 2x + \sec^2 2x)$

Minimum at $(0, 3)$, maximum at $\left(\dfrac{\pi}{2}, -3\right)$.

CHAPTER 15

EXERCISE 15.1

1 (a)

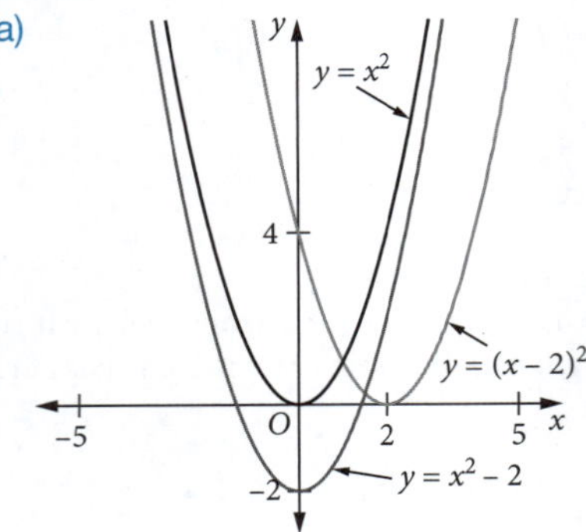

(b)

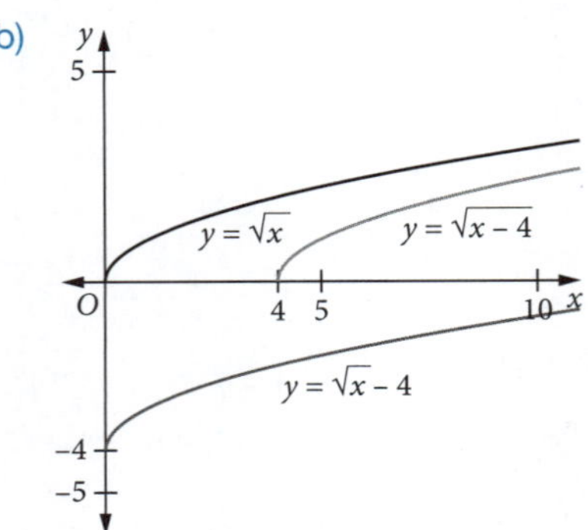

(c)

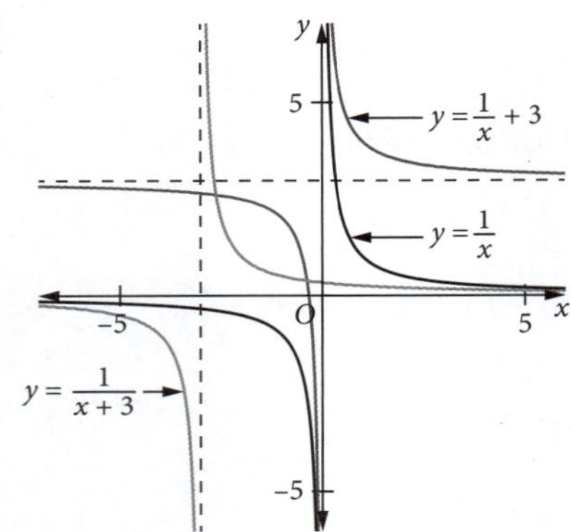

(d)

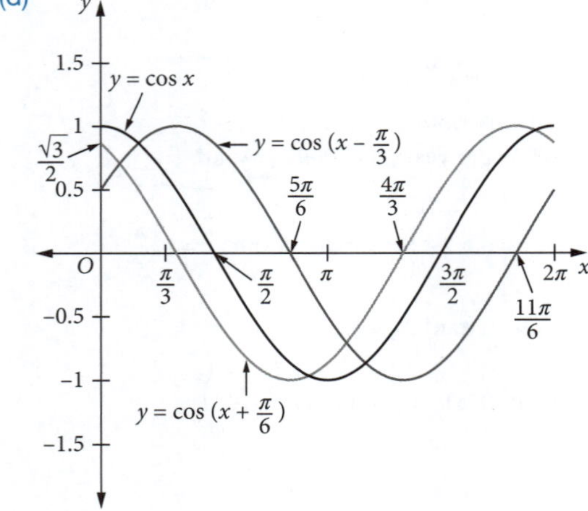

2 (a)

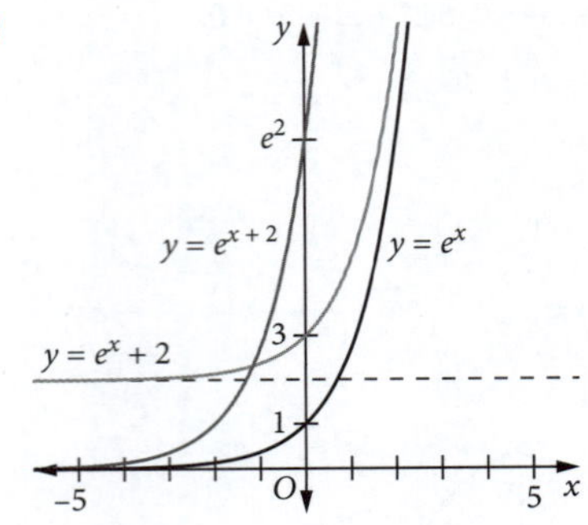

(b)

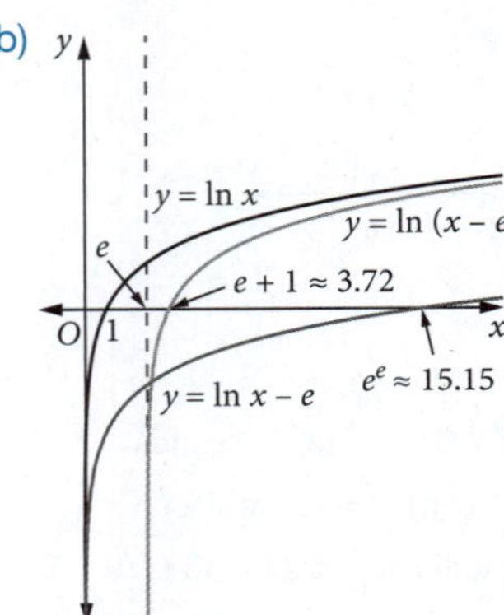

(c)

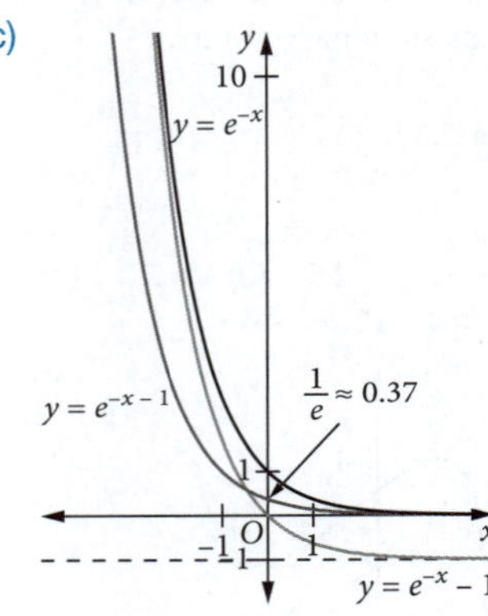

(d)

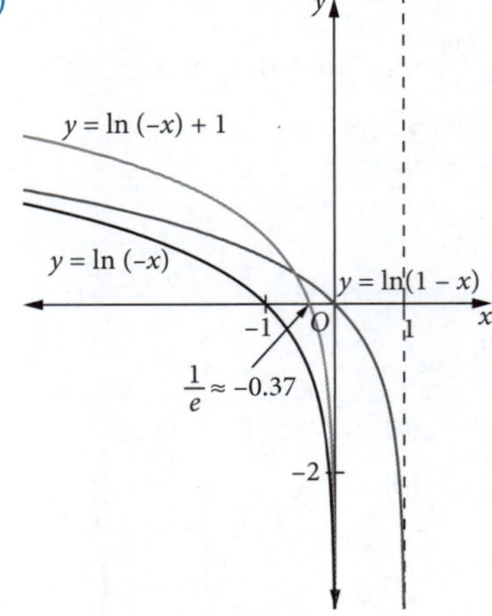

3 (a)

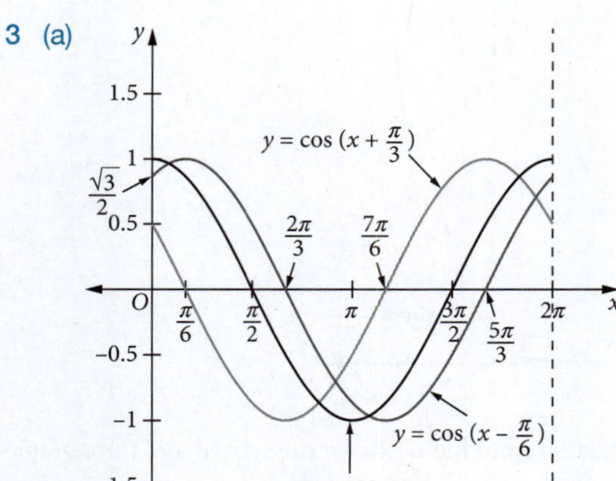

(b)

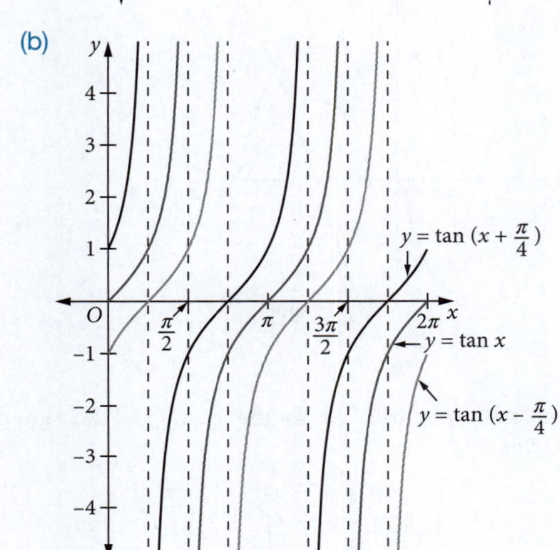

(c)

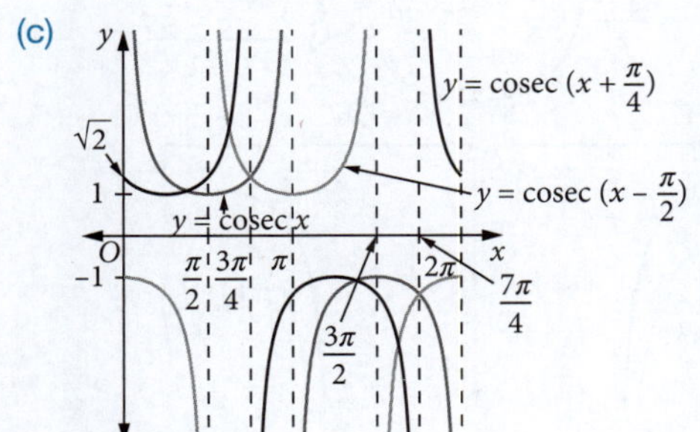

(d)

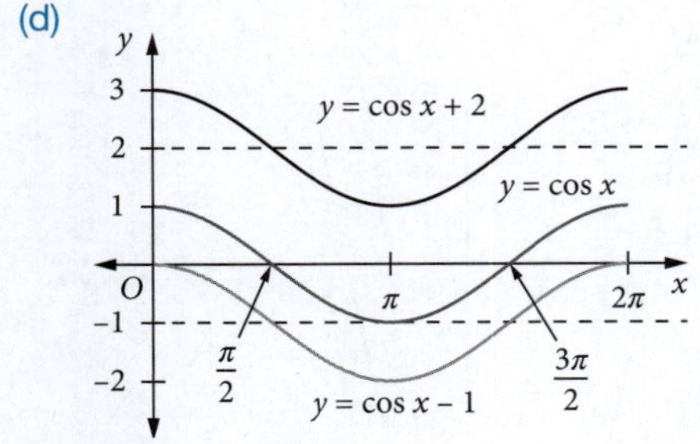

4 (a)

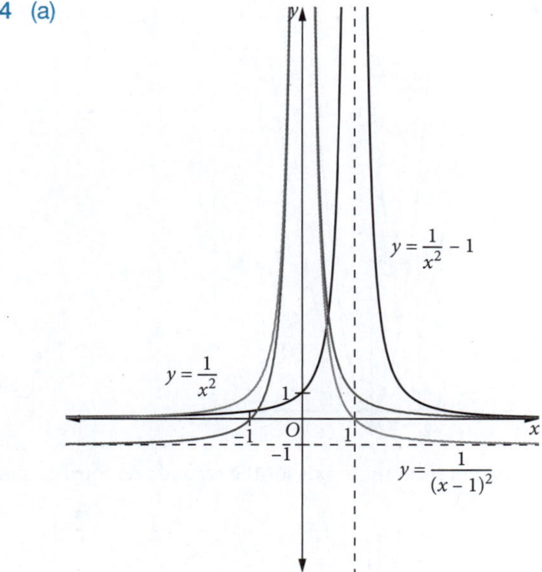

(b)

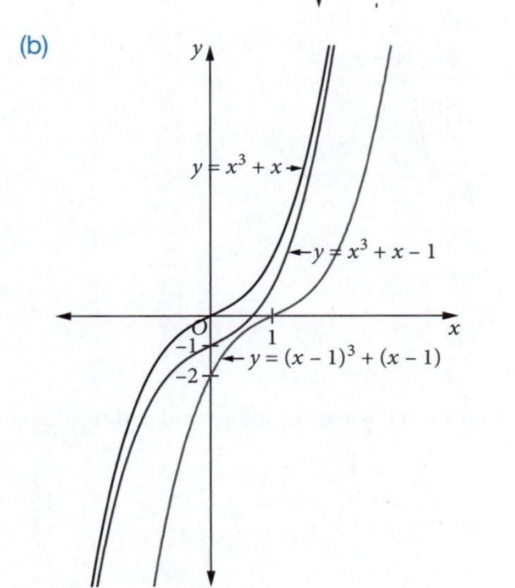

(c)

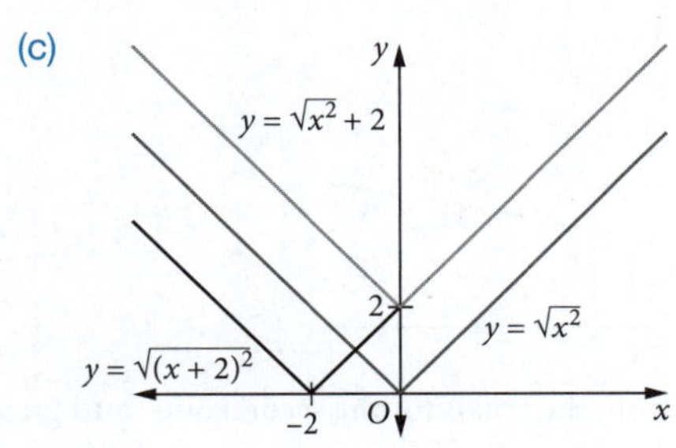

5 C.

EXERCISE 15.2

1 (a)

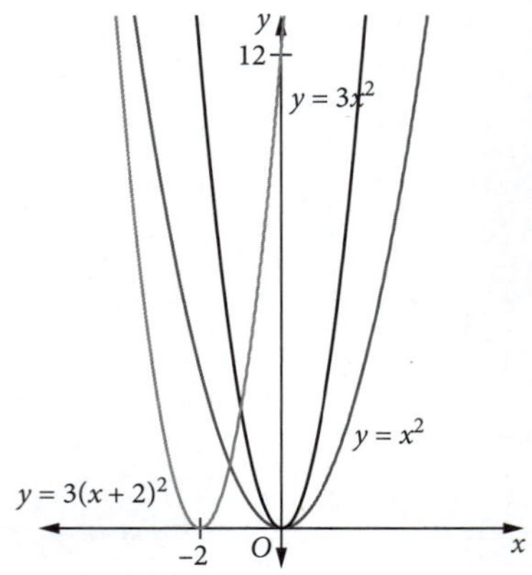

The dilation from the x-axis for the second and third graphs has factor 3.

(b)

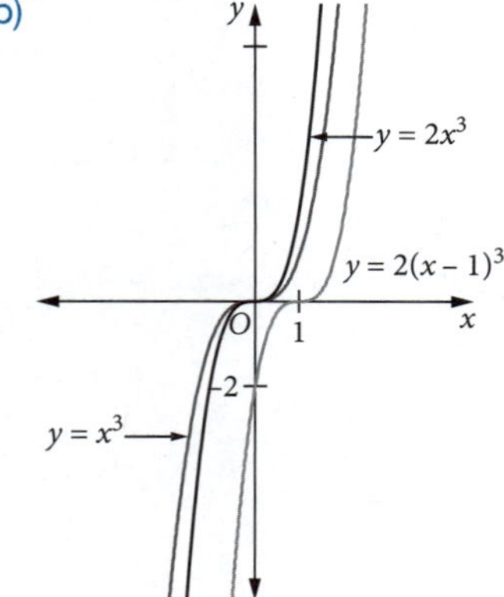

The dilation from the x-axis for the second and third graphs has factor 2.

(c)

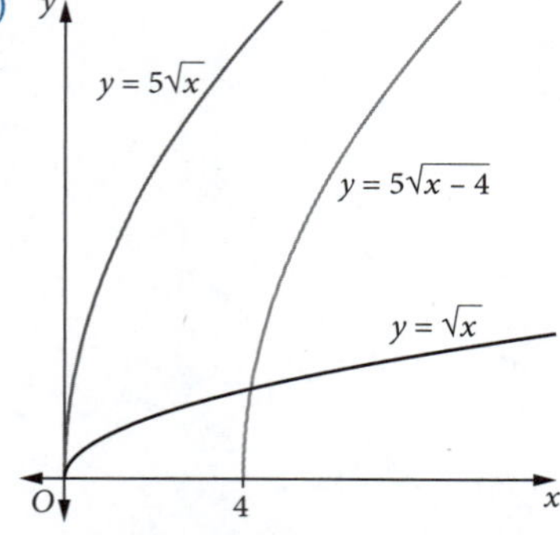

The dilation from the x-axis for the second and third graphs has factor 5.

(d)

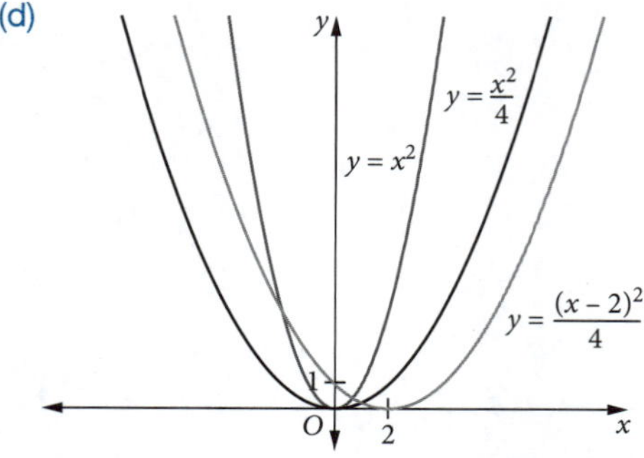

The dilation from the x-axis for the second and third graphs has factor $\frac{1}{4}$.

2 (a)

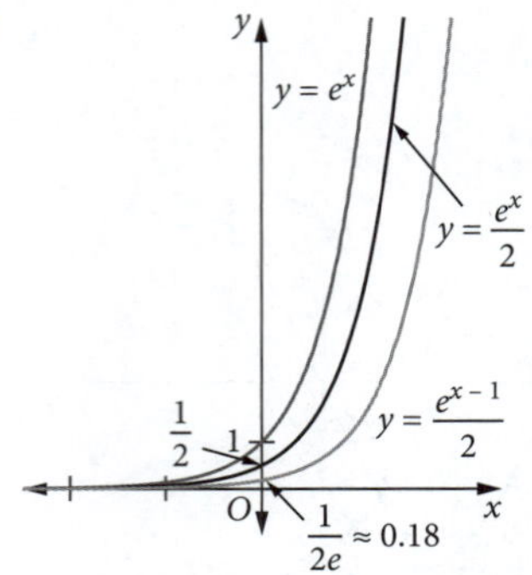

The dilation from the x-axis for the second and third graphs has factor $\frac{1}{2}$.

(b)

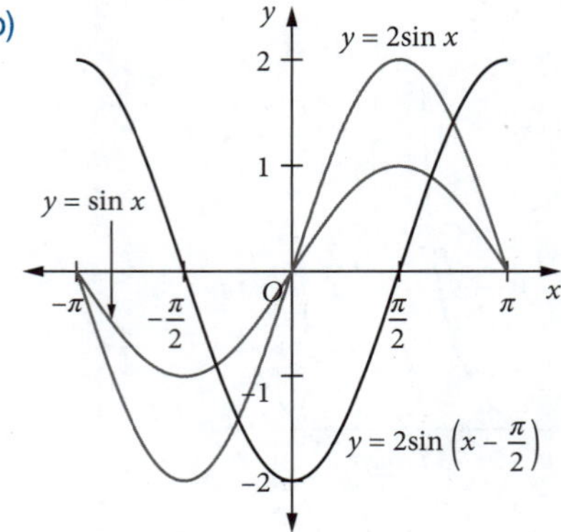

The dilation from the x-axis for the second and third graphs has factor 2.

(c)

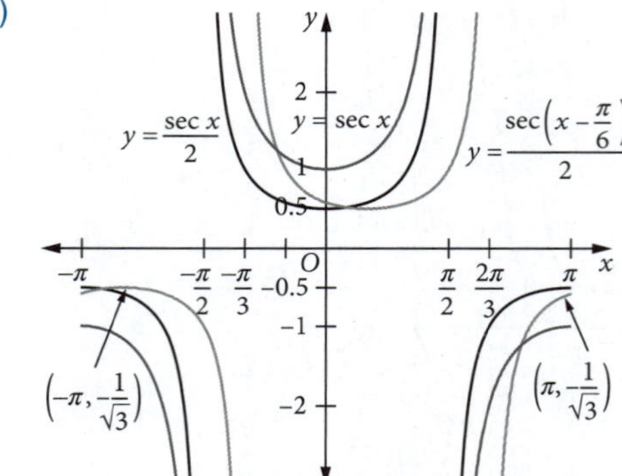

The dilation from the x-axis for the second and third graphs has factor $\frac{1}{2}$.

3 (a)

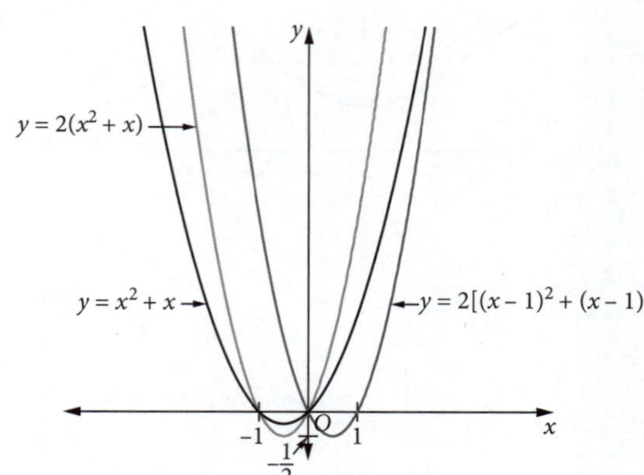

The dilation from the x-axis for the second and third graphs has factor 2.

(b) 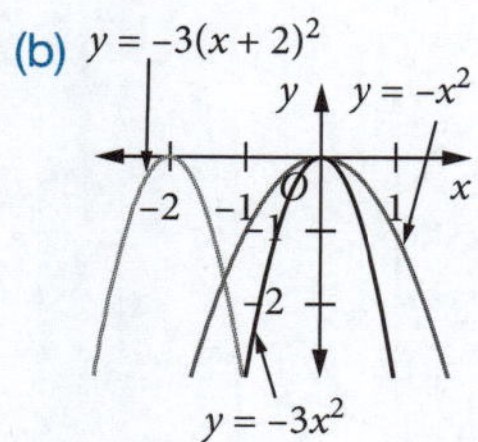

The dilation from the x-axis for the second and third graphs has factor 3.

(c) 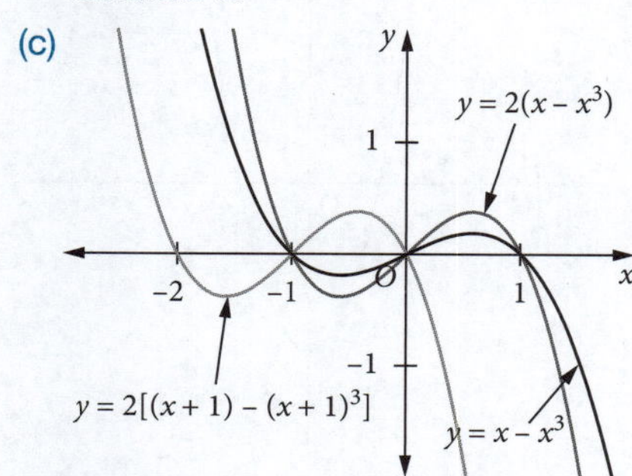

The dilation from the x-axis for the second and third graphs has factor 2.

4 B

EXERCISE 15.3

1 (a) 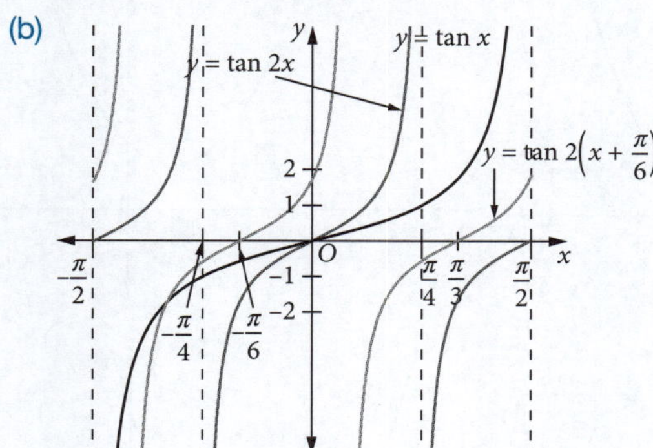

The dilation from the y-axis in the second and third graphs has a factor of 0.5. The third graph has also undergone a translation of 3 units to the right.

(b) 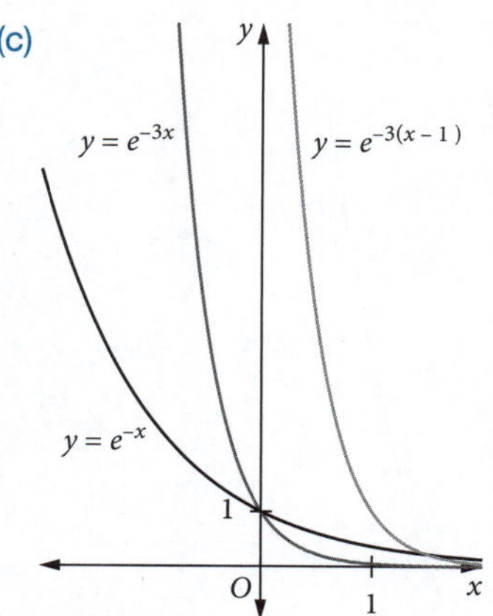

The dilation from the y-axis in the second and third graphs has a factor of 0.5. The third graph has also undergone a translation of $\frac{\pi}{6}$ units to the left.

(c) 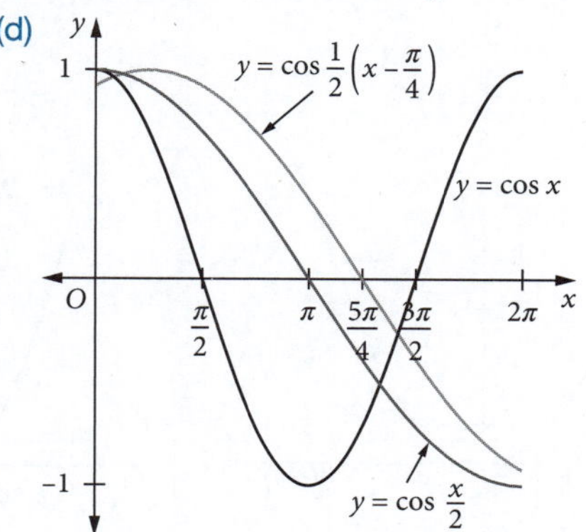

The dilation from the y-axis in the second and third graphs has a factor of $\frac{1}{3}$. The third graph has also undergone a translation of 1 unit to the right.

(d)

The dilation from the y-axis in the second and third graphs has a factor of 2. The third graph has also undergone a translation of $\frac{\pi}{4}$ units to the right.

2 (a) $f(2x) = (2x)^3 = 8x^3$ (b) $f(x-1) = (x-1)^3$
(c) $f(x) + 3 = x^3 + 3$ (d) $2f(x) + 1 = 2x^3 + 1$
(e) $3f(2(x+2)) - 4 = 3[2(x+2)]^3 - 4 = 24(x+2)^3 - 4$

3 (a) $f(2x) = \cos x$ (b) $f\left(x + \frac{\pi}{3}\right) = \cos \frac{1}{2}\left(x + \frac{\pi}{3}\right)$
(c) $2f(x) = 2\cos \frac{x}{2}$ (d) $f(x) - 1 = \cos \frac{x}{2} - 1$
(e) $2f\left(x + \frac{\pi}{6}\right) + 1 = 2\cos \frac{1}{2}\left(x + \frac{\pi}{6}\right) + 1$

4 (a)–(e) 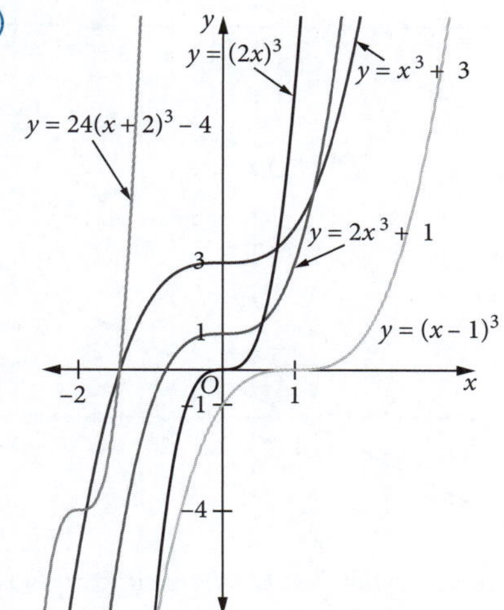

5 A

6 (a) 

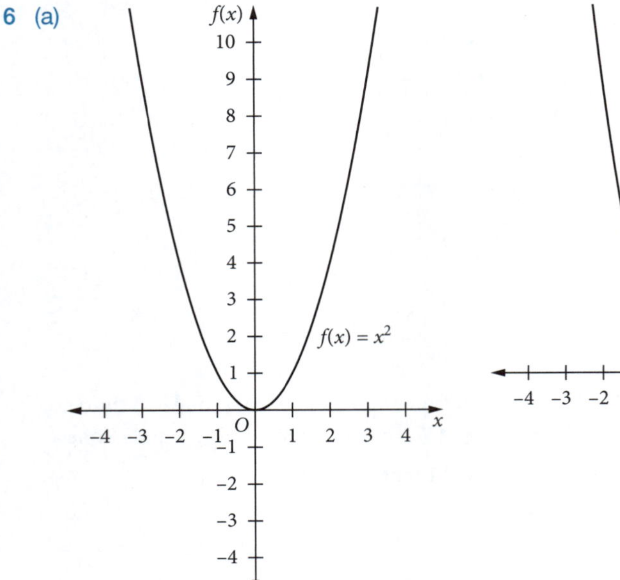

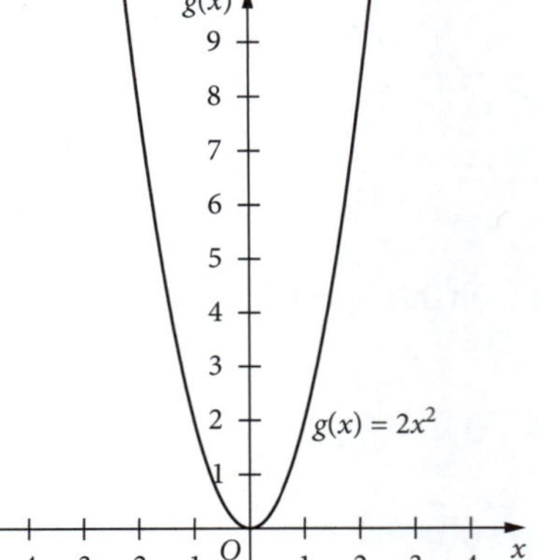

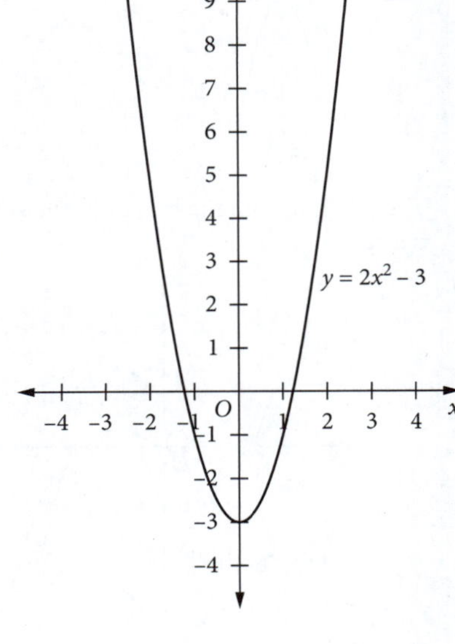

(b)

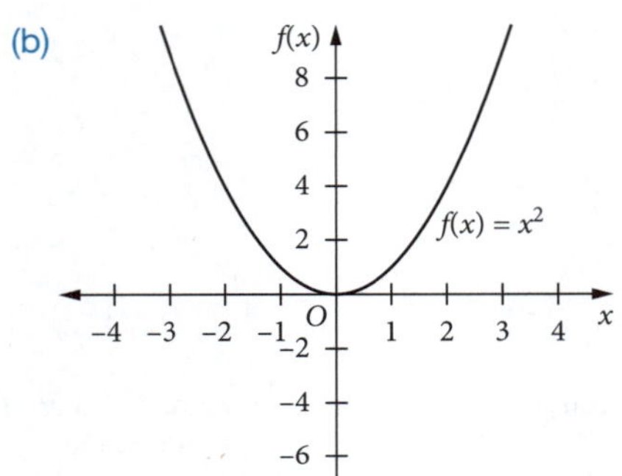

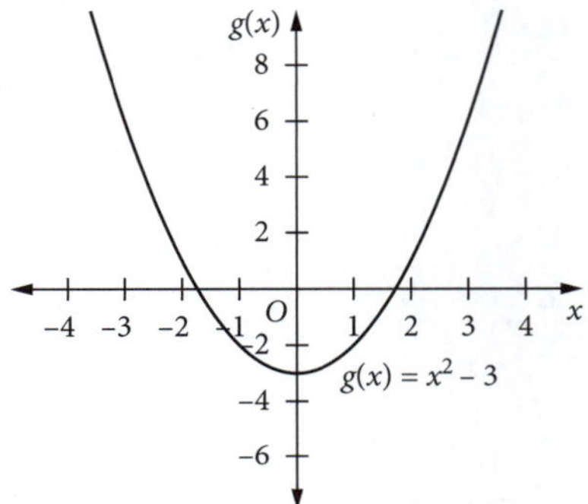

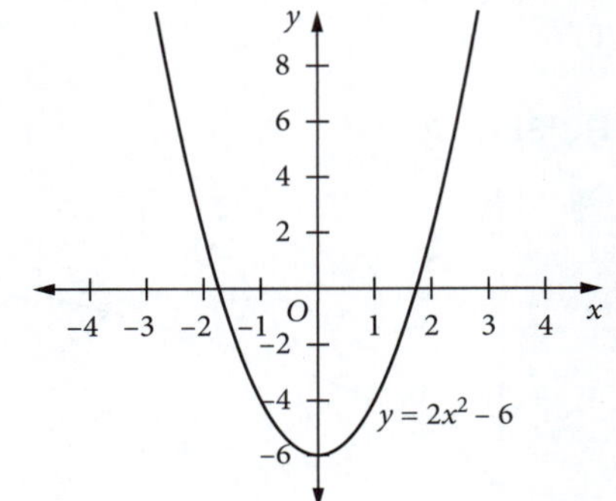

(c) The final graph in part (b) has been moved down 6 units, whereas part (a) was moved down 3 units. They have the same dilation.

7 (a)

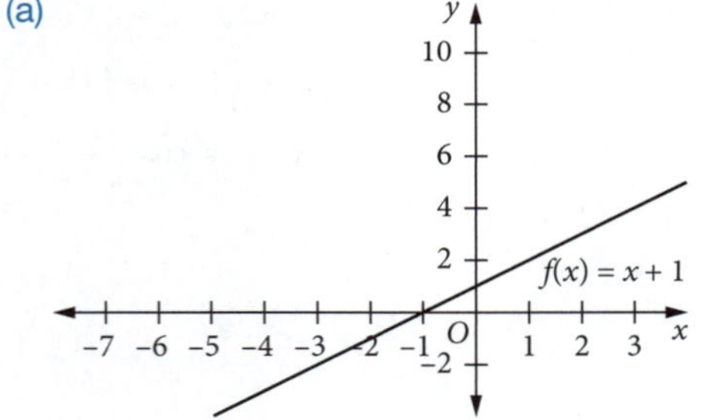

$g(x) = (x + 1)^2$

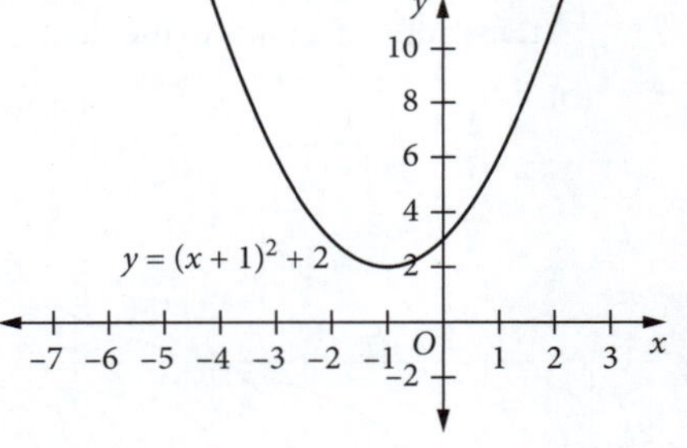

(b)

$f(x) = x + 1$

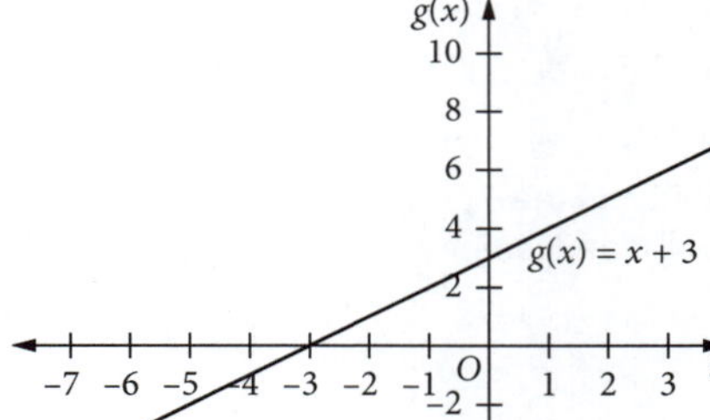

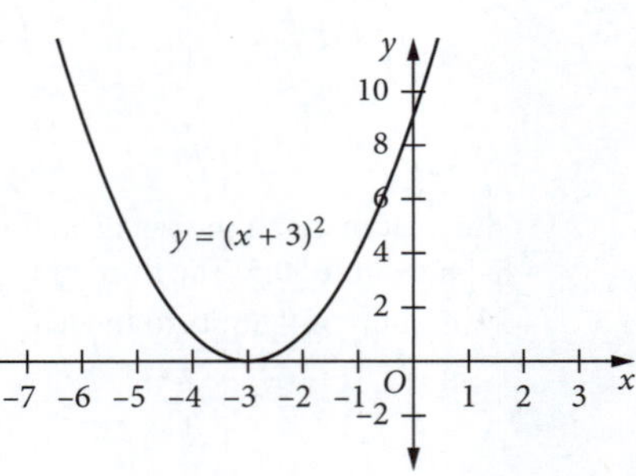

(c) Both curves are parabolas. The vertex in part (a) is at (−1, 2), whereas the vertex in part (b) is at (−3, 0).

8 (a)

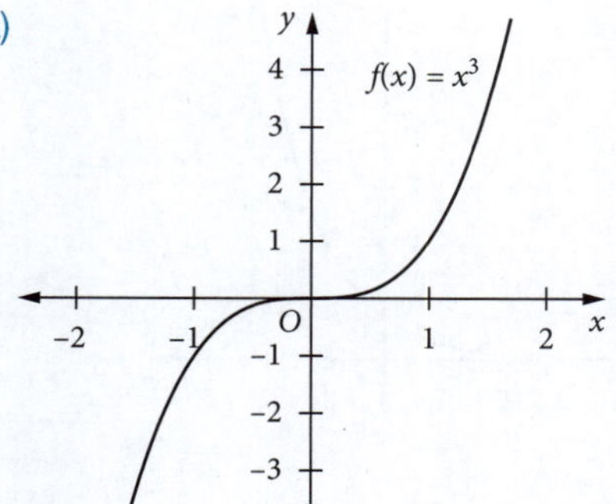

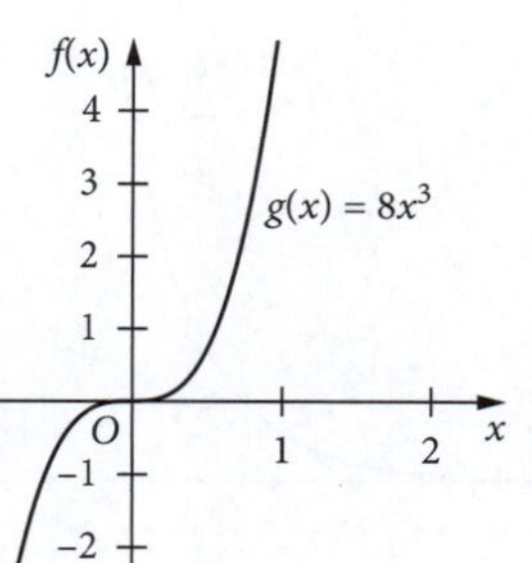
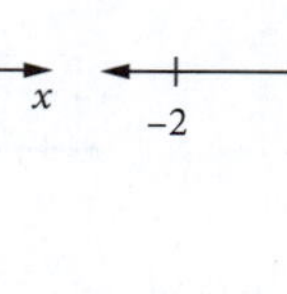

(b)

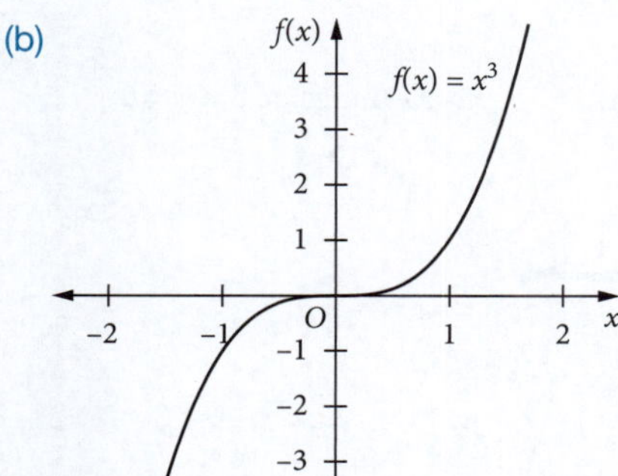

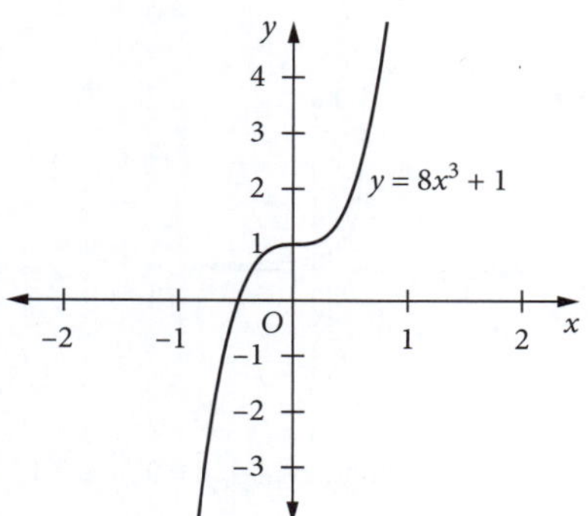

(c) The final graph is the same in each case.

9 (a)

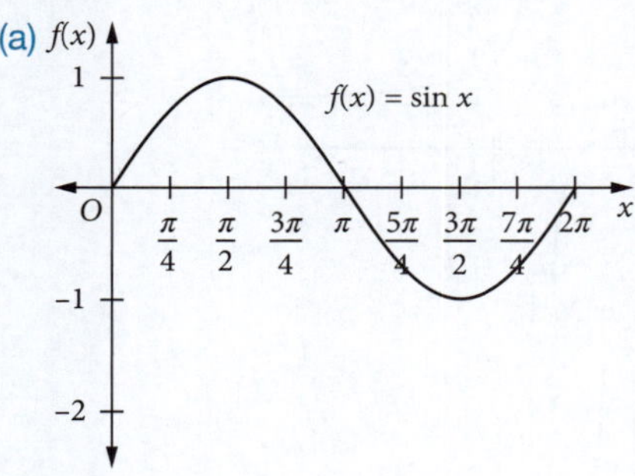

(b)

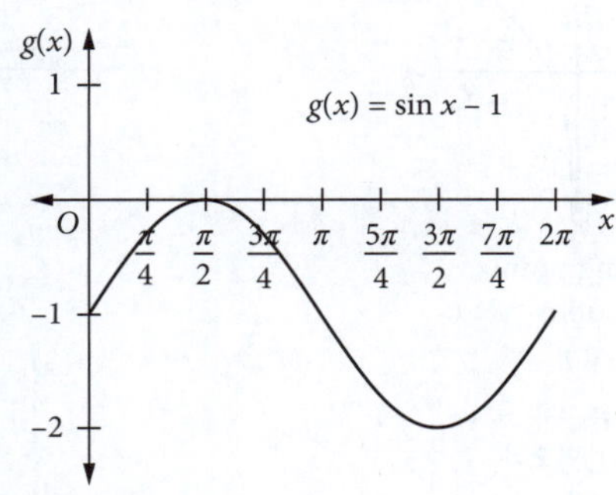

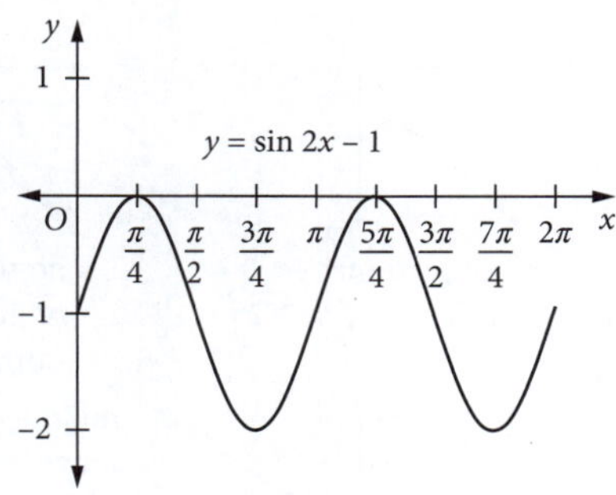

(c) The final graph is the same in each case.

10 (a)

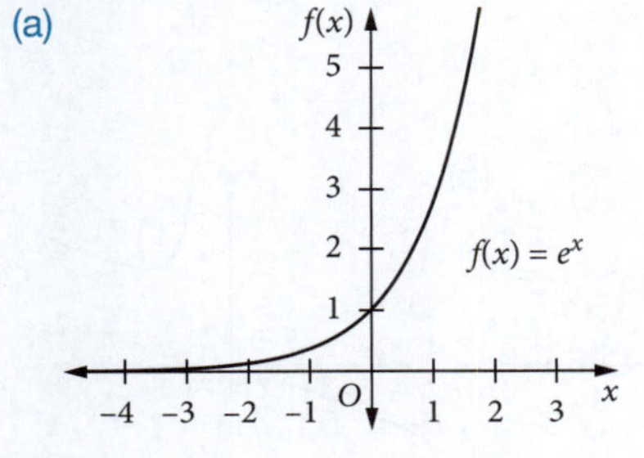

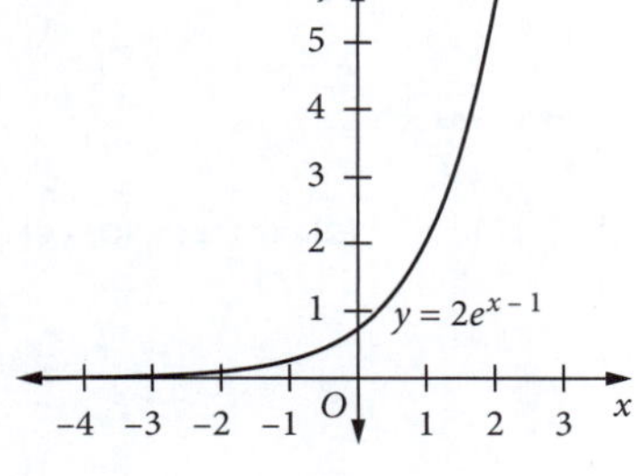

(b)

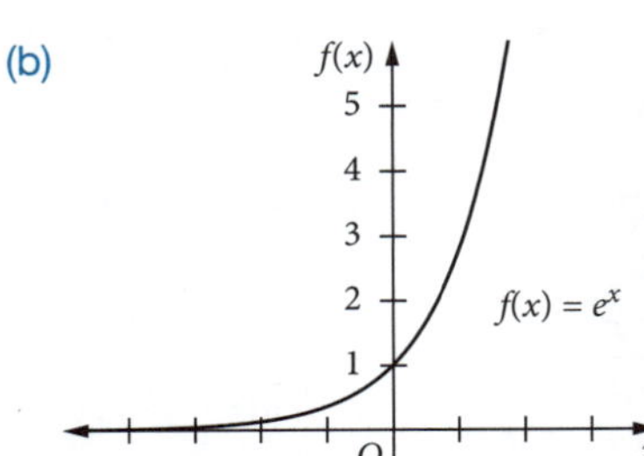

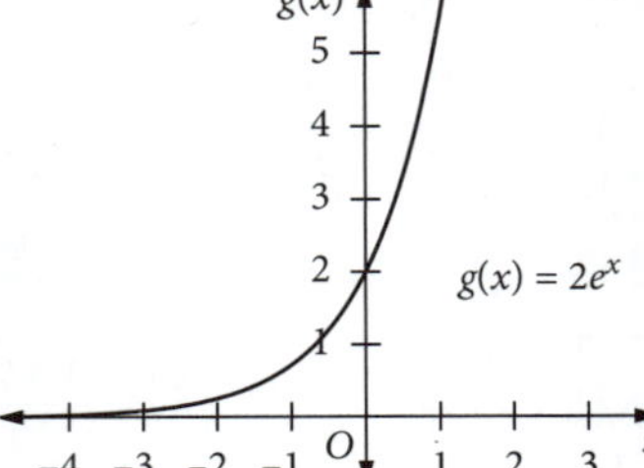

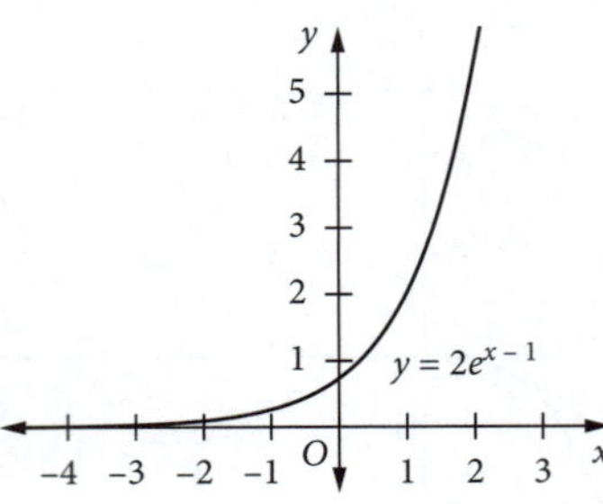

(c) The final graph is the same in each case.

EXERCISE 15.4

1 A

2 (a)

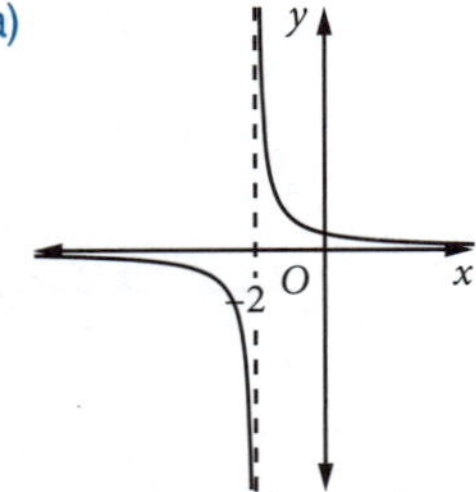

$x < -2$ range: real $y, y \neq 0$

(b)

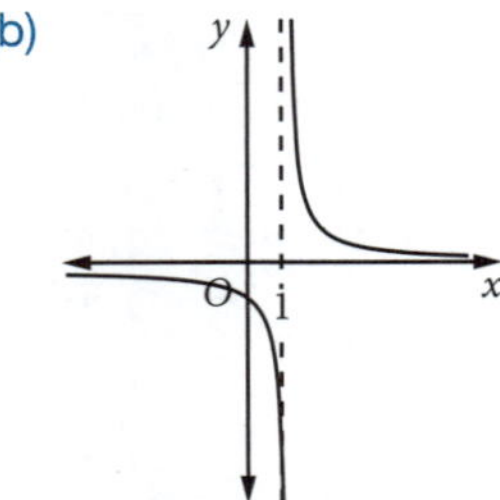

$x < 1$ range: real $y, y \neq 0$

(c)

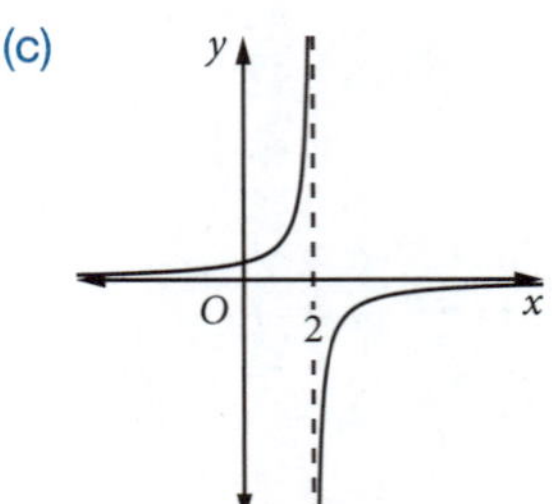

$x > 2$ range: real $y, y \neq 0$

3 (a) $1 + \frac{1}{x-2} = \frac{x-2+1}{x-2} = \frac{x-1}{x-2}$

(b)

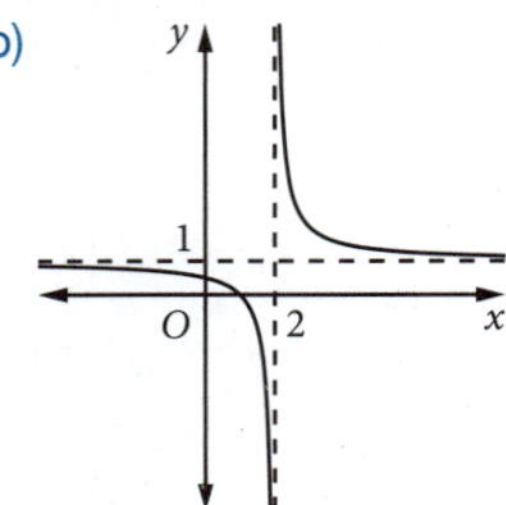

4 (a)

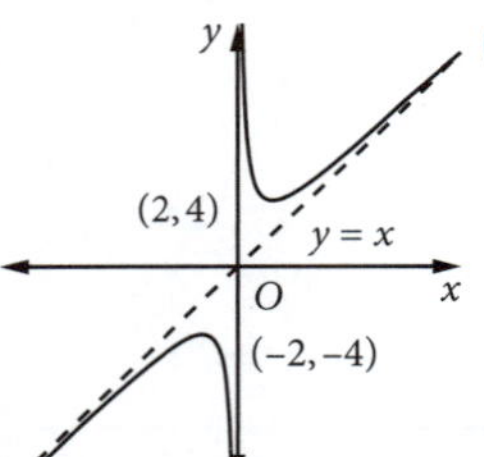

no inflections, $x > 0$
range: real $y, |y| \geq 2$

(b)

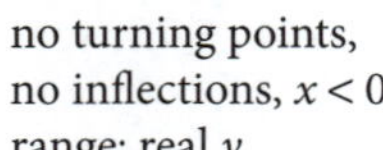

no turning points,
no inflections, $x < 0$
range: real y

(c)

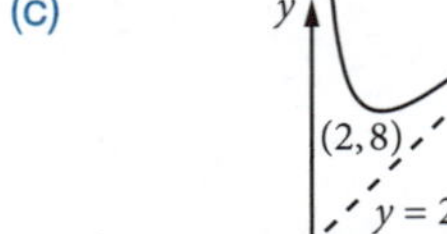

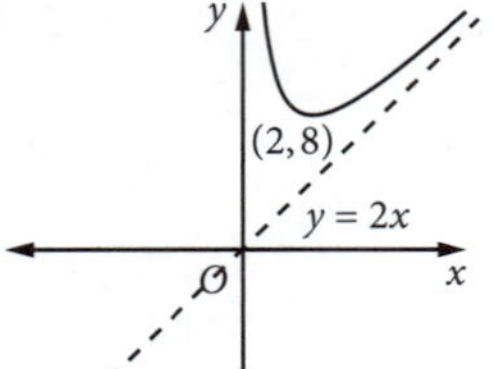

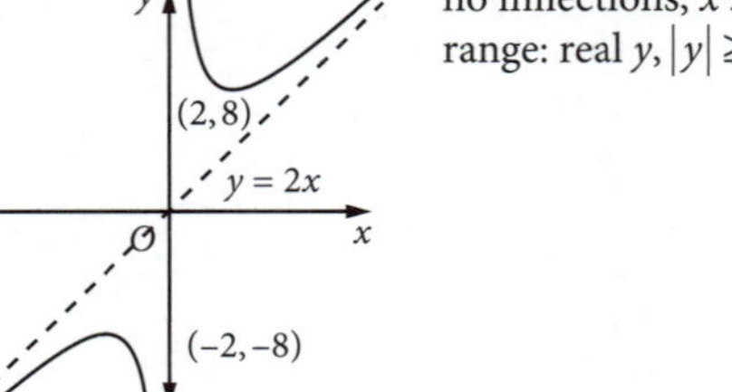

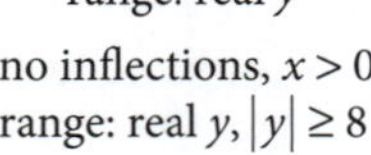

no inflections, $x > 0$
range: real $y, |y| \geq 8$

5 (a) correct (b) incorrect (c) correct (d) correct

6 (a)

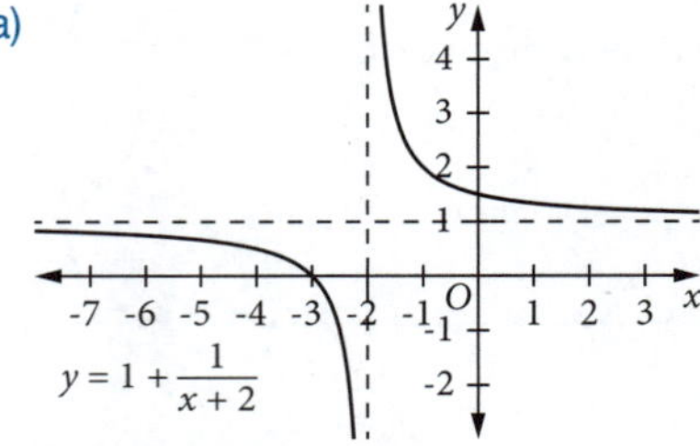

$x < -2$ range: real y

(b)

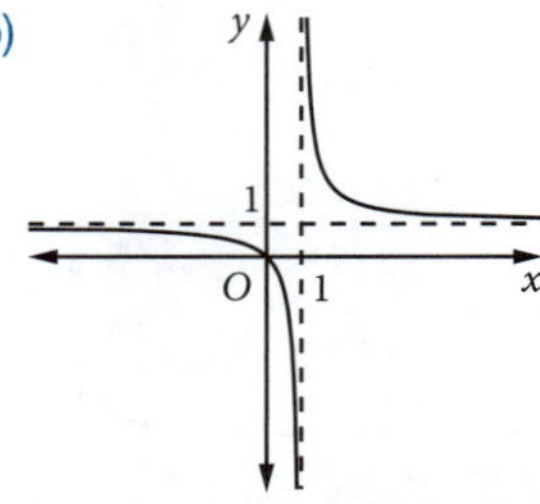

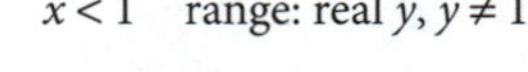

$x < 1$ range: real $y, y \neq 1$

(c)

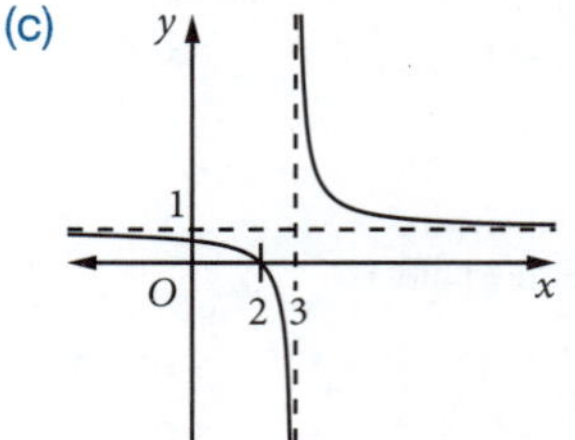

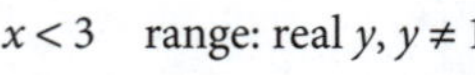

$x < 3$ range: real $y, y \neq 1$

7 (a)

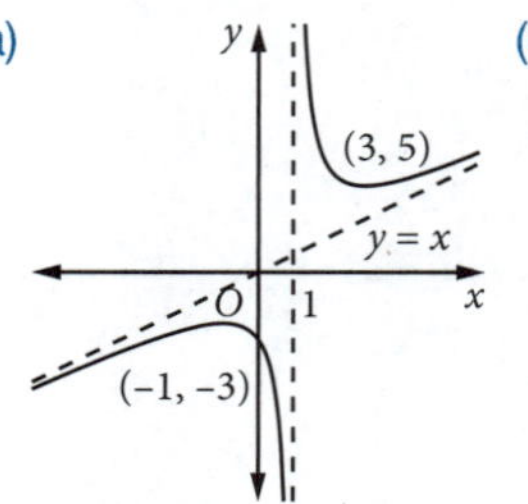

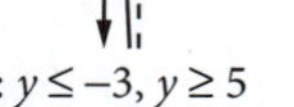

range: $y \leq -3, y \geq 5$

(b)

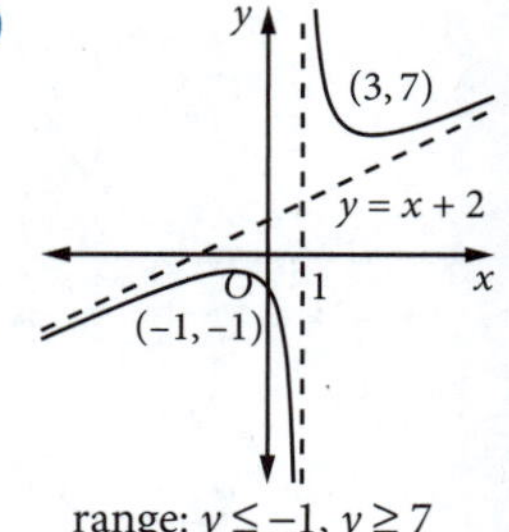

range: $y \leq -1, y \geq 7$

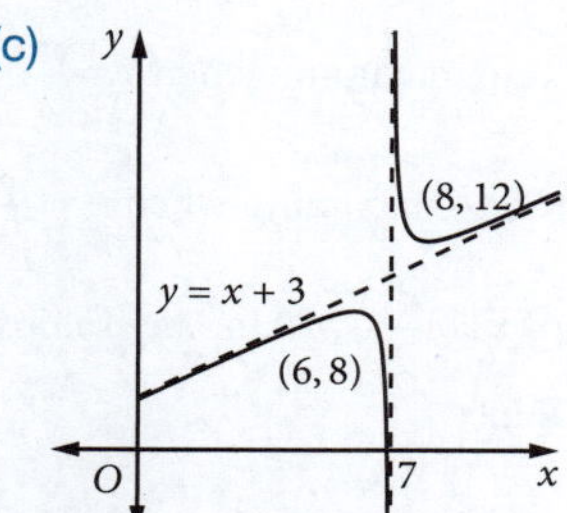

range: $y \leq -8, y \geq 12$

(d)

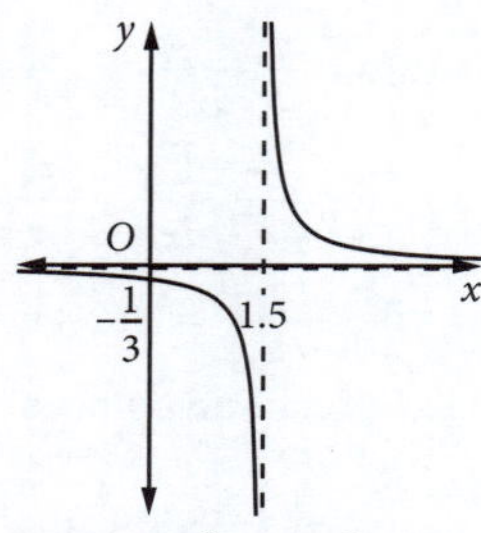

range: real $y, y \neq 0$

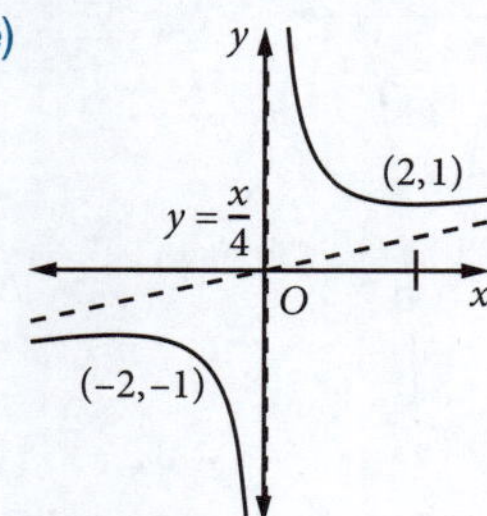

range: $y \leq -1, y \geq 1$

(f)

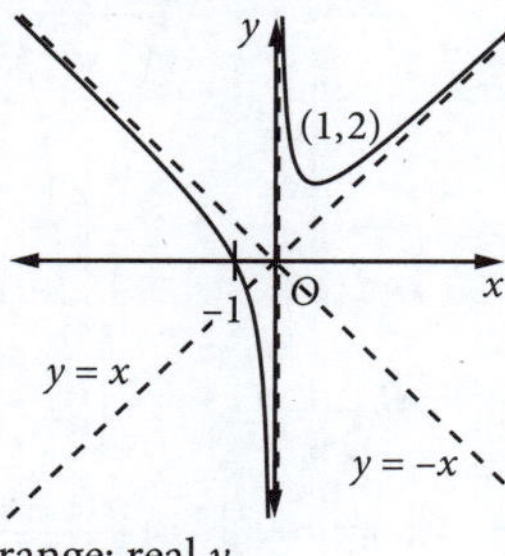

range: real y

EXERCISE 15.5

1 (a)

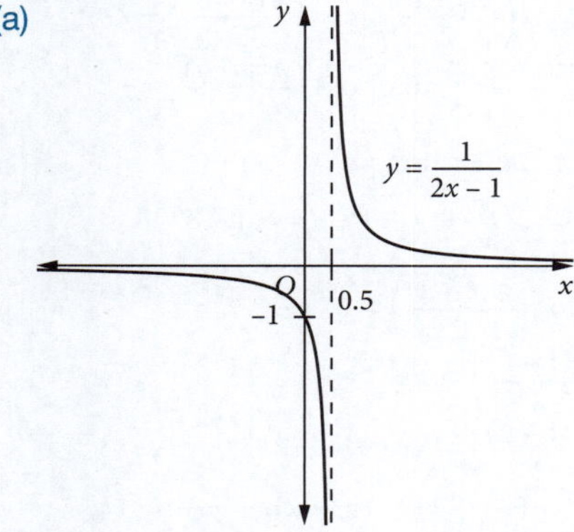

(b) $\frac{dy}{dx} = \frac{-2}{(2x-1)^2}$

$x = 1$: $\frac{dy}{dx} = -2$

$x = 1, y = 1$

Equation of tangent: $y - 1 = -2(x - 1)$

$2x + y - 3 = 0$

(c) $x = -1$: $\frac{dy}{dx} = -\frac{2}{9}$

Gradient of normal $= \frac{9}{2}$

$x = -1, y = -\frac{1}{3}$

Equation of normal: $y + \frac{1}{3} = \frac{9}{2}(x + 1)$

$6y + 2 = 27x + 27$

$27x - 6y + 25 = 0$

(d) $2x + y - 3 = 0$ [1]

$27x - 6y + 25 = 0$ [2]

[1] × 6: $12x + 6y - 18 = 0$ [3]

[2] + [3]: $39x + 7 = 0$

$x = -\frac{7}{39}$

Substitute into [1]: $-\frac{14}{39} + y - 3 = 0$

$y = \frac{131}{39}$

Point of intersection is $\left(-\frac{7}{39}, \frac{131}{39}\right)$

2 (a)

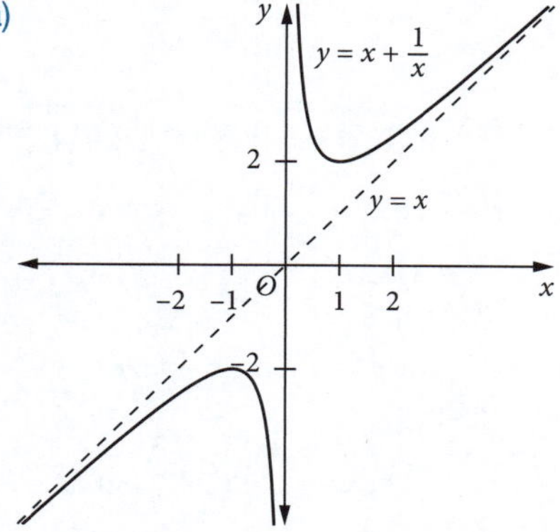

(b) $\frac{dy}{dx} = 1 - \frac{1}{x^2} = \frac{x^2 - 1}{x^2} = \frac{(x+1)(x-1)}{x^2}$

For stationary points, $\frac{dy}{dx} = 0$: $x = \pm 1$

$x = 1, y = 2$. $x = -1, y = -2$.

$\frac{d^2y}{dx^2} = \frac{2}{x^3}$

$x = 1$: $\frac{d^2y}{dx^2} = 2 > 0$

Minimum turning point at (1, 2)

$x = -1$: $\frac{d^2y}{dx^2} = -2 < 0$

Maximum turning point at (–1,–2)

(c) 2

3 (a)

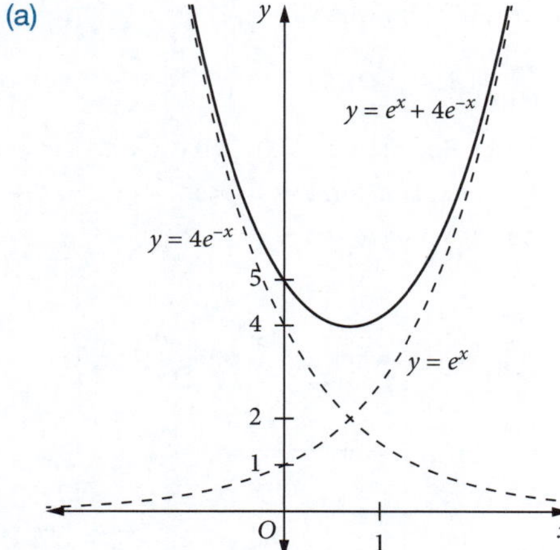

On the left the curve $y = 4e^{-x}$ is the asymptote, on the right the curve $y = e^x$ is the asymptote.

(b) $f'(x) = e^x - 4e^{-x} = \frac{e^{2x} - 4}{e^x}$

For stationary points, $\frac{dy}{dx} = 0$: $e^{2x} = 4$, $e^x = \pm 2$. Since $e^x > 0$, $e^x = 2$ is the only solution.

$e^x = 2 \rightarrow x = \ln 2$

$f'(x) > 0$ for $x > \ln 2$

(c) The minimum value of $f(x)$ is 4 and it occurs when $x = \ln 2$.

4 (a)

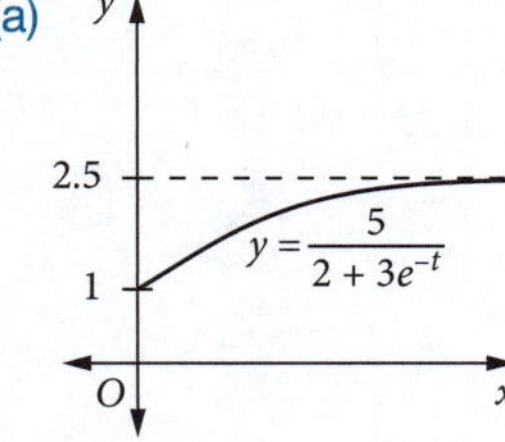

(b) $f'(t)=\frac{15e^{-t}}{\left(2+3e^{-t}\right)^2}$

$e^{-t}>0$ for all values of t, the denominator is always positive so $f'(t)>0$ for $t\geq 0$.

(c) As $t\to\infty$ then $f(t)\to 2.5$.

(d) $1\leq f(t)<2.5$

5 (a) $f(0)=1, f\left(\frac{\pi}{2}\right)=0, f(\pi)=-e^{-\pi}$

(b) $f'(x)=-e^{-x}\cos x-e^{-x}\sin x=-e^{-x}(\cos x+\sin x)$

(c) $f'(0)=-1\times 1-1\times 0=-1$

$$f'\left(\frac{3\pi}{4}\right)=-e^{-\frac{3\pi}{4}\left(\cos\frac{3\pi}{4}+\sin\frac{3\pi}{4}\right)}=-e^{-\frac{3\pi}{4}\left(\frac{-1}{\sqrt{2}}+\frac{1}{\sqrt{2}}\right)}=0$$

$$f'\left(-\frac{\pi}{4}\right)=-e^{\frac{\pi}{4}\left(\cos-\frac{\pi}{4}+\sin-\frac{\pi}{4}\right)}=-e^{-\frac{\pi}{4}\left(\frac{1}{\sqrt{2}}+\frac{-1}{\sqrt{2}}\right)}=0$$

(d)

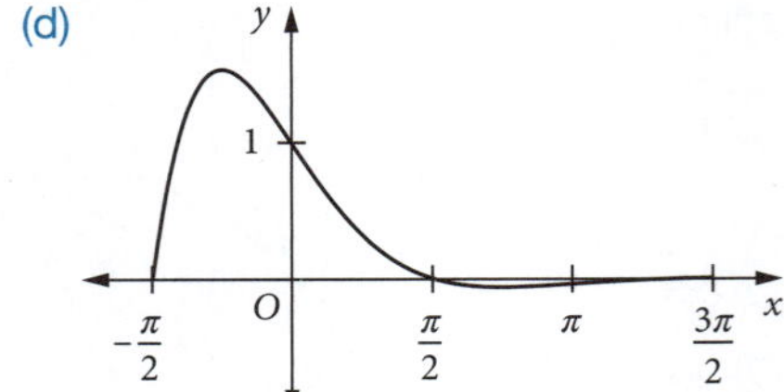

(e) $f''(x)=e^{-x}(\cos x+\sin x)-e^{-x}(-\sin x+\cos x)=2e^{-x}\sin x$

$x=-\frac{\pi}{4}, f''(x)=2e^{\frac{\pi}{4}\sin\left(-\frac{\pi}{4}\right)}=-\sqrt{2}e^{\frac{\pi}{4}}<0$

Maximum turning point when $x=-\frac{\pi}{4}$.

$$f\left(-\frac{\pi}{4}\right)=e^{\frac{\pi}{4}\cos\left(-\frac{\pi}{4}\right)}=\frac{e^{\frac{\pi}{4}}}{\sqrt{2}}\approx 1.55$$

6 (a) $f(x)=\log_e(\sin x)$. Require $\sin x>0$, so $0<x<\pi$.

(b) $f'(x)=\frac{\cos x}{\sin x}=\cot x$. Domain is $0<x<\pi$.

(c) $f'(x)=0$ when $\cot x=0$ so $x=\frac{\pi}{2}$.

$f''(x)=-\text{cosec}^2 x$

$f''\left(\frac{\pi}{2}\right)=-\text{cosec}^2\frac{\pi}{2}=-1<0$

Maximum value of $f(x)$ when $x=\frac{\pi}{2}$ is $\log_e\left(\sin\frac{\pi}{2}\right)=0$

(d)

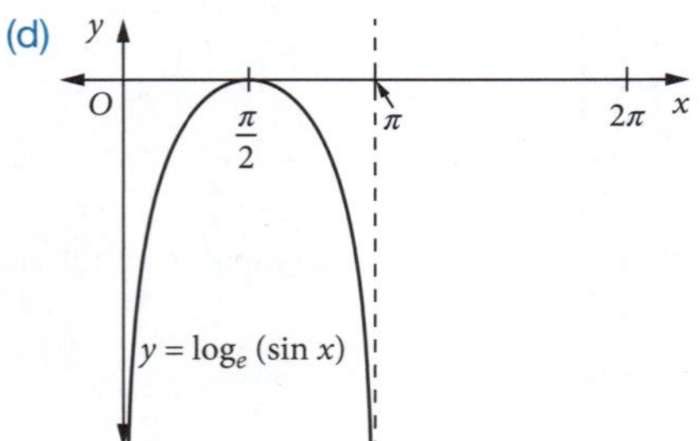

7 (a) $y=\log_e(1+\sin x)$

$$\frac{dy}{dx}=\frac{\cos x}{1+\sin x}$$

$$\frac{d^2y}{dx^2}=\frac{-\sin x(1+\sin x)-\cos x\times\cos x}{(1+\sin x)^2}$$

$$=\frac{-\sin x-\sin^2 x-\cos^2 x}{(1+\sin x)^2}=\frac{-(1+\sin x)}{(1+\sin x)^2}=\frac{-1}{1+\sin x}$$

(b) $\frac{dy}{dx}=0$ when $\cos x=0$, $x=\frac{\pi}{2},\frac{3\pi}{2},\frac{5\pi}{2},\frac{7\pi}{2}$ but y is undefined for $x=\frac{3\pi}{2},\frac{7\pi}{2}$.

Hence $x=\frac{\pi}{2},\frac{5\pi}{2}$

(c) $x=\frac{\pi}{2}$: $\frac{d^2y}{dx^2}=\frac{-1}{1+1}=-\frac{1}{2}<0$ so maximum when $x=\frac{\pi}{2}$.

$x=\frac{5\pi}{2}$: $\frac{d^2y}{dx^2}=\frac{-1}{1+1}=-\frac{1}{2}<0$ so maximum when $x=\frac{5\pi}{2}$.

(d) $\frac{-1}{1+\sin x}\neq 0$ for any value of x so $\frac{d^2y}{dx^2}\neq 0$ for any value of x.

Hence no points of inflection.

(e)

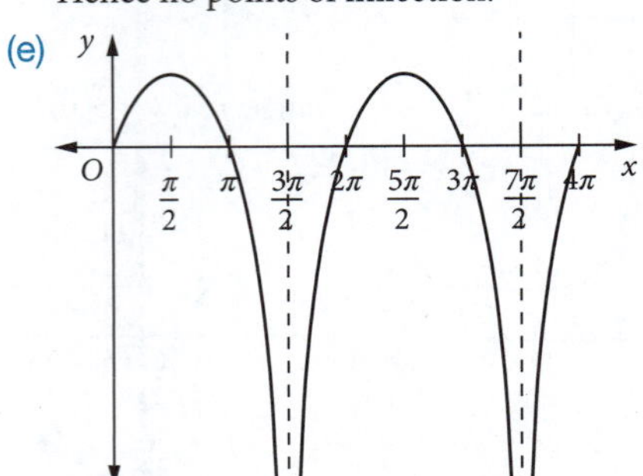

8 $C(t)=1000\left[\cos\left(\frac{\pi(t-8)}{2}\right)+2\right]^2-1000$, for $8\leq t\leq 16$

(a) $\frac{dC}{dt}=1000\times 2\left[\cos\left(\frac{\pi(t-8)}{2}\right)+2\right]\times\left[-\sin\left(\frac{\pi(t-8)}{2}\right)\right]\times\frac{\pi}{2}$

$=-1000\pi\sin\left(\frac{\pi(t-8)}{2}\right)\left[\cos\left(\frac{\pi(t-8)}{2}\right)+2\right]$

$\frac{dC}{dt}=0$: $\sin\left(\frac{\pi(t-8)}{2}\right)=0$ or $\cos\left(\frac{\pi(t-8)}{2}\right)=-2$

$\frac{\pi(t-8)}{2}=0,\pi,2\pi,3\pi,4\pi$

$t-8=0, t-8=2, t-8=4, t-8=6, t-8=8$

$t=8, 10, 12, 14, 16$

$t=8, 12, 16$: $\cos\left(\frac{\pi(t-8)}{2}\right)=1$

$t=10$: $\cos\left(\frac{\pi(t-8)}{2}\right)=\cos\pi=-1$

$t=14$: $\cos\left(\frac{\pi(t-8)}{2}\right)=\cos 3\pi=-1$

The least value of $C(t)$ will occur when $t=10, 14$.

Minimum concentration $=1000[\cos\pi+2]^2-1000=0$

(b)

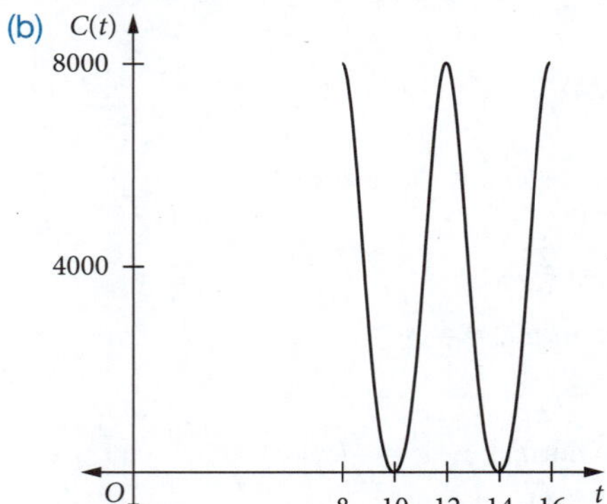

9 (a) $h^2+\left(\frac{x}{2}\right)^2=1.69x^2$

$h^2=1.69x^2-0.25x^2$

$h^2=1.44x^2$

$h=1.2x$

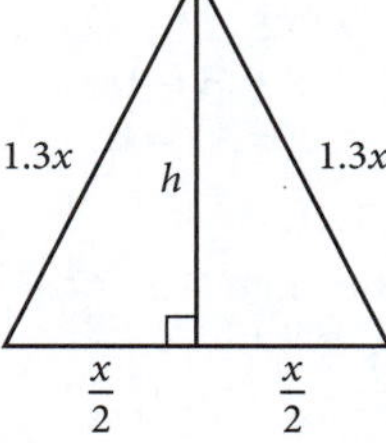

(b) $A=xy+\frac{1}{2}\times x\times 1.2x$

$=xy+0.6x^2$

(c) $2y+x+2.6x=48$

$2y+3.6x=48$

$y=24-1.8x$

(d) $A(x) = x \times (24 - 1.8x) + 0.6x^2$
$= 24x - 1.2x^2$

(e) 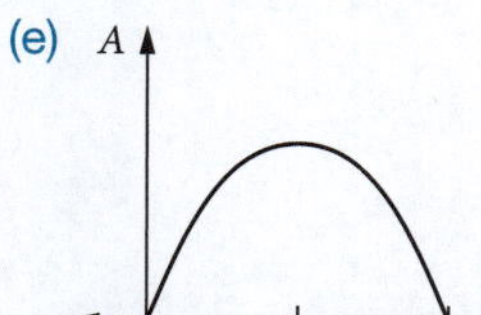

(f) Maximum area when $x = 10$ m, $y = 6$ m, equal sides of the isosceles triangle 13 m.
Maximum area = 120 m^2

10 $I(t) = 100(1 - e^{-5t})$

(a) $I(0) = 100(1 - 1) = 0$
$I(0.2) = 100(1 - e^{-1}) = 63.2$ amps
$I(1) = 100(1 - e^{-5}) = 99.3$ amps

(b) As t increases the current approaches 100 amps.

(c) $I'(t) = 100 \times 5e^{-5t} = 500e^{-5t}$

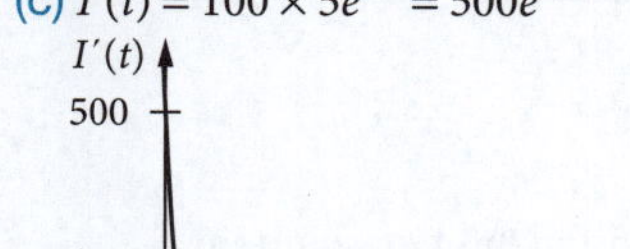
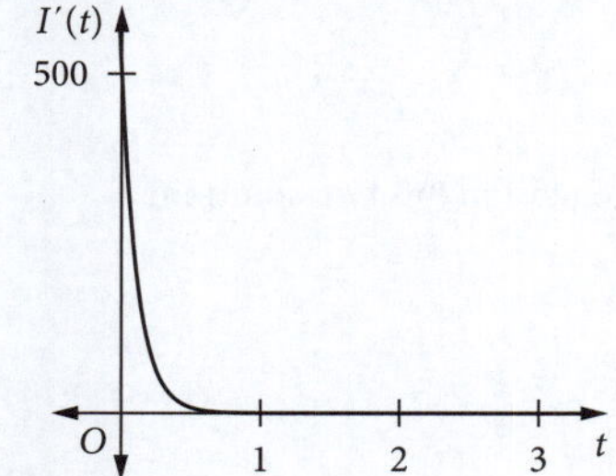

(d) 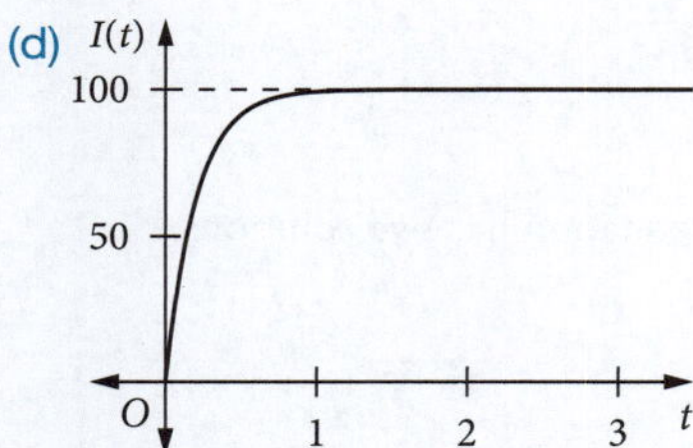

(e) The graph of $I'(t)$ shows the gradient of $I(t)$ over time. The gradient graph shows the current increasing rapidly at the start then staying practically the same.

EXERCISE 15.6

1 (a) 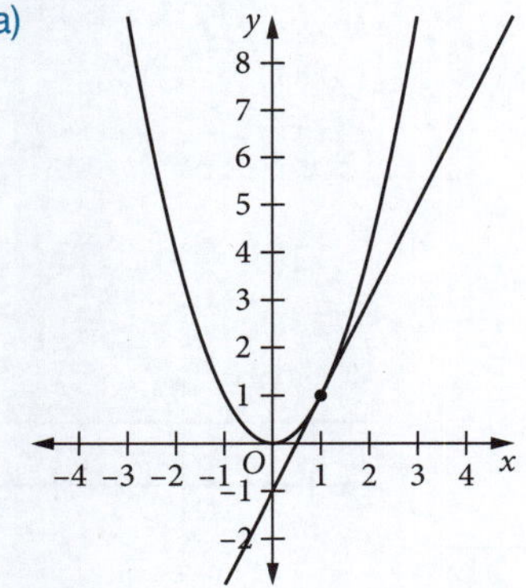

Graphs touch, equation has one solution.

(b) 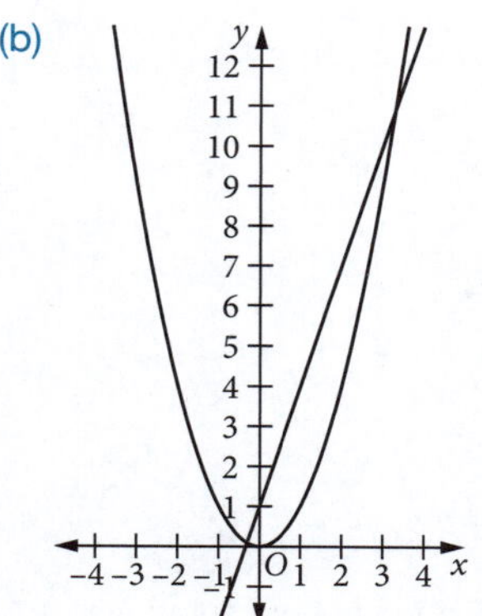

Graphs intersect twice, equation has two solutions.

(c) 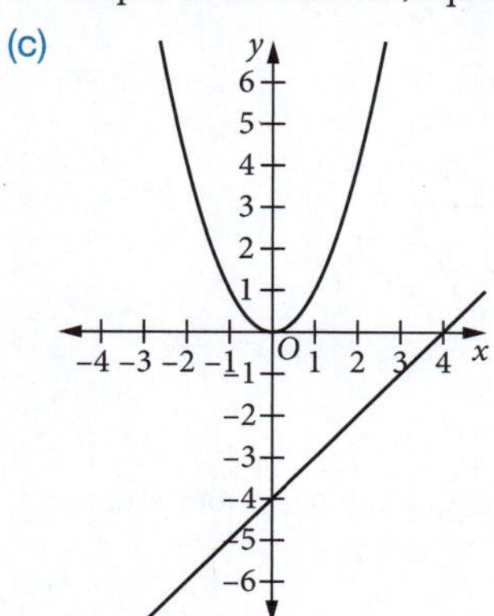

Graphs do not intersect so equation has no solutions.

(d) 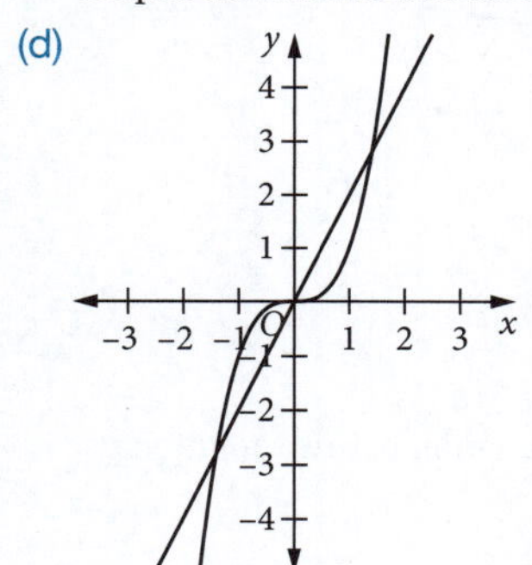

Graphs intersect 3 times. Equation has 3 solutions.

(e) 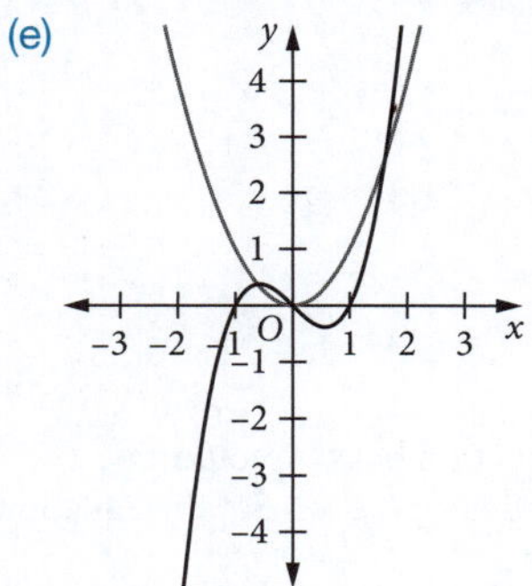

Graphs intersect 3 times. Equation has 3 solutions.

(f)

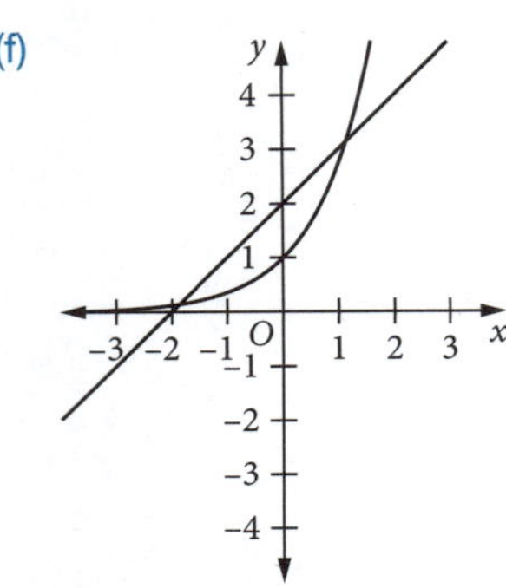

Graphs intersect twice, equation has two solutions.

2 (a)

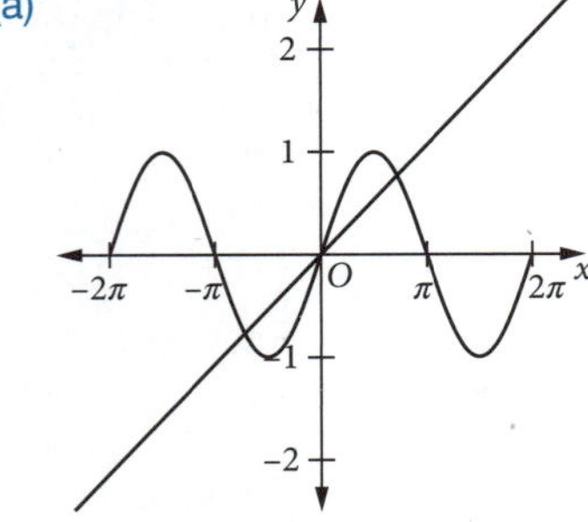

Graphs intersect 3 times. Equation has 3 solutions.

(b)

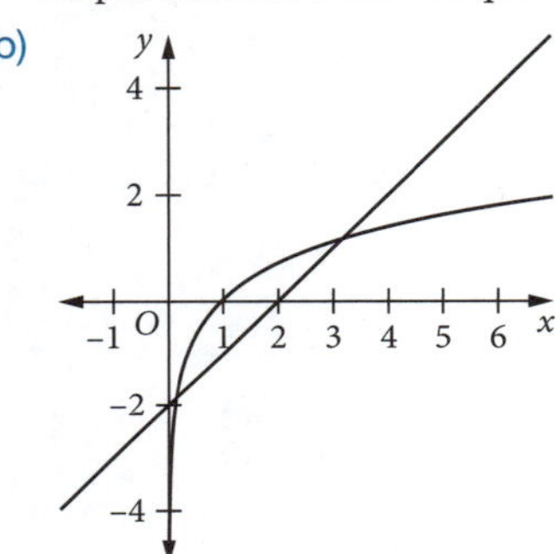

Graphs intersect twice, equation has two solutions.

(c)

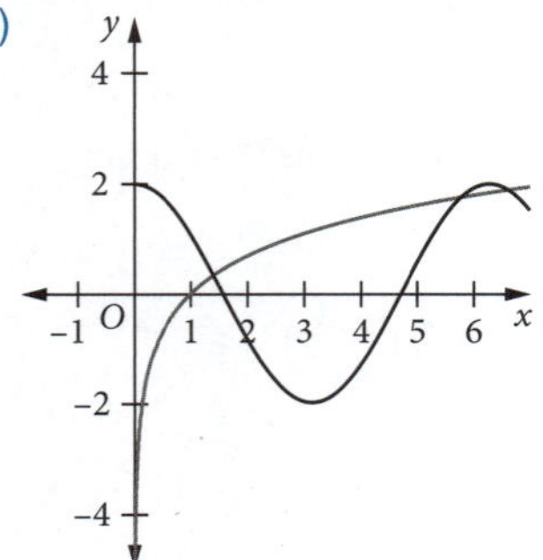

Graphs intersect twice, equation has two solutions.
If the domain of $2\cos x$ had been $0 \leq x \leq 3\pi$, the equation would have had 3 solutions.

(d)

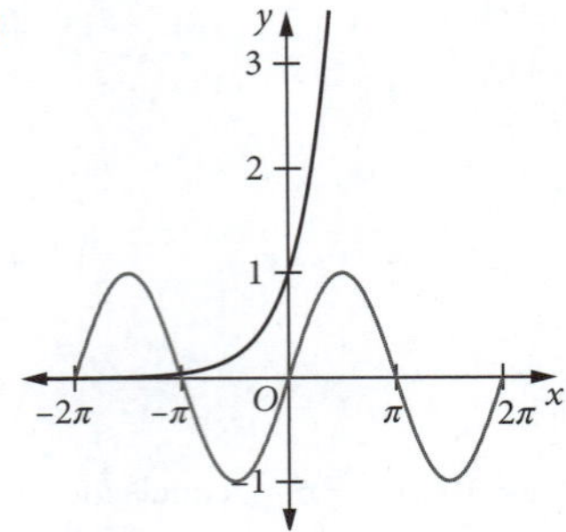

Graphs intersect 3 times, equation has 3 solutions.

(e)

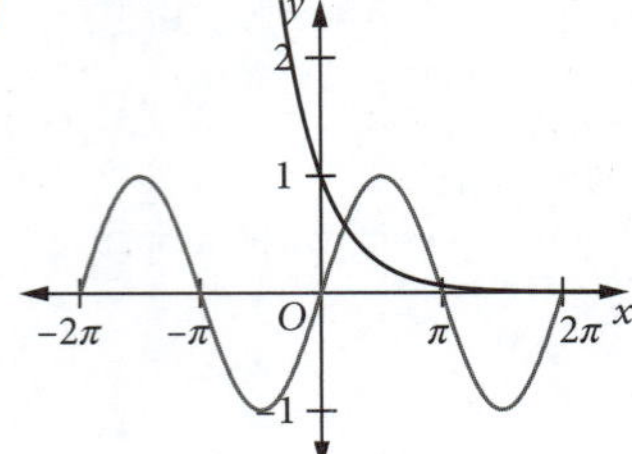

Graphs intersect twice, equation has two solutions.

(f)

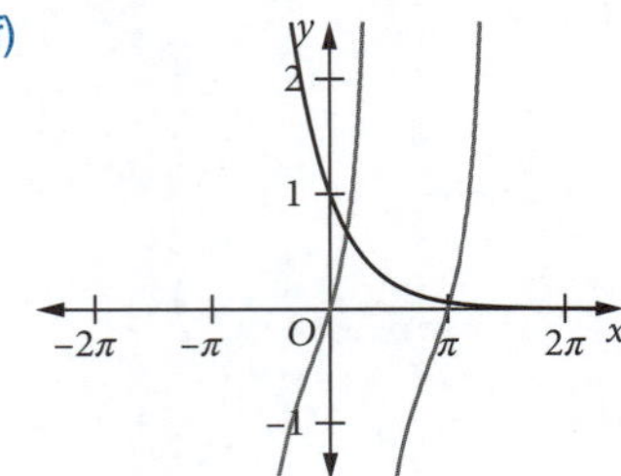

Graphs intersect twice, equation has two solutions.

3 2.84

EXERCISE 15.7

1

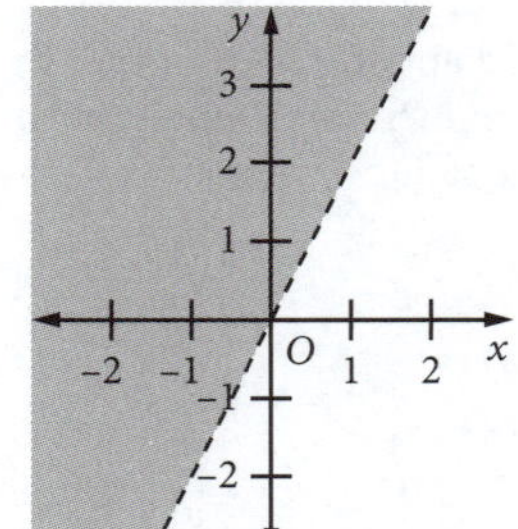

2

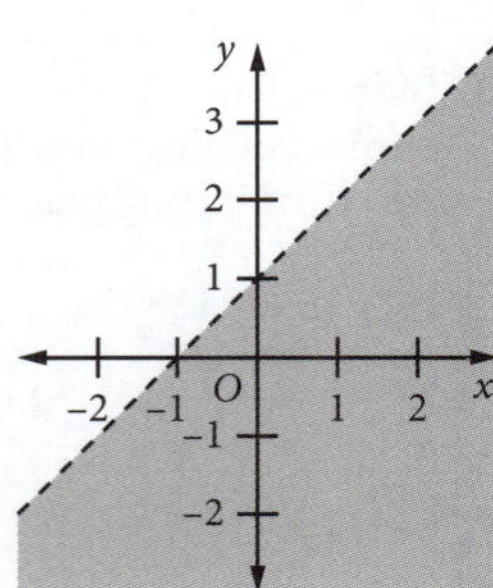

3

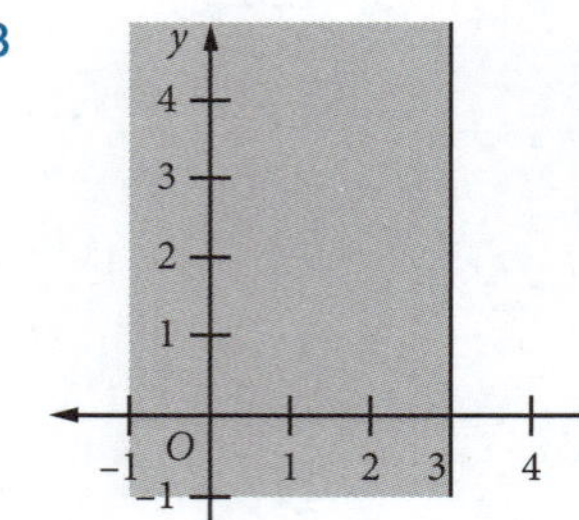

4

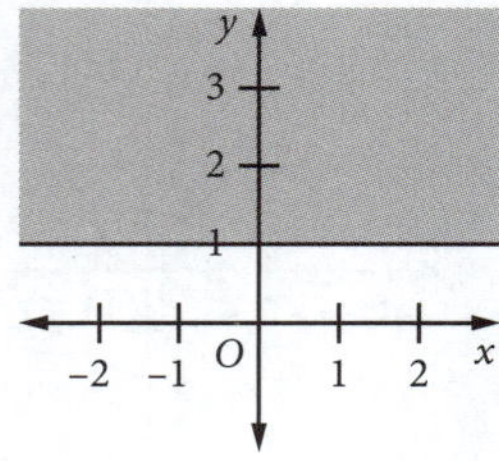

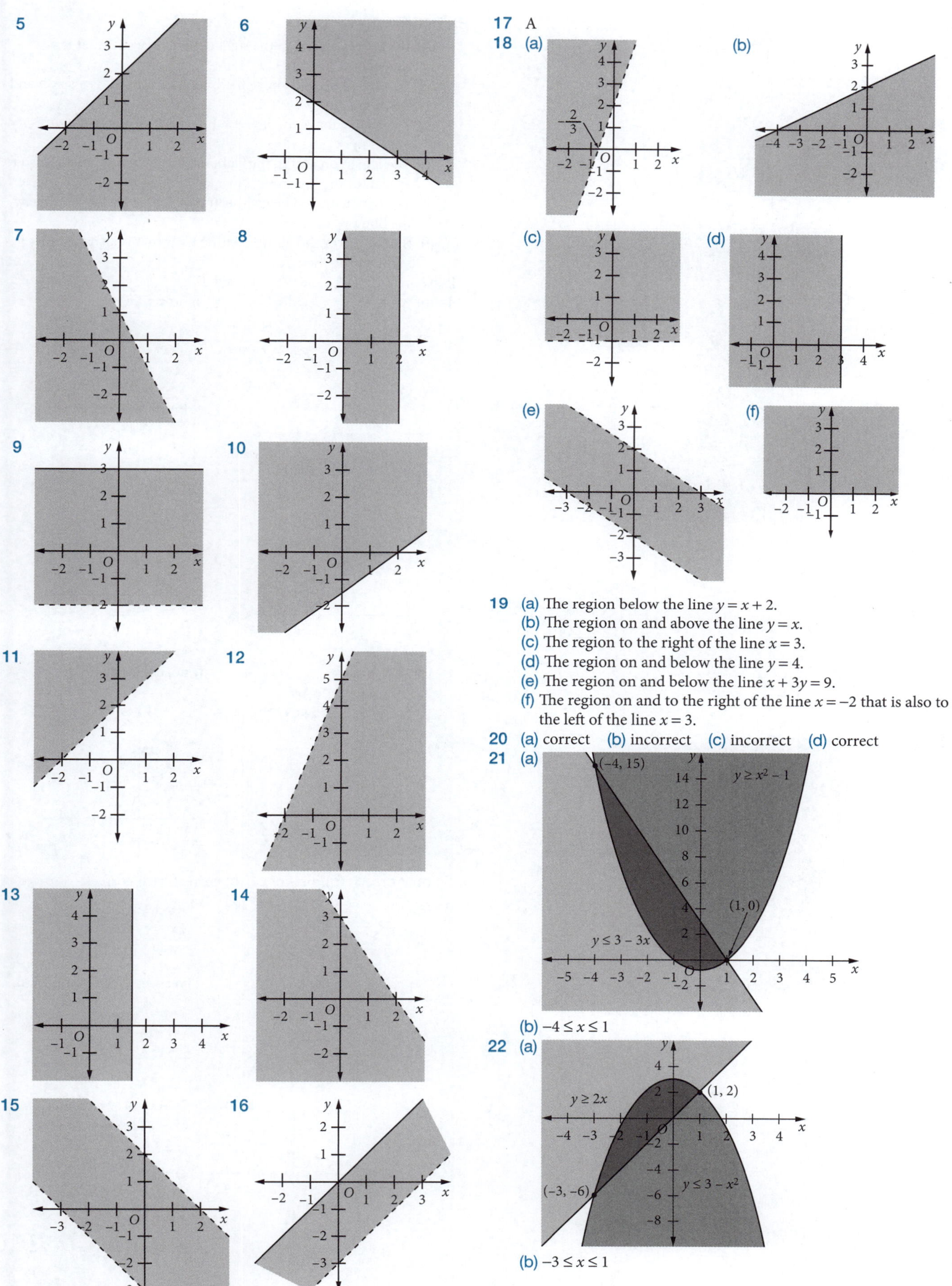

17 A

19 (a) The region below the line $y = x + 2$.
(b) The region on and above the line $y = x$.
(c) The region to the right of the line $x = 3$.
(d) The region on and below the line $y = 4$.
(e) The region on and below the line $x + 3y = 9$.
(f) The region on and to the right of the line $x = -2$ that is also to the left of the line $x = 3$.

20 (a) correct (b) incorrect (c) incorrect (d) correct

21 (b) $-4 \leq x \leq 1$

22 (b) $-3 \leq x \leq 1$

23  (a)

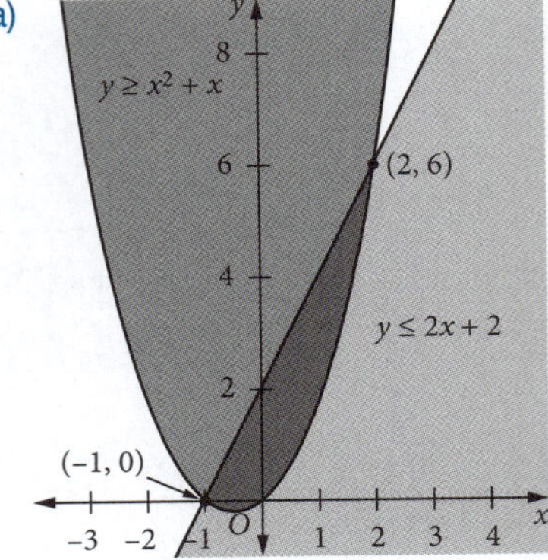

(b) $-1 \leq x \leq 2$

24 (a)

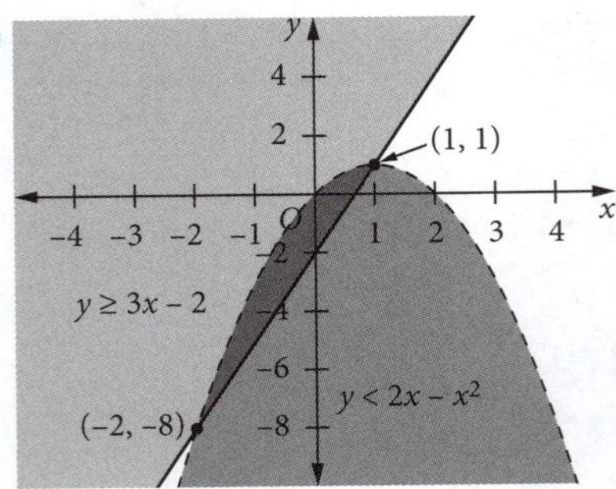

(b) $x^2 + x - 2 < 0$

$3x - 2 < 2x - x^2$

$-2 < x < 1$

(c) $x^2 + x - 2 < 0$

$(x + 2)(x - 1) < 0$

Roots: $x = -2, 1$

test using $x = 0$

$0 + 0 - 2 < 0$

$-2 < 0$: true

Solution is $-2 < x < 1$

25 (a)

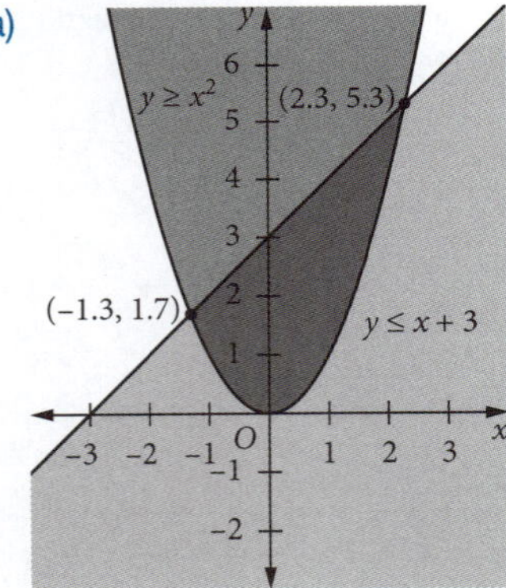

(b) $x^2 \leq x + 3$

$x^2 - x - 3 \leq 0$

$-1.3 \leq x \leq 2.3$

(c) Solve $x^2 - x - 3 = 0$

$x = \dfrac{1 \pm \sqrt{13}}{2}$

test using $x = 0$

$0 - 0 - 3 < 0$

$-3 < 0$: true

Solution is $\dfrac{1 - \sqrt{13}}{2} \leq x \leq \dfrac{1 + \sqrt{13}}{2}$

$-1.303 \leq x \leq 2.303$

EXERCISE 15.8

1 (a) The region on and to the right of the line $x = -1$ that is also below the line $y = 2$.

(b) The region on and above the line $x + y = 1$ that is also on and below the line $y = x + 1$.

(c) The region above the line $y = 1$ that is also above the line $x + y = 1$.

(d) The region on and to the right of the line $x = -1$ that is also to the left of the line $x = 2$.

(e) The region on and below the line $y = 2x + 2$ that is also below the line $x + y = 2$.

(f) The region on and above the line $y = x$ that is also on and below the line $y = 2x$.

2 C

3 $x + y \geq 3, y \leq x + 1$

4 (a) yes, no, yes

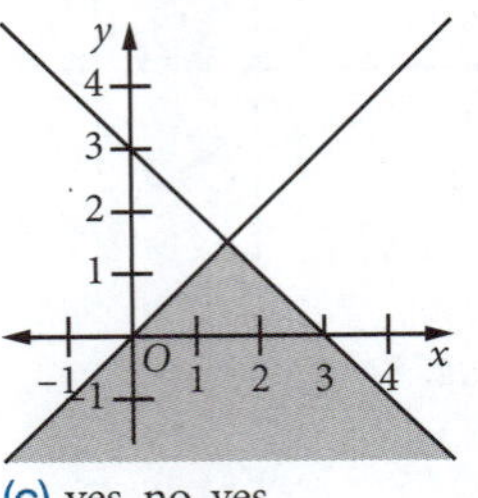

(b) no, no, yes

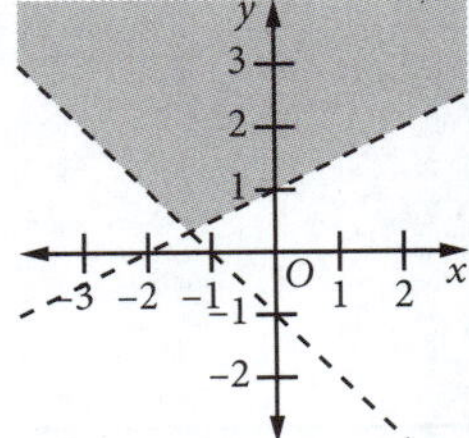

(c) yes, no, yes

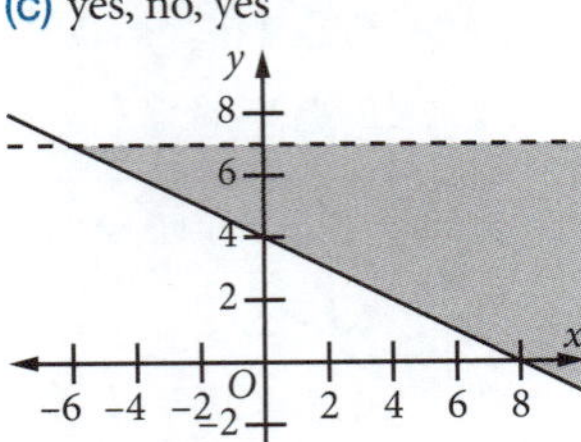

(d) no, yes, yes

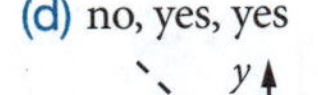

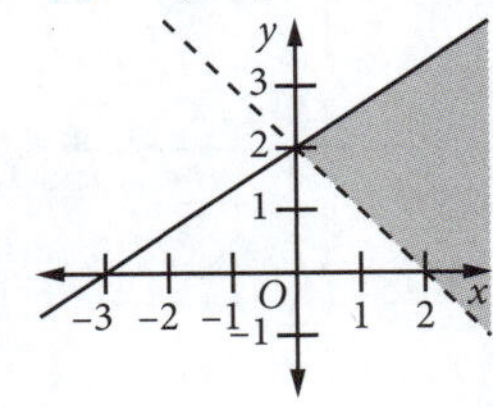

(e) yes, no, yes

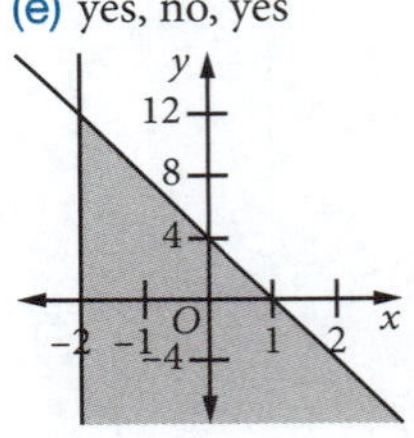

(f) no, no, no

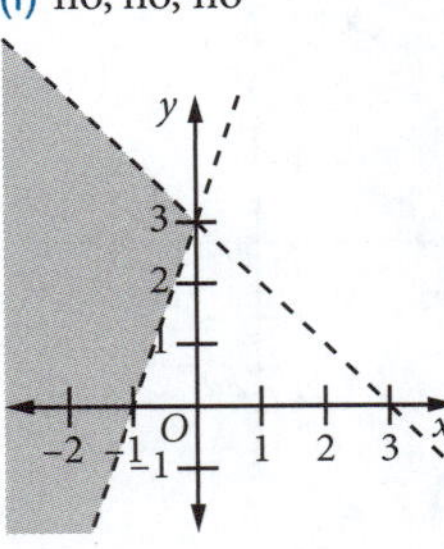

5 (a) correct (b) incorrect (c) correct (d) correct

6 (a)

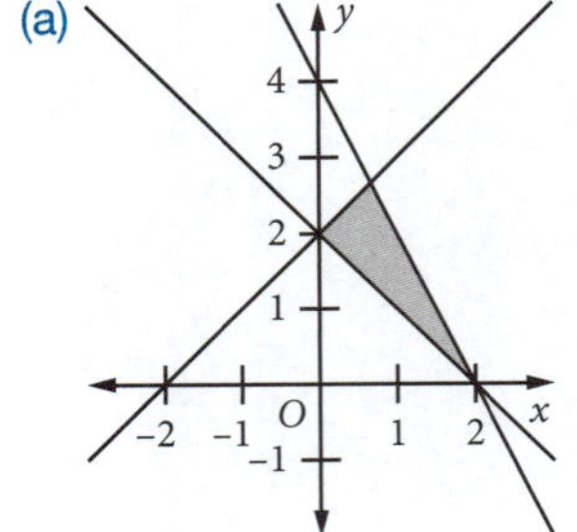

$y = x + 2, 2x + y = 4;$

$2x + x + 2 = 4, 3x = 2:$

$x = \frac{2}{3}, y = 2\frac{2}{3}$

vertices: $\left(\frac{2}{3}, 2\frac{2}{3}\right), (2, 0), (0, 2)$

(b)

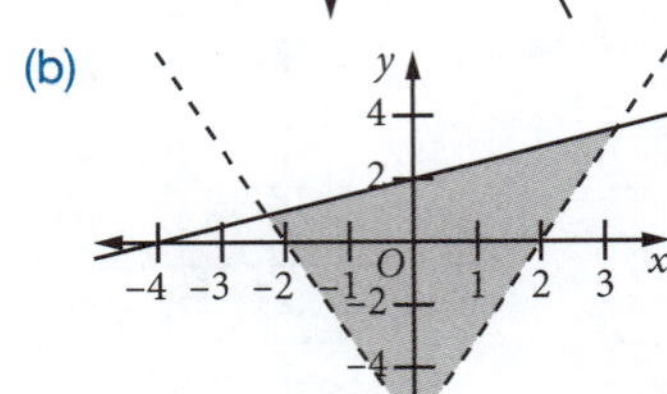

$2y - x = 4, y = 3x - 6;$

$6x - 12 - x = 4, 5x = 16:$

$x = 3\frac{1}{5}, y = 3\frac{3}{5}$

$2y - x = 4, 3x + y = -6;$

$6x - 12 - x = 4, 7x = -16:$

$x = -2\frac{2}{7}, y = \frac{6}{7}$

vertices: $\left(3\frac{1}{5}, 3\frac{3}{5}\right), \left(-2\frac{2}{7}, \frac{6}{7}\right), (0, -6)$

(c) vertices: $(-2, 0)$, $(0.5, 5)$, $\left(1\frac{1}{3}, 3\frac{1}{3}\right)$

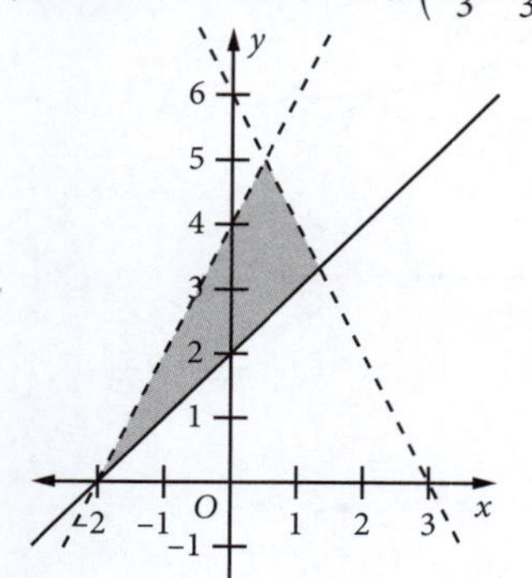

(d) vertices: $(-2, -3)$, $(4, 0)$, $(0, 3)$

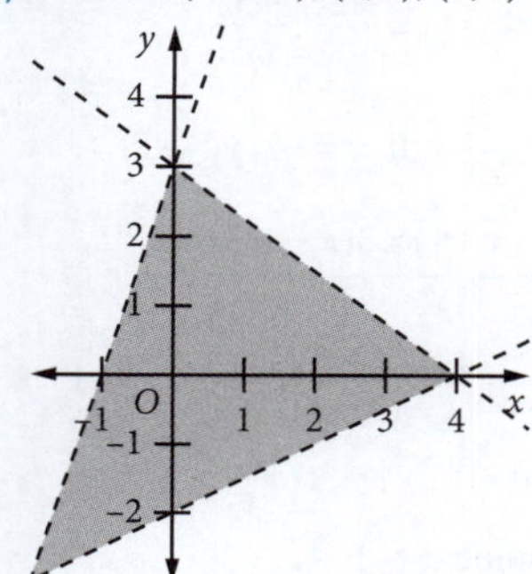

(e) vertices: $(0, 0)$, $(1, 0)$, $(1.5, 0.5)$, $(0, 2)$

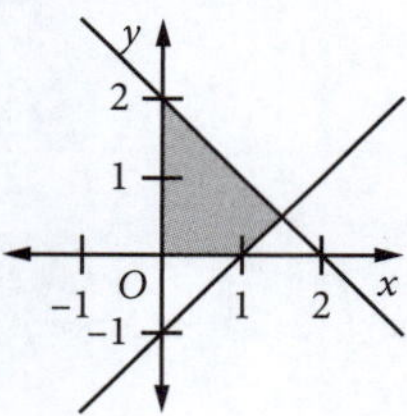

(f) vertices: $(-2, 0)$, $(0, 2)$, $(2, 0)$, $(0, -2)$

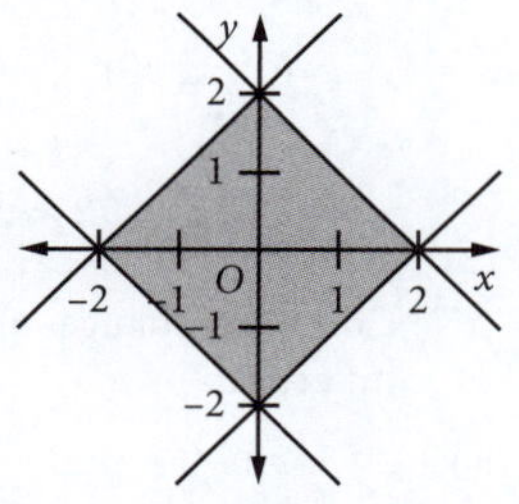

7 (a) the region bounded by the lines $y = x$, $x = 2$ and $y = 1$; $y \le x$, $x \le 2$, $y \ge 1$; $A(2, 2)$, $B(2, 1)$, $C(1, 1)$

(b) the region bounded by the lines $y = 2x$, $y = x$ and $x + y = 3$; $y \le 2x$, $y \ge x$, $x + y \le 3$; $A(0, 0)$, $B(1, 2)$, $C(1.5, 1.5)$

(c) the region bounded by the lines $y = \frac{x}{2} + 1$, $y = \frac{x}{6} + 1$ and $x + 2y = 6$; $y \le \frac{x}{2} + 1$, $y \ge \frac{x}{6} + 1$, $x + 2y \le 6$; $A(0, 1)$, $B(2, 2)$, $C(3, 1.5)$

CHAPTER REVIEW 15

1

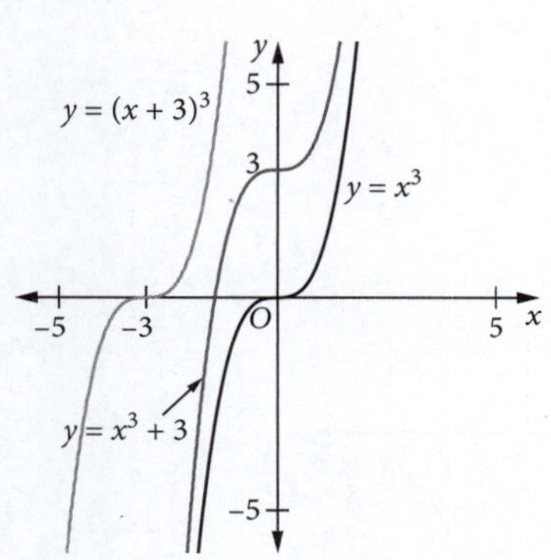

2

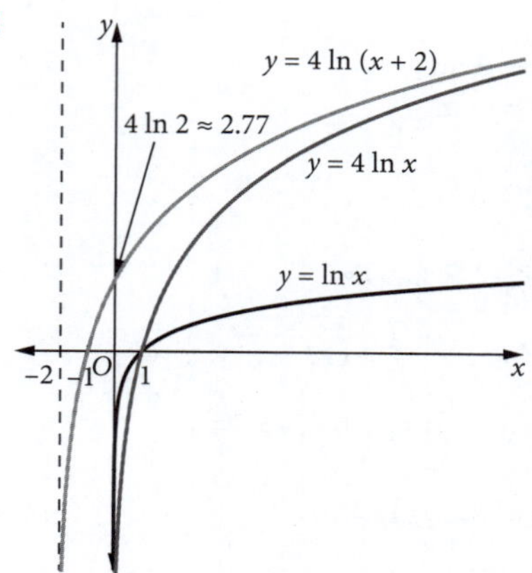

The dilation from the x-axis for the second and third graphs has factor 4.

3 (a)

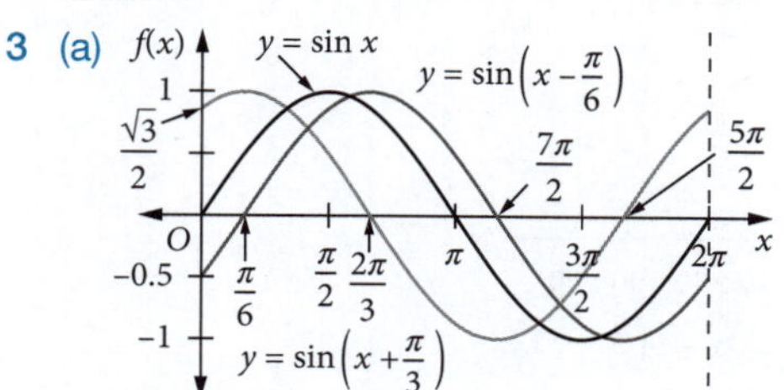

(b)

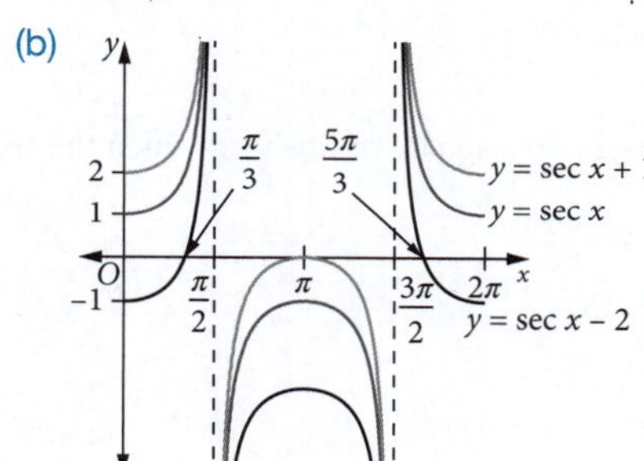

4 (a) $f(2x) = e^{2x}$ (b) $f(x-3) = e^{x-3}$ (c) $f(x) + 1 = e^{x} + 1$
(d) $2f(x) + 4 = 2e^{x} + 4$ (e) $f(2(x+2)) - 1 = e^{2(x+2)} - 1$

5 $C = v + \dfrac{6400}{v}$

(a) $0 < v \le 110$

(b)

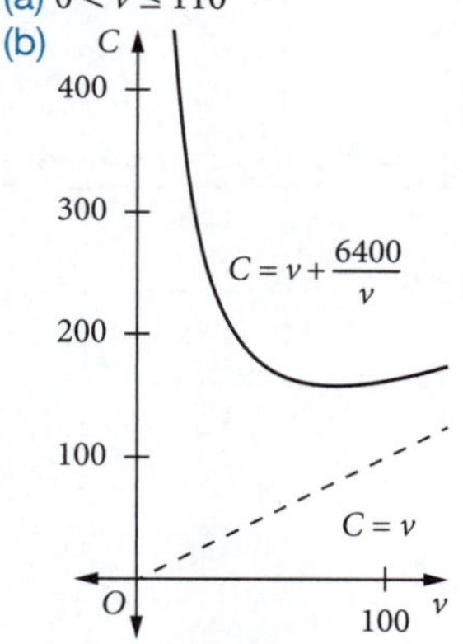

(c) $\dfrac{dC}{dt} = 1 - \dfrac{6400}{v^2}$

$\dfrac{dC}{dt} = 0 : 1 - \dfrac{6400}{v^2} = 0$

$v^2 = 6400$

$v = 80$

Average speed is $80\ \mathrm{km\,h^{-1}}$.

6 (a) Let $BC = y$ cm

$AB = AE = EB = DC = x$ cm

$BC = BF = FC = AD = y$ cm

$3x + 3y = 54$

$x + y = 18$

$y = 18 - x$

$A = xy + \frac{1}{2}x^2 \sin 60° + \frac{1}{2}y^2 \sin 60°$

$A(x) = x(18 - x) + x^2 \times \frac{1}{2} \times \frac{\sqrt{3}}{2} + (18 - x)^2 \times \frac{1}{2} \times \frac{\sqrt{3}}{2}$

$= 18x - x^2 + \frac{\sqrt{3}}{4}x^2 + 81\sqrt{3} - 9\sqrt{3}x + \frac{\sqrt{3}}{4}x^2$

$= \frac{(\sqrt{3} - 2)}{2}x^2 + 9(2 - \sqrt{3})x + 81\sqrt{3}$

(b) The domain is $0 < x < 18$.

(c)

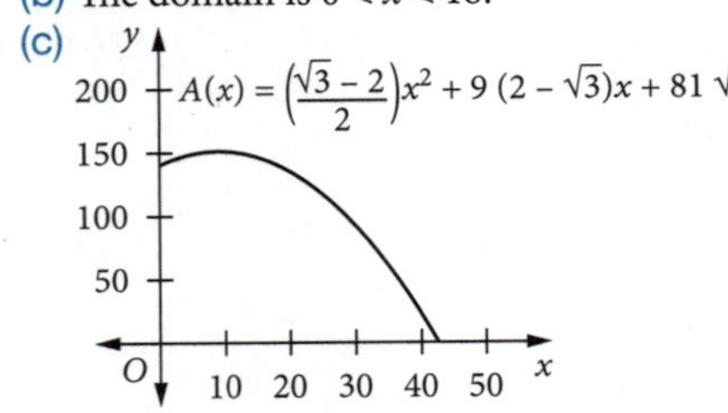

(d) $\frac{dA}{dx} = (\sqrt{3} - 2)x + 9(2 - \sqrt{3})$

$\frac{dA}{dx} = 0$: $(\sqrt{3} - 2)x + 9(2 - \sqrt{3}) = 0$

$x = 9$

The rectangle becomes a square of side 9 cm when the area is a maximum.

7 (a) (i) $f(x) = 4 + \frac{3x - 1}{x^2}$

$4 + \frac{3x - 1}{x^2} = 0$

$4x^2 + 3x - 1 = 0$

$(4x - 1)(x + 1) = 0$

$x = -1, \frac{1}{4}$

(ii) As $x \to \infty, f(x) \to 4$

(iii) $f(x)$ is undefined at $x = 0$, it approaches $-\infty$.

(iv) Asymptotes are $x = 0$ and $y = 4$

(v) $f'(x) = 0 + \frac{3x^2 - (3x - 1) \times 2x}{x^4} = \frac{3x - 6x + 2}{x^3} = \frac{2 - 3x}{x^3}$

$f'(x) = 0$: $x = \frac{2}{3}, y = 6\frac{1}{4}$

$f''(x) = \frac{-3x^3 - (2 - 3x) \times 3x^2}{x^6} = \frac{-3x - 6 + 9x}{x^4} = \frac{6(x - 1)}{x^4}$

$f''\left(\frac{2}{3}\right) = 6 \times \frac{3^4}{2^4} \times \left(-\frac{1}{3}\right) < 0$

$\left(\frac{2}{3}, 6\frac{1}{4}\right)$ is a maximum turning point.

(vi) $f''(x) = 0$ when $x = 1$.

$f''\left(\frac{2}{3}\right) < 0$

$f''(2) = 6 \times \frac{1}{16} > 0$

Concavity changes at $x = 1$ so $(1, 6)$ is a point of inflection.

(b)

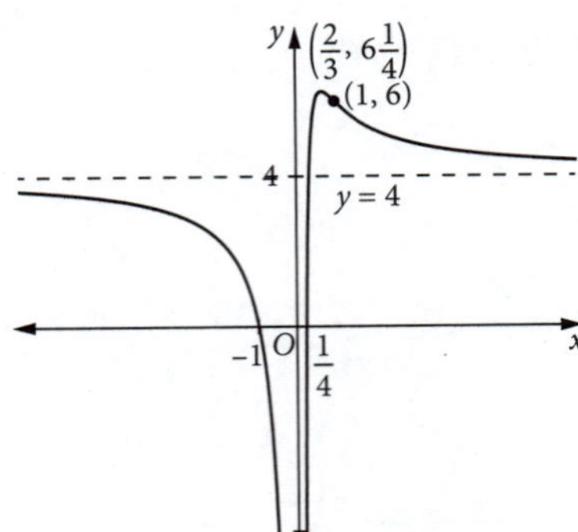

8 (a) If $y = \frac{4x}{(x - 1)^2}$

(i) $\frac{dy}{dx} = \frac{4(x - 1)^2 - 4x \times 2(x - 1)}{(x - 1)^4} = \frac{-4(x + 1)}{(x - 1)^3}$

For stationary points, $\frac{dy}{dx} = 0$: $x = -1, y = -1$

$\frac{d^2y}{dx^2} = \frac{-4\left((x - 1)^3 - (x + 1) \times 3(x - 1)^2\right)}{(x - 1)^6}$

$= \frac{-4(x - 1 - 3x - 3)}{(x - 1)^4} = \frac{8(x + 2)}{(x - 1)^4}$

$x = -1$: $\frac{d^2y}{dx^2} = \frac{8}{16} > 0$

Minimum turning point at $(-1, -1)$.

(ii) $\frac{d^2y}{dx^2} = 0$: $x = -2, y = -\frac{8}{9}$.

$x = -1$: $\frac{d^2y}{dx^2} > 0$

$x = -3$: $\frac{d^2y}{dx^2} = \frac{-8}{4^4} < 0$

Concavity changes at $x = -2$ so $\left(-2, \frac{-8}{9}\right)$ is a point of inflection.

(iii) $x = 1, y = 0$

(b)

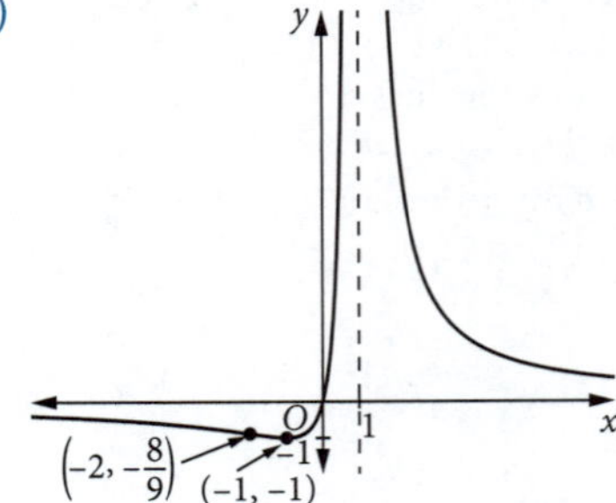

(c) $f'(x) = \frac{-4(x + 1)}{(x - 1)^3}$

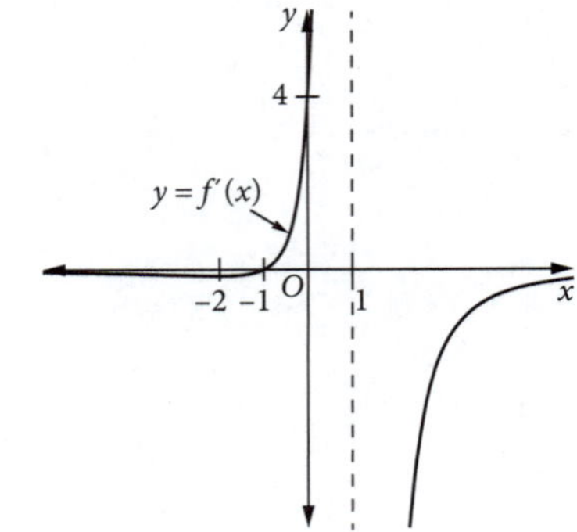

9 **(a)** $\dfrac{x(x+1)}{x-1}=\dfrac{x^2+x}{x-1}$, $px+q+\dfrac{r}{x-1}=\dfrac{px^2+(q-p)x+(r-q)}{x-1}$,

$p=1, q=2, r=2;$

$\dfrac{x(x+1)}{x-1}=x+2+\dfrac{2}{x-1}$

(b) $y=\dfrac{x(x+1)}{x-1}=x+2+\dfrac{2}{x-1}$,

$\dfrac{dy}{dx}=1-\dfrac{2}{(x-1)^2}$

$\dfrac{dy}{dx}=0$: $(x-1)^2=2$, $x-1=\pm\sqrt{2}$, $x=1\pm\sqrt{2}$

$x=1+\sqrt{2}$, $y=1+\sqrt{2}+2+\dfrac{2}{\sqrt{2}}=3+2\sqrt{2}$

$x=1-\sqrt{2}$, $y=1-\sqrt{2}+2+\dfrac{2}{-\sqrt{2}}=3-2\sqrt{2}$

$\dfrac{d^2y}{dx^2}=\dfrac{4}{(x-1)}$

$x=1+\sqrt{2}$: $\dfrac{d^2y}{dx^2}=\dfrac{4}{2\sqrt{2}}>0$

Minimum turning point at $(1+\sqrt{2}, 3+2\sqrt{2})$

$x=1-\sqrt{2}$: $\dfrac{d^2y}{dx^2}=\dfrac{4}{-2\sqrt{2}}<0$

Maximum turning point at $(1-\sqrt{2}, 3-2\sqrt{2})$

(c), (d)

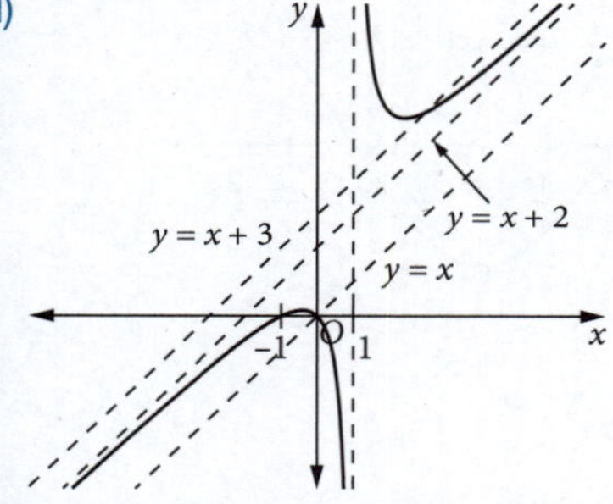

(e) $c=2$. $y=x+2$ is the sloping asymptote.

10

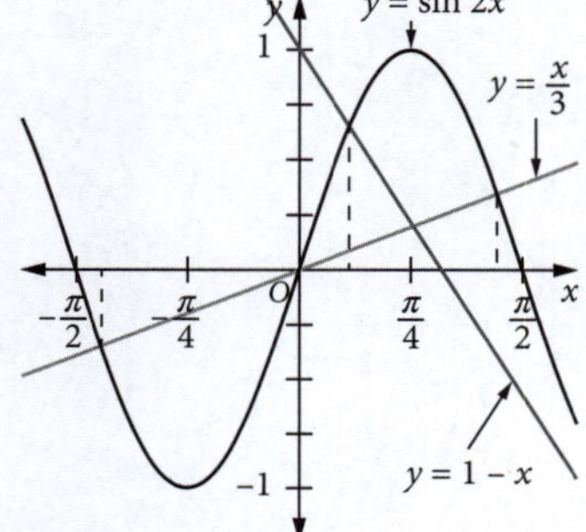

(a) The graphs $y=\sin 2x$ and $y=\dfrac{x}{3}$ intersect at three places: when $x=0$ and near $x=\pm0.4\pi$ or ±1.3.
More accurate solutions using technology give $x=0$, $\pm0.43\pi$ or ±1.34.

(b) The graphs $y=\sin 2x$ and $y=1-x$ intersect at one place: near $x=0.1\pi$ or 0.3.
A more accurate solution using technology gives $x=0.11\pi$ or 0.35.

11 **(a)**

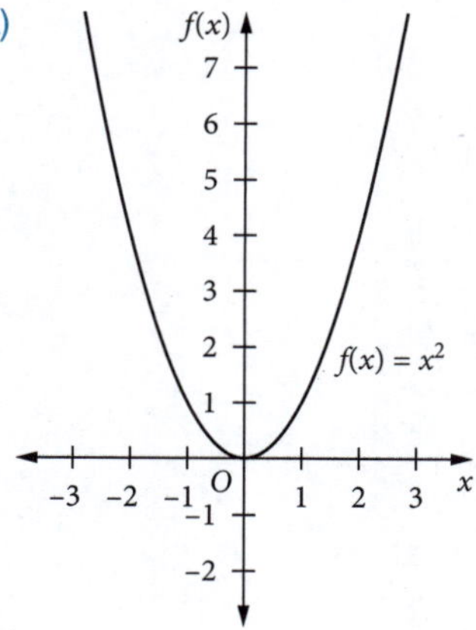

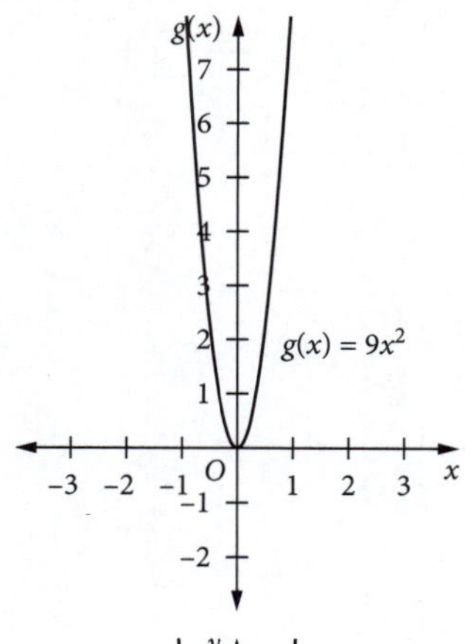

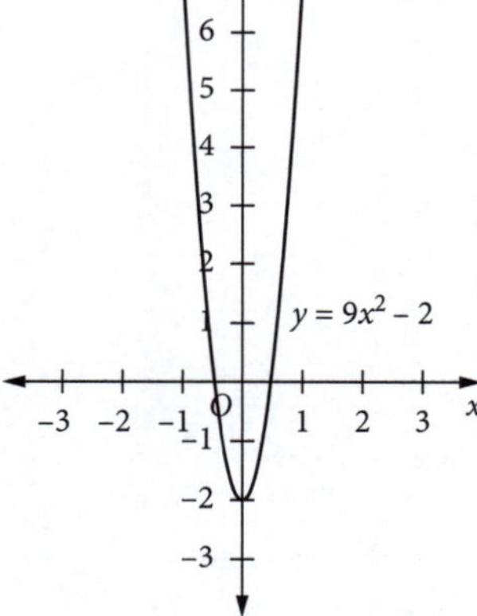

(b)

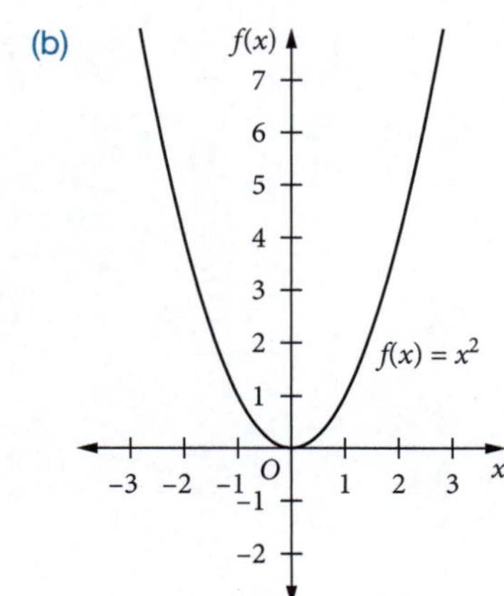

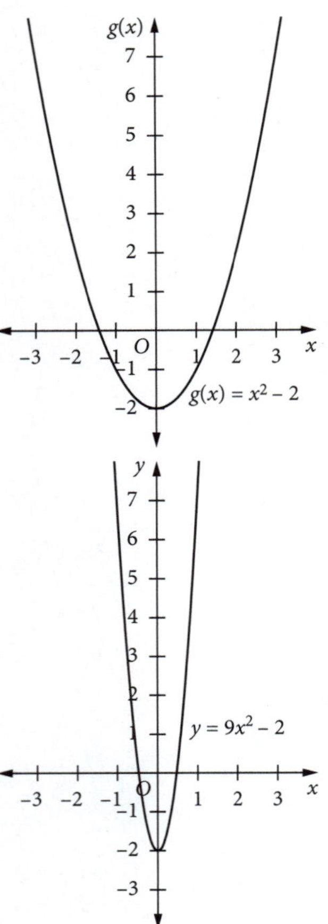

(c) The graphs are the same.

12 (a)

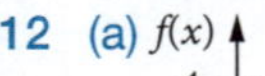

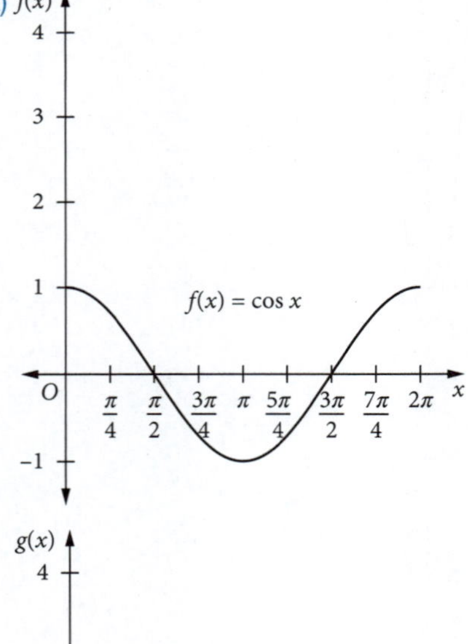

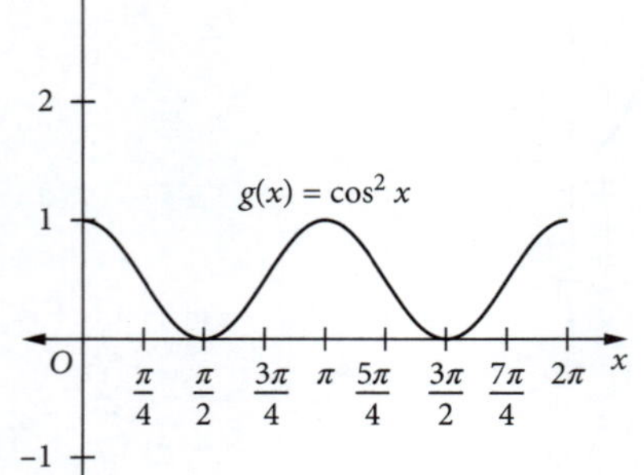

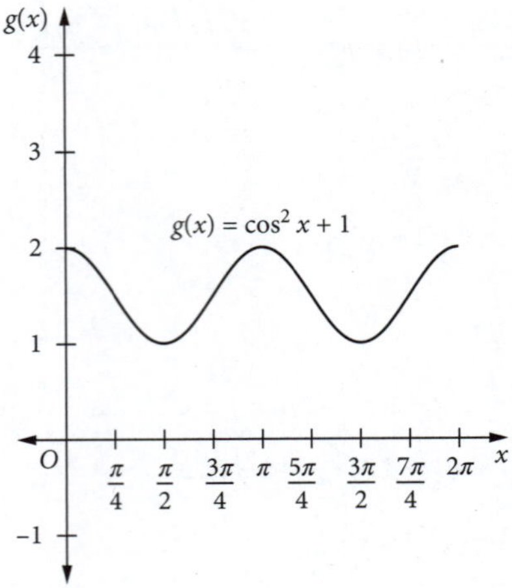

(b)

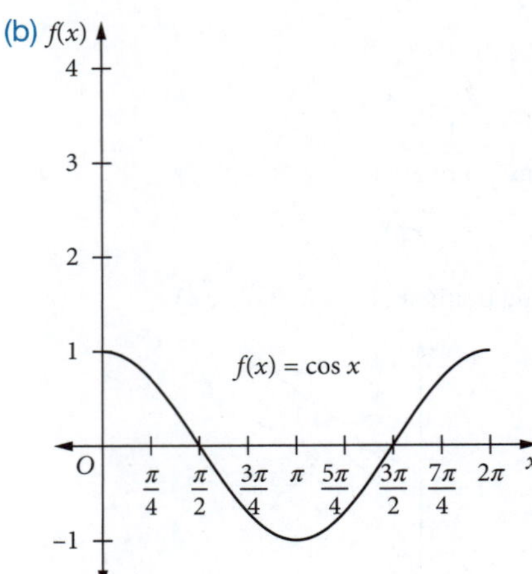

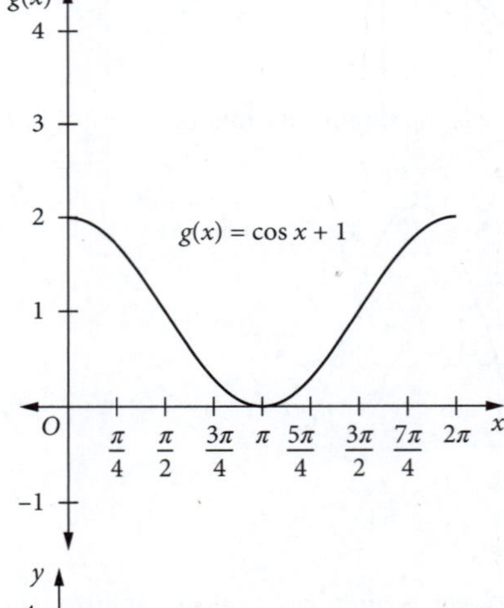

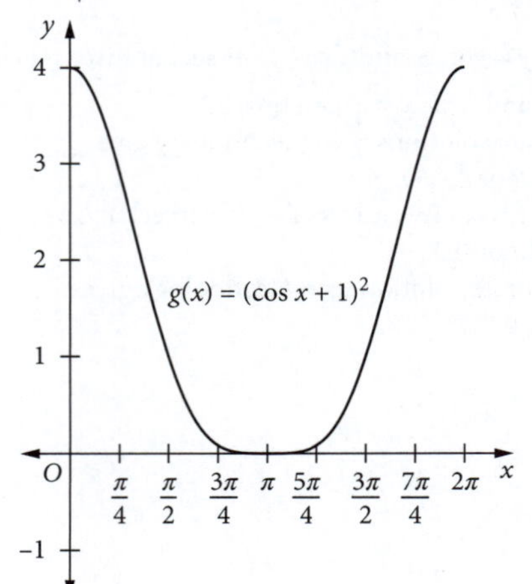

(c) The final graphs in each part are completely different due to the time the squaring is done.

13 (a)

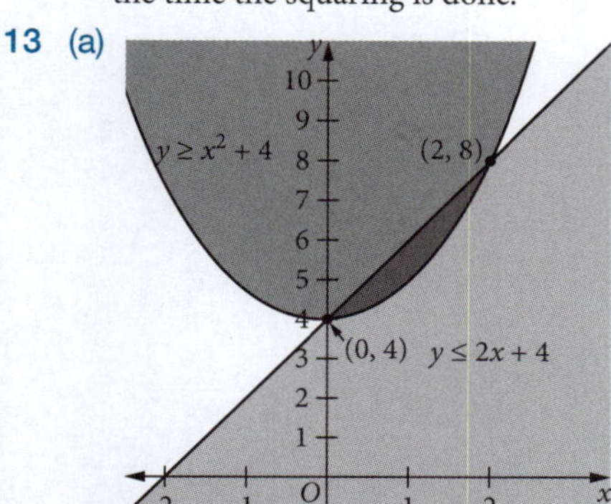

(b) $0 \leq x \leq 2$

CHAPTER 16

EXERCISE 16.1

1 (a) $2x^3 - 2x^2 + 5x + C$ (b) $3x + \frac{5x^2}{2} + \frac{x^3}{3} - \frac{3x^4}{4} + C$

(c) $\frac{x^3}{3} - x + C$ (d) $\frac{x^6}{6} - x^4 - \frac{2x^3}{3} + x + C$ (e) $\frac{3x^2}{2} - 5x + C$

(f) $\frac{2x^3}{3} - \frac{7x^2}{2} + 5x + C$

2 (a) correct (b) correct (c) incorrect (d) incorrect

3 (a) $\frac{4x^3}{3} + 2x^2 + x + C$ (b) $5x + C$ (c) $\frac{x^3}{3} + \frac{3x^2}{2} + C$

(d) $\frac{2x^3}{3} + \frac{3x^2}{2} - 2x + C$ (e) $x^4 - 2x^3 + \frac{x^2}{2} + C$

(f) $\frac{x^3}{3} + 3x^2 + 9x + C$

4 $\text{LHS} = \frac{8x^3 + 12x^2 + 6x + 1}{6} = \frac{4x^3}{2} + 2x^2 + x + 1$, which differs from RHS only by a constant

5 (a) $3x + x^2 - x^3 + C$ (b) $\frac{x^4}{4} + \frac{2x^3}{3} + C$ (c) $\frac{x^5}{5} - \frac{x^4}{4} + C$

(d) $\frac{x^3}{3} + \frac{x^2}{2} - 12x + C$ (e) $\frac{x^5}{5} - \frac{x^4}{12} + \frac{x^3}{6} + C$ (f) $\frac{2x^{\frac{3}{2}}}{3} + C$

6 (a) $\frac{x^4}{4} - x^3 + C$ (b) $\frac{x^3}{15} - \frac{x^4}{16} + C$ (c) $\frac{x^3}{3} - \frac{3x^4}{4} + C$

(d) $\frac{x^5}{5} - x + C$ (e) $-2x^{\frac{-1}{2}} - \frac{2x^{\frac{-3}{2}}}{3} + C$ (f) $\frac{2x^{\frac{3}{2}}}{3} + \frac{3x^{\frac{4}{3}}}{4} + C$

7 $x^2 - 2x + 5$ 8 C 9 $y = 2x^2 - 6x + 8$

10 $y = x^3 - x^2 + 3x - 24$ 11 $F(x) = \frac{(x+2)^3}{3} - 5$ 12 $y = x^2 - 4x$

13 $f(x) = \frac{x^3}{3} - x^2 + x + 2$ 14 $y = \frac{x^3}{3} + \frac{x^2}{2} - 2x + 4$

15 (a) $3x - \frac{1}{x} + \frac{1}{x^2} + C$ (b) $\frac{x^2 - 2}{x} + C = x - \frac{2}{x} + C$

(c) $\frac{3x^2}{2} - 2x - \frac{1}{2x^2} + C$ (d) $C - \frac{1}{2x^2}$ (e) $\frac{4x^{\frac{3}{2}}}{3} + 3x - \frac{5}{x} + C$

(f) $2\sqrt{x} - \frac{2}{\sqrt{x}} + C$

16 $s = 4t^3 - 3t^2 + t + 2$ 17 $N = 60t + 50, 410$ 18 $d = t^3 + 4t$

EXERCISE 16.2

1 (a) $\frac{x^2}{2} + C$ (b) $\frac{x^3}{3} + \frac{x^2}{2} + x + C$ (c) $3x - \frac{x^3}{3} + C$

(d) $x^6 - x^4 + x^2 + C$ (e) $x + C$ (f) $\frac{x^{n+1}}{n+1} + C$

2 (a) $\frac{2}{3}x^{\frac{3}{2}} + C$ (b) $C - \frac{1}{x}$ (c) $x + \frac{2x\sqrt{x}}{3} + \frac{x^2}{2} + C$

(d) $\frac{x^2}{2} - \frac{1}{x} + C$ (e) $\frac{x^3}{3} + 2x - \frac{1}{x} + C$ (f) $x - \frac{4}{3}x^{\frac{3}{2}} + \frac{x^2}{2} + C$

3 C 4 $y = x + \frac{x^2}{2} + x^3 - 6$

5 $y = x + \frac{2x\sqrt{x}}{3} + \frac{2}{3}$ OR $y = x + \frac{2}{3}x^{\frac{3}{2}} + \frac{2}{3}$

EXERCISE 16.3

1 (a) $C - \frac{\cos 2x}{2}$ (b) $\frac{\sin 3x}{3} + C$ (c) $\tan x + C$

(d) $\sin x - \cos x + C$ (e) $-2\cos x - 3\sin x + C$

(f) $C - \cos\left(x + \frac{\pi}{4}\right)$ (g) $2\sin\frac{x}{2} + C$ (h) $C - \cos 2x$

2 D

3 (a) $4\sin\frac{x}{4} - 4\cos\frac{x}{4} + C$ (b) $C - \cos x - \frac{\sin 2x}{2}$

(c) $-\frac{1}{2}\cos\left(2x + \frac{\pi}{2}\right) + C = \frac{\sin 2x}{2} + C$ (d) $\frac{1}{2}\sin\left(2x - \frac{\pi}{4}\right) + C$

(e) $\frac{\tan 3x}{3} + C$ (f) $C - \frac{1}{4}\cos 2x - \sin x$

(g) $2\sin\frac{x}{2} + \frac{1}{4}\cos 2x + C$ (h) $\frac{x^3}{3} - \frac{\cos 2x}{2} + C$

(i) $\frac{2}{3}x^{\frac{3}{2}} - \frac{\sin x}{2} + C$

4 (a) $\ln(1 - \cos x) + C$ (b) $\ln(\sin x) + C$

5 $y = \frac{7 - 2\cos 3x}{3}$

EXERCISE 16.4

1 (a) $\frac{e^{2x}}{2}$ (b) $\frac{e^{5x}}{5}$ (c) $\frac{-5e^{-0.4x}}{2}$

(d) $2e^{2.5x}$ (e) $e^x - \frac{e^{-3x}}{3}$ (f) $e^{-x} - \frac{e^{-2x}}{2}$

2 (a) $-e^{-x} + C$ (b) $2e^{\frac{x}{2}} + C$ (c) $C - \frac{e^{-3x}}{3}$

(d) $-e^{-t} - t + C$ (e) $\frac{e^{2u}}{2} + \frac{u^3}{3} + C$ (f) $\frac{-2e^{-2.5}}{5} + \frac{5e^{0.4x}}{2} + C$

EXERCISE 16.5

1 (a) $2\ln x + C$ (b) $\ln(x+1) + C$ (c) $\ln(2x+1) + C$

(d) $\frac{1}{2}\ln\left(x^2 - 4\right) + C$ (e) $\frac{1}{2}\ln(2x-1) + C$ (f) $\ln\left(4 + e^x\right) + C$

(g) $\frac{1}{4}\ln\left(x^4 + 1\right) + C$ (h) $-\frac{1}{2}\ln\left(4 - e^{2x}\right) + C$

2 C 3 $y = \log_e(2x+1)$

4 (a) correct (b) correct (c) incorrect (d) correct

5 $\frac{d}{dx}(\log_e(\cos x)) = \frac{-\sin x}{\cos x} = -\tan x$, $\int \tan x\, dx = C - \log_e(\cos x)$

$\text{area} = \int_0^{\frac{\pi}{3}} \tan x\, dx = \left[-\log_e(\cos x)\right]_0^{\frac{\pi}{3}} = -\log_e\frac{1}{2} + \log_e 1 = \log_e 2$

6 (a) $2^x \ln 2$ (b) $1 + 10^x \ln 10$ (c) $e^x + 5^x \ln 5$

(d) $\frac{d}{dx}\left(5^{x^2}\right) = 5^{x^2}\ln 5 \times 2x = 2x \times 5^{x^2}\ln 5$

(e) $\frac{d}{dx}\left(a^{\sqrt{x}}\right) = a^{\sqrt{x}}\ln a \times \frac{1}{2\sqrt{x}} = \frac{a^{\sqrt{x}}\ln a}{2\sqrt{x}}$

7 (a) $\frac{3^x}{\ln 3} + C$ (b) $\frac{x^2}{2} + \frac{10^x}{\ln 10} + C$ (c) $\ln|x| + x + e^x + \frac{a^x}{\ln a} + C$

CHAPTER REVIEW 16

1 (a) $\frac{x^2}{2} + 9x + C$ (b) $x^3 - x^2 + 4x + C$ (c) $\frac{x^5}{5} + \frac{x^4}{4} - 2x + C$

(d) $\frac{x^3}{3} + \frac{x^2}{2} - 6x + C$ (e) $\frac{(x+2)^3}{3} + C = \frac{x^3}{3} + 2x^2 + 4x + C$

(f) $7x + C$

2 (a) $\frac{5x^2}{2} + 4x + C$ (b) $5x - 2x^2 + x^3 + \frac{x^4}{4} + C$

(c) $x^2 + \frac{2x\sqrt{x}}{3} + 3x + C$

3 (a) $f(x)=\frac{x^3}{3}+\frac{x^4}{4}+x+2$ (b) $f(x)=3x-\frac{x^2}{2}+\frac{3x^4}{2}-1$
(c) $f(x)=x+\frac{1}{x}-2$

4 (a) $V=140t+\frac{13t^2}{2}-\frac{t^3}{3}$ (b) 2040 litres

5 (a) $\frac{d}{dx}\left(xe^x\right)=e^x+xe^x$
(b) $\int xe^x\,dx=\int\left(\frac{d}{dx}\left(xe^x\right)-e^x\right)dx=xe^x-e^x+C$

6 (a) $C-6\cos\frac{x}{2}$ (b) $\frac{x^2}{2}+\frac{1}{2}\tan 2x+C$ (c) $\ln(\sin t)+C$

7 (a) $5\log_e|x|+C$ (b) $3\log_e|x+4|+C$
(c) $2\log_e(x^2+1)+C$ (d) $\log_e(e^x+2)+C$

CHAPTER 17

EXERCISE 17.1

1 $0.2734<\text{area}<0.3984$

x	x^2
0	0
0.125	0.015 625
0.25	0.0625
0.375	0.140 625
0.5	0.25
0.625	0.390 625
0.75	0.5625
0.875	0.765 625
1	1

2 (a) 10 (b) 9 (c) 8.75
3 (a) 1.5 (b) 1.375 (c) 1.343 75
4 (a) 6 (b) 6 (c) 6 5 (a) 8 (b) 5 (c) 4.25

EXERCISE 17.2

1 $\int_0^3(x+2)dx=10.5$ 2 $\int_1^4(2x+1)dx=18$ 3 C

4 $\int_0^t 3x\,dx=\frac{3t^2}{2}$ 5 $\int_0^2 x^2\,dx$, 2.6 6 $\int_0^2(2-x)dx=2$

7 $\int_{-3}^3\sqrt{9-x^2}\,dx=\frac{9\pi}{2}$ 8 $\int_{-1}^5\sqrt{9-(x-2)^2}\,dx=\frac{9\pi}{2}$

EXERCISE 17.3

1 $9, -9; 9=-(-9)$ 2 $2\frac{1}{3}, 6\frac{1}{3}, 8\frac{2}{3}; 2\frac{1}{3}+6\frac{1}{3}=8\frac{2}{3}$

3 $8\frac{2}{3}, 8, 16\frac{2}{3}; 8\frac{2}{3}+8=16\frac{2}{3}$ 4 $72, 72; 72=72$

5 (a) -4 (b) 27.5 (c) $-10\frac{2}{3}$ (d) 0 (e) $6\frac{2}{3}$ (f) $61\frac{2}{3}$ 6 B

7 (a) 0 (b) $18\frac{2}{3}$ (c) 0 (d) $\frac{2}{3}$ (e) $\frac{16}{3}$ (f) $\frac{15}{64}$
(g) 38.4 (h) $\frac{1}{12}$ (i) $-\frac{1}{12}$

8 (a) a^3-b^3 (b) $\frac{8a}{3}+2b+2c$ (c) $\frac{a^3}{3}-a^2$

9 (a) correct (b) incorrect (c) correct (d) correct

10 4.5 11 $\frac{1}{12}$ 12 $14\frac{7}{24}$ 13 $a=2$

14 $a=1$ 15 $c=-\frac{1}{20}$ 16 6.75

EXERCISE 17.4

1 55.5 2 9 3 3 4 $10\frac{2}{3}$

5 (a) correct (b) incorrect (c) correct (d) correct

6 40.5 7 B 8 0.25 9 $k=0.75$

10 (a) $\frac{67}{192}$ (b) $1\frac{1}{3}$ (c) $\frac{5}{12}$

11 $85\frac{1}{3}$ 12 $1\frac{1}{3}$ 13 $3\frac{1}{2}$

14 Area of square = 1 unit^2
(a) $\int_0^1 x\,dx=\left[\frac{x^2}{2}\right]_0^1=\frac{1}{2}$
(b) $\int_0^1 x^2\,dx=\left[\frac{x^3}{3}\right]_0^1=\frac{1}{3}, \int_0^1\sqrt{x}\,dx=\left[\frac{2x^{\frac{3}{2}}}{3}\right]_0^1=\frac{2}{3};$
square is trisected
(c) $\int_0^1 x^3\,dx=\left[\frac{x^4}{4}\right]_0^1=\frac{1}{4}, \int_0^1 x\,dx=\left[\frac{x^2}{2}\right]_0^1=\frac{1}{2},$
$\int_0^1\sqrt[3]{x}\,dx=\left[\frac{3x^{\frac{4}{3}}}{4}\right]_0^1=\frac{3}{4}$; four equal parts

EXERCISE 17.5

1 $\frac{4}{3}$ 2 $21\frac{1}{3}$ 3 A 4 1

5 8 6 $10\frac{2}{3}$ 7 $6\frac{2}{3}$ 8 $\frac{4}{3}$

9 (a) correct (b) correct (c) incorrect (d) correct

10 $\frac{5}{3}$ 11 $\frac{4}{3}$ 12 $10\frac{2}{3}$

13 (a) $y=x$ (b) 4.5 (c) $6\frac{1}{6}$

14 (a) intersection at $\sqrt{x}=\frac{1}{x}$: $x^3=1, x=1$; for $x>1, \sqrt{x}>\frac{1}{x}$
(b) 1.61

15 (a) $x-20y+480=0$ (b) 16 m (c) $y=\frac{x^2}{200}$ (d) $1983\frac{1}{3}\text{ m}^2$

EXERCISE 17.6

1 4.5 2 9 3 C 4 8 5 $10\frac{2}{3}$

6 (a) $(2,5)$: $\frac{dy}{dx}=2x, m=4; y-5=4(x-2), y=4x-3$
(b) $2\frac{2}{3}$ (c) $1\frac{13}{24}$

7 $1\frac{11}{24}$

8 (a) $\frac{16\sqrt{2}}{3}$ (b) $\frac{16\left(2-\sqrt{2}\right)}{3}$

9 (a) correct (b) correct (c) correct (d) correct

10 4 11 $4\sqrt{6}$

12 (a)
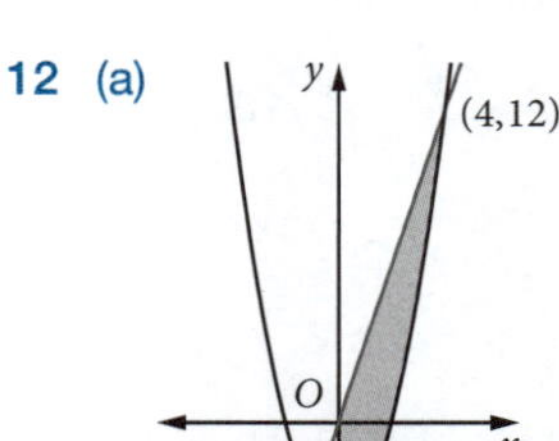

(b) $18\frac{2}{3}$

13 (a)
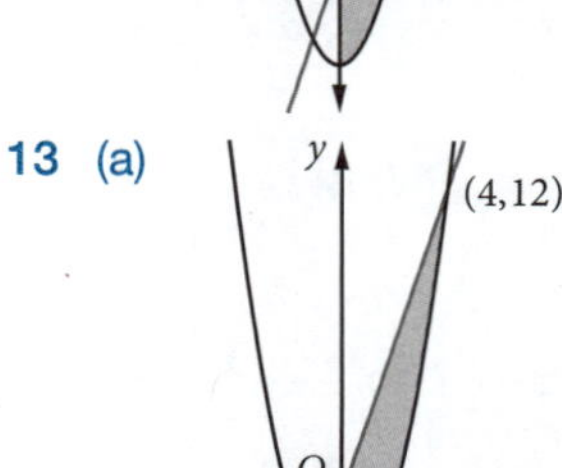

(b) intersection at $(-1,-3)$, $(4,12)$:

$$A=\int_{-4}^{-3}\sqrt{y+4}\,dy-\int_{-4}^{-3}-\sqrt{y+4}\,dy+\int_{-3}^{12}\left(\sqrt{y+4}-\frac{y}{3}\right)dy$$
$$=2\int_{-4}^{-3}\sqrt{y+4}\,dy+\int_{-3}^{12}\left(\sqrt{y+4}-\frac{y}{3}\right)dy$$

(c) $20\frac{5}{6}$ (d) $\int_{-1}^{4}\left(3x+4-x^2\right)dx=20\frac{5}{6}$

EXERCISE 17.7

1

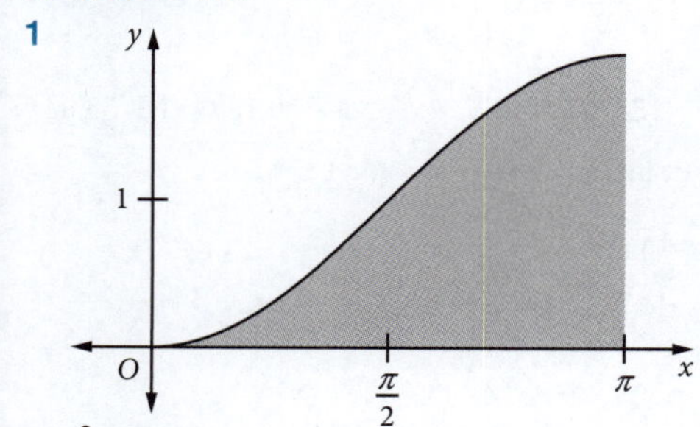

$\int$ area $=\pi$

2 (a) 2 (b) $\sqrt{3}$ (c) 1 (d) 1 (e) 2 (f) 0 (g) -0.25 (h) $\frac{\sqrt{2}-1}{2}$ (i) 0 (j) 4 (k) $\frac{-3}{8}$ (l) 0

3 (a) incorrect (b) correct (c) correct (d) correct

4

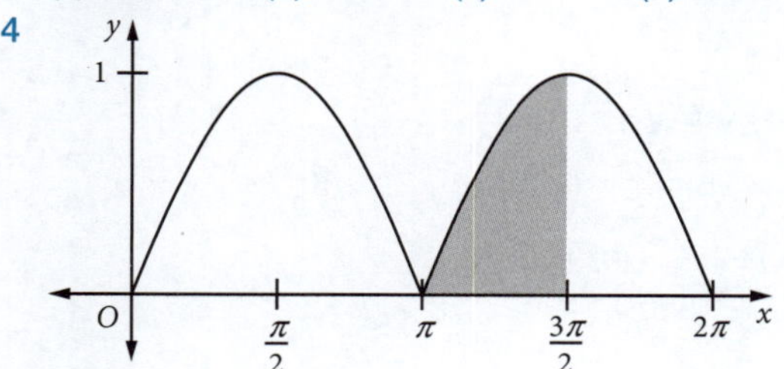

area $=1$

5 2

6 (a) $\log_e 1.5$ (b) $\ln 2$

7 $\frac{-9\sqrt{3}}{8}$ 8 0.5

9 (a) π (b) $\frac{\pi+1}{2}$ (c) $\frac{\pi+2\sqrt{2}}{4}$ (d) $\frac{ab}{\pi}$

10 $\left(\frac{\pi}{4},\frac{\sqrt{2}}{2}\right),\left(\frac{5\pi}{4},\frac{-\sqrt{2}}{2}\right)$

area $=4\sqrt{2}$

EXERCISE 17.8

1 (a) $\frac{e^2-1}{e}$ (b) $\frac{e^4-1}{2}$ (c) $\frac{2\left(e^2-1\right)}{e^{\frac{3}{2}}}$ (d) $\frac{2\left(e^{1.5}-1\right)}{3}$

(e) $\frac{e^2-1}{2e}$ (f) $\frac{e^3-1}{3}$ (g) $\frac{\left(e^2-1\right)^2}{2e^2}$ (h) 0 (i) 2.904

2 (a) correct (b) correct (c) correct (d) incorrect

3 e^2-e^{-2}

4 (a) e^2-1 (b) $y=e^2x-e^2$

(c) $\frac{e^2}{2}-1$ (d) e^2+1

5 $e^2+e^{-2}-2$ 6 0.511

7 (a) $\ln 2$ (b) $2\ln 2$ (c) $\frac{1}{2}\log_e 5$ (d) $\ln 5$ (e) $\ln 3$

(f) $3\ln 2.5$ (g) $\frac{1}{2}\log_e\frac{7}{3}$ (h) $3\log_e\frac{11}{3}$

8 (a) $\frac{26+3\ln 3}{3}$ (b) 0 (c) $6-\ln 2$ (d) $\ln 2+0.25$ (e) $\ln 2+2.5$

(f) $\int_1^2\left(x^2-\frac{2}{x}+\frac{1}{x^4}\right)dx=\left[\frac{x^3}{3}-2\log_e x-\frac{1}{3x^3}\right]_1^2=\frac{21}{8}-2\log_e 2$

(g) $e^3-e+\log_e 3$ (h) $\frac{14}{3}-2\ln 2$

9

area $=\int_1^4\frac{dx}{2x-1}=\frac{\ln 7}{2}$

10 $\int_2^4\frac{x\,dx}{x^2+1}=\frac{1}{2}\left[\ln\left(x^2+1\right)\right]_2^4=\frac{1}{2}\ln\frac{17}{5}$

11 (a) correct (b) correct (c) incorrect (d) correct

12

$y'=\frac{x^2-1}{x^2}$, $y''=\frac{2}{x^3}$.

turning point $(1,2)$

area $=1.5+\ln 2$

13 B 14 $\frac{1+e}{2}$ 15 $\frac{\log_e 2}{2}$

16

area $=a\ln a-a+1$

17 (a) $\frac{d}{dx}\left(\log_e(\cos x)\right)=-\tan x$ (b) Area $=\log_e 2$

EXERCISE 17.9

1 (a) $12.5\,\text{cm}^3\,\text{min}^{-1}$ (b) $V=1000-12.5t$, $0\le t\le 80$

2 (a) (i) 1 item/min (ii) 21 items/min (b) 110

3 (a) (i) $448\,\text{ML}\,\text{h}^{-1}$ (ii) 964 ML

(b) (i)

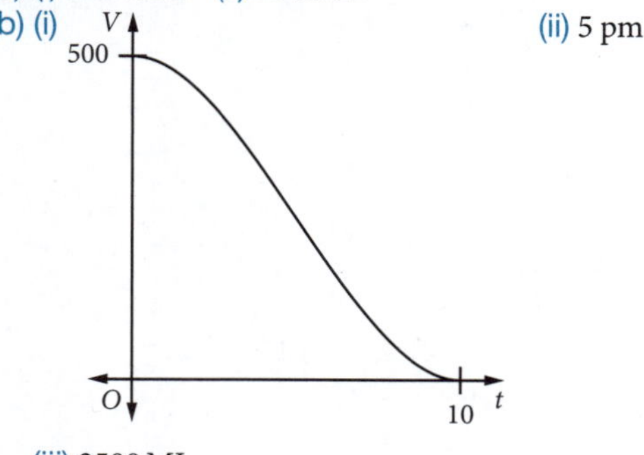

(ii) 5 pm

(iii) 2500 ML

4 (a) incorrect (b) correct (c) incorrect (d) correct

5 (a) correct (b) incorrect (c) incorrect (d) correct

6 (a) $x=\frac{t^3}{3}-t^4$ (b) $\ddot{x}=2t-12t^2$

7 (a) $x=2t^3+3t^2-12t+7$, $a=12t+6$

(b) $t=1$, $a=18\,\text{m}\,\text{s}^{-2}$ (c) $v=-12\,\text{m}\,\text{s}^{-1}$, $a=6\,\text{m}\,\text{s}^{-2}$

8 (a) $\dot{x}=11+10t-t^2$, $x=11t+5t^2-\frac{t^3}{3}$ (b) $t=11$, $x=282\frac{1}{3}$

9 (a) $v = 30 - 10t$ (b) $h = 30t - 5t^2$ (c) $t = 3, h = 45\,\text{m}$ (d) 6 s
10 (a) $v = 25 - 10t, x = 25t - 5t^2$ (b) 31.25 m (c) $t = 1.25\,\text{s}$
11 (a) $x = t^3 - t^2 - t + 1, a = 6t - 2$
(b) $t = 1\,\text{s}, v = 0\,\text{m s}^{-1}, a = 4\,\text{m s}^{-2}$ (c) 4 m
12 $s(t) = 4t^3 - 3t^2 + t + 2$
13 $t = 12\,\text{s}, x = 432\,\text{m}$
14 (a) $50\,\text{km h}^{-1}, 20\,\text{km h}^{-1}$ (b) 9 min (c) 13.2 km
(d) $x = \frac{112t}{3} + 2$
15 $x = -3$ 16 D
17 (a) $\frac{dx}{dt} = \frac{1}{1+t}, x = \int \frac{dt}{1+t} = \log_e(1+t) + C, C = 0, x = \log_e(1+t)$
(b) $\lim_{t\to\infty} \log_e(1+t) \to \infty$, moving away in positive direction
(c) -1
18 (a) A: $v = 100\,\text{km h}^{-1}$; B: $v = 62\,\text{km h}^{-1}$
(b) (i) $\ddot{x} = -40e^{-t}$ (ii) $\ddot{x} = 60 - v$
(c)

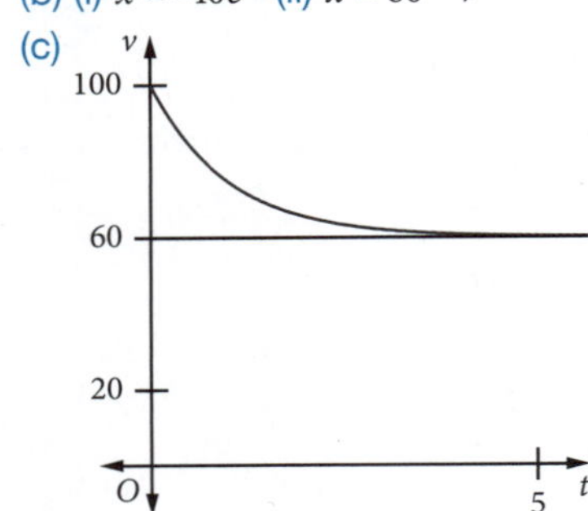

As t increases, the velocity decreases towards $60\,\text{km h}^{-1}$.
(d) 218 km
19 (a) $x = 2 - e^{-t}$

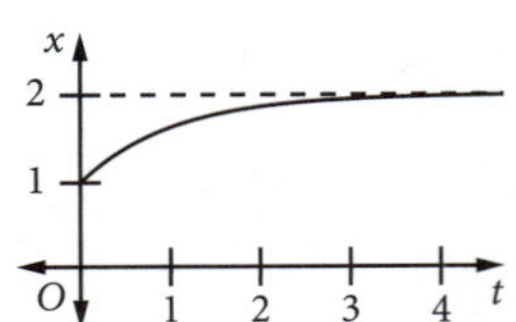

(b) $\frac{d^2x}{dt^2} = -e^{-t} = x - 2$
20 $a = 2\sin t, v = 3 - 2\cos t, x = 1 + 3t - 2\sin t$
21 (a) $x = 2 + t - 2\cos t$ (b) $t = \log_e 3$
22 (a) $x = 0$ (b) $\dot{x} = 2$
(c) $\ddot{x} = \frac{-4}{(1+t)^2}$; as $(1+t)^2 > 0$ for $t \geq 0$, $\ddot{x} < 0$ for all $t \geq 0$
(d) $t = 1$
23 $v = 6\sin 2t, x = 9 - 3\cos 2t$: 6 times

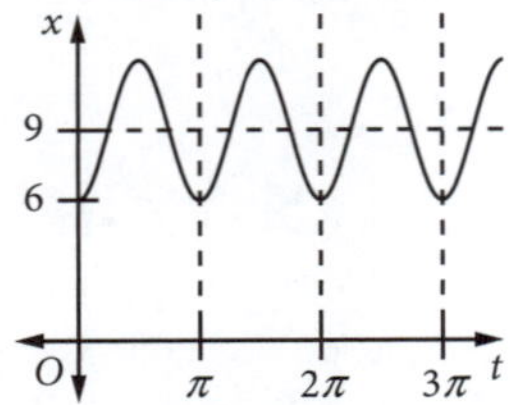

24 (a) $a = \frac{49}{2}e^{-0.5t}$ (b) 0 (c) 49 (d) 155

EXERCISE 17.10

1 $8 + 2\sqrt{13}$ 2 (a) $\frac{31}{40}$ (b) 0.783
3 (a) incorrect (b) correct (c) incorrect (d) correct
4 (a) $3\frac{3}{4}$ (b) $3\frac{15}{16}$
5

x	0	1	2	3	4
2^x	1	2	4	8	16

22.5
6 2.8125 7 (a) $\frac{2}{3}$ (b) 0.631 8 3.414
9 $2\frac{19}{60}$ 10 23.11
11 (a) $63\,750\,\text{m}^2$ (b) $159\,375\,000\,\text{m}^3$
12 (a) $\frac{\pi}{6}\left(\frac{1}{4} + 2 + \frac{1}{4}\right) = \frac{5\pi}{12} \approx 1.31$
(b) 1.31
13 1.57 14 3.1 15 1.54 16 3.65 17 0.188 18 53.44
19 (a) curves do not intersect; $\frac{1}{x^2} > \frac{-1}{x}$ for $1 \leq x \leq 4$
(b) $\frac{1}{2}\left(2 + 2\left(\frac{3}{4} + \frac{4}{9}\right) + \frac{5}{16}\right) \approx 2.35$

EXERCISE 17.11

1 $\frac{1}{3}$ 2 $3\frac{1}{5}$ 3 -2 4 D 5 2
6 $\frac{24}{7}(3 - \sqrt{2})$ 7 3
8 (a) $\frac{e^2 - 1}{2}$ (b) $\frac{e^2 - 1}{2e^2}$ (c) $\frac{e^{10} - e^2}{8}$ (d) $\frac{e^4 - 1}{2e^2}$
9 (a) $\frac{\log_e 3}{2}$ (b) $\frac{\log_e 5}{2}$ (c) $\frac{\log_e 3}{2}$ (d) 0
10 $\frac{4}{\pi}$ 11 $\frac{3}{\pi}$

CHAPTER REVIEW 17

1 (a) -20 (b) $\frac{512}{15}$ (c) 1820 2 $10\frac{2}{3}$
3 (a) 0 (b) $-26\frac{2}{3}$ (c) 48
4 $\frac{2}{3}$ 5 36 6 $42\frac{2}{3}$
7 (a) $\frac{\sqrt{2} - 1}{2}$ (b) $\frac{-\pi^2}{8}$ (c) $\log_e 3$
8 $f(x) = \frac{x^3}{3} + x + \frac{35}{24} - \frac{1}{4x}$
9 (a) 6.75 (b) $1\frac{17}{64}$
10 $\frac{19}{6}$ 11 $\frac{\sqrt{2} - \sqrt{3}}{2}$
12 (a) $0 < x < 4$ (b) $3\frac{2}{3}$ (c) $63°26'$
13 (a)

(b) $\frac{dy}{dx} = 6x - 3x^2$; at $x = 2$, $\frac{dy}{dx} = 0$; $x = 2, y = 4, A(2, 4)$
For B, $y = 4$; from the sketch $x = -1$.
Check by substitution: LHS $= 3 + 1 = 4 =$ RHS, so $B(-1, 4)$.
(c) 6.75
14 2.2

15 (a) $62\,\text{m}^2$
(b) Rotating about the y-axis, the subintervals for the trapezia would not be equal and the curve is not the same above and below the axis. Calculating the volume for each region and taking the average would give an approximate volume:

$$V = \pi\int_0^5 x^2\,dy \approx \pi\left[\frac{2.5}{2}\left(12^2+8^2\right)+\frac{4-2.5}{2}\left(8^2+4^2\right)+\frac{5-4}{2}\left(4^2+0^2\right)\right] \approx 288 \text{ for one side.}$$

16 $-\dfrac{2}{\pi}$

17 (a) 0 (b) 24.92 (c) $e^3+\ln 3-e$

18

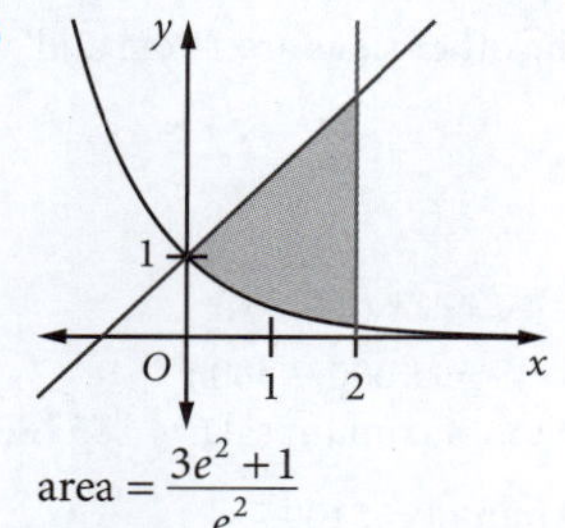

area $=\dfrac{3e^2+1}{e^2}$

19 (a) 0.4905 at $x=\dfrac{2}{3}$
(b)

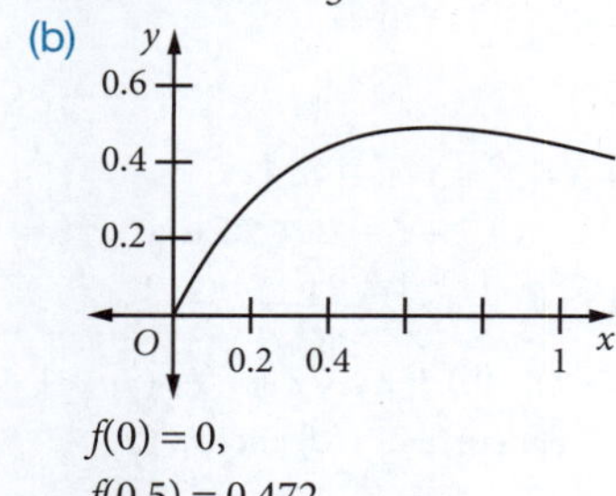

$f(0)=0$,
$f(0.5)=0.472$,
$f(1)=0.446$

(c) 0.348

20 0.7868

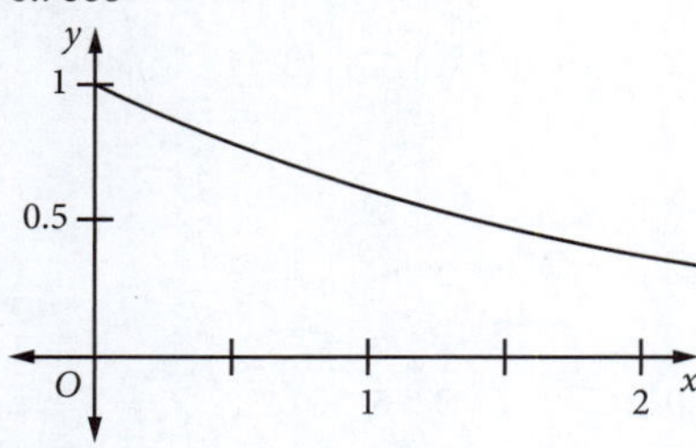

21 (a) $x=\dfrac{\pi}{6}, \dfrac{7\pi}{6}$ (b) 4

22 (a) $\dfrac{d}{dx}\left(\log_e(\sin x)\right)=\dfrac{1}{\sin x}\times\cos x=\cot x$
(b) $\dfrac{1}{2}\log_e 2-\dfrac{\pi^2}{32}$

23 (a) $x=1$: $y=\sin\dfrac{\pi x}{2}=\sin\dfrac{\pi}{2}=1$; $y=x^2=1^2=1$; curves meet at $x=1, y=1$
(b) $\dfrac{\pi+6}{3\pi}$

24 $t=2.5\,\text{s}, x=22\dfrac{11}{12}\,\text{m}$

25 (a) $14\,\text{m s}^{-2}$ (b) $275\,\text{m}$

26 (a) $x=2t^3-2t^2+t-10, a=12t-4$ (b) $8\,\text{m s}^{-2}$ (c) $17\,\text{m s}^{-1}$

27 (a) $t=20\,\text{min}$ (b) $V=110t+\dfrac{17t^2}{2}-\dfrac{t^3}{3}$ (c) 1968 L
(d) $t=22\,\text{min}$ (e) 1015 L

28 (a) $\dot{x}=\int 9\sin 3t\,dt$, $\dot{x}=-3\cos 3t+C$: $t=0, v=0, C=3$:
$\dot{x}=3-3\cos 3t$
(b) $t=\dfrac{2\pi}{3}$
(c) 2π (d) $x=3t-\sin 3t$

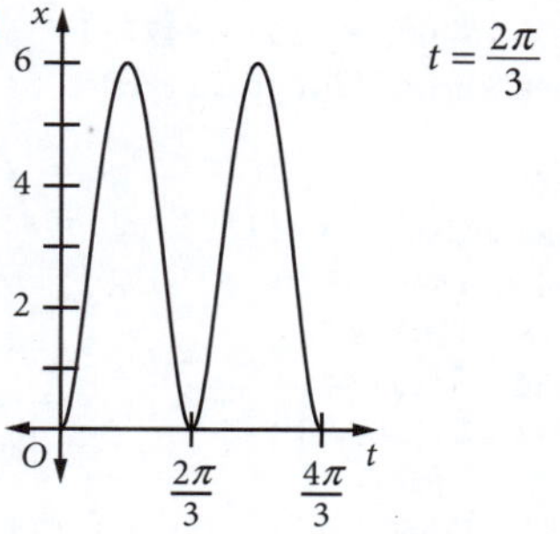

29 (a) 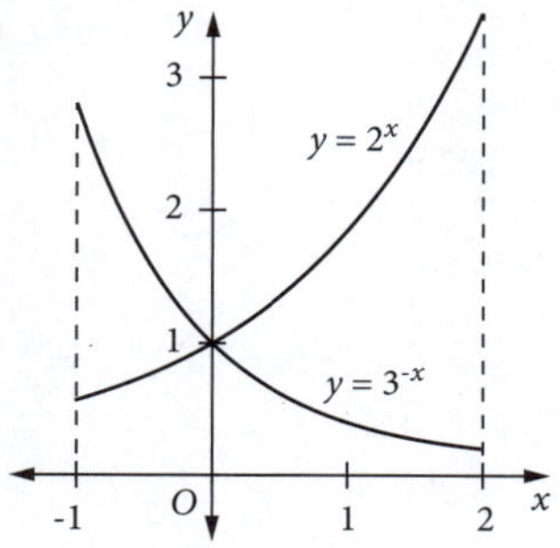
(b) Area $=\displaystyle\int_{-1}^{0}\left(3^{-x}-2^x\right)dx+\int_0^1\left(2^x-3^{-x}\right)dx$
$=\dfrac{10}{9\ln 3}+\dfrac{5}{2\ln 2}\approx 4.62 \text{ units}^2$

30 (a) 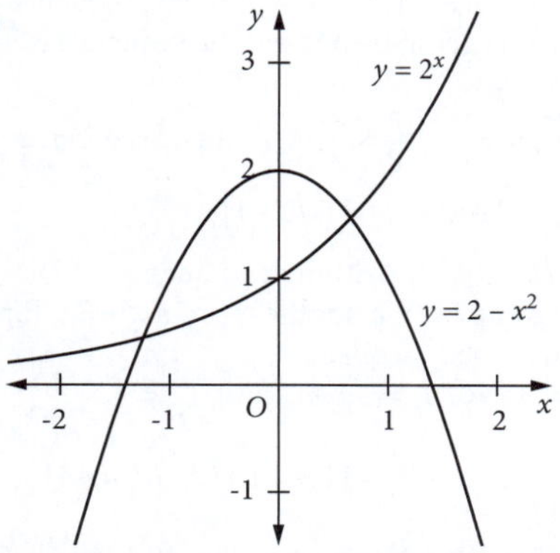
Points of intersection $(-1.26, 0.42)$, $(0.65, 1.57)$
(b) Area $=\displaystyle\int_{-1.26}^{0.65}\left(2-x^2-2^x\right)dx=1.40 \text{ units}^2$
(c) Area $=\displaystyle\int_{-\sqrt{2}}^{-1.26}\left(2-x^2\right)dx+\int_{-1.26}^{0.65}2^x\,dx+\int_{0.65}^{\sqrt{2}}\left(2-x^2\right)dx=2.37$

CHAPTER 18

EXERCISE 18.1

1 (a) $A=1000\times 1.003^{12}=\$1036.60$
(b) $A=2000\times 1.005^{10}=\$2102.28$
(c) $A=3400\times 1.0035^{20}=\$3646.08$

2 (a) B (b) D (c) A (d) C

3 $25\,000=P\times 1.0035^{120}$, $P=\dfrac{25\,000}{1.0035^{120}}=\$16\,439$

4 $P = \dfrac{50\,000}{1.005^{48}} = \$39{,}354.92$

5 (a) $A = \$2500 \times 1.267 = \$3167.50$
(b) $A = \$5600 \times 1.083 = \$6064.80$
(c) $A = \$3000 \times 1.276 = \$3828$
(d) First investment: $A = \$4000 \times 1.191 = \$4764$
Second investment: $A = \$4000 \times 1.194 = \$4776$
The second investment earns \$12 more interest
$(4000 \times 0.003 = 12)$

6 (a) A (b) D (c) C

7 (a) FVA = $\$1000 \times 6.4684 = \$6468.40$
(b) FVA = $\$1500 \times 25.6454 = \$38\,468.10$
(c) FVA = $\$200 \times 54.0978 = \$10\,819.56$
(d) FVA = $\$100 \times 51.9944 = \$5199.44$
(e) FVA = $\$400 \times 82.0232 = \$32\,809.28$

8 (a) B (b) C (c) A (d) D

9 (a) $P = \$100\,000 \times 0.0788 = \$7880$
(b) $P = \$3000 \times 0.0822 = \$246.60$
(c) $P = \$1\,000\,000 \times 0.0349 = \$34\,900$

10 (a) $A = \$3000 \times 1.093 = \$3279$
(b) $A = \$3279 \times 1.191 = \$3905.29$
Investment is worth \$3905.29
(c) $A_1 = \$3000 \times 1.126 = \$3378$
$A_2 = \$3378 \times 1.158 = \$3911.72$
Investment is worth \$3911.72
(d) The second bank earns \$6.43 more interest.

11 (a) 4.06% (b) 4.074% (c) 4.081%

12 B: $R = 5.221\%$. C: $R = 5.127\%$. B is best.

13 (a) 2.625% (b) 2.529% (c) 2.429%

14 N: $R = 5.536\%$. T: $R = 5.495\%$. T is best.

EXERCISE 18.2

1 (a) $17 - 7 = 27 - 17 = 37 - 27 = 10$. Arithmetic sequence.
(b) $2 - 5 = -1 - 2 = -4 - (-1) = -3$. Arithmetic sequence.
(c) The denominators go up by one but the sequence is not arithmetic.
(d) $1 - \frac{5}{8} = 1\frac{3}{8} - 1 = 1\frac{3}{4} - 1\frac{3}{8} = \frac{3}{8}$. Arithmetic sequence.
(e) $\sqrt{2} + 1 - \left(\sqrt{2} - 1\right) = \sqrt{2} + 3 - \left(\sqrt{2} + 1\right)$
$= \sqrt{2} + 5 - \left(\sqrt{2} + 3\right) = 2$. Arithmetic sequence.
(f) The constants go up by one but the term in π is multiplied by π. Not an arithmetic sequence.

2 (a) $a = 5$ (b) $d = 3$ (c) $T_n = 3n + 2$ (d) $T_{13} = 41$
(e) $98 = 3k + 2$, $k = 32$

3 (a) $a = 7$ (b) $d = -4$ (c) $T_n = 11 - 4n$ (d) $T_{13} = -41$
(e) $-101 = 11 - 4k$, $k = 28$

4 $a = 8, d = 6$. $T_n = 8 + 6(n - 1) = 6n + 2$. $T_8 = 48 + 2 = 50$, $T_{14} = 84 + 2 = 86$

5 $a = 17.2, d = -0.6$. $T_n = 17.2 - 0.6(n - 1) = 17.8 - 0.6n$.
$T_6 = 17.8 - 3.6 = 14.2$, $T_{11} = 17.8 - 6.6 = 11.2$

6 $a = p, d = q - p$. $T_{10} = p + 9 \times (q - p) = 9q - 8p$

7 $a = 14, d = -3$. $T_{n+2} = 14 + (n + 1) \times (-3) = 11 - 3n$

8 $a + 4d = 17$, $a + 11d = 52$, $7d = 35$, $d = 5$, $a = -3$.
Sequence is $-3, 2, 7, 12, \ldots$

9 $a + 2d = 5.6$, $a + 11d = -7$. $9d = -12.6$, $d = -1.4$. $a = 8.4$,
$T_6 = 8.4 - 7 = 1.4$

10 $a + 4d = m$, $a + 10d = n$. $6d = (n - m)$, $d = \dfrac{n - m}{6}$, $a = \dfrac{5m - 2n}{3}$.
$T_7 = \dfrac{5m - 2n}{3} + \dfrac{6(n - m)}{6} = \dfrac{2m + n}{3}$

11 $4p + 3 - (p + 5) = 8p - 2 - (4p + 3)$, $3p - 2 = 4p - 5$, $p = 3$

12 $a = 9, d = 3, T_n = 6p + 15$. $6p + 15 = 9 + 3(n - 1)$, $3n = 6p + 9$, $n = 2p + 3$ terms.

13 $a = -8$, $-8 + 6d = 22$, $d = 5$. Terms are $-3, 2, 7, 12, 17$.

14 $a = 6$. $6 + 4d = 2(6 + 3d)$, $2d = -6$, $d = -3$.

15 $a = 36, d = -5$. $36 - 5(n - 1) = -4$, $5(n - 1) = 40$, $n = 9$

16 $a = 3, d = 2, n = 12$. $T_{12} = 3 + 22 = 25$ cans of juice

17 $a^2 + b^2 = 25^2$. $a, b, 25$ is the sequence so $b - a = 25 - b$,
$a = 2b - 25$. $(2b - 25)^2 + b^2 = 25^2$.
$5b^2 - 100b = 0$, $b = 20$, $a = 15$. The other sides are 15 cm and 20 cm.

18 C
$a = 5\sqrt{2} - \sqrt{3}$, $d = \sqrt{3} - 2\sqrt{2}$.
$T_6 = 5\sqrt{2} - \sqrt{3} + 5\left(\sqrt{3} - 2\sqrt{2}\right) = 4\sqrt{3} - 5\sqrt{2}$

19 (a) $\$28\,000 + \$300 = \$31\,000$ (b) $S = 28\,000 + 300n$
(c) $n = 0, 1, 2, 3, \ldots, 16$ (d) $n = 16$, Maximum salary = \$32 800

20 (a) $I = 1000 \times \dfrac{0.5}{100} = \$5$ (b) Amount = \$1005
(c) Amount = $\$1000 + 12 \times \$5 = \$1060$
(d) Amount = $\$1000 + 24 \times \$5 = \$1120$

EXERCISE 18.3

1 (a) $1 + 4 + 9 + 16$ (b) $1 + 3 + 3^2 + 3^3 + 3^4 + 3^5$
(c) $1 + 3 + 5 + \ldots + (2p - 1)$ (d) $2 + 6 + 12 + 20 + 30$
(e) $x + 2x^2 + 3x^3 + \ldots + 8x^8$ (f) $\dfrac{1}{x} + \dfrac{1}{x^2} + \dfrac{1}{x^3} + \ldots + \dfrac{1}{x^k}$
(g) $9 + 25 + 49 + \ldots + (2p + 1)^2$ (h) $1 + 4 + 7 + \ldots + (3n + 1)$

2 (a) correct (b) incorrect (c) correct (d) incorrect

3 (a) $\sum_{k=1}^{9} k^2$ (b) $\sum_{k=1}^{10} k(k + 2)$ (c) $\sum_{k=1}^{p} (5k - 4)$
(d) $\sum_{k=2}^{p} \frac{1}{k(k+1)}$ (e) $\sum_{r=2}^{12} rx^r$ (f) $\sum_{k=1}^{n} ar^{k-1}$

4 (a) 30 (b) 48 (c) 22 (d) 30 (e) 55 (f) 24 (g) 40
(h) 9 (i) 288

EXERCISE 18.4

1 198 **2** 294 **3** 345

4 (a) 293 (b) 8730 **5** B

6 (a) 176.4 m (b) 53.9 m **7** 3275

8 157.5 **9** $1 + 3 + 5 + (2n - 1) = n^2$

10 (a) 95 (b) 152 (c) $n(2n + 1)$ (d) $n(3n + 5)$ (e) n^2 (f) 185

11 13 **12** 24

13 (a) correct (b) correct (c) correct (d) correct

14 (a) 2550 (b) 1050 (c) 550 (d) 2500

15 (a) $2n(n - 7)$ (b) $n = 16$

16 80°, 94°, 108°, 122°

17 115 **18** 15 **19** 16

20 $n = 12$, $S_{12} = 276$ **21** 297 m

22 $T_n = 6n - 14$: $-8, -2, 4, \ldots$
$a = -8, d = 6$

23 $T_n = 6n - 1$, $a = 5$, $d = 6$, $T_8 = 47$

24 $a = -3, d = 3$ **25** 9900

26 $n = \dfrac{b + c - 2a}{b - a}$, $S_n = \dfrac{(b + c - 2a)(a + c)}{2(b - a)}$

27 1, 3, 5 **28** (a) 12 (b) 26

29 9 cm, 12 cm **30** 20 **31** $n = 6, 10$

32 (a) n^2 (b) $n^2 + n$ (c) $\dfrac{n(n+1)}{2}$, 20 **33** 15

34 (a) 21 (b) 120 **35** $d = 2$

36 105°, 115°, 125°, 135°, 145°

EXERCISE 18.5

1 (a) $\frac{6}{3}=\frac{12}{6}=\frac{24}{12}=2$. This is a geometric sequence.

(b) $\frac{-2}{8}=\frac{\frac{1}{2}}{-2}=\frac{\frac{-1}{8}}{\frac{1}{2}}=-\frac{1}{4}$. This is a geometric sequence.

(c) $\frac{5}{2}\neq\frac{11}{5}$. This is not a geometric sequence.

(d) $\frac{\frac{1}{3}}{\frac{1}{9}}=\frac{1}{\frac{1}{3}}=\frac{3}{1}=3$. This is a geometric sequence.

(e) $\frac{\frac{1}{4}}{\frac{1}{2}}\neq\frac{\frac{1}{6}}{\frac{1}{4}}$. This is not a geometric sequence.

2 (a) $a=1$ (b) $r=3$ (c) $T_n=3^{n-1}$ (d) $T_{10}=3^9=19683$
(e) $3^{k-1}=6561=3^8, k=9$

3 (a) $a=144$ (b) $r=\frac{48}{144}=\frac{1}{3}$ (c) $T_n=144\times\frac{1}{3^{n-1}}=\frac{16}{3^{n-3}}$
(d) $T_{10}=\frac{16}{3^7}=\frac{16}{2187}$ (e) $\frac{16}{3^{k-3}}=\frac{16}{81}, 3^{k-3}=3^4, k=7$

4 $a=4, r=\frac{3}{2}, T_6=4\times\left(\frac{3}{2}\right)^5=\frac{243}{8}$

5 $ar^2=25, ar^4=156.25. r^2=\frac{156.25}{25}=6.25, r=\pm2.5$. There are two geometric sequences.

$a=4, r=2.5$ or $a=4, r=-2.5$.

6 $r=\frac{T_7}{T_6}=\frac{2}{3}. a\times\left(\frac{2}{3}\right)^5=\frac{32}{9}, a=27$

7 (a) $r=\frac{320}{160}=2. a\times2^5=160, a=5$. Sequence is 5, 10, 20
(b) $ar^4=4, ar^7=-\frac{1}{2}, r^3=-\frac{1}{8}, r=-\frac{1}{2}. a\times\frac{1}{16}=4,$

$a=64$. Sequence is 64, −32, 16.

8 $\frac{5p}{2p+1}=\frac{12p-4}{5p}, 25p^2=24p^2+4p-4, p^2-4p+4=0,$

$(p-2)^2=0, p=2.$

9 D

$\frac{5k+1}{2k+3}=\frac{8k-1}{5k+1}, (5k+1)^2=(2k+3)(8k-1),$

$25k^2+10k+1=16k^2+22k-3, 9k^2-12k+4=0.$

$(3k-2)^2=0, k=\frac{2}{3}.$

10 $x-1=y-x, \frac{y}{1}=\frac{x}{y}. y=2x-1, x=y^2. x=(2x-1)^2.$

$4x^2-5x+1=0. (4x-1)(x-1)=0, x=1,\frac{1}{4}.$

$y=1,-\frac{1}{2}. x=1$ and $y=1$ which gives both sequences as 1, 1, 1 or $x=\frac{1}{4}$ and $y=-\frac{1}{2}$.

11 $r=\frac{\sqrt{7}-\sqrt{5}}{\sqrt{7}+\sqrt{5}}=\frac{\left(\sqrt{7}-\sqrt{5}\right)^2}{7-5}=6-\sqrt{35}$

12 $a=10000, r=0.9. T_6=10000\times0.9^5\approx5905$

13 $a=40000. r=0.85. T_7=40000\times0.85^6=\$15086$

14 $\log ar=\log a+\log r, \log ar^2=\log a+2\log r,$
$\log ar^3=\log a+3\log r.$
The sequence becomes $\log a, \log a+\log r, \log a+2\log r, \log a+3\log r \ldots$ which is arithmetic with first term $\log a$ and $d=\log r$.

15 $\frac{6+x}{2+x}=\frac{13+x}{6+x}, (6+x)^2=(13+x)(2+x), 36+12x=26+15x,$

$x=\frac{10}{3}$ is the number.

16 (a) $a=200, r=2.$
After 4 hours, number present $=200\times2^4=3200$
(b) After 10 hours, number present $=200\times2^{10}=204800$
(c) $N=200\times2^t$

17 (a) 1 hour $=3\times20$ minutes. Fraction present $=\frac{1}{2^3}=\frac{1}{8}$
(b) $0.025\%=\frac{0.025}{100}=\frac{1}{4000}. \frac{1}{4000}=\frac{1}{2^t}, 2^t=4000,$

$t=\frac{\log_{10}4000}{\log_{10}2}\approx12.0.$

Length of time $=20\times12=240$ minutes $=4$ hours.

18 Birds $=30000, r=0.85$. Modelled by a geometric sequence with $T_1=30000\times0.85$
(a) $n=5: T_5=30000\times0.85^5=13311\approx13000$
(b) $T_n=9000: 9000=30000\times0.85n. 0.85n=0.3,$
$n=\frac{\log_{10}0.3}{\log_{10}0.85}=7.4$ years
(c) $P=3000\times0.85^t$ where $t\geq0$

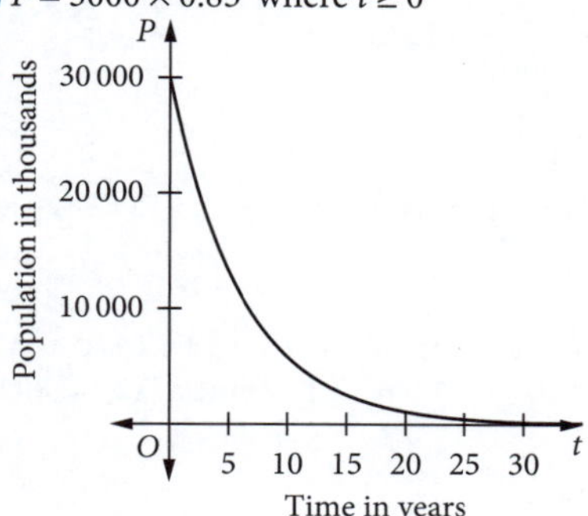

EXERCISE 18.6

1 geometric, $a=4, r=1.5: S_6=\frac{4\left(1.5^6-1\right)}{0.5}=83\frac{1}{8}$

2 geometric, $a=8, r=-0.5: 5\frac{21}{64}$ 3 A 4 $n=9; 10\frac{11}{16}$

5 176.37 6 $\frac{x\left[\left(\frac{2x}{3}\right)^n-1\right]}{\frac{2x}{3}-1}$, 189

7 (a) correct (b) correct (c) incorrect (d) correct

8 $10\log_{10}3+45\log_{10}2$ 9 $r=\pm2$

10 (a) 8 (b) (i) 1024 (ii) 2047 (iii) 22.5 cm 11 0

EXERCISE 18.7

1 (a) $5\frac{1}{3}$ (b) 16 (c) $17\frac{6}{7}$ (d) $\frac{4\sqrt{3}+6}{3}$ 2 $-\frac{1}{6}$

3 9, 6, 4; 45, −30, 20 4 D

5 16, 8, 4; 48, −24, 12 6 36 7 $\frac{a+1}{a}; a<-2$ or $a>0$

8 $r=4-\sqrt{15}; \frac{4\sqrt{3}+3\sqrt{5}}{3}$ 9 (a) $2\frac{7}{18}$ (b) $4\frac{62}{99}$ (c) $\frac{413}{990}$

10 $p=32, q=99$ 11 $\frac{58}{45}$ 12 $27\frac{49}{93}$

13 (a) $12\times\left(\frac{2}{3}\right)^3=3\frac{5}{9}$. The ball rebounds about 3.56 m.
(b) Downwards: $12+8+8\times\frac{2}{3}+\ldots=\frac{12}{1-\frac{2}{3}}=36,$

Upwards: $8+8\times\frac{2}{3}+8\times\left(\frac{2}{3}\right)^2+\ldots=\frac{8}{1-\frac{2}{3}}=24$

Total $=60$ m

14 (a) $1 \times \left(\frac{3}{4}\right)^2 = \frac{9}{16}$ square units

(b) $1 \times \left(\frac{3}{4}\right)^9 = \frac{19\,683}{262\,144} \approx 0.075$ square units

(c) The remaining shaded area becomes less, approaching zero, but never (theoretically) reaching it.

EXERCISE 18.8

1 C

2 (a) $A_0 = 1000$, $A_n = 1.031 \times A_{n-1}$

(b) $A_0 = 5000$, $r = \frac{6.1}{2} = 3.1\%$ per 6 months. $R = 1.031$.

$A_n = 1.031 \times A_{n-1}$

(c) $A_0 = 10\,000$, $r = \frac{2.7}{12} = 0.225\%$ p.m. $R = 1.002\,25$.

$A_n = 1.002\,25 \times A_{n-1}$

3 (a) $R = 1.027 = 1 + \frac{2.7}{100}$. $r = 2.7\%$ p.a.

(b) $R = 1.015 = 1 + \frac{1.5}{100}$.

Interest rate is 1.5% per six months $= 1.5 \times 2 = 3\%$ p.a.

(c) $R = 1 + \frac{0.5}{100}$.

Interest rate is 0.5% per month $= 0.5 \times 12 = 6\%$ p.a.

4 (a) 12442 (b) 49650 5 (a) 5117 tonnes (b) 41 809 tonnes

6 $27764 \approx 28000$ 7 5 to 6 years

8 (a) Megan: $348 600
Paul: $347 241.04
(b) the 3rd year

9 (a) $1 423 747 (b) $764 117 less

10 (a) Henry earns $46 000, Eleanor earns $44 994.56
(b) Henry earns $490 000, Eleanor earns $480 244.28
(c) the start of the 13th year

11 (a) $9853.23 (b) 9 years

12 (a) $200\,000 \times 0.007 = \$1400$
(b) to pay off more than the interest so that the balance owed is reduced
(c) $A_1 = 200\,000 \times 1.007 - 3000$;
$A_2 = (200\,000 \times 1.007 - 3000) \times 1.007 - 3000$
$= 200\,000 \times 1.007^2 - 3000(1 + 1.007)$
(d) $A_n = 200\,000 \times 1.007^n - \frac{3000(1.007^n - 1)}{0.007}$
(e) 91 months

13 (a) $252.15 (b) $32217.89

EXERCISE 18.9

1 $888.49 2 $1331.18

3 (a) $Y = \frac{400 \times 1.004(1.004^{360} - 1)}{0.004} = 322\,142.43$

(b) (i) $A_n = Y \times 1.004^n - 3000(1 + 1.004 + \ldots + 1.004^{n-1})$

$= Y \times 1.004^n - \frac{3000(1.004^n - 1)}{0.004}$

$= (Y - 75\,000) \times 1.004^n + 750\,000$

(ii) 141 months

4 (a) $399 422 (b) $3334.15
(c) 247 months (d) about $95 700

5 The lump sum investment is best.

6 (a) yes (b) $11 862.67 (c) $2455.80

7 (a) Amount deposited $= A_0 = \$100\,000$.

Interest rate per month $= r = \frac{2.7}{1200} = 0.00225$ so

$R = 1 + r = 1.00225$

Time period $= n$ months so that A_1 is the balance at the end of the first month after interest has been added and the first payment made.

Monthly payment: $M = \$300$

After 1 month: $A_1 = 100\,000 \times 1.00225 - 300$

This can be written as: $A_1 = A_0 \times R - M$

$= A_0 \times 1.00225 - 300$

After 2 months: $A_2 = A_1 \times 1.00225 - 300$

After n months (and n payments) this becomes:

$A_n = A_{n-1} \times 1.002\,25 - 300$

(b) $A_1 = 100\,000 \times 1.002\,25 - 300$
$= 99\,925$

$A_2 = 99\,925 \times 1.002\,25 - 300$
$= 99\,849.83$

After two withdrawals, the balance remaining is $99 849.83

(c) $r = 5\%$ p.a., $r = \frac{5}{1200}$ p.m. $= \frac{1}{240}$ p.m.

$R = 1 + \frac{1}{240} = \frac{241}{240}$

$A_1 = 100\,000 \times \frac{241}{240} - 300$

$= 100\,116.67$

After the first payment the balance remaining is larger than the initial deposit so the amount in the account will continue to grow.

(d) A monthly payment of $416.66 will leave $100 000 in the account if the annual interest rate remained at 5%. (Need to round down since the balance remaining has to be at least $100 000.)
If the interest rate is only 2.7%, then a monthly payment of $225 will leave $100 000 in the account.

CHAPTER REVIEW 18

1 (a) There is no common difference and no common ratio. The sequence is neither arithmetic nor geometric.

(b) The sequence is arithmetic with common difference $-3 + 2\sqrt{3}$.

(c) 1.6, 2.4, 3.6, 5.4, …
$T_2 - T_1 = 2.4 - 1.6 = 0.8$
$T_3 - T_2 = 3.6 - 2.4 = 1.2$
There is no common difference.
The sequence is not arithmetic.
$\frac{2.4}{1.6} = \frac{3.6}{2.4} = \frac{5.4}{3.6} = 1.5$. The sequence is geometric with common ratio 1.5.

2 (a) $a = 22, d = -7$. $T_{10} = 22 + 9 \times (-7) = -41$
(b) $22 + (k - 1)(-7) = -90$. $k = 17$

3 (a) $a = 36, r = \frac{126}{36} = 3.5$. $T_7 = 36 \times 3.5^6 = 66\,177.5625$

(b) $36 \times 3.5^{k-1} = 1\,000\,000$, $k - 1 = \frac{\log_{10}\left(\frac{1\,000\,000}{36}\right)}{\log_{10} 3.5}$,

$k \approx 9.2$. The smallest value is $k = 10$

4 (a) $a = 862.5, d = -87.5$. The arithmetic sequence is 862.5, 775, 687.5.

(b) $a = \pm 1200\sqrt{2}, r = \pm\frac{1}{\sqrt{2}}$. The two geometric sequences are

$1200\sqrt{2}, 1200, 600\sqrt{2}$ and $-1200\sqrt{2}, 1200, -600\sqrt{2}$

5 2250 6 12 7 380 m 8 2, 5, 8; 11, 5, −1

9 $\frac{3\sqrt{3}-5}{2}, \frac{7-4\sqrt{3}}{2}, \frac{2\sqrt{3}}{3}$ **10** $\frac{7}{30}$ **11** $r=3$

12 $T_n = 2n-4, a=-2, d=2$ **13** 93

14 2082000 **15** 0.8 **16** $2^{n+1}+n^2-2$

17 Working in centimetres: West: $100+\frac{100}{2^4}+\frac{100}{2^8}+\ldots=\frac{320}{3}$,

South: $50+\frac{50}{2^4}+\frac{50}{2^8}+\ldots=\frac{160}{3}$, East: $25+\frac{25}{2^4}+\frac{25}{2^8}+\ldots=\frac{80}{3}$,

North: $12.5+\frac{12.5}{2^4}+\frac{12.5}{2^8}+\ldots=\frac{40}{3}$.

The snail ends up 80 cm W, 40 cm S of its original position.

18 (a) 6 years at 4% gives a multiplication factor of 6.6330 from the table.
FVA = $4500 × 6.6330 = $29 848.50

(b) 5 years at 2% gives a multiplication factor of 5.2040 from the table.
FVA = $700 × 5.2040 = $3642.80

19 (a) 5 years at 3% gives a multiplication factor of 4.5797.
PVA = $6000 × 4.5797 = $27 478.20

(b) Let M be the yearly repayments.
Using the PVA table with $r = 5\%$ and the period 6 years, the present value of $1 is 5.0757.
Hence $5.0757 \times M = 10000$.
$M=\frac{10000}{5.0757}=\$1970.17$

20 (a) 74 months at 0.005 gives a multiplication factor of 61.725 71 from the table.
Present value of the annuity = $200 × 61.725 71 = $12 345.14

(b) 5y 11m = 5 × 12 + 11 = 71 months, $N = 71$
9% p.a. = 0.75% p.m. so $r = 0.0075$ as a decimal.
From the table, the value to be use is 54.892 93.
Hence $22000 = M \times 54.89293$
$M=\frac{19000}{54.89293}=346.13$
The repayments would be $347 per month, rounded to the next dollar.

21 $A = 3000 \times 1.004^{24} = \$3301.64$

22 $20000 = P \times 1.003^{60}, P=\frac{20000}{1.003^{60}}=\$16710$

23 $180.76

24 (a) $1788.21 (b) $1628.83

25 (a) $21 795 (b) $182 713.05 (c) 11 years, $10 623.86

26 (a) $3+\sqrt{3}$

(b) common ratio $r=\frac{1}{\sqrt{3}-1}=\frac{\sqrt{3}+1}{2}>1$, so there is no limiting sum

27 (a) $A_n = 200000\times1.005^n-\frac{M\left(1.005^n-1\right)}{0.005}$ (b) $1687.71

CHAPTER 19

EXERCISE 19.1

1

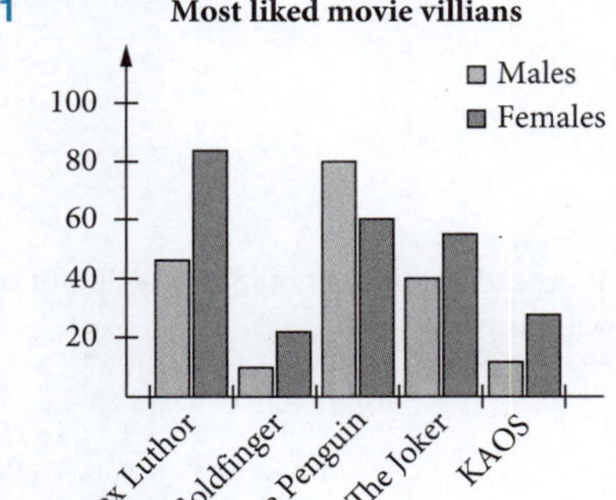

The most popular villain voted by teachers is Lex Luthor.
The most popular villain voted by students is The Penguin.

2

Percentage of Year 7 students	Travel time (minutes)	Percentage of Year 12 students
23.4	0–<10	1.7
37.4	10–<20	6.7
22.4	20–<30	20.8
11.2	30–<40	45.8
3.7	40–<50	20
1.9	50–<60	5

3 (a)

	Age group						
	18–24	25–34	35–44	45–54	55–64	65–74	>74
Male	7	13	19	29	33	37	42
Female	5	4	11	22	27	41	52

(b) The age group 65–74 years is when the percentage of women with high blood pressure starts to be higher than the percentage of men with high blood pressure.

(c) 7 out of 100 males, 5 out of 100 females, so a total of 12 out of 200. This gives 6% of the sample population with high blood pressure.

4 (a)

	Sugar tablet	Vitamin C tablet
Cold (%)	23	19
No cold (%)	77	81

(b) Of those who had taken the sugar tablet, 23% developed a cold. Of those who had taken the vitamin C tablet, 19% developed a cold. The number of students who developed a cold is quite small and a larger sample is needed. Four times as many students did not develop a cold and the result shows that 81% had taken the vitamin C tablet to 77%, showing no statistical difference. A much larger study would be needed to research the effectiveness of such products.

5 (a)

	Overweight to obese					
	males		females		Total	
	Number	%	Number	%	Number	%
Diabetic	14	14	8	6.7	22	10
Pre-diabetic	9	9	7	5.8	16	7.3
Healthy range	77	77	105	87.5	182	82.7
Total	100	100	120	100	220	100

(b)

	Normal to underweight					
	Males		Females		Total	
	Number	%	Number	%	Number	%
Diabetic	20	6.7	10	3.6	30	5.2
Pre-diabetic	13	4.3	5	1.8	18	3.1
Healthy range	267	89	265	94.6	532	91.7
Total	300	100	280	100	580	100

(c) 22 out of 220 or 10% of overweight or obese people have diabetes compared to normal weight or underweight people where 30 out of 580, or 5.2% have diabetes. Excess weight appears to significantly increase your chance of developing diabetes.

6 (a)

	0 aces	1 ace	2 aces	3 aces	4 aces	Total
Slow serve	7	8	8	4	0	27
Fast serve	1	8	14	30	20	73
Total	8	16	22	34	20	100

(b) 34 players served 3 aces.

(c) slow servers: 14.8%, fast servers: 68.5%

(d) Fast servers serve more aces than slow servers.

7 (a)

	Weekly salary May '10	Weekly salary Nov. '10	Weekly salary May '11	Weekly salary Nov. '11	Weekly salary May '12	Weekly salary Nov. '12	Weekly salary May '13
Men	1400	1400	1450	1450	1500	1550	1550
Women	1150	1200	1200	1200	1250	1300	1300
Difference	250	200	250	250	250	250	250

(b) For 2010 to 2013 men earned more, on average, than women by $250 per week.

8 (a)

Length of river (km)	100–<150	150–<200	200–<250	250–<300	>300	Total
North Island	6	7	1	1	1	16
South Island	5	3	2	1	1	12
Total	11	10	3	2	2	28

(b) There are 10 rivers in New Zealand in the group of 150–<200 km.
(c) There are 7 rivers in New Zealand which are longer than 200 km.
(d) The North Island has 16 rivers which are longer than 100 km. The South Island has 12 rivers which are longer than 100 km.
(e) A larger area does not indicate a greater number of substantial rivers; in fact, in this case the opposite is true.

EXERCISE 19.2

1

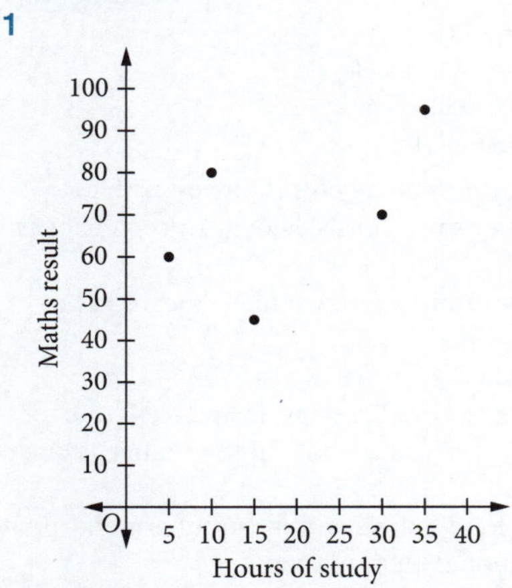

The maths results show a moderate, positive, linear association.

2 **(a)** A **(b)** A
(c) A **(d)** B

3 **(a)** no association **(b)** strong, positive, linear
(c) moderate, negative, linear **(d)** strong, positive, non-linear

4 **(a)**

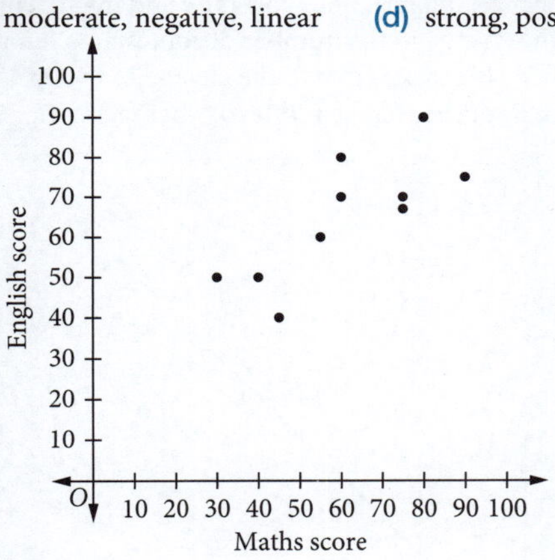

(b)

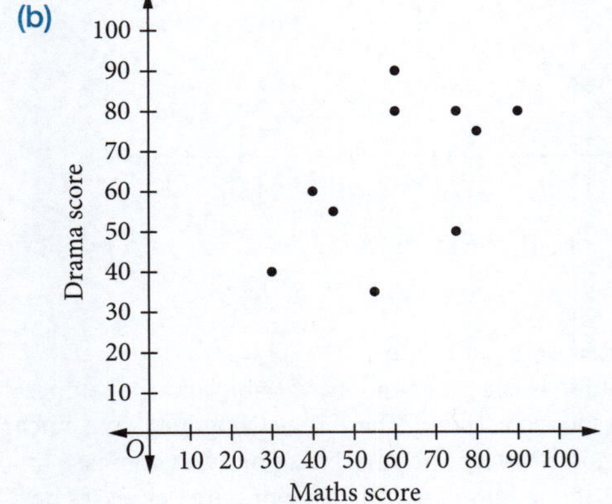

(c)

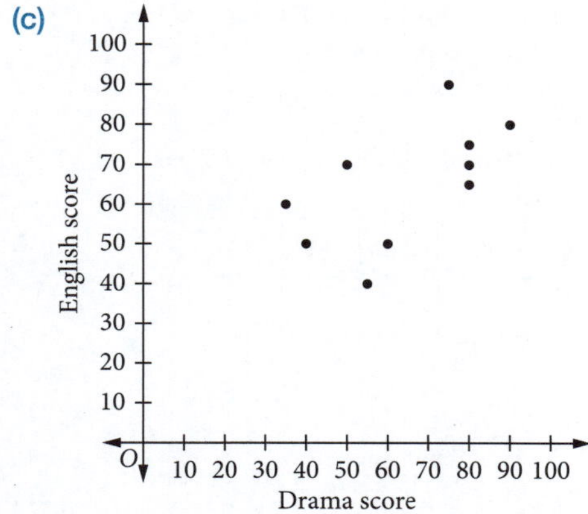

(d) The Maths versus English scores show the closest clustering around a positive line of best fit.

5 **(a)**

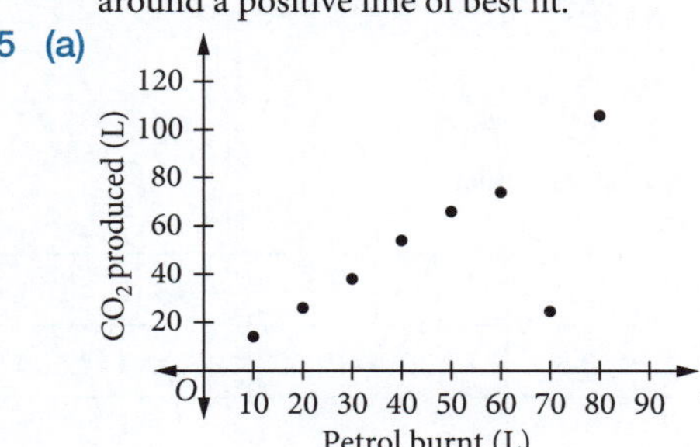

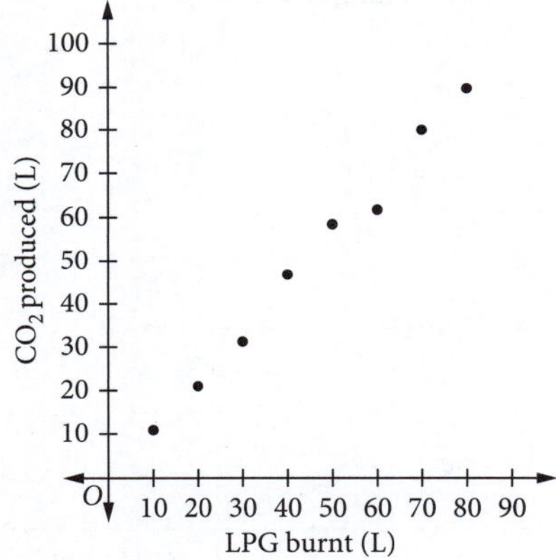

(b) Petrol (70, 25)
(c) Strong, positive, linear for both (if outlier is ignored)
(d) LPG; The gradient is roughly 1.1 L of CO_2 per litre of fuel while for petrol it is closer to 1.3 L of CO_2 per litre of fuel.

6 **(a)**

Year	1986	1991	1996	2001	2006	2011
Non-Indigenous	7	12	14	16	18	22
Indigenous	2	3	3	4	4	5

(b)

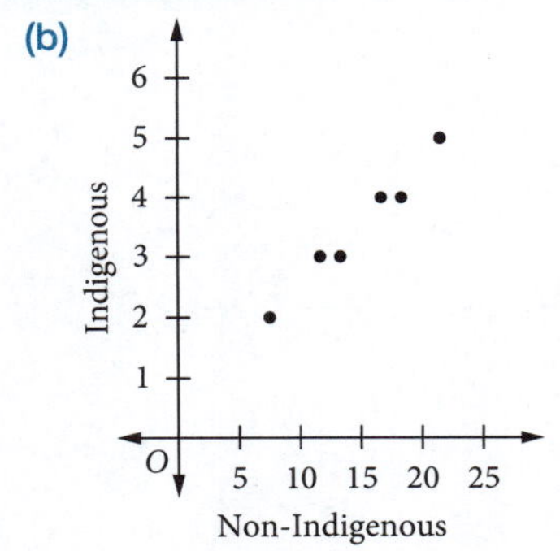

(c) The trend is strong, positive and linear.
(d) The non-Indigenous population.

EXERCISE 19.3

1 (a) $r = 0.98$ (b) $r = 0.94$ (c) $r = -0.13$ (d) $r = 0.90$
2 (a) D (b) A
3 (a) (i)

x	2	5	10	18	22
y	18	12	7	4	2

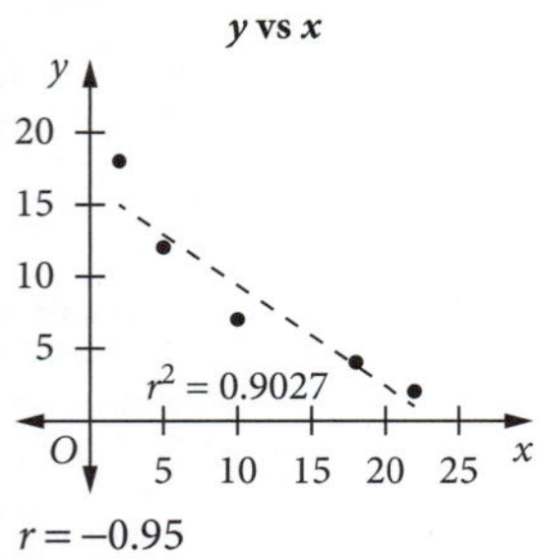

$r = -0.95$
(ii) strong, negative, linear
(b) (i)

Length (m)	1	2	4	7	9	15
Cost ($)	2.80	5.60	11.20	19.60	25.20	42.00

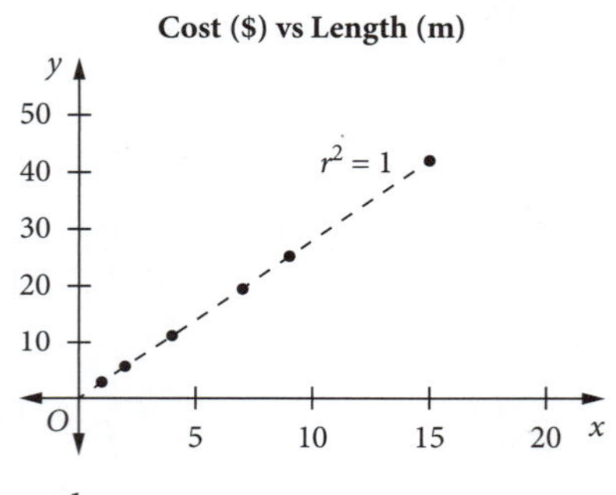

$r = 1$
(ii) strong, positive, linear
(c) (i)

Swimmers in the pool	10	25	45	80	100	120	180
Bacteria level (ppm)	0.2	0.5	0.8	0.6	2.3	3.0	2.8

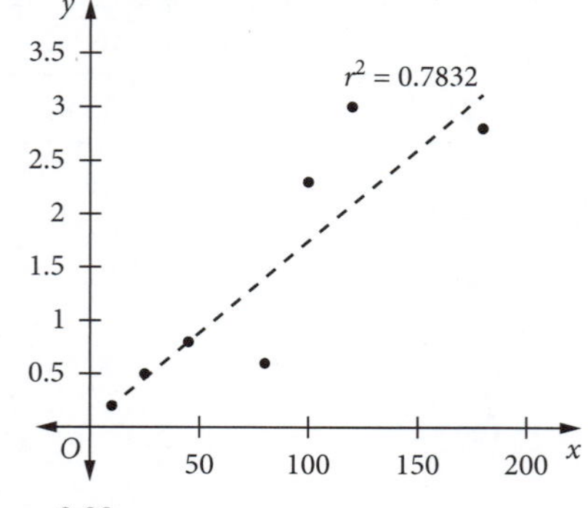

$r = 0.89$
(ii) strong, positive, linear

4 (a) $r = 0.46$ (b) (500, 182)
(c) A. Outliers are generally excluded from the data set, so calculations are an accurate representation of the study.
(d) Correlation, $r = 0.99$

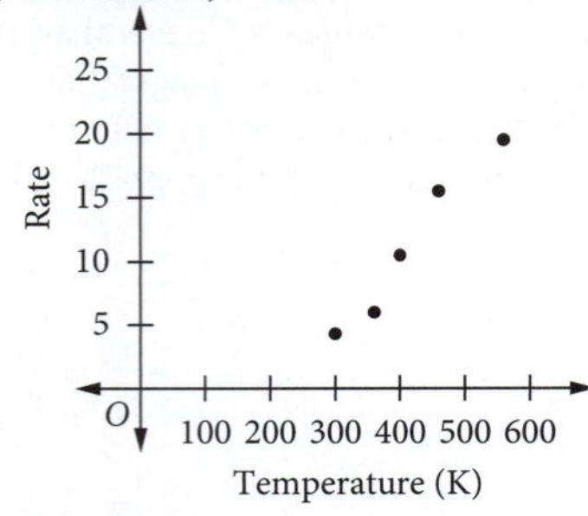

(e) Excluding the outlier means the points line up with a stronger linear relationship. This is evident since r changes from 0.46 to 0.99.
(f) The number 182 (mol/min) was most likely meant to be 18.2 (mol/min).
5 (a) Set B (b) Set A (c) Set C (d) Set B
6 (a) 64% of the change in the quality of performance is due to the change in hours of rehearsal. 36% of the change is due to other factors.
(b) The number of fast food outlets and the number of hospitals in a town depend upon the population.
(c) 36% of the change in the number of registered motorcycles is due to the change in the number of registered cars. 64% of the change is due to other factors.
(d) The results in maths and physics exams depend upon other factors such as interest, hours of study and intelligence.
(e) 96% of the change in the number of contestants remaining in a reality TV series is due to the change in the episode number. 4% of the change is due to other factors.

EXERCISE 19.4

1 (a)

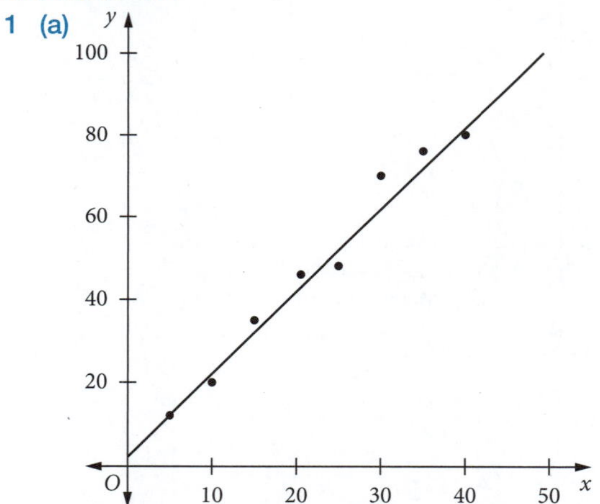

(b) y-intercept is 3, the gradient is 2.
(c) $y = 2x + 3$
2 (a) $r = -1$
temperature = $-6.5 \times$ altitude + 15
(b) The altitude is the independent variable and the temperature is the response variable. There is a strong, negative, linear association between the variables. For every increase in altitude of one kilometre, the temperature decreases by 6.5°C.

3

Length (m)	Diameter (mm)	Interpolation/Extrapolation
0.9	1	Interpolation
2.3	3	Interpolation
10	14	Extrapolation

4 (a) \$875 (b) \$150

5 (a) $r = 0.99$, $y = 2.7 + 0.91x$ (b) $r = 0.72$, $y = 11.31 + 0.62x$

(c) Set A shows a greater association than Set B. Set B has an outlier at (70, 32) which significantly affects the value of r.

6 (a) D (b) B

7 (a) (i) This is bivariate data, since there are two measurements for each person.

(ii) The data is ranked according to increasing arm span.

(iii) The other variable follows an increasing trend.

(iv) As both sets of data are increasing together, this would suggest a positive linear trend.

(b) To show any association you could draw a scatterplot.
As this is recorded data you need to check for outliers.
To find the equation for the line of best fit you find the least squares regression line.
To check on the strength of the association you calculate the correlation coefficient, r.
The independent variable is arm span.
The response variable is height.
The independent variable should be plotted on the the horizontal axis.
The response variable should be plotted on the vertical axis.

(c) (i) 128 cm to 172 cm (ii) 132 cm to 170 cm

(d) D (e) The association is strong, positive and linear.

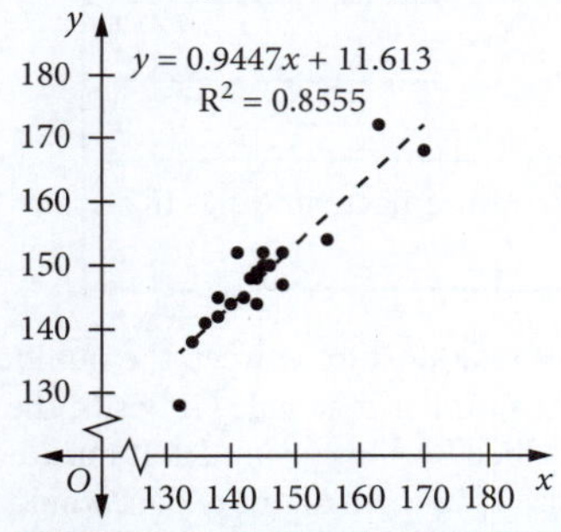

Arm Span, a (cm)	Height, h (cm)
132	128
134	138
136	141
138	142
138	145
140	144
141	152
142	145
143	148
144	144
144	148
144	149
145	152
145	150
146	150
148	147
148	152
155	154
164	172
170	168

(f) $r = 0.92$ (g) $h = 11.61 + 0.94a$

(h) A correlation coefficient of 0.92 shows a strong correlation, meaning that arm span is a good predictor of height.

8 (a) The independent variable is latitude.

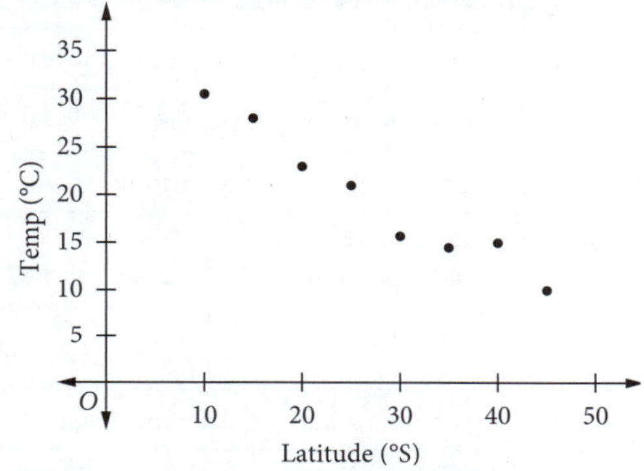

(b) There is a strong, negative, linear association. As the latitude increases the average maximum temperature decreases.

(c) $r = -0.98$

(d) $y = -0.6x + 36.0$
average maximum temperature (°C) = $-0.6 \times$ latitude (°S) + 36.0

(e) The independent variable is latitude.

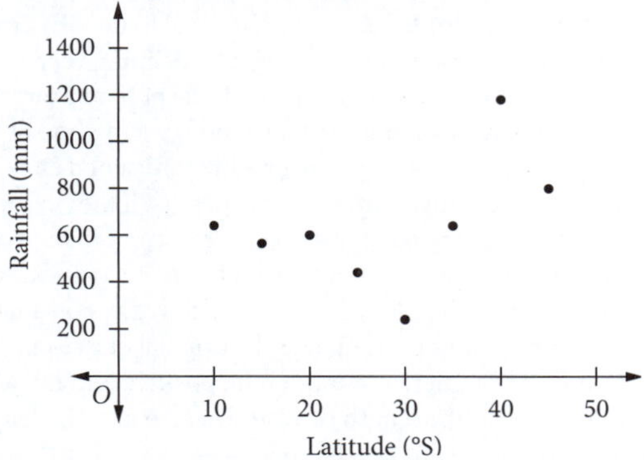

(f) There is a moderate, positive, linear association. As the latitude increases the average rainfall increases.

(g) $r = 0.46$

(h) $y = 10.1x + 360$
average annual rainfall (mm) = $10.1 \times$ latitude (°S) + 360

(i) The higher association exists between latitude and temperature. This is evident because the r value is furthest from 0.

(j) between 10°S and 45°S (k) $r = -0.29$

9 (a) Note: Answers for the correlation coefficient estimate will vary but should be similar to those in the table.

Association	Independent variable	Strength and direction of association	Correlation coefficient estimate
Area and population	area	weak, positive	0.2
Area and motorcar thefts	area	weak, positive	0.2
Area and average weekly earnings	area	moderate, positive	0.4
Population and motorcar thefts	population	strong, positive	0.9
Population and average weekly earnings	population	weak, negative	−0.2
Motorcar thefts and average weekly earnings	earnings	weak, negative	−0.2

(b)

Association	r
Population versus area	0.08
Motorcar thefts versus area	0.10
Average weekly earnings versus area	0.41
Motorcar thefts versus population	0.96
Average weekly earnings versus population	−0.17
Motorcar thefts versus average weekly earnings	−0.16

(c) population and motorcar thefts
(d) motorcar thefts = 2.64 × population (in millions) − 12.87

EXERCISE 19.5

1 D C F A

Step 1	Clarify the problem and formulate one or more questions that can be answered with data.
Step 2	Design and implement a plan to collect or obtain appropriate data.
Step 3	Select and apply appropriate graphical or numerical techniques to analyse the data.
Step 4	Interpret the results of this analysis and relate the interpretation to the original question; communicate findings in a systematic and concise manner.

2 (a) False (b) True (c) False (d) False (e) True (f) True

3 When researching an issue you must begin by defining very narrowly the question you wish to answer. If the potential target population is too large to handle you must design a method of randomly selecting a sample that will provide sufficient data, even if you need to eliminate any outliers. Identify, if they exist, the independent and dependent variables. Use a combination of appropriate statistical tools including scatterplots, back-to-back stem plots, frequency diagrams, line graphs and regression analysis. Summarise your findings in relation to the original question.

4 (Answers may vary.) There is a moderate, positive and linear relationship between the length of your forearm and the length of your right foot.

5 (a) While there is a minimal relationship between the number of male GP visits and the number of male specialist visits, the relationship is best described by a linear model. It must be noted that the sample size is very small; more data would be required for a more thorough analysis.
(b) There is a moderate positive relationship between the number of female GP visits and the number of female specialist visits; the relationship is best described by a linear model. It must be noted that the sample size is very small; more data would be required for a more thorough analysis.
(c) There is a strong positive relationship between the number of male GP visits and the number of female GP visits, the relationship is best described by a linear model. It must be noted that the sample size is very small; more data would be required for a more thorough analysis.
(d) There is minimal relationship between the average number of emergency department visits and the average number of hospital admissions; the relationship is best described by a linear model. It must be noted that the sample size is very small; more data would be required for a more thorough analysis.
(e) There is a strong positive relationship between the average number of GP visits and the average number of dental visits, the relationship is best described by a linear model. It must be noted that the sample size is very small; more data would be required for a more thorough analysis.

6 Answers will vary.

CHAPTER REVIEW 19

1 B 2 B 3 D 4 C 5 A 6 C
7 (a) C (b) C
8 B 9 C 10 D 11 C
12 (a) D (b) A

13 (a)

Age bracket	Enjoy	No opinion	Dislike	Total
10–20	4	18	2	24
20–30	5	6	5	16
30–40	3	20	27	50
>40	4	16	40	60
Total	16	60	74	150

(b) Age

(c)

Age bracket	Enjoy (%)	No opinion (%)	Dislike (%)
10–20	25	30	2.7
20–30	31.3	10	6.8
30–40	18.8	33.3	36.5
>40	25	26.7	54.1
Total	100	100	100

(d) Older people generally dislike hip-hop.

14 (a) weight

(b) This scatterplot shows a moderate, positive, linear association.

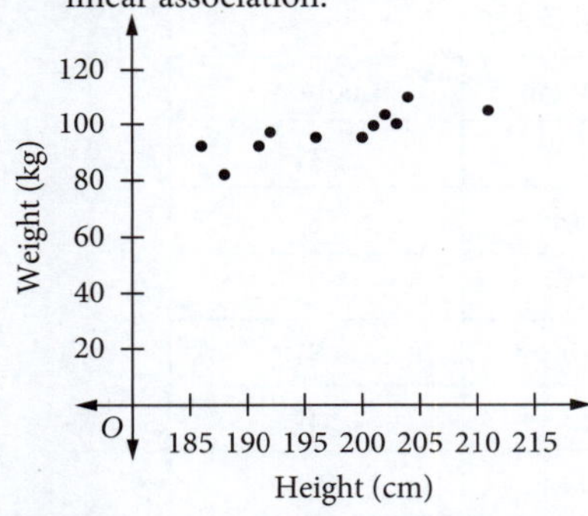

(c) $r = 0.83$ (d) $w = 0.8h - 59$

15 $r = 0.87$ for 'goals for' and 'points' on ladder, –0.79 for 'goals against' and 'points' on the ladder. Based on these statistics, the 'goals scored for' is the better predictor of ladder position.

16 (a) total hours of study

(b)

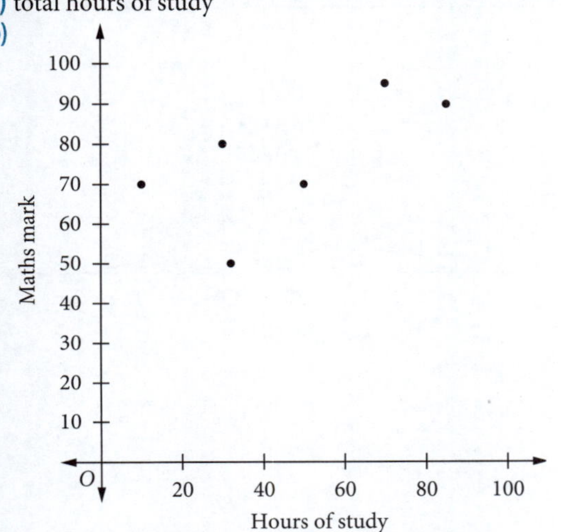

(c) The association is moderate, positive and linear.

(d) $r = 0.66$

(e)

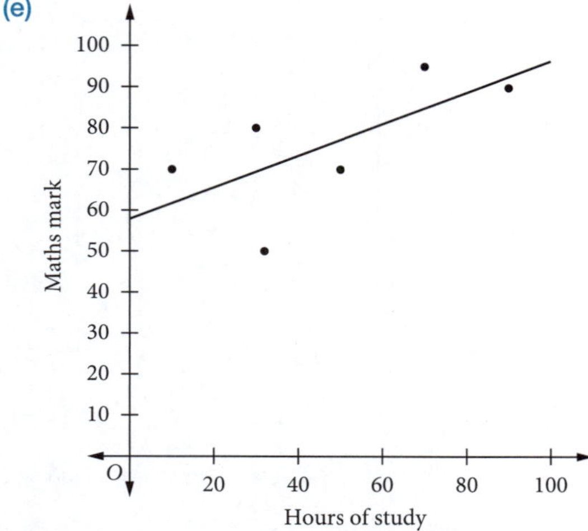

Maths mark = 0.4 × hours of study + 57.
(Answers may vary.)

(f) Inserting a line of best fit by eye will be slightly different by each individual.

(g) (i) English mark = 0.58 × hours of study + 48

(ii) $r = 0.83$

(h) The predicted English scores are 71.2 for 40 hours of study, 82.8 for 60 hours of study, and 117.6 for 120 hours of study.

(i) The prediction for 120 hours of study is outside the data set and is an extrapolation. In this case it is not possible to get a score of 117.6 out of 100.

(j) Both comparisons have a positive correlation; this indicates that more study improves your results.

(k) There is a clear association, but the association will have some confounding variables that are not controlled. For example, the effectiveness of the study time and the hours spent on each individual subject (only the total is given).

CHAPTER 20

EXERCISE 20.1

1 (a) 0.7248 (b) 0.8868 (c) 0.7680 (d) 0.3789

2 (a) $\frac{2}{27}$ (b) $\frac{7}{27}$ (c) $\frac{1900}{2197}$ (d) $\frac{492500}{970299}$

3 (a) $\mu = 0.5192$ (b) $\sigma = 0.2903$

4 $\mu = 150$

5 The median is 0.5296.

6 The median is 100.

7 C

8 (a) 0.16 (b) 0.65 (c) 0.2744

9 B

10 (a) $k = \frac{1}{6}$ (b) 0.5 (c) $\frac{5}{6}$

11 The median is $\frac{\pi}{2}$.

12 (a) 3.375 m (b) 2.19 m (c) 3.126 m

13 (a) $\frac{1}{10}$

(b) (i) 0.2212 (ii) 0.5273 (iii) 0.3221

(c) 10

(d) 100 (e) 0.6321

14 (a) $k = 12$ (b) 0.65 seconds (c) 0.636 seconds

(d) 0.949 (e) 0.649 (f) 0.313

15 (a)
$$f(x) = kx(4-x)^2 = kx(16-8x+x^2) = k(16x-8x^2+x^3)$$

$$\int_0^4 kx(4-x)^2\,dx = k\left[8x^2 - \frac{8x^3}{3} + \frac{x^4}{4}\right]_0^4$$
$$= k\left[\left(8\times 4^2 - \frac{8\times 4^3}{3} + \frac{4^4}{4}\right) - 0\right]$$
$$= k\left[128 - \frac{512}{3} + 64\right]$$
$$= k \times \frac{64}{3}$$
$$= 1$$
$$k = \frac{3}{64}$$

(b) $\mu = \frac{8}{5}$ and $\sigma = \frac{4}{5}$ (c) $\frac{13}{256}$ (d) \$489.84

(e) No, as they will effectively process only $\frac{2}{3}$ the number of trucks and so will have an effective return of \$433.33 per truck.

16 (a) $f(x)$ is a probability density function because it satisfies the two properties $f(x) \ge 0$ on R and $\int_{-\infty}^{\infty} f(x)dx = 1$.

(b) $f(x) = \begin{cases} 3x, & 0 \le x \le \frac{2}{3} \\ -6x+6, & \frac{2}{3} < x \le 1 \\ 0, & \text{otherwise} \end{cases}$

Alternative answers could include $x = 0, \frac{2}{3}, 1$ in any of the adjacent regions.

(c) $P\left(X < \frac{2}{3}\right) = \frac{2}{3}$, $P\left(X > \frac{2}{3}\right) = \frac{1}{3}$

(d) $P\left(X < \frac{2}{3}\right) = \frac{2}{3}$ and $P\left(X > \frac{2}{3}\right) = \frac{1}{3}$

Regardless of the method used (triangles or integration) the answers should be the same.

(e) $P\left(\frac{1}{2} < X < \frac{5}{6}\right) = \frac{13}{24}$

EXERCISE 20.2

1 (a) between 39 and 51 seconds. (b) 16% (c) 0.15%

2 (a) range 8 to 12 (b) range 6 to 14 (c) range 4 to 16

3 (a) 95% (b) 0.15% (c) 50% (d) 83.85%

4 C

5 B **6** $85 + 8 = 93\%$ **7** 16% **8** 2.5%

9 (a) (i) 68% of the packets produced by Machine A should weigh between 97 g and 103 g.
(ii) 68% of the packets produced by Machine B should weigh between 100 g and 108 g.
(b) (i) 95% of the packets produced by Machine A should weigh between 94 g and 106 g.
(ii) 95% of the packets produced by Machine B should weigh between 96 g and 112 g.
(c) (i) 99.7% of the packets produced by Machine A should weigh between 91 g and 109 g.
(ii) 99.7% of the packets produced by Machine B should weigh between 92 g and 116 g.
(d) Machine B is likely to produce packets with more chips.

10 (a)

(b) The histogram displays a distinct bell shape although it is not perfectly symmetrical. However, it would seem reasonable to at least entertain the notion that the sample does follow a normal distribution.

(c) Mean 1.765 m, standard deviation 0.074 m

(d) (i) 75.7% of the data lies within one standard deviation of the mean.
(ii) 94.3% of the data lies within two standard deviations of the mean.
(iii) 100% of the data lies within three standard deviations of the mean.

(e) The closest result was the two standard deviation range but this is not surprising as every distribution has approximately 95% of values within two standard deviations of the mean. The other results are not close enough to the expected values to support the hypothesis that the heights follow a normal distribution. Even although height is supposed to be normally distributed, that does not mean any particular sample (like this one) will necessarily be normally distributed as well.

11 (a) average height = 1.82 m

(b)

Class intervals (m)	Frequency
1.60–<1.65	1
1.65–<1.70	3
1.70–<1.75	7
1.75–<1.80	9
1.80–<1.85	10
1.85–<1.90	9
1.90–<1.95	7
1.95–<2.00	3
2.00–<2.05	1

(c)

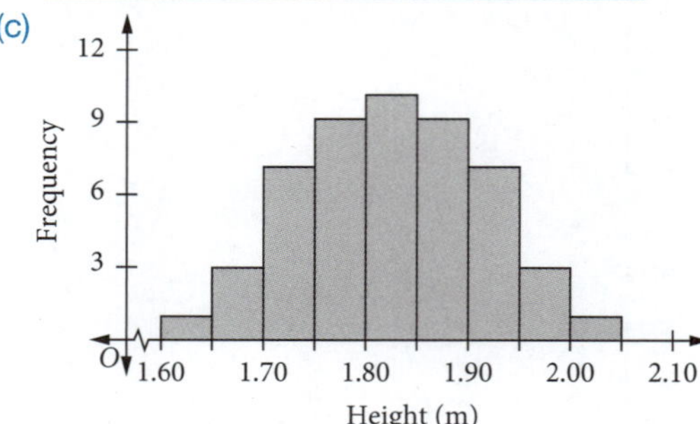

(d) $\sigma = 0.0884$

12 (a) $f(x) = e^{-x^2}$

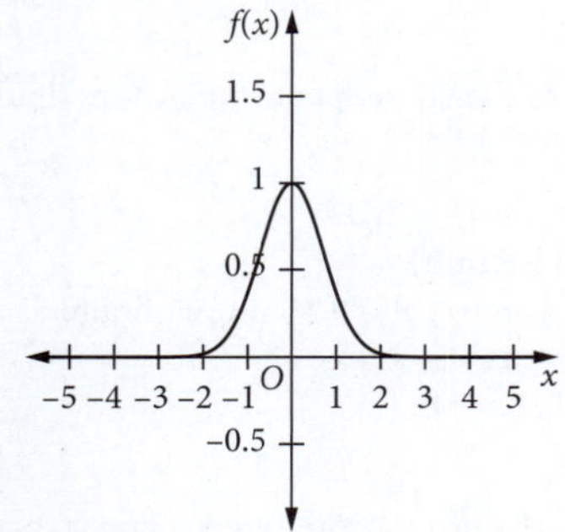

(b) $\int_{-4}^{4} e^{-x^2}\,dx = 1.772$

13 (a) (i) $f(x) = \frac{1}{3\sqrt{2\pi}} e^{-\frac{(x-10)^2}{18}}$

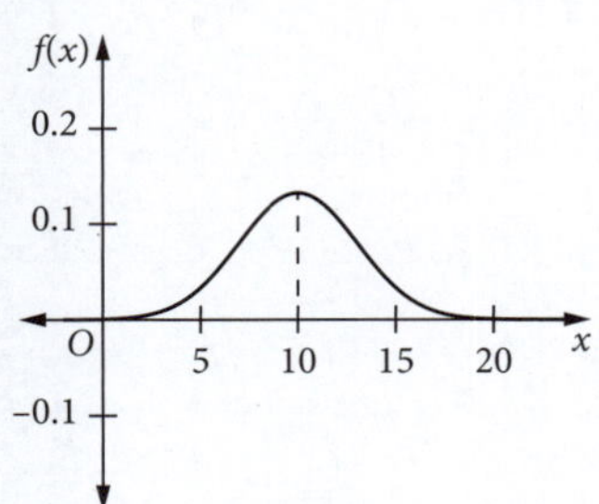

(ii) $f(x) = \frac{1}{\sqrt{2\pi}} e^{-\frac{x^2}{2}}$,

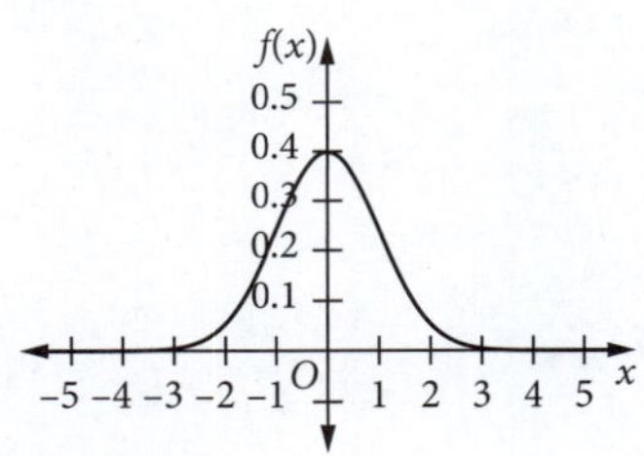

(b) (i) $\int_{1}^{19} \frac{1}{3\sqrt{2\pi}} e^{-\frac{(x-10)^2}{18}}\,dx = 0.9973$

(ii) $\int_{-3}^{3} \frac{1}{\sqrt{2\pi}} e^{-\frac{x^2}{2}}\,dx = 0.9973$

14 (a) $f(x) = \frac{1}{3\sqrt{2\pi}} e^{-\frac{(x-10)^2}{18}}$

x	1	4	7	10	13	16	19
$f(x)$	0.0015	0.0180	0.0807	0.1330	0.0807	0.0180	0.0015

$$\int_{1}^{19} \frac{1}{3\sqrt{2\pi}} e^{-\frac{(x-10)^2}{18}}\,dx$$

$$= \frac{3}{2}(0.0015 + 2 \times 0.0180 + 2 \times 0.0807 + 0.1330) \times 2 = 0.9957$$

(b) $f(x) = \frac{1}{\sqrt{2\pi}} e^{-\frac{x^2}{2}}$

x	−3	−3	−1	0	1	2	3
$f(x)$	0.0044	0.0540	0.2420	0.3989	0.2420	0.0540	0.0044

$$\int_{-3}^{3} \frac{1}{\sqrt{2\pi}} e^{-\frac{x^2}{2}}\,dx$$

$$= \frac{1}{2}(0.0044 + 2 \times 0.0540 + 2 \times 0.2420 + 0.3989) \times 2 = 0.9953$$

(c) Both are very close to the answer 0.9973 of part 13(b). They are the same, correct to 2 decimal places.

EXERCISE 20.3

1 (a) $z = 1$ (b) $z = 2$ (c) $z = 2\frac{1}{3}$ (d) $z = \frac{1}{3}$

2 (a) $z = 1$ (b) $z = 1.4$
(c) Felipe did better on the second test.

3 C

4 (a) 0.786 (b) 0.504 (c) 0.282

5 (a) 0.1977 (b) 0.1977 (c) 0.8023 (d) 0.6046

6 (a) 0.0568 (b) 0.9432 (c) 0.9432

7 B 8 D

9 (a) $z = -2.5$ (b) moderate (c) 7.9 kg (d) 8.7 kg

10 (a) $P(\text{mark at least } 50\%) = 0.9796$
50% is 2 standard deviations below the mean.

(b) $X = 45$, $\bar{x} = 72$, $\sigma = 11$. $z = \frac{45-72}{11} = -\frac{27}{11} = -2.45$

If $z = \frac{27}{11}$: $\frac{27}{11} = \frac{x-72}{11}$. $X = 99$

(c) To be in the top 2.5%, need to been 2 standard deviations above the mean, i.e. $72 + 2 \times 11 = 94\%$.
Minimum mark for A^{++} is 94%.

(d) $\bar{x} = 70$, $\sigma = 12$. $\bar{x} + 2\sigma = 70 + 24 = 94$. Yes, they would as they are in the top 2.5% of the group.

11 (a) $z = \frac{6.45-7.65}{1.2} = -1.1$ standard deviation below the mean,

$P(C > 6.45) = 1 - 0.16 = 0.84$

(b) \$8.05 is 0.33 SD above the mean.
\$6.65 is 0.83 SD below the mean.

(c)

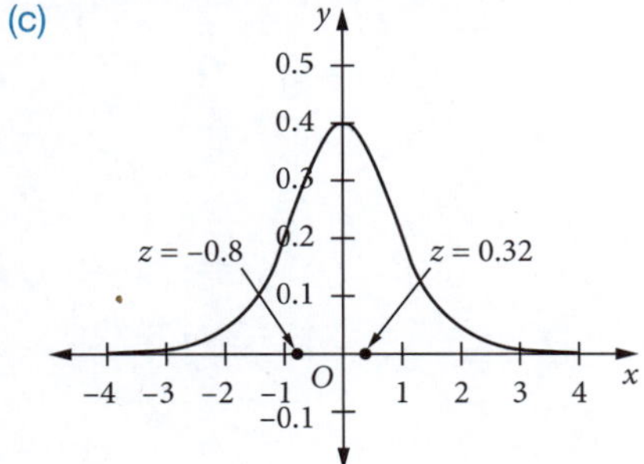

CHAPTER REVIEW 20

1 (a) $k = 0.1$ $\int_{0}^{10} k\,dx = 1$ (b) $P(X \le 3) = \int_{0}^{3} \frac{dx}{10} = 0.3$

(c) $P(X \le 5) = 0.5$ (d) $P(2 \le X \le 9) = 0.9 - 0.2 = 0.7$

2 No, $f(x)$ is negative for some values of x in the domain.

3 $\int_0^b \frac{x^2}{12} dx = 0.5$

$\frac{b^3}{36} = \frac{1}{2}$

Median is $\sqrt[3]{18}$

4 (a) 0.16

(b) $500 - 2 \times 5 = 490$ g

Bottom 2.5% rejected: number of jars $= 0.025 \times 1000 = 25$.

5 (a) $z = -1$ (b) $z = 2$

6 (a) $z = -1$ (b) $z = 3$

7 B

8 D 9 B 10 C 11 A

12 (a) It is reasonable because $f(x) \geq 0$ for the values in the domain.

(b) (i) 0.75 (ii) 0.1443 (iii) 0.5615

(c) 0.6619 (d) 0.2159

13 (a) $\frac{5}{7}$ hour = 43 minutes (b) ≈ 0.97 (c) 0.35

(d) 0.8232

So, expect about 8 of the 10 people to wait more than 30 minutes.

14 (a) Xandra: 1.75 hours (1 h 45 min)

Zack : 1.03 hours (1 h 2 min)

Xandra spends, on average, about 43 more minutes completing homework than Zack.

(b) 0.375 (c) 0.5375, 0.5730

15 (a) $54 + 2 \times 4 = 62$ cm.

(b) $z_R = -2$, $z_B = -1.75$. The proportion of red snapper that must be thrown back is smaller than the proportion of barramundi thrown back.

GLOSSARY

A

abscissa
The first member of an ordered pair on the number plane; for coordinates (x, y) the abscissa is x. Plural *abscissae*.

absolute value
The absolute value $|x|$ of a real number x is the magnitude or size of x without regard to its sign, i.e. on a number line, the distance of the number from the origin. Formally, $|x| = x$ if $x \geq 0$, $|x| = -x$ if $x < 0$.

acceleration
The rate of change of velocity with respect to time, $\frac{dv}{dt}$, standard units m s^{-2}.

ambiguous case (trigonometry)
In trigonometry, the ambiguous case refers to using the sine rule to calculate the size of an angle in a triangle where there are two possibilities for the angle, one obtuse and one acute, leading to two possible triangles.

amplitude
The amplitude of a wave function is the height from the horizontal centre line to the peak (or to the trough) of the graph of the function. Alternatively, it is half the distance between the maximum and minimum values.

angle of inclination
The angle of inclination of a straight line is the angle the line makes with the positive x-axis.

annuity
An annuity is a compound interest investment from which payments are made or received on a regular basis for a fixed period of time.

anti-derivative
An anti-derivative, primitive or indefinite integral of a function $f(x)$ is a function $F(x)$ whose derivative is $f(x)$, i.e. $F'(x) = f(x)$.

anti-differentiation
The reverse process to differentiation, also known as finding the primitive function.

arc of curve
A portion of a curve joining two points.

area under a curve
The region in the number plane bounded by a curve, the x-axis and two x values (abscissae).

arithmetic sequence
An arithmetic sequence is a sequence of numbers such that the difference of any two successive members of the sequence is a constant.

arithmetic series
An arithmetic series is a sum whose terms form an arithmetic sequence.

array
An array is an ordered rectangular collection of objects or numbers arranged in rows and columns.

asymptote
An asymptote is a line.

A horizontal asymptote is a horizontal line whose distance from the function $f(x)$ becomes as small as we please for all large values of x.

The line $x = a$ is a vertical asymptote if the function f is not defined at $x = a$ and values of $f(x)$ become as large as we please (positive or negative) as x approaches a.

B

break-even point
The break-even point is the point at which income and cost of production are equal.

C

Cartesian coordinates
An ordered pair of numbers (x, y) used to locate a point in the number plane.

chain rule
A rule used to find the derivative of a function of a function. If $y = f(u)$ and $u = g(x)$ then $\frac{dy}{dx} = \frac{dy}{du} \times \frac{du}{dx}$.

change of base
The function $\log_a n = \frac{\log_b n}{\log_b a}$ can be used to change the base of a logarithm from base b to base a.

chord
A straight line joining two points on a circle.

circular measure
The size of an angle, measured in radians.

common difference
The difference between pairs of successive terms of an arithmetic sequence.

common ratio
The ratio of successive terms of a geometric sequence.

complementary angles
Two angles whose sum is $90°$ or $\frac{\pi}{2}$ radians (i.e. a right angle).

complementary events
Two events such that one or the other event must occur, but not both. The sum of their probabilities is one.

composite events
Events that consist of two or more simple events.

composite functions
In a composite function, the output of one function becomes the input of a second function.

More formally, the composite of f and g, acting on x, can be written as $f(g(x))$ with $g(x)$ being performed first.

composite function rule
See *chain rule*.

concave downwards
If $f''(x) < 0$ then the curve of f is concave downwards.

concave upwards
If $f''(x) > 0$ then the curve of f is concave upwards.

concavity
If a function $f(x)$ is a differentiable function on a given interval, I, then:

- $f(x)$ is concave up on I if and only if $f'(x)$ is increasing on I. Graphically if the tangent at $x = a$ lies below the curve (locally), then the function is concave up at $x = a$.
- $f(x)$ is concave down on I if and only if $f'(x)$ is decreasing on I. Graphically if the tangent at $x = a$ lies above the curve (locally), then the function is concave down at $x = a$.

If $f(x)$ is doubly differentiable at $x = a$ then the second derivative can be used to identify concavity in the following way:

- If $f''(a)$ is positive then the curve is concave up at $x = a$
- If $f''(a)$ is negative then the curve is concave down at $x = a$

If $f''(a)$ is zero then the curve could be concave up, concave down or a point of inflection and further work is required to determine which one of these three cases applies at $x = a$.

conditional probability
The probability that an event A occurs can change if it becomes known that another event B has occurred. The new probability is known as conditional probability and is written as $P(A|B)$. If B has occurred, the sample space is reduced by discarding all outcomes that are not in the event B.

continuity
A function is said to be continuous (or that it 'has continuity') if for every x value in the domain there exists a corresponding y value, i.e. there are no gaps or holes.

continuous function
A function is continuous when sufficiently small changes in the input result in arbitrarily small changes in the output. Its graph is an unbroken curve.

continuous random variable
A continuous random variable is a numerical variable that can take any value along a continuum.

cosecant (cosec)
The reciprocal of the sine function.

cotangent (cot)
The reciprocal of the tangent function.

cumulative distribution function
Given a continuous random variable X, the cumulative distribution function $F(x)$ is the probability that $X \leq x$.

cumulative frequency
The cumulative frequency is the accumulating total of frequencies within an ordered dataset.

curve
A joined set of points representing a function or a relation.

D

decreasing function
A function that always has a negative gradient.

definite integral
An integral written $\int_a^b f(x)dx$, where a is the lower limit and b is the upper limit of the integration.

dependent variable
The variable that depends on the independent variable. In $y = f(x)$, x is the independent variable and y is the dependent variable.

derivative at a point
The value of the derivative at a particular point; the gradient of the tangent to the curve at that point.

derived function
The result of differentiation; gives the gradient of the tangent at any point on the curve. Also known as 'gradient function'.

differentiable function
A function that has a derivative at all points in the given domain.

differentiation
The process of finding the gradient function from a function.

differentiation from first principles
Finding the derivative using the formula $f'(x) = \lim_{h \to 0} \frac{f(x+h) - f(x)}{h}$.

dilation
A dilation stretches or compresses the graph of a function. This could happen either in the x or y direction or both.

direct variation
Two variables are in direct variation if one is a constant multiple of the other. This can be represented by the equation $y = kx$, where k is the constant of variation (or proportion). Also known as direct proportion, it produces a linear graph through the origin.

discontinuous function

If a function $f(x)$ is not continuous at $x = a$, then $f(x)$ is said to be discontinuous at $x = a$.

discrete random variable

A discrete random variable is a numerical variable whose values can be listed.

discriminant

For a quadratic equation $y = ax^2 + bx + c$, the discriminant is $\Delta = b^2 - 4ac$.

displacement

The signed distance (positive or negative) of a particle from the origin.

domain

The domain of a function is the set of x values of $y = f(x)$ for which the function is defined. Also known as the 'input' of a function.

E

empirical rule (normal distributions)

The empirical rule for normally distributed random variables is:

- approximately 68% of data will have-scores between −1 and 1
- approximately 95% of data will have-scores between −2 and 2
- approximately 99.7% of data will have-scores between −3 and 3.

equally likely outcomes

Outcomes that have the same chance of occurring.

Euler's number

The irrational number $e \approx 2.718\,28\ldots$ defined by $e = \lim\limits_{n \to \infty}\left(1 + \frac{1}{n}\right)^n$ such that the function $y = e^x$ is its own derivative, or $\frac{d}{dx}\left(e^x\right) = e^x$.

even function

Algebraically, a function is even if $f(-x) = f(x)$, for all values of x in the domain.

An even function has line symmetry about the y-axis.

expected value

In statistics, the expected value $E(X)$ of a random variable X is a measure of the central tendency of its distribution. Also known as the expectation or mean. $E(X)$ is calculated differently depending on whether the random variable is discrete or continuous.

exponential decay

Exponential decay (or 'exponential decline') occurs when a quantity decreases by a constant percentage over time.

exponential function

A function with its variable as an index, e.g. $y = a^x$.

exponential growth

Exponential growth occurs when a quantity increases by a constant percentage over time.

exponential growth and decay

Exponential growth occurs when the rate of change of a mathematical function is positive and proportional to the function's current value. Exponential decay occurs in the same way when the growth rate is negative.

extrapolate

To predict the value of a function outside the given domain.

extrapolation

Extrapolation occurs when the fitted model is used to make predictions using values that are outside the range of the original data upon which the fitted model was based. Extrapolation far beyond the range of the original data is not advisable as it can sometimes lead to quite erroneous predictions.

F

first derivative test

If $f'(x) = 0$, then the point at x is a stationary point. If the sign of $f'(x)$ changes from positive to negative through x, then it is a local maximum turning point; if the sign of $f'(x)$ changes from negative to positive through x, then it is a local minimum turning point.

function

A function f is a rule that associates each element x in a set S with a unique element $f(x)$ from a set T.

The set S is called the domain of f and the set T is called the co-domain of f. The subset of T consisting of those elements of T which occur as values of the function is called the range of f. The functions most commonly encountered in elementary mathematics are real functions of a real variable, for which both the domain and co-domain are subsets of the real numbers.

If we write $y = f(x)$, then we say that x is the independent variable and y is the dependent variable.

fundamental counting principle

If one event can occur in m ways and another event can occur in n ways, then the two-stage event (both events together) can occur in $m \times n$ ways. Also called 'multiplication principle'.

fundamental theorem of calculus

$\int_a^b f(x)\,dx = \left[F(x)\right]_a^b = F(b) - F(a)$, where $F(x)$ is the primitive function of $f(x)$.

Future value

The future value of an investment or annuity is the total value of the investment at the end of the term of the investment, including all contributions and interest earned.

future value interest factors
Future value interest factors are the values of an investment at a specific date. A table of these factors can be used to calculate the future value of different amounts of money that are invested at a certain interest rate for a specified period of time.

G

general term
An expression that allows you to calculate any of the terms of a series.

geometric sequence
A geometric sequence is a sequence of numbers where each term after the first is found by multiplying the previous one by a fixed number called the common ratio.

geometric series
A geometric series is a sum whose terms form a geometric sequence.

gradient of a secant
The slope of the line joining two points on a curve.

gradient of a tangent
The slope of the line that touches a point on a curve. This is the limiting position of the secant, where two points become the same point.

greatest value of a function
For a given domain, the value of a function at its highest maximum turning point or highest endpoint, whichever is the greatest.

H

half-life
The time taken for half of the atoms in a radioactive substance to decay.

horizontal line test
The horizontal line test is a method that can be used to determine whether a function is a one-to-one function. If any horizontal line intersects the graph of a function more than once then the function is not a one-to-one function.

horizontal point of inflection
The point on a curve where $\frac{dy}{dx} = 0$ and at which the gradient has the same sign on either side of the point.

I

identity
An identity is a statement involving a variable(s) that is true for all possible values of the variable(s).

increasing function
A function that always has a positive gradient.

indefinite integral
The primitive of a function, written $\int f(x)dx$.

independent events
Events are independent if the occurrence or non-occurrence of one event cannot change the probability of the occurrence of another event, i.e. the events have no effect on each other.

independent events (independence)
In probability, two events are independent of each other if the occurrence of one does not affect the probability of the occurrence of the other.

independent variable
The variable that we choose to substitute into a function to generate the dependent variable.

instantaneous rate of change
The instantaneous rate of change is the rate of change at a particular moment. For a differentiable function, the instantaneous rate of change at a point is the same as the gradient of the tangent to the curve at that point. This is defined to be the value of the derivative at that particular point.

integration
See *anti-differentiation*. The reverse process to differentiation, used to find the area under a curve.

interpolate
To predict the value of a function in the given domain using the data available when a given rule or function is not available.

interpolation
Interpolation occurs when a fitted model is used to make predictions using values that lie within the range of the original data.

interval notation
Interval notation is a notation for representing an interval by its endpoints. Parentheses and/or square brackets are used respectively to show whether the endpoints are excluded or included.

L

least-squares regression line
Least-squares regression is a method for finding a straight line that best summarises the relationship between two variables, within the range of the dataset.

The least-squares regression line is the line that minimises the sum of the squares of the residuals. Also known as the least-squares line of best fit.

limit
The limit of a function at a point a, if it exists, is the value the function approaches as the independent variable approaches a.

The notation used is: $\lim_{x \to a} f(x) = L$.

This is read as 'the limit of $f(x)$ as x approaches a is L'.

limiting sum
For a geometric series with a common ratio $|r| < 1$, the value that the sum of the series approaches as the number of terms increases to infinity.

line of best fit

A line of best fit is a line drawn through a scatterplot of data points that most closely represents the relationship between two variables.

local and global maximum and minimum

$f(x_0)$ is a local maximum of the function $f(x)$ if $f(x) \leq f(x_0)$ for all values of x near x_0. We say that $f(x_0)$ is a global maximum of the function $f(x)$ if $f(x) \leq f(x_0)$ for all values of x in the domain of f.

$f(x_0)$ is a local minimum of the function $f(x)$ if $f(x) \geq f(x_0)$ for all values of x near x_0. We say that $f(x_0)$ is a global minimum of the function $f(x)$ if $f(x) \geq f(x_0)$ for all values of x in the domain of f.

logarithmic function

The function $f(x) = \log_a x$, where $a^{\log_a x} = \log_a\left(a^x\right) = x$.

M

maximum turning point

A point on a curve where $\frac{dy}{dx} = 0$ and the gradient changes from positive to negative as x increases through the point. If $\frac{dy}{dx} = 0$ and $\frac{d^2y}{dx^2} < 0$ then it is a maximum turning point; if $\frac{d^2y}{dx^2} = 0$ then it may or may not be a maximum turning point.

measures of central tendency

Measures of central tendency are the values about which the set of data values for a particular variable are scattered. They are a measure of the centre or location of the data.

The two most common measures of central tendency are the mean and the median.

measures of spread

Measures of spread describe how similar or varied the set of data values are for a particular variable.

Common measures of spread include the range, combinations of quantiles (deciles, quartiles, percentiles), the interquartile range, variance and standard deviation.

minimum turning point

A point on a curve where $\frac{dy}{dx} = 0$ and the gradient changes from negative to positive as x increases through the point. If $\frac{dy}{dx} = 0$ and $\frac{d^2y}{dx^2} > 0$ then it is a minimum turning point; if $\frac{d^2y}{dx^2} = 0$ then it may or may not be a minimum turning point.

mutually exclusive events

Two events are mutually exclusive (or 'disjoint') if membership of one event excludes membership of the other, so that they cannot occur simultaneously.

N

natural logarithm

A logarithm to base e, written $\log_e x$ or $\ln x$. Also called 'Naperian' or 'Napierian' logarithm.

normal distribution

The normal distribution is a type of continuous distribution and is often called a 'bell curve' due to its shape. The mean, median and mode are equal and the scores are symmetrically arranged either side of the mean.

normal random variable

A normal random variable is a variable which varies according to the normal distribution.

normal to a curve

In calculus, the normal to a curve at a given point P is the straight line that is perpendicular to the tangent to the curve at that point.

O

odd function

Algebraically, a function is odd if $f(-x) = -f(x)$, for all values of in the domain.

An odd function has point symmetry about the origin.

one-to-one function

In a one-to-one function, every element in the range of a function corresponds to exactly one element of the domain.

ordinate

The second member of an ordered pair on the number plane; for coordinates (x, y) the ordinate is y.

oscillation

A movement from one extreme position to another and then back again, made by an object with a wave-like or vibrating motion.

P

Pareto chart

A Pareto chart is a type of chart that contains both a bar and a line graph, where individual values are represented in descending order by the bars and the cumulative total is represented by the line graph.

particle

A body that behaves such that all forces acting on the body can be regarded as acting through a single point. This means that the body can be represented as a single point, regardless of its actual size and shape.

Pearson's correlation coefficient

Pearson's correlation coefficient is a statistic that measures the strength of the linear relationship between a pair of variables or datasets. Its value lies between -1 and 1 (inclusive). Also known as simply the correlation coefficient. For a sample, it is denoted by r.

period
The period of a trigonometric function is the smallest interval for which the function repeats itself.

periodic phenomenon
Something that repeats itself in a regular (periodic) way.

phase
When a trigonometric function is translated horizontally, the phase (or phase shift) is the magnitude of this translation.

point of inflection
A point of inflection is a point on a curve where the tangent exists and crosses the curve.

Within these courses, points of inflection can be found by taking the zeros of the second derivative and checking whether the concavity changes around the point.

polynomial
A polynomial is an expression of the form $a_nx^n + a_{n-1}x^{n-1} + \ldots + a_2x^2 + a_1x + a_0$, where n is a non-negative integer.

population
The population in statistics is the entire dataset from which a statistical sample may be drawn.

power function
A power function is a function of the form $f(x) = kx^n$, where k and n are real numbers.

present value
The present value of an investment or annuity is the single sum of money (or principal) that could be initially invested to produce a future value over a given period of time.

primitive function
If $F'(x) = f(x)$, then $F(x)$ is the primitive function of $f(x)$.

product rule
A rule used to find the derivative of the product of two functions. If $y = uv$, where u and v are functions of x, then $\frac{dy}{dx} = v\frac{du}{dx} + u\frac{dv}{dx}$.

Q

quadratic formula
A formula that gives the roots of the general quadratic equation $ax^2 + bx + c = 0$: $x = \frac{-b \pm \sqrt{b^2 - 4ac}}{2a}$

quadratic inequality
A quadratic inequality is an inequality involving a quadratic expression.

quotient rule
A rule used to find the derivative of the quotient of two functions. If $y = \frac{u}{v}$, where u and v are functions of x, then $\frac{dy}{dx} = \frac{v\frac{du}{dx} - u\frac{dv}{dx}}{v^2}$.

R

radian
The angle at the centre of a unit circle subtended by an arc of unit length. 2π radians $= 360°$.

random variable
A random variable is a variable whose possible values are outcomes of a statistical experiment or a random phenomenon.

range (of function)
The range of a function is the set of values of the dependent variable for which the function is defined.

rate of change
A rate of change of a function, $y = f(x)$ is $\frac{\Delta y}{\Delta x}$ where Δx is the change in x and Δy is the corresponding change in y.

real number
The set of real numbers consists of the set of all rational and irrational numbers.

reducing balance loan
A reducing balance loan is a compound interest loan where the loan is repaid by making regular payments and the interest paid is calculated on the amount still owing (the reducing balance of the loan) after each payment is made.

relation
A set of points for which a given value of x may have more than one y value. A circle is a relation. Also called 'multi-valued function'.

S

sample space
All possible outcomes of an experiment.

scatterplot
A scatterplot is a two-dimensional data plot using Cartesian coordinates to display the values of two variables in a bivariate dataset. Also known as a scatter graph.

secant (sec)
The reciprocal of the cosine function.

secant (line)
A secant is the straight line passing through two points on the graph of a function.

second derivative
The second derivative is the derivative of the first derivative. It is denoted by: $f''(x)$ or $\frac{d^2y}{dx^2}$

sector
A portion of a circle bounded by two radii.

segment
A portion of a circle cut off by a chord.

sequence

In mathematics, a sequence is a set of numbers whose terms follow a prescribed pattern. Mathematical sequences include arithmetic sequences and geometric sequences.

series

A series is the sum of the terms of a particular sequence.

set language and notation

A set is a collection of distinct objects called elements.

The language and notation used in the study of sets includes:

- A set is a collection of objects, for example 'A is the set of the numbers 1, 3 and 5' is written as $A = \{1, 3, 5\}$.
- Each object is an element or member of a set, for example '1 is an element of set A' is written as $1 \in A$.
- The number of elements in set $A = \{1, 3, 5\}$ is written as $n(A) = 3$, or $|A| = 3$.
- The empty set is the set with no members and is written as {} or ϕ.
- The universal set contains all elements involved in a particular problem.
- B is a subset of A if every member of B is a member of A and is written as $B \subset A$, i.e. 'B is a subset of A'. B may also be equal to A in this scenario, and we can therefore write $B \subseteq A$.
- The complement of a set A is the set of all elements in the universal set that are not in A and is written as $\bar{A}$ or A^c.
- The intersection of sets A and B is the set of elements which are in both A and B and is written as $A \cap B$, i.e. 'A intersection B'.
- The union of sets A and B is the set of elements which are in A or B or both and is written as $A \cup B$, i.e. 'A union B'.

sigma notation (Σ)

The symbol Σ (the Greek capital letter sigma) means 'the sum of'. It is specifically used in series notation to mean 'all these terms added together, starting with the value underneath the Σ and continuing to the value above the Σ'.

sketch

A sketch is an approximate representation of a graph, including labelled axes, intercepts and any other important relevant features. Compared to the corresponding graph, a sketch should be recognisably similar but does not need to be precise.

standard deviation

Standard deviation is a measure of the spread of a dataset. It gives an indication of how far, on average, individual data values are spread from the mean.

stationary point

A stationary point on the graph $y = f(x)$ of a differentiable function is a point where $f'(x) = 0$.

A stationary point could be classified as a local or global maximum or minimum or a horizontal point of inflection.

subinterval

A section of an interval of a definite integral that has been divided into smaller parts, as when using the trapezoidal rule.

sum to *n* terms

To add together the terms of a series, from the first term to the nth term.

T

tangent

The tangent to a curve at a given point P can be described intuitively as the straight line that 'just touches' the curve at that point. At P the curve has 'the same direction' as the tangent. In this sense it is the best straight-line approximation to the curve at point P.

term

A single part of a sequence or series.

time rate of change

The rate at which something is changing over time.

trapezoidal rule

An approximation to the definite integral made by finding the area of a trapezoidal shape between two points on the curve:

$$\int_a^b f(x)\,dx \approx \frac{(b-a)}{2}\big(f(a) + f(b)\big)$$

tree diagram

A diagram that shows outcomes and probabilities of multiple and/or successive events by connecting possible outcomes along 'branches'.

trigonometric function

A function involving sin, cos, tan, cosec, sec or cot.

trigonometric identity

A general relation involving trigonometric functions that is true for all values, e.g. $\sin^2 x + \cos^2 x = 1$.

turning point

The point on a curve where $\frac{dy}{dx} = 0$ and at which the gradient changes sign on either side of the point.

V

variance

In statistics, the variance Var(X) of a random variable X is a measure of the spread of its distribution. Var(X) is calculated differently depending on whether the random variable is discrete or continuous.

velocity

The rate of change of displacement with respect to time, standard units m s^{-1}.

Venn diagram

A diagram that shows closed curves (e.g. circles or other shapes) around members of sets, using the shapes' intersections to show the common elements of different sets.

vertical line test

The vertical line test determines whether a relation or graph is a function. If a vertical line intersects or touches a graph at more than one point, then the graph is not a function.

Z

***z*-score**

A z-score is a statistical measurement of how many standard deviations a raw score is above or below the mean. A z-score can be positive or negative, indicating whether it is above or below the mean, or zero. Also known as a standardised score.